CHILTON'S™ GUIDE TO
SMALL ENGINE REPAIR—UP TO 20 HP

Repair, maintenance and service for gasoline engines up to and including 20 horsepower.

WITHDRAWN

President	Dean F. Morgantini, S.A.E.
Vice President–Finance	Barry L. Beck
Vice President–Sales	Glenn D. Potere
Executive Editor	Kevin M. G. Maher, A.S.E.
Manager–Consumer Automotive	Richard Schwartz, A.S.E.
Manager–Marine/Recreation	James R. Marotta, A.S.E.
Production Specialists	Brian Hollingsworth, Melinda Possinger
Project Managers	Will Kessler, A.S.E., S.A.E., Thomas A. Mellon, A.S.E., S.A.E., Richard Rivele, Todd W. Stidham, A.S.E., Ron Webb
Editor	Richard J. Rivele

CHILTON™ Automotive Books
PUBLISHED BY W. G. NICHOLS, INC.

Manufactured in USA
© 1994 Chilton Book Company
1020 Andrew Drive
West Chester, PA 19380
ISBN 0-8019-8325-8
Library of Congress Catalog Card No. 94-70513
7 8 9 0 06 05 04

™Chilton is a registered trademark of Cahners Business Information, a division of Reed Elsevier, Inc., and has been licensed to W. G. Nichols, Inc.

www.chiltononline.com

Contents

SAFETY NOTICE

Proper service and repair procedures are vital to the safe, reliable operation of all power equipment, as well as the personal safety of those performing repairs. This manual outlines procedures for servicing and repairing power equipment engines using safe, effective methods. The procedures contain many NOTES, CAUTIONS, and WARNINGS which should be followed along with standard procedures to eliminate the possibility of personal injury or improper service which could damage the equipment or compromise its safety.

It is important to note that the repair procedures and techniques, tools and parts for servicing power equipment, as well as the skill and experience of the individual performing the work vary widely. It is not possible to anticipate all of the conceivable ways or conditions under which engines may be serviced, or to provide cautions as to all of the possible hazards that may result. Standard and accepted safety precautions and equipment should be used when handling toxic or flammable fluids, and safety goggles or other protection should be used during cutting, grinding, chiseling, prying,or any other process that can cause material removal or projectiles.

Some procedures require the use of tools specially designed for a specific purpose. Before substituting another tool or procedure, you must be completely satisfied that neither your personal safety, nor the performance of the engine will be endangered.

Although information in this manual is based on industry sources and is complete as possible at the time of publication, the possibility exists that some vehicle manufacturers made later changes which could not be included here. While striving for total accuracy, W. G. Nichols, Inc. cannot assume responsibility for any errors, changes or omissions that may occur in the compilation of this data.

PART NUMBERS

Part numbers listed in this reference are not recommendations by Chilton for any product by brand name. They are references that can be used with interchange manuals and aftermarket supplier catalogs to locate each brand supplier's discrete part number.

SPECIAL TOOLS

Special tools are recommended by the engine manufacturer to perform their specific job. Use has been kept to a minimum, but where absolutely necessary, they are referred to in the text by the part number of the tool manufacturer. These tools can be purchased, under the appropriate part number, from your dealer or regional distributor, or an equivalent tool can be purchased locally from a tool supplier or parts outlet. Before substituting any tool for the one recommended, read the SAFETY NOTICE at the top of this page.

ACKNOWLEDGMENTS

W. G. Nichols, Inc. expresses appreciation to Briggs & Stratton Co., Milwaukee, WI; Kawasaki Engine Div., Shakopee, MN; Kohler Co., Kohler, WI; Tecumseh Power Products, Grafton, WI; Aviation Auto Supply, Bryn Mawr, PA, for their generous assistance.

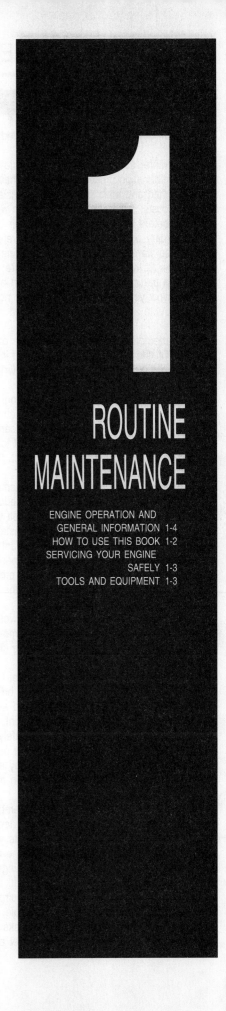

1

ROUTINE MAINTENANCE

HOW TO USE THIS BOOK

This book covers popular small, air-cooled engines of 2-20 hp.

Each Chapter contains maintenance, tune-up, repair and overhaul information procedures. Studies have shown that a properly tuned and maintained engine translates into lower operating costs, and periodic maintenance will catch minor problems before they turn into major repair bills.

A secondary purpose of this book is a reference guide for owners who want to understand their engine and/or their repair shop better. In this case, no tools at all are required. Knowing just what a particular repair job requires in parts and labor time will allow you to evaluate whether or not you're getting a fair price quote and help decipher itemized bills from a repair shop.

Before attempting any repairs or service on your engine, read through the entire procedure outlined in the appropriate Chapter. This will give you the overall view of what tools and supplies will be required.

Each operation should be approached logically and all procedures thoroughly understood before attempting any work. Some special tools that may be required can often be rented from local automotive or power equipment jobbers, or places specializing in renting tools and equipment. Check the yellow pages of your phone book.

All Chapters contain adjustments, maintenance, removal and installation procedures, and overhaul procedures. When overhaul is not considered practical, we tell you how to remove the failed part and then how to install the new or rebuilt replacement. In this way, you at least save the labor costs. Backyard overhaul of some components is just not practical, but the removal and installation procedure is often simple and well within the capabilities of the average do-it-yourselfer.

A basic mechanic's rule should be followed: all threaded fasteners (screws, nuts, and bolts) are removed by turning counterclockwise, and tightened by turning clockwise, unless otherwise noted.

Safety is always the most important rule. Constantly be aware of the dangers involved in working on or around an automobile and take proper precautions to avoid the risk of personal injury or damage to the equipment. See the section in this Chapter, Servicing Your Engine Safely, and the SAFETY NOTICE on the acknowledgment page before attempting any service procedures and pay attention to the instructions provided. There are 3 common mistakes in mechanical work:

1. Incorrect order of assembly, disassembly or adjustment. When taking something apart or putting it together, doing things in the wrong order usually just costs you extra time; however it CAN break something. Read the entire procedure before beginning disassembly. Do everything in the order in which the instructions say you should do it, even if you can't immediately see a reason for it. When you're taking apart something that is very intricate (for example, a carburetor), you might want to draw a picture of how it looks when assembled at one point in order to make sure you get everything back in

its proper position. We will supply exploded views whenever possible, but sometimes the job requires more attention to detail than an illustration provides. When making adjustments (especially tune-up adjustments), do them in order. One adjustment often affects another and you cannot expect satisfactory results unless each adjustment is made only when it cannot be changed by any other.

2. Overtorquing (or undertorquing) nuts and bolts. While it is more common for overtorquing to cause damage, undertorquing can cause a fastener to vibrate loose and cause serious damage, especially when dealing with aluminum parts. Pay attention to torque specifications and utilize a torque wrench in assembly. If a torque figure is not available, remember that, if you are using the right tool to do the job, you will probably not have to strain yourself to get a fastener tight enough. The pitch of most threads is so slight that the tension you put on the wrench will be multiplied many times in actual force on what you are tightening. A good example of how critical torque is can be seen in the case of spark plug installation, especially where you are putting the plug into an aluminum cylinder head. Too little torque can fail to crush the gasket, causing leakage of combustion gases and consequent overheating of the plug and engine parts. Too much torque can damage the threads or distort the plug, which changes the spark gap at the electrode. Since more and more manufacturers are using aluminum in their engine and chassis parts to save weight, a torque wrench should be in any serious do-it-yourselfer's tool box.

There are many commercial chemical products available for ensuring that fasteners won't come loose, even if they are not torqued just right (a very common brand is Loctite®). If you're worried about getting something together tight enough to hold, but loose enough to avoid mechanical damage during assembly, one of these products might offer substantial insurance. Read the label on the package and make sure the product is compatible with the materials, fluids, etc. involved before choosing one.

3. Crossthreading. This occurs when a part such as a bolt is screwed into a nut or casting at the wrong angle and forced, causing the threads to become damaged. Crossthreading is more likely to occur if access is difficult. It helps to clean and lubricate fasteners, and to start threading with the part to be installed going straight in, using your fingers. If you encounter resistance, unscrew the part and start over again at a different angle until it can be inserted and turned several times without much effort. Keep in mind that many parts, especially spark plugs, use tapered threads so that gentle turning will automatically bring the part you're threading to the proper angle if you don't force it or resist a change in angle. Don't put a wrench on the part until it's been turned in a couple of times by hand. If you suddenly encounter resistance and the part has not seated fully, don't force it. Pull it back out and make sure it's clean and threading properly.

Always take your time and be patient; once you have some experience, working on your piece of equipment will become an enjoyable hobby.

TOOLS AND EQUIPMENT

Naturally, without the proper tools and equipment it is impossible to properly service your engine. It would be impossible to catalog each tool that you would need to perform each or every operation in this book. It would also be unwise for the amateur to rush out and buy an expensive set of tools an the theory that he or she may need one or more of them at sometime.

The best approach is to proceed slowly, gathering together a good quality set of those tools that are used most frequently. Don't be misled by the low cost of bargain tools. It is far better to spend a little more for better quality. Forged wrenches, 6- or 12-point sockets and fine tooth ratchets are by far preferable to their less expensive counterparts. As any good mechanic can tell you, there are few worse experiences than trying to work on an engine with bad tools. Your monetary savings will be far outweighed by frustration and mangled knuckles.

Certain tools, plus a basic ability to handle tools, are required to get started.

Begin accumulating those tools that are used most frequently; those associated with routine maintenance and tune-up.

In addition to the normal assortment of screwdrivers and pliers you should have the following tools for routine maintenance jobs:

1. U.S. and metric wrenches, sockets and combination open end/box end wrenches in sizes from $\frac{1}{4}$ in. to $\frac{7}{8}$ in. and 3mm to 22mm; and a spark plug socket ($\frac{13}{16}$ in. and/or $\frac{5}{8}$ in.) If possible, buy various length socket drive extensions.
2. Wire spark plug gauge/adjusting tools
3. Set of feeler blades.
4. Hydrometer for checking the battery
5. A container for draining oil
6. Many rags for wiping up the inevitable mess.

In addition to the above items there are several others that are not absolutely necessary, but handy to have around. These include oil-dry (cat box litter works just as well and may be cheaper), a funnel, and the usual supply of lubricants and fluids, although these can be purchased as needed. This is a basic list for routine maintenance, but only your personal needs and desires can accurately determine your list of necessary tools.

In addition to these basic tools, there are several other tools and gauges you may find useful. These include:

7. A compression gauge. The screw-in type is slower to use, but eliminates the possibility of a faulty reading due to escaping pressure
8. A test light
9. An induction meter. This is used for determining whether or not there is current in a wire.

As a final note, you will probably find a torque wrench necessary for all but the most basic work. The beam type models are perfectly adequate, although the click (break-away) type are more precise, and you don't have to crane your neck to see a torque reading in awkward situations. The break-away torque wrenches are more expensive and should be recalibrated periodically.

Torque specification for each fastener will be given in the procedure in any case that a specific torque value is required. If no torque specifications are given, use the following values as a guide, based upon fastener size:

Bolts marked 6T
6mm bolt/nut — 5-7 ft. lbs.
8mm bolt/nut — 12-17 ft. lbs.
10mm bolt/nut — 23-34 ft. lbs.
12mm bolt/nut — 41-59 ft. lbs.
14mm bolt/nut — 56-76 ft. lbs.
Bolts marked 8T
6mm bolt/nut — 6-9 ft. lbs.
8mm bolt/nut — 13-20 ft. lbs.
10mm bolt/nut — 27-40 ft. lbs.
12mm bolt/nut — 46-69 ft. lbs.
14mm bolt/nut — 75-101 ft. lbs.

SERVICING YOUR ENGINE SAFELY

It is virtually impossible to anticipate all of the hazards involved with equipment maintenance and service but care and common sense will prevent most accidents.

The rules of safety for mechanics range from DON'T smoke around gasoline, to use the proper tool for the job. The trick to avoiding injuries is to develop safe work habits and take every possible precaution.

Do's

• Do keep a fire extinguisher and first aid kit within easy reach.
• Do wear safety glasses or goggles when cutting, drilling, grinding or prying. If you wear glasses for the sake of vision, then they should be made of hardened glass that can serve also as safety glasses, or wear safety goggles over your regular glasses.
• Do shield your eyes whenever you work around the battery. Batteries contain sulfuric acid; in case of contact with the eyes or skin, flush the area with water or a mixture of water and baking soda and get medical attention immediately.
• Do use adequate ventilation when working with any chemicals.
• Do disconnect the negative battery cable when working on anything electrical, or when work around or near the electrical system.
• Do follow manufacturer's directions whenever working with potentially hazardous materials.
• Do properly maintain your tools. Loose hammerheads, mushroomed punches and chisels, frayed or poorly grounded electrical cords, excessively worn screwdrivers, spread wrenches (open end), cracked sockets, slipping ratchets, or faulty droplight sockets can cause accidents.
• Do use the proper size and type of tool for the job being done.
• Do when possible, pull on a wrench handle rather than push on it, and adjust your stance to prevent a fall.

- Do be sure that adjustable wrenches are tightly adjusted on the nut or bolt and pulled so that the face is on the side of the fixed jaw.
- Do select a wrench or socket that fits the nut or bolt. The wrench or socket should sit straight, not cocked.
- Do strike squarely with a hammer to avoid glancing blows.

Don'ts

- Don't run an engine in a garage or anywhere else without proper ventilation--EVER! Carbon monoxide is poisonous; it is absorbed by the body 400 times faster than oxygen; it takes a long time to leave the human body and you can build up a deadly supply of it in your system by simply breathing in a little every day. You may not realize you are slowly poisoning yourself. Always use power vents, windows, fans or open the garage doors.
- Don't work around moving parts while wearing a necktie or other loose clothing. Short sleeves are much safer than long, loose sleeves. Hard-toed shoes with neoprene soles protect your toes and give a better grip on slippery surfaces. Jewelry such as watches, fancy belt buckles, beads or body adornment of any kind is not safe working around moving parts. Long hair should be hidden under a hat or cap.
- Don't use pockets for tool boxes. A fall or bump can drive a screwdriver deep into your body. Even a wiping cloth hanging from the back pocket can wrap around a spinning shaft.
- Don't smoke when working around gasoline, cleaning solvent or other flammable material.
- Don't smoke when working around the battery. When the battery is being charged, it gives off explosive hydrogen gas.
- Don't use gasoline to wash your hands; there are excellent soaps available. Aside from being flammable, gasoline also removes all the natural oils from the skin so that bone dry hands will suck up oil and grease.

ENGINE OPERATION AND GENERAL INFORMATION

How an Internal Combustion Engine Develops Power

The energy source that runs an internal combustion engine is heat created by the combustion of an air/fuel mixture. The combustion process takes place within a sealed cylinder containing a piston which is able to move up and down in the cylinder. The piston is connected to a crankshaft by a connecting rod. The lower end of the rod is connected to the crankshaft at a point which is offset from the centerline of the crankshaft, allowing it to turn a large circle. This is why the piston moves up and down, and why pressure on the top of the piston eventually becomes torque, or turning force, on the crankshaft.

The sealed cylinder, or 'combustion chamber' traps the air/fuel mixture. When burning takes place in a confined space such as this, the heat it produces becomes pressure which can be used mechanically to produce power. As the fuel burns and the piston goes down, the chamber becomes larger and larger, allowing for continued use of this pressure. The fact that the size of the chamber changes with the position of the piston also allows the air/fuel to be compressed, or packed into a confined space before it is burned, which has the effect of greatly increasing the amount of pressure made by the heat of burning, and makes the engine produce more power on less fuel. The variable size of the chamber also permits the engine to do its own breathing-to expel burnt gases and pull in fuel and fresh air (see the description of the four events in the operating cycle of a four-stroke engine below).

As you can see from how the engine produces power, leakage from the combustion chamber will have a tremendous effect on operating efficiency. One of the most important aspects of engine overhaul work involves repair or replacement of parts so that the combustion chamber will be as tightly sealed as possible.

FOUR-STROKE ENGINES

▶ See Figures 1, 2, 3 and 4

The entire series of four events that occur in order for an engine to operate may take place in one revolution of the crankshaft or it may take two revolutions of the crankshaft. The former is termed a two-cycle engine and the latter a four-cycle engine.

The four events that must occur in order for any internal combustion engine to operate are: intake, compression, expansion or power, and exhaust. When all of these take place in succession, this is considered one cycle.

In a four-cycle engine, the intake portion of the cycle takes place when the piston is traveling downward, creating a vacuum within the cylinder. Just as the piston starts to travel downward, a mechanically operated valve opens, allowing the fuel/air mixture to be drawn into the cylinder.

As the piston begins to travel upward, the valve closes and the fuel/air mixture becomes trapped in the cylinder. The piston travels upward and compresses the air/fuel mixture. This is the compression part of the cycle.

Just as the piston reaches the top of its stroke and starts back down the cylinder, an electric spark ignites the air/fuel

Fig. 1 Intake stroke of a 4-stroke engine

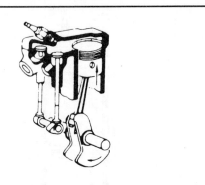

Fig. 2 Compression stroke of a 4-stroke engine

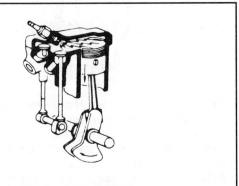

Fig. 3 Power stroke of a 4-stroke engine

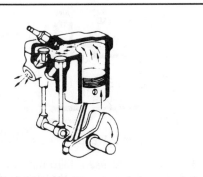

Fig. 4 Compression stroke of a 4-stroke engine

mixture and the resulting explosion and rapid expansion of the gases forces the piston downward in the cylinder. This is the expansion or power stroke of the cycle.

Just as the piston reaches the end of its downward travel on the compression stroke, another mechanically operated valve is opened. The next upward stroke of the piston forces the burned gases out the opened valve. This is the exhaust stroke.

When the piston reaches the top of the cylinder, thus ending the exhaust stroke, the exhaust valve is closed and the intake valve opened. The next downward stroke of the piston is the intake stroke that the whole series of events began with.

TWO-STROKE ENGINES

▶ See Figures 5, 6, 7 and 8

In a two-stroke engine, intake, compression, power, and exhaust take place in one downward stroke and one upward stroke of the piston. The spark plug fires every time the piston reaches the top of each stroke, not every other stroke as in a four-stroke engine. On some four-stroke engines, the magneto is operated by the crankshaft so that the spark plug actually fires once in every engine revolution. However, since the plug fires only into already burnt gases, this has no appreciable effect on engine operation.

The piston in a two-cycle engine is used as a sliding valve for the cylinder intake and exhaust ports. The crankcase is used as a pump in order to slightly compress new fuel/air mixture and force it through the cylinder. In some designs, the piston also opens and closes a third port connecting the intake tube and carburetor to the crankcase. The intake and exhaust ports are both open when the piston is at the end of its downward stroke, which is called bottom dead center or BDC. Since the exhaust port is opened to the outside atmospheric pressure, the exhaust gases, which are under a much higher pressure due to combustion, will escape to the outside through the exhaust port. After the pressure of the exhaust gases has been somewhat relieved, the piston uncovers the intake port, and a fresh charge of air/fuel is pumped through the intake port, forcing the exhaust gases out in front of it. We say that the air/fuel mixture is being pumped into the cylinder because it is under pressure caused by the downward movement of the piston. In the most common type of two-stroke engine, the air/fuel mixture is drawn through a one-way valve, known as a reed valve, and into the crankcase by the vacuum caused by upward movement of the piston. When the piston reaches top dead center or TDC and starts back down, the one-way valve in the crankcase is closed by its natural spring action and the building pressure caused by the downward movement of the piston. In the three port type of engine, the upward movement of the piston creates a vacuum in the crankcase. When the skirt (bottom) of the piston uncovers the third port as the piston nears the top of its travel, the vacuum in the crankcase draws in air/fuel mixture. As the piston descends, the third port is again covered by the piston skirt, and the crankcase is sealed for compression of the mixture. The air/fuel mixture is compressed until the piston moves past the intake port, thus allowing the compressed mixture to enter the cylinder. The piston starts its upward stroke, closing off the intake port and then the exhaust port, thus sealing the cylinder. The air/fuel mixture in the cylinder is compressed and at the same time, a fresh charge of air/fuel mixture is being drawn into the crankcase. When the piston reaches TDC, a spark ignites the compressed air/fuel mixture and the resulting expansion of the gases forces the piston back down the cylinder. The piston passes the exhaust port first, allowing the exhaust gases to begin to escape. The piston travels down the cylinder a little farther and past the intake port. The fresh air/fuel mixture in the crankcase, which was compressed by the downward stroke of the piston, is forced through the intake port and the whole cycle starts over.

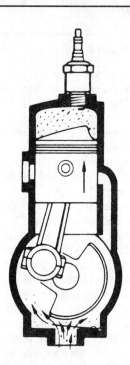

Fig. 5 The compression stroke of a 2-stroke engine. The intake port is open and the air/fuel mixture is entering the crankcase

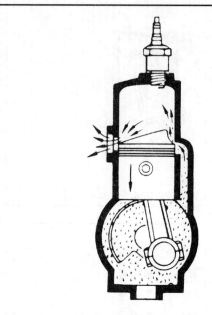

Fig. 7 The exhaust stroke pf a 2-stroke engine. The piston travels past the exhaust port, opening it, then, past the intake port opening that. As the exhaust gases flow out, the air/fuel mixture flows in, due to its being under pressure in the crankcase. the next stroke of the piston is the compression stroke and the series of events starts over again

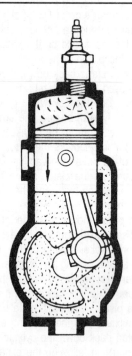

Fig. 6 The power stroke of a 2-stroke engine. The intake port is closed and, as the piston is being forced down by the expanding gases, the air/fuel mixture is being compressed

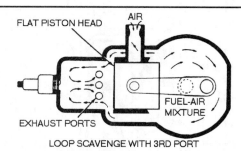

FLAT PISTON HEAD   AIR

FUEL-AIR MIXTURE

EXHAUST PORTS

LOOP SCAVENGE WITH 3RD PORT

Fig. 8 Some engines use a third port, which is opened and closed by the bottom of the piston, to control the admission of air/fuel mixture into the crankcase

Fuel Systems

The fuel system in a small engine consists of a fuel supply or storage vessel, a fuel pump, various fuel lines, and the carburetor. Fuel is stored in the fuel tank and pumped from the tank into the carburetor. Most small engines do not have an actual fuel pump. The carburetor receives gasoline by gravity feed or the fuel is drawn into the carburetor by venturi vacuum. The function of the carburetor is to mix the fuel with air in the proper proportions.

TYPES OF CARBURETORS

Small engine carburetors are categorized by the way in which the fuel is delivered to the carburetor fuel inlet passage in the venturi (fuel nozzle).

Plain Tube Carburetors
▶ See Figures 9 and 10

Carburetors used on engines that run at a constant speed, carrying the same load all of the time, can be relatively simple in design because they are only required to mix fuel and air at a constant ratio. Such is the case with plain tube carburetors.

All carburetors operate on the principle that a gas will flow from a large or wide volume passage through a smaller or narrower volume passage at an increased speed and a decreased pressure over that of the larger passage. This is called the venturi principle. The narrower passage in the carburetor where this acceleration takes place is therefore called the venturi.

A fuel inlet passage is placed at the carburetor venturi. Because of the reduced pressure (vacuum) and the high speed of the air rushing past at that point in the carburetor, the fuel is drawn out of the passage and atomized with the air.

The fuel inlet passage, in most carburetors, can regulate how much fuel is allowed to pass. This is done by a needle valve which consists of a tapered rod (needle) inserted in the opening (seat), partially blocking the flow of fuel out of the opening. The needle is movable in and out of the opening and, since it is tapered at that end, it can regulate the flow of fuel.

The choke of a plain tube carburetor is located before the venturi or on the atmospheric side. The purpose of the choke is to increase the vacuum within the carburetor during low cranking or starting speeds of the engine. During cranking speeds, there is not enough vacuum present to draw the fuel out through the inlet and into the carburetor to become mixed with the air. When the choke plate closes off the end of the carburetor open to the atmosphere, the vacuum condition in the venturi increases greatly, thus enabling the fuel to be drawn out of the opening.

In most cases the choke is manufactured with a small hole in it so that not all air is blocked off. When the engine starts, the choke is opened slightly to allow more air to pass. The choke usually is not opened all the way until the engine is warmed up and can operate on the leaner fuel/air mixture that comes into the engine when the choke is completely opened and not restricting the air flow at all. Thus, the choke also provides the richer mixtures (more fuel for the amount of air) required when the engine is cold.

Many of the chokes used on small engine carburetors are entirely automatic. In many cases, the choke is required to provide just sufficient vacuum to get the fuel moving and give rich mixture during cranking. Once the engine is running, the choke can be opened in just a few seconds without causing it to stall. It will run fairly well on a normal, lean mixture because the carburetor is close to the engine and puddling of fuel that occurs in engines with long manifolds while they are cold is no problem.

This type of choke is held shut by the pressure of a spring, and opened by the action of a diaphragm. Manifold vacuum is

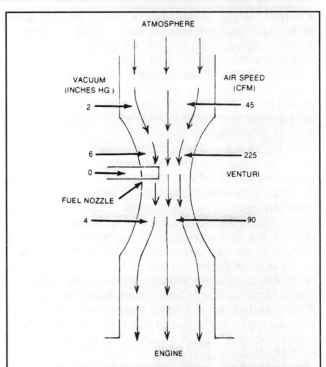

Fig. 9 All carburetors operate on the venturi priciple. The numbers represent hypothetical ratios of air speed and vacuum in relation to the venturi. Zero vacuum at the fuel nozzle is atmopheric pressure

fed to the side of the diaphragm on which the spring is located through a small orifice. There is no vacuum when the engine

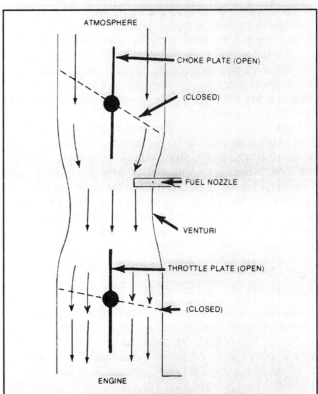

Fig. 10 The positioning of the choke and throttle plates in relation to the position of the venturi

is stopped, but upon starting, manifold vacuum gradually draws air out of the diaphragm chamber, and draws the diaphragm downward, against spring pressure. A rod linking the diaphragm to the choke then pulls it open.

This type of choke is also capable of keeping the mixture adequately rich during sudden increases in throttle opening; vacuum normally gets very low under these circumstances, and the choke tends to close slightly as the throttle opens, thus aiding fuel flow.

The throttle plate in a carburetor is installed on the engine side of the venturi. The purpose of the throttle plate is to regulate the flow of air/fuel mixture going into the engine. Thus the throttle plate regulates the speed and power output of an engine. By restricting the flow of air/fuel mixture going into the engine, the throttle plate is also restricting the combustion explosion and the energy created by the explosion.

These are the basic components required for any carburetor to operate on an engine. It is possible for such a carburetor to be installed on an engine and work. However, most engines operate at various speeds, under various load conditions, at different altitudes, and in a variety of temperatures. All of these variations require that the fuel/air mixture can be changed on command. In other words, the simple plain tube carburetor is not sufficient in most cases.

Suction Carburetor

Next to the plain tube carburetor, this is the simplest in design. With this type of carburetor, the fuel supply is located directly below the carburetor in a fuel tank. In fact, the carburetor and fuel tank are considered one assembly in most cases because a pipe extends from the carburetor venturi down into the fuel tank. When the engine is running, the partial vacuum in the venturi, and the relatively higher atmospheric pressure in the fuel tank, force the fuel up through the fuel pipe and into the carburetor venturi. There is a check ball located in the bottom of the pipe that prevents the fuel in the pipe from draining back into the fuel tank when the engine is shut down. In most engines there is a screen located at the end of the pipe to prevent dirt from entering and blocking the fuel nozzle.

Some suction carburetors are designed with an extra fuel inlet passage for when the engine is idling. This extra passage would be located on the engine side of the throttle plate. Since the vacuum condition in front of an idling engine might not be enough for the fuel to be drawn out at that point, the extra passage is installed behind the throttle plate where the vacuum is great enough to draw the fuel out. This extra fuel inlet passage would also have a regulating needle as does the main inlet.

Float Type Carburetors
▶ See Figure 12

The fuel tank used with float type carburetors is usually located on top of, or at least above, the level of the carburetor. Fuel is fed to the carburetor by gravity. If the fuel tank is located below the level of the carburetor, a fuel pump is employed to pump fuel to the carburetor. Fuel enters the carburetor through a valve and into a float bowl and, as it fills the bowl, a float rises on the surface of the fuel. The float is connected to the inlet valve and, as the level rises, a needle is inserted into the inlet valve and the flow coming into the float

bowl is checked. As the fuel is drawn into the main fuel nozzle in the venturi and the level in the bowl drops, the float drops and the needle opens the valve allowing more fuel to run into the bowl.

Although there are many different float carburetors, all operate in this manner.

Diaphragm Type Carburetors
▶ See Figure 13

Fuel is delivered to the diaphragm type carburetor in the same manner as to the float type carburetor.

A flexible diaphragm operates the fuel inlet valve, hence the description 'diaphragm carburetor.'

Atmospheric pressure is maintained on the under side of the diaphragm by a vent hole in the bottom of the carburetor. The other side of the diaphragm is acted upon by the varying vacuum conditions in the carburetor.

When the vacuum condition in the carburetor is increased by the opening of the throttle plate, and the demand for an increased flow of fuel, the center diaphragm is bellowed upward by the increased vacuum. The center of the diaphragm is connected by a lever to the fuel inlet valve needle. The needle is dropped down away from the inlet valve and fuel is allowed to drop in. When the throttle plate is closed and the need for fuel is reduced, the vacuum condition also decreases and the diaphragm is returned by a spring located on the atmospheric side to its normal flattened position. This closes the fuel inlet valve by raising the needle up into position against its seat.

This type of carburetor also has an idle orifice positioned behind the throttle plate to compensate for the low vacuum condition in front of the throttle plate during idle speeds.

Some diaphragm type carburetors incorporate an integral fuel pump. It consists of a second diaphragm having fuel on one side, and exposed to the pressure fluctuations of the crankcase or the intake manifold on the other side. The vibration of this diaphragm, in conjunction with the action of check valves in the passages leading to and from the fuel side of the diaphragm chamber work together to pump fuel, under pressure, to the fuel inlet valve.

REED VALVES

▶ See Figure 14

Reed valves are used on those two stroke engines in which the intake tube leading from the carburetor empties right into the crankcase. The reed serves as a kind of check valve, so that air/fuel cannot be forced back out of the crankcase. When the piston begins to rise, a slight vacuum is created under it, sucking the reed open against its spring pressure, as shown in the illustration. When the piston reaches the top of its travel, and the pressure in the crankcase gets near to outside (atmospheric) pressure, the valve's spring action pulls it flat against the intake opening (the reed is really a kind of flat spring). Then, as the piston descends, the build-up in pressure in the crankcase, in combination with the slight amount of oil that gets onto the reed, seals the intake very tightly, forcing practically the full charge into the cylinder.

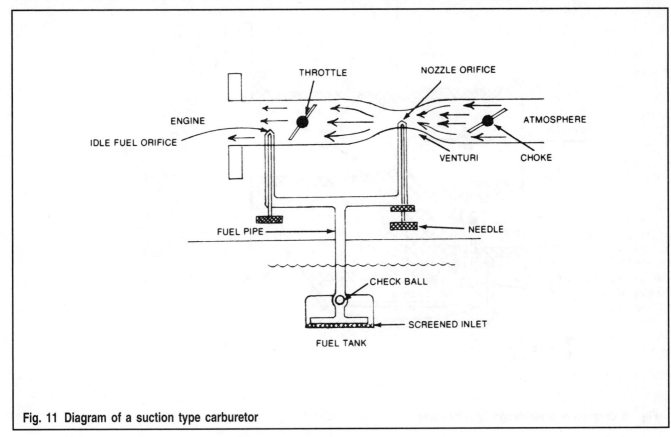

Fig. 11 Diagram of a suction type carburetor

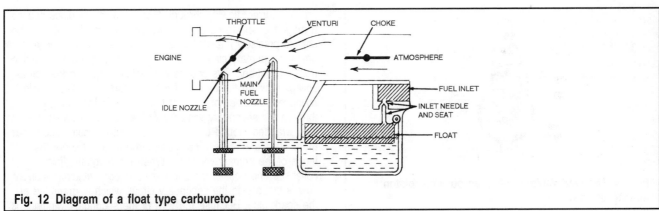

Fig. 12 Diagram of a float type carburetor

Ignition Systems

The function of an ignition system is to provide the electrical spark that ignites the air/fuel mixture in the cylinder at precisely the correct time.

There are three types of ignition systems used on small engines: either a battery ignition, a magneto ignition or a breakerless ignition system.

The battery ignition, as the name implies, gets its initial electrical charge from a storage battery. The magneto ignition actually generates its own electricity, thus eliminating the need for a battery as far as ignition is concerned.

Battery and magneto ignition systems are similar in one respect. Both systems take a relatively small amount of voltage, such as 12 volts in the case of a battery ignition, then

step up that to an extremely high voltage, in some systems as high as 20,000 volts.

The breakerless ignition system operates on the same general principle as the magneto system but does not use breaker points and conventional ignition condenser to time the spark. A trigger module containing solid state electronics performs the same function as the breaker points.

BASIC BATTERY IGNITION SYSTEM

▶ **See Figure 15**

Battery ignition systems are found mostly on larger one cylinder engines, such as those installed on lawn tractors, larger pumps and generators.

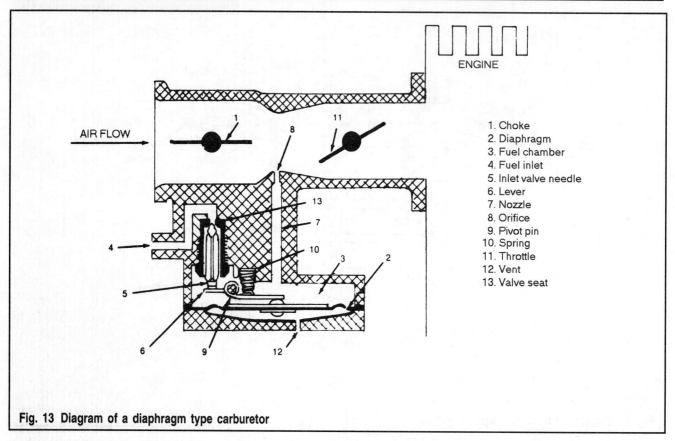

AIR FLOW

ENGINE

1. Choke
2. Diaphragm
3. Fuel chamber
4. Fuel inlet
5. Inlet valve needle
6. Lever
7. Nozzle
8. Orifice
9. Pivot pin
10. Spring
11. Throttle
12. Vent
13. Valve seat

Fig. 13 Diagram of a diaphragm type carburetor

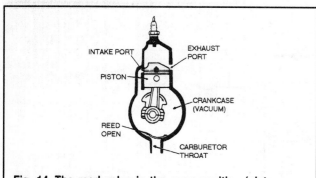

Fig. 14 The reed valve in the open position (piston moving upward)

As previously stated, the initial electrical charge comes from a storage battery. The entire ignition system is grounded so that current will flow from the battery throughout the primary circuit.

The ignition system is composed of two segments, the primary circuit and the secondary circuit. Current from the battery flows through the primary circuit and the current that fires the spark plug flows through the secondary circuit. The current in the primary circuit actually creates the secondary current magnetically.

When the engine is running, current flows from the battery, through the breaker points, and on to the ignition coil. The current then flows through the primary windings of the coil which are wrapped around a soft iron core and grounded. The primary current flowing through these primary windings causes a magnetic field to be created around the soft iron core of the coil. At precisely the right time, the breaker points open and break the primary circuit, cutting off the flow of current coming from the battery. This causes the magnetic field in the coil to collapse toward the center of the soft iron core. As the field collapses, the lines of force of the magnetic field must pass through the secondary windings of the coil. These windings are also wrapped around the soft iron core of the coil, except that there are many more windings and the wire is thinner than those of the primary windings. When the magnetic field collapses and passes through the secondary windings, current flow is induced in the secondary circuit which is grounded at the spark plug. Because the wire of the secondary windings is smaller and there are many more coils around the soft iron core, the current created or induced in the secondary side of the ignition is of much greater voltage than the primary side. This current in the secondary circuit flows toward the grounded end of the circuit which is the spark plug. The current flows down the center of the spark plug to the center electrode. In order to reach the ground, it must jump across a gap to the grounded electrode. When it does, a spark is created and it is this spark that ignites the air/fuel mixture in the cylinder.

MAGNETO TYPE IGNITION SYSTEMS

Magneto ignition systems create the initial primary circuit current, thereby eliminating the need for ignition batteries.

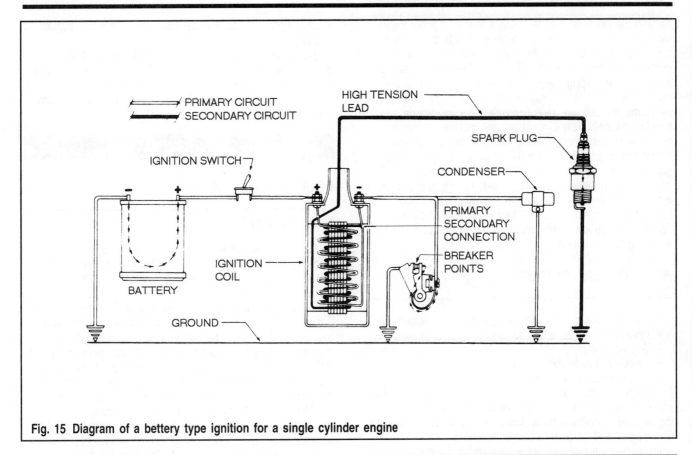

Fig. 15 Diagram of a bettery type ignition for a single cylinder engine

Flywheel Type Magnetos

▶ See Figure 16

On this type of magneto, the flywheel of the engine carries the permanent magnets that are used to create the primary current.

The magnets are arranged so that about 1/3 of the area enclosed by the flywheel is a magnetic field. In the center of the flywheel is a three pronged coil, the three prongs representing the core. The center prong has the primary and secondary windings and is set up just like a battery ignition coil.

As the magnets in the flywheel pass the two outside prongs, the primary current is induced in the coil and a magnetic field is set up around the center prong. At the point when the primary current is at its strongest which is also the time when the secondary current is needed at the spark plug, the breaker points interrupt the primary current, causing the magnetic field to collapse through the secondary windings. This induces secondary current which flows to the grounded spark plug.

Unit Type Magnetos

▶ See Figures 17, 18, 19, 20 and 21

Unit type magnetos operate in the same way as flywheel magnetos.

Permanent magnets are rotated through a mechanical connection to the crankshaft of the engine. The rotating magnets create the primary current which is routed through the breaker points and on to windings around a soft iron core which is also wrapped by the secondary windings. At precisely the right moment, the points interrupt the primary current,

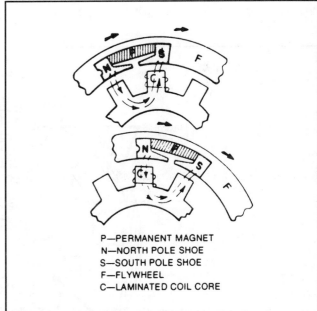

P—PERMANENT MAGNET
N—NORTH POLE SHOE
S—SOUTH POLE SHOE
F—FLYWHEEL
C—LAMINATED COIL CORE

Fig. 16 A cutaway view of a flywheel used in a flywheel magneto. The magnets are arranged so that there is a magnetic field covering about 1/3 of the area enclosed by the flywheel. On the flywheel magnetos, where the coil and core are mounted on the outside of the flywheel, the magnets would be arranged on the outside of the flywheel

causing the magnetic field to collapse through the secondary windings and inducing the secondary current to the spark plug.

Because the magnets in the unit magneto are driven by the crankshaft of the engine through a gear mechanism, starting is a problem. At starting speeds, the magnets in the magneto cannot be rotated fast enough to create a primary current. To overcome this difficulty, unit magnetos have an impulse coupling on their shaft which drives the magnets. When the engine is turned over at starting speed, a catch engages a coil spring that is wound up in much the same manner as those that propel wind-up toys. As the shaft rotates further, it releases the spring. The spring unwinds rapidly, spinning the magnets fast enough to cause the primary current to be induced in the primary circuit. When the engine starts to run, centrifugal force keeps the catch in the impulse coupling from engaging the wind-up spring.

BREAKERLESS IGNITION

The breakerless system consists of four major components:
- Ignition winding on alternator stator
- Trigger module
- Ignition coil assembly
- Flywheel-mounted trigger

The ignition winding is separate from other windings on the alternator stator. It functions like the magneto winding. The trigger module contains three diodes, a resistor, a sensing coil and magnet and an SCR, a sort of electronic switch. The ignition coil assembly includes a capacitor and a pulse transformer that serves the same purpose as the ignition coil in other systems. The flywheel has a projection that triggers ignition.

In some applications a 22 ohm, 1/2 watt resistor has been placed between the key switch and the ignition coil. This has been added to prevent current feedback through a dirty or wet switch. This feedback, if not held in check by a resistor, can damage the trigger unit.

Lubrication Systems

FOUR-STROKE ENGINES

▶ **See Figure 22**

Most small four stroke engines are lubricated by the splash system. All vital moving parts are splashed with lubricating oil that is stored in the crankcase. The connecting rod bearing gap usually has an arm extending down into the area in which the oil lies. As the crankshaft turns, the arm, commonly called a dipper, splashes oil up onto the cylinder walls, crankshaft bearings and camshaft bearings. In some cases, the connecting rod bearing cap screws are locked in place by lock plates which double as oil dippers. When the lock plate's tabs are bent up beside the cap screws, they extend a little past the tops of the screws and, when they enter the oil, they provide a sufficient splash to lubricate the engine.

In larger four stroke engines, a pressure system is used to lubricate the engine. This type of lubrication system pumps the oil through special passages in the block and to the components to be lubricated. In some cases, the camshaft and crankshaft have their center drilled so that oil can be pumped

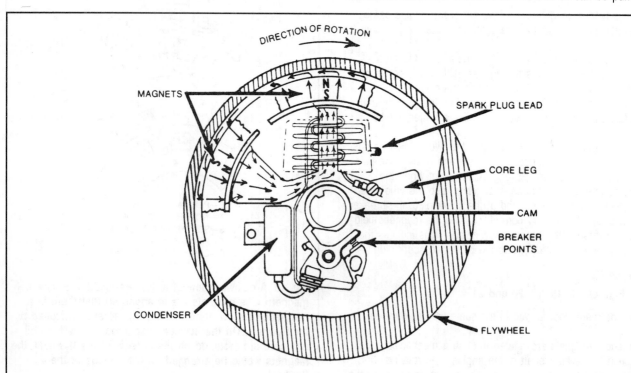

Fig. 17 The three legs of the core pass through the magnetic field created by the rotating magnets. In this position, the lines of force are concentrated in the left and center core legs and are interlocking the coil windings

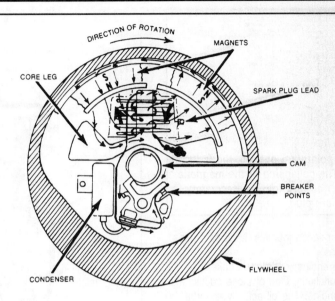

Fig. 18 The flywheel has turned to the point where the lines of force of the permanent magnets are being withdrawn from the left and center cores and are being attracted by the center and right cores. Since the center core is both drawing and attracting the lines of force, the lines are cutting up one side and down the other (indicated by the heavy black lines). The breaker points are now closed and a current is now indiced in the primary circuit by the lines of force cutting up and down the center core leg

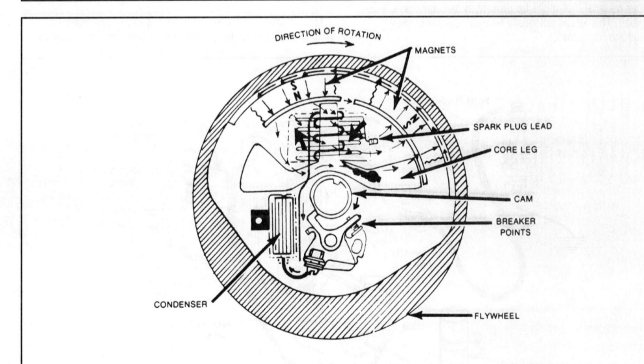

Fig. 19 With the induced current in the primary windings, a magnetic field is created around the center core leg (coil). When the field is created around the center leg, the points are opened and the condenser begins to absorb the reverse flow of current

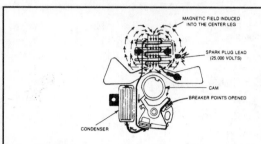

Fig. 20 When the breaker points are opened, the induced magnetic field collapses. The collapsing of the magnetic field induces the secondary current into the secondary windings leading to the spark plug

through the center and to the bearing journals. Most oil pumps in small engines are gear types.

In addition to lubricating the engine, oil in four stroke engines has other important functions. One of those additional functions is to help cool the engine. The oil actually absorbs heat from high temperature areas and dissipates it throughout other parts of the engine that are not directly exposed to the very high temperatures of combustion.

Oil in the crankcase also functions as a sealer. It helps to seal off the combustion chamber (top of the piston) from the crankcase, thus maintaining compression which is vital to satisfactory engine operation, while at the same time lubricating the cylinder walls and piston rings.

Engine oil also keeps harmful bits of dust, metal, carbon, or any other material that might be present in the crankcase, in suspension. This keeps potentially harmful abrasives away from vital moving parts.

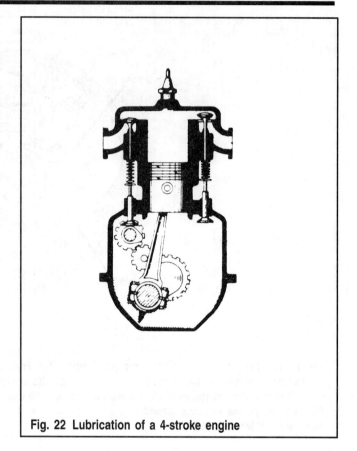

Fig. 22 Lubrication of a 4-stroke engine

Most manufacturers recommend a good grade, medium weight detergent oil for their engines. The detergent properties

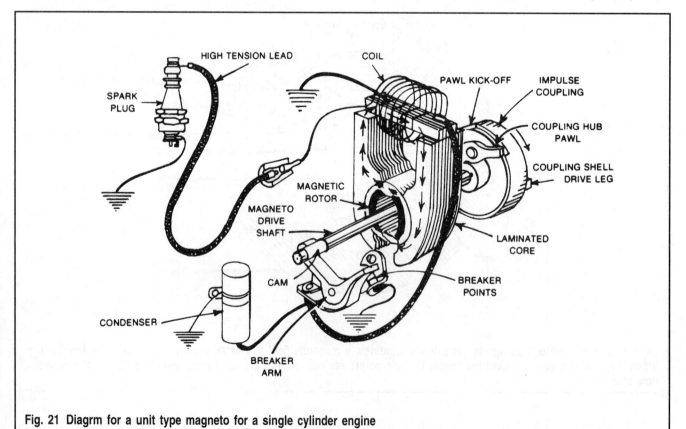

Fig. 21 Diagrm for a unit type magneto for a single cylinder engine

of engine oil do not mean that the oil is capable of cleaning away dirt or sludge deposits already present in the engine. It means that the oil will help fight the formation of such deposits. In other words, the oil keeps the dirt in suspension.

A certain amount of combustion leaks past the piston rings and into the crankcase. Raw gas, carbon, products of combustion, and other undesirable material still manage to make their way into the lubricating oil. The oil becomes diluted by the gas and loses its cooling properties. When it is filled with abrasives, it loses its detergent properties due to chemical reactions, constant heat, and pollutants. The oil can then actually become harmful to the engine. Thus, one can see the need for keeping oil as clean as possible by changing it frequently.

TWO-STROKE ENGINES

▶ **See Figure 23**

Since the two-stroke engine uses its crankcase to compress the air/fuel mixture so that it can be forced up into the combustion chamber, it cannot also be used as an oil sump. The oil would be splashed around and forced up into the combustion chamber, resulting in the loss of great amounts of lubricating oil while at the same time 'contaminating' the air/fuel mixture. In fact the air/fuel mixture would be so filled with oil that it would not ignite.

A two stroke engine is lubricated by mixing the lubricating oil in with the fuel. A mixture of fuel and lubricating oil is sucked into the crankcase and, while it is being compressed by the downward movement of the piston, enough oil attaches itself to the moving parts of the engine to sufficiently lubricate the engine. The rest of the oil is burned along with the fuel/air mixture. As can well be imagined, the mixture ratio of fuel and oil is critical to a two stroke engine. Too much oil will foul the spark plug and too little oil will not lubricate the engine sufficiently, causing excessive wear and possibly engine seizure. Follow the manufacturer's recommendations closely. Most recommend that detergent oils not be used in two-stroke engines.

Governors

▶ **See Figures 24 and 25**

The governor is a control that opens and closes the engine throttle to maintain its speed when the load on the engine changes. Consider an engine which is operating a generator. The generator must operate within a very narrow speed range in order to provide the stable output current electrical appliances require. Yet, as appliances in the circuit are turned on and off, the load on the generator changes, which tends to either allow the engine to speed up or force it to slow down.

The governor measures engine speed and responds accordingly. As the speed drops, the throttle is opened. As speed increases the throttle is closed. The governor has a 'speed droop,' which means that it does not try to keep the engine at exactly the same speed, but allows a gradual drop in speed as load increases. If a governor has a speed droop of 100 rpm, and is set so that the engine runs at 3650 rpm with no load, loading the engine will cause the throttle to be opened to about the half-way point at 3600 rpm, and all the way at 3550 rpm.

The governor is basically just a spring that pulls the throttle open while some other force, which varies with engine speed, tries to close it. The simplest method of providing the closing force is the air vane. The vane is hinged at the leading edge. When it turns on the hinge, it closes the throttle through a simple wire link. The vane is housed inside the cooling system, directly in the path of cooling air. When engine speed increases, air pressure on the vane increases. At a certain speed, the force on the vane is great enough to overcome the tension of the spring and begin stretching the spring, closing the throttle. The point at which the spring has stretched enough to nearly close the throttle, and keep the engine speed steady without any load, is somewhat above the point where the vane just begins to stretch the spring. This range is the speed droop. It accounts for the fact that as load is applied to an engine, it finds a slightly lower speed at which it operates.

The flyweight type governor works very similarly. Here, the increasing closing force comes from hinged weights which spin in a circle, driven through gearing from the engine's crankshaft. As engine speed increases, centrifugal force makes the weights turn outward on their hinges and force the throttle closed via a collar which the bottoms of the weights work against.

The spring and flyweights or wind vane are connected via a lever which contains various holes in which the end of the spring can be installed. If the governor is too sensitive (speed droop too small), it will hunt back and forth, causing the throttle to jump violently from full open to idle position and back. If the governor is not sensitive enough (speed drop too great), the engine will slow excessively before it adjusts to the load. The spring is moved from hole to hole in order to

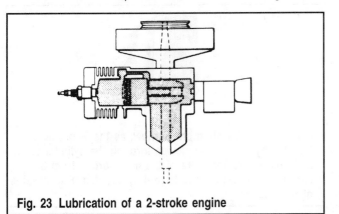

Fig. 23 Lubrication of a 2-stroke engine

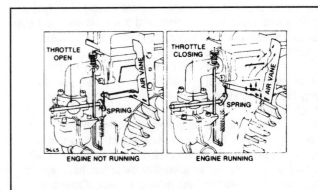

Fig. 24 Operation of an air vane governor

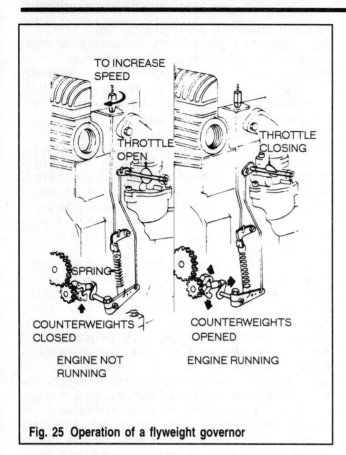

Fig. 25 Operation of a flyweight governor

change this sensitivity. If the spring is hooked near the shaft the lever turns on, a small change in force from the flyweights or air vane will stretch the spring a great deal, making the governor very sensitive. If the spring is hooked to a hole farther away from the pivot point of the lever, a large change in engine speed will be required to stretch it, making the governor slow to respond. The governor speed setting is changed by simply stretching the spring — pulling the end opposite the point where it fits into the lever away from the lever, utilizing any of various kinds of mechanical linkages.

Cooling Systems

▶ **See Figure 26**

Most small engines are cooled directly by the outside air because this type of system is much less expensive to manufacture, is more easily maintained and repaired, and; if the engine is operating correctly, is more reliable. This type of cooling also saves engine weight, and this is an important side benefit, as many small engines are used to operate hand carried or hand propelled machinery.

Air is a much less efficient coolant than water. It carries much less heat, so that a given volume of air will rise to a very high temperature without actually carrying away very much heat. And, it readily forms stagnant areas around solid objects. Water carries a lot of heat without much temperature rise; it grips tightly to whatever metal object it is cooling, and can easily be kept in uniform motion. For these reasons, air cooled engines are not as tolerant of inefficient, heat producing operation. They must be kept running at peak efficiency if maximum component life is to be realized.

The biggest key to making air cooling work is finning. The parts of the cylinder and cylinder head which are exposed to burning gases inside are cast in such a way as to very greatly increase the metal surface exposed to air. At a number of points, the metal of a cylinder, for example, is forced outward from the main structure into a thin sheet or fin that may increase the surface available to air for cooling as much as ten times. While the fins at first might seem to represent excess weight, they actually form an integral portion of the structure of the engine, increasing its rigidity, and decreasing the thickness required in the main structure. While the outer ends of the fins seem to be very far from the source of the heat they in fact carry lots of heat to the cooling air because metal parts are excellent heat conductors, moving energy from hot to cool portions in a process that is a little like the flow of electricity.

Since the air heats fast and can easily become stagnant, forming a blanket near the metal, the second secret of air cooling is high velocity. The air must be guided in a precise manner around every part of the engine coming into contact with the burning fuel. This is done by shrouding the outer edges of the fins where necessary with light metal ducts. Use of a very powerful, centrifugal fan which slings the air outward and through the engine (or draws the air through the engine at high speed) ensures that there will be enough air moving along all the hot parts to break up any stagnant pockets and prevent any of the air from being there long enough to reach excessively high temperatures.

Since air cooling is more sensitive than conventional water cooling systems, always observe the following precautions when operating any air-cooled engine:

1. You have to have a sharper eye with air cooled engines. Watch for sluggish operation and excessive oil consumption that may mean things are getting too hot.

2. Never overload the engine. If the engine runs full throttle below governed rpm, it will be running short of cooling air, and will usually overheat.

3. Use oil of the proper viscosity. Too low a viscosity (or even very dirty oil) will cause excess friction and overheating.

4. Keep all cooling system ducts, shrouds, or seals in top shape — replace any that become loose or even dented. Replace parts whose fins are broken off.

5. Keep fins clean of dust, oil and debris.

6. Keep the cooling air blower clean, and watch carefully for even partial clogging of the air intake screen. If any of the blower's blades break off, replace the blower or flywheel.

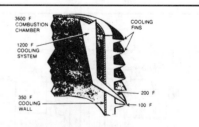

Fig. 26 The illustration shows how, as the ⅓ of energy handled by the engine passes through the cylinder walls and fins, the temperature of the metal drops from 350 deg. F on the cylinder wall, to only 100 deg. F at the tips of the fins

7. Some air cooled engines use thermostatically or throttle controlled air vane to restrict cooling air at low temperatures or under light loads; make sure it works freely and opens when it's supposed to.

8. Keep the engine in good tune. Lean fuel mixtures, late ignition timing, engine knock, or any other combustion problem will severely overheat the engine.

Battery

FLUID LEVEL (EXCEPT MAINTENANCE FREE BATTERIES)

▶ See Figures 27 and 28

Check the battery electrolyte level at least once a month, or more often in hot weather or during periods of extended operation. The level can be checked through the case on translucent polypropylene batteries; the cell caps must be removed on other models. The electrolyte level in each cell should be kept filled to the split ring inside, or the line marked on the outside of the case.

If the level is low, add only distilled water, or colorless, odorless drinking water, through the opening until the level is correct. Each cell is completely separate from the others, so each must be checked and filled individually.

If water is added in freezing weather, the engine should be operated to allow the water to mix with the electrolyte. Otherwise, the battery could freeze.

SPECIFIC GRAVITY (EXCEPT MAINTENANCE FREE BATTERIES)

At least once a year, check the specific gravity of the battery. It should be between 1.20 and 1.26 at room temperature.

The specific gravity can be checked with the use of an hydrometer, an inexpensive instrument available from many sources, including auto parts stores. The hydrometer has a squeeze bulb at one end and nozzle at the other. Battery electrolyte is sucked into the hydrometer until the float is lifted from its seat. The specific gravity is then read by noting the position of the float. Generally, if after charging, the specific gravity between any two cells varies more than 50 points (0.050), the battery is bad and should be replaced.

It is not possible to check the specific gravity in this manner on sealed (maintenance free) batteries. Instead, the indicator built into the top of the case must be relied on to display any signs of battery deterioration.

Cables and Clamps

▶ See Figures 29, 30, 31 and 32

Once a year, the battery terminals and the cable clamps should be cleaned.

1. Loosen the clamps and remove the cables, negative cable first. On batteries with posts on top, the use of a puller specially made for the purpose is recommended. These are inexpensive, and available in auto parts stores.

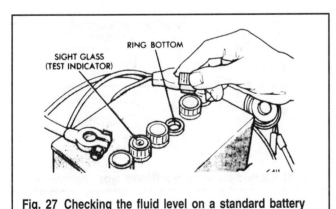

Fig. 27 Checking the fluid level on a standard battery

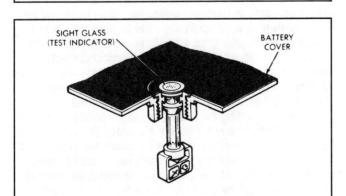

Fig. 28 Test indicator on a maintenance free battery

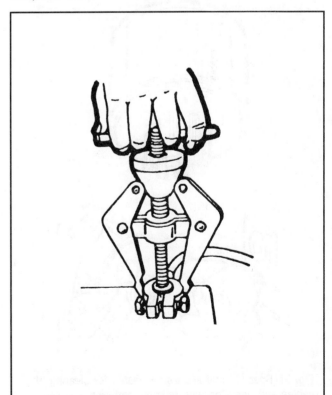

Fig. 29 Special pullers are available to remove the cable clamps

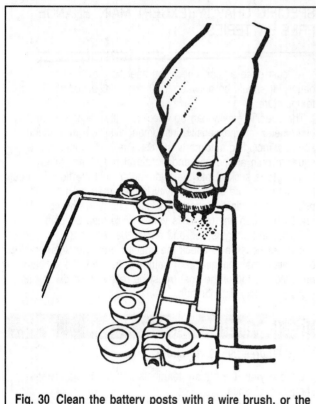

Fig. 30 Clean the battery posts with a wire brush, or the tool shown

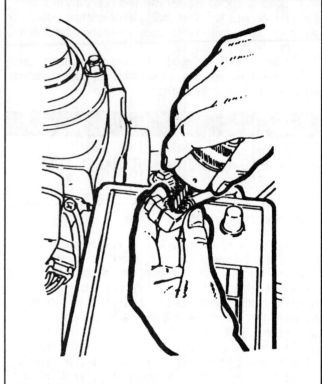

Fig. 32 Clean the inside of the clamp with a wire brush or the tool shown

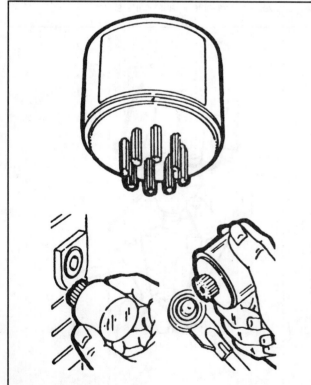

Fig. 31 Special tools are also available for cleaning the cables and posts on side terminal batteries

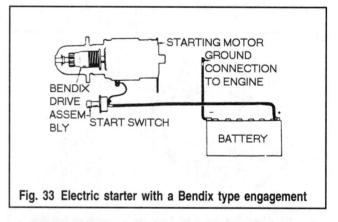

Fig. 33 Electric starter with a Bendix type engagement

2. Clean the cable clamps and the battery terminal with a wire brush, until all corrosion, grease, etc. is removed and the metal is shiny. It is especially important to clean the inside of the clamp thoroughly, since a small deposit of foreign material or oxidation there will prevent a sound electrical connection and inhibit either starting or charging. Special tools are available for cleaning these parts, one type for conventional batteries and another type for side terminal batteries.

3. Before installing the cables, loosen the battery hold-down clamp or strap, remove the battery and check the battery tray. Clear it of any debris, and check it for soundness. Rust should be wire brushed away, and the metal given a coat of anti-rust paint. Replace the battery and tighten the hold-down clamp or strap securely, but be careful not to overtighten, which will crack the battery case.

4. After the clamps and terminals are clean, install the cables, positive cable first; do not hammer on the clamps to install. Tighten the clamps securely, but do not distort them. Give the clamps and terminals a thin external coat of grease after installation, to retard corrosion.

5. Check the cables at the same time that the terminals are cleaned. If the cable insulation is cracked or broken, or if the ends are frayed, the cable should be replaced with a new cable of the same length and gauge.

Starting Systems

ELECTRIC STARTERS

▶ **See Figures 33 and 34**

Many small engines utilize direct current electric motors for starting purposes. These are run off a standard battery, and the entire system is quite similar to ordinary automotive equipment of this type, except for its smaller size. Direct current motors consist of a stationary coil of wire which is mounted on the inside of the housing, and a rotating coil, which is mounted on the shaft, and is surrounded by the 'stator,' or stationary coil. Direct current from the battery passes through both coils in series, being carried to the 'armature' or rotating coil through brushes. The brushes are mounted inside the housing and rub against commutator rings, which spin with the rotor shaft. The brushes are sprung against the rings so they stay in constant contact. The most common problems with these motors occur when the brushes wear and lose their spring tension.

Some starter motors turn the engine over by engaging a pinion — a gear on the end of the shaft — with a toothed flywheel. Some starter pinions are moved forward into mesh with the flywheel through the action of a magnetic solenoid, while others employ the Bendix drive. The Bendix type pinion spins forward, against the tension of a spring, via the action of spiral cut grooves in the armature shaft. When the engine starts, the pinion is spun back toward the starter, and then held by the spring. In the solenoid design, an overrunning clutch allows the pinion to spin faster than the starter armature, thus keeping the engine from over-speeding it as it starts. When operating engines with this type of drive, allow the starter to disengage as soon as the engine fires to avoid overheating the overrunning clutch. In both types, but especially the Bendix type, poor engagement may result from poor lubrication of the spiral grooves which guide the pinion along the armature shaft.

Some electric starters drive through belts and double as generators when the engine starts. This design incorporates a relay that switches the polarity (direction of current flow) through the unit when the engine comes up to speed. Because of the large amount of torque required to turn the engine over, it is important to keep the drive belt under proper tension, and to replace it if it becomes glazed (which can cause it to slip).

RECOIL STARTERS

▶ **See Figure 35**

Recoil starters employ a rope or cable wound around a pulley, but automatically rewind the rope when it is released. The rewind force is provided by a large clock type spring which connects the pulley to the engine housing. When the rope is pulled, a cam action engages dogs which cause the engine flywheel to turn with the pulley. When the engine starts, its motion disengages the dogs.

When working on recoil starters, it is absolutely necessary to completely release the tension of the spring before disassembling the unit. If spring tension is released suddenly or unexpectedly, the force released is tremendous, and is very likely to cause personal injury.

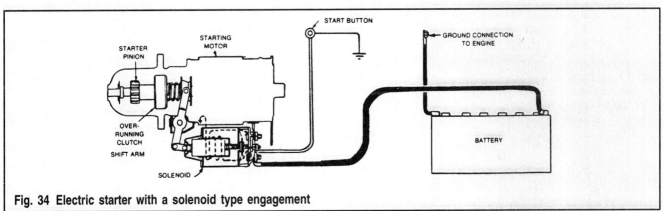

Fig. 34 Electric starter with a solenoid type engagement

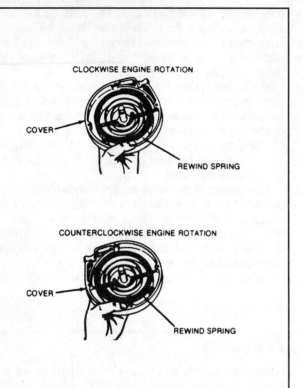

Fig. 35 A rewind spring is mounted inside the cover of most recoil starters

2

CARE AND MAINTENANCE

ROUTINE CARE, MAINTENANCE, STORAGE AND OPERATING PRECAUTIONS

Operating Precautions

GENERAL

1. Before starting the engine, make sure it is full of the proper grade of lubricating oil. Some sort of record of the viscosity of the oil in use should be kept so that the proper grade may be added when necessary and the oil may be changed to keep the viscosity in conformance with recommendations for the outside air temperature range that is prevalent. If the engine is run in hot weather with an oil of too low a viscosity in the crankcase, very serious damage can result. The same sort of damage (from overheating) can occur if the oil level is too low. Remember that all engines must consume some oil to run properly and that, even if the engine is in continuous service, it should be stopped every few hours for an oil level check, and the crankcase refilled.

In the case of 2-stroke engines, a recommended 2-stroke oil must be used, and it must be mixed in the proportions recommended for that engine by the manufacturer. If several 2-strokes using different fuel mixes are in use at the same location, each fuel container should be marked so that only the mix recommended for each engine is used in it. Do not simply use a pre-mixed batch of fuel — there is no standard mix. Make sure what you're using is in the exact proportion recommended by the manufacturer. Also, fuel and oil must be uniformly mixed. Usually, this is done by filling the container about 25% of the way with gas, pouring in the oil, and then completing the filling job. After this, the container should be shaken vigorously with the cap applied tightly to complete the mixing process. When using fuel that has been setting for several weeks, shake up the can before refilling the engine fuel tank to ensure uniform mixing.

2. Check that all cooling system air inlets are clear before starting the engine, and check occasionally to make sure nothing has clogged the air intake, especially if a screen is used. Check the condition of fins, fan blades, and thermostatic airflow controls as described above in the description of air cooling systems, periodically.

3. Watch the engine during operation to ensure it is running smoothly at governed speed. If operation is sluggish, stop it immediately and check for possible overloading, overheating, or lack of oil or two stroke lubrication. If the engine lacks power and overheats, check fuel air mixture adjustments (especially on 2-stroke engines) and ignition timing.

4. Check for exhaust smoke during operation. If the engine smokes, check first for excessive oil consumption. If this is not due to running too hot or with the crankcase too full, the crankcase breather or engine must be repaired. Smoke can also occur due to a very rich mixture. If necessary, service the air cleaner and check for any restrictions in the air intake. Make sure the choke is fully opened if it is a manual design, and that it is opening properly if it is an automatic design. Adjust the carburetor mixture, if necessary. Smoke may also accompany severe engine knock due to very advanced ignition timing. Check the timing if there is knock.

5. Make sure to use only clean, fresh fuel. Fuel can begin deteriorating only a month after it is purchased. Fuel containing water or rust or other dirt should be discarded. Use leaded or unleaded fuel as per manufacturer's recommendations. Make sure the fuel also meets octane requirements.

6. Set the throttle only slightly above idle in starting (unless this conflicts with specific recommendations for starting), and idle the engine for several minutes before putting it to work. This will allow oil to work. This will allow oil to become thin enough to reach all moving parts of the engine before they begin carrying too much load. It's also a good idea to idle the engine for a minute or two before shutdown, to cool the hottest parts more gradually.

7. Make sure the engine remains tightly mounted, as vibration can severely damage the engine or other parts, or cause potentially dangerous mechanical damage.

SAFETY PRECAUTIONS

1. Never refuel the engine unless it is stopped. Hot exhaust system parts or an electrical spark could ignite the fuel. Also, avoid spillage of fuel, especially when the engine is hot. If fuel spills, make sure it is completely removed before starting the engine. Check the engine occasionally for fuel leaks and repair them immediately. Also, keep ignition high tension wiring in top shape. Brittle insulation can crack, causing a spark which could ignite spilled fuel.

2. Keep all sources of ignition away from batteries, especially when they are being rapidly charged or caps are off. When installing jumper cables between batteries, remove caps and place a rag over open vent holes. Make positive connections first (make sure you connect positive to positive); make negative connections by first connecting the cable to the negative side of the good battery, and making the final connection to a bare spot on the frame of the piece of equipment with the dead battery.

3. Be careful not to come in contact with output terminals on an electric generator, which are often located externally. Remember that metal tools are excellent conductors, and that they too must be kept away from electrical terminals. Even if you do not become exposed to electrical shock via tools, they can serve as a conductor and become hot enough to cause serious burns. This can happen if they touch positive and negative terminals of a battery.

Many electric generators are used only in case of a failure in commercial power. In these cases, they are often connected into a circuit through a transfer switch. Remember that the transfer switch is energized even when the generator is not running. Remember, too, that the output terminals of a generator are hot once the generator is running and electrically excited — it does not need to be carrying a load.

4. Make sure that exhaust gases are properly vented. If you are working near an engine which is enclosed, leaks in the exhaust system can prove dangerous even when the bulk of the exhaust is being carried to the outside air.

5. If your small engine is utilized in a marine engine compartment, remember that the compartment must be

thoroughly aired out before starting the engine, or explosion of accumulated fumes may result.

6. Remember that many pieces of equipment driven by small engines are too heavy to be safely carried by a single person. Often, even though the total weight may be within reasonable limits, you may strain yourself because of the difficulty of handling the unit's bulk. Get help!

7. Keep the operating area clean. It should be wiped clean of fluids which could catch fire, and kept free of debris which might be drawn into a fan or blower and thrown out at high speed. This, of course, includes removing debris from a lawn which is to be moved with a rotary mower or blown clean by a rotary blower.

8. When working on any engine driven accessory, disconnect the spark plug wire to avoid possible accidental starting of the motor if it should be turned over.

9. Keep all safety guards tightly in place, and replace them should they become damaged. Always be fully aware of all moving parts and the possibility of coming into contact with them even if guards are in place.

10. Keep governors in good operating condition, and do not reset for a speed higher than the recommendations of the manufacturer of the equipment.

11. Periodically tighten mounting bolts of both the engine and any machinery driven off it, especially that which turns at crankshaft rpm, such as rotary blades.

HOT AND COLD WEATHER OPERATION

There are a few things that must be done when a small engine is to be operated in hot or cold weather, that is, when temperatures are either below 30°F (-1°C), or above 75°F (24°C).

Hot Weather

1. Keep the cooling fins of the engine block clean and free of all obstructions. Remove all dirt, built up oil and grease, flaking paint and grass.

2. Air should be able to flow to and from the engine with no obstructions. Keep all fairing and cover openings free from obstructions.

3. During hot weather service, heavier weight oil should be used in the crankcase. Follow the manufacturer's recommendations as to the heaviest weight oil allowed in the crankcase.

4. Check the oil level each time the fuel tank is filled. An engine will use more oil in extremely hot weather.

5. Check the battery water level more frequently since, in hot weather, the water in the battery will evaporate more quickly.

6. Be on the lookout for vapor lock, which occurs within the carburetor.

7. Use regular grade gasoline rather than premium.

8. Use unleaded gasoline if possible.

9. The most important thing to remember is to keep the engine as clean as possible. Blow it off with compressed air or wash it as often as possible.

❋❋WARNING

Wash the engine only after it has had sufficient time to cool down to ambient temperatures. Avoid getting water in or even near the carburetor intake opening.

Cold Weather

1. A lightweight oil should be installed in the crankcase when operating in cold weather. Consult the manufacturer's recommendations.

2. If the engine is filled with summer weight oil, the engine should be moved to a warm — above 60°F (16°C) — location and allowed to reach ambient temperature before starting. This is because a heavy summer weight oil will be even thicker at cold temperatures. So thick, in fact, that it will be unable to sufficiently lubricate the engine when it is first started and running. Damage could occur due to lack of lubrication.

3. Change the oil only after the engine has been operated long enough for operating temperatures to have been reached. Change the oil while the engine is still hot.

4. Use fresh gasoline. Fill the gas tank daily to prevent the formation of condensation in the tank and fuel lines.

5. Keep the battery in a fully charged condition, since cold weather infringes upon a battery's maximum current output capabilities.

6. If the engine is run only for short periods of time, have the battery charged every so often to ensure maximum power output when it is needed most.

Routine Maintenance

Care of a small engine is divided into the following five categories: lubrication; filter service; tune-up; carburetor overhaul and fuel pump repair/replacement; combustion chamber deposit removal and valve repair; complete overhaul. Routine maintenance consists of the first three categories. These are described below.

LUBRICATION

This category includes simple replenishment of lost fluids, and, in part, is the responsibility of the operator. Operators must be aware of not only the need to run the engine only when it is adequately lubricated, but of the need to cease operation and perform required maintenance, even if the actual work is done by a mechanic.

Before starting the engine, fill the crankcase and the air cleaner with the proper oil and fill the gasoline tank. Never try to fill the fuel tank of an engine that is running, and if the engine is still hot from running, allow it to cool down before refueling it.

Use a good grade, clean, fresh, lead free or leaded regular grade automotive gasoline. The use of highly leaded gasoline (high octane) should be avoided, as it causes deposits on the valves and valve seats, spark plugs, and the cylinder head, thus shortening engine life.

Any high quality detergent oil having the American Petroleum Institute classification "For Service SG" can be used. Detergent oils keep the engine cleaner by retarding the formation of gum and varnish deposits. Do not use any oil additives. In the summer — above 40°F (4°C) — use SAE 30 weight oil. If that is not available, use SAE 10W-30 or SAE 10W-40 weight oil. In the winter — under 40°F (4°C) — use SAE 5W-20 or SAE 5W-30 weight oil. If neither of these is available, use SAE 10W or SAE 10W-30 weight oil. If the engine is operated in ambient temperatures that are below 0°F (-18°C), use SAE 10W or SAE 10W-30 weight oil diluted 10% with kerosene.

The oil should be changed after each 25 hours of service or engine operation, and more often under dirty or dusty operating conditions, or as the manufacturer specifies. In normal running of any engine, small particles of metal from the cylinder walls, pistons and bearings will gradually work into the oil. Dust particles from the air also get into the oil. If the oil is not changed regularly, these foreign particles cause increased friction and a grinding action which shorten the life of the engine. Fresh oil also assists in cooling the engine, for old oil gradually becomes thick and cannot dissipate the heat fast enough. Oil oil will also gradually lose its lubrication properties.

In 2-stroke engines, lubrication consists of ensuring the engine runs on a mix of fuel and oil which is in the proper proportion, and that the oil used meets the specifications of the manufacturer. Since running a 2-stroke engine on straight gasoline or an improper mix is very much like operating a four stroke engine without oil, it must be seen that proper preparation of fuel/oil mix is literally a life and death matter for the engine — failure to provide the proper mix may result in immediate engine failure. Always observe the following points:

1. Use an oil specifically designed for 2-stroke engines, and of the viscosity recommended by the manufacturer. The wrong oil may solidify in many different parts of the engine, may leave ash deposits in the combustion chamber, or foul the spark plug. Don't forget that the oil must not only lubricate well, but burn well.

2. Measure the oil accurately into the fuel container in the exact proportion recommended. Mix thoroughly according to the directions on the can (see "2-Stroke Lubrication" above). Do not simply use a standard, pre-mixed fuel unless you can determine that it is in the correct proportion. Where engines requiring different mixes are used at a common site, label fuel cans with the fuel/oil mixture ratio contained.

3. Remember that available lubrication in a 2-stroke also depends on fuel/air ration. Ensure that carburetors are properly adjusted, and that there are no air leaks so that sufficient lubrication will always be available. Watch, too, for clogged air cleaners, partially closed chokes, or too rich an adjustment, as these will lead to plug fouling. Correct immediately any conditions causing 4-cycling or misfire, as gasoline may dilute oil lubing pistons and rings under these conditions.

FILTER SERVICE

Air filter service is usually performed at the time of oil change, but may be performed at a longer interval — check specific recommendations. The air cleaner must be serviced much more frequently if the engine is operated in dusty conditions. Check specific recommendations here, also.

Oil type air cleaners require draining of old oil; a thorough cleaning of oil bowl and element with solvent; oiling of the element; and refilling of bowl to the specified level with new oil of the type used in the engine.

Most dry element air cleaners require that the element be replaced — they usually cannot be cleaned with compressed air. Some also employ a swirl chamber to remove large dust particles before they reach the main element. This chamber must be thoroughly cleaned out and, in some cases, a dust catching bowl must be emptied and cleaned.

Other types of dry element type air cleaners may require cleaning in soap and water, thorough drying and, in some cases, oiling.

Fuel sediment bowls and strainers, or filters are used on many engines to ensure that the use of dirty fuel or the entrance of dirt into the gas tank will not cause dirt to get into the carburetor. Since only dirt that enters with the fuel or works its way into the tank reaches the filter, it should be obvious that the first step in fuel filter maintenance is the use of clean gas, and the second, proper maintenance of the tank filler cap and gasket. The filter is usually serviced at the same time the oil is changed or at twice that interval. Fuel tank valves are turned off, and the bowl or filter housing is removed and cleaned. Strainers are cleaned in solvent and dried, and pleated paper type elements are replaced.

Oil filters are replaced at every oil change or every other change — consult specific recommendations. Throw-away type filters usually require the use of a strap wrench for ready removal. Wipe the filter base clean, lubricate the seal on the filter with clean oil, and tighten only by hand, or the amount specified on the filter. In the case of cartridge type filters, clean the housing with solvent and dry. Make sure to replace seals both at the filter base and around the mounting bolt, as applicable.

Filters deserve the same consistency of attention to recommended service intervals as oil changes. Oil change intervals are determined by the ability of the filter to prolong the life of the oil directly in mind. If the filter is allowed to accumulate dirt to the point where it bypasses due to loss of oil pressure, the oil will be subjected to a much greater than normal amount of material to keep in suspension. This shortens the potential life of the oil drastically, greatly increases the changes of clogging engine oil passages, and may allow abrasive particles large enough to be trapped between moving parts to circulate with the oil. Remember, too, that operation in dusty areas can cause a filter to become clogged and by-passed very quickly. Follow manufacturer's recommendations for more frequent changes under these conditions.

Tune-Up

The following list of procedures is rather extensive for a simple tune-up. Normally one would just check the condition of the spark plug, points, condenser, and wiring, make the necessary adjustments to these components and the carburetor, maybe change the oil, and service the carburetor if needed.

However if the following is performed, you will either be sure that the engine is functioning properly or you will know what major repairs should be made. In other words the engine is going to run well or you will find the cause of any problems.

1. Remove the air cleaner and check for the proper servicing.

2. Check the oil level and drain the crankcase. Clean the fuel tank and lines if separate from the carburetor.

3. Remove the blower housing and inspect the rope, rewind assembly and starter clutch of the starter mechanism. Thoroughly clean the cooling fins with compressed air, if possible, and check that all control flaps operate freely.

4. Spin the flywheel to check compression. It should be spun in the direction opposite to normal rotation, and as rapidly as possible. A sharp rebound indicates good compression. 4-stroke engines may also be checked with a compression gauge in place of the spark plug — consult manufacturer's specifications in the individual repair section.

5. Remove the carburetor and disassemble and inspect it for wear or damage. Wash it in solvent, replace parts as necessary, and assemble. Set the initial adjustments.

6. Inspect the crossover tube or the intake elbow for damaged gaskets.

7. Check the governor blade, linkage, and spring for damage or wear; if it is mechanical, check the linkage adjustment.

8. Remove the flywheel and check for seal leakage, both on the flywheel and power take off sides. Check the flywheel key for wear and damage.

9. Remove the breaker cover and check for proper sealing.

10. Inspect the breaker points and condenser. Replace or clean and adjust them. Check the plunger or the cam. Lubricate the cam follower.

11. Check the coil and inspect all wires for breaks or damaged insulation. Be sure the lead wires do not touch the flywheel. Check the stop switch and the lead.

12. Replace the breaker cover, using sealer where the wires enter.

13. Install the flywheel and time the ignition if necessary. Set the air gap and check for ignition spark.

14. Remove the cylinder head, check the gasket, remove the spark plug, clean off the carbon, and inspect the valves for proper seating.

15. Replace the cylinder head, using a new gasket, torque it to the proper specification, and set the spark plug gap or replace the plug if necessary.

16. Replace the oil and fuel and check the muffler for restrictions or damage.

17. Adjust the remote control linkage and cable, if used, for correct operation.

18. Service the air cleaner and check the gaskets and element for damage.

19. Run the engine and adjust the idle mixture and high speed mixture of the carburetor.

Storage

If an engine is to be out of service for more than 30 days, the following steps should be performed:

1. Run the engine for 5-10 minutes until it is thoroughly warmed up to normal operating temperatures.

2. Turn off the fuel supply while the engine is still running, and continue running it until the engine stops from lack of fuel. This procedure removes all fuel from the carburetor.

3. Drain the oil from the crankcase while the engine is still warm.

4. Fill the crankcase with clean oil and tag the engine to indicate what weight oil was installed.

5. Remove the spark plug and squirt about an ounce of oil into the cylinder. Turn the engine over a few times to coat the cylinder walls, the top of the piston, and the head with a protective coating of oil. Reinstall the spark plug and tighten it to the proper torque.

6. Clean or replace the air cleaner. Refer to the manufacturer's recommendations.

7. Clean the governor linkage, making sure that it is in good working order and oiling all joints.

8. Plug the exhaust outlet and the fuel inlet openings. Use clean, lintless rags.

9. Remove the battery and store it in a cool place where there is no danger of freezing. Do not store any wet cell battery directly in contact with the ground or cement floor, as it will establish a ground and discharge itself. A completely discharged battery can never be brought back to its original output capacity. Store the battery on a work bench or on blocks of wood on the floor.

10. Wipe off or wash the engine. Wash only after the engine has had time to cool down to ambient temperature and avoid getting water in the carburetor intake port.

11. Coat all parts that might rust with a light coating of oil. Paint all non-operating parts with a rust inhibiting paint.

12. Provide the entire unit with a suitable covering. Plastic is good where the application and removal of sunlight will not promote the formation of condensation under the plastic covering. If this is the case, use a covering that is able to "breathe," such as a canvas.

Battery

▶ See Figures 1, 2, 3, 4, 5, 6 and 7

INSPECTION

Loose, dirty, or corroded battery terminals are a major cause of 'no-start.' Every 3 months or so, remove the battery terminals and clean them, giving them a light coating of petroleum jelly when you are finished. This will help to retard corrosion.

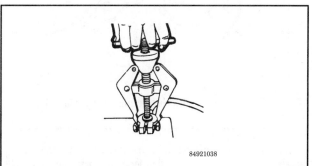

84921038

Fig. 1 Top terminal battery cables are easily removed with this inexpensive puller

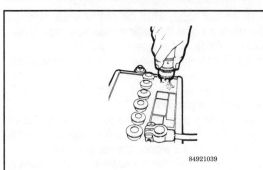

Fig. 2 Clean the battery posts with a wire terminal cleaner

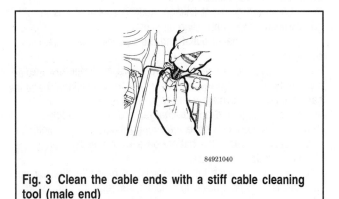

Fig. 3 Clean the cable ends with a stiff cable cleaning tool (male end)

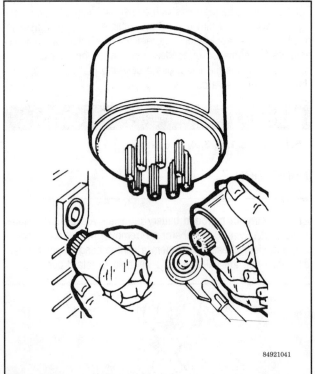

Fig. 4 Side terminal batteries require a special wire brush for cleaning

SPECIFIC GRAVITY (@ 80°F.) AND CHARGE	
Specific Gravity Reading (use the minimum figure for testing)	
Minimum	**Battery Charge**
1.260	100% Charged
1.230	75% Charged
1.200	50% Charged
1.170	25% Charged
1.140	Very Little Power Left
1.110	Completely Discharged

Fig. 5 Battery specific gravity chart

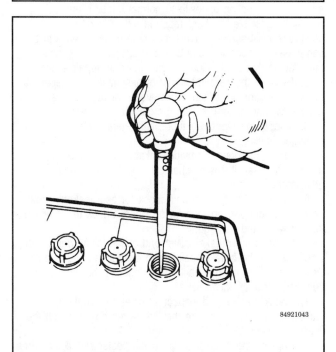

Fig. 6 The specific gravity of the battery can be checked with a simple float-type hydrometer. Some testers have colored balls which correspond to the values in the left column of the specific gravity chart

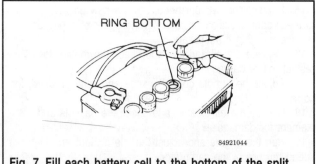

Fig. 7 Fill each battery cell to the bottom of the split ring with distilled water

Check the battery cables for signs of wear or chafing and replace any cable or terminal that looks marginal. Battery ter-minals can be easily cleaned and inexpensive terminal cleaning tools are an excellent investment that will pay for themselves many times over. They can usually be purchased from any well-equipped auto store or parts department. Side terminal batteries require a different tool to clean the threads in the battery case. The accumulated white powder and corrosion can be cleaned from the top of the battery with an old toothbrush and a solution of baking soda and water.

Unless you have a maintenance-free battery, check the electrolyte level (see Battery under Fluid Level Checks in this Section) and check the specific gravity of each cell. Be sure that the vent holes in each cell cap are not blocked by grease or dirt. The vent holes allow hydrogen gas, formed by the chemical reaction in the battery, to escape safely.

REPLACEMENT BATTERIES

The cold power rating of a battery measures battery starting performance and provides an approximate relationship between battery size and engine size. The cold power rating of a replacement battery should match or exceed your engine size in cubic inches.

FLUID LEVEL EXCEPT MAINTENANCE FREE BATTERIES

Check the battery electrolyte level at least once a month, or more often in hot weather or during periods of extended operation. The level can be checked through the case on translucent polypropylene batteries; the cell caps must be removed on other models. The electrolyte level in each cell should be kept filled to the split ring inside, or the line marked on the outside of the case.

If the level is low, add only distilled water, or colorless, odorless drinking water, through the opening until the level is correct. Each cell is completely separate from the others, so each must be checked and filled individually.

If water is added in freezing weather, the battery should be gently rocked to allow the water to mix with the electrolyte. DO NOT SPILL ANY ELECTROLYTE! Otherwise, the battery could freeze.

SPECIFIC GRAVITY EXCEPT MAINTENANCE FREE BATTERIES

At least once a year, check the specific gravity of the battery. It should be between 1.20 in. Hg and 1.26 in. Hg at room temperature.

The specific gravity can be checked with the use of an hydrometer, an inexpensive instrument available from many sources, including auto parts stores. The hydrometer has a squeeze bulb at one end and a nozzle at the other. Battery electrolyte is sucked into the hydrometer until the float is lifted from its seat. The specific gravity is then read by noting the position of the float. Generally, if after charging, the specific gravity between any two cells varies more than 50 points (0.50), the battery is bad and should be replaced.

It is not possible to check the specific gravity in this manner on sealed (maintenance free) batteries. Instead, the indicator built into the top of the case must be relied on to display any signs of battery deterioration. If the indicator is dark, the bat-

tery can be assumed to be OK. If the indicator is light, the specific gravity is low, and the battery should be charged or replaced.

CABLES AND CLAMPS

Once a year, the battery terminals and the cable clamps should be cleaned. Loosen the clamps and remove the cables, negative cable first. On batteries with posts on top, the use of a puller specially made for the purpose is recommended. These are inexpensive, and available in auto parts stores. Side terminal battery cables are secured with a bolt.

Clean the cable lamps and the battery terminal with a wire brush, until all corrosion, grease, etc., is removed and the metal is shiny. It is especially important to clean the inside of the clamp thoroughly, since a small deposit of foreign material or oxidation there will prevent a sound electrical connection and inhibit either starting or charging. Special tools are available for cleaning these parts, one type for conventional batteries and another type for side terminal batteries.

Before installing the cables, loosen the battery holddown clamp or strap, remove the battery and check the battery tray. Clear it of any debris, and check it for soundness. Rust should be wire brushed away, and the metal given a coat of anti-rust paint. Replace the battery and tighten the holddown clamp or strap securely, but be careful not to overtighten, which will crack the battery case.

After the clamps and terminals are clean, reinstall the cables, negative cable last; do not hammer on the clamps to install. Tighten the clamps securely, but do not distort them. Give the clamps and terminals a thin external coat of grease after installation, to retard corrosion.

Check the cables at the same time that the terminals are cleaned. If the cable insulation is cracked or broken, or if the ends are frayed, the cable should be replaced with a new cable of the same length and gauge.

CHARGING

If the battery becomes discharged because of an electrical system failure, or just leaving your lights on, it can be recharged using a battery charger.

A charger rated at 6 amps is more than sufficient for any need. Some battery chargers have a high amperage start feature of 50 amps or more. This will give the battery enough 'boost' to get the engine started when the battery is dead, however, a slow charge is far more beneficial. Follow the charger manufacturer's recommendation when using your charger. A charger with a charging rate gauge will be best, since it shows the rate of charge and can indicate a shorted battery.

✽✽CAUTION

Keep flame or sparks away from the battery; it gives off explosive hydrogen gas. Battery electrolyte contains sulfuric acid. If you should splash any on your skin or in your eyes, flush the affected area with plenty of clear water. If it lands in your eyes, get medical help immediately.

Fluid Disposal

Used fluids such as engine oil, transmission fluid, antifreeze and brake fluid are hazardous wastes and must be disposed of properly. Before draining any fluids, consult with the local authorities; in many areas, waste oil, etc. is being accepted as part of recycling programs. A number of service stations and auto parts stores are also accepting waste fluids for recycling.

Be sure of the recycling center's policies before draining any fluids, as many will not accept different fluids that have been mixed together, such as oil and antifreeze.

Engine Oil

▸ See Figure 8

The recommended oil viscosities for sustained temperatures ranging from below 0°F (-18°C) to above 32°F (0°C) are listed in this Section. They are broken down into multi-viscosity and single viscosities. Multi-viscosity oils are recommended because of their wider range of acceptable temperatures and driving conditions.

When adding oil to the crankcase or changing the oil or filter, it is important that oil of an equal quality to original equipment be used in your engine. The use of inferior oils may void the warranty, damage your engine, or both.

The SAE (Society of Automotive Engineers) grade number of oil indicates the viscosity of the oil (its ability to lubricate at a given temperature). The lower the SAE number, the lighter the oil; the lower the viscosity, the easier it is to crank the engine in cold weather but the less the oil will lubricate and protect the engine in high temperatures. This number is marked on every oil container.

Oil viscosities should be chosen from those oils recommended for the lowest anticipated temperatures during the oil change interval. Due to the need for an oil that embodies both good lubrication at high temperatures and easy cranking in cold weather, multigrade oils have been developed. Basically, a multigrade oil is thinner at low temperatures and thicker at high temperatures. For example, a 10W-40 oil (the W stands for winter) exhibits the characteristics of a 10 weight (SAE 10) oil when the engine is first started and the oil is cold. Its lighter weight allows it to travel to the lubricating surfaces quicker and offer less resistance to starter motor cranking than, say, a straight 30 weight (SAE 30) oil. But after the engine reaches operating temperature, the 10W-40 oil begins acting like straight 40 weight (SAE 40) oil, its heavier weight providing greater lubrication with less chance of foaming than a straight 30 weight oil.

The API (American Petroleum Institute) designations, also found on the oil container, indicates the classification of engine oil used under certain given operating conditions. Only oils designated for use Service SG heavy duty detergent should be used in your engine. Oils of the SG type perform may functions inside the engine besides their basic lubrication. Through a balanced system of metallic detergents and polymeric dispersants, the oil prevents high and low temperature deposits and also keeps sludge and dirt particles in suspension. Acids, particularly sulfuric acid, as well as other by-products of engine combustion are neutralized by the oil. If these acids are allowed to concentrate, they can cause corrosion and rapid wear of the internal engine parts.

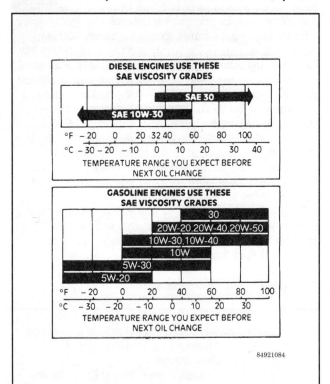

84921084

Fig. 8 Engine oil viscosities

❋❋CAUTION

Non-detergent motor oils or straight mineral oils should not be usedin your engine.

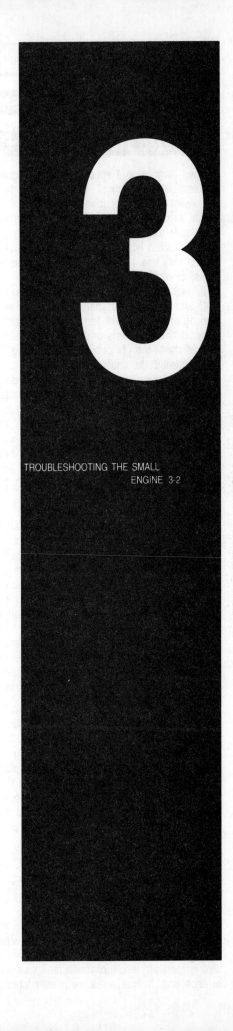

TROUBLESHOOTING THE SMALL
ENGINE 3-2

TROUBLESHOOTING THE SMALL ENGINE

Organized Troubleshooting

When diagnosing a specific problem, organized troubleshooting is a must. The complexity of modern equipment demands that you approach any problem in a logical, organized manner. There are certain troubleshooting techniques that are standard:

1. Establish when the problem occurs. Does the problem appear only under certain conditions? Were there any noises, odors, or other unusual symptoms?

2. Isolate the problem area. To do this, make some simple tests and observations; then eliminate the systems that are working properly. Check for obvious problems such as broken wires, dirty connections or split or disconnected vacuum hoses. Always check the obvious before assuming something complicated is the cause.

3. Test for problems systematically to determine the cause once the problem area is isolated. Are all the components functioning properly? Is there power going to electrical switches and motors? Is there vacuum at vacuum switches and/or actuators? Is there a mechanical problem such as bent linkage or loose mounting screws? Doing careful, systematic checks will often turn up most causes on the first inspection without wasting time checking components that have little or no relationship to the problem.

4. Test all repairs after the work is done to make sure that the problem is fixed. Some causes can be traced to more than one component, so a careful verification of repair work is important to pick up additional malfunctions that may cause a problem to reappear or a different problem to arise. A blown fuse, for example, is a simple problem that may require more than another fuse to repair. If you don't look for a problem that caused a fuse to blow, for example, a shorted wire may go undetected.

Experience has shown that most problems tend to be the result of a fairly simple and obvious cause, such as loose or corroded connectors or air leaks in the intake system; making careful inspection of components during testing essential to quick and accurate troubleshooting.

➡ Pinpointing the exact cause of trouble in an electrical system can sometimes only be accomplished by the use of special test equipment. The following describes commonly used test equipment and explains how to put it to best use in diagnosis. In addition to the information covered below, the manufacturer's instructions booklet provided with the tester should be read and clearly understood before attempting any test procedures.

TEST EQUIPMENT

Jumper Wires

Jumper wires are simple, yet extremely valuable, pieces of test equipment. Jumper wires are merely wires that are used to bypass sections of a circuit. The simplest type of jumper wire is merely a length of multistrand wire with an alligator clip at each end. Jumper wires are usually fabricated from lengths of standard automotive wire and whatever type of connector (alligator clip, spade connector or pin connector) that is required for the particular equipment being tested. The well equipped tool box will have several different styles of jumper wires in several different lengths. Some jumper wires are made with three or more terminals coming from a common splice for special purpose testing. In cramped, hard-to-reach areas it is advisable to have insulated boots over the jumper wire terminals in order to prevent accidental grounding, sparks, and possible fire, especially when testing fuel system components.

Jumper wires are used primarily to locate open electrical circuits, on either the ground (-) side of the circuit or on the hot (+) side. If an electrical component fails to operate, connect the jumper wire between the component and a good ground. If the component operates only with the jumper installed, the ground circuit is open. If the ground circuit is good, but the component does not operate, the circuit between the power feed and component is open. You can sometimes connect the jumper wire directly from the battery to the hot terminal of the component, but first make sure the component uses 12 volts in operation. Some electrical components, such as fuel injectors, are designed to operate on about 4 volts and running 12 volts directly to the injector terminals can burn out the wiring. By inserting an inline fuse holder between a set of test leads, a fused jumper wire can be used for bypassing open circuits. Use a 5 amp fuse to provide protection against voltage spikes. When in doubt, use a voltmeter to check the voltage input to the component and measure how much voltage is being applied normally. By moving the jumper wire successively back from the lamp toward the power source, you can isolate the area of the circuit where the open is located. When the component stops functioning, or the power is cut off, the open is in the segment of wire between the jumper and the point previously tested.

✳✳CAUTION

Never use jumpers made from wire that is of lighter gauge than used in the circuit under test. If the jumper wire is of too small gauge, it may overheat and possibly melt. Never use jumpers to bypass high resistance loads (Such as motors) in a circuit. Bypassing resistances, in effect, creates a short circuit which may, in turn, cause damage and fire. Never use a jumper for anything other than temporary bypassing of components in a circuit.

12 Volt Test Light

The 12 volt test light is used to check circuits and components while electrical current is flowing through them. It is used for voltage and ground tests. Twelve volt test lights come in different styles but all have three main parts; a ground clip, a probe, and a light. The most commonly used 12 volt test lights have pick-type probes. To use a 12 volt test light, connect the ground clip to a good ground and probe wherever necessary with the pick. The pick should be sharp so that it can penetrate wire insulation to make contact with the wire, without making a large hole in the insulation. The wrap-around light is handy in hard to reach areas or where it is difficult to support a wire to push a probe pick into it. To use the wrap around light, hook the wire to probed with the hook and pull the trigger. A small pick will be forced through the wire insulation into the wire core.

✳✳CAUTION

Do not use a test light to probe electronic ignition spark plug or coil wires. Never use a pick-type test light to probe wiring on computer controlled systems unless specifically instructed to do so. Any wire insulation that is pierced by the test light probe should be taped and sealed with silicone after testing.

Like the jumper wire, the 12 volt test light is used to isolate opens in circuits. But, whereas the jumper wire is used to bypass the open to operate the load, the 12 volt test light is used to locate the presence of voltage in a circuit. If the test light glows, you know that there is power up to that point; if the 12 volt test light does not glow when its probe is inserted into the wire or connector, you know that there is an open circuit (no power). Move the test light in successive steps back toward the power source until the light in the handle does glow. When it does glow, the open is between the probe and point previously probed.

➡**The test light does not detect that 12 volts (or any particular amount of voltage) is present; it only detects that some voltage is present. It is advisable before using the test light to touch its terminals across the battery posts to make sure the light is operating properly.**

Self-Powered Test Light

The self-powered test light usually contains a 1.5 volt penlight battery. One type of self-powered test light is similar in design to the 12 volt test light. This type has both the battery and the light in the handle and pick-type probe tip. The second type has the light toward the open tip, so that the light illuminates the contact point. The self-powered test light is dual purpose piece of test equipment. It can be used to test for either open or short circuits when power is isolated from the circuit (continuity test). A powered test light should not be used on any computer controlled system or component unless specifically instructed to do so. Many engine sensors can be destroyed by even this small amount of voltage applied directly to the terminals.

Open Circuit Testing

To use the self-powered test light to check for open circuits, first isolate the circuit from the equipment's 12 volt power source by disconnecting the battery or wiring harness connector. Connect the test light ground clip to a good ground and probe sections of the circuit sequentially with the test light. (start from either end of the circuit). If the light is out, the open is between the probe and the circuit ground. If the light is on, the open is between the probe and end of the circuit toward the power source.

Short Circuit Testing

By isolating the circuit both from power and from ground, and using a self-powered test light, you can check for shorts to ground in the circuit. Isolate the circuit from power and ground. Connect the test light ground clip to a good ground

and probe any easy-to-reach test point in the circuit. If the light comes on, there is a short somewhere in the circuit. To isolate the short, probe a test point at either end of the isolated circuit (the light should be on). Leave the test light probe connected and open connectors, switches, remove parts, etc., sequentially, until the light goes out. When the light goes out, the short is between the last circuit component opened and the previous circuit opened.

➡**The 1.5 volt battery in the test light does not provide much current. A weak battery may not provide enough power to illuminate the test light even when a complete circuit is made (especially if there are high resistances in the circuit). Always make sure that the test battery is strong. To check the battery, briefly touch the ground clip to the probe; if the light glows brightly the battery is strong enough for testing. Never use a self-powered test light to perform checks for opens or shorts when power is applied to the electrical system under test.**

Voltmeter

A voltmeter is used to measure voltage at any point in a circuit, or to measure the voltage drop across any part of a circuit. It can also be used to check continuity in a wire or circuit by indicating current flow from one end to the other. Voltmeters usually have various scales on the meter dial and a selector switch to allow the selection of different voltages. The voltmeter has a positive and a negative lead. To avoid damage to the meter, always connect the negative lead to the negative (-) side of circuit (to ground or nearest the ground side of the circuit) and connect the positive lead to the positive (+) side of the circuit (to the power source or the nearest power source). Note that the negative voltmeter lead will always be black and that the positive voltmeter will always be some color other than black (usually red). Depending on how the voltmeter is connected into the circuit, it has several uses.

A voltmeter can be connected either in parallel or in series with a circuit and it has a very high resistance to current flow. When connected in parallel, only a small amount of current will flow through the voltmeter current path; the rest will flow through the normal circuit current path and the circuit will work normally. When the voltmeter is connected in series with a circuit, only a small amount of current can flow through the circuit. The circuit will not work properly, but the voltmeter reading will show if the circuit is complete or not.

Available Voltage Measurement

Set the voltmeter selector switch to the 20V position and connect the meter negative lead to the negative post of the battery. Connect the positive meter lead to the positive post of the battery and turn the ignition switch ON to provide a load. Read the voltage on the meter or digital display. A well charged battery should register over 12 volts. If the meter reads below 11.5 volts, the battery power may be insufficient to operate the electrical system properly. This test determines voltage available from the battery and should be the first step in any electrical trouble diagnosis procedure. Many electrical problems, especially on computer controlled systems, can be caused by a low state of charge in the battery. Excessive corrosion at the battery cable terminals can cause a poor contact that will prevent proper charging and full battery current flow.

Normal battery voltage is 12 volts when fully charged. When the battery is supplying current to one or more circuits it is said to be 'under load'. When everything is off the electrical system is under a 'no-load' condition. A fully charged battery may show about 12.5 volts at no load; will drop to 12 volts under medium load; and will drop even lower under heavy load. If the battery is partially discharged the voltage decrease under heavy load may be excessive, even though the battery shows 12 volts or more at no load. When allowed to discharge further, the battery's available voltage under load will decrease more severely. For this reason, it is important that the battery be fully charged during all testing procedures to avoid errors in diagnosis and incorrect test results.

Voltage Drop

When current flows through a resistance, the voltage beyond the resistance is reduced (the larger the current, the greater the reduction in voltage). When no current is flowing, there is no voltage drop because there is no current flow. All points in the circuit which are connected to the power source are at the same voltage as the power source. The total voltage drop always equals the total source voltage. In a long circuit with many connectors, a series of small, unwanted voltage drops due to corrosion at the connectors can add up to a total loss of voltage which impairs the operation of the normal loads in the circuit.

INDIRECT COMPUTATION OF VOLTAGE DROPS

1. Set the voltmeter selector switch to the 20 volt position.
2. Connect the meter negative lead to a good ground.
3. Probe all resistances in the circuit with the positive meter lead.
4. Operate the circuit in all modes and observe the voltage readings.

DIRECT MEASUREMENT OF VOLTAGE DROPS

1. Set the voltmeter switch to the 20 volt position.
2. Connect the voltmeter negative lead to the ground side of the resistance load to be measured.
3. Connect the positive lead to the positive side of the resistance or load to be measured.
4. Read the voltage drop directly on the 20 volt scale.

Too high a voltage indicates too high a resistance. If, for example, a blower motor runs too slowly, you can determine if there is too high a resistance in the resistor pack. By taking voltage drop readings in all parts of the circuit, you can isolate the problem. Too low a voltage drop indicates too low a resistance. If, for example, a blower motor runs too fast in the MED and/or LOW position, the problem can be isolated in the resistor pack by taking voltage drop readings in all parts of the circuit to locate a possibly shorted resistor. The maximum allowable voltage drop under load is critical, especially if there is more than one high resistance problem in a circuit because all voltage drops are cumulative. A small drop is normal due to the resistance of the conductors.

HIGH RESISTANCE TESTING

1. Set the voltmeter selector switch to the 4 volt position.
2. Connect the voltmeter positive lead to the positive post of the battery.

3. Turn on the headlights and heater blower to provide a load.
4. Probe various points in the circuit with the negative voltmeter lead.
5. Read the voltage drop on the 4 volt scale. Some average maximum allowable voltage drops are:

    FUSE PANEL — 7 volts
    IGNITION SWITCH — 5volts
    HEADLIGHT SWITCH — 7 volts
    IGNITION COIL (+) — 5 volts
    ANY OTHER LOAD — 1.3 volts

➡**Voltage drops are all measured while a load is operating; without current flow, there will be no voltage drop.**

Ohmmeter

The ohmmeter is designed to read resistance (ohms) in a circuit or component. Although there are several different styles of ohmmeters, all will usually have a selector switch which permits the measurement of different ranges of resistance (usually the selector switch allows the multiplication of the meter reading by 10, 100, 1000, and 10,000). A calibration knob allows the meter to be set at zero for accurate measurement. Since all ohmmeters are powered by an internal battery (usually 9 volts), the ohmmeter can be used as a self-powered test light. When the ohmmeter is connected, current from the ohmmeter flows through the circuit or component being tested. Since the ohmmeter's internal resistance and voltage are known values, the amount of current flow through the meter depends on the resistance of the circuit or component being tested.

The ohmmeter can be used to perform continuity test for opens or shorts (either by observation of the meter needle or as a self-powered test light), and to read actual resistance in a circuit. It should be noted that the ohmmeter is used to check the resistance of a component or wire while there is no voltage applied to the circuit. Current flow from an outside voltage source (such as the battery) can damage the ohmmeter, so the circuit or component should be isolated from the electrical system before any testing is done. Since the ohmmeter uses its own voltage source, either lead can be connected to any test point.

➡**When checking diodes or other solid state components, the ohmmeter leads can only be connected one way in order to measure current flow in a single direction. Make sure the positive (+) and negative (-) terminal connections are as described in the test procedures to verify the one-way diode operation.**

In using the meter for making continuity checks, do not be concerned with the actual resistance readings. Zero resistance, or any resistance readings, indicate continuity in the circuit. Infinite resistance indicates an open in the circuit. A high resistance reading where there should be none indicates a problem in the circuit. Checks for short circuits are made in the same manner as checks for open circuits except that the circuit must be isolated from both power and normal ground. Infinite resistance indicates no continuity to ground, while zero resistance indicates a dead short to ground.

RESISTANCE MEASUREMENT

The batteries in an ohmmeter will weaken with age and temperature, so the ohmmeter must be calibrated or 'zeroed' before taking measurements. To zero the meter, place the selector switch in its lowest range and touch the two ohmmeter leads together. Turn the calibration knob until the meter needle is exactly on zero.

➡All analog (needle) type ohmmeters must be zeroed before use, but some digital ohmmeter models are automatically calibrated when the switch is turned on. Self-calibrating digital ohmmeters do not have an adjusting knob, but its a good idea to check for a zero readout before use by touching the leads together. All computer controlled systems require the use of a digital ohmmeter with at least 10 megohms impedance for testing. Before any test procedures are attempted, make sure the ohmmeter used is compatible with the electrical system or damage to the on-board computer could result.

To measure resistance, first isolate the circuit from the power source by disconnecting the battery cables or the harness connector. Make sure the key is OFF when disconnecting any components or the battery. Where necessary, also isolate at least one side of the circuit to be checked to avoid reading parallel resistances. Parallel circuit resistances will always give a lower reading than the actual resistance of either of the branches. When measuring the resistance of parallel circuits, the total resistance will always be lower than the smallest resistance in the circuit. Connect the meter leads to both sides of the circuit (wire or component) and read the actual measured ohms on the meter scale. Make sure the selector switch is set to the proper ohm scale for the circuit being tested to avoid misreading the ohmmeter test value.

✴✴CAUTION

Never use an ohmmeter with power applied to the circuit. Like the self-powered test light, the ohmmeter is designed to operate on its own power supply. The normal 12 volt automotive electrical system current could damage the meter.

Ammeters

An ammeter measures the amount of current flowing through a circuit in units called amperes or amps. Amperes are units of electron flow which indicate how fast the electrons are flowing through the circuit. Since Ohms Law dictates that current flow in a circuit is equal to the circuit voltage divided by the total circuit resistance, increasing voltage also increases the current level (amps). Likewise, any decrease in resistance will increase the amount of amps in a circuit. At normal operating voltage, most circuits have a characteristic amount of amperes, called 'current draw' which can be measured using an ammeter. By referring to a specified current draw rating, measuring the amperes, and comparing the two values, one can determine what is happening within the circuit to aid in diagnosis. An open circuit, for example, will not allow any current to flow so the ammeter reading will be zero. More current flows through a heavily loaded circuit or when the charging system is operating.

An ammeter is always connected in series with the circuit being tested. All of the current that normally flows through the circuit must also flow through the ammeter; if there is any other path for the current to follow, the ammeter reading will not be accurate. The ammeter itself has very little resistance to current flow and therefore will not affect the circuit, but it will measure current draw only when the circuit is closed and electricity is flowing. Excessive current draw can blow fuses and drain the battery, while a reduced current draw can cause motors to run slowly, lights to dim and other components to not operate properly. The ammeter can help diagnose these conditions by locating the cause of the high or low reading.

Multimeters

Different combinations of test meters can be built into a single unit designed for specific tests. Some of the more common combination test devices are known as Volt/Amp testers, Tach/Dwell meters, or Digital Multimeters. The Volt/Amp tester is used for charging system, starting system or battery tests and consists of a voltmeter, an ammeter and a variable resistance carbon pile. The voltmeter will usually have at least two ranges for use with 6, 12 and 24 volt systems. The ammeter also has more than one range for testing various levels of battery loads and starter current draw and the carbon pile can be adjusted to offer different amounts of resistance. The Volt/Amp tester has heavy leads to carry large amounts of current and many later models have an inductive ammeter pickup that clamps around the wire to simplify test connections. On some models, the ammeter also has a zero-center scale to allow testing of charging and starting systems without switching leads or polarity. A digital multimeter i s a voltmeter, ammeter and ohmmeter combined in an instrument which gives a digital readout. These are often used when testing solid state circuits because of their high input impedance (usually 10 megohms or more).

The tach/dwell meter combines a tachometer and a dwell (cam angle) meter and is a specialized kind of voltmeter. The tachometer scale is marked to show engine speed in rpm and the dwell scale is marked to show degrees of distributor shaft rotation. In most electronic ignition systems, dwell is determined by the control unit, but the dwell meter can also be used to check the duty cycle (operation) of some electronic engine control systems. Some tach/dwell meters are powered by an internal battery, while others take their power from the battery in use. The battery powered testers usually require calibration much like an ohmmeter before testing.

Wiring Harnesses

The average automobile contains about ½ mile of wiring, with hundreds of individual connections. To protect the many wires from damage and to keep them from becoming a confusing tangle, they are organized into bundles, enclosed in plastic or taped together and called wire harnesses. Individual wires are color coded to help trace them through a harness where sections are hidden from view.

A loose or corroded connection or a replacement wire that is too small for the circuit will add extra resistance and an additional voltage drop to the circuit. A ten percent voltage drop can result in slow or erratic motor operation, for example, even though the circuit is complete. Automotive wiring or circuit conductors can be in any one of three forms:

1. Single strand wire
2. Multistrand wire
3. Printed circuitry

Single strand wire has a solid metal core and is usually used inside such components as alternators, motors, relays and other devices. Multistrand wire has a core made of many small strands of wire twisted together into a single conductor. Most of the wiring in an automotive electrical system is made up of multistrand wire, either as a single conductor or grouped together in a harness. All wiring is color coded on the insulator, either as a solid color or as a colored wire with an identification stripe. A printed circuit is a thin film of copper or other conductor that is printed on an insulator backing. Occasionally, a printed circuit is sandwiched between two sheets of plastic for more protection and flexibility. A complete printed circuit, consisting of conductors, insulating material and connectors for lamps or other components is called a printed circuit board. Printed circuitry is used in place of individual wires or harnesses in places where space is limited, such as behind instrument panel.

WIRE GAUGE

Since computer controlled automotive electrical systems are very sensitive to changes in resistance, the selection of properly sized wires is critical when systems are repaired. The wire gauge number is an expression of the cross section area of the conductor. The most common system for expressing wire size is the American Wire Gauge (AWG) system.

Wire cross section area is measured in circular mils. A mil is 1/1000 in. (0.001 in.); a circular mil is the area of a circle one mil in diameter. For example, a conductor 1/4 in. in diameter is 0.250 in. or 250 mils. The circular mil cross section area of the wire is 250 squared (250^2)or 62,500 circular mils. Imported models usually use metric wire gauge designations, which is simply the cross section area of the conductor in square millimeters (mm^2).

Gauge numbers are assigned to conductors of various cross section areas. As gauge number increases, area decreases and the conductor becomes smaller. A 5 gauge conductor is smaller than a 1 gauge conductor and a 10 gauge is smaller than a 5 gauge. As the cross section area of a conductor decreases, resistance increases and so does the gauge number. A conductor with a higher gauge number will carry less current than a conductor with a lower gauge number.

➡Gauge wire size refers to the size of the conductor, not the size of the complete wire. It is possible to have two wires of the same gauge with different diameters because one may have thicker insulation than the other.

12 volt automotive electrical systems generally use 10, 12, 14, 16 and 18 gauge wire. Main power distribution circuits and larger accessories usually use 10 and 12 gauge wire. Battery cables are usually 4 or 6 gauge, although 1 and 2 gauge wires are occasionally used. Wire length must also be considered when making repairs to a circuit. As conductor length increases, so does resistance. An 18 gauge wire, for example, can carry a 10 amp load for 10 feet without excessive voltage drop; however if a 15 foot wire is required for the same 10 amp load, it must be a 16 gauge wire.

An electrical schematic shows the electrical current paths when a circuit is operating properly. It is essential to understand how a circuit works before trying to figure out why it doesn't. Schematics break the entire electrical system down into individual circuits and show only one particular circuit. In a schematic, no attempt is made to represent wiring and components as they physically appear on the equipment; switches and other components are shown as simply as possible. Face views of harness connectors show the cavity or terminal locations in all multi-pin connectors to help locate test points.

If you need to backprobe a connector while it is on the component, the order of the terminals must be mentally reversed. The wire color code can help in this situation, as well as a keyway, lock tab or other reference mark.

WIRING REPAIR

Soldering is a quick, efficient method of joining metals permanently. Everyone who has the occasion to make wiring repairs should know how to solder. Electrical connections that are soldered are far less likely to come apart and will conduct electricity much better than connections that are only 'pig-tailed' together. The most popular (and preferred) method of soldering is with an electrical soldering gun. Soldering irons are available in many sizes and wattage ratings. Irons with higher wattage ratings deliver higher temperatures and recover lost heat faster. A small soldering iron rated for no more than 50 watts is recommended, especially on electrical systems where excess heat can damage the components being soldered.

There are three ingredients necessary for successful soldering; proper flux, good solder and sufficient heat. A soldering flux is necessary to clean the metal of tarnish, prepare it for soldering and to enable the solder to spread into tiny crevices. When soldering, always use a resin flux or resin core solder which is non-corrosive and will not attract moisture once the job is finished. Other types of flux (acid core) will leave a residue that will attract moisture and cause the wires to corrode. Tin is a unique metal with a low melting point. In a molten state, it dissolves and alloys easily with many metals. Solder is made by mixing tin with lead. The most common proportions are 40/60, 50/50 and 60/40, with the percentage of tin listed first. Low priced solders usually contain less tin, making them very difficult for a beginner to use because more heat is required to melt the solder. A common solder is 40/60 which is well suited for all-around general use, but 60/40 melts easier, has more tin f or a better joint and is preferred for electrical work.

Soldering Techniques

Successful soldering requires that the metals to be joined be heated to a temperature that will melt the solder — usually 360-460°F (182-238°C). Contrary to popular belief, the purpose of the soldering iron is not to melt the solder itself, but to heat the parts being soldered to a temperature high enough to melt the solder when it is touched to the work. Melting flux-cored solder on the soldering iron will usually destroy the effectiveness of the flux.

→**Soldering tips are made of copper for good heat conductivity, but must be 'tinned' regularly for quick transference of heat to the project and to prevent the solder from sticking to the iron. To 'tin' the iron, simply heat it and touch the flux-cored solder to the tip; the solder will flow over the hot tip. Wipe the excess off with a clean rag, but be careful as the iron will be hot.**

After some use, the tip may become pitted. If so, simply dress the tip smooth with a smooth file and 'tin' the tip again. An old saying holds that 'metals well cleaned are half soldered.' Flux-cored solder will remove oxides but rust, bits of insulation and oil or grease must be removed with a wire brush or emery cloth. For maximum strength in soldered parts, the joint must start off clean and tight. Weak joints will result in gaps too wide for the solder to bridge.

If a separate soldering flux is used, it should be brushed or swabbed on only those areas that are to be soldered. Most solders contain a core of flux and separate fluxing is unnecessary. Hold the work to be soldered firmly. It is best to solder on a wooden board, because a metal vise will only rob the piece to be soldered of heat and make it difficult to melt the solder. Hold the soldering tip with the broadest face against the work to be soldered. Apply solder under the tip close to the work, using enough solder to give a heavy film between the iron and the piece being soldered, while moving slowly and making sure the solder melts properly. Keep the work level or the solder will run to the lowest part and favor the thicker parts, because these require more heat to melt the solder. If the soldering tip overheats (the solder coating on the face of the tip burns up), it should be retinned. Once the soldering is completed, let the soldered joint stand until cool. Tape and seal all soldered wire splices after the repair has cooled.

Wire Harness and Connectors

The on-board computer (ECM) wire harness electrically connects the control unit to the various solenoids, switches and sensors used by the control system. Most connectors in the engine compartment or otherwise exposed to the elements are protected against moisture and dirt which could create oxidation and deposits on the terminals. This protection is important because of the very low voltage and current levels used by the computer and sensors. All connectors have a lock which secures the male and female terminals together, with a secondary lock holding the seal and terminal into the connector. Both terminal locks must be released when disconnecting ECM connectors.

These special connectors are weather-proof and all repairs require the use of a special terminal and the tool required to service it. This tool is used to remove the pin and sleeve terminals. If removal is attempted with an ordinary pick, there is a good chance that the terminal will be bent or deformed. Unlike standard blade type terminals, these terminals cannot be straightened once they are bent. Make certain that the connectors are properly seated and all of the sealing rings in place when connecting leads. On some models, a hinge-type flap provides a backup or secondary locking feature for the terminals. Most secondary locks are used to improve the

connector reliability by retaining the terminals if the small terminal lock tangs are not positioned properly.

Molded-on connectors require complete replacement of the connection. This means splicing a new connector assembly into the harness. All splices in on-board computer systems should be soldered to insure proper contact. Use care when probing the connections or replacing terminals in them as it is possible to short between opposite terminals. If this happens to the wrong terminal pair, it is possible to damage certain components. Always use jumper wires between connectors for circuit checking and never probe through weatherproof seals.

Open circuits are often difficult to locate by sight because corrosion or terminal misalignment are hidden by the connectors. Merely wiggling a connector on a sensor or in the wiring harness may correct the open circuit condition. This should always be considered when an open circuit or a failed sensor is indicated. Intermittent problems may also be caused by oxidized or loose connections. When using a circuit tester for diagnosis, always probe connections from the wire side. Be careful not to damage sealed connectors with test probes.

All wiring harnesses should be replaced with identical parts, using the same gauge wire and connectors. When signal wires are spliced into a harness, use wire with high temperature insulation only. With the low voltage and current levels found in the system, it is important that the best possible connection at all wire splices be made by soldering the splices together. It is seldom necessary to replace a complete harness. If replacement is necessary, pay close attention to insure proper harness routing. Secure the harness with suitable plastic wire clamps to prevent vibrations from causing the harness to wear in spots or contact any hot components.

→**Weatherproof connectors cannot be replaced with standard connectors. Instructions are provided with replacement connector and terminal packages. Some wire harnesses have mounting indicators (usually pieces of colored tape) to mark where the harness is to be secured.**

In making wiring repairs, it's important that you always replace damaged wires with wires that are the same gauge as the wire being replaced. The heavier the wire, the smaller the gauge number. Wires are color-coded to aid in identification and whenever possible the same color coded wire should be used for replacement. A wire stripping and crimping tool is necessary to install solderless terminal connectors. Test all crimps by pulling on the wires; it should not be possible to pull the wires out of a good crimp.

Wires which are open, exposed or otherwise damaged are repaired by simple splicing. Where possible, if the wiring harness is accessible and the damaged place in the wire can be located, it is best to open the harness and check for all possible damage. In an inaccessible harness, the wire must be bypassed with a new insert, usually taped to the outside of the old harness.

When replacing fusible links, be sure to use fusible link wire, NOT ordinary automotive wire. Make sure the fusible segment is of the same gauge and construction as the one being replaced and double the stripped end when crimping the terminal connector for a good contact. The melted (open) fusible link segment of the wiring harness should be cut off as close to the harness as possible, then a new segment spliced in as described. In the case of a damaged fusible link that feeds two harness wires, the harness connections should be replaced with two fusible link wires so that each circuit will have its own separate protection.

➡**Most of the problems caused in the wiring harness are due to bad ground connections. Always check all ground connections for corrosion or looseness before performing any power feed checks to eliminate the chance of a bad ground affecting the circuit.**

Repairing Hard Shell Connectors

Unlike molded connectors, the terminal contacts in hard shell connectors can be replaced. Weatherproof hard-shell connectors with the leads molded into the shell have non-replaceable terminal ends. Replacement usually involves the use of a special terminal removal tool that depress the locking tangs (barbs) on the connector terminal and allow the connector to be removed from the rear of the shell. The connector shell should be replaced if it shows any evidence of burning, melting, cracks, or breaks. Replace individual terminals that are burnt, corroded, distorted or loose.

➡**The insulation crimp must be tight to prevent the insulation from sliding back on the wire when the wire is pulled. The insulation must be visibly compressed under the crimp tabs, and the ends of the crimp should be turned in for a firm grip on the insulation.**

The wire crimp must be made with all wire strands inside the crimp. The terminal must be fully compressed on the wire strands with the ends of the crimp tabs turned in to make a firm grip on the wire. Check all connections with an ohmmeter to insure a good contact. There should be no measurable resistance between the wire and the terminal when connected.

Mechanical Test Equipment

VACUUM GAUGE

Most gauges are graduated in inches of mercury (in. Hg), although a device called a manometer reads vacuum in inches of water (in. H2O). The normal vacuum reading usually varies between 18 and 22 in. Hg at sea level. To test engine vacuum, the vacuum gauge must be connected to a source of manifold vacuum. Many engines have a plug in the intake manifold which can be removed and replaced with an adapter fitting. Connect the vacuum gauge to the fitting with a suitable rubber hose or, if no manifold plug is available, connect the vacuum gauge to any device using manifold vacuum, such as EGR valves, etc. The vacuum gauge can be used to determine if enough vacuum is reaching a component to allow its actuation.

HAND VACUUM PUMP

Small, hand-held vacuum pumps come in a variety of designs. Most have a built-in vacuum gauge and allow the component to be tested without removing it. Operate the pump lever or plunger to apply the correct amount of vacuum required for the test specified in the diagnosis routines. The level of vacuum in inches of Mercury (in. Hg) is indicated on the pump gauge. For some testing, an additional vacuum gauge may be necessary.

To correctly diagnose and solve problems in vacuum control systems, a vacuum source is necessary for testing. In some cases, vacuum can be taken from the intake manifold when the engine is running, but vacuum is normally provided by a hand vacuum pump. These hand vacuum pumps have a built-in vacuum gauge that allow testing while the device is still attached to the component. For some tests, an additional vacuum gauge may be necessary.

How To Go About It

Start with the simplest, most obvious causes first — many engine mechanics and operators have difficulty identifying trouble because they start out assuming everything that is obvious has already been checked. Check to see that there is fuel in the tank, and that it is clean, that the tank is properly vented, and that the fuel filter or sediment bowl is not full of dirt. Check to see that the spark plug wire is connected and that the spark plug is not fouled. If the cause of the trouble is not immediately obvious, use your basic knowledge of how the engine works. For example, if the engine runs fine but is very hard to start, you might conclude that the choke does not close, since its function is confined, mainly, to engine starting.

The guide below will point out many possible causes of the most basic problems. Find the 'PROBLEM' which matches the engine's behavior, and then check out the possibilities listed under 'CAUSES AND REMEDIES.' Refer to the manufacturer's section which pertains to your engine, if necessary, in making repairs.

General Troubleshooting Guide

PROBLEM: The engine does not start or is hard to start.
CAUSES AND REMEDIES:
1. The fuel tank is empty.
2. The fuel shut-off valve is closed; open it.
3. The fuel line is clogged. Remove the fuel line and clean it. Clean the carburetor, if necessary.
4. The fuel tank is not vented properly. Check the fuel tank cap vent to see if it is open.
5. There is water in the fuel supply. Drain the tank, clean the fuel lines and the carburetor, and dry the spark plug. Fill the tank with fresh fuel. Check the fuel supply before pouring it into the engine's fuel tank. Chances are it might be the source of the water.
6. The engine is over-choked. Open the choke and throttle wide on manual choke engines. On engines with automatic chokes, close the throttle. Then, turn the engine over with several pulls of the starter rope. If engine does not start, set throttle to just above idle, close choke again, and again attempt to start the engine. If one or two pulls does not make engine fire, try cranking with the choke closed only half way. If engine still fails to start, remove the spark plug and dry it, and spin the engine over several times to clean excess fuel out of the engine. Replace the spark plug and perform the normal starting procedure. Over-choking is most often due to continued cranking with the choke fully shut.

7. The carburetor is improperly adjusted; adjust it to the standard recommended preliminary settings. See the carburetor section.

8. Magneto wiring is loose or defective. Check the magneto wiring for shorts or grounds and repair it, if necessary.

9. No spark. Check for spark, and if there is none, check and, if necessary, replace the contact points, and set contact gap and timing. If there is still no spark, replace further magneto parts (especially coil and high tension wire) as necessary.

10. The spark plug is fouled. Remove, clean, and regap the spark plug.

11. The spark plug is damaged (cracked porcelain, bent electrodes etc.). Replace the spark plug.

12. Compression is poor. The head is loose or the gasket is leaking. Sticking or burned valves or worn piston rings could also be the cause. In any case, the engine will have to be disassembled and the cause of the problem corrected.

PROBLEM: The engine misses under load (if a two-stroke, it may 'four-cycle.')
CAUSES AND REMEDIES:

13. The spark plug is fouled. Remove, clean, and regap the spark plug.

14. The spark plug is damaged. Replace the spark plug.

15. The spark plug is improperly gapped. Regap the spark plug to the proper gap.

16. The breaker points are pitted or improperly gapped. Replace the points, or set the gap.

17. The breaker point's breaker arm is sluggish. Clean and lubricate it.

18. The condenser is faulty. Replace it.

19. The carburetor is not adjusted properly. Adjust it.

20. The fuel system is partly clogged, or the fuel shut-off valve is partly closed. Open the valve and check the fuel filter/strainer, tank, lines, and carburetor for dirt. Clean all parts as necessary.

21. If the engine is a two-stroke, the exhaust ports may be clogged. Remove the exhaust manifold and inspect the ports. If they are clogged with carbon, clean them with a soft tool such as a wooden stick. Check also for bad crankshaft seals.

22. The valves are not adjusted properly. Adjust the valve clearance.

23. The valve springs are weak. Replace them.

PROBLEM: The engine knocks.
CAUSES AND REMEDIES:

24. The magneto is not timed correctly. Time the magneto.

25. The carburetor is not properly adjusted (may be too lean). Adjust the carburetor for best mixture.

26. The engine has overheated. Stop the engine and find the cause of overheating.

27. Carbon has built up in the combustion chamber, resulting in retention of excess heat and an increase in compression which causes pre-ignition. Remove the cylinder head, and remove the carbon from the head and the top of the piston.

28. The connecting rod is loose or worn. Replace it.

29. The flywheel is loose. Check the flywheel key and keyway and the end of the crankshaft. Replace any worn parts. Tighten the flywheel nut to the specified torque.

30. The cylinder is worn. Rebuild/replace parts as necessary.

PROBLEM: The engine vibrates excessively.
CAUSES AND REMEDIES:

31. The engine is not mounted securely to the equipment that it operates. Tighten any loose mounting bolts.

32. The equipment that the engine operates is not balanced. Check the equipment.

33. The crankshaft is bent. Replace the crankshaft.

34. The counter balance shaft is improperly timed (recent reassembly) or broken. Disassemble the crankcase, inspect, and replace or repair parts as necessary.

PROBLEM: The engine lacks power.
CAUSES AND REMEDIES:

35. The choke is partially closed. Open the choke.

36. The carburetor is not adjusted correctly. Adjust it.

37. The ignition is not timed correctly. Time the ignition.

38. There is a lack of lubrication or not enough oil in the crankcase. Fill the crankcase to the correct level.

39. The air cleaner is fouled. Clean it.

40. The valves are not sealing. Do a valve job.

41. Ring seal is poor. Repair/replace rings, piston, or cylinder/cylinder liner.

42. If the engine is a two stroke, the exhaust ports may be clogged with carbon. Remove the exhaust manifold and inspect. Clean with a soft instrument such as a wooden stick, if dirty. Ports may clog frequently if the carburetor mixture is adjusted too rich, or if there is excessive oil or oil of the wrong type in the fuel.

PROBLEM: The engine operates erratically, surges, and runs unevenly.
CAUSES AND REMEDIES:

43. The fuel line is clogged. Unclog it.

44. The fuel tank cap vent is clogged. Open the vent hole.

45. There is water in the fuel. Drain the tank, the carburetor, and the fuel lines and refill with fresh gasoline.

46. The fuel pump is faulty. Check the operation of the fuel pump if so equipped.

47. The governor is improperly set or parts are sticking or binding. Set the governor and check for binding parts and correct them.

48. The carburetor is not adjusted properly. Adjust it.

PROBLEM: Engine overheats.
CAUSES AND REMEDIES:

49. The ignition is not timed properly. Time the engine's ignition.

50. The fuel mixture is too lean. Adjust the carburetor.

51. The air intake screen or cooling fins are clogged. Clean away any obstructions.

52. The engine is being operated without the blower housing or shrouds in place. Install the blower housing and shrouds.

53. The engine is operating under an excessive load. Reduce the load and check associated equipment.

54. The oil level is too high. Check the oil level and drain some out if necessary.

55. There is not enough oil in the crankcase. Check the oil level and adjust accordingly.

56. The oil in the crankcase is of too low a viscosity or is excessively contaminated with fuel (four stroke). If the engine is a two-stroke, check for adequate fuel/oil mix — oil must be mixed with the fuel in proper proportions and be fully mixed. Check condition of crankcase oil (four-stroke) and if it appears very dirty, or there is doubt about proper viscosity, replace it.

57. The valve tappet clearance is too close. Adjust the valves to the proper specification.

58. Carbon has built up in the combustion chamber. Remove the cylinder and clean the head and piston of all carbon.

59. An improper amount of oil is mixed with the fuel (two stroke engines only). Drain the fuel tank and fill with correct mixture.

PROBLEM: The crankcase breather is passing oil (four stroke engines only).
CAUSES AND REMEDIES:

60. The crankcase is substantially over-filled with oil. Check oil level several minutes after engine has stopped. Wipe the dipstick clean before checking the level. If the crankcase is too full, drain oil as necessary until oil level is at or slightly below the upper mark.

61. The engine is being operated at two high rpm. Slow it down by adjusting the governor.

62. The oil fill cap or gasket is missing or damaged. Install a new cap and gasket and tighten it securely.

63. The breather mechanism is damaged. Replace the reed plate assembly.

64. The breather mechanism is dirty. Remove, clean, and replace it.

65. The drain hole in the breather is clogged. Clean the breather assembly and open the hole.

66. The piston ring gaps are aligned. Disassemble the engine and offset the ring gaps 90 degrees from each other.

67. The breather is loose or the gaskets are leaking. Tighten the breather to the crankcase.

68. The rings are not seated properly or they are worn. Install new rings.

PROBLEM: The engine backfires.
CAUSES AND REMEDIES:

69. The carburetor is adjusted so the air/fuel mixture is too lean. Adjust the carburetor.

70. The ignition is not timed correctly. Time the engine.

71. The valves are sticking. Do a valve job.

Charging System Troubleshooting

When performing charging system tests, turn off all lights and electrical components.

To ensure accurate meter indications, the battery terminal posts and battery cable clamps must be clean and tight.

✳✳WARNING

Do not make jumper wire connections except as instructed. Incorrect jumper wire connections can damage the regulator or fuse links.

PRELIMINARY INSPECTION

1. Make sure the battery cable connections are clean and tight.

2. Check all alternator and regulator wiring connections. Make sure all connections are clean and secure.

3. Check the alternator belt tension. Adjust, if necessary.

4. Check the fuse link between the starter relay and alternator. Replace if burned out.

5. Make sure the fuses/fuse links to the alternator are not burned or damaged. This could cause an open circuit or high resistance, resulting in erratic or intermittent charging problems.

6. If equipped with heated windshield, make sure the wiring connections to the alternator output control relay are correct and tight.

7. If equipped with heated windshield, make sure the connector to the heated windshield module is properly seated and there are no broken wires.

External Regulator Alternator

CHARGING SYSTEM INDICATOR LIGHT TEST

1. If the charging system indicator light does not come on with the ignition key in the **RUN** position and the engine not running, check the ignition switch-to-regulator I terminal wiring for an open circuit or burned out charging system indicator light. Replace the light, if necessary.

2. If the charging system indicator light does not come on, disconnect the electrical connector at the regulator and connect a jumper wire between the I terminal of the connector and the negative battery cable clamp.

3. The charging system indicator light should go on with the ignition switch in the **RUN** position.

4. If the light does not go on, check the light for continuity and replace, if necessary.

5. If the light is not burned out, there is an open circuit between the ignition switch and the regulator.

6. Check the 500 ohm resistor across the indicator light.

BASE VOLTAGE TEST

1. Connect the negative and positive leads of a voltmeter to the negative and positive battery cable clamps.

2. Make sure the ignition switch is in the **OFF** position and all electrical loads (lights, etc.) are OFF.

3. Record the battery voltage shown on the voltmeter; this is the base voltage.

NO-LOAD TEST

1. Connect a suitable tachometer to the engine.

2. Start the engine and bring the engine speed to 1500 rpm. With no other electrical loads (doors closed, foot off the

brake pedal), the reading on the voltmeter should increase, but no more than 2.5 volts above the base voltage.

➡**The voltage reading should be taken when the voltage stops rising. This may take a few minutes.**

3. If the voltage increases as in Step 2, perform the Load Test.

4. If the voltage continues to rise, perform the Over Voltage Tests.

5. If the voltage does not rise to the proper level, perform the Under Voltage Tests.

LOAD TEST

1. With the engine running, turn the blower speed switch to the high speed position and turn the headlights on to high beam.

2. Raise the engine speed to approximately 2000 rpm. The voltmeter reading should be a minimum of 0.5 volts above the base voltage. If not, perform the Under Voltage Tests.

➡**If the voltmeter readings in the No-Load Test and Load Test are as specified, the charging system is operating properly. Go to the following tests if one or more of the voltage readings differs, and also check for battery drain.**

OVER VOLTAGE TESTS

1. If the voltmeter reading was more than 2.5 volts above the base voltage in the No-Load Test, connect a jumper wire between the voltage regulator base and the alternator frame or housing. Repeat the No-Load Test.

2. If the over voltage condition disappears, check the ground connections on the alternator, regulator and from the engine to the dash panel and to the battery. Clean and securely tighten the connections.

3. If the over voltage condition still exists, disconnect the voltage regulator wiring connector from the voltage regulator. Repeat the No-Load Test.

4. If the over voltage condition disappears (voltmeter reads base voltage), replace the voltage regulator.

5. If the over voltage condition still exists with the voltage regulator wiring connector disconnected, check for a short between circuits A and F in the wiring harness and service, as necessary. Then reconnect the voltage regulator wiring connector.

UNDER VOLTAGE TESTS

1. If the voltage reading was not more than 0.5 volts above the base voltage, disconnect the wiring connector from the voltage regulator and connect an ohmmeter from the F terminal of the connector to ground. The ohmmeter should indicate more than 2.4 ohms.

2. If the ohmmeter reading is less than 2.4 ohms, service the grounded field circuit in the wiring harness or alternator and repeat the Load Test.

✳✳WARNING

Do not replace the voltage regulator before a shorted rotor coil or field circuit has been serviced. Damage to the regulator could result.

3. If the ohmmeter reading is more than 2.4 ohms, connect a jumper wire from the A to F terminals of the wiring connector and repeat the Load Test. If the voltmeter now indicates more than 0.5 volts above the base voltage, the regulator or wiring is damaged or worn. Perform the S and I Circuit Tests and service the wiring or regulator, as required.

4. If the voltmeter still indicates an under voltage problem, remove the jumper wire from the voltage regulator connector and leave the connector disconnected from the regulator.

5. Disconnect the FLD terminal on the alternator and pull back the protective cover from the BAT terminal. Connect a jumper wire between the FLD and BAT terminals and repeat the Load Test.

6. If the voltmeter indicates a 0.5 volts or more, increase above base voltage, perform the S and I Circuit Tests and service the wiring or regulator, as indicated.

7. If the voltmeter still indicates under voltage, shut the engine OFF and move the positive voltmeter lead to the BAT terminal of the alternator. If the voltmeter now indicates the base voltage, service the alternator. If the voltmeter indicates 0 volts, service the alternator-to-starter relay wire.

REGULATOR S AND I CIRCUIT TESTS

1. Disconnect the voltage regulator wiring connector and install a jumper wire between the A and F terminals.

2. With the engine idling and the negative voltmeter lead connected to the negative battery terminal, connect the positive voltmeter lead to the S terminal and then to the I terminal of the regulator wiring connector.

3. The S circuit voltage reading should be approximately ½ the I circuit reading. If the voltage readings are correct, remove the jumper wire. Replace the voltage regulator and repeat the Load Test.

4. If there is no voltage present, service the faulty wiring circuit. Connect the positive voltmeter lead to the positive battery terminal.

5. Remove the jumper wire from the regulator wiring connector and connect the connector to the regulator. Repeat the Load Test

FUSE LINK CONTINUITY

1. Make sure the battery is okay. (See Section 2)

2. Turn on any accessory. If it does not operate, the fuse link is probably burned out.

3. To test the fuse link that protects the alternator, check for voltage at the BAT terminal of the alternator, using a voltmeter. If there is no voltage, the fuse link is probably burned out.

Integral Regulator/External Fan Alternator

CHARGING SYSTEM INDICATOR LIGHT TEST

Two conditions can cause the charging system indicator light to come on when your engine is running: no alternator output, caused by a damaged alternator, regulator or wiring, or an over voltage condition, caused by a shorted alternator rotor, regulator or wiring.

In a normally functioning system, the charging system indicator light will be OFF when the ignition switch is in the **OFF** position, ON when the ignition switch is in the **RUN** position and the engine not running, and OFF when the ignition switch is in the **RUN** position and the engine is running.

1. If the charging system indicator light does not come on, disconnect the wiring connector from the regulator.

2. Connect a jumper wire between the connector I terminal and the negative battery cable clamp.

3. Turn the ignition switch to the **RUN** position, but leave the engine OFF. If the charging system indicator light does not come on, check for a light socket resistor. If there is a resistor, check the contact of the light socket leads to the flexible printed circuit. If they are good, check the indicator light for continuity and replace if burned out. If the light checks out good, perform the Regulator I Circuit Test.

4. If the indicator light comes on, remove the jumper wire and reconnect the wiring connector to the regulator. Connect the negative voltmeter lead to the negative battery cable clamp and connect the positive voltmeter lead to the regulator A terminal screw. Battery voltage should be indicated. If battery voltage is not indicated, service the A circuit wiring.

5. If battery voltage is indicated, clean and tighten the ground connections to the engine, alternator and regulator. Tighten loose regulator mounting screws to 15-26 inch lbs. (1.7-2.8 Nm).

6. Turn the ignition switch to the **RUN** position with the engine OFF. If the charging system indicator light still does not come on, replace the regulator.

BASE VOLTAGE TEST

1. Connect the negative and positive leads of a voltmeter to the negative and positive battery cable clamps.

2. Make sure the ignition switch is in the **OFF** position and all electrical loads (lights, radio, etc.) are OFF.

3. Record the battery voltage shown on the voltmeter; this is the base voltage.

NO-LOAD TEST

1. Connect a suitable tachometer to the engine.

2. Start the engine and bring the engine speed to 1500 rpm. With no other electrical loads (doors closed, foot off the brake pedal), the reading on the voltmeter should increase, but no more than 2.5 volts above the base voltage.

➡ **The voltage reading should be taken when the voltage stops rising. This may take a few minutes.**

3. If the voltage increases as in Step 2, perform the Load Test.

4. If the voltage continues to rise, perform the Over Voltage Tests.

5. If the voltage does not rise to the proper level, perform the Under Voltage Tests.

LOAD TEST

1. With the engine running, turn the blower speed switch to the high speed position and turn the headlights on to high beam.

2. Raise the engine speed to approximately 2000 rpm. The voltmeter reading should be a minimum of 0.5 volts above the base voltage. If not, perform the Under Voltage Tests.

➡ **If the voltmeter readings in the No-Load Test and Load Test are as specified, the charging system is operating properly. Go to the following tests if one or more of the voltage readings differs, and also check for battery drain.**

OVER VOLTAGE TESTS

If the voltmeter reading was more than 2.5 volts above base voltage in the No-Load Test, proceed as follows:

1. Turn the ignition switch to the **RUN** position, but do not start the engine.

2. Connect the negative voltmeter lead to the alternator rear housing. Connect the positive voltmeter lead first to the alternator output connection at the starter solenoid and then to the regulator A screw head.

3. If there is greater than 0.5 volts difference between the 2 locations, service the A wiring circuit to eliminate the high resistance condition indicated by excessive voltage drop.

4. If the over voltage condition still exists, check for loose regulator and alternator grounding screws. Tighten loose regulator grounding screws to 15-26 inch lbs. (1.7-2.8 Nm).

5. If the over voltage condition still exists, connect the negative voltmeter lead to the alternator rear housing. With the ignition switch in the **OFF** position, connect the positive voltmeter lead first to the regulator A screw head and then to the regulator F screw head. If there are different voltage readings at the 2 screw heads, a malfunctioning grounded brush lead or a grounded rotor coil is indicated; service or replace the entire alternator/regulator unit.

6. If the same voltage is obtained at both screw heads in Step 5 and there is no high resistance in the ground of the A+ circuit, replace the regulator.

UNDER VOLTAGE TESTS

If the voltmeter reading was not more than 0.5 volts above base voltage, proceed as follows:

1. Disconnect the electrical connector from the regulator. Connect an ohmmeter between the regulator A and F terminal screws. The ohmmeter reading should be more than 2.4 ohms. If it is less than 2.4 ohms, the regulator has failed. also check the alternator for a shorted rotor or field circuit. Perform the Load Test after servicing.

✳✳WARNING

Do not replace the voltage regulator before a shorted rotor coil or field circuit has been serviced. Damage to the regulator could result.

2. If the ohmmeter reading is greater than 2.4 ohms, connect the regulator wiring connector and connect the negative voltmeter lead to the alternator rear housing. Connect the positive voltmeter lead to the regulator A terminal screw. The voltmeter should indicate battery voltage. If there is no voltage, service the A wiring circuit and then perform the Load Test.

3. If the voltmeter indicates battery voltage, connect the negative voltmeter lead to the alternator rear housing. With the ignition switch in the **OFF** position, connect the positive

voltmeter lead to the regulator F terminal screw. The voltmeter should indicate battery voltage. If there is no voltage, there is an open field circuit in the alternator. Service or replace the alternator, then perform the Load Test after servicing.

4. If the voltmeter indicates battery voltage, connect the negative voltmeter lead to the alternator rear housing. Turn the ignition switch to the RUN position, leaving the engine off, and connect the positive voltmeter lead to the regulator F terminal screw. The voltmeter should read 1.5 volts or less. If more than 1.5 volts is indicated, perform the I circuit tests and service the I circuit if needed. If the I circuit is normal, replace the regulator, if needed, and perform the Load Test after servicing.

5. If 1.5 volts or less is indicated, disconnect the alternator wiring connector. Connect a set of 12 gauge jumper wires between the alternator B+ terminal blades and the mating wiring connector terminals. Perform the Load Test, but connect the positive voltmeter lead to one of the B+ jumper wire terminals. If the voltage increases more than 0.5 volts above base voltage, service the alternator-to-starter relay wiring. Repeat the Load Test, measuring voltage at the battery cable clamps after servicing.

6. If the voltage does not increase more than 0.5 volts above base voltage, connect a jumper wire from the alternator rear housing to the regulator F terminal. Repeat the Load Test with the positive voltmeter lead connected to one of the B+ jumper wire terminals. If the voltage increases more than 0.5 volts, replace the regulator. If the voltage does not increase more than 0.5 volts, service or replace the alternator.

REGULATOR S AND I CIRCUIT TEST

1. Disconnect the wiring connector from the regulator. Connect a jumper wire between the regulator A terminal and the wiring connector A lead and connect a jumper wire between the regulator F screw and the alternator rear housing.

2. With the engine idling and the negative voltmeter lead connected to the negative battery terminal, connect the positive voltmeter lead first to the S terminal and then to the I terminal of the regulator wiring connector.

3. The S circuit voltage should be approximately 1/2 that of the I circuit. If the voltage readings are correct, remove the jumper wire. Replace the regulator and connect the regulator wiring connector. Perform the Load Test.

4. If there is no voltage present, remove the jumper wire and service the faulty wiring circuit or alternator.

5. Connect the positive voltmeter lead to the positive battery terminal and connect the wiring connector to the regulator. Repeat the Load Test.

FUSE LINK CONTINUITY

1. Make sure the battery is okay (See Section 1).

2. Turn on the headlights or any accessory. If the headlights or accessory do not operate, the fuse link is probably burned out.

3. To test the fuse link that protects the alternator, check for voltage at the BAT terminal of the alternator and A terminal of the regulator, using a voltmeter. If there is no voltage, the fuse link is probably burned out.

FIELD CIRCUIT DRAIN

In all of the Field Circuit Drain test steps, connect the negative voltmeter lead to the alternator rear housing.

1. With the ignition switch in the OFF position, connect the positive voltmeter lead to the regulator F terminal screw. The voltmeter should read battery voltage if the system is operating normally. If less than battery voltage is indicated, go to Step 2.

2. Disconnect the wiring connector from the regulator and connect the positive voltmeter lead to the wiring connector I terminal. There should be no voltage indicated. If voltage is indicated, service the I lead from the ignition switch to identify and eliminate the voltage source.

3. If there was no voltage indicated in Step 2, connect the positive voltmeter lead to the wiring connector S terminal. No voltage should be indicated. If no voltage is indicated, replace the regulator.

4. If there was voltage indicated in Step 3, disconnect the wiring connector from the alternator rectifier connector. Connect the positive voltmeter lead to the regulator wiring connector S terminal. If voltage is indicated, service the S lead to the alternator connector to eliminate the voltage source. If no voltage is indicated, the alternator rectifier assembly is faulty.

Integral Regulator/Internal Fan Alternator

BASE VOLTAGE TEST

1. Connect the negative and positive leads of a voltmeter to the negative and positive battery cable clamps.

2. Make sure the ignition switch is in the OFF position and all electrical loads (lights, radio, etc.) are OFF.

3. Record the battery voltage shown on the voltmeter; this is the base voltage.

➡Turn the headlights ON for 10-15 seconds to remove any surface charge from the battery, then wait until the voltage stabilizes, before performing the base voltage test.

NO-LOAD TEST

1. Connect a suitable tachometer to the engine.

2. Start the engine and bring the engine speed to 1500 rpm. With no other electrical loads (doors closed, foot off the brake pedal), the reading on the voltmeter should increase, but no more than 3 volts above the base voltage.

➡The voltage reading should be taken when the voltage stops rising. This may take a few minutes.

3. If the voltage increases as in Step 2, perform the Load Test.

4. If the voltage continues to rise, perform the Over Voltage Tests.

5. If the voltage does not rise to the proper level, perform the Under Voltage Tests.

LOAD TEST

1. With the engine running, turn the blower speed switch to the high speed position and turn the headlights on to high beam.

2. Raise the engine speed to approximately 2000 rpm. The voltmeter reading should be a minimum of 0.5 volts above the base voltage. If not, perform the Under Voltage Tests.

➡If the voltmeter readings in the No-Load Test and Load Test are as specified, the charging system is operating properly. Go to the following tests if one or more of the voltage readings differs, and also check for battery drain.

OVER VOLTAGE TESTS

If the voltmeter reading was more than 3 volts above base voltage in the No-Load Test, proceed as follows:

1. Turn the ignition switch to the **RUN** position, but do not start the engine.

2. Connect the negative voltmeter lead to ground. Connect the positive voltmeter lead first to the alternator output connection at the starter solenoid (1992) or load distribution point (1993) and then to the regulator A screw head.

3. If there is greater than 0.5 volts difference between the 2 locations, service the A wiring circuit to eliminate the high resistance condition indicated by excessive voltage drop.

4. If the over voltage condition still exists, check for loose regulator and alternator grounding screws. Tighten loose regulator grounding screws to 16-24 inch lbs. (1.7-2.8 Nm).

5. If the over voltage condition still exists, connect the negative voltmeter lead to ground. Turn the ignition switch to the **OFF** position and connect the positive voltmeter lead first to the regulator A screw head and then to the regulator F screw head. If there are different voltage readings at the 2 screw heads, a malfunctioning regulator grounded brush lead or a grounded rotor coil is indicated; replace the regulator/brush set or the entire alternator.

6. If the same voltage reading, battery voltage, is obtained at both screw heads in Step 5, then there is no short to ground through the alternator field/brushes. Replace the regulator.

UNDER VOLTAGE TESTS

If the voltmeter reading was not more than 0.5 volts above base voltage, proceed as follows:

1. Disconnect the wiring connector from the regulator and connect an ohmmeter between the regulator A and F terminal screws. The ohmmeter should read more than 2.4 ohms. If the ohmmeter reads less than 2.4 ohms, check the alternator for shorted rotor to field coil or for shorted brushes. Replace the brush holder or the entire alternator assembly. Perform the Load Test after replacement.

✳✳WARNING

Do not replace the regulator if a shorted rotor coil or field circuit has been diagnosed, or regulator damage could result. Replace the alternator assembly.

2. If the ohmmeter reading is greater than 2.4 ohms, connect the regulator wiring connector and connect the negative voltmeter lead to ground. Connect the positive voltmeter lead to the regulator A terminal screw; battery voltage should be indicated. If there is no voltage, service the A wiring circuit and then perform the Load Test.

3. If battery voltage is indicated in Step 2, connect the negative voltmeter lead to ground. Turn the ignition switch to the **OFF** position, then connect the positive voltmeter lead to the regulator F terminal screw. Battery voltage should be indicated on the voltmeter. If there is no voltage, replace the alternator and then perform the Load Test.

4. If battery voltage is indicated in Step 3, connect the negative voltmeter lead to ground. Turn the ignition switch to the **RUN** position, but leave the engine OFF. Connect the positive voltmeter lead to the regulator F terminal screw; the voltmeter reading should be 2 volts or less. If more than 2 volts is indicated, perform the I circuit tests and service the I circuit, if needed. If the I circuit tests normal, replace the regulator, if needed, then perform the Load Test.

5. If 2 volts or less is indicated in Step 4, perform the Load Test, but connect the positive voltmeter lead to the alternator output stud. If the voltage increases more than 0.5 volts above base voltage, service the alternator-to-starter relay (1992) or alternator-to-load distribution point (1993) wiring. Repeat the Load Test, measuring the voltage at the battery cable clamps after servicing.

6. If the voltage does not increase more than 0.5 volts above base voltage in Step 5, perform the Load Test and measure the voltage drop from the battery to the A terminal of the regulator (regulator connected). If the voltage drop exceeds 0.5 volts, service the wiring from the A terminal to the starter relay (1992) or load distribution point (1993).

7. If the voltage drop does not exceed 0.5 volts, connect a jumper wire from the alternator rear housing to the regulator F terminal. Repeat the Load Test with the positive voltmeter lead connected to the alternator output stud. If the voltage increases more than 0.5 volts, replace the regulator. If voltage does not increase more than 0.5 volts, replace the alternator.

ALTERNATOR S CIRCUIT TEST

1. Disconnect the wiring connector from the regulator. Connect a jumper wire from the regulator A terminal to the wiring connector A lead. Connect a jumper wire from the regulator F screw to the alternator rear housing.

2. With the engine idling and the negative voltmeter lead connected to ground, connect the positive voltmeter lead first to the S terminal and then to the A terminal of the regulator wiring connector. The S circuit voltage should be approximately ½ the A circuit voltage. If the voltage readings are normal, remove the jumper wire, replace the regulator and connect the wiring connector. Repeat the Load Test.

3. If there is no voltage present, remove the jumper wire and service the damaged or worn wiring circuit or alternator.

4. Connect the positive voltmeter lead to the positive battery terminal. Connect the wiring connector to the regulator and repeat the Load Test.

FUSE LINK CONTINUITY

1. Make sure the battery is okay (See Section 1).

2. Turn on the headlights or any accessory. If the headlights or accessory do not operate, the fuse link is probably burned out.

3. To test the fuse link that protects the alternator, check for voltage at the BAT terminal of the alternator and A terminal of the regulator, using a voltmeter. If there is no voltage, the fuse link is probably burned out.

FIELD CIRCUIT DRAIN

In all of the Field Circuit Drain test steps, connect the negative voltmeter lead to the alternator rear housing.

1. With the ignition switch in the **OFF** position, connect the positive voltmeter lead to the regulator F terminal screw. The voltmeter should read battery voltage if the system is operating normally. If less than battery voltage is indicated, go to Step 2.

2. Disconnect the wiring connector from the regulator and connect the positive voltmeter lead to the wiring connector I terminal. There should be no voltage indicated. If voltage is indicated, service the I lead from the ignition switch to identify and eliminate the voltage source.

3. If there was no voltage indicated in Step 2, connect the positive voltmeter lead to the wiring connector S terminal. No voltage should be indicated. If no voltage is indicated, replace the regulator.

4. If there was voltage indicated in Step 3, disconnect the 1-pin S terminal connector. Again, connect the positive voltmeter lead to the regulator wiring connector S terminal. If voltage is indicated, service the S lead wiring to eliminate the voltage source. If no short is found, replace the alternator.

Starter Motor

TESTING

Place the transmission in **N** or **P**. Disconnect the vacuum line to the Thermactor® bypass valve, if equipped, before performing any cranking tests. After tests, run the engine for 3 minutes before connecting the vacuum line.

Starter Cranks Slowly

1. Connect jumper cables as shown in the Jump Starting procedure in Section 1. If, with the aid of the booster battery, the starter now cranks normally, check the condition of the battery. Recharge or replace the battery, as necessary. Clean the cables and battery posts and make sure connections are tight.

2. If Step 1 does not correct the problem, clean and tighten the connections at the starter relay and battery ground on the engine. You should not be able to rotate the eyelet terminals easily, by hand. Also make sure the positive cable is not shorted to ground.

3. If the starter still cranks slowly, it must be replaced.

Starter Relay Operates But Starter Doesn't Crank

1. Connect jumper cables as shown in the Jump Starting procedure in Section 1. If, with the aid of the booster battery, the starter now cranks normally, check the condition of the battery. Recharge or replace the battery, as necessary. Clean the cables and battery posts and make sure connections are tight.

2. If Step 1 does not correct the problem, clean and tighten the connections at the starter and relay. Make sure the wire strands are secure in the eyelets.

3. On models with a fender mounted solenoid, if the starter still doesn't crank, it must be replaced.

4. On equipment with starter mounted solenoid: Connect a jumper cable across terminals B and M of the starter solenoid.

If the starter does not operate, replace the starter. If the starter does operate, replace the solenoid.

✳✳CAUTION

Making the jumper connections could cause a spark. Battery jumper cables or equivalent, should be used due to the high current in the starting system.

Starter Doesn't Crank — Relay Chatters or Doesn't Click

1. Connect jumper cables as shown in the Jump Starting procedure in Section 1. If, with the aid of the booster battery, the starter now cranks normally, check the condition of the battery. Recharge or replace the battery, as necessary. Clean the cables and battery posts and make sure connections are tight.

2. If Step 1 does not correct the problem, remove the push-on connector from the relay (red with blue stripe wire). Make sure the connection is clean and secure and the relay bracket is grounded.

3. If the connections are good, check the relay operation with a jumper wire. Remove the push-on connector from the relay and, using a jumper wire, jump from the now exposed terminal on the starter relay to the main terminal (battery side or battery positive post). If this corrects the problem, check the ignition switch, neutral safety switch and the wiring in the starting circuit for open or loose connections.

4. If a jumper wire across the relay does not correct the problem, replace the relay.

Start Spins But Doesn't Crank Engine

1. Remove the starter.

2. Check the armature shaft for corrosion and clean or replace, as necessary.

3. If there is no corrosion, replace the starter drive.

Engine Overhaul Tips

Most engine overhaul procedures are fairly standard. In addition to specific parts replacement procedures and complete specifications for your individual engine, this section also is a guide to accept rebuilding procedures. Examples of standard rebuilding practice are shown and should be used along with specific details concerning your particular engine.

Competent and accurate machine shop services will ensure maximum performance, reliability and engine life.

In most instances it is more profitable for the do-it-yourself mechanic to remove, clean and inspect the component, buy the necessary parts and deliver these to a shop for actual machine work.

On the other hand, much of the rebuilding work (crankshaft, block, bearings, piston rods, and other components) is well within the scope of the do-it-yourself mechanic.

TOOLS

The tools required for an engine overhaul or parts replacement will depend on the depth of your involvement. With a few exceptions, they will be the tools found in a mechanic's tool kit. More in-depth work will require any or all of the following:

• A dial indicator (reading in thousandths) mounted on a universal base
• Micrometers and telescope gauges
• Jaw and screw-type pullers
• Scraper
• Valve spring compressor
• Ring groove cleaner
• Piston ring expander and compressor
• Ridge reamer
• Cylinder hone or glaze breaker
• Plastigage®
• Engine stand

The use of most of these tools is illustrated in this section. Many can be rented for a one-time use from a local parts jobber or tool supply house specializing in automotive work.

Occasionally, the use of special tools is called for. See the information on Special Tools and Safety Notice in the front of this book before substituting another tool.

INSPECTION TECHNIQUES

Procedures and specifications are given in this section for inspecting, cleaning and assessing the wear limits of most major components. Other procedures such as Magnaflux® and Zyglo® can be used to locate material flaws and stress cracks. Magnaflux® is a magnetic process applicable only to ferrous materials. The Zyglo® process coats the material with a fluorescent dye penetrant and can be used on any material Check for suspected surface cracks can be more readily made using spot check dye. The dye is sprayed onto the suspected area, wiped off and the area sprayed with a developer. Cracks will show up brightly.

OVERHAUL TIPS

Aluminum has become extremely popular for use in engines, due to its low weight. Observe the following precautions when handling aluminum parts:

• Never hot tank aluminum parts (the caustic hot tank solution will eat the aluminum.
• Remove all aluminum parts (identification tag, etc.) from engine parts prior to the tanking.
• Always coat threads lightly with engine oil or anti-seize compounds before installation, to prevent seizure.
• Never oveorque bolts or spark plugs especially in aluminum threads.

Stripped threads in any component can be repaired using any of several commercial repair kits (Heli-Coil® , Microdot® , Keenserts® , etc.).

When assembling the engine, any parts that will be frictional contact must be prelubed to provide lubrication at initial start-up. Any product specifically formulated for this purpose can be used, but engine oil is not recommended as a prelube.

When semi-permanent (locked, but removable) installation of bolts or nuts is desired, threads should be cleaned and coated with Loctite® or other similar, commercial non-hardening sealant.

REPAIRING DAMAGED THREADS

▶ **See Figures 1, 2, 3, 4 and 5**

Several methods of repairing damaged threads are available. Heli-Coil® (shown here), Keenserts® and Microdot® are among the most widely used. All involve basically the same principle — drilling out stripped threads, tapping the hole and installing a prewound insert — making welding, plugging and oversize fasteners unnecessary.

Two types of thread repair inserts are usually supplied: a standard type for most Inch Coarse, Inch Fine, Metric Course and Metric Fine thread sizes and a spark lug type to fit most spark plug port sizes. Consult the individual manufacturer's catalog to determine exact applications. Typical thread repair kits will contain a selection of prewound threaded inserts, a tap (corresponding to the outside diameter threads of the insert)

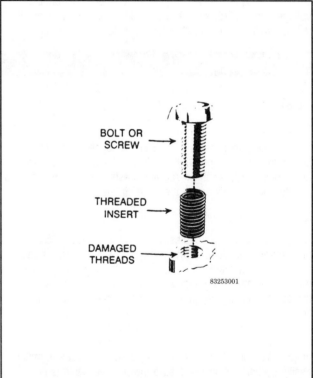

Fig. 1 Damaged bolt holes can be repaired with thread repair inserts

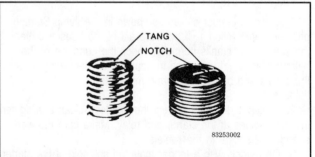

Fig. 2 Standard thread repair insert (left) and spark plug thread insert (right)

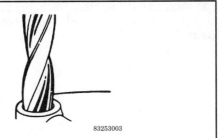

Fig. 3 Drill out the damaged threads with specified drill. Drill completely through the hole or to the bottom of a blind hole

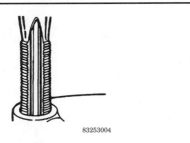

Fig. 4 With the tap supplied, tap the hole to receive the thread insert. Keep the tap well oiled and back it out frequently to avoid clogging the threads

Fig. 5 Screw the threaded insert onto the installation tool until the tang engages the slot. Screw the insert into the tapped hole until it is ¼-½ turn below the top surface, after installation break off the tang with a hammer and punch

and an installation tool. Spark plug inserts usually differ because they require a tap equipped with pilot threads and a combined reamer/tap section. Most manufacturers also supply blister-packed thread repair inserts separately in addition to a master kit containing a variety of taps and inserts plus installation tools.

Before effecting a repair to a threaded hole, remove any snapped, broken or damaged bolts or studs. Penetrating oil can be used to free frozen threads. The offending item can be removed with locking pliers or with a screw or stud extractor. After the hole is clear, the thread can be repaired, as shown in the series of accompanying illustrations.

Checking Engine Compression

▶ See Figure 6

A noticeable lack of engine power, excessive oil consumption and/or poor fuel mileage measured over an extended period are all indicators of internal engine war. Worn piston rings, scored or worn cylinder bores, blown head gaskets, sticking or burnt valves and worn valve seats are all possible culprits here. A check of each cylinder's compression will help you locate the problems.

As mentioned earlier, a screw-in type compression gauge is more accurate that the type you simply hold against the spark plug hole, although it takes slightly longer to use. It's worth it to obtain a more accurate reading. Follow the procedures below.

1. Warm up the engine to normal operating temperature.
2. Remove all the spark plugs.
3. Disconnect the high tension lead from the ignition coil.
4. On fully open the throttle either by operating the carburetor throttle linkage by hand or by having an assistant floor the accelerator pedal.
5. Screw the compression gauge into the No. 1 spark plug hole until the fitting is snug.

✲✲WARNING

Be careful not to crossthread the plug hole. On aluminum cylinder heads use extra care, as the threads in these heads are easily ruined.

6. Open the throttle fully. Then, while you read the compression gauge, crank the engine two or three times in short bursts.
7. Read the compression gauge at the end of each series of cranks, and record the highest of these readings. Compare the highest reading to the reading in each cylinder.

A cylinder's compression pressure is usually acceptable if it is not less than 80% of of the highest reading. For example, if the highest reading is 150 psi, the lowest should be no lower than 120 psi.

No cylinder should have a reading below 100 psi.

8. If a cylinder is unusually low, pour a tablespoon of clean engine oil into the cylinder through the spark plug hole and repeat the compression test. If the compression comes up after adding the oil, it appears that the cylinder's piston rings or bore are damaged or worn. If the pressure remains low, the valves may not be seating properly (a valve job is needed), or

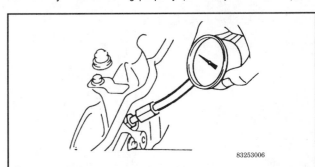

Fig. 6 The screw-in type compression gauge is more accurate

the head gasket may be blown near that cylinder. If compression in any two adjacent cylinders is low, and if the addition of oil doesn't help the compression, there is leakage past the head gasket. Oil and coolant water in the combustion chamber can result from this problem. There may be evidence of water droplets on the engine dipstick when a head gasket has blown.

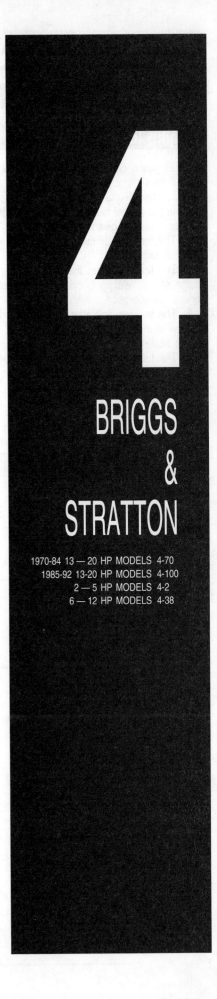

4

BRIGGS & STRATTON

2 — 5 HP MODELS

Engine Identification

The Briggs & Stratton model designation system consists of up to a six digit number. It is possible to determine most of the important mechanical features of the engine by merely knowing the model number. An explanation of what each number means is given below.

1. The first one or two digits indicate the cubic inch displacement (cid).

2. The first digit after the displacement indicates the basic design series, relating to cylinder construction, ignition and general configuration.

3. The second digit after the displacement indicates the position of the crankshaft and the type of carburetor the engine has.

4. The third digit after the displacement indicates the type of bearings and whether or not the engine is equipped with a reduction gear or auxiliary drive.

5. The last digit indicates the type of starter. The model identification plate is usually located on the air baffle surrounding the cylinder.

Air Cleaners

▶ **See Figures 1 and 2**

A properly serviced air cleaner protects the engine from dust particles that are in the air. When servicing an air cleaner,

check the air cleaner mounting and gaskets for worn or damaged mating surfaces. Replace any worn or damaged parts to prevent dirt and dust from entering the engine through openings caused by improper sealing. Straighten or replace any bent mounting studs.

SERVICING

Oil/Foam Air Cleaners

Clean and re-oil the air cleaner element every 25 hours of operation under normal operating conditions. The capacity of the oil/foam air cleaner is adequate for a full season's use without cleaning. Under very dusty conditions, clean the air cleaner every few hours of operation.

The oil/foam air cleaner is serviced in the following manner:

1. Remove the screw that holds the halves of the air cleaner shell together and retains it to the carburetor.

2. Remove the air cleaner carefully to prevent dirt from entering the carburetor.

3. Take the air cleaner apart (split the two halves).

4. Wash the foam in kerosene or liquid detergent and water to remove the dirt.

5. Wrap the foam in a clean cloth and squeeze it dry.

6. Saturate the foam in clean engine oil and squeeze it to remove the excess oil.

7. Assemble the air cleaner and fasten it to the carburetor with the attaching screw.

Briggs and Stratton Model Numbering System

Cubic Inch Displacement	First Digit After Displacement Basic Design Series	Second Digit After Displacement Crankshaft, Carburetor Governor	Third Digit After Displacement Bearings, Reduction Gears & Auxiliary Drives	Fourth Digit After Displacement Type of Starter
6	0	0-	0-Plain Bearing	0-Without Starter
8	1	1-Horizontal Vacu-Jet	1-Flange Mounting Plain Bearing	1-Rope Starter
9	2			
10	3	2-Horizontal Pulsa-Jet	2-Ball Bearing	2-Rewind Starter
13	4			
14	5	3-Horizontal (Pneumatic) Flo-Jet (Governor)	3-Flange Mounting Ball Bearing	3-Electric-110 Volt, Gear Drive
17	6			
19	7	4-Horizontal (Mechanical) Flo-Jet (Governor)	4-	4-Elec. Starter-Generator-12 Volt, Belt Drive
20	8			
23	9			
24		5-Vertical Vacu-Jet	5-Gear Reduction (6 to 1)	5-Electric Starter Only-12 Volt, Gear Drive
30				
32				
		6-	6-Gear Reduction (6 to 1) Reverse Rotation	6-Wind-up Starter
		7-Vertical Flo-Jet	7-	7-Electric Starter, 12 Volt Gear Drive, with Alternator
		8-	8-Auxiliary Drive Perpendicular to Crankshaft	8-Vertical-pull Starter
		9-Vertical Pulsa-Jet	9-Auxiliary Drive Parallel to Crankshaft	

832540C1

General Engine Specifications

Model	Bore Size (in.)	Horsepower
Aluminum Engines		
6B 60000	2.375	2
8B, 80000, 82000	2.375	3
92000	2.5625	3.5
100000	2.5	4
110000, 111000, 111,200	2.7812	4
130000	2.5625	5
Cast Iron Engines		
5, 6, N	2.000	2
8	2.250	3
9	2.250	3.5
14	2.625	5

832540C2

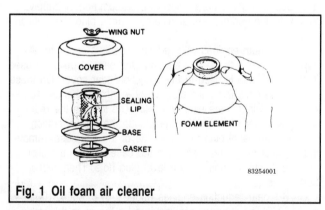

Fig. 1 Oil foam air cleaner

Fig. 2 Oil foam air cleaner

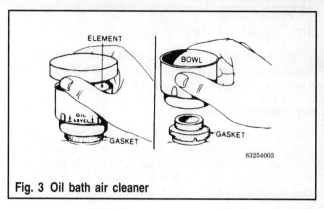

Fig. 3 Oil bath air cleaner

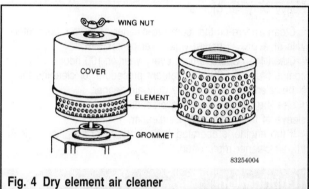

Fig. 4 Dry element air cleaner

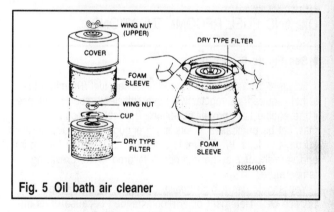

Fig. 5 Oil bath air cleaner

Oil Viscosity Recommendations

Winter (under 40°F.)	Summer (above 40°F.)
SAE 5W-20, or SAE 5W-30	SAE 30
If above are not available: SAE 10W or SAE 10W-30 (under 0°F.) Use SAE 10W or SAE 10W-30 in proportions of 90% motor oil/ 10% kerosene	If above is not available: SAE 10W-40 or SAE 10W-30

832540C3

Oil Bath Air Cleaner

♦ See Figures 3, 4 and 5

Pour the old oil out of the bowl. Wash the element thoroughly in solvent and squeeze it dry. Clean the bowl and refill it with the same type of oil used in the crankcase.

Dry Element Air Cleaner

Remove the element of the air cleaner and tap (top and bottom) it on a flat surface or wash it in non-sudsing detergent and flush it from the inside until the water coming out is clear. After washing, air dry the element thoroughly before reinstalling it on the engine. NEVER OIL A DRY ELEMENT.

Heavy Duty Air Cleaner

Clean and re-oil the foam pre-cleaner at three month intervals or every 25 hours, whichever comes first.

Clean the paper element every year or 100 hours, whichever comes first. Use the dry element procedure for cleaning the paper element of the heavy duty air cleaner.

Use the oil foam cleaning procedure to clean the foam sleeve of the heavy duty air cleaner.

If the engine is operated under very dusty conditions, clean the air cleaner more often.

Lubrication

OIL AND FUEL RECOMMENDATIONS

♦ See Figure 6

Briggs & Stratton recommends unleaded fuel. Premium fuel is not required, as regular will have sufficient knock resistance if the engine is in proper condition. The factory recommends that fuel be purchased in lots small enough to be used up in 30 days or less. When fuel is older than that, it can form gum and varnish, or may be improperly tailored to the prevailing temperature.

You should use a high quality detergent oil designated 'For Service SG'. Detergent oil is recommended because of its important ability to keep gum and varnish from clogging the lubrication system. Briggs & Stratton specifically recommends that no special oil additives be used.

Oil must be changed every 25 hours of operation. If the atmosphere in which the engine is operating is very dirty, oil

Engine Oil Capacity Chart

Basic Model Series	Capacity Pints
Aluminum	
6, 8, 9, 11 Cu. in. Vert. Crankshaft	1¼
6, 8, 9 Cu. in. Horiz. Crankshaft	1¼
10, 13 Cu. in. Vert. Crankshaft	1¾
10, 13 Cu. in. Horiz. Crankshaft	1¼
Cast Iron	
9, 14, 19, 20 Cu. in. Horiz. Crank.	3

832540C4

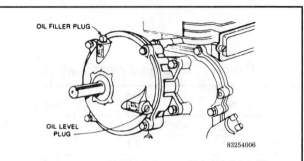

Fig. 6 Location of oil fill and level plugs on gear reduction engines

changes should be made more frequently, as often as every 12 hours, if necessary. Oil should be changed after 5 hours of operation in the case of brand new engines. Drain engine oil when hot.

In hot weather, when under heavy load, or when brand new, engines may consume oil at a rate which will require you to refill the crankcase several times between oil changes. Check the oil level every hour or so until you can accurately estimate how long the engine can go between refills. To check oil level, stop the engine and allow it to sit for a couple of minutes, then remove the dipstick or filler cap. Fill the crankcase to the top of the filler pipe when there is no dipstick, or wipe the dipstick clean, reinsert it, and add oil as necessary until the level reaches the upper mark.

On cast iron engines with a gear reduction unit, crankcase and reduction gears are lubricated by a common oil supply. When draining crankcase, also remove drain plug in reduction unit.

On aluminum engines with reduction gear, a separate oil supply lubricates the gears, although the same type of oil used in the crankcase is used in the reduction gear cover. On these engines, remove the drain plug every fourth oil change (100 hours), then install the plug and refill. The level in the reduction gear cover must be checked during the refill operation by removing the level plug from the side of the gearcase, removing the filler plug, and then filling the case through the filler plug hole until oil runs out the level plug hole. Then, install both plugs.

On 6-1 gear reduction engines (models 6, 8, 8000, 10000, and 13000), no changes are required for the oil in the reduction gear case, but level must be checked and the case refilled, as described in the paragraph above, every 100 hours. Make sure the oil level plug (with screwdriver slot and no vent) is installed in the hole on the side of the case.

Spark Plugs

♦ See Figure 7

Remove the spark plug with a ¾ in. (1½ in. plug) or a ¹⁵/₁₆ in. (2 in. plug) deep well socket wrench. Clean carbon deposits off the center and side electrodes with a sharp instrument. If possible, you should also attempt to remove deposits from the recess between the insulator and the threaded portion of the plug. If the electrodes are burned away or the insulator is cracked at any point, replace the plug. Using a wire type feeler gauge, adjust the gap by bending the side electrode where it is curved until the gap is 0.030 in. (0.8mm).

Tune-Up Specifications

Model	Plug Type	Plug Gap (in.)	Point Gap (in.)	Armature Gap		Idle Speed
				2 leg	3 leg	
Aluminum Block						
6B, 6000, 8B	①	.030	.020	.006–.010	.012–.016	1750
80000, 82000, 92000, 110900	①	.030	.020	.006–.010	.012–.016	1750
100000, 130000	①	.030	.020	.010–.014	.016–.019	1750
Cast Iron Block						
5, 6, N, 8	①	.030	.020	—	.022–.026	1750
9	①	.030	.020	—	—	1200
14	①	.030	.020	—	—	1200

① Manufacturer's Code		Manufacturer
1½ in. plug	2 in. plug	
CJ-8	J-8	Champion
RCJ-8	RJ-8	Champion (resistor)
A-7NX	A-71	Autolite
AR-7N	AR-80	Autolite (resistor)
CS-45	GC-46	A.C.
—	R-46	A.C. (resistor)

832540C5

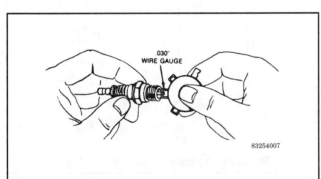

030″ WIRE GAUGE

83254007

Fig. 7 Checking spark plug gap with a wire feeler gauge

When installing the plug, make sure the threads of the plug and the threads in the cylinder head are clean. It is best to oil the plug threads very lightly. Be careful not to overtorque the plug, especially if the engine has an aluminum head. If you use a torque wrench, torque to about 15 ft. lbs.

Breaker Points

All Briggs & Stratton engines have magneto ignition systems. Three types are used: Flywheel Type — Internal Breaker, Flywheel Type — External Breaker, and Magna-Matic.

REMOVAL & INSTALLATION

Flywheel Type — Internal Breaker

▶ **See Figures 8, 9, 10, 11, 12, 13, 14, 15, 16 and 17**

This ignition system has the magneto located on the flywheel and the breaker points located under the flywheel.

The flywheel is located on the crankshaft with a soft metal key. It is held in place by a nut or starter clutch. The flywheel key must be in good condition to insure proper location of the flywheel for ignition timing. Do not use a steel key under any circumstances. Use only a soft metal key, as originally supplied.

The keyway in both flywheel and crankshaft should not be distorted. Flywheels are made of aluminum, zinc, or cast iron.

1. Place a block of wood under the flywheel fins to prevent the flywheel from turning while you are loosening the nut or

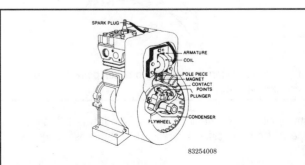

SPARK PLUG
ARMATURE
COIL
POLE PIECE
MAGNET
CONTACT POINTS
PLUNGER
FLYWHEEL
CONDENSER

83254008

Fig. 8 Flywheel magneto ignition with internal breaker points and external armature

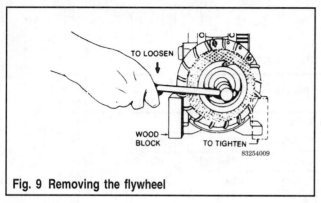

Fig. 9 Removing the flywheel

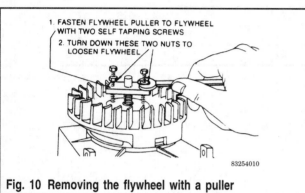

Fig. 10 Removing the flywheel with a puller

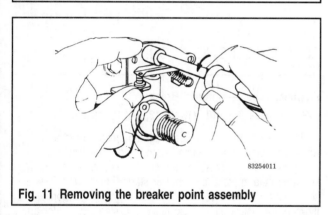

Fig. 11 Removing the breaker point assembly

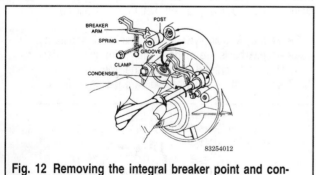

Fig. 12 Removing the integral breaker point and condenser assembly

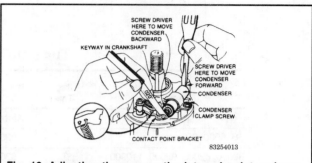

Fig. 13 Adjusting the gap on the integral point and condenser assembly

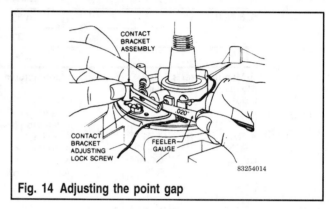

Fig. 14 Adjusting the point gap

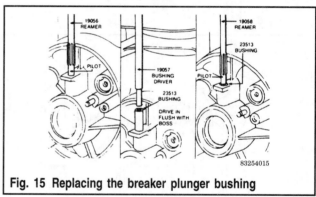

Fig. 15 Replacing the breaker plunger bushing

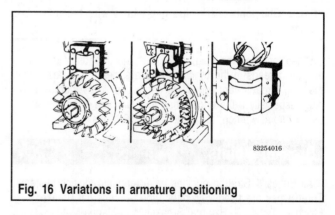

Fig. 16 Variations in armature positioning

starter clutch. Be careful not to bend the flywheel. There are special flywheel holders available for this purpose; Briggs &

Stratton recommends their use on flywheels of 6¾ in. 171.45mm) diameter or less.

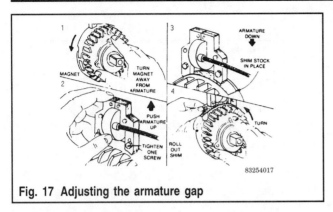

Fig. 17 Adjusting the armature gap

2. On rope starter engines, the ½ in. flywheel nut has a lefthand thread and the ⅝ in. nut has a righthand thread. The starter clutch used on rewind or wind-up starters has a right-hand thread. Some flywheels have two holes provided for the use of a flywheel puller. Use a small gear puller or auto-motive steering wheel puller to remove the flywheel if a fly-wheel puller is not available. Be careful not to bend the fly-wheel if a gear puller is used. On rope starter engines leave the nut on for the puller to bear against. Small cast iron flywheels do not require a puller.

3. Remove the breaker cover. Care should be taken when removing the cover, to avoid damaging it. If the cover is bent or damaged, it should be replaced to insure a proper seal. The breaker point gap on all models is 0.020 in. (0.5mm). Check the points for contact and for signs of burning or pitting. Points that are set too wide will advance the spark timing and may cause kickback when starting. Points that are set too close will retard the spark timing and decrease engine power. On mod-els that have a separate condenser, the point set is removed by first removing the condenser and armature wires from the breaker point clip. Loosen the adjusting lock screw and remove the breaker point assembly. On models where the condenser is incorporated with the breaker points, loosen the screw which holds the post. The condenser/point assembly is removed by loosening the screw which holds the condenser clamp.

4. When installing a point set with the separate condenser, be sure that the small boss on the magneto plate enters the hole in the point bracket. Mount the point set to the magneto plate or the cylinder with a lock screw. Fasten the armature lead wire to the breaker points with the clip and screw. If these lead wires do not have terminals, the bare end of the wires can be inserted into the clip and the screw tightened to make a good connection. Do not let the ends of the wire touch either the point bracket or the magneto plate, or the ignition will be grounded.

5. To install the integral condenser/point set, place the mounting post of the breaker arm into the recess in the cylin-der so that the groove in the post fits the notch in the recess. Tighten the mounting screw securely. Use a ¼ in. wrench. Slip the open loop of the breaker arm spring through the two holes in the arm, then hook the closed loop of the spring over the small post protruding from the cylinder. Push the flat end of the breaker arm into the groove in the mounting post. This places tension on the spring and pulls the arm against the plunger. If the condenser post is threaded, attach the soil pri-mary wire and the ground wire (if furnished) with the lock washer and nut. If the primary wire is fastened to the con-denser with a spring fastener, compress the spring and slip

the primary wire and ground wire into the hole in the con-denser post. Release the spring. Lay the condenser in place and tighten the condenser clamp securely. Install the spring in the breaker arm.

POINT GAP ADJUSTMENT

Turn the crankshaft until the points are open to the widest gap. When adjusting a breaker point assembly with an integral condenser, move the condenser forward or backward with a screwdriver until the proper gap is obtained — 0.020 in. (0.5mm). Point sets with a separate condenser are adjusted by moving the contact point bracket up and down after the lock screw has been loosened. The point gap is set to 0.020 in. (0.5mm).

BREAKER POINT PLUNGER

If the breaker point plunger hole becomes excessively worn, oil will leak past the plunger and may get on the points, causing them to burn. To check the hole, loosen the breaker point mounting screw and move the breaker points out of the way. Remove the plunger. If the flat end of the #19055 plug gauge will enter the plunger hole for a distance of ¼ in. (6mm) or more, the hole should be rebushed.

To install the bushing, it is necessary that the breaker points, armature, and crankshaft be removed. Use a #19056 reamer to ream out the old plunger hole. This should be done by hand. The reamer must be in alignment with the plunger hole. Drive the bushing, #23513, into the hole until the upper end of the bushing is flush with the top of the boss. Remove all metal chips and dirt from the engine.

If the breaker point plunger is worn to a length of 0.870 in. (22mm) or less, it should be replaced. Plungers must be in-serted with the groove at the top or oil will enter the breaker box. Insert the plunger into the hole in the cylinder.

ARMATURE AIR GAP ADJUSTMENT

Set the air gap between the flywheel and the armature as follows: With the armature up as far as possible and just one screw tightened, slip the proper gauge between the armature and flywheel. Turn the flywheel until the magnets are directly below the armature. Loosen the one mounting screw and the magnets should pull the armature down firmly against the thickness gauge. Tighten the mounting screws.

Flywheel Type — External Breaker
▶ See Figures 18, 19, 20, 21 and 22

1. Turn the crankshaft until the points open to their widest gap. This makes it easier to assemble and adjust the points later if the crankshaft is not removed.

2. Remove the condenser and upper and lower mounting screws.

3. Loosen the lock nut and back off the breaker point screw.

4. Install the points in the reverse order of removal.

To avoid the possibility of oil leaking past the breaker point plunger or moisture entering the crankcase between the plunger and the bushing, a plunger seal is installed on the engine models using this type of ignition system. To install a new seal on the plunger, remove the breaker point assembly and condenser. Remove the retainer and eyelet, remove the old seal, and install the new one. Use extreme care when

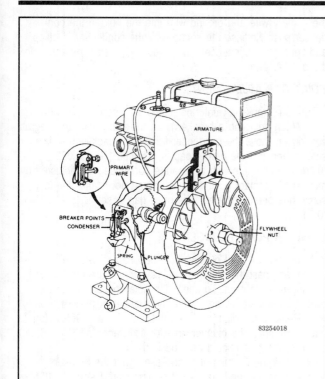

Fig. 18 Flywheel magneto ignition with an external breaker assembly

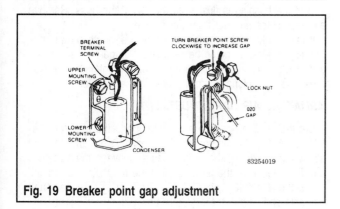

Fig. 19 Breaker point gap adjustment

Fig. 20 Removing an unthreaded plunger bushing

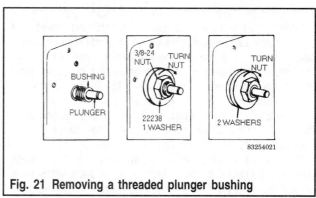

Fig. 21 Removing a threaded plunger bushing

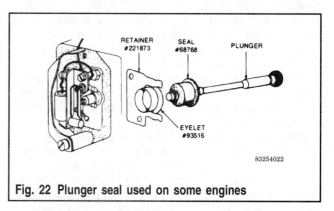

Fig. 22 Plunger seal used on some engines

installing the seal on the plunger to avoid damaging the seal. Replace the eyelet and retainer and replace the points and condenser.

➡Apply a small amount of sealer to the threads of both mounting screws and the adjustment screw. The sealer prevents oil from leaking into the breaker point area.

POINT GAP ADJUSTMENT

Turn the crankshaft until the points open to their widest gap. Turn the breaker point adjusting screw until the points open to 0.020 in. (0.5mm) and tighten the lock nut. When the cover is installed, seal the point where the primary wire passes under the cover. This area must be resealed to prevent the entry of dust and moisture.

REPLACING THREADED BREAKER PLUNGER AND BUSHING

Remove the breaker cover and the condenser and breaker point assembly.

Place a thick ⅜ in. (9.5mm) inside diameter washer over the end of the bushing and screw on the ⅜-24 nut. Tighten the nut to pull the bushing out of the hole. After the bushing has been moved about ⅛ in. (3mm), remove the nut and put on a second thick washer and repeat the procedure. A total stack of ⅜ in. (9.5mm) washers will be required to completely remove the bushing. Be sure the plunger does not fall out of the bushing as it is removed.

Place the new plunger in the bushing with the large end of the plunger opposite the threads on the bushing. Screw the ⅜-24 in. nut onto the threads to protect them and insert the bushing into the cylinder. Place a piece of tubing the same diameter as the nut and, using a hammer, drive the bushing into the cylinder until the square shoulder on the bushing is flush with the face of the cylinder. Check to be sure that the plunger operates freely.

REPLACING UNTHREADED BREAKER PLUNGER AND BUSHING

Pull the plunger out as far as possible and use a pair of pliers to break the plunger off as close as possible to the bushing. Use a ¼-20 in. tap or a #93029 self threading screw to thread the hole in the bushing to a depth of about ½-⅝ in. (13-15mm). Use a ¼-20 · ½ in. hex head screw and two spacer washers to pull the bushing out of the cylinder. The bushing will be free when it has been extracted ⁵⁄₁₆ in. (8mm). Carefully remove the bushing and the remainder of the broken plunger. Do not allow the plunger or metal chips to drop into the crankcase.

Correctly insert the new plunger into the new bushing. Insert the plunger and the bushing into the cylinder. Use a hammer and the old bushing to drive the new bushing into the cylinder until the new bushing is flush with the face of the cylinder. Make sure that the plunger operates freely.

PLUNGER SEAL

Later models with Flywheel Type-External Breaker Ignition feature a plunger seal. This seal keeps both oil and moisture from entering the breaker box. If the points have become contaminated on an engine manufactured without this feature, the seal may be installed. Parts, part numbers, and their locations are shown in the illustration. Install the seal onto the plunger very carefully to avoid fracturing it.

Magna-Matic Ignition System

▶ **See Figures 23, 24, 25, 26, 27 and 28**

1. Using pullers similar to factory designs numbered #19068 and 19203, screw the two bolts into the holes tapped into the flywheel. The bolts are turned until the flywheel is forced off the crankshaft. Only this type of device should be used to pull these flywheels.

2. Loosen the socket head screw in the rotor clamp which will allow the clamp to loosen. It may be necessary to use a puller to remove the rotor from the crankshaft. On older models, loosen the small lock screw, then the set screw.

3. Usually the coil and armature are not separated, but left assembled for convenience. However, if one or both need replacement, proceed as follows: the coil primary wire and the coil ground wire must be unfastened. Pry out the clips that hold the coil and coil core to the armature. The coil core is a slip fit in the coil and can be pushed out of the coil.

4. Turn the crankshaft until the points open to the widest gap. This makes it easier to assemble and adjust the points later if the crankshaft is not removed. With the terminal screw out, remove the spring screw. Loosen the breaker shaft nut until the nut is flush with the end of the shaft. Tap the nut to free the breaker arm from the tapered end of the breaker shaft. Remove the nut, lockwasher, and breaker arm. Remove the breaker plate screw, breaker plate, pivot, insulating plate, and eccentric. Pry out the breaker shaft seal with a sharp pointed tool.

5. Remove the two mounting screws, then remove the breaker box, turning it slightly to clean the arm at the inner

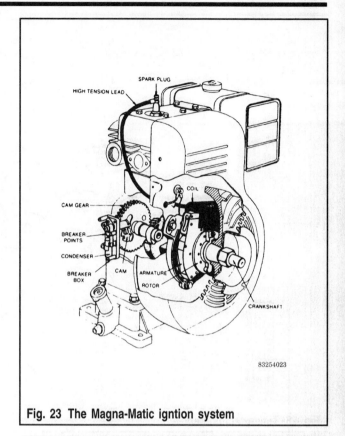

Fig. 23 The Magna-Matic igntion system

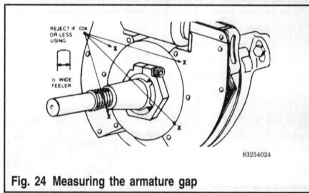

Fig. 24 Measuring the armature gap

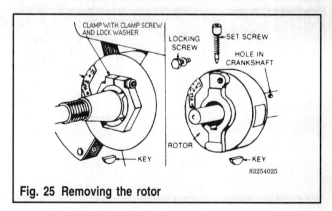

Fig. 25 Removing the rotor

end of the breaker shaft. The breaker points need not be removed to remove the breaker box.

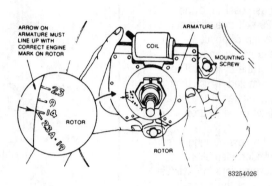

ARROW ON ARMATURE MUST LINE UP WITH CORRECT ENGINE MARK ON ROTOR

COIL

ARMATURE

MOUNTING SCREW

23
9
14
23A·19

ROTOR

ROTOR

83254026

Fig. 26 Timing adjustment

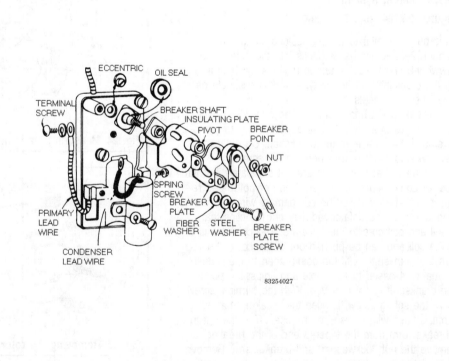

ECCENTRIC

OIL SEAL

TERMINAL SCREW

BREAKER SHAFT

INSULATING PLATE

PIVOT

BREAKER POINT

NUT

SPRING SCREW

BREAKER PLATE

PRIMARY LEAD WIRE

FIBER WASHER

STEEL WASHER

BREAKER PLATE SCREW

CONDENSER LEAD WIRE

83254027

Fig. 27 The breaker point assembly

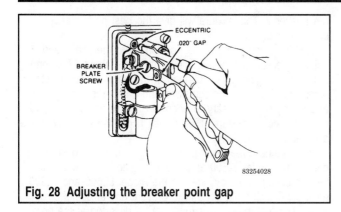

83254028

Fig. 28 Adjusting the breaker point gap

6. The breaker shaft can be removed, after the breaker points are removed, by turning the shaft one half turn to clear the retaining spur at the inside of the breaker box.

To install:

7. Install the breaker shaft with the arm upward so the arm will clear the retainer boss. Push the shaft all the way in, then turn the arm downward.

8. Pull the primary wire through the hole at the lower left corner of the breaker box. See that the primary wire rests in the groove at the top end of the box, then tighten the two mounting screws to hold the box in place.

9. To install the breaker points, press in the new oil seal with the metal side out. Put the new breaker plate on the top of the insulating plate, making sure that the detent in the breaker plate engages the hole in the insulating plate. Fasten the breaker plate screw enough to put a light tension on the plate. Adjust the eccentric so that the left edge of the insulating plate is parallel to the edge of the box and tighten the screw. This locates the breaker plate so that the proper gap adjustments may be made. Turn the breaker shaft clockwise as far as possible and hold it in this position. Place the new breaker points on the shaft, then the lockwasher, and tighten the nut down on the lockwasher. Replace the spring screw and terminal screw.

10. To adjust the breaker points, turn the crankshaft until the breaker points open to the widest gap. Loosen the breaker point plate screw slightly. Rotate the eccentric to obtain a point gap of 0.020 in. (0.5mm). Tighten the breaker plate screw.

11. Push the coil core into the coil with the rounded side toward the ignition cable. Place the coil and core on the armature with the coil retainer between the coil and the armature and with the rounded side toward the coil. Hook the lower end of the clips into the armature, then press the upper end onto the coil core.

12. Fasten the coil ground wire (bare double wires) to the armature support. Next, place the assembly against the cylinder and around the rotor and bearing support. Insert the three mounting screws together with the washer and lockwasher into the three long oval holes in the armature. Tighten them enough to hold the armature in place but loose enough so the armature can be moved for adjustment of the timing. Attach the primary wires from the coil and the breaker points to the terminal at the upper side of the backing plate. This terminal is insulated from the backing plate. Push the ignition cable through the louvered hole at the left side of the backing plate.

➡**On Model 9 engines, knot the ignition cable before inserting it through the backing plate. Be sure all wires are clean of the flywheel.**

13. The rotor and armature are correctly timed at the factory and require timing only if the armature has been removed from the engine, or if the cam gear or crankshaft has been replaced.   If it is necessary to adjust the rotor, proceed as follows: with the point gap set at 0.020 in. (0.5mm), turn the crankshaft in the normal direction of rotation until the breaker points close and just start to open. Use a timing light or insert a piece of tissue paper between the breaker points to determine when the points begin to open. With the three armature mounting screws slightly loose, rotate the armature until the arrow on the armature lines up with the arrow on the rotor. Align with the corresponding number of engine models, for example, on Model 9, align with #9. Retighten the armature mounting screws.

14. To install the set screw type rotor, place the woodruff key in the keyway on the crankshaft, then slide the rotor onto the crankshaft until the set screw hole in the rotor and the crankshaft are aligned. Be sure the key remains in place. Tighten the set screw securely, then tighten the lock screw to prevent the set screw from loosening. The lock screw is self-threading and the hole does not require tapping.

15. To install the clamp type rotor, place the woodruff key in place in the crankshaft and align the keyway in the rotor with the woodruff key. If necessary, use a short length of pipe and a hammer to drive the rotor onto the shaft until a 0.025 in. (0.6mm) feeler gauge can be inserted between the rotor and the bearing support. The split in the clamp must be between the slots in the rotor. Tighten the clamp screws to 60-70 inch lbs.

16. The armature air gap on engines equipped with Magna-Matic ignition system is fixed and can change only if wear occurs on the crankshaft journal and/or main bearing. Check for wear by inserting a ½ in. (12.7mm) wide feeler gauge at several points between the rotor and armature. Minimum feeler gauge thickness is 0.004 in. (0.1mm). Keep the feeler gauge away from the magnets on the rotor or you will have a false reading.

Mixture Adjustment

920000 Engines w/Automatic Choke

◆ **See Figure 29**

1. Start the engine and run it long enough to reach operating temperature. If the carburetor is so far out of adjustment that it will not start, close the needle valve by turning it clockwise. Then open the needle valve 1½ turns counterclockwise.

2. Move the control so that the engine runs at normal operating speed. Turn the needle valve clockwise until the engine starts to lose speed because of too lean a mixture. Then slowly turn the needle valve counterclockwise and out past the point of smoothest operation until the engine just begins to run unevenly because of too rich a mixture. Turn the needle back

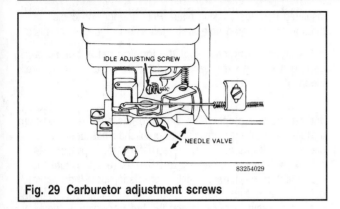

Fig. 29 Carburetor adjustment screws

clockwise to the midpoint between the rich and lean mixture extremes. This should be where the engine operates smoothest. The final adjustment of the needle valve should be slightly on the rich side (counterclockwise) of the mid-point.

3. Move the engine control to the slow position and turn the idle adjusting screw until a fast idle of about 1750 rpm is obtained. If the engine idles at a speed lower than 1750 rpm, it may not accelerate properly. It is not practical to attempt to obtain acceleration from speeds below 1750 rpm, because the mixture which would be required would be too rich for normal operating speeds.

4. To check the idle adjustment, move the engine control from slow to fast speed. The engine should accelerate smoothly. If the engine tends to stall or die out, increase the idle speed or readjust the carburetor, usually to a slightly richer mixture.

Flooding can occur if the engine is tipped at an angle for a prolonged period of time, if the engine is cranked repeatedly with the spark plug wire disconnected, or if the carburetor mixture is too rich.

In case of flooding, move the governor control to the stop position and pull the starter rope at least six times.

When the control is placed in the stop position, the governor spring holds the throttle in a closed idle position. Cranking the engine with a closed throttle creates a higher vacuum which opens the choke rapidly, permitting the engine to clean itself of excess fuel.

Then move the control to the fast position and start the engine. If the engine continues to flood, lean the carburetor needle valve by about 1/8-1/4 of a turn clockwise.

Pulsa-Jet and Vacu-Jet (Model Series 82000, 92000 Only)

Models 82500 and 92500 have a Vacu-Jet carburetor and Models 82900 and 92900 have a Pulsa-Jet carburetor.

Adjust the carburetor with the air cleaner installed and the fuel tank half full.

Turn the needle valve clockwise to close it. Then open it about 1½ turns. This will permit the engine to be started and warmed up before making the final adjustment.

With the engine running at normal operating speed (about 3000 rpm without a load) turn the needle valve clockwise until the engine starts to lose speed because of a too lean mixture.

Then slowly turn the needle valve counterclockwise past the point of smoothest operation, until the engine just begins to

run unevenly. This mixture will give the best performance under a load.

Hold the throttle in the idle position. Turn the idle speed adjusting screw until a fast idle is obtained (about 1750 rpm).

Test the engine under full load. If the engine tends to stall or die out, it usually indicates that the mixture is slightly lean and it may be necessary to open the needle valve slightly to provide a richer mixture. This slightly richer mixture may cause a slight unevenness in idling.

The breather tube and fuel intake tube thread into the cylinder on the model 82500 and 82900 engines. The fuel intake tube is bolted to the cylinder on the model 92500 and 92900 engines. Check for a good fit to prevent any air leaks or dirt entry. The fuel intake tube must not be distorted at the point where the carburetor O-ring fits or air leaks will occur.

Two Piece Flo-Jet

▶ **See Figure 30**

1. Start the engine and run it at 3000 rpm until it warms up.

2. Turn the needle valve (flat handle) to both extremes of operation noting the location of the valve at both points. That is, turn the valve inward until the mixture becomes too lean and the engine starts to slow, then note the position of the valve. Turn it outward slowly until the mixture becomes too rich and the engine begins to slow. Turn the valve back inward to the mid-point between the two extremes.

3. Install a tachometer on the engine. Pull the throttle to the idle position and hold it there through the rest of this step. Adjust the idle speed screw until the engine idles at 1750 rpm if it's an aluminum engine, or 1200 rpm, if it's a cast iron engine. Then, turn the idle valve in and out to adjust mixture, as described in Step 2. If idle valve adjustment changes idle speed, adjust speed to specification.

4. Release the throttle and observe the engine's response. The engine should accelerate without hesitation. If response is poor, one of the mixture adjustments is too lean. Readjust either or both as necessary. If idle speed was changed after idle valve was adjusted, readjust the idle valve first.

One Piece Flo-Jet

Follow the instructions for adjusting the Two Piece Flo-Jet carburetor (above). On the large, One Piece Flo-Jet, the nee-

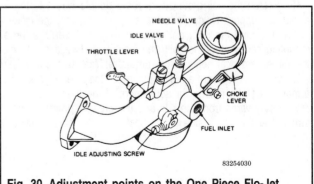

Fig. 30 Adjustment points on the One Piece Flo-Jet

dle valve is located under the float bowl, and the idle valve on top of the venturi passage. On the small One Piece Flo-Jet, both valves are adjusted by screws located on top of the venturi passage. The needle valve is located on the air horn side, is centered above the float bowl, and uses a larger screw head.

Governor Adjustments

SETTING MAXIMUM GOVERNED SPEED WITH ROTARY LAWNMOWER BLADES

➡Strict limits on engine rpm must be observed when setting top governed speed on rotary lawnmowers. This is done so that blade tip speeds will be kept to less than 19,000 feet per minute. Briggs & Stratton suggests setting the governor 200 rpm low to allow for possible error in the tachometer reading. These figures below, based on blade length, must be strictly adhered to, or a serious accident could result!

Models N, 6 and 8

There is no adjustment between the governor lever and the governor crank on these models. However, governor action can be changed by inserting the governor link or spring in different holes of the governor and throttle levers. In general, the closer to the pivot end of the lever, the smaller the difference between load and no-load engine speed. The engine will begin to 'hunt' if the spring is brought too close to the pivot point. The farther the spring is from the pivot end, the tendency to hunt will decrease, but the speed drop will be greater as the load increases. If the governor speed is lowered, the spring can usually be moved closer to the pivot. The standard setting is the 4th hole from the pivot point.

Models 6B, 8B, 60000 and 80000
▶ **See Figures 31, 32 and 33**

Loosen the screw which holds the governor lever to the governor shaft. Turn the governor lever counterclockwise until the carburetor throttle is wide open. With a screwdriver, turn the governor shaft counterclockwise as far as it will go. Tighten the screw which holds the governor lever to the governor shaft.

Blade Length (in.)	Max. Governed Speed (R.P.M.)
18	4032
19	3820
20	3629
21	3456
22	3299
23	3155
24	3024
25	2903
26	2791

832540C6

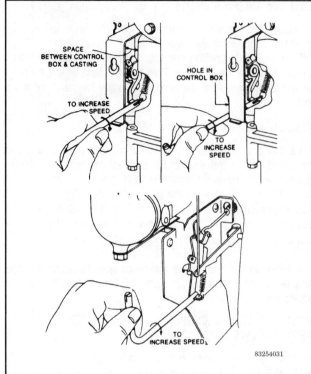

Fig. 31 Bending the spring anchor tang to get desired top speed

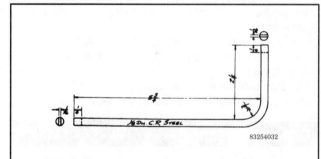

Fig. 32 You can make a tool like the one shown to adjust the spring anchor tang

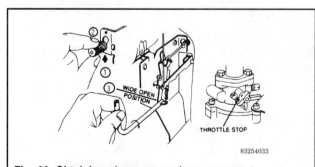

Fig. 33 Obtaining closer governing on generator applications

Cast Iron Models 9 and 14

Loosen the screw which holds the governor lever to the governor shaft. Push the lever counterclockwise as far as it will go. Hold it in position and turn the governor shaft counterclockwise as far as it will go. This can be done with a screwdriver. Securely tighten the screw that holds the governor lever to the shaft.

Aluminum Models 100000 and 130000

Vertical and horizontal shaft engine governors are adjusted by setting the control lever in the high speed position. Loosen the nut on the governor lever. Turn the governor shaft clockwise with a screwdriver to the end of its travel. Tighten the nut. The throttle must be wide open. Check to see if the throttle can be moved from idle to wide open without binding.

ADJUSTING TOP NO LOAD SPEED

Set the control lever to the maximum speed position with the engine running. Bend the spring anchor tang to get the desired top speed.

ADJUSTMENT FOR CLOSER GOVERNING (GENERATOR APPLICATIONS ONLY)

1. Snap knob upward to release adjusting nut.
2. Pull knob out against stop.
3. Then, bend the spring anchor tang to get top no-load speed as described below, depending upon the application.
   a. On models 100200 and 130200 with 3600 rpm generator, set the no load speed at 4,600 using the standard governor spring.
   b. On models 100200 and 130200 with 1800 rpm generator, set the no load speed at 2800 rpm, and set throttle stop at 1600 rpm.
4. Snap knob back into its normal position.
5. Adjust the knob for the desired generator speed.

Choke Adjustment

Choke-a-Matic, Pulsa-Jet and Vacu-Jet Carburetors

▶ **See Figure 34**

To check the operation of the choke linkage, move the speed adjustment lever to the choke position. If the choke slide does not fully close, bend the choke link. The speed

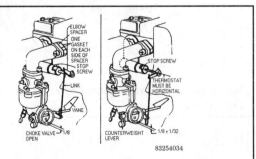

Fig. 34 Adjusting the automatic choke on the Two-Piece Flo-Jet

adjustment lever must make good contact against the top switch.

Install the carburetor and adjust it in the same manner as the Pulsa-Jet carburetor.

Two-piece Flo-Jet w/Automatic Choke

Hold the choke shaft so the thermostat lever is free. At room temperature (68°F) (20°C), the screw in the thermostat collar should be in the center of the stops. If not, loosen the stop screw and adjust the screw.

Loosen the set screw on the lever of the thermostat assembly. Slide the lever to the right or left on the shaft to ensure free movement of the choke link in any position. Rotate the thermostat shaft clockwise until the stop screw strikes the tube. Hold it in position and set the lever on the thermostat shaft so that the choke valve will be held open about 1/8 in. (3mm) from a closed position. Then tighten the set screw in the lever.

Rotate the thermostat shaft counterclockwise until the stop screw strikes the opposite side of the tube. Then open the choke valve manually until it stops against the top of the choke link opening. The choke valve should now be open approximately 1/8 in. (3mm) as before.

Check the position of the counterweight lever. With the choke valve in a wide open position (horizontal) the counterweight lever should also be in a horizontal position with the free end toward the right.

Operate the choke manually to be sure that all parts are free to move without binding or rubbing in any position.

Compression Checking

You can check the compression in any Briggs and Stratton engine by performing the following simple procedure: spin the flywheel counterclockwise (flywheel side) against the compression stroke. A sharp rebound indicates that there is satisfactory compression. A slight or no rebound indicates poor compression.

It has been determined that this test is an accurate indication of compression and is recommended by Briggs and Stratton. Briggs and Stratton does not supply compression pressures.

Loss of compression will usually be the result of one or a combination of the following:

1. The cylinder head gasket is blown or leaking.
2. The valves are sticking or not seating properly.
3. The piston rings are not sealing, which would also cause the engine to consume an excessive amount of oil.

Carbon deposits in the combustion chamber should be removed every 100 or 200 hours of use (more often when run at a steady load), or whenever the cylinder head is removed.

Carburetors

▶ **See Figures 35, 36 and 37**

Before removing any carburetor for repair, look for signs of air leakage or mounting gaskets that are loose, have deteriorated, or are otherwise damaged.

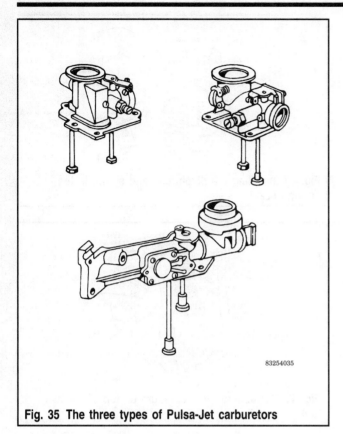

83254035

Fig. 35 The three types of Pulsa-Jet carburetors

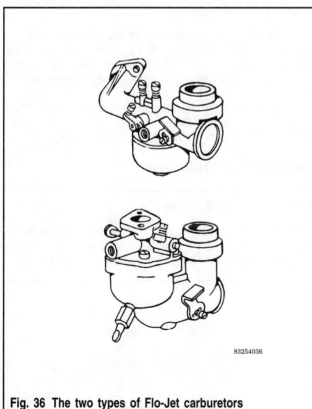

83254036

Fig. 36 The two types of Flo-Jet carburetors

Note the position of the governor springs, governor link, remote control, or other attachments to facilitate reassembly. Be careful not to bend the links or stretch the springs.

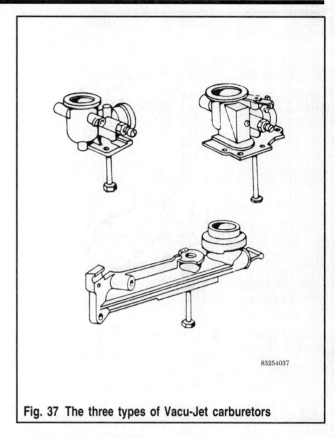

83254037

Fig. 37 The three types of Vacu-Jet carburetors

AUTOMATIC CHOKE

▶ **See Figure 38**

All 92000 model engines built since August 1968 have an automatic choke system.

The automatic choke operates in conjunction with engine vacuum, similar to the Pulsa-Jet fuel pump.

A diaphragm under the carburetor is connected to the choke shaft by a link. A calibrated spring under the diaphragm holds the choke closed when the engine is not running. Upon starting, vacuum created during the intake stroke is routed to the bottom of the diaphragm through a calibrated passage, thereby opening the choke.

This system also has the ability to respond in the same manner as an accelerator pump. As speed decreases during heavy loads, the choke valve partially closes, enriching the air/fuel mixture, thereby improving low speed performance and lugging power.

To check the automatic choke, remove the air cleaner and replace the stud. Observe the position of the choke valve; it should be fully closed. Move the speed control to the stop position; the governor spring should be holding the throttle in a closed position. Give the starter rope several quick pulls. The choke valve should alternately open and close.

If the choke valve does not react as stated in the previous paragraph, the carburetor will have to be disassembled to determine the problem. Before doing so, however, check the following items so you know what to look for:

Engine is Underchoked

1. Carburetor is adjusted too lean.

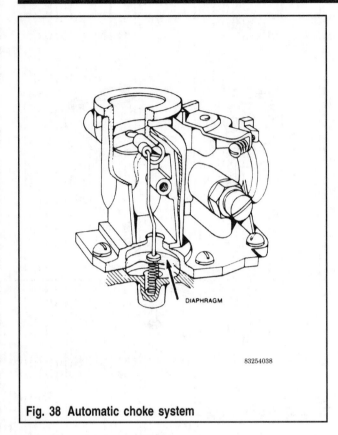

Fig. 38 Automatic choke system

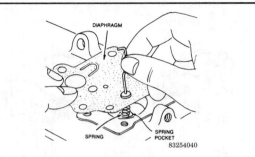

Fig. 40 Installing the diaphragm and spring in the spring pocket

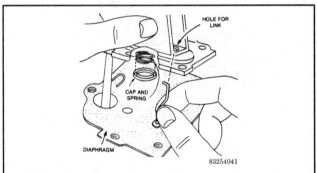

Fig. 41 Positioning the diaphragm on top of the fuel tank

2. The fuel pipe check valve is inoperative (Vacu-Jet only).
3. The air cleaner stud is bent.
4. The choke shaft is sticking due to dirt.
5. The choke spring is too short or damaged.
6. The diaphragm is not preloaded.

Engine is Overchoked

1. Carburetor is adjusted too rich.
2. The ari cleaner stud is bent.
3. The choke shaft is sticking due to dirt.
4. The diaphragm is ruptured.
5. The vacuum passage is restricted.
6. The choke spring is distorted or stretched.
7. There is gasoline or oil in the vacuum chamber.
8. There is a leak between the link and the diaphragm.
9. The diaphragm was folded during assembly, causing a vacuum leak.
10. The machined surface on the tank top is not flat.

Replacing the Automatic Choke
▶ See Figures 39, 40 and 41

Inspect the automatic choke for free operation. Any sticking problems should be corrected as proper choke operation depends on freedom of the choke to travel as dictated by engine vacuum.

Remove the carburetor and fuel tank assembly from the engine. The choke link cover may now be removed and the choke link disconnected from the choke shaft. Disassemble the carburetor from the tank top, being careful not to damage the diaphragm.

Checking the Diaphragm and Spring

The diaphragm can be reused, provided it has not developed wear spots or punctures. On the Pulsa-Jet models, make sure that the fuel pump valves are not damaged. Also check the choke spring length. The Pulsa-Jet spring minimum length is $1\frac{1}{8}$ in. (28.5mm) and the maximum is $1\frac{7}{32}$ in. (31mm). Vacu-Jet spring length minimum is $\frac{15}{16}$ in., (23.8mm) maximum length 1 in. (25.4mm). If the spring length is shorter or longer than specified, replace the diaphragm and the spring.

CHECKING THE TANK TOP

▶ See Figures 42, 43 and 44

The machined surface on the top of the tank must be flat in order for the diaphragm to provide an adequate seal between the carburetor and the tank. If the machined surface on the

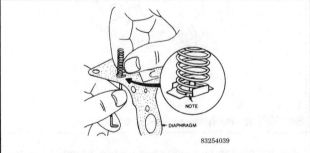

Fig. 39 Assembling the diaphragm spring in the new diaphragm

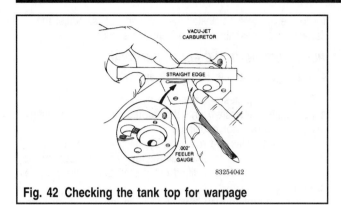

Fig. 42 Checking the tank top for warpage

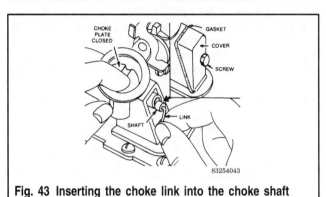

Fig. 43 Inserting the choke link into the choke shaft

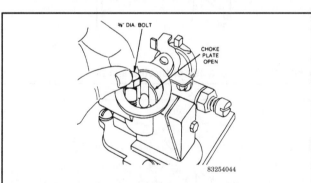

Fig. 44 Preloading the diaphragm to adjust the choke

tank is not flat, it is possible for gasoline to enter the vacuum chamber by passing between the machined surface and the diaphragm. Once fuel has entered the vacuum chamber, it can move through the vacuum passage and into the carburetor. The flatness of the machined surface on the tank top can be checked by using a straightedge and a feeler gauge. The surface should not vary more than 0.002 in. (0.05mm). Replace the tank if a 0.002 in. (0.05mm) feeler gauge can be passed under the straightedge.

If a new diaphragm is installed, assemble the spring to the replacement diaphragm, taking care not to bend or distort the spring.

Place the diaphragm on the tank surface, positioning the spring in the spring pocket.

Place the carburetor on the diaphragm ensuring that the choke link and diaphragm are properly aligned between the carburetor and the tank top. On Pulsa-Jet models, place the pump spring and cap on the diaphragm over the recess or

pump chamber in the fuel tank. Thread in the carburetor mounting screws to about two threads. Do not tighten them. Close the choke valve and insert the choke link into the choke shaft.

Remove the air cleaner gasket, if it is in place, before continuing. Insert a 3/8 in. bolt or rod into the carburetor air horn. With the bolt in position, tighten the carburetor mounting screws in a staggered sequence. Please note that the insertion of the 3/8 in. bolt opens the choke to an over-center position, which preloads the diaphragm.

Remove the 3/8 in. bolt. The choke valve should now move to a fully closed position. If the choke valve is not fully closed, make sure that the choke spring is properly assembled to the diaphragm, and also properly inserted in its pocket in the tank top.

ADJUSTMENTS

All carburetor adjustments should be made with the air cleaner on the engine. Adjustment is best made with the fuel tank half full.

Carburetors

REBUILDING

Pulsa-Jet and Vacu-Jet (Model Series 82000 and 92000 Only)

♦ See Figures 45, 46 and 47

Models 82500 and 92500 have a Vacu-Jet carburetor and Models 82900 and 92900 have a Pulsa-Jet carburetor.

1. Remove the carburetor and fuel tank assembly from the engine by removing the two attaching bolts.

2. Disconnect the governor link at the throttle, leaving the governor link and the governor spring hooked to the governor blade and control lever.

3. Slip the carburetor and tank assembly off of the engine.

4. Remove the carburetor from the tank. Always remove all nylon and rubber parts if the carburetor is soaked in solvent.

5. Remove the O-ring and discard it. Remove and inspect the needle valve, packing and seat.

6. Metering holes in the carburetor body should be cleaned with solvent and compressed air. Do not clean the holes with

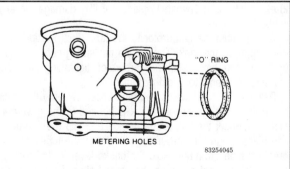

Fig. 45 Remove the O-ring and inspect the metering valve

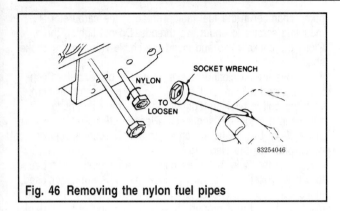

Fig. 46 Removing the nylon fuel pipes

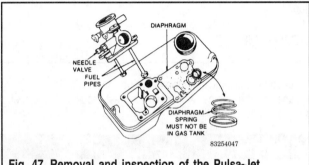

Fig. 47 Removal and inspection of the Pulsa-Jet diaphragm

a pin or a length of wire because of the danger of altering their size.

7. Remove the choke parts on models 82500 and 82900 by pulling the nylon choke shaft sideways to separate the choke shaft from the choke valve. On the 92500 and 92900, remove the choke parts by first disconnecting the choke return spring at the pin in the carburetor body. Then pull the nylon choke shaft sideways to separate the choke shaft from the choke valve.

8. If the choke valve is heat-sealed to the choke shaft, loosen it by sliding a sharp pointed tool along the edge of the choke shaft. Do not re-seal parts on assembly.

9. When replacing the choke valve and shaft, install the choke valve so the poppet valve spring is visible when the valve is in full choke position.

On these models, the nylon fuel pipe is threaded into the carburetor body. Use a socket to remove and replace it. Be careful not to overtighten it and do not use any sealer.

The Pulsa-Jet diaphragm also serves as a gasket between the carburetor and the tank. Inspect the diaphragm for punctures, wrinkles, and wear. Replace it if it is damaged in any way.

To assemble the carburetor to the tank, first position the diaphragm on the tank. Then place the spring cap and spring on the diaphragm. Install the carburetor, tightening the mounting screws evenly to avoid distortion.

To install the carburetor and tank assembly onto the engine, make sure that the governor link is hooked to the governor blade. Connect the link to the throttle and slip the carburetor into place. Align the carburetor with the intake tube and breather tube grommet. Hold the choke lever in the open position so it does not catch on the control plate. Be sure the O-ring in the carburetor does not distort when fitting the carbure-

tor to the intake tube. Install the mounting bolts. Adjust the carburetor as described in the Tune-Up section.

PULSA-JET THROTTLE PLATE REMOVAL
▶ See Figures 48, 49, 50, 51, 52 and 53

Cast throttle plates are removed by backing off the idle speed adjustment screw until the throttle clears the retaining lug on the carburetor housing.

Stamped throttles are removed by using a phillips screwdriver to remove the throttle valve screw. After removal of the valve, the throttle may be lifted out. Installation is the reverse of removal.

Some carburetors may have a spiral in the carburetor bore. To remove it, fasten the carburetor in a vise about ½ in. (13mm) below the top of the jaws. Grasp the spiral firmly with a pair of pliers. Place a screwdriver under the edge of the pliers. Using the edge of the vise, push down on the screwdriver to pry out the spiral. When installing the spiral, keep the

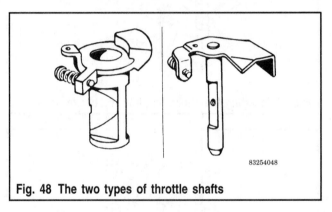

Fig. 48 The two types of throttle shafts

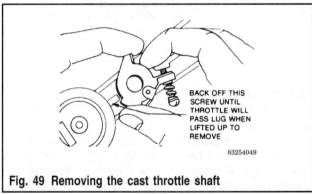

Fig. 49 Removing the cast throttle shaft

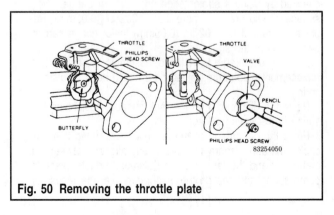

Fig. 50 Removing the throttle plate

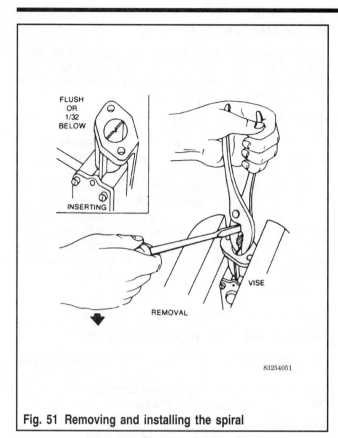

Fig. 51 Removing and installing the spiral

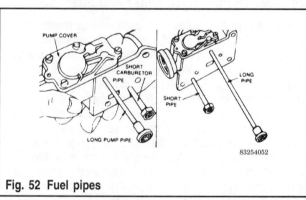

Fig. 52 Fuel pipes

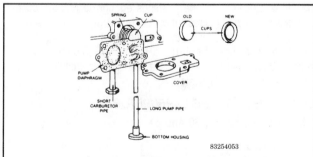

Fig. 53 Removal and inspection of the pump cover diaphragm from the Pulsa-Jet carburetor

top flush, or 1/32in. (0.8mm) below the carburetor flange, and parallel with the fuel tank mounting face.

FUEL PIPE

Check balls are not used in these fuel pipes. The screen housing or pipe must be replaced if the screen cannot be satisfactorily cleaned. The long pipe supplies fuel from the tank to the pump. The short pipe supplies fuel from the tank cup to the carburetor. Fuel pipes are nylon or brass. Nylon pipes are removed and installed by using a socket, or open-end wrench.

→Where brass pipes are used, replace only the screen housing. The housing is driven off the pipe with a screwdriver with the pipe held in a vise. The new housing is installed by lightly tapping it onto the pipe with a soft hammer.

NEEDLE VALVE AND SEAT

Remove the needle valve to inspect it. If the carburetor is gummy or dirty, remove the seat to allow better cleaning of the metering holes. Do not insert pins or wires in the metering holes. Use solvent or compressed air.

PUMP

Remove the fuel pump cover, diaphragm, spring, and cup. Inspect the diaphragm for punctures, cracks, and fatigue. Replace it if damaged. On early models, the spring cap is solid; on later models, the cap has a hole in it. The new style supersedes the old style. When installing the pump cover, tighten the screws evenly to insure a good seal.

CHOKE-A-MATIC (EXCEPT 100900 MODELS)
▶ See Figures 54, 55, 56 and 57

To remove the choke link, remove the speed adjustment lever and stop switch insulator plate. Remove the speed adjustment lever from the choke link, then pull out the choke link through the hole in the choke slide.

Replace worn or damaged parts. To assemble, slip the washers and spring over the choke link. Hook the choke link through the hole in the choke slide. Place the other end of the choke link through the hole in the speed adjustment lever and mount the lever and stop switch insulator plate to the carburetor.

Vacu-Jet Carburetors

Vacu-Jet carburetors are removed from the engine together with the fuel tank as one unit. The throttle plates are removed and installed in the same manner as the throttles in the Pulsa-Jet carburetors.

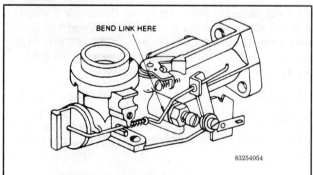

Fig. 54 Adjustment of the Choke-a-Matic choke linkage

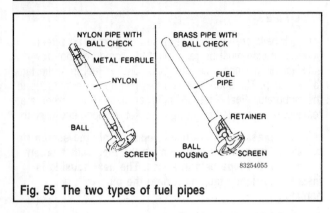

Fig. 55 The two types of fuel pipes

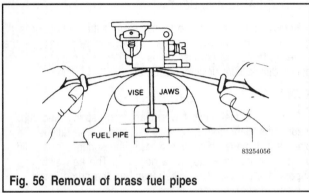

Fig. 56 Removal of brass fuel pipes

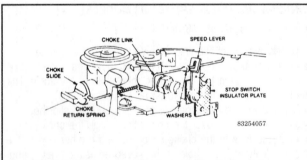

Fig. 57 Adjustment of a Choke-a-Matic choke linkage on a Vacu-Jet carburetor

FUEL PIPE

The fuel pipe contains a check ball and a fine mesh screen. To function properly, the screen must be clean and the check ball free. Replace the pipe if the screen and ball cannot be satisfactorily cleaned in carburetor cleaner.

➡**Do not leave the carburetor in the cleaner for more than ½ hour without removing all nylon parts. Nylon fuel pipes are removed and replaced with a 9/16 in. socket. Brass fuel pipes are removed by clamping the pipe in a vise and prying out the pipe with two screwdrivers.**

To install the brass fuel pipes, remove the throttle, if necessary, and place the carburetor and pipe in a vise. Press the pipe into the carburetor until it projects 2⁹/₃₂-2¹¹/₃₂ in. (57.9-58.7mm) from the carburetor face.

NEEDLE VALVE AND SEAT

Remove the needle valve assembly to inspect it. If the carburetor is gummy or dirty, remove the seat to allow better cleaning of the metering holes. Do not clean the metering holes with a pin or a length of wire.

CHOKE-A-MATIC LINKAGE

To remove the choke link, remove the speed adjustment lever and the top switch insulator plate. Work the link out through the hole in the choke slide.

Replace all worn or damaged parts. To assemble a carburetor using a choke slide, place the choke return spring and three washers on the choke link. Push the choke link through the hole in the carburetor body, turning the link to line up with the hole in the choke slide. The speed adjustment lever screw and the stop switch insulator plate should be installed as one assembly after placing the choke link through the end of the speed adjustment lever.

Two Piece Flo-Jet Carburetors (Large and Small Line)

CHECKING THE UPPER BODY FOR WARPAGE

With the carburetor assembled and the body gasket in place, try to insert a 0.002 in. (0.05mm) feeler gauge between the upper and lower bodies at the air vent boss, just below the idle valve. If the gauge can be inserted, the upper body is warped and should be replaced.

CHECKING THE THROTTLE SHAFT AND BUSHINGS

Wear between the throttle shaft and bushings should not exceed 0.010 in. (0.254mm). Check the wear by placing a short iron bar on the upper carburetor body so that it just fits under the throttle shaft. Measure the distance with a feeler gauge while holding the shaft down and then holding it up. If the difference is over 0.010 in. (0.254mm), either the upper body should be rebushed, the throttle shaft replaced, or both. Wear on the throttle shaft can be checked by comparing the worn and unworn portions of the shaft. To replace the bushings, remove the throttle shaft using a thin punch to drive out the pin which holds the throttle stop to the shaft; remove the throttle valve, then pull out the shaft. Place a ¼-20 tap or an E-Z Out in a vise. Turn the carburetor body so as to thread the tap or E-Z Out into the bushings enough to pull the bushings out of the body. Press the new bushings into the carburetor body with a vise. Insert the throttle shaft to be sure it is free in the bushings. If not, run a size ⁷/₃₂ in. (5.5mm) drill through both bushings to act as a line reamer. Install the throttle shaft, valve, and stop.

DISASSEMBLY OF THE CARBURETOR

▶ See Figures 58, 59, 60 and 61

1. Remove the idle valve.
2. Loosen the needle valve packing nut.
3. Remove the packing nut and needle valve together. To remove the nozzle, use a narrow, blunt screwdriver so as not to damage the threads in the lower carburetor body. The nozzle projects diagonally into a recess in the upper body and must be removed before the upper body is separated from the lower body, or it may be damaged.
4. Remove the screws which hold the upper and lower bodies together. A pin holds the float in place.

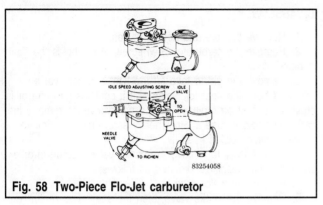

Fig. 58 Two-Piece Flo-Jet carburetor

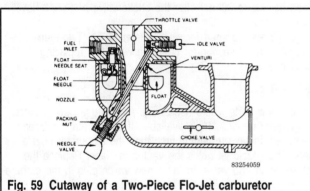

Fig. 59 Cutaway of a Two-Piece Flo-Jet carburetor

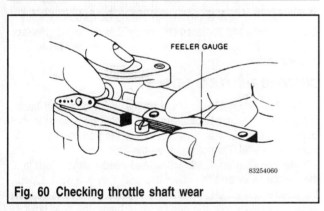

Fig. 60 Checking throttle shaft wear

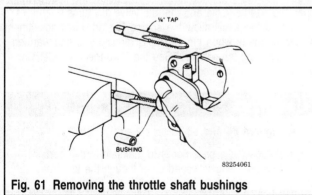

Fig. 61 Removing the throttle shaft bushings

5. Remove the pin to take out the float valve needle. Check the float for leakage. If it contains gasoline or is

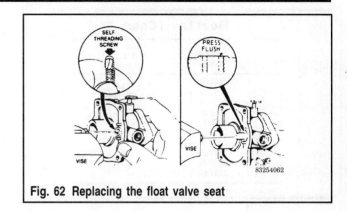

Fig. 62 Replacing the float valve seat

crushed, it must be replaced. Use a wide, proper fitting screwdriver to remove the float inlet seat.

6. Lift the venturi out of the lower body. Some carburetors have a welch plug. This should be removed only if necessary to remove the choke plate. Some carburetors have nylon choke shaft.

REPAIR

Use new parts where necessary. Always use new gaskets. Carburetor repair kits are available. Tighten the inlet seat with the gasket securely in place, if used. Some float valves have a spring clip to connect the float valve to the float tang. Others are nylon with a stirrup which fits over the float tang. Older float valves and engines with fuel pumps have neither a spring nor a stirrup.

A viton tip float valve is used in later models of the large, two-piece Flo-Jet carburetor. The seat is pressed into the upper body and does not need replacement unless it is damaged.

REPLACING THE PRESSED-IN FLOAT VALVE SEAT
▶ See Figure 62

Clamp the head of a #93029 self threading screw in a vise. Turn the carburetor body to thread the screw into the seat. Continue turning the carburetor body, drawing out the seat. Leave the seat fastened to the screw. Insert the new seat #230996 into the carburetor body. The seat has a starting lead.

➡**If the engine is equipped with a fuel pump, install a #231019 seat. Press the new seat flush with the body using the screw and old seat as a driver. Make sure that the seat is not pressed below the body surface or improper float-to-float valve contact will occur. Install the float valve.**

CHECKING THE FLOAT LEVEL

With the body gasket in place on the upper body and the float valve and float installed, the float should be parallel to the body mounting surface. If not, bend the tang on the float until they are parallel. Do not press on the flat to adjust it.

ASSEMBLY OF THE CARBURETOR
▶ See Figures 63 and 64

Assemble the venturi and the venturi gasket to the lower body. Be sure that the holes in the venturi and the venturi gasket are aligned. Some models do not have a removable venturi. Install the choke parts and welch plug if previously

Float Level Chart

Carburetor Number	Float Setting (in.)
2712-S	$^{19}/_{64}$
2713-S	$^{19}/_{64}$
2714-S	$^{1}/_{4}$
*2398-S	$^{1}/_{4}$
2336-S	$^{1}/_{4}$
2336-SA	$^{1}/_{4}$
2337-S	$^{1}/_{4}$
2337-SA	$^{1}/_{4}$
2230-S	$^{17}/_{64}$
2217-S	$^{11}/_{64}$

*When resilient seat is used, set float level at $^{9}/_{32}$ ± $^{1}/_{64}$.

832540C7

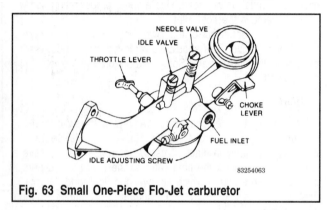

Fig. 63 Small One-Piece Flo-Jet carburetor

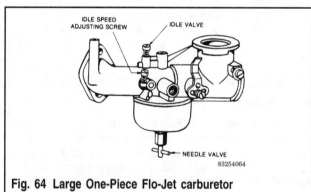

Fig. 64 Large One-Piece Flo-Jet carburetor

removed. Use a sealer around the welch plug to prevent entry of dirt.

Fasten the upper and lower bodies together with the mounting screws. Screw in the nozzle with a narrow, blunt screwdriver, making sure that the nozzle tip enters the recess in the upper body. Tighten the nozzle securely. Screw in the needle valve and idle valve until they just seat. Back off the needle valve 1½ turns. Do not tighten the packing nut. Back off the idle valve ¾ of a turn. These settings are about correct. Final adjustment will be made when the engine is running.

One-Piece Flo-Jet Carburetor

The large, one-piece Flo-Jet carburetor has its high speed needle valve below the float bowl. All other repair procedures are similar to the small, one-piece Flo-Jet carburetor.

DISASSEMBLY

1. Remove the idle and needle valves.
2. Remove the carburetor bowl screw. A pin holds the float in place.
3. Remove the pin to take off the float and float valve needle. Check the float for leakage. If it contains gasoline or is crushed, it must be replaced. Use a screwdriver to remove the carburetor nozzle. Use a wide, heavy screwdriver to remove the float valve seat, if used.

If it is necessary to remove the choke valve, venturi throttle shaft, or shaft bushings, proceed as follows:

4. Pry out the welch plug.
5. Remove the choke valve, then the shaft. The venturi will then be free to fall out after the choke valve and shaft have been removed.
6. Check the shaft for wear. (Refer to the 'Two-Piece Flo-Jet Carburetor' section for checking wear and replacing bushings.)

REPAIR

Use new parts where necessary. Always use new gaskets. Carburetor repair kits are available. If the venturi has been removed, install the venturi first, then the carburetor nozzle jets. The nozzle jet holds the venturi in place. Replace the choke shaft and valve. Install a new welch plug in the carburetor body. Use a sealer to prevent dirt from entering.

A viton tip float valve is used in the large, one-piece Flo-Jet carburetor. The seat is pressed in the upper carburetor body and does not need replacement unless it is damaged. Replace the seat in the same manner as for the two-piece Flo-Jet carburetor.

CHECKING THE FLOAT LEVEL

With the body gasket in place on the upper body and float valve and the float installed, the float should be parallel to the body mounting surface. If not, bend the tang on the float until they are parallel. Do not press on the float.

Install the float bowl, idle valve, and needle valve. Turn in the needle valve and the idle valve until they just seat. Open the needle valve 2½ turns and the idle valve 1½ turns. On the large carburetors with the needle valve below the float bowl, open the needle valve and the idle valve 1⅛ turns.

These settings will allow the engine to start. Final adjustment should be made when the engine is running and has warmed up to operating temperature. See the 'Two-Piece Flo-Jet Carburetor' adjustment procedure.

Governors

▶ **See Figures 65 and 66**

The purpose of a governor is to maintain, within certain limits, a desired engine speed even though the load may vary.

AIR VANE GOVERNORS

The governor spring tends to open the throttle. Air pressure against the air vane tends to close the throttle. The engine speed at which these two forces balance is called the gov-

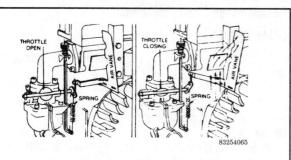

Fig. 65 Air vane governor installed on horizontal crankshaft engines

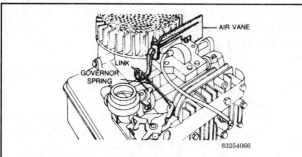

Fig. 66 Air vane governor installed on vertical crankshaft engines

erned speed. The governed speed can be varied by changing the governor spring tension.

Worn linkage or damaged governor springs should be replaced to insure proper governor operation. No adjustment is necessary.

MECHANICAL GOVERNORS

The governor spring tends to pull the throttle open. The force of the counterweights, which are operated by centrifugal force, tends to close the throttle. The engine speed at which these two forces balance is called the governed speed. The governed speed can be varied by changing the governor spring tension.

GOVERNOR REPAIR

▶ **See Figures 67, 68, 69, 70, 71, 72 and 73**

The procedures below describe disassembly and assembly of the various kinds of mechanical governors. Look for gears with worn or broken teeth, worn thrust washers, weight pins, cups, followers, etc.. Replace parts that are worn and reassemble.

Models N, 6 and 8

DISASSEMBLY

1. Remove the two governor housing mounting screws, and remove the housing.

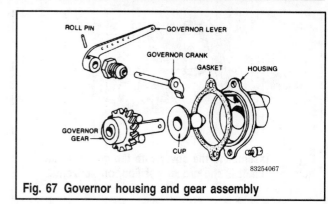

Fig. 67 Governor housing and gear assembly

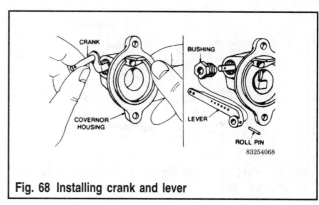

Fig. 68 Installing crank and lever

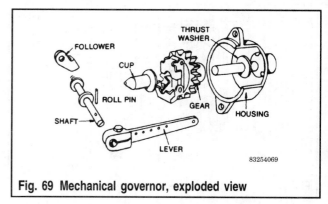

Fig. 69 Mechanical governor, exploded view

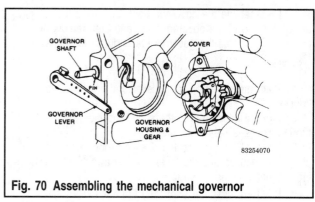

Fig. 70 Assembling the mechanical governor

2. Pull the cup off the governor gear, and then slide the gear off the shaft.

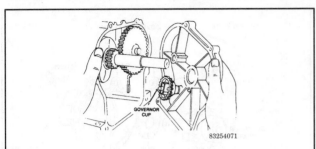

Fig. 71 Assembling the cover with the governor and governor shaft in the proper position, on horizontal engines

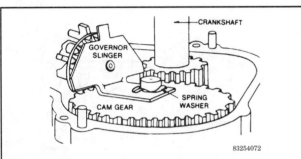

Fig. 72 Installing the spring on the camshaft on Models 100900 and 130900

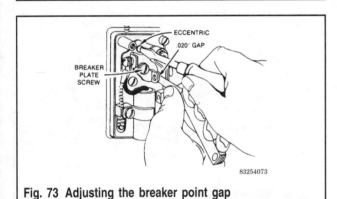

Fig. 73 Adjusting the breaker point gap

3. Disassemble the governor crank by driving the roll pin out of the end of the governor lever and then remove the crank bushing. Pull the governor crank out of the housing.

ASSEMBLY

1. Push the governor crank, lever end first, into the housing.
2. Slip the bushing onto the shaft, and then thread it into the housing and tighten securely.
3. Position the lever on the shaft with the lever and shaft pin holes lined up and the lever pointing away from the housing mounting flange. Push in the pin.
4. Push the governor gear onto the shaft in the engine block.
5. Position the gasket on the governor housing, put the housing into position on the block, and install the two housing mounting screws. Connect linkage.

Models 6B, 8B, 60000 and 80000

DISASSEMBLY

1. Loosen the governor lever mounting screw and pull the lever off the shaft.
2. Remove the two housing mounting screws. Carefully pull the housing off the block being careful to catch the governor gear, which will slip off the shaft. Pull the steel thrust washer off the shaft.
3. Remove the governor lever roll pin and washer. Unscrew the governor lever shaft by turning it clockwise and remove it.

ASSEMBLY

1. Push the governor lever shaft into the crankcase cover, threaded end first. Assemble the small washer onto the inner end of the shaft, and then screw the shaft into the governor crank follower by turning it counterclockwise. Tighten it securely.
2. Turn the shaft until the follower points down slightly, in a position where it would press against the cup when the housing is installed.
3. Place the washer on the outside end of the shaft. Install the rollpin, so the leading end just reaches the outside diameter of the shaft and the back end protrudes.
4. Install the thrust washer and the governor gear on the shaft in the housing (in that order).
5. Hold the crankcase cover in a vertical (the normal) position and install the housing with the gear in position so the point of the steel cup on the gear contacts the follower. Install and tighten the housing mounting screws.
6. Install the lever on the shaft pointing downward at an angle of about 30°. Adjust as described in the Tune-Up section.

Cast Iron Models

DISASSEMBLY

1. Remove the cotter key and washer from the outer end of the governor shaft. Remove the governor crank from inside the crankcase.
2. Slide the governor gear off the shaft.

ASSEMBLY

1. Install the governor gear onto the shaft inside the crankcase. Then, insert the governor shaft assembly through the bushing from inside the crankcase.
2. Install the governor lever to the shaft loosely, and then adjust it as described in the Tune-Up section.

Aluminum Models

DISASSEMBLY

On horizontal shaft models: Remove the governor assembly as a unit from the crankcase cover.

On vertical shaft models: Remove the entire assembly as part of the oil slinger (see the Overhaul Section).

ASSEMBLY

1. Assemble governors on horizontal crankshaft models with crankshaft in a horizontal position. The governor rides on a short stationary shaft which is integral with the crankcase cover. The governor shaft keeps the governor from sliding off

the shaft after the cover is installed. The governor shaft must hang straight down, or it may jam the governor assembly when the crankcase cover is installed, breaking it when the engine is started. The governor shaft adjustment should be made as soon as the crankcase cover is in place so that the governor lever will be clamped in the proper position.

2. On both horizontal and vertical crankshaft models, the governor is held together through normal operating forces. For this reason, the governor link and all other external linkages must be in place and properly adjusted whenever the engine is operated.

3. On vertical shaft models 100900 and 130900, be sure the spring washer is in place on the camshaft after the governor is in position.

Cylinder Head

▶ See Figure 74

REMOVAL & INSTALLATION

Always note the position of the different cylinder head screws so that they can be properly reinstalled. If a screw is used in the wrong position, it may be too short and not engage enough threads. If it is too long, it may bottom on a fin, either breaking the fin, or leaving the cylinder head loose.

1. Remove the cylinder screws and then the cylinder head. Be sure to remove the gasket and all remaining gasket material from the cylinder head and the block.

Basic Model Series	In. lbs. Torque
Aluminum Cylinder	
6B, 60000, 8B, 80000 82000, 92000, 110000, 100000, 130000	140
Cast Iron Cylinder	
5, 6, N, 8, 9	140
14	165

832540C8

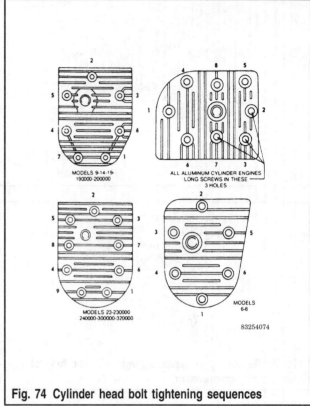

Fig. 74 Cylinder head bolt tightening sequences

2. Assemble the cylinder head with a new gasket, cylinder head shield, screws, and washers in their proper places. Graphite grease should be used on aluminum cylinder head screws. Do not use a sealer of any kind on the head gasket. Tighten the screws down evenly by hand. Use a torque wrench and tighten the head bolts in the correct sequence.

Valves

REMOVAL & INSTALLATION

▶ See Figures 75 and 76

1. Using a valve spring compressor, adjust the jaws so they touch the top and bottom of the valve chamber, and then place one of the jaws over the valve spring and the other underneath, between the spring and the valve chamber. This positioning of the valve spring compressor is for valves that have either pin or collar type retainers.

2. Tighten the jaws to compress the spring. Remove the collars or pin and lift out the valve. Pull out the compressor and the spring.

3. To remove valves with ring type retainers, position the compressor with the upper jaw over the top of the valve chamber and the lower jaw between the spring and the retainer. Compress the spring, remove the retainer, and pull out the valve. Remove the compressor and spring.

To install:

4. Before installing the valves, check the thickness of the valve springs. Some engines use the same spring for the intake and exhaust side, while others use a heavier spring on the exhaust side. Compare the springs before installing them.

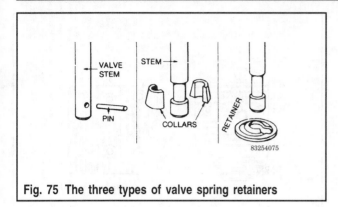

Fig. 75 The three types of valve spring retainers

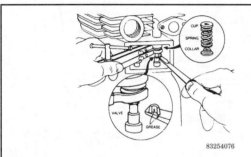

Fig. 76 Removing the valve springs with the help of a valve spring compressor

5. If the retainers are held by a pin or collars, place the valve spring and retainer and cup (Models 9-14-19-20-23-24-32) into the valve spring compressor. Compress the spring until it is solid. Insert the compressed spring and retainer into the valve chamber. Then drop the valve into place, pushing the stem through the retainer. Hold the spring up in the chamber, hold the valve down, and insert the retainer pin with needle nose pliers or place the collars in the groove in the valve stem. Loosen the spring until the retainer fits around the pin or collars, then pull out the spring compressor. Be sure the pin or collars are in place.

6. To install valves with ring type retainers, compress the retainer and spring with the compressor. The large diameter of the retainer should be toward the front of the valve chamber. Insert the compressed spring and retainer into the valve chamber. Drop the valve stem through the larger area of the retainer slot and move the compressor so as to center the small area of the valve retainer slot onto the valve stem shoulder. Release the spring tension and remove the compressor.

Valve Guides

REMOVAL & INSTALLATION

Aluminum Models
▶ See Figure 77

1. First check valve guide for wear with a plug gauge. If the flat end of the valve guide plug gauge can be inserted into the valve guide for a distance of 5/16 in. (8mm), the valve guide

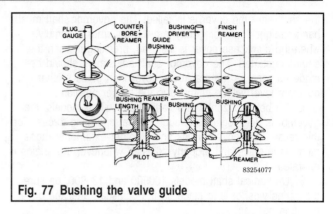

Fig. 77 Bushing the valve guide

is worn and should be rebushed in the following manner. See the illustration.

2. Procure a reamer and a reamer guide bushing. Lubricate the reamer with kerosene. Use reamer and reamer guide bushing to ream out the worn guide. Ream to only 1/16 in. (1.5mm) deeper than valve guide bushing #63709. BE CAREFUL NOT TO REAM THROUGH THE GUIDE!

3. Press in valve guide bushing #63709 until top end of bushing is flush with top end of valve guide. Use a soft metal driver (brass, copper, etc.) or driver #19065 so top end of bushing is not peened over.

4. Finish-ream the bushing. A standard valve can now be used.

➡**It is usually not necessary to bush factory installed brass valve guides. However, if bushing is required, DO NOT REMOVE ORIGINAL BUSHING, but follow standard procedure outlined.**

Cast Iron Models
▶ See Figure 78

1. First check valve guide for wear with a plug gauge, Briggs & Stratton part #19151 or equivalent. If the flat end of the valve guide plug gauge can be inserted into the valve guide for a distance of 5/16 in. (8mm) the guide is worn and should be rebushed in the following manner. See the illustration.

2. Procure a reamer #19183 and reamer guide bushing #19192, and lubricate the reamer with kerosene. Then, use reamer and reamer guide bushing to ream out the worn guide. Ream to only 1/16 in. (1.5mm) deeper than valve guide bushing #230655. BE CAREFUL NOT TO REAM THROUGH THE GUIDE!

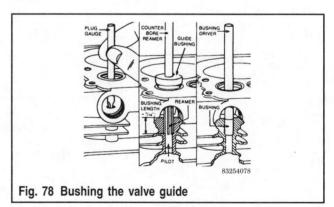

Fig. 78 Bushing the valve guide

3. Press in valve guide bushing #230655 until top end of bushing is flush with top end of valve guide. Use a soft metal driver (brass, copper, etc.) so top end of bushing is not peened over. The bushing #230655 is finish reamed to size at the factory, so no further reaming is necessary, and a standard valve can be used.

➡ **Valve seating should be checked after bushing the guide, and corrected if necessary by refacing the seat.**

REFACING VALVES AND SEATS

Faces on valves and valve seats should be resurfaced with a valve grinder or cutter to an angle of 45°.

➡ **Some engines have a 30° intake valve and seat.**

The valve and seat should then be lapped with a fine lapping compound to remove the grinding marks and ensure a good seat. The valve seat width should be 1.2-1.5mm. If the seat is wider, a narrowing stone or cutter should be used. If either the seat or valve is badly burned, it should be replaced. Replace the valve if the edge thickness (margin) is less than $1/64$. (0.4mm) after it has been resurfaced.

CHECK AND ADJUST TAPPET CLEARANCE

Valve Tappet Clearance Chart

Model Series	Intake		Exhaust	
	Max	Min	Max	Min
Aluminum Cylinder				
6B, 60000, 8B, 80000	.007	.005	.011	.009
82000, 92000, 100000, 110900	.007	.005	.011	.009
130000	.007	.005	.011	.009
Cast Iron Cylinder				
5, 6, 8, N, 9, 14	.009	.007	.016	.014

832540C9

Insert the valves in their respective positions in the cylinder. Turn the crankshaft until one of the valves is at its highest position. Turn the crankshaft one revolution. Check the clearance with a feeler gauge. Repeat for the other valve. Grind off the end of the valve stem if necessary to obtain proper clearance.

➡ **Check the valve tappet clearance with the engine cold.**

Valve Seat Inserts

▶ See Figure 79

Valve Seat Inserts Chart

Basic Model Series	Intake Standard	Exhaust Standard	Exhaust Stellite	Insert # Puller Assembly	Puller Nut
Aluminum Cylinder					
6B, 8B	211291	211291	210452	19138	19140 Ex. 19182 In.
60000, 80000	210879*	211291	210452	19138	19140 Ex. 19182 In.
82000, 92000, 110000	210879	211291	210452	19138	19140 Ex. 19182 In.
100000, 130000	211158	211172	211436	19138	19182 Ex. 19139 In.
Cast Iron Cylinder					
5, 6, N	63838	21865		19138	19140
8	210135	21865		19138	19140
9	63007	63007		19138	19139
14	21880	21880	21612	19138	19141

*21191 used before serial #5810060—210808 used from serial #5810060—6012010
Includes puller and #19182, 19141, 19140 and 19139 nuts

83254C10

Cast iron cylinder engines are equipped with an exhaust valve insert which can be removed and replaced with a new insert. The intake side must be counterbored to allow the installation of an intake valve seat insert (see below). Aluminum alloy cylinder models are equipped with inserts on both the exhaust and intake valves.

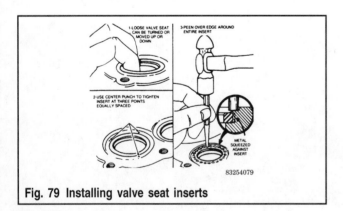

Fig. 79 Installing valve seat inserts

REMOVAL & INSTALLATION

➡**Valve seat inserts are removed with a special puller. On aluminum alloy cylinder models, it may be necessary to grind the puller nut until the edge is $1/32$. (0.8mm) thick in order to get the puller nut under the valve insert.**

When installing the valve seat insert, make sure that the side with the chamfered outer edge goes down into the cylinder. Install the seat insert and drive it into place with a driver. The seat should then be ground lightly and the valves and seats lapped lightly with grinding compound.

Aluminum alloy cylinder models use the old insert as a spacer between the driver and the new insert. Drive in the new insert until it bottoms. The top of the insert will be slightly below the cylinder head gasket surface. Peen around the insert using a punch and hammer.

➡**The intake valve seat on cast iron cylinder models has to be counterbored before installing the new valve seat insert.**

COUNTERBORING CYLINDER FOR INTAKE VALVE SEAT ON CAST IRON MODELS

1. Select the proper seat insert, cutter shank, counter bore cutter, pilot and driver from the table. These numbers refer to Briggs & Stratton parts — you may get equivalent parts from other sources if available.

2. With cylinder head resting on a flat surface, valve seats up, slide the pilot into the intake valve guide. Then, assemble the correct counterbore cutter to the shank with the cutting blades of the cutter downward.

3. Insert the cutter straight into the valve seat, over the pilot. Cut so as to avoid forcing the cutter to one side, and be sure to stop as soon as the stop on the cutter touches the cylinder head.

4. Blow out all cutting chips thoroughly.

Pistons, Piston Rings, and Connecting Rods

REMOVAL

▶ **See Figures 80, 81, 82, 83, 84, 85, 86 and 87**

To remove the piston and connecting rod from the engine, bend down the connecting rod lock. Remove the connecting rod cap. Remove any carbon or ridge at the top of the cylinder bore. This will prevent breaking the rings. Push the piston and rod out of the top of the cylinder.

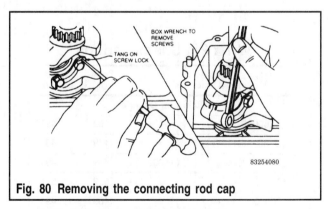

Fig. 80 Removing the connecting rod cap

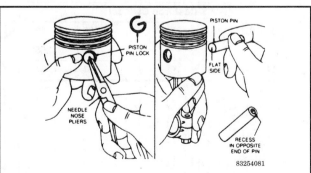

Fig. 81 Removing the piston pin and rod from the piston

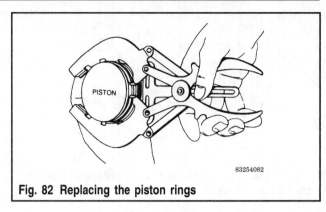

Fig. 82 Replacing the piston rings

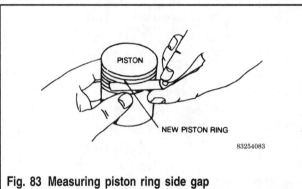

Fig. 83 Measuring piston ring side gap

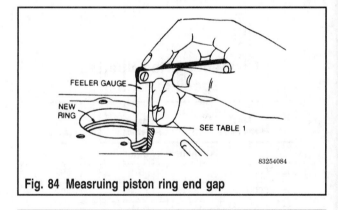

Fig. 84 Measruing piston ring end gap

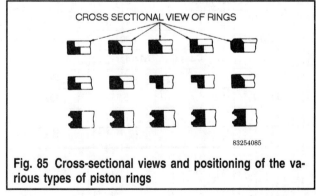

Fig. 85 Cross-sectional views and positioning of the various types of piston rings

Pistons used in sleeve bore, aluminum alloy engines are marked with an **L** on top of the piston. These pistons are tin plated and use an expander with the oil ring. This piston

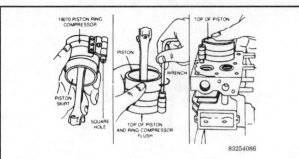

Fig. 86 Installing the piston and connecting rod assembly into the block

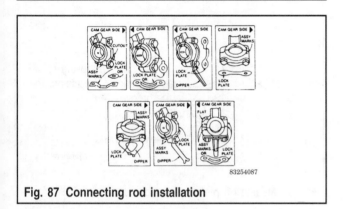

Fig. 87 Connecting rod installation

Piston Ring Gap Specifications

Basic Model Series	Comp. Ring	Oil Ring
Aluminum Cylinder		
6B, 60000, 8B, 80000		
82000, 92000, 110000, 111000	.035	.045
Cast Iron Cylinder		
5, 6, 8, N, 9	.035	.035

83254C12

Connecting Rod Bearing Specifications

Basic Model Series	Crank Pin Bearing	Piston Pin Bearing
Aluminum Cylinder		
6B, 60000	.876	.492
8B, 80000	1.001	.492
82000, 92000, 110000	1.001	.492
100000	1.001	.555
130000	1.001	.492
Cast Iron Cylinder		
5	.752	.492
6, 8, N	.751	.492
9	.876	.563
14	1.001	.674

83254C11

Wrist Pin Specifications

Basic Model Series	Piston Pin	Pin Bore
Aluminum Cylinder		
6B, 60000	.489	.491
8B, 80000	.489	.491
82000, 92000, 110000, 111000	.489	.491
100000	.552	.554
130000	.489	.491
Cast Iron Cylinder		
5, 6, 8, N	.489	.491
9	.561	.563
14	.671	.673

83254C13

assembly is not interchangeable with the piston used in the aluminum bore engines (Kool bore).

Pistons used in aluminum bore (Kool bore) engines are not marked on the top.

Connecting Rod Capscrew Torque

Basic Model Series	Inch lbs Avg. Torque
Aluminum Cylinder	
6B, 60000	100
8B, 80000	100
82000, 92000, 110000, 111000	100
100000, 130000	100
Cast Iron Cylinder	
5, 6, N, 8	100
9	140
14	190

83254C14

To remove the connecting rod from the piston, remove the piston pin lock with thin nose pliers. One end of the pin is drilled to facilitate removal of the lock.

Remove the rings one at a time, slipping them over the ring lands. Use a ring expander to remove the rings.

INSPECTION

Check the piston ring fit. Use a feeler gauge to check the side clearance of the top ring. Make sure that you remove all carbon from the top ring groove. Use a new piston ring to check the side clearance. If the cylinder is to be resized, there is no reason to check the piston, since a new oversized piston assembly will be installed. If the side clearance is more than 0.007 in. (0.18mm), the piston is excessively worn and should be replaced.

Check the piston ring end gap by cleaning all carbon from the ends of the rings and inserting them one at a time 1 in. (25mm) down into the cylinder. Check the end gap with a feeler gauge. If the gap is larger than recommended, the ring should be replaced.

➡**When checking the ring gap, do not deglaze the cylinder walls by installing piston rings in aluminum cylinder engines.**

Chrome ring sets are available fro all current aluminum and cast iron cylinder models. No honing or deglazing is required. The cylinder bore can be a maximum of 0.005 in. (0.127mm) oversize when using chrome rings.

If the crankpin bearing in the rod is scored, the rod must be replaced. 0.005 in. (0.127mm) oversize piston pins are availa-

ble in case the connecting rod and piston are worn at the piston pin bearing. If, however, the crankpin bearing in the connecting rod is worn, the rod should be replaced. Do not attempt to file or fit the rod.

If the piston pin is worn 0.0005 in. (0.0127mm) out of round or below the rejection sizes, it should be replaced.

INSTALLATION

The piston pin is a push fit into both the piston and the connecting rod. On models using a solid piston pin, one end is flat and the other end is recessed. Other models use a hollow piston pin.

1. Place a pin lock in the groove at one side of the piston. From the opposite side of the piston, insert the piston pin, flat end first for solid pins; with hollow pins, insert either end first until it stops against the pin lock. Use thin nose pliers to assemble the pin lock in the recessed end of the piston. Be sure the locks are firmly set in the groove.

2. Install the rings on the pistons, using a piston ring expander. Make sure that they are installed in the proper position. The scraper groove on the center compression ring should always be down toward the piston skirt. Be sure the oil return holes are clean and all carbon is removed from the grooves.

➡**Install the expander under the oil ring in sleeve bore aluminum alloy engines.**

3. Oil the rings and the piston skirt, then compress the rings with a ring compressor. On cast iron engines, install the compressor with the two projections downward; on aluminum engines, install the compressor with the two projections upward. These instructions refer to the piston in normal position — with skirt downward.

4. Turn the piston and compressor upside down on the bench and push downward so the piston head and the edge of the compressor band are even, all the while tightening the compressor. Draw the compressor up tight to fully compress the rings, then loosen the compressor very slightly.

✳✳WARNING

Do not attempt to install the piston and ring assembly without using a ring compressor.

5. Place the connecting rod and piston assembly, with the rings compressed, into the cylinder bore.

6. Push the piston and rod down into the cylinder. Oil the crankpin of the crankshaft.

7. Pull the connecting rod against the crankpin and assemble the rod cap so the assembly marks align.

Some rods do not have assembly marks, as the rod and cap will fit together only in one position. Use care to ensure proper installation. On the 251000 engine, the piston has a notch on the top surface. The notch must face the flywheel side of the block when installed. On models 300000 and 320000, the piston has an identification mark **F** located next to the piston pin bore. The mark must appear on the same side as the assembly mark on the rod. The assembly mark on the rod is also used to identify rod and cap alignment. Note, on these pistons, that the top ring has a beveled upper surface on the outside, while the center ring has a flat outer surface. The **F** mark or notch must face the flywheel when the piston is installed.

Where there are flat washers under the cap screws, remove and discard them prior to installing the rod. Assemble the cap screws and screw locks with the oil dippers (if used), and torque to the figure shown in the chart to avoid breakage or rod scoring later. Turn the crankshaft two revolutions to be sure the rod is correctly installed. If the rod strikes the camshaft, the connecting rod has been installed wrong or the cam gear is out of time. If the crankshaft operates freely, bend the cap screw locks against the screw heads. After tightening the rod screws, the rod should be able to move sideways on the crankpin of the shaft.

Crankshaft and Camshaft Gear

REMOVAL

Aluminum Cylinder Engines
▶ See Figures 88 and 89

To remove the crankshaft from aluminum alloy engines, remove any rust or burrs from the power take-off end of the crankshaft. Remove the crankcase cover or sump. If the sump or cover sticks, tap it lightly with a soft hammer on alternate sides near the dowel. Turn the crankshaft to align the crankshaft and camshaft timing marks, lift out the cam gear, then remove the crankshaft. On models that have ball bearings on the crankshaft, the crankshaft and the camshaft must be re-

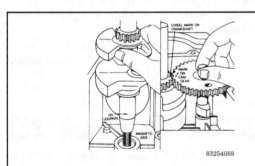

Fig. 88 Alignment of the camshaft and crankshaft timing marks

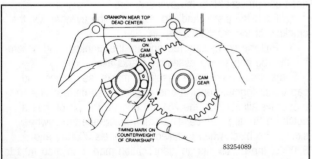

Fig. 89 Alignment of the camshaft and crankshaft timing marks on engines equipped with ball bearings

Crankshaft Specifications

Basic Model Series	PTO Journal	Mag. Journal	C Crankpin
Aluminum Cylinder			
6B, 60000	.873	.873	.870
8B, 80000*	.873	.873	.996
82000, 92000*, 110900*	.873	.873	.996
100000, 130000	.998	.873	.996
Cast Iron Cylinder			
5, 6, 8, N	.873	.873	.743
9	.983	.983	.873
14	1.179	1.179	.996

*Auxiliary drive models P.T.O. bearing reject size— 1.003

83254C15

Camshaft Specifications

Basic Model Series	Cam Gear or Shaft Journals	Cam Lobe
Aluminum Cylinder		
6B, 60000	.498	.883
8B, 80000*	.498	.883
82000, 92000	.498	.883
110900	.436 MAG. .498 PTO.	.870
100000, 130000	.498	.950
Cast Iron Cylinder		
5, 6, 8, N	.372	.875
9	.372	1.124
14	.497	1.115

*Auxiliary drive models P.T.O. .751

83254C16

moved together with the timing marks properly aligned — see illustration.

Cast Iron Cylinder Models

To remove the crankshaft from cast iron models, remove the crankcase cover. Revolve the crankshaft until the crankpin is pointing upward toward the breather at the rear of the engine (approximately a 45° angle). Pull the crankshaft out from the drive side, twisting it slightly if necessary. On models with ball bearings on the crankshaft, both the crankcase cover and bearing support should be removed.

On cast iron models with ball bearings on the drive side, first remove the magneto. Drive out the camshaft. Push the camshaft forward into the recess at the front of the engine. Then draw the crankshaft from the magneto side of the engine. Double thrust engines have cap screws inside the crankcase which hold the bearing in place. These must be removed before the crankshaft can be removed.

To remove the camshaft from all cast iron models, except the 300400 and 320400, use a long punch to drive the camshaft out toward the magneto side. Save the plug. Do not burr or peen the end of the shaft while driving it out. Hold the camshaft while driving it out. Hold the camshaft while removing the punch, so it will not drop and become damaged.

CHECKING THE CRANKSHAFT

Discard the crankshaft if it is worn beyond the allowable limit. Check the keyways for wear and make sure they are not spread. Remove all burrs from the keyway to prevent scratching the bearing. Check the three bearing journals, drive end, crankpin, and magneto end, for size and any wear or damage. Check the cam gear teeth for wear. They should not be worn at all. Check the threads at the magneto end for damage. Make sure that the crankshaft is straight.

➡**There are 0.020 in. (0.5mm) undersize connecting rods available for use on reground crankpin bearings.**

BALL BEARINGS REPLACEMENT

The ball bearings are pressed onto the crankshaft. If either the bearing or the crankshaft is to be removed, use an arbor press to remove them.

To install, heat the bearing in hot oil (325°F) (163°C) maximum. Don't let the bearing rest on the bottom of the pan in which it is heated. Place the crankshaft in a vise with the bearing side up. When the bearing is quite hot, it will slip fit onto the bearing journal. Grasp the bearing, with the shield down, and thrust it down onto the crankshaft. The bearing will tighten on the shaft while cooling. Do not quench the bearing (throw water on it to cool it).

CHECKING THE CAMSHAFT GEAR

Inspect the teeth for wear and nicks. Check the size of the camshaft and camshaft gear bearing journals. Check the size of the cam lobes. If the cam is worn beyond tolerance, discard it.

Check the automatic spark advance on models equipped with the Magna-Matic ignition system. Place the cam gear in the normal operating position with the movable weight down. Press the weight down and release it. The spring should lift the weight. If not, the spring is stretched or the weight is binding.

INSTALLATION

Aluminum Alloy Engines — Plain Bearing

In aluminum alloy engines, the tappets are inserted first, the crankshaft next, and then the cam gear. When inserting the cam gear, turn the crankshaft and the cam gear so that the timing marks on the gears align.

Aluminum Alloy Engines — Ball Bearing

On crankshafts with ball bearings, the gear teeth are not visible for alignment of the timing marks; therefore, the timing mark is on the counterweight. On ball bearing equipped engines, the tappets are installed first. The crankshaft and the cam gear must be inserted together and their timing marks aligned.

Crankshaft Cover and Crankshaft

INSTALLATION

Models 100900 and 130900

On these models, install the governor slinger onto the cam gear with the spring washer.

To protect the oil seal while assembling the crankcase cover, put oil or grease on the sealing edge of the oil seal. Wrap a piece of thin cardboard around the crankshaft so the seal will slide easily over the shoulder of the crankshaft. If the sharp edge of the oil seal is cut or bent under, the seal may leak.

Cast Iron Engines w/Plain Bearings

1. Assemble the tappets and cylinder, then insert the cam gear.
2. Push the camshaft into the camshaft hole in the cylinder, from the flywheel side, through the cam gear.
3. With a blunt punch, press or hammer the camshaft until the end is flush with the outside of the cylinder on the power takeoff side.
4. Place a small amount of sealer on the camshaft plug, then press or hammer it into the camshaft hole in the cylinder at the flywheel side.
5. Install the crankshaft so the timing marks on the teeth and on the cam gear align.

Cast Iron Engines w/Ball Bearings

1. Assemble the tappets, then insert the cam gear into the cylinder, pushing the cam gear forward into the recess in front of the cylinder.
2. Insert the crankshaft into the cylinder.
3. Turn the camshaft and crankshaft until the timing marks align, then push the cam gear back until it engages the gear on the crankshaft with the timing marks together.

4. Insert the camshaft.

5. Place a small amount of sealer on the camshaft plug and press or hammer it into the camshaft hole in the cylinder at the flywheel side.

Crankshaft End-Play Adjustment

The crankshaft end-play on all models, plain and ball bearing, should be 0.002-0.008 in. (0.05-0.20mm). The method of obtaining the correct end-play varies, however, between cast iron, aluminum, plain, and ball bearing models. New gasket sets include three crankcase cover or bearing support gaskets, 0.005 in. (0.127mm), 0.009 in. (0.228mm), and 0.015 in. (0.381mm) thick.

The end-play of the crankshaft may be checked by assembling a dial indicator on the crankshaft with the pointer against the crankcase. Move the crankshaft in and out. The indicator will show the end-play. Another way to measure the end-play is to assemble a pulley to the crankshaft and measure the end-play with a feeler gauge. Place the feeler gauge between the crankshaft thrust face and the bearing support. The feeler gauge method of measuring crankshaft end-play can only be used on cast iron plain bearing engines with removable bases.

On cast iron engines, the end-play should be 0.002-0.008 in. (0.05-0.20mm) with one 0.015 in. (0.381mm) gasket in place. If the end-play is less than 0.002 in. (0.05mm), which would be the case if a new crankcase or sump cover is used, additional gaskets of 0.005 in. (0.127mm), 0.009 in. (0.228mm), or 0.015 in. (0.381mm) may be added in various combinations to obtain the proper end-play.

ALUMINUM ENGINES ONLY

If the end-play is more than 0.008 in. (0.20mm) with one 0.015 in. (0.381mm) gasket in place, a thrust washer is available to be placed on the crankshaft power take-off end, between the gear and crankcase cover or sump on plain bearing engines. On ball bearing equipped aluminum engines, the thrust washer is added to the magneto end of the crankshaft instead of the power take-off end.

➡ **Aluminum engines never use less than the 0.015 in. (0.381mm) gasket.**

Cylinders

INSPECTION

▸ **See Figures 90 and 91**

Always inspect the cylinder after the engine has been disassembled. Visual inspection will show if there are any cracks, stripped bolt holes, broken fins, or if the cylinder wall is scored. Use an inside micrometer or telescoping gauge and micrometer to measure the size of the cylinder bore. Measure at right angles.

If the cylinder bore is more than 0.003 in. (0.076mm) oversize, or 0.0015 in. (0.038mm) out of round on lightweight (aluminum) cylinders, the cylinder must be resized (rebored).

➡ **Do not deglaze the cylinder walls when installing piston rings in aluminum cylinder engines. Also be aware that there are chrome ring sets available for most engines.**

Cylinder Bore Specifications

Basic Engine Model or Series	Std. Bore Size Diameter	
	Max	Min
Aluminum Cylinder		
6B		
60000 before Ser. #5810060	2.3125	2.3115
60000 after Ser. #5810030	2.375	2.374
8B, 80000, 82000	2.375	2.374
92000	2.5625	2.5615
100000	2.500	2.449
110000	2.7812	2.7802
130000	2.5625	2.5615
Cast Iron Cylinder		
5, 6, 5S, N	2.000	1.999
8	2.250	2.249
9	2.250	2.249
14	2.625	2.624

83254C17

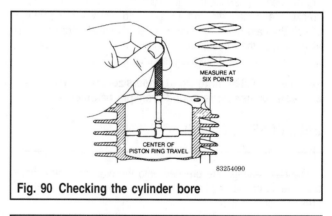

Fig. 90 Checking the cylinder bore

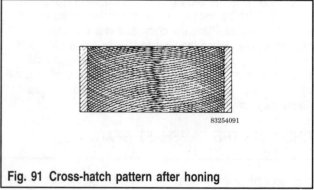

Fig. 91 Cross-hatch pattern after honing

These are used to control oil pumping in bores worn to 0.005 in. (0.127mm) over standard and do not require honing or glaze breaking to seat.

RESIZING

Always resize to exactly 0.010 in. (0.254mm), 0.020 in. (0.508mm), or 0.030 in. (0.762mm) over standard size. If this is done accurately, the stock oversize rings and pistons will fit perfectly and proper clearances will be maintained. Cylinders, either cast iron or lightweight, can be quickly resized with a good hone. Use the stones and lubrication recommended by the hone manufacturer to produce the correct cylinder wall finish for the various engine models.

If a boring bar is used, a hone must be used after the boring operation to produce the proper cylinder wall finish. Honing can be done with a portable electric drill, but it is easier to use a drill press.

1. Clean the cylinder at top and bottom to remove all burrs and pieces of base and head gaskets.

2. Fasten the cylinder to a heavy iron plate. Some cylinders require shims. Use a level to align the drill press spindle with the bore.

3. Oil the surface of the drill press table liberally. Set the iron plate and the cylinder on the drill press table. Do not anchor the cylinder to the drill press table. If you are using a portable drill, set the plate and the cylinder on the floor.

4. Place the hone driveshaft in the chuck of the drill.

5. Slip the hone into the cylinder. Connect the driveshaft to the hone and set the stop on the drill press so the hone can only extend ¾-1 in. (19-25mm) from the top or bottom of the cylinder. If you are using a portable drill, cut a piece of wood to place in the cylinder as a stop for the hone.

6. Place the hone in the middle of the cylinder bore. Tighten the adjusting knob with your finger or a small screwdriver until the stones fit snugly against the cylinder wall. Do not force the stones against the cylinder wall. The hone should operate at a speed of 300-700 rpm. Lubricate the hone as recommended by the manufacturer.

➡ **Be sure that the cylinder and the hone are centered and aligned with the driveshaft and the drill spindle.**

7. Start the drill and, as the hone spins, move it up and down at the lower end of the cylinder. The cylinder is not worn at the bottom but is round so it will act to guide the hone and straighten the cylinder bore. As the bottom of the cylinder increases in diameter, gradually increase your strokes until the hone travels the full length of the bore.

➡ **Do not extend the hone more than ¾-1 in. (19-25mm) past either end of the cylinder bore.**

8. As the cutting tension decreases, stop the hone and tighten the adjusting knob. Check the cylinder bore frequently with an accurate micrometer. Hone 0.0005 in. (0.0127mm) oversize to allow for shrinkage when the cylinder cools.

9. When the cylinder is within 0.0015 in. (0.0381mm) of the desired size, change from the rough stone to a finishing stone.

The finished resized cylinder should have a cross-hatched appearance. Proper stones, lubrication, and spindle speed along with rapid movement of the hone within the cylinder during the last few strokes, will produce this finish. Cross-hatching provides proper lubrication and ring break-in.

➡ **It is EXTREMELY important that the cylinder be thoroughly cleaned after honing to eliminate ALL grit. Wash the cylinder carefully in a solvent such as kerosene. The cylinder bore should be cleaned with a brush, soap, and water.**

Bearings

INSPECTION

Plain Type

▶ **See Figure 92**

Bearings should be replaced if they are scored or if a plug gauge will enter. Try the gauge at several points in the bearing.

REPLACING PLAIN BEARINGS

Models 9-14

The crankcase cover bearing support should be replaced if the bearing is worn or scored.

Crankshaft Bearing Specifications

Basic Engine Model or Series	PTO Bearing	Bearing Magneto
Aluminum Cylinder		
6B, 8B	.878	.878
60000, 80000	.878	.878
82000, 92000, 110900	.878	.878
100000, 130000	1.003	.878
Cast Iron Cylinder		
5, 6, N	.878	.878
9	.988	.988
14	1.185	1.185

83254C18

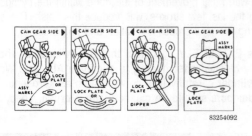

83254092

Fig. 92 Installing the connecting rod in a horizontal crankshaft engine

REPLACING THE MAGNETO BEARING

Aluminum Cylinder Engines

There are no removable bearings in these engines. The cylinder must be reamed out so a replacement bushing can be installed.

1. Place a pilot guide bushing in the sump bearing, with the flange of the guide bushing toward the inside of the sump.

2. Assemble the sump on the cylinder. Make sure that the pilot guide bushing does not fall out of place.

3. Place the guide bushing into the oil seal recess in the cylinder. This guide bushing will center the counterbore reamer even though the oil bearing surface might be badly worn.

4. Place the counterbore reamer on the pilot and insert them into the cylinder until the tip of the pilot enters the pilot guide bushing in the sump.

5. Turn the reamer clockwise with a steady, even pressure until it is completely through the bearing. Lubricate the reamer with kerosene or any other suitable solvent.

➡**Counterbore reaming may be performed without any lubrication. However, clean off shavings because aluminum material builds up on the reamer flutes causing eventual damage to the reamer and an oversize counterbore.**

6. Remove the sump and pull the reamer out without backing it through the bearing. Clean out the remaining chips. Remove the guide bushing from the oil seal recess.

7. Hold the new bushing against the outer end of the reamed out bearing, with the notch in the bushing aligned with the notch in the cylinder. Note the position of the split in the bushing. At a point in the outer edge of the reamed out bearing opposite to the split in the bushing, make a notch in the cylinder hub at a 45° angle to the bearing surface. Use a chisel or a screwdriver and hammer.

8. Press in the new bushing, being careful to align the oil notches with the driver and the support until the outer end of the bushing is flush with the end of the reamed cylinder hub.

9. With a blunt chisel or screwdriver, drive a portion of the bushing into the notch previously made in the cylinder. This is called staking and is done to prevent the bushing from turning.

10. Reassemble the sump to the cylinder with the pilot guide bushing in the sump bearing.

11. Place a finishing reamer on the pilot and insert the pilot into the cylinder bearing until the tip of the pilot enters the pilot guide bushings in the sump bearing.

12. Lubricate the reamer with kerosene, fuel oil, or other suitable solvent, then ream the bushing, turning the reamer clockwise with a steady even pressure until the reamer is completely through the bearing. Improper lubricants will produce a rough bearing surface.

13. Remove the sump, reamer, and the pilot guide bushing. Clean out all reaming chips.

REPLACING THE P.T.O. BEARING

Aluminum Cylinder Engines

The sump or crankcase bearing is repaired the same way as the magneto end bearing. Make sure to complete repair of one bearing before starting to repair the other. Press in new oil seals when bearing repair is completed.

➡**On Models 8B-HA, 80590, 81590, 82590, 80790, 81790, 82990, 92590 and 92990, the magneto bearing can be replaced as described above. However, if the sump bearing is worn, the sump must be replaced.**

REPLACING OIL SEALS

1. Assemble the seal with the sharp edge of leather or rubber toward the inside of the engine.

2. Lubricate the inside diameter of the seal with Lubriplate® or equivalent.

3. Press all seals but those listed below so they are flush with the hub. On models 60000, 80000, 100000 and 13000 with ball bearing which has a mounting flange, the seal must be pressed in until it is 3/16 (4.76mm) below the crankcase mounting flange.

Extended Oil Filler Tubes and Dipsticks

▶ **See Figures 93 and 94**

When installing the extended oil fill and dipstick assembly, the tube must be installed so the O-ring seal is firmly compressed. To do so, push the tube downward toward the sump, then tighten the blower housing screw, which is used to secure the tube and bracket. When the dipstick assembly is fully depressed, it seals the upper end of the tube.

A leak at the seal between the tube and the sump, or at the seal at the upper end of the dipstick can result in a loss of

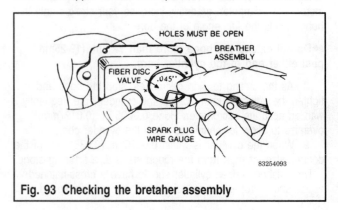

83254093

Fig. 93 Checking the bretaher assembly

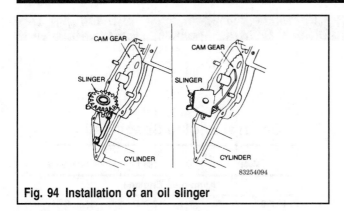

Fig. 94 Installation of an oil slinger

crankcase vacuum, and a discharge of smoke through the exhaust system.

Breathers

The function of the breather is to maintain a vacuum in the crankcase. The breather has a fiber disc valve which limits the direction of air flow caused by the piston moving back and forth in the cylinder. Air can flow out of the crankcase, but the one-way valve blocks the return flow, thus maintaining a vacuum in the crankcase. A partial vacuum must be maintained in the crankcase to prevent oil from being forced out of the engine at the piston rings, oil seals, breaker plunger, and gaskets.

INSPECTION OF THE BREATHER

If the fiber disc valve is stuck or binding, the breather cannot function properly and must be replaced. A 0.045 in. (1.14mm) wire gauge should not enter the space between the fiber disc valve and the body. Use a spark plug wire gauge to check the valve. The fiber disc valve is held in place by an internal bracket which will be distorted if pressure is applied to the fiber disc valve. Therefore, do not apply force when checking the valve with the wire gauge.

If the breather is removed for inspection or valve repair, a new gasket should be used when replacing the breather. Tighten the screws securely to prevent oil leakage.

Most breathers are now vented through the air cleaner, to prevent dirt from entering the crankcase. Check to be sure that the venting elbows or the tube are not damaged and that they are properly sealed.

Oil Dippers and Slingers

Oil dippers reach into the oil reservoir in the base of the engine and splash oil onto the internal engine parts. The oil dipper is installed on the connecting rod and has no pump or moving parts.

Oil slingers are driven by the cam gear. Old style slingers using a die cast bracket assembly have a steel bushing between the slinger and the bracket. Replace the bracket on which the oil slinger rides if it is worn to a diameter of 0.490 in. (12.4mm) or less. Replace the steel bushing if it is worn. Newer style oil slingers have a stamped steel bracket.

6 — 12 HP MODELS

Engine Identification

The Briggs and Stratton model designation system consists of up to a six digit number. It is possible to determine most of the important mechanical features of the engine by merely knowing the model number. An explanation of what each number means is given below.

1. The first one or two digits indicate the cubic inch displacement (cid).

2. The first digit after the displacement indicates the basic design series, relating to cylinder construction, ignition and general configuration.

3. The second digit after the displacement indicates the position of the crankshaft and the type of carburetor the engine has.

4. The third digit after the displacement indicates the type of bearings and whether or not the engine is equipped with a reduction gear or auxiliary drive.

5. The last digit indicates the type of starter. The model identification plate is usually located on the air baffle surrounding the cylinder.

Air Cleaners

▶ **See Figures 95, 96, 97, 98 and 99**

A properly serviced air cleaner protects the engine from dust particles that are in the air. When servicing an air cleaner, check the air cleaner mounting and gaskets for worn or dam-

General Engine Specifications

Model	Bore Size (In.)	Horsepower
Aluminum Engines		
140000	2.750	6
170000, 171700	3.000	7
190000, 191700	3.000	8
251000	3.4375	10
Cast Iron Engines		
19, 190000, 200000	3.000	8
23, 230000	3.000	9
243000	3.0625	10
300000	3.4375	13
320000	3.5625	16

832550C2

Briggs and Stratton Model Numbering System

Cubic Inch Displacement	First Digit After Displacement — Basic Design Series	Second Digit After Displacement — Crankshaft, Carburetor Governor	Third Digit After Displacement — Bearings, Reduction Gears & Auxiliary Drives	Fourth Digit After Displacement — Type of Starter
6	0	0-	0-Plain Bearing	0-Without Starter
8	1	1-Horizontal Vacu-Jet	1-Flange Mounting Plain Bearing	1-Rope Starter
9	2	2-Horizontal Pulsa-Jet	2-Ball Bearing	2-Rewind Starter
10	3			
13	4	3-Horizontal Flo-Jet (Pneumatic Governor)	3-Flange Mounting Ball Bearing	3-Electric-110 Volt, Gear Drive
14	5	4-Horizontal Flo-Jet (Mechanical Governor)	4-	4-Elec. Starter-Generator-12 Volt, Belt Drive
17	6			
19	7			
20	8			
23	9	5-Vertical Vacu-Jet	5-Gear Reduction (6 to 1)	5-Electric Starter Only-12 Volt, Gear Drive
24				
30		6-	6-Gear Reduction (6 to 1) Reverse Rotation	6-Wind-up Starter
32				
		7-Vertical Flo-Jet	7-	7-Electric Starter, 12 Volt Gear Drive, with Alternator
		8-	8-Auxiliary Drive Perpendicular to Crankshaft	8-Vertical-pull Starter
		9-Vertical Pulsa-Jet	9-Auxiliary Drive Parallel to Crankshaft	

832550C1

Fig. 95 Oil-foam air cleaner

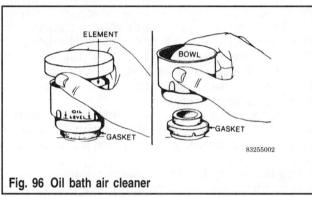

Fig. 96 Oil bath air cleaner

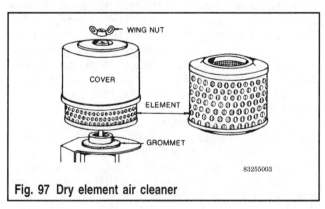

Fig. 97 Dry element air cleaner

Fig. 98 Heavy-duty oil bath air cleaner

aged mating surfaces. Replace any worn or damaged parts to prevent dirt and dust from entering the engine through open-

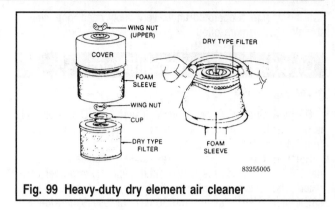

Fig. 99 Heavy-duty dry element air cleaner

ings caused by improper sealing. Straighten or replace any bent mounting studs.

SERVICING

Oil Foam Air Cleaners

Clean and re-oil the air cleaner element every 25 hours of operation under normal operating conditions. The capacity of the oil-foam air cleaner is adequate for a full season's use without cleaning. Under very dusty conditions, clean the air cleaner every few hours of operation.

The oil-foam air cleaner is serviced in the following manner:

1. Remove the screw that holds the halves of the air cleaner shell together and retains it to the carburetor.

2. Remove the air cleaner carefully to prevent dirt from entering the carburetor.

3. Take the air cleaner apart (split the two halves).

4. Wash the foam in kerosene or liquid detergent and water to remove the dirt.

5. Wrap the foam in a clean cloth and squeeze it dry.

6. Saturate the foam in clean engine oil and squeeze it to remove the excess oil.

7. Assemble the air cleaner and fasten it to the carburetor with the attaching screw.

Oil Bath Air Cleaner

Pour the old oil out of the bowl. Wash the element thoroughly in solvent and squeeze it dry. Clean the bowl and refill it with the same type of oil used in the crankcase.

Dry Element Air Cleaner

Remove the element of the air cleaner and tap (top and bottom) it on a flat surface or wash it in non-sudsing detergent and flush it from the inside until the water coming out is clear. After washing, air dry the element thoroughly before reinstalling it on the engine. NEVER OIL A DRY ELEMENT.

Heavy Duty Air Cleaner

Clean and re-oil the foam pre-cleaner at three month intervals or every 25 hours, whichever comes first.

Clean the paper element every year or 100 hours, whichever comes first. Use the dry element procedure for cleaning the paper element of the heavy duty air cleaner.

Use the oil foam cleaning procedure to clean the foam sleeve of the heavy duty air cleaner.

If the engine is operated under very dusty conditions, clean the air cleaner more often.

Oil and Fuel Recommendations

▶ **See Figure 100**

Briggs & Stratton recommends unleaded fuel. Unleaded fuel is preferable because of the reduction in deposits that results from its use, but its use is not required.

Premium fuel is not required, as regular or unleaded will have sufficient knock resistance if the engine is in proper condition. The factory recommends that fuel by purchased in lots small enough to be used up in 30 days or less. When fuel is older than that, it can form gum and varnish, or may be improperly tailored to the prevailing temperature.

Oil Viscosity Recommendations

Winter (under 40°F.)	Summer (above 40°F.)
SAE 5W-20, or SAE 5W-30	SAE 30
If above are not available:	If above is not available:
SAE 10W or SAE 10W-30 (under 0°F.)	SAE 10W-40 or SAE 10W-30
Use SAE 10W or SAE 10W-30 in proportions of 90% motor oil/ 10% kerosene	

832550C3

Engine Oil Capacity Chart

Basic Model Series	Capacity Pints
Aluminum	
14, 17 Cu. in. Vert. Crankshaft	2¼
14, 17, 19 Cu. in. Horiz. Crankshaft	2¾
25 Cu. in. Vert. Crankshaft	3
25 Cu. in. Horiz. Crankshaft	3
Cast Iron	
19, 20 Cu. in. Horiz. Crank.	3
23, 24, 30, 32 Cu. in. Horiz. Crank.	4

832550C4

You should use a high quality detergent oil designated 'For Service SG'. Detergent oil is recommended because of its important ability to keep gum and varnish from clogging the lubrication system. Briggs & Stratton specifically recommends that no special oil additives be used.

Oil must be changed every 25 hours of operation. If the atmosphere in which the engine is operating is very dirty, oil changes should be made more frequently, as often as every 12 hours, if necessary. Oil should be changed after 5 hours of operation in the case of brand new engines. Drain engine oil when hot.

In hot weather, when under heavy load, or when brand new, engines may consume oil at a rate which will require you to refill the crankcase several times between oil changes. Check the oil level every hour or so until you can accurately estimate how long the engine can go between refills. To check oil level, stop the engine and allow it to sit for a couple of minutes, then remove the dipstick or filler cap. Fill the crankcase to the top of the filler pipe when there is no dipstick, or wipe the dipstick clean, reinsert it, and add oil as necessary until the level reaches the upper mark.

On cast iron engines with a gear reduction unit, crankcase and reduction gears are lubricated by a common oil supply. When draining crankcase, also remove drain plug in reduction unit.

On aluminum engines with reduction gear, a separate oil supply lubricates the gears, although the same type of oil used in the crankcase is used in the reduction gear cover. On these engines, remove the drain plug every fourth oil change (100 hours), then install the plug and refill. The level in the reduction gear cover must be checked during the refill operation by removing the level plug from the side of the gearcase, removing the filler plug, and then filling the case through the filler plug hole until oil runs out the level plug hole. Then, install both plugs.

On 6-1 gear reduction engines (models 6, 8, 8000, 10000, and 13000), no changes are required for the oil in the reduction gear case, but level must be checked and the case refilled, as described in the paragraph above, every 100 hours. Make sure the oil level plug (with screwdriver slot and no vent) is installed in the hole on the side of the case.

Spark Plugs

▶ **See Figures 101, 102, 103 and 104**

Remove the spark plug with a ¾ in. (1½ in. plug) or a ¹³⁄₁₆ in. (2 in. plug) deep well socket wrench. Clean carbon deposits

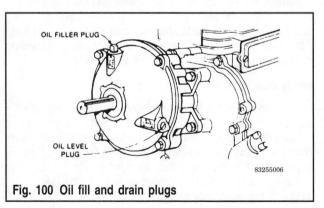

Fig. 100 Oil fill and drain plugs

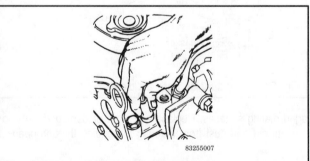

Fig. 101 Twist and pill on the rubber boot to remove the wire from the plug. Never pull on the wire itself

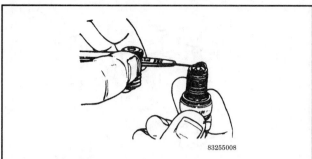

Fig. 102 Always use a wire-type gauge to check the spark plug electrode gap

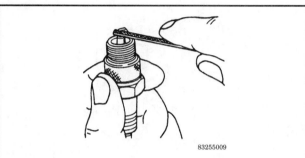

Fig. 103 Plugs that are in good condition can be filed and re-used

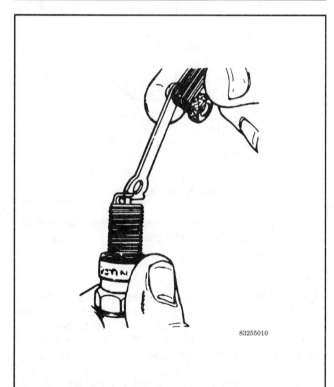

Fig. 104 Adjust the electrode gap by bending the side electrode

off the center and side electrodes with a sharp instrument. If possible, you should also attempt to remove deposits from the recess between the insulator and the threaded portion of the

plug. If the electrodes are burned away or the insulator is cracked at any point, replace the plug. Using a wire type feeler gauge, adjust the gap by bending the side electrode where it is curved until the gap is 0.030 in. (0.8mm).

When installing the plug, make sure the threads of the plug and the threads in the cylinder head are clean. It is best to oil the plug threads very lightly. Be careful not to overtorque the plug, especially if the engine has an aluminum head. If you use a torque wrench, torque to about 15 ft. lbs.

Breaker Points Ignition Systems

All Briggs and Stratton engines have magneto ignition systems. Three types are used: Flywheel Type — Internal Breaker Flywheel Type — External Breaker, and Magna-Matic.

FLYWHEEL TYPE — INTERNAL BREAKER

This ignition system has the magneto located on the flywheel and the breaker points located under the flywheel.

The flywheel is located on the crankshaft with a soft metal key. It is held in place by a nut or starter clutch. The flywheel key must be in good condition to insure proper location of the flywheel for ignition timing. Do not use a steel key under any circumstances. Use only a soft metal key, as originally supplied.

The keyway in both flywheel and crankshaft should not be distorted. Flywheels are made of aluminum, zinc, or cast iron.

Flywheel, Nut, and/or Starter Clutch
▶ **See Figures 105, 106 and 107**

REMOVAL & INSTALLATION

Place a block of wood under the flywheel fins to prevent the flywheel from turning while you are loosening the nut or starter clutch. Be careful not to bend the flywheel. There are special flywheel holders available for this purpose; Briggs & Stratton recommends their use on flywheels of 6¾ in. (171.45mm) diameter or less.

On rope starter engines, the ½ in. flywheel nut has a left-hand thread and the ⅝ in. nut has a right-hand thread. The starter clutch used on rewind or wind-up starters has a right-hand thread.

Some flywheels have two holes provided for the use of a flywheel puller. Use a small gear puller or automotive steering

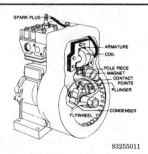

Fig. 105 Flywheel magneto ignition with internal breaker points and external armature

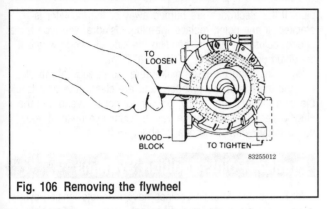

Fig. 106 Removing the flywheel

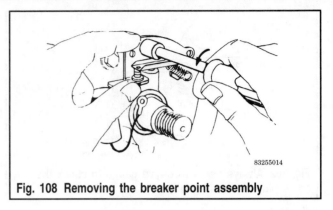

Fig. 108 Removing the breaker point assembly

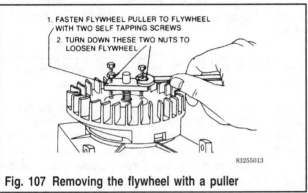

Fig. 107 Removing the flywheel with a puller

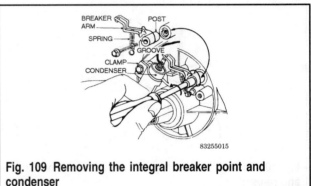

Fig. 109 Removing the integral breaker point and condenser

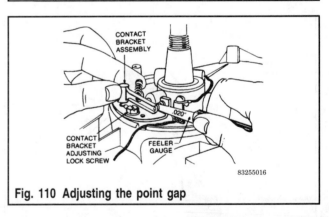

Fig. 110 Adjusting the point gap

wheel puller to remove the flywheel if a flywheel puller is not available. Be careful not to bend the flywheel if a gear puller is used. On rope starter engines leave the nut on for the puller to bear against. Small cast iron flywheels do not require a puller.

Install the flywheel in the reverse order of removal after inspecting the key and keyway for damage or wear.

Breaker Point Replacement
▶ **See Figures 108, 109 and 110**

Remove the breaker cover. Care should be taken when removing the cover, to avoid damaging it. If the cover is bent or damaged, it should be replaced to insure a proper seal.

The breaker point gap on all models is 0.020 in. (0.5mm). Check the points for contact and for signs of burning or pitting. Points that are set too wide will advance the spark timing and may cause kickback when starting. Points that are set too close will retard the spark timing and decrease engine power.

On models that have a separate condenser, the point set is removed by first removing the condenser and armature wires from the breaker point clip. Loosen the adjusting lock screw and remove the breaker point assembly.

On models where the condenser is incorporated with the breaker points, loosen the screw which holds the post. The condenser/point assembly is removed by loosening the screw which holds the condenser clamp.

When installing a point set with the separate condenser, be sure that the small boss on the magneto plate enters the hole in the point bracket. Mount the point set to the magneto plate or the cylinder with a lock screw. Fasten the armature lead wire to the breaker points with the clip and screw. If these lead wires do not have terminals, the bare end of the wires

can be inserted into the clip and the screw tightened to make a good connection. Do not let the ends of the wire touch either the point bracket or the magneto plate, or the ignition will be grounded.

To install the integral condenser/point set, place the mounting post of the breaker arm into the recess in the cylinder so that the groove in the post fits the notch in the recess. Tighten the mounting screw securely. Use a 1/4 in. wrench. Slip the open loop of the breaker arm spring through the two holes in the arm, then hook the closed loop of the spring over the small post protruding from the cylinder. Push the flat end of the breaker arm into the groove in the mounting post. This places tension on the spring and pulls the arm against the plunger. If the condenser post is threaded, attach the soil primary wire and the ground wire (if furnished) with the lock washer and nut. If the primary wire is fastened to the condenser with a spring fastener, compress the spring and slip the primary wire and ground wire into the hole in the condenser post. Release the spring. Lay the condenser in place and tighten the condenser clamp securely. Install the spring in the breaker arm.

Point Gap Adjustment

Turn the crankshaft until the points are open to the widest gap. When adjusting a breaker point assembly with an integral condenser, move the condenser forward or backward with a screwdriver until the proper gap is obtained — 0.020 in. (0.5mm). Point sets with a separate condenser are adjusted by moving the contact point bracket up and down after the lock screw has been loosened. The point gap is set to 0.020 in. (0.5mm).

Breaker Point Plunger

▶ See Figures 111 and 112

If the breaker point plunger hole becomes excessively worn, oil will leak past the plunger and may get on the points, causing them to burn. To check the hole, loosen the breaker point mounting screw and move the breaker points out of the way. Remove the plunger. If the flat end of the #19055 plug gauge will enter the plunger hole for a distance of ¼ in. (6mm) or more, the hole should be rebushed.

To install the bushing, it is necessary that the breaker points, armature, and crankshaft be removed. Use a #19056 reamer to ream out the old plunger hole. This should be done by hand. The reamer must be in alignment with the plunger hole. Drive the bushing, #23513, into the hole until the upper

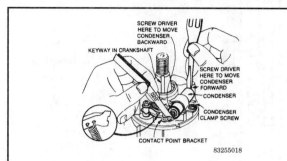

Fig. 112 Adjusting the point gap on integral point and condenser assembly

end of the bushing is flush with the top of the boss. Remove all metal chips and dirt from the engine.

If the breaker point plunger is worn to a length of 0.870 in. (22mm) or less, it should be replaced. Plungers must be inserted with the groove at the top or oil will enter the breaker box. Insert the plunger into the hole in the cylinder.

Armature Air Gap Adjustment

▶ See Figures 113 and 114

Set the air gap between the flywheel and the armature as follows: With the armature up as far as possible and just one screw tightened, slip the proper gauge between the armature and flywheel. Turn the flywheel until the magnets are directly below the armature. Loosen the one mounting screw and the magnets should pull the armature down firmly against the thickness gauge. Tighten the mounting screws.

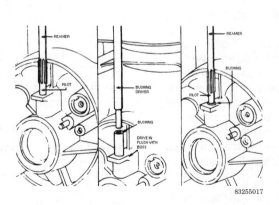

Fig. 111 Replacing the breaker plunger bushing

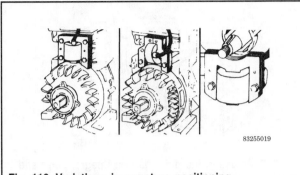

Fig. 113 Variations in armature positioning

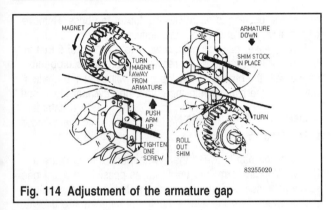

Fig. 114 Adjustment of the armature gap

FLYWHEEL TYPE — EXTERNAL BREAKER

Breaker Point Set Replacement
▶ **See Figure 115**

Turn the crankshaft until the points open to their widest gap. This makes it easier to assemble and adjust the points later if the crankshaft is not removed. Remove the condenser and upper and lower mounting screws. Loosen the lock nut and back off the breaker point screw. Install the points in the reverse order of removal.

To avoid the possibility of oil leaking past the breaker point plunger or moisture entering the crankcase between the plunger and the bushing, a plunger seal is installed on the engine models using this type of ignition system. To install a new seal on the plunger, remove the breaker point assembly and condenser. Remove the retainer and eyelet, remove the old seal, and install the new one. Use extreme care when installing the seal on the plunger to avoid damaging the seal. Replace the eyelet and retainer and replace the points and condenser.

➡**Apply a small amount of sealer to the threads of both mounting screws and the adjustment screw. The sealer prevents oil from leaking into the breaker point area.**

Point Gap Adjustment
▶ **See Figure 116**

Turn the crankshaft until the points open to their widest gap. Turn the breaker point adjusting screw until the points open to 0.020 in. (0.5mm) and tighten the lock nut. When the cover is installed, seal the point where the primary wire passes under

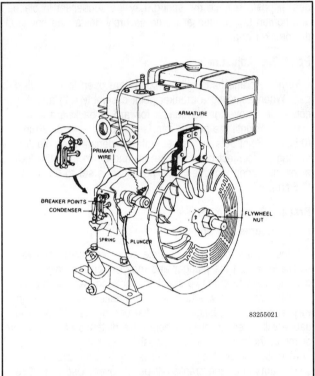

Fig. 115 Flywheel magneto ignition with an external breaker assembly

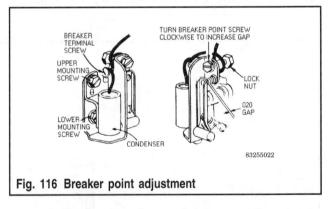

Fig. 116 Breaker point adjustment

the cover. This area must be resealed to prevent the entry of dust and moisture.

Armature Timing Adjustment
MODELS 193000, 200000, 230000, 243000

Using a puller, remove the flywheel. Set the point gap at 0.020 in. (0.5mm). Position the flywheel on the crankshaft taper. Slip the key in place. Install the flywheel and the crankshaft clockwise until the breaker points are just opening. Use a timing light. When the points just start to open, the arrow on the flywheel should line up with the arrow on the armature bracket.

If the arrows do not match, slip off the flywheel without disturbing the position of the crankshaft. Slightly loosen the mounting screw which holds the armature bracket to the cylinder. Slip the flywheel back onto the crankshaft. Insert the flywheel key. Install the flywheel nut finger tight. Move the armature and bracket assembly to align the arrows. Slip off the flywheel and tighten the armature bracket bolts. Install the key and flywheel. Tighten the flywheel nut to 110-118 ft. lbs. on the 193000 and 200000 series. On all the rest, tighten to 138-150 ft. lbs. Set the armature gap at 0.010-0.014 in. (0.25-0.35mm).

MODELS 19D AND 23D

With the points set at 0.020 in. (0.5mm) and the flywheel key screw finger tight together with the flywheel nut, rotate the flywheel clockwise until the breaker points are just opening. The flywheel key drives the crankshaft while doing this. Using a timing light, rotate the flywheel slightly counterclockwise until the edge of the armature lines up with the edge of the flywheel insert. The crankshaft must not turn while doing this. Tighten the key screw and the flywheel nut. Set the armature air gap at 0.022-0.026 in. (0.56-0.66mm).

Replacing Threaded Breaker Plunger and Bushing
▶ See Figure 117

Remove the breaker cover and the condenser and breaker point assembly.

Place a thick ³/₈ in. (9.5mm) inside diameter washer over the end of the bushing and screw on the ³/₈-24 nut. Tighten the nut to pull the bushing out of the hole. After the bushing has been moved about ¹/₈ in. (3mm), remove the nut and put on a second thick washer and repeat the procedure. A total stack of ³/₈ in. (9.5mm) washers will be required to completely remove the bushing. Be sure the plunger does not fall out of the bushing as it is removed.

Place the new plunger in the bushing with the large end of the plunger opposite the threads on the bushing. Screw the ³/₈-24 in. nut onto the threads to protect them and insert the bushing into the cylinder. Place a piece of tubing the same diameter as the nut and, using a hammer, drive the bushing into the cylinder until the square shoulder on the bushing is flush with the face of the cylinder. Check to be sure that the plunger operates freely.

Replacing Unthreaded Breaker Plunger and Bushing
▶ See Figure 118

Pull the plunger out as far as possible and use a pair of pliers to break the plunger off as close as possible to the bushing. Use a ¹/₄-20 in. tap or a #93029 self threading screw to thread the hole in the bushing to a depth of about ¹/₂-⁵/₈ in. (13-16mm). Use a ¹/₄-20 · ¹/₂ in. hex head screw and two

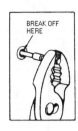
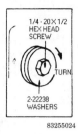

Fig. 118 Removing an unthreaded plunger bushing

spacer washers to pull the bushing out of the cylinder. The bushing will be free when it has been extracted ⁵/₁₆ in. (8mm). Carefully remove the bushing and the remainder of the broken plunger. Do not allow the plunger or metal chips to drop into the crankcase.

Correctly insert the new plunger into the new bushing. Insert the plunger and the bushing into the cylinder. Use a hammer and the old bushing to drive the new bushing into the cylinder until the new bushing is flush with the face of the cylinder. Make sure that the plunger operates freely.

Plunger Seal
▶ See Figure 119

Later models with Flywheel Type — External Breaker Ignition feature a plunger seal. This seal keeps both oil and moisture from entering the breaker box. If the points have become contaminated on an engine manufactured without this feature, the seal may be installed. Parts, part numbers, and their locations are shown in the illustration. Install the seal onto the plunger very carefully to avoid fracturing it.

MAGNA-MATIC IGNITION SYSTEM

Removing the Flywheel
▶ See Figure 120

Flywheels on engines with Magna-Matic ignition are removed with pullers similar to factory designs numbered #19068 and 19203. These pullers employ two bolts, which are screwed into holes tapped into the flywheel. The bolts are turned until the flywheel is forced off the crankshaft. Only this type of device should be used to pull these flywheels.

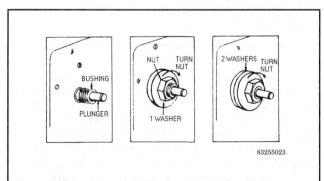

Fig. 117 Removing a threaded plunger bushing

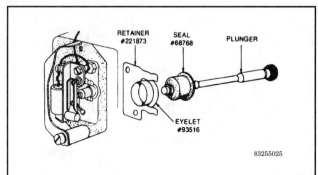

Fig. 119 Plunger seal used on later models

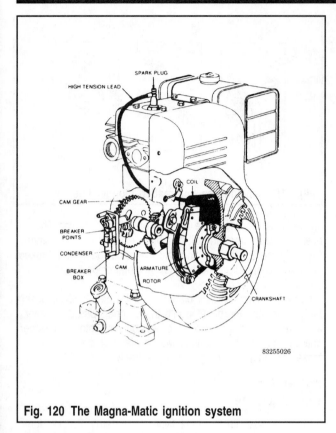

Fig. 120 The Magna-Matic ignition system

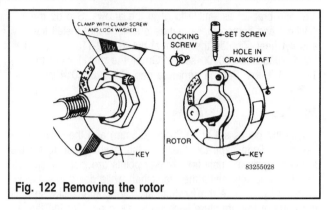

Fig. 122 Removing the rotor

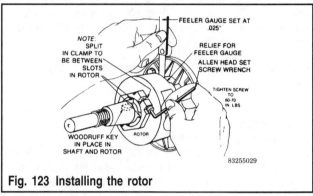

Fig. 123 Installing the rotor

Armature Air Gap

▶ See Figure 121

The armature air gap on engines equipped with Magna-Matic ignition system is fixed and can change only if wear occurs on the crankshaft journal and/or main bearing. Check for wear by inserting a ½ in. wide feeler gauge at several points between the rotor and armature. Minimum feeler gauge thickness is 0.004 in. (0.1mm). Keep the feeler gauge away from the magnets on the rotor or you will have a false reading.

Rotor Replacement

▶ See Figures 122 and 123

The rotor is held in place by a woodruff key and a clamp on later engines, and a woodruff key and set screw on older engines. The rotor clamp must always remain on the rotor, unless the rotor is in place on the crankshaft and within the armature, or a loss of magnetism will occur.

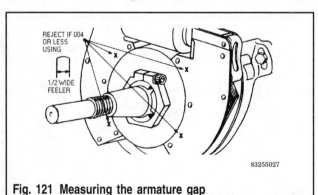

Fig. 121 Measuring the armature gap

Loosen the socket head screw in the rotor clamp which will allow the clamp to loosen. It may be necessary to use a puller to remove the rotor from the crankshaft. On older models, loosen the small lock screw, then the set screw.

To install the set screw type rotor, place the woodruff key in the keyway on the crankshaft, then slide the rotor onto the crankshaft until the set screw hole in the rotor and the crankshaft are aligned. Be sure the key remains in place. Tighten the set screw securely, then tighten the lock screw to prevent the set screw from loosening. The lock screw is self-threading and the hole does not require tapping.

To install the clamp type rotor, place the woodruff key in place in the crankshaft and align the keyway in the rotor with the woodruff key. If necessary, use a short length of pipe and a hammer to drive the rotor onto the shaft until a 0.025 in. (0.6mm) feeler gauge can be inserted between the rotor and the bearing support. The split in the clamp must be between the slots in the rotor. Tighten the clamp screws to 60-70 inch lbs.

Rotor Timing Adjustment

▶ See Figure 124

The rotor and armature are correctly timed at the factory and require timing only if the armature has been removed from the engine, or if the cam gear or crankshaft has been replaced.

If it is necessary to adjust the rotor, proceed as follows: with the point gap set at 0.020 in. (0.5mm), turn the crankshaft in the normal direction of rotation until the breaker points close and just start to open. Use a timing light or insert a piece of tissue paper between the breaker points to determine when the points begin to open. With the three armature mounting screws slightly loose, rotate the armature until the arrow on

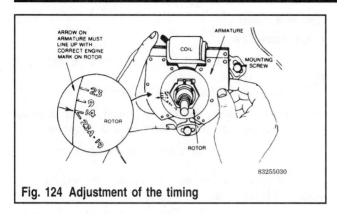

Fig. 124 Adjustment of the timing

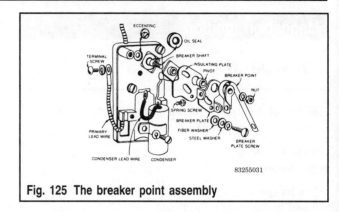

Fig. 125 The breaker point assembly

the armature lines up with the arrow on the rotor. Align with the corresponding number of engine models, for example, on Model 9, align with #9. Retighten the armature mounting screws.

Coil and/or Armature Replacement

Usually the coil and armature are not separated, but left assembled for convenience. However, if one or both need replacement, proceed as follows: the coil primary wire and the coil ground wire must be unfastened. Pry out the clips that hold the coil and coil core to the armature. The coil core is a slip fit in the coil and can be pushed out of the coil.

To reassemble, push the coil core into the coil with the rounded side toward the ignition cable. Place the coil and core on the armature with the coil retainer between the coil and the armature and with the rounded side toward the coil. Hook the lower end of the clips into the armature, then press the upper end onto the coil core.

Fasten the coil ground wire (bare double wires) to the armature support. Next, place the assembly against the cylinder and around the rotor and bearing support. Insert the three mounting screws together with the washer and lockwasher into the three long oval holes in the armature. Tighten them enough to hold the armature in place but loose enough so the armature can be moved for adjustment of the timing. Attach the primary wires from the coil and the breaker points to the terminal at the upper side of the backing plate. This terminal is insulated from the backing plate. Push the ignition cable through the louvered hole at the left side of the backing plate.

Breaker Point Replacement
▶ See Figure 125

Turn the crankshaft until the points open to the widest gap. This makes it easier to assemble and adjust the points later if the crankshaft is not removed. With the terminal screw out, remove the spring screw. Loosen the breaker shaft nut until the nut is flush with the end of the shaft. Tap the nut to free the breaker arm from the tapered end of the breaker shaft. Remove the nut, lockwasher, and breaker arm. Remove the breaker plate screw, breaker plate, pivot, insulating plate, and eccentric. Pry out the breaker shaft seal with a sharp pointed tool.

To install the breaker points, press in the new oil seal with the metal side out. Put the new breaker plate on the top of the insulating plate, making sure that the detent in the breaker plate engages the hole in the insulating plate. Fasten the breaker plate screw enough to put a light tension on the plate.

Adjust the eccentric so that the left edge of the insulating plate is parallel to the edge of the box and tighten the screw. This locates the breaker plate so that the proper gap adjustments may be made. Turn the breaker shaft clockwise as far as possible and hold it in this position. Place the new breaker points on the shaft, then the lockwasher, and tighten the nut down on the lockwasher. Replace the spring screw and terminal screw.

Breaker Box Removal & Installation

Remove the two mounting screws, then remove the breaker box, turning it slightly to clear the arm at the inner end of the breaker shaft. The breaker points need not be removed to remove the breaker box.

To install, pull the primary wire through the hole at the lower left corner of the breaker box. See that the primary wire rests in the groove at the top end of the box, then tighten the two mounting screws to hold the box in place.

Breaker Shaft Removal & Installation

The breaker shaft can be removed, after the breaker points are removed, by turning the shaft one half turn to clear the retaining spur at the inside of the breaker box.

Install by inserting the breaker shaft with the arm upward so the arm will clean the retainer boss. Push the shaft all the way in, then turn the arm downward.

Breaker Point Adjustment
▶ See Figure 126

To adjust the breaker points, turn the crankshaft until the breaker points open to the widest gap. Loosen the breaker point plate screw slightly. Rotate the eccentric to obtain a point gap of 0.020 in. (0.5mm). Tighten the breaker plate screw.

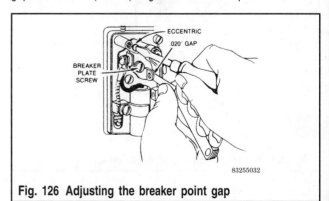

Fig. 126 Adjusting the breaker point gap

Magnetron Ignition

IDENTIFICATION

Magnetron has been produced in two versions, composite (Type I, Type II) and replaceable module.

Armature Testing
▶ **See Figure 127**

Use an approved tester to test armature. Specifications are supplied by the tester manufacturer.

Armature Removal
▶ **See Figure 128**

➡ **Removal of the flywheel is not required to remove Magnetron armatures except to inspect flywheel key and keyways on crankshaft and flywheel.**

1. Remove armature mounting screws and lift off armature.
2. Disconnect stop switch wire at spade terminal on composite armature. On armatures with replaceable Magnetron Modules, use breaker point condenser from Part #294628 point set or a 3/16 in. diameter pin punch to release wires from module.
3. Unsolder stop switch wire from module wire and armature primary wire.

Module Removal
▶ **See Figure 129**

1. Remove sealant and/or tape holding armature wires to armature.
2. Unsolder and separate remaining wires.

➡ **On some armatures, the module ground wire is soldered to the armature ground wire. Unsolder and disconnect.**

3. Move all wires so module will clear armature and laminations.
4. Pull module retainer away from laminations and push module off laminations.

Module Installation
▶ **See Figure 130**

The armature side identified by large rivet heads. The module is installed with the retainer on the back side, small rivet ends.

Stop Switch and Armature Primary Wire Installation
▶ **See Figures 131 and 132**

1. Be sure all insulating is removed from wires to ensure good contact.
2. Use a 3/16 in. diameter pin punch to compress wire retainer spring and insert stop switch and armature primary wire under hook of wire retainer.
3. Twist wires together and solder twisted section with 60/40 rosin core solder. Take care not to damage module case.
4. Install wires in module retainer.

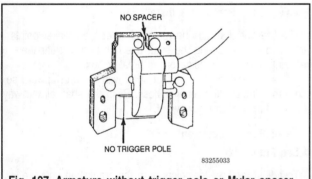

Fig. 127 Armature without trigger pole or Mylar spacer

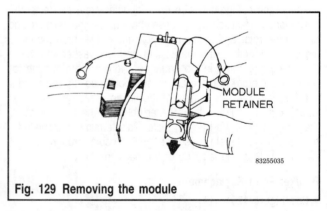

Fig. 129 Removing the module

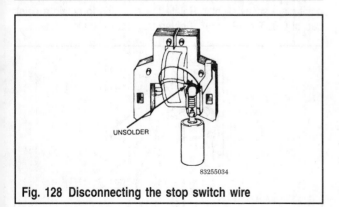

Fig. 128 Disconnecting the stop switch wire

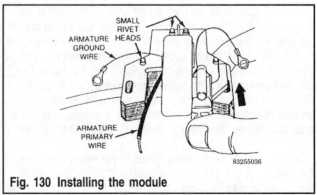

Fig. 130 Installing the module

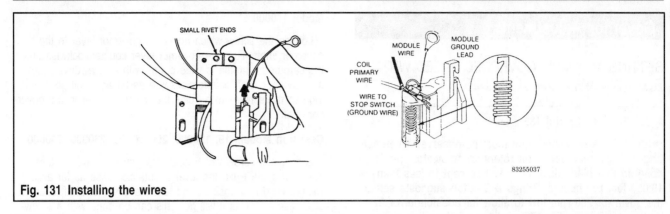

Fig. 131 Installing the wires

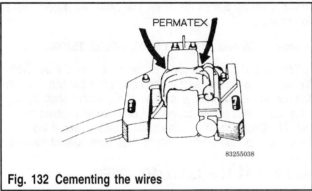

Fig. 132 Cementing the wires

5. Cement wires to armature with Permatex® No. 2 or similar sealer to prevent wires from vibrating and breaking.

Timing

Timing of Magnetron ignition is controlled by the flywheel key.

Mixture Adjustment

TWO PIECE FLO-JET

▶ See Figure 133

1. Start the engine and run it at 3000 rpm until it warms up.

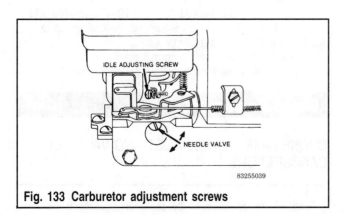

Fig. 133 Carburetor adjustment screws

2. Turn the needle valve (flat handle) to both extremes of operation noting the location of the valve at both points. That is, turn the valve inward until the mixture becomes too lean and the engine starts to slow, then note the position of the valve. Turn it outward slowly until the mixture becomes too rich and the engine begins to slow. Turn the valve back inward to the mid-point between the two extremes.

3. Install a tachometer on the engine. Pull the throttle to the idle position and hold it there through the rest of this step. Adjust the idle speed screw until the engine idles at 1750 rpm if it's an aluminum engine, or 1200 rpm, if it's a cast iron engine. Then, turn the idle valve in and out to adjust mixture, as described in Step 2. If idle valve adjustment changes idle speed, adjust speed to specification.

4. Release the throttle and observe the engine's response. The engine should accelerate without hesitation. If response is poor, one of the mixture adjustments is too lean. Readjust either or both as necessary. If idle speed was changed after idle valve was adjusted, readjust the idle valve first.

ONE PIECE FLO-JET

▶ See Figure 134

Follow the instructions for adjusting the Two Piece Flo-Jet carburetor (above). On the large, One Piece Flo-Jet, the needle valve is located under the float bowl, and the idle valve on top of the venturi passage. On the small One Piece Flo-Jet, both valves are adjusted by screws located on top of the venturi passage. The needle valve is located on the air horn side, is centered above the float bowl, and uses a larger screw head.

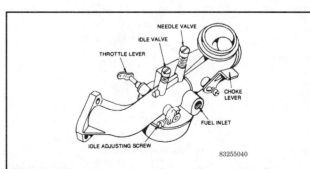

Fig. 134 Idle valve and needle locations on the One-Piece Flo-Jet

Governor Adjustments

SETTING MAXIMUM GOVERNED SPEED WITH ROTARY LAWNMOWER BLADES

▶ **See Figures 135 and 136**

➡ **Strict limits on engine rpm must be observed when setting top governed speed on rotary lawnmowers. This is done so that blade tip speeds will be kept to less than 19,000 feet per minute. Briggs & Stratton suggests setting the governor 200 rpm low to allow for possible error in the tachometer reading. These figures below, based on blade length, must be strictly adhered to, or a serious accident could result!**

Blade Length (in.)	Max. Governed Speed (R.P.M.)
18	4032
19	3820
20	3629
21	3456
22	3299
23	3155
24	3024
25	2903
26	2791

832550C7

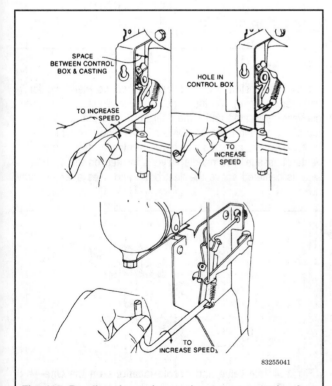

SPACE BETWEEN CONTROL BOX & CASTING

HOLE IN CONTROL BOX

TO INCREASE SPEED

TO INCREASE SPEED

TO INCREASE SPEED

83255041

Fig. 135 Bending the spring anchor tang to get the desired top speed

Model 140000

Loosen the screw which holds the governor lever to the governor shaft. Turn the governor lever counterclockwise until the carburetor throttle is wide open. With a screwdriver, turn the governor shaft counterclockwise as far as it will go. Tighten the screw which holds the governor lever to the governor shaft.

Cast Iron Models 19, 190000, 200000, 23, 230000, 240000

Loosen the screw which holds the governor lever to the governor shaft. Push the lever counterclockwise as far as it will go. Hold it in position and turn the governor shaft counterclockwise as far as it will go. This can be done with a screwdriver. Securely tighten the screw that holds the governor lever to the shaft.

Aluminum Models 140000, 170000, 190000, 251000

Vertical and horizontal shaft engine governors are adjusted by setting the control lever in the high speed position. Loosen the nut on the governor lever. Turn the governor shaft clockwise with a screwdriver to the end of its travel. Tighten the nut. The throttle must be wide open. Check to see if the throttle can be moved from idle to wide open without binding.

ADJUSTING TOP NO LOAD SPEED

Set the control lever to the maximum speed position with the engine running. Bend the spring anchor tang to get the desired top speed.

ADJUSTMENT FOR CLOSER GOVERNING

▶ **See Figure 137**

Generator Applications Only

1. Snap knob upward to release adjusting nut.
2. Pull knob out against stop.
3. Then, bend the spring anchor tang to get top no-load speed as described below, depending upon the application.
   On models 140400, 146400, 170400, 190400 and 251400 with an 1800 rpm generator: temporarily substituting a #260902 governor spring, set the no load speed at 2600 rpm, and then set throttle stop at 1600 rpm.
   On models 140400, 146400, 170400, 190400, 251400 and 251400 with an 3600 rpm generator: set no load speed to 4200 rpm with standard governor spring.
4. Snap knob back into its normal position.
5. Adjust the knob for the desired generator speed.

Choke Adjustment

CHOKE-A-MATIC, PULSA-JET AND VACU-JET CARBURETORS

To check the operation of the choke linkage, move the speed adjustment lever to the choke position. If the choke

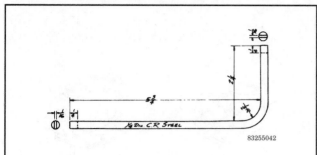

Fig. 136 You can make a tool like this one, to adjust the spring anchor tang

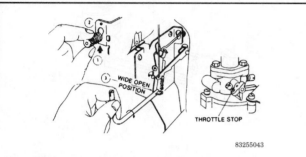

Fig. 137 Obtaining closer governing on generator applications

slide does not fully close, bend the choke link. The speed adjustment lever must make good contact against the top switch.

Install the carburetor and adjust it in the same manner as the Pulsa-Jet carburetor.

TWO-PIECE FLO-JET AUTOMATIC CHOKE

♦ **See Figure 138**

Hold the choke shaft so the thermostat lever is free. At room temperature — 68°F (20°C), the screw in the thermostat collar should be in the center of the stops. If not, loosen the stop screw and adjust the screw.

Loosen the set screw on the lever of the thermostat assembly. Slide the lever to the right or left on the shaft to ensure free movement of the choke link in any position. Rotate the

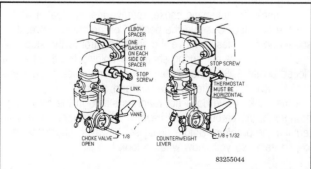

Fig. 138 Adjusting the choke on Two-Piece Flo-Jet models

thermostat shaft clockwise until the stop screw strikes the tube. Hold it in position and set the lever on the thermostat shaft so that the choke valve will be held open about 1/8 in. (3mm) from a closed position. Then tighten the set screw in the lever.

Rotate the thermostat shaft counterclockwise until the stop screw strikes the opposite side of the tube. Then open the choke valve manually until it stops against the top of the choke link opening. The choke valve should now be open approximately 1/8 in. (3mm) as before.

Check the position of the counterweight lever. With the choke valve in a wide open position (horizontal) the counterweight lever should also be in a horizontal position with the free end toward the right.

Operate the choke manually to be sure that all parts are free to move without binding or rubbing in any position.

Compression Checking

You can check the compression in any Briggs & Stratton engine by performing the following simple procedure: spin the flywheel counterclockwise (flywheel side) against the compression stroke. A sharp rebound indicates that there is satisfactory compression. A slight or no rebound indicates poor compression.

It has been determined that this test is an accurate indication of compression and is recommended by Briggs and Stratton. Briggs & Stratton does not supply compression pressures.

Loss of compression will usually be the result of one or a combination of the following:

1. The cylinder head gasket is blown or leaking.
2. The valves are sticking or not seating properly.
3. The piston rings are not sealing, which would also cause the engine to consume an excessive amount of oil.

Carbon deposits in the combustion chamber should be removed every 100 or 200 hours of use (more often when run at a steady load), or whenever the cylinder head is removed.

Carburetors

♦ **See Figures 139, 140 and 141**

There are three types of carburetors used on Briggs & Stratton engines. They are the Pulsa-Jet, Vacu-Jet and Flo-Jet. The first two types have three models each and the Flo-Jet has two versions.

Before removing any carburetor for repair, look for signs of air leakage or mounting gaskets that are loose, have deteriorated, or are otherwise damaged.

Note the position of the governor springs, governor link, remote control, or other attachments to facilitate reassembly. Be careful not to bend the links or stretch the springs.

AUTOMATIC CHOKE OPERATION

♦ **See Figure 142**

The automatic choke operates in conjunction with engine vacuum, similar to the Pulsa-Jet fuel pump.

A diaphragm under the carburetor is connected to the choke shaft by a link. A calibrated spring under the diaphragm holds the choke closed when the engine is not running. Upon start-

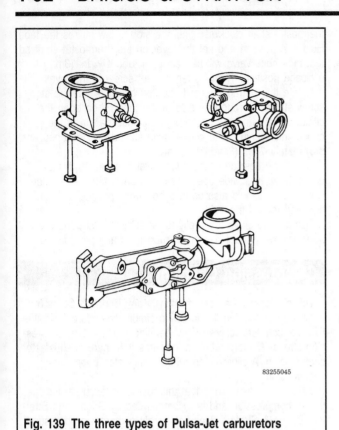

Fig. 139 The three types of Pulsa-Jet carburetors

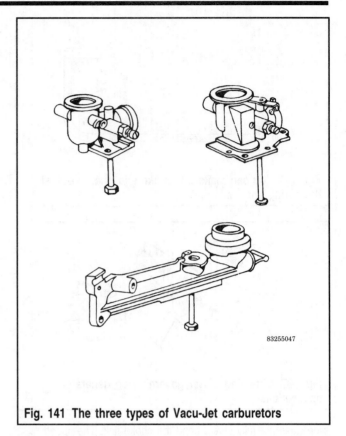

Fig. 141 The three types of Vacu-Jet carburetors

ing, vacuum created during the intake stroke is routed to the bottom of the diaphragm through a calibrated passage, thereby opening the choke.

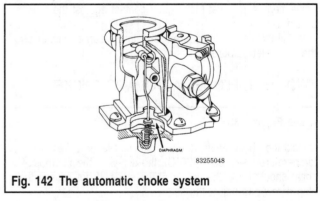

Fig. 142 The automatic choke system

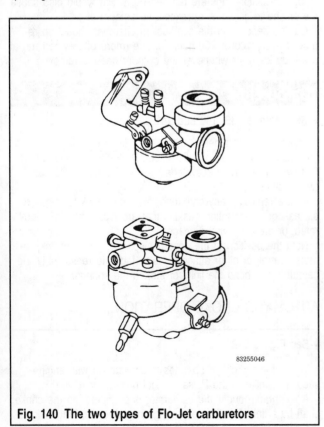

Fig. 140 The two types of Flo-Jet carburetors

This system also has the ability to respond in the same manner as an accelerator pump. As speed decreases during heavy loads, the choke valve partially closes, enriching the air/fuel mixture, thereby improving low speed performance and lugging power.

To check the automatic choke, remove the air cleaner and replace the stud. Observe the position of the choke valve; it should be fully closed. Move the speed control to the stop position; the governor spring should be holding the throttle in a closed position. Give the starter rope several quick pulls. The choke valve should alternately open and close.

If the choke valve does not react as stated in the previous paragraph, the carburetor will have to be disassembled to determine the problem. Before doing so, however, check the following items so you know what to look for:

Engine is Underchoked

1. Carburetor is adjusted too lean.

2. The fuel pipe check valve is inoperative (Vacu-Jet only).
3. The air cleaner stud is bent.
4. The choke shaft is sticking due to dirt.
5. The choke spring is too short or damaged.
6. The diaphragm is not preloaded.

Engine is Overchoked

1. Carburetor is adjusted too rich.
2. The air cleaner stud is bent.
3. The choke shaft is sticking due to dirt.
4. The diaphragm is ruptured.
5. The vacuum passage is restricted.
6. The choke spring is distorted or stretched.
7. There is gasoline or oil in the vacuum chamber.
8. There is a leak between the link and the diaphragm.
9. The diaphragm was folded during assembly, causing a vacuum leak.
10. The machined surface on the tank top is not flat.

REPLACING THE AUTOMATIC CHOKE

▶ **See Figures 143, 144 and 145**

Inspect the automatic choke for free operation. Any sticking problems should be corrected as proper choke operation depends on freedom of the choke to travel as dictated by engine vacuum.

Remove the carburetor and fuel tank assembly from the engine. The choke link cover may now be removed and the choke link disconnected from the choke shaft. Disassemble the carburetor from the tank top, being careful not to damage the diaphragm.

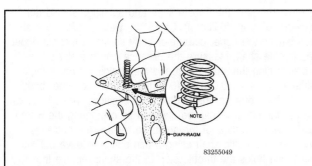

Fig. 143 Assembling the diaphragm spring on the new diaphragm

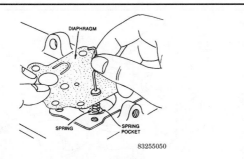

Fig. 144 Installing the diaphragm and spring into the spring pocket

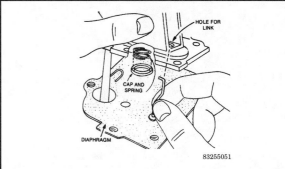

Fig. 145 Positioning the diaphragm on top of the fuel tank

Checking the Diaphragm and Spring

The diaphragm can be reused, provided it has not developed wear spots or punctures. On the Pulsa-Jet models, make sure that the fuel pump valves are not damaged. Also check the choke spring length. The Pulsa-Jet spring minimum length is $1\frac{1}{8}$ in. (28.5mm) and the maximum is $1\frac{7}{32}$ in. (31mm) Vacu-Jet spring length minimum is $\frac{15}{16}$ in. (24mm), maximum length 1 in. (25mm). If the spring length is shorter or longer than specified, replace the diaphragm and the spring.

CHECKING THE TANK TOP

▶ **See Figures 146, 147 and 148**

The machined surface on the top of the tank must be flat in order for the diaphragm to provide an adequate seal between the carburetor and the tank. If the machined surface on the

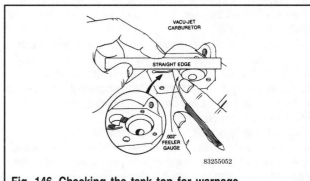

Fig. 146 Checking the tank top for warpage

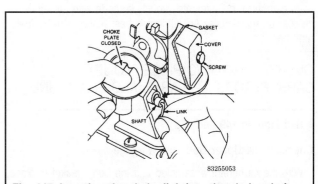

Fig. 147 Inserting the choke link into the choke shaft

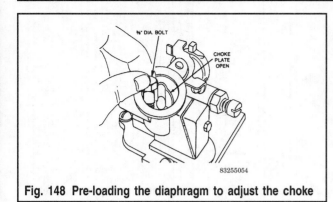

Fig. 148 Pre-loading the diaphragm to adjust the choke

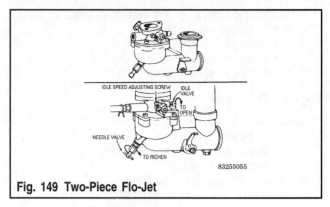

Fig. 149 Two-Piece Flo-Jet

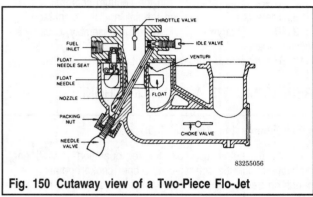

Fig. 150 Cutaway view of a Two-Piece Flo-Jet

tank is not flat, it is possible for gasoline to enter the vacuum chamber by passing between the machined surface and the diaphragm. Once fuel has entered the vacuum chamber, it can move through the vacuum passage and into the carburetor. The flatness of the machined surface on the tank top can be checked by using a straight-edge and a feeler gauge. The surface should not vary more than 0.002 in. (0.05mm). Replace the tank if a 0.002 in. (0.05mm) feeler gauge can be passed under the straight-edge.

If a new diaphragm is installed, assemble the spring to the replacement diaphragm, taking care not to bend or distort the spring.

Place the diaphragm on the tank surface, positioning the spring in the spring pocket.

Place the carburetor on the diaphragm ensuring that the choke link and diaphragm are properly aligned between the carburetor and the tank top. On Pulsa-Jet models, place the pump spring and cap on the diaphragm over the recess or pump chamber in the fuel tank. Thread in the carburetor mounting screws to about two threads. Do not tighten them. Close the choke valve and insert the choke link into the choke shaft.

Remove the air cleaner gasket, if it is in place, before continuing. Insert a 3/8 in. bolt or rod into the carburetor air horn. With the bolt in position, tighten the carburetor mounting screws in a staggered sequence. Please note that the insertion of the 3/8 in. bolt opens the choke to an over-center position, which preloads the diaphragm.

Remove the 3/8 in. bolt. The choke valve should now move to a fully closed position. If the choke valve is not fully closed, make sure that the choke spring is properly assembled to the diaphragm, and also properly inserted in its pocket in the tank top.

All carburetor adjustments should be made with the air cleaner on the engine. Adjustment is best made with the fuel tank half full.

OVERHAULING TWO PIECE FLO-JET CARBURETORS — LARGE AND SMALL LINE

▶ See Figures 149 and 150

Checking the Upper Body for Warpage

With the carburetor assembled and the body gasket in place, try to insert a 0.002 in. (0.05mm) feeler gauge between the upper and lower bodies at the air vent boss, just below the idle valve. If the gauge can be inserted, the upper body is warped and should be replaced.

Checking the Throttle Shaft and Bushings
▶ See Figures 151 and 152

Wear between the throttle shaft and bushings should not exceed 0.010 in. (0.25mm). Check the wear by placing a short iron bar on the upper carburetor body so that it just fits under the throttle shaft. Measure the distance with a feeler gauge while holding the shaft down and then holding it up. If the difference is over 0.010 in. (0.25mm), either the upper body should be rebushed, the throttle shaft replaced, or both. Wear on the throttle shaft can be checked by comparing the worn and unworn portions of the shaft. To replace the bushings, remove the throttle shaft using a thin punch to drive out the pin which holds the throttle stop to the shaft; remove the throttle valve, then pull out the shaft. Place a 1/4-20 tap or an E-Z Out in a vise. Turn the carburetor body so as to thread

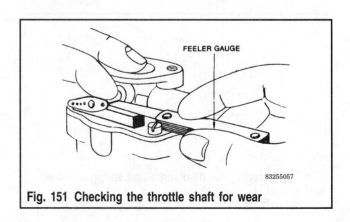

Fig. 151 Checking the throttle shaft for wear

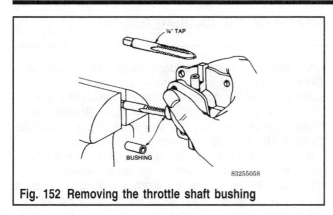

Fig. 152 Removing the throttle shaft bushing

the tap or E-Z Out into the bushings enough to pull the bushings out of the body. Press the new bushings into the carburetor body with a vise. Insert the throttle shaft to be sure it is free in the bushings. If not, run a size $7/32$ in. (5.5mm) drill through both bushings to act as a line reamer. Install the throttle shaft, valve, and stop.

DISASSEMBLY OF THE CARBURETOR

1. Remove the idle valve.
2. Loosen the needle valve packing nut.
3. Remove the packing nut and needle valve together. To remove the nozzle, use a narrow, blunt screwdriver so as not to damage the threads in the lower carburetor body. The nozzle projects diagonally into a recess in the upper body and must be removed before the upper body is separated from the lower body, or it may be damaged.
4. Remove the screws which hold the upper and lower bodies together. A pin holds the float in place.
5. Remove the pin to take out the float valve needle. Check the float for leakage. If it contains gasoline or is crushed, it must be replaced. Use a wide, proper fitting screwdriver to remove the float inlet seat.
6. Lift the venturi out of the lower body. Some carburetors have a welch plug. This should be removed only if necessary to remove the choke plate. Some carburetors have nylon choke shaft.

REPAIR

Use new parts where necessary. Always use new gaskets. Carburetor repair kits are available. Tighten the inlet seat with the gasket securely in place, if used. Some float valves have a spring clip to connect the float valve to the float tang. Others are nylon with a stirrup which fits over the float tang. Older float valves and engines with fuel pumps have neither a spring nor a stirrup.

A viton tip float valve is used in later models of the large, two-piece Flo-Jet carburetor. The seat is pressed into the upper body and does not need replacement unless it is damaged.

Replacing the Pressed-In Float Valve Seat
▶ **See Figure 153**

Clamp the head of a #93029 self threading screw in a vise. Turn the carburetor body to thread the screw into the seat. Continue turning the carburetor body, drawing out the seat. Leave the seat fastened to the screw. Insert the new seat #230996 into the carburetor body. The seat has a starting lead.

➡**If the engine is equipped with a fuel pump, install a #231019 seat. Press the new seat flush with the body using the screw and old seat as a driver. Make sure that the seat is not pressed below the body surface or improper float-to-float valve contact will occur. Install the float valve.**

Checking the Float Level

With the body gasket in place on the upper body and the float valve and float installed, the float should be parallel to the body mounting surface. If not, bend the tang on the float until they are parallel. Do not press on the flat to adjust it.

ASSEMBLY OF THE CARBURETOR

Assemble the venturi and the venturi gasket to the lower body. Be sure that the holes in the venturi and the venturi gasket are aligned. Some models do not have a removable venturi. Install the choke parts and welch plug if previously removed. Use a sealer around the welch plug to prevent entry of dirt.

Fasten the upper and lower bodies together with the mounting screws. Screw in the nozzle with a narrow, blunt screwdriver, making sure that the nozzle tip enters the recess in the upper body. Tighten the nozzle securely. Screw in the needle valve and idle valve until they just seat. Back off the needle valve $1\frac{1}{2}$ turns. Do not tighten the packing nut. Back off the idle valve $3/4$ of a turn. These settings are about correct. Final adjustment will be made when the engine is running.

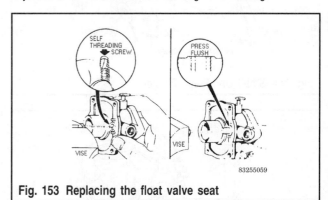

Fig. 153 Replacing the float valve seat

OVERHAULING ONE-PIECE FLO-JET CARBURETOR

▶ **See Figures 154 and 155**

The large, one-piece Flo-Jet carburetor has its high speed needle valve below the float bowl. All other repair procedures are similar to the small, one-piece Flo-Jet carburetor.

DISASSEMBLY

1. Remove the idle and needle valves.
2. Remove the carburetor bowl screw. A pin holds the float in place.
3. Remove the pin to take off the float and float valve needle. Check the float for leakage. If it contains gasoline or is crushed, it must be replaced. Use a screwdriver to remove the carburetor nozzle. Use a wide, heavy screwdriver to remove the float valve seat, if used.

If it is necessary to remove the choke valve, venturi throttle shaft, or shaft bushings, proceed as follows.

4. Pry out the welch plug.
5. Remove the choke valve, then the shaft. The venturi will then be free to fall out after the choke valve and shaft have been removed.
6. Check the shaft for wear. (Refer to the 'Two-Piece Flo-Jet Carburetor' section for checking wear and replacing bushings).

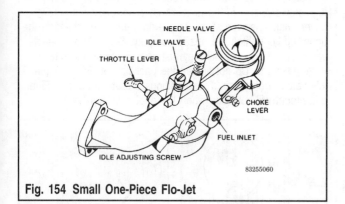

Fig. 154 Small One-Piece Flo-Jet

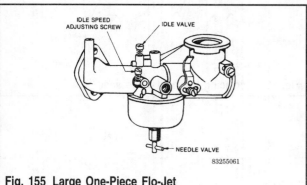

Fig. 155 Large One-Piece Flo-Jet

REPAIR OF THE CARBURETOR

Use new parts where necessary. Always use new gaskets. Carburetor repair kits are available. If the venturi has been removed, install the venturi first, then the carburetor nozzle jets. The nozzle jet holds the venturi in place. Replace the choke shaft and valve. Install a new welch plug in the carburetor body. Use a sealer to prevent dirt from entering.

A viton tip float valve is used in the large, one-piece Flo-Jet carburetor. The seat is pressed in the upper carburetor body and does not need replacement unless it is damaged. Replace the seat in the same manner as for the two-piece Flo-Jet carburetor.

Checking the Float Level

Float Level Chart

Carburetor Number	Float Setting (in.)
2712-S	$19/64$
2713-S	$19/64$
2714-S	$1/4$
*2398-S	$1/4$
2336-S	$1/4$
2336-SA	$1/4$
2337-S	$1/4$
2337-SA	$1/4$
2230-S	$17/64$
2217-S	$11/64$

*When resilient seat is used, set float level at $9/32 \pm 1/64$.

832550C8

With the body gasket in place on the upper body and float valve and the float installed, the float should be parallel to the body mounting surface. If not, bend the tang on the float until they are parallel. Do not press on the float.

Install the float bowl, idle valve, and needle valve. Turn in the needle valve and the idle valve until they just seat. Open the needle valve $2\frac{1}{2}$ turns and the idle valve $1\frac{1}{2}$ turns. On the large carburetors with the needle valve below the float bowl, open the needle valve and the idle valve $1\frac{1}{8}$ turns.

These settings will allow the engine to start. Final adjustment should be made when the engine is running and has warmed up to operating temperature.

Governors

The purpose of a governor is to maintain, within certain limits, a desired engine speed even though the load may vary.

AIR VANE GOVERNORS

▶ **See Figures 156 and 157**

The governor spring tends to open the throttle. Air pressure against the air vane tends to close the throttle. The engine speed at which these two forces balance is called the governed speed. The governed speed can be varied by changing the governor spring tension.

Worn linkage or damaged governor springs should be replaced to insure proper governor operation. No adjustment is necessary.

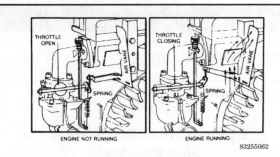

Fig. 156 Air vane governor installed on a horizontal crankshaft engine

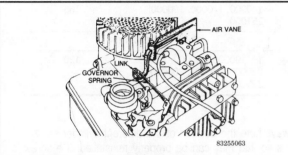

Fig. 157 Air vane governor installed on a vertical crankshaft engine

MECHANICAL GOVERNORS

The governor spring tends to pull the throttle open. The force of the counterweights, which are operated by centrifugal force, tends to close the throttle. The engine speed at which these two forces balance is called the governed speed. The governed speed can be varied by changing the governor spring tension.

GOVERNOR REPAIR

▶ See Figure 158

The procedures below describe disassembly and assembly of the various kinds of mechanical governors. Look for gears with worn or broken teeth, worn thrust washers, weight pins, cups, followers, etc. Replace parts that are worn and reassemble.

Model 140000

DISASSEMBLY

1. Loosen the governor lever mounting screw, and pull the lever off the shaft.
2. Remove the two housing mounting screws. Carefully pull the housing off the block, being careful to catch the governor gear, which will slip off the shaft. Pull the steel thrust washer off the shaft.
3. Remove the governor lever roll pin and washer. Unscrew the governor lever shaft by turning it clockwise and remove it.

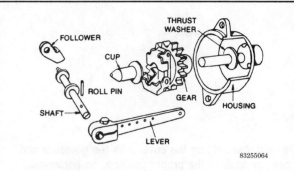

Fig. 158 Exploded view of a mechanical governor

ASSEMBLY

▶ See Figure 159

1. Push the governor lever shaft into the crankcase cover, threaded end first. Assemble the small washer onto the inner end of the shaft, and then screw the shaft into the governor crank follower by turning it counterclockwise. Tighten it securely.
2. Turn the shaft until the follower points down slightly, in a position where it would press against the cup when the housing is installed.
3. Place the washer on the outside end of the shaft. Install the rollpin, so the leading end just reaches the outside diameter of the shaft and the back end protrudes.
4. Install the thrust washer and the governor gear on the shaft in the housing (in that order).
5. Hold the crankcase cover in a vertical (the normal) position and install the housing with the gear in position so the point of the steel cup on the gear contacts the follower. Install and tighten the housing mounting screws.
6. Install the lever on the shaft pointing downward at an angle of about 30°. Adjust as described in the Tune-Up section.

Cast Iron Models 19, 190000, 20000, 23, 230000, 240000

▶ See Figures 160 and 161

DISASSEMBLY

1. Remove the cotter key and washer from the outer end of the governor shaft. Remove the governor crank from inside the crankcase.
2. Slide the governor gear off the shaft.

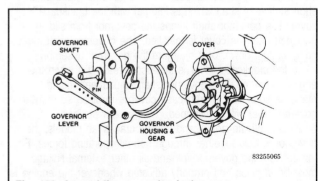

Fig. 159 Assembling a mechanical governor

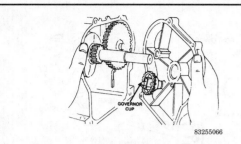

83255066

Fig. 160 Assembling the cover with the governor and governor shaft in the proper position, on horizontal shaft engines

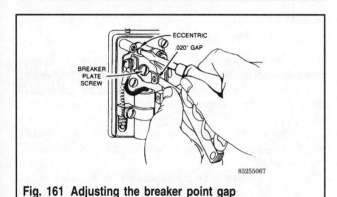

83255067

Fig. 161 Adjusting the breaker point gap

ASSEMBLY

1. Install the governor gear onto the shaft inside the crankcase. Then, insert the governor shaft assembly through the bushing from inside the crankcase.

2. Install the governor lever to the shaft loosely, and then adjust it as described in the Tune-Up section.

Aluminum Models 140000, 170000, 190000 and 251000

DISASSEMBLY

On horizontal shaft models: Remove the governor assembly as a unit from the crankcase cover.

On vertical shaft models: Remove the entire assembly as part of the oil slinger (see the Overhaul Section).

ASSEMBLY

1. Assemble governors on horizontal crankshaft models with crankshaft in a horizontal position. The governor rides on a short stationary shaft which is integral with the crankcase cover. The governor shaft keeps the governor from sliding off the shaft after the cover is installed. The governor shaft must hang straight down, or it may jam the governor assembly when the crankcase cover is installed, breaking it when the engine is started. The governor shaft adjustment should be made as soon as the crankcase cover is in place so that the governor lever will be clamped in the proper position.

2. On both horizontal and vertical crankshaft models, the governor is held together through normal operating forces. For this reason, the governor link and all other external linkages must be in place and properly adjusted whenever the engine is operated.

Cylinder Head

REMOVAL & INSTALLATION

▶ See Figure 162

Cylinder Head Bolt Torque Specifications

Basic Model Series	in. lbs. Torque
Aluminum Cylinder	
140000, 170000, 190000, 251000	165
Cast Iron Cylinder	
19, 190000, 200000, 23, 230000, 240000, 300000, 320000	190

832550C9

Always note the position of the different cylinder head screws so that they can be properly reinstalled. If a screw is used in the wrong position, it may be to short and not engage enough threads. If it is too long, it may bottom on a fin, either breaking the fin, or leaving the cylinder head loose.

Remove the cylinder screws and then the cylinder head. Be sure to remove the gasket and all remaining gasket material from the cylinder head and the block. Assemble the cylinder head with a new gasket, cylinder head shield, screws, and

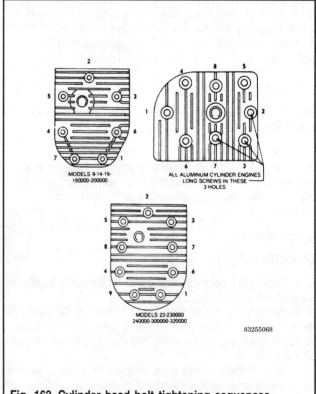

MODELS 9-14-19-190000-200000

ALL ALUMINUM CYLINDER ENGINES LONG SCREWS IN THESE 3 HOLES

MODELS 23-230000 240000-300000-320000

83255068

Fig. 162 Cylinder head bolt tightening sequences

washers in their proper places. Graphite grease should be used on aluminum cylinder head screws.

Do not use a sealer of any kind on the head gasket. Tighten the screws down evenly by hand. Use a torque wrench and tighten the head bolts in the correct sequence.

Valves

REMOVAL & INSTALLATION

▶ See Figures 163 and 164

1. Using a valve spring compressor, adjust the jaws so they touch the top and bottom of the valve chamber, and then place one of the jaws over the valve spring and the other underneath, between the spring and the valve chamber. This positioning of the valve spring compressor is for valves that have either pin or collar type retainers.

2. Tighten the jaws to compress the spring. Remove the collars or pin and lift out the valve.

3. Pull out the compressor and the spring.

4. To remove valves with ring type retainers, position the compressor with the upper jaw over the top of the valve chamber and the lower jaw between the spring and the retainer. Compress the spring, remove the retainer, and pull out the valve. Remove the compressor and spring.

5. If the retainers are held by a pin or collars, place the valve spring and retainer and cup (Models 14-19-20-23-24-32) into the valve spring compressor. Compress the spring until it is solid. Insert the compressed spring and retainer into the valve chamber. Then drop the valve into place, pushing the stem through the retainer. Hold the spring up in the chamber,

hold the valve down, and insert the retainer pin with needle nose pliers or place the collars in the groove in the valve stem. Loosen the spring until the retainer fits around the pin or collars, then pull out the spring compressor. Be sure the pin or collars are in place.

6. Before installing the valves, check the thickness of the valve springs. Some engines use the same spring for the intake and exhaust side, while others use a heavier spring on the exhaust side. Compare the springs before installing them.

7. To install valves with ring type retainers, compress the retainer and spring with the compressor. The large diameter of the retainer should be toward the front of the valve chamber. Insert the compressed spring and retainer into the valve chamber. Drop the valve stem through the larger area of the retainer slot and move the compressor so as to center the small area of the valve retainer slot onto the valve stem shoulder. Release the spring tension and remove the compressor.

Valve Guides

REMOVAL & INSTALLATION

▶ See Figures 165 and 166

First check valve guide for wear with a plug gauge, Briggs & Stratton part #19151 or equivalent. If the flat end of the valve guide plug gauge can be inserted into the valve guide for a distance of $5/16$ in. (8mm), the guide is worn and should be rebushed in the following manner.

1. Procure a reamer #19183 and reamer guide bushing #19192, and lubricate the reamer with kerosene. Then, use reamer and reamer guide bushing to ream out the worn guide.

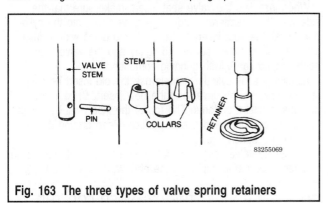

Fig. 163 The three types of valve spring retainers

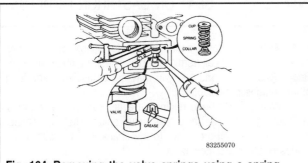

Fig. 164 Removing the valve springs using a spring compressor

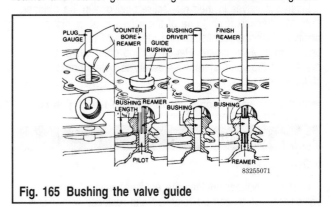

Fig. 165 Bushing the valve guide

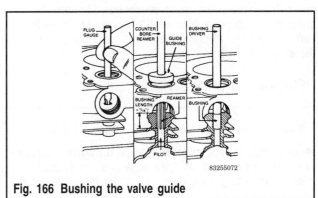

Fig. 166 Bushing the valve guide

Ream to only ¹/₁₆ in. (1.6mm) deeper than valve guide bushing #230655 . BE CAREFUL NOT TO REAM THROUGH THE GUIDE!

2. Press in valve guide bushing #230655 until top end of bushing is flush with top end of valve guide. Use a soft metal driver (brass, copper, etc.) so top end of bushing is not peened over.

➡ **The bushing #230655 is finish reamed to size at the factory, so no further reaming is necessary, and a standard valve can be used.**

Valve seating should be checked after bushing the guide, and corrected if necessary by refacing the seat.

REFACING VALVES AND SEATS

Faces on valves and valve seats should be resurfaced with a valve grinder or cutter to an angle of 45°.

➡ **Some engines have a 30° intake valve and seat.**

The valve and seat should then be lapped with a fine lapping compound to remove the grinding marks and ensure a good seat. The valve seat width should be 1.2-1.6mm. If the seat is wider, a narrowing stone or cutter should be used. If either the seat or valve is badly burned, it should be replaced. Replace the valve if the edge thickness (margin) is less than 0.4mm after it has been resurfaced.

CHECK AND ADJUST TAPPET CLEARANCE

Valve Tappet Clearance Chart

Model Series	Intake		Exhaust	
	Max	Min	Max	Min
Aluminum Cylinder				
140000, 170000, 190000, 251000	.007	.005	.011	.009
Cast Iron Cylinder				
19, 190000, 200000	.009	.007	.016	.014
23, 230000, 240000, 300000, 320000	.009	.007	.019	.017

83255C10

Insert the valves in their respective positions in the cylinder. Turn the crankshaft until one of the valves is at its highest position. Turn the crankshaft one revolution. Check the clearance with a feeler gauge. Repeat for the other valve. Grind off the end of the valve stem if necessary to obtain proper clearance.

➡ **Check the valve tappet clearance with the engine cold.**

Valve Seat Inserts

▶ **See Figure 167**

Cast iron cylinder engines are equipped with an exhaust valve insert which can be removed and replaced with a new insert. The intake side must be counter-bored to allow the installation of an intake valve seat insert (see below). Aluminum alloy cylinder models are equipped with inserts on both the exhaust and intake valves.

REMOVAL & INSTALLATION

Valve seat inserts are removed with a special puller.

➡ **On aluminum alloy cylinder models, it may be necessary to grind the puller nut until the edge is ³/₃₂ in. (0.8mm) thick in order to get the puller nut under the valve insert.**

When installing the valve seat insert, make sure that the side with the chamfered outer edge goes down into the cylinder. Install the seat insert and drive it into place with a driver. The seat should then be ground lightly and the valves and seats lapped lightly with grinding compound.

Aluminum alloy cylinder models use the old insert as a spacer between the driver and the new insert. Drive in the new insert until it bottoms. The top of the insert will be slightly below the cylinder head gasket surface. Peen around the insert using a punch and hammer.

The intake valve seat on cast iron cylinder models has to be counter-bored before installing the new valve seat insert.

COUNTER BORING THE CYLINDER FOR THE INTAKE VALVE SEAT ON CAST IRON MODELS

1. Select the proper seat insert, cutter shank, counter bore cutter, pilot and driver from the table. These numbers refer to Briggs & Stratton parts — you may get equivalent parts from other sources if available.

2. With cylinder head resting on a flat surface, valve seats up, slide the pilot into the intake valve guide. Then, assemble the correct counterbore cutter to the shank with the cutting blades of the cutter downward.

3. Insert the cutter straight into the valve seat, over the pilot. Cut so as to avoid forcing the cutter to one side, and be

Valve Seat Inserts Chart

Basic Model Series	Intake Standard	Exhaust Standard	Exhaust Stellite	Insert # Puller Assembly	Puller Nut
Aluminum Cylinder					
140000, 170000, 190000	211661	211661	210940*	19138	19141
250000	211661	211661	210940	19138	19141
Cast Iron Cylinder					
19, 190000	21880	21880	21612	19138	19141
200000, 23, 230000	21880	21880	21612	19138	19141
240000	21880	21612	21612	19138	19141
300000, 320000		21612	21612	19138	19141

* 21191 used before serial #5810060—210808 used from serial #5810060—6012010
\# Includes puller and #19182, 19141, 19140 and 19139 nuts

83255C11

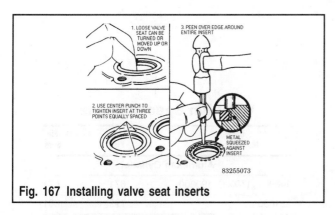

Fig. 167 Installing valve seat inserts

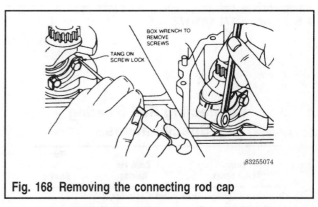

Fig. 168 Removing the connecting rod cap

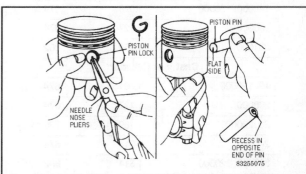

Fig. 169 Removing the piston pin and rod from the piston

sure to stop as soon as the stop on the cutter touches the cylinder head.

4. Blow out all cutting chips thoroughly.

Pistons, Piston Rings, and Connecting Rods

REMOVAL

▶ **See Figures 168, 169 and 170**

To remove the piston and connecting rod from the engine, bend down the connecting rod lock. Remove the connecting rod cap. Remove any carbon or ridge at the top of the cylinder bore. This will prevent breaking the rings. Push the piston and rod out of the top of the cylinder.

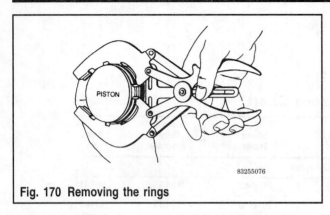

83255076

Fig. 170 Removing the rings

Pistons used in sleeve bore, aluminum alloy engines are marked with an **L** on top of the piston. These pistons are tin plated and use an expander with the oil ring. This piston assembly is not interchangeable with the piston used in the aluminum bore engines (Kool bore).

Pistons used in aluminum bore (Kool bore) engines are not marked on the top.

To remove the connecting rod from the piston, remove the piston pin lock with thin nose pliers. One end of the pin is drilled to facilitate removal of the lock.

Remove the rings one at a time, slipping them over the ring lands. Use a ring expander to remove the rings.

INSPECTION

▶ See Figures 171 and 172

Piston Ring Gap Specifications

Basic Model Series	Comp. Ring	Oil Ring
Aluminum Cylinder		
140000, 170000, 190000, 251000	.035	.045
Cast Iron Cylinder		
19, 190000, 200000, 23, 230000, 240000 300000, 320000	.035	.035

83255C13

Connecting Rod Bearing Specifications

Basic Model Series	Crank Pin Bearing	Piston Pin Bearing
Aluminum Cylinder		
140000, 170000	1.095	.674
190000	1.127	.674
251000	1.252	.802
Cast Iron Cylinder		
19, 190000	1.001	.674
200000	1.127	.674
23, 230000	1.189	.736
240000	1.314	.674
300000, 320000	1.314	.802

83255C12

Wrist Pin Specifications

Basic Model Series	Piston Pin	Pin Bore
Aluminum Cylinder		
140000, 170000, 190000	.671	.671
251000	.799	.801
Cast Iron Cylinder		
19, 190000	.671	.673
200000	.671	.673
23, 230000	.734	.736
240000	.671	.673
300000, 320000	.799	.801

83255C14

Check the piston ring fit. Use a feeler gauge to check the side clearance of the top ring. Make sure that you remove all carbon from the top ring groove. Use a new piston ring to

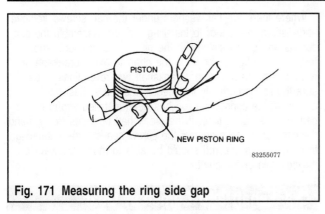

Fig. 171 Measuring the ring side gap

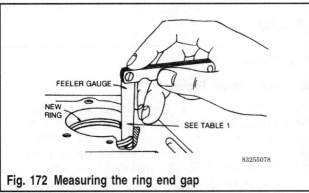

Fig. 172 Measuring the ring end gap

check the side clearance. If the cylinder is to be resized, there is no reason to check the piston, since a new oversized piston assembly will be installed. If the side clearance is more than 0.007 in. (0.178mm), the piston is excessively worn and should be replaced.

Check the piston ring end gap by cleaning all carbon from the ends of the rings and inserting them one at a time 1 in. (25mm) down into the cylinder. Check the end gap with a feeler gauge. If the gap is larger than recommended, the ring should be replaced.

➡**When checking the ring gap, do not deglaze the cylinder walls by installing piston rings in aluminum cylinder engines.**

Chrome ring sets are available for all current aluminum and cast iron cylinder models. No honing or deglazing is required. The cylinder bore can be a maximum of 0.005 in. (0.127mm) oversize when using chrome rings.

If the crankpin bearing in the rod is scored, the rod must be replaced. 0.005 in. (0.127mm) oversize piston pins are available in case the connecting rod and piston are worn at the piston pin bearing. If, however, the crankpin bearing in the connecting rod is worn, the rod should be replaced. Do not attempt to file or fit the rod.

If the piston pin is worn 0.0005 in out of round or below the rejection sizes, it should be replaced.

INSTALLATION

▶ **See Figures 173, 174, 175 and 176**

The piston pin is a push fit into both the piston and the connecting rod. On models using a solid piston pin, one end is flat and the other end is recessed. Other models use a hollow piston pin. Place a pin lock in the groove at one side of the piston. From the opposite side of the piston, insert the piston pin, flat end first for solid pins; with hollow pins, insert either end first until it stops against the pin lock. Use thin nose pliers to assemble the pin lock in the recessed end of the piston. Be sure the locks are firmly set in the groove.

Install the rings on the pistons, using a piston ring expander. Make sure that they are installed in the proper position. The scraper groove on the center compression ring should always

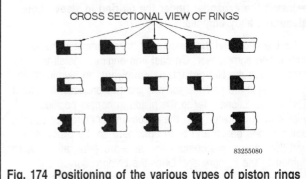

Fig. 173 Assembling the piston and connecting rod

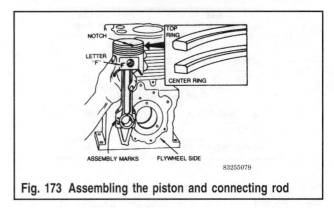

Fig. 174 Positioning of the various types of piston rings

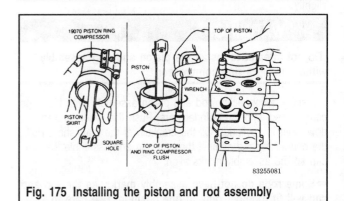

Fig. 175 Installing the piston and rod assembly

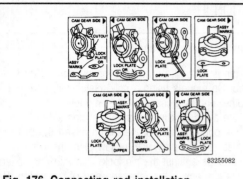

Fig. 176 Connecting rod installation

Connecting Rod Capscrew Torque

Basic Model Series	Inch lbs Avg. Torque
Aluminum Cylinder	
140000, 170000, 190000	165
251000	185
Cast Iron Cylinder	
19, 190000, 200000	190
23, 230000	190
240000, 300000, 320000	190

83255C15

be down toward the piston skirt. Be sure the oil return holes are clean and all carbon is removed from the grooves.

➡**Install the expander under the oil ring in sleeve bore aluminum alloy engines.**

Oil the rings and the piston skirt, then compress the rings with a ring compressor. On cast iron engines, install the compressor with the two projections downward; on aluminum engines, install the compressor with the two projections upward. These instructions refer to the piston in normal position — with skirt downward. Turn the piston and compressor upside down on the bench and push downward so the piston head and the edge of the compressor band are even, all the while tightening the compressor. Draw the compressor up tight to fully compress the rings, then loosen the compressor very slightly.

✳✳WARNING

Do not attempt to install the piston and ring assembly without using a ring compressor.

Place the connecting rod and piston assembly, with the rings compressed, into the cylinder bore. Push the piston and rod down into the cylinder. Oil the crankpin of the crankshaft. Pull the connecting rod against the crankpin and assemble the rod cap so the assembly marks align.

➡**Some rods do not have assembly marks, as the rod and cap will fit together only in one position. Use care to ensure proper installation. On the 251000 engine, the piston has a notch on the top surface. The notch must face the flywheel side of the block when installed.**

Where there are flat washers under the cap screws, remove and discard them prior to installing the rod. Assemble the cap screws and screw locks with the oil dippers (if used), and torque to the figure shown in the chart to avoid breakage or rod scoring later. Turn the crankshaft two revolutions to be sure the rod is correctly installed. If the rod strikes the camshaft, the connecting rod has been installed wrong or the cam gear is out of time. If the crankshaft operates freely, bend the cap screw locks against the screw heads. After tightening the rod screws, the rod should be able to move sideways on the crankpin of the shaft.

Crankshaft and Camshaft Gear

REMOVAL

▶ **See Figures 177 and 178**

Aluminum Cylinder Engines

To remove the crankshaft from aluminum alloy engines, remove any rust or burrs from the power take-off end of the crankshaft. Remove the crankcase cover or sump. If the sump or cover sticks, tap it lightly with a soft hammer on alternate sides near the dowel. Turn the crankshaft to align the crankshaft and camshaft timing marks, lift out the cam gear, then remove the crankshaft. On models that have ball bearings on the crankshaft, the crankshaft and the camshaft must be removed together with the timing marks properly aligned — see illustration.

Cast Iron Cylinder Models

To remove the crankshaft from cast iron models, remove the crankcase cover. Revolve the crankshaft until the crankpin is pointing upward toward the breather at the rear of the engine (approximately a 45° angle). Pull the crankshaft out from the drive side, twisting it slightly if necessary. On models with ball bearings on the crankshaft, both the crankcase cover and bearing support should be removed.

On cast iron models with ball bearings on the drive side, first remove the magneto. Drive into the recess at the front of the engine. Then draw the crankshaft from the magneto side of the engine. Double thrust engines have cap screws inside the crankcase which hold the bearing in place. These must be removed before the crankshaft can be removed.

To remove the camshaft from all cast iron models, except the 300400 and 320400, use a long punch to drive the

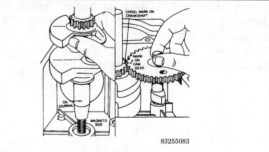

Fig. 177 Alignment of the camshaft and crankshaft without ball bearings

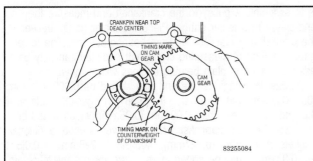

Fig. 178 Alignment of the camshaft and crankshaft with ball bearings

camshaft out toward the magneto side. Save the plug. Do not burr or peen the end of the shaft while driving it out. Hold the camshaft while driving it out. Hold the camshaft while removing the punch, so it will not drop and become damaged.

CHECKING THE CRANKSHAFT

Crankshaft Specifications

Basic Model Series	PTO Journal	Mag. Journal	C Crankpin
Aluminum Cylinder			
140000, 170000	1.179	.997 #	1.090
190000	1.179	.997 #	1.122
251000	1.376	1.376	1.247
Cast Iron Cylinder			
19, 190000	1.179	1.179	.996
200000	1.179	1.179	1.122
23, 230000 †	1.376	1.376	1.184
240000	Ball	Ball	1.309
300000, 320000	Ball	Ball	1.309

Synchro balanced magneto bearing reject size— 1.179
† Gear reduction P.T.O.—1.179

83255C16

Discard the crankshaft if it is worn beyond the allowable limit. Check the keyways for wear and make sure they are not spread. Remove all burrs from the keyway to prevent scratching the bearing. Check the three bearing journals, drive end, crankpin, and magneto end, for size and any wear or damage. Check the cam gear teeth for wear. They should not be worn at all. Check the threads at the magneto end for damage. Make sure that the crankshaft is straight.

➡ **There are 0.020 in. (0.5mm) undersize connecting rods available for use on reground crankpin bearings.**

REMOVAL & INSTALLATION OF THE BALL BEARINGS

The ball bearings are pressed onto the crankshaft. If either the bearing or the crankshaft is to be removed, use an arbor press to remove them.

To install, heat the bearing in hot oil — 325°F (163°C) maximum. Don't let the bearing rest on the bottom of the pan in which it is heated. Place the crankshaft in a vise with the bearing side up. When the bearing is quite hot, it will slip fit onto the bearing journal. Grasp the bearing, with the shield down, and thrust it down onto the crankshaft. The bearing will tighten on the shaft while cooling. Do not quench the bearing (throw water on it to cool it).

CHECKING THE CAMSHAFT GEAR

Camshaft Specifications

Basic Model Series	Cam Gear or Shaft Journals	Cam Lobe
Aluminum Cylinder		
140000, 170000, 190000	.498	.977
251000	.498	1.184
Cast Iron Cylinder		
19, 190000	.497	1.115
200000	.497	1.115
23, 230000	.497	1.184
240000	.497	1.184
300000	#	1.184
320000	#	1.215

Magneto side— .8105, P.T.O. side— .6145

83255C17

Inspect the teeth for wear and nicks. Check the size of the camshaft and camshaft gear bearing journals. Check the size of the cam lobes. If the cam is worn beyond tolerance, discard it.

Check the automatic spark advance on models equipped with the Magna-Matic ignition system. Place the cam gear in the normal operating position with the movable weight down. Press the weight down and release it. The spring should lift the weight. If not, the spring is stretched or the weight is binding.

INSTALLATION

Aluminum Alloy Engines — Plain Bearing

In aluminum alloy engines, the tappets are inserted first, the crankshaft next, and then the cam gear. When inserting the cam gear, turn the crankshaft and the cam gear so that the timing marks on the gears align.

Aluminum Alloy Engines — Ball Bearing

On crankshafts with ball bearings, the gear teeth are not visible for alignment of the timing marks; therefore, the timing mark is on the counterweight. On ball bearing equipped engines, the tappets are installed first. The crankshaft and the cam gear must be inserted together and their timing marks aligned.

Crankshaft Cover and Crankshaft

INSTALLATION

Cast Iron Engines w/Plain Bearings

Assemble the tappets to the cylinder, then insert the cam gear. Push the camshaft into the camshaft hole in the cylinder from the flywheel side through the cam gear. With a blunt punch, press or hammer the camshaft until the end is flush with the outside of the cylinder on the power takeoff side. Place a small amount of sealer on the camshaft plug, then press or hammer it into the camshaft hole in the cylinder at the flywheel side. Install the crankshaft so the timing marks on the teeth and on the cam gear align.

Cast Iron w/Ball Bearings

Assemble the tappets, then insert the cam gear into the cylinder, pushing the cam gear forward into the recess in front of the cylinder. Insert the crankshaft into the cylinder. Turn the camshaft and crankshaft until the timing marks align, then push the cam gear back until it engages the gear on the crankshaft with the timing marks together. Insert the camshaft. Place a small amount of sealer on the camshaft plug and press or hammer it into the camshaft hole in the cylinder at the flywheel side.

CRANKSHAFT END-PLAY ADJUSTMENT

The crankshaft end-play on all models, plain and ball bearing, should be 0.002-0.008 in. (0.2mm). The method of obtaining the correct end-play varies, however, between cast iron, aluminum, plain, and ball bearing models. New gasket sets include three crankcase cover or bearing support gaskets, 0.005 in. (0.127mm), 0.009 in. (0.228mm), and 0.015 in. (0.381mm) thick.

The end-play of the crankshaft may be checked by assembling a dial indicator on the crankshaft with the pointer against the crankcase. Move the crankshaft in and out. The indicator will show the end-play. Another way to measure the end-play is to assemble a pulley to the crankshaft and measure the end-play with a feeler gauge. Place the feeler gauge between the crankshaft thrust face and the bearing support. The feeler gauge method of measuring crankshaft end-play can only be used on cast iron plain bearing engines with removable bases.

On cast iron engines, the end-play should be 0.002-0.008 in. (0.05-0.20mm) with one 0.015 in. (0.381mm) gasket in place. If the end-play is less than 0.002 in. (0.05mm), which would be the case if a new crankcase or sump cover is used, additional gaskets of 0.005 in. (0.127mm), 0.009 in. (0.228mm) or 0.015 in. (0.381mm) may be added in various combinations to obtain the proper end-play.

Aluminum Engines Only

If the end-play is more than 0.008 in. (0.20mm) with one 0.015 in. (0.381mm) gasket in place, a thrust washer is available to be placed on the crankshaft power take-off end, between the gear and crankcase cover or sump on plain bearing engines. On ball bearing equipped aluminum engines, the thrust washer is added to the magneto end of the crankshaft instead of the power take-off end.

➡Aluminum engines never use less than the 0.015 in. (0.381mm) gasket.

Cylinders

INSPECTION

▶ See Figure 179

Always inspect the cylinder after the engine has been disassembled. Visual inspection will show if there are any cracks, stripped bolt holes, broken fins, or if the cylinder wall is scored. Use an inside micrometer or telescoping gauge and micrometer to measure the size of the cylinder bore. Measure at right angles.

If the cylinder bore is more than 0.003 in. (0.076mm) oversize, or 0.0015 in. (0.381mm) out of round on light-weight (aluminum) cylinders, the cylinder must be resized (rebored).

➡Do not deglaze the cylinder walls when installing piston rings in aluminum cylinder engines. Also be aware that there are chrome ring sets available for most engines. These are used to control oil pumping in bores worn to 0.005 in. (0.127mm) over standard and do not require honing or glaze breaking to seat.

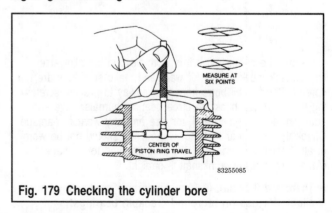

Fig. 179 Checking the cylinder bore

RESIZING

▶ **See Figure 180**

Cylinder Bore Specifications

Basic Engine Model or Series	Std. Bore Size Diameter	
	Max	Min
Aluminum Cylinder		
140000	2.750	2.749
170000, 190000	3.000	2.999
251000	3.4375	3.4365
Cast Iron Cylinder		
19, 23, 190000, 200000	3.000	2.999
230000	3.000	2.999
243400	3.0625	3.0615
300000	3.4375	3.4365
320000	3.5625	3.5615

83255C18

Always resize to exactly 0.010 in. (0.254mm), 0.020 in. (0.50mm) or 0.030 in. (0.762mm) over standard size. If this is done accurately, the stock oversize rings and pistons will fit perfectly and proper clearances will be maintained. Cylinders, either cast iron or lightweight, can be quickly resized with a good hone. Use the stones and lubrication recommended by the hone manufacturer to produce the correct cylinder wall finish for the various engine models.

If a boring bar is used, a hone must be used after the boring operation to produce the proper cylinder wall finish.

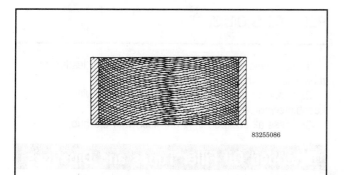

83255086

Fig. 180 Cross-hatch pattern after honing

Honing can be done with a portable electric drill, but it is easier to use a drill press.

1. Clean the cylinder at top and bottom to remove all burrs and pieces of base and head gaskets.
2. Fasten the cylinder to a heavy iron plate. Some cylinders require shims. Use a level to align the drill press spindle with the bore.
3. Oil the surface of the drill press table liberally. Set the iron plate and the cylinder on the drill press table. Do not anchor the cylinder to the drill press table. If you are using a portable drill, set the plate and the cylinder on the floor.
4. Place the hone driveshaft in the chuck of the drill.
5. Slip the hone into the cylinder. Connect the driveshaft to the hone and set the stop on the drill press so the hone can only extend ¾-1 in. (19-25mm) from the top or bottom of the cylinder. If you are using a portable drill, cut a piece of wood to place in the cylinder as a stop for the hone.
6. Place the hone in the middle of the cylinder bore. Tighten the adjusting knob with your finger or a small screwdriver until the stones fit snugly against the cylinder wall. Do not force the stones against the cylinder wall. The hone should operate at a speed of 300-700 rpm. Lubricate the hone as recommended by the manufacturer.

➡**Be sure that the cylinder and the hone are centered and aligned with the driveshaft and the drill spindle.**

7. Start the drill and, as the hone spins, move it up and down at the lower end of the cylinder. The cylinder is not worn at the bottom but is round so it will act to guide the hone and straighten the cylinder bore. As the bottom of the cylinder increases in diameter, gradually increase your strokes until the hone travels the full length of the bore.

➡**Do not extend the hone more than ¾-1 in. (19-25mm) past either end of the cylinder bore.**

8. As the cutting tension decreases, stop the hone and tighten the adjusting knob. Check the cylinder bore frequently with an accurate micrometer. Hone 0.0005 in. (0.0127mm) oversize to allow for shrinkage when the cylinder cools.
9. When the cylinder is within 0.0015 in. (0.038mm) of the desired size, change from the rough stone to a finishing stone.

The finished resized cylinder should have a cross-hatched appearance. Proper stones, lubrication, and spindle speed along with rapid movement of the hone within the cylinder during the last few strokes, will produce this finish. Cross-hatching provides proper lubrication and ring break-in.

➡**It is EXTREMELY important that the cylinder be thoroughly cleaned after honing to eliminate ALL grit. Wash the cylinder carefully in a solvent such as kerosene. The cylinder bore should be cleaned with a brush, soap, and water.**

Bearings

INSPECTION

Plain Type

Crankshaft Bearing Specifications

Basic Engine Model or Series	PTO Bearing	Bearing Magneto
Aluminum Cylinder		
140000, 170000	1.185	1.004
190000	1.185	1.004
251000	1.383	1.383
Cast Iron Cylinder		
19, 190000, 200000	1.185	1.185
23, 230000	1.382	1.382
240000, 300000	Ball	Ball
320000	Ball	Ball

83255C19

Bearings should be replaced if they are scored or if a plug gauge will enter. Try the gauge at several points in the bearings.

REPLACING PLAIN BEARINGS

Models 19-20-23

The crankcase cover bearing support should be replaced if the bearing is worn or scored.

REPLACING THE MAGNETO BEARING

Aluminum Cylinder Engines

There are no removable bearings in these engines. The cylinder must be reamed out so a replacement bushing can be installed.

1. Place a pilot guide bushing in the sump bearing, with the flange of the guide bushing toward the inside of the sump.
2. Assemble the sump on the cylinder. Make sure that the pilot guide bushing does not fall out of place.
3. Place the guide bushing into the oil seal recess in the cylinder. This guide bushing will center the counterbore reamer even though the oil bearing surface might be badly worn.
4. Place the counterbore reamer on the pilot and insert them into the cylinder until the tip of the pilot enters the pilot guide bushing in the sump.

5. Turn the reamer clockwise with a steady, even pressure until it is completely through the bearing. Lubricate the reamer with kerosene or any other suitable solvent.

➡ Counterbore reaming may be performed without any lubrication. However, clean off shavings because aluminum material builds up on the reamer flutes causing eventual damage to the reamer and an oversize counterbore.

6. Remove the sump and pull the reamer out without backing it through the bearing. Clean out the remaining chips. Remove the guide bushing from the oil seal recess.
7. Hold the new bushing against the outer end of the reamed out bearing, with the notch in the bushing aligned with the notch in the cylinder. Note the position of the split in the bushing. At a point in the outer edge of the reamed out bearing opposite to the split in the bushing, make a notch in the cylinder hub at a 45° angle to the bearing surface. Use a chisel or a screwdriver and hammer.
8. Press in the new bushing, being careful to align the oil notches with the driver and the support until the outer end of the bushing is flush with the end of the reamed cylinder hub.
9. With a blunt chisel or screwdriver, drive a portion of the bushing into the notch previously made in the cylinder. This is called staking and is done to prevent the bushing from turning.
10. Reassemble the sump to the cylinder with the pilot guide bushing in the sump bearing.
11. Place a finishing reamer on the pilot and insert the pilot into the cylinder bearing until the tip of the pilot enters the pilot guide bushings in the sump bearing.
12. Lubricate the reamer with kerosene, fuel oil, or other suitable solvent, then ream the bushing, turning the reamer clockwise with a steady even pressure until the reamer is completely through the bearing. Improper lubricants will produce a rough bearing surface.
13. Remove the sump, reamer, and the pilot guide bushing. Clean out all reaming chips.

REPLACING THE P.T.O. BEARING

Aluminum Cylinder Engines

The sump or crankcase bearing is repaired the same way as the magneto end bearing. Make sure to complete repair of one bearing before starting to repair the other. Press in new oil seals when bearing repair is completed.

REPLACING OIL SEALS

1. Assemble the seal with the sharp edge of leather or rubber toward the inside of the engine.
2. Lubricate the inside diameter of the seal with Lubriplate® or equivalent.
3. Press all seals so they are flush with the hub.

Extended Oil Filler Tubes and Dipsticks

When installing the extended oil fill and dipstick assembly, the tube must be installed so the O-ring seal is firmly compressed. To do so, push the tube downward toward the sump, then tighten the blower housing screw, which is used to secure the tube and bracket. When the dipstick assembly is fully depressed, it seals the upper end of the tube.

A leak at the seal between the tube and the sump, or at the seal at the upper end of the dipstick can result in a loss of crankcase vacuum, and a discharge of smoke through the exhaust system.

Breathers

▶ See Figure 181

The function of the breather is to maintain a vacuum in the crankcase. The breather has a fiber disc valve which limits the direction of air flow caused by the piston moving back and forth in the cylinder. Air can flow out of the crankcase, but the one-way valve blocks the return flow, thus maintaining a vacuum in the crankcase. A partial vacuum must be maintained in the crankcase to prevent oil from being forced out of the engine at the piston rings, oil seals, breaker plunger, and gaskets.

INSPECTION OF THE BREATHER

If the fiber disc valve is stuck or binding, the breather cannot function properly and must be replaced. A 0.045 in. (1.14mm)

wire gauge should not enter the space between the fiber disc valve and the body. Use a spark plug wire gauge to check the valve. The fiber disc valve is held in place by an internal bracket which will be distorted if pressure is applied to the fiber disc valve. Therefore, do not apply force when checking the valve with the wire gauge.

If the breather is removed for inspection or valve repair, a new gasket should be used when replacing the breather. Tighten the screws securely to prevent oil leakage.

Most breathers are now vented through the air cleaner, to prevent dirt from entering the crankcase. Check to be sure that the venting elbows or the tube are not damaged and that they are properly sealed.

Oil Dippers and Slingers

▶ See Figure 182

Oil dippers reach into the oil reservoir in the base of the engine and splash oil onto the internal engine parts. The oil dipper is installed on the connecting rod and has no pump or moving parts.

Oil slingers are driven by the cam gear. Old style slingers using a die cast bracket assembly have a steel bushing between the slinger and the bracket. Replace the bracket on which the oil slinger rides if it is worn to a diameter of 0.490 in. (12.4mm) or less. Replace the steel bushing if it is worn. Newer style oil slingers have a stamped steel brackets.

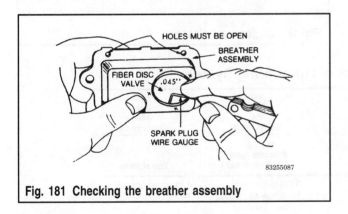

Fig. 181 Checking the breather assembly

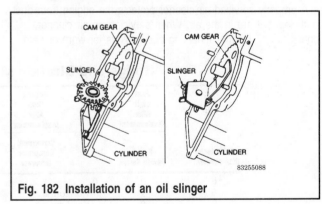

Fig. 182 Installation of an oil slinger

1970-84 13 — 20 HP Models

Engine Identification

The Briggs and Stratton model designation system consists of up to a six digit number. It is possible to determine most of the important mechanical features of the engine by merely knowing the model number. An explanation of what each number means is given below.

1. The first one or two digits indicate the cubic inch displacement (cid).

2. The first digit after the displacement indicates the basic design series, relating to cylinder construction, ignition and general configuration.

3. The second digit after the displacement indicates the position of the crankshaft and the type of carburetor the engine has.

4. The third digit after the displacement indicates the type of bearings and whether or not the engine is equipped with a reduction gear or auxiliary drive.

5. The last digit indicates the type of starter. The model identification plate is usually located on the air baffle surrounding the cylinder.

Air Cleaners

A properly serviced air cleaner protects the engine from dust particles that are in the air. When servicing an air cleaner, check the air cleaner mounting and gaskets for worn or damaged mating surfaces. Replace any worn or damaged parts to prevent dirt and dust from entering the engine through openings caused by improper sealing. Straighten or replace any bent mounting studs.

SERVICING

Oil Foam Air Cleaners
▶ See Figure 183

Clean and re-oil the air cleaner element every 25 hours of operation under normal operating conditions. The capacity of the oil-foam air cleaner is adequate for a full season's use

83256001

Fig. 183 Oil foam air cleaner

Briggs and Stratton Model Numbering System

Cubic Inch Displacement	First Digit After Displacement — Basic Design Series	Second Digit After Displacement — Crankshaft, Carburetor Governor	Third Digit After Displacement — Bearings, Reduction Gears & Auxiliary Drives	Fourth Digit After Displacement — Type of Starter
6	0	0-	0-Plain Bearing	0-Without Starter
8	1	1-Horizontal Vacu-Jet	1-Flange Mounting Plain Bearing	1-Rope Starter
9	2			
10	3	2-Horizontal Pulsa-Jet	2-Ball Bearing	2-Rewind Starter
13	4			
14	5	3-Horizontal Flo-Jet (Pneumatic Governor)	3-Flange Mounting Ball Bearing	3-Electric-110 Volt. Gear Drive
17	6			
19	7	4-Horizontal Flo-Jet (Mechanical Governor)	4-	4-Elec. Starter-Generator-12 Volt. Belt Drive
20	8			
23	9			
24		5-Vertical Vacu-Jet	5-Gear Reduction (6 to 1)	5-Electric Starter Only-12 Volt. Gear Drive
30				
32				
		6-	6-Gear Reduction (6 to 1) Reverse Rotation	6-Wind-up Starter
		7-Vertical Flo-Jet	7-	7-Electric Starter, 12 Volt Gear Drive, with Alternator
		8-	8-Auxiliary Drive Perpendicular to Crankshaft	8-Vertical-pull Starter
		9-Vertical Pulsa-Jet	9-Auxiliary Drive Parallel to Crankshaft	

without cleaning. Under very dusty conditions, clean the air cleaner every few hours of operation.

The oil-foam air cleaner is serviced in the following manner:

1. Remove the screw that holds the halves of the air cleaner shell together and retains it to the carburetor.

2. Remove the air cleaner carefully to prevent dirt from entering the carburetor.

3. Take the air cleaner apart (split the two halves).

4. Wash the foam in kerosene or liquid detergent and water to remove the dirt.

5. Wrap the foam in a clean cloth and squeeze it dry.

6. Saturate the foam in clean engine oil and squeeze it to remove the excess oil.

7. Assemble the air cleaner and fasten it to the carburetor with the attaching screw.

Oil Bath Air Cleaner
▶ **See Figure 184**

Pour the old oil out of the bowl. Wash the element thoroughly in solvent and squeeze it dry. Clean the bowl and refill it with the same type of oil used in the crankcase.

Dry Element Air Cleaner
▶ **See Figure 185**

Remove the element of the air cleaner and tap (top and bottom) it on a flat surface or wash it in non-sudsing detergent and flush it from the inside until the water coming out is clear. After washing, air dry the element thoroughly before reinstalling it on the engine. NEVER OIL A DRY ELEMENT.

Heavy Duty Air Cleaner
▶ **See Figure 186**

Clean and re-oil the foam pre-cleaner at three month intervals or every 25 hours, whichever comes first.

Clean the paper element every year or 100 hours, whichever comes first. Use the dry element procedure for cleaning the paper element of the heavy duty air cleaner.

Use the oil foam cleaning procedure to clean the foam sleeve of the heavy duty air cleaner.

If the engine is operated under very dusty conditions, clean the air cleaner more often.

Oil and Fuel Recommendations

▶ **See Figure 187**

Briggs & Stratton recommends unleaded fuel. Unleaded fuel is preferable because of the reduction in deposits that results from its use, but its use is not required.

Premium fuel is not required, as regular or unleaded will have sufficient knock resistance if the engine is in proper condition. The factory recommends that fuel by purchased in lots small enough to be used up in 30 days or less. When fuel is older than that, it can form gum and varnish, or may be improperly tailored to the prevailing temperature.

You should use a high quality detergent oil designated 'For Service SG'. Detergent oil is recommended because of its important ability to keep gum and varnish from clogging the lubrication system. Briggs & Stratton specifically recommends that no special oil additives be used.

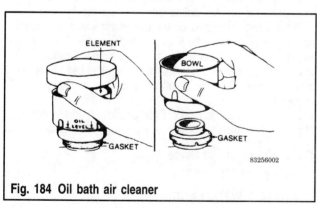

Fig. 184 Oil bath air cleaner

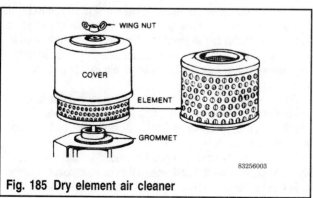

Fig. 185 Dry element air cleaner

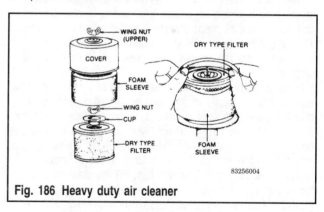

Fig. 186 Heavy duty air cleaner

Oil Viscosity Recommendations

Winter (under 40°F.)	Summer (above 40°F.)
SAE 5W-20, or SAE 5W-30	SAE 30
If above are not available:	If above is not available:
SAE 10W or SAE 10W-30 (under 0°F.)	SAE 10W-40 or SAE 10W-30
Use SAE 10W or SAE 10W-30 in proportions of 90% motor oil/ 10% kerosene	

832560C2

Engine Oil Capacity Chart

Basic Model Series	Capacity Pints
Aluminum	
25 Cu. in. Vert. Crankshaft	3
25 Cu. in. Horiz. Crankshaft	3
Cast Iron	
23, 24, 30, 32 Cu. in. Horiz. Crank.	4

832560C3

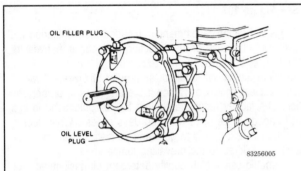

Fig. 187 Oil fill and drain plugs on gear reduction engines

Oil must be changed every 25 hours of operation. If the atmosphere in which the engine is operating is very dirty, oil changes should be made more frequently, as often as every 12 hours, if necessary. Oil should be changed after 5 hours of operation in the case of brand new engines. Drain engine oil when hot.

In hot weather, when under heavy load, or when brand new, engines may consume oil at a rate which will require you to refill the crankcase several times between oil changes. Check the oil level every hour or so until you can accurately estimate how long the engine can go between refills. To check oil level, stop the engine and allow it to sit for a couple of minutes, then remove the dipstick or filler cap. Fill the crankcase to the top of the filler pipe when there is no dipstick, or wipe the dipstick clean, reinsert it, and add oil as necessary until the level reaches the upper mark.

On cast iron engines with a gear reduction unit, crankcase and reduction gears are lubricated by a common oil supply. When draining crankcase, also remove drain plug in reduction unit.

On aluminum engines with reduction gear, a separate oil supply lubricates the gears, although the same type of oil used in the crankcase is used in the reduction gear cover. On these engines, remove the drain plug every fourth oil change (100 hours), then install the plug and refill. The level in the reduction gear cover must be checked during the refill operation by removing the level plug from the side of the gearcase, removing the filler plug, and then filling the case through the filler plug hole until oil runs out the level plug hole. Then, install both plugs.

Spark Plugs

▶ **See Figures 188, 189, 190, 191, 192 and 193**

Remove the spark plug with a ¾ in. (1½ in. plug) or a ¹³⁄₁₆ in. (2 in. plug) deep well socket wrench. Clean carbon deposits off the center and side electrodes with a sharp instrument. If possible, you should also attempt to remove deposits from the recess between the insulator and the threaded portion of the plug. If the electrodes are burned away or the insulator is cracked at any point, replace the plug. Using a wire type feeler gauge, adjust the gap by bending the side electrode where it is curved until the gap is 0.030 in. (0.8mm).

When installing the plug, make sure the threads of the plug and the threads in the cylinder head are clean. It is best to oil the plug threads very lightly. Be careful not to overtorque the plug, especially if the engine has an aluminum head. If you use a torque wrench, torque to about 15 ft. lbs.

Breaker Points

All Briggs and Stratton engines have magneto ignition systems. Three types are used: Flywheel Type — Internal Breaker Flywheel Type — External Breaker, and Magna-Matic.

FLYWHEEL TYPE — INTERNAL BREAKER

▶ **See Figure 194**

This ignition system has the magneto located on the flywheel and the breaker points located under the flywheel.

The flywheel is located on the crankshaft with a soft metal key. It is held in place by a nut or starter clutch. The flywheel key must be in good condition to insure proper location of the flywheel for ignition timing. Do not use a steel key under any circumstances. Use only a soft metal key, as originally supplied.

The keyway in both flywheel and crankshaft should not be distorted. Flywheels are made of aluminum, zinc, or cast iron.

Flywheel, Nut, and/or Starter Clutch

▶ **See Figures 195 and 196**

REMOVAL & INSTALLATION

Place a block of wood under the flywheel fins to prevent the flywheel from turning while you are loosening the nut or starter clutch. Be careful not to bend the flywheel. There are special flywheel holders available for this purpose; Briggs & Stratton recommends their use on flywheels of 6¾ in. (171.45mm) diameter or less.

On rope starter engines, the ½ in. flywheel nut has a left-hand thread and the ⅝ in. nut has a right-hand thread. The starter clutch used on rewind or wind-up starters has a right-hand thread.

Some flywheels have two holes provided for the use of a flywheel puller. Use a small gear puller or automotive steering wheel puller to remove the flywheel if a flywheel puller is not available. Be careful not to bend the flywheel if a gear puller is used. On rope starter engines leave the nut on for the puller

Tune-Up Specifications

Model	Plug Type	Plug Gap (in.)	Point Gap (in.)	Armature Gap		Idle Speed
				2 leg	3 leg	
Aluminum Block						
All	①	.030	.020	.010–.014	.016–.019	1750
Cast Iron Block						
23, 230000	①	.030	.020	.010–.014	.022–.026	1200
243400, 300000, 320000	①	.030	.020	.010–.014	—	1200

① Manufacturer's Code		Manufacturer
1½ in. plug	2 in. plug	
CJ-8	J-8	Champion
RCJ-8	RJ-8	Champion (resistor)
A-7NX	A-71	Autolite
AR-7N	AR-80	Autolite (resistor)
CS-45	GC-46	A.C.
—	R-46	A.C. (resistor)

832560C4

Fig. 188 Twist and pull on the boot to remove the spark plug wire; never pull on the wire itself

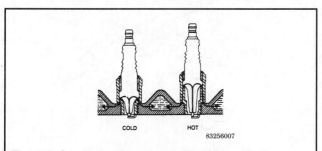

COLD   HOT
83256007

Fig. 189 Spark plug heat range. The plug with the higher heat range is on the right. Higher heat ranges operate better in 2-cycle engines

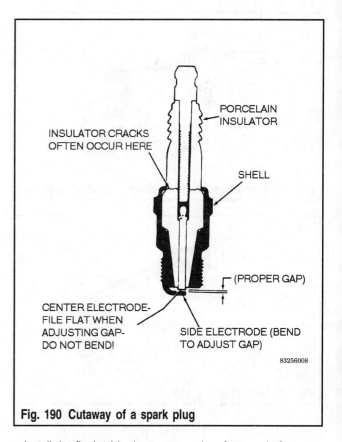

PORCELAIN INSULATOR

INSULATOR CRACKS OFTEN OCCUR HERE

SHELL

(PROPER GAP)

CENTER ELECTRODE-FILE FLAT WHEN ADJUSTING GAP-DO NOT BEND!

SIDE ELECTRODE (BEND TO ADJUST GAP)

83256008

Fig. 190 Cutaway of a spark plug

to bear against. Small cast iron flywheels do not require a puller.

Install the flywheel in the reverse order of removal after inspecting the key and keyway for damage or wear.

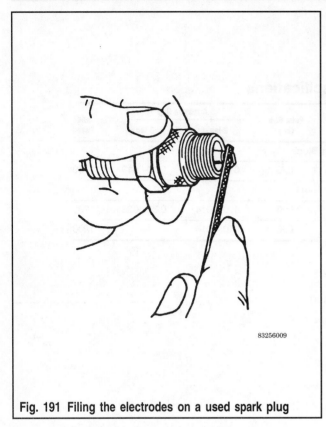

Fig. 191 Filing the electrodes on a used spark plug

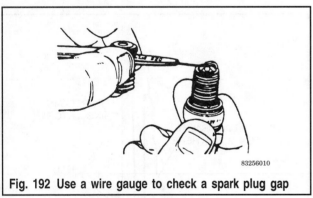

Fig. 192 Use a wire gauge to check a spark plug gap

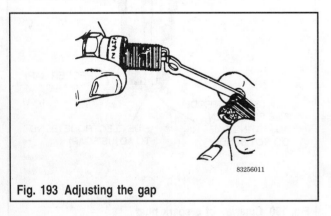

Fig. 193 Adjusting the gap

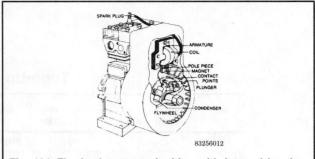

Fig. 194 Flywheel magneto ignition with internal breaker points and external armature

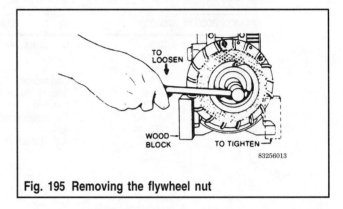

Fig. 195 Removing the flywheel nut

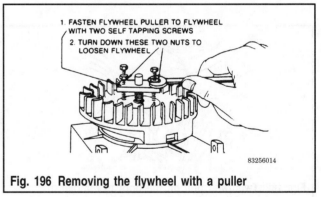

Fig. 196 Removing the flywheel with a puller

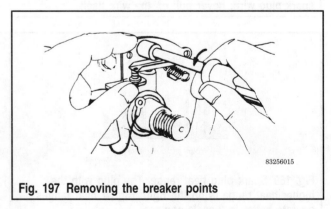

Fig. 197 Removing the breaker points

Breaker Point Replacement

▶ **See Figures 197 and 198**

Remove the breaker cover. Care should be taken when removing the cover, to avoid damaging it. If the cover is bent or damaged, it should be replaced to insure a proper seal.

The breaker point gap on all models is 0.020 in. (0.5mm). Check the points for contact and for signs of burning or pitting. Points that are set too wide will advance the spark timing and may cause kickback when starting. Points that are set too close will retard the spark timing and decrease engine power.

On models that have a separate condenser, the point set is removed by first removing the condenser and armature wires from the breaker point clip. Loosen the adjusting lock screw and remove the breaker point assembly.

On models where the condenser is incorporated with the breaker points, loosen the screw which holds the post. The condenser/point assembly is removed by loosening the screw which holds the condenser clamp.

When installing a point set with the separate condenser, be sure that the small boss on the magneto plate enters the hole in the point bracket. Mount the point set to the magneto plate or the cylinder with a lock screw. Fasten the armature lead wire to the breaker points with the clip and screw. If these lead wires do not have terminals, the bare end of the wires can be inserted into the clip and the screw tightened to make a good connection. Do not let the ends of the wire touch either the point bracket or the magneto plate, or the ignition will be grounded.

To install the integral condenser/point set, place the mounting post of the breaker arm into the recess in the cylinder so that the groove in the post fits the notch in the recess. Tighten the mounting screw securely. Use a ¼ in. wrench. Slip the open loop of the breaker arm spring through the two holes in the arm, then hook the closed loop of the spring over the small post protruding from the cylinder. Push the flat end of the breaker arm into the groove in the mounting post. This places tension on the spring and pulls the arm against the plunger. If the condenser post is threaded, attach the soil primary wire and the ground wire (if furnished) with the lock washer and nut. If the primary wire is fastened to the condenser with a spring fastener, compress the spring and slip the primary wire and ground wire into the hole in the condenser post. Release the spring. Lay the condenser in place and tighten the condenser clamp securely. Install the spring in the breaker arm.

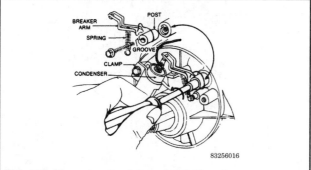

Fig. 198 Removing the integral point/condenser assembly

Point Gap Adjustment

▶ **See Figures 199 and 200**

Turn the crankshaft until the points are open to the widest gap. When adjusting a breaker point assembly with an integral condenser, move the condenser forward or backward with a screwdriver until the proper gap is obtained — 0.020 in. (0.5mm). Point sets with a separate condenser are adjusted by moving the contact point bracket up and down after the lock screw has been loosened. The point gap is set to 0.020 in. (0.5mm).

Breaker Point Plunger

▶ **See Figure 201**

If the breaker point plunger hole becomes excessively worn, oil will leak past the plunger and may get on the points, causing them to burn. To check the hole, loosen the breaker

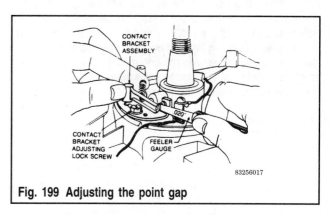

Fig. 199 Adjusting the point gap

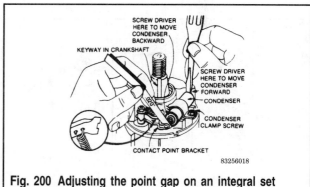

Fig. 200 Adjusting the point gap on an integral set

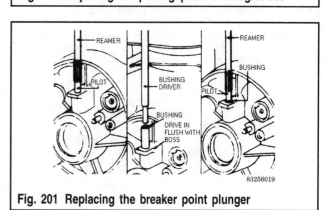

Fig. 201 Replacing the breaker point plunger

point mounting screw and move the breaker points out of the way. Remove the plunger. If the flat end of the #19055 plug gauge will enter the plunger hole for a distance of ¼ in. (6mm) or more, the hole should be rebushed.

To install the bushing, it is necessary that the breaker points, armature, and crankshaft be removed. Use a #19056 reamer to ream out the old plunger hole. This should be done by hand. The reamer must be in alignment with the plunger hole. Drive the bushing, #23513, into the hole until the upper end of the bushing is flush with the top of the boss. Remove all metal chips and dirt from the engine.

If the breaker point plunger is worn to a length of 0.870 in. (22mm) or less, it should be replaced. Plungers must be inserted with the groove at the top or oil will enter the breaker box. Insert the plunger into the hole in the cylinder.

Armature Air Gap Adjustment
▶ See Figures 202 and 203

Set the air gap between the flywheel and the armature as follows: With the armature up as far as possible and just one screw tightened, slip the proper gauge between the armature and flywheel. Turn the flywheel until the magnets are directly below the armature. Loosen the one mounting screw and the magnets should pull the armature down firmly against the thickness gauge. Tighten the mounting screws.

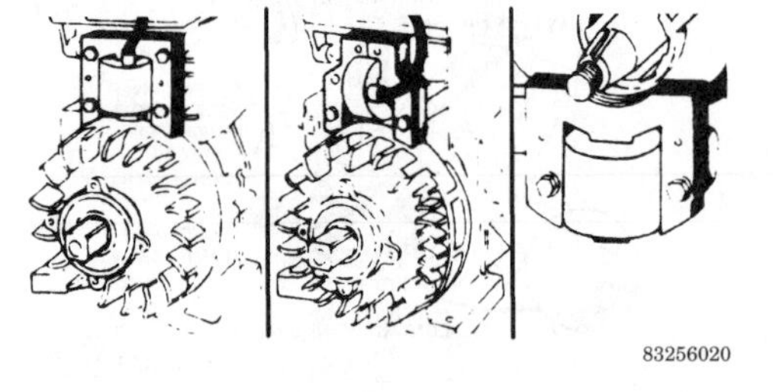

FLYWHEEL TYPE — EXTERNAL BREAKER

Breaker Point Set Replacement
▶ See Figure 204

Turn the crankshaft until the points open to their widest gap. This makes it easier to assemble and adjust the points later if the crankshaft is not removed. Remove the condenser and upper and lower mounting screws. Loosen the lock nut and back off the breaker point screw. Install the points in the reverse order of removal.

To avoid the possibility of oil leaking past the breaker point plunger or moisture entering the crankcase between the plunger and the bushing, a plunger seal is installed on the engine models using this type of ignition system. To install a new seal on the plunger, remove the breaker point assembly and condenser. Remove the retainer and eyelet, remove the old seal, and install the new one. Use extreme care when installing the seal on the plunger to avoid damaging the seal. Replace the eyelet and retainer and replace the points and condenser.

➡**Apply a small amount of sealer to the threads of both mounting screws and the adjustment screw. The sealer prevents oil from leaking into the breaker point area.**

Point Gap Adjustment
▶ See Figure 205

Turn the crankshaft until the points open to their widest gap. Turn the breaker point adjusting screw until the points open to 0.020 in. (0.5mm) and tighten the lock nut. When the cover is installed, seal the point where the primary wire passes under

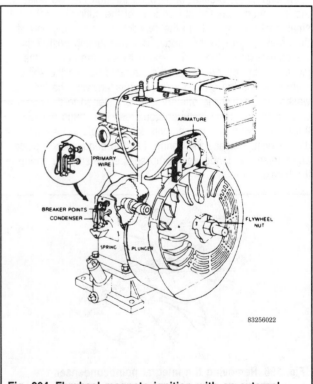

Fig. 204 Flywheel magneto ignition with an external breaker assembly

Fig. 202 Variations in armature positioning

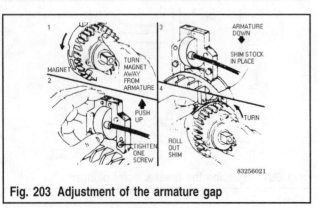

Fig. 203 Adjustment of the armature gap

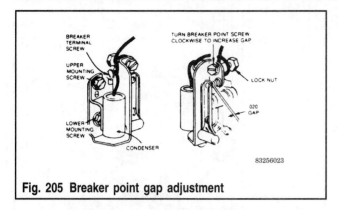

Fig. 205 Breaker point gap adjustment

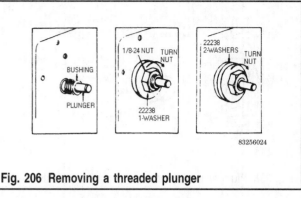

Fig. 206 Removing a threaded plunger

the cover. This area must be resealed to prevent the entry of dust and moisture.

Armature Timing Adjustment

MODELS 230000, 243000

Using a puller, remove the flywheel. Set the point gap at 0.020 in. (0.5mm). Position the flywheel on the crankshaft taper. Slip the key in place. Install the flywheel and the crankshaft clockwise until the breaker points are just opening. Use a timing light. When the points just start to open, the arrow on the flywheel should line up with the arrow on the armature bracket.

If the arrows do not match, slip off the flywheel without disturbing the position of the crankshaft. Slightly loosen the mounting screw which holds the armature bracket to the cylinder. Slip the flywheel back onto the crankshaft. Insert the flywheel key. Install the flywheel nut finger tight. Move the armature and bracket assembly to align the arrows. Slip off the flywheel and tighten the armature bracket bolts. Install the key and flywheel. Tighten the flywheel nut to 138-150 ft. lbs. Set the armature gap at 0.010-0.014 in. (0.25-0.35mm).

MODEL 23D

With the points set at 0.020 in. (0.5mm) and the flywheel key screw finger tight together with the flywheel nut, rotate the flywheel clockwise until the breaker points are just opening. The flywheel key drives the crankshaft while doing this. Using a timing light, rotate the flywheel slightly counterclockwise until the edge of the armature lines up with the edge of the flywheel insert. The crankshaft must not turn while doing this. Tighten the key screw and the flywheel nut. Set the armature air gap at 0.022-0.026 in. (0.56-0.66mm).

Replacing Threaded Breaker Plunger and Bushing
♦ **See Figure 206**

Remove the breaker cover and the condenser and breaker point assembly.

Place a thick ⅜ in. (9.5mm) inside diameter washer over the end of the bushing and screw on the ⅜-24 nut. Tighten the nut to pull the bushing out of the hole. After the bushing has been moved about ⅛ in. (3mm), remove the nut and put on a second thick washer and repeat the procedure. A total stack of ⅜ in. (9.5mm) washers will be required to completely remove the bushing. Be sure the plunger does not fall out of the bushing as it is removed.

Place the new plunger in the bushing with the large end of the plunger opposite the threads on the bushing. Screw the ⅜-24 in. nut onto the threads to protect them and insert the bushing into the cylinder. Place a piece of tubing the same diameter as the nut and, using a hammer, drive the bushing into the cylinder until the square shoulder on the bushing is flush with the face of the cylinder. Check to be sure that the plunger operates freely.

Replacing Unthreaded Breaker Plunger and Bushing
♦ **See Figure 207**

Pull the plunger out as far as possible and use a pair of pliers to break the plunger off as close as possible to the bushing. Use a ¼-20 in. tap or a #93029 self threading screw to thread the hole in the bushing to a depth of about ½-⅝ in. (13-16mm). Use a ¼-20 · ½ in. hex head screw and two spacer washers to pull the bushing out of the cylinder. The bushing will be free when it has been extracted ⁵⁄₁₆ in. (8mm). Carefully remove the bushing and the remainder of the broken plunger. Do not allow the plunger or metal chips to drop into the crankcase.

Correctly insert the new plunger into the new bushing. Insert the plunger and the bushing into the cylinder. Use a hammer and the old bushing to drive the new bushing into the cylinder until the new bushing is flush with the face of the cylinder. Make sure that the plunger operates freely.

Plunger Seal
♦ **See Figure 208**

Later models with Flywheel Type — External Breaker Ignition feature a plunger seal. This seal keeps both oil and moisture from entering the breaker box. If the points have become contaminated on an engine manufactured without this feature,

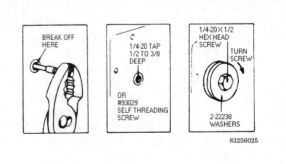

Fig. 207 Removing an unthreaded plunger

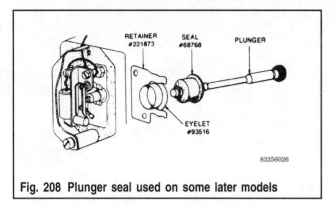

Fig. 208 Plunger seal used on some later models

the seal may be installed. Parts, part numbers, and their locations are shown in the illustration. Install the seal onto the plunger very carefully to avoid fracturing it.

MAGNA — MATIC IGNITION SYSTEM

Removing the Flywheel
▶ See Figure 209

Flywheels on engines with Magna-Matic ignition are removed with pullers similar to factory designs numbered #19068 and 19203. These pullers employ two bolts, which are screwed into holes tapped into the flywheel. The bolts are turned until the flywheel is forced off the crankshaft. Only this type of device should be used to pull these flywheels.

Armature Air Gap
▶ See Figure 210

The armature air gap on engines equipped with Magna-Matic ignition system is fixed and can change only if wear occurs on the crankshaft journal and/or main bearing. Check for wear by inserting a ½ in. wide feeler gauge at several points between the rotor and armature. Minimum feeler gauge thickness is 0.004 in. (0.1mm). Keep the feeler gauge away from the magnets on the rotor or you will have a false reading.

Rotor Replacement
▶ See Figures 211 and 212

The rotor is held in place by a woodruff key and a clamp on later engines, and a woodruff key and set screw on older engines. The rotor clamp must always remain on the rotor, unless the rotor is in place on the crankshaft and within the armature, or a loss of magnetism will occur.

Loosen the socket head screw in the rotor clamp which will allow the clamp to loosen. It may be necessary to use a puller to remove the rotor from the crankshaft. On older models, loosen the small lock screw, then the set screw.

To install the set screw type rotor, place the woodruff key in the keyway on the crankshaft, then slide the rotor onto the crankshaft until the set screw hole in the rotor and the crankshaft are aligned. Be sure the key remains in place. Tighten the set screw securely, then tighten the lock screw to prevent the set screw from loosening. The lock screw is self-threading and the hole does not require tapping.

To install the clamp type rotor, place the woodruff key in place in the crankshaft and align the keyway in the rotor with

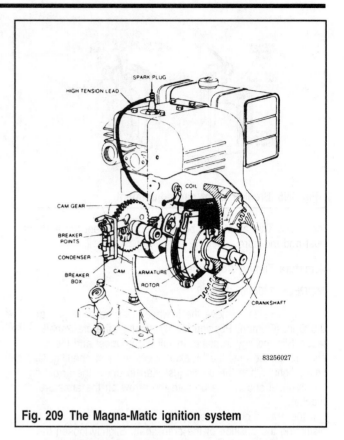

Fig. 209 The Magna-Matic ignition system

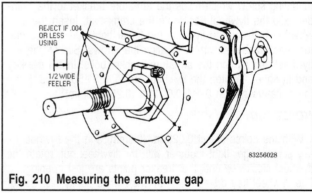

Fig. 210 Measuring the armature gap

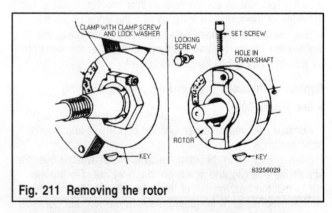

Fig. 211 Removing the rotor

the woodruff key. If necessary, use a short length of pipe and a hammer to drive the rotor onto the shaft until a 0.025 in. (0.6mm) feeler gauge can be inserted between the rotor and

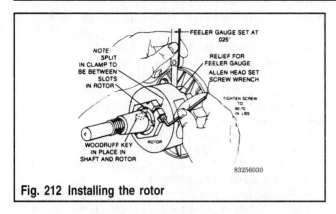

Fig. 212 Installing the rotor

the bearing support. The split in the clamp must be between the slots in the rotor. Tighten the clamp screws to 60-70 inch lbs.

Rotor Timing Adjustment
▶ See Figure 213

The rotor and armature are correctly timed at the factory and require timing only if the armature has been removed from the engine, or if the cam gear or crankshaft has been replaced.

If it is necessary to adjust the rotor, proceed as follows: with the point gap set at 0.020 in. (0.5mm), turn the crankshaft in the normal direction of rotation until the breaker points close and just start to open. Use a timing light or insert a piece of tissue paper between the breaker points to determine when the points begin to open. With the three armature mounting screws slightly loose, rotate the armature until the arrow on the armature lines up with the arrow on the rotor. Align with the corresponding number of engine models, for example, on Model 23, align with #23. Retighten the armature mounting screws.

Coil and/or Armature Replacement

Usually the coil and armature are not separated, but left assembled for convenience. However, if one or both need replacement, proceed as follows: the coil primary wire and the coil ground wire must be unfastened. Pry out the clips that hold the coil and coil core to the armature. The coil core is a slip fit in the coil and can be pushed out of the coil.

To reassemble, push the coil core into the coil with the rounded side toward the ignition cable. Place the coil and core on the armature with the coil retainer between the coil and the armature and with the rounded side toward the coil. Hook the

lower end of the clips into the armature, then press the upper end onto the coil core.

Fasten the coil ground wire (bare double wires) to the armature support. Next, place the assembly against the cylinder and around the rotor and bearing support. Insert the three mounting screws together with the washer and lockwasher into the three long oval holes in the armature. Tighten them enough to hold the armature in place but loose enough so the armature can be moved for adjustment of the timing. Attach the primary wires from the coil and the breaker points to the terminal at the upper side of the backing plate. This terminal is insulated from the backing plate. Push the ignition cable through the louvered hole at the left side of the backing plate.

Breaker Point Replacement
▶ See Figure 214

Turn the crankshaft until the points open to the widest gap. This makes it easier to assemble and adjust the points later if the crankshaft is not removed. With the terminal screw out, remove the spring screw. Loosen the breaker shaft nut until the nut is flush with the end of the shaft. Tap the nut to free the breaker arm from the tapered end of the breaker shaft. Remove the nut, lockwasher, and breaker arm. Remove the breaker plate screw, breaker plate, pivot, insulating plate, and eccentric. Pry out the breaker shaft seal with a sharp pointed tool.

To install the breaker points, press in the new oil seal with the metal side out. Put the new breaker plate on the top of the insulating plate, making sure that the detent in the breaker plate engages the hole in the insulating plate. Fasten the breaker plate screw enough to put a light tension on the plate. Adjust the eccentric so that the left edge of the insulating plate is parallel to the edge of the box and tighten the screw. This locates the breaker plate so that the proper gap adjustments may be made. Turn the breaker shaft clockwise as far as possible and hold it in this position. Place the new breaker points on the shaft, then the lockwasher, and tighten the nut down on the lockwasher. Replace the spring screw and terminal screw.

Breaker Box Replacement

Remove the two mounting screws, then remove the breaker box, turning it slightly to clear the arm at the inner end of the breaker shaft. The breaker points need not be removed to remove the breaker box.

To install, pull the primary wire through the hole at the lower left corner of the breaker box. See that the primary wire rests

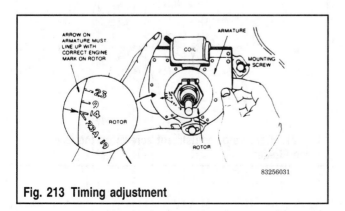

Fig. 213 Timing adjustment

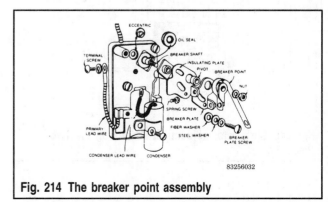

Fig. 214 The breaker point assembly

in the groove at the top end of the box, then tighten the two mounting screws to hold the box in place.

Breaker Shaft Replacement

The breaker shaft can be removed, after the breaker points are removed, by turning the shaft one half turn to clear the retaining spur at the inside of the breaker box.

Install by inserting the breaker shaft with the arm upward so the arm will clean the retainer boss. Push the shaft all the way in, then turn the arm downward.

Breaker Point Adjustment

▶ **See Figure 215**

To adjust the breaker points, turn the crankshaft until the breaker points open to the widest gap. Loosen the breaker point plate screw slightly. Rotate the eccentric to obtain a point gap of 0.020 in. (0.5mm). Tighten the breaker plate screw.

Mixture Adjustment

TWO PIECE FLO-JET

▶ **See Figure 216**

1. Start the engine and run it at 3000 rpm until it warms up.

2. Turn the needle valve (flat handle) to both extremes of operation noting the location of the valve at both points. That is, turn the valve inward until the mixture becomes too lean and the engine starts to slow, then note the position of the valve. Turn it outward slowly until the mixture becomes too rich and the engine begins to slow. Turn the valve back inward to the mid-point between the two extremes.

3. Install a tachometer on the engine. Pull the throttle to the idle position and hold it there through the rest of this step. Adjust the idle speed screw until the engine idles at 1750 rpm if it's an aluminum engine, or 1200 rpm, if it's a cast iron engine. Then, turn the idle valve in and out to adjust mixture, as described in Step 2. If idle valve adjustment changes idle speed, adjust speed to specification.

4. Release the throttle and observe the engine's response. The engine should accelerate without hesitation. If response is poor, one of the mixture adjustments is too lean. Readjust either or both as necessary. If idle speed was changed after idle valve was adjusted, readjust the idle valve first.

ONE PIECE FLO-JET

▶ **See Figure 217**

Follow the instructions for adjusting the Two Piece Flo-Jet carburetor (above). On the large, One Piece Flo-Jet, the needle valve is located under the float bowl, and the idle valve on top of the venturi passage. On the small One Piece Flo-Jet, both valves are adjusted by screws located on top of the venturi passage. The needle valve is located on the air horn side, is centered above the float bowl, and uses a larger screw head.

Governor Adjustments

SETTING MAXIMUM GOVERNED SPEED

➡**Strict limits on engine rpm must be observed when setting top governed speed on rotary lawnmowers. This is done so that blade tip speeds will be kept to less than 19,000 feet per minute. Briggs & Stratton suggests setting the governor 200 rpm low to allow for possible error in the tachometer reading. These figures below, based on blade length, must be strictly adhered to, or a serious accident could result!**

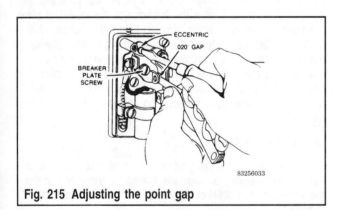

Fig. 215 Adjusting the point gap

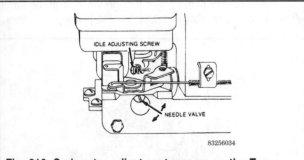

Fig. 216 Carburetor adjustment screws on the Two-Piece Flo-Jet

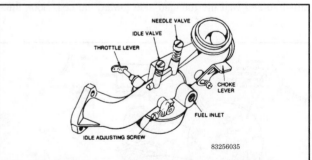

Fig. 217 Carburetor adjustment screws on the One-Piece Flo-Jet

Blade Length (in.)	Max. Governed Speed (R.P.M.)
18	4032
19	3820
20	3629
21	3456
22	3299
23	3155
24	3024
25	2903
26	2791

832560C5

Aluminum Models

Vertical and horizontal shaft engine governors are adjusted by setting the control lever in the high speed position. Loosen the nut on the governor lever. Turn the governor shaft clockwise with a screwdriver to the end of its travel. Tighten the nut. The throttle must be wide open. Check to see if the throttle can be moved from idle to wide open without binding.

Cast Iron Models

Loosen the screw which holds the governor lever to the governor shaft. Push the lever counterclockwise as far as it will go. Hold it in position and turn the governor shaft counterclockwise as far as it will go. This can be done with a screwdriver. Securely tighten the screw that holds the governor lever to the shaft.

ADJUSTING TOP NO LOAD SPEED

▶ **See Figures 218 and 219**

Set the control lever to the maximum speed position with the engine running. Bend the spring anchor tang to get the desired top speed.

ADJUSTMENT FOR CLOSER GOVERNING

▶ **See Figure 220**

Generator Applications Only

1. Snap knob upward to release adjusting nut.
2. Pull knob out against stop.
3. Then, bend the spring anchor tang to get top no-load speed as described below, depending upon the application.

On models with an 1800 rpm generator: temporarily substituting a #260902 governor spring, set the no load speed at 2600 rpm, and then set throttle stop at 1600 rpm.

On models with a 3600 rpm generator: set no load speed to 4200 rpm with standard governor spring.

4. Snap knob back into its normal position.
5. Adjust the knob for the desired generator speed.

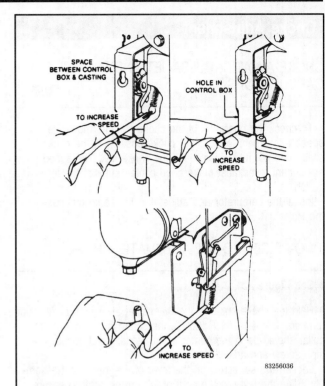

Fig. 218 Bending the spring anchor to get the desired top speed

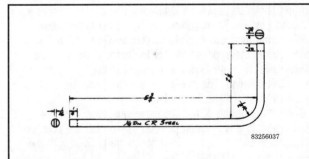

Fig. 219 The dimensions and diagram for making a home-made adjusting tool

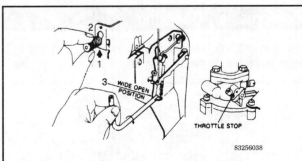

Fig. 220 Obtaining closer governing on generator applications

Choke Adjustment

CHOKE-A-MATIC, PULSA-JET AND VACU-JET CARBURETORS

To check the operation of the choke linkage, move the speed adjustment lever to the choke position. If the choke slide does not fully close, bend the choke link. The speed adjustment lever must make good contact against the top switch.

Install the carburetor and adjust it in the same manner as the Pulsa-Jet carburetor.

TWO-PIECE FLO-JET AUTOMATIC CHOKE

▶ **See Figure 221**

Hold the choke shaft so the thermostat lever is free. At room temperature — 68°F (20°C), the screw in the thermostat collar should be in the center of the stops. If not, loosen the stop screw and adjust the screw.

Loosen the set screw on the lever of the thermostat assembly. Slide the lever to the right or left on the shaft to ensure free movement of the choke link in any position. Rotate the thermostat shaft clockwise until the stop screw strikes the tube. Hold it in position and set the lever on the thermostat shaft so that the choke valve will be held open about 1/8 in. (3mm) from a closed position. Then tighten the set screw in the lever.

Rotate the thermostat shaft counterclockwise until the stop screw strikes the opposite side of the tube. Then open the choke valve manually until it stops against the top of the choke link opening. The choke valve should now be open approximately 1/8 in. (3mm) as before.

Check the position of the counterweight lever. With the choke valve in a wide open position (horizontal) the counterweight lever should also be in a horizontal position with the free end toward the right.

Operate the choke manually to be sure that all parts are free to move without binding or rubbing in any position.

Compression Checking

You can check the compression in any Briggs & Stratton engine by performing the following simple procedure: spin the flywheel counterclockwise (flywheel side) against the compression stroke. A sharp rebound indicates that there is satisfactory compression. A slight or no rebound indicates poor compression.

It has been determined that this test is an accurate indication of compression and is recommended by Briggs and Stratton. Briggs & Stratton does not supply compression pressures.

Loss of compression will usually be the result of one or a combination of the following:
1. The cylinder head gasket is blown or leaking.
2. The valves are sticking or not seating properly.
3. The piston rings are not sealing, which would also cause the engine to consume an excessive amount of oil.

Carbon deposits in the combustion chamber should be removed every 100 or 200 hours of use (more often when run at a steady load), or whenever the cylinder head is removed.

Carburetors

▶ **See Figures 222, 223 and 224**

There are three types of carburetors used on Briggs & Stratton engines. They are the Pulsa-Jet, Vacu-Jet and Flo-Jet. The first two types have three models each and the Flo-Jet has two versions.

Before removing any carburetor for repair, look for signs of air leakage or mounting gaskets that are loose, have deteriorated, or are otherwise damaged.

Note the position of the governor springs, governor link, remote control, or other attachments to facilitate reassembly. Be careful not to bend the links or stretch the springs.

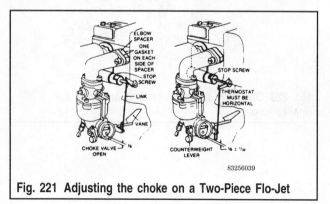

Fig. 221 Adjusting the choke on a Two-Piece Flo-Jet

Fig. 222 The two types of Flo-Jet carburetors

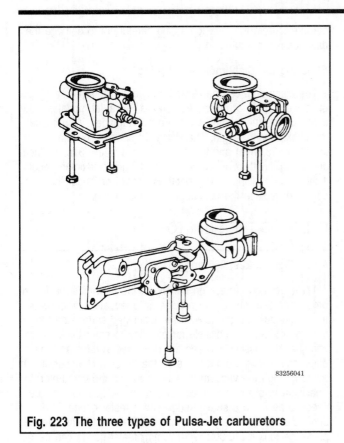

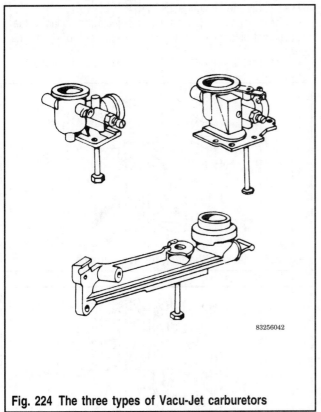

Fig. 223 The three types of Pulsa-Jet carburetors

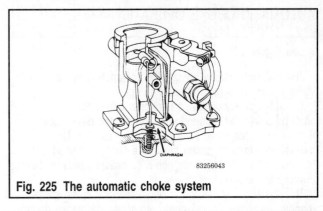

Fig. 225 The automatic choke system

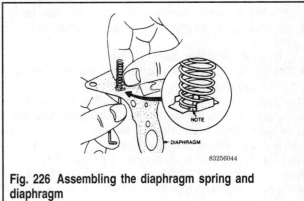

Fig. 226 Assembling the diaphragm spring and diaphragm

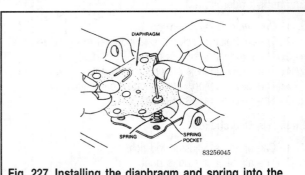

Fig. 227 Installing the diaphragm and spring into the spring pocket

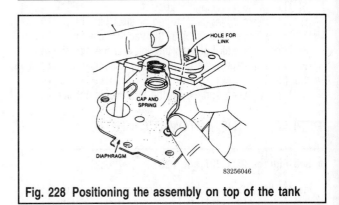

Fig. 228 Positioning the assembly on top of the tank

Fig. 224 The three types of Vacu-Jet carburetors

AUTOMATIC CHOKE OPERATION

▶ **See Figure 225**

The automatic choke operates in conjunction with engine vacuum, similar to the Pulsa-Jet fuel pump.

A diaphragm under the carburetor is connected to the choke shaft by a link. A calibrated spring under the diaphragm holds the choke closed when the engine is not running. Upon starting, vacuum created during the intake stroke is routed to the bottom of the diaphragm through a calibrated passage, thereby opening the choke.

This system also has the ability to respond in the same manner as an accelerator pump. As speed decreases during heavy loads, the choke valve partially closes, enriching the air/fuel mixture, thereby improving low speed performance and lugging power.

To check the automatic choke, remove the air cleaner and replace the stud. Observe the position of the choke valve; it should be fully closed. Move the speed control to the stop position; the governor spring should be holding the throttle in a closed position. Give the starter rope several quick pulls. The choke valve should alternately open and close.

If the choke valve does not react as stated in the previous paragraph, the carburetor will have to be disassembled to determine the problem. Before doing so, however, check the following items so you know what to look for:

Engine is Underchoked

1. Carburetor is adjusted too lean.
2. The fuel pipe check valve is inoperative (Vacu-Jet only).
3. The air cleaner stud is bent.
4. The choke shaft is sticking due to dirt.
5. The choke spring is too short or damaged.
6. The diaphragm is not preloaded.

Engine is Overchoked

1. Carburetor is adjusted too rich.
2. The air cleaner stud is bent.
3. The choke shaft is sticking due to dirt.
4. The diaphragm is ruptured.
5. The vacuum passage is restricted.
6. The choke spring is distorted or stretched.
7. There is gasoline or oil in the vacuum chamber.
8. There is a leak between the link and the diaphragm.
9. The diaphragm was folded during assembly, causing a vacuum leak.
10. The machined surface on the tank top is not flat.

REPLACING THE AUTOMATIC CHOKE

▶ **See Figures 226, 227 and 228**

Inspect the automatic choke for free operation. Any sticking problems should be corrected as proper choke operation depends on freedom of the choke to travel as dictated by engine vacuum.

Remove the carburetor and fuel tank assembly from the engine. The choke link cover may now be removed and the choke link disconnected from the choke shaft. Disassemble the

carburetor from the tank top, being careful not to damage the diaphragm.

Checking the Diaphragm and Spring

The diaphragm can be reused, provided it has not developed wear spots or punctures. On the Pulsa-Jet models, make sure that the fuel pump valves are not damaged. Also check the choke spring length. The Pulsa-Jet spring minimum length is $1\frac{1}{8}$ in. (28.5mm) and the maximum is $1\frac{7}{32}$ in. (31mm) Vacu-Jet spring length minimum is $\frac{15}{16}$ in. (24mm), maximum length 1 in. (25mm). If the spring length is shorter or longer than specified, replace the diaphragm and the spring.

CHECKING THE TANK TOP

▶ **See Figures 229, 230 and 231**

The machined surface on the top of the tank must be flat in order for the diaphragm to provide an adequate seal between the carburetor and the tank. If the machined surface on the tank is not flat, it is possible for gasoline to enter the vacuum chamber by passing between the machined surface and the diaphragm. Once fuel has entered the vacuum chamber, it can move through the vacuum passage and into the carburetor. The flatness of the machined surface on the tank top can be checked by using a straight-edge and a feeler gauge. The surface should not vary more than 0.002 in. (0.05mm). Replace the tank if a 0.002 in. (0.05mm) feeler gauge can be passed under the straight-edge.

If a new diaphragm is installed, assemble the spring to the replacement diaphragm, taking care not to bend or distort the spring.

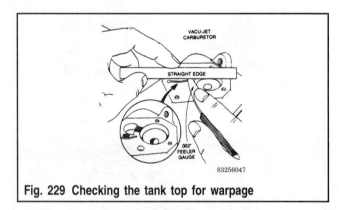

Fig. 229 Checking the tank top for warpage

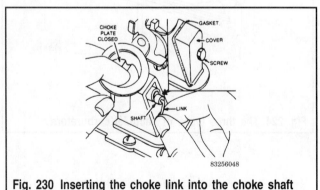

Fig. 230 Inserting the choke link into the choke shaft

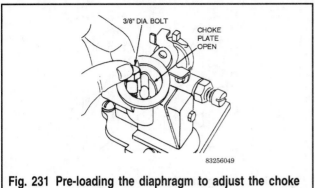

Fig. 231 Pre-loading the diaphragm to adjust the choke

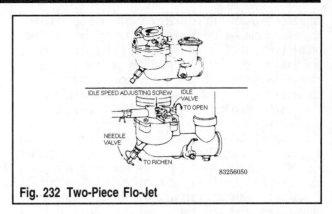

Fig. 232 Two-Piece Flo-Jet

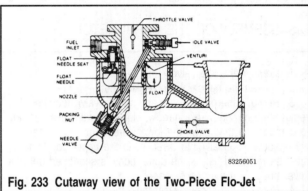

Fig. 233 Cutaway view of the Two-Piece Flo-Jet

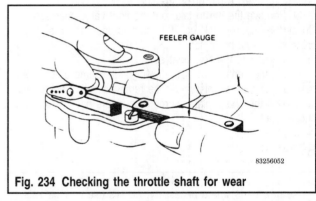

Fig. 234 Checking the throttle shaft for wear

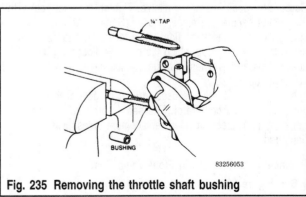

Fig. 235 Removing the throttle shaft bushing

Place the diaphragm on the tank surface, positioning the spring in the spring pocket.

Place the carburetor on the diaphragm ensuring that the choke link and diaphragm are properly aligned between the carburetor and the tank top. On Pulsa-Jet models, place the pump spring and cap on the diaphragm over the recess or pump chamber in the fuel tank. Thread in the carburetor mounting screws to about two threads. Do not tighten them. Close the choke valve and insert the choke link into the choke shaft.

Remove the air cleaner gasket, if it is in place, before continuing. Insert a ⅜ in. bolt or rod into the carburetor air horn. With the bolt in position, tighten the carburetor mounting screws in a staggered sequence. Please note that the insertion of the ⅜ in. bolt opens the choke to an over-center position, which preloads the diaphragm.

Remove the ⅜ in. bolt. The choke valve should now move to a fully closed position. If the choke valve is not fully closed, make sure that the choke spring is properly assembled to the diaphragm, and also properly inserted in its pocket in the tank top.

All carburetor adjustments should be made with the air cleaner on the engine. Adjustment is best made with the fuel tank half full.

OVERHAULING TWO PIECE FLO-JET CARBURETORS — LARGE AND SMALL LINE

▶ See Figures 232 and 233

CHECKING THE UPPER BODY FOR WARPAGE

With the carburetor assembled and the body gasket in place, try to insert a 0.002 in. (0.05mm) feeler gauge between the upper and lower bodies at the air vent boss, just below the idle valve. If the gauge can be inserted, the upper body is warped and should be replaced.

CHECKING THE THROTTLE SHAFT AND BUSHINGS

▶ See Figures 234 and 235

Wear between the throttle shaft and bushings should not exceed 0.010 in. (0.25mm). Check the wear by placing a short iron bar on the upper carburetor body so that it just fits under the throttle shaft. Measure the distance with a feeler gauge while holding the shaft down and then holding it up. If the difference is over 0.010 in. (0.25mm), either the upper body should be rebushed, the throttle shaft replaced, or both. Wear on the throttle shaft can be checked by comparing the worn and unworn portions of the shaft. To replace the bushings, remove the throttle shaft using a thin punch to drive out the pin which holds the throttle stop to the shaft; remove the throttle valve, then pull out the shaft. Place a $\frac{1}{4}$-20 tap or an E-Z Out in a vise. Turn the carburetor body so as to thread the tap or E-Z Out into the bushings enough to pull the bushings out of the body. Press the new bushings into the carburetor body with a vise. Insert the throttle shaft to be sure it is free in the bushings. If not, run a size $\frac{7}{32}$ in. (5.5mm) drill through both bushings to act as a line reamer. Install the throttle shaft, valve, and stop.

DISASSEMBLY OF THE CARBURETOR

1. Remove the idle valve.
2. Loosen the needle valve packing nut.
3. Remove the packing nut and needle valve together. To remove the nozzle, use a narrow, blunt screwdriver so as not to damage the threads in the lower carburetor body. The nozzle projects diagonally into a recess in the upper body and must be removed before the upper body is separated from the lower body, or it may be damaged.
4. Remove the screws which hold the upper and lower bodies together. A pin holds the float in place.
5. Remove the pin to take out the float valve needle. Check the float for leakage. If it contains gasoline or is crushed, it must be replaced. Use a wide, proper fitting screwdriver to remove the float inlet seat.
6. Lift the venturi out of the lower body. Some carburetors have a welch plug. This should be removed only if necessary to remove the choke plate. Some carburetors have nylon choke shaft.

REPAIR

Use new parts where necessary. Always use new gaskets. Carburetor repair kits are available. Tighten the inlet seat with the gasket securely in place, if used. Some float valves have a spring clip to connect the float valve to the float tang. Others are nylon with a stirrup which fits over the float tang. Older float valves and engines with fuel pumps have neither a spring nor a stirrup.

A viton tip float valve is used in later models of the large, two-piece Flo-Jet carburetor. The seat is pressed into the upper body and does not need replacement unless it is damaged.

Replacing the Pressed-In Float Valve Seat
▶ See Figure 236

Clamp the head of a #93029 self threading screw in a vise. Turn the carburetor body to thread the screw into the seat. Continue turning the carburetor body, drawing out the seat. Leave the seat fastened to the screw. Insert the new seat

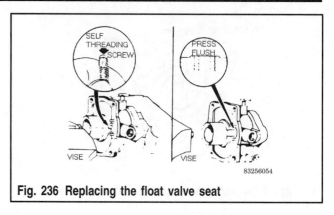

Fig. 236 Replacing the float valve seat

#230996 into the carburetor body. The seat has a starting lead.

➡**If the engine is equipped with a fuel pump, install a #231019 seat. Press the new seat flush with the body using the screw and old seat as a driver. Make sure that the seat is not pressed below the body surface or improper float-to-float valve contact will occur. Install the float valve.**

Checking the Float Level

With the body gasket in place on the upper body and the float valve and float installed, the float should be parallel to the body mounting surface. If not, bend the tang on the float until they are parallel. Do not press on the flat to adjust it.

ASSEMBLY OF THE CARBURETOR

Assemble the venturi and the venturi gasket to the lower body. Be sure that the holes in the venturi and the venturi gasket are aligned. Some models do not have a removable venturi. Install the choke parts and welch plug if previously removed. Use a sealer around the welch plug to prevent entry of dirt.

Fasten the upper and lower bodies together with the mounting screws. Screw in the nozzle with a narrow, blunt screwdriver, making sure that the nozzle tip enters the recess in the upper body. Tighten the nozzle securely. Screw in the needle valve and idle valve until they just seat. Back off the needle valve $1\frac{1}{2}$ turns. Do not tighten the packing nut. Back off the idle valve $\frac{3}{4}$ of a turn. These settings are about correct. Final adjustment will be made when the engine is running.

OVERHAULING ONE-PIECE FLO-JET CARBURETORS

▶ **See Figures 237 and 238**

The large, one-piece Flo-Jet carburetor has its high speed needle valve below the float bowl. All other repair procedures are similar to the small, one-piece Flo-Jet carburetor.

DISASSEMBLY

1. Remove the idle and needle valves.

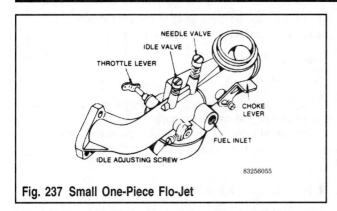

Fig. 237 Small One-Piece Flo-Jet

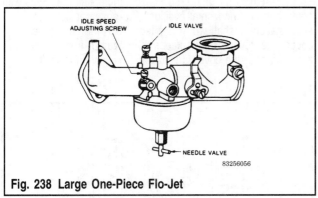

Fig. 238 Large One-Piece Flo-Jet

2. Remove the carburetor bowl screw. A pin holds the float in place.

3. Remove the pin to take off the float and float valve needle. Check the float for leakage. If it contains gasoline or is crushed, it must be replaced. Use a screwdriver to remove the carburetor nozzle. Use a wide, heavy screwdriver to remove the float valve seat, if used.

If it is necessary to remove the choke valve, venturi throttle shaft, or shaft bushings, proceed as follows.

4. Pry out the welch plug.

5. Remove the choke valve, then the shaft. The venturi will then be free to fall out after the choke valve and shaft have been removed.

6. Check the shaft for wear. (Refer to the 'Two-Piece Flo-Jet Carburetor' section for checking wear and replacing bushings).

REPAIR OF THE CARBURETOR

Use new parts where necessary. Always use new gaskets. Carburetor repair kits are available. If the venturi has been removed, install the venturi first, then the carburetor nozzle jets. The nozzle jet holds the venturi in place. Replace the choke shaft and valve. Install a new welch plug in the carburetor body. Use a sealer to prevent dirt from entering.

A viton tip float valve is used in the large, one-piece Flo-Jet carburetor. The seat is pressed in the upper carburetor body and does not need replacement unless it is damaged. Replace the seat in the same manner as for the two-piece Flo-Jet carburetor.

Checking the Float Level

Float Level Chart

Carburetor Number	Float Setting (in.)
2712-S	$^{19}/_{64}$
2713-S	$^{19}/_{64}$
2714-S	$^1/_4$
*2398-S	$^1/_4$
2336-S	$^1/_4$
2336-SA	$^1/_4$
2337-S	$^1/_4$
2337-SA	$^1/_4$
2230-S	$^{17}/_{64}$
2217-S	$^{11}/_{64}$

*When resilient seat is used, set float level at $^9/_{32} \pm {}^1/_{64}$.

With the body gasket in place on the upper body and float valve and the float installed, the float should be parallel to the body mounting surface. If not, bend the tang on the float until they are parallel. Do not press on the float.

Install the float bowl, idle valve, and needle valve. Turn in the needle valve and the idle valve until they just seat. Open the needle valve 2½ turns and the idle valve 1½ turns. On the large carburetors with the needle valve below the float bowl, open the needle valve and the idle valve 1⅛ turns.

These settings will allow the engine to start. Final adjustment should be made when the engine is running and has warmed up to operating temperature.

Governors

The purpose of a governor is to maintain, within certain limits, a desired engine speed even though the load may vary.

AIR VANE GOVERNORS

▶ See Figures 239 and 240

The governor spring tends to open the throttle. Air pressure against the air vane tends to close the throttle. The engine speed at which these two forces balance is called the governed speed. The governed speed can be varied by changing the governor spring tension.

Worn linkage or damaged governor springs should be replaced to insure proper governor operation. No adjustment is necessary.

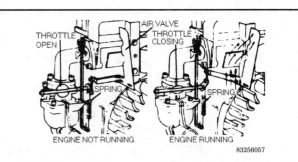

Fig. 239 Air vane governor installed on a horizontal shaft engine

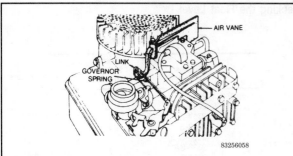

Fig. 240 Air vane governor installed on a vertical shaft engine

MECHANICAL GOVERNORS

▶ **See Figure 241**

The governor spring tends to pull the throttle open. The force of the counterweights, which are operated by centrifugal force, tends to close the throttle. The engine speed at which these two forces balance is called the governed speed. The governed speed can be varied by changing the governor spring tension.

GOVERNOR REPAIR

The procedures below describe disassembly and assembly of the various kinds of mechanical governors. Look for gears with worn or broken teeth, worn thrust washers, weight pins, cups, followers, etc. Replace parts that are worn and reassemble.

Cast Iron Models

DISASSEMBLY

1. Remove the cotter key and washer from the outer end of the governor shaft. Remove the governor crank from inside the crankcase.
2. Slide the governor gear off the shaft.

ASSEMBLY

▶ **See Figure 242**

1. Install the governor gear onto the shaft inside the crankcase. Then, insert the governor shaft assembly through the bushing from inside the crankcase.

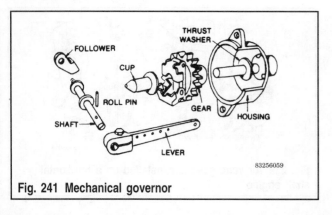

Fig. 241 Mechanical governor

2. Install the governor lever to the shaft loosely, and then adjust it as described in the Tune-Up section.

Aluminum Models

DISASSEMBLY

On horizontal shaft models: Remove the governor assembly as a unit from the crankcase cover.

On vertical shaft models: Remove the entire assembly as part of the oil slinger (see the Overhaul Section).

ASSEMBLY

▶ **See Figures 243 and 244**

1. Assemble governors on horizontal crankshaft models with crankshaft in a horizontal position. The governor rides on a short stationary shaft which is integral with the crankcase cover. The governor shaft keeps the governor from sliding off the shaft after the cover is installed. The governor shaft must

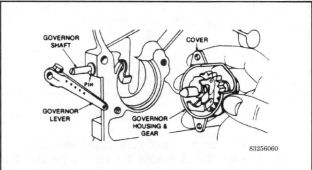

Fig. 242 Assembling the governor on a cast iron engine

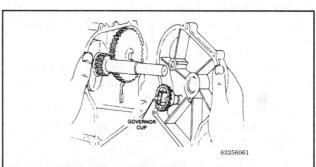

Fig. 243 Assembling the governor on aluminum, horizontal shaft engines

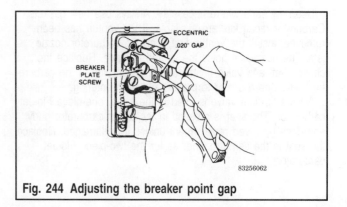

Fig. 244 Adjusting the breaker point gap

hang straight down, or it may jam the governor assembly when the crankcase cover is installed, breaking it when the engine is started . The governor shaft adjustment should be made as soon as the crankcase cover is in place so that the governor lever will be clamped in the proper position.

2. On both horizontal and vertical crankshaft models, the governor is held together through normal operating forces. For this reason, the governor link and all other external linkages must be in place and properly adjusted whenever the engine is operated.

Cylinder Head

REMOVAL & INSTALLATION

▶ **See Figure 245**

Always note the position of the different cylinder head screws so that they can be properly reinstalled. If a screw is used in the wrong position, it may be to short and not engage enough threads. If it is too long, it may bottom on a fin, either breaking the fin, or leaving the cylinder head loose.

Cylinder Head Bolt Torque Specifications

Basic Model Series	In. lbs. Torque
Aluminum Cylinder	165
Cast Iron Cylinder	190

832560C7

Remove the cylinder screws and then the cylinder head. Be sure to remove the gasket and all remaining gasket material from the cylinder head and the block. Assemble the cylinder head with a new gasket, cylinder head shield, screws, and washers in their proper places. Graphite grease should be used on aluminum cylinder head screws.

Do not use a sealer of any kind on the head gasket. Tighten the screws down evenly by hand. Use a torque wrench and tighten the head bolts in the correct sequence.

Valves

REMOVAL & INSTALLATION

▶ **See Figures 246 and 247**

1. Using a valve spring compressor, adjust the jaws so they touch the top and bottom of the valve chamber, and then place one of the jaws over the valve spring and the other underneath, between the spring and the valve chamber. This positioning of the valve spring compressor is for valves that have either pin or collar type retainers.

2. Tighten the jaws to compress the spring. Remove the collars or pin and lift out the valve.

3. Pull out the compressor and the spring.

4. To remove valves with ring type retainers, position the compressor with the upper jaw over the top of the valve chamber and the lower jaw between the spring and the retainer. Compress the spring, remove the retainer, and pull out the valve. Remove the compressor and spring.

5. If the retainers are held by a pin or collars, place the valve spring and retainer and cup into the valve spring com-

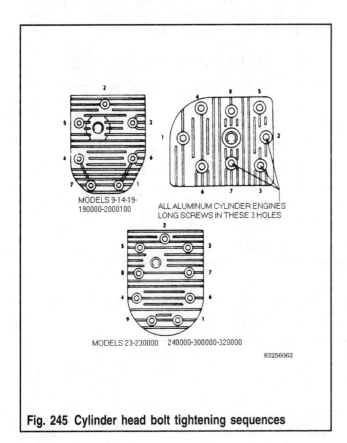

MODELS 9-14-19-
190000-2000100

ALL ALUMINUM CYLINDER ENGINES
LONG SCREWS IN THESE 3 HOLES

MODELS 23-230000   240000-300000-320000

83256063

Fig. 245 Cylinder head bolt tightening sequences

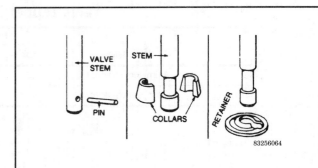

Fig. 246 The three types of valve spring retainers

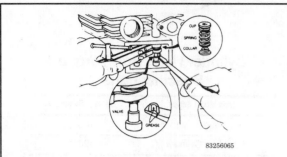

Fig. 247 Removing the valve springs using a spring compressor

pressor. Compress the spring until it is solid. Insert the compressed spring and retainer into the valve chamber. Then drop the valve into place, pushing the stem through the retainer. Hold the spring up in the chamber, hold the valve down, and insert the retainer pin with needle nose pliers or place the collars in the groove in the valve stem. Loosen the spring until the retainer fits around the pin or collars, then pull out the spring compressor. Be sure the pin or collars are in place.

6. Before installing the valves, check the thickness of the valve springs. Some engines use the same spring for the intake and exhaust side, while others use a heavier spring on the exhaust side. Compare the springs before installing them.

7. To install valves with ring type retainers, compress the retainer and spring with the compressor. The large diameter of the retainer should be toward the front of the valve chamber. Insert the compressed spring and retainer into the valve chamber. Drop the valve stem through the larger area of the retainer slot and move the compressor so as to center the small area of the valve retainer slot onto the valve stem shoulder. Release the spring tension and remove the compressor.

Valve Guides

REMOVAL & INSTALLATION

▶ **See Figures 248 and 249**

First check valve guide for wear with a plug gauge, Briggs & Stratton part #19151 or equivalent. If the flat end of the valve

guide plug gauge can be inserted into the valve guide for a distance of 5/16 in. (8mm), the guide is worn and should be rebushed in the following manner.

1. Procure a reamer #19183 and reamer guide bushing #19192, and lubricate the reamer with kerosene. Then, use reamer and reamer guide bushing to ream out the worn guide. Ream to only 1/16 in. (1.6mm) deeper than valve guide bushing #230655 . BE CAREFUL NOT TO REAM THROUGH THE GUIDE!

2. Press in valve guide bushing #230655 until top end of bushing is flush with top end of valve guide. Use a soft metal driver (brass, copper, etc.) so top end of bushing is not peened over.

➡**The bushing #230655 is finish reamed to size at the factory, so no further reaming is necessary, and a standard valve can be used.**

Valve seating should be checked after bushing the guide, and corrected if necessary by refacing the seat.

REFACING VALVES AND SEATS

Faces on valves and valve seats should be resurfaced with a valve grinder or cutter to an angle of 45°.

➡**Some engines have a 30° intake valve and seat.**

The valve and seat should then be lapped with a fine lapping compound to remove the grinding marks and ensure a good seat. The valve seat width should be 1.2-1.6mm. If the seat is wider, a narrowing stone or cutter should be used. If either the seat or valve is badly burned, it should be replaced. Replace the valve if the edge thickness (margin) is less than 0.4mm after it has been resurfaced.

CHECK AND ADJUST TAPPET CLEARANCE

Valve Tappet Clearance Chart

Model Series	Intake		Exhaust	
	Max	Min	Max	Min
Aluminum Cylinder	.007	.005	.011	.009
Cast Iron Cylinder	.009	.007	.019	.017

832560C8

Insert the valves in their respective positions in the cylinder. Turn the crankshaft until one of the valves is at its highest

position. Turn the crankshaft one revolution. Check the clearance with a feeler gauge. Repeat for the other valve. Grind off

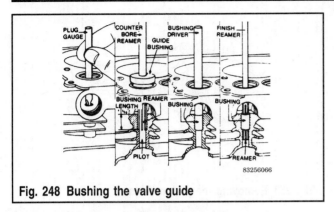

Fig. 248 Bushing the valve guide

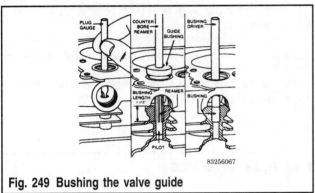

Fig. 249 Bushing the valve guide

the end of the valve stem if necessary to obtain proper clearance.

➡**Check the valve tappet clearance with the engine cold.**

Valve Seat Inserts

▶ **See Figure 250**

Cast iron cylinder engines are equipped with an exhaust valve insert which can be removed and replaced with a new insert. The intake side must be counter-bored to allow the installation of an intake valve seat insert (see below). Aluminum alloy cylinder models are equipped with inserts on both the exhaust and intake valves.

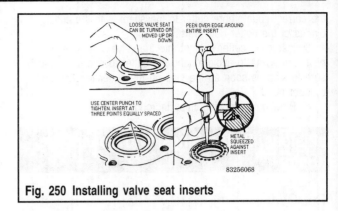

Fig. 250 Installing valve seat inserts

REMOVAL & INSTALLATION

Valve seat inserts are removed with a special puller.

➡**On aluminum alloy cylinder models, it may be necessary to grind the puller nut until the edge is 3/32 in. (0.8mm) thick in order to get the puller nut under the valve insert.**

When installing the valve seat insert, make sure that the side with the chamfered outer edge goes down into the cylinder. Install the seat insert and drive it into place with a driver. The seat should then be ground lightly and the valves and seats lapped lightly with grinding compound.

Aluminum alloy cylinder models use the old insert as a spacer between the driver and the new insert. Drive in the new insert until it bottoms. The top of the insert will be slightly below the cylinder head gasket surface. Peen around the insert using a punch and hammer.

The intake valve seat on cast iron cylinder models has to be counter-bored before installing the new valve seat insert.

COUNTERBORING THE CYLINDER FOR THE INTAKE VALVE SEAT ON CAST IRON MODELS

1. Select the proper seat insert, cutter shank, counter bore cutter, pilot and driver from the table. These numbers refer to Briggs & Stratton parts — you may get equivalent parts from other sources if available.
2. With cylinder head resting on a flat surface, valve seats up, slide the pilot into the intake valve guide. Then, assemble

Valve Seat Inserts Chart

Basic Model Series	Intake Standard	Exhaust Standard	Exhaust Stellite	Insert # Puller Assembly	Puller Nut
Aluminum Cylinder					
250000	211661	211661	210940	19138	19141
Cast Iron Cylinder					
23, 230000	21880	21880	21612	19138	19141
240000	21880	21612	21612	19138	19141
300000, 320000		21612	21612	19138	19141

Includes puller and #19182, 19141, 19140 and 19139 nuts

the correct counterbore cutter to the shank with the cutting blades of the cutter downward.

3. Insert the cutter straight into the valve seat, over the pilot. Cut so as to avoid forcing the cutter to one side, and be sure to stop as soon as the stop on the cutter touches the cylinder head.

4. Blow out all cutting chips thoroughly.

Pistons, Piston Rings, and Connecting Rods

REMOVAL

▶ **See Figures 251, 252 and 253**

To remove the piston and connecting rod from the engine, bend down the connecting rod lock. Remove the connecting rod cap. Remove any carbon or ridge at the top of the cylinder bore. This will prevent breaking the rings. Push the piston and rod out of the top of the cylinder.

Pistons used in sleeve bore, aluminum alloy engines are marked with an **L** on top of the piston. These pistons are tin plated and use an expander with the oil ring. This piston assembly is not interchangeable with the piston used in the aluminum bore engines (Kool bore).

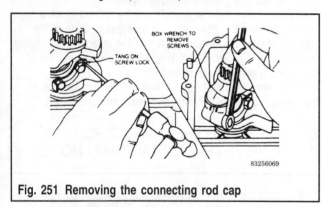

Fig. 251 Removing the connecting rod cap

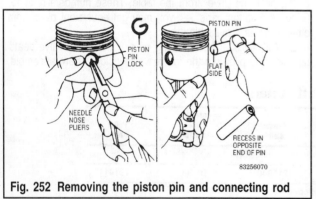

Fig. 252 Removing the piston pin and connecting rod

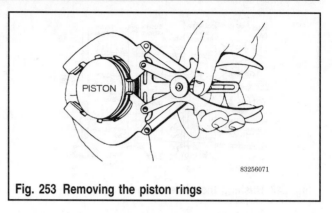

Fig. 253 Removing the piston rings

Pistons used in aluminum bore (Kool bore) engines are not marked on the top.

To remove the connecting rod from the piston, remove the piston pin lock with thin nose pliers. One end of the pin is drilled to facilitate removal of the lock.

Remove the rings one at a time, slipping them over the ring lands. Use a ring expander to remove the rings.

INSPECTION

▶ **See Figures 254 and 255**

Connecting Rod Bearing Specifications

Basic Model Series	Crank Pin Bearing	Piston Pin Bearing
Aluminum Cylinder		
251000	1.252	.802
Cast Iron Cylinder		
23, 230000	1.189	.736
240000	1.314	.674
300000, 320000	1.314	.802

83256C10

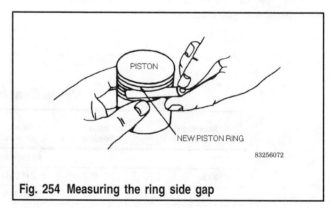

Fig. 254 Measuring the ring side gap

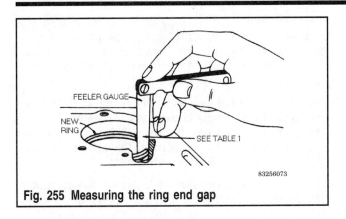

Fig. 255 Measuring the ring end gap

Piston Ring Gap Specifications

Basic Model Series	Comp. Ring	Oil Ring
Aluminum Cylinder	.035	.045
Cast Iron Cylinder	.035	.035

83256C11

Wrist Pin Specifications

Basic Model Series	Piston Pin	Pin Bore
Aluminum Cylinder		
251000	.799	.801
Cast Iron Cylinder		
23, 230000	.734	.736
240000	.671	.673
300000, 320000	.799	.801

83256C12

Check the piston ring fit. Use a feeler gauge to check the side clearance of the top ring. Make sure that you remove all carbon from the top ring groove. Use a new piston ring to check the side clearance. If the cylinder is to be resized, there is no reason to check the piston, since a new oversized piston assembly will be installed. If the side clearance is more than 0.007 in. (0.178mm), the piston is excessively worn and should be replaced.

Check the piston ring end gap by cleaning all carbon from the ends of the rings and inserting them one at a time 1 in. (25mm) down into the cylinder. Check the end gap with a feeler gauge. If the gap is larger than recommended, the ring should be replaced.

➡When checking the ring gap, do not deglaze the cylinder walls by installing piston rings in aluminum cylinder engines.

Chrome ring sets are available for all current aluminum and cast iron cylinder models. No honing or deglazing is required. The cylinder bore can be a maximum of 0.005 in. (0.127mm) oversize when using chrome rings.

If the crankpin bearing in the rod is scored, the rod must be replaced. 0.005 in. (0.127mm) oversize piston pins are available in case the connecting rod and piston are worn at the piston pin bearing. If, however, the crankpin bearing in the connecting rod is worn, the rod should be replaced. Do not attempt to file or fit the rod.

If the piston pin is worn 0.0005 in out of round or below the rejection sizes, it should be replaced.

INSTALLATION

◗ See Figures 256, 257, 258 and 259

Connecting Rod Capscrew Torque

Basic Model Series	Inch lbs Avg. Torque
Aluminum Cylinder	
251000	185
Cast Iron Cylinder	
23, 230000	190
240000, 300000, 320000	190

83256C13

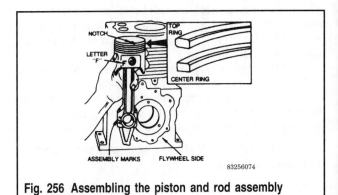

Fig. 256 Assembling the piston and rod assembly

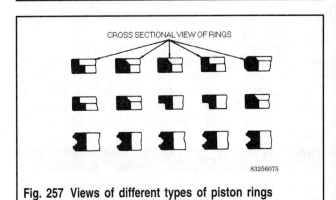

Fig. 257 Views of different types of piston rings

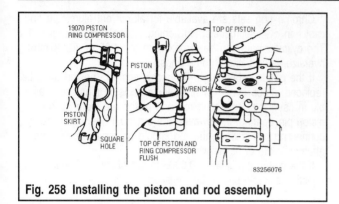

Fig. 258 Installing the piston and rod assembly

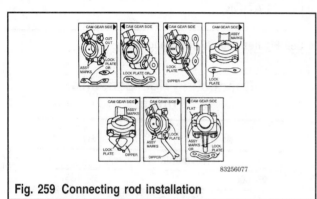

Fig. 259 Connecting rod installation

The piston pin is a push fit into both the piston and the connecting rod. On models using a solid piston pin, one end is flat and the other end is recessed. Other models use a hollow piston pin. Place a pin lock in the groove at one side of the piston. From the opposite side of the piston, insert the piston pin, flat end first for solid pins; with hollow pins, insert either end first until it stops against the pin lock. Use thin nose pliers to assemble the pin lock in the recessed end of the piston. Be sure the locks are firmly set in the groove.

Install the rings on the pistons, using a piston ring expander. Make sure that they are installed in the proper position. The scraper groove on the center compression ring should always be down toward the piston skirt. Be sure the oil return holes are clean and all carbon is removed from the grooves.

➥Install the expander under the oil ring in sleeve bore aluminum alloy engines.

Oil the rings and the piston skirt, then compress the rings with a ring compressor. On cast iron engines, install the compressor with the two projections downward; on aluminum engines, install the compressor with the two projections upward. These instructions refer to the piston in normal position — with skirt downward. Turn the piston and compressor upside down on the bench and push downward so the piston head and the edge of the compressor band are even, all the while tightening the compressor. Draw the compressor up tight to fully compress the rings, then loosen the compressor very slightly.

✳✳WARNING

Do not attempt to install the piston and ring assembly without using a ring compressor.

Place the connecting rod and piston assembly, with the rings compressed, into the cylinder bore. Push the piston and rod down into the cylinder. Oil the crankpin of the crankshaft. Pull the connecting rod against the crankpin and assemble the rod cap so the assembly marks align.

➥Some rods do not have assembly marks, as the rod and cap will fit together only in one position. Use care to ensure proper installation. On the 251000 engine, the piston has a notch on the top surface. The notch must face the flywheel side of the block when installed.

Where there are flat washers under the cap screws, remove and discard them prior to installing the rod. Assemble the cap screws and screw locks with the oil dippers (if used), and torque to the figure shown in the chart to avoid breakage or rod scoring later. Turn the crankshaft two revolutions to be sure the rod is correctly installed. If the rod strikes the camshaft, the connecting rod has been installed wrong or the cam gear is out of time. If the crankshaft operates freely, bend the cap screw locks against the screw heads. After tightening the rod screws, the rod should be able to move sideways on the crankpin of the shaft.

Crankshaft and Camshaft Gear

REMOVAL

▶ **See Figures 260 and 261**

Aluminum Cylinder Engines

To remove the crankshaft from aluminum alloy engines, remove any rust or burrs from the power take-off end of the crankshaft. Remove the crankcase cover or sump. If the sump or cover sticks, tap it lightly with a soft hammer on alternate sides near the dowel. Turn the crankshaft to align the crankshaft and camshaft timing marks, lift out the cam gear, then remove the crankshaft. On models that have ball bearings on the crankshaft, the crankshaft and the camshaft must be removed together with the timing marks properly aligned — see illustration.

Cast Iron Cylinder Models

To remove the crankshaft from cast iron models, remove the crankcase cover. Revolve the crankshaft until the crankpin is pointing upward toward the breather at the rear of the engine (approximately a 45° angle). Pull the crankshaft out from the

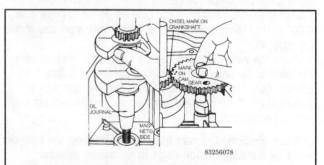

Fig. 260 Alignment of the camshaft and crankshaft timing marks, without ball bearings

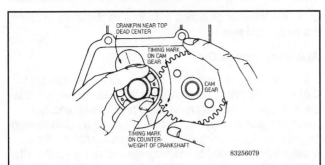

Fig. 261 Alignment of the camshaft and crankshaft timing marks, with ball bearings

drive side, twisting it slightly if necessary. On models with ball bearings on the crankshaft, both the crankcase cover and bearing support should be removed.

On cast iron models with ball bearings on the drive side, first remove the magneto. Drive into the recess at the front of the engine. Then draw the crankshaft from the magneto side of the engine. Double thrust engines have cap screws inside the crankcase which hold the bearing in place. These must be removed before the crankshaft can be removed.

To remove the camshaft from all cast iron models, except the 300400 and 320400, use a long punch to drive the camshaft out toward the magneto side. Save the plug. Do not burr or peen the end of the shaft while driving it out. Hold the camshaft while driving it out. Hold the camshaft while removing the punch, so it will not drop and become damaged.

CHECKING THE CRANKSHAFT

Crankshaft Specifications

Basic Model Series	PTO Journal	Mag. Journal	C Crankpin
Aluminum Cylinder			
251000	1.376	1.376	1.247
Cast Iron Cylinder			
23, 230000†	1.376	1.376	1.184
240000	Ball	Ball	1.309
300000, 320000	Ball	Ball	1.309

†Gear reduction P.T.O.—1.179

83256C14

Discard the crankshaft if it is worn beyond the allowable limit. Check the keyways for wear and make sure they are not spread. Remove all burrs from the keyway to prevent scratching the bearing. Check the three bearing journals, drive end, crankpin, and magneto end, for size and any wear or damage. Check the cam gear teeth for wear. They should not be worn at all. Check the threads at the magneto end for damage. Make sure that the crankshaft is straight.

➡**There are 0.020 in. (0.5mm) undersize connecting rods available for use on reground crankpin bearings.**

REMOVAL & INSTALLATION OF THE BALL BEARINGS

The ball bearings are pressed onto the crankshaft. If either the bearing or the crankshaft is to be removed, use an arbor press to remove them.

To install, heat the bearing in hot oil — 325°F (163°C) maximum. Don't let the bearing rest on the bottom of the pan in which it is heated. Place the crankshaft in a vise with the bearing side up. When the bearing is quite hot, it will slip fit onto the bearing journal. Grasp the bearing, with the shield down, and thrust it down onto the crankshaft. The bearing will tighten on the shaft while cooling. Do not quench the bearing (throw water on it to cool it).

CHECKING THE CAMSHAFT GEAR

Camshaft Specifications

Basic Model Series	Cam Gear or Shaft Journals	Cam Lobe
Aluminum Cylinder		
251000	.498	1.184
Cast Iron Cylinder		
23, 230000	.497	1.184
240000	.497	1.184
300000	#	1.184
320000	#	1.215

Magneto side—.8105, P.T.O. side—.6145

83256C15

Inspect the teeth for wear and nicks. Check the size of the camshaft and camshaft gear bearing journals. Check the size of the cam lobes. If the cam is worn beyond tolerance, discard it.

Check the automatic spark advance on models equipped with the Magna-Matic ignition system. Place the cam gear in the normal operating position with the movable weight down. Press the weight down and release it. The spring should lift the weight. If not, the spring is stretched or the weight is binding.

INSTALLATION

Aluminum Alloy Engines — Plain Bearing

In aluminum alloy engines, the tappets are inserted first, the crankshaft next, and then the cam gear. When inserting the cam gear, turn the crankshaft and the cam gear so that the timing marks on the gears align.

Aluminum Alloy Engines — Ball Bearing

On crankshafts with ball bearings, the gear teeth are not visible for alignment of the timing marks; therefore, the timing mark is on the counterweight. On ball bearing equipped engines, the tappets are installed first. The crankshaft and the cam gear must be inserted together and their timing marks aligned.

Crankshaft Cover and Crankshaft

INSTALLATION

Cast Iron Engines w/Plain Bearings

Assemble the tappets to the cylinder, then insert the cam gear. Push the camshaft into the camshaft hole in the cylinder from the flywheel side through the cam gear. With a blunt punch, press or hammer the camshaft until the end is flush with the outside of the cylinder on the power takeoff side. Place a small amount of sealer on the camshaft plug, then press or hammer it into the camshaft hole in the cylinder at the flywheel side. Install the crankshaft so the timing marks on the teeth and on the cam gear align.

Cast Iron w/Ball Bearings

Assemble the tappets, then insert the cam gear into the cylinder, pushing the cam gear forward into the recess in front of the cylinder. Insert the crankshaft into the cylinder. Turn the camshaft and crankshaft until the timing marks align, then push the cam gear back until it engages the gear on the crankshaft with the timing marks together. Insert the camshaft. Place a small amount of sealer on the camshaft plug and press or hammer it into the camshaft hole in the cylinder at the flywheel side.

CRANKSHAFT END-PLAY ADJUSTMENT

The crankshaft end-play on all models, plain and ball bearing, should be 0.002-0.008 in. (0.2mm). The method of obtaining the correct end-play varies, however, between cast iron, aluminum, plain, and ball bearing models. New gasket sets include three crankcase cover or bearing support gaskets, 0.005 in. (0.127mm), 0.009 in. (0.228mm), and 0.015 in. (0.381mm) thick.

The end-play of the crankshaft may be checked by assembling a dial indicator on the crankshaft with the pointer against the crankcase. Move the crankshaft in and out. The indicator will show the end-play. Another way to measure the end-play is to assemble a pulley to the crankshaft and measure the end-play with a feeler gauge. Place the feeler gauge between the crankshaft thrust face and the bearing support. The feeler gauge method of measuring crankshaft end-play can only be used on cast iron plain bearing engines with removable bases.

On cast iron engines, the end-play should be 0.002-0.008 in. (0.05-0.20mm) with one 0.015 in. (0.381mm) gasket in place. If the end-play is less than 0.002 in. (0.05mm), which would be the case if a new crankcase or sump cover is used, additional gaskets of 0.005 in. (0.127mm), 0.009 in. (0.228mm) or 0.0l5

in. (0.381mm) may be added in various combinations to obtain the proper end-play.

Aluminum Engines Only

If the end-play is more than 0.008 in. (0.20mm) with one 0.015 in. (0.381mm) gasket in place, a thrust washer is available to be placed on the crankshaft power take-off end, between the gear and crankcase cover or sump on plain bearing engines. On ball bearing equipped aluminum engines, the thrust washer is added to the magneto end of the crankshaft instead of the power take-off end.

➡**Aluminum engines never use less than the 0.015 in. (0.381mm) gasket.**

Cylinders

INSPECTION

▶ **See Figure 262**

Always inspect the cylinder after the engine has been disassembled. Visual inspection will show if there are any cracks, stripped bolt holes, broken fins, or if the cylinder wall is scored. Use an inside micrometer or telescoping gauge and micrometer to measure the size of the cylinder bore. Measure at right angles.

If the cylinder bore is more than 0.003 in. (0.076mm) oversize, or 0.0015 in. (0.381mm) out of round on light-weight (aluminum) cylinders, the cylinder must be resized (rebored).

➡**Do not deglaze the cylinder walls when installing piston rings in aluminum cylinder engines. Also be aware that there are chrome ring sets available for most engines. These are used to control oil pumping in bores worn to 0.005 in. (0.127mm) over standard and do not require honing or glaze breaking to seat.**

RESIZING

▶ **See Figure 263**

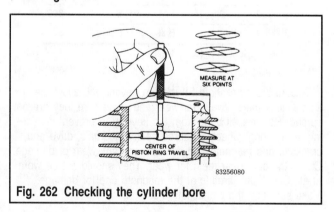

MEASURE AT SIX POINTS

CENTER OF PISTON RING TRAVEL

83256080

Fig. 262 Checking the cylinder bore

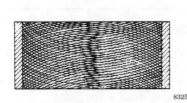

83256081

Fig. 263 Correct honing pattern. Your cylinder should look like this when you're done

Cylinder Bore Specifications

Basic Engine Model or Series	Std. Bore Size Diameter	
	Max	Min
Aluminum Cylinder		
251000	3.4375	3.4365
Cast Iron Cylinder		
23,	3.000	2.999
230000	3.000	2.999
243400	3.0625	3.0615
300000	3.4375	3.4365
320000	3.5625	3.5615

83256C16

Always resize to exactly 0.010 in. (0.254mm), 0.020 in. (0.50mm) or 0.030 in. (0.762mm) over standard size. If this is done accurately, the stock oversize rings and pistons will fit perfectly and proper clearances will be maintained. Cylinders, either cast iron or lightweight, can be quickly resized with a good hone. Use the stones and lubrication recommended by the hone manufacturer to produce the correct cylinder wall finish for the various engine models.

If a boring bar is used, a hone must be used after the boring operation to produce the proper cylinder wall finish. Honing can be done with a portable electric drill, but it is easier to use a drill press.

1. Clean the cylinder at top and bottom to remove all burrs and pieces of base and head gaskets.

2. Fasten the cylinder to a heavy iron plate. Some cylinders require shims. Use a level to align the drill press spindle with the bore.

3. Oil the surface of the drill press table liberally. Set the iron plate and the cylinder on the drill press table. Do not anchor the cylinder to the drill press table. If you are using a portable drill, set the plate and the cylinder on the floor.

4. Place the hone driveshaft in the chuck of the drill.

5. Slip the hone into the cylinder. Connect the driveshaft to the hone and set the stop on the drill press so the hone can only extend ¾-1 in. (19-25mm) from the top or bottom of the cylinder. If you are using a portable drill, cut a piece of wood to place in the cylinder as a stop for the hone.

6. Place the hone in the middle of the cylinder bore. Tighten the adjusting knob with your finger or a small screwdriver until the stones fit snugly against the cylinder wall. Do

not force the stones against the cylinder wall. The hone should operate at a speed of 300-700 rpm. Lubricate the hone as recommended by the manufacturer.

➡ **Be sure that the cylinder and the hone are centered and aligned with the driveshaft and the drill spindle.**

7. Start the drill and, as the hone spins, move it up and down at the lower end of the cylinder. The cylinder is not worn at the bottom but is round so it will act to guide the hone and straighten the cylinder bore. As the bottom of the cylinder increases in diameter, gradually increase your strokes until the hone travels the full length of the bore.

➡ **Do not extend the hone more than ¾-1 in. (19-25mm) past either end of the cylinder bore.**

8. As the cutting tension decreases, stop the hone and tighten the adjusting knob. Check the cylinder bore frequently with an accurate micrometer. Hone 0.0005 in. (0.0127mm) oversize to allow for shrinkage when the cylinder cools.

9. When the cylinder is within 0.0015 in. (0.038mm) of the desired size, change from the rough stone to a finishing stone.

The finished resized cylinder should have a cross-hatched appearance. Proper stones, lubrication, and spindle speed along with rapid movement of the hone within the cylinder during the last few strokes, will produce this finish. Cross-hatching provides proper lubrication and ring break-in.

➡ **It is EXTREMELY important that the cylinder be thoroughly cleaned after honing to eliminate ALL grit. Wash the cylinder carefully in a solvent such as kerosene. The cylinder bore should be cleaned with a brush, soap, and water.**

Bearings

INSPECTION

Plain Type

Crankshaft Bearing Specifications

Basic Engine Model or Series	PTO Bearing	Bearing Magneto
Aluminum Cylinder		
251000	1.383	1.383
Cast Iron Cylinder		
23, 230000	1.382	1.382
240000, 3000000	Ball	Ball
320000	Ball	Ball

83256C17

Bearings should be replaced if they are scored or if a plug gauge will enter. Try the gauge at several points in the bearings.

REPLACING PLAIN BEARINGS

The crankcase cover bearing support should be replaced if the bearing is worn or scored.

REPLACING THE MAGNETO BEARING

Aluminum Cylinder Engines

There are no removable bearings in these engines. The cylinder must be reamed out so a replacement bushing can be installed.

1. Place a pilot guide bushing in the sump bearing, with the flange of the guide bushing toward the inside of the sump.
2. Assemble the sump on the cylinder. Make sure that the pilot guide bushing does not fall out of place.
3. Place the guide bushing into the oil seal recess in the cylinder. This guide bushing will center the counterbore reamer even though the oil bearing surface might be badly worn.
4. Place the counterbore reamer on the pilot and insert them into the cylinder until the tip of the pilot enters the pilot guide bushing in the sump.
5. Turn the reamer clockwise with a steady, even pressure until it is completely through the bearing. Lubricate the reamer with kerosene or any other suitable solvent.

➡**Counterbore reaming may be performed without any lubrication. However, clean off shavings because aluminum material builds up on the reamer flutes causing eventual damage to the reamer and an oversize counterbore.**

6. Remove the sump and pull the reamer out without backing it through the bearing. Clean out the remaining chips. Remove the guide bushing from the oil seal recess.
7. Hold the new bushing against the outer end of the reamed out bearing, with the notch in the bushing aligned with the notch in the cylinder. Note the position of the split in the bushing. At a point in the outer edge of the reamed out bearing opposite to the split in the bushing, make a notch in the cylinder hub at a 45° angle to the bearing surface. Use a chisel or a screwdriver and hammer.
8. Press in the new bushing, being careful to align the oil notches with the driver and the support until the outer end of the bushing is flush with the end of the reamed cylinder hub.
9. With a blunt chisel or screwdriver, drive a portion of the bushing into the notch previously made in the cylinder. This is called staking and is done to prevent the bushing from turning.
10. Reassemble the sump to the cylinder with the pilot guide bushing in the sump bearing.
11. Place a finishing reamer on the pilot and insert the pilot into the cylinder bearing until the tip of the pilot enters the pilot guide bushings in the sump bearing.
12. Lubricate the reamer with kerosene, fuel oil, or other suitable solvent, then ream the bushing, turning the reamer clockwise with a steady even pressure until the reamer is completely through the bearing. Improper lubricants will produce a rough bearing surface.
13. Remove the sump, reamer, and the pilot guide bushing. Clean out all reaming chips.

REPLACING THE P.T.O. BEARING

Aluminum Cylinder Engines

The sump or crankcase bearing is repaired the same way as the magneto end bearing. Make sure to complete repair of one bearing before starting to repair the other. Press in new oil seals when bearing repair is completed.

REPLACING OIL SEALS

1. Assemble the seal with the sharp edge of leather or rubber toward the inside of the engine.
2. Lubricate the inside diameter of the seal with Lubriplate® or equivalent.
3. Press all seals so they are flush with the hub.

Extended Oil Filler Tubes and Dipsticks

When installing the extended oil fill and dipstick assembly, the tube must be installed so the O-ring seal is firmly compressed. To do so, push the tube downward toward the sump, then tighten the blower housing screw, which is used to secure the tube and bracket. When the dipstick assembly is fully depressed, it seals the upper end of the tube.

A leak at the seal between the tube and the sump, or at the seal at the upper end of the dipstick can result in a loss of crankcase vacuum, and a discharge of smoke through the exhaust system.

Breathers

▶ **See Figure 264**

The function of the breather is to maintain a vacuum in the crankcase. The breather has a fiber disc valve which limits the direction of air flow caused by the piston moving back and forth in the cylinder. Air can flow out of the crankcase, but the one-way valve blocks the return flow, thus maintaining a vacuum in the crankcase. A partial vacuum must be maintained in the crankcase to prevent oil from being forced out of the engine at the piston rings, oil seals, breaker plunger, and gaskets.

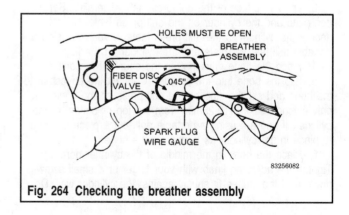

Fig. 264 Checking the breather assembly

INSPECTION OF THE BREATHER

If the fiber disc valve is stuck or binding, the breather cannot function properly and must be replaced. A 0.045 in. (1.14mm) wire gauge should not enter the space between the fiber disc valve and the body. Use a spark plug wire gauge to check the valve. The fiber disc valve is held in place by an internal bracket which will be distorted if pressure is applied to the fiber disc valve. Therefore, do not apply force when checking the valve with the wire gauge.

If the breather is removed for inspection or valve repair, a new gasket should be used when replacing the breather. Tighten the screws securely to prevent oil leakage.

Most breathers are now vented through the air cleaner, to prevent dirt from entering the crankcase. Check to be sure that the venting elbows or the tube are not damaged and that they are properly sealed.

Oil Dippers and Slingers

▶ **See Figure 265**

Oil dippers reach into the oil reservoir in the base of the engine and splash oil onto the internal engine parts. The oil dipper is installed on the connecting rod and has no pump or moving parts.

Oil slingers are driven by the cam gear. Old style slingers using a die cast bracket assembly have a steel bushing between the slinger and the bracket. Replace the bracket on which the oil slinger rides if it is worn to a diameter of 0.490 in. (12.4mm) or less. Replace the steel bushing if it is worn. Newer style oil slingers have a stamped steel .

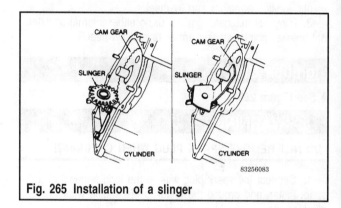

Fig. 265 Installation of a slinger

1985-92 13-20 HP Models

Ignition System

▶ See Figures 266, 267 and 268

There are three basic types of ignition systems:

1. Magnetron ignition, a self-contained transistor module, ignition armature and flywheel.

2. Magnevac ignition, using magnetically operated sealed points, ignition armature and flywheel.

3. Flywheel magneto ignition, using either internal or external breaker points, ignition armature and flywheel.

Ignition Check

▶ See Figure 269

✳✳WARNING

DO NOT REMOVE SPARK PLUG WHEN CHECKING!

1. Connect the spark plug wire to the long terminal of a spark tester, and ground the tester to the engine with an alligator clip.

2. Operate starter and observe spark gap in tester. If spark jumps tester gap, you can assume ignition is good.

➡**Flywheel must rotate at 350 rpm minimum on engines equipped with Magnetron ignition.**

If engine runs but misses during operation, a quick check can determine if the miss is ignition or not.

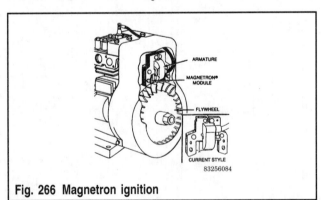

Fig. 266 Magnetron ignition

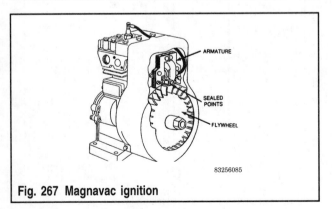

Fig. 267 Magnavac ignition

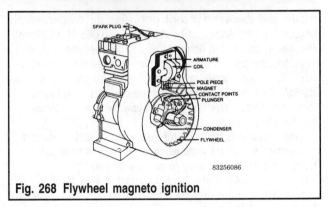

Fig. 268 Flywheel magneto ignition

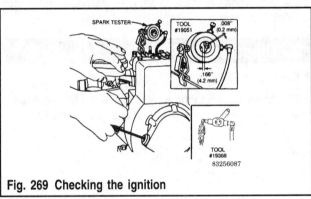

Fig. 269 Checking the ignition

CHECK FOR SPARK MISS

▶ See Figure 270

Place spark tester in series with engine's spark plug and spark plug wire. An ignition miss will be readily apparent when the engine is started and run.

Spark Plugs

➡**In some areas, local laws requires the use of a resistor spark plug to suppress ignition signals. If an engine was originally equipped with a resistor spark plug, be sure to use the same type of spark plug for replacement.**

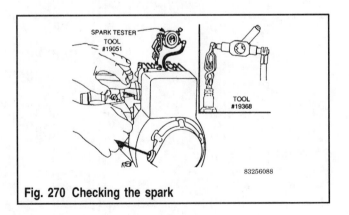

Fig. 270 Checking the spark

SPARK PLUG SERVICE

♦ **See Figure 271**

Gap spark plug to 0.030 in. gap. Replace spark plug if electrodes are burned away or porcelain is cracked. DO NOT USE ABRASIVE CLEANING MACHINES.

COIL AND CONDENSER TESTING

All Models

Use an approved tester to test coils and condensers. Specificat0ons are supplied by the tester manufacturer.

Flywheel

REMOVAL

All Aluminum Series except 280000
♦ **See Figures 272, 273 and 274**

1. Use a flywheel holder to hold flywheel from turning. Use Starter Clutch Wrench, Tool #19244, to remove rewind starter clutch.

➡**DO NOT use fins on magnet insert to prevent flywheel from turning.**

2. Use flywheel nut to protect crankshaft threads and for puller to bear against. Thread flywheel nut onto crankshaft until

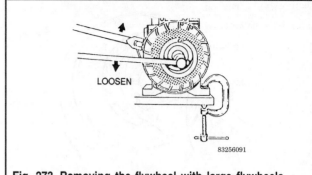

Fig. 273 Removing the flywheel with large flywheels

top of nut is flush with crankshaft threads or slightly above end threads.

➡**Care is required not to damage flywheel fins, magnets or ring gear.**

Model Series 280000
♦ **See Figures 275, 276 and 277**

1. Remove blower and rotating screen, when so equipped.
2. Place a flywheel holder on fan retainer with lugs of flywheel holder engaging the slots of the fan retainer.
3. Loosen flywheel nut or rewind starter clutch with socket and wrench.
4. Remove two screws and fan retainer. Use flywheel nut to protect crankshaft threads and for puller to bear on.

➡**If puller screws on flywheel puller are too short, use two head bolts from Model Series 280000, Part #93723.**

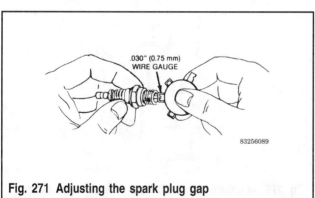

Fig. 271 Adjusting the spark plug gap

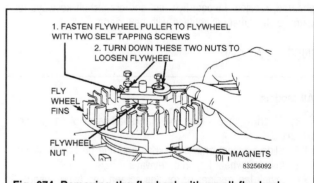

Fig. 274 Removing the flywheel with small flywheels

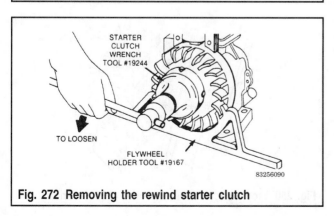

Fig. 272 Removing the rewind starter clutch

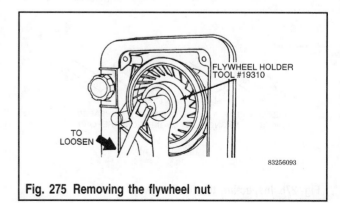

Fig. 275 Removing the flywheel nut

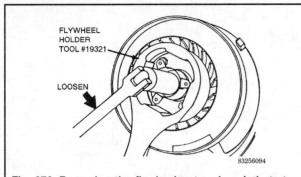

Fig. 276 Removing the flywheel nut and rewind starter

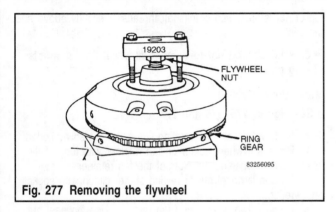

Fig. 277 Removing the flywheel

INSPECTION

▶ **See Figure 278**

Inspect flywheel key for partial or complete shearing. If sheared, replace. Inspect flywheel and crankshaft keyways for damage. If damaged, replace with new parts.

INSTALLATION

All Models

1. Clean flywheel taper and crankshaft taper of all grease, oil and dirt.
2. Slide flywheel onto crankshaft and line up both keyways.
3. Insert flywheel key into keyways.
4. Install fan retainer or rotating screen cup (when used), then flat or Belleville washer, and flywheel nut or rewind starter

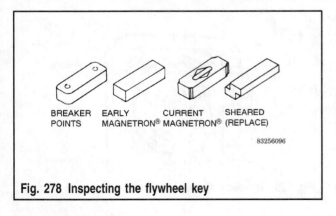

Fig. 278 Inspecting the flywheel key

clutch. When installing Bellville washer have hollow side towards flywheel.

➡ **Use only the ORIGINAL FLYWHEEL KEYS SUPPLIED WITH THE ENGINE. DO NOT use a steel key under any circumstances!**

Magnetron Ignition

Magnetron has been produced in two versions, composite (Type I, Type II) and replaceable module.

➡ **Magnetron armatures used on Model Series 280000 do not have visible trigger pole and do not have a Mylar spacer.**

MAGNETRON ARMATURE TESTING

▶ **See Figure 279**

Use an approved tester to test armature. Specifications are supplied by the tester manufacturer.

MAGNETRON ARMATURE REMOVAL

▶ **See Figures 280, 281, 282, 283 and 284**

➡ **Removal of the flywheel is not required to remove Magnetron armatures except to inspect flywheel key and keyways on crankshaft and flywheel.**

1. Remove armature mounting screws and lift off armature.

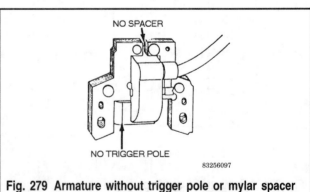

Fig. 279 Armature without trigger pole or mylar spacer

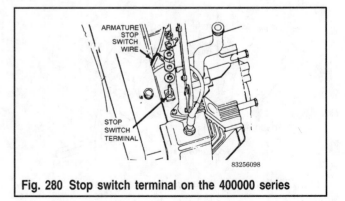

Fig. 280 Stop switch terminal on the 400000 series

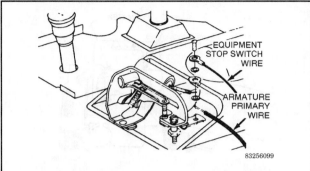

Fig. 281 Removing the primary wire on the 400000 series

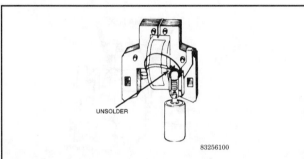

Fig. 282 Disconnecting the stop switch wire on Magnetron ignition

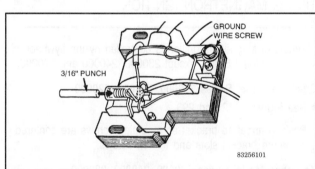

Fig. 283 Removing the armature stop wire on 400000 series

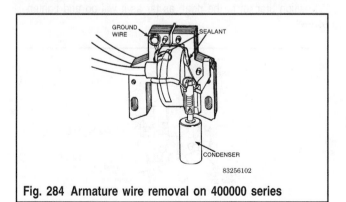

Fig. 284 Armature wire removal on 400000 series

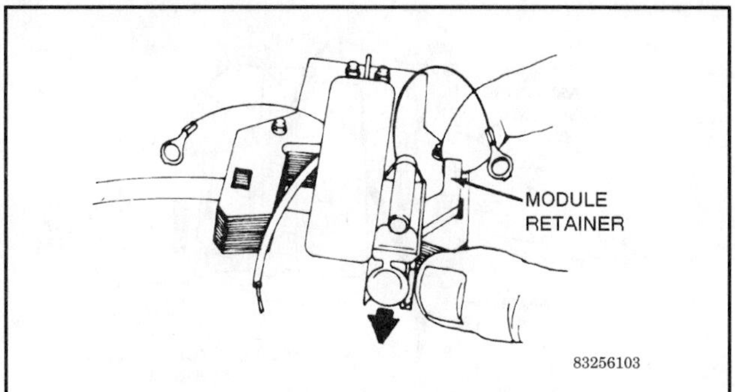

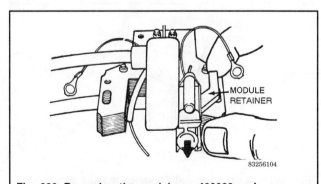

Fig. 285 Removing the Magnetron module

2. Disconnect stop switch wire at spade terminal on composite armature. On armatures with replaceable Magnetron Modules, use breaker point condenser from Part #294628 point set or a ⁹⁄₁₆ in. diameter pin punch to release wires from module.

3. Unsolder stop switch wire from module wire and armature primary wire.

MODULE REMOVAL

▶ **See Figures 285 and 286**

1. Remove sealant and/or tape holding armature wires to armature.

2. Unsolder and separate remaining wires.

➡**On some armatures, the module ground wire is soldered to the armature ground wire. Unsolder and disconnect.**

3. Move all wires so module will clear armature and laminations.

4. Pull module retainer away from laminations and push module off laminations.

MODULE INSTALLATION

▶ **See Figure 287**

The armature side identified by large rivet heads. The module is installed with the retainer on the back side, small rivet ends.

Fig. 286 Removing the module on 400000 series

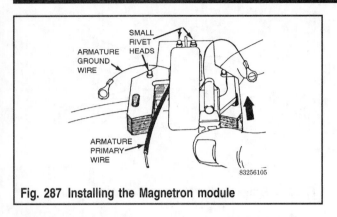

Fig. 287 Installing the Magnetron module

STOP SWITCH AND ARMATURE PRIMARY WIRE INSTALLATION

▶ **See Figures 288, 289, 290 and 291**

1. Be sure all insulating is removed from wires to ensure good contact.
2. Use a ³⁄₁₆ in. diameter pin punch to compress wire retainer spring and insert stop switch and armature primary wire under hook of wire retainer.
3. Twist wires together and solder twisted section with 60/40 rosin core solder. Take care not to damage module case.
4. Install wires in module retainer.
5. Cement wires to armature with Permatex® No. 2 or similar sealer to prevent wires from vibrating and breaking.

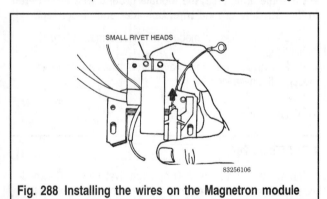

Fig. 288 Installing the wires on the Magnetron module

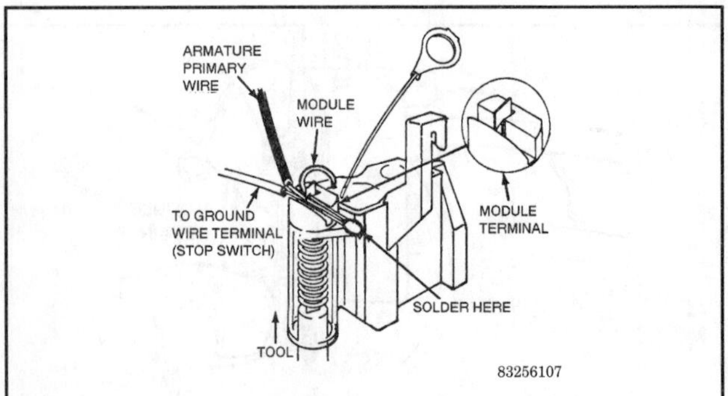

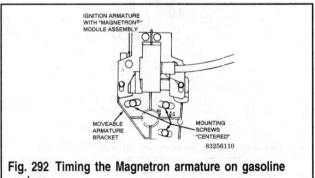

Fig. 289 Installing the 400000 series module wire

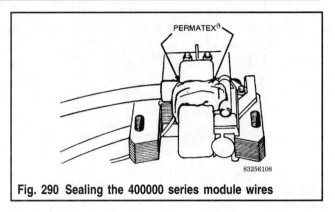

Fig. 290 Sealing the 400000 series module wires

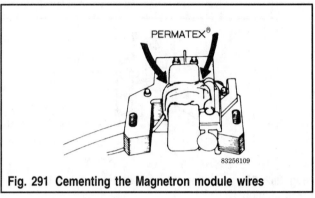

Fig. 291 Cementing the Magnetron module wires

TIMING MAGNETRON IGNITION

Timing of Magnetron ignition is controlled by the flywheel key on all Model Series except 230000, 240000 and 320000.

Gasoline Model Series 230000, 240000, 320000
▶ **See Figures 292 and 293**

Position armature bracket so mounting screws are centered in armature bracket slots and tighten screws.

Kerosene Model Series 230000, 240000, 320000
▶ **See Figure 294**

Position bracket to the right, as far as it will go and tighten screws.

Fig. 292 Timing the Magnetron armature on gasoline engines

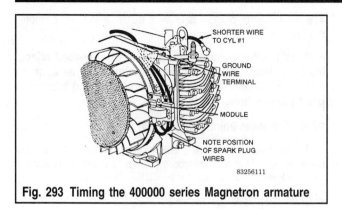

Fig. 293 Timing the 400000 series Magnetron armature

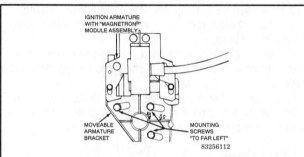

Fig. 294 Timing the Magnetron armature on kerosene engines

Magnevac Ignition

ARMATURE TESTING

Use an approved tester to test armature. Specifications are supplied by the tester manufacturer.

ARMATURE REMOVAL

▶ **See Figures 295 and 296**

➡**Removal of the flywheel is not required to remove Magnevac armatures except to inspect flywheel key and keyways on crankshaft and flywheel.**

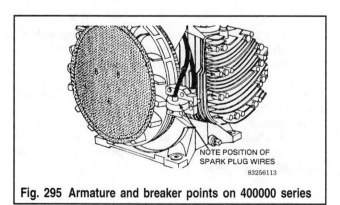

Fig. 295 Armature and breaker points on 400000 series

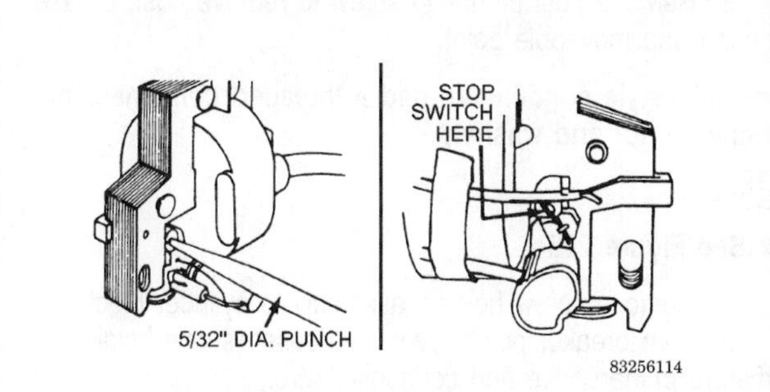

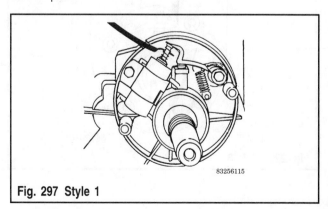

Fig. 296 Removing the stop switch wire

1. Remove armature mounting screws and lift off armature.
2. Disconnect stop switch wire by using a $^5/_{32}$ in. diameter pin punch to depress spring and retainer.

Flywheel Magneto Breaker Points Internal Type

BREAKER COVER REMOVAL

Care should be taken when removing breaker cover, to avoid damaging cover. If cover is bent or damage, it should be replaced to ensure a proper dust seal.

➡**Breaker point gap is 0.20 in. on all models. Breaker points should be checked for contact and for signs of burning or pitting. Points gapped too wide will advance spark timing and may cause kickback when starting. Points gapped too close will retard spark timing and decrease engine power.**

REMOVAL

Style 1
▶ **See Figure 297**

1. Remove condenser clamp screw and clamp.
2. Lift condenser and wires away from cylinder and compress condenser spring to remove stop switch wire and armature primary wire. The tip of the condenser is one-half of the breaker points.

Fig. 297 Style 1

3. Remove post mounting screw to remove post, breaker spring and moveable point.

➡️**Early style condensers had a threaded condenser post. Remove nut and washer.**

Style II

▶ **See Figure 298**

1. Remove screw holding assembly to cylinder block.
2. Turn breaker points over and loosen screw holding armature primary wire and condenser wire.
3. Remove wires.

Style III

▶ **See Figure 299**

1. Loosen screw on breaker point assembly and remove armature primary wire and condenser wire.
2. Remove screw holding breaker point to armature plate and remove points.

CHECK BREAKER POINT PLUNGER HOLE

A worn breaker point plunger hole can cause oil to leak past the plunger and contaminate the breaker points causing the points to burn. To check for plunger hole wear, remove breaker points and plunger. If flat end of Plug Gauge, Tool

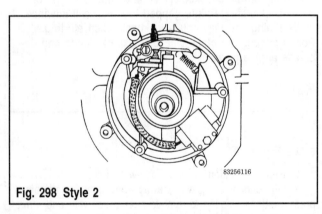

Fig. 298 Style 2

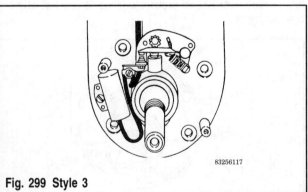

Fig. 299 Style 3

#19055 will enter plunger hole for a distance of ¼ in. or more, the hole should be rebushed.

➡️**When breaker point plunger hole is worn beyond reject, installing Magnetron on two leg armatures can be done on aluminum cylinder engines instead of rebushing breaker point plunger hole.**

Rebush Breaker Point Plunger Hole

1. Remove breaker point, armature, crankshaft, and breaker point plunger. Use a counterbore reamer to hand ream worn plunger hole.

➡️**Crankshaft must be removed**

2. Drive service bushing, Part No. 23513, with Bushing Driver, Tool #19057, until upper end of bushing is flush with the top of the boss.
3. Use a finish reamer to hand ream the new bushing. Keep reamer in alignment with bushing and plunger hole. Remove all reaming chips and dirt.
4. Check breaker point plunger if worn to 0.870 in. or less. Insert plunger with groove towards breaker points or oil will enter breaker point box.

INSTALLATION

Style I

1. Install new breaker point plunger.
2. Install post into recess of cylinder with groove of post in notch of recess. Note position of braided wire.
3. Tighten mounting screw securely.
4. Hook open loop of breaker sprig into two holes of breaker arm, and then hook closed loop of spring over spring post and into groove of post.
5. Push flat of breaker arm toward groove in mounting post until flat engages groove.
6. Compress spring on condenser and slip armature primary wire (and stop switch wire, if used) into hole of condenser post.
7. Release spring to clamp wire(s).
8. Lay condenser into cylinder recess and install clamp and screw securely.

➡️**On early style threaded post condensers install wire(s), eyelet(s), washer and nut.**

Style II

1. Install armature primary wire in slot of insulation with end of wire under clamp and wire from condenser under clamp under breaker point set. Note position of wires.
2. Place screw through eyelet of stop switch wire and install screw and eyelet on breaker point set terminal. Note position of eyelet.
3. Tighten screw while holding wires in correct position.
4. Install breaker point plunger hole.
5. Place breaker set on cylinder with pin in hole of breaker set. Tighten screw finger tight.

Style III

1. Place armature wire and condenser wire under wire terminal. Note position of wires.

2. Install breaker point set on engine with cast boss in plate entering hole in point set.

ADJUSTING BREAKER POINT GAP

Turn crankshaft until breaker points open to their widest gap.

Style I

With a screwdriver, move condenser back and forth until breaker points are gapped 0.020 in. wide.

Style II

Mounting screw should be finger tight. With a screwdriver in the adjusting slot, move point bracket until breaker points are gapped 0.020 in. wide. Tighten mounting screw and recheck gap.

Style III

With mounting screw finger tight, move point bracket until breaker points are gapped 0.020 in. wide. Tighten mounting screw and recheck point gap.

➡ **Always clean breaker points after adjustment. Open breaker points and insert a piece of lint-free paper between points. Rotate paper using breaker points as a pivot. Open breaker points to remove paper so it will not tear or leave dirt on breaker points. Continue to clean breaker points until paper comes out clean.**

Breaker point Cover

The breaker point cover protects the breaker points from dirt and moisture. The opening for the armature primary wire (and stop switch wire, when used) should be sealed with Permatex® or similar sealant to prevent dirt and moisture from entering breaker box. Distorted covers will not seal around the outer edge and should be replaced.

➡ **Engines used in winter applications use vented breaker points.**

Magneto Ignition Breaker Points External Type

REMOVAL

Model Series 233000, 243000, 300000, 320000

▶ **See Figures 300, 301, 302 and 303**

1. Remove breaker cover.
2. For ease of assembly and point adjustment, if crankshaft was not removed, turn crankshaft until breaker points are at their widest gap.
3. Remove condenser, upper, and lower mounting screws.

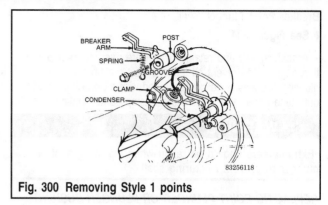

Fig. 300 Removing Style 1 points

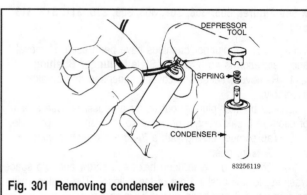

Fig. 301 Removing condenser wires

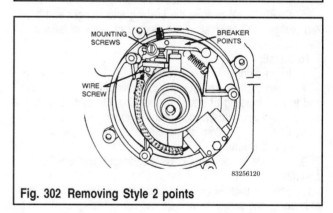

Fig. 302 Removing Style 2 points

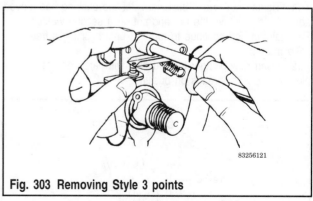

Fig. 303 Removing Style 3 points

4. Loosen lock nut and turn point adjustment screw counterclockwise to remove breaker points.

Breaker Point Plunger Seal

▶ **See Figure 304**

A seal, eyelet, and retainer wire were used on later production engines to prevent oil leakage past the breaker point plunger. If points were contaminated with oil on engines without these parts, add these parts to stop contamination.

✳✳WARNING

Extreme care should be taken when installing seal on plunger to prevent fracturing seal!

REPLACING POINT PLUNGER OR PLUNGER BUSHING

▶ **See Figures 305, 306, 307, 308, 309, 310, 311, 312, 313 and 314**

Two styles of plunger bushings have been used. Removal and installation is as follows: **Style I Plunger Bushing**

1. Remove breaker box cover, condenser, and breaker assembly.

2. Pull breaker plunger out as far as possible. Use a pair of pliers to break plunger off as close to bushing as possible.

3. Tap plunger busing with a ¼-20 tap or self-tapping screw, ½-⅝ in. deep.

4. Use a ¼-20 · ½ in. long hex head screw and two spacer washers, to pull bushing out of the cylinder. The bushing will be free when it has been pulled 5/16 in.

5. CAREFULLY remove the bushing and plunger and broken plunger. DO NOT allow the plunger or chips to fall into the crankcase.

To install:

6. Insert plunger in new bushing.

7. Insert plunger and bushing into cylinder. Use a hammer and the old bushing to drive new bushing into cylinder until bushing is flush with the face of the cylinder.

8. Check for freedom of movement of the plunger.

Style II Bushing

9. Remove breaker box cover, condenser, and breaker assembly.

10. Place a thick ⅜ in. inside diameter washer over the end of bushing and screw on a ⅜-24 nut.

11. Tighten nut to pull bushing. After the bushing has moved about ⅛ in., remove the nut and put on a second washer. Reinstall nut and continue to turn nut until bushing is free.

To install:

12. Insert new plunger into bushing with large end of plunger opposite threads on bushing.

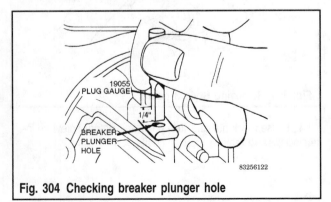

Fig. 304 Checking breaker plunger hole

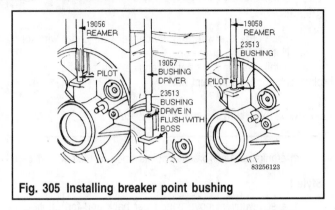

Fig. 305 Installing breaker point bushing

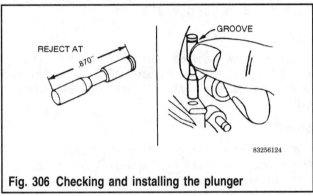

Fig. 306 Checking and installing the plunger

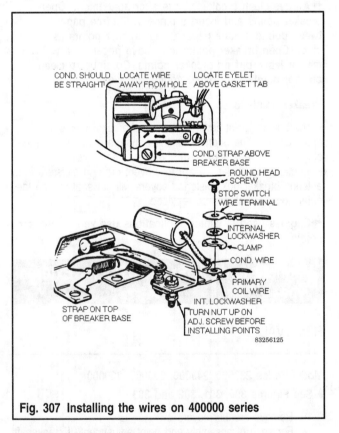

Fig. 307 Installing the wires on 400000 series

13. Screw ⅜-24 nut onto threaded end of bushing to protect threads.

14. Insert bushing and plunger into cylinder.

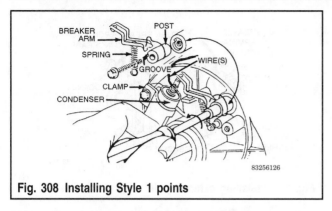

Fig. 308 Installing Style 1 points

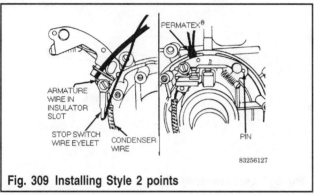

Fig. 309 Installing Style 2 points

15. Use a piece of metal tubing or Part #295840 piston pin to drive bushing into cylinder. Use a piece of metal tubing or Part #295840 piston pin to drive bushing into cylinder.

16. Drive bushing until square shoulder of bushing is flush with the face of cylinder. Check to be sure that the plunger moves freely.

CONDENSER AND BREAKER POINTS ASSEMBLY

1. Install armature primary wire and condenser lead under terminal clamp and tighten screw. Note position of armature primary and condenser lead.

2. Apply sealer, such as Permatex® or similar sealant, to adjusting screw and two mounting screws. Sealant prevents engine oil from leaking into breaker cover and onto points.

3. On new breaker points, turn lock nut back until it contacts ferrule on breaker point bracket. On breaker points that are being reused, turn lock nut back until it contacts ferrule on

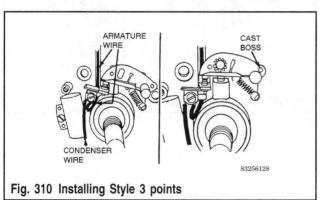

Fig. 310 Installing Style 3 points

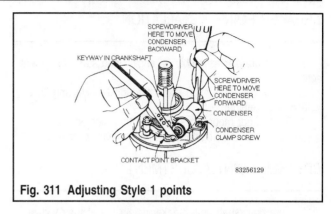

Fig. 311 Adjusting Style 1 points

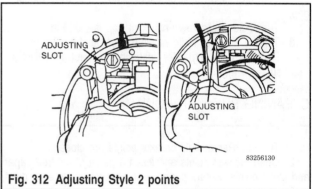

Fig. 312 Adjusting Style 2 points

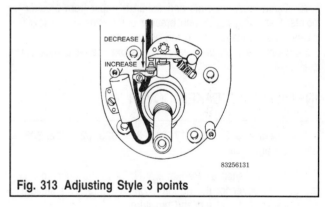

Fig. 313 Adjusting Style 3 points

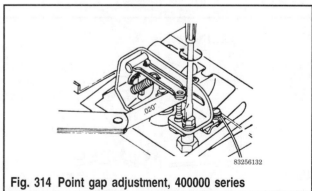

Fig. 314 Point gap adjustment, 400000 series

breaker point bracket. While holding adjusting screw, tighten nut against ferrule. This secures the adjustment screw to the breaker point bracket.

BREAKER POINT INSTALLATION

1. Place breaker assembly on engine and start adjustment screw.
2. Install lower mounting screw through bracket and lower hole of seal retainer.
3. Start upper mounting screw and then tighten lower mounting screw. Now tighten the upper mounting screw.

BREAKER POINT ADJUSTMENT

1. Turn crankshaft until breaker points are at their widest gap.
2. Turn adjusting screw until point gap is 0.020 in.
3. Tighten lock nut while holding adjustment screw.
4. Recheck point gap after tightening lock nut. Readjust as required.

CLEANING BREAKER POINTS

1. Turn crankshaft until breaker points are closed.
2. Open breaker points and insert a piece of lint-free paper and close points. Rotate paper using breaker points as a pivot point.
3. Open breaker points and withdraw paper from breaker points. Removing paper with breaker points closed can tear paper and will leave dirt on the breaker points.
4. Continue to clean breaker points until paper comes out clean.

INSTALLING BREAKER COVER

▶ **See Figures 315, 316, 317, 318, 319, 320, 321, 322, 323, 324, 325 and 326**

Apply sealer such as Permatex® No. 2 at the opening on the breaker cover for the armature primary wire. This sealant is to prevent entry of dirt and moisture.

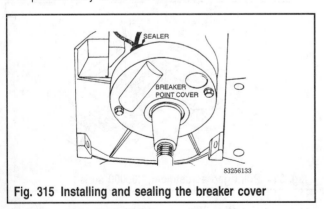

Fig. 315 Installing and sealing the breaker cover

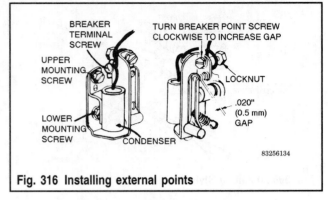

Fig. 316 Installing external points

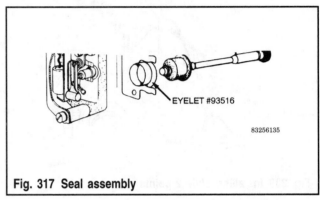

Fig. 317 Seal assembly

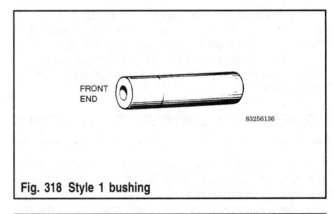

Fig. 318 Style 1 bushing

Fig. 319 Removing plunger and bushing

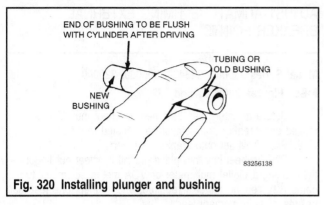

Fig. 320 Installing plunger and bushing

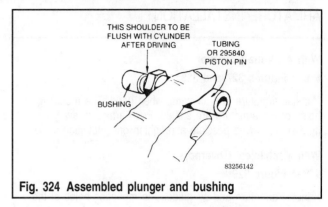

Fig. 324 Assembled plunger and bushing

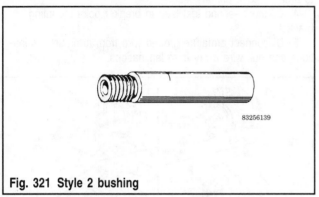

Fig. 321 Style 2 bushing

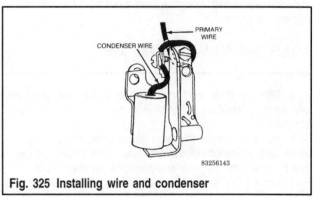

Fig. 325 Installing wire and condenser

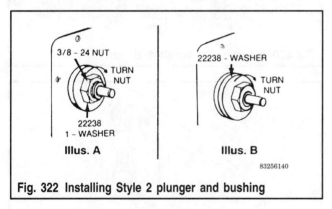

Fig. 322 Installing Style 2 plunger and bushing

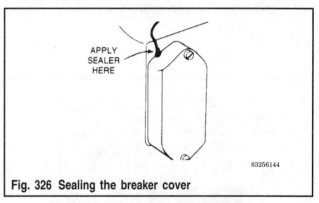

Fig. 326 Sealing the breaker cover

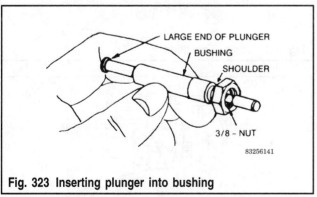

Fig. 323 Inserting plunger into bushing

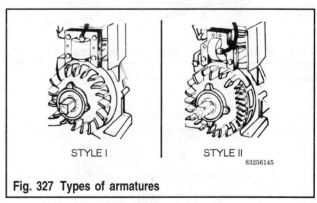

Fig. 327 Types of armatures

ARMATURE INSTALLATION

With Air Vane Governor
▶ **See Figures 327 and 328**

Install armature and air vane, when used. The mounting holes of the armature are slotted. Push armature away from flywheel as far as possible and tighten one mounting screw.

With Mechanical Governors
▶ **See Figure 329**

Install armature and air guide. The mounting holes of the armature are slotted. Push armature away from flywheel and tighten one screw.

ADJUSTING ARMATURE AIR GAP

1. With armature away from flywheel as far as possible and one screw tightened, turn flywheel so magnets are away from armature.
2. Place the proper thickness gauge between rim of flywheel and laminations of the armature. While holding gauge, turn flywheel until magnets are directly under laminations.
3. Loosen the one screw holding armature and let magnets pull armature down against flywheel. Tighten both mounting screws. Rotate flywheel until gauge is free.

ADJUST ARMATURE TIMING EXTERNAL BREAKER POINTS

Model Series 230000, 243400, 300000, 320000
▶ **See Figures 330, 331 and 332**

1. Before armature can be timed, flywheel must be removed and breaker points must be adjusted to 0.020 in.
2. Slide flywheel onto crankshaft taper.
3. Slip flywheel key into place. Install flywheel nut finger tight. Using a digital multimeter or VOA meter, set meter to ohms (Ω), zeroing meter if required, and connect one test lead to breaker point primary lead.
4. Connect second test lead to breaker point mounting bracket.
5. Disconnect armature ground wire from armature laminations and pull wire away from laminations.

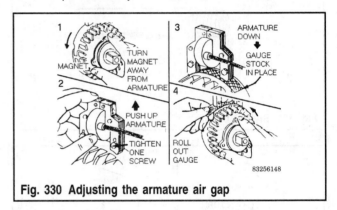

Fig. 330 Adjusting the armature air gap

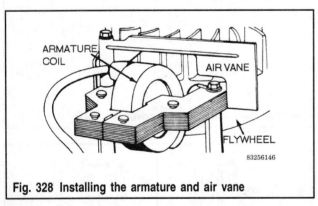

Fig. 328 Installing the armature and air vane

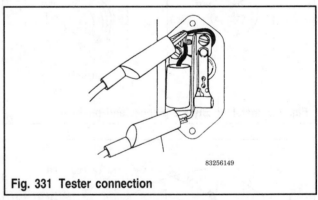

Fig. 331 Tester connection

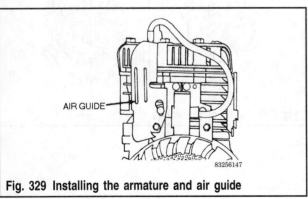

Fig. 329 Installing the armature and air guide

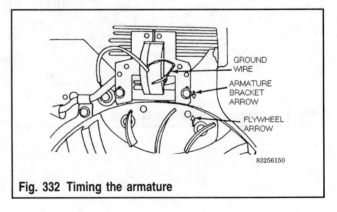

Fig. 332 Timing the armature

6. Turn flywheel clockwise until meter shows points are closed (low ohms reading).

7. Turn flywheel slowly clockwise until points open (high ohms reading). Arrow on flywheel should be in line with arrow on armature bracket when points just open.

8. If arrows do not line up, remove flywheel without moving crankshaft.

9. Loosen screws holding armature bracket until bracket can be moved with a slight drag.

10. Slip flywheel back on crankshaft without crankshaft. Insert flywheel key.

11. Install nut finger tight.

12. Move armature bracket assembly until arrows line up.

13. Remove flywheel and tighten armature bracket screws.

14. Tighten flywheel and adjust armature air gap.

Stop Switches

▶ See Figure 333

Stationary, rotary, toggle, and key stop switches are used to meet various equipment needs.

STATIONARY STOP SWITCH

▶ See Figures 334 and 335

Stationary stop switches are located on fuel tank brackets, governor control brackets, cylinder head brackets, System 2 and System 4 band brake control brackets.

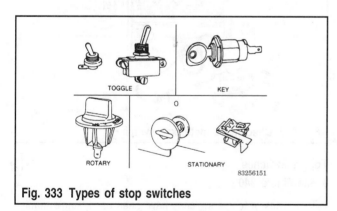

Fig. 333 Types of stop switches

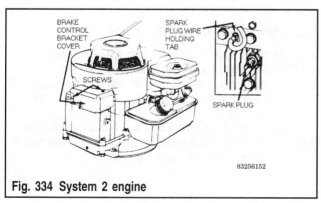

Fig. 334 System 2 engine

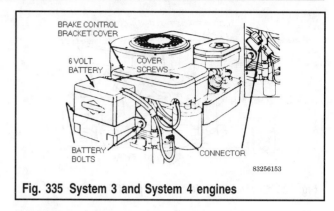

Fig. 335 System 3 and System 4 engines

Mechanical Check

▶ See Figures 336, 337 and 338

Move control lever away from stop switch by moving control lever at engine, in direction shown. Then release control completely. Control lever at engine must contact stop switch at tang shown.

Electrical Check

Push down on wire retainer and remove stop switch wire. On System 2 and System 4 the band brake control cover must be removed.

➡**On System 3 & System 4 engines with battery mounted on the engine, the battery must be removed from the battery holder before the cover can be removed. Disconnect the battery to prevent accidental starter operation.**

1. Using a digital multimeter or VOA meter, set meter to ohms (Ω) zeroing meter if required, and connect test leads to engine ground and other test lead to wire retainer.

2. Move control lever to run position. On System 2 and System 4 engines operate safety control (operator presence control) to move control lever away from stop switch.

3. With control lever in run position, VOA meter should show no continuity (high ohms reading).

4. Move control lever to stop position or release safety control (operator presence control) to move control lever. Meter should show continuity (low ohms reading). If switch shows continuity in both run and stop positions or no continuity in both positions, replace stop switch.

Rotary Stop Switch

CHECKING

▶ See Figure 339

1. Remove blower housing from engine and disconnect stop switch wire from switch.

2. Using a digital multimeter or VOA meter, set meter to ohms (Ω) zeroing meter if required, and connect test leads to blower housing and to stop switch terminal. With switch in **OFF** position there should be continuity (low ohms reading).

3. Turn switch to **ON** position and there should be no continuity (high ohms reading). Replace switch if there is no continuity in both **ON** and **OFF** positions or there is continuity in both positions.

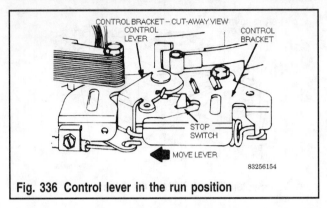

Fig. 336 Control lever in the run position

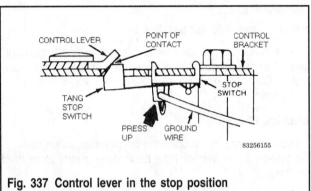

Fig. 337 Control lever in the stop position

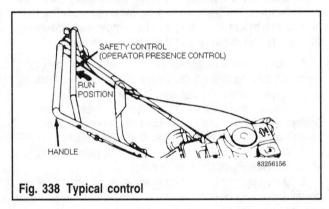

Fig. 338 Typical control

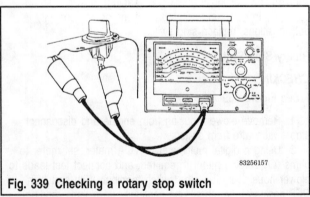

Fig. 339 Checking a rotary stop switch

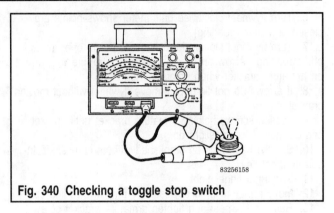

Fig. 340 Checking a toggle stop switch

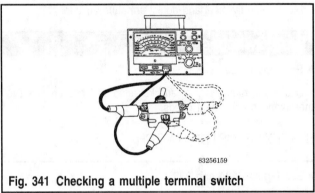

Fig. 341 Checking a multiple terminal switch

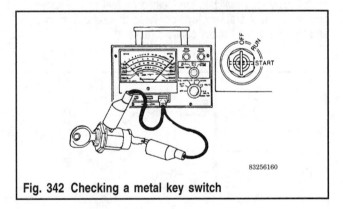

Fig. 342 Checking a metal key switch

Toggle Switches

▶ See Figure 340

Two styles of toggle switches have been used, single terminal and multiple terminals.

CHECKING SINGLE TERMINAL SWITCH

1. Disconnect stop switch wire from spade terminal. Using a digital multimeter or VOA meter, set meter to ohms (Ω) zeroing meter if required, and connect test leads to spade terminal and to switch mounting surface. Mounting surface must be free of paint, rust or dirt.
2. With switch in **OFF** position there should be continuity (low ohms reading). Move switch to **ON** position. There should be no continuity (high ohms reading).

CHECKING MULTIPLE TERMINAL SWITCH

▶ **See Figure 341**

1. Disconnect all wires from switch marking each wire for correct installation. Using a digital multimeter or VOA meter, set meter to ohms (Ω) zeroing meter if required, and connect test leads to either center terminal and a terminal on either end of switch on the same side as the center terminal. If meter shows continuity (low ohms reading) move switch to other position and the reading should not show continuity (high ohms reading).

2. Move test leads from end terminal to other end terminal and repeat tests. Test results should be the opposite of first tests.

3. Repeat tests for terminals on other side of switch. If there is continuity in either switch position, replace switch.

Key Switch

Two styles of key switches have been used, metal key and plastic key.

CHECKING METAL KEY SWITCH

▶ **See Figure 342**

1. Disconnect stop switch wire from spade terminal. Using a digital multimeter or VOA meter, set meter to ohms (Ω) zeroing meter if required, and connect test leads to spade terminal and to switch mounting surface. Surface must be free of paint, rust and dirt.

2. With key in **OFF** position there should be continuity (low ohms reading). Turn key to **ON** position. There should not be continuity (high ohms reading).

CHECK SWITCH WITH PLASTIC KEY

▶ **See Figure 343**

1. Disconnect stop switch wire from spade terminal. Using a digital multimeter or VOA meter, set meter to ohms (Ω) zeroing meter if required, and connect test leads to spade terminal and to switch mounting surface. Mounting surface must be free of paint, rust and dirt.

2. With key pushed all the way in there should be no continuity (high ohms reading). Pull key out and there should be continuity (low ohms reading).

Stop Switch Wire, All Models

CHECKING STOP SWITCH WIRE

▶ **See Figure 344**

Using a digital multimeter or VOA meter, set meter to ohms (Ω) zeroing meter if required, and connect one test lead to end of stop switch wire. Connect other test lead to engine ground. There should be continuity (low ohms reading). Move wire back and forth and up and down. If readings change, repair or replace damaged wire.

STOP SWITCH WIRE CONTINUITY TEST

▶ **See Figures 345, 346, 347 and 348**

1. Place control lever, and safety control, if engine is mounted on equipment, in run position.

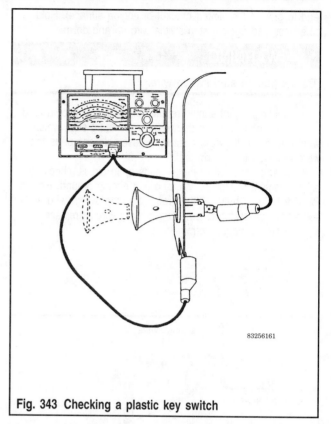

Fig. 343 Checking a plastic key switch

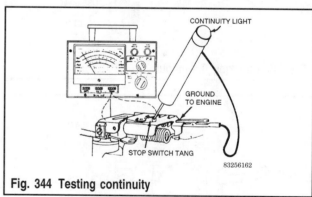

Fig. 344 Testing continuity

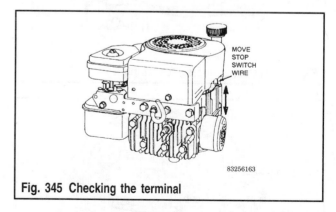

Fig. 345 Checking the terminal

2. Using a continuity light, digital multimeter, or VOA meter, set meter to ohms (Ω) zeroing meter, and connect one test

lead to ground (unpainted bracket or engine surface). Hold other test lead against stop switch wire up and down.

✳✳WARNING

Do not pull on stop switch wire.

3. Continuity light should remain ON or meter should read less than 1 ohm or more than 0.3 ohms during stop switch wire movement. If test is positive, reassemble any parts that were removed to perform test.

4. If light goes out or meter reads open circuit, check for proper contact at stop switch tang and engine ground. Retest. Poor or no continuity requires replacing stop switch wire and/or soldering stop switch to armature primary wire at replaceable MAGNETRON module terminal.

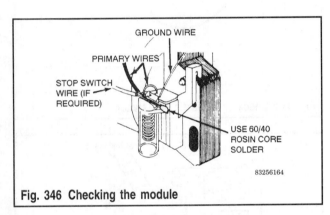

Fig. 346 Checking the module

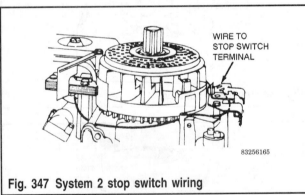

Fig. 347 System 2 stop switch wiring

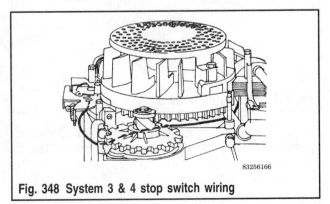

Fig. 348 System 3 & 4 stop switch wiring

Air Cleaners

A properly serviced air cleaner protects internal parts of engine from dust particles in the air. If air cleaner instructions are not carefully followed, dirt and dust which should be collected in cleaner, will be drawn into engine. It will become a part of oil film and is very detrimental to engine life.; dirt in oil forms an abrasive mixture which wears moving parts, instead of protecting them.

No engine can stand up under the grinding action which takes place when this occurs. The air cleaner on every engine brought in for a check up or repair should be examined and serviced. If cleaner shows sign of neglect, show it to customer before cleaning, and instruct him on proper care to ensure long engine life.

➡**Air cleaner element and/or cartridge should be replaced if damaged or restricted. Replace air cleaner gaskets and mounting gaskets that are worn or damaged to prevent dirt and dust entering engine through improper sealing. Straighten or replace bent mounting studs.**

SERVICE

Oil-Foam Air Cleaner
▶ **See Figures 349, 350, 351, 352, 353, 354 and 355**

Clean and re-oil air cleaner element every 25 hours or at three month intervals under normal conditions. Capacity of the air cleaner is adequate for a full season's use, without cleaning, in average homeowner's lawn mower service. Clean every few hours under extremely dusty condition.

1. Remove screw or wing nut.

2. Remove air cleaner carefully to prevent dirt from entering carburetor.

3. Take air cleaner apart and clean.

  a. WASH foam element in kerosene or liquid detergent and water to remove dirt.

  b. Wrap foam in cloth and squeeze dry.

  c. Saturate foam with engine oil. Squeeze to remove excess oil.

4. Reassemble parts and fasten to carburetor securely with screw or wing nut.

Dual Element Air Cleaner
▶ **See Figure 356**

Clean and re-oil foam precleaner every 25 hours or at three month intervals, whichever occurs first.

➡**Service more often under dusty conditions.**

1. Remove knob and cover.

2. Remove foam precleaner by sliding it off of paper cartridge.

  a. Wash foam precleaner in kerosene or liquid detergent and water.

  b. Wrap foam precleaner in cloth and squeeze dry.

  c. Saturate foam precleaner in engine oil. Squeeze to remove excess oil.

3. Install foam precleaner over paper cartridge. Reassemble cover and screw knob down tight.

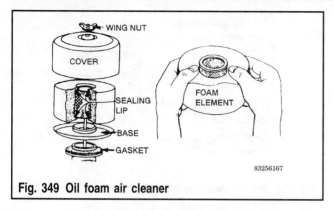

Fig. 349 Oil foam air cleaner

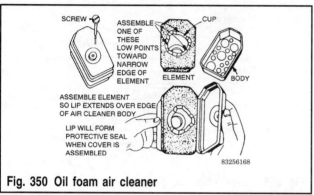

Fig. 350 Oil foam air cleaner

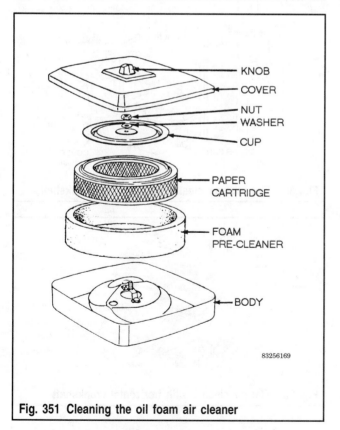

Fig. 351 Cleaning the oil foam air cleaner

Yearly or every 100 hours, whichever occurs first, remove paper cartridge. Service more often if necessary. Clean by

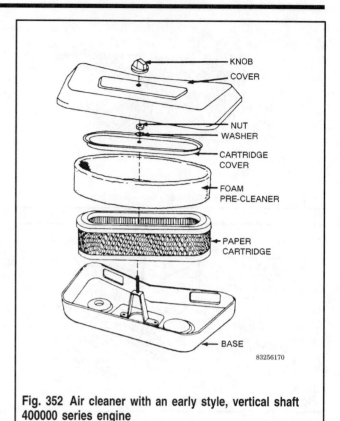

Fig. 352 Air cleaner with an early style, vertical shaft 400000 series engine

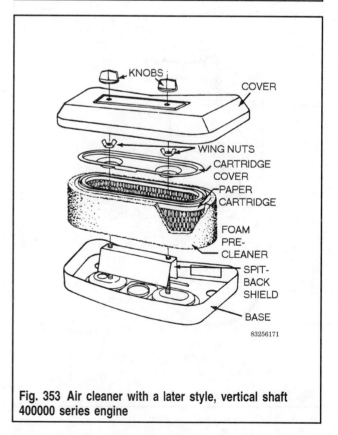

Fig. 353 Air cleaner with a later style, vertical shaft 400000 series engine

tapping gently on flat surface. If very dirty, replace cartridge, or wash in a low or non-sudsing detergent and warm water solution. Rinse thoroughly with flowing water from inside until water

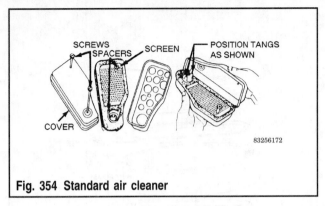

Fig. 354 Standard air cleaner

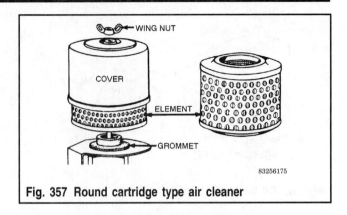

Fig. 357 Round cartridge type air cleaner

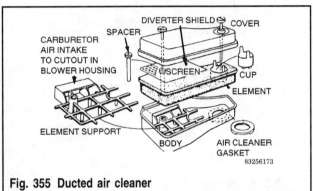

Fig. 355 Ducted air cleaner

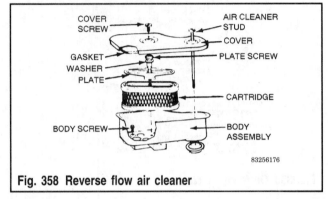

Fig. 358 Reverse flow air cleaner

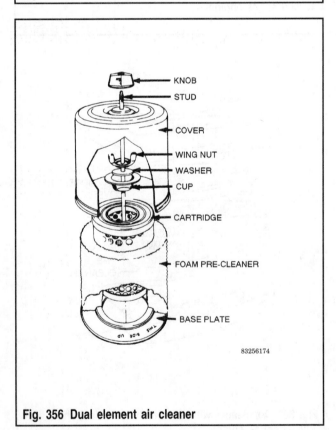

Fig. 356 Dual element air cleaner

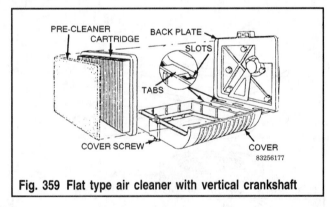

Fig. 359 Flat type air cleaner with vertical crankshaft

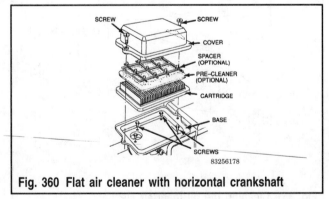

Fig. 360 Flat air cleaner with horizontal crankshaft

is clear. Cartridge must be allowed to stand and air dry thoroughly before using.

✳✳CAUTION

Petroleum solvents, such as kerosene, are not to be used to clean cartridge. They may cause deterioration of cartridge. DO NOT USE PRESSURIZED AIR TO CLEAN OR DRY CARTRIDGE.

Cartridge Type (Round)

▶ **See Figure 357**

To clean, tap cartridge (top or bottom) on flat surface or wash in non-sudsing detergent and flush from inside until water is clear. After washing, air dry thoroughly before using.

✳✳CAUTION

Petroleum solvents, such as kerosene, are not to be used to clean cartridge. They may cause deterioration of cartridge. DO NOT USE PRESSURIZED AIR TO CLEAN OR DRY CARTRIDGE!

Cartridge Type Reverse Air Flow, Vertical Crankshaft

▶ **See Figure 358**

1. Remove air cleaner stud, cover screw, cover and gasket. Replace gasket if damaged.
2. Remove plate screw, washer and plate.
3. Remove cartridge and clean air cleaner body carefully to prevent dirt from body through holes into duct.
4. Clean cartridge by tapping gently on flat surface.
   a. If very dirty, replace cartridge or wash in a low or non-sudsing detergent and warm water solution.
   b. Rinse thoroughly from OUTSIDE IN until water is clear.
   c. Cartridge must be allowed to stand and air dry thoroughly before using.
5. Reassemble air cleaner.

Cartridge Type Flat Cartridge, Vertical Crankshaft

▶ **See Figure 359**

1. Loosen screw and tilt cover.
2. Carefully remove cartridge and foam precleaner when so equipped.
3. Clean by tapping gently on a flat surface. If very dirty, replace cartridge and precleaner, or clean as follows:
   a. Wash cartridge and precleaner in a low or non-sudsing detergent and warm water solution.

➡**Do not use petroleum solvents such as kerosene, to clean cartridge or precleaner.**

   b. Rinse thoroughly with flowing water from inside out until water is clear.
   c. Allow cartridge and precleaner to stand and air dry thoroughly before using. DO NOT OIL CARTRIDGE. DO NOT USE PRESSURIZED AIR TO CLEAN OR DRY CARTRIDGE.

4. Install cartridge and foam precleaner, when so equipped. Then close cover and fasten screw securely. Tabs in cover must be in slots of back plate.

✳✳CAUTION

Petroleum solvents, such as kerosene, are not to be used to clean cartridge. They may cause deterioration of cartridge. DO NOT USE PRESSURIZED AIR TO CLEAN OR DRY CARTRIDGE!

Cartridge Type Flat Cartridge, Horizontal Crankshaft

▶ **See Figure 360**

1. Loosen screws and remove cover.
2. Carefully remove precleaner (when so equipped) and cartridge.
3. Clean cartridge by tapping gently on a flat surface. If very dirty, replace cartridge and precleaner or clean as follows:
   a. Wash cartridge and precleaner in a non-sudsing detergent and warm water solution.
   b. Rinse thoroughly with flowing water from mesh side until water is clear.
   c. Wrap foam precleaner in cloth and squeeze dry.
   d. Saturate foam precleaner in engine oil. Squeeze to remove excess oil.
   e. Allow cartridge to stand and dry thoroughly before using.

✳✳CAUTION

Petroleum solvents, such as kerosene, are not to be used to clean cartridge. They may cause deterioration of cartridge. DO NOT USE PRESSURIZED AIR TO CLEAN OR DRY CARTRIDGE!

4. Install cartridge and precleaner.
5. Reinstall air cleaner cover and tighten screws.

Carburetors

Before removing any carburetor for repair, look for signs of air leakage, or mounting gaskets that are loose, have deteriorated, or are otherwise damaged.

Note position of governor springs, governor link, remote control of other attachments to facilitate reassembly. Do not bend links or stretch springs.

BRIGGS & STRATTON/WALBRO LARGE ONE-PIECE FLO-JET MODEL SERIES 254700, 257700, 283700, 286700 WITH VERTICAL CRANKSHAFTS

▶ **See Figure 361**

These carburetors have a FIXED HIGH SPEED MAIN jet with ADJUSTABLE IDLE. The different carburetors are identified as LMT 1 and up. The letters LMT are cast into the body of the carburetor while the numbers are stamped into carburetor mounting flange next to idle mixture.

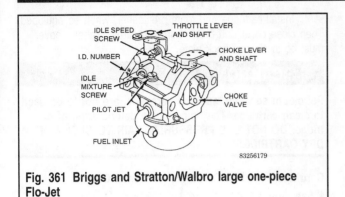

Fig. 361 Briggs and Stratton/Walbro large one-piece Flo-Jet

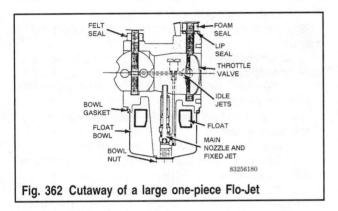

Fig. 362 Cutaway of a large one-piece Flo-Jet

Disassembly

▶ **See Figures 363, 364 and 365**

1. Remove bowl screw and fuel bowl washer.
2. Remove float bowl and float bowl gasket from carburetor.
3. Remove float hinge pin, float and inlet valve.
4. With a small hook or crochet hook remove inlet valve seat.
5. Remove idle mixture screw with spring and idle speed screw with spring.
6. Rotate throttle shaft and lever to closed position and remove two (2) throttle screws.
7. Remove throttle valve and throttle shaft and lever with foam seal.
8. Remove throttle shaft seal from carburetor body.
9. On models with plastic choke shaft and valve: Pull choke valve out of choke shaft and lever. Remove choke shaft and lever, return spring, and foam washer.

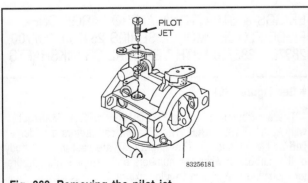

Fig. 363 Removing the pilot jet

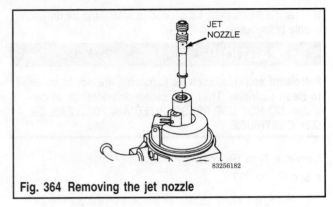

Fig. 364 Removing the jet nozzle

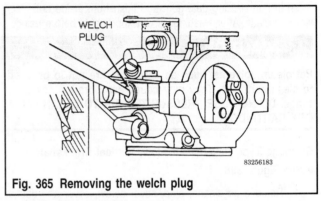

Fig. 365 Removing the welch plug

10. On models with metal choke shaft and valve: Rotate choke shaft and lever to close position. Service replacement carburetors may have a spring detent that will hold choke in closed position. Hold choke closed and remove two (2) screws holding choke valve. Remove valve. Release tension on choke shaft and remove choke shaft lever, return spring and seal assembly.
11. Remove brass pilot jet from carburetor body.
12. Using a screwdriver, remove main carburetor jet nozzle.
13. With a modified $5/32$ in. pin punch, remove welch plug from carburetor body.

➡**If poor performance is experienced at higher altitudes, a high altitude kit is available as Part #494386 (Model Series 286700 only). This kit consists of a new jet nozzle, pilot jet and float bowl gasket.**

CLEANING AND INSPECTION

Carburetor can be cleaned in any commercially available carburetor cleaner. After cleaning, inspect for wear, damage, cracks, or plugged openings. Replace body if any of the above conditions exist. Use only compressed air to clear plugged openings.

Inspect idle mixture for bent needle point or a groove in tip of needle. Replace if bent or grooved.

ASSEMBLY

▶ **See Figures 366, 367, 368, 369, 370, 371, 372 and 373**

1. Install welch plug with pin punch slightly smaller than outside diameter of plug. Press in until plug is flat. DO NOT cave in plug. After plug is installed, seal outside edge of plug with fingernail polish or non-hardening sealant.

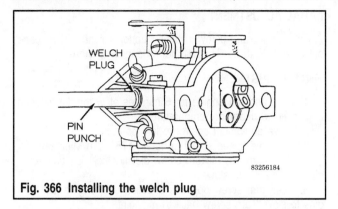

Fig. 366 Installing the welch plug

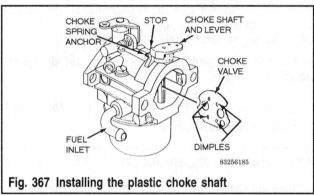

Fig. 367 Installing the plastic choke shaft

2. With plastic choke shaft: Insert spring inside large foam seal and slide seal and spring onto choke shaft with straight end of spring up toward choke shaft lever. Insert choke shaft into carburetor body until hook of spring hooks on spring anchor. Lift choke shaft and lever up slightly and turn counterclockwise until stop on lever clears spring anchor and push shaft down. Insert Choke valve into choke shaft and lever with dimples toward fuel inlet side of carburetor. Dimples help to hold and align choke valve n shaft.

3. With metal choke shaft: Install foam seal and return spring n choke shaft hooking small hook in notch on choke lever inset. Insert choke shaft assembly into carburetor body and engage large end of return spring on boss.   If carburetor has spring detent, guide detent spring into slot on choke shaft lever. Place choke valve on shaft with single notch on edge toward fuel inlet. Two (2) half moon dimples will help to position valve on shaft.

4. Install throttle shaft seal with sealing lip down in carburetor body until top of seal is flush with top of carburetor. install

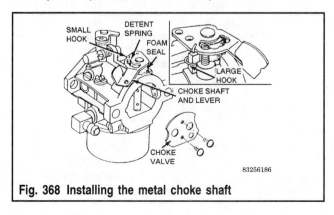

Fig. 368 Installing the metal choke shaft

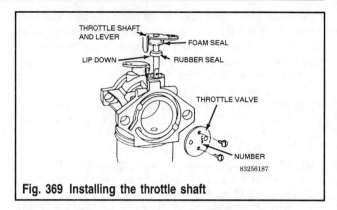

Fig. 369 Installing the throttle shaft

throttle shaft and small foam washer. Turn shaft until flat is facing out. Lay throttle valve on shaft with numbers toward idle mixture screw and dimples facing in resting on edge of shaft. install two (2) screws.

➡ These carburetors are equipped with a tan or black inlet valve seat for gravity feed systems and a brown inlet valve seat for fuel pump feed systems. Both seats are installed the same way.

5. Be sure inlet valve seat area is clean. Install inlet needle seat with GROOVE DOWN using a bushing driver, until seated. After installing seat use compressed air to remove any debris that may be in carburetor. Blow from seat out to fuel inlet.

6. Install jet nozzle using a screwdriver, until nozzle seats. After installing jet nozzle, use compressed air to blow out any chips of debris that may have been loosened while installing jet nozzle.

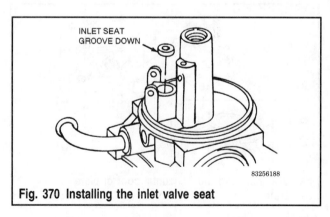

Fig. 370 Installing the inlet valve seat

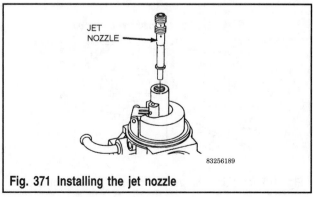

Fig. 371 Installing the jet nozzle

7. Insert inlet needle valve in slot on float. Place float and needle on carburetor and install float hinge pin centering hinge pin. Place bowl gasket on carburetor body. Place bowl on carburetor and install bowl screw and float bowl washer torquing bowl screw to 40 inch lbs.

8. Install pilot jet until it seats securely. Install idle mixture screw with spring and turn until head of screw touches spring.

9. If carburetor elbow was removed, place new gasket and elbow on intake port and torque screws to 100 inch lbs.

10. Place two (2) studs on carburetor and new gasket on studs with long edge on side of gasket opposite fuel inlet. Hook governor link spring in throttle lever hole with grommet with link on top of lever.

11. With Choke-A-Matic Link, Horizontal Control Plate: Insert 'Z' bend of Choke-A-Matic link in outer hole of choke lever from bottom of lever. Slide 'U' bend of link into slot on governor control bracket and place carburetor on intake elbow. Torque studs to 65 inch lbs.

12. With Choke-A-Matic Link, Vertical Control Plate: Insert 'Z' bend of Choke-A-Matic link in inner hole of choke lever from bottom of lever. Slide 'U' bend of link into slot on governor control bracket and place carburetor on intake elbow. Torque studs to 65 inch lbs.

13. Install air cleaner body and torque nuts to 55 inch lbs. be sure breather tube is on air cleaner body opening. Install air cleaner brace.

14. Install air cleaner cartridge and precleaner with small air cleaner nut. install air cleaner cover and large nut.

15. Place foam element into air cleaner body, making sure lip of foam extends over edges of body to form protective seal. Inserts slots of cover into tabs on body and press down on cover until latch snaps into place.

INITIAL ADJUSTMENT

Turn idle mixture screw in until it bottoms lightly. Then back out screw 1½ turns. This will permit engine to start.

FINAL ADJUSTMENT

1. Start and run engine for 5 minutes at ½ throttle to bring engine up to operating temperature.

2. Move equipment control to idle position.

3. Turn idle speed screw to obtain 1750 rpm Minimum.

4. Then turn idle mixture screw slowly clockwise until engine just begins to slow.

5. Then turn screw opposite direction until engine just begins to slow. Turn screw back to midpoint.

6. Move equipment speed control from idle to high speed position. Engine should accelerate smoothly. If it does not, open idle mixture needle screw ⅛ turn open.

➡**If engine is adjusted for governed idle, reset idle speed screw to 1200 rpm.**

ONE PIECE FLO-JET CARBURETOR, SMALL AND LARGE, VERTICAL CRANKSHAFT

◆ **See Figures 374, 375, 376 and 377**

These are float feed carburetors with high speed and idle needle valve adjustments.

The large One-Piece Flo-Jet carburetor is similar to the small One-Piece Flo-Jet. The main difference is that high speed needle is below float bowl.

Repair procedures for small and large One-Piece Flo-Jet carburetors are similar except for location of adjusting needles.

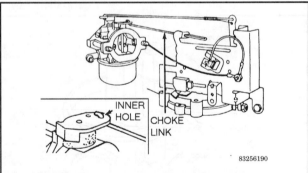

Fig. 372 Installing the horizontal control plate carburetor

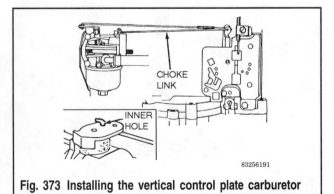

Fig. 373 Installing the vertical control plate carburetor

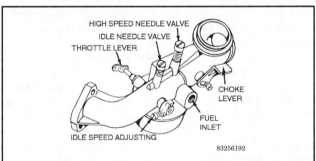

Fig. 374 Small, one-piece Flo-Jet; small and large vertical crankshaft

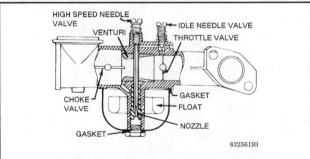

Fig. 375 Cross-section of a small, one-piece Flo-Jet; small and large vertical crankshaft

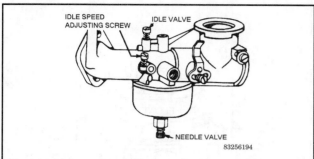

Fig. 376 Large, one-piece Flo-Jet; small and large vertical crankshaft

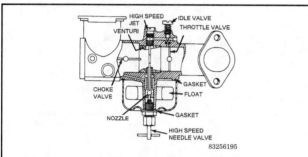

Fig. 377 Cross-section of a large, one-piece Flo-Jet; small and large vertical crankshaft

Disassembly

▶ **See Figures 378 and 379**

1. Small One-Piece Flo-Jet: Remove idle and high speed adjusting needles. Remove bowl nut and float bowl. Use a screwdriver to remove nozzle. Remove float pin to remove float and float needle. Use a large wide screwdriver to remove float valve seat.

2. Large One-Piece Flo-Jet: Remove idle needle valve. Remove high speed needle valve assembly from float bowl and remove float bowl. Use Tool #19280, carburetor Nozzle Screwdriver, to remove nozzle, then remove jet from top of carburetor. remove float pin to remove float and float needle.

3. Pry out welch plug.

4. Remove choke valve.

5. On carburetors with nylon choke shaft, remove choke valve. Venturi can now be removed.

➡**Large One-Piece Choke-A-Matic carburetors have a plate stop pin which must be pressed out to remove venturi.**

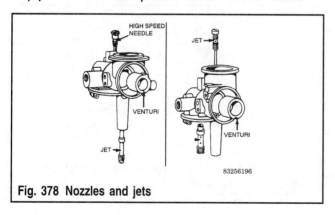

Fig. 378 Nozzles and jets

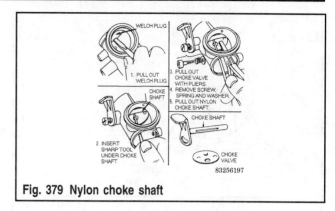

Fig. 379 Nylon choke shaft

6. Remove throttle shaft seals, when so equipped.

Inspection

▶ **See Figures 380, 381 and 382**

1. Discard idle and high speed mixture needles if damaged.

2. Check float for leakage. If it contains fuel or is crushed, it must be replaced.

3. Replace float needle, if worn.

4. If carburetor leaks with new float needle on carburetors with pressed in float needle seat:

   a. Use a Part #93029 self-threading screw or remove one self-threading screw from Tool #19069, Flywheel Puller, and clamp head of screw in a vise.

➡**On carburetors with removable viton inlet seat, use a ¼-20 tap and screw Part #93029 or a screw extractor to remove pressed in seat.**

   b. Turn carburetor body to thread screw or screw extractor into seat. Continue turning carburetor body drawing seat out. Leave screw or screw extractor fastened to seat.

   c. Insert new seat from repair kit Part #394682 into carburetor body (seat has chamfer).

➡**If engine is equipped with a fuel pump, install repair kit Part #394683.**

   d. Press new seat flush with body using screw/screw extractor and old seat as driver. Use care to ensure seat is not pressed below body surface or improper float to float valve contact will occur.

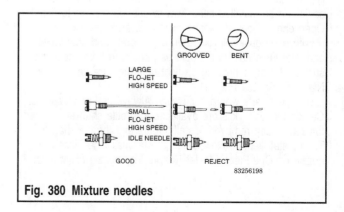

Fig. 380 Mixture needles

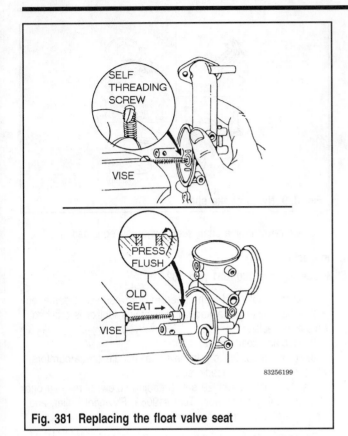

Fig. 381 Replacing the float valve seat

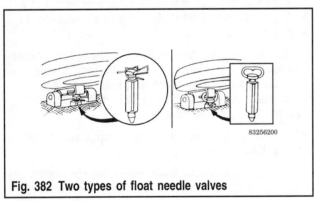

Fig. 382 Two types of float needle valves

Assembly
▶ See Figure 383

On carburetors equipped with throttle shaft seals, rubber lipped seals are installed with lip out on both sides. Foam seals can be installed either way. Install float needle to float. Open end of hook on spring must face away from venturi.

With body gasket in place on upper body and float valve and float installed, float should be parallel to body mounting surface. If not, bend tang on float until they are parallel. DO NOT PRESS ON FLOAT!

Use new parts where necessary. Always use new gaskets. Carburetor repair kits are available. If throttle shaft and/or venturi have been removed, install throttle and throttle shaft first. Then install venturi. Now install jet on small One-Piece or nozzle on One-Piece Flo-Jet Nozzle. The nozzle or jet holds

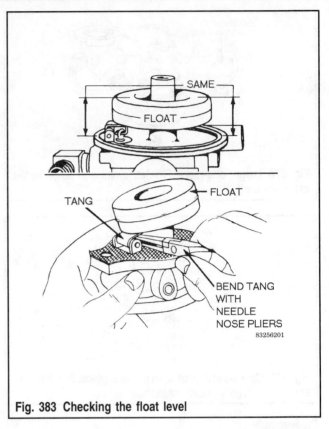

Fig. 383 Checking the float level

venturi in place. Replace choke shaft and valve. Install new welch plug using sealer around edge of plug. Stake plug in eight places. Sealer is to prevent entry of dirt into engine. Install float bowl, idle and high speed adjustment needles.

INITIAL ADJUSTMENT
▶ See Figures 384, 385 and 386

On One-Piece Flo-Jets, turn both idle ad high speed needles in until they just bottom. Then turn both valves 1½ turns open.

These settings will allow engine to start. Final adjustment should be made when engine is running and has warmed up. See carburetor adjustment (two piece Flo-Jet carburetor).

Choke-A-Matic Remote Control Adjustment

On Choke-A-Matic carburetors, remote control must be correctly adjusted in order to obtain proper operation of choke and stop switch.

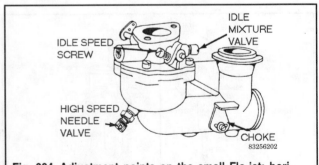

Fig. 384 Adjustment points on the small Flo-jet; horizontal and vertical shaft

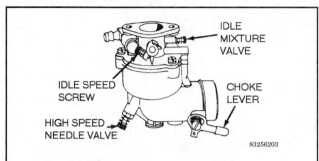

Fig. 385 Adjustment points on the medium Flo-jet; horizontal and vertical shaft

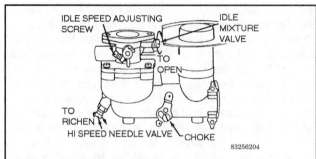

Fig. 386 Adjustment points on the large Flo-jet; horizontal and vertical shaft

TWO-PIECE FLO-JET CARBURETOR, SMALL, MEDIUM AND LARGE FLO-JET

▶ **See Figures 387, 388 and 389**

Disassembly

1. Remove idle needle valve.
2. On early small Flo-Jet loosen high speed packing nut. Remove packing nut and high speed needle valve together.
3. On current small, medium and large Flo-jets remove high speed needle valve assembly. Remove nozzle a screwdriver. Take care to avoid damage to threads in lower carburetor body.

➡ **If threads have been damaged in lower carburetor body, Tool #19245 Tap Set, can be used to clean damaged threads.**

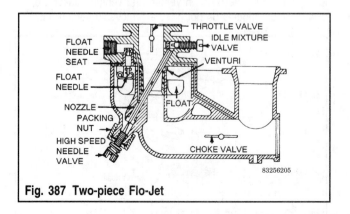

Fig. 387 Two-piece Flo-Jet

4. Because nozzle projects diagonally into a recess in upper body, it must be removed before separating upper and lower bodies.
5. Remove screw holding upper and lower bodies together and separate two bodies.
6. Remove float pin and remove float and float needle as an assembly.
7. With a wide blade screwdriver that completely fills lot, remove float inlet seat.   On carburetors with pressed in seats, see 'Replacing Pressed in Float Seats.'   On small Flo-Jet carburetors venturi is a separate part and can be slipped out of lower body.
8. Some two piece Flo-Jet carburetors have a welch plug and it should be removed only if choke shaft or choke valve is going to be removed. Some carburetors have a nylon choke shaft.

➡ **Throttle shaft should be removed only when necessary to replace throttle shaft and/or bushings.**

9. Use a thin punch to drive out pin holding throttle stop to shaft, remove throttle valve, then pull out shaft.

Inspection

▶ **See Figures 390 and 391**

With carburetor assembled and body gasket in place, if a 0.002 in. feeler gauge can be inserted between upper and lower bodies at air vent boss, just below idle valve, upper body is warped or gasket surfaces are damaged and should be replaced.

Wear between throttle shaft and bushings should not exceed 0.010 in. Check wear by placing a short iron bar on upper carburetor body. Measure distance between bar and shaft with

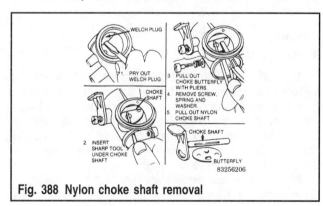

Fig. 388 Nylon choke shaft removal

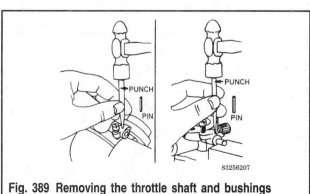

Fig. 389 Removing the throttle shaft and bushings

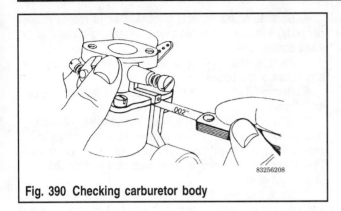

Fig. 390 Checking carburetor body

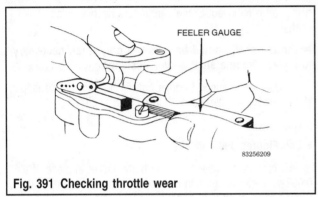

Fig. 391 Checking throttle wear

feeler gauge while holding shaft down and then holding shaft up.

If difference is over 0.010 in., either upper body should be rebushed, throttle shaft replaced, or both. Wear on throttle shaft can be checked by comparing worn and unworn portions of shaft.

Assembly

▶ **See Figures 392, 393 and 394**

1. To replace throttle shaft bushings: Place a 1/4 in. · 20 tap or screw extractor in a vise. Turn carburetor body so as to thread tap into bushings enough to pull bushings out of body. Press new bushings into carburetor body with a vise. Insert throttle shaft to be sure it is free in bushings. If not, run a size 7/32 in. drill through both bushings to act as a line reamer. Install throttle shaft, valve and stop.

2. To replace float needle bushing or seat: Use new parts when necessary. Always use new gaskets. Old gaskets take a

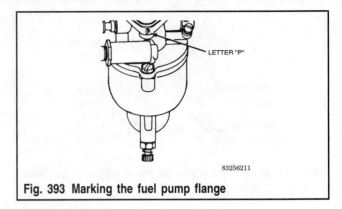

Fig. 393 Marking the fuel pump flange

set or harden and may leak. Carburetor repair kits are available. Tighten inlet seat with gasket securely in place, if used. Some float valves have a spring clip to connect float valve to float tang. Others are nylon with a stirrup which fits over float tang. Older type float valves and earlier engine with fuel pumps have neither spring or stirrup. A viton tip float valve is used on later models of Flo-Jet carburetors. These needles are used with inlet needle seat pressed into upper carburetor body and does not need replacement unless damaged. Use a Part #93029 self-threading screw or remove one self-threading screw from Tool #19069, Flywheel Puller, and clamp head of screw in a vise.

➡**On carburetors with removable viton inlet seat, use a 1/4-20 tap and screw Part #93029 or a screw extractor to remove pressed in seat.**

Turn Carburetor body to thread screw or screw extractor into seat. Continue turning carburetor body drawing seat out. Leave screw or screw extractor fastened to seat. Insert new seat from repair kit Part #394682 into carburetor body (seat has chamfer).

➡**If engine is equipped with a fuel pump, install repair kit Part #394683.**

Press new seat flush with body using screw and old seat as a driver. Use care to ensure seat is not pressed below body surface or improper float to float valve contact will occur.

3. With body gasket in place on upper body and float valve and float installed, float should be parallel to body mounting surface. If not, bend tang on float until they are parallel. DO NOT PRESS ON FLOAT TO ADJUST!

4. Assemble venturi and venturi gasket, when used, to lower body. Be sure holes in venturi and venturi gasket are aligned. Most models do not have a removable venturi. Install

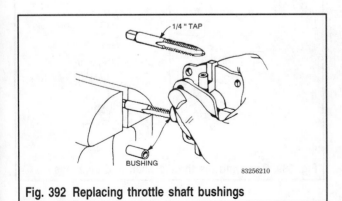

Fig. 392 Replacing throttle shaft bushings

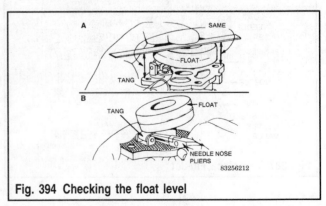

Fig. 394 Checking the float level

choke parts and welch plug if previously removed. Use a sealer around welch plug to prevent entry of dirt. Stake welch plug at least twice on small two-piece Flo-Jets and eight places on large two-piece Flo-Jets.

5. Fasten upper and lower bodies together with mounting screws. Screw in nozzle, being careful that nozzle tip enters recess in upper body. Tighten nozzle securely. Screw in needle valve and idle valve until they just seat. Back off high speed needle valve 1½ turns. Do not tighten packing nut. Back off idle needle valve 1¼ turn. These settings are approximately correct. Final adjustment will be made with air cleaner installed.

Adjustments

▶ See Figure 395

1. Start engine and run to warm up.
2. Place governor speed control lever in **FAST** position.
3. Turn high speed needle valve in until engine slows (clockwise — lean mixture). Then turn it out past smooth operating point (rich mixture).
4. Turn high speed needle valve to midpoint between rich and lean.
5. Rotate throttle counterclockwise and hold against stop. Adjust idle speed adjusting screw to obtain 1750 rpm, aluminum engines; 1200 rpm, cast iron engines.
6. Holding throttle against idle stop, turn idle valve (lean) and out (rich). Set at midpoint between rich and lean.
7. Recheck idle rpm. Release throttle. If engine will not accelerate properly, carburetor should be readjusted, usually to a slightly richer mixture.

Choke-A-Matic Remote Control Adjustment

On Choke-A-Matic carburetors, remote control must be correctly adjusted in order to obtain proper operation of choke and stop switch.

Idling Device and Throttle Control (Two-Piece Flo-Jet)

▶ See Figure 396

A manual friction control may be used to limit throttle movement, to any pre-set position. It is commonly used for two purposes.
1. To return throttle to a 'no-load' position on a pump, generator, etc.
2. For cold weather starting on governed idle engines. Throttle can easily be kept in a 'near-closed' position, while starting, which is most favorable for cold weather starts.

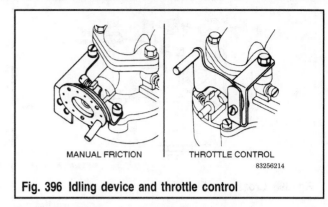

MANUAL FRICTION   THROTTLE CONTROL

83256214

Fig. 396 Idling device and throttle control

Remote Throttle Control (Two-Piece Flo-Jet)

▶ See Figure 397

The remote throttle control opens carburetor throttle until full governed speed is obtained, at which point governor takes over control of throttle. At any point below governed speed, throttle is held in fixed position and engine speed will vary with load.

Cross-over Flo-Jet, Horizontal Crankshaft

▶ See Figure 398

The cross-over Flo-Jet carburetor is used on Model series 253400, 255400 engines and is a float type carburetor with idle and high speed adjustment needles. This carburetor also has an integral fuel pump. All adjustments can be made from top of carburetor.

Disassembly

1. Remove idle and high speed needle adjustment valves.
2. Remove float bowl mounting screw, washer and float bowl.
3. Using a large blunt screwdriver, remove nozzle screw.
4. Remove float hinge pin, float and float inlet needle.
5. Use screwdriver to remove two screws from choke shaft. Then remove choke plate and choke shaft.
6. Use screwdriver to remove screw from throttle shaft.
7. Remove throttle plate and throttle shaft.
8. Remove three screws from fuel pump body. Remove fuel pump from carburetor taking care not to lose pump valve springs.

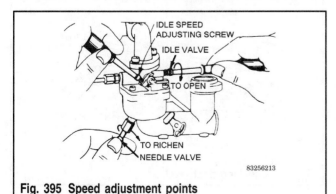

IDLE SPEED ADJUSTING SCREW

IDLE VALVE

TO OPEN

TO RICHEN

NEEDLE VALVE

83256213

Fig. 395 Speed adjustment points

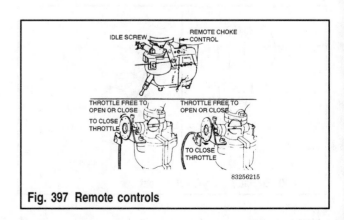

IDLE SCREW   REMOTE CHOKE CONTROL

THROTTLE FREE TO OPEN OR CLOSE   THROTTLE FREE TO OPEN OR CLOSE

TO CLOSE THROTTLE   TO CLOSE THROTTLE

83256215

Fig. 397 Remote controls

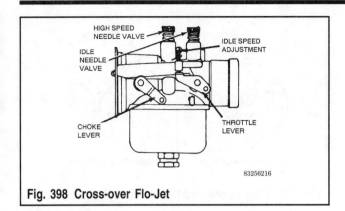

Fig. 398 Cross-over Flo-Jet

Inspection

▶ **See Figures 399, 400, 401 and 402**

1. Check idle and high speed needle valves for burrs, grooves or bent needle tips. Replace if damaged.

2. Check float for fuel in float, damage or leaks. If it contains fuel or is crushed, it must be replaced.

3. If carburetor leaks with new inlet needle valve, replace inlet needle seat:

a. Use a #93029 self-threading screw or remove one self-threading screw from Tool # 19069, Flywheel Puller, and clamp head of screw in a vise.

b. Turn carburetor body to thread screw into seat. Continue turning carburetor body drawing seat out. leave seat fastened to screw.

c. Insert new seat from repair kit Part #394683 into carburetor body (seat has staring chamfer).

d. Press new seat flush with body using screw and old seat as a driver. Use care to ensure seat is not pressed below body surface or improper float to float needle valve contact will occur. Install float valve. Hook on spring must face away from venturi.

4. With float needle valve, valve and float hinge pin installed, hold carburetor upside down. Float should be parallel to bowl mounting surface. If not, bend tang on float until they are parallel. DO NOT PRESS ON FLOAT TO ADJUST!

Assembly

▶ **See Figures 403, 404 and 405**

Use new parts where necessary. Always use new gaskets. Old gaskets take a set or harden and may leak. Carburetor repair kits are available. These carburetors use a viton tip float needle and a pressed-in needle seat. Seat does not need replacement unless seat is damaged or leaks with a new float needle.

1. Install main nozzle using blunt screwdriver to prevent damage to slot and metering hole. Place bowl on carburetor and install bowl nut and washer. Install one (1) pump valve spring on spring boss, and then place diaphragm on carburetor.

2. Place a pump valve spring on spring boss in pump body, and place pump body on carburetor. Place damping diaphragm, pump gasket and pump cover on pump body and install three screws. A fuel pump repair kit is available.

3. Place choke shaft in carburetor body and slide in choke valve with notch out and dimple down toward float bowl. Install two (2) screws using a screwdriver. Slide in throttle shaft and then slide in throttle plate with (2) dimples facing toward idle valve. When valve is installed correctly, dimples will be down

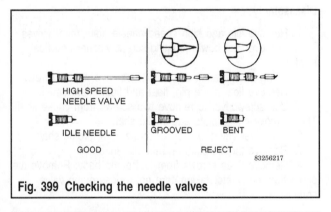

Fig. 399 Checking the needle valves

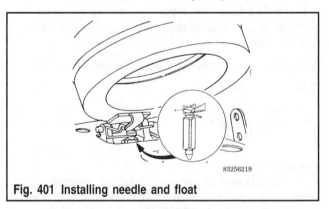

Fig. 401 Installing needle and float

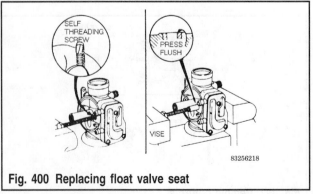

Fig. 400 Replacing float valve seat

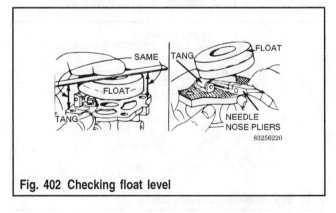

Fig. 402 Checking float level

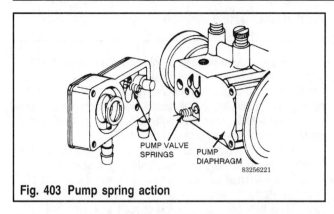

Fig. 403 Pump spring action

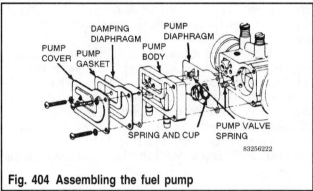

Fig. 404 Assembling the fuel pump

and number on plate is visible with throttle in closed or idle position. install idle and high speed needle valves.

➡️**Carburetor adjustments should be made with air cleaner installed on engine.**

Initial Adjustment
▶ **See Figure 406**

1. Turn idle and high speed needle valves clockwise until they just close.

➡️**Valves may be damaged by turning them into too far.**

2. Now open high speed needle valve 1½ turns counterclockwise and idle needle valve one turn. This initial adjustment will permit engine to be started and warmed up prior to final adjustment.

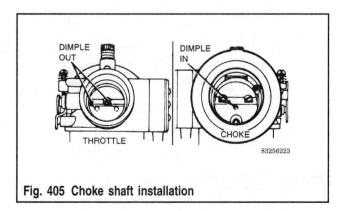

Fig. 405 Choke shaft installation

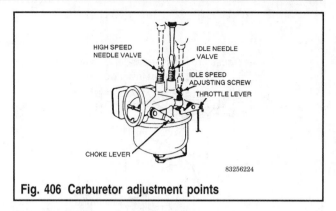

Fig. 406 Carburetor adjustment points

Final Adjustment

1. Place governor speed control lever in **IDLE** position.
2. Set idle speed adjusting screw to obtain 1750 rpm minimum while holding throttle lever against screw.
3. Turn idle needle valve in until rpm slows or misses (clockwise — lean mixture), then turn it out past smooth idling point until engine runs unevenly (rich mixture). Now turn idle needle valve to midpoint between rich and lean so engine runs smoothly. Release throttle lever.
4. Move governor speed control lever to **FAST** position.
5. Turn high speed needle valve in until engine slows or misses (lean mixture), then turn it out past smooth operating point until engine runs unevenly (rich mixture). Now turn high speed needle valve to midpoint between rich and lean so engine runs smoothly.
6. Engine should accelerate smoothly. If engine does not accelerate properly, carburetor should be readjusted usually to a slightly richer mixture.

LP Gas Fuel System

The following information is provided to assist you in servicing LP gas fuel systems. This information applies only to Garretson Equipment Company systems installed by Briggs & Stratton.

For LP fuel systems not covered in this section contact manufacturer of fuel system.

✳✳CAUTION

Lp gas fuel system should only be worked in in a very well insulated area. Many state, county and city governments require that service be performed only outdoors. Before loosening any fuel line connections, have a fan blowing directly across engine.

CHECKING AND ADJUSTING THE FUEL SYSTEM

▶ **See Figure 407**

1. Loosen fuel line at primary regulator. Open valve on cylinder for an instant, to be sure there is pressure in fuel cylinder. Escaping gas can be heard. Shut off valve at cylinder.
2. Remove fuel line between primary and secondary regulator (fuel controller). Attach pressure gauge to outlet of primary

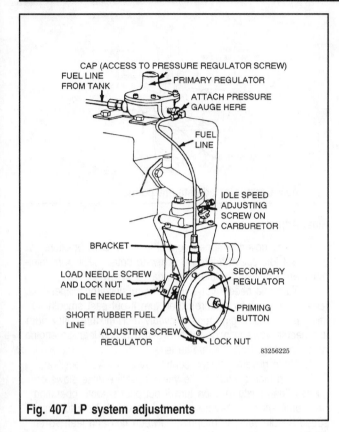

Fig. 407 LP system adjustments

regulator, leaving gauge connection loose enough to permit a slight leakage of gas. (This will permit adjustment of regulator under conditions of actual gas flow.) Remove cap or top of primary regulator.

3. Open fuel cylinder valve. Turn pressure regulator screw in the psi primary regulator, until a pressure of 1½ psi is obtained at pressure gauge. Shut off fuel cylinder valve. Reassemble cap, Remove pressure gauge. Loosen secondary regulator bracket from carburetor. Pull secondary regulator away from carburetor so that short rubber fuel line is disconnected. Assemble fuel line between primary regulator and secondary regulator (fuel controller). Secondary regulator must remain mounted so diaphragm is in a vertical plane.

4. Open fuel cylinder valve. Apply soap suds to outlet at center of secondary regulator to which rubber fuel line has been attached. If a bubble forms, it indicates that valve is leaking or not locking off. If no bubble appears, press primer button. A bubble should appear, indicating fuel is flowing into regulator. Put soap suds on outlet again, then slowly turn adjusting screw at bottom of secondary regulator counterclockwise until a bubble forms at outlet. Turn adjusting screw in (clockwise) slowly until soap bubbles on outlet no longer form.

5. Hold adjusting screw at this point and tighten locknut. Press primer button to allow fuel to flow. Release and again put soap suds on outlet to make certain fuel shuts off. Repeat several times. If bubble should form after primer button is released, adjusting screw should be turned in until flow stops and soap bubble does not break or enlarge. Loosen fuel line between regulators. Re-assemble secondary regulator to carburetor with short rubber fuel line in place. Retighten fuel line connections.

ADJUSTING THE CARBURETOR

1. Loosen locknut on load needle screw and turn needle screw in until it seats. Do not force; open 2½ turns. Turn idle needle in until it seats, then open one turn. If engine will not be required to idle, leave idle needle closed. Depress primer button momentarily, then start engine, run engine to allow it to warm up before final adjustment.

2. With engine running at normal operating speed, turn load needle screw in slowly (clockwise) until engine starts to miss (lean Mixture). Then turn load screw out slowly past point of best operation until engine begins to run unevenly (rich mixture). Then turn load screw in just enough so engine will run smoothly. Hold load screw and tighten locknut. Hold throttle at idle position, then release throttle. Engine should accelerate quickly and smoothly.

3. If engine will be required to run at idle, turn idle speed adjusting screw on throttle until engine runs at proper idle speed for engine model. Hold throttle at this point and turn idle needle slowly in or out until engine runs at maximum idle speed. Then readjust idle speed screw until proper idle speed is obtained. Allow throttle to open. Engine should accelerate quickly and smoothly. If not, readjust load screw, usually to a richer mixture. To stop engine, turn off fuel supply valve at fuel cylinder.

CLEANING THE LP GAS FILTER

▶ **See Figure 408**

1. Unscrew filter head from filter body.
2. Remove element assembly from head.
3. Wash element in commercial solvent cleaner or gasoline. If accumulated dirt is gummy, we suggest a short soaking period in solvent cleaner. Element should then be rinsed in clean gasoline and blown with compressed air.

➡**Always use reverse flow from inside out, never use compressed air on outside surface of element. Never dip element in Bright Dip® or other acid solution.**

4. To re-assemble filter, insert element into filter head with round washer entering first. Gasket is put on filter body. Spring is located in filter body so that when filter body and head are put together, spring will hold element against head.
5. Tighten body and head with 75 ft. lbs. torque.

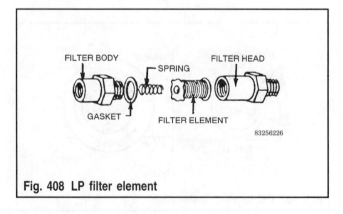

Fig. 408 LP filter element

6. After filter has been re-assembled to engine, point at gasket and other line connections should be checked with soap suds, with fuel turned on, to be sure there are no leaks.

Kerosene System

▶ **See Figures 409 and 410**

Efficient engine performance will be obtained only when following changes are accomplished:

1. A low compression cylinder head is required for models 233000, and 243000. Other models may use tow cylinder head gaskets.

2. A special spark plug — #291835 — must be used on models 233000, and 243000. Spark plug gap 0.030 in. all models.

3. A reduced breaker gap 0.015 in. is used on models 233000 and 243000. Engine must be retimed using breaker gap.

4. Power loss will vary between 15% to 25% and fuel consumption will be approximately 15% less while running on kerosene.

5. Due to the low volatility of kerosene, engines operated on kerosene/gasoline fuel systems can be started on kerosene only, when engine is at operating temperature. Cold engines must be started on gasoline and switched over to kerosene operation only after warmed up.

6. After warm-up and while operating on kerosene, adjust carburetor needle valves to a point where engine runs smoothly, and accelerates without hesitation when throttle is quickly opened. When shutting down engine, carburetor must be emptied of kerosene so engine can be started on gasoline when cold. Refer to Flo-Jet carburetor for adjustment of carburetor and adjust carburetor while running on kerosene.

FUEL PUMPS (CRANKCASE VACUUM OPERATED)

Some models are factory or field equipped with fuel pumps operated by crankcase vacuum. Fuel pumps may be mounted directly to carburetor or on a mounting bracket. Crankcase vacuum is obtained by a fitting on dipstick tube, and hollow bolt and fitting, or from crankcase breather valve.

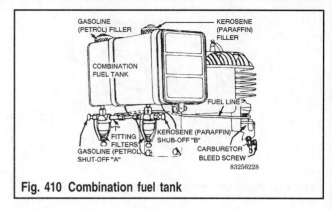

Fig. 410 Combination fuel tank

OPERATION

▶ **See Figures 411, 412, 413, 414, 415, 416 and 417**

Any restriction in fuel or vacuum lines will affect operation. Also any leaks that cause air to get into fuel line or reduce vacuum in vacuum line will reduce performance.

To service fuel pump, remove pump from carburetor or mounting bracket. When removing fuel supply line from tank to pump, be sure to plug fuel line or turn off fuel valve, if so equipped.

Disassemble fuel pump by removing four ¼ in. head cap screws from pump cover. Separate pump cover, pumping chamfer and impulse chamber. Discard old gaskets, diaphragms and spring. Clean pump parts in carburetor solvent or lacquer thinner.

A repair kit is available. Kit includes all parts needed. Install chamber gasket using locator pins. Place springs in spring recesses and install pump diaphragm on locator pins. Place

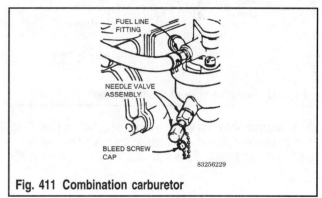

Fig. 411 Combination carburetor

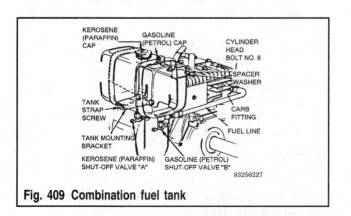

Fig. 409 Combination fuel tank

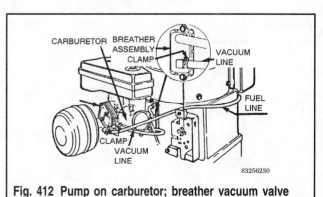

Fig. 412 Pump on carburetor; breather vacuum valve

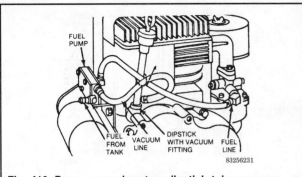

Fig. 413 Pump on carburetor; dipstick tube vacuum

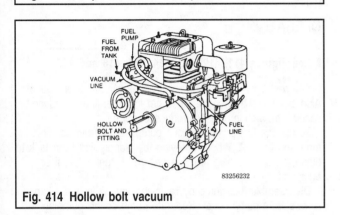

Fig. 414 Hollow bolt vacuum

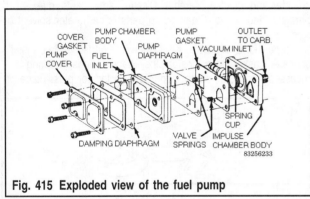

Fig. 415 Exploded view of the fuel pump

pump chamber body on impulse body using locator pins. Place damping diaphragm and cover gasket on pump body. Install cover and four screws. Torque screws to 10-15 inch lbs.

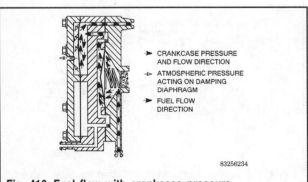

Fig. 416 Fuel flow with crankcase pressure

Governor Controls & Carburetor Linkage

REMOTE CONTROLS

In general, there are three types of remote controls:
1. Governor control
2. Throttle control
3. Choke-A-Matic control

Governor Control
▶ **See Figure 418**

The remote governor control regulates the engine speed by changing the governor spring tension, thus allowing the governor to control the carburetor throttle at all times and maintain any desired speed.

Throttle Control
▶ **See Figures 419 and 420**

The remote throttle control is used on an engine having a fixed no load governor speed setting such as 3600 or 4000 rpm.

This control enables an operator to control the speed of an engine, similar to an accelerator used on an automobile. However, when full governed speed is obtained, the governor prevents overspeeding and possible damage to the engine. At any point below the governed speed, the throttle is held in a fixed position and the engine speed will vary with the load.

Choke-A-Matic Control
▶ **See Figure 421**

On Choke-A-Matic carburetors, the remote control must be correctly adjusted in order to obtain proper operation of the choke and stop switch.

➡ **Remote control system must be mounted on powered equipment in normal operating position before adjustments are made.**

ADJUSTMENTS

Choke-A-Matic

1. Remove remote control lever to **FAST** position. Choke actuating lever **A** should just contact choke shaft **B** or link **B** as shown.
2. If not, loosen screw **C** slightly and move casing and wire **D** in or out to obtain this condition.
3. Check operation by moving remote control lever to **START** or **CHOKE** position. Choke valve should be completely closed.
4. Move remote control lever to **STOP** position. Control must contact stop switch blade.

Choke-A-Matic Dial Control
▶ **See Figure 422**

➡ **Dial controls seldom require adjustments unless blower housing has been removed.**

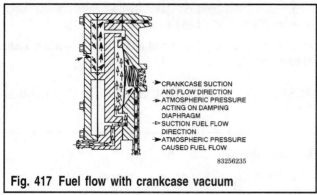

Fig. 417 Fuel flow with crankcase vacuum

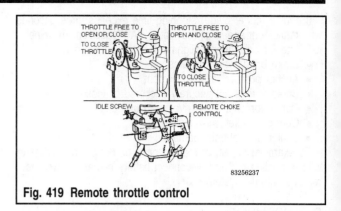

Fig. 419 Remote throttle control

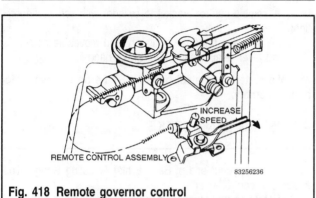

Fig. 418 Remote governor control

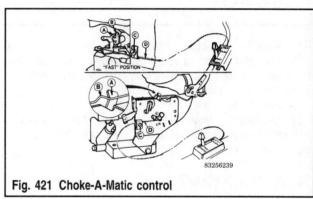

Fig. 421 Choke-A-Matic control

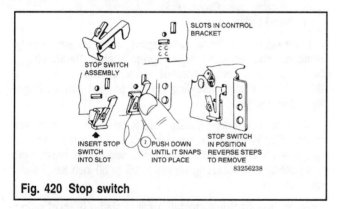

Fig. 420 Stop switch

1. Place dial control knob in **START** position.

2. Loosen control wire screw **A** and move lever **C** to full choke position. Allow a 1/8 in. gap between lever and bracket as shown.

3. While holding lever, tighten screw **A**.

Band Brake Controls

System 2 and 4 engines are equipped with the band brake feature. The band brake MUST STOP the engine (cutter blade) within three seconds after operator releases equipment safety control. Cutter blade stopping time can be checked using BLADE MONITOR, Tool #19255.

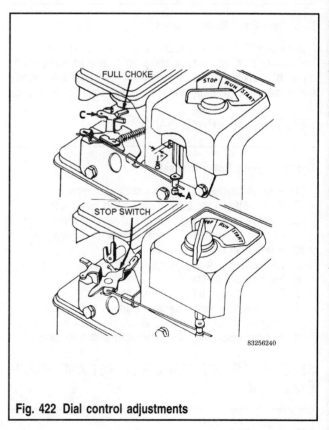

Fig. 422 Dial control adjustments

If engine (cutter blade) stopping time exceeds three seconds with equipment speed control set in **FAST** position, examine following for adjustment, alignment, alignment, or damage:

- Brake band, worn or damaged
- Anchor post, misaligned or bent
- Brake spring, not securely anchored or loose
- Control bracket lever, rivet worn or loose.
- Control bracket, misadjusted
- Equipment controls, damaged

To examine, adjust, or replace band brake, place spark plug wire in holding tab and disconnect battery holder screws and remove battery, System 4 only.

DISASSEMBLY

1. Remove brake control bracket cover.
2. Loosen cable clamp screw and remove cable from control lever. Contact equipment manufacturer for control cable specifications or replacement.
3. Remove two switch cover screws.
4. On System 4 engines, move cover as shown. Handle with care to prevent damage to lever caused by link when moving switch cover.

REMOVAL OF BRAKE BAND

New Style Band Brake Bracket

Remove blower housing and rotating screen. On new style band brake brackets, release brake spring and lift brake band up off both stationary and moving posts. Replace brake band if brake material is damaged or worn to less than 0.030 in. thick.

REMOVAL OF BRAKE BAND

Old Style Band Brake Bracket

1. Remove blower housing and rotating screen.
2. Bend control lever tang to clear band brake loop.
3. Release brake spring tension and remove brake band.
4. Replace brake band if brake material is damaged or worn to less than 0.030 in. thick.
5. Loosen screw **A** (System 4 engines only).
6. Remove two screws **B**,.
7. Remove control bracket from cylinder.
8. Disconnect control lever from starter link (System 4 engines only) using care to prevent switch cover lever damage.
9. Remove stop switch wire from stop switch terminal.

ASSEMBLY OF CONTROL BRACKET AND BRAKE BAND

New Style Bracket

1. Install stop switch wire on control bracket.
2. Assemble control bracket to cylinder with screws finger tight.

3. Place brake band on stationary post and hook over end of moveable post until band bottoms.

➡**Brake material on steel band MUST be on flywheel side after assembly.**

ASSEMBLY OF CONTROL BRACKET AND BRAKE BAND

Old Style Bracket

1. Install stop switch wire on control bracket stop switch terminal.
2. Assemble control bracket to cylinder with screws finger tight.
3. Install brake band on stationary and moveable posts.
4. Bend lever retainer tang over brake band loop.

➡**Brake material on steel band MUST be on flywheel side after assembly.**

ADJUSTING BAND BRAKE

1. Place bayonet end of Band Brake Adjusting Gauge, Tool #19256, in control lever.
2. Rotate control lever far enough to install other end of gauge in cable clamp screw hole.
3. Install brake spring.

➡**For ease of assembly, brake spring must be temporarily removed from control bracket spring anchor. Re-attach brake spring to control bracket spring anchor IMMEDIATELY after installing control bracket screws finger tight and gauge.**

4. With brake spring installed, apply pressure to the control bracket ONLY. Move it in direction indicated by arrow, until gauge link tension is JUST eliminated. Hold control bracket in this position while torquing screws to 25 to 30 inch lbs. Remove gauge.

➡**Some manufacturers install a cable clamp bracket using a pop rivet in the control bracket cable clamp screw hole. Place bayonet end of gauge in control lever as shown.**

5. Rotate control lever sufficiently to install other end of gauge into pop rivet hole. Check as noted above for tension at bayonet end of gauge.

TESTING THE BAND BRAKE

To test brake adjustment, use Torque Wrench, Tool #19197, Starter Clutch Wrench, Tool #19244, and/or a 7/8 in. socket. With band brake engaged, rotate flywheel clockwise. If less than 45 inch lbs. of torque is required to rotate flywheel, check the following for damage, misalignment, or misadjustment:

1. Brake Band Lining
2. Contour or Wear
3. Brake Band Anchors
4. Control Bracket
5. Brake Spring

6. Brake Spring Anchor, Correct, readjust, and repeat band brake test.

When band brake is released, engine must turn freely. If brake band drags against flywheel restricting movement, check for damaged brake band or anchors.

FINAL ASSEMBLY

Install rotating screen and blower housing on engine. Tighten screw. Note location of blower housing guard.

INSTALLING ELECTRIC STARTER CONTROLS, SYSTEM 4

1. Install starter link into control lever.
2. Carefully insert other end of link in switch cover lever.
3. Rotate switch cover into position on starter motor.
4. Fasten screws securely.

➥**If equipped with key switch, ignition link may be omitted.**

5. Install equipment safety control cable to control lever.
6. Tighten cable clamp screw securely.
7. Conduct Stop Switch and Stop Switch Wire Tests.
8. Install brake control bracket cover and tighten screws.
9. Place battery in holder and tighten screws (System 4 engines only).
10. After engine is installed on equipment, connect battery wires to connector and place wire on spark plug.
11. Service engine. Test blade stopping time, using BLADE MONITOR, Tool #19255.

Governor Control & Linkages

GOVERNED IDLE ADJUSTMENTS

Models 220700, 252700, 253700, 280000 Vertical Crankshaft

▶ **See Figure 423**

1. First make final carburetor mixture adjustment.
2. Then place remote control in idle position.
3. Hold throttle in closed position with finger, adjusting idle speed screw to 1550 rpm. Release throttle.

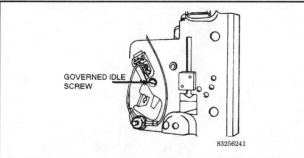

Fig. 423 Setting the governed idle on 220700, 252700, 253700, 280000 vertical crankshaft

4. Set remote control to 1750 rpm. Turn screw in until it contacts remote control lever.

Models 254700, 257700, 283700, 286700 Vertical Crankshaft with Horizontal Control

▶ **See Figure 424**

1. First make final carburetor mixture adjustment.
2. Place remote control in idle position.
3. Hold throttle in closed position with finger, adjusting idle speed screw to 1200 rpm. Release throttle.
4. Set remote control to 1750 rpm and turn screw in until it contacts remote control lever.

Models 254700, 257700, 283700, 286700 Vertical Crankshaft with Vertical Control

▶ **See Figure 425**

1. First make final carburetor mixture adjustment.
2. Place remote control in idle position.
3. Hold throttle in closed position with finger, adjusting idle speed screw to 1200 rpm. Release throttle.
4. Set remote control to 1750 rpm and bend tang until it contact remote control slide.

REMOTE GOVERNOR CONTROL ADJUSTMENT

Attach remote control casing and wire. Do not change the position of the small elastic stop nuts. They provide for a governed idle speed and protection against overspeeding.

THUMB ADJUSTMENT

Remove thumb nut and upper elastic stop nut. Replace thumb nut and adjust to desired operating speed. Do not

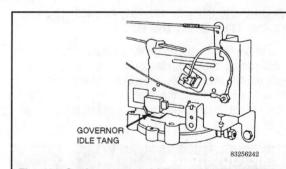

Fig. 424 Setting governed idle on 254700, 257700, 283700, 286700 vertical crankshaft with horizontal control

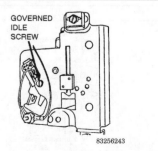

Fig. 425 Setting governed idle on 254700, 257700, 283700, 286700 vertical crankshaft with vertical control

change the position of the lower elastic stop nut. it provides protection against overspeeding.

Models 230000 Horizontal Crankshaft

▶ **See Figure 426**

All engine in Model series 243400, 300400, 320400 and some Model Series 23D and 233400 engine use two governor springs. The shorter spring keeps the engine on governor, even at idle speed. If moderate loads are applied at idle, the engine will not stall.

Models 220000, 250000 Horizontal Crankshaft, except 253400, 255400

1. First make final carburetor mixture adjustment.
2. Place remote control in idle position.
3. Hold throttle in closed position with finger, adjusting idle speed screw to 1550 rpm. Release throttle.
4. Set remote control to 1750 rpm and turn screw in until it contacts remote control lever.
5. Always set desired TOP NO LOAD speed at power test by bending end of control lever at the spring anchor.

Models 253400, 255400 Horizontal Crankshaft

1. First make final carburetor mixture adjustment.
2. Place remote control in idle position.
3. Hold throttle in closed position with finger, adjusting idle speed screw to 1550 rpm. Release throttle.
4. Set remote control to 1750 rpm. Using Tang Bender, Tool #19229, bend tang to obtain 1750 rpm.

Models 230000, 240000, 300000, 320000 Horizontal Crankshaft

▶ **See Figures 427, 428, 429, 430, 431, 432, 433, 434, 435, 436, 437, 438, 439, 440, 441 and 442**

1. First make final carburetor mixture adjustments.
2. Place remote control in idle position.
3. Hold throttle shaft in closed position and adjust idle screw to 1000 rpm. Release the throttle.
4. With remote control in idle position, adjust upper elastic stop nut to 1200 rpm.

Air Vane Governors

The purpose of the governor is to maintain within certain limits. A desired engine speed, even though the load may vary.

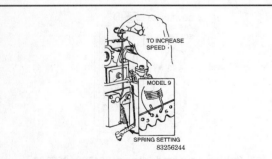
Fig. 426 Remote governor control spring setting on 230000

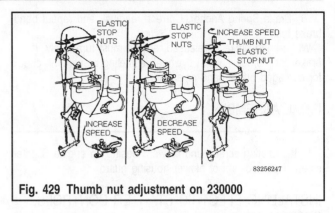
Fig. 429 Thumb nut adjustment on 230000

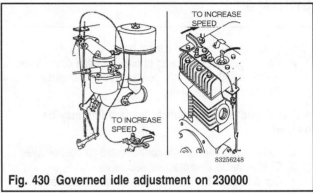
Fig. 430 Governed idle adjustment on 230000

The governor spring tends to open the throttle. Air pressure against the air vane tends to close the throttle. The engine speed at which these two forces balance is called the gov-

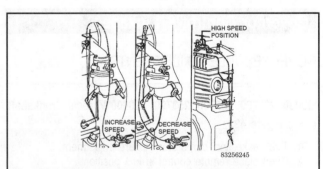
Fig. 427 Remote governor control casing and wire attachments on 230000

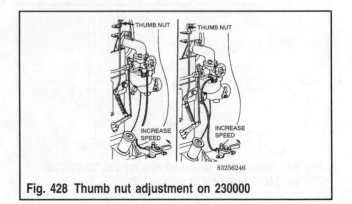
Fig. 428 Thumb nut adjustment on 230000

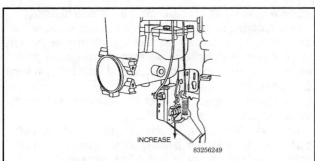

Fig. 431 220000 and 250000 horizontal crankshaft rack and pinion control adjustment

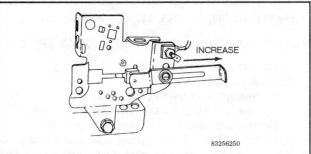

Fig. 432 220000 and 250000 horizontal crankshaft rack and pinion control adjustment

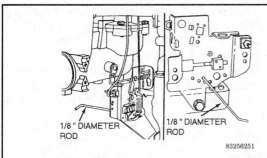

Fig. 433 220000 and 250000 horizontal crankshaft rack and pinion control adjustment

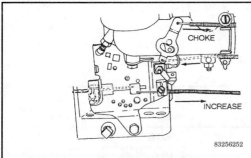

Fig. 434 220000 and 250000 horizontal crankshaft rack and pinion control adjustment

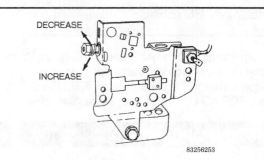

Fig. 435 220000 and 250000 horizontal crankshaft rack and pinion control adjustment

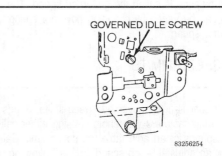

Fig. 436 Governed idle speed adjustments on 220000 and 250000 horizontal crankshaft

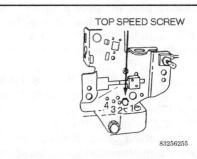

Fig. 437 Top speed screw adjustment on 220000 and 250000 horizontal crankshaft

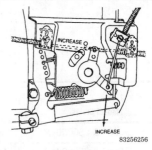

Fig. 438 Governor controls and carburetor linkage adjustments on 253400 and 255400

erned speed. The governed speed can be varied by changing governor spring tension, or changing governor spring.

CHECKING GOVERNOR

Worn linkage or damaged governor springs should be replaced to ensure proper governor operation. If spring or linkage is changed, check and adjust TOP NO LOAD rpm, or check TOP NO LOAD rpm, with engine assembled.

Mechanical Governor

The governor spring tends to pull the throttle open. The force of the counterweights, which are operated by centrifugal force, tends to close the throttle. The engine speed at which these two forces balance is called the governed speed. The governed speed can be varied by changing governor spring tension or governor spring.

GOVERNED SPEED LIMITS

To comply with specified top governed speed limits, Briggs & Stratton supplies manufacturers with engines using either calibrated governor springs or an adjustable top speed limit. Calibrated springs or an adjustable top speed limit will allow no more than a desired top governed speed when the engine is operated on a rigid test stand at our own Factory. However, the design of the cutter blade, deck, etc., can affect engine speeds. Therefore, the top governed speed should be checked with a tachometer when the engine is operated on a completely assembled machine. If on a lawn mower, it should be operated on a hard surface to eliminate cutting load on the blade.

If a service replacement engine is used, check the top governed speed using a tachometer, with the engine operating on a completely assembled mower, to be sure the blade tip speed will not exceed 19,000 feet per minute. If necessary, change the governor spring or adjust the top speed limit device, so the engine will not exceed the recommended speed, based on blade length as shown.

➡After a new governor spring is installed, check engine top governed speed with an accurate tachometer.

Run engine at half throttle to allow the engine to reach normal operating temperature before measuring speed with a tachometer. To account for tolerances, which may be required by tachometer manufacturers, we suggest that the top governed speed of the engine be adjusted at least 200 rpm lower than the maximum speeds shown.

Since blade tip speed is a function of engine rpm, lower tip speeds require lower engine speeds.

Table No.1 lists various lengths of rotary lawn mower cutter blades, and the maximum blade rotational speeds, which will produce blade tip speeds of 19,000 feet per minute.

DISASSEMBLY & ASSEMBLY

▶ See Figures 443, 444, 445, 446, 447, 448 and 449

Cast Iron Model Series 230000, 240000, 300000, 320000

1. Remove engine base.
2. Loosen governor bolt and nut.
3. Remove governor lever from governor crank assembly.
4. Remove hair pin and washer from governor crank.
5. Remove any paint or burrs from governor crank.
6. Remove governor crank. Current production engines have a spacer on the governor crank. Earlier production engine have a long bushing without spacer.
7. Slide governor gear assembly off governor shaft.
8. Press old bushing out of cylinder.
9. Press new bushing into cylinder until busing is flush with outside surface of cylinder. Ream new bushing with Finish Reamer Tool #19333, governor bushing reamer, using Stanisol® or kerosene as lubricant.
10. Assemble governor gear and cup assembly on governor shaft in cylinder. Slide governor crank (and spacer, when used) through bushing from inside cylinder. Install lever, governor spring, and links.

ADJUSTMENT

▶ See Figure 450

Table No. 4

Model Series	Governor Type	Governor Pre-Set RPM	Notes
220000 & 250000	Mechanical	4200	50 & 60 Cycle, 3000 & 3600 RPM
220000 & 250000	Mechanical	2400	60 Cycle, 1800 RPM

83256C18

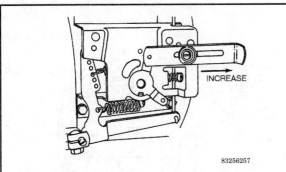

83256257

Fig. 439 Governed idle adjustments on 253400 and 255400

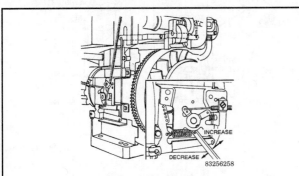

83256258

Fig. 440 Governed idle adjustments on 253400 and 255400

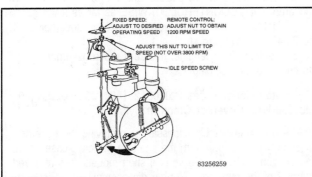

83256259

Fig. 441 Governed idle adjustments on 253400 and 255400

Blade Length	Maximum Rotational RPM
18"	3800
19"	3600
20"	3400
21"	3250
22"	3100
23"	2950
24"	2800
25"	2700

83256260

Fig. 442 Blade length table

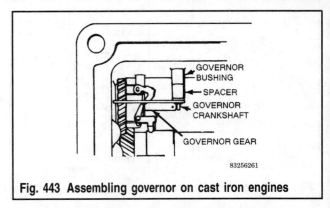

83256261

Fig. 443 Assembling governor on cast iron engines

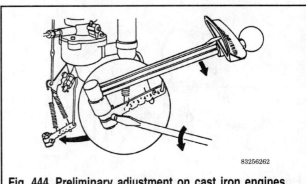

83256262

Fig. 444 Preliminary adjustment on cast iron engines

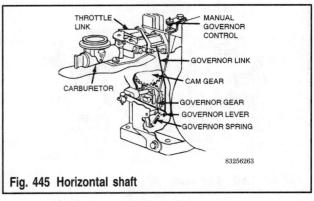

83256263

Fig. 445 Horizontal shaft

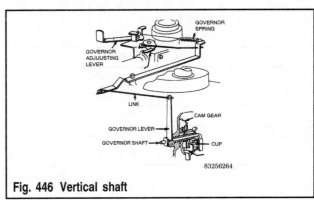

83256264

Fig. 446 Vertical shaft

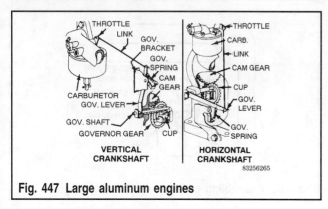

Fig. 447 Large aluminum engines

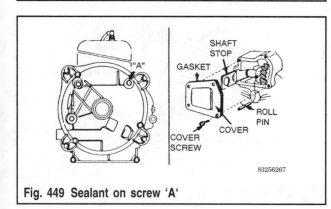

Fig. 448 Proper positioning of the governor crank

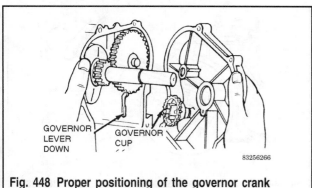

Fig. 449 Sealant on screw 'A'

Hole No.	RPM Range
3	2800-3100
4	3200-3400
5	3500-3700
6	3800-4000
7	4100-4200

83256C19

1. Loosen screw holding governor lever to governor shaft.
2. Place throttle in high speed position.

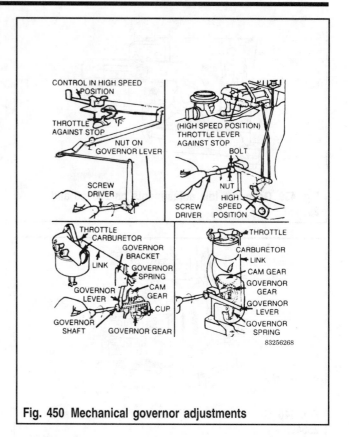

Fig. 450 Mechanical governor adjustments

3. Hold throttle in this position and with a screwdriver turn governor shaft COUNTERCLOCKWISE as far as it will go.
4. Tighten screw holding governor lever to governor shaft to 35-45 inch lbs. torque.
5. Before starting engine, manually move governor linkage to check for any binding.

Aluminum Model Series 220000, 251400, 252400, 254400, 280000 With Governor Crank in Cylinder

The only disassembly necessary is removing the governor assembly as one unit from the shaft on the crankcase cover on horizontal models. On vertical shaft models, it is removed as part of the oil slinger. Further disassembly is unnecessary.

To Assemble:

1. On horizontal crankshaft models, the governor rides on a short stationary shaft and is retained by the governor shaft, with which it comes in contact after the crankcase cover is secured in place.
2. Press governor cup against crankcase cover to seat retaining ring on shaft, prior to installing crankcase cover. It is suggested that the assembly of the crankcase cover be made with the crankshaft in a horizontal position.
3. The governor shaft should hang straight down parallel to the cylinder axis. If the governor shaft is clamped in an angular position, pointing toward the crankcase cover, it is possible for the end of the shaft to be jammed into the inside of the governor assembly, resulting in broken parts when the engine is started.
4. After the crankcase cover and gasket are in place, install cover screws. be sure that screw **A,** has non-hardening sealant, such as Permatex® No.2 on threads of screw.

5. Complete installation of remaining governor linkages and carburetor and then adjust governor shaft and lever. See 'Adjustment.'

6. On vertical crankshaft models the governor is part of the oil slinger and is installed as shown. Before installing sump be sure that governor cup is in line with governor shaft paddle. install sump and gasket being sure **A,** has non-hardening sealant on threads such as Permatex® No.2.

Complete installation of remaining governor linkages and carburetor. Then adjust governor shaft and lever. See 'Adjustments.'

➡**If governor shaft bushing is replaced, it must be finished reamed with Governor Bushing Reamer, Tool #19333, for ¼ in. governor crank or with Governor Bushing Reamer, Tool #19058, for ³/₁₆ in. governor crank.**

ADJUSTMENT

Loosen screw holding governor lever to governor shaft. Place throttle in high speed position. While holding throttle in this position with a screwdriver, turn governor shaft clockwise as far as it will go. Tighten screw holding governor lever to governor shaft. Torque to 35-45 inch lbs.

Before starting engine, manually move governor linkage to check for any binding.

ADJUST MANUAL FRICTION CONTROL

For fixed speed place speed control lever in maximum rpm position and tighten wing nut down until lever cannot be moved.

For manual friction, tighten wing nut until lever will stay in any position, without moving while engine is running.

For remote control, loosen wing nut until speed control lever drops of its own weight down to idle.

Generator Engines

Place governor spring in governor lever hole number 2 for 50 Cycle or hole number 4 for 60 Cycle generators.

Start and run engine at half throttle for five minutes to bring engine to operating temperature. then move speed control lever to maximum rpm position. If tab on lever is touching head of TOP NO LOAD rpm Adjusting screw, back out screw until tab no longer touches screw when control lever is in maximum rpm position.

Bend spring anchor tang using Tang Bender, Tool #19229, to 33000 for 50 cycle or 3800 for 60 cycle.

Turn TOP NO LOAD rpm screw clockwise until 3150 rpm, 50 Cycle or 3750 rpm, 60 Cycles is obtained, No load.

If available, use a load bank to load engine to full generator rated output. With generator at full rate output, turn screw to obtain 3000 rpm, 50 Cycle or 3600 rpm, 60 Cycle.

Aluminum Model Series 220000, 250000, 280000 (Except 253400, 255400)

GOVERNOR ADJUSTMENT

▶ **See Figure 451**

1. Set control lever to maximum speed position, with engine running.

2. Use Tang Bender Tool, Tool #19229, to bend spring anchor tang to obtain the proper top no load rpm. If Tool #19229 is not available, make tool.

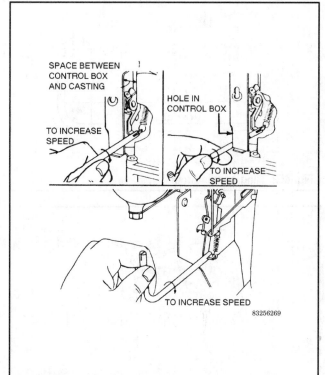

Fig. 451 Adjusting top no-load speed on aluminum 220000, 250000 and 280000

GENERATOR APPLICATIONS ONLY

Governor regulation to within two cycles of either 60 or 50 cycles can be obtained is the procedures indicated below are followed:

1. Push speed adjusting nut in and up to release spring tension on nut.

2. Start engine and pull out on speed adjusting nut to maximum length of travel.

3. Set engine speed per Table No.3 by bending governor tang. With engine still running, return speed adjusting nut down into slot.

4. Then turn speed adjusting nut to obtain:

• 1600 rpm Top No Load for 1500 rpm 50 cycle generator

• 1875 rpm Top No Load for 1800 rpm 60 Cycle generator

• 3100 rpm Top No Load for 3000 rpm 50 cycle generator

• 3700 rpm Top No Load for 3600 rpm 60 cycle generator

Model Series 253400, 255400

GOVERNED IDLE

▶ **See Figures 452, 453 and 454**

Turn carburetor idle speed adjusting screw to obtain 1600 rpm while holding throttle lever against screw. Release throttle lever. Align holes in control bracket and inside lever with ⅛ in. diameter rod. Governor speed control lever of equipment should be in **IDLE** position. Adjust if necessary. bend spring tang to obtain 1750 rpm. Remove ⅛ in. diameter rod.

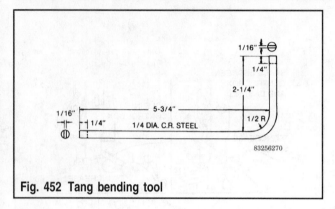

Fig. 452 Tang bending tool

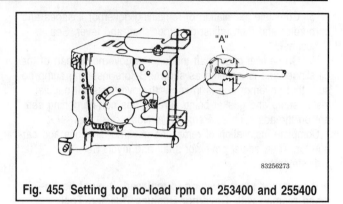

Fig. 455 Setting top no-load rpm on 253400 and 255400

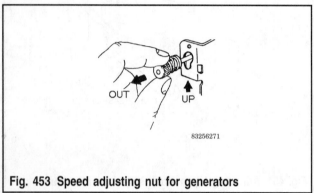

Fig. 453 Speed adjusting nut for generators

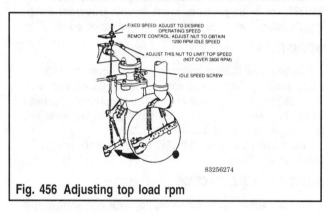

Fig. 456 Adjusting top load rpm

Adjust lower stop nut to obtain top no load rpm.

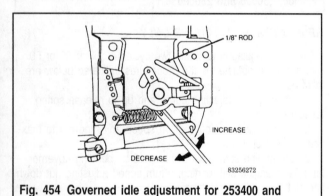

Fig. 454 Governed idle adjustment for 253400 and 255400

ADJUSTING TOP NO LOAD SPEED
▶ See Figure 455

On Model series 253400, 255400 with speed control at fast position, turn screw **A** to set top no load rpm. Turn clockwise to increase or counterclockwise to decrease speed.

Cast Iron Model Series 230000, 240000, 300000, 320000w/Mechanical Governor

ADJUSTING TOP NO LOAD SPEED
▶ See Figure 456

Fixed Speed Operation
1. Loosen lower stop nut.
2. Align top stop nut to obtain top no load rpm.
3. After speed is set, tighten lower stop nut.

Remote Control Operation

Cylinder Head

REMOVAL & INSTALLATION

▶ See Figures 457 and 458

Cylinder Head Torque

Basic Model Series	
Aluminum Cylinder	Inch Pounds
220000, 250000, 280000	165
Cast Iron Cylinder	Inch Pounds
230000, 240000, 300000, 320000	190

Always note the position of the different cylinder head screws so that they can be properly reinstalled. If a screw is used in the wrong position, it may be to short and not engage enough threads. If it is too long, it may bottom on a fin, either breaking the fin, or leaving the cylinder head loose.

Remove the cylinder screws and then the cylinder head. Be sure to remove the gasket and all remaining gasket material from the cylinder head and the block. Assemble the cylinder head with a new gasket, cylinder head shield, screws, and

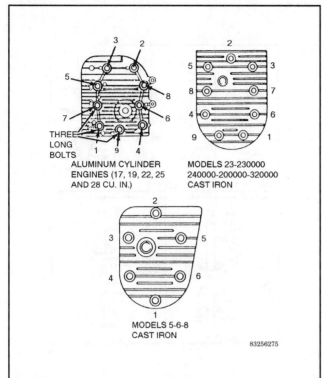

Fig. 457 Head bolt torque patterns for single cylinder engines

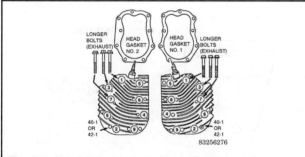

Fig. 458 Head bolt torque patterns for twin cylinder engines

washers in their proper places. Graphite grease should be used on aluminum cylinder head screws.

Do not use a sealer of any kind on the head gasket. Tighten the screws down evenly by hand. Use a torque wrench and tighten the head bolts in the correct sequence.

Valves

REMOVAL AND INSTALLATION

▶ See Figures 459, 460, 461, 462 and 463

1. Using a valve spring compressor, adjust the jaws so they touch the top and bottom of the valve chamber, and then place one of the jaws over the valve spring and the other underneath, between the spring and the valve chamber. This

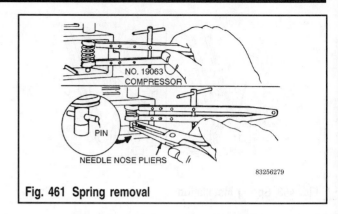

Fig. 461 Spring removal

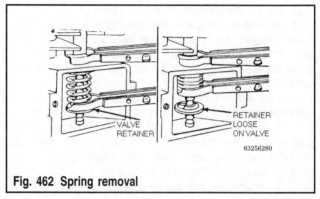

Fig. 462 Spring removal

positioning of the valve spring compressor is for valves that have either pin or collar type retainers.

2. Tighten the jaws to compress the spring. Remove the collars or pin and lift out the valve.

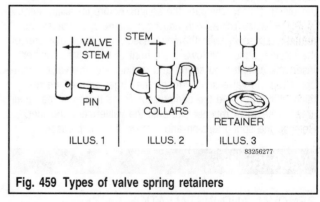

Fig. 459 Types of valve spring retainers

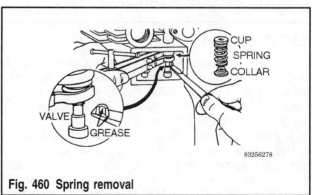

Fig. 460 Spring removal

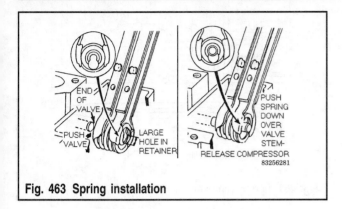

Fig. 463 Spring installation

3. Pull out the compressor and the spring.

4. To remove valves with ring type retainers, position the compressor with the upper jaw over the top of the valve chamber and the lower jaw between the spring and the retainer. Compress the spring, remove the retainer, and pull out the valve. Remove the compressor and spring.

5. If the retainers are held by a pin or collars, place the valve spring and retainer and cup into the valve spring compressor. Compress the spring until it is solid. Insert the compressed spring and retainer into the valve chamber. Then drop the valve into place, pushing the stem through the retainer. Hold the spring up in the chamber, hold the valve down, and insert the retainer pin with needle nose pliers or place the collars in the groove in the valve stem. Loosen the spring until the retainer fits around the pin or collars, then pull out the spring compressor. Be sure the pin or collars are in place.

6. Before installing the valves, check the thickness of the valve springs. Some engines use the same spring for the intake and exhaust side, while others use a heavier spring on the exhaust side. Compare the springs before installing them.

7. To install valves with ring type retainers, compress the retainer and spring with the compressor. The large diameter of the retainer should be toward the front of the valve chamber. Insert the compressed spring and retainer into the valve chamber. Drop the valve stem through the larger area of the retainer slot and move the compressor so as to center the small area of the valve retainer slot onto the valve stem shoulder. Release the spring tension and remove the compressor.

Valve Guides

REMOVAL AND INSTALLATION

▶ **See Figures 464, 465, 466, 467, 468 and 469**

First check valve guide for wear with a plug gauge, Briggs & Stratton part #19151 or equivalent. If the flat end of the valve guide plug gauge can be inserted into the valve guide for a distance of 5/16 in. (8mm), the guide is worn and should be rebushed in the following manner.

1. Procure a reamer #19183 and reamer guide bushing #19192, and lubricate the reamer with kerosene. Then, use reamer and reamer guide bushing to ream out the worn guide. Ream to only 1/16 in. (1.6mm) deeper than valve guide bushing #230655 . BE CAREFUL NOT TO REAM THROUGH THE GUIDE!

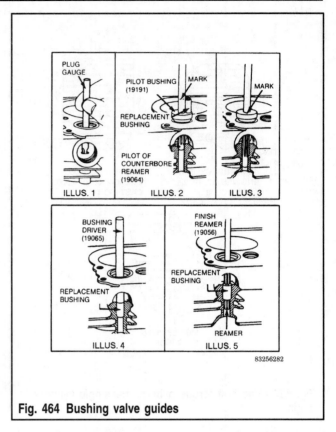

Fig. 464 Bushing valve guides

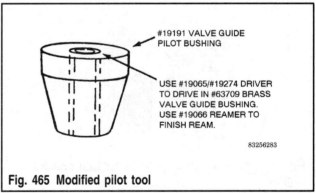

Fig. 465 Modified pilot tool

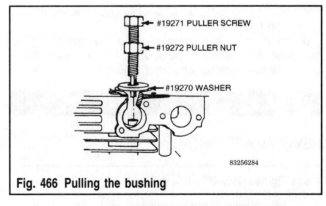

Fig. 466 Pulling the bushing

2. Press in valve guide bushing #230655 until top end of bushing is flush with top end of valve guide. Use a soft metal

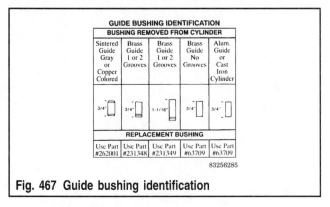

Fig. 467 Guide bushing identification

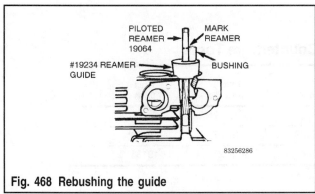

Fig. 468 Rebushing the guide

driver (brass, copper, etc.) so top end of bushing is not peened over.

➡**The bushing #230655 is finish reamed to size at the factory, so no further reaming is necessary, and a standard valve can be used.**

Valve seating should be checked after bushing the guide, and corrected if necessary by refacing the seat.

REFACING VALVES AND SEATS

Faces on valves and valve seats should be resurfaced with a valve grinder or cutter to an angle of 45°.

➡**Some engines have a 30° intake valve and seat.**

The valve and seat should then be lapped with a fine lapping compound to remove the grinding marks and ensure a good seat. The valve seat width should be 1.2-1.6mm. If the seat is wider, a narrowing stone or cutter should be used. If either the seat or valve is badly burned, it should be replaced. Replace the valve if the edge thickness (margin) is less than 0.4mm after it has been resurfaced.

CHECK AND ADJUST TAPPET CLEARANCE

Cylinder Head Torque

Basic Model Series	
Aluminum Cylinder	**Inch Pounds**
220000, 250000, 280000	165
Cast Iron Cylinder	**Inch Pounds**
230000, 240000, 300000, 320000	190

83256C20

Insert the valves in their respective positions in the cylinder. Turn the crankshaft until one of the valves is at its highest position. Turn the crankshaft one revolution. Check the clearance with a feeler gauge. Repeat for the other valve. Grind off the end of the valve stem if necessary to obtain proper clearance.

➡**Check the valve tappet clearance with the engine cold.**

Valve Seat Inserts

Cast iron cylinder engines are equipped with an exhaust valve insert which can be removed and replaced with a new insert. The intake side must be counter-bored to allow the installation of an intake valve seat insert (see below). Aluminum alloy cylinder models are equipped with inserts on both the exhaust and intake valves.

REMOVAL AND INSTALLATION

▶ **See Figures 470, 471, 472, 473, 474, 475 and 476**

Valve Seat Inserts

Basic Model Series	Intake Standard	Exhaust Standard	Exhaust Cobalite™	Insert* Puller Assembly	Puller Nut
Aluminum Cylinder					
220000, 250000, 280000	261463	211661	210940	19138	19141 Ex.
Cast Iron Cylinder					
230000	21880	21880	21612	19138	19141
240000	21880	None	21612	19138	19141
300000, 320000	None	None	21612	19138	19141 Ex.

* Includes puller and No. 19182, 19140 and 19139 nuts.

83256C21

GUIDE BUSHING IDENTIFICATION

BUSHING REMOVED FROM CYLINDER			
Sintered Guide Gray or Copper Colored	Brass Guide 1 or 2 Grooves	Brass Guide No Grooves	Alum. Guide or Cast Iron Cylinder
REPLACEMENT BUSHING			
Use Part #261961	Use Part #231218	Use Part #230655	Use Part #231218

83256287

Fig. 469 Guide bushing identification

Valve Seat Insert and Counterbore Tools

Basic Model Series	Counter-Bore Cutter	Shank	Cutter & Driver Pilot	Insert Driver
Cast Iron Cylinder				
230000, 240000	19131	19129	19127	19136
300000, 320000	—	—	19127	19136

83256C22

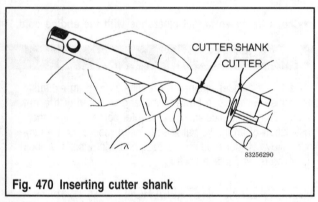

Fig. 470 Inserting cutter shank

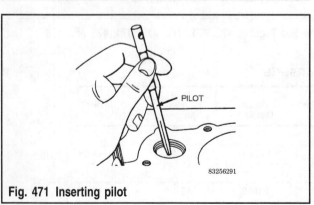

Fig. 471 Inserting pilot

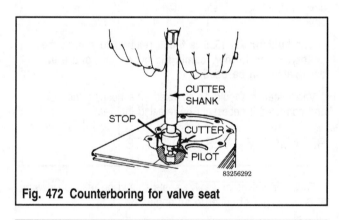

Fig. 472 Counterboring for valve seat

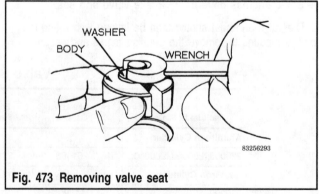

Fig. 473 Removing valve seat

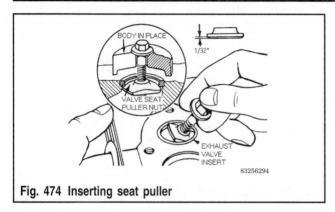

Fig. 474 Inserting seat puller

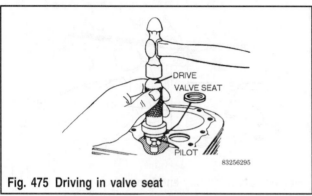

Fig. 475 Driving in valve seat

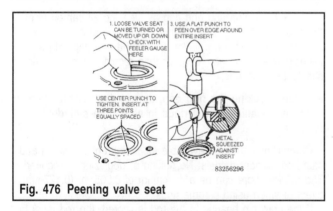

Fig. 476 Peening valve seat

Valve seat inserts are removed with a special puller.

➡**On aluminum alloy cylinder models, it may be necessary to grind the puller nut until the edge is ³⁄₃₂ in. (0.8mm) thick in order to get the puller nut under the valve insert.**

When installing the valve seat insert, make sure that the side with the chamfered outer edge goes down into the cylinder. Install the seat insert and drive it into place with a driver. The seat should then be ground lightly and the valves and seats lapped lightly with grinding compound.

Aluminum alloy cylinder models use the old insert as a spacer between the driver and the new insert. Drive in the new insert until it bottoms. The top of the insert will be slightly below the cylinder head gasket surface. Peen around the insert using a punch and hammer.

The intake valve seat on cast iron cylinder models has to be counter-bored before installing the new valve seat insert.

COUNTERBORING THE CYLINDER FOR THE INTAKE VALVE SEAT ON CAST IRON MODELS

1. Select the proper seat insert, cutter shank, counter bore cutter, pilot and driver from the table. These numbers refer to Briggs & Stratton parts — you may get equivalent parts from other sources if available.

2. With cylinder head resting on a flat surface, valve seats up, slide the pilot into the intake valve guide. Then, assemble the correct counterbore cutter to the shank with the cutting blades of the cutter downward.

3. Insert the cutter straight into the valve seat, over the pilot. Cut so as to avoid forcing the cutter to one side, and be sure to stop as soon as the stop on the cutter touches the cylinder head.

4. Blow out all cutting chips thoroughly.

Pistons, Piston Rings, and Connecting Rods

REMOVAL

▶ **See Figures 477, 478, 479 and 480**

To remove the piston and connecting rod from the engine, bend down the connecting rod lock. Remove the connecting rod cap. Remove any carbon or ridge at the top of the cylinder bore. This will prevent breaking the rings. Push the piston and rod out of the top of the cylinder.

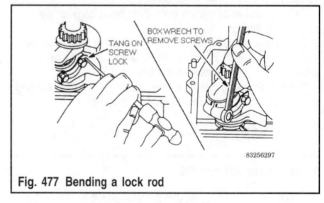

Fig. 477 Bending a lock rod

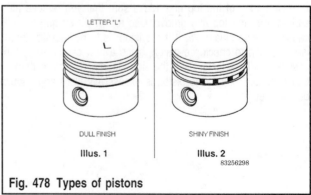

Fig. 478 Types of pistons

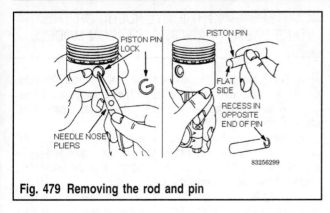

Fig. 479 Removing the rod and pin

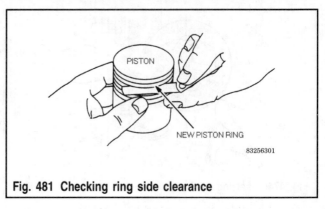

Fig. 481 Checking ring side clearance

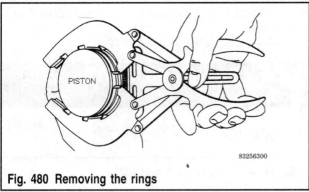

Fig. 480 Removing the rings

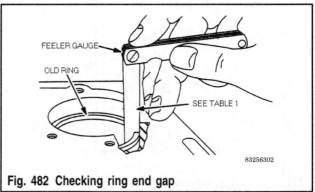

Fig. 482 Checking ring end gap

Pistons used in sleeve bore, aluminum alloy engines are marked with an **L** on top of the piston. These pistons are tin plated and use an expander with the oil ring. This piston assembly is not interchangeable with the piston used in the aluminum bore engines (Kool bore).

Pistons used in aluminum bore (Kool bore) engines are not marked on the top.

To remove the connecting rod from the piston, remove the piston pin lock with thin nose pliers. One end of the pin is drilled to facilitate removal of the lock.

Remove the rings one at a time, slipping them over the ring lands. Use a ring expander to remove the rings.

INSPECTION

♦ See Figures 481 and 482

Check the piston ring fit. Use a feeler gauge to check the side clearance of the top ring. Make sure that you remove all carbon from the top ring groove. Use a new piston ring to check the side clearance. If the cylinder is to be resized, there is no reason to check the piston, since a new oversized piston assembly will be installed. If the side clearance is more than 0.007 in. (0.178mm), the piston is excessively worn and should be replaced.

Check the piston ring end gap by cleaning all carbon from the ends of the rings and inserting them one at a time 1 in. (25mm) down into the cylinder. Check the end gap with a feeler gauge. If the gap is larger than recommended, the ring should be replaced.

➡**When checking the ring gap, do not deglaze the cylinder walls by installing piston rings in aluminum cylinder engines.**

Chrome ring sets are available for all current aluminum and cast iron cylinder models. No honing or deglazing is required. The cylinder bore can be a maximum of 0.005 in. (0.127mm) oversize when using chrome rings.

If the crankpin bearing in the rod is scored, the rod must be replaced. 0.005 in. (0.127mm) oversize piston pins are available in case the connecting rod and piston are worn at the piston pin bearing. If, however, the crankpin bearing in the connecting rod is worn, the rod should be replaced. Do not attempt to file or fit the rod.

If the piston pin is worn 0.0005 in out of round or below the rejection sizes, it should be replaced.

INSTALLATION

♦ See Figures 483, 484, 485, 486, 487, 488, 489 and 490

Connecting Rod Reject Size

Basic Model Series	Crank Pin Bearing	Piston Pin Bearing
Aluminum Cylinder		
220000, 250000, 280000	1.252"	.802"
Cast Iron Cylinder		
230000	1.189"	.736"
240000	1.314"	.674"
300000, 320000	1.314"	.802"

83256C24

Piston Pin Reject Sizes

Basic Model Series	Piston Pin O.D.	Pin Bore I.D.
Aluminum Cylinder		
220000, 250000, 280000	.799"	.801"
Cast Iron Cylinder		
230000	.734"	.736"
240000	.671"	.673"
300000, 320000	.799"	.801"

83256C25

The piston pin is a push fit into both the piston and the connecting rod. On models using a solid piston pin, one end is flat and the other end is recessed. Other models use a hollow piston pin. Place a pin lock in the groove at one side of the piston. From the opposite side of the piston, insert the piston pin, flat end first for solid pins; with hollow pins, insert either end first until it stops against the pin lock. Use thin nose pliers to assemble the pin lock in the recessed end of the piston. Be sure the locks are firmly set in the groove.

Install the rings on the pistons, using a piston ring expander. Make sure that they are installed in the proper position. The scraper groove on the center compression ring should always be down toward the piston skirt. Be sure the oil return holes are clean and all carbon is removed from the grooves.

➡**Install the expander under the oil ring in sleeve bore aluminum alloy engines.**

Oil the rings and the piston skirt, then compress the rings with a ring compressor. On cast iron engines, install the compressor with the two projections downward; on aluminum engines, install the compressor with the two projections upward. These instructions refer to the piston in normal position — with skirt downward. Turn the piston and compressor upside down on the bench and push downward so the piston head and the edge of the compressor band are even, all the while tightening the compressor. Draw the compressor up tight to fully compress the rings, then loosen the compressor very slightly.

✳✳WARNING

Do not attempt to install the piston and ring assembly without using a ring compressor.

Place the connecting rod and piston assembly, with the rings compressed, into the cylinder bore. Push the piston and rod down into the cylinder. Oil the crankpin of the crankshaft. Pull the connecting rod against the crankpin and assemble the rod cap so the assembly marks align.

➡**Some rods do not have assembly marks, as the rod and cap will fit together only in one position. Use care to ensure proper installation. On the 251000 engine, the piston has a notch on the top surface. The notch must face the flywheel side of the block when installed.**

Where there are flat washers under the cap screws, remove and discard them prior to installing the rod. Assemble the cap screws and screw locks with the oil dippers (if used), and torque to the figure shown in the chart to avoid breakage or rod scoring later. Turn the crankshaft two revolutions to be sure the rod is correctly installed. If the rod strikes the camshaft, the connecting rod has been installed wrong or the cam gear is out of time. If the crankshaft operates freely, bend the cap screw locks against the screw heads. After tightening the rod screws, the rod should be able to move sideways on the crankpin of the shaft.

Crankshaft and Camshaft Gear

REMOVAL

Aluminum Cylinder Engines

Crankshaft Reject Sizes

Model Series	PTO Journal (Inches)	Crankpin Journal (Inches)	Magneto Journal (Inches)
Aluminum Cylinder			
220000, 250000, 280000	1.376	1.247	1.376
Cast Iron Cylinder			
230000 ♦	1.376 ♦	1.184	1.376
240000 ♦	Ball ♦	1.309	Ball
300000, 320000	Ball	1.309	Ball

♦ Gear Reduction PTO 1.179 in.

83256C26

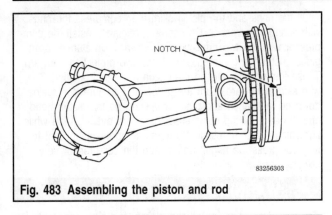

Fig. 483 Assembling the piston and rod

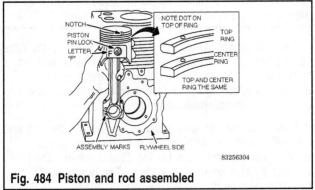

Fig. 484 Piston and rod assembled

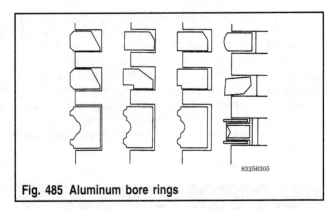

Fig. 485 Aluminum bore rings

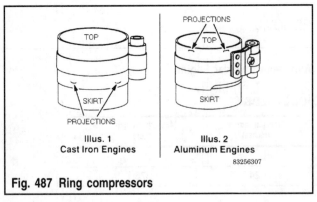

Fig. 487 Ring compressors

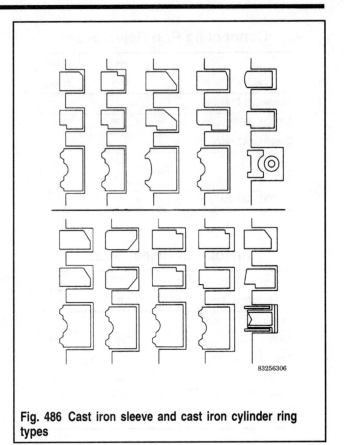

Fig. 486 Cast iron sleeve and cast iron cylinder ring types

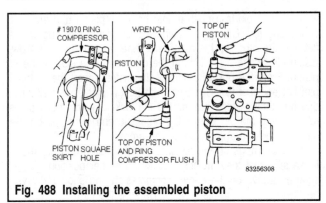

Fig. 488 Installing the assembled piston

Crankshaft Reject Sizes

Model Series	PTO Journal (Inches)	Crankpin Journal (Inches)	Magneto Journal (Inches)
Aluminum Cylinder			
220000, 250000, 280000	1.376	1.247	1.376
Cast Iron Cylinder			
230000 ♦	1.376 ♦	1.184	1.376
240000 ♦	Ball ♦	1.309	Ball
300000, 320000	Ball	1.309	Ball

♦ Gear Reduction PTO 1.179 in.

83256C26

Crankshaft Reject Sizes

Model Series	PTO Journal (Inches)	Crankpin Journal (Inches)	Magneto Journal (Inches)
Aluminum Cylinder			
220000, 250000, 280000	1.376	1.247	1.376
Cast Iron Cylinder			
230000 ♦	1.376 ♦	1.184	1.376
240000 ♦	Ball ♦	1.309	Ball
300000, 320000	Ball	1.309	Ball

♦ Gear Reduction PTO 1.179 in.

83256C26

Crankshaft Reject Sizes

Model Series	PTO Journal (Inches)	Crankpin Journal (Inches)	Magneto Journal (Inches)
Aluminum Cylinder			
220000, 250000, 280000	1.376	1.247	1.376
Cast Iron Cylinder			
230000 ♦	1.376 ♦	1.184	1.376
240000 ♦	Ball ♦	1.309	Ball
300000, 320000	Ball	1.309	Ball

♦ Gear Reduction PTO 1.179 in.

83256C26

To remove the crankshaft from aluminum alloy engines, remove any rust or burrs from the power take-off end of the crankshaft. Remove the crankcase cover or sump. If the sump or cover sticks, tap it lightly with a soft hammer on alternate sides near the dowel. Turn the crankshaft to align the crankshaft and camshaft timing marks, lift out the cam gear, then remove the crankshaft. On models that have ball bearings on the crankshaft, the crankshaft and the camshaft must be removed together with the timing marks properly aligned — see illustration.

Cast Iron Cylinder Models
▶ See Figures 491, 492, 493, 494, 495 and 496

To remove the crankshaft from cast iron models, remove the crankcase cover. Revolve the crankshaft until the crankpin is pointing upward toward the breather at the rear of the engine (approximately a 45° angle). Pull the crankshaft out from the drive side, twisting it slightly if necessary. On models with ball bearings on the crankshaft, both the crankcase cover and bearing support should be removed.

On cast iron models with ball bearings on the drive side, first remove the magneto. Drive into the recess at the front of the engine. Then draw the crankshaft from the magneto side of the engine. Double thrust engines have cap screws inside the crankcase which hold the bearing in place. These must be removed before the crankshaft can be removed.

To remove the camshaft from all cast iron models, except the 300400 and 320400, use a long punch to drive the camshaft out toward the magneto side. Save the plug. Do not burr or peen the end of the shaft while driving it out. Hold the camshaft while driving it out. Hold the camshaft while removing the punch, so it will not drop and become damaged.

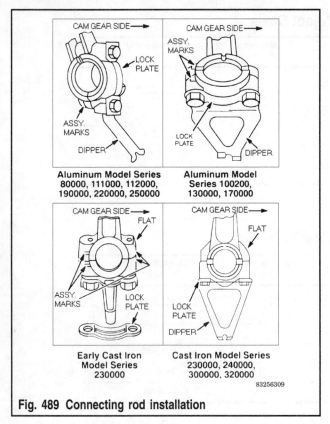

Fig. 489 Connecting rod installation

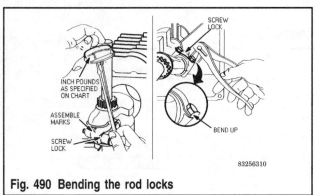

Fig. 490 Bending the rod locks

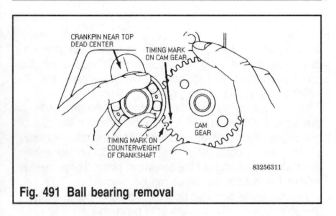

Fig. 491 Ball bearing removal

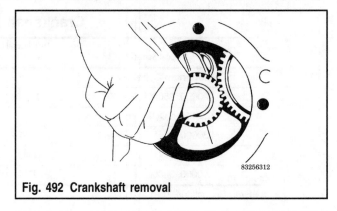

Fig. 492 Crankshaft removal

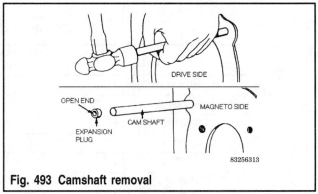

Fig. 493 Camshaft removal

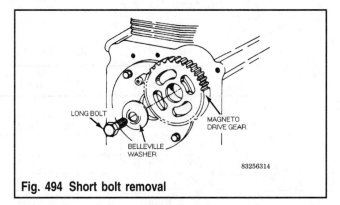

Fig. 494 Short bolt removal

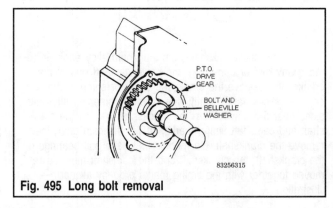

Fig. 495 Long bolt removal

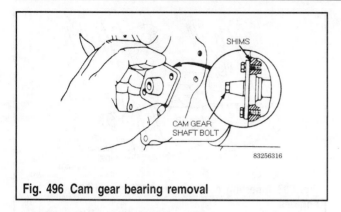

Fig. 496 Cam gear bearing removal

CHECKING THE CRANKSHAFT

▶ **See Figure 497**

Discard the crankshaft if it is worn beyond the allowable limit. Check the keyways for wear and make sure they are not spread. Remove all burrs from the keyway to prevent scratching the bearing. Check the three bearing journals, drive end, crankpin, and magneto end, for size and any wear or damage. Check the cam gear teeth for wear. They should not be worn at all. Check the threads at the magneto end for damage. Make sure that the crankshaft is straight.

➡**There are 0.020 in. (0.5mm) undersize connecting rods available for use on reground crankpin bearings.**

REMOVAL AND INSTALLATION OF THE BALL BEARINGS

▶ **See Figures 498 and 499**

The ball bearings are pressed onto the crankshaft. If either the bearing or the crankshaft is to be removed, use an arbor press to remove them.

To install, heat the bearing in hot oil — 325°F (163°C) maximum. Don't let the bearing rest on the bottom of the pan in which it is heated. Place the crankshaft in a vise with the bearing side up. When the bearing is quite hot, it will slip fit onto the bearing journal. Grasp the bearing, with the shield down, and thrust it down onto the crankshaft. The bearing will tighten on the shaft while cooling. Do not quench the bearing (throw water on it to cool it).

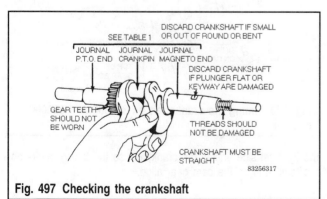

Fig. 497 Checking the crankshaft

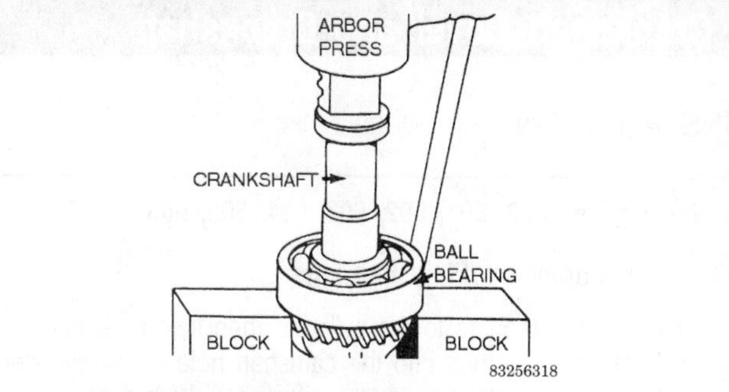

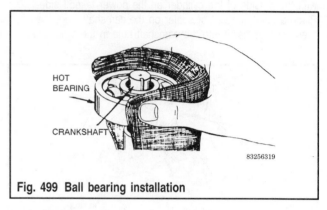

Fig. 498 Ball bearing removal

Fig. 499 Ball bearing installation

CHECKING THE CAMSHAFT GEAR

Inspect the teeth for wear and nicks. Check the size of the camshaft and camshaft gear bearing journals. Check the size of the cam lobes. If the cam is worn beyond tolerance, discard it.

Check the automatic spark advance on models equipped with the Magna-Matic ignition system. Place the cam gear in the normal operating position with the movable weight down. Press the weight down and release it. The spring should lift the weight. If not, the spring is stretched or the weight is binding.

INSTALLATION

Aluminum Alloy Engines — Plain Bearing

In aluminum alloy engines, the tappets are inserted first, the crankshaft next, and then the cam gear. When inserting the cam gear, turn the crankshaft and the cam gear so that the timing marks on the gears align.

Aluminum Alloy Engines — Ball Bearing

On crankshafts with ball bearings, the gear teeth are not visible for alignment of the timing marks; therefore, the timing mark is on the counterweight. On ball bearing equipped engines, the tappets are installed first. The crankshaft and the cam gear must be inserted together and their timing marks aligned.

Crankshaft Cover and Crankshaft

INSTALLATION

▶ **See Figures 500, 501, 502, 503, 504, 505, 506 and 507**

Cast Iron Engines w/Plain Bearings

Assemble the tappets to the cylinder, then insert the cam gear. Push the camshaft into the camshaft hole in the cylinder from the flywheel side through the cam gear. With a blunt punch, press or hammer the camshaft until the end is flush with the outside of the cylinder on the power takeoff side. Place a small amount of sealer on the camshaft plug, then press or hammer it into the camshaft hole in the cylinder at

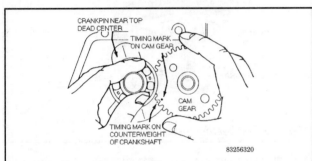

Fig. 500 Installing the crankshaft and cam gear on ball bearing engines

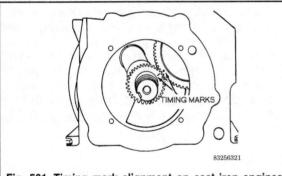

Fig. 501 Timing mark alignment on cast iron engines

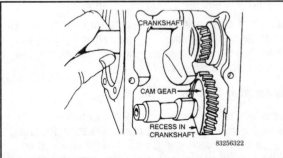

Fig. 502 Installing the crankshaft and cam gear on abll bearing cast iron engines, exc. 300000 and 320000

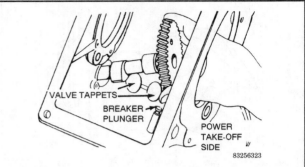

Fig. 503 Inserting can gear on 300000 and 320000 engines

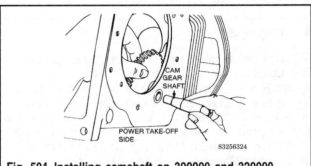

Fig. 504 Installing camshaft on 300000 and 320000 engines

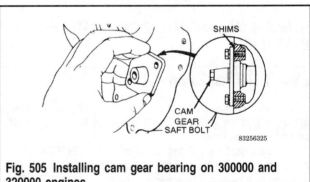

Fig. 505 Installing cam gear bearing on 300000 and 320000 engines

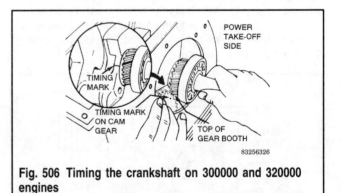

Fig. 506 Timing the crankshaft on 300000 and 320000 engines

the flywheel side. Install the crankshaft so the timing marks on the teeth and on the cam gear align.

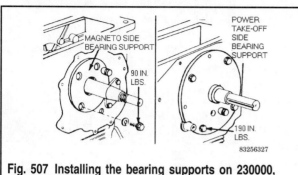

Fig. 507 Installing the bearing supports on 230000, 240000, 300000 and 320000 engines

Cast Iron w/Ball Bearings

Assemble the tappets, then insert the cam gear into the cylinder, pushing the cam gear forward into the recess in front of the cylinder. Insert the crankshaft into the cylinder. Turn the camshaft and crankshaft until the timing marks align, then push the cam gear back until it engages the gear on the crankshaft with the timing marks together. Insert the camshaft. Place a small amount of sealer on the camshaft plug and press or hammer it into the camshaft hole in the cylinder at the flywheel side.

CRANKSHAFT END-PLAY ADJUSTMENT

▶ **See Figures 508, 509 and 510**

The crankshaft end-play on all models, plain and ball bearing, should be 0.002-0.008 in. (0.2mm). The method of obtaining the correct end-play varies, however, between cast iron, aluminum, plain, and ball bearing models. New gasket sets include three crankcase cover or bearing support gaskets, 0.005 in. (0.127mm), 0.009 in. (0.228mm), and 0.015 in. (0.381mm) thick.

The end-play of the crankshaft may be checked by assembling a dial indicator on the crankshaft with the pointer against the crankcase. Move the crankshaft in and out. The indicator will show the end-play. Another way to measure the end-play is to assemble a pulley to the crankshaft and measure the end-play with a feeler gauge. Place the feeler gauge between the crankshaft thrust face and the bearing support. The feeler

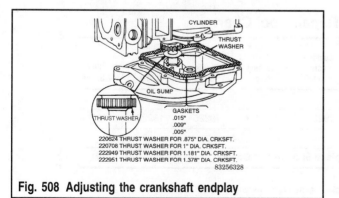

Fig. 508 Adjusting the crankshaft endplay

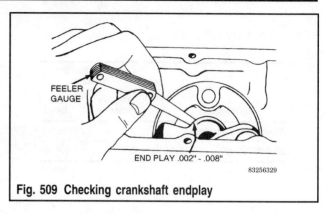

Fig. 509 Checking crankshaft endplay

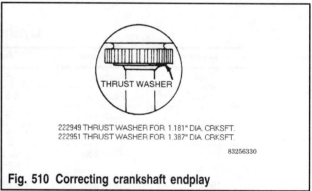

Fig. 510 Correcting crankshaft endplay

gauge method of measuring crankshaft end-play can only be used on cast iron plain bearing engines with removable bases.

On cast iron engines, the end-play should be 0.002-0.008 in. (0.05-0.20mm) with one 0.015 in. (0.381mm) gasket in place. If the end-play is less than 0.002 in. (0.05mm), which would be the case if a new crankcase or sump cover is used, additional gaskets of 0.005 in. (0.127mm), 0.009 in. (0.228mm) or 0.015 in. (0.381mm) may be added in various combinations to obtain the proper end-play.

Aluminum Engines Only

If the end-play is more than 0.008 in. (0.20mm) with one 0.015 in. (0.381mm) gasket in place, a thrust washer is available to be placed on the crankshaft power take-off end, between the gear and crankcase cover or sump on plain bearing engines. On ball bearing equipped aluminum engines, the thrust washer is added to the magneto end of the crankshaft instead of the power take-off end.

➡ **Aluminum engines never use less than the 0.015 in. (0.381mm) gasket.**

Cylinders

INSPECTION

▶ **See Figures 511 and 512**

Model Series	Standard Bore Size Diameter	
	Max. Inches	Min. Inches
Aluminum Cylinder		
220000, 250000, 280000	3.4375	3.4365

83256C30

Cylinder Hones

Hone Set #	Bore Material	Bore Size	Stone Set #	Carrier Set #
19205	Aluminum	1-7/8 to 2-3/4″	19206	19205
19205	Aluminum	2-5/8 to 3-1/2″	19207	19205
	Cast Iron	1-7/8 to 2-3/4″	19303 (60 grit)	19205
	Cast Iron	1-7/8 to 2-3/4″	19304 (220 grit)	19205
19211	Cast Iron	2-1/2 to 3-5/16″ 2-1/2 to 3-5/16″	19212 (60 grit) 19213 (220 grit)	19214 19214
19211	Cast Iron	3-5/16 to 4-1/8″ 3-5/16 to 4-1/8″	19212 (60 grit) 19213 (220 grit)	19215 19215

83256C31

Model Series	Standard Bore Size Diameter	
	Max. Inches	Min. Inches
Cast Iron Cylinder		
230000	3.0000	2.9990
240000	3.0625	3.0615
300000	3.4375	3.4365
320000	3.5625	3.5615

83256C32

Cylinder Bearing Reject Size Chart

Model Series	Magneto Bearing (Inches)	PTO Bearing (Inches)
Aluminum Cylinder		
220000, 250000, 280000	1.383	1.383
Cast Iron Cylinder		
230000 ♦	1.382	1.382
240000, 300000, 320000	Ball	Ball

♦ Gear Reduction PTO—1.185″

83256C33

Magneto Bearing Repair Tool Chart

Model Series	Cylinder Support	Pilot	Counter-bore Reamer	Reamer Guide Bushing Mag.	Bushing Driver	Pilot Guide Bushing PTO	Plug Gauge
Aluminum Model Series							
220000, 250000, 280000	19227	19220•	19224•	19222•	19226•	19220•	19219
Cast Iron Model Series							
230000			Replace Support and Cover				19117

•Tools for DU™ Bushing only, in positions shown.
NOTE: Tools listed may be used to install either steel backed aluminum bushing or DU™ bushing except as noted above.

83256C34

PTO Bearing Repair Tool Chart

	Cylinder Support	Pilot	Counter-bore Reamer	Reamer Guide Bushing Mag.	Bushing Driver	Pilot Guide Bushing PTO	Plug Gauge
220000, 250000, 280000	19227	19220●	19224●	19222●	19226●	19220●	19219
Cast Iron Model Series							
230000			**Replace Support and Cover**				19117

●Tools for DU™ Bushing only, in positions shown.
NOTE: Tools listed may be used to install either steel backed aluminum bushing or DU™ bushing except as noted above.

83256C35

Model Series	Depth Mag.	Depth P.T.O.
220000, 250000, 280000	7/64″	1/8″

83256C36

Seal Protectors

Tool #1	Color	Crankshaft Journal Size
19334/1	White	.787
19334/2	Red	.875
19334/3	Blue	.984
19334/4	Orange	1.000
19334/5	Brown	1.062
19334/6	Green	1.181
19334/7	Yellow	1.378
19356/8	Purple	1.317
19356/9	Black	1.503

83256C37

Always inspect the cylinder after the engine has been disassembled. Visual inspection will show if there are any cracks,

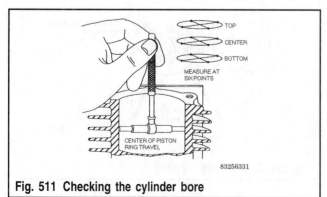

Fig. 511 Checking the cylinder bore

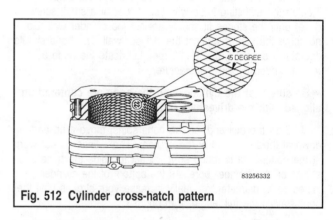

Fig. 512 Cylinder cross-hatch pattern

stripped bolt holes, broken fins, or if the cylinder wall is scored. Use an inside micrometer or telescoping gauge and micrometer to measure the size of the cylinder bore. Measure at right angles.

If the cylinder bore is more than 0.003 in. (0.076mm) oversize, or 0.0015 in. (0.381mm) out of round on light-weight (aluminum) cylinders, the cylinder must be resized (rebored).

➡**Do not deglaze the cylinder walls when installing piston rings in aluminum cylinder engines. Also be aware that there are chrome ring sets available for most engines. These are used to control oil pumping in bores worn to 0.005 in. (0.127mm) over standard and do not require honing or glaze breaking to seat.**

RESIZING

Always resize to exactly 0.010 in. (0.254mm), 0.020 in. (0.50mm) or 0.030 in. (0.762mm) over standard size. If this is done accurately, the stock oversize rings and pistons will fit perfectly and proper clearances will be maintained. Cylinders, either cast iron or lightweight, can be quickly resized with a good hone. Use the stones and lubrication recommended by the hone manufacturer to produce the correct cylinder wall finish for the various engine models.

If a boring bar is used, a hone must be used after the boring operation to produce the proper cylinder wall finish. Honing can be done with a portable electric drill, but it is easier to use a drill press.

1. Clean the cylinder at top and bottom to remove all burrs and pieces of base and head gaskets.

2. Fasten the cylinder to a heavy iron plate. Some cylinders require shims. Use a level to align the drill press spindle with the bore.

3. Oil the surface of the drill press table liberally. Set the iron plate and the cylinder on the drill press table. Do not anchor the cylinder to the drill press table. If you are using a portable drill, set the plate and the cylinder on the floor.

4. Place the hone driveshaft in the chuck of the drill.

5. Slip the hone into the cylinder. Connect the driveshaft to the hone and set the stop on the drill press so the hone can only extend ¾-1 in. (19-25mm) from the top or bottom of the cylinder. If you are using a portable drill, cut a piece of wood to place in the cylinder as a stop for the hone.

6. Place the hone in the middle of the cylinder bore. Tighten the adjusting knob with your finger or a small screwdriver until the stones fit snugly against the cylinder wall. Do not force the stones against the cylinder wall. The hone should operate at a speed of 300-700 rpm. Lubricate the hone as recommended by the manufacturer.

➡**Be sure that the cylinder and the hone are centered and aligned with the driveshaft and the drill spindle.**

7. Start the drill and, as the hone spins, move it up and down at the lower end of the cylinder. The cylinder is not worn at the bottom but is round so it will act to guide the hone and straighten the cylinder bore. As the bottom of the cylinder increases in diameter, gradually increase your strokes until the hone travels the full length of the bore.

➡**Do not extend the hone more than ¾-1 in. (19-25mm) past either end of the cylinder bore.**

8. As the cutting tension decreases, stop the hone and tighten the adjusting knob. Check the cylinder bore frequently with an accurate micrometer. Hone 0.0005 in. (0.0127mm) oversize to allow for shrinkage when the cylinder cools.

9. When the cylinder is within 0.0015 in. (0.038mm) of the desired size, change from the rough stone to a finishing stone.

The finished resized cylinder should have a cross-hatched appearance. Proper stones, lubrication, and spindle speed along with rapid movement of the hone within the cylinder during the last few strokes, will produce this finish. Cross-hatching provides proper lubrication and ring break-in.

➡**It is EXTREMELY important that the cylinder be thoroughly cleaned after honing to eliminate ALL grit. Wash the cylinder carefully in a solvent such as kerosene. The cylinder bore should be cleaned with a brush, soap, and water.**

Bearings

INSPECTION

Plain Type

Bearings should be replaced if they are scored or if a plug gauge will enter. Try the gauge at several points in the bearings.

REPLACING PLAIN BEARINGS

The crankcase cover bearing support should be replaced if the bearing is worn or scored.

REPLACING THE MAGNETO BEARING

Aluminum Cylinder Engines
▶ **See Figures 513, 514, 515, 516 and 517**

There are no removable bearings in these engines. The cylinder must be reamed out so a replacement bushing can be installed.

1. Place a pilot guide bushing in the sump bearing, with the flange of the guide bushing toward the inside of the sump.

2. Assemble the sump on the cylinder. Make sure that the pilot guide bushing does not fall out of place.

3. Place the guide bushing into the oil seal recess in the cylinder. This guide bushing will center the counterbore reamer even though the oil bearing surface might be badly worn.

4. Place the counterbore reamer on the pilot and insert them into the cylinder until the tip of the pilot enters the pilot guide bushing in the sump.

5. Turn the reamer clockwise with a steady, even pressure until it is completely through the bearing. Lubricate the reamer with kerosene or any other suitable solvent.

➡**Counterbore reaming may be performed without any lubrication. However, clean off shavings because aluminum material builds up on the reamer flutes causing eventual damage to the reamer and an oversize counterbore.**

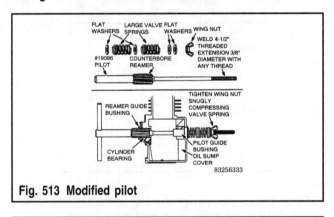

Fig. 513 Modified pilot

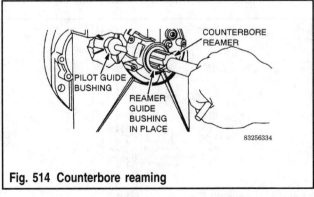

Fig. 514 Counterbore reaming

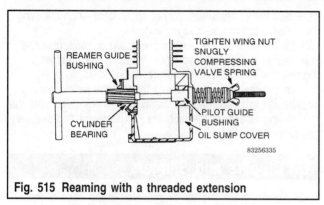

Fig. 515 Reaming with a threaded extension

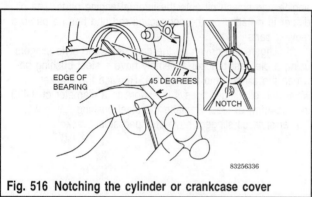

Fig. 516 Notching the cylinder or crankcase cover

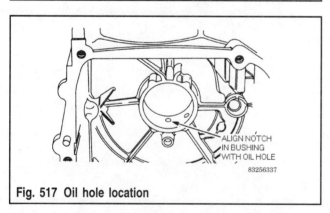

Fig. 517 Oil hole location

6. Remove the sump and pull the reamer out without backing it through the bearing. Clean out the remaining chips. Remove the guide bushing from the oil seal recess.

7. Hold the new bushing against the outer end of the reamed out bearing, with the notch in the bushing aligned with the notch in the cylinder. Note the position of the split in the bushing. At a point in the outer edge of the reamed out bearing opposite to the split in the bushing, make a notch in the cylinder hub at a 45° angle to the bearing surface. Use a chisel or a screwdriver and hammer.

8. Press in the new bushing, being careful to align the oil notches with the driver and the support until the outer end of the bushing is flush with the end of the reamed cylinder hub.

9. With a blunt chisel or screwdriver, drive a portion of the bushing into the notch previously made in the cylinder. This is called staking and is done to prevent the bushing from turning.

10. Reassemble the sump to the cylinder with the pilot guide bushing in the sump bearing.

11. Place a finishing reamer on the pilot and insert the pilot into the cylinder bearing until the tip of the pilot enters the pilot guide bushings in the sump bearing.

12. Lubricate the reamer with kerosene, fuel oil, or other suitable solvent, then ream the bushing, turning the reamer clockwise with a steady even pressure until the reamer is completely through the bearing. Improper lubricants will produce a rough bearing surface.

13. Remove the sump, reamer, and the pilot guide bushing. Clean out all reaming chips.

REPLACING THE P.T.O. BEARING

Aluminum Cylinder Engines

The sump or crankcase bearing is repaired the same way as the magneto end bearing. Make sure to complete repair of one bearing before starting to repair the other. Press in new oil seals when bearing repair is completed.

REPLACING OIL SEALS

1. Assemble the seal with the sharp edge of leather or rubber toward the inside of the engine.

2. Lubricate the inside diameter of the seal with Lubriplate® or equivalent.

3. Press all seals so they are flush with the hub.

Extended Oil Filler Tubes and Dipsticks

▶ See Figure 518

When installing the extended oil fill and dipstick assembly, the tube must be installed so the O-ring seal is firmly compressed. To do so, push the tube downward toward the sump, then tighten the blower housing screw, which is used to secure the tube and bracket. When the dipstick assembly is fully depressed, it seals the upper end of the tube.

A leak at the seal between the tube and the sump, or at the seal at the upper end of the dipstick can result in a loss of crankcase vacuum, and a discharge of smoke through the exhaust system.

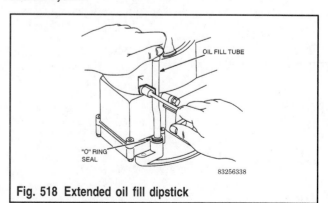

Fig. 518 Extended oil fill dipstick

Breathers

The function of the breather is to maintain a vacuum in the crankcase. The breather has a fiber disc valve which limits the direction of air flow caused by the piston moving back and forth in the cylinder. Air can flow out of the crankcase, but the one-way valve blocks the return flow, thus maintaining a vacuum in the crankcase. A partial vacuum must be maintained in the crankcase to prevent oil from being forced out of the engine at the piston rings, oil seals, breaker plunger, and gaskets.

INSPECTION OF THE BREATHER

If the fiber disc valve is stuck or binding, the breather cannot function properly and must be replaced. A 0.045 in. (1.14mm) wire gauge should not enter the space between the fiber disc valve and the body. Use a spark plug wire gauge to check the valve. The fiber disc valve is held in place by an internal bracket which will be distorted if pressure is applied to the fiber disc valve. Therefore, do not apply force when checking the valve with the wire gauge.

If the breather is removed for inspection or valve repair, a new gasket should be used when replacing the breather. Tighten the screws securely to prevent oil leakage.

Most breathers are now vented through the air cleaner, to prevent dirt from entering the crankcase. Check to be sure that the venting elbows or the tube are not damaged and that they are properly sealed.

Oil Dippers and Slingers

Oil dippers reach into the oil reservoir in the base of the engine and splash oil onto the internal engine parts. The oil dipper is installed on the connecting rod and has no pump or moving parts.

Oil slingers are driven by the cam gear. Old style slingers using a die cast bracket assembly have a steel bushing between the slinger and the bracket. Replace the bracket on which the oil slinger rides if it is worn to a diameter of 0.490 in. (12.4mm) or less. Replace the steel bushing if it is worn. Newer style oil slingers have stamped steel brackets.

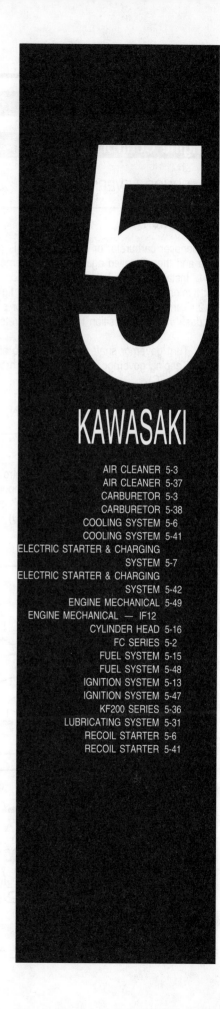

FC SERIES

Engine Control System

GOVERNOR LEVER SETTING

▶ **See Figure 1**

Whenever carburetor or governor lever is removed from engine and then installed again, adjust governor lever position.

1. Install governor lever (A) on governor shaft (B) but do not tighten nut (C). Loosen nut (C) if it is tight.
2. Turn governor lever (A) clockwise or place throttle lever on dash in 'FAST' position to open carburetor throttle valve fully.
3. Turn governor shaft (B) clockwise to end of travel.
4. Keeping governor lever position of throttle fully open, tighten nut (C).

THROTTLE CABLE INSTALLATION

▶ **See Figure 2**

1. Link throttle cable (G) to speed control lever (C) and clamp throttle cable outer housing (F) temporarily.
2. With throttle lever on dash in 'FAST' position, align hole (B) of speed control lever (C) with hole (D) of control plate (E)

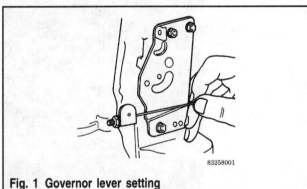

Fig. 1 Governor lever setting

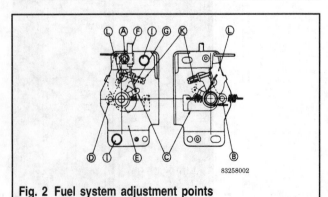

Fig. 2 Fuel system adjustment points

and insert 6mm (0.24 in.) diameter pin or 6mm bolt through two holes.
3. Pull out throttle cable outer housing (F) to remove any slack and tighten cable clamp bolt (A).
4. Remove 6mm pin and set throttle lever on dash in 'CHOKE' position. Make sure carburetor choke valve is completely closed.

FAST IDLE SPEED ADJUSTMENT

▶ **See Figure 2**

➡ **Air cleaner must be installed to engine before starting.**

1. Start and warm up engine without load.
2. Loosen two control plate bolts (I).
3. Align hole (B) of speed control lever (C) with hole (D) of control plate (E) and insert 6mm (0.24 in.) diameter pin or 6mm bolt through two holes.

➡ **Make sure choke valve is fully opened.**

4. Adjust fast idle speed for specified rpm by moving control plate (E).
5. Tighten two bolts (I) securely in a manner to avoid changing specified speed.
6. Stop engine, remove 6mm pin, and set throttle lever on dash in 'CHOKE' position. Make sure carburetor choke valve is closed completely.

CHOKE ADJUSTMENT

▶ **See Figure 2**

1. Align hole (B) of speed control lever (C) with hole (D) of control plate (E) and insert 6mm (0.24 in.) diameter pin or 6mm bolt through two holes.
2. Turn choke setting screw (K) counterclockwise until it is clear of choke control lever (L) and then turn choke setting screw clockwise until it just contacts choke control lever.
3. Remove 6mm pin and set throttle lever on dash in 'CHOKE' position. Make sure carburetor choke valve is closed completely.

SLOW IDLE SPEED ADJUSTMENT

▶ **See Figure 3**

1. Turn carburetor pilot screw (B) in until it just seats and then back out 1½ turns.
2. Start and warm up engine without load.
3. Move throttle lever on dash to 'SLOW' position.
4. Adjust slow idle speed to specified rpm by moving throttle stop screw (A).
5. Adjust pilot screw (B) until engine idles at maximum speed and then turn pilot screw out additional ¼ turn.
6. Re-adjust slow idle speed, to specified rpm.

AIR CLEANER

K-KLEEN System

➧ See Figures 4 and 5

Intake air is induced through rotary screen (A), fan housing (B), and air cleaner (C) to remove grass and rubbish from air. Therefore, the condition of air passages affects volume of intake air and carburetor functions.

1. Assemble related parts neatly to minimize air leakage.
2. Keep clearance between rotary screen and fan housing as shown.
3. Do not remove any parts constructing air passages when running engine.
4. Keep air passages free from grass and rubbish.

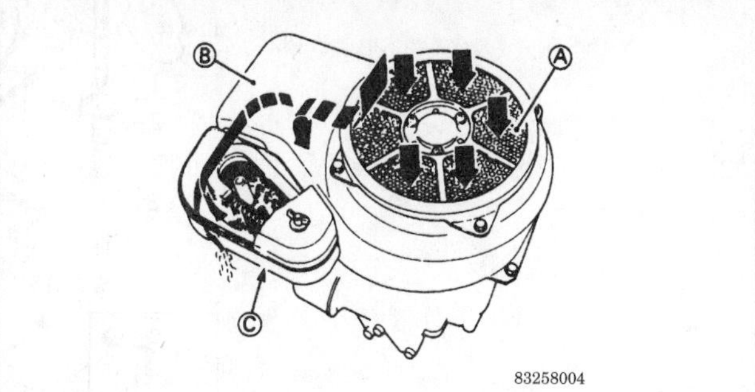
A
B
C
83258004

Fig. 4 K-Kleen system

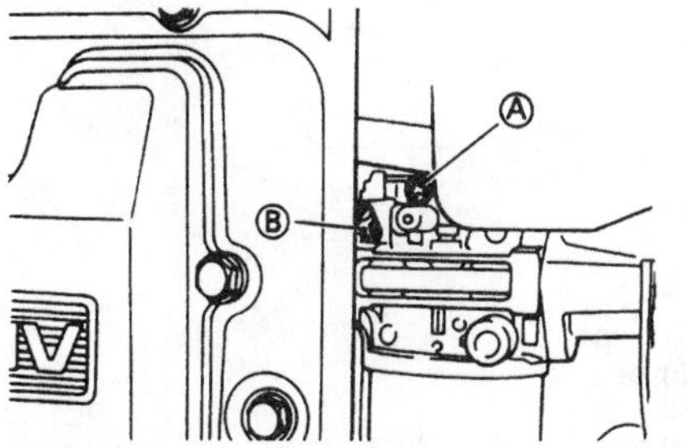
A
B
83258003

Fig. 3 Slow idle speed adjustment

MAINTENANCE

Air cleaner maintenance is one of the most important items to keep engine performing well.

1. The foam element must be lightly oiled to perform as intended. Make sure to soak element in engine oil and squeeze excessive oil, after washed.
2. The paper element is cleaned by gentle tapping or washing in detergent and water.

****WARNING**

Do not use pressurized air to paper element to avoid breakage. Do not oil paper element.

CARBURETOR

Components

➧ See Figure 6

This carburetor is float type with adjustable pilot screw (idle mixture), fixed main jet, and float chamber drain screw.

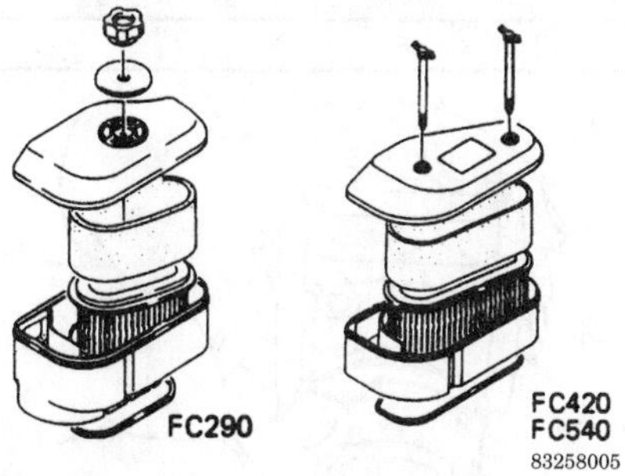
FC290
FC420
FC540
83258005

Fig. 5 Air filter element replacement

REMOVAL

➧ See Figure 7

****CAUTION**

Gasoline is extremely flammable. Avoid fires due to smoking or careless practices.

Before removing carburetor from engine, drain fuel in float chamber to suitable container loosening drain screw (A).

Float Chamber

➧ See Figure 8

Before removing float chamber, rotate float chamber clockwise and counterclockwise $1/6$ turn either way, 2 or 3 times, pushing float chamber to carburetor body to release sticking of float chamber and rubber gasket.

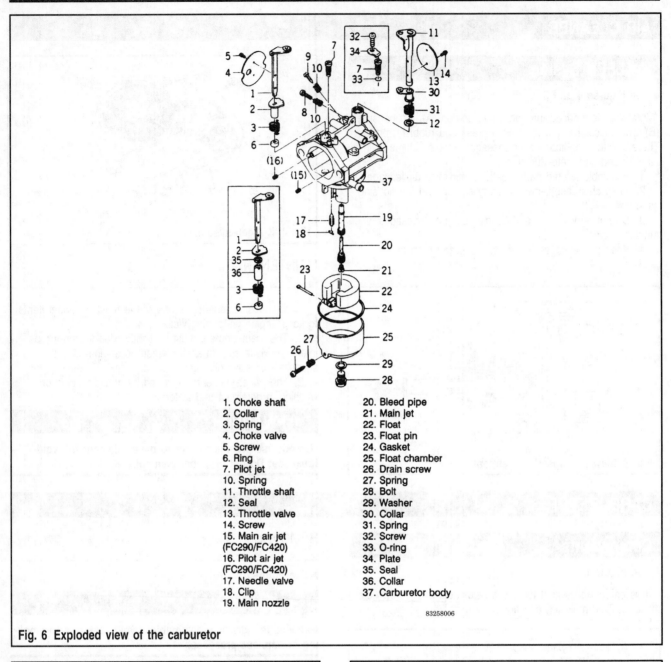

1. Choke shaft
2. Collar
3. Spring
4. Choke valve
5. Screw
6. Ring
7. Pilot jet
10. Spring
11. Throttle shaft
12. Seal
13. Throttle valve
14. Screw
15. Main air jet
(FC290/FC420)
16. Pilot air jet
(FC290/FC420)
17. Needle valve
18. Clip
19. Main nozzle
20. Bleed pipe
21. Main jet
22. Float
23. Float pin
24. Gasket
25. Float chamber
26. Drain screw
27. Spring
28. Bolt
29. Washer
30. Collar
31. Spring
32. Screw
33. O-ring
34. Plate
35. Seal
36. Collar
37. Carburetor body

83258006

Fig. 6 Exploded view of the carburetor

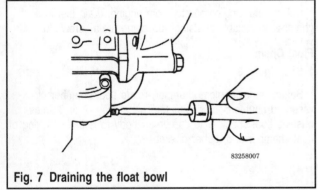

83258007

Fig. 7 Draining the float bowl

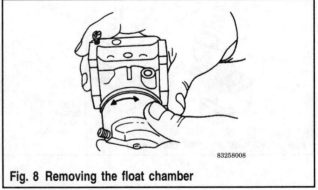

83258008

Fig. 8 Removing the float chamber

CLEANING

> ❄❄**WARNING**
>
> Do not clean jet orifices with a hard object. Follow instructions prepared by cleaner manufacturer when using cleaner.

1. Dip carburetor components except non-metallic parts such as gasket into carburetor cleaner until dirt is removed and rinse them out with solvent.

➡Rinse carburetor aluminum components in hot water to neutralize corrosive action of cleaner, if so instructed by the manufacturer.

2. Dry components with compressed air. Make sure all orifices and passages are free from dirt or foreign object.

> ❄❄**WARNING**
>
> Do not use rags or paper to dry components to avoid plugging orifices by lint.

FLOAT ADJUSTMENT

▶ See Figures 9 and 10

> ❄❄**WARNING**
>
> Do not strike float pin to remove or install it, to avoid breakage of pin holder. To remove float pin, pull the transformed end with pliers.

To install float pin, push float pin transformed end until the other end flushes with pin holder outer outer surface.

➡The white float (made of polyacetal) does not require check and adjustment.

> ❄❄**WARNING**
>
> Do not push down float or needle valve when checking parallelism.

1. Check float angle and correct it as follows; Place carburetor upside down and check parallelism of float surface (A) and carburetor body (B).

2. If not parallel, adjust float surface angle bending tang (C) with needle-nose pliers.

3. Put carburetor float side down and check dimension (D).

4. If dimension is out of standard, adjust it by bending stopper (E).

Standard Float Stroke
- FC290 10.5-12.5mm (0.413-0.492 in.)
- FC540 13.5-15.5mm (0.531-0.610 in.)

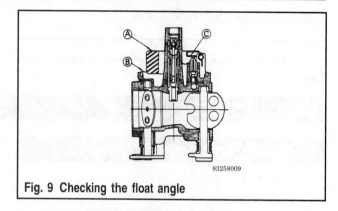

Fig. 9 Checking the float angle

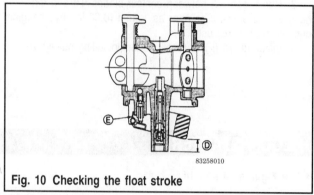

Fig. 10 Checking the float stroke

INSTALLATION

Precautions

▶ See Figure 11

1. Do not over-tighten small carburetor components. Finger-tighten pilot screw.

2. Do not bend throttle and choke shafts when assembling.

3. Apply screw locking agent to screws of throttle valve or choke valve. Do not allow agent to flow into shaft bearing surfaces.

4. Make sure movement of throttle and choke valves is smooth.

5. Before carburetor installation, install throttle linkage (S) and choke linkage (B) on carburetor.

6. Make sure to fasten fuel line at carburetor inlet with clamp.

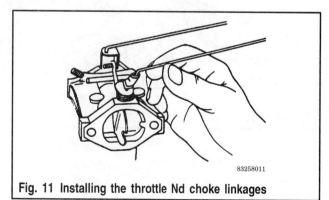

Fig. 11 Installing the throttle Nd choke linkages

CARBURETOR IDENTIFICATION

A portion of carburetor, part number is marked on carburetor body.

COOLING SYSTEM

Blade Gap Adjustment

▶ **See Figure 12**

1. Install fan housing (A) and rotary screen (B).
2. Check gap (C) between contour blades (D) and fan housing and adjust it as near as 1.5mm (0.06 in.) by changing number of shim (E) used.
3. Install guard (F) or recoil starter (including pulley).

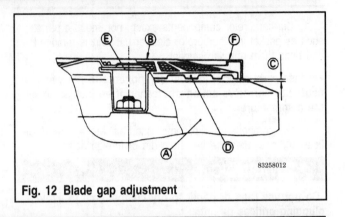

Fig. 12 Blade gap adjustment

RECOIL STARTER

▶ **See Figures 13 and 14**

Disassembly

✳✳WARNING

Do not wedge rope between reel and case.

1. Pull handle out about 250mm (10 in.). Then hold rope in place with locking pliers or knot (A).
2. Pull knot in handle (B) out and untie it.

✳✳WARNING

Wear gloves during disassembling to avoid injury.

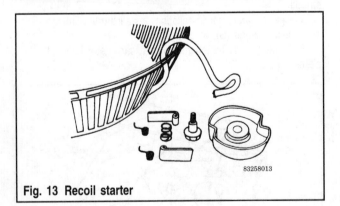

Fig. 13 Recoil starter

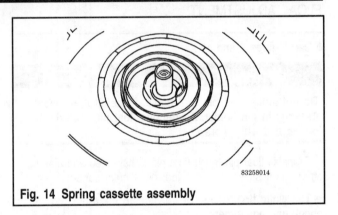

Fig. 14 Spring cassette assembly

3. While carefully holding reel (C) and case (D), remove locking pliers or untie knot.
4. Unwind spring tension slowly.
5. Remove bolt (G) at center part and then retainer (H), pawls (J), pawl springs (K), and center spring (L) but not including reel (C).

✳✳CAUTION

When removing reel (C), be careful that recoil spring under the reel does not fly loose and cause injury! The spring is under great pressure! There should be no spring tension on reel when removing reel. If tension is felt, push reel back into place and gently 'wiggle' it until reel can be easily removed.

6. Rotate reel (C) one-quarter turn clockwise from rest position where no tension can be felt. Then, slowly lift reel straight up out of case.

✳✳WARNING

If recoil spring cassette (N) is sticking with removed reel (C), be careful that the cassette does not drop.

✳✳CAUTION

Be careful that recoil spring (M) does not fly loose from cassette (N) and causes injury. The spring is under great pressure.

7. Slowly lift recoil spring cassette (N), straight up out of case (D) or reel (C).
8. If recoil (M) must be removed from cassette (N), hold the cassette with opening side downward in suitable container and tap cassette to remove recoil spring.

INSPECTION

1. Dip metal parts in bath of high flash-point solvent, if necessary.

✳✳WARNING

Do not clean any non-metallic parts in solvent. They may be damaged by the solvent.

2. Check starter pawls for chips or excessive wear.
3. Check starter rope for excessive wear or fraying.

4. Check springs for break, rust, distortion, or weakened condition. If damage is found, replace the part.

✳✳CAUTION

Do not throw away recoil spring as installed in cassette. Recoil spring may fly loose from cassette and cause injury.

REASSEMBLY

✳✳CAUTION

Wear gloves during recoil spring (M) installation to avoid injury. The recoil spring must be assembled with great pressure.

1. Lightly grease recoil spring.
2. Set recoil spring in spring cassette (N) and install cassette so that hook (P) of spring catches tab (R) on case.
3. If rope is unwound from reel (C), wind rope i reel counterclockwise facing pawl groove(s).
4. Aligning guide pin (T) on reel with recoil spring hook (U) on cassette, install reel into case.
5. Install pawl springs (K), pawls (J), center spring and retainer, then tighten bolt.

➡Pawl springs (K) and pawls (J) are assembled into reel as shown.

6. Rotate reel two turns counterclockwise to preload recoil spring.
7. Pull rope out of case through rope hole and install handle.

ELECTRIC STARTER & CHARGING SYSTEM

Troubleshooting

1. Disconnect spark plug cap, and ground the cap terminal.
2. Turn engine switch to 'START' position and check condition.

✳✳CAUTION

Engine may be cranked in this test. Do not touch any rotating parts of engine and equipment during test. If starter does not stop by engine switch 'OFF', disconnect negative (-) lead from battery as soon as possible.

SOLENOID AND CIRCUIT CHECK

◆ **See Figures 15, 16, 17 and 18**

This procedure is for shift type starter, but can be used for Bendix type by replacing word 'SOLENOID' with 'SOLENOID SWITCH.'

➡Before this test, make sure battery is fully charged.

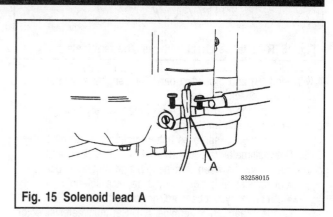

Fig. 15 Solenoid lead A

1. Disconnect lead (A) between starter motor and solenoid, and keep the lead away from terminal of solenoid to avoid accidental cranking in test.

➡Painted part is not grounded.

2. Disconnect lead (B) between solenoid and engine switch. Set multimeter selector switch to 25V DC position and connect

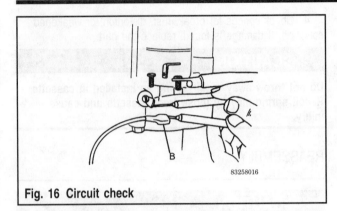

Fig. 16 Circuit check

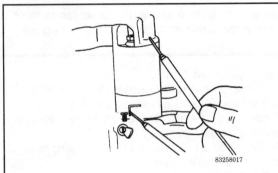

Fig. 17 Resistance check between terminal and ground

Fig. 18 Resistance check between the terminals

it with lead (B) and grounded part. Turn engine switch to 'START' and check voltage.
- If voltage is 0 or much less than battery voltage, check engine switch and/or circuit.
- If voltage is same as battery voltage, go to next step.

3. Set multimeter selector switch R · 1Ω position and check resistance between terminal (C) and grounded part.
- If resistance is not nearly 0, replace solenoid.
- If continuity is observed, go to next step.

4. Connect lead (B) to terminal (C). Set multimeter selector switch to R x 1 ohm position and connect it with two terminals on solenoid as shown.. Turn engine switch to 'START' and check resistance.
- If resistance is not nearly 0, or solenoid does not click, replace solenoid.
- If solenoid is normal, go to next step.

STARTER MOTOR CHECK

▶ See Figures 19, 20 and 21

❊❊WARNING

Disconnect negative (-) lead first and then positive (+) lead to prevent spark at terminal.

1. Disconnect battery before removing electric starter from engine to avoid accidental running of starter in handling.
2. Remove electric starter from engine.

❊❊CAUTION

The test room must be free from any flammable object. Keep away your body from pinion.

❊❊WARNING

Be careful not to deform electric starter body by holding.

3. Hold electric starter with vise.
4. Connect first jumper cable with positive (+) battery terminal and starter motor terminal (D) on solenoid.
5. Connect second jumper cable with negative (-) battery terminal.
6. Touch starter body (not painted part) with the other end of second starter jumper cable intermittently (within one second). If pinion does not turn, repair or replace starter motor.

BRUSH SERVICE LIMIT

Check overall length of each brush with vernier calipers. If length is less than minimum, replace brush.
- Brush length minimum: FC290 8.5mm (0.34 in.); FC420 6.0mm (0.24 in.); FC540 10.5mm (0.41 in.)

BRUSH SPRING CHECK

Check each brush spring for breakage or distortion. If brush spring does not snap brush firmly into place, replace.

BRUSH HOLDER CHECK

▶ See Figure 22

Check multimeter selector switch to R · 1kΩ position and check resistance between brush holder (A) and its base (B). If resistance is not infinite (°), replace brush holder.

ARMATURE CHECK

▶ See Figure 23

1. Check surface of commutator and grooves. If it has a scratch or is dirty, polish it with very fine emery paper and clean the grooves.

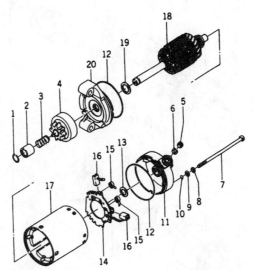

Fig. 19 FC290 starter exploded view

1. Snapring
2. Collar
3. Spring
4. Pinion assembly
5. Nut
6. Spring washer
7. Through bolt
8. Spring washer
9. Washer
10. O-ring
11. Rear cover
12. O-ring
13. Washer
14. Brush holder
15. Brush spring
16. Brush
17. Yoke
18. Armature
19. Washer
20. Front cover

83258019

2. Set multimeter selector switch to R · 1Ω position and check resistance between segments. If resistance is not nearly 0Ω, replace armature (open circuit).

3. Set multimeter selector switch to R · 1kΩ position and check resistance between commutator and armature shaft. If resistance is not infinite (°), replace armature (short circuit).

4. Check armature windings for shorts.

a. Place armature on growler.

b. Hold thin metal strip (e.g. hack saw blade) on top of armature.

c. Turn on growler and rotate armature one complete turn. If metal strip vibrates at any position, replace armature (short circuit).

ARMATURE SERVICE LIMIT

1. Check commutator outside at several points with vernier caliper. If diameter is less than 27mm (1.06 in.), replace armature.

2. Check groove depths between commutator segments. If depths are less than 0.2mm (0.01 in), cut insulating material to standard depth with thin file and clean grooves.

• Standard groove depth: 0.5-0.8mm (0.02-0.03 in.)

3. Support armature in alignment jig at each ends of shaft as shown. Position dial indicator perpendicular to commutator. Rotate armature slowly and check run-out. If run-out is more than maximum, turn down commutator or replace armature.Commutator run-out — maximum:

FC290/FC420: 0.4mm (0.016 in.)
FC540: 0.5mm (0.019 in.)

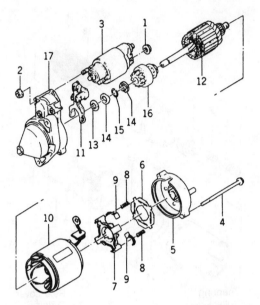

1. Nut
2. Nut
3. Solenoid
4. Bolt
5. Rear cover
6. Insulator
7. Brush holder
8. Brush spring
9. Brush
10. Yoke
11. Shift lever
12. Armature
13. Washer
14. Stopper
15. Snapring
16. Pinion
17. Front cover

83258020

Fig. 20 FC420 starter exploded view

YOKE CHECK

▶ **See Figures 24, 25, 26 and 27**

1. Set multimeter selector switch to R · 1Ω position and check resistance between negative terminals (FC420/FC540). If resistance is not nearly 0Ω, replace yoke assembly.

2. Set multimeter selector switch to R · 1Ω position and check resistance between negative terminals and yoke case (FC420/FC540). If resistance is not nearly 0Ω, replace yoke assembly.

3. Set multimeter selector switch to R · 1Ω position and check resistance between positive terminals (FC420). If resistance is not nearly 0Ω, replace yoke assembly.

4. Set multimeter selector switch to R · 1kΩ position and check resistance between positive terminal and yoke case (FC290/FC420). If resistance is not infinite (°), replace armature.

PINION CLUTCH CHECK

1. Check that pinion rotates with armature shaft when turned clockwise.

2. Check that pinion rotates freely when turned counterclockwise.

3. If pinion does not move normally, replace pinion assembly.

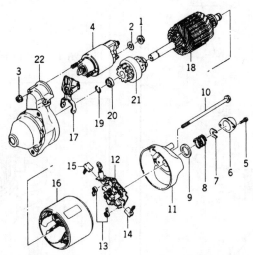

1. Nut
2. Washer
3. Nut
4. Solenoid assembly
5. Screw
6. Cover
7. Stopper
8. Spring
9. Seal
10. Through bolt
11. Rear cover

12. Brush holder
13. Brush spring
14. Brush
15. Brush
16. Yoke
17. Shift lever
18. Armature
19. Snapring
20. Pinion stopper
21. Pinion
22. Front cover

83258021

Fig. 21 FC540 starter exploded view

STARTER MOTOR REASSEMBLY

1. Coat multi-purpose grease to following parts.
   • Sliding surface or pinion and spline shaft
   • Bearings
   • Dust seal lip
   • Contact surface of shift lever
2. Do not use removed snap ring again.

REGULATOR CHECK

▶ **See Figure 28**

1. Remove all leads from regulator.
2. Set multimeter selector switch to R · 1kΩ position.

3. Check resistance between terminals as shown on table. A: Key switch   B: Charging monitor   C: Multimeter terminal

If resistance is not as specified, replace regulator.

➡**Resistance value may vary with individual meters.**

UNREGULATED STATOR OUTPUT

▶ **See Figures 29, 30 and 31**

(13-15A TYPE)
1. Disconnect regulator.
2. Start and warm up engine.
3. Set multimeter selector switch to 250V AC position.
4. Connect meter with stator lead terminals (G,H) in 3P coupler.

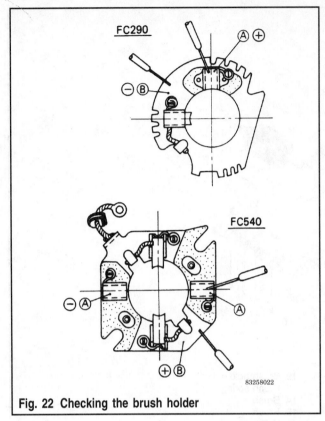

Fig. 22 Checking the brush holder

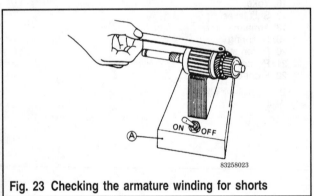

Fig. 23 Checking the armature winding for shorts

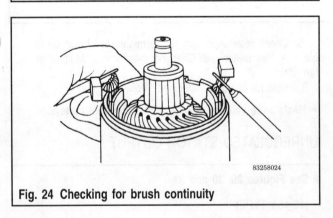

Fig. 24 Checking for brush continuity

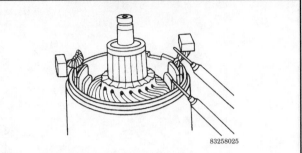

Fig. 25 Checking for continuity between brush ground and case

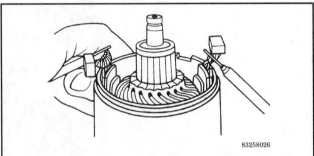

Fig. 26 Checking for continuity between brush positive leads

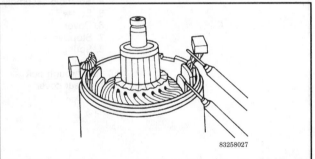

Fig. 27 Checking for continuity between brush positive leads and case

5. Run engine at 3,350 rpm and check voltage. If voltage is less than minimum, replace stator. Stator output — minimum:
FC290/FC420: 26 V
FC540: 24 V

IGNITION SYSTEM

Types of Systems

▶ **See Figure 32**

Transistor controlled ignition system is used for these engines and this system consists of following components.

1. Ignition coil
2. Control unit
3. Flywheel (with permanent magnet)

These components do not mechanically contact and periodic maintenance is not required.

- F: Flywheel
- L_1: Primary coil
- L_2: Secondary coil
- CU: Control unit

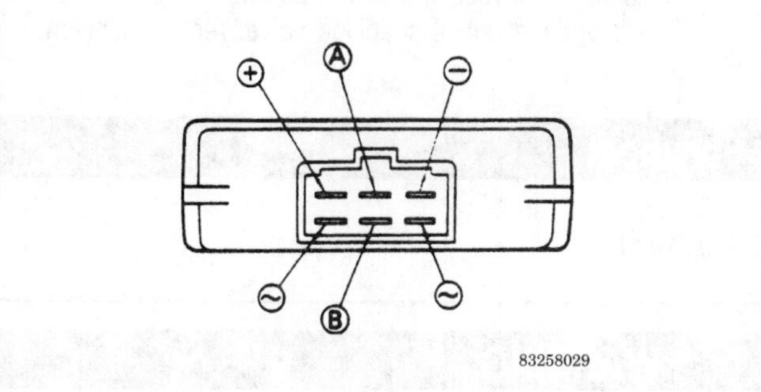

83258029

Fig. 29 Regulator connector

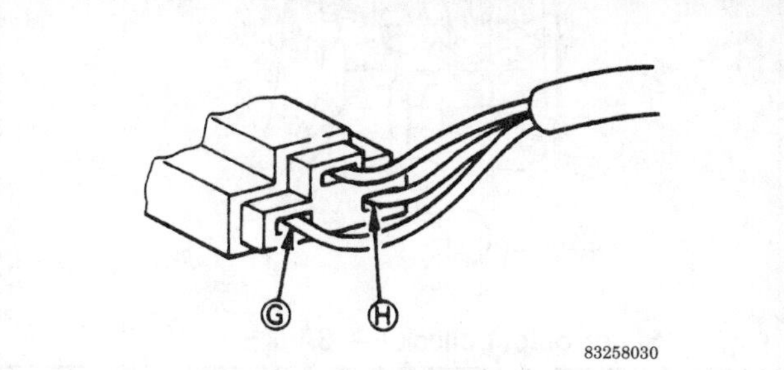

83258030

Fig. 30 Stator output check — 13A-15A

- R_1: Control resistor
- R_2: Control resistor
- SP: Spark plug
- SW: Engine switch
- TR_1: Transistor
- TR_2: Transistor

SPARK CHECK

To check ignition system, check spark as follows:

1. Remove spark plug and connect plug cap with the removed spark plug.

⊕ / ⊖ ©	+	Ⓐ	−	~	~	Ⓑ
+		∞	∞	∞	∞	∞
Ⓐ	4kΩ ~ 20kΩ		200Ω ~ 1kΩ	1kΩ ~ 5kΩ	1kΩ ~ 5kΩ	200kΩ ~ ∞
−	3kΩ ~ 15kΩ	200Ω ~ 1kΩ		1kΩ ~ 5kΩ	1kΩ ~ 5kΩ	200kΩ ~ ∞
~	1kΩ ~ 5kΩ	∞	∞			∞	∞
~	1kΩ ~ 5kΩ	∞	∞	∞		∞
Ⓑ	10kΩ ~ 50kΩ	1kΩ ~ 5kΩ	1.5kΩ ~ 7.5kΩ	4kΩ ~ 20kΩ	4kΩ ~ 20kΩ	

83258028

Fig. 28 Regulator lead resistance check

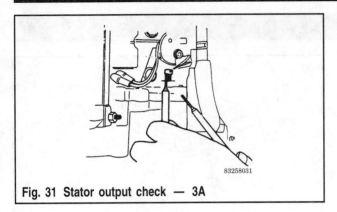

Fig. 31 Stator output check — 3A

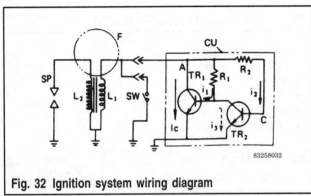

Fig. 32 Ignition system wiring diagram

2. Install spare plug to plug hole to avoid fuel spitting from hole.

❊❊CAUTION

To avoid electric shock, hold plug cap, but not spark plug.

3. Keeping contact with spark plug metal part (not center electrode) and engine block, crank engine.

❊❊WARNING

Do not clean spark plug with bead or sand cleaner.

- If no or very weak spark is observed, clean spark plug and regap it to 0.7-.08mm (0.028-0.031 in.) and try engine cranking again.
- If spark is not improved by cleaning, try checking again with new spark plug.
- If spark is not improved yet, check ignition system.

CONTROL UNIT CHECK

▶ **See Figure 33**

1. Set multimeter selector switch to R · 10Ω position.
2. Check resistance between terminal (A) and case (B). If resistance is out of specified value, replace control unit.

➡**This check may not cover every defect.**

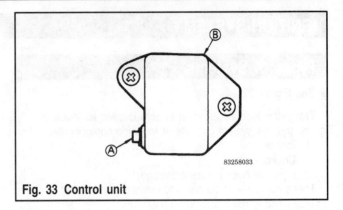

Fig. 33 Control unit

IGNITION COIL CHECK

▶ **See Figure 34**

1. Resistance between the points as specified.
2. If resistance is out of specified valve, replace ignition coil.

Flywheel

REMOVAL

❊❊WARNING

Remove plug cap from spark plug to avoid engine starting. Do not insert and tool in flywheel fins or ring gear to avoid rotation. The tool will damage flywheel.

1. Loosen nut by turning it counterclockwise. Use ½ in. air impact wrench or strap wrench to avoid flywheel rotation.

❊❊WARNING

Keep shoulder of nut flush with end of crankshaft until taper engagement is released.

2. Use flywheel puller (A) to remove flywheel.

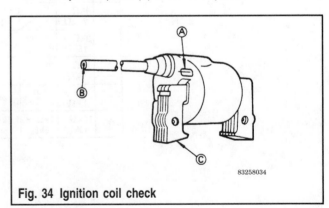

Fig. 34 Ignition coil check

INSTALLATION

1. Before installing flywheel, remove grease and oil from taper part of crankshaft and taper hole of flywheel.
2. Make sure key is in place when installing flywheel
3. Tighten the nut.

IGNITION COIL AIR GAP ADJUSTMENT

1. If ignition coil is removed or replaced, adjust the air gap when installing coil.
2. If ignition coil is removed or replaced, adjust the air gap when installing coil.
3. Insert 0.3mm (0.012 in.) feeler gauge between coil legs (A) and flywheel rim (B).
4. Pushing coil to flywheel, tighten coil mounting screws firmly.

FUEL SYSTEM

Fuel Pump

REMOVAL AND DISASSEMBLY

▶ **See Figure 35**

✳✳CAUTION

Gasoline is extremely flammable. Avoid fires due to smoking or careless practices. Avoid spilling of gasoline in removing and disassembling fuel pump. Plug fuel line disconnected from pump intake joint, immediately.

INSPECTION

1. Check vent hole and screen for plugging or clogging. If vent hole and screen are plugged or clogged, remove dirt from them.
2. Check diaphragms for crack, tear or hole. If defect is found, replace it.
3. Check valves for crack, tear or wear. If defect is found, replace them.

✳✳WARNING

Make sure rubber gasket and rubber diaphragm are placed as shown.

INSTALLATION

▶ **See Figure 36**

1. Make sure to use heat shield washers (A) as shown on FC420 and FC540.
2. Make sure to fasten fuel and pulse lines at pump with clamps.

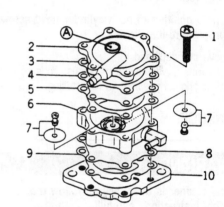

1. Screw	6. Body
2. Cover	7. Inlet valve
3. Gasket	8. Outlet valve
4. Diaphragm	9. Gasket
5. Gasket (rubber)	10. Base

83258036

Fig. 35 Fuel pump exploded view

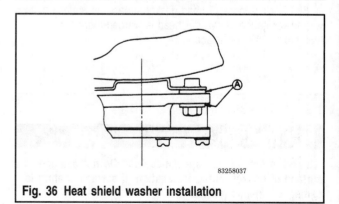

83258037

Fig. 36 Heat shield washer installation

ENGINE MECHANICAL — IF12 Cylinder Head

Compression Check

1. Remove spark plug and set compression gauge to plug hole.
2. Crank engine with recoil or electric starter several times and check highest reading. If highest reading is less than 483 kPa (71 pi.), excessive wear or engine damage is indicated.

➡Battery should be fully charged for this test.

REMOVAL

1. Disconnect plug cap from spark plug.
2. Remove rocker cover.

➡Always note position of each cylinder head screws so they are properly re-installed.

3. Remove cylinder head.
4. If push rods are removed, mark push rods so they are placed in their original positions in re-installing.

MAINTENANCE

❋❋WARNING

If chemical cleaner is used, always follow the manufacturer's safety instructions carefully.

1. Remove coating material stuck on surfaces, with oil stone.

❋❋WARNING

If deposit on combustion chamber is removed, remove valve, from cylinder head. Do not damage valve seat and gasket surface of cylinder head in deposit cleaning.

2. Remove deposit from cylinder head.
3. Check flatness of head gasket surface on surface plate with feeler gauge. If cylinder head is warped more than 0.05mm (0.03 in.), replace it.

INSTALLATION

▶ See Figures 37 and 38

❋❋WARNING

Gasket is coated with special sealant. Do not damage surface of gasket during installation. If surface coating is damaged, replace gasket.

1. Rotate crankshaft until piston comes up at highest position in compression stroke.
2. Install push rods in their respective position in cylinder.

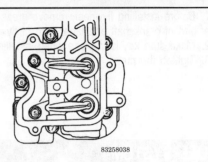

Fig. 37 Cylinder head bolt installation sequence — FC290

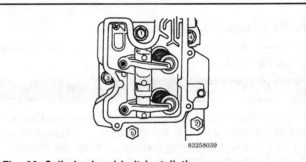

Fig. 38 Cylinder head bolt installation sequence — FC420 & 540

3. Place head gasket and cylinder head assembly on cylinder block.
4. Align push rod ends with recess of rocker arms adjust screws.
5. Install the head bolts hand tight, then, using the numbers in the accompanying illustrations, tighten them evenly in the following sequence:
- FC290: 1-2-3-4-5
- FC420/FC540: 1-3-4-2-5
6. Tighten screws down evenly by hand and then tighten them in sequence as specified.

Valves

VALVE CLEARANCE ADJUSTMENT

▶ See Figure 39

When any part related to valve clearance is changed or modified for defect correction, or after engine has been used for a long period, adjust valve clearance.

➡Do adjustment while engine is cold.

1. Turn crankshaft until piston comes up at highest position in compression stroke.
2. Check clearance between valve stem and rocker arm with 0.15mm (0.006 in.) feeler gauge (A). — for both valves. If clearance is out of specified value, adjust it turning adjuster (B), and then lock the adjuster with the nut.

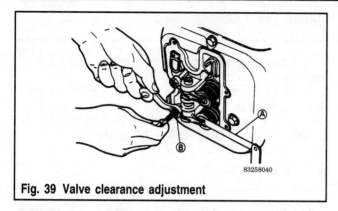

Fig. 39 Valve clearance adjustment

AUTOMATIC COMPRESSION RELEASE CHECK

▶ **See Figure 40**

ACR is device to release compression during engine start, for easy cranking.
1. Remove rocker cover.
2. Remove spark plug to ease hand cranking.
3. Make sure valve clearance is as specified.
4. Rotate crankshaft slowly in usual direction observing movement of exhaust valve (A) and rocker arm (B). If exhaust valve does not open more than 0.25mm (0.01 in.) briefly after intake valve closes, ACR mechanism on camshaft is faulty.

REMOVAL

▶ **See Figures 41, 42 and 43**

➡**Mark push rods so they are placed in their original positions in re-installing.**

1. Remove push rods.
2. Pull out rocker shaft (A) with pliers and remove rocker arms (B).
3. Remove valve rotator cap (C) on exhaust valve stem end.
4. Place screw head of valve spring compressor (D) on the valve head and slip jaw of compressor between spring (E) and retainer.
5. Compress spring and remove retainer (F) and collets (G) — for intake valve — with needle nose pliers.

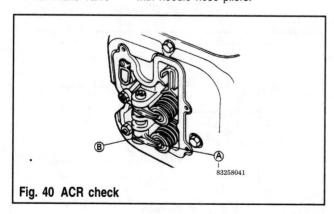

Fig. 40 ACR check

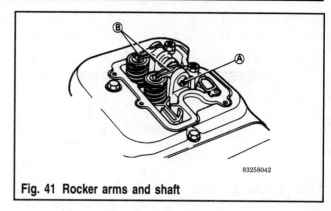

Fig. 41 Rocker arms and shaft

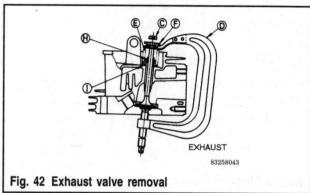

Fig. 42 Exhaust valve removal

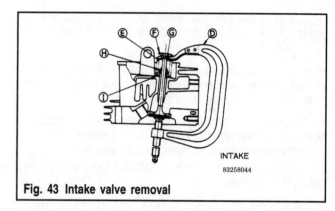

Fig. 43 Intake valve removal

6. Remove compressor and valve spring.

➡**Before pulling valve out of guide, remove all burrs from valve stem, and oil to stem to avoid damaging valve stem sea (H). If stem seal is not damaged, do not remove seal and bottom spring retainer (I).**

7. If stem seal (H) must be replaced, remove it with screw driver.

INSPECTION AND MAINTENANCE

▶ **See Figures 44 and 45**

1. Check valve head for excessive deposit and gas leakage.
2. Remove carbon from valve head with wire brush.

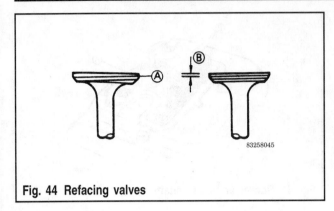

Fig. 44 Refacing valves

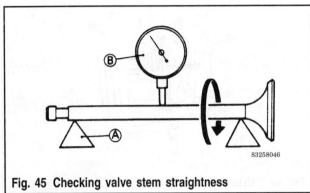

Fig. 45 Checking valve stem straightness

3. Check valve head for warped face (A), dent on face and margin (B) of less than 0.6mm (0.24 in.). If valve head has above defect, replace valve.

➡**Excessive deposit is caused by leaded gasoline, and deposit triggers gas leakage causing valve defects. Therefore unleaded gasoline is recommended.**

4. Check valve stem for sticking, gummy deposit, discoloration at area covered by valve guide, and excessive corrosion.

5. Remove carbon from valve stem as well as head. If valve stem is worn excessively or does not move smoothly in guide, replace valve.

➡**Sticking and discoloration are caused by overheating of engine, or gas leakage from valve face. Therefore such causes must be corrected as well as valve maintenance. Gummy deposit is caused by old or stale gasoline. Clean fuel system and use fresh gasoline. Remove gasoline from fuel system before long storage.**

6. Check diameter of valve stem in area covered by valve guide at several points with micrometer. If diameter is less than minimum, replace valve.
   • Valve stem diameter minimum: Intake: 6.930mm (0.2728 in.); Exhaust: 6.915mm (0.2722 in.)

7. Check bend of valve stem at center part with V blocks (A) and dial indicator (B). If bend (dial gauge reading) of valve stem is more than 0.03mm (0.0012 in.), replace valve.

✳✳WARNING

Do not try to grind or recondition valve face. If valve face is worn or damaged, replace it.

VALVE SPRINGS

1. Check valve spring for any damage and replace it if necessary.
2. Check free length of valve spring with vernier calipers. If length is less than minimum, replace spring.
   • Free length minimum (Intake & Exhaust): FC290: 31.0mm (1.22 in.); FC420/FC540: 37.5mm (1.47 in.)

VALVE LAPPING

If valve does not contact all way around with seat, lap valve into seat.
1. Coat fine lapping compound sparingly on valve face.
2. Rotate valve in circular motion with valve lapper (A).

➡**Lapping mark should appear on or near center of valve face.**

3. Check valve every 8 to 10 strokes and continue lapping until uniform ring appears on valve seat all way around.
4. After lapping, wash parts in solvent to remove compound. Dry
ts thoroughly.

VALVE SEAT RECONDITIONING

⬦ **See Figures 46 and 47**

1. Reface valve seat with 45° cutter, removing only enough material to make smooth and concentric seat.
2. Use 30° cutter to narrow seat to standard with as specified below.
   • Standard valve seat width (Intake & Exhaust): FC290: 0.5-1.10mm (0.020-0.043 in.); FC420/FC540: 1.1-1.46mm (0.043-0.057 in.)
3. Make a light pass with 45° cutter to remove any burr at edge of seat.
4. Coat marker and check contact of valve face and seat. Contact should be at center part of valve face as shown and all way around.
5. Lap valve into seat.

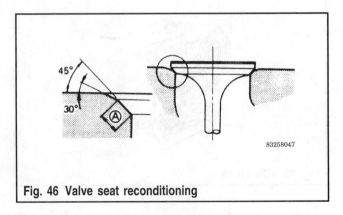

Fig. 46 Valve seat reconditioning

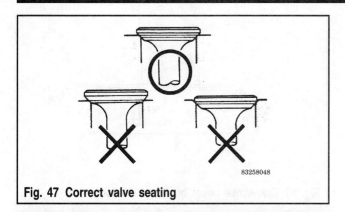

Fig. 47 Correct valve seating

VALVE GUIDES

INSPECTION

1. Use valve guide cleaner to clean inside of valve guides.
2. Check inside diameter of valve guide at several points with inside micrometer. If diameter is more than 7.065mm (0.2781 in.) replace valve guide.

REPLACEMENT

▶ **See Figures 48, 49 and 50**

1. Place cylinder head on support plate with combustion chamber upward.

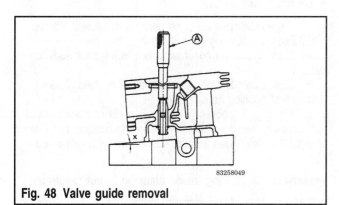

Fig. 48 Valve guide removal

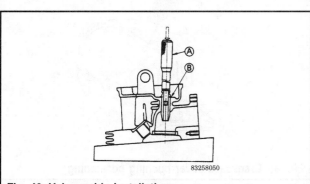

Fig. 49 Valve guide installation

2. Remove valve guide with driver (A).
3. Place cylinder head on support plate with combustion chamber downward.
4. Install snap ring (B) on valve guide.
5. Coat a film of oil on outer surface of valve guide.
6. Install valve guide with driver (A) until snap ring (B) just seats on head.

✳✳WARNING

Be careful not to damage head gasket surface while installing valve guide.

7. Finish valve guide to 7.00-7.015mm (0.2756-0.2762 in.) — for all valves — with valve guide reamer.
8. Lubricate reamer with kerosene or proper lubricant and turn reamer clockwise.
9. Clean parts thoroughly before assembly.
10. Check valve seating and reface valve seat, if necessary.

Rocker Arms

INSPECTION

1. Check rocker arm for pitted or worn contact surface with valve stem.
2. Check inside diameter of rocker arm at several points with inside micrometer. If diameter is more than 13.068mm (0.5145 in.), replace rocker arm.
3. Check outside diameter of rocker shaft at several points with micrometer. If diameter is less than 12.936mm (0.5093 in.), replace rocker shaft.

Pushrods

▶ **See Figure 51**

INSPECTION

Check bend of push rod at center part with V blocks (A) and dial indicator (B). If bend (diameter gauge reading) is more than 0.3mm (0.012 in.), replace push rod.

Crankcase Cover

REMOVAL

▶ **See Figures 52 and 53**

✳✳WARNING

Be careful not to burn yourself by hot oil.

1. Drain engine oil to suitable container.
2. Remove rust and burr from edge of PTO shaft step.
3. Loosen screws, and tap parts near dowel alternately with soft mallet.

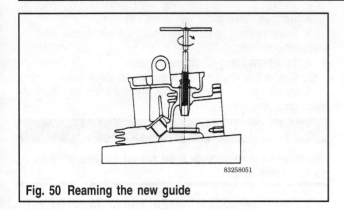

Fig. 50 Reaming the new guide

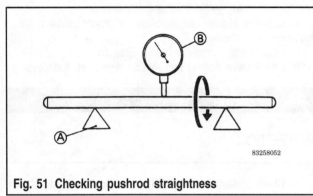

Fig. 51 Checking pushrod straightness

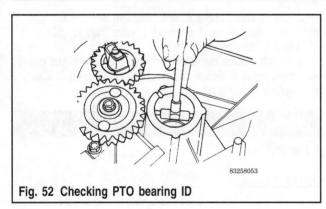

Fig. 52 Checking PTO bearing ID

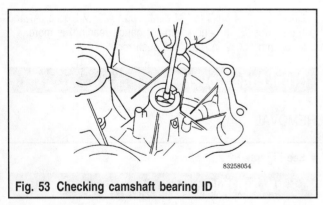

Fig. 53 Checking camshaft bearing ID

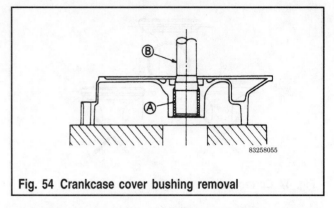

Fig. 54 Crankcase cover bushing removal

4. Check inside diameter of PTO bearing (A) at several points with inside micrometer. If diameter is more than maximum, replace crankcase cover (FC420/FC540) or bushing (FC290).
   • PTO bearing inside diameter maximum: FC290: 30.125mm (1.1860 in); FC420: 35.069mm (1.3807 in.); FC540: 38.056mm (1.493 in.)

5. Check inside diameter of camshaft bearing (B) at several points with inside micrometer. If diameter is more than maximum, replace crankcase cover.
   • Camshaft bearing inside diameter maximum: FC290: 14.054mm (0.5533 in.); FC420/FC540: 21.076mm (0.8298 in.)

BUSHING REPLACEMENT — FC290

▶ **See Figures 54, 55 and 56**

1. Remove oil seal from crankcase cover.

➡**Do not re-use oil seal.**

2. Place crankcase cover on bench with oil seal side up.
3. Push out bushing (A) with bushing tool (B).
4. Place crankcase cover on bench with gasket surface side up.
5. Coat a light film of oil on outside of bushing (A) and entrance of bushing housing (C).
6. Aligning oil grooves (D) and seam (E) as shown, install new bushing (A) into crankcase cover with bushing tool (F) until bushing end goes 1mm (0.039 in.) below housing end (G).

➡**Finishing of bushing inside diameter is not required.**

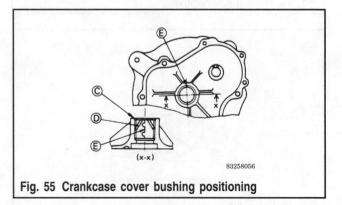

Fig. 55 Crankcase cover bushing positioning

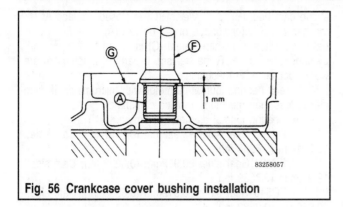

Fig. 56 Crankcase cover bushing installation

OIL SEAL REPLACEMENT

▶ **See Figures 57, 58, 59 and 60**

If oil leakage through oil seal is observed or seal lip is damaged, replace oil seal.

1. Remove oil seal by tapping it out with screw driver or punch.

2. Placing spring held seal lip (A) inside, push oil seal to be flush with housing end (C).

3. Before final assembly, pack some amount of grease for high temperature application into space between seal lip (A) and dust lip (B).

4. Clean gasket surface and place new gasket on crankcase cover.

5. Pack grease into oil seal.

6. Coat a light film of oil on bearings.

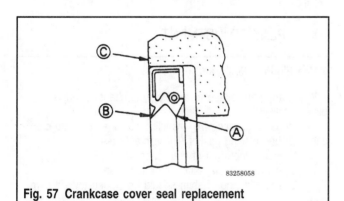

Fig. 57 Crankcase cover seal replacement

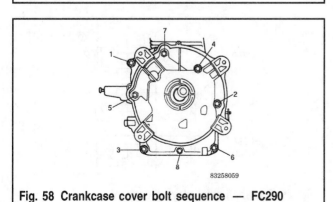

Fig. 58 Crankcase cover bolt sequence — FC290

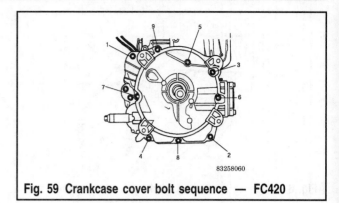

Fig. 59 Crankcase cover bolt sequence — FC420

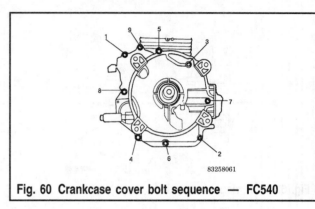

Fig. 60 Crankcase cover bolt sequence — FC540

7. Make sure governor weights are closed.

➡**Make sure to place shim on crankshaft.**

8. Make sure governor gear and oil pump gear are properly aligned to mesh with their respective drive gears when installing crankcase cover. Do not force into position.

9. Install crankcase cover and tighten bolts down evenly by hand. Tighten bolts in the sequence shown and to the specified torque. Do not tighten one bolt completely before the others. It may cause warped crankcase cover.

Camshaft

REMOVAL

▶ **See Figure 61**

1. Place cylinder block upside down on bench.

2. Rotate crankshaft until timing marks on crankshaft gear and camshaft gear align, to avoid interference between tappets and camshaft in removal.

3. Remove tappets, and mark them so they can be placed in their original position in re-installing.

INSPECTION

▶ **See Figures 62, 63 and 64**

1. Check cam gear for worn or broken teeth. If excessively worn or broken teeth are observed, replace camshaft.

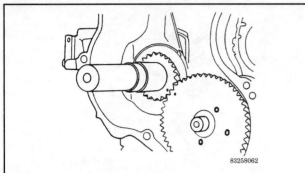

Fig. 61 Aligning the timing marks prior to disassembly

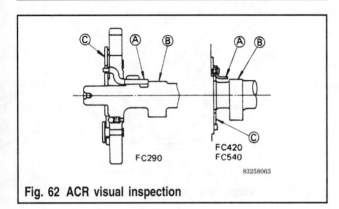

Fig. 62 ACR visual inspection

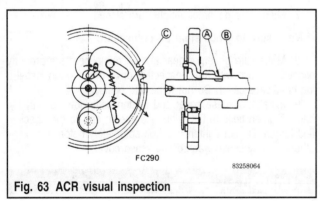

Fig. 63 ACR visual inspection

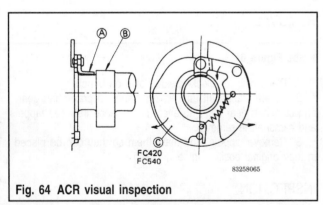

Fig. 64 ACR visual inspection

2. Check movement and damage of ACR mechanism on camshaft. If outer surface of tab or arm (A) is not placed

above cam heel (B) when weight (C) is closed, replace ACR mechanism. If outer surface of tab or arm (A) does not lower below cam heel (B) when weight (C) is pulled toward outside by finger, replace ACR mechanism. If parts of mechanism are worn, replace ACR mechanism.

3. Check bearing journal diameter with micrometer. If diameter is less than minimum, replace camshaft.
- Bearing journal diameter minimum:
- FC290 Gear side: 13.922mm (0.5481 in.); Cam side: 15.921mm (0.6268 in.)
- FC420 Gear side: 20.912mm (0.8233 in.); Cam side: 19.912mm (0.7839 in.)
- FC540 Gear side: 20.912mm (0.8233 in.); Cam side: 20.912mm (0.8233 in.)

4. Check cam lobe height with vernier calipers. If lobe height is less then minimum, replace camshaft.
- Lobe height minimum
- Intake and Exhaust
- FC290: 27.08mm (1.066 in.)
- FC420: 36.75mm (1.447 in.)
- FC540: 37.10mm (1.461 in.)

INSTALLATION

1. Place cylinder block upside down on bench.
2. Install tappets in their respective positions and push them all the way into guide to avoid interference with camshaft in assembling.
3. Rotate crankshaft until piston is at highest position.
4. Aligning timing marks and install camshaft into crankcase.

AXIAL PLAY ADJUSTMENT

▶ **See Figures 65, 66 and 67**

When any part related to axial play is changed, adjust axial play as follows:
1. Measure distance (A), installing camshaft (C) into crankcase (D) and placing gasket (E) on crankcase.
2. Measure depth (B) of crankcase cover (F).
3. Select appropriate shim with each chart.
4. Chamfered side of shim should be placed toward cam gear.

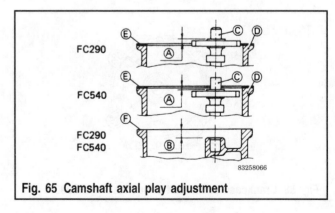

Fig. 65 Camshaft axial play adjustment

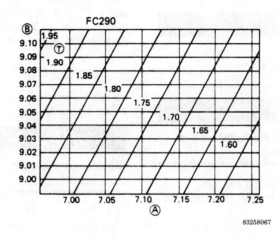

Fig. 66 Shim selection chart — FC290

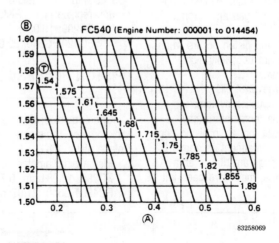

Fig. 67 Shim selection chart — FC540

5. Play after adjustment is 0.07-0.19mm (0.0028-0.0075 in.) for FC290 and FC540.

➡**FC420 and FC540 with serial number higher than 014455 do not require camshaft axial play adjustment. The camshaft is hardened to avoid out of adjustment.**

Piston & Connecting Rod

REMOVAL

> ✳✳**WARNING**
>
> **Remove any carbon or ridge at top of cylinder bore to avoid piston ring breakage in removing.**

1. Rotate crankshaft to expose connecting rod bolts.
2. Loosen connecting rod bolts and remove connecting rod cap.
3. Push piston and connecting rod out through top of cylinder.
4. Remove piston from connecting rod.
5. Remove piston rings from piston with ring expander (C).

PISTON & PISTON RING INSPECTION

▶ **See Figure 68**

Appearance of piston and piston rings shows condition of engine in running. If excessive damage is observed, replace piston and/or piston rings and remove cause of such damage.

1. Rings of wrong size or rings having improper end gap will not fit to shape of cylinder. This causes high oil consumption and excessive blowby. Check ring end gap and arrange end gap as shown.
2. Scuffing or scoring of both rings and piston occurs when friction and/or combustion temperature are unusually high.
   - Check and clean cooling system.
   - Check and correct quality and level of oil.
   - Check and adjust fuel and combustion systems.
3. Engine running at abnormally high temperature may cause varnish, lacquer, or carbon deposit formed in piston ring grooves making rings stick. Apply same treatment as above 2.
4. Vertical scratches across piston rings are due to abrasive in engine. Abrasive may be airborne, may have been left

in engine during overhaul, or may be loose lead and carbon deposit.
   - Check air cleaner and clean or replace damaged one.
   - Check any air intake through abnormal route.
   - Clean engine inside and change oil.
5. Scratches across oil side rails (A) are due to abrasive in engine oil, and other rings will also be worn in this condition, increased deposit in combustion chamber, and ring sticking. Clean engine inside and change oil.

PISTON CLEANING

▶ **See Figure 69**

> ✳✳**WARNING**
>
> **Do not use caustic cleaning solution or wire brush to clean piston.**

1. Remove all deposits from piston.
2. Clean carbon from piston ring grooves with ring groove cleaner. If cleaning tool is not available, use old piston ring breaking into suitable size.
3. Make sure oil return passages in ring groove are open.

PISTON SERVICE LIMIT

▶ **See Figures 70 and 71**

1. Check clearance between ring groove and ring using new ring and feeler gauge. If clearance is more than maximum, replace piston.
2. Check inside diameter of piston pin hole at several points with inside micrometer. If diameter is more than maximum, replace piston. Piston pin hole diameter — maximum:
   FC290: 19.031mm (0.7493 in.)
   FC420/FC540: 22.037mm (0.8676 in.)

PISTON PIN SERVICE LIMIT

Check outside diameter of piston pin at several points with micrometer. If piston pin diameter is less than minimum, replace piston pin.
   Piston pin diameter — minimum
   FC290: 18.981mm (0.7473 in.)
   FC420/FC540: 21.977mm (0.8652 in.)

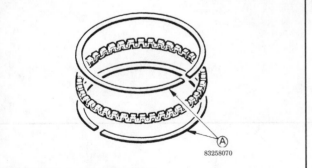

Fig. 68 Oil ring configuration

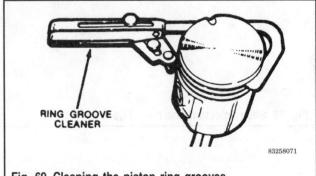

Fig. 69 Cleaning the piston ring grooves

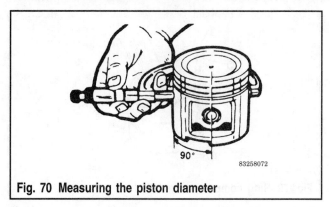

Fig. 70 Measuring the piston diameter

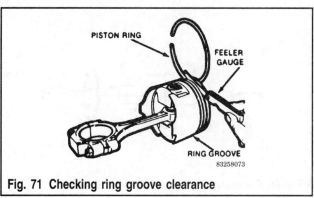

Fig. 71 Checking ring groove clearance

PISTON RING SERVICE LIMIT

▶ **See Figures 72, 73, 74 and 75**

Check thickness of piston ring at several points with microm-
eter. If thickness is less than minimum, replace ring.
- Piston ring thickness minimum
  Top and Second
    FC290: 1.44mm (0.056 in.)
    FC420/FC540: 1.94mm (0.076 in.)
Check piston ring end gap with feeler gauge, installing each
ring squarely in cylinder at approximately 25mm (1 in.) from
top. If gap is more than maximum replace piston ring.
- Piston ring end gap maximum
- Top and Second
- FC290: 0.7mm (0.028 in.)
- FC420/FC540: 0.9mm (0.035 in.)

➡**replace oil ring together with compression rings.**

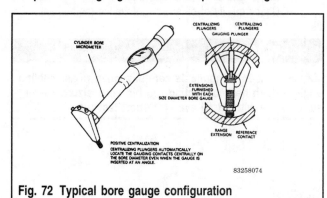

Fig. 72 Typical bore gauge configuration

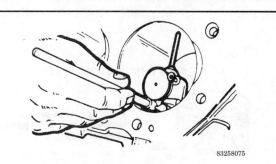

Fig. 73 Measuring the cylinder bore diameter with a
bore gauge

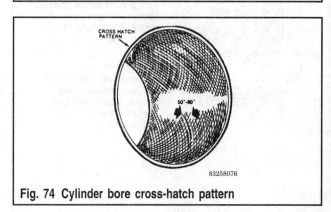

Fig. 74 Cylinder bore cross-hatch pattern

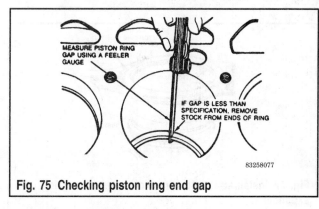

Fig. 75 Checking piston ring end gap

PISTON RING INSTALLATION

▶ **See Figures 76, 77 and 78**

1. Use ring expander.

➡**Face up 'R' mark on top and second rings.**

2. Install rings in following sequence:
   a. Lower side rail (A1)
   b. Spacer (B)
   c. Upper side rail (A2)
   d. Second ring (C)
   e. Top ring (D)
3. Place end gaps as follows:
   - I: C, A1
   - II: D, A2

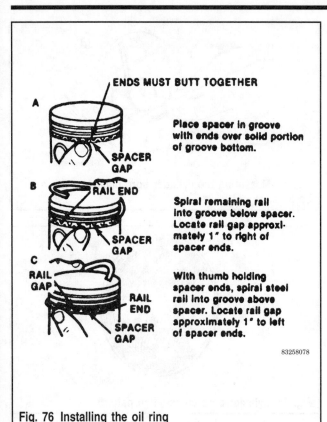

Fig. 76 Installing the oil ring

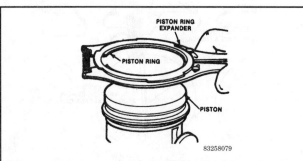

Fig. 77 Installing the compression rings with a ring expander

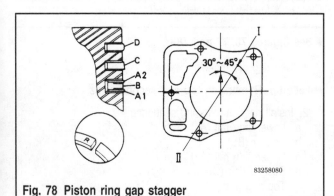

Fig. 78 Piston ring gap stagger

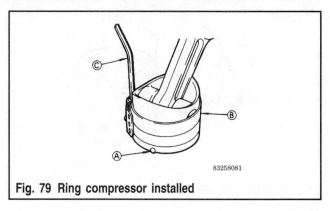

Fig. 79 Ring compressor installed

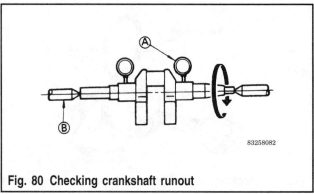

Fig. 80 Checking crankshaft runout

CONNECTING ROD INSPECTION

Check connecting rod especially big end for wearing, scratching, scoring, and.or discoloring.

1. Abnormal wearing and scratching are caused by foreign particle(s) in oil.
- Clean engine inside and change oil.
- Check oil filter and air cleaner and clean or replace damaged one.
- If connecting rod and/or crankshaft are used again, remove ridges on their surface carefully.

2. Scoring and discoloring are symptom of poor lubricating and/or overheating.
- If crankpin surface is damaged by melted connecting rod metal, replace connecting rod and crankshaft.
- Check lubricating system including oil pump, oil filter, and oil passages and repair or replace damaged part.
- Check and clean cooling system.
- Check quality of oil and maintenance method with user.

✳✳WARNING

Check re-used components carefully when re-assembling engine which had connecting rod big end seizure. Never start engine without oil even for short test run.

CONNECTING ROD SERVICE LIMIT

1. Check inner diameter of small end at several points with inside micrometer. If inside diameter is more than maximum, replace connecting rod.
- Small end inside diameter maximum
- FC290: 19.050mm (0.7540 in.)
- FC420/FC540: 22.059mm (0.8685 in.)

2. Assemble connecting rod big end aligning pilot grooves (A), and tighten connecting rod bolts.

3. Check inner diameter of big end at several points with inside micrometer. If inside diameter is more than maximum, replace connecting rod.
- Big end inside diameter maximum
- FC290: 35.567mm (1.4003 in.)
- FC420/FC540: 41.068mm (1.6169 in.)

PISTON AND CONNECTING ROD ASSEMBLY

1. Aligning 'square' mark on piston head with 'MADE IN JAPAN' on connecting rod, assemble piston over connecting rod.

2. Coat a light film of oil on piston pin and insert pin through piston and connecting rod.

3. Install retaining rings in each grooves firmly.

✷✷WARNING

Do not re-use retaining ring removed. Removal may deform or weaken the ring allowing it to come out during operation causing damage to cylinder wall.

PISTON/CONNECTING ROD INSTALLATION

▶ **See Figure 79**

1. Set ring compressor (B) over piston, flushing with piston top, with projection (A) on compressor toward top of piston.

2. Tighten compressor with wrench (C), then loosen it slightly.

3. Coat a light film of oil on cylinder wall.

4. Rotate crankshaft with crank pin in lowest position.

5. Aligning 'square' mark on piston head toward flywheel side, install piston/connecting rod assembly into cylinder.

6. Leading big end of connecting rod to crank pin, push piston down further.

7. Coat a light film of oil on crank pin, connecting rod big end, cap, and connecting rod bolts.

8. Aligning pilot groove, install cap to big end and tighten cap bolts (D).

9. Make sure connecting rod moves sideways lightly on crank pin.

✷✷WARNING

connecting rod bolt tightening is one of the most important items in assembling. Always use torque wrench.

Crankshaft

REMOVAL

1. Remove balancer support shaft from crankcase.
2. Pull crankshaft and balancer assembly out of crankcase.
3. Remove crank gear and balancer link rods from crankshaft.

INSPECTION

▶ **See Figure 80**

1. Check crank pin part and bearing journals for score, wear, or corrosion. If crank pin part shows any damage, carefully check connecting rod big end and repair or replace connecting rod and/or crankshaft.

2. Check crank gear for worn or broken teeth. If excessively worn or broken teeth are observed, replace crank gear.

3. Check outside diameter of both main bearing journals at several points with micrometer. If outside diameter is less than minimum, replace crankshaft. Journal outside diameter — minimum:

  FC290
  PTO side: 29.922mm (1.1780 in.)
  Magneto side: 29.940mm (1.1784 in.)
  FC420
  PTO side: 34.919mm (1.3747 in.)
  Magneto side: 34.945mm (1.3757 in.)
  FC540
  PTO side: 37.904mm (1.4923 in.)
  Magneto side: 34.945mm (1.3757 in.)

4. Check outside diameter of crank pin at several points with micrometer. If outside diameter is less than minimum, repair or replace crankshaft. Crankpin outside diameter — minimum:

  FC290: 35.428mm (1.3948 in.)
  FC420/FC540: 40.928mm (1.6113 in.)

5. Check outside diameter of balancer link rod journal at several points with micrometer. If outside is less than minimum, replace crankshaft. Balancer link rod journal outside diameter — minimum:

  FC290: 46.953mm (1.8453 in.)
  FC420: 53.950mm (2.1249 in.)
  FC540: 57.941mm (2.2811 in.)

6. Check run out of crankshaft at both bearing journals with dial indicator (A), setting crankshaft to alignment jig (B). If total reading of run is more than 0.05mm (0.002 in.), replace crankshaft.

AXIAL PLAY ADJUSTMENT

▶ **See Figures 81, 82, 83**

When any part related to axial play is changed, adjust axial play as follows:

1. Measure distance (A), installing crankshaft (C) into crankcase (D) and placing gasket (E) on crankcase.

2. Measure depth (B) of crankcase cover (F).

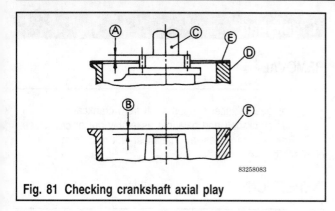

Fig. 81 Checking crankshaft axial play

3. Select appropriate shim with each chart.
Axial play after adjustment is 0.009-0.22mm (0.0035-0.0087 in.)

INSTALLATION

▶ **See Figure 85**

1. Coat a film of oil on bearing surfaces of parts assembled.

2. Assemble balance weight (A), balancer link rods (B), collar (C), spacer or governor drive gear for FC290 (D), and crank gear (E) to crankshaft.

3. Install the following parts in the order listed:
Balance weight (A): Oil hole (F)-to-flywheel side Link rods (B) of FC420 and FC540: Side oil grooves-to-collar (C) and spacer (D) side Collar (C): Conic face-to-ball bearing side Governor drive gear of FC290 and spacer of FC420 and FC540 (D): Chamfered face to link rod (B) side

4. Tape key way at taper of crankshaft to avoid cutting of oil seal lips.

5. Pack some amount of grease for high temperature application into oil seal.

6. Placing crank pin for crankcase bottom, install crankshaft and assembled parts into crankcase.

7. Install balancer support shaft (H) with 0-ring (I) through hole (J) of crankcase.

CRANK PIN RE-GRINDING

▶ **See Figure 86**

Re-grind crank pin to specified under size.
- FC290
- A: 29.950-30.000mm (1.1791-1.1811 in.)
- B: 34.980-34.990mm (1.3772-1.3776 in.)
- C: 2.80-3.20mm (0.110-0.126 in.)
- D: 28.4mm (1.12 in.) maximum
- FC420
- A: 33.950-34.000mm (1.3366-1.3386 in.)
- B: 40.467-40.480mm (1.5932-1.5937 in.)
- C: 3.30-3.70mm (0.130-0.146 in.)
- D: 32.5mm (1.28 in.) maximum
- FC540
- A: 42.950-43.000mm (1.6909-1.6929 in.)
- B: 40.467-40.480mm (1.5932-1.5937 in.)
- C: 3.30-3.70mm (0.130-0.146 in.)
- D: 32.5mm (1.28 in.) maximum

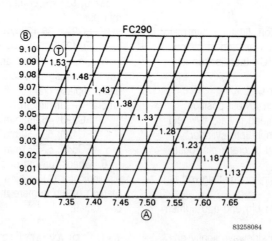

Fig. 82 Crankshaft axial play shim chart — FC290

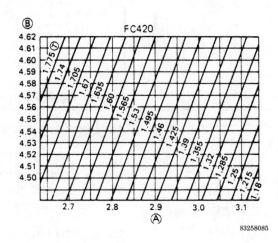

Fig. 83 Crankshaft axial play shim chart — FC420

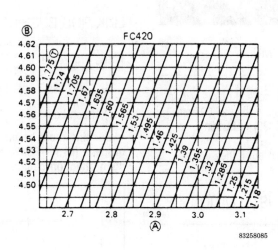

Fig. 84 Crankshaft axial play shim chart — FC540

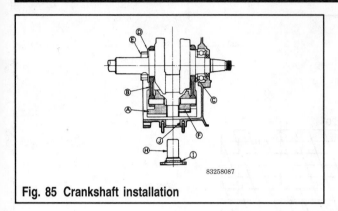

Fig. 85 Crankshaft installation

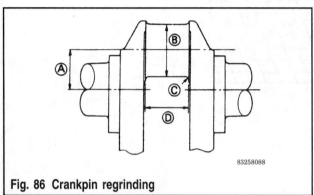

Fig. 86 Crankpin regrinding

➡Crank pin must be concentric and parallel within
**0.005mm (0.0002 in.) full indicator reading, and surface
must be finished very smooth with super finishing stone.**

If crank pin is re-ground, under-size connecting rod must
be used to keep specified clearance.

Balancer

BALANCE WEIGHT SERVICE LIMIT

1. Check bearing inside diameter at several points with inside micrometer. If inside diameter is more than 26.097mm
(1.024 in.), replace balance weight (FC290) or bushing
(FC420/FC540).
2. Wrist pins require no maintenance but if they are severely damaged, replace balance weight.

BALANCE WEIGHT BUSHING INSTALLATION

▶ **See Figure 87**

FC420 & FC5450

1. Align oil hole (A) in bushing and oil passage (B) in
balance weight, when installing new bushing.
2. Push in bushing as shown.

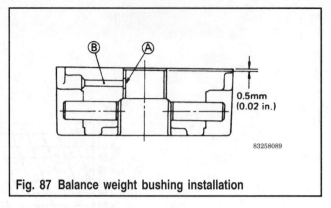

Fig. 87 Balance weight bushing installation

BALANCER SUPPORT SHAFT SERVICE LIMIT

Check outside diameter of shaft at several points with micrometer. If outside diameter is less than 25.927mm (1.0208
in.), replace balancer support shaft.

BALANCER LINK ROD SERVICE LIMIT

1. Check inside diameter of small end bearing at several
points with inside micrometer. If inside diameter is more than
12.60mm (0.4748 in.), replace balancer link rod.
2. Check inside diameter of big end bearing at several
points with inside micrometer. If inside diameter is more than
maximum, replace bushing.
   Big end inside diameter maximum
      FC290: 47.121mm (1.8552 in.)
      FC420: 54.121mm (2.1307 in.)
      FC540: 58.153mm (2.2895 in.)

LINK ROD BUSHING ASSEMBLY

▶ **See Figure 88**

1. Place seam (A) of bushing to right angle with lengthwise
center (C).
2. Push bushing into link rod as shown.

➡**Bushing must be pushed from opposite side of oil
grooves (B) for FC420 and FC540. FC290 does not need
such attention because link rod is symmetric.**

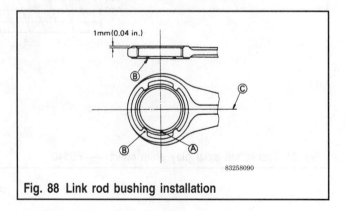

Fig. 88 Link rod bushing installation

Governor

GOVERNOR GEAR CHECK AND REMOVAL

▶ **See Figure 89**

1. Check governor gear assembly for wear and damage, as installed in crankcase cover.

✳✳WARNING

Do not remove governor gear assembly from crankcase cover except to replace it. If once removed, it cannot be re-used.

2. If governor gear assembly (A) must be replaced, remove it with proper size screw drivers (B).

✳✳WARNING

Do not damage gasket surface by screw drivers.

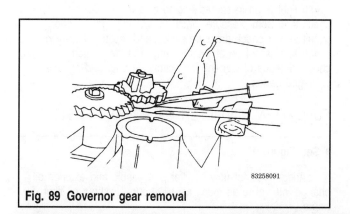

Fig. 89 Governor gear removal

GOVERNOR GEAR INSTALLATION

▶ **See Figure 90**

1. Place sleeve (A) into governor gear assembly (B).

✳✳WARNING

Sleeve can not be assembled after governor gear assembly installed in crankcase cover.

2. Place thrust washer (C) on boss of shaft (D) and then install governor gear assembly (B) with sleeve (A) to shaft until step (E) fits into groove (F).
3. Check free rotation of governor assembly after installation.

GOVERNOR SHAFT INSTALLATION

▶ **See Figure 91**

Install governor shaft (A) into crankcase and set locking pin (B) to governor shaft positioning as shown.

➡**Be careful for position of locking pin end and projection (C) which is stopper of governor shaft (A).**

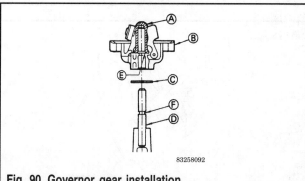

Fig. 90 Governor gear installation

LUBRICATING SYSTEM

Engine Oil

▶ **See Figure 92**

Use high quality detergent engine oil classified 'API Service SG' or equivalent.

➡**Detergent engine oil delays formation of gum, and varnish. Do not add any additives to detergent oil.**

Select oil viscosity depending on expected environmental temperature as shown.

Oil Slinger

INSPECTION

FC290

▶ **See Figure 93**

Lubricating system of FC290 is oil splash type in which oil sling we with paddles is driven by governor gear.

1. Remove bolt (A) and washer (B).
2. Remove shaft (C), washer (D) and slinger (E).
3. Check slinger for worn or broken teeth and paddles. If slinger is damaged, replace it.

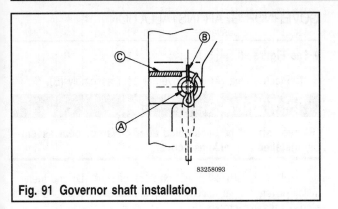

Fig. 91 Governor shaft installation

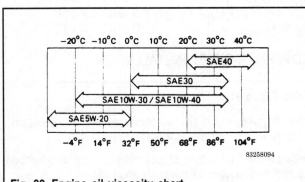

Fig. 92 Engine oil viscosity chart

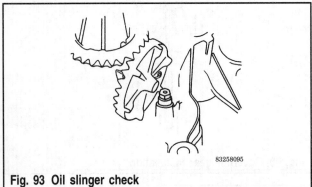

Fig. 93 Oil slinger check

Oil Warning System

FC420 & FC540

▶ **See Figure 94**

Oil pressure switch (A) is available for FC420 and FC549. It turns on warning light on dash if oil pressure gauge falls below 29.4 kPa (4.2 psi).

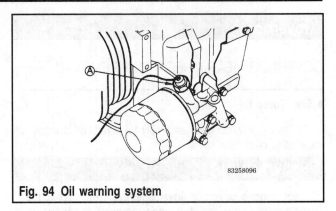

Fig. 94 Oil warning system

This pressure switch is installed on oil filter adapter (with oil filter) or oil passage cover (without oil filter).

❊❊WARNING

Whenever you start engine, make sure warning light is not on in started engine. If warning light comes on, stop engine immediately and check oil level.

Check oil warning system as follows:
When you start engine, observe warning light on dash carefully.
• If light is on in started engine in spite of adequate oil, check lead from pressure switch to warning light for short circuit and/or check pressure switch and replace damaged
• If light is not on at the moment of starter switch operation, check all leads of warning light circuit or bulb and replace damaged parts.

FULL FLOW OIL FILTER — FC420 & FC540

▶ **See Figure 95**

Cartridge type full-flow oil filter is available and extends oil change interval to as much as 100 hours. (without oil filter: 50 hours)
Oil flows through inlet (A), element (B), and outlet (C). If element is clogged, oil flows through bypass (D) to avoid oil shortage.

OIL FILTER CHANGE

❊❊CAUTION

Be careful not to burn yourself by hot oil.

❊❊WARNING

Use only KAWASAKI genuine replacement filter. Others may be fit but may be inferior in quality.

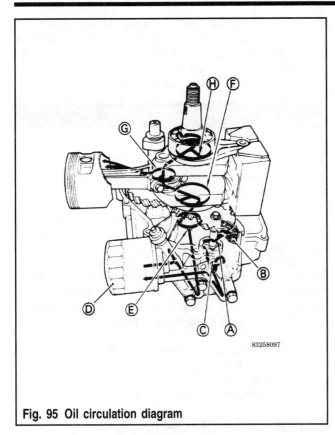

Fig. 95 Oil circulation diagram

1. Drain engine oil to suitable container.

✳✳WARNING

Before removing oil filter, place suitable pan under filter connection.

2. Turn filter counterclockwise to remove it.

3. Coat a film of engine oil on seal (E) of new filter.

4. Install new filter turning it clockwise until seal contacts mounting surface. Then turn filter ¾ turn more by hand.

5. Supply engine oil as specified.

6. Run engine for about 3 minutes, stop engine, and check oil leakage around filter.

7. Add oil to compensate oil level down due to oil filter capacity.

OIL PASSAGE CHECK

FC240 & FC540

FC420 and FC540 are equipped with crank gear driven trochoid pump to pressurize lubricating oil.

Oil is drawn into pump chamber (A) trough oilscreen (B) and pressurized by pump. Oil pressure is controlled by relief valve (C) to 294 kPa (42.7 psi).

Then oil flows through oil filter (D — if used), PTO side main journal (E), and crankshaft lubricating lower balancer link rod (F), crank pin (G), and upper balancer link rod (H). A portion of oil at crank pin passes through orifice in connecting rod and spreads on piston and cylinder.

Another portion of oil at upper balancer link rod spreads on magneto side ball bearing.

If sufficient lubrication is observed in spite of adequate oil in oil pan, check oil passages as shown above.

Oil Pump Components — FC420 & FC540

- A: Bolt
- B: Washer
- C: Pump gear
- D: Bolt
- E: Relief valve bolt
- F1: Pump cover
- F2: Pump housing
- G: Relief valve spring
- H: Check ball
- I: Oil screen
- J: Rotors (inner and outer)
- K: Pump shaft bearing

OIL PUMP CHECK

▶ **See Figures 96 and 97**

FC420 & FC540

1. Check oil screen (I) for clogging and clean or replace it if necessary.

2. Check components such as relief valve seat in crankcase cover, rotors (J), and pump gear for any damage, and repair or replace faulty part.

3. Check free length of relief valve spring (G) with vernier calipers. If free length is less than 19.0mm (0.75 in.), replace valve spring.

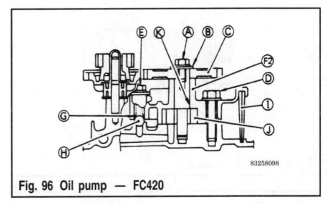

Fig. 96 Oil pump — FC420

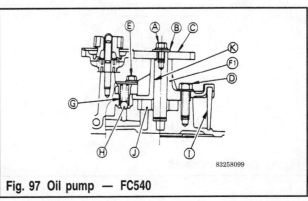

Fig. 97 Oil pump — FC540

4. Check outside diameter of pump shaft (L) at several points with micrometer. If outside diameter is less than 12.627mm (0.4971 in.), replace pump shaft including rotors (J).

5. Check inside diameter of pumpshaft bearing (K) at several points with inside micrometer. If inside diameter is more than 12.760mm (0.5024 in.), replace pump housing (FC420) or pump cover (FC540).

Breather System

▶ **See Figure 98**

Function of breather is to keep vacuum in crankcase avoiding oil being forced out of engine.

Reed valve controls direction of air flow caused by piston movement so that air flow from inside of crankcase to outside can pass reed valve but not from outside to crankcase. Rocker chamber works as oil separator of oil laden air with space for expansion and maze.

Finally, air from crankcase flows to air cleaner and mixed with intake air.

➡**Reed valve is installed on cylinder top surface in FC290 and on rocker chamber in FC420 and FC540.**

BREATHER REED VALVE CHECK

▶ **See Figure 99**

1. Check valve and valve seat for any damage such as crack or wear.

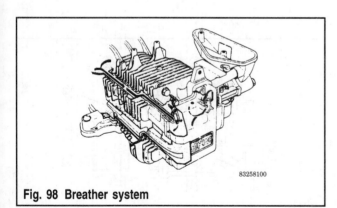

Fig. 98 Breather system

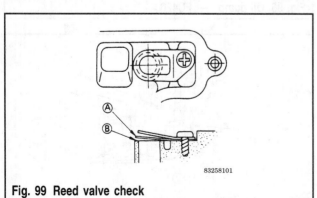

Fig. 99 Reed valve check

2. Check gap between reed valve (A) and seat (B) with feeler gauge. If gap is more than maximum, turn over valve or replace valve. REED VALVE GAP MIX:
- FC290: 0.2mm (0.008 in.)
- FC420/FC540: 1-2mm (0.04-0.08 in.)

Cylinder/Crankcase

CYLINDER SERVICE LIMIT

▶ **See Figure 100**

1. Visually check cylinder block for scored bore, broken fin, or other damages. If unrepairable damage is observed, replace cylinder block.

2. Check inside diameter at 10mm (0.4 in.) from top in directions of parallel and right angle to crankshaft, with inside micrometer. If inside diameter and/or out of round are more than maximum, resize cylinder bore or replace cylinder block.
- Cylinder bore inside diameter maximum
- FC290: 78.067mm (3.0735 in.)
- FC420/FC540: 89.076mm (3.5069 in.)
- Cylinder bore out of round maximum
- FC290: 0.056mm (0.0022 in.)
- FC420/FC540: 0.063mm (0.0025 in.)

CYLINDER BORE RESIZING

Oversize piston and piston rings for standard plus 0.25mm (0.01 in.), 0.50mm (0.02 in.), and 0.75mm (0.03 in.) are available. Select suitable size depending on condition of cylinder bore to be resized.

1. Bore cylinder finely to size as shown on right table before honing.

2. Hone cylinder to final a=size as shown on table following procedure shown below.

✲✲WARNING

Consider shrinkage of cylinder bore for 0.006-0.008mm (0.0002-0.0003 in.) after cooling down from honing heat.

➡**Use honing stone recommended by hone manufacturer.**

HONING

▶ **See Figures 101 and 102**

1. Align centers of cylinder bore and drill press carefully and set cylinder block on drill press table.

2. Install hone to drive shaft and set stopper of drill press so that hone can only extend 20-25mm (¾-1 in.) from top to bottom of cylinder bore.

3. Adjust honing stone to contact snuggly against cylinder wall at narrowest point.

➡**DO NOT FORCE!**

4. Rotate hone by hand. If it cannot be rotated, adjust hone until it can be rotated by hand.

5. Set drill press rpm to 200-250.

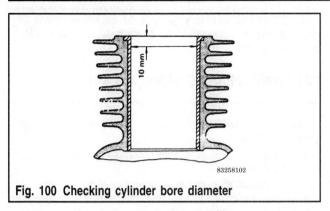

Fig. 100 Checking cylinder bore diameter

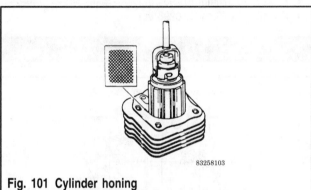

Fig. 101 Cylinder honing

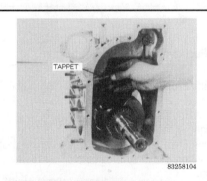

Fig. 102 Cross-hatch pattern after honing

6. Coat honing oil on cylinder bore.

✳✳WARNING

Do not use solvent or gasoline.

7. Drive drill press and move hone up and down in cylinder bore about 20 cycles/minute.

✳✳WARNING

Stop drill press when checking or measuring cylinder bore.

8. Measure inside diameter in suitable periods with inside micrometer, and check finishing pattern which should be 40 to 60 degrees crosshatch.

9. Clean cylinder block thoroughly with soap and warm water for 'white glove inspection'.

✳✳WARNING

Thoroughly wash honing grit from cylinder. Grit is extremely abrasive to engine components.

10. Dry cylinder block and coat engine oil to cylinder bore.

BALL BEARING CHECK

✳✳WARNING

Do not remove ball bearing from housing except replacing.

1. Clean ball bearing with high flash point solvent.
2. Pour engine oil to bearing.
3. Rotate bearing inner race slowly by hand. If any roughness is felt and it can not be removed by re-cleaning, replace ball bearing.

BALL BEARING REPLACEMENT

▶ **See Figure 103**

1. Remove oil seal preceding ball bearing replacement. Do not re-use removed oil seal.
2. Push out ball bearing with bearing driver.
3. Clean bearing housing with high flash point solvent and dry it.
4. Coat a light film of oil on bearing.
5. Push new ball bearing into housing to end.

OIL SEAL REPLACEMENT

▶ **See Figure 104**

If oil leakage through oil seal is observed or seal lip is damaged, replace oil seal.
1. Remove oil seal by tapping it out with a screw driver or punch.
2. Placing spring held seal lip (A) inside, push oil seal to be flush with housing end (C).

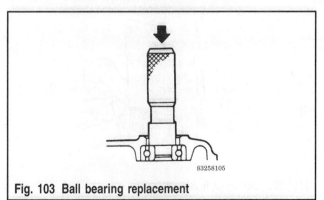

Fig. 103 Ball bearing replacement

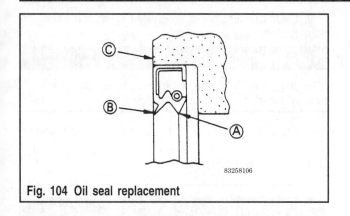

Fig. 104 Oil seal replacement

3. Before final assembly, pack some amount of grease for high temperature application into space between seal lip (A) and dust lip (B).

CAMSHAFT BEARING SERVICE LIMIT

Check inside diameter of camshaft bearing in crankcase at several points with inside micrometer.

If inside diameter is more than maximum, replace cylinder/crankcase.
- Camshaft bearing inside diameter maximum
- FC290: 16.055mm (0.6321 in.)
- FC420: 20.076mm (0.7904 in.)
- FC540: 21.076mm (0.8298 in.)

KF200 SERIES

Engine Control System

GOVERNOR LEVER SETTING

▶ **See Figure 105**

Whenever carburetor or governor lever is removed from engine and then installed again, adjust governor lever position.

1. Install governor lever (A) on governor shaft (B) but do not tighten nut (C). Loosen nut (C) if it is tight.

2. Turn governor lever (A) clockwise or place throttle lever on dash in 'FAST" position to open carburetor throttle valve fully.

3. Turn governor shaft (B) clockwise to end of travel.

4. Keeping governor lever position of throttle fully open, tighten nut (C).

THROTTLE CABLE INSTALLATION

▶ **See Figure 106**

1. Link throttle cable (G) to speed control lever (C) and clamp throttle cable outer housing (F) temporarily.

2. With throttle lever on dash in 'FAST" position, align hole (B) of speed control lever (C) with hole (D) of control plate (E)

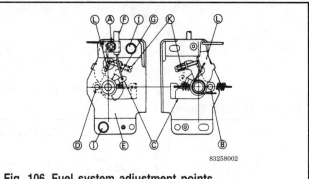

Fig. 106 Fuel system adjustment points

and insert 6mm (0.24 in.) diameter pin or 6mm bolt through two holes.

3. Pull out throttle cable outer housing (F) to remove any slack and tighten cable clamp bolt (A).

4. Remove 6mm pin and set throttle lever on dash in 'CHOKE" position. Make sure carburetor choke valve is completely closed.

FAST IDLE SPEED ADJUSTMENT

▶ **See Figure 106**

➡**Air cleaner must be installed to engine before starting.**

1. Start and warm up engine without load.

2. Loosen two control plate bolts (I).

3. Align hole (B) of speed control lever (C) with hole (D) of control plate (E) and insert 6mm (0.24 in.) diameter pin or 6mm bolt through two holes.

➡**Make sure choke valve is fully opened.**

4. Adjust fast idle speed for specified rpm by moving control plate (E).

5. Tighten two bolts (I) securely in a manner to avoid changing specified speed.

6. Stop engine, remove 6mm pin, and set throttle lever on dash in 'CHOKE" position. Make sure carburetor choke valve is closed completely.

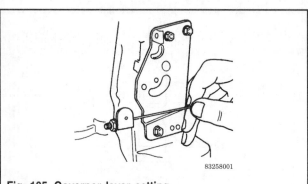

Fig. 105 Governor lever setting

CHOKE ADJUSTMENT

▶ **See Figure 106**

1. Align hole (B) of speed control lever (C) with hole (D) of control plate (E) and insert 6mm (0.24 in.) diameter pin or 6mm bolt through two holes.

2. Turn choke setting screw (K) counterclockwise until it is clear of choke control lever (L) and then turn choke setting screw clockwise until it just contacts choke control lever.

3. Remove 6mm pin and set throttle lever on dash in 'CHOKE" position. Make sure carburetor choke valve is closed completely.

SLOW IDLE SPEED ADJUSTMENT

▶ **See Figure 107**

1. Turn carburetor pilot screw (B) in until it just seats and then back out 1½ turns.

AIR CLEANER

K-KLEEN System

▶ **See Figures 108 and 109**

Intake air is induced through rotary screen (A), fan housing (B), and air cleaner (C) to remove grass and rubbish from air. Therefore, the condition of air passages affects volume of intake air and carburetor functions.

1. Assemble related parts neatly to minimize air leakage.

2. Keep clearance between rotary screen and fan housing as shown.

3. Do not remove any parts constructing air passages when running engine.

4. Keep air passages free from grass and rubbish.

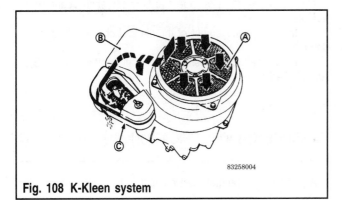

Fig. 108 K-Kleen system

2. Start and warm up engine without load.

3. Move throttle lever on dash to 'SLOW" position.

4. Adjust slow idle speed to specified rpm by moving throttle stop screw (A).

5. Adjust pilot screw (B) until engine idles at maximum speed and then turn pilot screw out additional ¼ turn.

6. Re-adjust slow idle speed, to specified rpm.

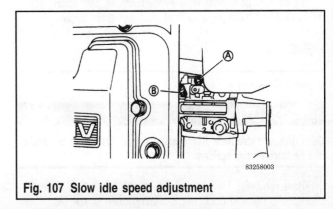

Fig. 107 Slow idle speed adjustment

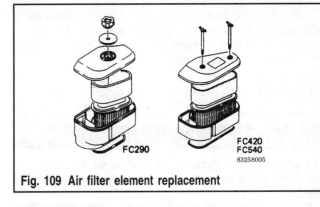

Fig. 109 Air filter element replacement

MAINTENANCE

Air cleaner maintenance is one of the most important items to keep engine performing well.

1. The foam element must be lightly oiled to perform as intended. Make sure to soak element in engine oil and squeeze excessive oil, after washed.

2. The paper element is cleaned by gentle tapping or washing in detergent and water.

✳✳WARNING

Do not use pressurized air to paper element to avoid breakage. Do not oil paper element.

CARBURETOR

Components

▶ **See Figure 110**

This carburetor is float type with adjustable pilot screw (idle mixture), fixed main jet, and float chamber drain screw.

REMOVAL

▶ **See Figure 111**

❊❊CAUTION

Gasoline is extremely flammable. Avoid fires due to smoking or careless practices.

Before removing carburetor from engine, drain fuel in float chamber to suitable container loosening drain screw (A).

Float Chamber

▶ **See Figure 112**

Before removing float chamber, rotate float chamber clockwise and counterclockwise 1/6 turn either way, 2 or 3 times, pushing float chamber to carburetor body to release sticking of float chamber and rubber gasket.

CLEANING

❊❊WARNING

Do not clean jet orifices with a hard object. Follow instructions prepared by cleaner manufacturer when using cleaner.

1. Dip carburetor components except non-metallic parts such as gasket into carburetor cleaner until dirt is removed and rinse them out with solvent.

➡**Rinse carburetor aluminum components in hot water to neutralize corrosive action of cleaner, if so instructed by the manufacturer.**

2. Dry components with compressed air. Make sure all orifices and passages are free from dirt or foreign object.

❊❊WARNING

Do not use rags or paper to dry components to avoid plugging orifices by lint.

FLOAT ADJUSTMENT

▶ **See Figures 113 and 114**

❊❊WARNING

Do not strike float pin to remove or install it, to avoid breakage of pin holder. To remove float pin, pull the transformed end with pliers.

To install float pin, push float pin transformed end until the other end flushes with pin holder outer outer surface.

➡**The white float (made of polyacetal) does not require check and adjustment.**

❊❊WARNING

Do not push down float or needle valve when checking parallelism.

1. Check float angle and correct it as follows; Place carburetor upside down and check parallelism of float surface and carburetor body.
2. If not parallel, adjust float surface angle bending tang with needle-nose pliers.

INSTALLATION

▶ **See Figure 115**

Precautions

1. Do not over-tighten small carburetor components. Finger-tighten pilot screw.
2. Do not bend throttle and choke shafts when assembling.
3. Apply screw locking agent to screws of throttle valve or choke valve. Do not allow agent to flow into shaft bearing surfaces.
4. Make sure movement of throttle and choke valves is smooth.
5. Before carburetor installation, install throttle linkage (S) and choke linkage (B) on carburetor.
6. Make sure to fasten fuel line at carburetor inlet with clamp.

CARBURETOR IDENTIFICATION

A portion of carburetor, part number is marked on carburetor body.

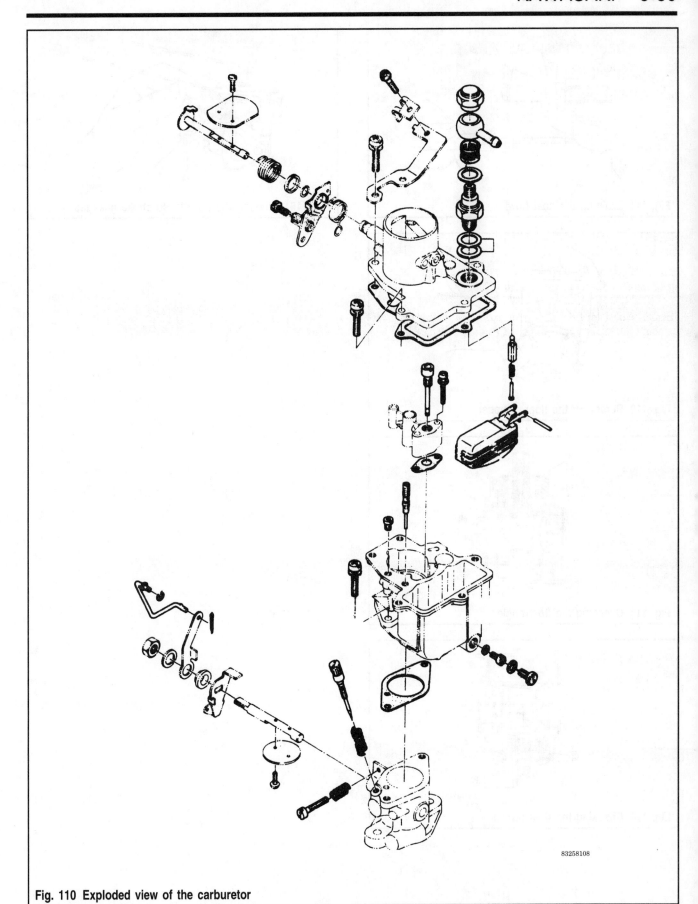

83258108

Fig. 110 Exploded view of the carburetor

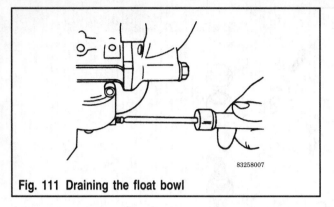

Fig. 111 Draining the float bowl

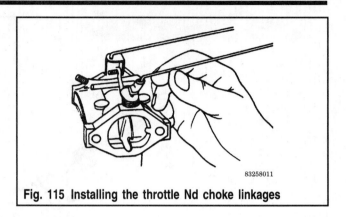

Fig. 115 Installing the throttle Nd choke linkages

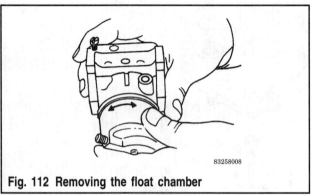

Fig. 112 Removing the float chamber

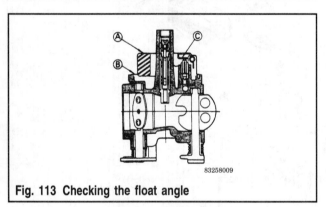

Fig. 113 Checking the float angle

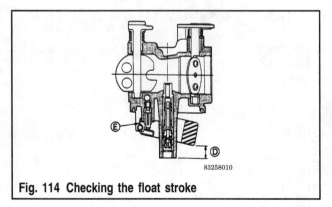

Fig. 114 Checking the float stroke

COOLING SYSTEM

Blade Gap Adjustment

▶ **See Figure 116**

1. Install fan housing (A) and rotary screen (B).
2. Check gap (C) between contour blades (D) and fan housing and adjust it as near as 1.5mm (0.06 in.) by changing number of shim (E) used.
3. Install guard (F) or recoil starter (including pulley).

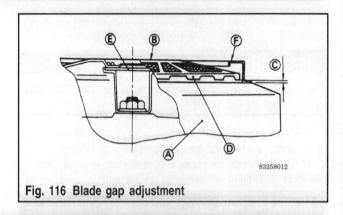

Fig. 116 Blade gap adjustment

RECOIL STARTER

▶ **See Figures 117 and 118**

Disassembly

✴✴WARNING

Do not wedge rope between reel and case.

1. Pull handle out about 250mm (10 in.). Then hold rope in place with locking pliers or knot (A).

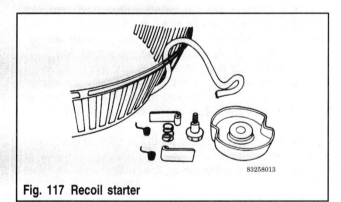

Fig. 117 Recoil starter

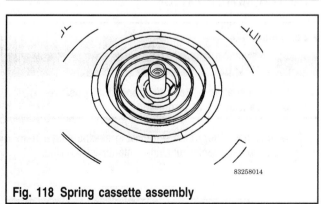

Fig. 118 Spring cassette assembly

2. Pull knot in handle (B) out and untie it.

✴✴WARNING

Wear gloves during disassembling to avoid injury.

3. While carefully holding reel (C) and case (D), remove locking pliers or untie knot.
4. Unwind spring tension slowly.
5. Remove bolt (G) at center part and then retainer (H), pawls (J), pawl springs (K), and center spring (L) but not including reel (C).

✴✴CAUTION

When removing reel (C), be careful that recoil spring under the reel does not fly loose and cause injury! The spring is under great pressure! There should be no spring tension on reel when removing reel. If tension is felt, push reel back into place and gently 'wiggle'' it until reel can be easily removed.

6. Rotate reel (C) one-quarter turn clockwise from rest position where no tension can be felt. Then, slowly lift reel straight up out of case.

✴✴WARNING

If recoil spring cassette (N) is sticking with removed reel (C), be careful that the cassette does not drop.

✴✴CAUTION

Be careful that recoil spring (M) does not fly loose from cassette (N) and causes injury. The spring is under great pressure.

7. Slowly lift recoil spring cassette (N), straight up out of case (D) or reel (C).
8. If recoil (M) must be removed from cassette (N), hold the cassette with opening side downward in suitable container and tap cassette to remove recoil spring.

INSPECTION

1. Dip metal parts in bath of high flash-point solvent, if necessary.

✳✳WARNING

Do not clean any non-metallic parts in solvent. They may be damaged by the solvent.

2. Check starter pawls for chips or excessive wear.
3. Check starter rope for excessive wear or fraying.
4. Check springs for break, rust, distortion, or weakened condition. If damage is found, replace the part.

✳✳CAUTION

Do not throw away recoil spring as installed in cassette. Recoil spring may fly loose from cassette and cause injury.

ELECTRIC STARTER & CHARGING SYSTEM

Troubleshooting

1. Disconnect spark plug cap, and ground the cap terminal.
2. Turn engine switch to 'START' position and check condition.

✳✳CAUTION

Engine may be cranked in this test. Do not touch any rotating parts of engine and equipment during test. If starter does not stop by engine switch 'OFF', disconnect negative (-) lead from battery as soon as possible.

SOLENOID AND CIRCUIT CHECK

▶ **See Figures 120, 121, 122 and 123**

This procedure is for shift type starter, but can be used for Bendix type by replacing word 'SOLENOID' with 'SOLENOID SWITCH.'

➡**Before this test, make sure battery is fully charged.**

1. Disconnect lead (A) between starter motor and solenoid, and keep the lead away from terminal of solenoid to avoid accidental cranking in test.

➡**Painted part is not grounded.**

2. Disconnect lead (B) between solenoid and engine switch. Set multimeter selector switch to 25V DC position and connect

REASSEMBLY

✳✳CAUTION

Wear gloves during recoil spring (M) installation to avoid injury. The recoil spring must be assembled with great pressure.

1. Lightly grease recoil spring.
2. Set recoil spring in spring cassette (N) and install cassette so that hook (P) of spring catches tab (R) on case.
3. If rope is unwound from reel (C), wind rope i reel counterclockwise facing pawl groove(s).
4. Aligning guide pin (T) on reel with recoil spring hook (U) on cassette, install reel into case.
5. Install pawl springs (K), pawls (J), center spring and retainer, then tighten bolt.

➡**Pawl springs (K) and pawls (J) are assembled into reel as shown.**

6. Rotate reel two turns counterclockwise to preload recoil spring.
7. Pull rope out of case through rope hole and install handle.

it with lead (B) and grounded part. Turn engine switch to 'START' and check voltage.

- If voltage is 0 or much less than battery voltage, check engine switch and/or circuit.
- If voltage is same as battery voltage, go to next step.

3. Set multimeter selector switch R · 1Ω position and check resistance between terminal (C) and grounded part.

- If resistance is not nearly 0Ω, replace solenoid.
- If continuity is observed, go to next step.

4. Connect lead (B) to terminal (C). Set multimeter selector switch to R x 1 ohm position and connect it with two terminals on solenoid as shown.. Turn engine switch to 'START' and check resistance.

- If resistance is not nearly 0Ω or solenoid does not click, replace solenoid.
- If solenoid is normal, go to next step.

STARTER MOTOR CHECK

▶ **See Figure 124**

✳✳WARNING

Disconnect negative (-) lead first and then positive (+) lead to prevent spark at terminal.

1. Disconnect battery before removing electric starter from engine to avoid accidental running of starter in handling.

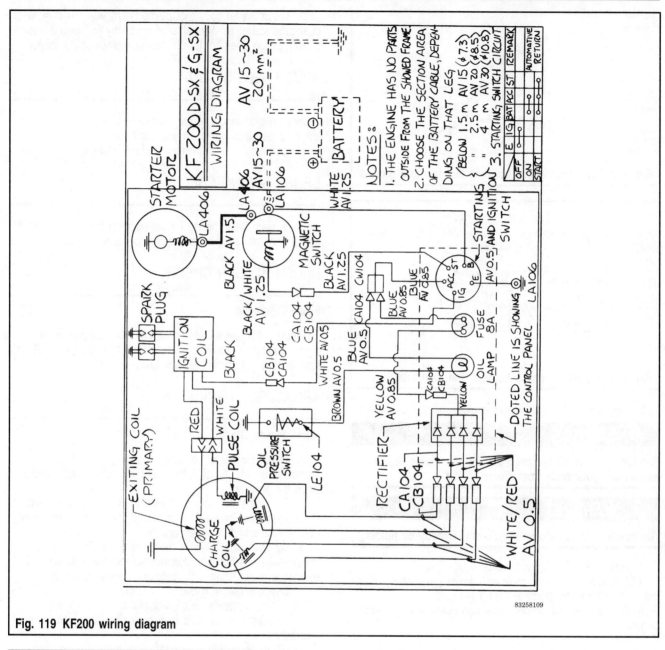

Fig. 119 KF200 wiring diagram

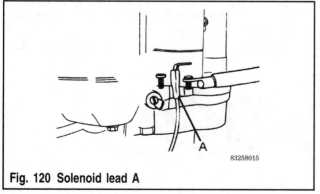

Fig. 120 Solenoid lead A

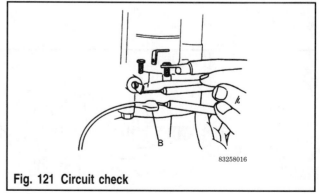

Fig. 121 Circuit check

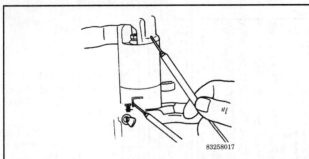

Fig. 122 Resistance check between terminal and ground

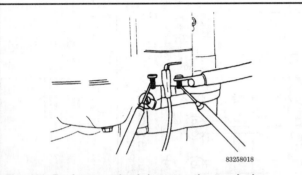

Fig. 123 Resistance check between the terminals

2. Remove electric starter from engine.

✳✳CAUTION

The test room must be free from any flammable object. Keep away your body from pinion.

✳✳WARNING

Be careful not to deform electric starter body by holding.

3. Hold electric starter with vise.
4. Connect first jumper cable with positive (+) battery terminal and starter motor terminal (D) on solenoid.
5. Connect second jumper cable with negative (-) battery terminal.

6. Touch starter body (not painted part) with the other end of second starter jumper cable intermittently (within one second). If pinion does not turn, repair or replace starter motor.

BRUSH SERVICE LIMIT

Check overall length of each brush with vernier calipers. If length is less than minimum, replace brush.
• Brush length — minimum: 10.5mm (0.41 in.)

BRUSH SPRING CHECK

Check each brush spring for breakage or distortion. If brush spring does not snap brush firmly into place, replace.

BRUSH HOLDER CHECK

▶ **See Figure 125**

Check multimeter selector switch to R · 1kΩ position and check resistance between brush holder (A) and its base (B). If resistance is not infinite, replace brush holder.

ARMATURE CHECK

▶ **See Figure 126**

1. Check surface of commutator and grooves. If it has a scratch or is dirty, polish it with very fine emery paper and clean the grooves.
2. Set multimeter selector switch to R · 1Ω position and check resistance between segments. If resistance is not nearly 0Ω, replace armature (open circuit).
3. Set multimeter selector switch to R · 1kΩ position and check resistance between commutator and armature shaft. If resistance is not infinite (°), replace armature (short circuit).
4. Check armature windings for shorts.
   a. Place armature on growler (A).
   b. Hold thin metal strip (e.g. hack saw blade) on top of armature.
   c. Turn on growler and rotate armature one complete turn. If metal strip vibrates at any position, replace armature (short circuit).

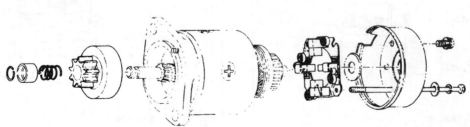

Fig. 124 Starter exploded view

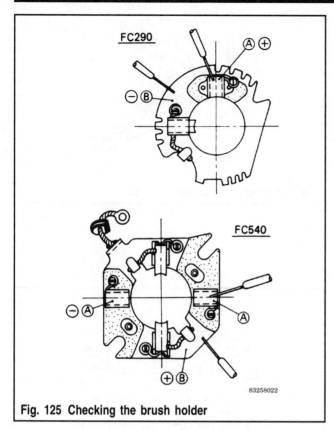

Fig. 125 Checking the brush holder

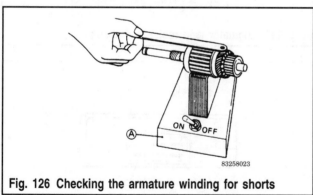

Fig. 126 Checking the armature winding for shorts

ARMATURE SERVICE LIMIT

1. Check commutator outside at several points with vernier caliper. If diameter is less than 27mm (1.06 in.), replace armature.

2. Check groove depths between commutator segments. If depths are less than 0.2mm (0.01 in), cut insulating material to standard depth with thin file and clean grooves.
   • Standard groove depth: 0.5-0.8mm (0.02-0.03 in.)

3. Support armature in alignment jig at each ends of shaft as shown. Position dial indicator perpendicular to commutator. Rotate armature slowly and check run-out. If run-out is more than maximum, turn down commutator or replace armature.
   • Commutator run-out — maximum: 0.5mm (0.019 in.)

YOKE CHECK

▶ **See Figures 127, 128, 129 and 130**

1. Set multimeter selector switch to R · 1Ω position and check resistance between negative terminals (E).If resistance is not nearly 0Ω, replace yoke assembly.

2. Set multimeter selector switch to R · 1Ω position and check resistance between negative terminals (E) and yoke case (F). If resistance is not nearly 0Ω, replace yoke assembly.

3. Set multimeter selector switch to R · 1Ω position and check resistance between positive terminals (G). If resistance is not nearly 0Ω, replace yoke assembly.

4. Set multimeter selector switch to R · 1kΩ position and check resistance between positive terminal (G) and yoke case (F). If resistance is not infinite (°), replace armature.

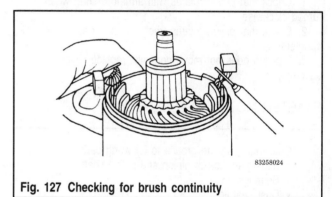

Fig. 127 Checking for brush continuity

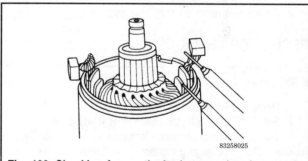

Fig. 128 Checking for continuity between brush ground and case

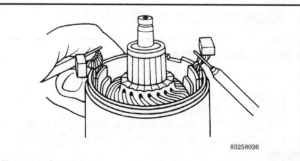

Fig. 129 Checking for continuity between brush positive leads

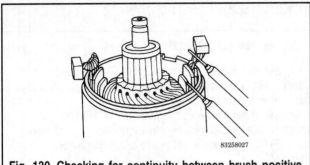

Fig. 130 Checking for continuity between brush positive leads and case

PINION CLUTCH CHECK

1. Check that pinion rotates with armature shaft when turned clockwise.
2. Check that pinion rotates freely when turned counterclockwise.
3. If pinion does not move normally, replace pinion assembly.

STARTER MOTOR REASSEMBLY

1. Coat multi-purpose grease to following parts.
   - Sliding surface or pinion and spline shaft
   - Bearings
   - Dust seal lip
   - Contact surface of shift lever
2. Do not use removed snap ring again.

REGULATOR CHECK

▶ **See Figure 131**

1. Remove all leads from regulator.
2. Set multimeter selector switch to R · 1kΩ position.
3. Check resistance between terminals as shown on table.
A: Key switch B: Charging monitor C: Multimeter terminal
   If resistance is not as specified, replace regulator.

➡**Resistance value may vary with individual meters.**

UNREGULATED STATOR OUTPUT

▶ **See Figures 132, 133 and 134**

(13-15A TYPE)
1. Disconnect regulator.
2. Start and warm up engine.
3. Set multimeter selector switch to 250V AC position.
4. Connect meter with stator lead terminals (G,H) in 3P coupler.
5. Run engine at 3,350 rpm and check voltage. If voltage is less than minimum, replace stator.
   - Stator output — minimum: 24 V

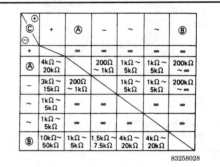

Fig. 134 Stator output check — 3A

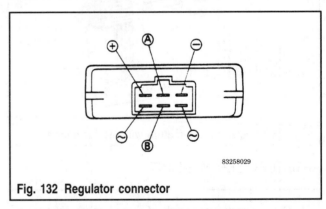

$\overset{\oplus}{\underset{\ominus}{\text{C}}}$	+	Ⓐ	−	~	~	Ⓑ
+		∞	∞	∞	∞	∞
Ⓐ	4kΩ ~ 20kΩ		200Ω ~ 1kΩ	1kΩ 5kΩ	1kΩ 5kΩ	200kΩ ~ ∞
−	3kΩ ~ 15kΩ	200Ω ~ 1kΩ		1kΩ 5kΩ	1kΩ 5kΩ	200kΩ ~ ∞
~	1kΩ ~ 5kΩ	∞	∞		∞	∞
~	1kΩ ~ 5kΩ	∞	∞	∞		∞
Ⓑ	10kΩ ~ 50kΩ	1kΩ ~ 5kΩ	1.5kΩ ~ 7.5kΩ	4kΩ ~ 20kΩ	4kΩ ~ 20kΩ	

Fig. 131 Regulator lead resistance check

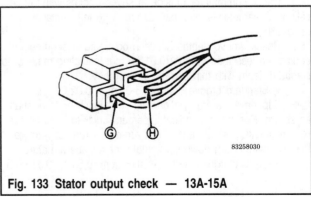

Fig. 132 Regulator connector

Fig. 133 Stator output check — 13A-15A

IGNITION SYSTEM

Types of Systems

▶ **See Figure 135**

Transistor controlled ignition system is used for these engines and this system consists of following components.
1. Ignition coil
2. Control unit
3. Flywheel (with permanent magnet)

These components do not mechanically contact and periodic maintenance is not required.
- F: Flywheel
- L_1: Primary coil
- L_2: Secondary coil
- CU: Control unit
- R_1: Control resistor
- R_2: Control resistor
- SP: Spark plug
- SW: Engine switch
- TR_1: Transistor
- TR_2: Transistor

SPARK CHECK

To check ignition system, check spark as follows:
1. Remove spark plug and connect plug cap with the removed spark plug.
2. Install spare plug to plug hole to avoid fuel spitting from hole.

☀☀CAUTION

To avoid electric shock, hold plug cap, but not spark plug.

3. Keeping contact with spark plug metal part (not center electrode) and engine block, crank engine.

☀☀WARNING

Do not clean spark plug with bead or sand cleaner.

- If no or very weak spark is observed, clean spark plug and regap it to 0.7-.08mm (0.028-0.031 in.) and try engine cranking again.
- If spark is not improved by cleaning, try checking again with new spark plug.
- If spark is not improved yet, check ignition system.

CONTROL UNIT CHECK

▶ **See Figure 136**

1. Set multimeter selector switch to R · 10Ω position.
2. Check resistance between terminal (A) and case (B). If resistance is out of specified value, replace control unit.

➡**This check may not cover every defect.**

IGNITION COIL CHECK

▶ **See Figure 137**

1. Resistance between the points as specified.
2. If resistance is out of specified valve, replace ignition coil.

Flywheel

REMOVAL

☀☀WARNING

Remove plug cap from spark plug to avoid engine starting. Do not insert and tool in flywheel fins or ring gear to avoid rotation. The tool will damage flywheel.

1. Loosen nut by turning it counterclockwise. Use ½ in. air impact wrench or strap wrench to avoid flywheel rotation.

☀☀WARNING

Keep shoulder of nut flush with end of crankshaft until taper engagement is released.

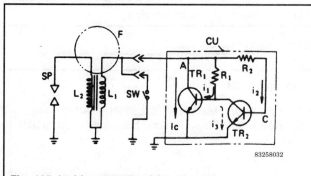

Fig. 135 Ignition system wiring diagram

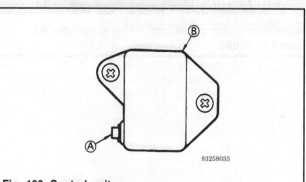

Fig. 136 Control unit

2. Use flywheel puller (A) to remove flywheel.

INSTALLATION

1. Before installing flywheel, remove grease and oil from taper part of crankshaft and taper hole of flywheel.
2. Make sure key is in place when installing flywheel
3. Tighten the nut.

IGNITION COIL AIR GAP ADJUSTMENT

1. If ignition coil is removed or replaced, adjust the air gap when installing coil.
2. If ignition coil is removed or replaced, adjust the air gap when installing coil.

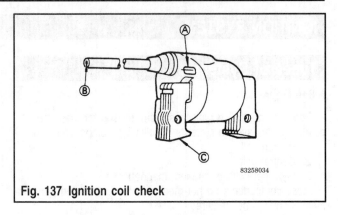

Fig. 137 Ignition coil check

3. Insert 0.3mm (0.012 in.) feeler gauge or solid sheet (A) between coil legs (B) and flywheel rim (C).
4. Pushing coil to flywheel, tighten coil mounting screws firmly.

FUEL SYSTEM

Fuel Pump

REMOVAL AND DISASSEMBLY

▶ See Figure 138

✳✳CAUTION

Gasoline is extremely flammable. Avoid fires due to smoking or careless practices. Avoid spilling of gasoline in removing and disassembling fuel pump. Plug fuel line disconnected from pump intake joint, immediately.

INSPECTION

1. Check vent hole and screen for plugging or clogging. If vent hole and screen are plugged or clogged, remove dirt from them.
2. Check diaphragms for crack, tear or hole. If defect is found, replace it.
3. Check valves for crack, tear or wear. If defect is found, replace them.

✳✳WARNING

Make sure rubber gasket and rubber diaphragm are placed as shown.

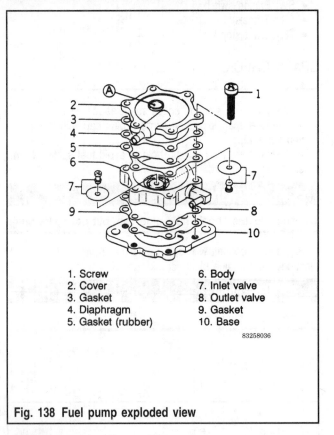

1. Screw
2. Cover
3. Gasket
4. Diaphragm
5. Gasket (rubber)
6. Body
7. Inlet valve
8. Outlet valve
9. Gasket
10. Base

Fig. 138 Fuel pump exploded view

INSTALLATION

▶ See Figure 139

1. Make sure to use heat shield washers (A) as shown.
2. Make sure to fasten fuel and pulse lines at pump with clamps.

ENGINE MECHANICAL

Engine

▶ **See Figures 140, 141, 142, 143, 144, 145, 146, 147, 148 and 149**

ENGINE DISASSEMBLY

1. Remove the oil level gauge.
2. Remove the oil drain plug and allow the oil to drain.
3. Remove the air cleaner wing nut.
4. Loosen the clamp at the top side of the carburetor.
5. Remove the bolt on the air cleaner mount bracket.
6. Lift up to remove the air cleaner assembly.
7. Remove the muffler bracket bolt.
8. Remove the bolts holding the muffler to the exhaust manifold.
9. Remove the muffler assembly using care not to damage the exhaust gasket.
10. Turn the fuel valve to the 'OFF" position.
11. Disconnect the fuel line at the fuel valve.
12. Remove the bolts holding the fuel tank assembly to the engine.
13. Remove the tank and fuel shut-off as an assembly.
14. Remove the top cover.
15. Disconnect the throttle control link at the throttle lever.
16. Disconnect the wire from the oil pressure shut-off switch.
17. Disconnect the wire at the starter solenoid.
18. Disconnect the choke control cable at the carburetor.
19. Disconnect the wire at the starter motor.
20. Disconnect the ignition coil wire.
21. Remove the bolts holding the control panel to the engine and remove the panel assembly.
22. Remove the ignition coil.
23. Remove the governor arm from the governor shaft.
24. Remove the intake manifold bolts and remove the carburetor and intake manifold as an assembly.
25. Remove the exhaust manifold cover, then remove the nuts holding the manifold to the engine.
26. Remove the bolts holding the fuel pump to the crankcase.
27. Disconnect the fuel lines at the fuel pump.
28. Carefully lift the fuel pump from the crankcase.

29. Remove the bolts holding the starter to the crankcase and remove the starter motor.
30. Remove all the bolts holding the spiral case and lift it off. It should not be necessary to pry on the spiral case. If it is, check for bolts that may have been missed.
31. Remove the spark plugs from both cylinders.
32. Loosen the pulley bolts approximately $\frac{3}{4}$-1 turn. Do not remove the pulley as this becomes the flywheel puller.
33. Loosen the flywheel nut and turn it clockwise until the shoulder of the nut makes contact with the pulley, then apply sufficient force to loosen the flywheel.
34. After the flywheel is loose, continue turning the flywheel nut clockwise until nut is completely loose from the crankshaft.
35. Remove the flywheel, nut and pulley as an assembly. Before attempting to assemble the engine, these parts (flywheel, nut and pulley) should be separated.
36. Remove the outer case cover.
37. Remove the inner case cover.
38. Remove the woodruff key from the crankshaft.
39. Remove the magneto stator assembly.
40. Loosen and remove the cylinder head bolts and lift off the cylinder head.
41. Remove the nuts holding the cylinder to the crankcase.
42. Gently pull the cylinder free of the piston. Care must be taken not to damage the piston by letting it fall against the crankcase. Repeat for other side.
43. Place the cylinder top side down on a work bench.
44. Remove the valve collets by applying pressure against the valve spring until the collets are exposed. Remove them with a magnet or needlenose pliers.

Fig. 140 Removing the flywheel

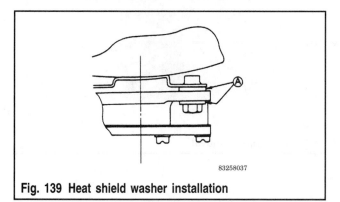

Fig. 139 Heat shield washer installation

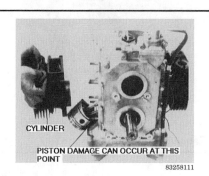

Fig. 141 Cylinder removed

45. The valve and valve spring are now free to be removed. Repeat for other side.

46. Place the engine on blocks with the magneto side facing down and remove the oil pan bolts and oil pan.

47. Remove the oil pump drive gear.

48. Remove the oil pump bolts and lift out the oil pump assembly.

49. Bend the connecting rod bolt locking tab to expose the bolts and remove the bolts and connecting rod cap.

50. Pull the piston and connecting rod free from the engine. Repeat for other side.

➡**The connecting rod cap and bolt should be replaced as an assembly.**

51. Place the crankcase assembly in the upright position.

52. Remove the side base bolts and pull the side base off the crankcase.

✳✳WARNING

If normal pulling pressure is not enough to remove the side base, use a puller. DO NOT pry between crankcase and side base. This can cause damage to the sealing surface.

53. Remove the governor assembly.

54. Gently remove the camshaft assembly.

➡**Be careful not to damage cam the lobe surfaces.**

55. Remove the tappets from the crankcase.

56. Remove the crankshaft from the crankcase.

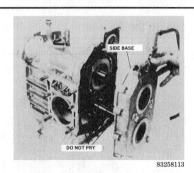

Fig. 142 Oil pan removed

Fig. 143 Exploded view of the oil pump

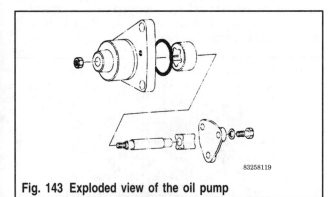

Fig. 144 Removing the side base

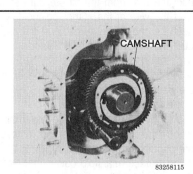

Fig. 145 Removing the governor

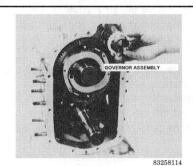

Fig. 146 Removing the camshaft

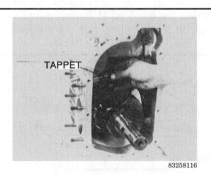

Fig. 147 Tappet removal

Fig. 148 Crankshaft removal

ASSEMBLY

▶ **See Figures 150, 151, 152, 153, 154, 155, 156, 157 and 158**

1. Install the crankshaft in the crankcase.

➡**Be careful not to damage the oil seal.**

2. Apply a light coat of clean oil to the tappets and install them in the crankcase.
3. Rotate the crankshaft until the timing mark is topside.
4. Install the camshaft so that the timing mark on the crankshaft is perfectly aligned with that of the camshaft.
5. Install the governor assembly.
6. Install the side base gasket and side base.

➡**Be careful not to damage the oil seal. Torque the side base bolts to 10 ft. lbs.**

7. If the piston was removed from the connecting rod, assemble the piston to the connecting rod in such a way that the mark on top of the piston is facing the same direction as the Japanese letters on the connecting rod. These letters translated mean 'magneto side". So, when the piston and the connecting rod are assembled, they would be installed on the crankshaft in such a way that the arrow on the piston and the Japanese letters face the magneto side of the crankcase.
8. Apply a light coat of oil to the crankshaft and connecting rod surfaces and install as explained above. Torque the connecting rod bolts to 29-30 ft. lbs.
9. Carefully bend the tabs of the lock plate until contact is made with the connecting rod bolts. Repeat for the other side.
10. Install the oil pump in the crankcase.
11. Install the oil pimp drive gear on the oil pump shaft.
12. Install the oil pan gasket and oil pan on the crankcase. Torque the bolts to 10 ft. lbs.

13. Apply a light coat of oil to each valve stem and insert the proper valve in the cylinder.
14. Assemble the valve spring components according to the accompanying diagram.
15. Compress the valve spring and install the collets.
16. Place the cylinder gasket on the crankcase and align the piston ring so there is 180° difference among rings.
17. Compress the piston rings and install the cylinder in the crankcase. Torque the cylinder nuts to 10-12 ft. lbs.
18. Install the head gasket and cylinder head. Torque the head bolts to 25-29 ft. lbs. Repeat for the other side.
19. To adjust the valve, rotate the crankshaft so that the piston is at TDC on the compression stroke. (Both valves closed). Clearance should be 0.007-0.010 in. Adjustment is made by adding or subtracting shims on the tappet stem.
20. Install the woodruff key in the crankshaft.
21. Install the inner case cover.
22. Install the inner spiral case and magneto stator assembly.
23. Install the flywheel on the crankshaft and torque the flywheel nut to 60-65 ft. lbs.
24. Install the flywheel pulley and spiral case.
25. Install the starter motor in the crankcase.
26. Install the fuel pump in the crankcase and connect the fuel lines.
27. Install the exhaust gaskets, exhaust manifold and manifold cover.
28. Install the intake gaskets, intake manifold and carburetor assembly.
29. Install the governor link and link spring on the carburetor and attach the governor spring from the carburetor to the governor arm.
30. Install the governor arm on the governor shaft. Do not tighten.
31. Turn the governor shaft counterclockwise until it stops and open the carburetor throttle valve to the full open position. At this point tighten the governor arm on the governor shaft.
32. Install the ignition coil.
33. Install the control panel assembly.
34. Connect the wire from the ignition switch to the ignition coil white wire.
35. Install the fuel tank and fuel shut-off assembly.
36. Install the exhaust gasket and muffler.
37. Install the air cleaner assembly.
38. Connect the fuel line to the fuel shut-off and check it for fuel flow.
39. Install the oil drain plug and fill with good grade oil. Capacity is 3800cc — approximately 4 qts.
40. Visually inspect the engine and make a test run.

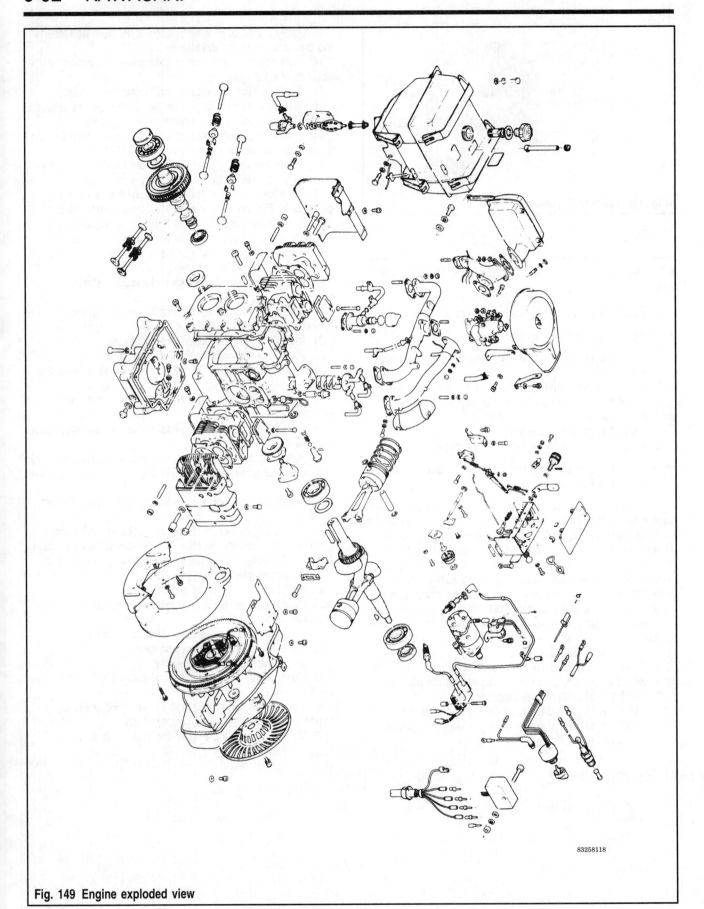

83258118

Fig. 149 Engine exploded view

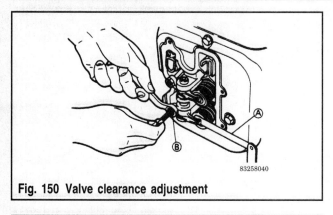

Fig. 150 Valve clearance adjustment

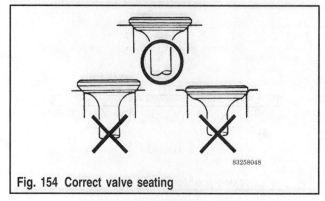

Fig. 154 Correct valve seating

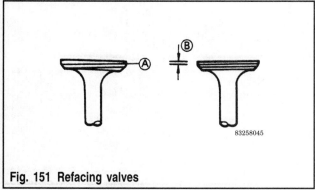

Fig. 151 Refacing valves

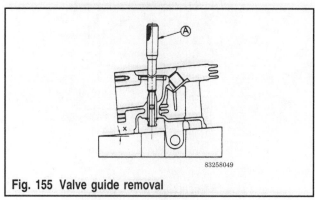

Fig. 155 Valve guide removal

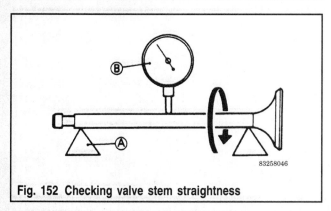

Fig. 152 Checking valve stem straightness

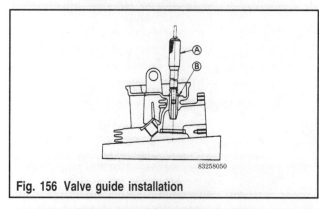

Fig. 156 Valve guide installation

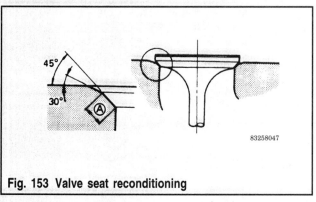

Fig. 153 Valve seat reconditioning

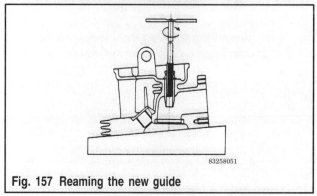

Fig. 157 Reaming the new guide

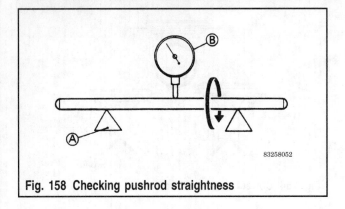

Fig. 158 Checking pushrod straightness

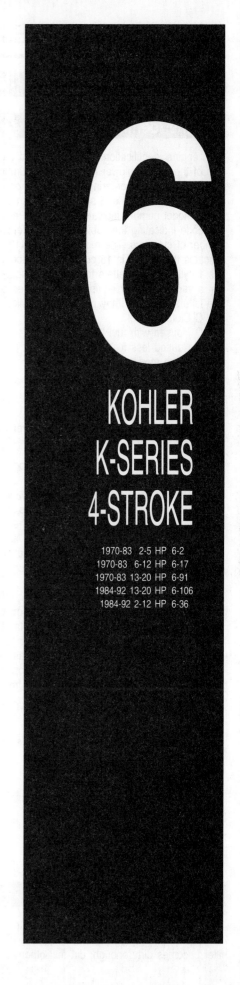

6

KOHLER
K-SERIES
4-STROKE

1970-83 2-5 HP

Engine Identification

An engine identification plate is mounted on the carburetor side of the engine blower housing. The numbers that are important, as far as ordering replacement parts is concerned, are the model, serial, and specification numbers.

The model number indicates the engine model series. It also is a code indicating the cubic inch displacement and the number of cylinders. The model number K181, for instance, indicates the engine is 18 cu. in. in displacement and that it has 1 cylinder. The letters following the model number indicate that a variety of other equipment is installed on the engine. The letters and what they mean are as follows:

- C Clutch model
- G Housed with fuel tank
- H Housed less fuel tank
- P Pump model
- R Reduction gear
- S Electric start
- T Retractable start

➡A model number without a suffix letter indicates a basic rope start version.

The specification number indicates a model variation. It indicates a combination of various groups used to build the engine. It may have a letter preceding it which is sometimes important in determining superseding parts. The first two numbers of the specifications number is the code designation the engine model: the remaining numbers are issued in numerical sequence as each new specification is released, for example, 2899, 28100, 28101, etc. The current specification number model code is K91.

The serial number lists the order in which the engine was built. If a change takes place to a model or a specification, the serial number is used to indicate the points at which the change takes place. The first letter or number in the serial number indicates what year the engine was built. The letter prefix to the engine serial number was dropped in 1969 and thereafter the prefix is a number. Engines made in 1969 have either the letter **E** or the number **1**. The code is as follows:

- A — 1965
- B — 1966
- C — 1967
- D — 1968
- E — 1969 First Digit Numbers:
- 1 — 1969
- 2 — 1970
- 3 — 1971
- 4 — 1972
- 5 — 1973
- 6 — 1974

Air Cleaners

A dirty air cleaner can cause rich fuel/air mixture and consequent poor engine operation and sludge deposits. If the filter becomes dirty enough, dirt that otherwise would be

trapped can pass through and may wear the engine's moving parts prematurely. It is therefore necessary that all maintenance work be performed precisely as specified.

DRY AIR CLEANERS

▶ **See Figure 1**

Clean dry element air cleaners every 50 hours of operation, or every 6 months (whichever comes first) under good operating conditions. Service more frequently if the operating area is dusty. Remove the element and tap it lightly against a hard surface to remove the bulk of the dirt. If dirt will not drop off easily, replace the element. Do not use compressed air or solvents. Replace the air cleaner every 100-200 hours, under good conditions, and more frequently if the air is dusty.

Observe the following precautions:

1. Handle the element carefully-do not allow the gasket surfaces to become bent or twisted.

2. Make sure gasket surfaces seal against back plate and cover.

3. Tighten wing nut only finger tight — if it is too tight, cleaner may not seal properly.

If the dry type air cleaner is equipped with a precleaner, service this unit when cleaning the paper element. Servicing consists of cleaning the precleaner in soap and water, squeezing the excess out, and then allowing it to air dry before installation. Do not oil!

OIL BATH AIR CLEANERS

This type of unit may be used to replace the dry type in applications where very frequent replacement of the element is required. The conversion is simple and requires the use of an elbow to fit the oil bath unit onto the engine in a vertical position.

Service the unit every 25 hours of operation under good conditions and, under dusty conditions, as often as every 8 hours of operation.

Service as follows:

1. Remove cover and lift element out of bowl.

2. Drain dirty oil from bowl, and then wash thoroughly in clean solvent.

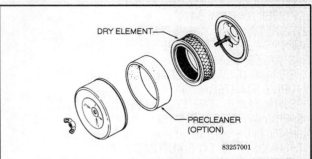

83257001

Fig. 1 Exploded view of a dry type air cleaner with precleaner

3. Swish the element in the solvent and then allow it to drip dry. Do not dry with compressed air. Lightly oil the element with engine oil.

4. Inspect air horn, filter bowl, and cover gaskets, and replace as necessary (if grooved or cracked).

5. Install filter bowl gasket on air horn, then put the bowl into position. Fill bowl to indicated level with engine oil.

6. Install element, put the cover in position, and then install copper gasket (if used) and wingnut. Tighten wingnut with fingers only to avoid distorting housing. Make sure all joints in the unit seal tightly.

Lubrication

CRANKCASE

Oil level must be maintained between **F** and **L** marks — do not overfill. Check every day and add as necessary. On new engines, be especially careful to stop engine and check level frequently. When checking, make sure regular type dipstick is inserted fully. On screw type dipstick, check level with dipstick inserted fully but not screwed in. On this type, however, make sure to screw dipstick back in tightly when oil level check is completed.

Use SG type oils meeting viscosity specifications according to the prevailing temperature as shown in the chart below.

Change initial fill of oil on new engines after five hours of operation. Then, change oil every 25 hours of operation. Change oil when engine is hot. Change more frequently in dusty areas. If the engine has just been overhauled, it is best to fill it initially with a non-detergent oil. Then, after 5 hours, refill with SG type oil.

Oil capacity is about ½ qt. (473mL)

REDUCTION GEAR UNITS

Every 50 hours, remove the oil plug on the lower part of the reduction unit cover to check level. If oil does not reach the level of the oil plug, remove the vented fill plug from the top of the cover and refill with engine oil until level is correct. This oil need not be changed unless unit has been out of service for several months. In this situation, remove the drain plug, drain

oil, then replace plug and fill to proper level as described above.

FUEL RECOMMENDATIONS

Use either leaded or unleaded regular grade fuel of at least 90 octane. Unleaded fuel produces fewer combustion chamber deposits, so its use is preferred.

Purchase fuel from a reputable dealer, and make sure to use only fresh fuel (fuel less than 30 days old). If the engine is stored, drain the fuel system or use a fuel stabilizer that is compatible with the type of fuel tank the engine is equipped with.

Spark Plugs

SERVICE

▶ **See Figures 2, 3, 4, 5 and 6**

The spark plug should be removed and serviced every 100 hours of engine operation. The plug should have a light coating of light gray colored deposits. If deposits are black, fuel/air mixture could be too rich due to improper carburetor adjustment or a dirty air cleaner. If deposits are white, the engine may be overheating or a spark plug of too high a heat range could be in use.

Kohler recommends that the plug be replaced rather than sandblasted or scraped if there are excessive deposits. Torque plugs to 18-22 ft. lbs.

TESTING

To test a plug for adequate performance, remove it from the engine, attach the ignition wire, and then rest the side electrode against the cylinder head. Crank the engine vigorously. If there is a sharp spark, the plug and ignition system are all right, although ignition timing should be checked if the engine fires irregularly.

Breaker Points

Tune-Up Specifications

Plug Gap (in.)	Breaker Point Gap (in.)	Trigger Air Gap (in.)	Normal Timing (deg)	Retard Timing (deg)
.025①	.020	—	20B	—

B—Before
① Shielded plug gap—.020 in.

Spark Plug Specifications

Plug Size	Hex Size	Plug Reach	Standard Plugs		Resistor Plugs	
			Solid Post	Knurled Nut	Non-Shielded	Shielded
14 mm	$^{13}/_{16}$"	$^{3}/_{8}$"	J-8 270321-S	J-8 220040-S	XJ-8 232604-S	XEJ-8 220258-S

832570C2

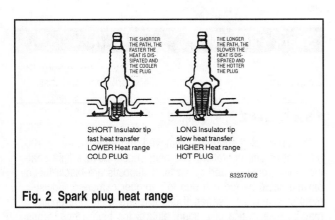

Fig. 2 Spark plug heat range

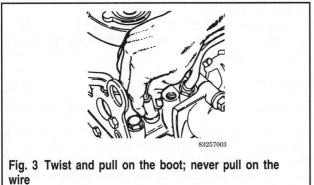

Fig. 3 Twist and pull on the boot; never pull on the wire

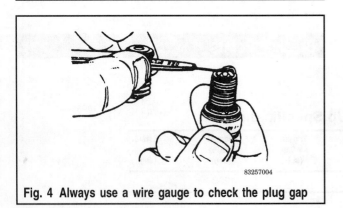

Fig. 4 Always use a wire gauge to check the plug gap

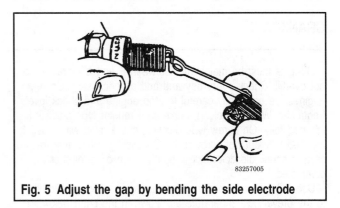

Fig. 5 Adjust the gap by bending the side electrode

Fig. 6 Plugs that are in good condition can be cleaned with a wire brush and file, then re-gapped

INSPECTION

Remove the breaker cover and inspect the points for pitting or buildup of metal on either the movable or stationary contact every 100 hours of operation. Replace the points if they are badly burned. If there is a great deal of metal buildup on either contact, the condenser may be faulty and should be replaced.

To replace points, remove the primary wiring connector screw and pull off the primary wire. Then, remove the contact set mounting screws and remove the contact set. Install the new set of points in reverse order, leaving upper mounting screw slightly loose. Then set point gap and timing as described below.

SETTING BREAKER GAP AND TIMING

♦ See Figures 7 and 8

1. Remove the breaker cover and disconnect the spark plug lead. Rotate the engine in direction of normal rotation until the points reach the maximum opening.

2. Using a clean, flat feeler gauge of 0.020 in. (0.50mm) size, check the gap between the points. Gauge should just slide between the contacts without opening them when flat between them. If the gap is incorrect, loosen the upper mounting screw (if necessary), and shift the breaker base with the blade of the screwdriver until gap is correct.

3. There is a timing sight hole in either the bearing plate or the blower housing. If there is a snap button in the hole, pry it out with a screwdriver.

4. While observing the sight hole, turn the engine slowly in normal direction of rotation. When the T mark appears in the hole, the points should just be beginning to open. If timing is incorrect, breaker gap will have to be reset slightly — 0.018-0.022 in. (0.46-0.56mm). If the points are not yet opening when the timing mark is centered in the hole, make the point gap wider. If points open too early, narrow it. Recheck the setting after tightening the upper breaker mounting screw by turning the engine in normal direction of rotation past the firing point and checking that the points open at just the right time.

➡This procedure may be performed with the engine running at 1200-1800 rpm if a timing light is available. Connect the timing light according to manufacturer's instructions. You may have to chalk the timing mark to see it adequately.

TRIGGER AIR GAP

Trigger air gap is set within the range 0.005-0.010 in. (0.13-0.25mm). As long as the gap falls within this range, the ignition system should perform adequately. Optimum ignition performance during cold weather starting is provided if the gap is adjusted to 0.005 in. (0.13mm). If you wish to adjust this or to ensure that the gap falls within the proper range, rotate the flywheel until the flywheel projection is lined up with the trigger assembly. Then, loosen the trigger bracket capscrews and slide the trigger back and forth to get the proper gap, as measured with a flat feeler gauge. Then, retighten capscrews.

IGNITION COILS

Coils do not require regular service, except to make sure they are kept clean, that the connections are tight, and that rubber insulators are in good condition (replace if cracked). If you suspect poor performance of a breakerless type ignition system and trigger air gap is correct, check resistance with an ohmmeter. To do this, disconnect the high tension lead at the coil and connect the meter between coil terminal and coil mounting bracket. If resistance is not about 11,500 ohms, replace the coil. Also, check the reading with the meter lead going to the coil terminal pulled off and connected to the spark plug connector of the high tension lead. If there is continuity here, replace the coil.

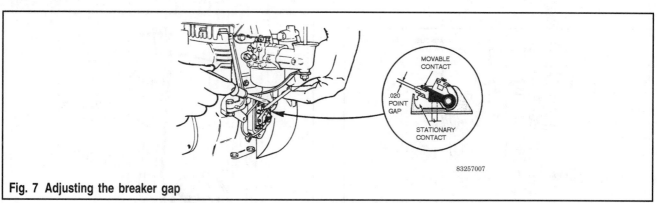

Fig. 7 Adjusting the breaker gap

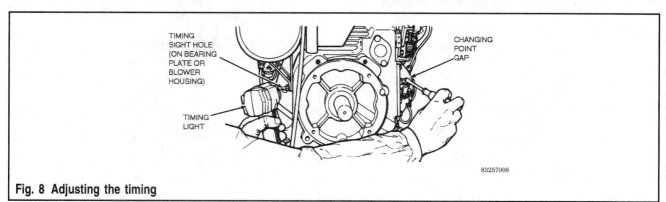

Fig. 8 Adjusting the timing

PERMANENT MAGNETS

These may be checked for magnet strength by holding a screwdriver (non-magnetic) blade within one inch of the magnet. If the magnetic field is good, the blade will be attracted to the magnet. Otherwise, replace it.

Mixture Adjustments

▶ **See Figures 9 and 10**

➡**Before making any adjustments, be sure that the carburetor air cleaner is not clogged. A clogged air cleaner will cause an over-rich mixture, black exhaust smoke, and may lead you to believe that the carburetor is out of adjustment when, in reality, it is not. The carburetor is set at the factory and rarely needs adjustment unless, of course, it has been disassembled or rebuilt.**

1. With the engine stopped, turn the main and idle fuel adjusting screws all the way in until they bottom lightly. Do not force the screws or you will damage the needles.

2. For a preliminary setting, turn the main fuel screw out 2 full turns and the idle screw out 1¼ turns.

3. Start the engine and allow it to reach operating temperatures; then operate the engine at full throttle and under a load, if possible.

4. For final adjustment, turn the main fuel adjustment screw in until the engine slows down (lean mixture), then out until it slows down again (rich mixture). Note the positions of the screw at both settings, then set it about halfway between the two positions.

5. Set the idle mixture adjustment screw in the same manner. The idle speed (no-load) on most engines is 1200 rpm; however, on engines with a parasitic load (hydrostatic drives) the engine idle speed may have to be increased to as much as 1700 rpm for best no-load idle.

Governor Adjustment

▶ **See Figure 11**

All Kohler engines use mechanical, camshaft driven governors.

INITIAL ADJUSTMENT

1. Loosen, but do not remove, the nut that holds the governor arm to the governor cross shaft.

2. Grasp the end of the cross shaft with a pair of pliers and turn it in counterclockwise as far as it will go. The tab on the cross shaft will stop against the rod on the governor gear assembly.

3. Pull the governor arm away from the carburetor, then retighten the nut which holds the governor arm to the shaft. With updraft carburetors, lift the arm as far as possible, then retighten the arm nut.

FINAL ADJUSTMENT

After making the initial adjustment and connecting the throttle wire on the variable speed applications, start the engine and check the maximum operating speed with a tachometer.

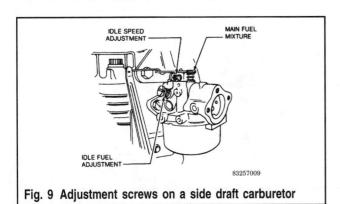

Fig. 9 Adjustment screws on a side draft carburetor

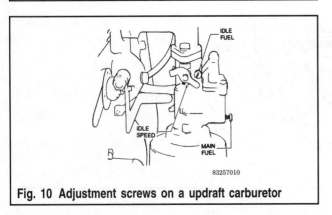

Fig. 10 Adjustment screws on a updraft carburetor

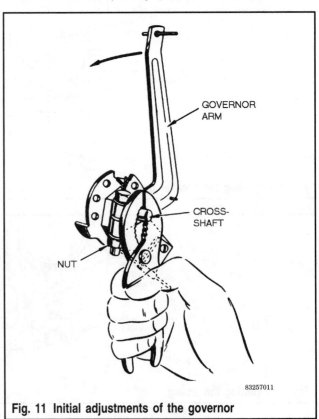

Fig. 11 Initial adjustments of the governor

If adjustment is necessary:

1. Loosen the bushing nut slightly.
2. Move the throttle bracket in a counterclockwise direction to increase speed, or in a clockwise direction to decrease engine speed. Maximum speed is 4000 rpm.
3. With the speed set to the proper range, tighten the bushing nut to lock the throttle bracket in position.

Choke Adjustment

THERMOSTATIC TYPE

If the engine does not start when cranked, continue cranking and move the choke lever first to one side and then to the other to determine whether the setting is too lean or too rich. Once the direction in which lever must be moved has been determined, loosen the adjusting screw on the choke body. Then, move the bracket downward to increase choking or upward to decrease it. Then, tighten the lockscrew. Try starting it again and readjust as necessary.

ELECTRIC-THERMOSTATIC TYPE

▶ See Figure 12

Remove the air cleaner from the carburetor and check the position of the choke plate. The choke should be fully closed when engine is at outside temperature and the temperature is very low. In milder temperatures, slightly less closure is required.

If adjustment is required, move the choke arm until the hole in the brass shaft lines up with the slot in the bearings. Insert a #43 (0.089 in.) drill through the shaft and push it downward so it engages the notch in the base of the choke unit. Then, loosen the clamp bolt on the choke lever and push the arm upward to move the choke plate toward the closed position. When the desired position is obtained, tighten the clamp bolt. Then, remove the drill.

Remount the air cleaner, and then check for any binding in the choke linkage. Correct as necessary. Finally, run the engine until hot, and make sure the choke opens fully. If not, readjust it toward the open position as necessary.

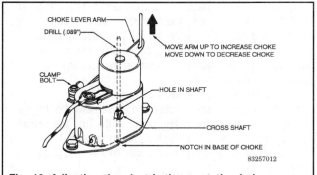

CHOKE LEVER ARM
DRILL (.089")
MOVE ARM UP TO INCREASE CHOKE
MOVE DOWN TO DECREASE CHOKE
CLAMP BOLT
HOLE IN SHAFT
CROSS SHAFT
NOTCH IN BASE OF CHOKE
83257012

Fig. 12 Adjusting the electric-thermostatic choke

Compression Check

Compression is checked by removing the spark plug lead and spinning the flywheel forward against compression. If the piston does not bounce backward with considerable force, checking with a gauge may be necessary. On Automatic Compression Release engines, rotate the flywheel backward against power stroke-if little resistance is felt, check compression with a gauge.

The compression gauge check requires rapid motoring (spinning) of the crankshaft, at about 1000 rpm. Install the gauge in the spark plug hole and motor the engine. Gauge should read 110-120 psi. If reading is less than 100 psi, the engine requires major repair to piston rings or valves.

Carburetor

If a carburetor will not respond to mixture screw adjustments, then you can assume that there are dirt, gum, or varnish deposits in the carburetor or worn/damaged parts. To remedy these problems, the carburetor will have to be completely disassembled, cleaned, and worn parts replaced and reassembled.

Parts should be cleaned with solvent to remove all deposits. Replace worn parts and use all new gaskets. Carburetor rebuilding kits are available.

REBUILDING

Side Draft Carburetors

▶ See Figure 13

1. Remove the carburetor from the engine.
2. Remove the bowl nut, gasket, and bowl. If the carburetor has a bowl drain, remove the drain spring, spacer and plug, and gasket from inside the bowl.
3. Remove the float pin, float, needle, and needle seat. Check the float for dents, leaks, and wear on the float lip or in the float pin holes.
4. Remove the bowl ring gasket.
5. Remove the idle fuel adjusting needle, main fuel adjusting needle, and springs.
6. Do not remove the choke and throttle plates or shafts. If these parts are worn, replace the entire carburetor assembly.

To assemble:

7. Install the needle seat, needle, float, and float pin.
8. Set the float level. With the carburetor casting inverted and the float resting against the needle in its seat, there should be $^{11}/_{64}$ in. $\pm$ $^{1}/_{32}$ in. (4.4mm $\pm$ 0.8mm) clearance between the machined surface of the casting and the free end of the float.
9. Adjust the float level by bending the lip of the float with a small screwdriver.
10. Install the new bowl ring gasket, new bowl nut gasket, and bowl nut. Tighten the nut securely.
11. Install the main fuel adjustment needle. Turn it in until the needle seats in the nozzle and then back out two turns.
12. Install the idle fuel adjustment needle. Back it out about $1^{1}/_{4}$ turns after seating it lightly against the jet.
13. Install the carburetor on the engine.

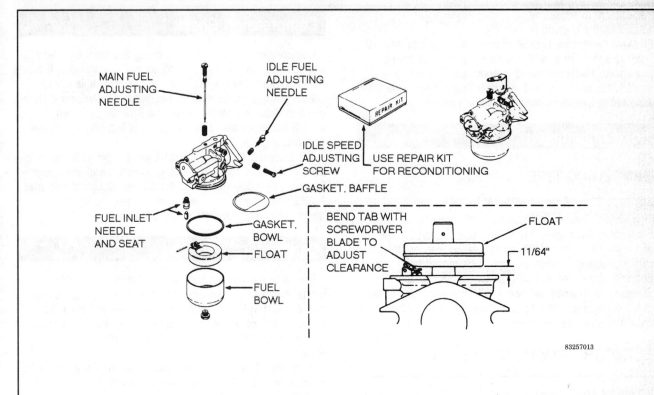

Fig. 13 Exploded view of a side draft carburetor. The inset shows the float adjustment procedure

Updraft Carburetors

▶ **See Figure 14**

1. Remove the carburetor from the engine.
2. Remove the bowl cover and the gasket.
3. Remove the float pin, float, needle and needle seat. Check the float pin for wear.
4. Remove the idle fuel adjustment needle, main fuel adjustment needle, and the springs. Do not remove the choke plate or the shaft unless the replacement of these parts is necessary.

To assemble:

5. Install the throttle shaft and plate. The elongated side of the valve must be toward the top.
6. Install the needle seat. A 5/16 in. socket should be used. Do not overtighten.
7. Install the needle, float, and float pins.
8. Set the float level. With the bowl cover casting inverted and the float resting lightly against the needle in its seat, there should be 7/16 in. ± 1/32 in. (11mm ± 0.8mm) clearance between the machined surface casting and the free end of the float.
9. Adjust the float level by bending the lip of the float with a small screwdriver.
10. Install the new carburetor bowl gasket, bowl cover, and bowl cover screws. Tighten the screws securely.
11. Install the main fuel adjustment needle. Turn it in until the screw seats in the nozzle and then back it out 2 turns.
12. Install the idle fuel adjustment needle. Back it out about 1½ turns after seating the screw lightly against the jet. Install the idle speed screw and spring. Adjust the idle to the desired speed with the engine running.
13. Install the carburetor on the engine.

Fuel Pump

▶ **See Figure 15**

Fuel pumps used on single cylinder Kohler engines are either the mechanical or vacuum actuated type. The mechanical type is operated by an eccentric on the camshaft and the vacuum type is operated by the pulsating negative pressures in the crankcase. The K91 vacuum type pump is not serviceable and must be replaced when faulty. The mechanical pump is serviceable and rebuilding kits are available.

1. Disconnect fuel lines, remove mounting screws, and pull the pump off the engine.
2. File a mark across some point at the union of pump body and cover. Remove the screws and remove the cover.
3. Turn the cover upside down and remove the valve plate screw and washer. Remove the valve retainer, valves, valve springs, and valve gasket, after noting the position of each part. Discard the valve springs, valves and valve retainer gasket.
4. Clean the fuel head with solvent and a soft wire brush. Hold the pump cover with the diaphragm surface upward; position a new gasket into the cavity. Put the valve spring and valves into position in the cavity and reassemble the valve retainer. Lock the retainer into position by installing the fuel pump valve retainer screw.
5. Rebuild the lower diaphragm section.
6. Hold the mounting bracket and press down on the diaphragm to compress the spring underneath. Turn the bracket 90 degrees to unhook the diaphragm and remove it.
7. Clean the mounting bracket with solvent and a wire brush.

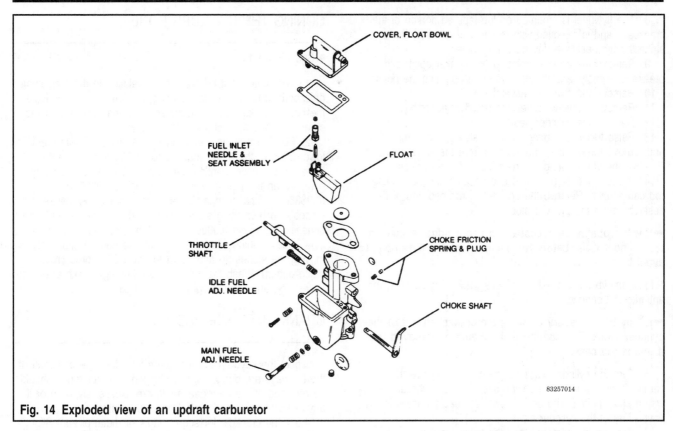

Fig. 14 Exploded view of an updraft carburetor

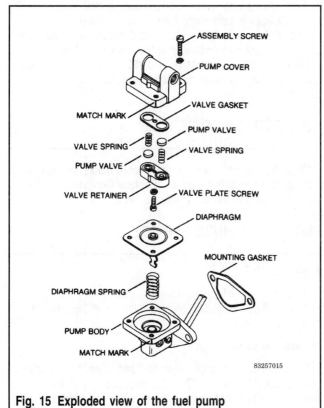

Fig. 15 Exploded view of the fuel pump

8. Stand a new diaphragm spring in the casting, put the diaphragm into position, and push downward to compress the spring. Turn the diaphragm 90 degrees to reconnect it.

9. Position the pump cover on top of the mounting bracket with the indicating marks lined up. Install the screws loosely on mechanical pumps; on vacuum pumps, tighten the screws.

10. Holding only the mounting bracket, push the pump lever to the limit of its travel, hold it there, and then tighten the four screws.

11. Remount the fuel pump on the engine with a new gasket, tighten the mounting bolts, and reconnect the fuel lines.

Engine Overhaul

DISASSEMBLY

The following procedure is designed to be a general guide rather than a specific and all-inclusive disassembly procedure. The sequence may have to be varied slightly to allow for the removal of special equipment or accessory items such as motor/generators, starters, instrument panels, etc.

1. Disconnect the high tension spark plug lead and remove the spark plug.

2. Close the valve on the fuel sediment bowl and remove the fuel line at the carburetor.

3. Remove the air cleaner from the carburetor intake.

4. Remove the carburetor.

5. Remove the fuel tank. The sediment bowl and brackets remain attached to the fuel tank.

6. Remove the blower housing, cylinder baffle, and head baffle.

7. Remove the rotating screen and the starter pulley.

8. The flywheel is mounted on the tapered portion of the crankcase and is removed with the help of a puller. Do not strike the flywheel with any type of hammer.

9. Remove the breaker point cover, breaker point lead, breaker assembly, and the push-rod that operates the points.

10. Remove the magneto assembly.

11. Remove the valve cover and breather assembly.

12. Remove the cylinder head.

13. Raise the valve springs with a valve spring compressor and remove the valve spring keepers from the valve stems. Remove the valve spring retainers, springs, and valves.

14. Remove the oil pan base and unscrew the connecting rod can screws. Remove the connecting rod cap and piston assembly from the cylinder block.

➡ **It will probably be necessary to use a ridge reamer on the cylinder walls before removing the piston assembly, to avoid breaking the piston rings.**

15. Remove the crankshaft, oil seals and, if necessary, the anti-friction bearings.

➡ **It may be necessary to press the crankshaft out of the cylinder block. The bearing plate should be removed first, if this is the case.**

16. Turn the cylinder block upside down and drive the camshaft pin out from the power take-off side of the engine with a small punch. The pin will slide out easily once it is driven free of the cylinder block.

17. Remove the camshaft and the valve tappets.

18. Loosen and remove the governor arm from the governor shaft.

19. Unscrew the governor bushing nut and remove the governor shaft from the inside of the cylinder block.

20. Loosen, but do not remove, the screw located at the lower right of the governor bushing nut until the governor gear is free to slide off of the stub shaft.

CYLINDER BLOCK SERVICE

Make sure that all surfaces are free of gasket fragments and sealer materials. The crankshaft bearings are not to be removed unless replacement is necessary. One bearing is pressed into the cylinder block and the other is located in the bearing plate. If there is no evidence of scoring or grooving and the bearings turn easily and quietly it is not necessary to replace them.

The cylinder bore must not be worn, tapered, or out-of-round more than 0.005 in. (0.13mm). Check at two locations 90 degrees apart and compare with specifications. If it is, the cylinder must be rebored. If the cylinder is very badly scored or damaged it may have to be replaced, since the cylinder can only be rebored to either 0.010 in. (0.25mm) or 0.020 in. (0.50mm) and 0.030 in. (0.76mm) maximum. Select the nearest suitable oversize and bore it to that dimension. On the other hand, if the cylinder bore is only slightly damaged, only a light deglazing may be necessary.

HONING THE CYLINDER BORE

▸ **See Figure 16**

1. The hone must be centered in relation to the crankshaft crossbore. It is best to use a low speed drill press. Lubricate the hone with kerosene and lower it into the bore. Adjust the stones so they contact the cylinder walls.

2. Position the lower edge of the stones even with the lower edge of the bore, hone at about 600 rpm. Move the hone up and down continuously. Check bore size frequently.

3. When the bore reaches a dimension 0.0025 in. (0.0635mm) smaller than desired size, replace the coarse stones with burnishing stones. Use burnishing stones until the dimension is within 0.0005 in. (0.013mm) of desired size.

4. Use finishing stones and polish the bore to final size, moving the stones up and down to get a 60 degree cross-hatch pattern. Wash the cylinder wall thoroughly with soap and water, dry, and apply a light coating of oil.

CRANKSHAFT SERVICE

Inspect the keyway and the gears that drive the camshaft. If the keyways are badly worn or chipped, the crankshaft should be replaced. If the cam gear teeth are excessively worn or if any are broken, the crankshaft must be replaced.

Check the crankpin for score marks or metal pickup. Slight score marks can be removed with a crocus cloth soaked in oil. If the crankpin is worn more than 0.002 in. (0.05mm), the crankshaft is to be either replaced or the crankpin reground to 0.010 in. (0.254mm) undersize. If the crankpin is reground to 0.010 in. (0.254mm) undersize, a 0.010 in. (0.254mm) undersize connecting rod must be used to achieve proper running clearance.

CONNECTING ROD SERVICE

Check the bearing area for wear, score marks, and excessive running and side clearance. Replace the rod and cap if they are worn beyond the limits allowed.

PISTON AND RINGS SERVICE

▸ **See Figure 17**

Rings are available in the standard size as well as 0.010 in. (0.25mm), 0.020 in. (0.50mm), and 0.030 in. (0.76mm) oversize sets.

➡ **Never reuse old rings.**

The standard size rings are to be used when the cylinder is not worn or out-of-round. Oversize rings are only to be used when the cylinder has been rebored to the corresponding oversize. Service type rings are used only when the cylinder is worn but within the wear and out-of-round limitations; wear limit is 0.005 in. (0.127mm) oversize and out-of-round limit is 0.004 in. (0.10mm).

The old piston may be reused if the block does not need reboring and the piston is within wear limits. Never reuse old

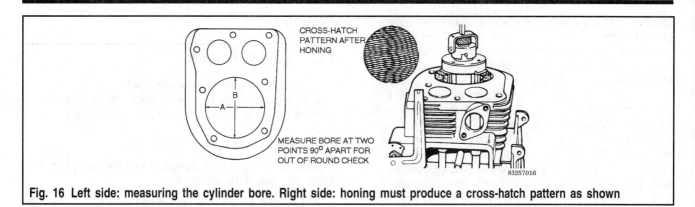

Fig. 16 Left side: measuring the cylinder bore. Right side: honing must produce a cross-hatch pattern as shown

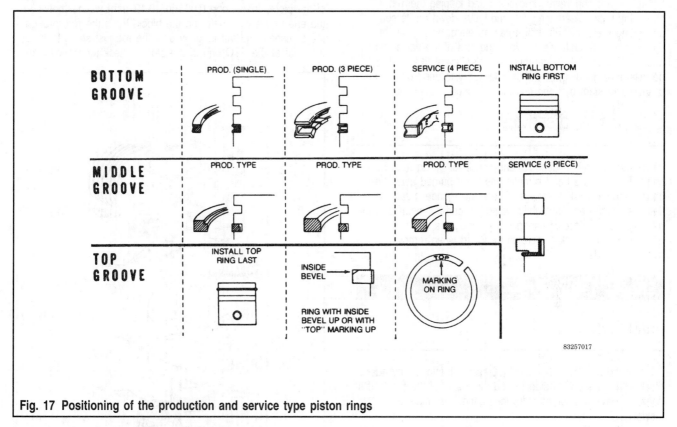

Fig. 17 Positioning of the production and service type piston rings

rings. After removing old rings, thoroughly remove deposits from ring grooves. New rings must each be positioned in its running area of the cylinder bore for an end clearance check, and each must meet specifications.

The cylinder must be deglazed before replacing the rings. If chrome plated rings are used, the chrome plated ring must be installed in the top groove. Make sure that the ring grooves are free from all carbon deposits. Use a ring expander to install the rings. Then check side clearance.

PISTON AND ROD SERVICE

Normally very little wear will take place at the piston boss and piston pin. If the original piston and connecting rod can be used after rebuilding, the piston pin may also be used. However if a new piston or connecting rod or both have to be used, a new piston pin must also be installed. Lubricate the

pin before installing it with a loose to light interference fit. Use new piston pin retainers whether or not the pin is new. Make sure they're properly engaged.

VALVES AND VALVE MECHANISM SERVICE

Inspect the valve mechanism, valves, and valve seats or inserts for evidence of wear, deep pitting, cracks or distortion. Check the clearance between the valve stems and the valve guides.

Valve guides must be replaced if they are worn beyond the limit allowed. K91 model engines do not use valve guides. To remove valve guides, press the guide down into the valve chamber and carefully break off the protruding end until the guide is completely removed. Be careful not to damage the block when removing the old guides. Use an arbor press to install the new guides. Press the new guides to the depth specified, then use a valve guide reamer to gain the proper inside diameter.

Make sure that replacement valves are the correct type (special hard faced valves are needed in some cases). Exhaust valves are always hard faced.

Intake valve seats are usually machined into the block, although inserts are used in some engines. Exhaust valve seats are made of special hardened material. The seating surfaces should be held as close to $\frac{1}{32}$ in. (0.8mm) in width as possible. Seats more than $\frac{1}{16}$ in. (1.6mm) wide must be reground with 45 degree and 15 degree cutters to obtain the proper width. Reground or new valves and seats must be lapped in for a proper fit.

After resurfacing valves and seats and lapping them in, check the valve clearance. Hold the valve down on its seat and rotate the camshaft until it has no effect on the tappet, then check the clearance between the end of the valve stem and the tappet. If the clearance is not sufficient (it will always be less after grinding), it will be necessary to grind off the end of the valve stem until the correct clearance is obtained.

CYLINDER HEAD SERVICE

Remove all carbon deposits and check for pitting from hot spots. Replace the head if metal has been burned away because of head gasket leakage. Check the cylinder head for flatness. If the head is slightly warped, it can be resurfaced by rubbing it on a piece of sandpaper placed on a flat surface. Be careful not to nick or scratch the head when removing carbon deposits.

Engine Assembly

REAR MAIN BEARING

Install the rear main bearing by pressing it into the cylinder block with the shielded side toward the inside of the block. If it does not have a shielded side, then either side may face inside.

GOVERNOR SHAFT

1. Place the cylinder block on its side and slide the governor shaft into place from the inside of the block. Place the speed control disc on the governor bushing nut and thread the nut into the block, clamping the throttle bracket into place.
2. There should be a slight end-play in the governor shaft and that can be adjusted by moving the needle bearing in the block.
3. Place a space washer on the stub shaft and slide the governor gear assembly into place.
4. Tighten the holding screw from outside the cylinder block.
5. Rotate the governor gear assembly to be sure that the holding screw does not contact the weight section of the gear.

CAMSHAFT

▶ **See Figures 18, 19 and 20**

1. Turn the cylinder block upside down.
2. The tappets must be installed before the camshaft is installed. Lubricate and install the tappets into the valve guides.
3. Position the camshaft inside the block.

➡**Align the marks on the camshaft and the automatic spark advance, if so equipped.**

4. Lubricate the rod and insert it into the bearing plate side of the block. Install one 0.005 in. (0.127mm) washer between the end of the camshaft and the block. Push the rod through the camshaft and tap it lightly until the rod just starts to enter the bore at the PTO end of the block. Check the endplay and

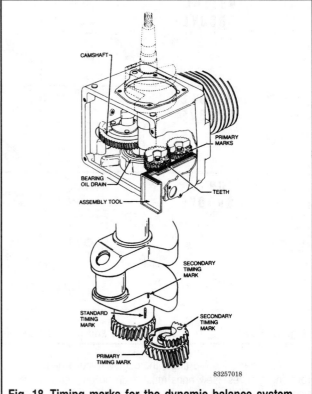

Fig. 18 Timing marks for the dynamic balance system

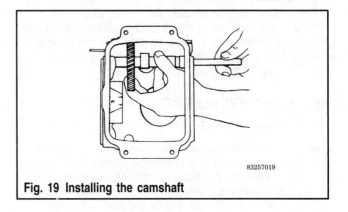

Fig. 19 Installing the camshaft

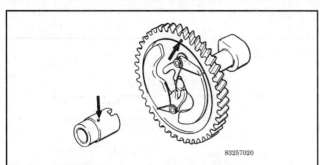

Fig. 20 Timing marks for the automatic spark advance system

adjust it with additional washers if necessary. Press the rod into its final position.

5. The fit at the bearing plate for the camshaft rod is a light to loose fit to allow oil that might leak past to drain back into the block.

CRANKSHAFT

▶ See Figure 21

1. Place the block on the base of an arbor press and carefully insert the tapered end of the crankshaft through the inner race of the anti-friction bearing.

2. Turn the crankshaft and camshaft until the timing mark in the shoulder of the crankshaft lines up with the mark on the cam gear.

3. When the marks are aligned, press the crankshaft into the bearing, making sure that the gears mesh as it is being pressed in. Recheck the alignment of the timing marks on the crankshaft and the camshaft.

4. The end-play of the crankshaft is controlled by the application of various thickness gaskets between the bearing plate and the block. Normal end-play is achieved by installing 0.020 in. (0.50mm) and 0.010 in. (0.25mm) gaskets, with the thicker gaskets on the inside.

BEARING PLATE

1. Press the front main bearing into the bearing plate. Make sure that the bearing is straight.

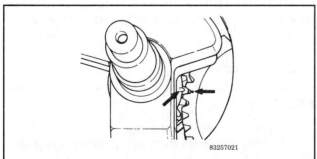

Fig. 21 Alignment of the timing marks for the crankshaft and camshaft

2. Press the bearing plate onto the crankshaft and into position on the block. Install the cap screws and secure the plate to the block. Draw up evenly on the screws.

3. Measure the crankshaft end-play, which is very critical on gear reduction engines.

PISTON AND ROD ASSEMBLY

▶ See Figure 22

1. Lubricate the pin and assemble it to the connecting rod and piston. Install the wrist pin retaining ring. Use new retaining rings.

2. Lubricate the entire assembly, stagger the ring gaps and, using a ring compressor, slide the piston and rod assembly into the cylinder bore with the connecting rod marks on the flywheel side of the engine.

3. Place the block on its end and oil the connecting rod end and the crankpin.

4. Attach the rod cap, lock or lock washers, and the cap screws. Tighten the screws to the correct torque.

➡**Align the marks on the cap and the connecting rod.**

5. Bend the lock tabs to lock the screws.

CRANKSHAFT OIL SEALS

Apply a coat of grease to the lip and guide the oil seals onto the crankshaft. Make sure no foreign material gets onto the knife edges of seal, and make sure the seal does not

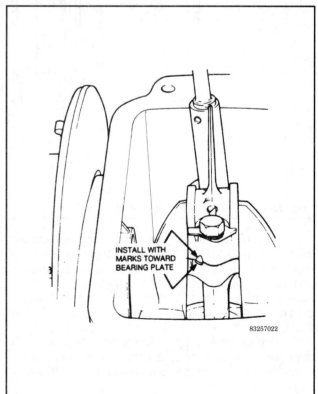

INSTALL WITH MARKS TOWARD BEARING PLATE

Fig. 22 Connecting rod and cap alignment

bend. Place the block on its side and drive the seals squarely into the bearing plate and block.

OIL PAN BASE

Using a new gasket on the base, install pilot studs to align the cylinder block, gasket, and base. Tighten the four attaching screws to the correct torque.

VALVES

▶ **See Figure 23**

Valve Specifications

Dimension		Model K91	
		Intake	Exhaust
A	Seat Angle	89°	89°
B	Seat Width	.037/.045	.037/.045
C	Insert OD	—	.972/.973
D	Guide Depth	None	None
E	Guide ID	None	None
F	Valve Head Diameter	.979/.989	.807/.817
G	Valve Face Angle	45°	45°
H	Valve Stem Diameter	.2480/.2485	.2460/.2465

832570C4

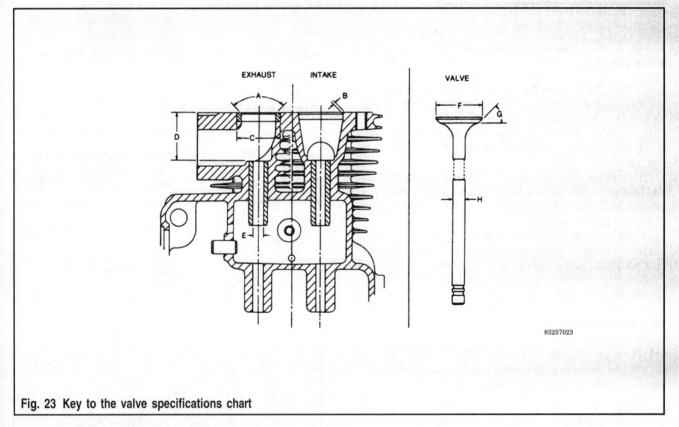

Fig. 23 Key to the valve specifications chart

1. Clean the valves, seats, and parts thoroughly. Grind and lap-in the valves and seats for proper seating. Valve seat width must be $\frac{1}{32}$-$\frac{1}{16}$ in. (0.8-1.6mm). After grinding and lapping, slide the valves into position and check the clearance between stem and tappet. If the clearance is too small, grind the stem ends square and remove all burrs. On engines with adjustable valves, make the adjustment at this time.

2. Place the valve springs, retainers, and rotators under the valve guides. Lubricate the valve stems, and then install the valves down through the guides, compress the springs, and place the locking keys or pins in the grooves of the valve stems.

CYLINDER HEAD

▶ **See Figure 24**

1. Use a new cylinder head gasket.
2. Lubricate and tighten the head bolts evenly, and in sequence, to the proper torque.
3. Install the spark plug.

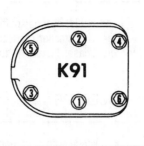

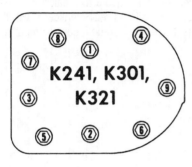
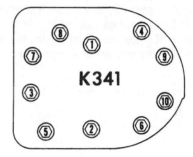

83257024

Fig. 24 Cylinder head torque sequences

BREATHER ASSEMBLY

▶ **See Figure 25**

Assemble the breather assembly, making sure that all parts are clean and the cover is securely tightened to prevent oil leakage.

MAGNETO

On flywheel magneto systems, the coil-core assembly is secured onto the bearing plate. On magneto-alternator systems, the coil is part of the stator assembly, which is secured to the bearing plate. On rotor type magneto systems, the rotor has a keyway and is press fitted onto the crankshaft. The magnet rotor is marked 'engine-side' for proper assembly. Run all leads through the hole provided at the 11 o'clock position on the bearing plate.

FLYWHEEL

1. Place the washer in place on the crankshaft and place the flywheel in position. Install the key.
2. Install the starter pulley, lock washer, and retaining nut. Tighten the retaining nut to the specified torque.

BREAKER POINTS

1. Install the pushrod.
2. Position the breaker points and fasten them with the two screws.
3. Place the cover gasket into position and attach the magneto lead.
4. Set the gap and install the cover.

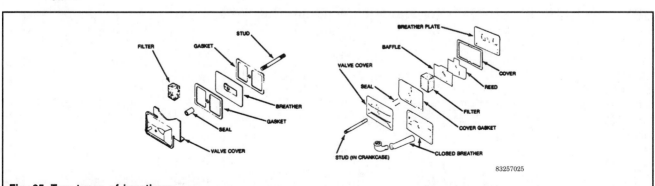

83257025

Fig. 25 Two types of breathers

CARBURETOR

Insert a new gasket and assemble the carburetor to the intake port with the two attaching screws.

GOVERNOR ARM AND LINKAGE

1. Insert the carburetor linkage in the throttle arm.
2. Connect the governor arm to the carburetor linkage and slide the governor arm into the governor shaft.
3. Position the governor spring in the speed control disc.
4. Before tightening the clamp bolt, turn the shaft counterclockwise with pliers as far as it will go; pull the arm as far as it will go to the left (away from the carburetor), tighten the nut, and check for freedom of movement. Adjust the governor.

BLOWER HOUSING AND FUEL TANK

Install the head baffle, cylinder baffle, and the blower housing, in that order. The smaller cap screws are used on the bottom of the crankcase. Install the fuel tank and connect the fuel line.

RUN-IN PROCEDURE

1. Fill the crankcase with a non-detergent oil and run it under load for 5 hours to break it in.
2. Drain oil and refill crankcase with the recommended detergent type oil. Non-detergent oil must not be used except for break-in.

1970-83 6-12 HP

Engine Identification

General Engine Specifications

Model	Bore & Stroke (in.)	Displacement	Horsepower
K141 (—29355)	$2^7/_8 \times 2^1/_2$	16.22	6.25
K141 (29356—)	$2^{15}/_{16} \times 2^1/_2$	16.9	6.25
K161 (—281161)	$2^7/_8 \times 2^1/_2$	16.22	6.25
K161 (281162—)	$2^{15}/_{16} \times 2^1/_2$	16.9	6.25
K181	$2^{15}/_{16} \times 2^3/_4$	18.6	8.0
K241	$3^1/_4 \times 2^7/_8$	23.9	10.0
K241A	$3^1/_4 \times 2^7/_8$	23.9	8.0
K301	$3^3/_8 \times 3^1/_4$	29.07	12.0
K301A	$3^3/_8 \times 3^1/_4$	29.07	12.0
K321	$3^1/_2 \times 3^1/_4$	31.27	14.0
K321A	$3^1/_2 \times 3^1/_4$	31.27	14.0
K341	$3^3/_4 \times 3^1/_4$	35.89	16.0
K341A	$3^3/_4 \times 3^1/_4$	35.89	16.0

832570C5

An engine identification plate is mounted on the carburetor side of the engine blower housing. The numbers that are important, as far as ordering replacement parts is concerned, are the model, serial, and specification numbers.

The model number indicates the engine model series. It also is a code indicating the cubic inch displacement and the number of cylinders. The model number K181, for instance, indicates the engine is 18 cu. in. in displacement and that it has 1 cylinder. The letters following the model number indicate that a variety of other equipment is installed on the engine. The letters and what they mean are as follows:
- C Clutch model
- G Housed with fuel tank
- H Housed less fuel tank
- P Pump model
- R Reduction gear
- S Electric start
- T Retractable start

➡A model number without a suffix letter indicates a basic rope start version.

The specification number indicates a model variation. It indicates a combination of various groups used to build the engine. It may have a letter preceding it which is sometimes important in determining superseding parts. The first two numbers of the specifications number is the code designation the

engine model: the remaining numbers are issued in numerical sequence as each new specification is released, for example, 2899, 28100, 28101, etc. The current specification number model code is as follows:
- K141 — 29
- K161 — 28
- K181 — 30
- K241 — 46
- K301 — 47

The serial number lists the order in which the engine was built. If a change takes place to a model or a specification, the serial number is used to indicate the points at which the change takes place. The first letter or number in the serial number indicates what year the engine was built. The letter prefix to the engine serial number was dropped in 1969 and thereafter the prefix is a number. Engines made in 1969 have either the letter **E** or the number **1**. The code is as follows:
- A — 1965
- B — 1966
- C — 1967
- D — 1968
- E — 1969 First Digit Numbers:
- 1 — 1969
- 2 — 1970
- 3 — 1971
- 4 — 1972
- 5 — 1973
- 6 — 1974

Air Cleaners

A dirty air cleaner can cause rich fuel/air mixture and consequent poor engine operation and sludge deposits. If the filter becomes dirty enough, dirt that otherwise would be trapped can pass through and may wear the engine's moving parts prematurely. It is therefore necessary that all maintenance work be performed precisely as specified.

DRY AIR CLEANERS

▶ See Figure 26

Clean dry element air cleaners every 50 hours of operation, or every 6 months (whichever comes first) under good operating conditions. Service more frequently if the operating area is dusty. Remove the element and tap it lightly against a hard surface to remove the bulk of the dirt. If dirt will not drop off easily, replace the element. Do not use compressed air or solvents. Replace the air cleaner every 100-200 hours, under good conditions, and more frequently if the air is dusty.

Observe the following precautions:

1. Handle the element carefully-do not allow the gasket surfaces to become bent or twisted.

2. Make sure gasket surfaces seal against back plate and cover.

3. Tighten wing nut only finger tight — if it is too tight, cleaner may not seal properly.

Engine Rebuilding Specifications

Specification	K91	Specification	K91
Displacement		**Valve-Intake**	
Cubic Inches	8.86	Valve-Tappet Cold Clear	.005/.009
Cubic Centimeters	145.19	Valve Lift (Zero Lash)	.2095
Horsepower (Max RPM)	4.0	Stem to Guide Max Wear	
		Clear	.004
Cylinder Bore		**Valve-Exhaust**	
New Diameter	2.375	Valve-Tappet Cold Clear	.011/.015
Maximum Wear Diameter	2.378	Valve Lift (Zero Lash)	.1828
Maximum Taper	.0025	Stem to Guide Max Wear	
Maximum out of Round	.005	Clear	.006
Crankshaft		**Tappet**	
End Play (Free)	.0228/.0038	Clearance in Guide	.0005/.002
Crankpin		**Ignition**	
New Diameter	.936	Spark Plug Gap-Gasoline	.025
Maximum Out of Round	.0005	Spark Plug Gap-Gas	.018
Maximum Taper	.001	Spark Plug Gap (Shielded)	.020
Camshaft		Breaker Point Gap	.020
Run Clearance on Pin	.001/.0025	Trigger Air Gap (Breakerless)	NOT USED
End Play	.005/.020	SparkRun ° BTDC	20°
Connecting Rod		Spark Retard	NO RETARD
Big End Maximum Diameter	.9385	**Torque Valves**	
Rod-Crankpin Max Clear	.0035	**(Also See Pages 15.3)**	
Small (Pin) End-New Dia	.56315	Spark Plug (foot lbs)	18–22
Rod to Pin Clearance	.0007/.0008	Cylinder Head	200 in. lbs
Piston		Connecting Rod	140 in. lbs
Thrust Face-Max Wear Dia*	2.359	Flywheel Nut	40–50 ft lbs
Thrust Face*-Bore Clear	.0035/.006		
Ring-Max Slide Clearance	.006		
Ring-End Gap in New Bore	.007/.017		
Ring-End Gap in Used Bore	.027		

* Measured just below oil ring and at right angles to piston pin

832570C3

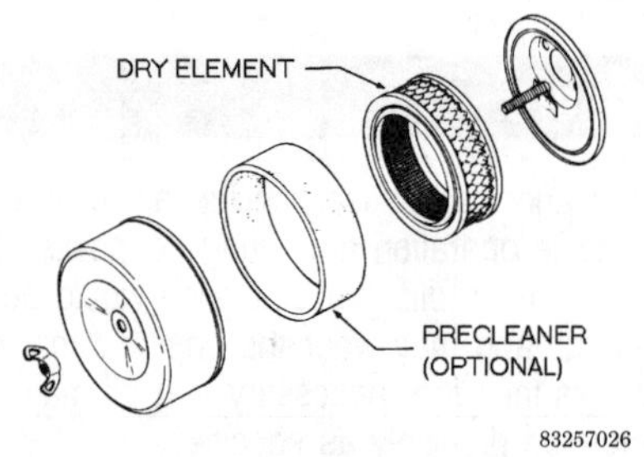

Fig. 26 Exploded view of a dry type air cleaner with pre-cleaner

If the dry type air cleaner is equipped with a precleaner, service this unit when cleaning the paper element. Servicing consists of cleaning the precleaner in soap and water, squeezing the excess out, and then allowing it to air dry before installation. Do not oil!

OIL BATH AIR CLEANERS

This type of unit may be used to replace the dry type in applications where very frequent replacement of the element is required. The conversion is simple and requires the use of an elbow to fit the oil bath unit onto the engine in a vertical position.

Service the unit every 25 hours of operation under good conditions and, under dusty conditions, as often as every 8 hours of operation.

Service as follows:

1. Remove cover and lift element out of bowl.

2. Drain dirty oil from bowl, and then wash thoroughly in clean solvent.

3. Swish the element in the solvent and then allow it to drip dry. Do not dry with compressed air. Lightly oil the element with engine oil.

4. Inspect air horn, filter bowl, and cover gaskets, and replace as necessary (if grooved or cracked).

5. Install filter bowl gasket on air horn, then put the bowl into position. Fill bowl to indicated level with engine oil.

6. Install element, put the cover in position, and then install copper gasket (if used) and wingnut. Tighten wingnut with fingers only to avoid distorting housing. Make sure all joints in the unit seal tightly.

Lubrication

CRANKCASE

Oil Viscosity Chart

Air Temperature	Oil Viscosity	Oil Type
Above 30° F	SAE 30	API Service SC*
30° to 0° F	SAE 10W-30	API Service SC*
Below 0° F	SAE 5W-20	API Service SC*

*SC standard recommendation—CC (MIL-2104B) and SD class oils may also be used.

832570C6

Oil level must be maintained between **F** and **L** marks — do not overfill. Check every day and add as necessary. On new engines, be especially careful to stop engine and check level frequently. When checking, make sure regular type dipstick is inserted fully. On screw type dipstick, check level with dipstick inserted fully but not screwed in. On this type, however, make sure to screw dipstick back in tightly when oil level check is completed.

Use SG type oils meeting viscosity specifications according to the prevailing temperature as shown in the chart below.

Change initial fill of oil on new engines after five hours of operation. Then, change oil every 25 hours of operation. Change oil when engine is hot. Change more frequently in dusty areas. If the engine has just been overhauled, it is best to fill it initially with a non-detergent oil. Then, after 5 hours, refill with SG type oil.

Oil capacities are: K141, 161 and 181 — 1 qt. (0.9L); K241 and 301 — 2 qts. (1.9L).

On K241A or 301A: add 1 qt., then fill to the F mark on the dipstick.

REDUCTION GEAR UNITS

Every 50 hours, remove the oil plug on the lower part of the reduction unit cover to check level. If oil does not reach the level of the oil plug, remove the vented fill plug from the top of the cover and refill with engine oil until level is correct. This oil need not be changed unless unit has been out of service for several months. In this situation, remove the drain plug, drain oil, then replace plug and fill to proper level as described above.

FUEL RECOMMENDATIONS

Use either leaded or unleaded regular grade fuel of at least 90 octane. Unleaded fuel produces fewer combustion chamber deposits, so its use is preferred.

Purchase fuel from a reputable dealer, and make sure to use only fresh fuel (fuel less than 30 days old). If the engine is stored, drain the fuel system or use a fuel stabilizer that is compatible with the type of fuel tank the engine is equipped with.

Spark Plugs

SERVICE

The spark plug should be removed and serviced every 100 hours of engine operation. The plug should have a light coating of light gray colored deposits. If deposits are black, fuel/air mixture could be too rich due to improper carburetor adjustment or a dirty air cleaner. If deposits are white, the engine may be overheating or a spark plug of too high a heat range could be in use.

Kohler recommends that the plug be replaced rather than sandblasted or scraped if there are excessive deposits. Torque plugs to 18-22 ft. lbs.

TESTING

To test a plug for adequate performance, remove it from the engine, attach the ignition wire, and then rest the side electrode against the cylinder head. Crank the engine vigorously. If there is a sharp spark, the plug and ignition system are all right, although ignition timing should be checked if the engine fires irregularly.

Breaker Points

Tune-Up Specifications

Model	Plug Gap (in.)	Breaker Point Gap (in.)	Trigger Air Gap (in.)	Normal Timing (deg)	Retard Timing (deg)
K141 (small bore)	.025 ①	.020	.005 – .010	20	3B
K141 (large bore)	.025 ①	.020	.005 – .010	20	—
K161 (small bore)	.025 ①	.020	.005 – .010	20	3B
K161 (large bore)	.025 ①	.020	.005 – .010	20	—
K181	.025 ①	.020	.005 – .010	20	3B
K241	.025 ①	.020	.005 – .010	20	3A
K301	.025 ①	.020	.005 – .010	20	3A
K321	.025 ①	.020	.005 – .010	20	—
K341	.025 ①	.020	.005 – .010	20	—

B—Before
A—After
① Shielded plug gap—.020 in.

832570C7

Spark Plug Specifications

Engine Model	Plug Size	Hex Size	Plug Reach	Standard Plugs		Resistor Plugs	
				Solid Post	Knurled Nut	Non-Shielded	Shielded
K141	14 mm	$^{13}/_{16}$″	$^{3}/_{8}$″	J-8 270321-S	J-8 220040-S	XJ-8 232604-S	XEJ-8 220258-S
K161	14 mm	$^{13}/_{16}$″	$^{3}/_{8}$″	J-8 270321-S	J-8 220040-S	XJ-8 232604-S	XEJ-8 220258-S
K181	14 mm	$^{13}/_{16}$″	$^{3}/_{8}$″	J-8 270321-S	J-8 220040-S	XJ-8 232604-S	XEJ-8 220258-S
K241	14 mm	$^{13}/_{16}$″	$^{7}/_{16}$″	H-10 235040-S	Not Available	XH-10 235041-S	XEH-10 235259-S
K301	14 mm	$^{13}/_{16}$″	$^{7}/_{16}$″	H-10 235040-S	Not Available	XH-10 235041-S	XEH-10 235259-S
K321	14 mm	$^{13}/_{16}$″	$^{7}/_{16}$″	H-10 235040-S	Not Available	XH-10 235041-S	XEH-10 235259-S
K341	14 mm	$^{13}/_{16}$″	$^{7}/_{16}$″	H-10 235040-S	Not Available	XH-10 235041-S	XEH-10 235259-S

Gap Setting—Gasoline .025″ (Shielded .020″) Tightening Torque—All plugs 18 to 22 foot lbs.
(Champion plugs listed—use Champion or equivalent plugs.)

832570C8

INSPECTION

Remove the breaker cover and inspect the points for pitting or buildup of metal on either the movable or stationary contact every 100 hours of operation. Replace the points if they are badly burned. If there is a great deal of metal buildup on either contact, the condenser may be faulty and should be replaced.

To replace points, remove the primary wiring connector screw and pull off the primary wire. Then, remove the contact set mounting screws and remove the contact set. Install the new set of points in reverse order, leaving upper mounting screw slightly loose. Then set point gap and timing as described below.

SETTING BREAKER GAP AND TIMING

▶ **See Figures 27 and 28**

1. Remove the breaker cover and disconnect the spark plug lead. Rotate the engine in direction of normal rotation until the points reach the maximum opening.

2. Using a clean, flat feeler gauge of 0.020 in. (0.50mm) size, check the gap between the points. Gauge should just slide between the contacts without opening them when flat between them. If the gap is incorrect, loosen the upper mounting screw (if necessary), and shift the breaker base with the blade of the screwdriver until gap is correct.

3. There is a timing sight hole in either the bearing plate or the blower housing. If there is a snap button in the hole, pry it out with a screwdriver.

4. While observing the sight hole, turn the engine slowly in normal direction of rotation. When the T mark appears in the hole, the points should just be beginning to open. If timing is incorrect, breaker gap will have to be reset slightly — 0.018-0.022 in. (0.46-0.56mm). If the points are not yet opening when the timing mark is centered in the hole, make the point gap wider. If points open too early, narrow it. Recheck the setting after tightening the upper breaker mounting screw by turning the engine in normal direction of rotation past the firing point and checking that the points open at just the right time.

➡ **This procedure may be performed with the engine running at 1200-1800 rpm if a timing light is available. Connect the timing light according to manufacturer's instructions. You may have to chalk the timing mark to see it adequately.**

TRIGGER AIR GAP

Trigger air gap is set within the range 0.005-0.010 in. (0.13-0.25mm). As long as the gap falls within this range, the ignition system should perform adequately. Optimum ignition performance during cold weather starting is provided if the gap is adjusted to 0.005 in. (0.13mm). If you wish to adjust this or to ensure that the gap falls within the proper range, rotate the flywheel until the flywheel projection is lined up with the trigger assembly. Then, loosen the trigger bracket capscrews and slide the trigger back and forth to get the proper gap, as measured with a flat feeler gauge. Then, retighten capscrews.

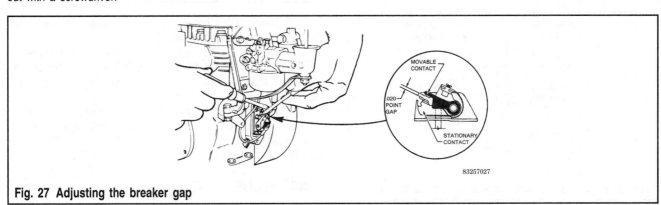

Fig. 27 Adjusting the breaker gap

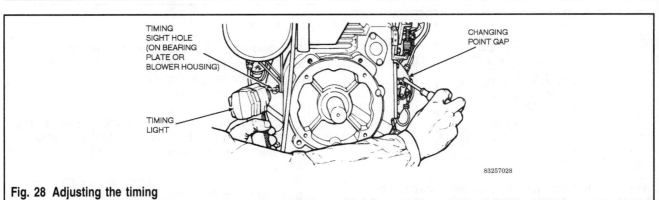

Fig. 28 Adjusting the timing

IGNITION COILS

Coils do not require regular service, except to make sure they are kept clean, that the connections are tight, and that rubber insulators are in good condition (replace if cracked). If you suspect poor performance of a breakerless type ignition system and trigger air gap is correct, check resistance with an ohmmeter. To do this, disconnect the high tension lead at the coil and connect the meter between coil terminal and coil mounting bracket. If resistance is not about 11,500 ohms, replace the coil. Also, check the reading with the meter lead going to the coil terminal pulled off and connected to the spark plug connector of the high tension lead. If there is continuity here, replace the coil.

PERMANENT MAGNETS

These may be checked for magnet strength by holding a screwdriver (non-magnetic) blade within one inch of the magnet. If the magnetic field is good, the blade will be attracted to the magnet. Otherwise, replace it.

Mixture Adjustments

▶ See Figures 29 and 30

➡Before making any adjustments, be sure that the carburetor air cleaner is not clogged. A clogged air cleaner will cause an over-rich mixture, black exhaust smoke, and may lead you to believe that the carburetor is out of adjustment when, in reality, it is not. The carburetor is set at the factory and rarely needs adjustment unless, of course, it has been disassembled or rebuilt.

1. With the engine stopped, turn the main and idle fuel adjusting screws all the way in until they bottom lightly. Do not force the screws or you will damage the needles.
2. For a preliminary setting, turn the main fuel screw out 2 full turns and the idle screw out 1¼ turns.
3. Start the engine and allow it to reach operating temperatures; then operate the engine at full throttle and under a load, if possible.
4. For final adjustment, turn the main fuel adjustment screw in until the engine slows down (lean mixture), then out until it slows down again (rich mixture). Note the positions of the screw at both settings, then set it about halfway between the two positions.
5. Set the idle mixture adjustment screw in the same manner. The idle speed (no-load) on most engines is 1200 rpm; however, on engines with a parasitic load (hydrostatic drives) the engine idle speed may have to be increased to as much as 1700 rpm for best no-load idle.

Governor Adjustment

▶ See Figure 31

All Kohler engines use mechanical, camshaft driven governors.

INITIAL ADJUSTMENT

1. Loosen, but do not remove, the nut that holds the governor arm to the governor cross shaft.
2. Grasp the end of the cross shaft with a pair of pliers and turn it in counterclockwise as far as it will go. The tab on the cross shaft will stop against the rod on the governor gear assembly.
3. Pull the governor arm away from the carburetor, then retighten the nut which holds the governor arm to the shaft. With updraft carburetors, lift the arm as far as possible, then retighten the arm nut.

FINAL ADJUSTMENT

K141, 161, 181

After making the initial adjustment and connecting the throttle wire on the variable speed applications, start the engine and check the maximum operating speed with a tachometer.
If adjustment is necessary:
1. Loosen the bushing nut slightly.
2. Move the throttle bracket in a counterclockwise direction to increase speed, or in a clockwise direction to decrease engine speed. Maximum speed is 4000 rpm for the K161; 3600 rpm for the K141 and 181.
3. With the speed set to the proper range, tighten the bushing nut to lock the throttle bracket in position.

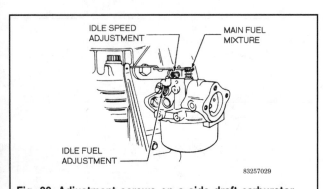

Fig. 29 Adjustment screws on a side draft carburetor

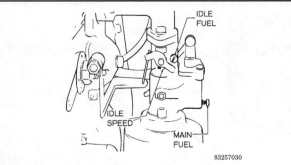

Fig. 30 Adjustment screws on a updraft carburetor

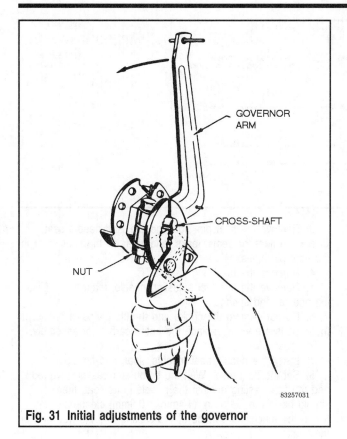

Fig. 31 Initial adjustments of the governor

GOVERNOR ARM

CROSS-SHAFT

NUT

83257031

K241

1. Start the engine and check the speed with a tachometer
2. If the speed is not correct, adjust it as follows:
3. With a constant speed governor — Tighten the governor adjusting screw to increase speed; loosen it to decrease speed.
4. With a variable speed governor — Loosen the capscrew, move the high speed stop bracket until the correct speed is obtained, then, tighten the capscrew.

If the governor is too sensitive, causing hesitation and/or surging, or not sensitive enough, causing a speed drop under load, adjust the sensitivity by moving the governor spring to a different hole. Standard setting is the 3rd hole from the bottom on the governor arm and 2nd hole from the top on the speed control bracket.

Choke Adjustment

THERMOSTATIC TYPE

If the engine does not start when cranked, continue cranking and move the choke lever first to one side and then to the other to determine whether the setting is too lean or too rich. Once the direction in which lever must be moved has been determined, loosen the adjusting screw on the choke body. Then, move the bracket downward to increase choking or upward to decrease it. Then, tighten the lockscrew. Try starting it again and readjust as necessary.

ELECTRIC-THERMOSTATIC TYPE

♦ See Figure 32

Remove the air cleaner from the carburetor and check the position of the choke plate. The choke should be fully closed when engine is at outside temperature and the temperature is very low. In milder temperatures, slightly less closure is required.

If adjustment is required, move the choke arm until the hole in the brass shaft lines up with the slot in the bearings. Insert a #43 (0.089 in.) drill through the shaft and push it downward so it engages the notch in the base of the choke unit. Then, loosen the clamp bolt on the choke lever and push the arm upward to move the choke plate toward the closed position. When the desired position is obtained, tighten the clamp bolt. Then, remove the drill.

Remount the air cleaner, and then check for any binding in the choke linkage. Correct as necessary. Finally, run the engine until hot, and make sure the choke opens fully. If not, readjust it toward the open position as necessary.

Valve Adjustment

K241 and 301 engines

1. The engine should be cold for this adjustment. Turn the crankshaft until the piston reaches TDC and the timing marks indicates that position.
2. If the valves are slightly open, turn the crankshaft and additional revolution to TDC. The valves should both be closed.
3. Using a flat feeler gauge, check the valve clearances. Intake clearance should be 0.008-0.010 in.; exhaust clearance should be 0.017-0.020 in. A slight drag on the gauge indicates a proper clearance.
4. If adjustment is needed, loosen the locknut and turn the adjusting nut to obtain the proper clearance. Hold the adjusting nut while tightening the locknut.

Compression Check

Compression is checked by removing the spark plug lead and spinning the flywheel forward against compression. If the piston does not bounce backward with considerable force, checking with a gauge may be necessary. On Automatic Compression Release engines, rotate the flywheel backward against power stroke-if little resistance is felt, check compression with a gauge.

The compression gauge check requires rapid motoring (spinning) of the crankshaft, at about 1000 rpm. Install the gauge in the spark plug hole and motor the engine. Gauge should read 110-120 psi. If reading is less than 100 psi, the engine requires major repair to piston rings or valves.

Carburetor

If a carburetor will not respond to mixture screw adjustments, then you can assume that there are dirt, gum, or var-

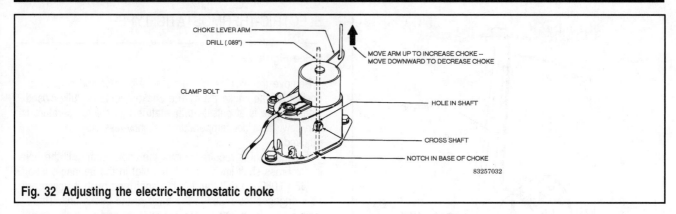

Fig. 32 Adjusting the electric-thermostatic choke

nish deposits in the carburetor or worn/damaged parts. To remedy these problems, the carburetor will have to be completely disassembled, cleaned, and worn parts replaced and reassembled.

Parts should be cleaned with solvent to remove all deposits. Replace worn parts and use all new gaskets. Carburetor rebuilding kits are available.

REBUILDING

Side Draft Carburetors

▶ **See Figure 33**

1. Remove the carburetor from the engine.
2. Remove the bowl nut, gasket, and bowl. If the carburetor has a bowl drain, remove the drain spring, spacer and plug, and gasket from inside the bowl.

3. Remove the float pin, float, needle, and needle seat. Check the float for dents, leaks, and wear on the float lip or in the float pin holes.
4. Remove the bowl ring gasket.
5. Remove the idle fuel adjusting needle, main fuel adjusting needle, and springs.
6. Do not remove the choke and throttle plates or shafts. If these parts are worn, replace the entire carburetor assembly.

To assemble:

7. Install the needle seat, needle, float, and float pin.
8. Set the float level. With the carburetor casting inverted and the float resting against the needle in its seat, there should be $^{11}/_{64}$ in. $\pm$ $^{1}/_{32}$ in. (4.4mm $\pm$ 0.8mm) clearance between the machined surface of the casting and the free end of the float.
9. Adjust the float level by bending the lip of the float with a small screwdriver.
10. Install the new bowl ring gasket, new bowl nut gasket, and bowl nut. Tighten the nut securely.

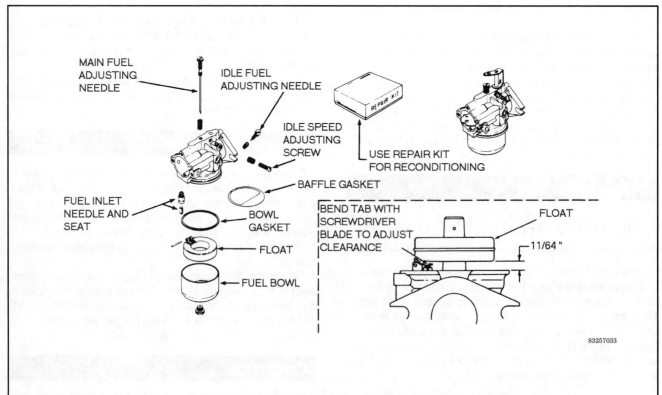

Fig. 33 Exploded view of a side draft carburetor. The inset shows the float adjustment procedure

11. Install the main fuel adjustment needle. Turn it in until the needle seats in the nozzle and then back out two turns.

12. Install the idle fuel adjustment needle. Back it out about 1¼ turns after seating it lightly against the jet.

13. Install the carburetor on the engine.

Updraft Carburetors

▶ See Figure 34

1. Remove the carburetor from the engine.

2. Remove the bowl cover and the gasket.

3. Remove the float pin, float, needle and needle seat. Check the float pin for wear.

4. Remove the idle fuel adjustment needle, main fuel adjustment needle, and the springs. Do not remove the choke plate or the shaft unless the replacement of these parts is necessary.

To assemble:

5. Install the throttle shaft and plate. The elongated side of the valve must be toward the top.

6. Install the needle seat. A ⁵⁄₁₆ in. socket should be used. Do not overtighten.

7. Install the needle, float, and float pins.

8. Set the float level. With the bowl cover casting inverted and the float resting lightly against the needle in its seat, there should be ⁷⁄₁₆ in. ± ¹⁄₃₂ in. (11mm ± 0.8mm) clearance between the machined surface casting and the free end of the float.

9. Adjust the float level by bending the lip of the float with a small screwdriver.

10. Install the new carburetor bowl gasket, bowl cover, and bowl cover screws. Tighten the screws securely.

11. Install the main fuel adjustment needle. Turn it in until the screw seats in the nozzle and then back it out 2 turns.

12. Install the idle fuel adjustment needle. Back it out about 1½ turns after seating the screw lightly against the jet. Install the idle speed screw and spring. Adjust the idle to the desired speed with the engine running.

13. Install the carburetor on the engine.

Fuel Pump

▶ See Figure 35

Fuel pumps used on single cylinder Kohler engines are either the mechanical or vacuum actuated type. The mechanical type is operated by an eccentric on the camshaft and the vacuum type is operated by the pulsating negative pressures in the crankcase. The K91 vacuum type pump is not serviceable and must be replaced when faulty. The mechanical pump is serviceable and rebuilding kits are available.

1. Disconnect fuel lines, remove mounting screws, and pull the pump off the engine.

2. File a mark across some point at the union of pump body and cover. Remove the screws and remove the cover.

3. Turn the cover upside down and remove the valve plate screw and washer. Remove the valve retainer, valves, valve springs, and valve gasket, after noting the position of each part. Discard the valve springs, valves and valve retainer gasket.

4. Clean the fuel head with solvent and a soft wire brush. Hold the pump cover with the diaphragm surface upward; position a new gasket into the cavity. Put the valve spring and valves into position in the cavity and reassemble the valve retainer. Lock the retainer into position by installing the fuel pump valve retainer screw.

5. Rebuild the lower diaphragm section.

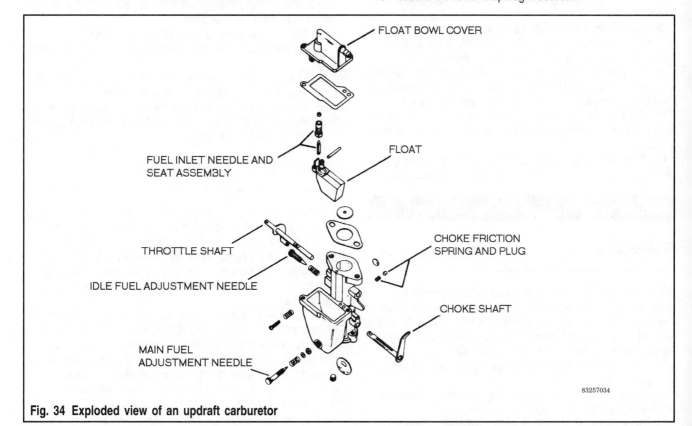

FLOAT BOWL COVER

FUEL INLET NEEDLE AND SEAT ASSEMBLY

FLOAT

THROTTLE SHAFT

CHOKE FRICTION SPRING AND PLUG

IDLE FUEL ADJUSTMENT NEEDLE

CHOKE SHAFT

MAIN FUEL ADJUSTMENT NEEDLE

83257034

Fig. 34 Exploded view of an updraft carburetor

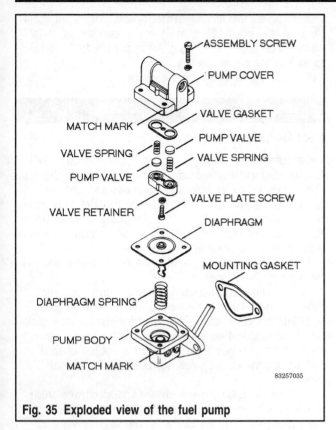

ASSEMBLY SCREW

PUMP COVER

VALVE GASKET

MATCH MARK

PUMP VALVE

VALVE SPRING

VALVE SPRING

PUMP VALVE

VALVE PLATE SCREW

VALVE RETAINER

DIAPHRAGM

MOUNTING GASKET

DIAPHRAGM SPRING

PUMP BODY

MATCH MARK

83257035

Fig. 35 Exploded view of the fuel pump

6. Hold the mounting bracket and press down on the diaphragm to compress the spring underneath. Turn the bracket 90 degrees to unhook the diaphragm and remove it.

7. Clean the mounting bracket with solvent and a wire brush.

8. Stand a new diaphragm spring in the casting, put the diaphragm into position, and push downward to compress the spring. Turn the diaphragm 90 degrees to reconnect it.

9. Position the pump cover on top of the mounting bracket with the indicating marks lined up. Install the screws loosely on mechanical pumps; on vacuum pumps, tighten the screws.

10. Holding only the mounting bracket, push the pump lever to the limit of its travel, hold it there, and then tighten the four screws.

11. Remount the fuel pump on the engine with a new gasket, tighten the mounting bolts, and reconnect the fuel lines.

Engine Overhaul

DISASSEMBLY

The following procedure is designed to be a general guide rather than a specific and all-inclusive disassembly procedure. The sequence may have to be varied slightly to allow for the removal of special equipment or accessory items such as motor/generators, starters, instrument panels, etc.

1. Disconnect the high tension spark plug lead and remove the spark plug.

2. Close the valve on the fuel sediment bowl and remove the fuel line at the carburetor.

3. Remove the air cleaner from the carburetor intake.

4. Remove the carburetor.

5. Remove the fuel tank. The sediment bowl and brackets remain attached to the fuel tank.

6. Remove the blower housing, cylinder baffle, and head baffle.

7. Remove the rotating screen and the starter pulley.

8. The flywheel is mounted on the tapered portion of the crankcase and is removed with the help of a puller. Do not strike the flywheel with any type of hammer.

9. Remove the breaker point cover, breaker point lead, breaker assembly, and the push-rod that operates the points.

10. Remove the magneto assembly.

11. Remove the valve cover and breather assembly.

12. Remove the cylinder head.

13. Raise the valve springs with a valve spring compressor and remove the valve spring keepers from the valve stems. Remove the valve spring retainers, springs, and valves.

14. Remove the oil pan base and unscrew the connecting rod can screws. Remove the connecting rod cap and piston assembly from the cylinder block.

➡**It will probably be necessary to use a ridge reamer on the cylinder walls before removing the piston assembly, to avoid breaking the piston rings.**

15. Remove the crankshaft, oil seals and, if necessary, the anti-friction bearings.

➡**It may be necessary to press the crankshaft out of the cylinder block. The bearing plate should be removed first, if this is the case.**

16. Turn the cylinder block upside down and drive the camshaft pin out from the power take-off side of the engine with a small punch. The pin will slide out easily once it is driven free of the cylinder block.

17. Remove the camshaft and the valve tappets.

18. Loosen and remove the governor arm from the governor shaft.

19. Unscrew the governor bushing nut and remove the governor shaft from the inside of the cylinder block.

20. Loosen, but do not remove, the screw located at the lower right of the governor bushing nut until the governor gear is free to slide off of the stub shaft.

CYLINDER BLOCK SERVICE

Make sure that all surfaces are free of gasket fragments and sealer materials. The crankshaft bearings are not to be removed unless replacement is necessary. One bearing is pressed into the cylinder block and the other is located in the bearing plate. If there is no evidence of scoring or grooving and the bearings turn easily and quietly it is not necessary to replace them.

The cylinder bore must not be worn, tapered, or out-of-round more than 0.005 in. (0.13mm). Check at two locations 90 degrees apart and compare with specifications. If it is, the cylinder must be rebored. If the cylinder is very badly scored or damaged it may have to be replaced, since the cylinder can only be rebored to either 0.010 in. (0.25mm) or 0.020 in. (0.50mm) and 0.030 in. (0.76mm) maximum. Select the nearest suitable oversize and bore it to that dimension. On the other hand, if the cylinder bore is only slightly damaged, only a light deglazing may be necessary.

HONING THE CYLINDER BORE

▶ **See Figure 36**

1. The hone must be centered in relation to the crankshaft crossbore. It is best to use a low speed drill press. Lubricate the hone with kerosene and lower it into the bore. Adjust the stones so they contact the cylinder walls.

2. Position the lower edge of the stones even with the lower edge of the bore, hone at about 600 rpm. Move the hone up and down continuously. Check bore size frequently.

3. When the bore reaches a dimension 0.0025 in. (0.0635mm) smaller than desired size, replace the coarse stones with burnishing stones. Use burnishing stones until the dimension is within 0.0005 in. (0.013mm) of desired size.

4. Use finishing stones and polish the bore to final size, moving the stones up and down to get a 60 degree cross-hatch pattern. Wash the cylinder wall thoroughly with soap and water, dry, and apply a light coating of oil.

CRANKSHAFT SERVICE

Inspect the keyway and the gears that drive the camshaft. If the keyways are badly worn or chipped, the crankshaft should be replaced. If the cam gear teeth are excessively worn or if any are broken, the crankshaft must be replaced.

Check the crankpin for score marks or metal pickup. Slight score marks can be removed with a crocus cloth soaked in oil. If the crankpin is worn more than 0.002 in. (0.05mm), the crankshaft is to be either replaced or the crankpin reground to 0.010 in. (0.254mm) undersize. If the crankpin is reground to 0.010 in. (0.254mm) undersize, a 0.010 in. (0.254mm) under-size connecting rod must be used to achieve proper running clearance.

CONNECTING ROD SERVICE

Check the bearing area for wear, score marks, and excessive running and side clearance. Replace the rod and cap if they are worn beyond the limits allowed.

PISTON AND RINGS SERVICE

▶ **See Figure 37**

Rings are available in the standard size as well as 0.010 in. (0.25mm), 0.020 in. (0.50mm), and 0.030 in. (0.76mm) over-size sets.

➡**Never reuse old rings.**

The standard size rings are to be used when the cylinder is not worn or out-of-round. Oversize rings are only to be used when the cylinder has been rebored to the corresponding over-size. Service type rings are used only when the cylinder is worn but within the wear and out-of-round limitations; wear limit is 0.005 in. (0.127mm) oversize and out-of-round limit is 0.004 in. (0.10mm).

The old piston may be reused if the block does not need reboring and the piston is within wear limits. Never reuse old rings. After removing old rings, thoroughly remove deposits from ring grooves. New rings must each be positioned in its running area of the cylinder bore for an end clearance check, and each must meet specifications.

The cylinder must be deglazed before replacing the rings. If chrome plated rings are used, the chrome plated ring must be installed in the top groove. Make sure that the ring grooves are free from all carbon deposits. Use a ring expander to install the rings. Then check side clearance.

PISTON AND ROD SERVICE

Normally very little wear will take place at the piston boss and piston pin. If the original piston and connecting rod can be used after rebuilding, the piston pin may also be used. However if a new piston or connecting rod or both have to be used, a new piston pin must also be installed. Lubricate the pin before installing it with a loose to light interference fit. Use new piston pin retainers whether or not the pin is new. Make sure they're properly engaged.

VALVES AND VALVE MECHANISM SERVICE

All exc. K241 and K301

Inspect the valve mechanism, valves, and valve seats or inserts for evidence of wear, deep pitting, cracks or distortion.

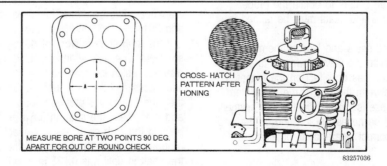

MEASURE BORE AT TWO POINTS 90 DEG. APART FOR OUT OF ROUND CHECK

CROSS-HATCH PATTERN AFTER HONING

83257036

Fig. 36 Left side: measuring the cylinder bore. Right side: honing must produce a cross-hatch pattern as shown

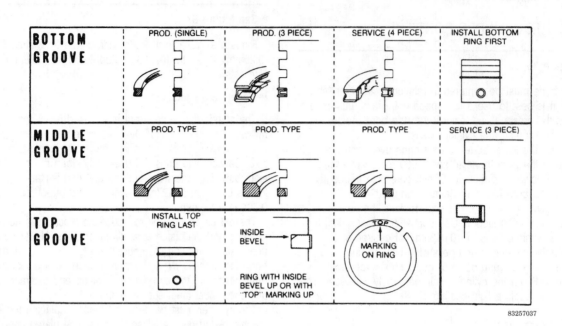

Fig. 37 Positioning of the production and service type piston rings

Check the clearance between the valve stems and the valve guides.

Valve guides must be replaced if they are worn beyond the limit allowed. K91 model engines do not use valve guides. To remove valve guides, press the guide down into the valve chamber and carefully break off the protruding end until the guide is completely removed. Be careful not to damage the block when removing the old guides. Use an arbor press to install the new guides. Press the new guides to the depth specified, then use a valve guide reamer to gain the proper inside diameter.

Make sure that replacement valves are the correct type (special hard faced valves are needed in some cases). Exhaust valves are always hard faced.

Intake valve seats are usually machined into the block, although inserts are used in some engines. Exhaust valve seats are made of special hardened material. The seating surfaces should be held as close to $1/32$ in. (0.8mm) in width as possible. Seats more than $1/16$ in. (1.6mm) wide must be reground with 45 degree and 15 degree cutters to obtain the proper width. Reground or new valves and seats must be lapped in for a proper fit.

After resurfacing valves and seats and lapping them in, check the valve clearance. Hold the valve down on its seat and rotate the camshaft until it has no effect on the tappet, then check the clearance between the end of the valve stem and the tappet. If the clearance is not sufficient (it will always be less after grinding), it will be necessary to grind off the end of the valve stem until the correct clearance is obtained.

CYLINDER HEAD SERVICE

Remove all carbon deposits and check for pitting from hot spots. Replace the head if metal has been burned away because of head gasket leakage. Check the cylinder head for flatness. If the head is slightly warped, it can be resurfaced by rubbing it on a piece of sandpaper placed on a flat surface. Be careful not to nick or scratch the head when removing carbon deposits.

DYNAMIC BALANCE SYSTEM

◆ **See Figure 38**

The Dynamic balance system consists of two balance gears which ride in needle bearings. The gears are assembled on two stub shafts that are pressed into special bosses in the crankcase. Snaprings hold the gears, and spacer washers are used to control endplay. The gears are driven off the crankgear. The system is found on special versions of the 241 and 301.

If the stub shaft is worn or damaged, press the old stub shaft out. The new shafts must be pressed in a specified distance which is dependent on the distance between the stub shaft boss and the main bearing boss. Measure the distance the stub shaft boss protrudes above the main bearing boss and then press the shaft in for a protrusion of the shaft end beyond the stub shaft boss as specified. If the stub shaft boss protrudes $7/16$ in. (11mm) beyond the main bearing boss, press the shaft in until it is 0.735 in. (18.7mm) above the stub shaft boss. If the protrusion is about $1/16$ in. (1.6mm), press the stub

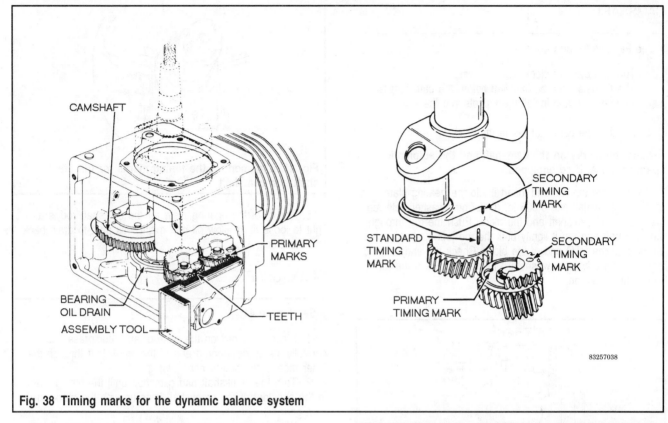

Fig. 38 Timing marks for the dynamic balance system

shaft in until it is 1.11 in. (28mm) above the stub shaft boss, and then, use a ³/₈ in. (9.5mm) spacer.

When installing the balance gears, slip one 0.010 in. (0.254mm) spacer on the stub shaft, then install the gear/bearing assembly on the stub shaft with the timing marks facing out. Proper endplay of 0.002-0.010 in. (0.05-0.25mm) is attained with one 0.005 in. (0.127mm) spacer, one 0.010 in. (0.254mm) spacer and one 0.020 in. (0.50mm) spacer which are all installed on the snapring retainer end of the shaft. Install the thickest spacer next to the retainer. Check the endplay and adjust it by adding or subtracting spacers.

To time the balance gears, first press the crankshaft into the block and align the primary timing mark on top of the balance gear with the standard timing mark next to the crankgear. Press the shaft in until the crankgear is engaged ¹/₁₆ in. (1.6mm) into the top gear (narrow side). Rotate the crankshaft to align the timing marks on the crankgear and cam gear. Press the crankshaft the remainder of the way into the block.

Rotate the crankshaft until it is about 15 degrees past BDC and slip one 0.010 in. (0.25mm) spacer over the stub shaft before installing the bottom gear/bearing assembly.

Align the secondary timing mark on this gear with the secondary timing mark on the counterweight of the crankshaft and then install the gear on the shaft. The secondary timing mark will also be aligned with the standard timing mark on the crankshaft after installation. Use one 0.005 in. (0.127mm) spacer and one 0.020 in. (0.50mm) spacer with the larger spacer next to the retainer to get the proper endplay of 0.002-0.010 in. (0.05-0.25mm). Install the snapring retainer, then check and adjust the endplay.

Engine Assembly

REAR MAIN BEARING

Install the rear main bearing by pressing it into the cylinder block with the shielded side toward the inside of the block. If it does not have a shielded side, then either side may face inside.

GOVERNOR SHAFT

1. Place the cylinder block on its side and slide the governor shaft into place from the inside of the block. Place the speed control disc on the governor bushing nut and thread the nut into the block, clamping the throttle bracket into place.
2. There should be a slight end-play in the governor shaft and that can be adjusted by moving the needle bearing in the block.
3. Place a space washer on the stub shaft and slide the governor gear assembly into place.
4. Tighten the holding screw from outside the cylinder block.
5. Rotate the governor gear assembly to be sure that the holding screw does not contact the weight section of the gear.

CAMSHAFT

▶ **See Figures 39 and 40**

1. Turn the cylinder block upside down.
2. The tappets must be installed before the camshaft is installed. Lubricate and install the tappets into the valve guides.
3. Position the camshaft inside the block.

➡**Align the marks on the camshaft and the automatic spark advance, if so equipped.**

4. Lubricate the rod and insert it into the bearing plate side of the block. Install one 0.005 in. (0.127mm) washer between the end of the camshaft and the block. Push the rod through the camshaft and tap it lightly until the rod just starts to enter the bore at the PTO end of the block. Check the endplay and adjust it with additional washers if necessary. Press the rod into its final position.

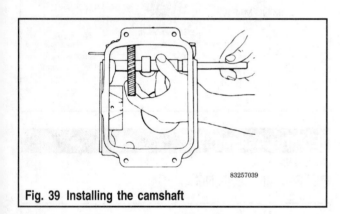

Fig. 39 Installing the camshaft

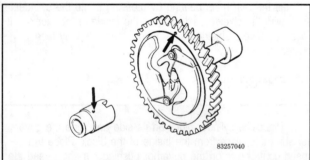

Fig. 40 Timing marks for the automatic spark advance system

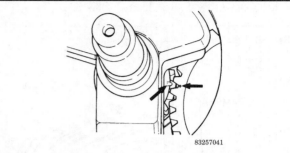

Fig. 41 Alignment of the timing marks for the crankshaft and camshaft

5. The fit at the bearing plate for the camshaft rod is a light to loose fit to allow oil that might leak past to drain back into the block.

CRANKSHAFT

▶ **See Figure 41**

1. Place the block on the base of an arbor press and carefully insert the tapered end of the crankshaft through the inner race of the anti-friction bearing.
2. Turn the crankshaft and camshaft until the timing mark in the shoulder of the crankshaft lines up with the mark on the cam gear.
3. When the marks are aligned, press the crankshaft into the bearing, making sure that the gears mesh as it is being pressed in. Recheck the alignment of the timing marks on the crankshaft and the camshaft.
4. The end-play of the crankshaft is controlled by the application of various thickness gaskets between the bearing plate and the block. Normal end-play is achieved by installing 0.020 in. (0.50mm) and 0.010 in. (0.25mm) gaskets, with the thicker gaskets on the inside.

BEARING PLATE

1. Press the front main bearing into the bearing plate. Make sure that the bearing is straight.
2. Press the bearing plate onto the crankshaft and into position on the block. Install the cap screws and secure the plate to the block. Draw up evenly on the screws.
3. Measure the crankshaft end-play, which is very critical on gear reduction engines.

PISTON AND ROD ASSEMBLY

▶ **See Figure 42**

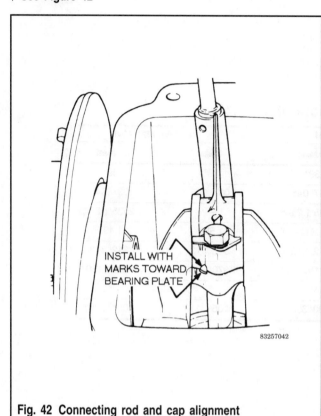

INSTALL WITH
MARKS TOWARD
BEARING PLATE

83257042

Fig. 42 Connecting rod and cap alignment

1. Lubricate the pin and assemble it to the connecting rod and piston. Install the wrist pin retaining ring. Use new retaining rings.

2. Lubricate the entire assembly, stagger the ring gaps and, using a ring compressor, slide the piston and rod assembly into the cylinder bore with the connecting rod marks on the flywheel side of the engine.

3. Place the block on its end and oil the connecting rod end and the crankpin.

4. Attach the rod cap, lock or lock washers, and the cap screws. Tighten the screws to the correct torque.

➡**Align the marks on the cap and the connecting rod.**

5. Bend the lock tabs to lock the screws.

CRANKSHAFT OIL SEALS

Apply a coat of grease to the lip and guide the oil seals onto the crankshaft. Make sure no foreign material gets onto the knife edges of seal, and make sure the seal does not bend. Place the block on its side and drive the seals squarely into the bearing plate and block.

OIL PAN BASE

Using a new gasket on the base, install pilot studs to align the cylinder block, gasket, and base. Tighten the four attaching screws to the correct torque.

VALVES

▶ **See Figure 43**

Valve Specifications

Dimension	Model K141, K161, K181		K241, K301, K321, K341	
	Intake	Exhaust	Intake	Exhaust
A Seat Angle	89°	89°	89°	89°
B Seat Width	.037/.045	.037/.045	.037/.045	.037/.045
C Insert OD	—	1.2535/1.2545	—	1.2535/1.2545
D Guide Depth	1.312	1.312	1.586	1.497
E Guide ID	.312/.313	.312/.313	.312/.313	.312/.313
F Valve Head Diameter	$1^3/_8$	$1^1/_8$	1.370/1.380	1.120/1.130*
G Valve Face Angle	45°	45°	45°	45°
H Valve Stem Diameter	.3105/.3110	.3090/.3095	.3105/.3110	.3084/.3091

*2.125″ on all K341 and K321 engines with spec suffix "D" and later.

832570C9

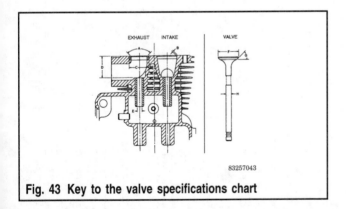

EXHAUST INTAKE   VALVE

83257043

Fig. 43 Key to the valve specifications chart

1. Clean the valves, seats, and parts thoroughly. Grind and lap-in the valves and seats for proper seating. Valve seat width must be $^1/_{32}$-$^1/_{16}$ in. (0.8-1.6mm). After grinding and lapping, slide the valves into position and check the clearance between stem and tappet. If the clearance is too small, grind the stem ends square and remove all burrs. On engines with adjustable valves, make the adjustment at this time.

2. Place the valve springs, retainers, and rotators under the valve guides. Lubricate the valve stems, and then install the valves down through the guides, compress the springs, and place the locking keys or pins in the grooves of the valve stems.

CYLINDER HEAD

▶ **See Figure 44**

1. Use a new cylinder head gasket.
2. Lubricate and tighten the head bolts evenly, and in sequence, to the proper torque.
3. Install the spark plug.

BREATHER ASSEMBLY

▶ **See Figure 45**

Assemble the breather assembly, making sure that all parts are clean and the cover is securely tightened to prevent oil leakage.

MAGNETO

On flywheel magneto systems, the coil-core assembly is secured onto the bearing plate. On magneto-alternator systems, the coil is part of the stator assembly, which is secured to the bearing plate. On rotor type magneto systems, the rotor has a keyway and is press fitted onto the crankshaft. The magnet rotor is marked 'engine-side' for proper assembly. Run all

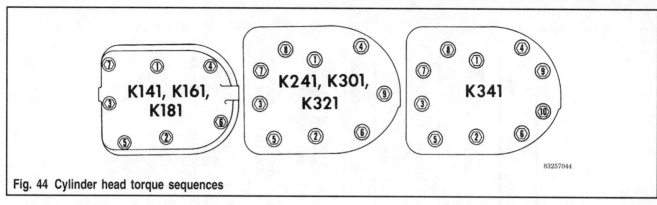

Fig. 44 Cylinder head torque sequences

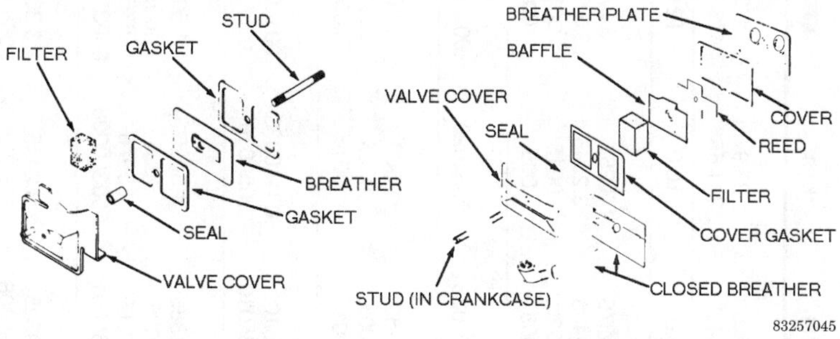

Fig. 45 Two types of breathers

leads through the hole provided at the 11 o'clock position on the bearing plate.

FLYWHEEL

1. Place the washer in place on the crankshaft and place the flywheel in position. Install the key.
2. Install the starter pulley, lock washer, and retaining nut. Tighten the retaining nut to the specified torque.

BREAKER POINTS

1. Install the pushrod.
2. Position the breaker points and fasten them with the two screws.
3. Place the cover gasket into position and attach the magneto lead.
4. Set the gap and install the cover.

CARBURETOR

Insert a new gasket and assemble the carburetor to the intake port with the two attaching screws.

GOVERNOR ARM AND LINKAGE

1. Insert the carburetor linkage in the throttle arm.
2. Connect the governor arm to the carburetor linkage and slide the governor arm into the governor shaft.
3. Position the governor spring in the speed control disc on the 141, 161 and 181.
4. Before tightening the clamp bolt, turn the shaft counterclockwise with pliers as far as it will go; pull the arm as far as it will go to the left (away from the carburetor), tighten the nut, and check for freedom of movement. Adjust the governor.

BLOWER HOUSING AND FUEL TANK

Install the head baffle, cylinder baffle, and the blower housing, in that order. The smaller cap screws are used on the bottom of the crankcase. Install the fuel tank and connect the fuel line.

RUN-IN PROCEDURE

1. Fill the crankcase with a non-detergent oil and run it under load for 5 hours to break it in.
2. Drain oil and refill crankcase with the recommended detergent type oil. Non-detergent oil must not be used except for break-in.

Engine Rebuilding Specifications

Specifications	K141 (2⁷/₁₆" Bore)	K141 (2¹⁵/₁₆" Bore)	K161 (2⁷/₈" Bore)	K161 (2¹⁵/₁₆" Bore)	K181	K241	K301	K321	K341
Displacement									
Cubic Inches	16.22	16.9	16.22	16.9	18.6	23.9	29.07	31.27	35.89
Cubic Centimeter	265.8	276.99	265.8	276.99	304.8	391.65	476.37	528.46	588.24
Horsepower (Max RPM)	6.25	6.25	7.0	7.0	8.0	10.0	12.0	14.0	16.0
Cylinder Bore									
New Diameter	2.875	2.9375	2.875	2.9375	2.9375	3.251	3.375	3.500	3.750
Maximum Wear Diameter	2.878	2.9405	2.878	2.9405	2.9405	3.2545	3.3785	3.503	3.753
Maximum Taper	.0025	.0025	.0025	.0025	.0025	.0015	.0015	.0015	.0015
Maximum out of Round	.005	.005	.005	.005	.005	.005	.005	.005	.005
Crankshaft									
End Play (Free)	.002/.023	.002/.023	.002/.023	.002/.023	.002/.023	.003/.020	.003/.020	.003/.020	.003/.020
Crankpin									
New Diameter	1.186	1.186	1.186	1.186	1.186	1.500	1.500	1.500	1.500
Maximum Out of Round	.0005	.0005	.0005	.0005	.0005	.0005	.0005	.0005	.0005
Maximum Taper	.001	.001	.001	.001	.001	.001	.001	.001	.001
Camshaft									
Run Clearance on Pin	.0005/.003	.0005/.003	.0005/.003	.0005/.003	.0005/.003	.001/.0035	.001/.0035	.001/.0035	.001/.0035
End Play	.005/.010	.005/.010	.005/.010	.005/.010	.005/.010	.005/.010	.005/.010	.005/.010	.005/.010
Connecting Rod									
Big End Maximum Diameter	1.1885	1.1885	1.1885	1.1885	1.1885	1.5025	1.5025	1.5025	1.5025
Rod-Crankpin Max Clear	.0035	.0035	.0035	.0035	.0035	.0035	.0035	.0035	.0035
Small (Pin) End-New Dia	.62565	.62565	.62565	.62565	.62565	.85975	.87585	.87585	.87585
Rod to Pin Clearance	.0006/.0011	.0006/.0011	.0006/.0011	.0006/.0011	.0006/.0011	.0003/.0008	.0003/.0008	.0003/.0008	.0003/.0008
Piston									
Thrust Face-Max Wear Dia*	2.866	2.9305	2.866	2.9305	2.9305	3.2445	3.3625	3.4945	3.7425
Thrust Face*-Bore Clear	.006/.0075	.006/.008	.006/.0075	.006/.008	.006/.008	.0075/.0085	.0065/.0095	.007/.010	.007/.010
Ring-Max Side Clearance	.006	.006	.006	.006	.006	.006	.006	.006	.006
Ring-End Gap in New Bore	.007/.017	.007/.017	.007/.017	.007/.017	.007/.017	.010/.020	.010/.020	.010/.020	.010/.020
Ring-End Gap in Used Bore	.027	.027	.027	.027	.027	.027	.030	.030	.030

83257C10

	1	2	3	4	5	6	7	8	9	10
Valve-Intake										
Valve-Tappet Cold Clear	.008/.010	.008/.010	.008/.010	.008/.010	.008/.010	.006/.008	.006/.008	.006/.008	.006/.008	.006/.008
Valve Lift (Zero Lash)	.324	.324	.324	.324	.324	.2778	.2778	.2778	.2778	.2778
Stem to Guide Max Wear Clear	.0045	.0045	.0045	.0045	.0045	.0045	.0045	.0045	.0045	.0045
Valve-Exhaust										
Valve-Tappet Cold Clear	.017/.020	.017/.020	.017/.020	.017/.020	.017/.020	.015/.017	.015/.017	.015/.017	.015/.017	.015/.017
Valve Lift (Zero Lash)	.324	.324	.324	.324	.324	.2542	.2542	.2542	.2542	.2542
Stem to Guide Max Wear Clear	.0065**	.0065**	.0065**	.0065**	.0065**	.006	.006	.006	.006	.006
Tappet										
Clearance in Guide	.0008/.0023	.0008/.0023	.0008/.0023	.0008/.0023	.0008/.0023	.0005/.002	.0005/.002	.0005/.002	.0005/.002	.0005/.002
Ignition										
Spark Plug Gap-Gasoline	.025	.025	.025	.025	.025	.025	.025	.025	.025	.025
Spark Plug Gap-Gas	.018	.018	.018	.018	.018	.018	.018	.018	.018	.018
Spark Plug Gap (Shielded)	.020	.020	.020	.020	.020	.020	.020	.020	.020	.020
Breaker Point Gap	.020	.020	.020	.020	.020	.020	.020	.020	.020	.020
Trigger Air Gap (Breakerless)										
Spark Run ° BTDC	.005/.010 20°	.005/.010 20°	.005/.010 20°	.005/.010 20°	.005/.010 20°	.005/.010 20°	.005/.010 20°	.005/.010 20°	.005/.010 20°	.005/.010 20°
Spark Retard	ACR ONLY (No Retard)	3° ATDC*** (ACR-NONE)	3° ATDC*** (ACR-NONE)	3° ATDC*** (ACR-NONE)	3° ATDC*** (ACR-NONE)	3° BTDC*** (ACR-NONE)	ACR ONLY (No Retard)	3° BTDC*** (ACR-NONE)	ACR ONLY (No Retard)	3° BTDC*** (ACR-NONE)
Torque Values (Also See Page 15.3)										
Spark Plug (ft. lbs.)	18–22	18–22	18–22	18–22	18–22	18–22	18–22	18–22	18–22	18–22
Cylinder Head	25–30 ft. lbs.	25–30 ft. lbs.	25–30 ft. lbs.	25–30 ft. lbs.	25–30 ft. lbs.	15–20 ft. lbs.	15–20 ft. lbs.	15–20 ft. lbs.	15–20 ft. lbs.	15–20 ft. lbs.
Connecting Rod	300 in. lbs.	300 in. lbs.	300 in. lbs.	300 in. lbs.	300 in. lbs.	200 in. lbs.	200 in. lbs.	200 in. lbs.	200 in. lbs.	200 in. lbs.
Flywheel Nut	60–70 ft. lbs.	60–70 ft. lbs.	60–70 ft. lbs.	60–70 ft. lbs.	60–70 ft. lbs.	50–60 ft. lbs.	50–60 ft. lbs.	50–60 ft. lbs.	50–60 ft. lbs.	50–60 ft. lbs.

*Measured just below oil ring and at right angles to piston pin

**Measured at top of guide with valve closed

***Engines built before automatic compression release (ACR)

83257C1a

1984-92 2-12 HP

Air Cleaner and Air Intake System

▶ **See Figures 46 and 47**

K series engine are equipped with a high density paper air cleaner element. Engines of some specifications are also equipped with an oiled foam precleaner that surrounds the paper element.

AIR CLEANER DISASSEMBLY

1. Remove the wing nut and air cleaner cover.
2. Remove the air precleaner (if so equipped), paper element and seal.

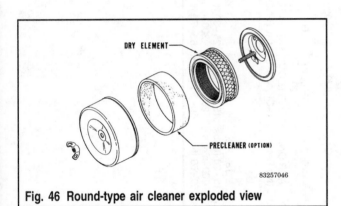

Fig. 46 Round-type air cleaner exploded view

3. Remove the base screws, air cleaner base, gasket and hose.

AIR CLEANER SERVICE

If so equipped, wash and re-oil the precleaner every 25 operating hours (more often under extremely dusty or dirty conditions).

1. Wash the precleaner in warm eater and detergent.
2. Rinse the precleaner thoroughly until all traces of detergent are eliminated. Squeeze out excess water (do not wring). Allow precleaner to dry.
3. Saturate the precleaner with clean, fresh engine oil. Squeeze out excess oil.
4. Reinstall the precleaner over the paper element.

Paper Element

Every 100 operating hours (more often under extremely dusty or dirty conditions) check the paper element. Replace the element as follows:

1. Remove the precleaner (if so equipped), element cover nut, element cover and paper element.
2. Replace a dirty, bent or damaged element with a new genuine Kohler element. Handle new elements carefully; do not use if surfaces are bent or damaged.

➡**Do not wash the paper element or use compressed air as this will damage the element.**

3. Reinstall the paper element.

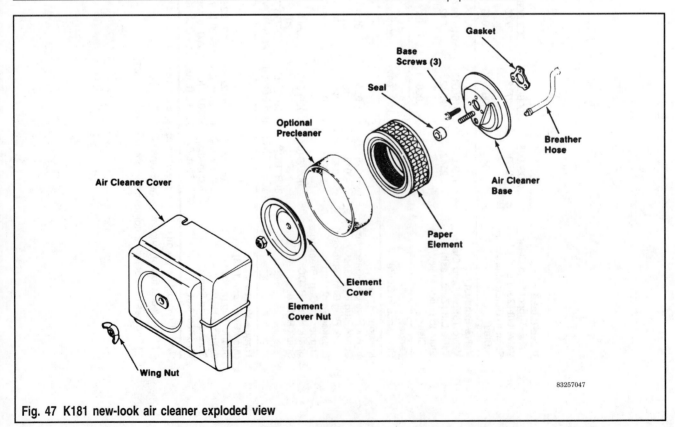

Fig. 47 K181 new-look air cleaner exploded view

4. Install the precleaner (cleaned and oiled) over the paper element.

5. Install the air cleaner cover and wing nut. Tighten wing nut. Make sure element is sealed tightly against air cleaner base.

Inspect Air Cleaner Components

Whenever the air cleaner cover is removed, or when servicing the paper element or precleaner, check the following components:

1. Air cleaner Base - Make sure it is secured tightly to carburetor and is not bent or damaged.

2. Element Cover and Element Cover Nut - On K181 New Look engines only, make sure element cover is not bent or damaged. Check that element cover nut is secured tightly to seal element between air cleaner base and element cover. Tighten nut to 50 inch lbs. torque.

3. Breather Tube - Make sure it is sealed tightly in the air cleaner base and breather cover.

➡**On Model K181 New Look engines of certain specifications, the element cover may contact the breather tube, making it impossible to maintain crankcase vacuum. To prevent this problem, cut the end of the breather tube that protrudes through the air cleaner base at approximately a 45 degree angle.**

❋❋CAUTION

WARNING: Damaged, worn or loose air cleaner components could allow unfiltered air into the engine causing premature wear and failure. Replace any damaged or worn components.

OPTIONAL OIL BATH AIR CLEANER

▶ **See Figure 48**

If the engine has an oil bath type air cleaner, clean and service it after every 25 hours of operation or more frequently if conditions warrant.

1. Remove the cover, lift the element out of the bowl and drain the oil from the bowl.

2. Thoroughly wash bowl and cover in clean solvent. Swish the element in the solvent and allow it to dry.

➡**Do not use compressed air to dry the element. The filtering material could be damaged.**

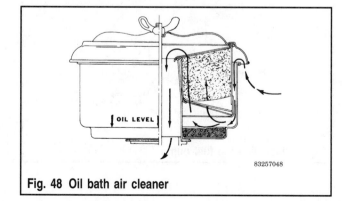

83257048

Fig. 48 Oil bath air cleaner

3. Lightly re-oil the element with engine oil.
4. Inspect base and cover gaskets. Replace if damaged.
5. Install base gasket and place filter on air horn.
6. Add engine oil to filter and fill to the OIL LEVEL mark.
7. Install filter element, cover gasket and cover. Secure with wing nut finger tight only.

COOLING AIR INTAKE SYSTEM

Effective cooling of an air cooled engine depends on an unobstructed flow of air over the cooling fins. Air is drawn into the cooling shroud by fins located on the flywheel. The blower housing, cooling shroud, air screen covering the flywheel and cooling fins on the cylinder and cylinder head must be kept clean and unobstructed at all time.

Never operate the engine with the blower housing or cooling shroud - removed. These devices direct air flow over the cooling fins.

➡**Some engines use a plastic grass screen and some use metal. The two are not interchangeable unless other modifications are made to the engine.**

FUEL SYSTEM

Fuel System

The typical gasoline fuel system and related components include the fuel tank with vented cap, shut-off valve screen, in-line fuel filter, fuel pump (some models), carburetor and interconnecting fuel line.

Operation

The fuel from the tank is moved through the screen and shut-off valve, in-line filter and fuel lines by the fuel pump (if so equipped) or gravity. Fuel enters the carburetor float bowl and is moved into the carburetor body where it is mixed with air. The fuel-air mixture is drawn into the combustion chamber where it is compressed, then ignited by the spark plug.

❋❋CAUTION

CAUTION: Gasoline may be present in the carburetor and fuel system. Gasoline is extremely flammable and it can explode if ignited. Keep sparks, open flames, and other sources of ignition away from the engine. Disconnect an ground the spark plug lead to prevent the possibility of sparks from the ignition system.

Fuel Tank

▶ **See Figure 49**

Engine-mounted fuel tanks on K series engines are constructed of steel. They are fitted with a vented cap. The venting properties of the cap should be checked regularly. A clogged vent can cause pressure buildup in the tank, which could result in fuel spraying from the filler when the cap is

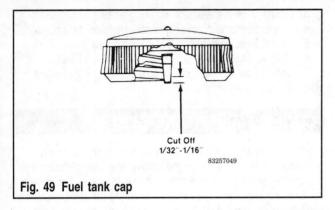

Fig. 49 Fuel tank cap

loosened. It can also cause a partial vacuum in the tank, stopping the engine.

Fuel Shut-off Valve

Some engines are equipped with a fuel shut-off valve with a wire mesh screen. On engines without a shut-off valve, a straight outlet fitting is used. The wire mesh prevents relatively large particles in the tank from reaching the carburetor. The shut-off valve permits work on the fuel system without the need for draining the tank.

Fuel Filter

Some engines covered by this manual may be equipped with a see-through inline fuel filter. When the interior of the filter appears to be dirty, it should be replaced.

Fuel Pump

▶ **See Figure 50**

All K series except the K91 have provisions for mounting a mechanically operated fuel pump. If not fuel pump is mounted on these engines, a cover is placed over the pump mounting pad on the crankcase.Older fuel pumps have a metal body. Later models have a body made of plastic. The plastic body better insulates the fuel from the hot engine, minimizing the chance of vapor lock.

OPERATION

The mechanical fuel pump is operated by a lever that rides on the engine camshaft. The lever transmits a pumping action to the flexible diaphragm inside the pump body. The pumping action draws fuel in through the inlet check valve on the downward stroke of the diaphragm. On the upward stroke, the fuel is forced out through the outlet check valve.

REMOVAL

1. Disconnect the fuel lines from the inlet and outlet fittings of the pump.

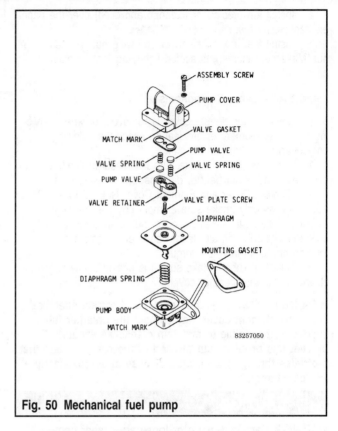

Fig. 50 Mechanical fuel pump

2. Remove the fillister head screws, flat washers, fuel pump and gasket.
3. If required, remove the fittings from the pump body.

REPAIR

Plastic bodied fuel pumps are not serviceable and must be replaced when faulty. Replacement pumps are available in kits which include the pump, mounting gasket and plain washers.

INSTALLATION

▶ **See Figure 51**

1. Fittings - Apply a small amount of Permatex Aviation Perm A Gasket (or equivalent gasoline resistant thread sealant) to fittings. Turn fittings into pump six full turns; continue turning fittings in the same direction until desired direction is reached.

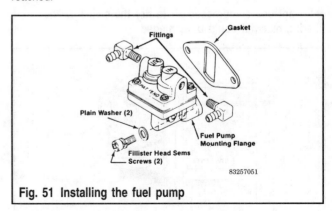

Fig. 51 Installing the fuel pump

2. Install new gasket, fuel pump, flat washers, lock washers and fillister head screws.

➡Make sure that the fuel pump lever is positioned above the camshaft. Damage to the engine could result if the lever is positioned below the camshaft.Make sure that the flat washers are installed next to the mounting flange to prevent damage from the lock washers.If a metal bodied pump was replaced by a plastic bodied pump, make sure that the old thick gasket is discarded and the new thin gasket is used.

3. Torque screws 37-45 inch lbs.
4. Connect fuel lines to inlet and outlet fittings.

Kohler-Built Carburetor

▶ See Figure 52

✳✳CAUTION

CAUTION: Gasoline may be present in the carburetor and fuel system. Gasoline is extremely flammable and it can explode if ignited. Keep sparks, open flames, and other sources of ignition away from the engine. Disconnect and ground the spark plug lead to prevent the possibility of sparks from the ignition system.

ADJUSTMENT

▶ See Figure 53

The carburetor is designed to deliver the correct fuel/air mixture to the engine under all operating conditions. Carburetors are set at the factory and normally do not need adjustment. If the engine exhibits conditions like those found in the table that follows, it may be necessary to adjust the carburetor.

In general, turning the adjusting needles in (clockwise) decreases the supply of fuel to the carburetor. This gives a leaner fuel-to-air mixture. Turning the adjusting needles out (counterclockwise) increases the supply of fuel to the carburetor. This gives a richer fuel-to-air mixture. Setting the needles

midway between the lean and rich positions will usually give the best results. Adjust the carburetor as follows:

1. With the engine stopped, turn the low idle fuel adjusting needle in (clockwise) until it bottoms lightly.

➡The tip of the low idle fuel and high idle fuel adjusting needles are tapered to critical dimensions. Damage to the needles and the seats in carburetor body will result if the needles are forced.

2. Preliminary Settings: Turn the adjusting needles out (counterclockwise) from lightly bottomed according to the table shown.

3. Start the engine and run at half throttle for five to ten minutes to warm up. The engine must be warm before making final settings (Steps 4, 5, 6, and 7).

4. High Idle Fuel Needle Setting: This adjustment is required only for adjustable high idle (main) jet carburetors. If the carburetor is a fixed main jet type, go to step 5.

   a. Place the throttle into the 'fast' position. If possible, place the engine under load.

   b. Turn the high idle fuel adjusting needle out - counterclockwise - from the preliminary setting until the engine speed decreases (rich). Note the position of the needle.

   c. Now turn the adjusting needle in (clockwise). The engine speed may increase, then it will decrease as the needle id turned in (lean). Note the position of the needle.

   d. Set the adjusting needle midway between the rich and lean settings.

5. Low Idle Speed Setting: Place the throttle control into the 'idle' or 'slow' position. Set the low idle speed to 1200 rpm + or - 75 rpm by turning the low idle speed adjusting screw in or out. Check the speed using a tachometer.

➡The actual low idle speed depends on the application. Refer to the equipment manufacturer's instructions for specific low idle speed settings. The recommended low idle speed for Basic Engines is 1200 rpm. To ensure best results when setting the low idle fuel needle, the low idle speed must not exceed 1500 rpm.

6. Low Idle Fuel Setting: Place the throttle into the 'idle' or 'slow' position.

   a. Turn the low idle fuel adjusting needle out (counterclockwise) from the preliminary setting until the engine speed decreases (rich). Note the position of the needle.

   b. Now turn the adjusting needle in (clockwise). The engine speed may increase, then it will decrease as the needle is turned in (lean). Note the position of the needle.

   c. Set the adjusting needle midway between the rich and lean settings.

7. Recheck the low idle speed using a tachometer. Readjust the speed as necessary.

DISASSEMBLY

▶ See Figure 54

1. Remove the bowl retaining screw, retaining screw gasket and fuel bowl.

2. Remove the float pin, float, fuel inlet needle, baffle gasket and bowl gasket.

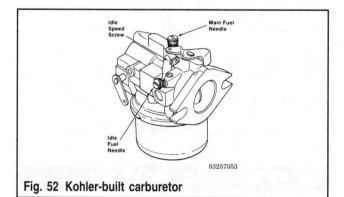

Fig. 52 Kohler-built carburetor

PRELIMINARY SETTINGS – K-SERIES MODELS

| | KOHLER ADJUSTABLE JET | | WALBRO FIXED JET | |
	Low Idle	High Idle	Low Idle	
K91	1–1/2 turns	2 turns	NOT APPL.	NOTE: Refer to publication TP2377B Carburetor Reference Manual for additional information.
K141	1–1/2 turns	3 turns	NOT APPL.	
K161*	1–1/2 turns	3 turns	NOT APPL.	
K181*	1–1/4 turns	2 turns	2–1/2 turns	
K241	2–1/2 turns	2 turns	1–1/4 turns	
K301	2–1/2 turns	2 turns	1–1/4 turns	
K321	2–1/2 turns	3–1/4 turns	1–1/2 turns	
K341	2–1/2 turns	3–1/2 turns	1 turn	

* Includes "New Look" Models

83257054

Fig. 53 Preliminary low and high idle fuel needle settings

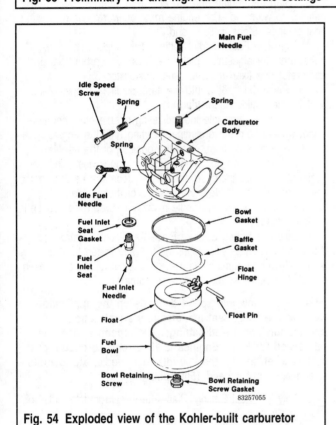

Fig. 54 Exploded view of the Kohler-built carburetor

3. Remove the fuel inlet seat and inlet seat gasket. Remove the idle fuel and main fuel adjusting needles and springs. Remove the idle speed adjusting screw and spring.

4. Further disassembly to remove the throttle and choke shafts is recommended only if these parts are to be replaced. Refer to 'Throttle and Choke Shaft Replacement' later in this section.

Cleaning

❊❊CAUTION

CAUTION: Carburetor cleaners and solvents are extremely flammable. Keep sparks, flames and other sources of ignition away from the area. Follow the cleaner manufacturer's warnings and instructions on its proper and safe use. Never use gasoline as a cleaning agent.

All parts should be carefully cleaned using a carburetor cleaner (such as acetone). Be sure all gum deposits are removed from the following areas:
• Carburetor body and bore; especially the areas where throttle plate, choke plate and shafts are seated.
• Float and hinge.
• Fuel bowl.
• Idle fuel and 'off-idle' ports in carburetor bore, ports in main fuel adjusting needle and main fuel seat.

➡These areas can be cleaned using a piece of fine wire in addition to cleaners. Be careful not to enlarge the ports or break the cleaning wire within the ports.Blow out all passages with compressed air.

➡Do not submerge carburetor in cleaner or solvent when fiber or rubber seals are installed. The cleaner may damage these seals.

INSPECTION

1. Carefully inspect all components and replace those that are worn or damaged.

2. Inspect the carburetor body for cracks, holes and other wear or damage.

3. Inspect the float for dents or holes. Check the float hinge for wear and missing or damaged float tabs.

4. Inspect the inlet needle and seat for wear or grooves.

5. Inspect the tips of the main and idle adjusting needles for wear or grooves.

6. Inspect the throttle and choke shafts and plate assemblies for wear or excessive play.

CHOKE PLATE MODIFICATION

The choke action has been changed on production carburetors to reduce the chances of over choking. On production carburetors now used on the K241 and K301, one relief hole is now 11/32 in. and the other is 3/16 in. If you find that the relief holes are smaller than this, enlarge them to these dimensions.

➡**When redrilling the holes, take the necessary precautions to prevent chips from entering the engine.**

REPAIR

Always use new gaskets when servicing and reinstalling carburetors. Several repair kits, which include the gaskets and other components, are available.

✳✳CAUTION

CAUTION: Suitable eye protection (safety glasses, goggles, or face hood) should be worn for any procedure involving the use of compressed air, punches, hammers, chisels, drills or grinding tools.

Throttle and Choke Shaft Replacement

▶ **See Figure 55**

Two kits are available that allow replacement of the carburetor throttle and choke shafts.

1. To ensure correct reassembly, mark choke plate and carburetor body with a marking pen. Also take note of choke plate position in bore and choke lever position.

2. Carefully and slowly remove the screws securing choke plate to choke shaft, remove and save the choke plate as is will be reused.

3. File off any burrs which may have been left on the choke shaft when the screws were removed. Place carburetor on workbench with choke side down. Remove choke shaft; the detent ball and spring will fall out.

4. Note the position of the choke lever with respect to the cut out portion of the choke shaft.

5. Carefully grind or file away the riveted portion of the shaft, remove and save the choke lever; discard the old choke shaft.

6. Attach the choke lever to the new choke shaft from the kit. Make sure the lever is installed correctly as noted in step 9.

Secure lever to choke shaft as follows:
- Models K91 - K181; Apply Loctite to threads of 1 #2-56x 7/32 in. brass screw. Secure lever to shaft.
- Models K241 - K341; Apply Loctite to thread of 1 #3-48x 7/32 in. brass screw. Secure lever to shaft.

Throttle Place and Throttle Shaft; Transfer Throttle Lever

1. To ensure correct reassembly, mark throttle plate and carburetor body with a marking pen. Also take note of the throttle plate position in the bore and the throttle lever position.

2. Carefully and slowly remove the screws securing the throttle plate to throttle shaft. Remove and save the throttle plate for reuse.

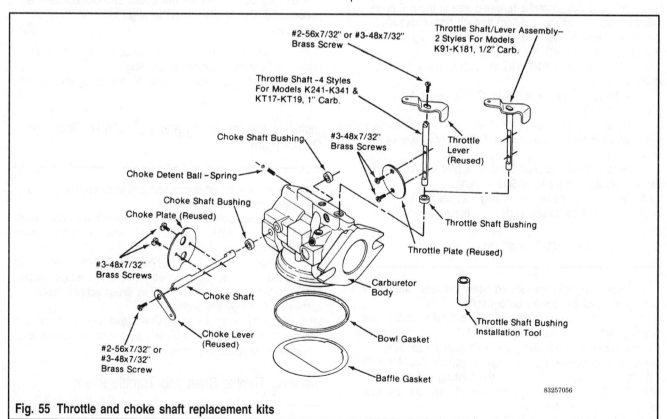

Fig. 55 Throttle and choke shaft replacement kits

3. File off any burrs that may have been left on the throttle shaft when screws were removed.

➡**Failure to remove burrs from the throttle shaft may cause permanent damage to carburetor body when shaft is removed.**

4. Remove throttle shaft from carburetor body. Remove and discard the foam rubber dust seal from the throttle shaft.

5. Remove and transfer the throttle lever as follows:
   - Models K91 - K181 (1/2 in Carb.): Carefully grind or file away the riveted portion of the throttle shaft. Save the throttle shaft as it will be used to install the new throttle shaft bushing. Discard the throttle lever.
   - Models K241 - K341 (1 in. Carb.):
     a. Note the position of the throttle lever with respect to the cutout portion of the throttle shaft.
     b. Carefully grind or file away the riveted portion of the shaft. Remove the throttle lever.
     c. Compare the old shaft with the new shafts in the kit. Select the appropriate new shaft and discard the old shaft.
     d. Attach throttle lever to throttle shaft. Make sure lever is installed correctly as noted in step a.
     e. Apply Loctite to threads of 1 #2-56x 7/32 in. brass screw (use #3-48x 7/32 in. screw if shaft is 2 49/64 in. long. Secure lever to shaft.

Drilling Choke Shaft Bores Using a Drill Press

1. Mount the carburetor body in a drill press vise. Keep the vise jaws slightly loose.

2. Install a drill bit of the following size in the drill press chuck. Lower the bit (not rotating) through both choke shaft bores; then tighten vise. This ensures accurate alignment of the carburetor body with the drill press chuck.
   - Models K91 - K181 (1/2 in. Carb.); Use a 7/32 in. diameter drill bit.
   - Models K241 - K341 (1 in. Carb.): Use a 1/4 in. diameter drill bit.

3. Install a 19/64 in. drill bit in the chuck. Set drill press to a low speed suitable for aluminum. Drill slowly to ensure a good finish.

4. Ream the choke shaft bores to a final size of 5/16 in. for best results use a piloted 5/16 in. reamer.

5. Blow out all metal chips using compressed air. Thoroughly clean the carburetor body in carburetor cleaner.

Installing Choke Shaft Bushings

1. Install screws in the tapered holes that enter the choke shaft bores until the screws bottom tightly.

2. Coat the outside surface of the kit-supplied choke shaft bushings with Loctite from the kit. Carefully press the bushings into the carburetor body using a smooth-jawed vise. Stop pressing when bushings bottom against screws. On Model K91 - K181 (1/2 in. Carb.); Make sure the bushing is pressed below the surface of the large choke shaft boss until the bushing bottoms against the screw.

3. Allow Loctite to 'set' for 5 to 10 minutes, then remove screws.

4. Install new choke shaft in bushings. Rotate shaft and check that it does not bind.

➡**If binding occurs, locate and correct the cause before proceeding. Use choke shaft to align bushings if necessary.**

5. Remove choke shaft and allow Loctite to 'set' for an additional 30 minutes before proceeding.

6. Wipe away any excess Loctite from bushings and choke shaft.

Installing Throttle Shaft Bushing

1. Make sure the dust seal counterbore in the carburetor body is thoroughly clean and free of chips and burrs.

2. Install a throttle shaft (without throttle lever) in carburetor body to use as a pilot:
   - Models K91 - K181 (1/2 in. Carb.); Use the old throttle shaft removed previously.
   - Models K141 - K341 (1 in. Carb.); Use one of the remaining new throttle shafts from the kit.

3. Coat the outside surface of the throttle shaft bushing with Loctite from the kit. Slip the bushing over that shaft. Using a vise and the installation tool from the kit, press the bushing into the counterbore until it bottoms in the carburetor body.

4. Allow the Loctite to 'set' for 5 to 10 minutes, then remove the throttle shaft.

5. Install the new throttle shaft and lever in carburetor body. Rotate the shaft and check that it does not bind.

➡**If binding occurs, locate the cause and correct before proceeding. Use throttle shaft to align bushing if necessary.**

6. Remove the shaft and allow the Loctite to 'set' for an additional 30 minutes before proceeding.

7. Wipe away all excess Loctite from bushing and throttle shaft.

Installing Detent Spring and Ball, Choke Shaft and Choke Plate

1. Install new detent spring and ball in carburetor body in the side opposite the choke lever.

2. Compress detent ball and spring and insert choke shaft through bushings. Make sure the choke lever is on the correct side of the carburetor body.

3. Compress choke plate to choke shaft. Make sure marks are aligned and plate is positioned properly in the bore. Apply Loctite to threads of 2 #3-48x 7/32 in. brass screws. Install screws so that they are slightly loose.

4. Operate the choke lever. Check that there is no binding between choke plate and carburetor bore. Loosen screws and adjust plate as necessary; then tighten screws.

Installing Throttle Shaft and Throttle Plate

1. Install throttle shaft in carburetor with cutout portion of the shaft facing out.

2. Attach throttle plate to throttle shaft. Make sure marks are aligned and plate is positioned properly in the bore. Apply Loctite to threads of 2 #3-48 x 7/32 in. brass screws. Install screws so that they are slightly loose.

3. Apply finger pressure to throttle shaft to keep it firmly seated against pivot in carburetor body. Rotate the throttle shaft until the throttle plate fully closes the bore around its perimeter; then tighten screws.

4. Operate the throttle lever and check that the throttle plate does not bind in the bore. Loosen screws and adjust plate if necessary; then tighten screws securely.

CARBURETOR ASSEMBLY

▶ **See Figures 56 and 57**

1. Install the fuel inlet seat gasket and fuel inlet seat into the carburetor body. Torque seat to 34-45 inch lbs.

2. Install the fuel inlet needle into inlet seat. Install float and slide float pin through float hinge and float hinge towers on carburetor body.

3. Set float level: Invert carburetor so the float tab rests on the fuel inlet needle. There should be 11/64 in. + or - 1/32 in. clearance between the machined surface of the body and the free end of the float. Bend the float tab with a small screwdriver to adjust.

4. Set float drop: Turn the carburetor over to its normal operating position and allow float to drop to its lowest level. The float drop should be limited to 1 1/32 in. between the machined surface of body and the bottom of the free end of float. Bend the float tab with a small screwdriver to adjust.

5. Check float-to-float hinge tower clearance: Invert the carburetor so the float tab rests on the fuel inlet needle. Insert a 0.010 in. feeler gauge between the float and float hinge towers. If the feeler gauge cannot be inserted, or there is interference between the float and towers, file the towers to obtain the proper clearance.

6. Install the bowl so it is centered on the baffle gasket. Make sure the baffle gasket and bowl are positioned properly to ensure a good seal.

7. Install the fuel bowl so it is centered on the baffle gasket. Make sure the baffle gasket and bowl are positioned properly to ensure a good seal.

8. Install the bowl retaining screw gasket and bowl retaining screw Torque screw to 50-60 inch lbs.

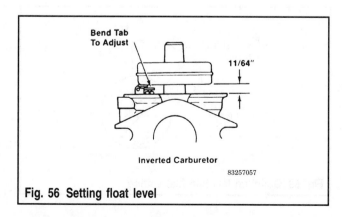

Fig. 56 Setting float level

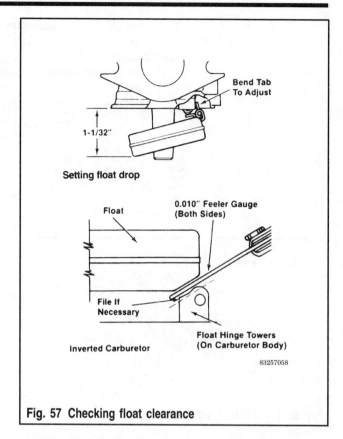

Fig. 57 Checking float clearance

9. Install the idle speed adjusting screw and spring. Install the idle fuel and main fuel adjusting needles and springs. Turn the adjusting needles clockwise until they are bottomed lightly.

➡**The ends of the adjusting needles are tapered to critical dimensions. Damage to the needles and seats will result if needles are forced.**

10. Reinstall the carburetor to the engine using a new gasket.

11. Adjust the carburetor as outlined under the 'Adjustment' portion of this section.

Walbro Fixed/Adjustable Carburetor

▶ **See Figure 58**

This section covers the idle adjustment, disassembly, cleaning, inspection, repair, and assembly of the Walbro-built, side draft, fixed/adjustable main jet carburetors.

✳✳CAUTION

CAUTION: Before servicing the carburetor, engine, or equipment, always remove the spark plug leads to prevent the engine from starting accidentally. Ground the leads to prevent sparks that could cause fires.

Gasoline may be present in the carburetor and fuel system. Gasoline is extremely flammable and its vapors can explode if ignited. Keep sparks, open flame, and other sources of ignition away from the area to prevent the possibility of fires or explosions.

Suitable eye protection (safety glasses, goggles, or face shield) should be worn for any procedure involving the use of

PRELIMINARY SETTINGS

K-SERIES MODELS	KOHLER ADJUSTABLE JET		WALBRO FIXED JET	WALBRO ADJUSTABLE JET	
	Low Idle	High Idle	Low Idle	Low Idle	High Idle
K91	1–1/2 turns	2 turns	NOT APPL.	2 B DETERMD	2 B DETERMD
K141	1–1/2 turns	3 turns	NOT APPL.	NOT APPL.	NOT APPL.
K161*	1–1/2 turns	3 turns	NOT APPL.	2 B DETERMD	2 B DETERMD
K181*	1–1/4 turns	2 turns	2–1/2 turns	2–1/2 turns	3/4 turn
K241	2–1/2 turns	2 turns	1–1/4 turns	1–3/4 turns	1–1/8 turns
K301	2–1/2 turns	2 turns	1–1/4 turns	1–3/4 turns	1–1/8 turns
K321	2–1/2 turns	3–1/4 turns	1–1/2 turns	2 B DETERMD	2 B DETERMD
K341	2–1/2 turns	3–1/2 turns	1 turn	2 B DETERMD	2 B DETERMD
KT17	1 turn	2–1/2 turns	1–1/4 turns	1–1/4 turns	1–1/4 turns
KT19	1 turn	2–1/2 turns	1–1/4 turns	1–1/4 turns	1 turn
K582	1–1/4 turns	3 turns	NOT APPL.	2 B DETERMD	2 B DETERMD

*Includes "New Look" Models

83257C12

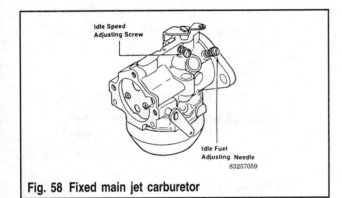

Fig. 58 Fixed main jet carburetor

compressed air, punches, hammers, chisels, drills, or grinding tools.

TROUBLESHOOTING

If engine troubles are experienced that appear to be fuel system related, check the following areas before adjusting or disassembling the carburetor.
- Make sure the fuel tank is filled with clean, fresh gasoline.
- Make sure the fuel tank cap vent is not blocked and that it is operating properly.
- Make sure fuel is reaching the carburetor. This includes checking the fuel shut-off valve, fuel tank filler screen, in-line fuel filter, fuel lines, and fuel pump for restrictions or faulty components as necessary.
- Make sure the carburetor is securely fastened to the engine using gaskets in good condition.
- Make sure the air cleaner element is clean and all air cleaner components are fastened securely.
- Make sure the ignition system, governor system, exhaust system, and throttle and choke controls are operating properly.
- If, after checking the items listed above, starting problems or other conditions similar to those listed in the following table exist, it may be necessary to adjust or service the carburetor.

CARBURETOR ADJUSTMENT

▶ See Figure 59

➡The tip of the low idle fuel and high idle fuel adjusting needles are tapered to critical dimensions. Damage to the needles and the seats in carburetor body will result if the needles are forced.

In general, turning the adjusting needles in (clockwise) decreases the supply of fuel to the carburetor. This gives a leaner fuel-to-air mixture. Turning the adjusting needles out (counterclockwise) increases the supply of fuel to the carburetor. This gives richer fuel-to-air mixture. Setting the needles midway between the lean and rich positions will usually give the best results.

Adjust the Carburetor as follows:

1. With the engine stopped, turn the low idle fuel adjusting needles in (clockwise) until it bottoms lightly.
2. Preliminary Settings: Turn the adjusting needles out (counterclockwise) from lightly bottomed according to the table shown.
3. Start the engine and run at half throttle for five to ten minutes to warm up. The engine must be warm, before making final settings (Steps 4, 5, 6 and 7).
4. High Idle Fuel Needle Setting: This adjustment is required only for adjustable high idle (main) jet carburetors. If the carburetor is fixed main jet type, go to step 5.
   a. Place the throttle into the 'fast' position. If possible, place the engine under load.

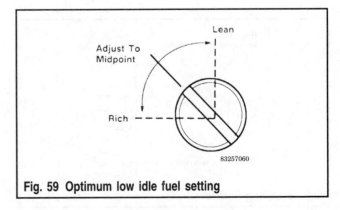

Fig. 59 Optimum low idle fuel setting

b. Turn the high idle fuel adjusting needle out (counter-clockwise) from the preliminary setting until the engine speed decreases (rich). Note the position of the needle.

c. Now turn the adjusting needle in (clockwise). The engine speed may increase, then it will decrease as the needle is turned in (lean). Note the position of the needle.

d. Set the adjusting needle midway between the rich and lean settings.

5. Low Idle Speed Setting: Place the throttle control into the 'idle' or 'slow' position. Set the low idle speed to 1200 rpm + or - 75 rpm by turning the low idle speed adjusting screw in or out. Check the speed using a tachometer.

➡**The actual low idle speed depends on the application. The recommended low idle speed for Basic Engines is 1200 rpm. To ensure best results when setting the low idle fuel needle, the low idle speed must not exceed 1500 rpm.**

6. Low Idle Fuel Needle setting: Place the throttle into the 'idle' or 'slow' position.

a. Turn the low idle fuel adjusting needle out (counter-clockwise) from the preliminary setting until the engine speed decreases (rich). Note the position of the needle.

b. Now turn the adjusting needle in (clockwise). The engine speed may increase, then it will decrease as the needle is turned in (lean). Note the position of the needle.

c. Set the adjusting needle midway between the rich and lean settings.

7. Recheck the low idle speed using a tachometer. Readjust the speed as necessary.

DISASSEMBLY

◆ **See Figures 60 and 61**

1. Remove the bowl retaining screw, retaining screw gasket, and fuel bowl.
2. Remove the bowl gasket, float pin, float, and fuel inlet needle.

❈❈CAUTION

CAUTION: To prevent damage to the carburetor, do not attempt to remove the fuel inlet seat as it is not serviceable. Replace the carburetor if the fuel inlet seat is damaged.

3. Remove the idle fuel adjusting needle and spring. Remove the idle speed adjusting screw and spring.
4. Remove the main fuel jet.
5. In order to clean the 'off-idle' ports and the bowl vent channel thoroughly, the welch plugs covering these areas must be removed. Use tool No. K01018 and the following procedure to remove the welch plugs.

a. Pierce the welch plug with the tip of the tool.

❈❈CAUTION

CAUTION: To prevent damage to the carburetor, do not allow the tool to strike the carburetor body.

b. Pry out the welch plug using the tool.

Throttle and Choke Shaft Removal

Further disassembly to remove the throttle shaft and choke shaft is recommended only if these parts are to be cleaned or replaced.

THROTTLE SHAFT REMOVAL

1. Because the edges of the throttle plate are beveled, mark the throttle plate and carburetor body with a marking pen to ensure correct reassembly. Also take note of the throttle plate position in bore, and the position of the throttle lever.
2. Carefully and slowly remove the screws securing the throttle plate to throttle shaft. Remove the throttle plate.
3. File off any burrs which may have been left on the throttle shaft when the screws were removed. Do this before removing the throttle shaft from carburetor body.
4. Remove the throttle lever/shaft assembly with foam dust seal from carburetor body.

CHOKE SHAFT REMOVAL

1. Because the edges of choke plate are beveled, mark the choke plate and carburetor body with a marking pen to ensure correct reassembly. Also take note of the choke plate position in bore, and the position of the choke lever.
2. Carefully and slowly remove the screws securing the choke plate to choke shaft. Remove the choke plate.
3. File off any burrs which may have been left on the choke shaft when the screws were removed. Do this before removing the choke shaft from carburetor body.
4. Rotate the choke shaft until the cutout portion of shaft is facing the air cleaner mounting surface. Place the carburetor body on the work bench with choke side down. Remove the choke lever/shaft assembly from carburetor body; the detent ball and spring will drop out.

CLEANING

❈❈CAUTION

CAUTION: Carburetor cleaners and solvents are extremely flammable. Keep sparks, flames, and other sources of ignition away from the area. Follow the cleaner manufacturer's warnings and instructions on its proper and safe use. Never use gasoline as a cleaning agent.

All parts should be carefully cleaned using a carburetor cleaner (such as acetone). Be sure all gum deposits are removed from the following areas:

• Carburetor body and bore; especially the areas where the throttle plate, choke plate, and shafts are seated.

• Idle fuel and 'off-idle' ports in carburetor bore, main jet, bowl vent, and fuel inlet seat.

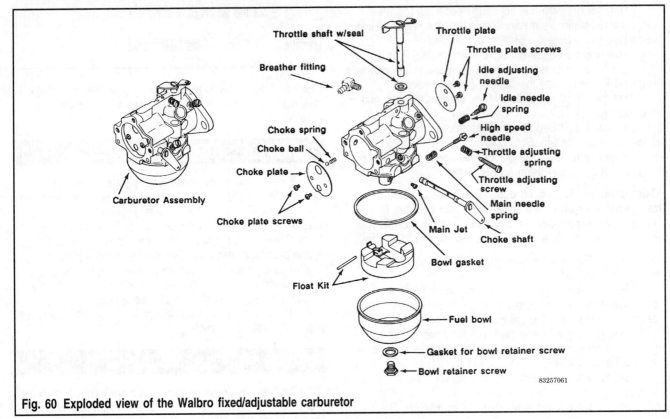

Fig. 60 Exploded view of the Walbro fixed/adjustable carburetor

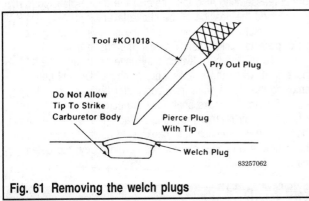

Fig. 61 Removing the welch plugs

➡These areas can be cleaned using a piece of fine wire in addition to cleaners. Be careful not to enlarge the ports, or break the cleaning wire within ports. Blow out all passages with compressed air.

- Float and Float hinge.
- Fuel Bowl.
- Throttle plate, choke plate, throttle shaft and choke shaft.

✳✳CAUTION

CAUTION: Do not submerge the carburetor in cleaner or solvents when fiber, rubber, or foam seals or gaskets, or the fuel inlet needle are installed. The cleaner may damage these parts.

- Carefully inspect all components and replace those that are worn or damaged.
- Inspect the carburetor body for cracks, holes, and other wear or damage.
- Inspect the float for cracks or holes. Check the float hinge for wear, and missing or damaged float tabs.
- Inspect the fuel inlet needle for wear or grooves.
- Inspect the tip of the idle fuel adjusting needle for wear or grooves.
- Inspect the throttle and choke shaft and plate assemblies for wear or excessive play.

REPAIR

Always use new gaskets when servicing and reinstalling carburetors. Repair kits are available which include new gaskets and other components. These kits are described below.

Components such as the throttle and choke shaft assemblies, throttle plate, choke plate, idle fuel needle, main jet, and others, are available separately.

REASSEMBLY

Throttle Shaft Installation

1. Install the foam dust seal on throttle shaft. Insert the throttle lever/shaft assembly into carburetor body with the cut-out portion of shaft facing the carburetor mounting flange.

2. Install the throttle plate to throttle shaft. Make sure the plate is positioned properly in bore as marked and noted during disassembly (the number stamped on plate should face the

carburetor mounting flange). Apply Loctite #609 to threads of 2 plate retaining screws. Install screws so they are slightly loose.

3. Apply finger pressure to the throttle lever/shaft to keep it firmly seated against pivot in carburetor body. Rotate the throttle shaft until the throttle plate fully closes the bore around its entire perimeter; then tighten screws.

4. Operate the throttle lever; check for binding between the throttle plate and carburetor bore. Loosen screws and adjust throttle plate as necessary; then torque screws to 8-12 inch lbs.

Choke Shaft Installation

1. Install the detent spring and ball into the carburetor body.

✳✳CAUTION

CAUTION: If the detent ball does not drop through the tapped air cleaner base screw hole by its own weight, do not force it. Forcing the ball could permanently lodge it in the hole. Install the ball through the choke shaft bore instead.

2. Compress the detent ball and spring. Insert the choke lever/shaft assembly into carburetor body with the cutout portion of shaft facing the air cleaner mounting surface. Make sure the choke lever is on the correct side of carburetor body.

3. Install the choke plate to choke shaft. Make sure the plate is positioned properly in bore as marked and noted during disassembly. (The numbers stamped on plate should face the air cleaner mounting surface and be upright). Apply Loctite #609 to threads of 2 plate retaining screws. Install the screws so they are slightly loose.

4. Operate the choke lever; check for binding between the choke plate and carburetor bore. Adjust plate as necessary; then torque screws to 8-12 inch lbs.

Carburetor Reassembly

▶ **See Figure 62**

1. If the welch plugs have been removed for cleaning, new welch plugs must be installed. Use tool No. K01017 and the following procedure to install the welch plugs.
   a. Position the carburetor body securely with the welch plug cavities to the top.
   b. Place a new welch plug into the cavity with the raised portion up. Use the end of the tool that is about the same size as plug and flatten the plug. Do not force the plug below the top surface.
   c. After welch plugs are installed, seal the exposed surface with sealant. Allow the sealant to dry.

➡**If a commercial sealant is not available, fingernail polish can be used.**

2. Install the main fuel jet.

3. Install fuel inlet needle into inlet seat. Install float and slide float pin through float hinge and float hinge towers on carburetor body.

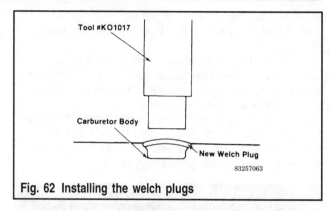

Fig. 62 Installing the welch plugs

4. Set Float Level: Invert the carburetor so the float tab rests on the fuel inlet needle. The exposed surface of float should be parallel with the bowl gasket surface of the carburetor body (exposed, free end of float 0.690-0.720 in. from bowl gasket surface). Bend the float tab with a small screwdriver to adjust.

5. Install a new gasket and the fuel bowl. Make sure the bowl gasket and bowl are centered and positioned properly to ensure a good seal.

6. Install a new bowl retaining screw gasket and the bowl retaining screw. Torque screw to 45-55 inch lbs.

7. Install the idle speed adjusting screw and spring.

8. Install the idle fuel adjusting needle and spring. Turn the adjusting needle in (clockwise) until it bottoms lightly.

✳✳CAUTION

CAUTION: The tip of the idle fuel adjusting needle is tapered to critical dimensions. Damage to the needle and the seat in carburetor body will result if the needle is forced.

9. Turn the idle fuel needle out (counterclockwise) from lightly bottomed according to the instructions in the 'Adjustment' section of this Bulletin.

HIGH ALTITUDE OPERATION (FIXED JET)

When operating the engine at high altitudes, the main fuel mixture tends to get over-rich. An over-rich mixture can cause conditions such as black, sooty exhaust smoke, misfiring, loss of speed and power, poor fuel economy, and poor or slow governor response.

To compensate for this, a special high altitude main fuel jet is available for each carburetor. The high altitude main fuel jet is sold in a kit which includes the jet and necessary gaskets.

High Altitude Jet Installation (Fixed Jet)

1. Remove the fuel bowl retaining screw, retaining screw gasket, fuel bowl, and bowl gasket.

➡**If necessary, remove the air cleaner and carburetor from engine to make fuel bowl removal easier.**

2. Remove the float pin, float, and fuel inlet needle.

3. Remove the existing main fuel jet.

4. Install the new high altitude main fuel jet and torque to 12-16 inch lbs.

5. Reinstall the fuel inlet needle, float, and float pin.

6. Install the new bowl gasket from kit and the fuel bowl. Make sure the bowl gasket and bowl are centered and positioned properly to ensure a good seal.

7. Install the new bowl retaining screw gasket from kit and the bowl retaining screw. Torque screw to 45-55 inch lbs.

8. Reinstall the carburetor and air cleaner to engine as necessary using the new gaskets from kit.

IDLE ADJUSTMENT PROCEDURE FOR K341QS ENGINES WITH ANTI-DIESELING SOLENOID

▶ **See Figure 63**

The idle speed of some vibro-mounted K341AQS engines has been increased to allow smoother operation at low idle and an anti-dieseling solenoid has been added to prevent dieseling during shut-down at the higher idle speed. If called upon to adjust the idle on any K341AQS engine with this solenoid, use the following procedure.

STEP 1 - IDLE FUEL MIXTURE ADJUSTMENT: With the engine stopped, turn the idle fuel adjusting screw all the way in (clockwise) until it bottoms lightly then back out 1/2 turn.

STEP 2 - IDLE SPEED ADJUSTMENT: Start engine and check idle speed with a hand tachometer. Idle, no load, speed should be 2100 rpm. To set the idle speed, loosen the jam nut on the anti-dieseling solenoid and turn the solenoid in or out until 2100 rpm idle speed is attained - re-tighten jam nut to lock solenoid in position.

Thermostatic Type Automatic Chokes

The automatic choke is a heat sensitive thermostatic unit. At room temperature, choke lever will be set in a vertical position. If engine should fail to start when cranked, adjust choke lever by hand to determine if choke setting is too lean or too rich. Once this has been established, adjustment can be made to remedy the situation.

ADJUSTMENT

1. Loosen adjustment lock screw on choke body. This allows the position of adjustment to be changed.

2. Moving adjustment bracket downward will increase the amount of choking. Upward movement will result in less choking.

3. After adjustment is made, tighten adjustment lock screw.

Electric-Thermostatic Type Automatic Chokes

▶ **See Figure 64**

Remove air cleaner from carburetor to observe position of choke plate. Choke adjustment must be made on cold engine. If starting in extreme cold, choke should be in full closed position before engine is started. A lesser degree of choking is needed in milder temperatures.

ADJUSTMENT

1. Move choke arm until hole in brass shaft lines up with slot in bearings.

2. Insert #43 drill (0.089 in.) and push all the way down to engine manifold to engage in notch in base of choke unit.

3. Loosen clamp bolt choke lever, push arm upward to move choke plate toward closed position. After desired position is attained, tighten clamp bolt then remove drill.

4. After replacing air cleaner, check for evidence of binding in linkage, adjust as needed. Be sure chokes are fully open when engine is at normal operating temperature.

TROUBLESHOOTING

Check resistance of heater terminal using an ohmmeter. Resistance should be 3 ohms or more. If resistance is less than 3 ohms, replace the choke.

CHOKE REPLACEMENT AND ADJUSTMENT

1. Position the choke unit on the two mounting screws so that it is slightly loose.

2. Hold the choke plate in the wide open position.

3. Rotate the choke unit clockwise on the carburetor (viewed from the choke side) with a slight pressure until it can no longer be rotated.

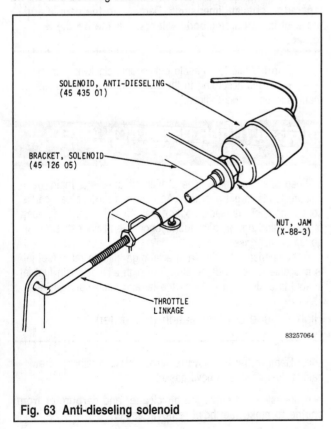

SOLENOID, ANTI-DIESELING
(45 435 01)

BRACKET, SOLENOID
(45 126 05)

NUT, JAM
(X-88-3)

THROTTLE
LINKAGE

83257064

Fig. 63 Anti-dieseling solenoid

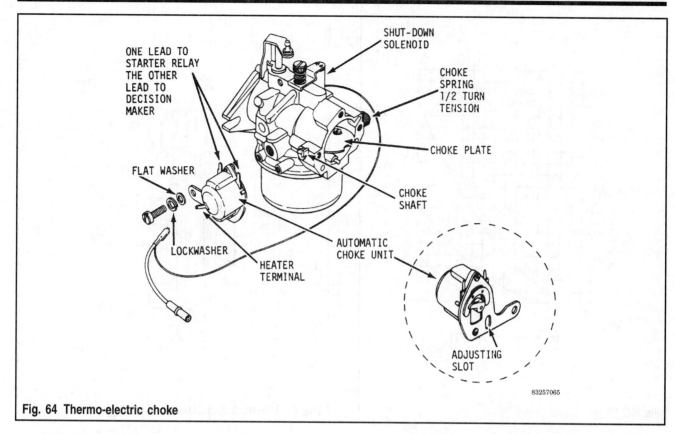

Fig. 64 Thermo-electric choke

4. While holding the choke unit in the above position, tighten the two mounting screws.

➡**With the engine not running and before any cranking, the choke plate will be closed 5-10 degrees at a temperature of about 75 degrees F. As the temperature decreases the choke plate will close even more.**

5. Check choke function by removing the spark plug lead and cranking the engine. The choke plate should close a minimum of 45 degrees at temperatures above 75 degrees F. The plate will close more at lower temperature.

➡**During cranking, the choke will remain closed only 5 to 10 seconds, as choke closing time is controlled by the Decision Maker.**

CHOKE SHAFT SPRING ADJUSTMENT

To adjust the choke spring, hold the plate in the wide open position. Windup the spring 1/2 turn and then place the straight end of the spring through the hole in the shaft.

LP Fuel System

▶ **See Figure 65**

The main components of the LP fuel system as used on K-series engines are:
- Fuel tank
- Primary regulator
- Secondary regulator
- Carburetor

In some applications, the primary and secondary regulators are combined in one two-stage unit. The gas carburetor and secondary regulator (or two stage regulator) are normally furnished with the engine. Other components are furnished with the engine. Other components are furnished by the fuel supplier.

There are some isolated instances in which the equipment manufacturer supplies the entire fuel system for operation with gas. Information on servicing these systems must be obtained from the equipment manufacturer. Depending on the air temperature and the mixture of gasses in the tank, pressure at the outlet of the tank can be as high as 180 to 200 psi.

Secondary Regulator

▶ **See Figures 66, 67 and 68**

The secondary regulators used on Kohler engines are compact single diaphragm types. This type regulator accurately regulates the flow of gas to the carburetor and shuts the gas off automatically when the demand for gas ceases. If the regulator fails, it must be replaced or reconditioned by an authorized gas equipment repair shop. Do not attempt to repair a faulty regulator.

Secondary regulators used on Kohler engines require only one adjustment. This adjustment should be performed while the engine is running.

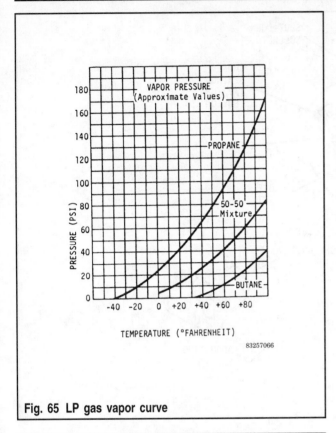

Fig. 65 LP gas vapor curve

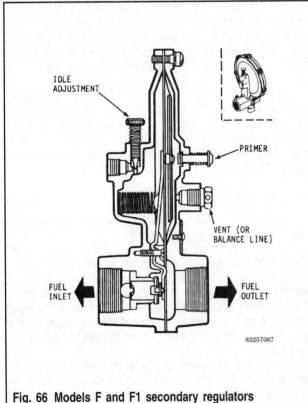

Fig. 66 Models F and F1 secondary regulators

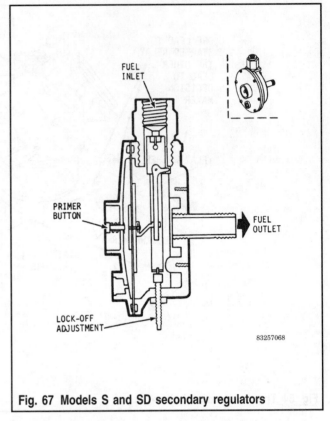

Fig. 67 Models S and SD secondary regulators

Garretson Model S, SD and KN regulators have a lockoff or fuel control adjustment. Use the following procedure to make this adjustment.

➡**The regulator should be mounted as close to vertical as possible, and adjusted in the position in which it will be mounted on the engine.**

1. Connect regulator inlet to a source of clean compressed air, not over 10 psi. Do not connect to a gas supply.
2. Turn air supply on.
3. If the regulator being adjusted is a Model KN, open the lock off adjusting screw until air just starts flowing through the regulator.
4. Turn the lock off adjusting screw in slowly until air flow stops.

➡**A soap bubble test is a good way to check for complete shut-off. If bubbles indicate that air is still flowing, turn the screw in one more full turn.**

The lock off adjusting screw may be used to adjust fuel flow while the engine is idling. Never adjust at any speed above idle.

5. If the regulator being adjusted is a Model S or SD, depress the primer button for an instant. This will allow air to flow through the regulator.
6. Check the air flow stops when the primer is released.
7. If air flow does not stop completely, loosen the adjustment screw lock nut and turn the adjustment screw in until air flow stops, then one more full turn.
8. Repeat steps 5 through 7 until air flow stops every time.
9. Tighten adjustment screw lock nut.

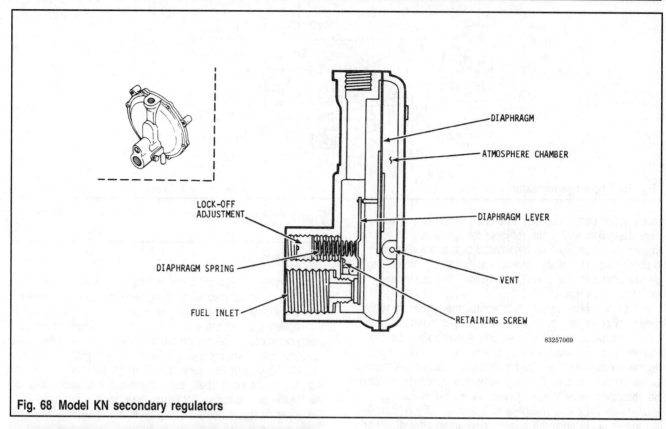

Fig. 68 Model KN secondary regulators

PRIMARY REGULATOR

▶ **See Figure 69**

The primary regulator provides initial control of the fuel under pressure as it comes from the fuel supply tank. The inlet pressure for primary regulators should never exceed 250 psi. The primary regulator is adjusted for outlet pressure of approximately 6 ounces per square inch (11 in. W.C.). If the regulator does not function properly, replace it or have it serviced by an authorized gas equipment shop. Never attempt to service a faulty primary regulator.

Upon demand for fuel, pressure drops on the outside of the regulator diaphragm. The gas inlet valve then begins to open, allowing fuel to pass through the regulator to the secondary regulator. As the need for more fuel increases, the fuel inlet valve opens further, allowing more fuel to pass. Pressure may be adjusted by removing the bonnet cap and turning the spring tension adjustment with a large screwdriver. Turning clockwise increases the pressure; turning counterclockwise decreases it.

TWO STAGE REGULATOR

▶ **See Figure 70**

The two stage regulator used on Kohler engines is a double diaphragm type regulator designed for use with air cooled engines. It combines primary and secondary regulation in one unit. The regulator fuel inlet is connected to the fuel tank. Its outlet is connected to the carburetor. If the regulator fails to operate properly, replace it or have it serviced by an authorized gas equipment shop. Never attempt to service a faulty regulator.

Vaporized fuel is admitted to the regulator at fuel tank pressure (up to 250 psi). Because the secondary valve is closed (engine not running), the pressure on the internal side of the primary diaphragm builds until the pressure overcomes the

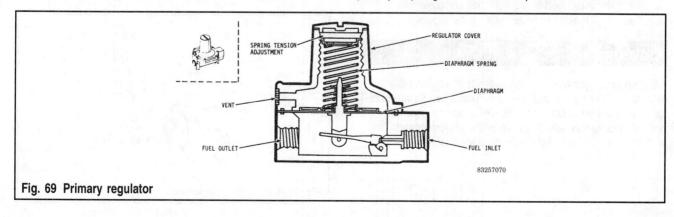

Fig. 69 Primary regulator

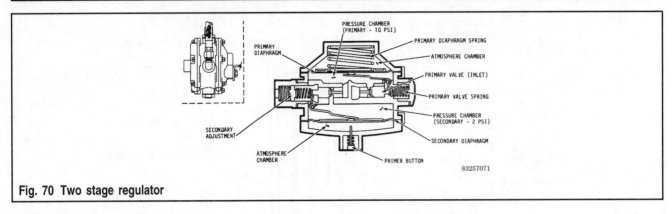

Fig. 70 Two stage regulator

spring action on the opposite side of the diaphragm. This primary diaphragm spring has sufficient tension to require approximately 10 psi pressure on the internal side of the diaphragm to counteract the opening force due to the spring. When the pressure reaches this level, the valve is closed, preventing further pressure rise.

The secondary diaphragm acts against the secondary valve spring. Its action results from vacuum caused by the carburetor. As the vacuum begins acting on the diaphragm, the diaphragm is moved nearer to the center of the regulator, opening the secondary valve until equilibrium is reached. As more fuel is needed, vacuum from the carburetor increases, causing the secondary valve to open further. When fuel is flowing, pressure on the primary diaphragm is lowered slightly, permitting the spring to open the primary valve in an attempt to bring the pressure back to 10 psi.

ADJUSTMENT

1. Turn the secondary adjustment counter clockwise as far as it will go. Then turn it clockwise 3 turns.

2. Connect a source of clean compressed air of at least 25 psi to the regulator inlet and depress primer button 3 times.

3. Connect a 0 to 15 psi pressure gauge to the fuel outlet and press and hold the primer button. The pressure gauge should read approximately 2 psi and hold steady at this reading. If pressure rises slowly, the primary valve is leaking and the regulator must be replaced. If pressure remains constant, proceed.

4. Remove pressure gauge and cover outlet with a film of soap solution. If a bubble forms, the secondary valve is leading.

5. Slowly turn the secondary adjustment to the left until the bubble expands, then to the right one complete turn to stop the leak. If leaking persists, replace the regulator.

Governor

Engine speed governors in the K-series of engine (with the exception of the K91) are of the centrifugal flyweight mechanical type. The K91 utilizes a flyball. The governor gear and flyweight mechanism are contained within the crankcase. The governor gear is driven by a gear on the camshaft.

OPERATION

▶ **See Figure 71**

In operation, centrifugal force causes the flyweights (or flyball) to move outward with an increase in speed and inward with a decrease. As the flyweights move outward, they force the regulating pin of the assembly to move outward. The regulating pin contacts the tab on the cross shaft, causing the shaft to rotate with changing speed. One end of the cross protrudes through the side of the crankcase. Through external linkage attached to the cross shaft, the rotating action is transmitted to the throttle on the carburetor.When the engine is not running, the governor spring holds the throttle in the open position. When a normal load is applied to an operating engine, the speed tends to decrease. The resulting rotation of the cross shaft acts against the governor spring, opening the throttle wider. This action admits more fuel, restoring engine speed. As speed again reaches the governed setting, the shaft rotates to close the throttle valve enough to maintain governed speed.Governed speed may be at a fixed point as constant speed applications or variable as determined by a throttle control setting.

INITIAL ADJUSTMENT K91, K141, K161, K181

▶ **See Figure 72**

Governors are adjusted at the factory. Further adjustment should not be necessary unless the governor arm or linkage work loose or become disconnected. The need for governor adjustment may be indicated by engine speed surges or hunting with changes in load or by a considerable drop in engine

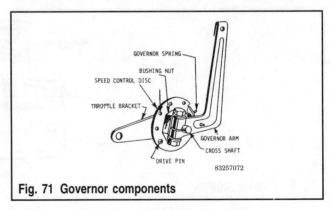

Fig. 71 Governor components

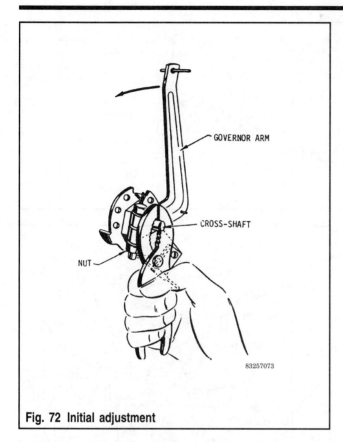

GOVERNOR ARM

CROSS-SHAFT

NUT

83257073

Fig. 72 Initial adjustment

speed when a normal speed is applied. The internal governor mechanism is different on the K241 through K341 models. Be sure to follow the adjustment procedure for the model engine being serviced.

1. Loosen, but do not remove, the nut that holds the governor arm to the governor cross shaft.
2. Grasp the end of the cross shaft with pliers and turn counterclockwise as far as possible. The tab on the cross shaft will touch the rod on the governor gear assembly.
3. Pull the governor arm away from the carburetor as far as it will go, then tighten nut holding governor arm to cross shaft.

THROTTLE WIRE INSTALLATION

▶ **See Figure 73**

In those applications where a throttle body is to be connected to the engine, connect it as follows.

1. Bend the end of the throttle wire.
2. Place throttle control in open position. Insert throttle wire in speed control disc hole nearest the throttle bracket.
3. Install throttle cable clamp and bolt it to the throttle bracket.
4. Remove drive pin from speed control disc and operate the throttle control, rotating the disc from idle to full speed.

SPEED ADJUSTMENT

▶ **See Figure 74**

❊❊CAUTION

CAUTION: The maximum allowable speed for Model K91 is 4000 rpm. Model K161 and K181 are restricted to 3600 rpm maximum. Never tamper with the governor setting to increase engine speed above these limits. Severe personal injury and damage to the engine or equipment can result if the engine is operated at speeds above these maximums.

After making an initial adjustment or connecting a throttle wire, set speed adjustment as follows.

1. Start the engine and allow a few minutes for warmup.
2. Open the throttle to full speed and check engine speed with a tachometer. Speed should be approximately 4000 rpm for Model K91 and 3600 rpm for Model K161 or K181.
3. If speed is not as required, slightly loosen the bushing nut at the speed control disc.
4. Move the throttle bracket counterclockwise to increase engine speed or clockwise to decrease speed.
5. When proper speed is set, tighten the bushing nut.

➡**Do not use excessive force in tightening the bushing nut. Excessive force could cause binding or stripping of threads.**

HIGH SPEED ADJUSTMENT

▶ **See Figures 75, 76 and 77**

The maximum allowable speed is 3600 rpm, no load. The actual high speed setting depends on the application. Refer to the equipment manufacturer's instructions for specific high speed settings. Check the operating speed with a tachometer; do not exceed the maximum. To adjust high speed stop:

1. Loosen the lock nut on high speed adjustment screw.
2. Turn the adjusting screw in or out until desired speed is reached. Tighten the lock nut.

Recheck the speed with the tachometer; readjust if necessary.

SENSITIVITY ADJUSTMENT

Governor sensitivity is adjusted by repositioning the governor spring in the holes in governor arm. If set too sensitive, speed surging will occur with a change in load. If a big drop in speed occurs when normal load is applied, the governor should be set for greater sensitivity.

The standard spring position is in the third hole from the cross shaft. The position can vary, depending on the engine application. Therefore, make a note of (or mark) the spring position before removing it from the governor arm.

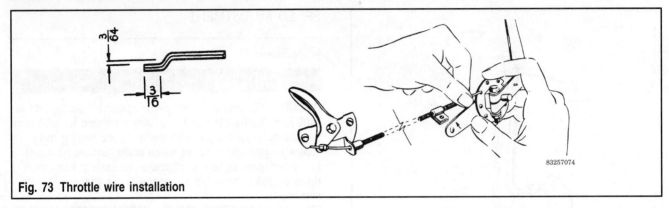

Fig. 73 Throttle wire installation

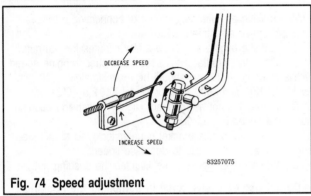

Fig. 74 Speed adjustment

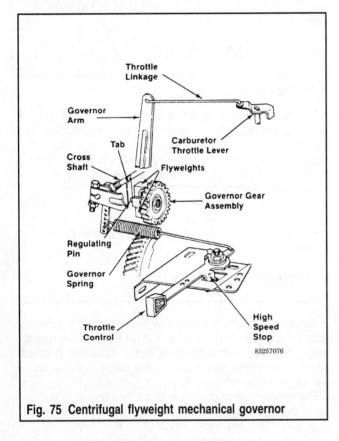

Fig. 75 Centrifugal flyweight mechanical governor

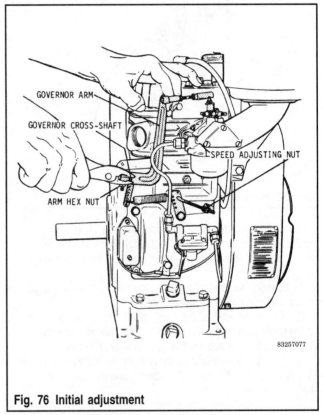

Fig. 76 Initial adjustment

Model K91 w/Fairbanks-Morse Starters

▶ See Figure 78

Retractable starters are lubricated during manufacturer and should require no further lubrication until disassembly for cord or rewind spring replacement or for other repair.

Frequently check mounting screws to make sure starter is securely tightened on blower housing of engine. If screws are loose, starter realignment may be necessary. Also make sure that the air intake screen is maintained in clean condition at all times.

Starters have die cast aluminum housings. A friction shoe assembly under spring tension is used and engages in the drive cup when the starter hand is pulled. The drive cup is held in place on the engine with flywheel nut. A pin on the cup is engaged in crankshaft keyway to prevent slippage of the drive cup.

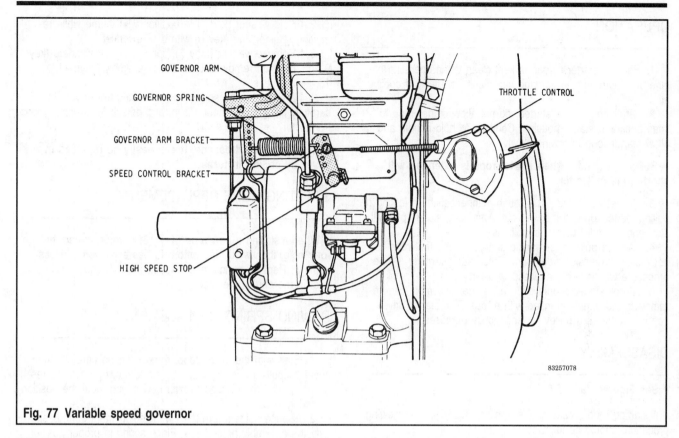

Fig. 77 Variable speed governor

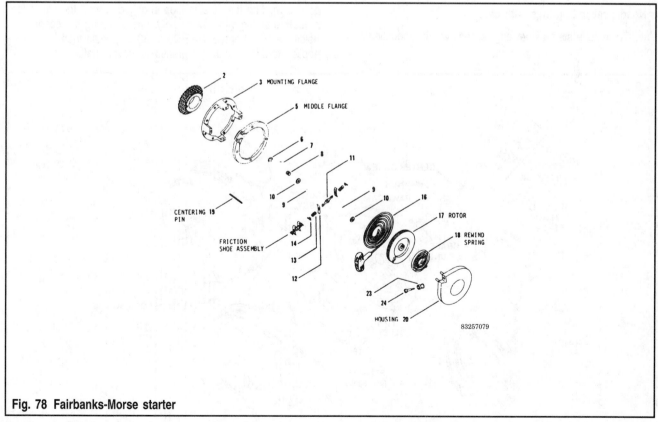

Fig. 78 Fairbanks-Morse starter

OPERATION

1. Be sure starter screen is kept clean when operating engine or serious engine damage can result from lack of cooling air.

2. After engine has started, do not allow starter rope to snap back into starter housing. Continue to hold handle and allow starter rope to rewind slowly.

➡**Releasing handle when starter rope is extended will shorten life of starter.**

3. Do not use starter in a rough manner, such as jerking or pulling starter rope all the way out. A smooth, steady pull will start engine under normal conditions.

4. Always pull starter handle straight out so that rope will not receive excessive wear from friction against guide. Proper procedure will prevent unnecessary wear.

5. If coil starter should ever fail, starter assembly can be removed and engine cranked with a rope. The starter drive cup will serve as a pulley for emergency purposes.

DISASSEMBLY

▶ **See Figures 79 and 80**

If starting rope breaks or if starting spring fails, the following procedure should be followed.

➡**Handle rewind springs with caution.**

1. To remove starter from engine, remove four mounting bolts.

2. Hold washer (key 7) in position with thumb while removing retainer ring (key 6) with a screwdriver.

3. Remove washer ((key 7), spring (key 8), washers (keys 9 and 10) then remove friction shoe assembly (keys 11, 12, 13, and 14).

4. Prevent rewind spring from escaping from cover by carefully lifting rotor about 1/2 in. and detach inside spring loop from rotor.

➡**If spring should escape, it can easily be replaced in cover by coiling in turns.**

STARTING ROPE REPLACEMENT

When installing a new rope (Key 16) in rotor, thread through rotor hole, then wind rope onto rotor, as explained in 'Reassembly'. Replace handle and washer, if used, and tie a double knot in the end of the rope.

REWIND SPRING REPLACEMENT

1. Start with the inside loop, remove spring carefully from cover by pulling out one loop at a time, holding back the rest of the turns. When replacing with new spring, note the position of spring loop.

2. Spring holders furnished with replacement springs simplify the assembly procedures. Place spring in proper position as shown, with the outside loop engaged around the pin. Then press the spring into cover cavity thus releasing the spring holder. A few drops of SAE 20 or 30 oil should then be applied to spring and light grease on cover shaft.

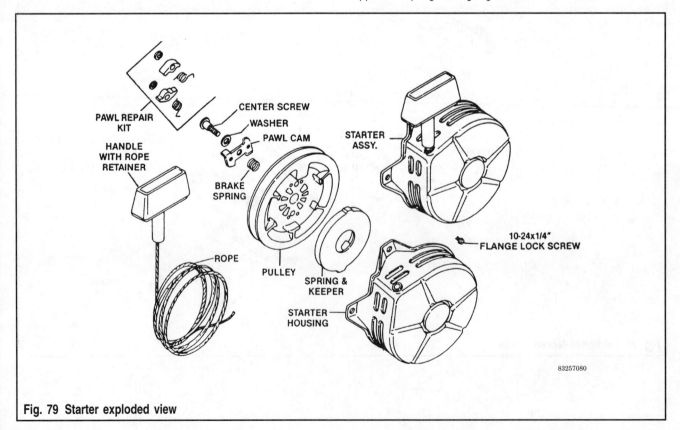

PAWL REPAIR KIT
CENTER SCREW
WASHER
PAWL CAM
STARTER ASSY.
HANDLE WITH ROPE RETAINER
BRAKE SPRING
ROPE
PULLEY
SPRING & KEEPER
STARTER HOUSING
10-24x1/4" FLANGE LOCK SCREW
83257080

Fig. 79 Starter exploded view

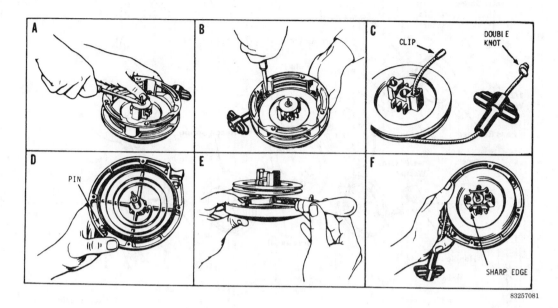

Fig. 80 Starter disassembly

REASSEMBLY

1. Replace washers (keys 9 and 10), friction shoe assembly, washers (keys 9 and 10), spring (key 8), washer (key 7), and retaining ring (key 6).
2. Starter rope is now completely wound on rotor in the direction shown.
3. The starter will be damaged if not centered properly. To ensure the proper centering of the starter, pull out the centering pin (Key 19) about 1/8 in.. Place the starter on the four screws, make sure the centering pin engages the center hole in the crankshaft and press into position. Hold the starter with one hand and place the lock washers and nuts on the screws and tighten securely.

INSTALLING STARTER

1. To align the starter, place it on the blower housing in the desired position, with the centering pin engaged in the center hole of the crankshaft. (If the centering pin is too short to reach the crankshaft, use a pair of pliers and pull the pin out to the correct length).
2. Press the starter assembly in this centered position and securely tighten the four screws with lock washers and flat washers.
3. Hold the starter assembly in this centered position and securely tighten the four screws.

Stamped Housing Models

▶ See Figure 81

✳✳CAUTION

CAUTION: Retractable starters contain a powerful wire recoil spring that is under tension. Do not remove the center screw from the starter until the tension is released. Removing the center screw before releasing spring tension, or improper starter disassembly, can cause the sudden and potentially dangerous release of the spring.

Always wear safety goggles when servicing retractable starters - full face protection is recommended. To ensure personal safety and proper starter assembly, the following procedures must be followed carefully.

REMOVAL

Remove the five screws securing the starter assembly to blower housing.

INSTALLATION

1. Install starter to blower housing using the five mounting screws. Leave screws slightly loose.
2. Pull the handle out approximately 8 in. to 10 in. until the pawls engage in the drive cup. Hold the handle in this position and tighten the screws securely.

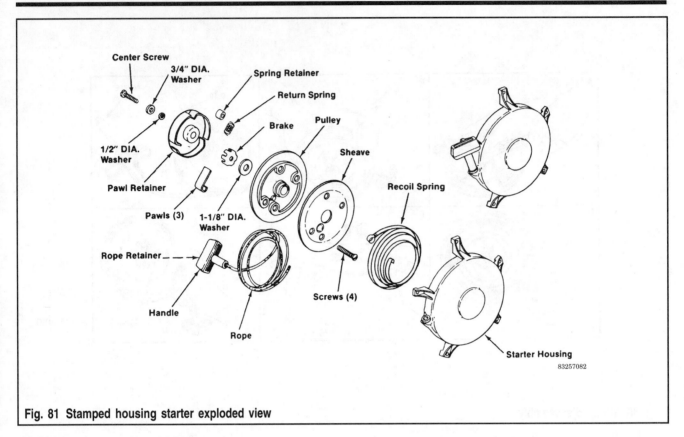

Fig. 81 Stamped housing starter exploded view

STARTER PAWLS REPLACEMENT

A pawl repair kit, No. 4175702 is available. This kit includes two starter pawls, two pawl springs, two retaining rings, and installation instructions.

1. Remove starter from engine.

✳✳CAUTION

CAUTION: Do not remove the center screw of the starter when replacing pawls. Removal of the center screw can cause the sudden and potentially dangerous release of the recoil spring. It is not necessary to remove the center screw when making this repair.

2. Carefully note position of the pawls, pawl springs, and retaining ring before disassembly. (Components must be assembled correctly for proper operation).

3. Remove the retaining rings, pawls, and pawl springs from pawl pins on pulley.

4. Clean pins and lubricate with any commercially available bearing grease.

5. Install new pawl springs, pawls, and retaining rings. When properly installed, the pawl springs will hold the pawls against the pawl cam.

➡**Make sure the snap rings are securely seated in grooves of pawl pins. Failure to seat the snap ring can cause pawls to dislodge during operation.**

6. Pull rope to make sure pawls operate properly.

7. Install starter to engine as instructed under 'To Install Starter'.

ROPE REPLACEMENT

The rope can be replaced without complete starter disassembly.

1. Remove the starter from engine.

2. Pull the rope out approximately 12 in. and tie a temporary (slip) knot in to keep it from retracting into starter.

3. Remove the rope retainer from inside handle. Untie the knot and remove the retainer and handle.

4. Hold the pulley firmly with thumb and untie the slip knot. Allow the pulley to rotate slowly as the spring tension is released.

5. When all spring tension on the starter pulley is released, remove old rope from pulley.

6. Tie a single knot in one end of new rope. 9.
Rotate the pulley counterclockwise (when viewed from pawl side of pulley) until the spring is tight. (Approx. 6 full turns of pulley).

7. Rotate the pulley clockwise until the rope pocket is aligned with the rope guide bushing of housing.

➡**Do not allow pulley/spring to unwind. Enlist the aid of a helper if necessary, or use a c-clamp to hold the pulley in position.**

8. Insert the new rope into the rope pocket of pulley and through rope guide bushing in housing.

9. Tie a slip knot approximately 12 in. from the free end of rope. Hold pulley firmly with thumb and allow pulley to rotate slowly until the temporary knot reached the rope guide bushing in housing.

10. Slip the handle and rope retainer onto rope. Tie a single knot at the end of rope and install rope retainer into handle.

11. Untie the slip knot in rope and pull the handle out until the rope is fully extended. Slowly retract the rope into the starter. If the spring has been properly tensioned, the rope will fully retract until the handle hits the housing.

DISASSEMBLY

1. Remove starter from engine.

❄❄CAUTION

CAUTION: Do not remove the center screw of that starter until the tension of recoil spring has been released. Removing the center screw before releasing spring tension, or improper starter disassembly can cause the sudden and potentially dangerous release of the recoil spring. Follow these instructions carefully to ensure personal safety and proper starter disassembly. Make sure adequate face protection is worn by all persons in the area.

2. Pull the rope out approximately 12 in. and tie a temporary (slip) knot in it to keep it from retracting into starter.

3. Remove the rope retainer from inside handle and untie the knot to remove retainer and handle.

4. Hold the pulley firmly with thumb and untie the slip knot. Allow pulley to rotate slowly as the spring tension is released.

5. When all spring tension on the starter pulley has been released, remove the rope from the pulley.

6. Remove the center screw, washer, pawl cam, and brake spring.

7. Rotate the pulley clockwise 2 full turns. This will ensure the pulley is disengaged from the spring.

8. Hold the pulley into starter housing and invert starter so the pulley is away from the your face, and away from others in the area.

9. Rotate the pulley slightly from side to side and carefully separate the pulley from the starter housing.

➡**If the pulley and housing do not separate easily, the spring could be engaged with the pulley, or there is still tension on the spring. Return the pulley to the housing and repeat step 7 before separating the pulley and housing.**

10. Note the position of the spring and keeper assembly on the pulley. (The spring and keeper assembly must be correctly positioned on pulley for proper operation). Remove the spring and keeper assembly from the pulley as a package.

❄❄CAUTION

CAUTION: Do not remove the spring from the keeper. Severe personal injury could result from sudden uncoiling of the spring.

11. Remove the rope from pulley. If necessary, remove the starter pawl components from pulley as instructed under 'To Replace Starter Pawls'.

INSPECTION AND SERVICE

1. Carefully inspect rope, starter paws, housing, center screw, and other components for wear or damage.

2. Replace all worn or damaged components.

3. Do not attempt to rewind a spring that has come out of the keeper. Order and install a new spring and keeper assembly.

4. Clean away all old grease and dirt from starter components. Generously lubricate the spring and center shaft of starter housing with any commercially available bearing grease.

REASSEMBLY

1. Make sure spring is well lubricated with grease. Position the spring and keeper assembly to pulley (side opposite pawls). The outside spring tail must be positioned opposite rope pocket.

2. Install the pulley with spring and keeper assembly into starter housing.

➡**The pulley is in position when the center shaft is extending slightly above the face of the pulley. Do not wind the pulley and recoil spring at this time.**

3. Lubricate the brake spring sparingly with grease. Install the brake spring into the recess in center shaft of starter housing. (Make sure the threads in center shaft remain clean, dry and free of grease or oil).

4. Apply a small amount of Loctite #271 to the threads to center screw. Install the center screw with washer and cam to the center shaft. Torque screw to 65-75 inch lbs.

5. If necessary, install the pawl springs, pawls, and retaining rings to pins on starter pulley. Refer to 'To Replace Starter Pawls'.

6. Tension the spring and install the rope and handle as instructed in steps 5 through 12 under 'To Replace Rope'.

7. Install the starter to engine.

Cast Housing Models

DISASSEMBLY

1. Remove the starter from engine.

❄❄CAUTION

CAUTION: Do not remove the center screw of the starter until the tension of recoil spring has been released. Removing the center screw before releasing spring tension, or improper starter disassembly can cause the sudden and potentially dangerous release of the recoil spring. Follow these instructions carefully to ensure personal safety and proper starter disassembly. Make sure adequate face protection is worn by all persons in the area.

2. Pull the rope out approximately 12 in. and tie a temporary (slip) knot in it to keep it from retracting into starter.

3. Remove the rope retainer from inside handle. Untie the knot and remove the retainer and handle.

4. Rotate the pulley counterclockwise until the notch in pulley is next to the rope guide bushing.

5. Hold the pulley firmly to keep it from turning. Untie the slip knot and pull the rope through the bushing.

6. Place the rope into the notch in pulley. This will keep the rope from interfering with the starter housing leg reinforcements as the pulley is rotated (step 7).

7. Hold the housing and pulley with both hands. Release pressure on the pulley and allow it to rotate slowly as the spring tension is released. Be sure to keep the rope in the notch.

8. Make sure the spring tension is fully released. (The pulley should rotate easily in either direction).

9. When all spring tension on the pulley is released, remove the center screw, 3/4 in. DIA washer, and 1/2 in. DIA. washer.

10. Carefully lift the pawl retainer from pulley.

➡**A small return spring and nylon spring retainer (spacer) are located under the pawl retainer. These parts are fragile and can be easily lost or damaged. If necessary, use a small screwdriver to loosen the spring retainer from the post on pulley. Replace the spring if it is broken, stretched, or shows other signs of damage.**

11. Remove the 1 1/8 in. DIA. thrust washer, brake, return spring, nylon spring retainer, and pawls.

12. Rotate the pulley clockwise 2 full turns. There should be no resistance to this rotation. This will ensure the pulley is disengaged from the recoil spring.

13. Hold the pulley into starter housing and invert starter so the pulley is away from your face and others in the area.

14. Rotate the pulley slightly from side to side and carefully separate the pulley from the starter housing.

➡**Pulley and housing do not separate easily, the spring could be engage with pulley, or there is still tension on the spring. Return the pulley to the housing and repeat step 12 before separating the pulley and housing.**

15. Only if it is necessary for the repair of starter, remove the spring from the starter housing as instructed under 'To Replace Recoil Spring'. Do not remove the spring unless it is absolutely necessary.

INSPECTION AND SERVICE

1. Carefully inspect the rope, starter pawls, housing, center screw, center shaft, spring and other components for wear or damage.

2. Replace all worn or damaged components.

3. Carefully clean all old grease and dirt from starter components. Lubricate the spring, center shaft, and certain other components as specified in these instructions with any commercial available bearing grease.

ROPE REPLACEMENT

The starter must be completely disassembled to replace the rope.

1. Remove the starter from engine.

2. Disassemble starter as instructed in steps 2 through 14 under 'Disassembly'.

3. Remove the 4 Phillips head screws securing the pulley and sheave. Separate the pulley and sheave and remove the old rope.

4. Position the new rope in the notch in the pulley and around the rope lock post.

➡**Use only a genuine Kohler replacement rope which is designed for this starter. Using rope of the incorrect diameter and/or type will not lock properly in the pulley.**

5. Install the sheave to the pulley and install the 4 Phillips head screws. Use care not to strip or cross-thread the threads in pulley.

6. Inspect the pulley to make sure the sheave is securely joined to the pulley. Pull firmly on the rope to make sure it is securely retained in the pulley.

RECOIL SPRING REPLACEMENT

❋❋CAUTION

CAUTION: Do not attempt to pull or pry the recoil spring from the housing. Doing so can cause the sudden and potentially dangerous release of the spring from the housing. Follow these instructions carefully to ensure personal safety and proper spring replacement. Make sure adequate face protection is worn throughout the following procedure.

1. Carefully note the position of the spring in the housing. The new spring must be installed in the proper position - it is possible to install it backwards in the housing.

2. Place the housing on a flat wooden surface with the recoil spring and center shaft down and away from you.

3. Grasp the housing by the top so that your fingers are protected. Do not wrap your fingers around the edge of the housing.

4. Lift the housing and rap it firmly against the wooded surface. Repeat this procedure until the spring is released from the spring pocket in housing.

5. Discard the old spring.

❋❋CAUTION

CAUTION: Do not attempt to rewind or reinstall a spring once it has been removed from the starter housing. Severe personal injury could result from the sudden uncoiling of the spring. Always order and install a new spring which is held in a specially designed 'C-ring' spring retainer.

6. Thoroughly clean the starter housing removing all old grease and dirt.

7. Carefully remove the masking tape surrounding the new spring/C-ring.

8. Position the spring/C-ring to the housing to the spring hook is over the post in the housing. Make sure the spring is coiled on the correct direction.

9. Obtain installer #11791 and Handle #11799.

Hook the spring hook over the post in housing. Make sure the spring/C-ring is centered over the spring pocket in housing. Drive the spring out of the c-ring and into the spring pocket using the seal installer and hand

10. Make sure all of the spring coils are bottomed against ribs in spring pocket. Use the seal installer and handle to bottom the coils, as necessary.

11. Lubricate the spring moderately with wheel bearing grease before reassembling the starter.

ASSEMBLY

1. Install the recoil spring into the starter housing as instructed under Replace Recoil Spring.

2. Sparingly lubricate the center shaft of starter with wheel bearing grease.

3. Make sure the rope is in good condition. If necessary, replace the old rope as instructed under 'To Replace Rope'.

➡**Ready the pulley and rope for assembly by unwinding all of the rope from the pulley. Place the rope in the notch in the pulley. This will keep the rope from interfering with the starter leg reinforcements as the pulley is rotated later during reassembly.**

4. Install the pulley onto the center shaft.

➡**If the pulley does not fully seat, it is resting on the inner center spring coil. Rotate the pulley slightly from side to side while exerting slight downward pressure. This should move the inner spring coil out of the way and allow the pulley to drop onto position.**

The pulley is in position when the center shaft is flush with the face of the pulley. Do not wind the pulley and recoil spring at this time.

5. Install the starter pawls into the appropriate pockets in the pulley.

6. Sparingly lubricate the underside of the 1 1/8 in. DIA. washer with grease and install it over the center shaft. Make sure the threads in center shaft remain clean, dry, and free of grease or oil.

7. Sparingly lubricate the inside of the 'legs' of the brake spider with grease. Install the brake to the retainer.

8. Install the small return ring to the pawl retainer. Make sure it is positioned properly.

9. Position the pawl retainer and return spring next to the small post on pulley. Install the free loop of the return spring over the post. Install the free loop of the return spring over the post.

10. Invert the pawl retainer over the pawls and center hub of pulley. Take great care not to damage or unhook the return spring. Make sure the pawls are positioned in the slots of pawl retainer.

11. As a test, rotate the pawl retainer slightly clockwise. Pressure from the return spring should be felt. In addition, the

pawl retainer should return to its original position when released. If no spring pressure is felt or the retainer does not return, the spring is damaged, unhooked, or improperly assembled. Repeat steps 8, 9, and 10 to correct the problem.

12. Sparingly lubricate the 1/2 in. DIA. washer and 3/4 in. DIA. washer in the center of pawl retainer. Make sure the threads in center shaft remain clean, dry and free of grease or oil.

13. Apply a small amount of Loctite #271 to the threads of center screw. Install the center screw to center shaft. Torque screw to 55-70 inch lbs.

14. Rotate the pulley counterclockwise (when viewed from the pawl side of pulley) until the spring is tight. (Approximately 4 full turns of pulley). Make sure the fully extended rope is held in the notch in pulley to prevent interference with the housing leg reinforcements.

15. Rotate the pulley clockwise until the notch is aligned with the rope guide bushing of housing.

➡**Do not allow the pulley/spring to unwind. Enlist the aid of a helper, or use a c-clamp to hold pulley in position.**

16. Insert the free end of rope through rope guide bushing. Tie a temporary (slip) knot approximately 12 in. from the free end of the rope.

17. Hold the pulley firmly with thumbs and allow the pulley to rotate slowly until the slip knot reaches the rope guide bushing of housing.

18. Slip the handle and rope retainer onto rope. Tie a single knot at the end of rope and install retainer into handle.

19. Untie the slip knot and pull the handle out until the rope is fully extended. Slowly retract the rope into the starter. If the spring has been properly tensioned, the rope will fully retract until the handle hits the housing.

Magneto Ignition System

OPERATION

▶ **See Figures 82 and 83**

In all magneto ignition systems, high strength permanent magnets provide the energy for ignition. In rotor type systems, the magnet is pressed onto the crankshaft and is rotated inside a coil-core assembly (stator) mounted on the breaking plat. In the other systems, a permanent magnet ring on the inside of the flywheel revolves around the stator. Movement of

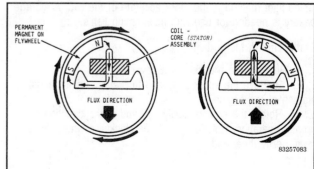

Fig. 82 Magneto cycle showing flux reversal

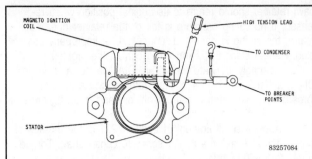

Fig. 83 Typical flywheel magneto ignition coil and starter

the magnets past the stator induces electric current flow in the stator coil (and in alternator and lighting coils if provided). The magnets are mounted with alternate North and South poles so that the direction of magnetic flux constantly changes, producing an alternating current (AC) in the stator coil windings.The stator windings are connected to the magneto ignition coil. Current flow in the ignition coil reaches its highest peak at the instant the magnetic flux reverses direction. This is the point at which the system is timed to provide a spark at the spark plug.

The ignition coil has a low tension primary winding and a high tension secondary winding. The secondary winding has approximately 100 turns of wire for every 1 turn in the primary. This relationship causes the voltage induced in the secondary winding to be about 100 times higher than in the primary. If the magneto produces 250 volts in the primary winding, the secondary winding voltage will be 25,000 volts.

When ignition is required, the breaker points open to break the primary circuit. The resultant sudden collapse of the field around the primary winding causes sufficient energy to be produced in the secondary winding, but the condenser shunts this energy to ground, preventing it from bridging the breaker point gap.

TIMING

Engines are equipped with a timing sight hole either in the bearing plate or in the blower housing. If a snap button covers the hole, pry it out with a screwdriver or similar tool so that the timing marks may be seen. Two marks will be present on the flywheel; T for top dead center, and S or SP for the firing point (20 deg. before top dead center).There are two ways to time a magneto ignition system, static and timing light. The timing light method is the more accurate of the two. A storage battery is needed for use with most timing lights.

Static Timing Method

1. Remove the breaker point cover.
2. Remove the spark plug lead to prevent unintentional starting of the engine.
3. Rotate the engine slowly by hand in the direction of normal operation. Rotation should be clockwise when viewed from the flywheel end.

4. The breaker points should just begin to open when the S or SP mark (Y mark on Model K91) appears in the center of the timing sight hole. Continue rotating the engine until the breaker points are fully opened.
5. Measure the breaker point gap with a feeler gauge. The gap should be 0.020 in..
6. If the gap is not 0.020 in., loosen the gap adjustment screw and adjust the gap.
7. Tighten the gap adjustment screw.
8. Replace the breaker point cover.

Timing Light Timing Method

Several different type of timing lights are available. Follow the manufacturer's directions for use. Perform timing with a timing light as follows.
1. Remove the lead from the spark plug.
2. Wrap one end of a short piece of fine bare wire around the spark plug terminal and replace the lead. The free end of the wire must protrude from beneath the rubber boot on the lead.

➥**The preceding is for timing lights using an alligator clip to connect the spark plug. If the light in use has a sharp prong on the spark plug lead, simply penetrate the rubber boot with the prong and make contact with spark plug lead metal connector.**

3. Connect one timing light lead to the wire wrapped around the spark plug terminal.
4. Connect on timing light lead to the hot (ungrounded) terminal of the battery.
5. Connect the third timing light lead to engine ground.
6. Start the engine and run it a 1200 to 1800 RPM.
7. Aim the timing light at the timing sight hole. The light should flash just as the S or SP mark is centered in the sight hole or is in line with the center mark on the bearing plate or blower housing.
8. If timing is not as specified, carefully remove the breaker point and slightly loosen the gap adjusting screw, shift the breaker point plate until the timing mark is properly positioned, and tighten the screw.
9. Shut off the engine and replace the breaker point cover.

Battery Ignition System

OPERATION

▸ **See Figures 84 and 85**

The battery ignition system operates in a manner similar to the magneto system. The major difference is that, in the battery system, energy is provided by a battery. The battery is maintained at full charge by an engine mounted motor-generator or alternator.

The coil in a battery ignition system is connected as follows:a. The positive (+) terminal is connected to the positive terminal of the battery.b. The negative (-) terminal is connected to the breaker points.c. The high tension (center) terminal is connected to the spark plug.

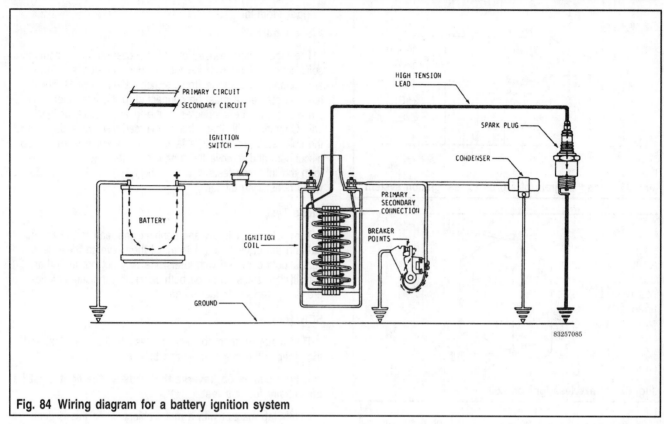

Fig. 84 Wiring diagram for a battery ignition system

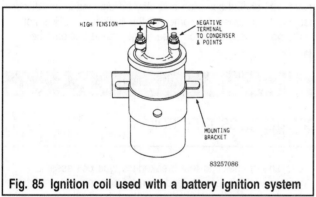

Fig. 85 Ignition coil used with a battery ignition system

TIMING

The timing procedure for the battery ignition system is the same for the magneto system. When using a timing light, refer to the manufacturer's instructions.

➡**The model k341QS Specification 71276A engine is unique in that it is timed slightly differently then other K series engines. These engines operate at lower speed, so the timing is set at 16 degrees before top dead center to improve running smoothness. Instead of having an S or SP at the timing mark on the flywheel, these engines have a 1 above and a 6 below the mark. When timing these engines, the timing mark is centered as with other engine.**

IGNITION SERVICE

Single Cylinder Models

Ignition problems and poor performance on these models are often the result of using an incorrect ignition coil, spark plug, or plug gap setting. When replacing an ignition coil always use the genuine Kohler replacement. Use of the correct spark plug and gap setting is also important. The specified plug is a Champion H10/RH10, or equivalent, gapped at 0.035 in. (0.9mm). Failure to follow these recommendations will result in erratic high speed ignition misfire or cutting-out under load.

Breakerless Ignition System

OPERATION

◆ **See Figures 86 and 87**

The breakerless ignition system operates on the same general principle as the magneto system but does not use breaker points and conventional ignition condenser to time the spark. A trigger module containing solid state electronics performs the same function as the breaker points. The breakerless system consists of four major components:
- Ignition winding on alternator stator
- Trigger module
- Ignition coil assembly
- Flywheel-mounted trigger

The ignition winding is separate from other windings on the alternator stator. If functions like the magneto winding. The

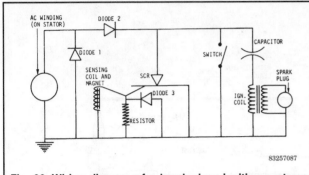

Fig. 86 Wiring diagram of a breakerless ignition system

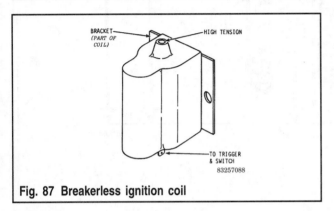

Fig. 87 Breakerless ignition coil

trigger module contains three diodes, a resistor, a sensing coil and magnet and an SCR, a type of electronic switch. The ignition coil assembly includes a capacitor and a pulse transformer that serves the same purpose as the ignition coil in other systems. The flywheel has a projection that triggers ignition.In some applications a 22 ohm, 1/2 watt resistor has been placed between the key switch and the ignition coil. This has been added to prevent current feedback through a dirty or wet switch. This feedback, if not held in check by a resistor, can damage the trigger unit.

TIMING

Because there are no breaker points in this system, there is not requirement for timing. However, there is a requirement for positioning the trigger module for proper relationship with the flywheel projection. The gap between the projection and trigger module is normally set between 0.005 in. and 0.010 in.. This setting is not critical, but selecting a 0.005 in. gap promotes better cold weather starting. Set the gap as follows.

1. Remove the spark plug lead to prevent starting.
2. Rotate the flywheel so that the projection is aligned with the trigger module.
3. Loosen the cap screws on the trigger module bracket and insert a 0.005 in. feeler gauge in the gap.
4. Move the trigger module until it touched the feeler gauge, making sure that the flat surfaces of module and projection are parallel.
5. Tighten the cap screws and replace the spark plug lead.

Trigger Module
▶ **See Figure 88**

The trigger module used on breakerless ignition systems is a solid state device which includes diodes, resistor, sensing coil and magnet plus an electronic switch called an SCR. The terminal marked A must be connected to the alternator while terminal I must be connected to the ignition switch or ignition coil. Operating with these leads reversed will cause damage to the solid state devices. If a faulty trigger module is suspected, disconnect and remove the trigger from the engine and perform the following tests with a flashlight tester. Reset air gap when reinstalling trigger.

Diode Test

Turn tester switch ON and connect one lead to the I terminal and the other to the A terminal then reverse these leads-- light should come on with leads one way but not the other way. If light stays on or off both ways, this indicates diodes are faulty--replace trigger module.

SCR Test

Turn tester on then connect one lead to the I terminal and the other to the trigger mounting bracket.

➡**If light comes on, reverse the leads as the light must be off initially for this test.**

Lightly tap magnet with a metal object--when this is done, tester light should come on and stay on until leads are disconnected. If light does not come on, this indicates SCR is not switching properly in which case trigger module should be replaced.

Ignition Coils

BREAKERLESS TYPE COIL

Use an ohmmeter to test breakerless type coil assembly. (A)--Remove high tension lead from terminal on coil. Insert one ohmmeter lead in coil terminal and the other to the coil mounting bracket. A resistance of about 11,500 ohms should be indicated here. (B)--Connect one tester lead to the coil mounting bracket and the other to the ignition switch wire. Continuity should not be indicated here. Replace ignition coil assembly if wrong results are obtained from either of these tests.

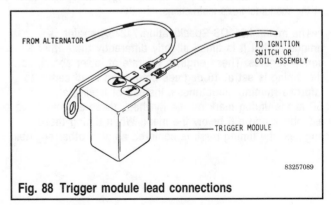

Fig. 88 Trigger module lead connections

MAGNETO AND BATTERY SYSTEM BREAKER POINTS

▶ See Figure 89

Engine operation is greatly affected by breaker point condition and adjustment of the gap. If points are burned or badly oxidized, little or not current will pass. The engine may not operate at all or may miss at high speed. Size of the breaker point gap affects the amount of time the points are open and closed. If the gap is set too wide, they will open too early and close too late. A definite period of time is required for the field to build in the ignition coil. If the points are closed for too long or too short a period, a weak spark will be produced by the coil.

Severe metal buildup on either contact indicates that the condenser is not properly matched to the rest of the system and should be replace.

Spark Plugs

Engine misfire and starting difficulty are often caused by the spark plug's being in poor condition or being improperly gapped. The spark plug should be removed after every 100 hours of operation for a check of its condition. At this time the gap should be reset or the spark plug replaced as necessary.

SERVICE

▶ See Figures 90, 91, 92, 93 and 94

1. Clean the area around the base of the spark plug to keep dirt out of the engine upon removal.
2. Remove the spark plug and check its condition. Replace it if it is badly worn or if re-use is questionable. Clean it if it is re-usable.

➡ Do not clean the spark plug in a machine that uses abrasive grit. Some grit could remain in recesses and enter the engine, causing extensive wear and damage.

3. Check the gap with a wire type feeler gauge. Set the gap as shown in the following table by carefully bending the side electrode.
4. Install the spark plug and torque it to 18 to 20 ft. lbs.

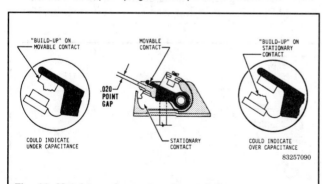

Fig. 89 Metal transfer on breaker points

Fig. 90 Twist and pull on the boot, never on the plug wire

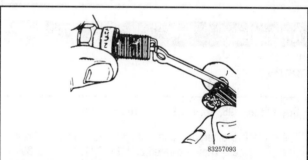

Fig. 91 Plugs that are in good condition can be filed and reused

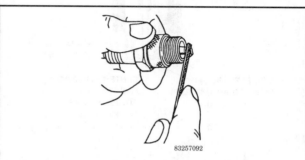

Fig. 92 Adjust the electrode gap by bending the side electrode

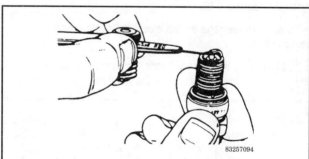

Fig. 93 Always used a wire gauge to check the plug gap

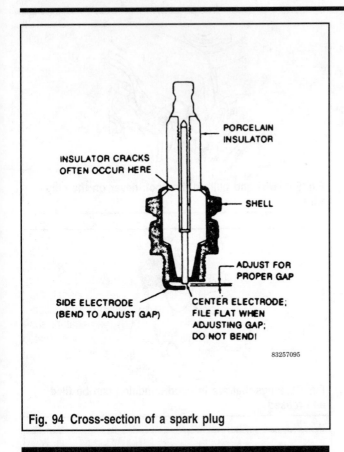

Fig. 94 Cross-section of a spark plug

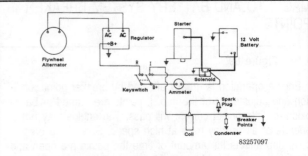

Fig. 96 Wiring diagram for electric start engines with 15 amp or 16 amp regulated battery charging system

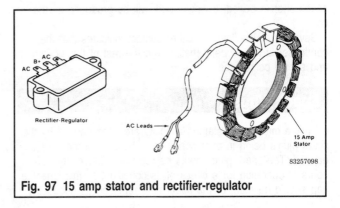

Fig. 97 15 amp stator and rectifier-regulator

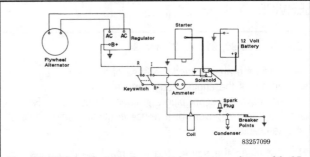

Fig. 98 Wiring diagram for electric start engines with 25 amp regulated battery charging system

Alternator

OPERATION

◆ **See Figures 95, 96, 97, 98, 99, 100 and 101**

There are five different models of alternators used in the K-series of engines. They are rated at 1.25, 3, 10, 15 and 30 amperes. The 1.25 amp system is intended for battery charging only. The 3 amp device is intended for battery charging and lighting. There are no adjustments provided for in these systems. Replace if faulty.

➡**To prevent damage to the electrical system and components:**

(a) Make sure the battery polarity is correct. A negative (-) ground system is used with K series engines. (b) Disconnect the rectifier-regulator leads and/or wiring harness plug if electric (arc) welding is to be done on the equipment powered by the engine. Disconnect any other electrical accessories that share a common ground with the engine. (c) Make sure the

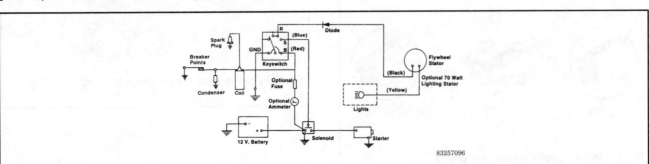

Fig. 95 Wiring diagram for electric start engines with 1.25 amp or 3 amp unregulated battery charging system and 70 watt lighting

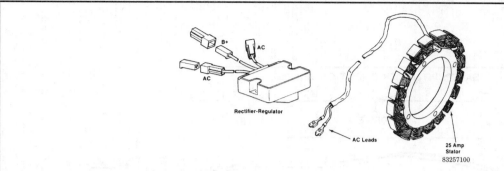

Fig. 99 25 amp stator and rectifier-regulator

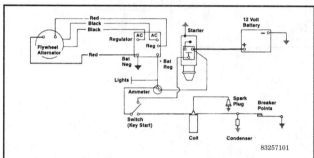

Fig. 100 Wiring diagram for electric start engines with 25 amp regulated battery charging system

stator (AC) leads do not touch. Shorting them together could permanently damage the stator. (d) Do not operate the engine with the battery disconnected.

➡️**If a battery has discharged to less than 4 volts, there may not be sufficient power to activate the rectifier-regulator. If the battery fails to accept a charge from the alternator, charge it on a battery charger and reinstall.**

Electric Starters

There are three types of electric starters used in the K-series of engines. The three types are: Motor-Generator - This starter also functions as a DC generator. In the starting mode, it turns the crankshaft through a V belt arrangement. The V belt transmits turning force from a small pulley on the motor-generator to a large pulley on the crankshaft.

Wound-Field Bendix Drive Starter - In the field-wound starters, electrical current flows through coils to built up a strong

magnetic field to turn the armature. When the armature starts to rotate, a drive pinion moves forward on the armature shaft and meshes with a ring gear on the flywheel. The armature and ring gear remain engaged until the engine starts to run. When the flywheel begins to turn faster than the starter, the pinion is thrown from the ring gear and returns to the disengaged position. A small anti-drift spring on the armature shaft holds the pinion in this position as the starter slows to a stop.

Permanent Magnet Bendix Drive Starter - Operation of this type starter is the same as that of the wound-field starter. The major difference between the two is in the method of generating the magnetic field to turn the armature. This starter uses strong permanent magnets in place of field coils.

SAFETY INTERLOCKS

◆ **See Figures 102 and 103**

In an effort to enhance safe operation of their equipment, many manufacturers install safety interlocks to prevent engine start before certain safety requirements are met. These interlocks are usually incorporated in the starter circuit. Unless all interlock switches are closed, the starter will not function.

Before servicing a starter that has failed, always check the safety interlock system first. This is done by bypassing the interlock switches with a temporary jumper wire.

❋❋CAUTION

CAUTION: Other than interlock testing, never operate an engine with the safety interlock system removed or bypassed. Great bodily harm or equipment damage could result!

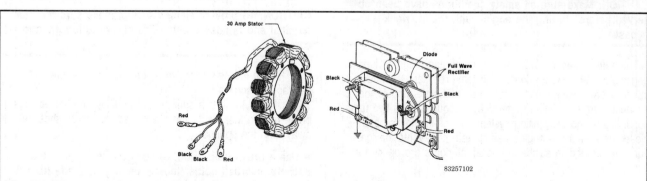

Fig. 101 30 amp stator and rectifier-regulator

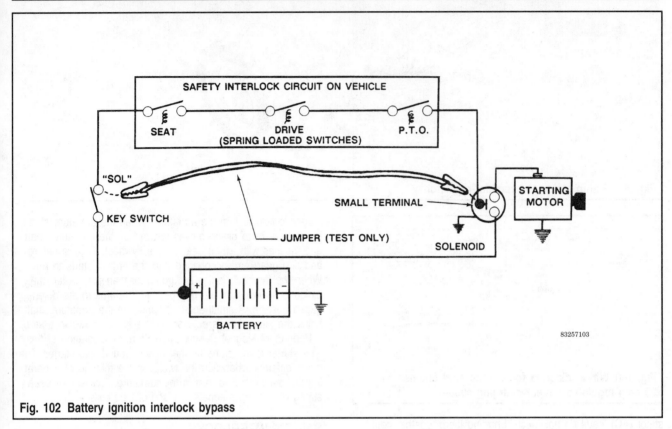

Fig. 102 Battery ignition interlock bypass

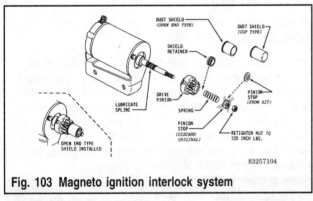

Fig. 103 Magneto ignition interlock system

Interlocks connected to an engine with a battery ignition system are bypassed simply by placing a jumper wire as shown.

✳✳CAUTION

CAUTION: Make sure all safety conditions have been observed before starting the engine with the interlock bypassed.

The safety interlock system on manual start magneto ignition engines is placed in the ignition system. The series connected interlock switches are connected to a solid state module that is connected to the ignition system. The module serves two functions. It grounds the ignition system until all interlocks have closed and, after the engine has started, it prevents the ignition from grounding as the individual interlocks are opened in

normal operation (transmission placed in Drive, PTO engaged, etc.).

Motor-Generator Type Starter

BRUSH REPLACEMENT

1. Remove the brush springs from the pockets in brush holder.
2. Remove the self-tapping screws and negative (-) brushes.
3. Remove the stud terminal with positive (+) brushes, and plastic brush holder from end cap.
4. Reinstall the brush holder and new holder and new stud terminal with positive (+) brushes into end cap. Secure with the fiber washer, plain washer, split lock washer, and hex nut.

✳✳CAUTION

CAUTION: To prevent electric arcing, make sure the stud terminal and braided brush leads do not touch the end cap.

5. Install the new negative (-) brushes and secure with the self-tapping screws.
6. Install the brush springs and brushes into the pockets in brush holder. Make sure the chamfered side of brushes are away from the springs.

➡Use a brush holder tool to keep the brushes in the pockets. A brush holder tool can easily be made from thin sheet metal.

COMMUTATOR SERVICE

Clean the commutator with a coarse, lint free cloth. Do not use emery cloth. If the commutator is badly worn or grooved, turn down on a lathe, or replace the armature.

REASSEMBLY

1. Insert the armature into the starter frame. Make sure the magnets are closer to the drive shaft end of armature. The magnets will hold the armature inside the frame.
2. Install the thrust washer and drive end cap. Make sure the match marks on end cap and frame are aligned.
3. Install the brush holder tool to keep the brushes in the pockets of commutator end cap.
4. Install the commutator end cap to armature and starter frame. Firmly hold the drive end cap and commutator end cap to the starter frame. Remove the brush holder tool.
5. Make sure the match marks on end cap and frame are aligned. Install the through-bolts.
6. Install the drive pinion, dust cover spacer, anti-drifting spring, stop gear spacer, stop nut, and dust cover. Refer to 'Starter Drive Service'.

➡**If the engine being serviced is equipped with special shouldered cap screws and lock washers for mounting, make sure these same parts are used for reinstalling the starter. These special parts ensure alignment of the pinion and ring gear.**

Wound-Field Bendix Drive Starter

▶ **See Figure 104**

❈❈CAUTION

WARNING: In the event of a false start (engine starts but fails to keep running) the engine must be allowed to come to a complete stop before the starter is re-engaged. If the flywheel is still rotating when the starter is engaged, the pinion and ring gear may be damaged.Do not crank the engine for longer than 10 seconds. A 60-second cool-down period must be allowed between starting attempts. Failure to follow this procedure could result in starter burnout.

➡**If the engine being serviced is a Model K161 or K181 and has special shouldered cap screws and lock washers for mounting. Make sure these same parts are used for reinstalling the starter. These special parts ensure alignment of the pinion and ring gear.**

SERVICE

1. Remove the end cap assembly by taking out the two through bolts and carefully slipping the end cap off the armature.
2. Lift the spring holding the positive brush and remove the brush.

3. Carefully remove the armature.
4. Inspect both brushes (positive on frame; negative on end cap). If brushes are worn unevenly or are shorter than 5/16 in., replace them.
5. Remove the negative brush by drilling out the rivet holding it to the end cap. Install the replacement brush and rivet.
6. Remove the positive brush by peeling back insulating material on the field winding and unclipping or unsoldering. Install the replacement brush and clip or solder in place.
7. Use a coarse cloth to clean the commutator. If the commutator is grooved or extremely dirty, use a commutator stone or fine sandpaper.

➡**Never use emery cloth to clean a commutator.**

8. Carefully insert the armature.
9. Lightly coat the end cap bushing and armature shaft with light engine oil.
10. Hold the positive brush spring back and carefully place end in position on armature shaft. Release spring after brushes are contacting commutator.
11. Insert two through bolts and torque to 40 to 55 inch lbs.
12. Inspect pinion and splined shaft. If any damage is noted, replace the Bendix drive.
13. If the Bendix drive is in good condition, wipe everything clean and apply a very thin coat of special silicone grease (Kohler Part No. 52 357 01) to the splined portion of the armature shaft.

Permanent Magnet Bendix Drive Starter

▶ **See Figure 105**

SERVICE

❈❈CAUTION

WARNING: In the event of a false start (engine starts but fails to keep running) the engine must be allowed to come to a complete stop before the starter is re-engaged. If the flywheel is still rotating when the starter is engaged, the pinion and ring gear may be damaged.

Do not crank the engine for longer than 10 seconds. A 60-second cool-down period must be allowed before starting attempts. Failure to follow this procedure could result in starter burnout.

1. Remove the stop nut and the remainder of the Bendix drive.
2. Remove both through bolts.
3. Remove the end bracket capscrew from the end cap.
4. Remove mounting bracket and frame by rotating the end bracket and slipping the mounting bracket and frame off of the drive end of the armature.
5. Separate the end cap from the armature, being careful to restrain the brushes in the end cap.
6. Inspect the commutator. If dirty, clean it with a coarse, lint-free cloth. If grooved, dress it with a commutator stone or turn it down on a lathe and undercut the mica.

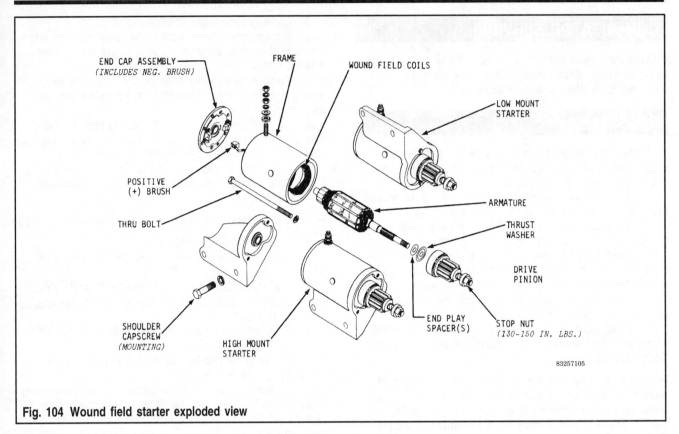

Fig. 104 Wound field starter exploded view

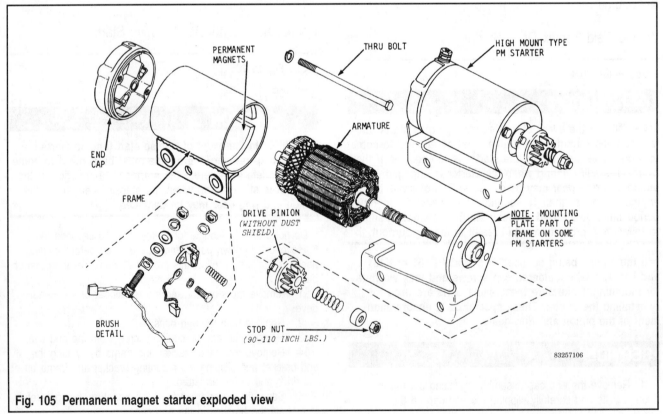

Fig. 105 Permanent magnet starter exploded view

TROUBLESHOOTING

Starter failures from overcranking or cranking with an abnormal parasitic load on the engine, will display one or a number of the following signs:

1. The armature wire insulation or coating will appear discolored and may be swollen. In many cases, you may be able to detect an odor from the burnt wire coating or see it oozing from the starter hosing.

2. One or a number of the armature windings may have wires or wire connections that have burnt in tow. Wires may have insulation missing or be partially fused together.

3. The starter brushes will show heavy surface galling and brush material transfer. Additionally, in many instances the starter brushes will be welded or stuck in the brush holders.Some of the frequent causes of abnormal parasitic load at cranking are:

- Improper viscosity engine crankcase oil.
- Incorrect fluid in a direct coupled hydrostatic unit-remember, even in the idle or neutral position, a direct coupled hydrostatic pump will place a parasitic load on the engine at cranking.
- Malfunctioning or inoperative direct coupled clutch assembly.
- Engaged accessory or drive clutch assembly.
- Overcranking - cranking the starter continuously for more than the recommended period and/or not allowing a sufficient cool down period between starting attempts.
- Parasitic Load at cranking - a load or force on the engine at cranking that opposes normal engine rotation.

Battery

BATTERY TEST

If the battery doe not have enough charge to crank the engine, recharge it.

➡ **Do not attempt to jump-start the engine with another battery. Starting with a battery larger than recommended can burn out the starter motor.**

The battery is tested by connecting a DC voltmeter across the battery terminals and cranking the engine. If the battery voltage drops below 9 while cranking, the battery is in need of a charge or replacement.

BATTERY CHARGING

✳✳CAUTION

CAUTION: Batteries contain sulfuric acid. To prevent acid burns, avoid contact with skin, eyes and clothing.

Batteries produce explosive hydrogen gas while being charged. Charge the battery only in a well ventilated area. Keep cigarettes, sparks, open flame and other sources of ignition away from the battery at all times.To prevent accidental shorting and the resultant sparks, remove all jewelry before servicing the battery.

When disconnecting battery cables, always disconnect the negative (-) cable first. When connecting battery cables, always connect the negative cable last.Before disconnecting the negative (-) cable, make sure all switches are OFF. If any switch is ON, a spark will occur at the ground terminal. This could result in an explosion if hydrogen gas or gasoline vapors are present. Keep batteries and acid out of the reach of children.

BATTERY MAINTENANCE

Regular maintenance will ensure that the battery will accept and hold a charge.

✳✳CAUTION

CAUTION: Always turn the ignition switch OFF or disconnect the battery cables before charging the battery. Failure to do this could result in overheating and explosion of the ignition coil.

1. Check the level of the electrolyte regularly. Add distilled water to maintain it at its recommended level.

➡ **Do not overfill the battery. Poor performance or early failure will result.**

2. Keep the cables, terminals and external battery surfaces clean. A buildup of corrosive acid or dirt on the surfaces can cause the battery to self-discharge. Wash the cables, terminals and external surfaces with a baking soda and water solution. Rinse thoroughly with clean water.

➡ **Do not allow the baking soda solution to enter the battery cells. The solution will chemically destroy the electrolyte.**

Automatic Compression Release

All K-series cylinder engines, except the K91, are equipped with Automatic Compression Release (ACR). The mechanism lowers compression at cranking speeds to make starting easier.

OPERATION

◆ **See Figures 106 and 107**

The ACR mechanism consists of two flyweights and a spring attached to the gear on camshaft. When the engine is rotating at low cranking speeds (600 RPM or lower) the flyweights are held by the spring in the position shown. In this position, the tab on the larger flyweight protrudes above the exhaust cam lobe. This lifts the exhaust valve off its seat during the first part of the compression stroke. The reduced compression results in an effective compression ratio of about 2:1 during cranking.

After the engine speed increases to about 600 RPM, centrifugal force moves the flyweights to the position shown. In this position the tab on the larger flyweight drops into the recess in the exhaust cam lobe. When in the recess, the tab

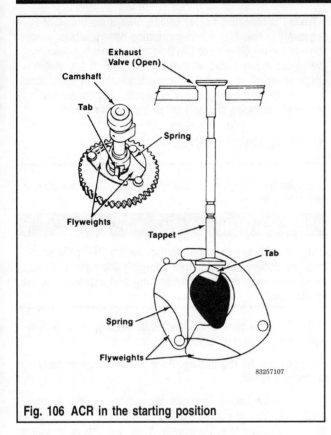

Fig. 106 ACR in the starting position

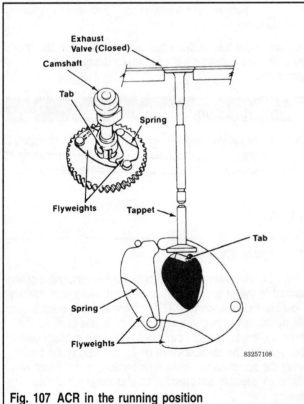

Fig. 107 ACR in the running position

has no effect on the exhaust valve and the engine operates at full compression and full power.

When the engine is stopped, the spring returns the fly-weights to the position shown, ready for the next start.

INSPECTION AND SERVICE

The ACR mechanism is extremely rugged and virtually trouble-free. If hard starting is experienced, check the exhaust valve for lift as follows:

1. Check exhaust valve to tappet clearance and adjust as necessary to specification.
2. Remove cylinder head and turn the crankshaft clockwise by hand and observe the exhaust valve carefully. When the piston is approximately 2/3 of the way up the cylinder during the compression stroke, the exhaust valve should lift off the seat slightly.

➡️**If the exhaust valve does not lift, the ACR spring may be unhooked or broken. To service the spring, remove the oil pan and rehook spring or replace it. The camshaft does not have to be removed.**

➡️**The flyweights are not serviceable. If they are stuck or worn excessively, the camshaft must be replaced.**

➡️**The tab on the flyweights is hardened and is not adjustable. Do not attempt to bend the tab - it will break and a new camshaft will be required.**

COMPRESSION TESTING

Because of the ACR mechanism, it is difficult to obtain an accurate compression reading. To check the condition of the combustion chamber, and related mechanisms, physical inspection and a crankcase vacuum test are recommended.

AUTOMATIC COMPRESSION RELEASE (ACR) CHANGES

New ACR Tests

Engines with serial No. 9006118 and after have hardened and ground steel ACR tabs on the camshaft assemblies. These new assemblies are manufactured with improved techniques, which permanently set the ACR mechanism, making adjustments to the mechanism unnecessary and impossible.

➡️**Do not attempt to bend these hardened steel ACR tabs. These tabs will break if bent.**

Procedures For Checking And Adjusting ACR On Engines Prior To Serial No. 9006118

On engines manufactured before serial No. 9006118 the ACR can still be checked and reset using the procedure described below.

ACR is set according to the amount of valve lift on the exhaust valve. The correct amount of lift is established by the height of the lifting tab in relation to the camshaft. If improper lift is suspected, the setting can be checked and adjusted as follows:

1. Check valve tappet clearances and adjust as necessary to specification.

2. Remove cylinder head and turn the engine over by hand until you reach BDC of the intake stroke (intake valve will be closing).

3. Mount a dial indicator on the top of the exhaust valve and set a 0.

4. Slowly turn the flywheel clockwise and watch the dial indicator. When the piston is about 2/3 of the way up the cylinder, the exhaust valve should open for ACR. Exhaust valve opening as indicated on the dial indicator should be 0.031-0.042 in.

If the exhaust valve does not open to the specified amount, adjust the ACR according to Step 5.

➡**Caution must be exercised in the bending of the tab as it is hardened and may crack or break if bent back and forth more than 3 or 4 times.**

5. If the valve list was above 0.042 in., hold a wooden dowel or peg on the top of the valve and tap it down carefully to within the 0.031-0.042 in. range. If the valve lift was below 0.031, remove the camshaft cover on the side of the engine exposing the cam gear and bend the ACR tab carefully upward until the valve lift is within the specified range.

Engine Mechanical

DISASSEMBLY

◆ **See Figures 108 and 109**

✳✳CAUTION

CAUTION: Before servicing the engine or equipment, always remove the spark plug lead to prevent the engine from starting accidentally. Ground the lead to prevent sparks that could cause fires.

Clean all parts thoroughly as the engine is disassembled. Only clean parts can be accurately inspected and gauged for wear of damage. There are many commercially available cleaners that quickly remove grease, oil, and grime rpm engine parts. When such a cleaner is used, follow the manufacturer's instructions carefully. Make sure all the engine is reassembled and placed in operation - even small amounts of these cleaners quickly break down the lubricating properties of engine oil. Check all parts for evidence of:

- Excessive sludge and varnish
- Scoring of the cylinder wall
- Piston damage
- Evidence of external oil leaks
- Evidence of overheating

Any of the listed problems could be the result of improper engine servicing or maintenance. The owner should be made aware of the benefits of proper servicing and maintenance.

1. Disconnect the spark plug lead and position it away from the spark plug terminal.

2. Unscrew the oil drain plug(s) and drain the crankcase oil into a suitable container for disposal.

3. Remove the wing nut, air cleaner cover, precleaner (if so equipped), paper element, three base screws, base, and base gasket.

4. If the engine is equipped with a flat muffler, remove muffler and gasket by unscrewing cap screws. If equipped with a round muffler remove by turning the threaded exhaust pipe between the muffler and engine with a pipe wench.

✳✳CAUTION

CAUTION: Gasoline may be present in the carburetor and fuel system. Gasoline is extremely flammable and it can explode if ignited. Keep sparks, open flames, and other sources of ignition away from the engine. Disconnect and ground the spark plug lead to prevent the possibility of sparks from the ignition system.

5. Close the fuel shut-off valve at fuel tank (if so equipped) or drain fuel from tank.

6. Loosen the hose clamp and remove fuel line from the carburetor inlet.

7. Remove two slotted hex cap screws, the carburetor, the gasket.

8. Remove the throttle linkage from the carburetor throttle lever.

9. Note the position of the governor spring in governor arm.

10. Loosen pawl nut. Remove governor arm and space from cross shaft.

➡**Loosening pawl nut of removing governor arm will disrupt governor arm to cross shaft adjustment. Readjustment will be required upon reassembly.**

11. Remove the governor spring from the governor arm.

12. Remove the governor hex cap screw, plain washer, spacer, bracket and throttle lever.

✳✳CAUTION

CAUTION: Gasoline may be present in the carburetor and fuel system. Gasoline is extremely flammable and can explode if ignited. Keep sparks, open flames, and other sources of ignition away from the engine. Disconnect and ground the spark plug lead to prevent the possibility of sparks from the ignition system.

13. Disconnect the fuel line from the fuel pump inlet fitting.

14. Disconnect the fuel line from the fuel pump outlet fitting.

15. Remove the fillister head screws, plain washers fuel pump, and gasket.

16. With retractable starter: Remove screws, washers and the retractable starter assembly.

17. With electric starter:

  a. Disconnect electrical connector(s) from back of key switch.

  b. Disconnect lead from electrical starter.

  c. Remove key switch panel.

  d. Remove hex cap screws which mount electric starter to engine.

e. Remove electric starter.

❋❋CAUTION

CAUTION: Gasoline may be present in the carburetor and fuel system. Gasoline in extremely flammable and it can explode if ignited. Keep sparks, open flames, and other sources of ignition away from the engine. Disconnect and ground the spark plug lead to prevent the possibility of sparks from the ignition system.

18. Remove fuel line from fuel tank outlet fitting.
19. Remove tank with bracket(s).
20. Remove the dipstick.
21. Remove the cylinder head baffle.
22. Remove the carburetor side air baffle.
23. Remove the starter side air baffle.
24. Remove pawl nut, breather cover, and gasket.
25. Remove the filter, seal, reed stop, reed, breather plate, gasket, and stud.
26. Remove the spark plug, cylinder head, and gasket.
27. Remove breaker point cover, gasket, breaker point lead, breaker assembly and push rod.

➡**Always use a flywheel strap wrench to hold the flywheel when loosening or tightening flywheel and fan retaining fasteners. Do not use any type of bar or wedge between fins of cooling fan as the fins could become cracked or damaged. Always use a puller to remove flywheel from crankshaft. Do not strike the crankshaft or flywheel, as these parts could become cracked or damaged.**

28. On rope start models:
   a. Remove the grass screen retainer and wire mesh grass screen from rope pulley.
   b. Hold the flywheel with a strap wrench and loosen the hex cap screw, Remove the hex cap screw, plain washer, rope pulley, and spacer. Remove the nylon grass screen from the fan.
29. On retractable start models:
   a. Hold the flywheel with a strap wrench and loosen hex cap screw securing flywheel to crankshaft. Remove the hex cap screw, plain washer, and drive cup.
   b. Remove the grass screen from the drive cup.
30. On electric start models:
   a. Remove the grass screen from the fan.
   b. Hold the flywheel with a strap wrench and loosen hex cap screw or hex nut securing flywheel to crankshaft. Remove the hex cap screw or hex nut. Remove plain washer.

31. On all models, the flywheel is mounted on tapered portion of crankshaft. Use of a puller is recommended for removing flywheel. Bumping end of crankshaft with a hammer to loosen flywheel should be avoided as this can damage crankshaft.

➡**Ignition magnet is not removable or serviceable!**

Do not attempt to remove ignition magnet from flywheel. Loosening or removing magnet mounting screws could cause the magnet to come off during engine operation and be thrown from the engine causing severe injury. Replace the flywheel if magnet is damaged.

32. Remove the screws and stator.
33. Rotate the crankshaft until the piston is at top dead center of compression stroke (both valves closed and piston flush with top of bore).
34. Compress the valve springs with a valve spring compressor and remove the keepers.
35. Remove the valve spring compressor, then remove the valves, intake valve spring lower retainer, exhaust valve rotator, valve springs, and valve spring upper retainers.

➡**Some models use a valve rotator on both valves.**

Make sure the piston is at top dead center in bore to prevent damage to oil dipper on connecting rod.

36. Remove the hex cap screws, oil pan, and gasket.
37. Remove the connecting rod cap.

➡**If a carbon ridge is present at top of bore, use a ridge reamer tool to remove it before attempting to remove piston.**

38. Carefully push the connecting rod and piston out top of bore.
39. Remove the retainer and wrist pin. Separate the piston from the connecting rod.
40. Remove the top and center compression rings and the oil control ring spacer using a ring expander tool.
41. Remove the rails and expander spring(s).
42. Remove the hex cap screws securing the bearing plate to crankcase.
43. Remove the bearing plate from the crankshaft using a puller.

➡**The front bearing may remain either in the bearing plate or on the crankshaft when the bearing plate is removed.**

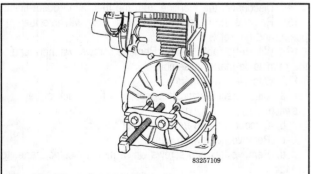

83257109

Fig. 108 Removing the flywheel

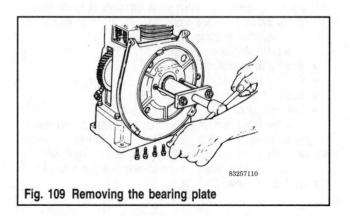

83257110

Fig. 109 Removing the bearing plate

44. Press the crankshaft out of the crankcase from the PTO side. It may be necessary to press crankshaft out of cylinder block. Bearing plate should be removed first if this is done.

➡If the repair does not require separating the bearing plate from crankshaft, the crankshaft and bearing plate can be pressed out as necessary.

45. Drive the camshaft pin (and cup plug on bearing side plate) out of the crankcase from the PTO side.
46. Remove the camshaft pin, camshaft, and shim(s) on bearing plate side of camshaft.
47. Mark the tappets as being either intake or exhaust. Remove the tappets from the crankcase.

➡The intake valve tappet is closest to the bearing plate side of crankcase. The exhaust valve tappet is closest to the PTO side of crankcase.

48. Remove the retaining rings, shims, balance gears with needle bearings, shims and spacers.

➡Extreme care must be taken when handling the new needle bearings or when removing balance gears containing the new bearings. The needles are no longer caged and will drop out. If this should occur, the bearing case should be greased and the needles reset. There are 27 individual needles in each bearing.

49. Remove the stop pin, copper washer, governor gear, and thrust washer.
50. Remove bushing nut and sleeve. Remove cross shaft from inside crankcase.
51. Remove the oil seals from the crankcase and bearing plate.
52. Press the bearings out of the bearing plate and crankcase.

➡If the bearings have remained on the crankshaft, remove bearing by using a puller.

INSPECTION AND REPAIR

All parts should be thoroughly cleaned. Dirty parts cannot be accurately gauged or inspected properly for wear or damage. There are many commercially available cleaners that quickly remove grease, oil and grime accumulation from engine parts. If such a cleaner is used, follow the manufacturer's instructions carefully, and make sure that all of the cleaner is removed before the engine is reassembled and placed in operation. Even small amounts of these cleaners quickly break down the lubricating properties of engine oils.

Flywheel Inspection

Inspect the flywheel for cracks, and the flywheel keyway for damage. Replace flywheel if cracked. Replace the flywheel, the crankshaft, and the key if the flywheel key is sheared or the keyway is damaged.

Inspect ring gear for cracks or damage. Kohler no longer provides ring gears as a serviceable part. Replace flywheel if the ring gear is damaged.

Flywheel Key Inspection

Shearing is possible on engines with flywheel drives and battery ignition systems. Check conditions such as overload, ignition timing and spark plug gap when flywheel key shearing occurs. Spark plus gap on battery ignition engines must be set as specified. If improperly gaped, a maverick spark can occur, which can cause improper ignition of unburned gases and can create a force causing the flywheel key to shear.

When repairing this type of failure, replace the flywheel, crankshaft, key, flywheel washer and nut or bolt.

Cylinder Head Inspection

Blocked cooling fins often cause localized 'hot spots' which can result in a 'blown' cylinder head gasket. If the gasket fails, high temperature gases can burn away portions of the aluminum alloy head. A cylinder head in this condition must be replaced.

If the cylinder head appears in good condition, use a block of wood or plastic scraper to scrape away carbon deposits. Be careful not to nick or scratch the aluminum, especially in gasket seating area.

The cylinder head should be checked for flatness. Use a feeler gauge and a surface plate or a piece of plate glass to make check. Cylinder head flatness should not vary more than 0.003 in.; if it does, replace the cylinder head.

NOTE: Measure cylinder head flatness between each cap screw hole.

In cases where the head is warped or burned, it will also be necessary to replace the head screws. The high temperatures that warped or burned the head could have made the screws ductile which will cause them to stretch when tightened.

Cylinder Block Inspection and Reconditioning

Check all gasket surfaces to make sure they are free of gasket fragments. Gasket surfaces must also be free of deep scratches or nicks. Scoring of the Cylinder Wall: Unburned fuel, in severe case, can cause scuffing and scoring of the cylinder wall, it washes the necessary lubricating oils off the piston and cylinder wall so that the piston rings make metal to metal contact with the wall. Scoring of the cylinder wall can also be caused by localized hot spots resulting from blocked fins or from inadequate or contaminated lubrication.

If the cylinder bore is badly scored, excessively worn, tapered, or out of round, resizing is necessary. Use an inside micrometer to determine the amount of wear, then select the nearest suitable oversizes of either 0.003 in., 0.010 in., 0.020 in., or 0.030 in.. Resizing to one of these oversizes will allow usage of the available oversize piston and ring assemblies. Initially, resize using a boring bar, then use the following procedures for honing the cylinder:

HONING

While most commercially available cylinder hones can be used with either portable drills or drill presses, the use of a low speed drill press is preferred as it facilitates more accurate alignment of the bore in relation to the crankshaft crossbore. Honing is best accomplished at a drill speed of about 250 RPM and 60 strokes per minute. After installing coarse stones in hone, proceed as follows:

1. Lower hone into bore and after centering, adjust so that stones are in contact with the cylinder wall. Use of a commercial cutting-cooling agent is recommended.

2. With the lower edge of each stone positioned even with the lowest edge of the bore, start drill and honing process. Move hone up and down while resizing to prevent formation of cutting ridges. Check size frequently.

➡**Keep in mind the temperatures caused by honing may cause inaccurate measurements. Make sure the block is cool when measuring.**

3. When bore is within 0.0025 in. of desired size, remove coarse stones and replace with burnishing stones. Continue with burnishing stones until within 0.0025 in. of desired size and then use finish stones (220-280 grit) and polish to final size. A crosshatch should be observed if honing is done correctly. The crosshatch should intersect at approximately 23 degrees to 33 degrees from the horizontal. Too flat an angle could cause the rings to skip and wear excessively, too high an angle will result in high oil consumption.

4. After resizing, check the bore for roundness, taper, and size. Use an inside micrometer, telescoping gauge, or bore gauge to take measurements. The measurements should be taken at three locations in the cylinder - at the top, middle, and bottom. Two measurements should be taken (perpendicular to each other) at each of the three locations.

5. Thoroughly clean cylinder wall with soap and hot water. Use a scrub brush to remove all traces or boring/honing process. Dry thoroughly and apply a light coat of SAE 10 oil to prevent rust.

MEASURING PISTON-TO-BORE CLEARANCE

▶ **See Figure 110**

Before installing the piston into the cylinder bore, it is necessary that the clearance be accurately checked. This step is often overlooked, and if the clearances are not within specifications, generally engine failure will result.

➡**Do not use a feeler gauge to measure piston-to-bore clearance, it will yield inaccurate measurements, use a micrometer.**

The following procedures should be used to measure the piston-to-bore clearance:

1. Use a micrometer and measure the diameter of the piston as shown.

2. Use an inside micrometer and measure the diameter of the piston as shown.

3. Use an inside micrometer, telescoping gauge, or bore gauge and measure the cylinder bore. Take the measurement approximately 2 1/2 in. below the top of the bore and perpendicular to the piston pin.

4. Piston-to-bore clearance is the difference between the bore and the piston diameter (step 2 minus step 1). For style A pistons only, clearance should be: 0.0035-0.006 in. for K91, 0.007-0.010 in. for K161-K349.

For piston styles C and D, clearance should be: 0.0034-0.0051 in. for K181, 0.0045-0.0062 in. for K301.

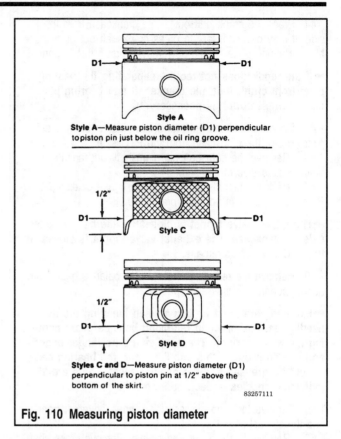

Style A—Measure piston diameter (D1) perpendicular to piston pin just below the oil ring groove.

Styles C and D—Measure piston diameter (D1) perpendicular to piston pin at 1/2" above the bottom of the skirt.

83257111

Fig. 110 Measuring piston diameter

VALVE INSPECTION AND SERVICE

▶ **See Figure 111**

Carefully inspect valve mechanism parts. Inspect valve springs and related hardware for excessive wear or distortion. Valve spring free height should be approximately the dimension given in the chart below. Check valves and valve seat area or inserts for evidence of deep pitting, cracks or distortion. Check clearance of valve stems in guides.

Hard starting, or loss of power accompanied by high fuel consumption may be symptoms of faulty valve. Although these symptoms could also be attributed to worn rings, remove and check valves first. After removal, clean valve head, face and stem with power wire brush and then carefully inspect for defects such as warped valve head, excessive corrosion or worn stem end. Replace valves found to be in bad condition. A normal valve and valves in bad condition are shown in the accompanying illustrations.

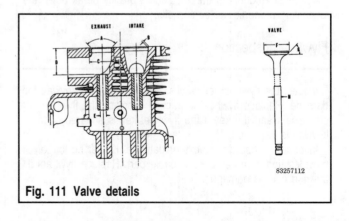

83257112

Fig. 111 Valve details

Valve Guides

If a valve guide is worn beyond specifications, it will not guide the valve in a straight line. This may result in a burnt valve face or seat, loss of compression, and excessive oil consumption.

To check valve guide to valve stem clearance, thoroughly clean the valve guide and, using a split-ball gauge, measure the inside diameter. Then, using an outside micrometer, measure the diameter of the valve stem at several points on the stem where it moves in the valve guide. Use the largest stem diameter to calculate the clearance. On models K91, K161, and K181, the clearance should not exceed 0.005 in. for intake and 0.007 in. for exhaust valves. On model K241, K301, the clearance should not exceed 0.006 in. for intake and 0.008 in. for exhaust valves. If the clearance exceeds these specifications, determine whether the valve stem or the guide is responsible for the excessive clearance.

➡**The exhaust valves on these engines have a slightly tapered valve stem to help prevent sticking. Because of the taper, the valve stem must be measured in two places to determine if the valve stem is worn. If the valve stem diameter is within specifications, replace the valve guide.**

VALVE GUIDE REMOVAL

The valve guides are a tight press fit in the cylinder block. A valve guide removal tool is recommended to remove valve guides. To remove valve guide, proceed as follows:

1. Install 5/16-18 NC nut on coarse threaded end of 2 1/2 in. long stud (K161 and K181) or 3 1/2 in. long stud (K241, K301).
2. Insert other end of stud through valve guide bore and install 5/16-24 NF nut. Tighten both nuts securely.

➡**Valve guide must be held firmly by the stud assembly so that all slide hammer force will act on the guide.**

3. Assemble the valve guide removal adapter to the stud and then slide hammer to the adapter.
4. Use the slide hammer to pull the guide out.

VALVE GUIDE INSTALLATION

1. Make sure valve guide bore is clean and free of nicks or burns.
2. Using valve guide driver, align and then press guide in until valve guide driver bottoms on valve guide counterbore.
3. Valve guides are often slightly compressed during insertion. Use a piloted reamer and then a finishing reamer to resize the guide bore to 0.3125 in. for K161, K181, K241, K301.

Valve Seat Inserts

The intake valve seat is usually machined into the cylinder block, however, certain applications may specify a hard allow insert. If the seat becomes badly pitted, cracked, or distorted, the insert must be replaced.

The insert is a tight press fit in the cylinder block. A valve seat removal tool is recommended for this job. Since insert removal causes loss of metal in the insert bore area, use only Kohler service replacement inserts, which are slightly larger to provide proper retention in the cylinder block. Make sure new insert is properly started and pressed into bore to prevent cocking of the insert.

VALVE SEAL INSERT REMOVAL

▶ **See Figure 112**

1. Install valve seat puller on forcing screw and lightly secure with washer and nut.
2. Center the puller assembly on valve seat insert.
3. Hold forcing screw with a hexing wrench to prevent turning and slowly tighten nut.

➡**Make sure sharp lip on puller engages in joint between bottom of valve seat insert and cylinder block counterbore, all the way around.**

4. Continue to tighten nut until puller is tight against valve seat insert.
5. Assemble adapter to valve seat puller forcing screw and slide hammer to adapter.
6. Use slide hammer to remove valve seat insert.

VALVE SEAT INSERT INSTALLATION

▶ **See Figure 113**

1. Make sure valve seat insert bore is clean and free of nicks and burns.
2. Align valve seat insert in counterbore and using valve seat installer and driver, press seat in until bottomed.
3. Use a standard valve seat cutter and cut seat to dimensions shown.

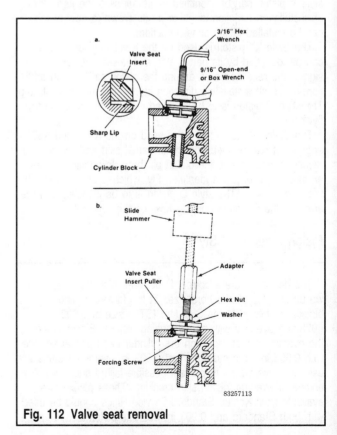

Fig. 112 Valve seat removal

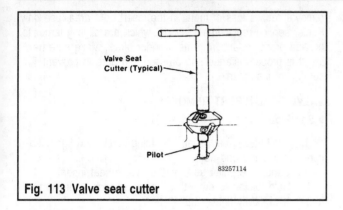

Fig. 113 Valve seat cutter

Reground or new valves must be lapped in to provide proper fit. Use a hand valve grinder with suction for final lapping. Lightly coat valve face with 'fine' grade of grinding compound, then rotate valve on seat with grinder. Continue grinding until smooth surface is obtained on seat and on valve face. Thoroughly clean cylinder block in soap and hot water to remove all traces of grinding compound. After dying cylinder block apply a light coating of SAE 10 oil to prevent rusting.

PISTON AND RINGS

Identification

Three different styles of pistons are currently being used in Kohler K-series engines.

Style 'A' pistons can be used in all K-series engines. The style A piston can be identified by its full skirt and its lack of an installation direction identifier on its crown (a new piston can be installed facing either direction).

The Style 'C' piston is used on the K341 engines only. It can be identified by its partial skirt and raised criss-cross design in the recessed area around the piston pin bore. In addition, it has an installation direction identifier (a notch) at its top. The style C piston is to be installed with the notch facing the flywheel.

The Style 'D' piston has been used on the K181 and K301 engines. It can be identified by its partial skirt and rectangular recessed area around the piston pin bore. In addition, it has an installation direction identifier, Fly, which is stamped into the top of the piston. The style D piston is to be installed with the arrow of the Fly mark pointing towards the flywheel.

Piston Sizes - All Styles

In order to ensure a correct fit between piston and cylinder we utilize two cylinder bore sizes at the factory. Cylinder blocks are honed to the Standard (STD) size or 0.003: (.075mm) oversize with corresponding pistons. Blocks using the oversize are stamped on the cylinder head gasket surface with 0.003 in.. It is essential that 0.003 in. oversize pistons are used in these blocks to prevent possible failure such as noisy engine or eventual piston skirt cracking. These pistons are available from Kohler. Standard Service Rings should be used with both Standard and 0.003 in. oversize pistons. Ring end

gap will increase slightly when installed on 0.003 in. oversize pistons; however, sealing is maintained due to the ring design.

Inspection

◗ **See Figures 114, 115 and 116**

Scuffing and scoring of piston and cylinder wall occur when internal temperatures approach the melting point of the piston. Temperatures high enough to do this are created by friction, which is usually attributed to improper lubrication, and/or overheating of the engine.

Normally, very little wear takes place in the piston boss-piston area. If the original piston and connecting rod can be reused after new rings are installed, the original pin can also be reused but new piston pin retainers are required. The piston pin is included as part of the piston assembly - if the pin

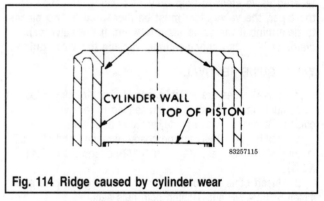

Fig. 114 Ridge caused by cylinder wear

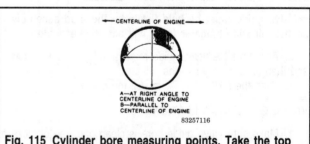

Fig. 115 Cylinder bore measuring points. Take the top measurement ½ inch below the top; the bottom measurement ½ inch above the top of the piston at BDC

Fig. 116 Measuring the bore with a dial gauge

boss in piston, or the pin are worn or damaged, a new piston assembly is required.

Ring failure is usually indicated by excessive oil consumption and blue exhaust smoke. When rings fail, oil is allowed to enter the combustion chamber where it is burned along with the fuel. High oil consumption can also occur when the piston ring end gap is incorrect because the ring cannot properly conform to the cylinder wall under this condition. Oil control is also lost when ring gaps are not staggered during installation.

When cylinder temperatures get too high, lacquer and varnish collect on piston causing rings to stick which results in rapid wear. A worn ring usually takes on a shiny or bright appearance. Scratches on rings and piston are caused by abrasive material such as carbon, dirt, or pieces of hard metal.

Detonation damage occurs when a portion of the fuel charge ignites spontaneously from heat and pressure shortly after ignition. This creates two frame fronts which meet and explode to create extreme hammering pressures on a specific area of the piston. Detonation generally occurs from using fuels with too low of an octane rating.

Pre-ignition of the fuel charge before the timed spark can cause damage similar to detonation. Pre-ignition damage is often more ever than detonation damage - often, a hole is quickly burned right through the piston dome. Pre-ignition is caused by a hot spot in the combustion chamber from sources such as: glowing carbon deposits, blocked fins, improperly seated valves or wrong spark plug.

Service

▶ **See Figures 117, 118, 119, 120, 121, 122 and 123**

K-series replacement pistons are available in STD bore size, and in 0.003 in., 0.010 in., 0.020 in., and 0.030 in. oversizes. Replacement pistons include new piston ring sets and new piston pins.

Service replacement piston ring sets are also available separately for STD-0.003 in. (same ring set for both sizes), 0.010 in., 0.020 in., and 0.030 in., oversized pistons. Always use new piston rings when installing pistons. Never reuse old rings.

The cylinder bore must be deglazed before service ring sets are used. Some important points to remember when servicing piston rings:

1. If the cylinder block does not need reboring and if the old piston is within wear limits and free of score or scuff marks, the old piston may be reused.

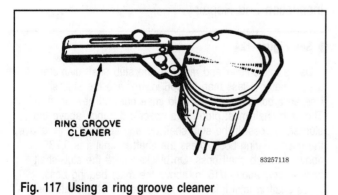

RING GROOVE CLEANER

83257118

Fig. 117 Using a ring groove cleaner

90°

83257119

Fig. 118 Using a micrometer to check the piston diameter

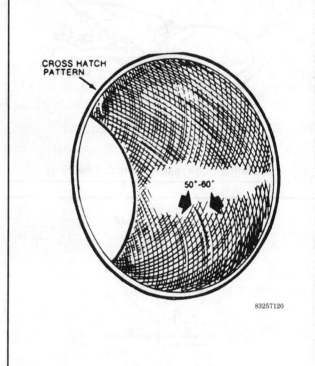

CROSS HATCH PATTERN

50°-60°

83257120

Fig. 119 Proper cylinder bore cross-hatching after honing

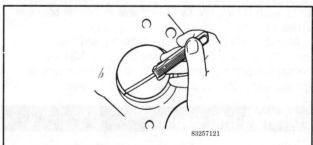

83257121

Fig. 120 Check the piston ring end gap with a feeler gauge, with the ring positioned in the bore, one inch below the deck of the block

2. Remove old rings and clean up grooves. Never reuse old rings.

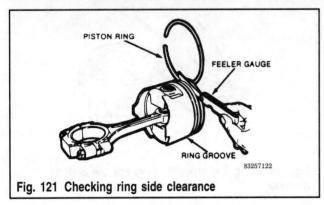

Fig. 121 Checking ring side clearance

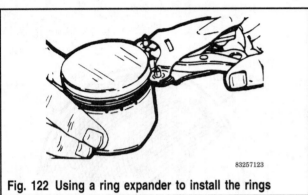

Fig. 122 Using a ring expander to install the rings

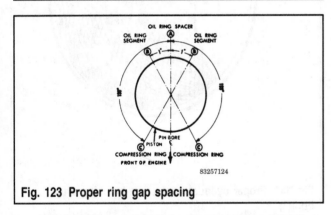

Fig. 123 Proper ring gap spacing

3. Before installing new rings on piston, place top two rings, each in turn, into its running area in cylinder bore and check end gap.

After installing the new compression (top and middle) rings on piston, check piston-to-ring side clearance. Maximum recommended side clearance is 0.006 in. If side clearance is greater than 0.006 in., a new piston must be used.

Oil Ring End Gaps

Although 4 sizes of service ring sets are available (Std., +/- 0.010 in., +/- 0.020 in, +/- 0.030 in.), only two sizes of oil rings are supplied (Std. and 0.020 in. oversize). When using 0.010 in. and 0.030 in. oversize ring sets, the oil rings appear to have excessive end gap. This is not detrimental and proper

sealing will be achieved due to the additional scraper rings and expander.

➡**Scraper and main ring end gaps should be staggered around the groove to prevent combustion blow-by.**

Piston Ring Installation

➡**Rings must be installed correctly. Ring installation instructions are usually included with new ring sets. Follow instructions carefully. Use a piston ring expander to install rings. Install the bottom (oil control) ring first and the top compression last.**

POSI-LOCK CONNECTION RODS

Posi-Lock connecting rods are used in some K-series engines. On model K181 engines with the style D pistons (refer to 'Piston and Rings, Identification' earlier in this section), the connecting rods have a narrower piston pin end than on the earlier (style A) Posi-Lock connecting rods. Therefore, the Posi-Lock connecting rods used with the style D pistons are not interchangeable with the Posi-Lock connecting rods used with style A pistons.

Inspection and Service

Check bearing area (big end) for excessive wear, score marks, running and side clearances. Replace rod and cap if scored or excessively worn.

Service replacement connecting rods are available in STD crank pin size and 0.010 in. undersize. The 0.010 in. undersize can be identified by the drilled hole located in the lower end of the rod shank.

BALANCE GEARS AND STUB SHAFTS

Some K241 and K301 engines are equipped with a balance gear system.The system consists of two gears and spacer (used to control end play) mounted on stub shafts which are pressed into the crankcase. The gears and spacers are held on the shafts with snap-ring retainers. The gears are timed with and driven by the engine crankshaft.

Inspection and Repair

▶ See Figure 124

Use a micrometer and measure the stub shaft diameter. If the diameter is less than 0.4996 in., replace the stub shaft. Use an arbor press to push old shaft out and new shaft in. The stub shaft must protrude a specific distance above the stub shaft boss. If the stub shaft boss is about 7/16 in. above the main bearing boss, press the shaft in until it is 0.735 in. above the stub shaft boss. On blocks where the stub shaft boss is only about 1/16 in. above the main bearing boss, press shaft in until it is 1.110 in. above the stub shaft boss. A 3/8 in. spacer must be used with the shaft which protrudes

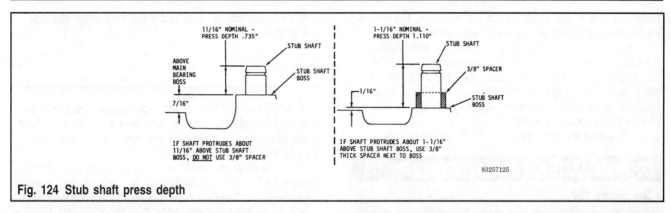

Fig. 124 Stub shaft press depth

1.110 in.Inspect the gears for worn or chipped teeth and for worn needle bearings, if required.

BALANCE GEAR BEARING AND BALANCE GEAR ASSEMBLY

A new needle bearing for the Dynamic Balance System is now being used on 10-16 HP Kohler engines. The new bearing (part number 47030 01) has been in use beginning with serial number 9641319.

It is not interchangeable with the old needle bearing, part number 236506.

Complete balance gear assemblies are interchangeable - and old style gear assemblies have been superseded with a new gear assembly, part number 47 042 01.Critical consideration is required when only the needle bearing is to be replaced. The engine serial number alone can correctly determine which needle bearing is involved on original equipment engines. However should a bearing replacement be required after a complete balance gear has been replaced on an engine with a serial number prior to 9641311, the following methods will assist in identifying the correct bearing:

Method #1 - I.D. of Balance Gear Bore
0.6825 - 236506
0.6821
0.6865 - 47 030 01
0.6869
Method #2 - O.D. of Old Bearing
0.6825 - 236506
0.6828
0.6870 - 47 030 01
0.6875

GOVERNOR GEAR

Inspection

Inspect the governor gear teeth. Look for any evidence of worn, chipped or cracked teeth. If one or more of these problems is noted, replace the governor gear.

CAMSHAFT AND CRANKSHAFT

Inspection and Service

▶ **See Figure 125**

Inspect the gear teeth on both the crankshaft and camshaft. If the teeth are badly worn, chipped or some are missing, replacement of the damaged components will be necessary.

Also, inspect the crankshaft bearings for scoring, grooving, etc. Do not replace bearings unless they show signs of damage or are out of running clearance specifications. If crankshaft turns easily and noiselessly, and there is not evidence of scoring, grooving, etc., on the races of bearing surfaces, the bearings can be reused.

Check crankshaft keyways. If worn or chipped, replacement of the crankshaft will be necessary. Also inspect the crank pin for score marks or metallic pickup. Slight score marks can be cleaned with crocus cloth soaked in oil. If wear limits, as stated in Section, 'General Information', are exceeded, it will be necessary to either replace the crankshaft or regrind the crank pin to 0.010 in. undersize. If reground, a 0.010 in. undersize connecting rod (big end) must then be used to achieve proper running clearance. Measure the crank pin for size, taper and out-of-round.

➡**If the crank pin is reground, visually check to ensure that the fillet blends smoothly with the crank pin surface.**

When replacing a crankshaft with external threads on the flywheel end with one that has internal threads, different mounting hardware is required. The internally threaded crank-

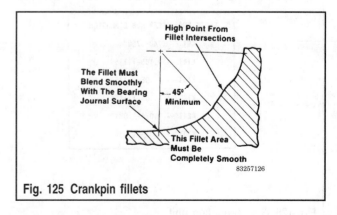

Fig. 125 Crankpin fillets

shafts are sold in kits which include the hardware. The mounting hardware can also be purchased separately.

1. Use a 13/32 I.D. x 1 1/4 O.D. x 1/8 TH. plain washer (Part No. 52114 01) when installing drive cups with a 1 1/4 in. Dia. spot face (machined, recessed area around mounting hole). These drive cups are primarily used on International Harvester applications. Therefore, use the drive cup to identify which washer should be used.

2. Required for drive cups with 5/8 in. mounting hole.

OPTIONAL GEAR REDUCTION UNIT

▶ **See Figure 126**

The reduction unit consists of a driven gear which is pressed on the power take off (PTO) shaft. The drive gear is an integral part of the engine crankshaft. The gear reduction on the K91 and K181 units is 6:9.

The gear reduction on the K301 engines is 4:

1. The PTO shaft is supported by two bearings, one in the cover and the other in the housing. Oil seals are provided at both ends of the shaft.

Removal

1. Drain lubricating oil from unit.
2. Remove four cap screws from gear housing and slide cover off along drive gear.
3. Remove four cap screws holding gear housing to engine.
4. Wash all parts and inspect shaft, bushing and gear for wear. Replace worn parts.

5. Remove old oil seals and install new seals (flat side out) in gear housing and cover.

Installation

1. Wrap piece of tape or roll paper around crankshaft gear to protect the oil seal, slide housing over the shaft and attach to the block. Two lock washers are used on the outside of housing and copper washers inside.
2. Tape or paper should be wrapped around the shaft to prevent the keyway from damaging the cover oil seal. Install the gasket(s) and reduction gear cover and tighten cap screws.
3. Adjust shaft end clearance to 0.001-0.006 in. by varying the total gasket thickness, adding or removing gaskets as required.
4. Remove oil fill plug and level plug, fill unit to the oil level hole. Use the same grade of oil as used in the engine.

ENGINE REASSEMBLY

▶ **See Figures 127, 128 and 129**

The following sequence is suggested for complete engine reassembly. This procedure assumes that all components are new or have been reconditioned, and all component subassembly work has been completed. This procedure may have to vary slightly to accommodate options or special equipment.

➡**Make sure the engine is assembled using all specified torque valves, tightening sequences, and clearances. Failure to observe specifications could cause severe engine wear or damage. Always use new gaskets.**

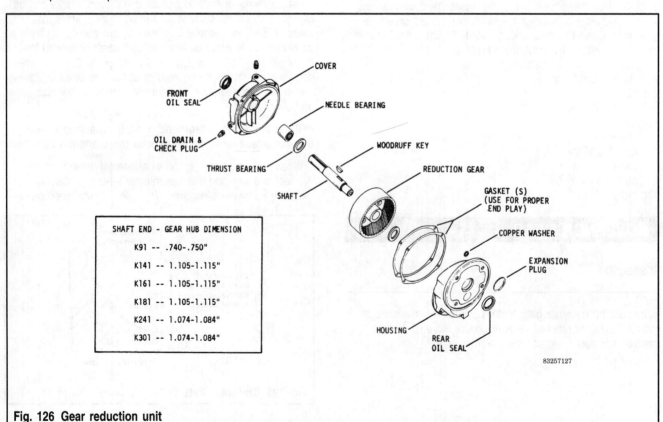

SHAFT END - GEAR HUB DIMENSION
K91 -- .740-.750"
K141 -- 1.105-1.115"
K161 -- 1.105-1.115"
K181 -- 1.105-1.115"
K241 -- 1.074-1.084"
K301 -- 1.074-1.084"

Fig. 126 Gear reduction unit

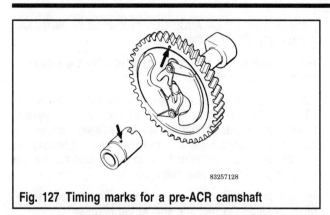

Fig. 127 Timing marks for a pre-ACR camshaft

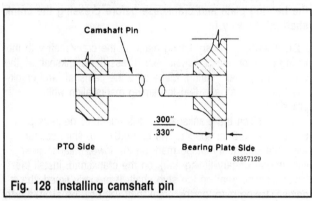

Fig. 128 Installing camshaft pin

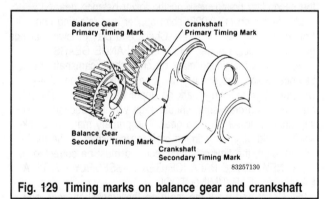

Fig. 129 Timing marks on balance gear and crankshaft

1. Install the rear bearing into crankcase using the #4747 handle and appropriate bearing installer. Make sure the bearing is bottomed fully, and is straight and true in bore. Install the rear main bearing by pressing it into cylinder block. If using a shielded type bearing, install with shielded side facing toward inside of block.

2. slide cross shaft into place from inside of block.

3. Place speed control disc on governor bushing nut and thread bushing nut into block. On earlier models, the cross shaft has an extension riveted in place to line up with governor gear. Torque bushing nut as follows:
- K91 70-90 inch lbs.
- K161, K181 - 130-150 inch lbs.
- K241, K301 - 100-120 inch lbs.

4. Install the thrust washer, governor gear, copper washer, and stop pin.

5. Rotate governor gear assembly to be sure stop pin does not contact weight section governor gear.

6. Install the intake valve tappet and exhaust valve tappet into crankcase. (Intake valve tappet towards bearing plate side; exhaust valve tappet towards PTO side of crankcase).

➡ **On K161 and K181 ACR engines, install the shorter tappet in the exhaust bore guide. Intake and exhaust tappets are interchangeable on other models.**

7. Install the camshaft, one 0.005 in. shim spacer, and the camshaft pin (from bearing plate side). Do not drive the camshaft pin into its final position at this time.

➡ **On pre-ACR models with the automatic spark advance camshaft, spread actuators and insert cam. Align the timing marks on cam and gear as shown.**

8. Measure the camshaft end play between the spacer and crankcase boss using a flat feeler gauge. Recommended camshaft end play is 0.005-0.020 in. for model K91 and 0.005-0.010 in. - for all other K-series models. Add or subtract 0.005 in. and/or 0.010 in. shim spacers as necessary to obtain the proper end play.

9. The K-series engines now use a new camshaft pin, the new camshaft pins are shorter than the old pins originally used in K-series engines.

10. To install the new (shorter) camshaft pin, drive the camshaft pin from the bearing plate side of crankshaft into the PTO side of crankcase:
   a. For Models K161 and K181 - drive the camshaft pin to a depth of 0.275-0.285 in. from the machined baring plate gasket surface.
   b. For Models K241, K301 - drive the camshaft pin to a depth of 0.300-0.330 in. from the machined bearing plate gasket surface. To install the old (longer) camshaft pin, drive the camshaft pin into the crankcase pin into the crankcase until the PTO end of camshaft pin is flush with the mounting surface on PTO side of crankcase.

11. On Engines So Equipped: The balance gears must be timed to the crankshaft whenever the crankshaft is installed. Use a balance gear timing tool to simplify this procedure. If the balance gears must be timed without using the tool, do not install the lower balance gear (closest to the oil pan) until after the crankshaft has been installed.

12. Install the 3/8 in. spacer, one 0.010 in. shim spacer, balance gear, one 0.020 in. shim spacer, and retaining ring (rounded edge towards balance gear). A new style needle bearing is now being used on the K-series balance gear assembly.

➡ **Extreme care must be taken when handling the new needle bearings. The needles are not longer caged and will drop out. If this should occur, the bearing case should be greased and the needles reset. There are 27 individual needles in each bearing.**

13. Check end play with a flat feeler gauge. Recommended end play is 0.002-0.010 in.. If end play is not within range, install or remove 0.005 in. and 0.010 in. spacers, as necessary.

14. On Engines Without Balance Gears.
   a. Lubricate the crankshaft rear bearing surface. Insert the crankshaft through the rear bearing.

➥If the crankshaft and bearing plate have not been separated, position the fuel line and wiring harness between the bearing plate and crankcase before pressing the crankshaft all the way in.

   b. Align the primary timing mark on crankshaft with the timing mark on camshaft. Press the crankshaft into rear bearing. Make sure the camshaft and crankshaft gears mesh and that the timing marks remain aligned while pressing.

15. On Engines With Balance Gears:K-series have two styles of balance gear assemblies. To provide improved vibration reducing characteristics, redesigned balance gear assemblies are being used in the K241 and K301 single cylinder engines. These new balance gear assemblies (Part No. 45 043 03) are being used in engines with a Serial No. of 1613600013 and later, and for service replacement.Because of the physical differences of the gear, new procedures for installing the crankshaft, and timing the balance gears, crankshaft, and camshaft are required.

The following crankshaft installation procedures are broken down into four sections:

1A) OLD STYLE BALANCE GEAR ASSEMBLY - WITH A BALANCE GEAR TIMING TOOL1B) OLD STYLE BALANCE GEAR ASSEMBLY - WITHOUT A BALANCE GEAR TIMING TOOL

2A) NEW STYLE BALANCE GEAR ASSEMBLY - WITH A BALANCE GEAR TIMING TOOL

2B) NEW STYLE BALANCE GEAR ASSEMBLY - WITHOUT A BALANCE GEAR TIMING TOOL
   METHOD
1A) OLD STYLE BALANCE GEAR ASSEMBLY - WITH A BALANCE GEAR TIMING TOOL

16. Align the primary timing marks of balance gears with the teeth on timing tool. Insert tool so it meshes with gears. Hold or clamp tool against oil pan gasket surface.

17. Lubricate the crankshaft rear bearing surface. Insert the PTO end of crankshaft through rear bearing. 'Straddle' the primary and secondary timing marks on crankshaft over the rear bearing oil drain. Press the crankshaft into rear bearing until the crankgear is just above the camshaft gear but not in mesh with it.

➥If the crankshaft and bearing plate have not been separated, position the fuel line and wiring harness between the bearing plate and crankcase before pressing the crankshaft all the way in.

18. Remove the balance gear timing tool and align the primary timing mark on the camshaft gear. Press the crankshaft all the way in to the rear bearing. Make sure the camshaft and crankshaft gears mesh and that the timing marks align while pressing.

19. Check the timing of the crankshaft, camshaft, and balance gears:
   a. The primary timing mark on crankshaft should align with the secondary timing mark on lower balance gear.
   b. The primary timing mark on crankshaft should align with the timing mark on camshaft.
   c. If the marks do not align, the timing is incorrect and must be corrected.

1B) OLD STYLE BALANCE GEAR ASSEMBLY - WITHOUT A BALANCE GEAR TIMING TOOL

➥The balance gear should be installed after the crankshaft has been installed.

20. Lubricate the crankshaft rear bearing surface. Insert the PTO end of crankshaft through rear bearing. Align the primary timing mark on crankshaft with the primary timing mark on upper balance gear. Press the crankshaft into rear bearing until the crankgear just starts to mesh (about 1/16 in.) with the center ring of balance gear teeth.

➥If the crankshaft and bearing plate have not been separated, position the fuel line and wiring harness between the bearing plate and crankcase before pressing the crankshaft all the way in.

21. Align the primary timing mark on the crankshaft with the timing mark on the camshaft gear. Press the crankshaft all the way into the rear bearing. Make sure the camshaft and crankshaft gears mesh and that the timing marks align while pressing.

22. Position the crankshaft so it is about 15 degrees past BDC. Install 3/8 in. spacer, and one 0.010 in. shim spacer. Align the secondary timing mark on the lower balance gear with the secondary timing mark on the crankshaft. Install the lower balance gear on the stub shaft. If properly timed, the primary timing mark on the crankshaft will now be aligned with the secondary timing mark on the lower balance gear.

23. Install on (1) 0.020: shim spacer and retaining ring (rounded edge towards gear). Check end play of lower balance gear as instructed under 'INSTALL BALANCE GEARS'.

24. Check the timing mark on crankshaft, camshaft and balance gears. The primary mark on crankshaft should align with the primary timing mark on upper balance gear. The primary mark on crankshaft should align with the secondary timing mark on lower balance gear. The primary mark on crankshaft should align with timing mark on camshaft. If the marks do not align, the timing is incorrect and must be corrected.

2A) NEW STYLE BALANCE GEAR ASSEMBLY - WITH A BALANCE GEAR TIMING TOOL

25. Count and mark the teeth on the crankshaft gear, and the lands (notches between teeth) on the camshaft gear as follows:
   a. Crankshaft - Locate the primary timing mark on crankshaft. While looking at the PTO end of crankshaft, start with the tooth directly below timing mark and count five (5) teeth in a counterclockwise direction. Mark the fifth tooth.
   b. Camshaft - Locate the timing mark on camshaft. Starting with the land next to the timing mark, count five (5) lands in a counterclockwise direction. Mark the fifth land.

26. Align the primary timing marks on balance gears with the teeth on timing tool. Insert the tool so it meshes with the gears. Hold or clamp the tool against oil pan gasket surface of crankcase.

27. Lubricate the rear bearing surface of crankshaft, Insert the PTO end of crankshaft through the rear bearing. 'Straddle' the primary and secondary timing marks on crankshaft over the rear bearing oil drain. Press the crankshaft into the rear bearing until the crankshaft gear is just above the camshaft gear, but not in mesh with it. Do not remove the balance gear timing tool at this time.

28. Align the fifth (5th) land marked on camshaft gear with the fifth (5th) tooth marked on camshaft gear. Press the crankshaft all the way into the rear bearing. Make sure the camshaft and crankshaft gears mesh and the marks align while pressing.

29. Remove the balance gear timing tool. Check the timing of the crankshaft, camshaft, and balance gears. The primary timing mark on crankshaft should align with the secondary timing mark on lower balance gear. The primary timing mark on crankshaft should align with the timing mark on camshaft.

2B) NEW STYLE BALANCE GEAR ASSEMBLY - WITHOUT A BALANCE GEAR TIMING TOOL

➡**The lower balance gear should be installed after the crankshaft has been installed.**

30. Count and mark the teeth on the crankshaft gear, and the land (notches between teeth) on the upper balance gear as follows: a. Crankshaft - Locate the primary timing mark on crankshaft. While looking at the PTO end of crankshaft, start with the tooth directly below timing mark and count twelve (12) teeth in a counterclockwise direction. Mark the twelfth tooth.   b. Upper Balance Gear - Locate the secondary timing mark on balance gear. Starting with the land next to the timing mark, count seven (7) lands in a clockwise direction. Mark the seventh land.

31. Lubricate the rear bearing surface of crankshaft. Insert the PTO end of crankshaft through the rear bearing. Align the twelfth (12th) tooth marked on crankshaft gear with the seventh (7th) land marked on upper balance gear. Press the crankshaft gear is just above the camshaft gear, but not in mesh with it.

32. Align the timing mark on camshaft with the primary timing mark on crankshaft.

➡ **Align the timing mark on camshaft with the primary timing mark on crankshaft. To align the marks, rotate the camshaft only - do not rotate the crankshaft. Rotating the crankshaft could cause the crankshaft gear to come out of mesh and the marks align while pressing.**

33. Install the 3/8 in. spacer and one (1) 0.010 in. shim spacer to the stub shaft for the lower balance gear.

34. Position the crankshaft so it is about 15 degrees past bottom dead center(BDC). Align the secondary timing mark on crankshaft. Install the lower balance gear to the stub shaft. If properly timed, the secondary timing mark on lower balance gear will now be aligned with the primary timing mark on crankshaft.

35. Secure the lower balance gear to stub shaft using on (1) 0.020 in. shim spacer and retaining ring (rounded edge towards gear). Check end play of lower balance gear as instructed in 'INSTALL BALANCE GEARS'.

36. Check the timing of the crankshaft, camshaft, and balance gears. The primary timing mark on crankshaft should align with the secondary timing mark on lower balance gear. The primary timing mark on crankshaft should align with the primary timing mark on upper balance gear. If the marks do not align, the timing is incorrect and must be corrected.

Front Bearing Installation

➧ **See Figure 130**

Install the front bearing into the bearing plate using the #4747 handle and appropriate bearing installer. Make sure the bearing is bottomed fully, and straight and true in the bore.

CONTINUE ENGINE ASSEMBLY

➧ **See Figures 131, 132, 133, 134 and 135**

1. Position the fuel line and wiring harness (if so equipped) to crankcase.

2. Adjust the fuel line and wiring harness to their final positions just before securing the bearing plate to the crankcase.

3. The installation of bearing plate and gaskets can be made considerably easier with the use of two simple, easy to make alignment guides. Using 2 1/2 in. long bolts with the hexagon heads removed and screwdriver slots cut in the stem, screw the two headless bolts into the cylinder block diagonally from each other. Bolt thread sizes are 1/4-20 U.N.C. for K91-K181; 3/8-16 U.N.C. for K241-K361.

4. Lubricate the bearing surface of crankshaft and bearing. Install the gasket, two or three 0.005 in. shims (as required), and bearing plate over studs.

➡**Crankshaft end play is determined by the thickness of the gasket and shims between crankcase and bearing plate. Check the end play after the bearing plate is installed.**

5. Install two hex cap screws and hand tighten. Remove the locating studs, and install the remaining two hex cap screws and hand tighten.

6. Tighten the screws evenly, drawing bearing plate to crankcase. Torque K91-K181 to 115 inch lbs. Torque K241-K341 to 35 ft. lbs.

7. Check crankshaft end play between the inner bearing race and should of crankshaft using a flat feeler gauge. Recommended total end play is:
- K91 - 0.004-0.023 in.
- K161, K181 - 0.002-0.023 in.
- K241, K301 - 0.003-0.020 in.

If measured end play is not within limits, remove the bearing end plate and, remove or install shims as necessary.

➡**Crankshaft end play is especially critical on gear reduction engines.**

8. Slide the appropriate seal sleeves over the crankshaft. Generously lubricate the lips of the oil seals with light grease. Slide the oil seals over the sleeves.

9. Use the #11795 handle and appropriate seal drivers to install the front oil seals to the following depths. Note that the front oil seal depth varies with engine model and type of bearing plate used-bearing plate configuration differs with type of ignition system used.

➡**For detailed piston inspection and piston ring installation procedures, refer to the 'Inspection And Repair/Reconditioning' section.**

10. Style A pistons: Install wrist pin and retainers.

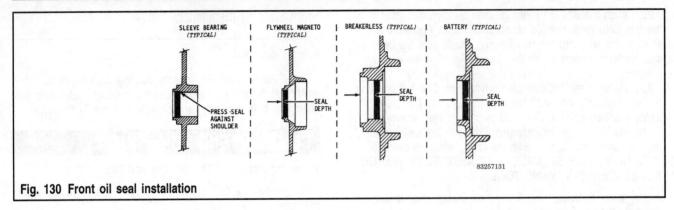

Fig. 130 Front oil seal installation

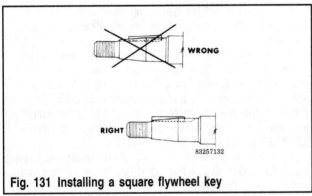

Fig. 131 Installing a square flywheel key

11. Style C and D pistons:

➡**Proper orientation of the piston to the connecting rod is extremely important. Improper orientation may cause extensive wear or damage.**

12. Orient piston and connecting rod so that the notch (Style 'C' piston) or Fly symbol (Style 'D' piston) on piston and the match mark on connecting rod are facing the same direction.

13. Install wrist pin and retainers.

➡**Proper orientation of the piston to the connecting rod is extremely important. Improper orientation may cause extensive wear or damage.**

14. Stagger the piston rings in their grooves until end gaps are 120 degrees apart.

15. Lubricate the piston and rings with engine oil. Install the piston ring compressor around piston.

16. Orient the notch (on style 'C' piston) or Fly symbol (Style 'D' piston) and match marks on connecting rod towards the flywheel end of crankshaft. Gently push the piston/connecting rod into bore - do not pound on piston.

17. Lubricate the crankshaft and connecting rod journal surfaces with engine oil. Install the connecting rod and cap line up and face flywheel end of engine.

18. Torque the capscrew to 20% over the nominal torque value listed below. Loosen cap screws to below the nominal value--do not leave overtorqued. Retorque bolts to nominal value.

➡**To prevent damage to connecting rod and engine, do not overtorque-loosen--and retorque the hex nuts on Posi-Lock connecting rods. Torque nuts, in increments, directly to the specified value.**

19. Rotate the crankshaft until the piston is at top dead center in bore to protect the dipper on the connecting rod. If locking tabs are used, bend tabs to lock cap screws.

20. Install screws and oil drain plug as specified in 'General Information'.

21. Rotate the crankshaft until piston is at top dead center of compression stroke.

22. Install the valves and measure the valve-to-tappet clearance using a flat feeler gauge.

➡**Valve faces and seats must be lapped-in before checking/adjusting valve clearance.**

23. Adjust valve-to-tappet clearance, as necessary.

   a. On models K91, K141, K161, and K181: If clearance is too small, grind end of valve stems until correct clearance is obtained. Make sure stems are ground perfectly flat and smooth. If clearance is too large, replace the valves and recheck clearance.

➡**Large clearances can also be reduced by grinding the valves and/or valve seats.**

   b. On Models K241 and K301, adjust valve-to-tappet clearance by turning the adjusting screw on tappets.

   c. On Models K91, K161, and K181, install the valve springs (close coils to top), intake valve spring retainer, exhaust valve rotator or retainer, and valves.

   d. On Models K241 and K301, install the valve spring upper retainers, valve springs (close coils to top), intake valve spring lower retainer, exhaust valve rotator or retainer, and valves.

➡**Some models use a valve rotator on both valves.**

24. Compress springs using a valve spring compressor and install keepers.

25. On flywheel-magneto ignition systems, the magneto coil-core assembly ignition systems, the magneto coil-core assembly is secured in stationary position on the bearing plate. On the magneto-alternator systems, the coil is part of the stator assembly which is also secured to the bearing plate. Permanent magnets are affixed to the inside rim of the flywheel except in rotor type magneto systems. On these the magnet or rotor has a keyway and is press fitted on crankshaft. The magnet rotor is marked 'engine-side' for proper assembly.

26. After installing magneto components, run all leads out through holes provided (in 11 O'clock position) on bearing plate.

✳✳CAUTION

CAUTION: Damaging Crankshaft and Flywheel Could Cause Personal Injury!Using improper procedures to install the flywheel can crack or damage the crankshaft and/or flywheel. This not only causes extensive engine damage, but also is a serious threat to the safety of persons nearby, since broken fragments could be thrown from the engine. Always observe and use the following precautions and procedures when installing the flywheel.

➡**Before installing the flywheel, make sure the crankshaft taper and flywheel hub are clean, dry and completely free of lubricants. The presence of lubricants can cause the flywheel to be overstressed and damaged when the cap screw is torqued to specification.**

Make sure square flywheel key is installed only in the flat area of keyway, not in the rounded are. The flywheel can become cracked or damaged if the key is installed in the rounded area of keyway.Always use a flywheel strap wrench to hold flywheel when tightening flywheel fastener. Do not use any type of bar or wedge between the cooling fins or flywheel ring gear, as these parts could become cracked or damaged. Do not use impact wrenches to install the flywheel retaining nut as this may overstress the nut and crack the flywheel hub.

Do not reuse a flywheel if it has been dropped or damaged in any way. Do make a thorough visual inspection of the flywheel and crankshaft before installation to make sure they are in good condition and free of cracks. The old crankshaft design has a externally threaded end and uses a square key, plain washer, and marsden nut to align and secure the flywheel. The new crankshaft design had an internally threaded end and uses a woodruff key, washer, and/or brushing, hex cap screw or hexnut.

27. Position key properly in keyway as shown, and carefully guide key slot in flywheel hub over the key while installing to avoid pushing the key inward.

28. On models K91, K161, and K181 w/Rope Start: Install rope pulley, plain washer, and hex nut (lubricate threads with oil). Hold flywheel with strap wrench and torque hex nut to 40-50 ft. lbs for K91 and 85-90 ft. lbs for K161 and K189.

If a hex head screw is used, torque screw to 250 inch lbs.

29. On models K91, K161, and K181 w/Retractable Start:

   a. Install the grass screen.

   b. Install the drive cup, plain washer, and hex nut (lubricate threads with oil). Hold the flywheel with strap wrench and torque hex nut to 40-50 ft. lbs. for K91 and 85-90 ft. lbs. for K161 and K1

30. If a hex head screw is used, torque screw to 250 inch lbs.

31. On models K91, K161, and K181 w/Electric Start:

   a. Install the plain washer and hex nut (lubricate threads with oil). Hold the flywheel with strap wrench and torque hex nut to 40-50 ft. lbs. for K91 and 85-90 ft. lbs. for K161 and K189.

   b. If a hex head screw is used, torque screw to 250 inch lbs.

32. On models K241, K301 w/ Rope Start:

   a. Install the nylon grass screen.

   b. Install the spacer, rope pulley, plain washer, and hex cap screw (lubricate threads with oil). Hold the flywheel with a strap wrench and torque hex cap screw to 35-40 ft. lbs. If a hex nut is used, torque to 50-60 ft. lbs.

   c. Install the wire mesh grass screen and grass screen retainer to rope pulley.

33. On models K241, K301 w/Retractable Start:

   a. Install the grass screen.

   b. Install the drive cup, plain washer, and hex cap screw (lubricate threads with oil). Hold the flywheel with a strap wrench and torque hex cap screw to 35-40 ft. lbs. If a hex nut is used, torque to 50-60 ft. lbs.

34. On models K241, K301 w/Electric Start:

   a. Install the plain washer and hex cap screw (lubricate threads with oil). Hold the flywheel with a strap wrench and torque hex cap screw to 35-40 ft. lbs. If a hex nut is used, torque to 50-60 ft. lbs.

   b. Install the grass screen.

35. For all models, torque the grass screen fasteners to 70-140 ft. lbs. for a metal grass screen and 20-30 inch lbs. for a plastic grass screen.

36. Install the spark plug lead and kill lead into the slots in the baffle.

37. Install the remaining self-tapping screws and the blower housing.

➡**On some models, the grass screen must be installed before installing the blower housing.**

38. Install push rod, breaker assembly, and breaker point lead.

39. Set breaker point gap at 0.020 in. full open.

40. Install gasket and breaker point cover.

41. Install the gasket and cylinder head. Always use a new gasket when head has been removed for service work.

42. Torque the hex cap screws and hex nuts (in increments) in the sequence and torques shown.

➡**The importance of torquing cylinder head bolts to specified values and following the recommended sequences cannot be overemphasized. Blown head gaskets and cylinder head distortion may result from improper torquing. Following is the recommended torquing procedure:**

43. Lubricate cylinder head bolts with oil before installation.

44. Initially torque each bolt to 10 ft. lbs. following the recommended torque sequence.

45. Sequentially tighten each bolt in 10 ft. lbs. increments until the specified torque values are reached.

After reaching the final torque value, run the engine for 15 minutes, stop, and allow to cool. Then, sequentially retorque the head bolts to the specified torque value.

46. Make sure the spark plug is properly gapped.

47. Install the spark plug and torque it to 18-22 ft. lbs.

48. Install the stud, gasket, breather plate, reed, reed stop, seal, and filter. The accompanying illustrations show the correct order of assembly for two types of breather assemblies. Make sure reed valve is installed properly and that oil drain hole on breather plate is down.

➡**All K181 Specifications have been changed to call for 2 pieces of breather filter 231419 instead of 9.**

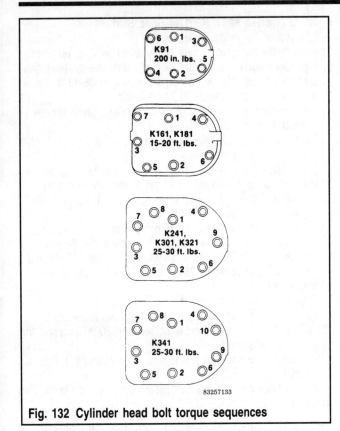

Fig. 132 Cylinder head bolt torque sequences

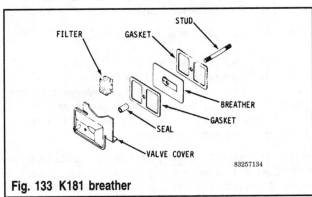

Fig. 133 K181 breather

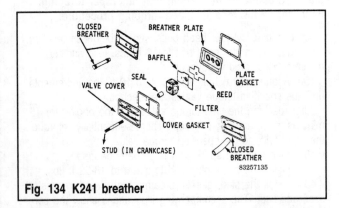

Fig. 134 K241 breather

Testing has revealed the use of two filters prevents oil droplets from being expelled through the breather system. All K181 engines are now being built with two filters and when serviced, two should always be used.

49. Install the gasket, breather cover, and pawlnut.

50. Install the starter side air baffle, plain washer, and hex cap screws. Leave the screws loose.

51. Install the carburetor side air baffle, plain washer, and hex cap screws. Leave the screws loose.

52. Install the cylinder head baffle, plain washer, and hex cap screws. Leave the screws loose. 69. Tighten the screws securely when all pieces are in position.

➡ **Shorter screws go into lower portion of blower housing.**

53. Install dipstick.

✳✳CAUTION

CAUTION: Gasoline may be present in the carburetor and fuel system. Gasoline is extremely flammable and it can explode if ignited. Keep sparks, open flames, and other sources of ignition away from the engine. Disconnect and ground the spark plug lead to prevent the possibility of sparks from the ignition system.

54. Install fuel tank with brackets.

55. Install fuel line on fuel tank outlet fitting.

56. Install electric starter.

57. Install hex cap screws which mount electric starter to engine.

58. Install key switch panel.

59. Connect lead to electrical starter.

60. Connect electrical connector(s).

61. Install retractable starter and hex cap screws. Leave the screws slightly loose.

62. Pull the starter handle out 8-10 in. until the pawls engage in the drive cup. Hold the handle in this position and tighten screws securely.

63. Install the gasket, fuel pump, plain washers, and fillister head screws. Torque the screws to 37-45 inch lbs.

➡ **Make sure the fuel pump lever is positioned above the camshaft. Damage to the fuel pump, and subsequent severe engine damage could result it the lever is positioned below the camshaft.**

64. Connect the fuel lines to fuel pump inlet and outlet fittings.

65. Install the throttle lever, bracket, spacer, plain washer and hex cap screw.

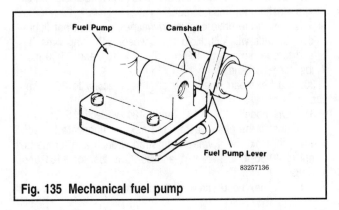

Fig. 135 Mechanical fuel pump

66.   Install the governor spring to the governor arm. Install the governor arm to the cross shaft. Leave the pawlnut slightly loose as the governor arm and cross shaft will be adjusted after the carburetor and throttle linkage are installed.

67.   Install the fuel line and hose clamps.

68.   Install the gasket, carburetor, and slotted hex cap screws.

69.   Install the throttle linkage into the nylon inserts in the governor arm and carburetor throttle lever.

70.   Adjust the governor as instructed below.

71.   Refer to 'Fuel System And Governor' section for carburetor adjustment procedure.

GOVERNOR ADJUSTMENT

The governor cross shaft/governor arm must be adjusted every time the governor arm is loosened or removed from cross shaft. a. Pull the governor arm away from the carburetor as far as it will go. b. Grasp end of cross shaft with pliers and turn counterclockwise as far as it will go. The governor shaft can be adjusted for end clearance by moving needle bearing in block. Set bearing to allow a slight back-and-forth movement of shaft. c. Torque the pawlnut on governor arm to 15 inch lbs.

➡**Make sure there is at least 1/16 in. clearance between the governor arm and the upper-left cam gear cover fastener to prevent interference.**

1. If the engine is equipped with a flat muffler, install muffler and gasket using cap screws. If equipped with a round muffler, install muffler and threaded exhaust pipe between the muffler and engine using a pipe wrench.

2. Install the base gasket, base, and air cleaner.

PREPARE THE ENGINE FOR OPERATION

Before operating the engine, be sure to do the following:

• Make sure all hardware is tightened securely and oil drain plugs are installed.

• Fill the crankcase with the right amount, weight, and type of oil.

• Fill the fuel tank with the proper type of gasoline and open fuel shut-off valve (if equipped).

• Adjust the carburetor main fuel needle, idle fuel needle, or idle speed adjusting screw as necessary. Refer to the 'Fuel System And Governor' section.

Run-in Procedures (Reconditioned Engines)

For overhauled engines or those rebuilt with a new short block or miniblock, use straight SAE 30, SG-quality oil for the first 5 hours of operation. Change the oil after this initial run-in period. Refill with SG-quality oil as specified under 'Oil Types'.

HORSEPOWER (Maximum RPM) Engine Model			4 K91	7 K161	8 K181	10 K241	12 K301	14 K321	16 K341
GENERAL	Bore x Stroke		2.375x2.000	2.938x2.500	2.938x2.750	3.251x2.875	3.375x3.250	3.500x3.250	3.750x3.250
	Displacement Cu. In.		8.86	16.94	18.64	23.85	29.07	31.27	35.90
	Max. Operating RPM		4000	3600	3600	3600	3600	3600	3600
BALANCE GEAR	Shaft O.D.	New	—	—	—	.4998/.5001	.4998/.5001	.4998/.5001	.4998/.5001
		Minimum Wear Limit	—	—	—	.4996	.4996	.4996	.4996
	End Play		—	—	—	.002/.010	.002/.010	.002/.010	.002/.010
CAMSHAFT	Sleeve I.D. Installed		—	—	—	—	—	—	—
	End Play		.005/.020	.005/.010	.005/.010	.005/.010	.005/.010	.005/.010	.005/.010
CONNECTING ROD	Running Clearance	Rod To Crank-Pin (New)	.001/.0025	.001/.002	.001/.002	.001/.002	.001/.002	.001/.002	.001/.002
		Rod To Crank-Pin Wear Limit	.003	.0025	.0025	.0025	.0025	.0025	.0025
		Rod To Piston Pin (New)	.0007/.0000	.0006/.0011	.0006/.0011	.0003/.0006	.0003/.0006	.0003/.0006	.0003/.0006
	Small End I.D. (New)		.5630/.5633	.6255/.6250	.6255/.6250	.8596/.8599	.8757/.8760	.8757/.8760	.8757/.8760
CRANKSHAFT MAINS	PTO & Flywheel End O.D.	New	.9041/.9044	1.1811/1.1814	1.1811/1.1814	1.5745/1.5749	1.5745/1.5749	1.5745/1.5749	1.5745/1.5749
		Maximum Wear Limit	.9041	1.1811	1.1811	1.5745	1.5745	1.5745	1.5745
	Max. Out of Round (Sleeve)		—	—	—	—	—	—	—
	Max. Taper (Sleeve)		—	—	—	—	—	—	—
	Running Clearance (Sleeve)	Minimum Bore	—	—	—	—	—	—	—
		Wear Limit ①	—	—	—	—	—	—	—
	New Sleeve Bearing I.D. Installed		—	—	—	—	—	—	—
CRANKSHAFT CRANKPIN	New		.9360/.9355	1.1860/1.1855	1.1860/1.1855	1.5000/1.4995	1.5000/1.4995	1.5000/1.4995	1.5000/1.4995
	Max. Wear Limit		.9350	1.1850	1.1850	1.4990	1.4990	1.4990	1.4990
	Max. Out of Round		.0005	.0005	.0005	.0005	.0005	.0005	.0005
	Max. Taper		.001	.001	.001	.001	.001	.001	.001
	End Play		.004/.023	.002/.023	.002/.023	.003/.020	.003/.020	.003/.020	.003/.020
CYLINDER BORE	Inside Diameter	New	2.3755/2.3745	2.9380/2.9370	2.9380/2.9370	3.2515/3.2505	3.3755/3.3745	3.5005/3.4995	3.7505/3.7495
		Maximum Wear Limit	2.376	2.941	2.941	3.254	3.376	3.503	3.753
	Max. Out of Round		.003	.003	.003	.003	.003	.003	.003
	Max. Taper		.003	.003	.003	.002	.002	.002	.002
CYLINDER HD.	Max. Out of Flatness		.003	.003	.003	.003	.003	.003	.003
IGNITION	Spark Plug Type & Gap	Type⑥	RCJ-8	RCJ-8	RCJ-8	RH-10	RH-10	RH-10	RH-10
		Battery	.025	.025	.025	.035	.035	.035	.035
		Magneto	.025	.025	.025	.025	.025	.025	.025
		Gaseous Fuels	.018	.018	.018	.018	.018	.018	.018
	Nominal Point Gap		.020	.020	.020	.020	.020	.020	.020
PISTON ⑦	Service Replacement Sizes					.003 — .010 — .020 — .030			
	Thrust Face O.D.⑦	New	2.371/2.369	2.9297/2.9281	2.9297/2.9281	3.2432/3.2413	3.364/3.365	3.4941/3.4925	3.7425/3.7410③
		Maximum Wear Limit	2.366	2.925	2.925	3.238	3.363	3.491	3.738③
	Thrust Face To Bore Clearance (New) ⑦②		.0035/.006	.007/.010	.007/.010	.007/.010	.007/.010	.007/.010	.007/.010③
	Ring End Gap	New Bore	.007/.017	.007/.017	.007/.017	.010/.020	.010/.020	.010/.020	.010/.020
		Used Bore (Max.)	.027	.027	.027	.030	.030	.030	.030
	Max. Ring Side Clearance		.006	.006	.006	.006	.006	.006	.006
PISTON ⑨	Service Replacement Sizes					.003 — .010 — .020 — .030			
	Thrust Face O.D.⑨	New	—	—	2.9329/2.9336	—	3.3700/3.3693	3.4945/3.4938	3.7465/3.7455
		Maximum Wear Limit	—	—	2.931	—	3.367	3.492	3.744
	Thrust Face To Bore Clearance (New)⑨②		—	—	.0034/.0051	—	.0045/.0062	.0050/.0067	.0030/.0050
	Ring End Gap	New Bore⑧	—	—	.010/.023	—	.010/.020	.010/.020	.010/.020
		Used Bore⑧ (Max.)	—	—	.032	—	.030	.030	.030
	Max. Ring Side Clearance		—	—	.006	—	.006	.006	.006
PISTON PIN	Outside Diameter		.5623/.5625	.6247/.6249	.6247/.6249	.8591/.8593	.8752/.8754	.8752/.8754	.8752/.8754
	Guide Reamer Size		.250	.3125	.3125	.3125	.3125	.3125	.3125
VALVES	Tappet Clearance (Cold)	Intake	.005/.009	.006/.008	.006/.008	.008/.010	.008/.010	.008/.010	.008/.010④
		Exhaust	.011/.015	.017/.019	.017/.019	.017/.019	.017/.019	.017/.019	.017/.019
	Minimum Lift (Zero Lash)	Intake	.2035	.2718	.2718	.318	.318	.318	.318
		Exhaust	.1768	.2482	.2482	.318	.318	.318	.318
	Minimum Valve Stem O.D.	Intake	.2470	.3183	.3183	.3183	.3183	.3183	.3183
		Exhaust	.2450	.3068	.3068	.3074	.3074	.3074	.3074
	Nominal Angle Valve Seat		45°	45°	45°	45°	45°	45°	45°
	Guide I.D. Maximum Wear Limit①	Intake	.005	.005	.005	.006	.006	.006	.006
		Exhaust	.007	.007	.007	.008	.008	.008	.008

① Maximum limits combination of I.D. and O.D. measurements
② Ball bearing 1.3779/1.3784, Maximum Wear 1.3779
③ Ball bearing 1.7716/1.7721, Maximum Wear 1.7716
④ Pre Series II 1.3733/1.3738, Maximum Wear 1.3728
⑤ Ball bearing .002/.023
⑥ Champion spark plugs or equivalent
⑦ Measure just below oil ring groove and at right angles to piston pin
⑧ 1800 RPM generator sets .005/.007
⑨ Measure ½ above the bottom of the piston skirt.
⑩ Top and center compression rings.

* Includes K141

83257C11

1970-83 13-20 HP

Engine Identification

An engine identification plate is mounted on the carburetor side of the engine blower housing. The numbers that are important, as far as ordering replacement parts is concerned, are the model, serial, and specification numbers.

The model number indicates the engine model series. It also is a code indicating the cubic inch displacement and the number of cylinders. The model number K341, for instance, indicates the engine is 41 cu in. in displacement and that it has 1 cylinder. The letters following the model number indicate that a variety of other equipment is installed on the engine. The letters and what they mean are as follows:
- C Clutch model
- G Housed with fuel tank
- H Housed less fuel tank
- P Pump model
- R Reduction gear
- S Electric Start
- T Retractable start

➡A model number without a suffix letter indicates a basic rope start version.

The specification number indicates a model variation. It indicates a combination of various groups used to build the engine. It may have a letter preceding it which is sometimes important in determining superseding parts. The first two numbers of the specifications number is the code designating the engine model; the remaining numbers are issued in numerical sequence as each new specification is released, for example, 2899, 28100, 28101, etc.

The serial number lists the order in which the engine was built. If a change takes place to a model or a specification, the serial number is used to indicate the points at which the change takes place. The first letter or number in the serial number indicates what year the engine was built. The letter prefix to the engine serial number was dropped in 1969 and thereafter the prefix is a number. Engines made in 1969 have either the letter **E** or the number **1**. The code is as follows:
- 2 — 1970
- 3 — 1971
- 4 — 1972
- 5 — 1973
- 6 — 1974

Air Cleaners

A dirty air cleaner can cause rich fuel/air mixture and consequent poor engine operation and sludge deposits. If the filter becomes dirty enough, dirt that otherwise would be trapped can pass through and may wear the engine's moving parts prematurely. It is therefore necessary that all maintenance work be performed precisely as specified.

DRY AIR CLEANERS

◆ See Figure 136

Clean dry element air cleaners every 50 hours of operation, or every 6 months (whichever comes first) under good operating conditions. Service more frequently if the operating area is dusty. Remove the element and tap it lightly against a hard surface to remove the bulk of the dirt. If dirt will not drop off easily, replace the element. Do not use compressed air or solvents. Replace the air cleaner every 100-200 hours, under good conditions, and more frequently if the air is dusty.

Observe the following precautions:

1. Handle the element carefully — do not allow the gasket surfaces to become bent or twisted.

2. Make sure gasket surfaces seal against back plate and cover.

3. Tighten wing nut only finger tight — if it is too tight, cleaner may not seal properly.

If the dry type air cleaner is equipped with a precleaner, service this unit when cleaning the paper element. Servicing consists of cleaning the precleaner in soap and water, squeezing the excess out, and then allowing it to air dry before installation. Do not oil!

OIL BATH AIR CLEANERS

This type of unit may be used to replace the dry type in applications where very frequent replacement of the element is required. The conversion is simple and requires the use of an elbow to fit the oil bath unit onto the engine in a vertical position.

Service the unit every 25 hours of operation under good conditions and, under dusty conditions, as often as every 8 hours of operation.

Service as follows:

1. Remove cover and lift element out of bowl.

2. Drain dirty oil from bowl, and then wash thoroughly in clean solvent.

3. Swish the element in the solvent and then allow it to drip dry. Do not dry with compressed air. Lightly oil the element with engine oil.

4. Inspect air horn, filter bowl, and cover gaskets, and replace as necessary (if grooved or cracked).

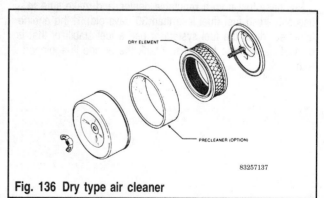

Fig. 136 Dry type air cleaner

5. Install filter bowl gasket on air horn, then put the bowl into position. Fill bowl to indicated level with engine oil.

6. Install element, put the cover in position, and then install copper gasket (if used) and wingnut. Tighten wingnut with fingers only to avoid distorting housing. Make sure all joints in the unit seal tightly.

Lubrication

CRANKCASE

Oil level must be maintained between **F** and **L** marks — do not overfill. Check every day and add as necessary. On new engines, be especially careful to stop engine and check level frequently. When checking, make sure regular type dipstick is inserted fully. On screw type dipstick, check level with dipstick inserted fully but not screwed in. On this type, however, make sure to screw dipstick back in tightly when oil level check is completed.

Use SG type oils meeting viscosity specifications according to the prevailing temperature as shown in the chart below.

Change initial fill of oil on new engines after five hours of operation. Then, change oil every 25 hours of operation. Change oil when engine is hot. Change more frequently in dusty areas. If the engine has just been overhauled, it is best to fill it initially with a non-detergent oil. Then, after 5 hours, refill with SG type oil.

REDUCTION GEAR UNITS

Every 50 hours, remove the oil plug on the lower part of the reduction unit cover to check level. If oil does not reach the level of the oil plug, remove the vented fill plug from the top of the cover and refill with engine oil until level is correct. This oil need not be changed unless unit has been out of service for several months. In this situation, remove the drain plug, drain oil, then replace plug and fill to proper level as described above.

FUEL RECOMMENDATIONS

Use either leaded or unleaded regular grade fuel of at least 90 octane. Unleaded fuel produces fewer combustion chamber deposits, so its use is preferred.

Purchase fuel from a reputable dealer, and make sure to use only fresh fuel (fuel less than 30 days old). If the engine is stored, drain the fuel system or use a fuel stabilizer that is compatible with the type of fuel tank the engine is equipped with.

Spark Plugs

SERVICE

▶ **See Figures 137, 138, 139, 140 and 141**

The spark plug should be removed and serviced every 100 hours of engine operation. The plug should have a light coating of light gray colored deposits. If deposits are black, fuel/air mixture could be too rich due to improper carburetor adjustment or a dirty air cleaner. If deposits are white, the engine may be overheating or a spark plug of too high a heat range could be in use.

Kohler recommends that the plug be replaced rather than sandblasted or scraped if there are excessive deposits. Torque plugs to 18-22 ft. lbs.

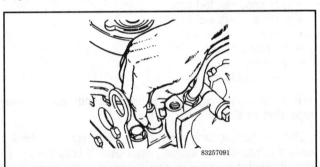

Fig. 137 Twist and pull on the boot, never on the plug wire

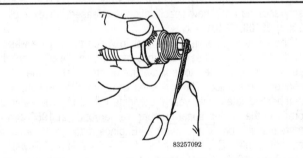

Fig. 138 Plugs that are in good condition can be filed and reused

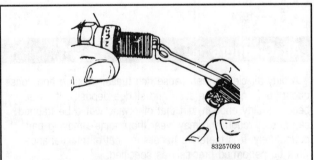

Fig. 139 Adjust the electrode gap by bending the side electrode

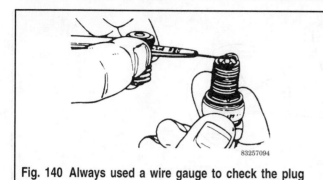

Fig. 140 Always used a wire gauge to check the plug gap

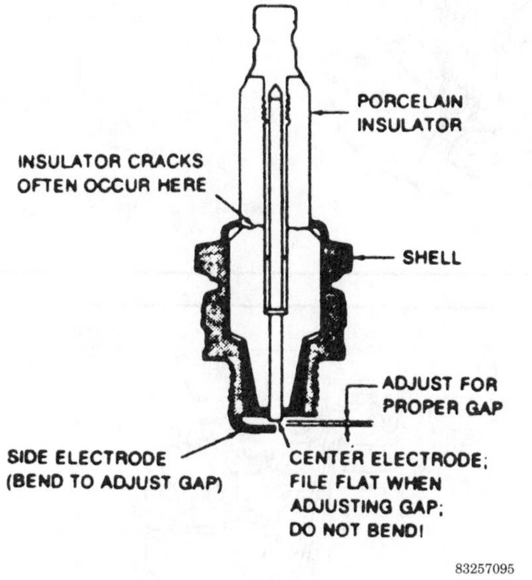

every 100 hours of operation. Replace the points if they are badly burned. If there is a great deal of metal buildup on either contact, the condenser may be faulty and should be replaced.

To replace points, remove the primary wiring connector screw and pull off the primary wire. Then, remove the contact set mounting screws and remove the contact set. Install the new set of points in reverse order, leaving upper mounting screw slightly loose. Then set point gap and timing as described below.

SETTING BREAKER GAP AND TIMING

▶ **See Figures 142 and 143**

1. Remove the breaker cover and disconnect the spark plug lead. Rotate the engine in direction of normal rotation until the points reach the maximum opening.

2. Using a clean, flat feeler gauge of 0.020 in. (0.50mm) size, check the gap between the points. Gauge should just slide between the contacts without opening them when flat between them. If the gap is incorrect, loosen the upper mounting screw (if necessary), and shift the breaker base with the blade of the screwdriver until gap is correct.

3. There is a timing sight hole in either the bearing plate or the blower housing. If there is a snap button in the hole, pry it out with a screwdriver.

4. While observing the sight hole, turn the engine slowly in normal direction of rotation. When the **S** or **SP** mark (engines with Automatic Compression Release) or the **T** mark on engines without ACR, appears in the hole, the points should just be beginning to open. If timing is incorrect, breaker gap will have to be reset slightly — 0.018-0.022 in. (0.46-0.56mm). If the points are not yet opening when the timing mark is centered in the hole, make the point gap wider. If points open too early, narrow it. Recheck the setting after tightening the upper breaker mounting screw by turning the engine in normal direction of rotation past the firing point and checking that the points open at just the right time.

➡**This procedure may be performed with the engine running at 1200-1800 rpm if a timing light is available. Connect the timing light according to manufacturer's instructions. You may have to chalk the timing mark to see it adequately.**

TRIGGER AIR GAP

Trigger air gap is set within the range 0.005-0.010 in. (0.127-0.254mm). As long as the gap falls within this range, the ignition system should perform adequately. Optimum ignition performance during cold weather starting is provided if the gap is adjusted to 0.005 in. (0.127mm). If you wish to adjust this or to ensure that the gap falls within the proper range, rotate the flywheel until the flywheel projection is lined up with the trigger assembly. Then, loosen the trigger bracket capscrews and slide the trigger back and forth to get the proper gap, as measured with a flat feeler gauge. Then, retighten capscrews.

PORCELAIN
INSULATOR

INSULATOR CRACKS
OFTEN OCCUR HERE

SHELL

ADJUST FOR
PROPER GAP

SIDE ELECTRODE
(BEND TO ADJUST GAP)

CENTER ELECTRODE;
FILE FLAT WHEN
ADJUSTING GAP;
DO NOT BEND!

Fig. 141 Cross-section of a spark plug

TESTING

To test a plug for adequate performance, remove it from the engine, attach the ignition wire, and then rest the side electrode against the cylinder head. Crank the engine vigorously. If there is a sharp spark, the plug and ignition system are all right, although ignition timing should be checked if the engine fires irregularly.

Breaker Points

INSPECTION

Remove the breaker cover and inspect the points for pitting or buildup of metal on either the movable or stationary contact

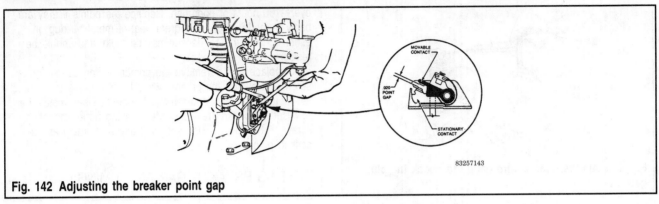

Fig. 142 Adjusting the breaker point gap

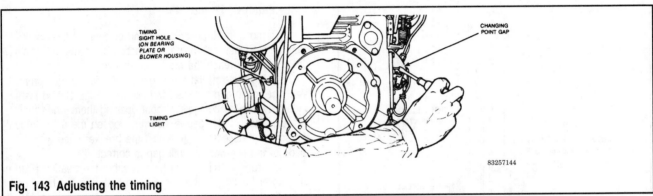

Fig. 143 Adjusting the timing

IGNITION COILS

Coils do not require regular service, except to make sure they are kept clean, that the connections are tight, and that rubber insulators are in good condition (replace if cracked). If you suspect poor performance of a breakerless type ignition system and trigger air gap is correct, check resistance with an ohmmeter. To do this, disconnect the high tension lead at the coil and connect the meter between coil terminal and coil mounting bracket. If resistance is not about 11,500 ohms, replace the coil. Also, check the reading with the meter lead going to the coil terminal pulled off and connected to the spark plug connector of the high tension lead. If there is continuity here, replace the coil.

PERMANENT MAGNETS

These may be checked for magnet strength by holding a screwdriver (non-magnetic) blade within one inch of the magnet. If the magnetic field is good, the blade will be attracted to the magnet. Otherwise, replace it.

Mixture Adjustments

▶ **See Figures 144 and 145**

➡Before making any adjustments, be sure that the carburetor air cleaner is not clogged. A clogged air cleaner will cause an over-rich mixture, black exhaust smoke, and may lead you to believe that the carburetor is out of adjust-

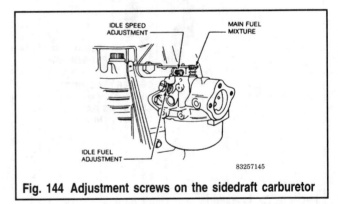

Fig. 144 Adjustment screws on the sidedraft carburetor

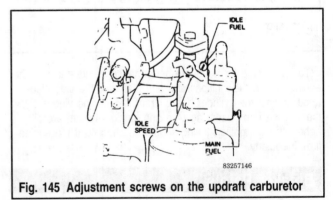

Fig. 145 Adjustment screws on the updraft carburetor

ment when, in reality, it is not. The carburetor is set at the factory and rarely needs adjustment unless, of course, it has been disassembled or rebuilt.

1. With the engine stopped, turn the main and idle fuel adjusting screws all the way in until they bottom lightly. Do not force the screws or you will damage the needles.

2. For a preliminary setting, turn the main fuel screw out 2 full turns and the idle screw out 1¼ turns.

3. Start the engine and allow it to reach operating temperatures; then operate the engine at full throttle and under a load, if possible.

4. For final adjustment, turn the main fuel adjustment screw in until the engine slows down (lean mixture), then out until it slows down again (rich mixture). Note the positions of the screw at both settings, then set it about halfway between the two positions.

5. Set the idle mixture adjustment screw in the same manner. The idle speed (no-load) on most engines is 1200 rpm; however, on engines with a parasitic load (hydrostatic drives) the engine idle speed may have to be increased to as much as 1700 rpm for best no-load idle.

Governor Adjustment

All Kohler engines use mechanical, camshaft driven governors.

INITIAL ADJUSTMENT

▶ **See Figure 146**

1. Loosen, but do not remove, the nut that holds the governor arm to the governor cross shaft.

2. Grasp the end of the cross shaft with a pair of pliers and turn it in counterclockwise as far as it will go. The tab on

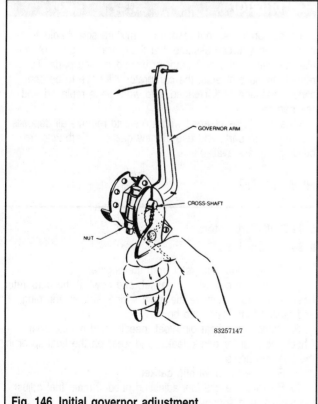

Fig. 146 Initial governor adjustment

the cross shaft will stop against the rod on the governor gear assembly.

3. Pull the governor arm away from the carburetor, then retighten the nut which holds the governor arm to the shaft. With updraft carburetors, lift the arm as far as possible, then retighten the arm nut.

FINAL ADJUSTMENT

Engine must be adjusted to 3600 rpm.

1. Start the engine and measure the speed with a tachometer.

2. If the speed is incorrect, adjust as follows:

a. Constant Speed Governor — Tighten the governor adjusting screw to increase speed, or loosen to decrease speed until the correct speed is attained.

b. Variable Speed Governor — Loosen the capscrew, move the high speed stop bracket until the correct speed is attained, and then retighten the capscrew.

If the governor is too sensitive (causing hunting or surging), or not sensitive enough (causing too great a drop in speed when load is applied), the governor sensitivity should be adjusted. Make the governor more sensitive by moving the spring to holes further apart. Make it less sensitive by moving it to holes that are closer together. Standard setting is the third hole from the bottom on the governor arm and second hole from the top on the speed control bracket.

Choke Adjustment

THERMOSTATIC TYPE

If the engine does not start when cranked, continue cranking and move the choke lever first to one side and then to the other to determine whether the setting is too lean or too rich. Once the direction in which lever must be moved has been determined, loosen the adjusting screw on the choke body. Then, move the bracket downward to increase choking or upward to decrease it. Then, tighten the lockscrew. Try starting it again and readjust as necessary.

ELECTRIC-THERMOSTATIC TYPE

▶ **See Figure 147**

Remove the air cleaner from the carburetor and check the position of the choke plate. The choke should be fully closed when engine is at outside temperature and the temperature is very low. In milder temperatures, slightly less closure is required.

If adjustment is required, move the choke arm until the hole in the brass shaft lines up with the slot in the bearings. Insert a #43 (0.089 in.) drill through the shaft and push it downward so it engages the notch in the base of the choke unit. Then, loosen the clamp bolt on the choke lever and push the arm upward to move the choke plate toward the closed position. When the desired position is obtained, tighten the clamp bolt. Then, remove the drill.

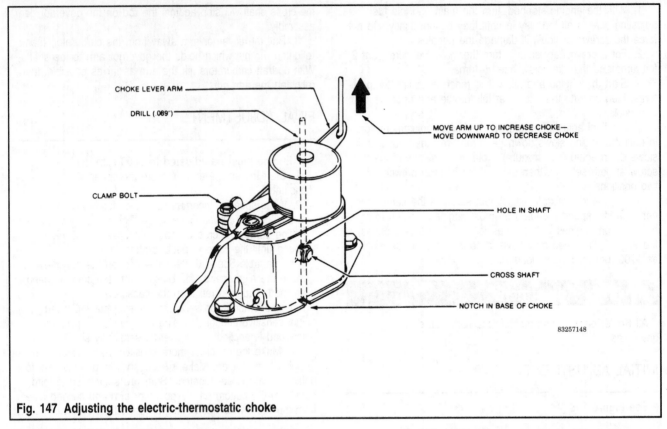

Fig. 147 Adjusting the electric-thermostatic choke

Remount the air cleaner, and then check for any binding in the choke linkage. Correct as necessary. Finally, run the engine until hot, and make sure the choke opens fully. If not, readjust it toward the open position as necessary.

Valve Adjustment

Adjustable valve tappets are provided. With the engine cold, turn crankshaft until it reaches Top Dead Center timing mark. If valves are slightly open, turn the crankshaft another turn until valves are closed and engine is again at Top Center. Check valve clearances with a flat feeler gauge. Note that exhaust and intake clearances are different, and make sure you're using the right gauge for each valve. If the valve clearance is correct, a gauge can just be inserted between tappet and valve stem. A slight pull is required to bring it back out. If clearance is incorrect, loosen the locking nut and turn the adjusting nut in or out to get the proper clearance. Hold the adjusting nut while tightening the locknut and recheck clearance.

Compression Check

Compression is checked by removing the spark plug lead and spinning the flywheel forward against compression. If the piston does not bounce backward with considerable force, checking with a gauge may be necessary. On Automatic Compression Release engines, rotate the flywheel backward against power stroke — if little resistance is felt, check compression with a gauge.

The compression gauge check requires rapid motoring (spinning) of the crankshaft, at about 1000 rpm. Install the gauge in the spark plug hole and motor the engine. Gauge should read 110-120 psi. If reading is less than 100 psi, the engine requires major repair to piston rings or valves.

Carburetor

If a carburetor will not respond to mixture screw adjustments, then you can assume that there are dirt, gum, or varnish deposits in the carburetor or worn/damaged parts. To remedy these problems, the carburetor will have to be completely disassembled, cleaned, and worn parts replaced and reassembled.

Parts should be cleaned with solvent to remove all deposits. Replace worn parts and use all new gaskets. Carburetor rebuilding kits are available.

REBUILDING

Side Draft Carburetors
▶ **See Figure 148**

1. Remove the carburetor from the engine.
2. Remove the bowl nut, gasket, and bowl. If the carburetor has a bowl drain, remove the drain spring, spacer and plug, and gasket from inside the bowl.
3. Remove the float pin, float, needle, and needle seat. Check the float for dents, leaks, and wear on the float lip or in the float pin holes.
4. Remove the bowl ring gasket.
5. Remove the idle fuel adjusting needle, main fuel adjusting needle, and springs.

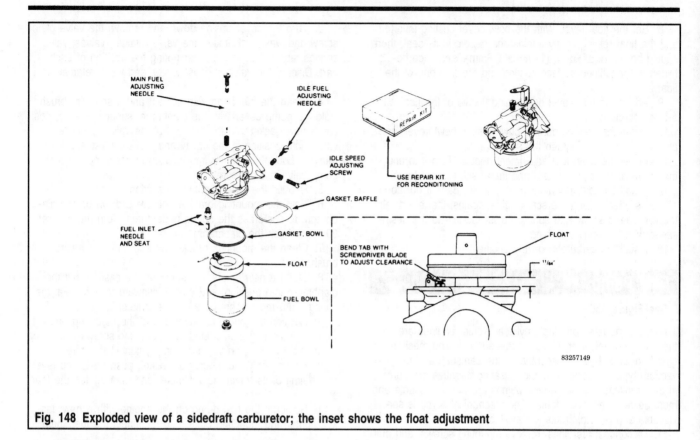

Fig. 148 Exploded view of a sidedraft carburetor; the inset shows the float adjustment

6. Do not remove the choke and throttle plates or shafts. If these parts are worn, replace the entire carburetor assembly.

To assemble:

7. Install the needle seat, needle, float, and float pin.

8. Set the float level. With the carburetor casting inverted and the float resting against the needle in its seat, there should be $^{11}/_{64} \pm ^{1}/_{32}$ in. (4.4mm ± 0.8mm) clearance between the machined surface of the casting and the free end of the float.

9. Adjust the float level by bending the lip of the float with a small screwdriver.

10. Install the new bowl ring gasket, new bowl nut gasket, and bowl nut. Tighten the nut securely.

11. Install the main fuel adjustment needle. Turn it in until the needle seats in the nozzle and then back out two turns.

12. Install the idle fuel adjustment needle. Back it out about $1^{1}/_{4}$ turns after seating it lightly against the jet.

13. Install the carburetor on the engine.

Updraft Carburetors

▶ **See Figure 149**

1. Remove the carburetor from the engine.
2. Remove the bowl cover and the gasket.
3. Remove the float pin, float, needle and needle seat. Check the float pin for wear.
4. Remove the idle fuel adjustment needle, main fuel adjustment needle, and the springs. Do not remove the choke plate or the shaft unless the replacement of these parts is necessary.

To assemble:

5. Install the throttle shaft and plate. The elongated side of the valve must be toward the top.

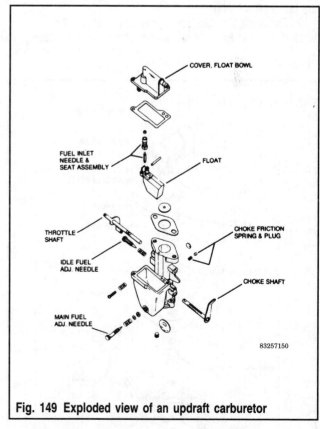

Fig. 149 Exploded view of an updraft carburetor

6. Install the needle seat. A $^{5}/_{16}$ in. socket should be used. Do not over-tighten.

7. Install the needle, float, and float pins.

8. Set the float level. With the bowl cover casting inverted and the float resting lightly against the needle in its seat, there should be $7/16$ in. $\pm$ $1/32$ in. (11mm $\pm$ 0.8mm). clearance between the machined surface casting and the free end of the float.

9. Adjust the float level by bending the lip of the float with a small screwdriver.

10. Install the new carburetor bowl gasket, bowl cover, and bowl cover screws. Tighten the screws securely.

11. Install the main fuel adjustment needle. Turn it in until the screw seats in the nozzle and then back it out 2 turns.

12. Install the idle fuel adjustment needle. Back it out about $1\frac{1}{2}$ turns after seating the screw lightly against the jet. Install the idle speed screw and spring. Adjust the idle to the desired speed with the engine running.

13. Install the carburetor on the engine.

Fuel Pump

▶ **See Figure 150**

Fuel pumps used on single cylinder Kohler engines are either the mechanical or vacuum actuated type. The mechanical type is operated by an eccentric on the camshaft and the vacuum type is operated by the pulsating negative pressures in the crankcase. Vacuum type pumps are not serviceable and must be replaced when faulty. The mechanical pump is serviceable and rebuilding kits are available.

1. Disconnect fuel lines, remove mounting screws, and pull the pump off the engine.

2. File a mark across some point at the union of pump body and cover. Remove the screws and remove the cover.

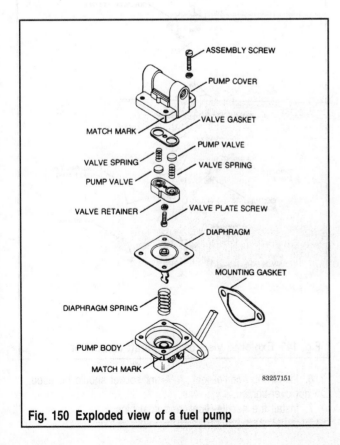

Fig. 150 Exploded view of a fuel pump

3. Turn the cover upside down and remove the valve plate screw and washer. Remove the valve retainer, valves, valve springs, and valve gasket, after noting the position of each part. Discard the valve springs, valves and valve retainer gasket.

4. Clean the fuel head with solvent and a soft wire brush. Hold the pump cover with the diaphragm surface upward; position a new gasket into the cavity. Put the valve spring and valves into position in the cavity and reassemble the valve retainer. Lock the retainer into position by installing the fuel pump valve retainer screw.

5. Rebuild the lower diaphragm section.

6. Hold the mounting bracket and press down on the diaphragm to compress the spring underneath. Turn the bracket 90° to unhook the diaphragm and remove it.

7. Clean the mounting bracket with solvent and a wire brush.

8. Stand a new diaphragm spring in the casting, put the diaphragm into position, and push downward to compress the spring. Turn the diaphragm 90° to reconnect it.

9. Position the pump cover on top of the mounting bracket with the indicating marks lined up. Install the screws loosely on mechanical pumps; on vacuum pumps, tighten the screws.

10. Holding only the mounting bracket, push the pump lever to the limit of its travel, hold it there, and then tighten the four screws.

11. Remount the fuel pump on the engine with a new gasket, tighten the mounting bolts, and reconnect the fuel lines.

Engine Overhaul

DISASSEMBLY

The following procedure is designed to be a general guide rather than a specific and all inclusive disassembly procedure. The sequence may have to be varied slightly to allow for the removal of special equipment or accessory items such as motor/generators, starters, instrument panels, etc.

1. Disconnect the high tension spark plug lead and remove the spark plug.

2. Close the valve on the fuel sediment bowl and remove the fuel line at the carburetor.

3. Remove the air cleaner from the carburetor intake.

4. Remove the carburetor.

5. Remove the fuel tank. The sediment bowl and brackets remain attached to the fuel tank.

6. Remove the blower housing, cylinder baffle, and head baffle.

7. Remove the rotating screen and the starter pulley.

8. The flywheel is mounted on the tapered portion of the crankcase and is removed with the help of a puller. Do not strike the flywheel with any type of hammer.

9. Remove the breaker point cover, breaker point lead, breaker assembly, and the pushrod that operates the points.

10. Remove the magneto assembly.

11. Remove the valve cover and breather assembly.

12. Remove the cylinder head.

13. Raise the valve springs with a valve spring compressor and remove the valve spring keepers from the valve stems. Remove the valve spring retainers, springs, and valves.

14. Remove the oil pan base and unscrew the connecting rod can screws. Remove the connecting rod cap and piston assembly from the cylinder block.

➡**It will probably be necessary to use a ridge reamer on the cylinder walls before removing the piston assembly, to avoid breaking the piston rings.**

15. Remove the crankshaft, oil seals and, if necessary, the anti-friction bearings.

➡**It may be necessary to press the crankshaft out of the cylinder block. The bearing plate should be removed first, if this is the case.**

16. Turn the cylinder block upside down and drive the camshaft pin out from the power take-off side of the engine with a small punch. The pin will slide out easily once it is driven free of the cylinder block.

17. Remove the camshaft and the valve tappets.

18. Loosen and remove the governor arm from the governor shaft.

19. Unscrew the governor bushing nut and remove the governor shaft from the inside of the cylinder block.

20. Loosen, but do not remove, the screw located at the lower right of the governor bushing nut until the governor gear is free to slide off of the stub shaft.

CYLINDER BLOCK SERVICE

▶ **See Figure 151**

Make sure that all surfaces are free of gasket fragments and sealer materials. The crankshaft bearings are not to be removed unless replacement is necessary. One bearing is pressed into the cylinder block and the other is located in the bearing plate. If there is no evidence of scoring or grooving and the bearings turn easily and quietly it is not necessary to replace them.

The cylinder bore must not be worn, tapered, or out-of-round more than 0.005 in. (0.127mm). Check at two locations 90 degrees apart and compare with specifications. If it is, the cylinder must be rebored. If the cylinder is very badly scored or damaged it may have to be replaced, since the cylinder can only be rebored to either 0.010 in. (0.254mm) or 0.020 in. (0.500mm) and 0.030 in. (0.762mm) maximum. Select the

nearest suitable oversize and bore it to that dimension. On the other hand, if the cylinder bore is only slightly damaged, only a light deglazing may be necessary.

HONING THE CYLINDER BORE

1. The hone must be centered in relation to the crankshaft crossbore. It is best to use a low speed drill press. Lubricate the hone with kerosene and lower it into the bore. Adjust the stones so they contact the cylinder walls.

2. Position the lower edge of the stones even with the lower edge of the bore, hone at about 600 rpm. Move the hone up and down continuously. Check bore size frequently.

3. When the bore reaches a dimension 0.0025 in. (0.0635mm) smaller than desired size, replace the coarse stones with burnishing stones. Use burnishing stones until the dimension is within 0.0005 in. (0.0127mm) of desired size.

4. Use finishing stones and polish the bore to final size, moving the stones up and down to get a 60 degree cross-hatch pattern. Wash the cylinder wall thoroughly with soap and water, dry, and apply a light coating of oil.

CRANKSHAFT SERVICE

Inspect the keyway and the gears that drive the camshaft. If the keyways are badly worn or chipped, the crankshaft should be replaced. If the cam gear teeth are excessively worn or if any are broken, the crankshaft must be replaced.

Check the crankpin for score marks or metal pickup. Slight score marks can be removed with a crocus cloth soaked in oil. If the crankpin is worn more than 0.002 in. (0.05mm), the crankshaft is to be either replaced or the crankpin reground to 0.010 in. (0.254mm) undersize. If the crankpin is reground to 0.010 in. (0.254mm) undersize, a 0.010 in. (0.254mm) undersize connecting rod must be used to achieve proper running clearance.

CONNECTING ROD SERVICE

Check the bearing area for wear, score marks, and excessive running and side clearance. Replace the rod and cap if they are worn beyond the limits allowed.

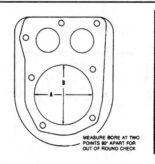
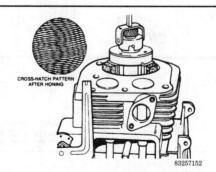

Fig. 151 Left side: measuring the cylinder bore; right side: honing must be a cross-hatch pattern as shown

PISTON AND RINGS SERVICE

Production and Service Type

▶ **See Figure 152**

Rings are available in the standard size as well as 0.010 in. (0.254mm), 0.020 in. (0.500mm), and 0.030 in. (0.762mm) oversize sets.

➡**Never reuse old rings.**

The standard size rings are to be used when the cylinder is not worn or out-of-round. Oversize rings are only to be used when the cylinder has been rebored to the corresponding oversize. Service type rings are used only when the cylinder is worn but within the wear and out-of-round limitations; wear limit is 0.005 in. (0.127mm) oversize and out-of-round limit is 0.004 in. (0.10mm).

➡**Never reuse old rings.**

The standard size rings are to be used when the cylinder is not worn or out-of-round. Oversize rings are only to be used when the cylinder has been rebored to the corresponding oversize. Service type rings are used only when the cylinder is worn but within the wear and out-of-round limitations; wear limit is 0.005 in. (0.127mm) oversize and out-of-round limit is 0.004 in. (0.10mm).

The old piston may be reused if the block does not need reboring and the piston is within wear limits. Never reuse old rings. After removing old rings, thoroughly remove deposits from ring grooves. New rings must each be positioned in its running area of the cylinder bore for an end clearance check, and each must meet specifications.

The cylinder must be deglazed before replacing the rings. If chrome plated rings are used, the chrome plated ring must be installed in the top groove. Make sure that the ring grooves are free from all carbon deposits. Use a ring expander to install the rings. Then check side clearance.

PISTON AND ROD SERVICE

Normally very little wear will take place at the piston boss and piston pin. If the original piston and connecting rod can be used after rebuilding, the piston pin may also be used. However if a new piston or connecting rod or both have to be used, a new piston pin must also be installed. Lubricate the pin before installing it with a loose to light interference fit. Use new piston pin retainers whether or not the pin is new. Make sure they're properly engaged.

VALVES AND VALVE MECHANISM SERVICE

Inspect the valve mechanism, valves, and valve seats or inserts for evidence of wear, deep pitting, cracks or distortion. Check the clearance between the valve stems and the valve guides.

Valve guides must be replaced if they are worn beyond the limit allowed.

To remove valve guides, press the guide down into the valve chamber and carefully break off the protruding end until the guide is completely removed. Be careful not to damage the block when removing the old guides. Use an arbor press to install the new guides. Press the new guides to the depth

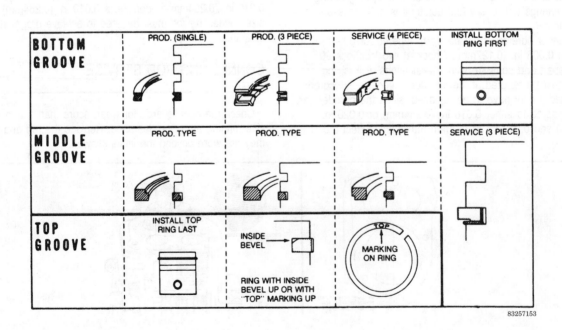

Fig. 152 Piston ring positioning for production-type rings

specified, then use a valve guide reamer to gain the proper inside diameter.

Make sure that replacement valves are the correct type (special hard faced valves are needed in some cases). Exhaust valves are always hard faced.

Intake valve seats are usually machined into the block, although inserts are used in some engines. Exhaust valve seats are made of special hardened material. The seating surfaces should be held as close to ¹⁄₃₂ in. (0.8mm) in width as possible. Seats more than ¹⁄₁₆ in. (1.6mm) wide must be reground with 45 degree and 15 degree cutters to obtain the proper width. Reground or new valves and seats must be lapped in for a proper fit.

After resurfacing valves and seats and lapping them in, check the valve clearance. Hold the valve down on its seat and rotate the camshaft until it has no effect on the tappet, then check the clearance between the end of the valve stem and the tappet. If the clearance is not sufficient (it will always be less after grinding), it will be necessary to grind off the end of the valve stem until the correct clearance is obtained. This is necessary on all engines except those which all have adjustable tappets.

CYLINDER HEAD SERVICE

Remove all carbon deposits and check for pitting from hot spots. Replace the head if metal has been burned away because of head gasket leakage. Check the cylinder head for flatness. If the head is slightly warped, it can be resurfaced by rubbing it on a piece of sandpaper placed on a flat surface. Be careful not to nick or scratch the head when removing carbon deposits.

DYNAMIC BALANCE SYSTEM SERVICE

▶ **See Figure 153**

The dynamic balance system consists of two balance gears which run on needle bearings. The gears are assembled on two stub shafts that are pressed into special bosses in the crankcase. Snap-rings hold the gears and spacer washers are used to control end-play. The gears are driven off of the crankgear. The dynamic balance system is found on some special versions.

If the stub shaft is worn or damaged, press the old shaft out. The new shafts must be pressed in a specified distance which depends upon the distance between the stub shaft boss and main bearing boss. Measure the distance the stub shaft boss protrudes above the main bearing boss and then press the shaft in for a protrusion of the shaft end beyond stub shaft boss as specified. If stub shaft boss protrudes about ⁷⁄₁₆ in. (11mm) beyond main bearing boss, press the shaft in until it is 0.735 in. (18.7mm) above stub shaft boss. If protrusion is about ¹⁄₁₆ in. (1.6mm), press the stub shaft in until it is 1.110 in. (28mm) above the stub shaft boss, and then use a ³⁄₈ in. (9.5mm) spacer.

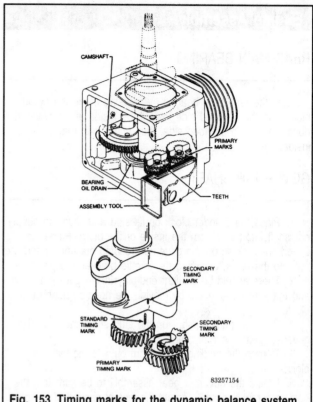

Fig. 153 Timing marks for the dynamic balance system

When installing the balance gears, slip one 0.010 in. (0.254mm) spacer onto the stub shaft, then install the gear/bearing assembly onto the stub shaft with the timing marks facing out. Proper end-play of 0.002-0.010 in. (0.05-0.25mm) is attained with one 0.005 in. (0.127mm) spacer, one 0.010 in. (0.254mm) spacer, and one 0.020 in. (0.50mm) spacer which are all installed on the snap-ring retainer end of the shaft. Install the thickest spacer next to the retainer. Check the end-play and adjust it by adding or subtracting 0.005 in. (0.127mm) spacers.

To time the balance gears, first press the crankshaft into the block and align the primary timing mark on the top of the balance gear with the standard timing mark next to the crankgear. Press the shaft in until the crankgear is engaged ¹⁄₁₆ in. (1.6mm) into the top gear (narrow side). Rotate the crankshaft to align the timing marks on the crankgear and cam gear. Press the crankshaft the remainder of the way into the block.

Rotate the crankshaft until it is about 15 degrees past BDC and slip one 0.010 in. (0.25mm) spacer over the stub shaft before installing the bottom gear/bearing assembly.

Align the secondary timing mark on this gear with the secondary timing mark on the counterweight of the crankshaft and then install the gear on the shaft. The secondary timing mark will also be aligned with the standard timing mark on the crankshaft after installation. Use one 0.005 in. (0.127mm) spacer and one 0.020 in. (0.50mm) spacer (with larger spacer next to retainer) to get proper end play of 0.002-0.010 in. (0.05-0.25mm). Install the snap-ring retainer, then check and adjust the endplay.

Engine Assembly

REAR MAIN BEARING

Install the rear main bearing by pressing it into the cylinder block with the shielded side toward the inside of the block. If it does not have a shielded side, then either side may face inside.

GOVERNOR SHAFT

1. Place the cylinder block on its side and slide the governor shaft into place from the inside of the block. Place the speed control disc on the governor bushing nut and thread the nut into the block, clamping the throttle bracket into place.
2. There should be a slight end-play in the governor shaft and that can be adjusted by moving the needle bearing in the block.
3. Place a space washer on the stub shaft and slide the governor gear assembly into place.
4. Tighten the holding screw from outside the cylinder block.
5. Rotate the governor gear assembly to be sure that the holding screw does not contact the weight section of the gear.

CAMSHAFT

▶ **See Figures 154 and 155**

1. Turn the cylinder block upside down.
2. The tappets must be installed before the camshaft is installed. Lubricate and install the tappets into the valve guides.
3. Position the camshaft inside the block.

➡**Align the marks on the camshaft and the automatic spark advance, if so equipped.**

4. Lubricate the rod and insert it into the bearing plate side of the block. Install one 0.005 in. (0.127mm) washer between the end of the camshaft and the block. Push the rod through the camshaft and tap it lightly until the rod just starts to enter the bore at the PTO end of the block. Check the endplay and adjust it with additional washers if necessary. Press the rod into its final position.

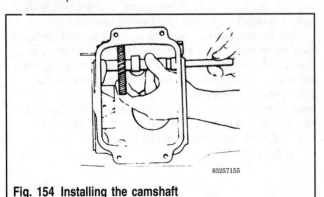

Fig. 154 Installing the camshaft

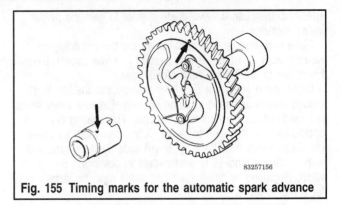

Fig. 155 Timing marks for the automatic spark advance

5. The fit at the bearing plate for the camshaft rod is a light to loose fit to allow oil that might leak past to drain back into the block.

CRANKSHAFT

▶ **See Figure 156**

1. Place the block on the base of an arbor press and carefully insert the tapered end of the crankshaft through the inner race of the anti-friction bearing.
2. Turn the crankshaft and camshaft until the timing mark in the shoulder of the crankshaft lines up with the mark on the cam gear.
3. When the marks are aligned, press the crankshaft into the bearing, making sure that the gears mesh as it is being pressed in. Recheck the alignment of the timing marks on the crankshaft and the camshaft.
4. The end-play of the crankshaft is controlled by the application of various thickness gaskets between the bearing plate and the block. Normal end-play is achieved by installing 0.020 in. (0.50mm) and 0.010 in. (0.25mm) gaskets, with the thicker gaskets on the inside.

BEARING PLATE

1. Press the front main bearing into the bearing plate. Make sure that the bearing is straight.
2. Press the bearing plate onto the crankshaft and into position on the block. Install the cap screws and secure the plate to the block. Draw up evenly on the screws.

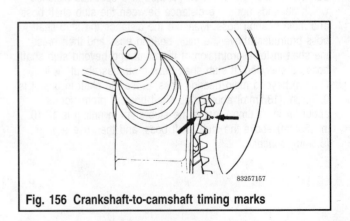

Fig. 156 Crankshaft-to-camshaft timing marks

3. Measure the crankshaft end-play, which is very critical on gear reduction engines.

PISTON AND ROD ASSEMBLY

▶ **See Figure 157**

1. Lubricate the pin and assemble it to the connecting rod and piston. Install the wrist pin retaining ring. Use new retaining rings.

2. Lubricate the entire assembly, stagger the ring gaps and, using a ring compressor, slide the piston and rod assembly into the cylinder bore with the connecting rod marks on the flywheel side of the engine.

3. Place the block on its end and oil the connecting rod end and the crankpin.

4. Attach the rod cap, lock or lock washers, and the cap screws. Tighten the screws to the correct torque.

➡ **Align the marks on the cap and the connecting rod.**

5. Bend the lock tabs to lock the screws.

CRANKSHAFT OIL SEALS

Apply a coat of grease to the lip and guide the oil seals onto the crankshaft. Make sure no foreign material gets onto the knife edges of seal, and make sure the seal does not bend. Place the block on its side and drive the seals squarely into the bearing plate and block.

OIL PAN BASE

Using a new gasket on the base, install pilot studs to align the cylinder block, gasket, and base. Tighten the four attaching screws to the correct torque.

VALVES

1. Clean the valves, seats, and parts thoroughly. Grind and lap-in the valves and seats for proper seating. Valve seat width must be $1/32$-$1/16$ in. (0.8-1.6mm). After grinding and lapping, slide the valves into position and check the clearance

between stem and tappet. If the clearance is too small, grind the stem ends square and remove all burrs. On engines with adjustable valves, make the adjustment at this time.

2. Place the valve springs, retainers, and rotators under the valve guides. Lubricate the valve stems, and then install the valves down through the guides, compress the springs, and place the locking keys or pins in the grooves of the valve stems.

CYLINDER HEAD

▶ **See Figure 158**

1. Use a new cylinder head gasket.

2. Lubricate and tighten the head bolts evenly, and in sequence, to the proper torque.

3. Install the spark plug.

BREATHER ASSEMBLY

▶ **See Figure 159**

Assemble the breather assembly, making sure that all parts are clean and the cover is securely tightened to prevent oil leakage.

MAGNETO

On flywheel magneto systems, the coil-core assembly is secured onto the bearing plate. On magneto-alternator systems, the coil is part of the stator assembly, which is secured to the bearing plate. On rotor type magneto systems, the rotor has a keyway and is press fitted onto the crankshaft. The magnet rotor is marked 'engine-side' for proper assembly. Run all leads through the hole provided at the 11 o'clock position on the bearing plate.

FLYWHEEL

1. Place the washer in place on the crankshaft and place the flywheel in position. Install the key.

2. Install the starter pulley, lock washer, and retaining nut. Tighten the retaining nut to the specified torque.

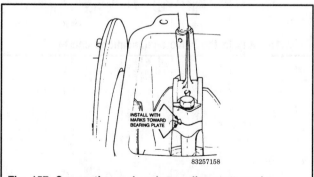

Fig. 157 Connecting rod and cap alignment marks

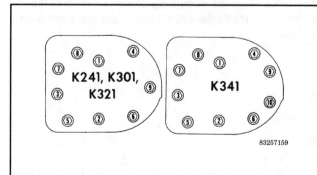

Fig. 158 Cylinder head torque sequences

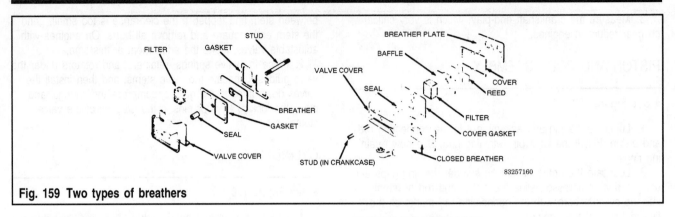

Fig. 159 Two types of breathers

BREAKER POINTS

1. Install the pushrod.
2. Position the breaker points and fasten them with the two screws.
3. Place the cover gasket into position and attach the magneto lead.
4. Set the gap and install the cover.

CARBURETOR

Insert a new gasket and assemble the carburetor to the intake port with the two attaching screws.

GOVERNOR ARM AND LINKAGE

1. Insert the carburetor linkage in the throttle arm.
2. Connect the governor arm to the carburetor linkage and slide the governor arm into the governor shaft.
3. Position the governor spring in the speed control disc.
4. Before tightening the clamp bolt, turn the shaft counterclockwise with pliers as far as it will go; pull the arm as far as it will go to the left (away from the carburetor), tighten the nut, and check for freedom of movement. Adjust the governor.

BLOWER HOUSING AND FUEL TANK

Install the head baffle, cylinder baffle, and the blower housing, in that order. The smaller cap screws are used on the bottom of the crankcase. Install the fuel tank and connect the fuel line.

RUN-IN PROCEDURE

1. Fill the crankcase with a non-detergent oil and run it under load for 5 hours to break it in.
2. Drain oil and refill crankcase with the recommended detergent type oil. Non-detergent oil must not be used except for break-in.

Valve Specifications

Dimension	Intake	Exhaust
A Seat Angle	89°	89°
B Seat Width	.037/.045	.037/.045
C Insert OD	—	1.2535/1.2545
D Guide Depth	1.586	1.497
E Guide ID	.312/.313	.312/.313
F Valve Head Diameter	1.370/1.380	1.120/1.130*
G Valve Face Angle	45°	45°
H Valve Stem Diameter	.3105/.3110	.3084/.3091

*2.125" on all K341 and K321 engines with spec suffix "D" and later.

83257C13

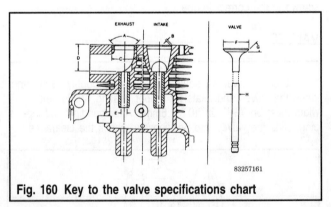

Fig. 160 Key to the valve specifications chart

Engine Rebuilding Specifications

Specifications	K321	K341
Displacement		
Cubic Inches	31.27	35.89
Cubic Centimeter	528.46	588.24
Horsepower (Max RPM)	14.0	16.0
Cylinder Bore		
New Diameter	3.500	3.750
Maximum Wear Diameter	3.503	3.753
Maximum Taper	.0015	.0015
Maximum out of Round	.005	.005
Crankshaft		
End Play (Free)	.003/.020	.003/.020
Crankpin		
New Diameter	1.500	1.500
Maximum Out of Round	.0005	.0005
Maximum Taper	.001	.001
Camshaft		
Run Clearance on Pin	.001/.0035	.001/.0035
End Play	.005/.010	.005/.010
Connecting Rod		
Big End Maximum Diameter	1.5025	1.5025
Rod-Crankpin Max Clear	.0035	.0035
Small (Pin) End-New Dia	.87585	.87585
Rod to Pin Clearance	.0003/.0008	.0003/.0008
Piston		
Thrust Face-Max Wear Dia*	3.4945	3.7425
Thrust Face*-Bore Clear	.007/.010	.007/.010
Ring-Max Side Clearance	.006	.006
Ring-End Gap in New Bore	.010/.020	.010/.020
Ring-End Gap in Used Bore	.030	.030

	K321	K341
Valve-Intake		
Valve-Tappet Cold Clear	.008/.010	.008/.010
Valve Lift (Zero Lash)	.324	.324
Stem to Guide Max Wear Clear	.0045	.0045
Valve-Exhaust		
Valve-Tappet Cold Clear	.017/.020	.017/.020
Valve Lift (Zero Lash)	.324	.324
Stem to Guide Max Wear Clear	.0065**	.0065**
Tappet		
Clearance in Guide	.0008/.0023	.0008/.0023
Ignition		
Spark Plug Gap-Gasoline	.025	.025
Spark Plug Gap-Gas	.018	.018
Spark Plug Gap (Shielded)	.020	.020
Breaker Point Gap	.020	.020
Trigger Air Gap (Breakerless)		
Spark Run ° BTDC	.005/.010	.005/.010
Spark Retard	20°	20°
	ACR ONLY	ACR ONLY
	(No Retard)	(No Retard)
Torque Values		
(Also See Page 15.3)		
Spark Plug (ft. lbs.)	18–22	18–22
Cylinder Head	25–30 ft. lbs.	25–30 ft. lbs.
Connecting Rod	300 in. lbs.	300 in. lbs.
Flywheel Nut	60–70 ft. lbs.	60–70 ft. lbs.

*Measured just below oil ring and at right angles to piston pin

**Measured at top of guide with valve closed

83257C14

A. MODEL NO.

K 32 1 PT

K-Series Engine

Approximate Displacement (Cu. In.)

Single Cylinder

	Version Code
A –	Special Oil Pan
C –	Clutch Model
G –	Generator Application
P –	Pump Model
Q –	Quiet Model
R –	Reduction Gear
S –	Electric Start
T –	Retractable Start
ST –	Electric Start And Retractable Start
EP –	Electric Plant

B. SPEC NO.

Engine Model Code

Code	Model
26, 27, 31	K91
28	K161
29	K141
30	K181
46	K241
47	K301
60	K321
71	K341

60 124B

Variation of Basic Engine

C. SERIAL NO.

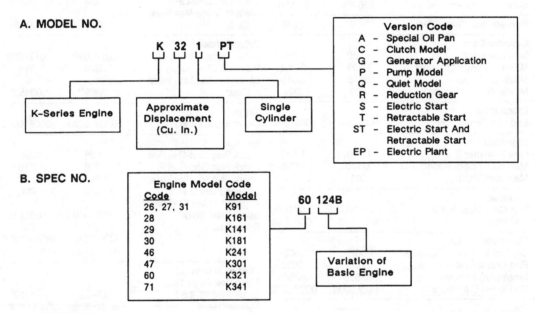

E – 172452	9076430	10026692	1501897591
A Letter	First Two Digits / If Seven Digit Number	First Three Digits / If Eight Digit Number	First Two Digits / If Ten Digit Number

A Letter		First Two Digits		First Three Digits		First Two Digits	
A	1965	10–19	1969	100–109	1980	15	1985
B	1966	20–29	1970	110–119	1981	16	1986
C	1967	30–39	1971	120–129	1982	17	1987
D	1968	40–49	1972	130–139	1983	18	1988
E	1969	50–59	1973	140–149	1984	19	1989
		60–69	1974	150–159	1985	20	1990
		70–72	1975			21	1991
		73–79	1976			22	1992
		80–89	1977			23	1993
		90–94	1978			24	1994
		95–99	1979			25	1995

Remaining digits are a factory code.

83257162

Fig. 161 Engine identification

Engine Model		12 K301	14 K321	16 K341
CONNECTING RODS①	Posi-lock②	New 260 in. lbs. Used 200 in. lbs.		
	Capscrew③	285 in. lbs.		
SPARK PLUGS		18-22 ft. lbs.		
CYLINDER HEAD①		K241, K301, K321 25-30 ft. lbs.	K341 25-30 ft. lbs.	
FLYWHEEL RETAINING	NUT	50-60 ft. lbs.		
	SCREW	22-27 ft. lbs.		
GOVERNOR BUSHING		100-120 in. lbs.		
GRASS SCREEN	Metal	70-140 in. lbs.		
	Plastic	20-30 in. lbs.		
OIL PAN	Aluminum	—		
	Cast Iron	35 ft. lbs.		
	Sheet Metal④	200 in. lbs.		
MANIFOLD SCREW/NUT		—		
CAMSHAFT NUT		—		
NON METALLIC FUEL PUMP MOUNTING SCREWS		37-45 in. lbs.		

① Lubricate with engine oil
② DO NOT overtorque — loosen — and retorque the hex nuts on Posi-Lock connecting rods.
  NEW — Component directly from stock.
  USED — Component that was in a running engine.
③ Overtorque 20%, loosen below torque value and retorque to final torque value
④ Torque twice with minimum of one minute interval
⑤ 3/8-16 thread with hex head nut and fibre gasket
⑥ Prior to Ser. #23209632   45-55 ft. lbs.

* Includes K141

83257C15

Air Cleaner and Air Intake System

K series engines are equipped with a high-density paper air cleaner element. Engines of some specifications are also equipped with an oiled foam precleaner that surrounds the paper element.

AIR CLEANER DISASSEMBLY

▶ **See Figure 161**

1. Remove the wing nut and air cleaner cover.
2. Remove the precleaner (if so equipped), paper element and seal.
3. Remove the base screws, air cleaner base, gasket and hose.

AIR CLEANER SERVICE

Precleaner

If so equipped, wash and re-oil the precleaner every 25 operating hours (more often under extremely dusty or dirty conditions).

1. Wash the precleaner in warm eater and detergent.
2. Rinse the precleaner thoroughly until all traces of detergent are eliminated. Squeeze out excess water (do not wring). Allow precleaner to dry.
3. Saturate the precleaner with clean, fresh engine oil. Squeeze out excess oil.

4. Reinstall the precleaner over the paper element.

Paper Element

Every 100 operating hours (more often under extremely dusty or dirty conditions) check the paper element. Replace the element as follows:

1. Remove the precleaner (if so equipped), element cover nut, element cover and paper element.
2. Replace a dirty, bent or damaged element with a new genuine Kohler element. Handle new elements carefully; do not use if surfaces are bent or damaged.

➡ **Do not wash the paper element or use compressed air as this will damage the element.**

3. Reinstall the paper element.
4. Install the precleaner (cleaned and oiled) over the paper element.
5. Install the air cleaner cover and wing nut. Tighten wing nut. Make sure element is sealed tightly against air cleaner base.

Inspect Air Cleaner Components

Whenever the air cleaner cover is removed, or when servicing the paper element or precleaner, check the following components:

1. Air cleaner Base - Make sure it is secured tightly to carburetor and is not bent or damaged.
2. Element Cover and Element Cover Nut - Check that element cover nut is secured tightly to seal element between air cleaner base and element cover. Tighten nut to 50 inch lbs. torque.

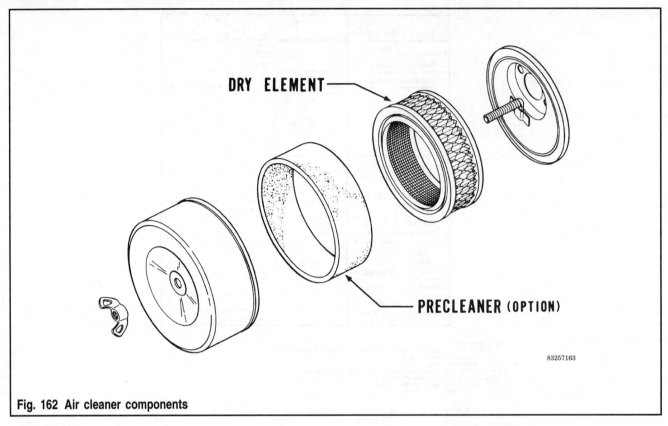

Fig. 162 Air cleaner components

3. Breather Tube - Make sure it is sealed tightly in the air cleaner base and breather cover.

✳✳WARNING

Damaged, worn or loose air cleaner components could allow unfiltered air into the engine causing premature wear and failure. Replace any damaged or worn components.

OPTIONAL OIL BATH AIR CLEANER

If the engine has an oil bath type air cleaner, clean and service it after every 25 hours of operation or more frequently if conditions warrant.

1. Remove the cover, lift the element out of the bowl and drain the oil from the bowl.

2. Thoroughly wash bowl and cover in clean solvent. Swish the element in the solvent and allow it to dry.

➡**Do not use compressed air to dry the element. The filtering material could be damaged.**

3. Lightly re-oil the element with engine oil.

4. Inspect base and cover gaskets. Replace if damaged.

5. Install base gasket and place filter on air horn.

6. Add engine oil to filter and fill to the OIL LEVEL mark.

7. Install filter element, cover gasket and cover. Secure with wing nut finger tight only.

COOLING AIR INTAKE SYSTEM

Effective cooling of an air cooled engine depends on an unobstructed flow of air over the cooling fins. Air is drawn into the cooling shroud by fins located on the flywheel. The blower housing, cooling shroud, air screen covering the flywheel and cooling fins on the cylinder and cylinder head must be kept clean and unobstructed at all times.

Never operate the engine with the blower housing or cooling shroud - removed. These devices direct air flow over the cooling fins.

➡**Some engines use a plastic grass screen and some use metal. The two are not interchangeable unless other modifications are made to the engine.**
Fuel Tank

✳✳CAUTION

Gasoline may be present in the carburetor and fuel system. Gasoline is extremely flammable and it can explode if ignited. Keep sparks, open flames, and other sources of ignition away from the engine. Disconnect and ground the spark plug lead to prevent the possibility of sparks from the ignition system.

Engine-mounted fuel tanks on K series engines are constructed of steel. They are fitted with a vented cap. The venting properties of the cap should be checked regularly. A clogged vent can cause pressure buildup in the tank, which could result in fuel spraying from the filler when the cap is loosened. It can also cause a partial vacuum in the tank, stopping the engine.

Fuel Shutoff Valve

Some engines are equipped with a fuel shutoff valve with a wire mesh screen. On engines without a shutoff valve, a straight outlet fitting is used. The wire mesh prevents relatively large particles in the tank from reaching the carburetor. The shutoff valve permits work on the fuel system without the need for draining the tank.

Fuel Filter

Some engines covered by this manual may be equipped with a see-through inline fuel filter. When the interior of the filter appears to be dirty, it should be replaced.

Fuel Pump

All K series have provisions for mounting a mechanically operated fuel pump. If no fuel pump is mounted on these engines, a cover is placed over the pump mounting pad on the crankcase.

Older fuel pumps have a metal body. Later models have a body made of plastic. The plastic body better insulates the fuel from the hot engine, minimizing the chance of vapor lock.

OPERATION

The mechanical fuel pump is operated by a lever that rides on the engine camshaft. The lever transmits a pumping action to the flexible diaphragm inside the pump body. The pumping action draws fuel in through the inlet check valve on the downward stroke of the diaphragm. On the upward stroke, the fuel is forced out through the outlet check valve.

REMOVAL

1. Disconnect the fuel lines from the inlet and outlet fittings of the pump.
2. Remove the fillister head screws, flat washers, fuel pump and gasket.
3. If required, remove the fittings from the pump body.

REPAIR

▶ See Figure 163

Plastic bodied fuel pumps are not serviceable and must be replaced when faulty. Replacement pumps are available in kits which include the pump, mounting gasket and plain washers.

INSTALLATION

▶ See Figure 164

1. Fittings - Apply a small amount of Permatex Aviation Perm-A-Gasket® (or equivalent gasoline resistant thread seal-

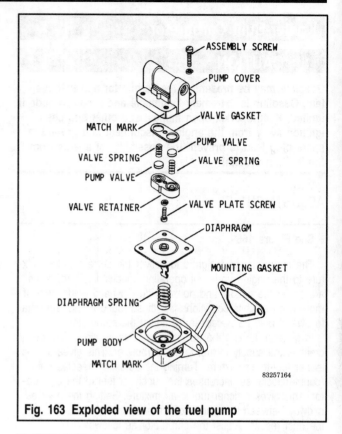

Fig. 163 Exploded view of the fuel pump

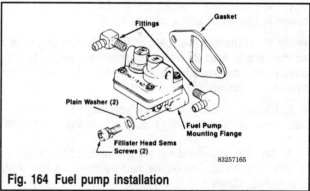

Fig. 164 Fuel pump installation

ant) to fittings. Turn fittings into pump six full turns; continue turning fittings in the same direction until desired direction is reached.

2. Install new gasket, fuel pump, flat washers, lock washers and fillister head screws.

➡**Make sure that the fuel pump lever is positioned above the camshaft. Damage to the fuel pump ad severe damage to the engine could result if the lever is positioned below the camshaft. Make sure that the flat washers are installed next to the mounting flange to prevent damage from the lock washers.**

3. If a metal bodied pump was replaced by a plastic bodied pump, make sure that the old thick gasket is discarded and the new thin gasket is used.
4. Torque screws 37-45 inch lbs.
5. Connect fuel lines to inlet and outlet fittings.

Kohler-Built Carburetor

✻✻CAUTION

Gasoline may be present in the carburetor and fuel system. Gasoline is extremely flammable and it can explode if ignited. Keep sparks, open flames, and other sources of ignition away from the engine. Disconnect and ground the spark plug lead to prevent the possibility of sparks from the ignition system.

ADJUSTMENT

▶ **See Figure 165**

The carburetor is designed to deliver the correct fuel/air mixture to the engine under all operating conditions. Carburetors are set at the factory and normally do not need adjustment. If the engine exhibits conditions like those found in the table that follows, it may be necessary to adjust the carburetor.

In general, turning the adjusting needles in (clockwise) decreases the supply of fuel to the carburetor. This gives a leaner fuel-to-air mixture. Turning the adjusting needles out (counterclockwise) increases the supply of fuel to the carburetor. This gives a richer fuel-to-air mixture. Setting the needles midway between the lean and rich positions will usually give the best results. Adjust the carburetor as follows:

1. With the engine stopped, turn the low idle fuel adjusting needle in (clockwise) until it bottoms lightly.

➡**The tip of the low idle fuel and high idle fuel adjusting needles are tapered to critical dimensions. Damage to the needles and the seats in carburetor body will result if the needles are forced.**

2. Preliminary Settings: Turn the adjusting needles out (counterclockwise) from lightly bottomed according to the table shown.

3. Start the engine and run at half throttle for five to ten minutes to warm up. The engine must be warm before making final settings (Steps 4, 5, 6, and 7).

4. High Idle Fuel Needle Setting: This adjustment is required only for adjustable high idle (main) jet carburetors. If the carburetor is a fixed main jet type, go to step 5.

   a. Place the throttle into the 'fast' position. If possible, place the engine under load.

   b. Turn the high idle fuel adjusting needle out — counterclockwise — from the preliminary setting until the engine speed decreases (rich). Note the position of the needle.

   c. Now turn the adjusting needle in (clockwise). The engine sped may increase, then it will decrease as the needle id turned in (lean). Note the position of the needle.

   d. Set the adjusting needle midway between the rich and lean settings.

5. Low Idle Speed Setting: Place the throttle control into the 'idle' or 'slow' position. Set the low idle speed to 1200 rpm ± 75rpm by turning the low idle speed adjusting screw in or out. Check the speed using a tachometer.

➡**The actual low idle speed depends on the application. The recommended low idle speed for Basic Engines is 1200 rpm. To ensure best results when setting the low idle fuel needle, the low idle speed must not exceed 1500 rpm.**

6. Low Idle Fuel Setting: Place the throttle into the 'idle' or 'slow' position.

   a. Turn the low idle fuel adjusting needle out (counterclockwise) from the preliminary setting until the engine speed decreases (rich). Note the position of the needle.

   b. Now turn the adjusting needle in (clockwise). The engine speed may increase, then it will decrease as the needle is turned in (lean). Note the position of the needle.

   c. Set the adjusting needle midway between the rich and lean settings.

7. Recheck the low idle speed using a tachometer. Readjust the speed as necessary.

DISASSEMBLY

▶ **See Figures 166 and 167**

1. Remove the bowl retaining screw, retaining screw gasket and fuel bowl.

2. Remove the float pin, float, fuel inlet needle, baffle gasket and bowl gasket.

3. Remove the fuel inlet seat and inlet seat gasket. Remove the idle fuel and main fuel adjusting needles and springs. Remove the idle speed adjusting screw and spring.

4. Further disassembly to remove the throttle and choke shafts is recommended only if these parts are to be replaced. Refer to 'Throttle and Choke Shaft Replacement' later in this section.

CLEANING

✻✻CAUTION

Carburetor cleaners and solvents are extremely flammable. Keep sparks, flames and other sources of ignition away from the area. Follow the cleaner manufacturer's warnings and instructions on its proper and safe use. Never use gasoline as a cleaning agent.

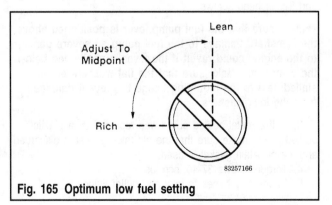

Fig. 165 Optimum low fuel setting

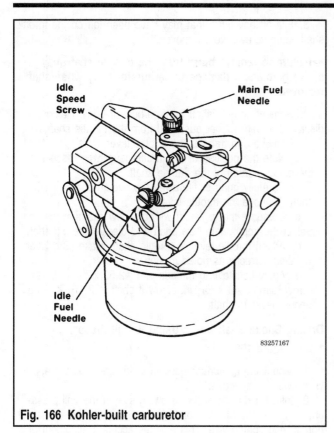

Fig. 166 Kohler-built carburetor

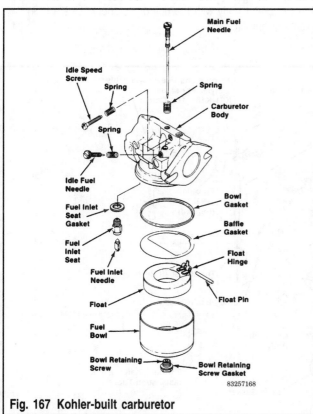

Fig. 167 Kohler-built carburetor

All parts should be carefully cleaned using a carburetor cleaner (such as acetone). Be sure all gum deposits are removed from the following areas:

- Carburetor body and bore; especially the areas where throttle plate, choke plate and shafts are seated.
- Float and hinge.
- Fuel bowl.
- Idle fuel and 'off-idle' ports in carburetor bore, ports in main fuel adjusting needle and main fuel seat.

➡**These areas can be cleaned using a piece of fine wire in addition to cleaners. Be careful not to enlarge the ports or break the cleaning wire within the ports.**

Blow out all passages with compressed air.

➡**Do not submerge carburetor in cleaner or solvent when fiber or rubber seals are installed. The cleaner may damage these seals.**

INSPECTION

1. Carefully inspect all components and replace those that are worn or damaged.
2. Inspect the carburetor body for cracks, holes and other wear or damage.
3. Inspect the float for dents or holes. Check the float hinge for wear and missing or damaged float tabs.
4. Inspect the inlet needle and seat for wear or grooves.
5. Inspect the tips of the main and idle adjusting needles for wear or grooves.
6. Inspect the throttle and choke shafts and plate assemblies for wear or excessive play.

CHOKE PLATE MODIFICATION

The choke action has been changed on production carburetors to reduce the chances of over choking. On production carburetors now used on the K241 and K301, one relief hole is now $11/16$ in. and the other is $3/16$ in. If you find that the relief holes are smaller than this, enlarge them to these dimensions.

➡**When redrilling the holes, take the necessary precautions to prevent chips from entering the engine.**

REPAIR

Always use new gaskets when servicing and reinstalling carburetors. Several repair kits, which include the gaskets and other components, are available.

✳✳CAUTION

Suitable eye protection (safety glasses, goggles, or face hood) should be worn for any procedure involving the use of compressed air, punches, hammers, chisels, drills, or grinding tools.

Throttle And Choke Shaft Replacement

▶ **See Figure 168**

Two kits are available that allow replacement of the carburetor throttle and choke shafts.

1. To ensure correct reassembly, mark choke plate and carburetor body with a marking pen. Also take note of choke plate position in bore and choke lever position.

2. Carefully and slowly remove the screws securing choke plate to choke shaft. remove and save the choke plate as it will be reused.

3. File off any burrs which may have been left on the choke shaft when the screws were removed. Place carburetor on workbench with choke side down. Remove choke shaft; the detent ball and spring will fall out.

4. Note the position of the choke lever with respect to the cut out portion of the choke shaft.

5. Carefully grind or file away the riveted portion of the shaft. remove and save the choke lever; discard the old choke shaft.

6. Attach the choke lever to the new choke shaft from the kit. Make sure the lever is installed correctly as noted in step 4. Secure lever to choke shaft. Apply Loctite® to threads of 1 #3-48 x $^7/_{32}$ in. brass screw. Secure lever to shaft.

Throttle Plate and Throttle Shaft; Transfer Throttle Lever

1. To ensure correct reassembly, mark throttle plate and carburetor body with a marking pen. Also take note of the throttle plate position in the bore and the throttle lever position.

2. Carefully and slowly remove the screws securing the throttle plate to throttle shaft. Remove and save the throttle plate for reuse.

3. File off any burrs that may have been left on the throttle shaft when screws were removed.

➡**Failure to remove burrs from the throttle shaft may cause permanent damage to carburetor body when shaft is removed.**

4. Remove throttle shaft from carburetor body. Remove and discard the foam rubber dust seal from the throttle shaft.

5. Remove and transfer the throttle lever.

a. Note the position of the throttle lever with respect to the cutout portion of the throttle shaft.

b. Carefully grind or file away the riveted portion of the shaft. Remove the throttle lever.

c. Compare the old shaft with the new shafts in the kit. Select the appropriate new shaft and discard the old shaft.

d. Attach throttle lever to throttle shaft. Make sure lever is installed correctly as noted in step a.

e. Apply Loctite® to threads of 1 #2-56 x $^7/_{32}$ in. brass screw (use #3-48 x $^7/_{32}$ in. screw if shaft is 2 $^{49}/_{64}$ in. long. Secure lever to shaft.

Drilling Choke Shaft Bores Using A Drill Press

▶ **See Figure 169**

1. Mount the carburetor body in a drill press vise. Keep the vise jaws slightly loose.

2. Install a drill bit of the following size in the drill press chuck. Lower the bit (not rotating) through both choke shaft bores; then tighten vise. This ensures accurate alignment of the carburetor body with the drill press chuck. Use a ¼ in. diameter drill bit.

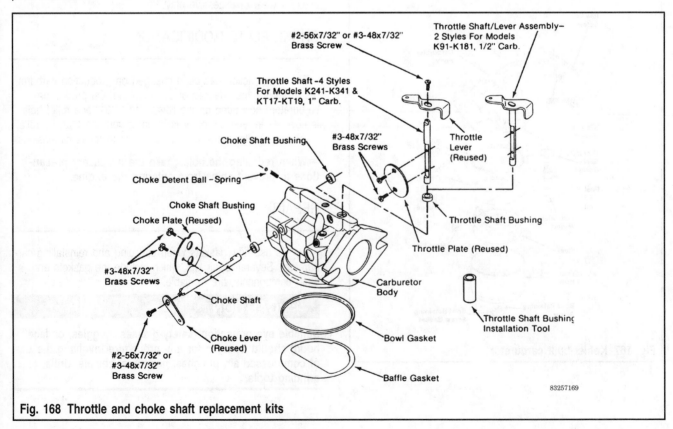

Fig. 168 Throttle and choke shaft replacement kits

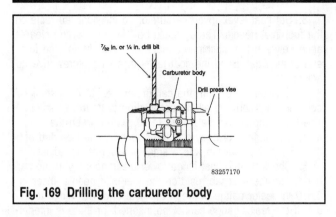

Fig. 169 Drilling the carburetor body

3. install a $^{19}/_{64}$ in. drill bit in the chuck. Set drill press to a low speed suitable for aluminum. Drill slowly to ensure a good finish.

4. Ream the choke shaft bores to a final size of $^5/_{16}$ in.. For best results use a piloted $^5/_{16}$ in. reamer.

5. Blow out all metal chips using compressed air. Thoroughly clean the carburetor body in carburetor cleaner.

Installing Choke Shaft Bushings

▶ **See Figure 170**

1. Install screws in the tapered holes that enter the choke shaft bores until the screws bottom lightly.

2. Coat the outside surface of the kit-supplied choke shaft bushings with Loctite® from the kit. Carefully press the bushings into the carburetor body using a smooth-jawed vise. Stop pressing when bushings bottom against screws.

3. Allow Loctite® to 'set' for 5 to 10 minutes, then remove screws.

4. Install new choke shaft in bushings. Rotate shaft and check that it does not bind.

➡**If binding occurs, locate and correct the cause before proceeding. Use choke shaft to align bushings if necessary.**

5. Remove choke shaft and allow Loctite® to 'set' for an additional 30 minutes before proceeding.

6. Wipe away any excess Loctite® from bushings and choke shaft.

Installing throttle Shaft Bushing

▶ **See Figure 171**

1. Make sure the dust seal counterbore in the carburetor body is thoroughly clean and free of chips and burrs.

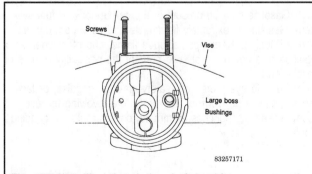

Fig. 170 Installing the choke shaft bushings

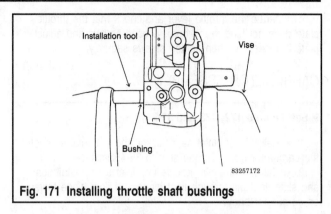

Fig. 171 Installing throttle shaft bushings

2. Install a throttle shaft (without throttle lever) in carburetor body to use as a pilot. Use one of the remaining new throttle shafts from the kit.

3. Coat the outside surface of the throttle shaft bushing with Loctite® from the kit. Slip the bushing over the shaft. Using a vise and the installation tool from the kit, press the bushing into the counterbore until it bottoms in the carburetor body.

4. Allow the Loctite® to 'set' for 5 to 10 minutes, then remove the throttle shaft.

5. Install the new throttle shaft and lever in carburetor body. Rotate the shaft and check that it does not bind.

➡**If binding occurs, locate the cause and correct before proceeding. use throttle shaft to align bushing if necessary.**

6. Remove the shaft and allow the Loctite® to 'set' for an additional 30 minutes before proceeding.

7. Wipe away all excess Loctite® from bushing and throttle shaft.

Installing Detent Spring and Ball, Choke Shaft and Choke Plate

1. Install new detent spring and ball in carburetor body in the side opposite the choke lever.

2. Compress detent ball and spring and insert choke shaft through bushings. Make sure the choke lever is on the correct side of the carburetor body.

3. Compress choke plate to choke shaft. Make sure marks are aligned and plate is positioned properly in the bore. Apply Loctite® to threads of 2 #3-48 x $^7/_{32}$ in. brass screws. Install screws so that they are slightly loose.

4. Operate the choke lever. Check that there is no binding between choke plate and carburetor bore. Loosen screws and adjust plate as necessary; then tighten screws.

Installing Throttle Shaft and Throttle Plate

1. Install throttle shaft in carburetor with cutout portion of the shaft facing out.

2. Attach throttle plate to throttle shaft. Make sure marks are aligned and plate is positioned properly in the bore. Apply Loctite® to threads of 2 #3-48 x $^7/_{32}$ in. brass screws. Install screws so that they are slightly loose.

3. Apply finger pressure to throttle shaft to keep it firmly seated against pivot in carburetor body. Rotate the throttle shaft until the throttle plate fully closes the bore around its perimeter; then tighten screws.

4. Operate the throttle lever and check that the throttle plate does not bind in the bore. Loosen screws and adjust plate if necessary; then tighten screws securely.

CARBURETOR ASSEMBLY

▶ **See Figures 172, 173 and 174**

1. Install the fuel inlet seat gasket and fuel inlet seat into the carburetor body. Torque seat to 35-45 inch lbs.

2. Install the fuel inlet needle into inlet seat. Install float and slide float pin through float hinge and float hinge towers on carburetor body.

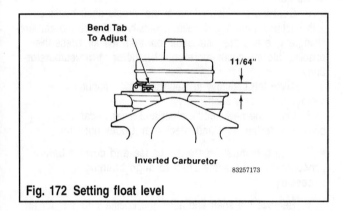

Fig. 172 Setting float level

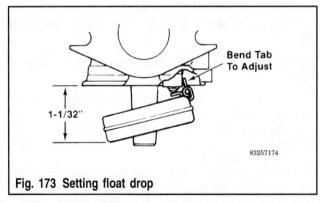

Fig. 173 Setting float drop

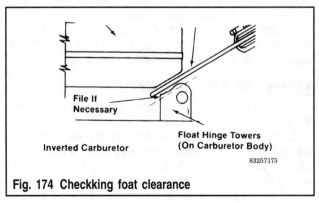

Fig. 174 Checkking foat clearance

3. Set float level: Invert carburetor so the float tab rests on the fuel inlet needle. There should be 11/64 in. ± 1/32 in.) clearance between the machined surface of the body and the free end of the float. Bend the float tab with a small screwdriver to adjust.

4. Set float drop: Turn the carburetor over to its normal operating position and allow float to drop to its lowest level. The float drop should be limited to 1 1/32 in. between the machined surface of body and the bottom of the free end of float. Bend the float tab with a small screwdriver to adjust.

5. Check float-to-float hinge tower clearance: Invert the carburetor so the float tab rests on the fuel inlet needle. Insert a 0.010 in. feeler gauge between the float and float hinge towers. If the feeler gauge cannot be inserted, or there is interference between the float and towers, file the towers to obtain the proper clearance.

6. Install the bowl and baffle gasket. Position baffle gasket so the inner edge is against the float hinge towers.

7. Install the fuel bowl so it is centered on the baffle gasket. Make sure the baffle gasket and bowl are positioned properly to ensure a good seal.

8. Install the bowl retaining screw gasket and bowl retaining screw Torque screw to 50-60 inch lbs.

9. Install the idle speed adjusting screw and spring. Install the idle fuel and main fuel adjusting needles and springs. Turn the adjusting needles clockwise until they are bottomed lightly.

➡**The ends of adjusting needles are tapered to critical dimensions. Damage to needles and seats will result if needles are forced.**

10. Reinstall the carburetor to the engine using a new gasket.

11. Adjust the carburetor as outlined under the 'Adjustment' portion of this section.

Walbro Fixed/Adjustable Carburetor

▶ **See Figures 175 and 176**

This section covers the idle adjustment, disassembly, cleaning, inspection, repair, and reassembly of the Walbro-built, side draft, fixed/adjustable main jet carburetors.

❉❉CAUTION

Before servicing the carburetor, engine, or equipment, always remove the spark plug leads to prevent the engine from starting accidentally. Ground the leads to prevent sparks that could cause fires.

Gasoline may be present in the carburetor and fuel system. Gasoline is extremely flammable and its vapors can explode if ignited. Keep sparks, open flame, and other sources of ignition away from the area to prevent the possibility of fires or explosions.

Suitable eye protection (safety glasses, goggles, or face shield) should be worn for any procedure involving the use of compressed air, punches, hammers, chisels, drills, or grinding tools.

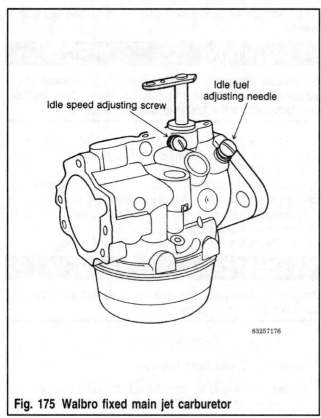

Fig. 175 Walbro fixed main jet carburetor

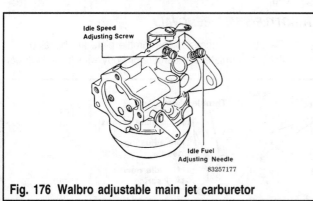

Fig. 176 Walbro adjustable main jet carburetor

TROUBLESHOOTING

If engine troubles are experienced that appear to be fuel system related, check the following areas before adjusting or disassembling the carburetor.
• Make sure the fuel tank is filled with clean, fresh gasoline.
• Make sure the fuel tank cap vent is not blocked and that it is operating properly.
• Make sure fuel is reaching the carburetor. This includes checking the fuel shut-off valve, fuel tank filler screen, in-line fuel filter, fuel lines, and fuel pump for restrictions or faulty components as necessary.
• Make sure the carburetor is securely fastened to the engine using gaskets in good condition.
• Make sure the air cleaner element is clean and all air cleaner components are fastened securely.

• Make sure the ignition system, governor system, exhaust system, and throttle and choke controls are operating properly.
• If, after checking the items listed above, starting problems or other conditions similar to those listed in the following table exist, it may be necessary to adjust or service the carburetor.

CARBURETOR ADJUSTMENT

▶ **See Figure 177**

➥**The tip of the low idle fuel and high idle fuel adjusting needles are tapered to critical dimensions. Damage to the needles and the seats in carburetor body will result if the needles are forced.**

In general, turning the adjusting needles in (clockwise) decreases the supply of fuel to the carburetor. This gives a leaner fuel-to-air mixture. Turning the adjusting needles out (counterclockwise) increases the supply of fuel to the carburetor. This gives richer fuel-to-air mixture. Setting the needles midway between the lean and rich positions will usually give the best results.

Adjust the carburetor as follows:
1. With the engine stopped, turn the low idle fuel adjusting needles in (clockwise) until it bottoms lightly.
2. Preliminary Settings: Turn the adjusting needles out (counterclockwise) from lightly bottomed according to the table shown.
3. Start the engine and run at half throttle for five to ten minutes to warm up. The engine must be war, before making final settings (Steps 4, 5, 6, and 7).
4. High Idle Fuel Needle Setting: This adjustment is required only for adjustable high idle (main) jet carburetors. If the carburetor is fixed main jet type, go to step 5.
   a. Place the throttle into the 'fast' position. If possible, place the engine under load.
   b. Turn the high idle fuel adjusting needle out (counterclockwise) from the preliminary setting until the engine speed decreases (rich). Note the position of the needle.
   c. Now turn the adjusting needle in (clockwise). The engine speed may increase, then it will decrease as the needle is turned in (lean). Note the position of the needle.
   d. Set the adjusting needle midway between the rich and lean settings.
5. Low Idle Speed Setting: Place the throttle control control into the 'idle' or 'slow' position. Set the low idle speed to 1200

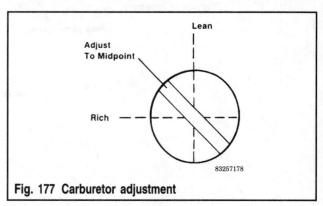

Fig. 177 Carburetor adjustment

rpm ± 75 rpm by turning the low idle speed adjusting screw in or out. Check the speed using a tachometer.

➡The actual low idle speed depends on the application. The recommended low idle speed for Basic Engines is 1200 rpm. To ensure best results when setting the low idle fuel needle, the low idle speed must not exceed 1500 rpm.

6. Low Idle Fuel Needle setting: Place the throttle into the 'idle' or 'slow' position.

a. Turn the low idle fuel adjusting needle out (counterclockwise) from the preliminary setting until the engine speed decreases (rich). Note the position of the needle.

b. Now turn the adjusting needle in (clockwise). The engine speed may increase, then it will decrease as the needle is turned in (lean). Note the position of the needle.

c. Set the adjusting needle midway between the rich and lean settings.

7. Recheck the low idle speed using a tachometer. Readjust the speed as necessary.

DISASSEMBLY

▶ **See Figures 178 and 179**

1. Remove the bowl retaining screw, retaining screw gasket, and fuel bowl.

2. Remove the bowl gasket, float pin, float, and fuel inlet needle.

✳✳CAUTION

To prevent damage to the carburetor, do not attempt to remove the fuel inlet seat as it is not serviceable. Replace the carburetor if the fuel inlet seat is damaged.

3. Remove the idle fuel adjusting needle and spring. Remove the idle speed adjusting screw and spring.

4. Remove the main fuel jet.

5. In order to clean the 'off-idle' ports and the bowl vent channel thoroughly, the welch plugs covering these areas must be removed. Use tool No. K01018 and the following procedure to remove the welch plugs.

a. Pierce the welch plug with the tip of the tool.

✳✳CAUTION

To prevent damage to the carburetor, do not allow the tool to strike the carburetor body.

b. Pry out the welch plug using the tool.

Throttle And Choke Shaft Removal

Further disassembly to remove the throttle shaft and choke shaft is recommended only if these parts are to be cleaned or replaced.

THROTTLE SHAFT REMOVAL

1. Because the edges of the throttle plate are beveled, mark the throttle plate and carburetor body with a marking pen

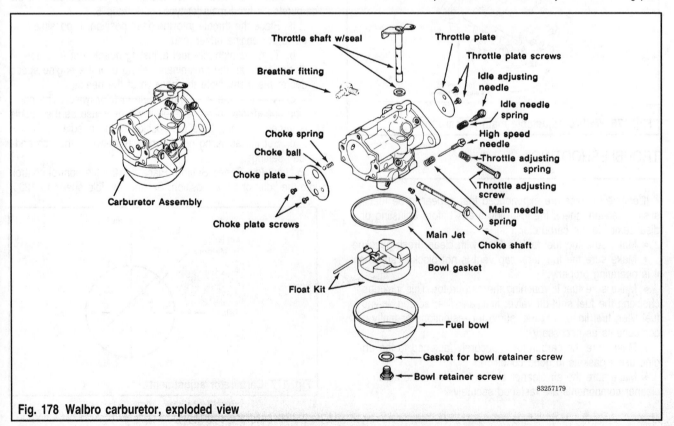

Fig. 178 Walbro carburetor, exploded view

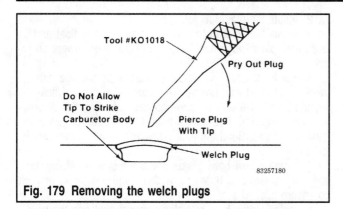

Fig. 179 Removing the welch plugs

to ensure correct reassembly. Also take note of the throttle plate position in bore, and the position of the throttle lever.

2. Carefully and slowly remove the screws securing the throttle plate to throttle shaft. Remove the throttle plate.

3. File off any burrs which may have been left on the throttle shaft when the screws were removed. Do this before removing the throttle shaft from carburetor body.

4. Remove the throttle lever/shaft assembly with foam dust seal from carburetor body.

CHOKE SHAFT REMOVAL

1. Because the edges of choke plate are beveled, mark the choke pate and carburetor body with a marking pen to ensure correct reassembly. Also take note of the choke plate position in bore, and the position of the choke lever.

2. Carefully and slowly remove the screws securing the choke plate to choke shaft. Remove the choke plate.

3. file off any burrs which may have been left on the choke shaft when the screws were removed. Do this before removing the choke shaft from carburetor body.

4. Rotate the choke shaft until the cutout portion of shaft is facing the air cleaner mounting surface. Place the carburetor body on the work bench with choke side down. Remove the choke lever/shaft assembly from carburetor body; the detent ball and spring will drop out.

CLEANING

✶✶CAUTION

Carburetor cleaners and solvents are extremely flammable. Keep sparks, flames, and other sources of ignition away from the area. Follow the cleaner manufacturer's warnings and instructions on its proper and safe use. Never use gasoline as a cleaning agent.

All parts should be carefully cleaned using a carburetor cleaner (such as acetone). Be sure all gum deposits are removed form the following areas:

• Carburetor body and bore; especially the areas where the throttle plate, choke plate, and shafts are seated.

• Idle fuel and 'off-idle' ports in carburetor bore, main jet, bowl vent, and fuel inlet seat.

➡These areas can be cleaned using a piece of fine wire in addition to cleaners. Be careful not to enlarge the ports, or break the cleaning wire within ports. Blow out all passages with compressed air.

• Float and Float hinge.
• Fuel Bowl.
• Throttle plate, choke plate, throttle shaft, and choke shaft.

✶✶CAUTION

Do not submerge the carburetor in cleaner or solvents when fiber, rubber, or foam seals or gaskets, or the fuel inlet needle are installed. The cleaner may damage these parts.

INSPECTION

• Carefully inspect all components and replace those that are worn or damaged.

• Inspect the carburetor body for cracks, holes, and other wear or damage.

• Inspect the float for cracks or holes. Check the float hinge for wear, and missing or damaged float tabs.

• Inspect the fuel inlet needle for wear or grooves.

• Inspect the tip of the idle fuel adjusting needle for wear or grooves.

• Inspect the throttle and choke shaft and plate assemblies for wear or excessive play.

REPAIR

Always use new gaskets when servicing and reinstalling carburetors. Repair kits are available which include new gaskets and other components. These kits are described below.

Components such as the throttle and choke shaft assemblies, throttle plate, choke plate, idle fuel needle, main jet, and others, are available separately.

REASSEMBLY

Throttle Shaft Installation

1. Install the foam dust seal on throttle shaft. Insert the throttle lever/shaft assembly into carburetor body with the cut-out portion of shaft facing the carburetor mounting flange.

2. Install the throttle plate to throttle shaft. Make sure the plate is positioned properly in bore as marked and noted during disassembly (the numbers stamped on plate should face the carburetor mounting flange). Apply Loctite® #609 to threads of 2 plate retaining screws. Install screws so they are slightly loose.

3. Apply finger pressure to the throttle lever/shaft to keep it firmly seated against pivot in carburetor body. Rotate the throttle shaft until the throttle plate fully closes the bore around its entire perimeter; then tighten screws.

4. Operate the throttle lever; check for binding between the throttle plate and carburetor bore. Loosen screws and adjust

throttle plate as necessary; then torque screws to 8-12 inch lbs.

Choke Shaft Installation

1. Install the detent spring and ball into the carburetor body.

❋❋CAUTION

If the detent ball does not drop through the tapped air cleaner base screw hole by its own weight, do not force it. Forcing the ball could permanently lodge it in the hole. Install the ball through the choke shaft bore instead.

2. Compress the detent ball and spring. Insert the choke lever/shaft assembly into carburetor body with the cutout portion of shaft facing the air cleaner mounting surface. Make sure the choke lever is on the correct side of carburetor body.

3. Install the choke plate to choke shaft. Make sure the plate is positioned properly in bore as marked and noted during disassembly. (The numbers stamped on plate should face the air cleaner mounting surface and be upright.) Apply Loctite® #609 to threads of 2 plate retaining screws. Install the screws so they are slightly loose.

4. Operate the choke lever; check for binding between the choke plate and carburetor bore. Adjust plate as necessary; then torque screws to 8-12 inch lbs.

Carburetor Reassembly

▶ **See Figure 180**

1. If the welch plugs have been removed for cleaning, new welch plugs must be installed. Use tool No. K01017 and the following procedure to install the welch plugs.

   a. Position the carburetor body securely with the welch plug cavities to the top.

   b. Place a new welch plug into the cavity with the raised portion up. Use the end of the tool that is about the same size as plug and flatten the plug. Do not force the plug below the top surface.

   c. After welch plugs are installed, seal the exposed surface with sealant. Allow the sealant to dry.

➡ **If a commercial sealant is not available, fingernail polish can be used.**

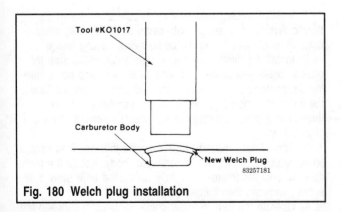

Fig. 180 Welch plug installation

2. Install the main fuel jet.

3. Install fuel inlet needle into inlet seat. Install float and slide float pin through float hinge and float hinge towers on carburetor body.

4. Set Float Level: Invert the carburetor so the float tab rests on the fuel inlet needle. The exposed surface of float should be parallel with the bowl gasket surface of the carburetor body (exposed, free end of float 0.690-0.720 in. from bowl gasket surface). Bend the float tab with a small screwdriver to adjust.

5. Install a new bowl gasket and the fuel bowl. Make sure the bowl gasket and bowl are centered and positioned properly to ensure a good seal.

6. Install a new bowl retaining screw gasket and the bowl retaining screw. Torque screw to 45-55 inch lbs.

7. Install the idle speed adjusting screw and spring.

8. Install the idle fuel adjusting needle and spring. turn the adjusting needle in (clockwise) until it bottoms lightly.

❋❋CAUTION

The tip of the idle fuel adjusting needle is tapered to critical dimensions. Damage to the needle and the seat in carburetor body will result if the needle is forced.

9. Turn the idle fuel needle out (counterclockwise) from lightly bottomed according to the instructions in the 'Adjustment' section of this Bulletin.

HIGH ALTITUDE OPERATION (FIXED JET)

When operating the engine at high altitudes, the main fuel mixture tends to get over-rich. An over-rich mixture can cause conditions such as black, sooty exhaust smoke, misfiring, loss of speed and power, poor fuel economy, and poor or slow governor response.

To compensate for this, a special high altitude main fuel jet is available for each carburetor. The high altitude main fuel jet is sold in a kit which includes the jet and necessary gaskets.

High Altitude Jet Installation (Fixed Jet)

1. Remove the fuel bowl retaining screw, retaining screw gasket, fuel bowl, and bowl gasket.

➡ **If necessary, remove the air cleaner and carburetor from engine to make fuel bowl removal easier.**

2. Remove the float pin, float, and fuel inlet needle.

3. Remove the existing main fuel jet.

4. Install the new high altitude main fuel jet and torque to 12-16 inch lbs.

5. Reinstall the fuel inlet needle, float, and float pin.

6. Install the new bowl gasket from kit and the fuel bowl. Make sure the bowl gasket and bowl are centered and positioned properly to ensure a good seal.

7. Install the new bowl retaining screw gasket from kit and the bowl retaining screw. Torque screw to 45-55 inch lbs.

8. Reinstall the carburetor and air cleaner to engine as necessary using the new gaskets from kit.

IDLE ADJUSTMENT PROCEDURE FOR ENGINES WITH ANTI-DIESELING SOLENOID

▶ **See Figure 181**

The idle speed of some vibro-mounted engines has been increased to allow smoother operation at low idle and an anti-dieseling solenoid has been added to prevent dieseling during shut-down at the higher idle speed. If called upon to adjust the idle on any engine with this solenoid, use the following procedure.

STEP 1 - IDLE FUEL MIXTURE ADJUSTMENT: With engine stopped, turn the idle fuel adjusting screw all the way in (clockwise) until it bottoms lightly then back out ½ turn.

STEP 2 - IDLE SPEED ADJUSTMENT: Start engine and check idle speed with a hand tachometer. Idle, no load, speed should be 2100 RPM. To set the idle speed, loosen the jam nut on the anti-dieseling solenoid and turn the solenoid in or out until 2100 RPM idle speed is attained - retighten jam nut to lock solenoid in position.

Thermostatic Type Automatic Chokes

The automatic choke is a heat sensitive thermostatic unit. At room temperature, choke lever will be set in a vertical position. If engine should fail to start when cranked, adjust choke lever by hand to determine if choke setting is too lean or too rich. Once this has been established, adjustment can be made to remedy the situation.

ADJUSTMENT

1. Loosen adjustment lock screw on choke body. This allows the position of adjustment to be changed.
2. Moving adjustment bracket downward will increase the amount of choking. Upward movement will result in less choking.
3. After adjustment is made, tighten adjustment lock screw.

Electric-Thermostatic Type Automatic Chokes

▶ **See Figure 182**

Remove air cleaner from carburetor to observe position of choke plate. Choke adjustment must be made on cold engine. If starting in extreme cold, choke should be in full closed position before engine is started. A lesser degree of choking is needed in milder temperatures.

ADJUSTMENT

1. Move choke arm until hole in brass shaft lines up with slot in bearings.
2. Insert #43 drill (0.089 in.) and push all the way down to engine manifold to engage in notch in base of choke unit.
3. Loosen clamp bolt choke lever, push arm upward to move choke plate toward closed position. After desired position is attained, tighten clamp bolt then remove drill.

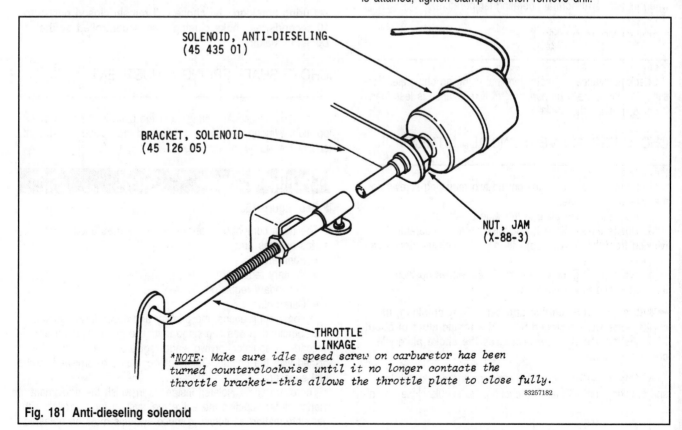

SOLENOID, ANTI-DIESELING
(45 435 01)

BRACKET, SOLENOID
(45 126 05)

NUT, JAM
(X-88-3)

THROTTLE
LINKAGE

NOTE: Make sure idle speed screw on carburetor has been turned counterclockwise until it no longer contacts the throttle bracket--this allows the throttle plate to close fully.

83257182

Fig. 181 Anti-dieseling solenoid

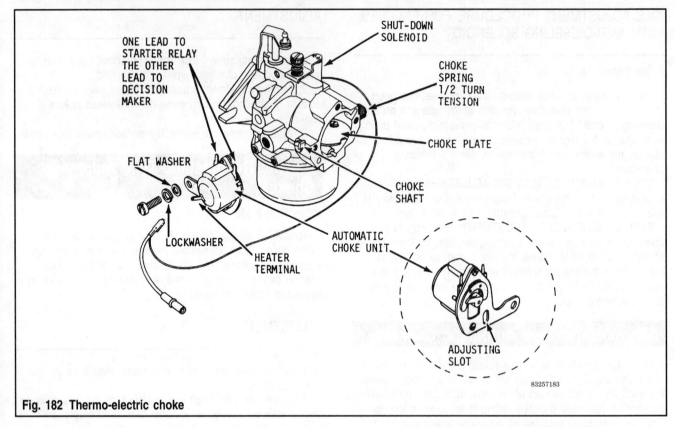

Fig. 182 Thermo-electric choke

4. After replacing air cleaner, check for evidence of binding in linkage, adjust as needed. Be sure chokes are fully open when engine is at normal operating temperature.

TROUBLESHOOTING

Check resistance of heater terminal using an ohmmeter. Resistance should be 3 ohms or more. If resistance is less than 3 ohms, replace the choke.

CHOKE REPLACEMENT AND ADJUSTMENT

1. Position the choke unit on the two mounting screws so that it is slightly loose.
2. Hold the choke plate in the wide open position.
3. Rotate the choke unit clockwise on the carburetor (viewed from the choke side) with a slight pressure until it can no longer be rotated.
4. While holding the choke unit in the above position, tighten the two mounting screws.

➡**With engine not running and before any cranking, the choke plate will be closed 5-10° at a temperature of about 75°F. As the temperature decreases the choke plate will close even more.**

5. Check choke function by removing the spark plug lead and cranking the engine. The choke plate should close a mini-

mum of 45 degrees at temperatures above 75 degrees F. The plate will close more at lower temperatures.

➡**During cranking, the choke will remain closed only 5 to 10 seconds, as choke closing time is controlled by the Decision Maker.**

CHOKE SHAFT SPRING ADJUSTMENT

To adjust the choke spring, hold the plate in the wide open position. Windup the spring ½ turn and then place the straight end of the spring through the hole in the shaft.

LP Fuel System

▶ **See Figure 183**

The main components of the fuel system as used on K series engines are:

• Fuel tank
• Primary regulator
• Secondary regulator
• Carburetor

In some applications, the primary and secondary regulators are combined in one two-stage unit. The gas carburetor and secondary regulator (or two stage regulator) are normally furnished with the engine. Other components are furnished by the fuel supplier.

There are some isolated instances in which the equipment manufacturer supplies the entire fuel system for operation with gas. Information on servicing these systems must be obtained from the equipment manufacturer.

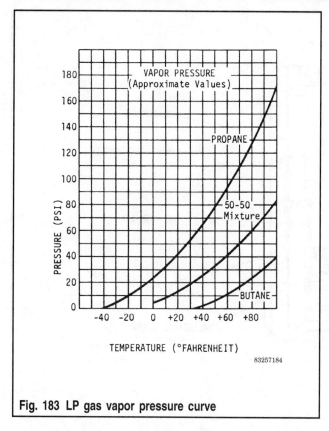

Fig. 183 LP gas vapor pressure curve

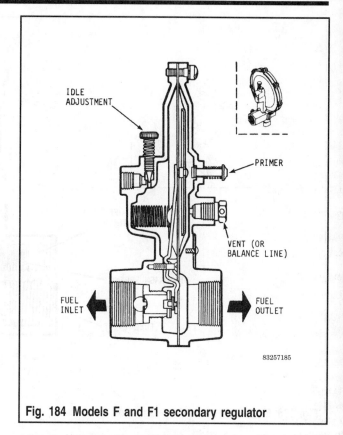

Fig. 184 Models F and F1 secondary regulator

Depending on the air temperature and the mixture of gasses in the tank, pressure at the outlet of the tank can be as high as 180 to 200 psi.

SECONDARY REGULATOR

▶ **See Figures 184, 185 and 186**

The secondary regulators used on Kohler engines are compact single diaphragm types. This type regulator accurately regulates the flow of gas to the carburetor and shuts the gas off automatically when the demand for gas ceases. If the regulator fails, it must be replaced or reconditioned by an authorized gas equipment repair shop. Do not attempt to repair a faulty regulator.

Secondary regulators used on Kohler engines require only one adjustment. This adjustment should be performed while the engine is running.

Garretson Model S, SD and KN regulators have a lock-off or fuel control adjustment. Use the following procedure to make this adjustment.

➡**The regulator should be mounted as close to vertical as possible, and adjusted in the position in which it will be mounted on the engine.**

1. Connect regulator inlet to a source of clean compressed air, not over 10 psi. Do not connect to a gas supply.
2. Turn air supply on.
3. If the regulator being adjusted is a Model KN, open the lock off adjusting screw until air just starts flowing through the regulator.

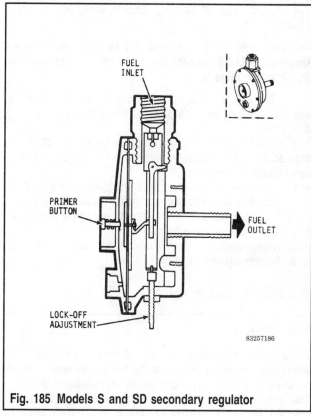

Fig. 185 Models S and SD secondary regulator

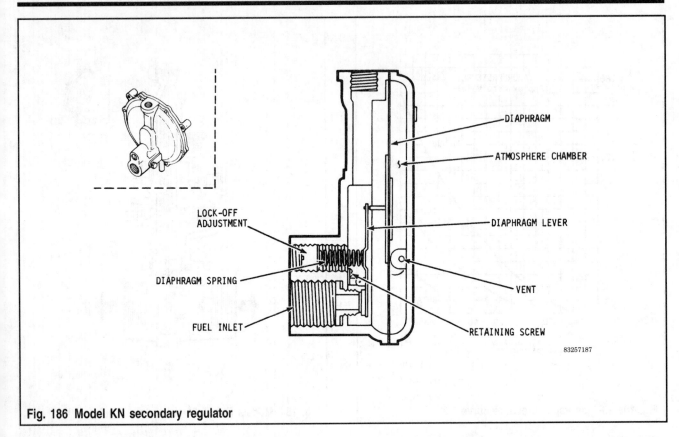

Fig. 186 Model KN secondary regulator

4. Turn the lock off adjusting screw in slowly until air flow stops.

➡**A soap bubble test is a good way to check for complete shutoff. If bubbles indicate that air is still flowing, turn the screw in one more full turn.**

The lock off adjusting screw may be used to adjust fuel flow while the engine is idling. Never adjust at any speed above idle.

5. If the regulator being adjusted is a Model S or SD, depress the primer button for an instant. This will allow air to flow through the regulator.

6. Check the air flow stops when the primer is released.

7. If air flow does not stop completely, loosen the adjustment screw lock nut and turn the adjustment screw in until air flow stops, then one more full turn.

8. Repeat steps 5 through 7 until air flow sops every time.

9. Tighten adjustment screw lock nut.

PRIMARY REGULATOR

◗ **See Figure 187**

The primary regulator provides initial control of the fuel under pressure as it comes from the fuel supply tank. The inlet pressure for primary regulators should never exceed 250 psi. The primary regulator is adjusted for outlet pressure of approximately 6 ounces per square inch (11 in. W.C.). If the regulator does not function properly, replace it or have it serviced by an authorized gas equipment shop. Never attempt to service a faulty primary regulator.

Upon demand for fuel, pressure drops on the outside of the regulator diaphragm. The gas inlet valve then begins to open, allowing fuel to pass through the regulator to the secondary regulator. As the need for more fuel increases, the fuel inlet valve opens further, allowing more fuel to pass.

Pressure may be adjusted by removing the bonnet cap and turning the spring tension adjustment with a large screwdriver. Turning clockwise increases the pressure; turning counterclockwise decreases it.

TWO STAGE REGULATOR

◗ **See Figure 188**

The two stage regulator used on Kohler engines is a double diaphragm type regulator designed for use with air-cooled engines. It combines primary and secondary regulation in one unit. The regulator fuel inlet is connected to the fuel tank. Its outlet is connected to the carburetor. If the regulator fails to operate properly, replace it or have it serviced by an authorized gas equipment shop. Never attempt to service a faulty regulator.

Vaporized fuel is admitted to the regulator at fuel tank pressure (up to 250 psi). Because the secondary valve is closed (engine not running), the pressure on the internal side of the primary diaphragm builds until the pressure overcomes the spring action on the opposite side of the diaphragm. This primary diaphragm spring has sufficient tension to require approximately 10 psi pressure on the internal side of the diaphragm to counteract the opening force due to the spring. When the pressure reaches this level, the valve is closed, preventing further pressure rise.

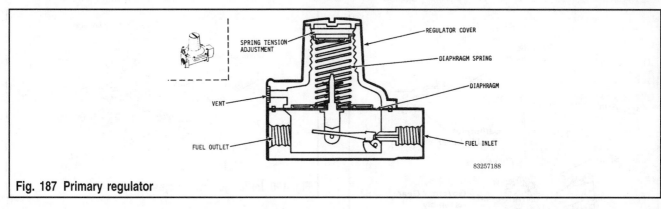

Fig. 187 Primary regulator

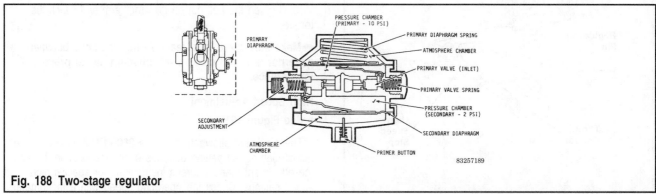

Fig. 188 Two-stage regulator

The secondary diaphragm acts against the secondary valve spring. Its action results from vacuum caused by the carburetor. As the vacuum begins acting on the diaphragm, the diaphragm is moved nearer to the center of the regulator, opening the secondary valve until equilibrium is reached. As more fuel is needed, vacuum from the carburetor increases, causing the secondary valve to open further. When fuel is flowing, pressure on the primary diaphragm is lowered slightly, permitting the spring to open the primary valve in an attempt to bring the pressure back to 10 psi.

ADJUSTMENT

1. Turn the secondary adjustment counterclockwise as far as it will go. Then turn it clockwise 3 turns.
2. Connect a source of clean compressed air of at least 25 psi to the regulator inlet an depress primer button 3 times.
3. Connect a 0 to 15 psi pressure gauge to the fuel outlet and press and hold the primer button. The pressure gauge should read approximately 2 psi and hold steady at this reading. If pressure rises slowly, the primary valve is leaking and the regulator must be replaced. If pressure remains constant, proceed.
4. Remove pressure gauge and cover outlet with a film of soap solution. If a bubble forms, the secondary valve is leaking.
5. Slowly turn the secondary adjustment to the left until the bubble expands, then to the right one complete turn to stop the leak. If leaking persists, replace the regulator.

Governor

▶ **See Figure 189**

Engine speed governors in the K series of engines are of the centrifugal flyweight mechanical type. The governor gear and flyweight mechanism are contained within the crankcase. The governor gear is driven by a gear on the camshaft.

OPERATION

Centrifugal force acting on the rotating governor gear assembly causes the flyweights to move outward as speed increases and inward as speed decreases. As the flyweights move outward they force the regulating pin of the assembly to move outward. The regulating pin contacts the tab on the cross shaft, causing the shaft to rotate with changing speed. One end of the cross shaft protrudes through the side of the crankcase. Through external attached to the cross shaft, the rotating action is transmitted to the throttle plate of carburetor.

When the engine is at rest and the throttle is in the 'fast' position, the tension of the governor spring holds the throttle valve open. When the engine is operating (governor gear assembly is rotating), the force applied by the regulating pin against the cross shaft tends to close the throttle valve. The governor spring tension and the force applied by the regulating pin are in 'equilibrium' during operation, holding the engine speed constant.

When a load is applied and the engine speed (and governor speed) decreases, the governor spring tension moves the governor arm to open the throttle plate wider. This admits more fuel and restores engine speed. (This action takes place very

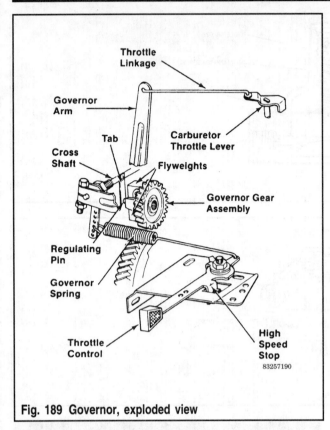

Fig. 189 Governor, exploded view

rapidly, so a reduction in speed is hardly noticed.) As the speed reaches the governed setting, the governor spring tension and the force applied by the regulating pin will again be in equilibrium. This maintains engine speed at a relatively constant level.

Governed speed may be at a fixed point as on constant speed applications, or variable as determined by a throttle control lever.

ADJUSTMENT

❊❊CAUTION

The maximum allowable speed for these engines is 3600 RPM, no load. Never tamper with the governor setting to increase the maximum speed. Severe personal injury and damage to the engine or equipment can result if operated at speeds above maximum.

Initial Adjustment
▶ See Figure 190

Make this initial adjustment whenever the governor arm is loosened or removed from cross shaft. Make sure the throttle linkage is connected to governor arm and throttle lever on carburetor to ensure proper setting.

1. Pull the governor arm away from the carburetor as far as it will go.

2. Grasp the end of cross shaft with pliers and turn counterclockwise as far as it will go.

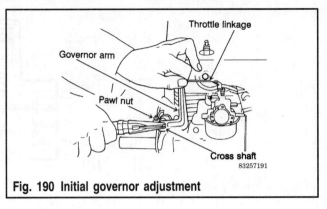

Fig. 190 Initial governor adjustment

3. Tighten the pall nut on governor arm to 15 inch lbs. torque.

➡**Make sure there is at least ¹⁄₁₆ in. clearance between governor arm and cross shaft bushing nut to prevent interference.**

High Speed Adjustment
▶ See Figure 191

The maximum allowable speed is 3600 RPM, no load. The actual high speed setting depends on the application. Refer to the equipment manufacturer's instructions for specific high speed settings. Check the operating speed with a tachometer; do not exceed the maximum. To adjust high speed stop:

1. Loosen the lock nut on high speed adjusting screw.

2. Turn the adjusting screw in or out until desired speed is reached. Tighten the lock nut.

3. Recheck the speed with the tachometer; readjust if necessary.

Sensitivity Adjustment
▶ See Figure 192

Governor sensitivity is adjusted by repositioning the governor spring in the holes in governor arm. If set too sensitive, speed surging will occur with a change in load. If a big drop in speed occurs when normal load is applied, the governor should be set for greater sensitivity.

The standard spring position is in the third hole from the cross shaft. the position can vary, depending on the engine application. Therefore, make a note of (or mark) the spring position before removing it from the governor arm.

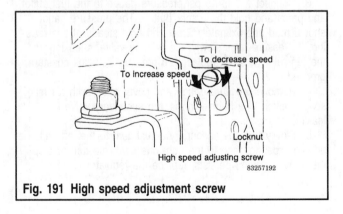

Fig. 191 High speed adjustment screw

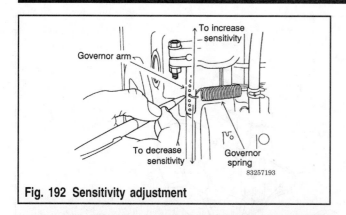

Fig. 192 Sensitivity adjustment

Retractable Starters

Retractable starters are lubricated during manufacturer and should require no further lubrication until disassembly for cord or rewind spring replacement or for other repair.

Frequently check mounting screws to make sure starter is securely tightened on blower housing of engine. If screws are loose, starter realignment may be necessary. Also make sure that the air intake screen is maintained in clean condition at all times.

STAMPED HOUSING MODELS

▶ See Figure 193

❋❋CAUTION

Retractable starters contain a powerful wire recoil spring that is under tension. Do not remove the center screw from the starter until the tension is released. Removing the center screw before releasing spring tension, or improper starter disassembly, can cause the sudden and potentially dangerous release of the spring.

Always wear safety goggles when servicing retractable starters - full face protection is recommended.

To ensure personal safety and proper starter disassembly, the following procedures must be followed carefully.

Removal

Remove the five screws securing the starter assembly to blower housing.

Installation

1. Install starter to blower housing using the five mounting screws. Leave screws slightly loose.
2. Pull the handle out approximately 8 in. to 10 in. until the pawls engage in the drive cup. Hold the handle in this position and tighten the screws securely.

STARTER PAWLS REPLACEMENT

A pawl repair kit, No. 41 757 02 is available. This kit includes two starter pawls, two pawl springs, two retaining rings, and installation instructions.
1. Remove starter from engine.

❋❋CAUTION

Do not remove the center screw of the starter when replacing pawls. Removal of the center screw can cause the sudden and potentially dangerous release of the recoil spring. It is not necessary to remove the center screw when making this repair.

2. Carefully not position of the pawls, pawl springs, and retaining ring before disassembly. (Components must be assembled correctly for proper operation.).
3. Remove the retaining rings, pawls, and pawl spring springs from pawl pins on pulley.
4. Clean pins and lubricate with any commercially available bearing bearing grease.
5. Install new pawl springs, pawls, and retaining rings. When properly installed, the pawl springs will hold the pawls against the pawl cam.

➡**Make sure the snap rings are securely seated in grooves of pawl pins. Failure to seat the snap ring can cause pawls to dislodge during operation.**

6. Pull rope to make sure pawls operate properly.
7. Install starter to engine as instructed under 'To Install Starter.'

ROPE REPLACEMENT

The rope can be replaced without complete starter disassembly.
1. Remove the starter from engine.
2. Pull the rope out approximately 12 in. and tie a temporary (slip) knot in it to keep it from retracting into starter.
3. Remove the rope retainer from inside handle. Untie the knot and remove the retainer and handle.
4. Hold the pulley firmly with thumb and untie the slip knot. Allow the pulley to rotate slowly as the spring tension is released.
5. When all spring tension on the starter pulley is released, remove old rope from pulley.
6. Tie a single knot in one end of new rope.
7. Rotate the pulley counterclockwise (when viewed from pawl side of pulley) until the spring is tight. (Approx. 6 full turns of pulley).
8. Rotate the pulley clockwise until the rope pocket is aligned with the rope guide bushing of housing.

➡**Do not allow pulley/spring to unwind. Enlist the aid of a helper if necessary, or use a c-clamp to hold the pulley in position.**

9. Insert the new rope into the rope pocket of pulley and through rope guide bushing in housing.

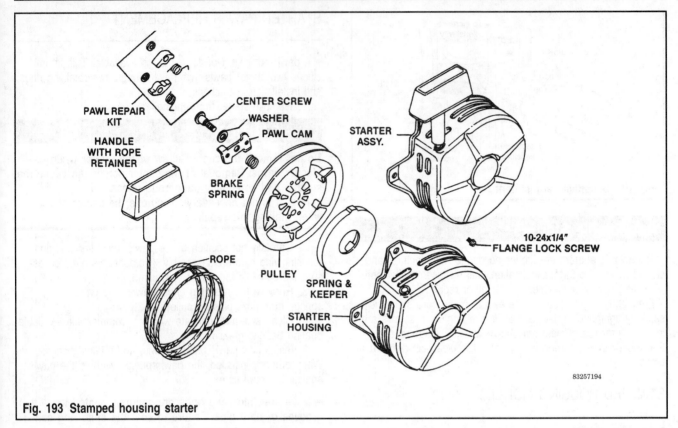

Fig. 193 Stamped housing starter

10. Tie a slip knot approximately 12 in. from the free end of rope. Hold pulley firmly with thumb and allow pulley to rotate slowly until the temporary knot reaches the rope guide bushing in housing.

11. Slip the handle and rope retainer onto rope. Tie a single knot at the end of rope and install rope retainer into handle.

12. Untie the slip knot in rope and pull the handle out until the rope is fully extended. Slowly retract the rope into the starter. If the spring has been properly tensioned, the rope will fully retract until the handle hits the housing.

DISASSEMBLY

1. Remove starter from engine.

✳✳CAUTION

Do not remove the center screw of the starter until the tension of recoil spring has been released. Removing the center screw before releasing spring tension, or improper starter disassembly can cause the sudden and potentially dangerous release of the recoil spring. Follow these instructions carefully to ensure personal safety and proper starter disassembly. Make sure adequate face protection is worn by all persons in the area.

2. Pull the rope out approximately 12 in. and tie a temporary (slip) knot in it to keep it from retracting into starter.

3. Remove the rope retainer from inside handle and untie the knot to remove retainer and handle.

4. Hold the pulley firmly with thumb and untie the slip knot. Allow pulley to rotate slowly as the spring tension is released.

5. When all spring tension on the starter pulley has been released, remove the rope from the pulley.

6. Remove the center screw, washer, pawl cam, and brake spring.

7. Rotate the pulley clockwise 2 full turns. This will ensure the pulley is disengaged from the spring.

8. Hold the pulley into starter housing and invert starter so the pulley is away from your face, and away from others in the area.

9. Rotate the pulley slightly from side to side and carefully separate the pulley from the starter housing.

➥**If the pulley and housing do not separate easily, the spring could be engaged with the pulley, or there is still tension on the spring. Return the pulley to the housing and repeat step 7 before separating the pulley and housing.**

10. Note the position of the spring and keeper assembly on the pulley. (The spring and keeper assembly must be correctly positioned on pulley for proper operation.) Remove the spring and keeper assembly from the pulley as a package.

✳✳CAUTION

Do not remove the spring from the keeper. Severe personal injury could result from sudden uncoiling of the spring.

11. Remove the rope from pulley. If necessary, remove the starter pawl components from pulley as instructed under 'To Replace Starter Pawls.'

INSPECTION AND SERVICE

1. Carefully inspect rope, starter paws, housing, center screw, and other components for wear or damage.

2. Replace all worn or damaged components.

3. Do not attempt to rewind a spring that has come out of the keeper. Order and install a new spring and keeper assembly.

4. Clean away all old grease and dirt from starter components. Generously lubricate the spring and center shaft of starter housing with any commercially available bearing grease.

REASSEMBLY

1. Make sure spring is well lubricated with grease. Position the spring and keeper assembly to pulley (side opposite pawls). The outside spring tail must be positioned opposite rope pocket.

2. Install the pulley with spring and keeper assembly into starter housing.

➡️The pulley is in position when the center shaft is extending slightly above the face of the pulley. Do not wind the pulley and recoil spring at this time.

3. Lubricate the brake spring sparingly with grease. Install the brake spring into the recess in center shaft of starter housing. (Make sure the threads in center shaft remain clean, dry, and free of grease or oil.).

4. Apply a small amount of Loctite® #271 to the threads to center screw. Install the center screw with washer and cam to the center shaft. Torque screw to 65-75 inch lbs.

5. If necessary, install the pawl springs, pawls, and retaining rings to pins on starter pulley. Refer to 'To Replace Starter Pawls.'

6. Tension the spring and install the rope and handle as instructed in steps 5 through 12 under 'To Replace Rope.'

7. Install the starter to engine.

Cast Housing Models

▶ **See Figure 194**

DISASSEMBLY

1. Remove the starter from engine.

❊❊CAUTION

Do not remove the center screw of the starter until the tension of recoil spring has been released. Removing the center screw before releasing spring tension, or improper starter disassembly can cause the sudden and potentially dangerous release of the recoil spring. Follow these instructions carefully to ensure personal safety and proper starter disassembly. Make sure adequate face protection is worn by all persons in the area.

2. Pull the rope out approximately 12 in. and tie a temporary (slip) knot in it to keep it from retracting into starter.

3. Remove the rope retainer from inside handle. Untie the knot and remove the retainer and handle.

4. Rotate the pulley counterclockwise until the notch in pulley is next to the rope guide bushing.

5. Hold the pulley firmly to keep it from turning. Untie the slip knot and pull the rope through the bushing.

6. Place the rope into the notch in pulley. This will keep the rope from interfering with the starter housing leg reinforcements as the pulley is rotated (step 7).

7. Hold the housing and pulley with both hands. Release pressure on the pulley and allow it to rotate slowly as the spring tension is released. Be sure to keep the rope in the notch.

8. Make sure the spring tension is fully released. (the pulley should rotate easily in either direction.).

9. When all spring tension on the pulley is released, remove the center screw, ¾ in. DIA. washer, and ½ in. DIA. washer.

10. Carefully lift the pawl retainer from pulley.

➡️**A small return spring and nylon spring retainer (spacer) are located under the pawl retainer. These parts are fragile and can be easily lost or damaged. If necessary, use a small screwdriver to loosen the spring retainer from the post on pulley. Replace the spring if it is broken, stretched, or shows other signs of damage.**

11. Remove the 1⅛ in. DIA. thrust washer, brake, return spring, nylon spring retainer, and pawls.

12. Rotate the pulley clockwise 2 full turns. There should be no resistance to this rotation. This will ensure the pulley is disengaged from the recoil spring.

13. Hold the pulley into starter housing and invert starter so the pulley is away from your face and others in the area.

14. Rotate the pulley slightly from side to side and carefully separate the pulley from the starter housing.

➡️**Pulley and housing do not separate easily, the spring could be engaged with pulley, or there is still tension on the spring. Return the pulley to the housing and repeat step 12 before separating the pulley and housing.**

15. Only if it is necessary for the repair of starter, remove the spring from the starter housing as instructed under 'To Replace Recoil Spring.' Do not remove the spring unless it is absolutely necessary.

INSPECTION AND SERVICE

1. Carefully inspect the rope, starter pawls, housing, center screw, center shaft, spring and other components for wear or damage.

2. Replace all worn or damaged components.

3. Carefully clean all old grease and dirt from starter components. Lubricate the spring, center shaft, and certain other components as specified in these instructions with any commercially available bearing grease.

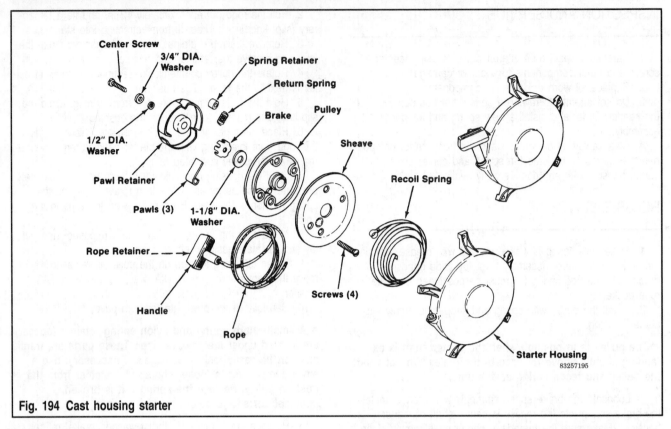

Fig. 194 Cast housing starter

ROPE REPLACEMENT

The starter must be completely disassembled to replace the rope.

1. Remove the starter from engine.
2. Disassemble starter as instructed in steps 2 through 14 under 'Disassembly.'
3. Remove the 4 Phillips head screw securing the pulley and sheave. Separate the pulley and sheave and remove the old rope.
4. Position the new rope in the notch in the pulley and around the rope lock post.

➡**Using rope of the incorrect diameter and/or type will not lock properly in the pulley.**

5. Install the sheave to the pulley and install the 4 Phillips head screws. Use care not to strip or cross-thread the threads in pulley.
6. Inspect the pulley to make sure the sheave is securely joined to the pulley. Pull firmly on the rope to make sure it is securely retained in the pulley.

RECOIL SPRING REPLACEMENT

❋❋CAUTION

Do not attempt to pull or pry the recoil spring from the housing. Doing so can cause the sudden and potentially dangerous release of the spring from the housing. Follow

these instructions carefully to ensure personal safety and proper spring replacement. Make sure adequate face protection is worn throughout the following procedure.

1. Carefully not the position of the spring in the housing. The new spring must be installed in the proper position - it is possible to install it backwards in the housing.
2. Place the housing on a flat wooden surface with the recoil spring and center shaft down and away from you.
3. Grasp the housing by the top so that your fingers are protected. Do not wrap your fingers around the edge of the housing.
4. Lift the housing and rap it firmly against the wooden surface. repeat this procedure until the spring is released from the spring pocket in housing.
5. Discard the old spring.

❋❋CAUTION

Do not attempt to rewind or reinstall a spring once it has been removed from the starter housing. Severe personal injury could result from the sudden uncoiling of the spring. Always order and install a new spring which is held in a specially designed 'C-ring' spring retainer.

6. Thoroughly clean the starter housing removing all old grease and dirt.
7. Carefully remove the masking tape surrounding the new spring/C-ring.
8. Position the spring/C-ring to the housing so the spring hook is over the post in the housing. Make sure the spring is coiled on the correct direction.

9. Obtain Deal Installer #11791 and Handle #11795. Hook the spring hook over the post in housing. Make sure the spring/C-ring is centered over the spring pocket in housing. Drive the spring out of the C-ring and into the spring pocket using the seal installer and handle.

10. Make sure all of the spring coils are bottomed against ribs in spring pocket. Use the seal installer and handle to bottom the coils, as necessary.

11. Lubricate the spring moderately with wheel bearing grease before reassembling the starter.

ASSEMBLY

1. Install the recoil spring into the starter housing as instructed under Replace Recoil Spring.

2. Sparingly lubricate the center shaft of starter with wheel bearing grease.

3. Make sure the rope is in good condition. If necessary, replace the old rope as instructed under 'To Replace Rope.'

➡**Ready the pulley and rope for assembly by unwinding all of the rope from the pulley. Place the rope in the notch in the pulley. This will keep the rope from interfering with the starter leg reinforcements as the pulley is rotated later during reassembly.**

4. Install the pulley onto the center shaft.

➡**If the pulley does not fully seat, it is resting on the inner center spring coil. Rotate the pulley slightly from side to side while exerting slight downward pressure. This should move the inner spring coil out of the way and allow the pulley to drop into position.**

The pulley is in position when the center shaft is flush with the face of the pulley. Do not wind the pulley and recoil spring at this time.

5. Install the starter pawls into the appropriate pockets in the pulley.

6. Sparingly lubricate the underside of the 1⅛ in. DIA. washer with grease and install it over the center shaft. Make sure the threads in center shaft remain clean, dry, and free of grease or oil.

7. Sparingly lubricate the insides of the 'legs' of the brake spider with grease. Instal the brake to the retainer.

8. Install the small return ring to the pawl retainer. Make sure it is positioned properly.

9. Position the pawl retainer and return spring next to the small post on pulley. Install the free loop of the return spring over the post. Install the nylon spring retainer over the post.

10. Invert the pawl retainer over the pawls and center hub of pulley. Take great care not to damage or unhook the return spring. Make sure the pawls are positioned in the slots of pawl retainer.

11. As a test, rotate the pawl retainer slightly clockwise. Pressure from the return spring should be felt. In addition, the pawl retainer should return to its original position when released. If no spring pressure is felt or the retainer does not return, the spring is damaged, unhooked, or improperly assembled. Repeat steps 8, 9, and 10 to correct the problem.

12. Sparingly lubricate the ½ in. DIA. washer and ¾ in. DIA. washer in the center of pawl retainer. Make sure the threads in center shaft remain clan, dry, and free of grease or oil.

13. Apply a small amount of Loctite® #271 to the threads of center screw. Install the center screw to center shaft. Torque screw to 55-70 inch lbs.

14. Rotate the pulley counterclockwise (when viewed from the pawl side of pulley) until the spring is tight. (Approximately 4 full turns of pulley.) Make sure the fully extended rope is held in the notch in pulley to prevent interference with the housing leg reinforcements.

15. Rotate the pulley clockwise until the notch is aligned with the rope guide bushing of housing.

➡**Do not allow the pulley/spring to unwind. Enlist the aid of a helper, or use a c-clamp to hold pulley in position.**

16. Insert the free end of rope through rope guide bushing. Tie a temporary (slip) knot approximately 12 in. from the free end of the rope.

17. Hold the pulley firmly with thumbs and allow the pulley to rotate slowly until the slip knot reaches the rope guide bushing of housing.

18. Slip the handle and rope retainer onto rope. Tie a single knot at the end of rope and install retainer into handle.

19. Untie the slip knot and pull the handle out until the rope is fully extended. Slowly retract the rope into the starter. If the spring has been properly tensioned, the rope will fully retract until the handle hits the housing.

Magneto Ignition System

OPERATION

▶ **See Figures 195 and 196**

In all magneto ignition systems, high-strength permanent magnets provide the energy for ignition. In rotor type systems, the magnet is pressed onto the crankshaft and is rotated inside a coil-core assembly (stator) mounted on the breaking plate. In the other systems, a permanent magnet ring on the inside of the flywheel revolves around the stator. Movement of the magnets past the stator induces electric current flow in the stator coil (and in alternator and lighting coils if provided). The magnets are mounted with alternate North and South poles so that the direction of magnetic flux constantly changes, producing an alternating current (AC) in the stator coil windings.

The stator windings are connected to the magneto ignition coil. Current flow in the ignition coil reaches its highest peak at the instant the magnetic flux reverses direction. This is the point at which the system is timed to provide a spark at the spark plug.

The ignition coil has a low tension primary winding and a high tension secondary winding. The secondary winding has approximately 100 turns of wire for evert 1 turn in the primary. This relationship causes the voltage induced in the secondary winding to be about 100 times higher than in the primary. If the magneto produces 250 volts in the primary winding, the secondary winding voltage will be 25,000 volts.

When ignition is required, the breaker points open to break the primary circuit. The resultant sudden collapse of the field

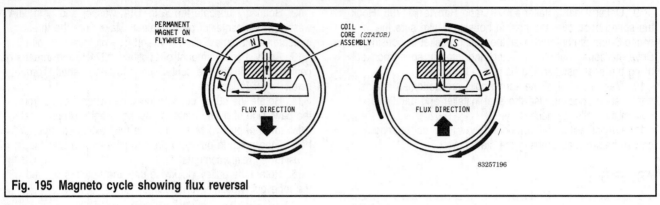

Fig. 195 Magneto cycle showing flux reversal

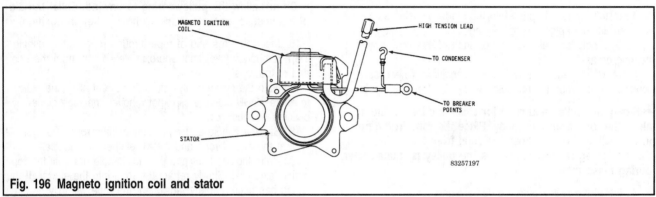

Fig. 196 Magneto ignition coil and stator

around the primary winding causes sufficient energy to be produced in the secondary winding to bridge the spark plug gap. The collapsing field also induces energy in the primary winding, but the condenser shunts this energy to ground, preventing it from bridging the breaker point gap.

TIMING

Engines are equipped with a timing sight hole either in the bearing plate or in the blower housing. If a snap button covers the hole, pry it out with a screwdriver or similar tool so that the timing marks may be seen. Two marks will be present on the flywheel; T for top dead center, and S or SP for the firing point (20 deg. before top dead center).

There are two ways to time a magneto ignition system, static and timing light. The timing light method is the more accurate of the two. A storage battery is needed for use with most timing lights.

Static Timing Method

1. Remove the breaker point cover.
2. Remove the spark plug lead to prevent unintentional starting of the engine.
3. Rotate the engine slowly by hand in the direction of normal operation. Rotation should be clockwise when viewed from the flywheel end.
4. The breaker points should just begin to open when the S or SP mark appears in the center of the timing sight hole. Continue rotating the engine until the breaker points are fully opened.

5. Measure the breaker point gap with a feeler gauge. The gap should be 0.020 in..
6. If the gap is not 0.020 in., loosen the gap adjustment screw and adjust the gap.
7. Tighten the gap adjustment screw.
8. Replace the breaker point cover.

Timing Light Timing Method

Several different types of timing lights are available. Follow the manufacturer's directions for use. Perform timing with a timing light as follows.

1. Remove the lead from the spark plug.
2. Wrap one end of a short piece of fine bare wire around the spark plug terminal and replace the lead. The free end of the wire must protrude from beneath the rubber boot on the lead.

➡The preceding is for timing lights using an alligator clip to connect the spark plug. If the light in use has a sharp prong on the spark plug lead, simply penetrate the rubber boot with the prong and make contact with spark plug lead metal connector.

3. Connect one timing light lead to the wire wrapped around the spark plug terminal.
4. Connect one timing light lead to the hot (ungrounded) terminal of the battery.
5. Connect the third timing light lead to engine ground.
6. Start the engine and run it at 1200 to 1800 RPM.
7. Aim the timing light at the timing sight hole. The light should flash just as the S or SP mark is centered in the sight hole or is in line with the center mark on the bearing plate or blower housing.

8. If timing is not as specified, carefully remove the breaker point and slightly loosen the gap adjusting screw, shift the breaker point plate until the timing mark is properly positioned, and tighten the screw.

9. Shut off the engine and replace the breaker point cover.

Battery Ignition System

OPERATION

▶ **See Figures 197 and 198**

The battery ignition system operates in a manner similar to the magneto system. the major difference is that, in the battery system, energy is provided by a battery. The battery is maintained at full charge by an engine mounted motor-generator or alternator.

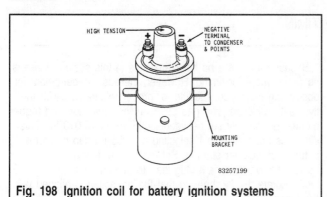

Fig. 198 Ignition coil for battery ignition systems

The coil in a battery ignition system is connected as follows:

1. The positive (+) terminal is connected to the positive terminal of the battery.

2. The negative (-) terminal is connected to the breaker points.

3. The high tension (center) terminal is connected to the spark plug.

TIMING

The timing procedure for the battery ignition system is the same as for the magneto system. When using a timing light, refer to the manufacturer's instructions.

➡**The model K341QS Specification 71276A engine is unique in that it is timed slightly differently then other K series engines. These engines operate at lower speed, so the timing is set at 16 degrees before top dead center to improve running smoothness. Instead of having an S or SP at the timing mark on the flywheel, these engines have a 1 above and a 6 below the mark. When timing these engines, the timing mark is centered as with other engines.**

IGNITION SERVICE

Ignition problems and poor performance on these models are often the result of using an incorrect ignition coil, spark plug, or plug gap setting. Use of the correct spark plug and gap setting is also important. Failure to follow the correct parts will

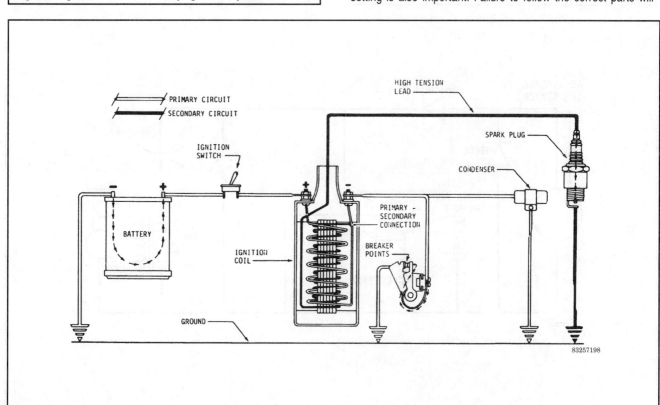

Fig. 197 Wiring diagram for battery ignition systems

result in erratic high speed ignition misfire or cutting-out under load.

Breakerless Ignition System

OPERATION

♦ **See Figures 199 and 200**

The breakerless ignition system operates on the same general principle as the magneto system but does not use breaker points and conventional ignition condenser to time the spark. A trigger module containing solid state electronics performs the same function as the breaker points.

The breakerless system consists of four major components:
- Ignition winding on alternator stator
- Trigger module
- Ignition coil assembly
- Flywheel-mounted trigger

The ignition winding is separate from other windings on the alternator stator. It functions like the magneto winding. The trigger module contains three diodes, a resistor, a sensing coil and magnet and an SCR, a sort of electronic switch. The ignition coil assembly includes a capacitor and a pulse transformer that serves the same purpose as the ignition coil in other systems. The flywheel has a projection that triggers ignition.

In some applications a 22 ohm, ½ watt resistor has been placed between the key switch and the ignition coil. This has been added to prevent current feedback through a dirty or wet switch. This feedback, if not held in check by a resistor, can damage the trigger unit.

TIMING

Because there are no breaker points in this system, there is no requirement for timing. However, there is a requirement for positioning the trigger module for proper relationship with the flywheel projection. The gap between the projection and trigger module is normally set between 0.005 in. and 0.010 in.. This setting is not critical, but selecting a 0.005 in. gap promotes better cold weather starting. Set the gap as follows.

1. Remove the spark plug lead to prevent starting.
2. Rotate the flywheel so that the projection is aligned with the trigger module.

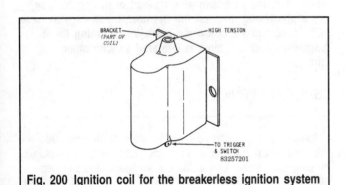

Fig. 200 Ignition coil for the breakerless ignition system

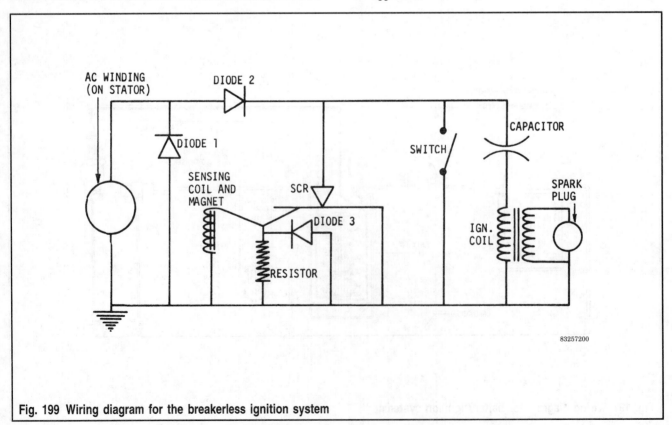

Fig. 199 Wiring diagram for the breakerless ignition system

3. Loosen the cap screws on the trigger module bracket and insert a 0.005 in. feeler gauge in the gap.

4. Move the trigger module until it touches the feeler gauge, making sure that the flat surfaces of module and projection are parallel.

5. Tighten the cap screws and replace the spark plug lead.

Trigger Module
▶ **See Figure 201**

The trigger module used on breakerless ignition systems is a solid state device which includes diodes, resistor, sensing coil and magnet plus an electronic switch called an SCR. The terminal marked A must be connected to the alternator while terminal I must be connected to the ignition switch or ignition coil. Operating with these leads reversed will cause damage to the solid state devices. If a faulty trigger module is suspected, disconnect and remove the trigger from the engine and perform the following tests with a flashlight tester. Reset air gap when reinstalling trigger.

Diode Test

Turn tester switch ON and connect one lead to the I terminal and the other to the A terminal then reverse these leads-- light should come on with leads one way but not the other way. If light stays on or off both ways, this indicates diodes are faulty--replace trigger module.

SCR Test

Turn tester on then connect one lead to the I terminal and the other to the trigger mounting bracket.

➡ **If light comes on, reverse the leads as the light must be off initially for this test.**

Lightly tap magnet with a metal object--when this is done, tester light should come on and stay on until leads are disconnected. If light does not come on, this indicates SCR is not switching properly in which case trigger module should be replaced.

Ignition Coils

BREAKERLESS TYPE IGNITION COIL

Use an ohmmeter to test breakerless type coil assembly. (A) -- Remove high tension lead from terminal on coil. Insert one ohmmeter lead in coil terminal and the other to the coil mounting bracket. A resistance of about 11,500 ohms should be indicated here. (B) -- Connect one tester lead to the coil mounting bracket and the other to the ignition switch wire. Continuity should not be indicated here. Replace ignition coil assembly if wrong results are obtained from either of these tests.

MAGNETO AND BATTERY SYSTEM BREAKER POINTS

▶ **See Figure 202**

Engine operation is greatly affected by breaker point condition and adjustment of the gap. If points are burned or badly oxidized, little or no current will pass. The engine may not operate at all or may miss at high speed. Size of the breaker point gap affects the amount of time the points are open and closed. If the gap is set too wide, they will open too early and close too late. A definite period of time is required for the field to build in the ignition coil. If the points are closed for too long or too short a period, a weak spark will be produced by the coil.

Severe metal buildup on either contact indicates that the condenser is not properly matched to the rest of the system and should be replaced.

Spark Plugs

Engine misfire and starting difficulty are often caused by the spark plug's being in poor condition or being improperly gapped. The spark plug should be removed after every 100 hours of operation for a check of its condition. At this time the gap should be reset or the spark plug replaced as necessary.

SERVICE

▶ **See Figures 203, 204, 205, 206 and 207**

1. Clean the area around the base of the spark plug to keep dirt out of the engine upon removal.

2. Remove the spark plug and check its condition. Replace it if it is badly worn or if re-use is questionable. Clean it if it re-usable.

➡ **Do not clean the spark plug in a machine that uses abrasive grit. Some grit could remain in recesses and enter the engine, causing extensive wear and damage.**

3. Check the gap with a wire type feeler gauge. set the gap a s shown in the following table by carefully bending the side electrode.

4. Install the spark plug and torque it to 18 to 20 ft. lbs.

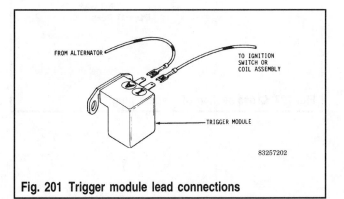

FROM ALTERNATOR

TO IGNITION
SWITCH OR
COIL ASSEMBLY

TRIGGER MODULE

83257202

Fig. 201 Trigger module lead connections

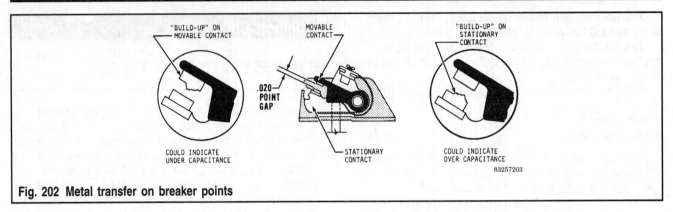

Fig. 202 Metal transfer on breaker points

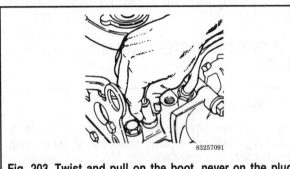

Fig. 203 Twist and pull on the boot, never on the plug wire

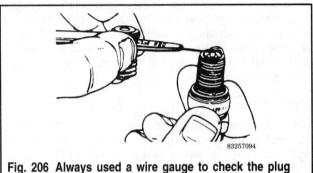

Fig. 206 Always used a wire gauge to check the plug gap

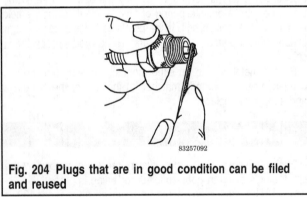

Fig. 204 Plugs that are in good condition can be filed and reused

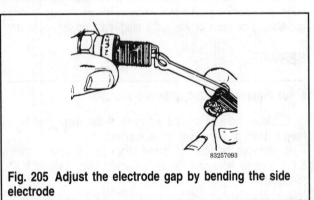

Fig. 205 Adjust the electrode gap by bending the side electrode

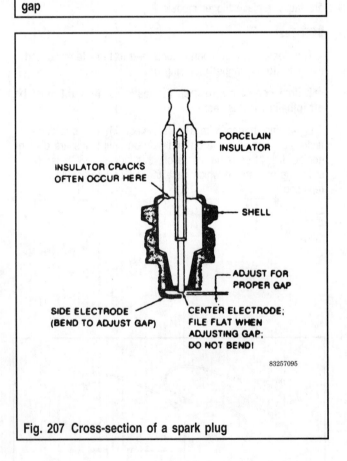

Fig. 207 Cross-section of a spark plug

Alternator

OPERATION

▶ **See Figures 208, 209, 210, 211, 212, 213 and 214**

There are five different models of alternators used in the K series of engines. They are rated at 1.25, 3, 10, 15 and 30 amperes. the 1.25 amp system is intended for battery charging only. The 3 amp device is intended for battery charging and lighting.

There are no adjustments provided for in these systems. Replace if faulty.

➡**To prevent damage to the electrical system and components:**

1. Make sure the battery polarity is correct. A negative (-) ground system is used with K series engines.

2. Disconnect the rectifier-regulator leads and/or wiring harness plug if electric (arc) welding is to be done on the equipment powered by the engine. Disconnect any other electrical accessories that share a common ground with the engine.

3. Make sure the stator (AC) leads do not touch. Shorting them together could permanently damage the stator.

4. Do not operate the engine with the battery disconnected.

➡**If a battery has discharged to less than 4 volts, there may not be sufficient power to activate the rectifier-regulator. If the battery fails to accept a charge from the alternator, charge it on a battery charger and reinstall.**

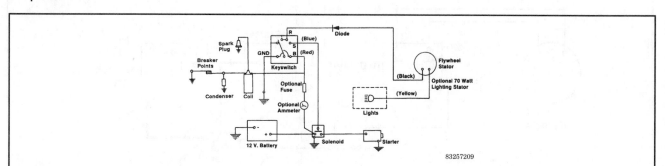

Fig. 208 Wiring diagram for electric start engines w/1.25 amp or 3 amp unregulated battery charging systems w/70 watt lighting

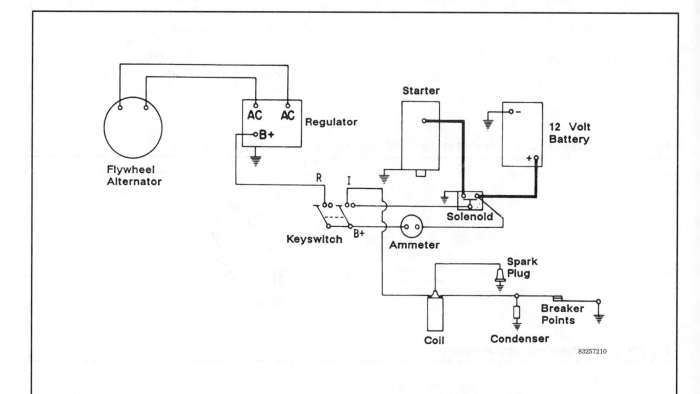

Fig. 209 Wiring diagram for electric start engines w/15 amp regulated battery charging systems

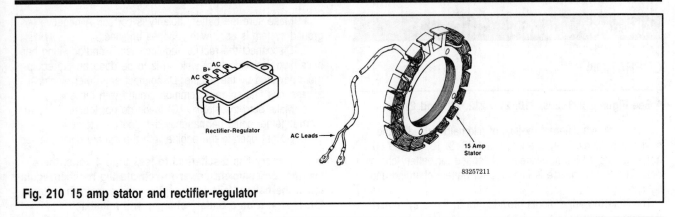

Fig. 210 15 amp stator and rectifier-regulator

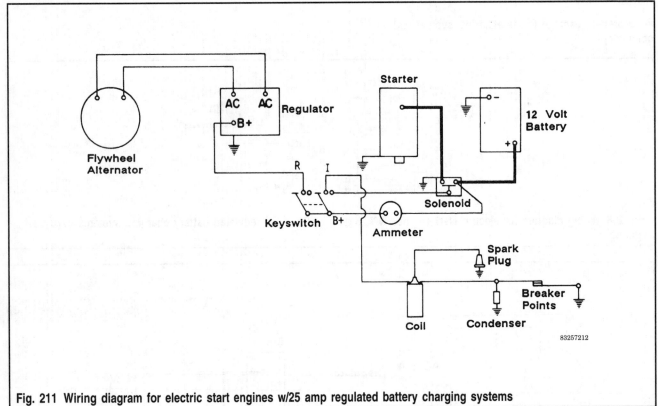

Fig. 211 Wiring diagram for electric start engines w/25 amp regulated battery charging systems

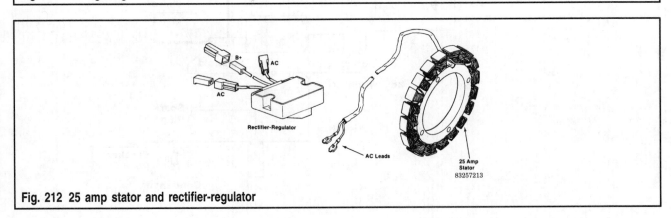

Fig. 212 25 amp stator and rectifier-regulator

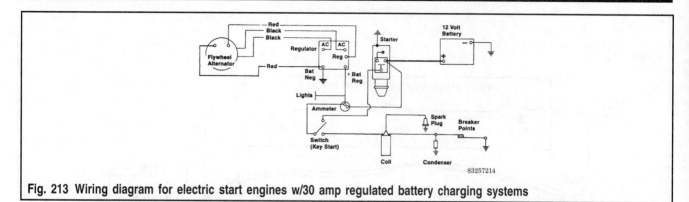

Fig. 213 Wiring diagram for electric start engines w/30 amp regulated battery charging systems

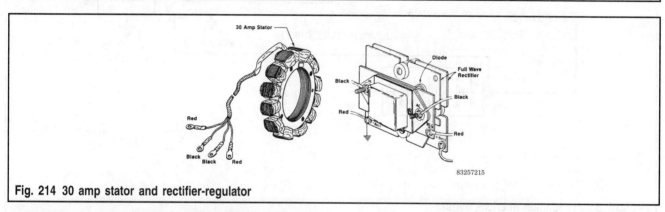

Fig. 214 30 amp stator and rectifier-regulator

Electric Starters

There are three types of electric starters used in the K series of engines. The three types are:

Motor-Generator - This starter also functions as a DC generator. In the starting mode, it turns the crankshaft through a V belt arrangement. The V belt transmits turning force from a small pulley on the motor-generator to a large pulley on the crankshaft.

Wound-Field Bendix Drive Starter - In the field-wound starters, electrical current flows through coils to build up a strong magnetic field to turn the armature. When the armature starts to rotate, a drive pinion moves forward on the armature shaft and meshes with a ring gear on the flywheel. The armature and ring gear remain engaged until the engine starts to run. When the flywheel begins to turn faster than the starter, the pinion is thrown from the ring gear and returns to the disengaged position. A small anti-drift spring on the armature shaft holds the pinion in this position as the starter slows to a stop.

Permanent Magnet Bendix Drive Starter - Operation of this type starter is the same as that of the wound-filed starter. the major difference between the two is in the method of generating the magnetic field to turn the armature. This starter uses strong permanent magnets in place of field coils.

Safety Interlocks

▶ **See Figure 215**

In an effort to enhance safe operation of their equipment, many manufacturers install safety interlocks to prevent engine start before certain safety requirements are met. These interlocks are usually incorporated in the starter circuit. Unless all interlock switches are closed, the starter will not function.

Before servicing a starter that has failed, always check the safety interlock system first. This is done by bypassing the interlock switches with a temporary jumper wire.

❋❋CAUTION

Other than interlock testing, never operate an engine with the safety interlock system removed or bypassed. Great bodily harm or equipment damage could result!

Interlocks connected to an engine with a battery ignition system are bypassed simply by placing a jumper wire as shown.

❋❋CAUTION

Make sure all safety conditions have been observed before starting as engine with the interlocks bypassed.

The safety interlock system on manual start magneto ignition engines is placed in the ignition system. The series connected interlock switches are connected to a solid state module that is connected to the ignition system. The module serves two functions. It grounds the ignition system until all interlocks have closed and, after the engine has started, it prevents the ignition from grounding as the individual interlocks are opened in normal operation (transmission placed in Drive, PTO engaged, etc.).

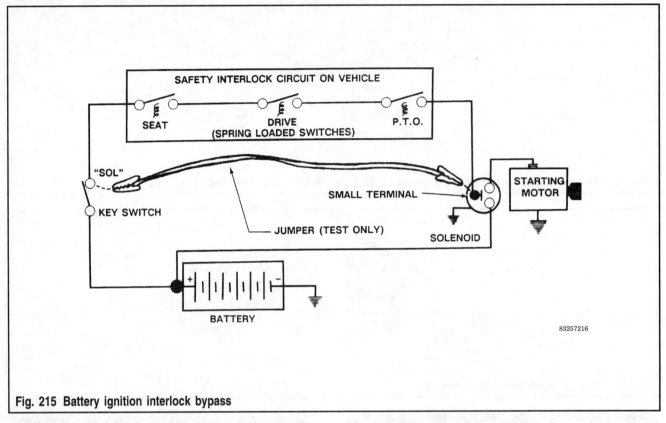

Fig. 215 Battery ignition interlock bypass

Motor-Generator Type Starter

▶ See Figure 216

BRUSH REPLACEMENT

▶ See Figure 217

1. Remove the brush springs from the pockets in brush holder.
2. Remove the self-tapping screws and negative (-) brushes.
3. Remove the stud terminal with positive (+) brushes, and plastic brush holder from end cap.
4. Reinstall the brush holder and new holder and new stud terminal with positive (+) brushes into end cap. Secure with the fiber washer, plain washer, split lock washer, and hex nut.

✳✳CAUTION

To prevent electric arcing, make sure the stud terminal and braided brush leads do not touch the end cap.

5. Install the new negative (-) brushes and secure with the self-tapping screws.
6. Install the brush springs and brushes into the pockets in brush holder. Make sure the chamfered sides of brushes are away from the springs.

➡Use a brush holder tool to keep the brushes in the pockets. A brush holder tool can easily be made from thin sheet metal.

COMMUTATOR SERVICE

Clean the commutator with a coarse, lint free cloth. Do not use emery cloth. if the commutator is badly worn or grooved, turn down on a lathe, or replace the armature.

REASSEMBLY

1. Insert the armature into the starter frame. Make sure the magnets are closer to the drive shaft end of armature. the magnets will hold the armature inside the frame.
2. Install the thrust washer and drive end cap. Make sure the match marks on end cap and frame are aligned.
3. Install the brush holder tool to keep the brushes in the pockets of commutator end cap.
4. Install the commutator end cap to armature and starter frame. Firmly hold the drive end cap and commutator end cap to the starter frame. Remove the brush holder tool.
5. Make sure the match marks on end cap and frame are aligned. Install the through-bolts.
6. Install the drive pinion, dust cover spacer, anti-drifting spring, stop gear spacer, stop nut, and dust cover. Refer to 'Starter Drive Service.'

➡If the engine being serviced is equipped with special shouldered cap screws and lock washers for mounting, make sure these same parts are used for reinstalling the starter. these special parts ensure alignment of the pinion and ring gear.

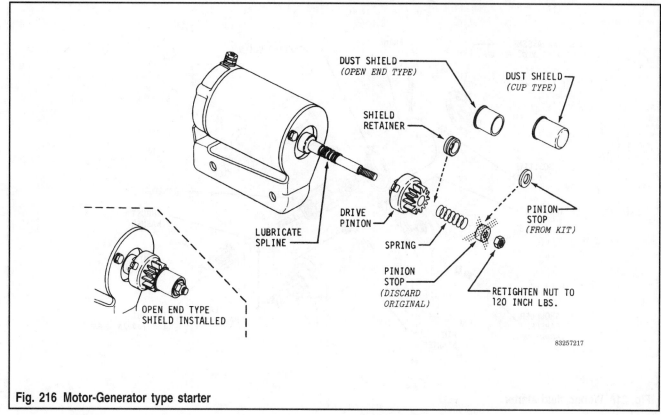

Fig. 216 Motor-Generator type starter

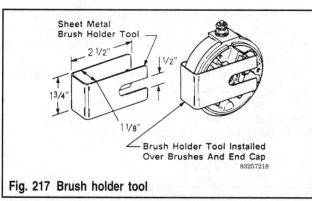

Fig. 217 Brush holder tool

Wound-Field Bendix Drive Starter

▶ **See Figure 218**

✳✳WARNING

In the event of a false start (engine starts but fails to keep running) the engine must be allowed to come to a complete stop before the starter is re-engaged. If the flywheel is still rotating when the starter is engaged, the pinion and ring gear may be damaged.

Do not crank the engine for longer than 10 seconds. A 60-second cool-down period must be allowed between starting attempts. Failure to follow this procedure could result in starter burnout.

➡**If the engine being serviced has special shouldered cap screws and lock washers for mounting, make sure these same parts are used for reinstalling the starter. These special parts ensure alignment of the pinion and ring gear.**

SERVICE

1. Remove the end cap assembly by taking out the two through bolts and carefully slipping the end cap off the armature.
2. Lift the spring holding the positive brush and remove the brush.
3. Carefully remove the armature.
4. Inspect both brushes (positive on frame; negative on end cap). If brushes are worn unevenly or are shorter than $5/16$ in., replace them.
5. Remove the negative brush by drilling out the rivet holding it to the end cap. Install the replacement brush and rivet.
6. Remove the positive brush by peeling back insulating material on the field winding and unclipping or unsoldering. Install the replacement brush and clip or solder in place.
7. Use a coarse cloth to clean the commutator. If the commutator is grooved or extremely dirty, use a commutator stone or fine sandpaper.

➡**Never use emery cloth to clean a commutator.**

8. Carefully insert the armature.
9. Lightly coat the end cap bushing and armature shaft with light engine oil.

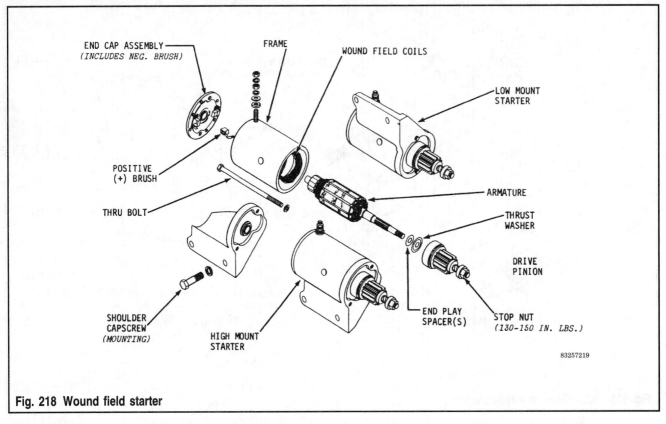

Fig. 218 Wound field starter

10. hold the positive brush spring back and carefully place end in position on armature shaft. Release spring after brushes are contacting commutator.

11. insert two through bolts and torque to 40 to 55 inch lbs.

12. inspect pinion and splined shaft. If any damage is noted, replace the Bendix drive.

13. If the Bendix drive is in good condition, wipe everything clean and apply a very thin coat of special silicone grease (Kohler Part No. 52 357 01) to the splined portion of the armature shaft.

Permanent Magnet Bendix Drive Starter

▶ **See Figure 219**

SERVICE

✳✳WARNING

In the event of a false start (engine starts but fails to keep running) the engine must be allowed to come to a complete stop before the starter is re-engaged. If the flywheel is still rotating when the starter is engaged, the pinion and ring gear may be damaged.

Do not crank the engine for longer than 10 seconds. A 60-second cool-down period must be allowed before starting attempts. Failure to follow this procedure could result in starter burnout.

1. Remove the stop nut and the remainder of the Bendix drive.

2. Remove both through bolts.

3. Remove the end bracket capscrew from the end cap.

4. Remove mounting bracket and frame by rotating the end bracket and slipping the mounting bracket and frame off of the drive end of the armature.

5. Separate the end cap from the armature, being careful to restrain the brushes in the end cap.

6. Inspect the commutator. If dirty, clean it with a coarse, lint-free cloth. If grooved, dress it with a commutator stone or turn it down on a lathe and undercut the mica.

TROUBLESHOOTING

Starter failures from overcranking or cranking with an abnormal parasitic load on the engine, will display one or a number of the following signs:

1. The armature wire insulation or coating will appear discolored and may be swollen. In many cases, you may be able to detect an odor from the burnt wire coating or see it oozing from the starter housing.

2. One or a number of the armature windings may have wires or wire connections that have burnt in tow. Wires may have insulation missing or be partially fused together.

3. The starter brushes will show heavy surface galling and brush material transfer. Additionally, in many instances the starter brushes will be welded or stuck in the brush holders.

Some of the frequent causes of abnormal parasitic load at cranking are:

- Improper viscosity engine crankcase oil.
- Incorrect fluid in a direct coupled hydrostatic unit - remember, even in the idle or neutral position, a direct coupled hydrostatic pump will place a parasitic load on the engine at cranking.

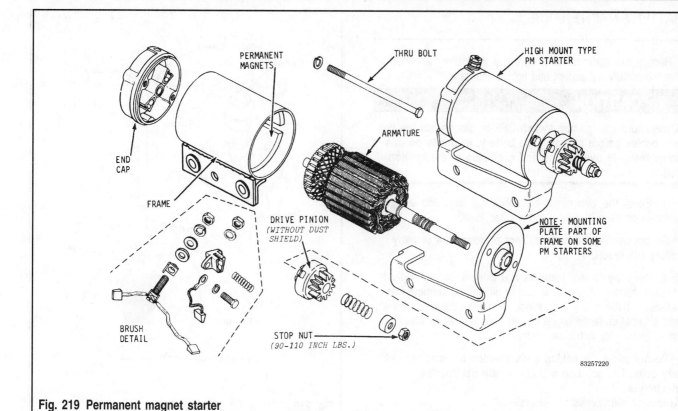

Fig. 219 Permanent magnet starter

• Malfunctioning or inoperative direct coupled clutch assembly.

• Engaged accessory or drive clutch assembly.

• Overcranking - cranking the starter continuously for more than the recommended period and/or not allowing a sufficient cool down period between starting attempts.

• Parasitic Load at cranking - a load or force on the engine at cranking that opposes normal engine rotation.

Battery

BATTERY TEST

If the battery does not have enough charge to crank the engine, recharge it.

➡**Do not attempt to jump-start the engine with another battery. Starting with a battery larger than recommended can burn out the starter motor.**

The battery is tested by connecting a DC voltmeter across the battery terminals and cranking the engine. If the battery voltage drops below 9 while cranking, the batter is in need of a charge or replacement.

BATTERY CHARGING

✳✳CAUTION

Batteries contain sulfuric acid. To prevent acid burns, avoid contact with skin, eyes and clothing.

Batteries produce explosive hydrogen gas while being charged. Charge the battery only in a well ventilated area. Keep cigarettes, sparks, open flame and other sources of ignition away from the battery at all times.

To prevent accidental shorting and the resultant sparks, remove all jewelry before servicing the battery.

When disconnecting battery cables, always disconnect the negative (-) cable first. When connecting battery cables, always connect the negative cable last.

Before disconnecting the negative (-) cable, make sure all switches are OFF. If any switch is ON, a spark will occur at the ground terminal. This could result in an explosion if hydrogen gas or gasoline vapors are present.

Keep batteries and acid out of the reach of children.

BATTERY MAINTENANCE

Regular maintenance will ensure that the battery will ensure that the battery will accept and hold a charge.

✳✳CAUTION

Always turn the ignition switch OFF or disconnect the battery cables before charging the battery. failure to do this could result in overheating and explosion of the ignition coil.

1. Check the level of the electrolyte regularly. Add distilled water to maintain it at its recommended level.

➡**Do not overfill the battery. Poor performance or early failure will result.**

2. Keep the cables, terminals and external battery surfaces clean. A buildup of corrosive acid or dirt on the surfaces can cause the battery to self-discharge. Wash the cables, terminals and external surfaces with a baking soda and water solution. Rinse thoroughly with clean water.

➡**Do not allow the baking soda solution to enter the battery cells. The solution will chemically destroy the electrolyte.**

Automatic Compression Release

▶ **See Figures 220 and 221**

All K-series cylinder engines are equipped with Automatic Compression Release (ACR). The ACR mechanism lowers compression at cranking speeds to make starting easier.

OPERATION

The ACR mechanism consists of two flyweights and a spring attached to the gear on camshaft. When the engine is rotating at low cranking speeds (600 RPM or lower) the flyweights are held by the spring in the position shown. In this position, the tab on the larger flyweight protrudes above the exhaust cam lobe. This lifts the exhaust valve off its seat during the first part of the compression stroke. The reduced compression results in an effective compression ratio of about 2:1 during cranking.

After the engine speed increases to about 600 RPM, centrifugal force moves the flyweights to the position shown. In this position the tab on the larger flyweight drops into the recess in the exhaust cam lobe. When in the recess, the tab has no effect on the exhaust valve and the engine operates at full compression and full power.

When the engine is stopped, the spring returns the flyweights to the position shown, ready for the next start.

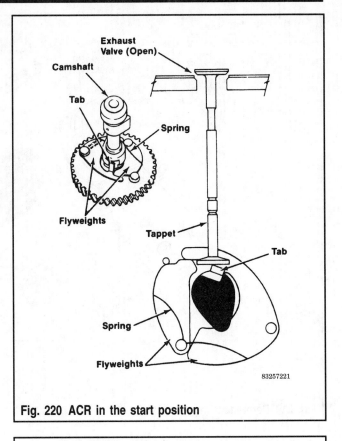

Fig. 220 ACR in the start position

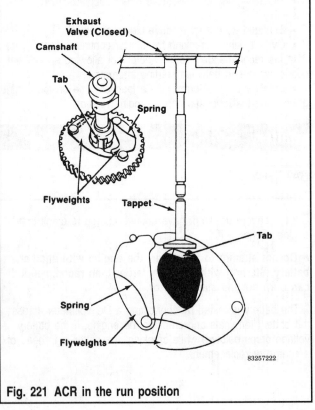

Fig. 221 ACR in the run position

INSPECTION AND SERVICE

The ACR mechanism is extremely rugged and virtually trouble-free. If hard starting is experienced, check the exhaust valve for lift as follows:

1. Check exhaust valve to tappet clearance and adjust as necessary to specification.
2. Remove cylinder head and turn the crankshaft clockwise by hand and observe the exhaust valve carefully. When the piston is approximately ⅔ of the way up the cylinder during the compression stroke, the exhaust valve should lift off the seat slightly.

➡️**If the exhaust valve does not lift, the ACR spring may be unhooked or broken. To service the spring, remove the oil pan and rehook spring or replace it. The camshaft does not have to be removed.**

The flyweights are not serviceable. If they are stuck or worn excessively, the camshaft must be replaced.

➡️**The tab on the flyweights is hardened and is not adjustable. Do not attempt to bend the tab - it will break and a new camshaft will be required.**

COMPRESSION TESTING

Because of the ACR mechanism, it is difficult to obtain an accurate compression reading.

To check the condition of the combustion chamber, and related mechanisms, physical inspection and a crankcase vacuum test are recommended.

AUTOMATIC COMPRESSION RELEASE (ACR) CHANGES

New ACR Tests

Engines with serial no. 9006118 and after have hardened and ground steel ACR tabs on the camshaft assemblies. These new assemblies are manufactured with improved techniques, which permanently set the ACR mechanism, making adjustments to the mechanism unnecessary and impossible.

➡️**Do not attempt to bend these hardened steel ACR tabs. These tabs will break if bent.**

Procedure For Checking And Adjusting ACR On Engines Prior To Serial No. 9006118

On engines manufactured before serial no. 9006118 the ACR can still be checked and reset using the procedure described below.

ACR is set according to the amount of valve lift on the exhaust valve. The correct amount of lift is established by the height of the lifting tab in relation to the camshaft. if improper lift is suspected, the setting can be checked and adjusted as follows:

1. Check valve tappet clearances and adjust as necessary to specification.

2. Remove cylinder head and turn the engine over by hand until you reach BDC of the intake stroke (intake valve will be closing).
3. Mount a dial indicator on the top of the exhaust valve and set a 0.
4. Slowly turn the flywheel clockwise and watch the dial indicator. When the piston is about ⅔ of the way up the cylinder, the exhaust valve should open for ACR. Exhaust valve opening as indicated on the dial indicator should be 0.031-0.042 in.

If the exhaust valve does not open to the specified amount, adjust the ACR according to STEP 5.

➡️**Caution must be exercised in the bending of the tab as it is hardened and may crack or break if bent back and forth more than 3 or 4 times.**

5. If the valve lift was above 0.042 in., hold a wooden dowel or peg on the top of the valve and tap it down carefully to within the 0.031-0.042 in. range. If the valve lift was below 0.031, remove the camshaft cover on the side of the engine exposing the cam gear and bend the ACR tab carefully upward until the valve lift is within the specified range.

Engine Overhaul

DISASSEMBLY

▶ **See Figures 222 and 223**

✳✳CAUTION

Before servicing the engine or equipment, always remove the spark plug lead to prevent the engine from starting accidentally. Ground the lead to prevent sparks that could cause fires.

Clean all parts thoroughly as the engine is disassembled. Only clean parts can be accurately inspected and gauged for wear or damage. There are many commercially available cleaners that quickly remove grease, oil, and grime rpm engine parts. When such a cleaner is used, follow the manufacturer's instructions carefully. Make sure all traces of the cleaner are removed before the engine is reassembled and placed in operation - even small amounts of these cleaners quickly break down the lubricating properties of engine oil. Check all parts for evidence of:

- Excessive sludge and varnish
- Scoring of the cylinder wall
- Piston damage
- Evidence of external oil leaks
- Evidence of overheating

Any of the listed problems could be the result of improper engine servicing or maintenance. The owner should be made aware of the benefits of proper servicing and maintenance.

1. Disconnect the spark plug lead and position it away from the spark plug terminal.
2. Unscrew the oil drain plug(s) and drain the crankcase oil into a suitable container for disposal.

3. Remove the wing nut, air cleaner cover, precleaner (if so equipped), paper element, three base screws, base, and base gasket.

4. If the engine is equipped with a flat muffler, remove muffler and gasket by unscrewing cap screws. If equipped with a round muffler remove by turning the threaded exhaust pipe between the muffler and engine with a pipe wench.

✳✳CAUTION

Gasoline may be present in the carburetor and fuel system. Gasoline is extremely flammable and it can explode if ignited. Keep sparks, open flames, and other sources of ignition away from the engine. Disconnect and ground the spark plug lead to prevent the possibility of sparks from the ignition system.

5. Close the fuel shut-off valve at fuel tank (if so equipped) or drain fuel from tank.

6. Loosen the hose clamp and remove fuel line from the carburetor inlet.

7. Remove two slotted hex cap screws, the carburetor, and gasket.

8. Remove the throttle linkage from the carburetor throttle lever.

9. Note the position of the governor spring in governor arm.

10. Loosen pawl nut. Remove governor arm and space from cross shaft.

➡Loosening pawl nut or removing governor arm will disrupt governor arm to cross shaft adjustment. Readjustment will be required upon reassembly.

11. Remove the governor spring from the governor arm.

12. Remove the hex cap screw, plain washer, spacer, bracket and throttle lever.

✳✳CAUTION

Gasoline may be present in the carburetor and fuel system. Gasoline is extremely flammable and can explode if ignited. Keep sparks, open flames, and other sources of ignition away from the engine. Disconnect and ground the spark plug lead to prevent the possibility of sparks from the ignition system.

13. Disconnect the fuel line from the fuel pump inlet fitting.

14. Disconnect the fuel line from the fuel pump outlet fitting.

15. Remove the fillister head screws, plain washers, fuel pump, and gasket.

16. With retractable starter: Remove screws, washers and the retractable starter assembly.

17. With electric starter:

a. Disconnect electrical connector(s) from back of key switch.

b. Disconnect lead from electrical starter.

c. Remove key switch panel.

d. Remove hex cap screws which mount electric starter to engine.

e. Remove electric starter.

✳✳CAUTION

Gasoline may be present in the carburetor and fuel system. Gasoline is extremely flammable and it can explode if ignited. Keep sparks, open flames, and other sources of ignition away from the engine. Disconnect and ground the spark plug lead to prevent the possibility of sparks from the ignition system.

18. Remove fuel line from fuel tank outlet fitting.

19. Remove tank with bracket(s).

20. Remove the dipstick.

21. Remove the cylinder head baffle.

22. Remove the carburetor side air baffle.

23. Remove the starter side air baffle.

24. Remove pawl nut, breather cover, and gasket.

25. Remove the filter, seal, reed stop, reed, breather plate, gasket, and stud.

26. Remove the spark plug, cylinder head, and gasket.

27. Remove breaker point cover, gasket, breaker point lead, breaker assembly and push rod.

➡Always use a flywheel strap wrench to hold the flywheel when loosening or tightening flywheel and fan retaining fasteners. Do not use any type of bar or wedge between fins of cooling fan as the fins could become cracked or damaged. Always use a puller to remove flywheel from crankshaft. Do not strike the crankshaft or flywheel, as these parts could become cracked or damaged.

28. On rope start models:

a. Remove the grass screen retainer and wire mesh grass screen from rope pulley.

b. Hold the flywheel with a strap wrench and loosen the hex cap screw. Remove the hex cap screw, Remove the hex cap screw, plain washer, rope pulley, and spacer. Remove the nylon grass screen from the fan.

29. On retractable start models:

a. Hold the flywheel with a strap wrench and loosen hex cap screw securing flywheel to crankshaft. Remove the hex cap screw, plain washer, and drive cup.

b. Remove the grass screen from the drive cup.

30. On electric start models:

a. remove the grass screen from the fan.

b. Hold the flywheel with a strap wrench and loosen hex cap screw or hex nut securing flywheel to crankshaft. Remove the hex cap screw or hex nut. remove plain washer.

31. On all models

Flywheel is mounted on tapered portion of crankshaft. Use of a puller is recommended for removing flywheel. Bumping end of crankshaft with a hammer to loosen flywheel should be avoided as this can damage crankshaft.

➡Ignition magnet is not removable or serviceable!

Do not attempt to remove ignition magnet from flywheel. Loosening or removing magnet mounting screws could cause the magnet to come off during engine operation and be thrown from the engine causing severe injury. Replace the flywheel if magnet is damaged.

32. remove the screws and stator.

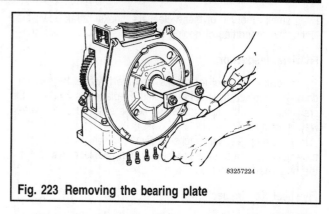

Fig. 223 Removing the bearing plate

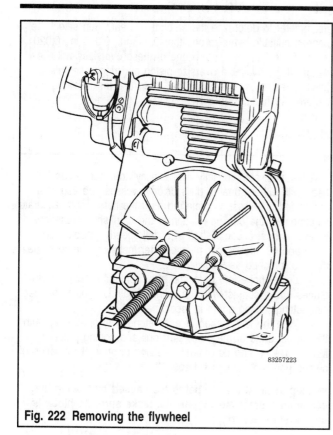

Fig. 222 Removing the flywheel

33. Rotate the crankshaft until the piston is at top dead center of compression stroke (both valves closed and piston flush with top of bore).

34. Compress the valve springs with a valve spring compressor and remove the keepers.

35. Remove the valve spring compressor, then remove the valves, intake valve spring lower retainer, exhaust valve rotator, valve springs, and valve spring upper retainers.

➡**Some models use a valve rotator on both valves.**

Make sure the piston is at top dead center in bore to prevent damage to oil dipper on connecting rod.

36. Remove the hex cap screws, oil pan, and gasket.

37. Remove the connecting rod cap.

➡**If a carbon ridge is present at top of bore, use a ridge reamer tool to remove it before attempting to remove piston.**

38. Carefully push the connecting rod and piston out top of bore.

39. Remove the retainer and wrist pin. Separate the piston from the connecting rod.

40. remove the top and center compression rings and the oil control ring spacer using a ring expander tool.

41. Remove the rails and expander spring(s).

42. Remove the hex cap screws securing the bearing plate to crankcase.

43. Remove the bearing plate from the crankshaft using a puller.

➡**The front bearing may remain either in the bearing plate or on the crankshaft when the bearing plate is removed.**

44. Press the crankshaft out of the crankcase from the PTO side. It may be necessary to press crankshaft out of cylinder block. Bearing plate should be removed first if this is done.

➡**If the repair does not require separating the bearing plate from crankshaft, the crankshaft and bearing plate can be pressed out as necessary.**

45. Drive the camshaft pin (and cup plug on bearing side plate) out of the crankcase from the PTO side.

46. Remove the camshaft pin, camshaft, and shim(s) on bearing plate side of camshaft.

47. Mark the tappets as being either intake or exhaust. Remove the tappets from the crankcase.

➡**The intake valve tappet is closest to the bearing plate side of crankcase. The exhaust valve tappet is closest to the PTO side of crankcase.**

48. Remove the retaining rings, shims, balance gears with needle bearings, shims and spacers.

➡**Extreme care must be taken when handling the new needle bearings or when removing balance gears containing the new bearings. The needles are no longer caged and will drop out. If this should occur, the bearing case should be greased and the needles reset. There are 27 individual needles in each bearing.**

49. Remove the stop pin, copper washer, governor gear, and thrust washer.

50. Remove bushing nut and sleeve. Remove cross shaft from inside crankcase.

51. Remove the oil seals from the crankcase and bearing plate.

52. Press the bearings out of the bearing plate and crankcase.

➡**If the bearings have remained on the crankshaft, remove bearing by using a puller.**

INSPECTION AND REPAIR

All parts should be thoroughly cleaned. Dirty parts cannot be accurately gauged or inspected properly for wear or damage. There are many commercially available cleaners that quickly remove grease, oil and grime accumulation from engine parts. If such a cleaner is used, follow the manufacturer's instructions carefully, and make sure that all of the cleaner is removed before the engine is reassembled and placed in operation.

Even small amounts of these cleaners quickly break down the lubricating properties of engine oils.

Flywheel Inspection

Inspect the flywheel for cracks, and the flywheel keyway for damage. Replace flywheel if cracked. Replace the flywheel, the crankshaft, and the key if flywheel key is sheared or the keyway is damaged.

Inspect ring gear for cracks or damage. Kohler no longer provides ring gears as a serviceable part. Replace flywheel if the ring gear is damaged.

Flywheel Key Inspection

Shearing is possible on engines with flywheel drives and battery ignition systems. Check conditions such as overload, ignition timing and spark plug gap when flywheel key shearing occurs.

Spark plug gap on battery ignition engines must be set as specified. If improperly gaped, a maverick spark can occur, which can cause improper ignition of unburned gases and can create a force causing the flywheel key to shear.

When repairing this type of failure, replace the flywheel, crankshaft, key, flywheel washer and nut or bolt.

Cylinder Head Inspection

Blocked cooling fins often cause localized 'hot spots' which can result in a 'blown' cylinder head gasket. If the gasket fails, high temperature gases can burn away portions of the aluminum alloy head. A cylinder head in this condition must be replaced.

If the cylinder head appears in good condition, use a block of wood or plastic scraper to scrape away carbon deposits. Be careful not to nick or scratch the aluminum, especially in gasket seating area.

The cylinder head should be checked for flatness. Use a feeler gauge and a surface plate or a piece of plate glass to make this check. Cylinder head flatness should not vary more than 0.003 in.; if it does, replace the cylinder head.

➡**Measure cylinder head flatness between each cap screw hole.**

In cases where the head is warped or burned, it will also be necessary to replace the head screws. The high temperatures that warped or burned the head could have made the screws ductile which will cause them to stretch when tightened.

Cylinder Block Inspection and Reconditioning

Check all gasket surfaces to make sure they are free of gasket fragments. Gasket surfaces must also be free of deep scratches or nicks.

Scoring of the Cylinder Wall: Unburned fuel, in severe case, can cause scuffing and scoring of the cylinder wall. As raw fuel seeps down the cylinder wall, it washes the necessary lubricating oils off the piston and cylinder wall so that the piston rings make metal to metal contact with the wall. Scoring of the cylinder wall can also be caused by localized hot spots resulting from blocked fins or from inadequate or contaminated lubrication.

If the cylinder bore is badly scored, excessively worn, tapered, or out of round, resizing is necessary. Use an inside

micrometer to determine the amount of wear, then select the nearest suitable oversize of either 0.003 in., 0.010 in., 0.020 in., or 0.030 in.. Resizing to one of these oversizes will allow usage of the available oversize piston and ring assemblies. Initially, resize using a boring bar, then use the following procedures for honing the cylinder:

HONING

While most commercially available cylinder hones can be used with either portable drills or drill presses, the use of a low speed drill press is preferred as it facilitates more accurate alignment of the bore in relation to the crankshaft crossbore. Honing is best accomplished at a drill speed of about 250 RPM and 60 strokes per minute. After installing coarse stones in hone, proceed as follows:

1. Lower hone into bore and after centering, adjust so that stones are in contact with the cylinder wall. Use of a commercial cutting-cooling agent is recommended.
2. With the lower edge of each stone positioned even with the lowest edge of the bore, start drill and honing process. Move hone up and down while resizing to prevent formation of cutting ridges. Check size frequently.

➡**Keep in mind the temperatures caused by honing may cause inaccurate measurements. Make sure the block is cool when measuring.**

3. When bore is within 0.0025 in. of desired size, remove coarse stones and replace with burnishing stones. Continue with burnishing stones until within 0.0025 in. of desired size and then use finish stones (220-280 grit) and polish to final size. A crosshatch should be observed if honing is done correctly. The crosshatch should intersect at approximately 23 degrees to 33 degrees off the horizontal. Too flat an angle could cause the rings to skip and wear excessively, too high an angle will result in high oil consumption.
4. After resizing, check the bore for roundness, taper, and size. Use an inside micrometer, telescoping gauge, or bore gauge to take measurements. The measurements should be taken at three locations in the cylinder - at the top, middle, and bottom. Two measurements should be taken (perpendicular to each other) at each of the three locations.
5. Thoroughly clean cylinder wall with soap and hot water. Use a scrub brush to remove all traces or boring/honing process. Dry thoroughly and apply a light coat of SAE 10 oil to prevent rust.

MEASURING PISTON-TO-BORE CLEARANCE

◆ **See Figure 224**

Before installing the piston into the cylinder bore, it is necessary that the clearance be accurately checked. This step is often overlooked, and if the clearances are not within specifications, generally engine failure will result.

➡**Do not use a feeler gauge to measure piston-to-bore clearance, it will yield inaccurate measurements. use a micrometer.**

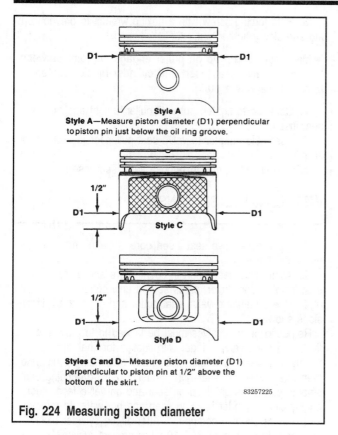

Style A—Measure piston diameter (D1) perpendicular to piston pin just below the oil ring groove.

Styles C and D—Measure piston diameter (D1) perpendicular to piston pin at 1/2" above the bottom of the skirt.

83257225

Fig. 224 Measuring piston diameter

The following procedures should be used to measure the piston-to-bore clearance:

1. Use a micrometer and measure the diameter of the piston as shown.

2. Use an inside micrometer and measure the diameter of the piston as shown.

3. Use an inside micrometer, telescoping gauge, or bore gauge and measure the cylinder bore. Take the measurement approximately 2½ in. below the top of the bore and perpendicular to the piston pin.

4. Piston-to-bore clearance is the difference between the bore and the piston diameter (step 2 minus step 1). For style A pistons only, clearance should be 0.007-0.010 in. For piston styles C and D, clearance should be 0.0045-0.0062 in.

VALVE INSPECTION AND SERVICE

Carefully inspect valve mechanism parts. Inspect valve springs and related hardware for excessive wear or distortion. Valve spring free height should be approximately the dimension given in the chart below. Check valves and valve seat area or inserts for evidence of deep pitting, cracks or distortion. Check clearance of valve stems in guides.

Hard starting, or loss of power accompanied by high fuel consumption may be symptoms of faulty valve. Although these symptoms could also be attributed to worn rings, remove and check valves first. After removal, clean valve head, face and stem with power wire brush and then carefully inspect for defects such as warped valve head, excessive corrosion or worn stem end. Replace valves found to be in bad condition. A normal valve and valves in bad condition are shown in the accompanying illustrations.

Valve Guides

If a valve guide is worn beyond specifications, it will not guide the valve in a straight line. This may result in a burnt valve face or seat, loss of compression, and excessive oil consumption.

To check valve guide to valve stem clearance, thoroughly clean the valve guide and, using a split-ball gauge, measure the inside diameter. Then, using an outside micrometer, measure the diameter of the valve stem at several points on the stem where it moves in the valve guide. Use the largest stem diameter to calculate the clearance. The clearance should not exceed 0.006 in. for intake and 0.008 in. for exhaust valves. If the clearance exceeds these specifications, determine whether the valve stem or the guide is responsible for the excessive clearance.

➡**The exhaust valves on these engines have a slightly tapered valve stem to help prevent sticking. Because of the taper, the valve stem must be measured in two places to determine if the valve stem is worn. If the valve stem diameter is within specifications, replace the valve guide.**

VALVE GUIDE REMOVAL

▶ **See Figure 225**

The valve guides are a tight press fit in the cylinder block. A valve guide removal tool is recommended to remove valve guides. To remove valve guide, proceed as follows:

1. Install 5/16-18 NC nut on coarse threaded end of 3½ in. long stud.

2. Insert other end of stud through valve guide bore and install 5/16-24 NF nut. Tighten both nuts securely.

➡**Valve guide must be held firmly by the stud assembly so that all slide hammer force will act on the guide.**

3. Assemble the valve guide removal adapter to the stud and then slide hammer to the adapter.

4. Use the slide hammer to pull the guide out.

VALVE GUIDE INSTALLATION

1. Make sure valve guide bore is clean and free of nicks or burrs.

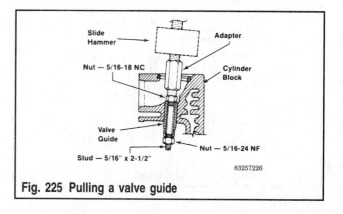

83257226

Fig. 225 Pulling a valve guide

2. Using valve guide driver align and then press guide in until valve guide driver bottoms on valve guide counterbore.

3. Valve guides are often slightly compressed during insertion. Use a piloted reamer and then a finishing reamer to resize the guide bore.

VALVE SEAT INSERTS

The intake valve seat is usually machined into the cylinder block, however, certain applications may specify a hard alloy insert. If the seat becomes badly pitted, cracked, or distorted, the insert must be replaced.

The insert is a tight press fit in the cylinder block. A valve seat removal tool is recommended for this job. Since insert removal causes loss of metal in the insert bore area, use only Kohler service replacement inserts, which are slightly larger to provide proper retention in the cylinder block. Make sure new insert is properly started and pressed into bore to prevent cocking of the insert.

VALVE SEAL INSERT REMOVAL

▶ **See Figure 226**

1. Install valve seat puller on forcing screw and lightly secure with washer and nut.
2. Center the puller assembly on valve seat insert.

3. Hold forcing screw with a hexing wrench to prevent turning and slowly tighten nut.

➡**Make sure sharp lip on puller engages in joint between bottom of valve seat insert and cylinder block counterbore, all the way around.**

4. Continue to tighten nut until puller is tight against valve seat insert.
5. Assemble adapter to valve seat puller forcing screw and slide hammer to adapter.
6. Use slide hammer to remove valve seat insert.

INSTALL VALVE SEAT INSERT

1. Make sure valve seat insert bore is clean and free of nicks and burrs.
2. Align valve seat insert in counterbore and using valve seat installer and driver, press seat in until bottomed.
3. Use a standard valve seat cutter and cut seat to dimensions shown.

Reground or new valves must be lapped in to provide proper fit. Use a hand valve grinder with suction for final lapping. Lightly coat valve face with 'fine' grade of grinding compound, then rotate valve on seat with grinder. Continue grinding until smooth surface is obtained on seat and on valve face. Thoroughly clean cylinder block in soap and hot water to remove all traces of grinding compound. After drying cylinder block apply a light coating of SAE 10 oil to prevent rusting.

PISTON AND RINGS

Identification

▶ **See Figures 227, 228 and 229**

Three different styles of pistons are currently being used in Kohler K-series engines.

Style 'A' pistons can be used in all K-series engines. The style A piston can be identified by its full skirt and its lack of an installation direction identifier on its crown (a new piston can be installed facing either direction).

The Style 'C' piston can be identified by its partial skirt and raised criss-cross design in the recessed area around the piston pin bore. In addition, it has an installation direction identifier (a notch) at its top. the style C piston is to be installed with the notch facing the flywheel.

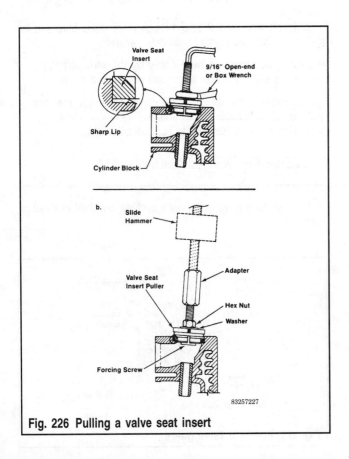

Fig. 226 Pulling a valve seat insert

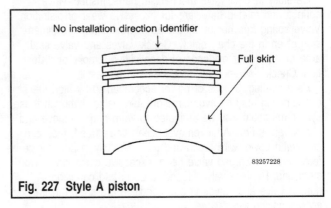

Fig. 227 Style A piston

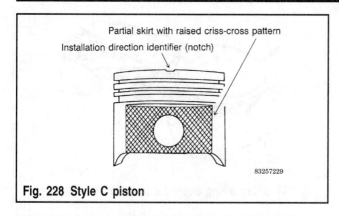

Fig. 228 Style C piston

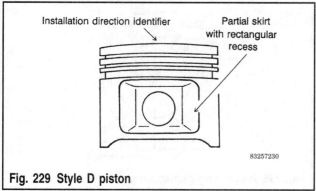

Fig. 229 Style D piston

The Style 'D' piston can be identified by its partial skirt and rectangular recessed area around the piston pin bore. In addition, it has an installation direction identifier, Fly, which is stamped into the top of the piston. the style D piston is to be installed with the arrow of the Fly mark pointing towards the flywheel.

Piston Sizes — All Styles

In order to ensure a correct fit between piston and cylinder we utilize two cylinder bore sizes at the factory. Cylinder blocks are honed to the Standard (STD) size or 0.003: (.075mm) oversize with corresponding pistons. Blocks using the oversize are stamped on the cylinder head gasket surface with 0.003 in.. It is essential that 0.003 in. oversize pistons are used in these blocks to prevent possible failure such as noisy engine or eventual piston skirt cracking. These pistons are available from Kohler. Standard Service Rings should be used with both Standard and 0.003 in. oversize pistons. ring end gap will increase slightly when installed on 0.003 in. oversize pistons; however, sealing is maintained due to the ring design.

Inspection

Scuffing and scoring of piston and cylinder wall occur when internal temperatures approach the melting point of the piston. Temperatures high enough to do this are created by friction, which is usually attributed to improper lubrication, and/or over-heating of the engine.

Normally, very little wear takes place in the piston boss-piston pin area. If the original piston and connecting rod can be reused after new rings are installed, the original pin can also be reused but new piston pin retainers are required. The piston pin is included as part of the piston assembly - if the

pin boss in piston, or the pin are worn or damaged, a new piston assembly is required.

Ring failure is usually indicated by excessive oil consumption and blue exhaust smoke. When rings fail, oil is allowed to enter the combustion chamber where it is burned along with the fuel. High oil consumption can also occur when the piston ring end gap is incorrect because the ring cannot properly conform to the cylinder wall under this condition. Oil control is also lost when ring gaps are not staggered during installation.

When cylinder temperatures get too high, lacquer and varnish collect on piston causing rings to stick which results in rapid wear. A worn ring usually takes on a shiny or bright appearance. Scratches on rings and piston are caused by abrasive material such as carbon, dirt, or pieces of hard metal.

Detonation damage occurs when a portion of the fuel charge ignites spontaneously from heat and pressure shortly after ignition. This creates two frame fronts which meet and explode to create extreme hammering pressures on a specific area of the piston. detonation generally occurs from using fuels with too low of an octane rating.

Pre-ignition of the fuel charge before the timed spark can cause damage similar to detonation. Pre-ignition damage is often more sever than detonation damage - often, a hole is quickly burned right through the piston dome. Pre-ignition is caused by a hot spot in the combustion chamber from sources such as: glowing carbon deposits, blocked fins, improperly seated valves or wrong spark park plug.

Service

▶ **See Figures 230, 231, 232, 233, 234, 235 and 236**

K-series replacement pistons are available in STD bore size, and in 0.003 in., 0.010 in., 0.020 in. and 0.030 in. oversizes.

Fig. 230 Using a ring groove cleaner

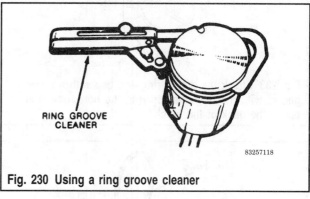

Fig. 231 Using a micrometer to check the piston diameter

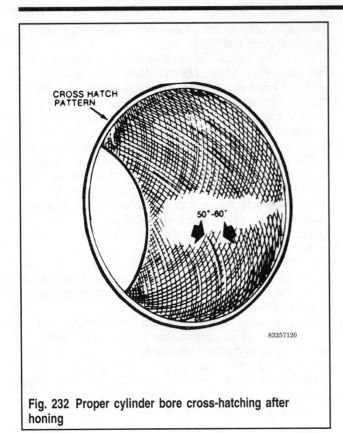

Fig. 232 Proper cylinder bore cross-hatching after honing

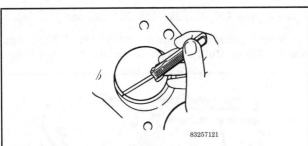

Fig. 233 Check the piston ring end gap with a feeler gauge, with the ring positioned in the bore, one inch below the deck of the block

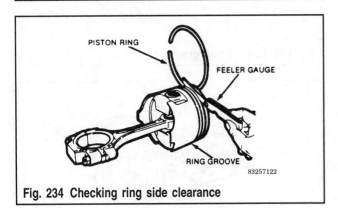

Fig. 234 Checking ring side clearance

Replacement pistons include new piston ring sets and new piston pins.

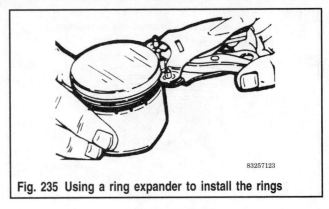

Fig. 235 Using a ring expander to install the rings

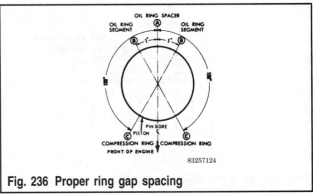

Fig. 236 Proper ring gap spacing

Service replacement piston ring sets are also available separately for STD-0.003 in. (same ring set for both sizes), 0.010 in., 0.020 in. and 0.030 in. oversized pistons. Always use new piston rings when installing pistons. Never reuse old rings.

The cylinder bore must be deglazed before service ring sets are used.

Some important points to remember when servicing piston rings:

1. If the cylinder block does not need reboring and if the old piston is within wear limits and free of score or scuff marks, the old piston may be reused.

2. Remove old rings and clean up grooves. Never reuse old rings.

3. Before installing new rings on piston, place top two rings, each in turn, into its running area in cylinder bore and check end gap.

4. After installing the new compression (top and middle) rings on piston, check piston-to-ring side clearance. Maximum recommended side clearance is 0.006 in.. If side clearance is greater than 0.006 in., a new piston must be used.

Oil Ring End Gaps

Although 4 sizes of service ring sets are available (Std., ± 0.010 in., ± 0.020, ± 0.030 in.), only two sizes of oil rings are supplied (Std. and 0.020 in. oversize). When using 0.010 in. and 0.030 in. oversize ring sets, the oil rings appear to have excessive end gap. This is not detrimental and proper sealing will be achieved due to the additional scraper rings and expander.

➡Scraper and main ring end gaps should be staggered around the groove to prevent combustion blow-by.

Piston Ring Installation

Rings must be installed correctly. Ring installation instructions are usually included with new ring sets. Follow instructions carefully. Use a piston ring expander to install rings. Install the bottom (oil control) ring first and the top compression last.

POSI-LOCK CONNECTING RODS

▶ **See Figure 237**

Posi-Lock connecting rods are used in some K-series engines. On models with the style D pistons (refer to 'Piston and Rings, Identification' earlier in this section), the connecting rods have a narrower piston pin end than on the earlier (style A) Posi-Lock connecting rods. Therefore, the Posi-Lock connecting rods used with the style D pistons are not interchangeable with the Posi-Lock connecting rods used with style A pistons.

Inspection and Service
▶ **See Figure 238**

Check bearing area (big end) for excessive wear, score marks, running and side clearances. Replace rod and cap if scored or excessively worn.

Service replacement connecting rods are available in STD crank pin size and 0.010 in. undersize. The 0.010 in. undersize can be identified by the drilled hole located in the lower end of the rod shank.

BALANCE GEARS AND STUB SHAFTS

Some engines are equipped with a balance gear system.

The system consists of two gears and spacer (used to control end play) mounted on stub shafts which are pressed into the crankcase. The gears and spacer s are held on the shafts with snap-ring retainers. The gears are timed with and driven by the engine crankshaft.

Inspection and Repair
▶ **See Figure 239**

Use a micrometer and measure the stub shaft diameter. If the diameter is less then 0.4996 in., replace the stub shaft.

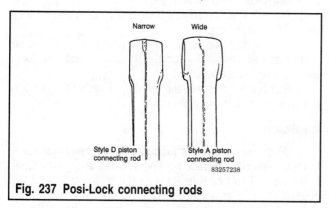

Fig. 237 Posi-Lock connecting rods

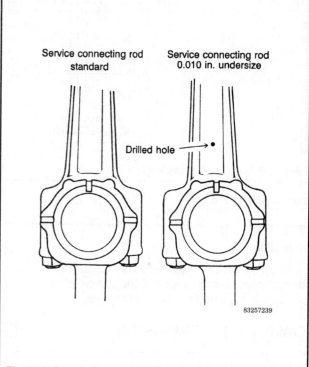

Fig. 238 Standard and 0.010 in. undersize connecting rods

Use an arbor press to push old shaft out and new shaft in. The stub shaft must protrude a specific distance above the stub shaft boss. If the stub shaft boss is about $7/16$ in. above the main bearing boss, press the shaft in until it is 0.735 in. above the stub shaft boss. On blocks where the stub shaft boss is only about $1/16$ in. above the main bearing boss, press shaft in until it is 1.110 in. above the stub shaft boss. A $3/8$ in. spacer must be used with the shaft which protrudes 1.110 in..

Inspect the gears for worn or chipped teeth and for worn needle bearings, if required.

BALANCE GEAR BEARING AND BALANCE GEAR ASSEMBLY

A new needle bearing for the Dynamic Balance System is now being used on some engines. the new bearing (part number 47030 01) has been in use beginning with serial number 9641311. It is not interchangeable with the old needle bearing, part number 236506.

Complete balance gear assemblies are interchangeable - and old style gear assemblies have been superceded with a new gear assembly, part number 47 042 01.

Critical consideration is required when only the needle bearing is to be replaced. The engine serial number alone can correctly determine which needle bearing is involved on original equipment engines. However should a bearing replacement be required after a complete balance gear has been replaced on an engine with a serial number prior to 9641311, the following methods will assist in identifying the correct bearing:

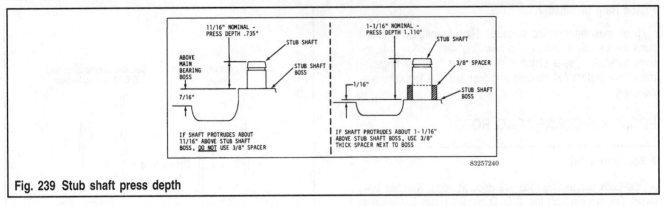

Fig. 239 Stub shaft press depth

GOVERNOR GEAR

Inspection

Inspect the governor gear teeth. Look for any evidence of worn, chipped or cracked teeth. If one or more of these problems is noted, replace the governor gear.

CAMSHAFT AND CRANKSHAFT

Inspection and Service
▶ **See Figure 240**

Inspect the gear teeth on both the crankshaft and camshaft. If the teeth are badly worn, chipped or some are missing, replacement of the damaged components will be necessary.

Also, inspect the crankshaft bearings for scoring, grooving, etc. Do not replace bearings unless they show signs of damage or are out of running clearance specifications. If crankshaft turns easily and noiselessly, and there is not evidence of scoring, grooving, etc., on the races of bearing surfaces, the bearings can be reused.

Check crankshaft keyways. If worn or chipped, replacement of the crankshaft will be necessary. Also inspect the crank pin for score marks or metallic pickup. Slight score marks can be cleaned with crocus cloth soaked in oil. If wear limits are exceeded, it will be necessary to either replace the crankshaft or regrind the crank pin to 0.010 in. undersize. If reground, a 0.010 in. undersize connecting rod (big end) must then be used to achieve proper running clearance. measure the crank pin for size, taper and out-of-round.

➡**If the crank pin is reground, visually check to ensure that the fillet blends smoothly with the crank pin surface.**

When replacing a crankshaft with external threads on the flywheel end with one that has internal threads, different mounting hardware is required. The internally threaded crankshafts are solid in kits which include the hardware. The mounting hardware can also be purchased separately.

1. Use a $^{13}/_{32}$ I.D. x $1^1/_4$ O.D. x $^1/_8$ TH. plain washer (Part No. 52114 01) when installing drive cups with a $1^1/_4$ in. Dia. spot face (machined, recessed area around mounting hole). These drive cups are primarily used on International Harvester applications, but may be found on other applications. Therefore, use the drive cup to identify which washer should be used.

2. Required for drive cups with $^5/_8$ in. mounting hole.

OPTIONAL GEAR REDUCTION UNIT

▶ **See Figure 241**

The reduction unit consists of a driven gear which is pressed on the power take off (PTO) shaft. The drive gear is an integral part of the engine crankshaft. The PTO shaft is supported by two bearings, one in the cover and the other in the housing. Oil seals are provided at both ends of the shaft.

Removal

1. Drain lubricating oil from unit.
2. Remove four cap screws from gear housing and slide cover off along drive gear.
3. Remove four cap screws holding gear housing to engine.
4. Wash all parts and inspect shaft, bushing and gear for wear. Replace worn parts.
5. Remove old oil seals and install new seals (flat side out) in the gear housing and cover.

Installation

1. Wrap piece of tape or roll paper around crankshaft gear to protect the oil seal, slide housing over the shaft and attach to the block. Two lock washers are used on the outside of housing and copper washers inside.
2. Tape or paper should be wrapped around the shaft to prevent the keyway from damaging the cover oil seal. Install the gasket(s) and reduction gear cover and tighten cap screws.

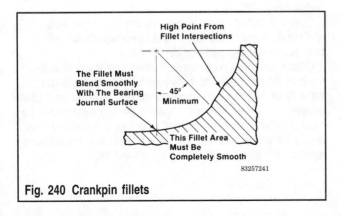

Fig. 240 Crankpin fillets

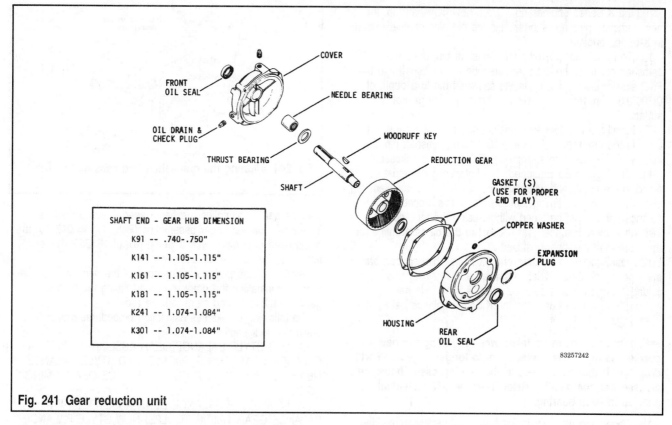

SHAFT END - GEAR HUB DIMENSION

K91 -- .740-.750"

K141 -- 1.105-1.115"

K161 -- 1.105-1.115"

K181 -- 1.105-1.115"

K241 -- 1.074-1.084"

K301 -- 1.074-1.084"

Fig. 241 Gear reduction unit

3. Adjust shaft end clearance to 0.001-0.006 in. by varying the total gasket thickness, adding or removing gaskets as required.

4. Remove oil fill plug and level plug, fill unit to the oil level hole. Use the same grade of oil as used in the engine.

ENGINE REASSEMBLY

▶ **See Figures 242, 243, 244, 245, 246, 247, 248, 249, 250, 251, 252, 253, 254, 255, 256, 257 and 258**

The following sequence is suggested for complete engine reassembly. This procedure assumes that all components are new or have been reconditioned, and all component subassembly work has been completed. This procedure may have to varied slightly to accommodate options or special equipment.

➡**Make sure the engine is assembled using all specified torque values, tightening sequences, and clearances. failure to observe specifications could cause severe engine wear or damage. Always use new gaskets.**

1. Install the rear bearing into crankcase using the #4747 handle and appropriate bearing installer. Make sure the bearing is bottomed fully, and is straight and true in bore. Install the rear main bearing by pressing it into cylinder block. If using a shielded type bearing, install with shielded side facing toward inside of block.

2. Slide cross shaft into place from inside of block.

3. Place speed control disc on governor bushing nut and thread bushing nut into block. On earlier models, the cross shaft has an extension riveted in place to line up with governor gear. Torque bushing nut as follows 100-120 inch lbs.

4. Install the thrust washer, governor gear, copper washer, and stop pin.

5. Rotate governor gear assembly to be sure stop pin does not contact weight section governor gear.

6. Install the intake valve tappet and exhaust valve tappet into crankcase. (Intake valve tappet towards bearing plate side; exhaust valve tappet towards PTO side of crankcase.).

7. Install the camshaft, one 0.005 in. shim spacer, and the camshaft pin (from bearing plate side). Do not driver the camshaft pin into its final position at this time.

➡**On pre-ACR models with the automatic spark advance camshaft, spread actuators and insert cam. Align the timing marks on cam and gear as shown.**

8. Measure the camshaft end play between the spacer and crankcase boss using a flat feeler gauge. Recommended camshaft end play is 0.005-0.010 in. for all models. Add or subtract 0.005 in. and/or 0.010 in. shim spacers as necessary to obtain the proper end play.

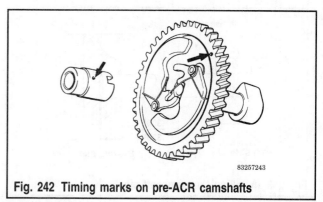

Fig. 242 Timing marks on pre-ACR camshafts

9. the K-Series engines now use a new camshaft pin, the new camshaft pins are shorter than the old pins originally used in K-Series engines.

10. To install the new (shorter) camshaft pin, drive the camshaft pin from the bearing plate side of crankshaft into the PTO side of crankcase. Drive the camshaft pin to a depth of 0.300-0.330 in. from the machined bearing plate gasket surface.

To install the old (longer) camshaft pin, drive the camshaft pin into the crankcase until the PTO end of camshaft pin is flush with the mounting surface on PTO side of crankcase.

11. On Engines So Equipped: The balance gears must be timed to the crankshaft whenever the crankshaft is installed. Use a balance gear timing tool to simplify this procedure. if the balance gears must be timed without using the tool, do not install the lower balance gear (closest to the oil pan) until after the crankshaft has been installed.

12. Install the ⅜ in. spacer, one 0.010 in. shim spacer, balance gear, one 0.020 in. shim spacer, and retaining ring (rounded edge towards balance gear). A new style needle bearing is now being used on the K-Series balance gear assembly.

➡**Extreme care must be taken when handling the new needle bearings. The needles are no longer caged and will drop out. If this should occur, the bearing case should be greased and the needles reset. There are 27 individual needles in each bearing.**

13. Check end play with a flat feeler gauge. Recommended end play is 0.002-0.010 in.. If end play is not within range, install or remove 0.005 in. and 0.010 in. spacers, as necessary.

14. On Engines Without Balance Gears

a. Lubricate the crankshaft rear bearing surface. Insert the crankshaft through the rear bearing.

➡**If the crankshaft and bearing plate have not been separated, position the fuel line and wiring harness between the bearing plate and crankcase before pressing the crankshaft all the way in.**

b. Align the primary timing mark on crankshaft with the timing mark on camshaft. Press the crankshaft into rear bearing. Make sure the camshaft and crankshaft gears mesh and that the timing marks remain aligned while pressing.

15. On Engines With Balance Gears:

K-Series have two styles of balance gear assemblies. To provide improved vibration reducing characteristics, redesigned

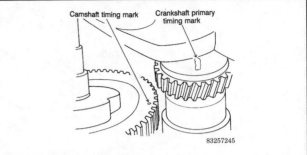

Fig. 244 Aligning the crankshaft and camshaft timing marks

balance gear assemblies are being used in some engines. These new balance gear assemblies (Part No. 45 043 03) are being used in engines with a Serial No. of 1613600013 and later, and for service replacement.

Because of the physical differences of the gear, new procedures for installing the crankshaft, and timing the balance gears, crankshaft, and camshaft are required.

The following crankshaft installation procedures are broken down into four sections:

1A) OLD STYLE BALANCE GEAR ASSEMBLY - WITH A BALANCE GEAR TIMING TOOL 1B) OLD STYLE BALANCE GEAR ASSEMBLY - WITHOUT A BALANCE GEAR TIMING TOOL

2A) NEW STYLE BALANCE GEAR ASSEMBLY - WITH A BALANCE GEAR TIMING TOOL 2B) NEW STYLE BALANCE GEAR ASSEMBLY - WITHOUT A BALANCE GEAR TIMING TOOL METHOD

1A) Old Style Balance Gear Assembly - with a Balance Gear Timing Tool

1. Align the primary timing marks of balance gears with the teeth on timing tool. Insert tool so it meshes with gears. Hold or clamp tool against oil pan gasket surface.

2. lubricate the crankshaft rear bearing surface. Insert the PTO end of crankshaft through rear bearing. 'Straddle' the primary and secondary timing marks on crankshaft over the rear bearing oil drain. Press the crankshaft into rear bearing until the crankgear is just above the camshaft gear but not in mesh with it.

➡**If the crankshaft and bearing plate have not been separated, position the fuel line and wiring harness between the bearing plate and crankcase before pressing the crankshaft all the way in.**

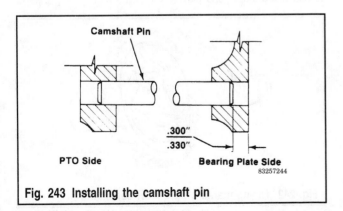

Fig. 243 Installing the camshaft pin

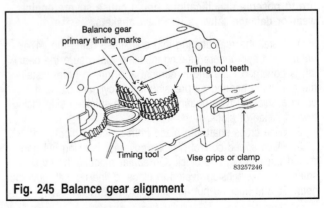

Fig. 245 Balance gear alignment

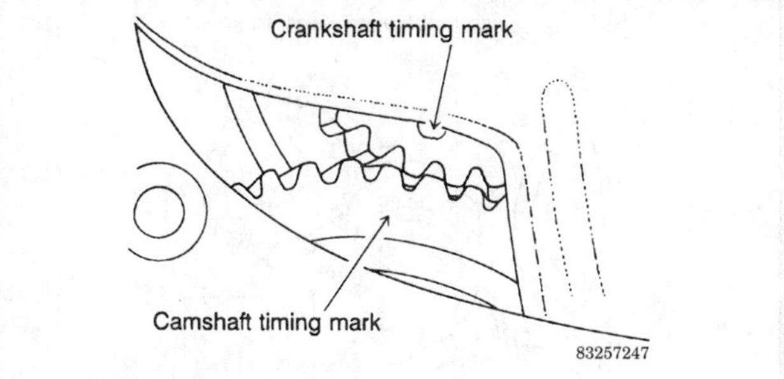

Fig. 246 Aligning the crankshaft and camshaft gear timing marks

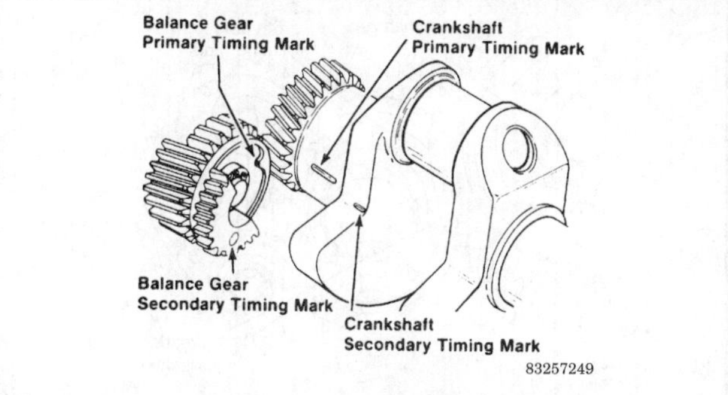

Fig. 248 Balance gear and crankshaft timing marks

3. Remove the balance gear timing tool and align the primary timing mark on the crankshaft with the timing mark on the camshaft gear. Press the crankshaft all the way in to the rear bearing. Make sure the camshaft and crankshaft gears mesh and that the timing marks align while pressing.

4. Check the timing of the crankshaft, camshaft, and balance gears:

a. The primary timing mark on crankshaft should align with the secondary timing mark on lower balance gear.

b. The primary timing mark should mark on crankshaft should align with the timing mark on camshaft.

c. If the mark do not align, the timing is incorrect and must be corrected.

1B) Old Style Balance Gear Assembly - without a Balance Gear Timing Tool

➡The balance gear should be installed after the crankshaft has been installed.

1. Lubricate the crankshaft rear bearing surface. Insert the PTO end of crankshaft through rear bearing. Align the primary timing mark on crankshaft with the primary timing mark on upper balance gear. Press the crankshaft into rear bearing until the crankgear just starts to mesh (about $^1/_{16}$ in.) with the center ring of balance gear teeth.

➡If the crankshaft and bearing plate have not been separated, position the fuel line and wiring harness between the bearing plate and crankcase before pressing the crankshaft all the way in.

2. Align the primary timing mark on the crankshaft with the timing mark on the camshaft gear. Press the crankshaft all the way into the rear bearing. Make sure the camshaft and crankshaft gears mesh and that the timing marks align while pressing.

3. Position the crankshaft so it is about 15 degrees past BDC. Install $^3/_8$ in. spacer, and one 0.010 in. shim spacer. Align the secondary timing mark on the lower balance gear with the secondary timing mark on the crankshaft. Install the lower balance gear on the stub shaft. If properly timed, the primary timing mark on the crankshaft will now be aligned with the secondary timing mark on the lower balance gear.

4. Install one (1) 0.020: shim spacer and retaining ring (rounded edge towards gear). Check end play of lower balance gear as instructed under 'INSTALL BALANCE GEARS.'

5. Check the timing mark on crankshaft, camshaft, and balance gears. The primary mark on crankshaft should align with the primary timing mark on upper balance gear. The primary mark on crankshaft should align with the secondary timing mark on lower balance gear. The primary mark on crankshaft should align with timing mark on camshaft. If the marks do not align, the timing is incorrect and must be corrected.

2A) New Style Balance Gear Assembly - with a Balance Gear Timing Tool

1. Count and mark the teeth on the crankshaft gear, and the lands (notches between teeth) on the camshaft gear as follows:

a. Crankshaft - Locate the primary timing mark on crankshaft. While looking at the PTO end of crankshaft, start with the tooth directly below timing mark and count five (5) teeth in a counterclockwise direction. Mark the fifth tooth.

b. Camshaft - Locate the timing mark on camshaft. starting with the land next to the timing mark, count five (5) lands in a counterclockwise direction. Mark the fifth land.

2. Align the primary timing marks on balance gears with the teeth on timing tool. Insert the tool so it meshes with the

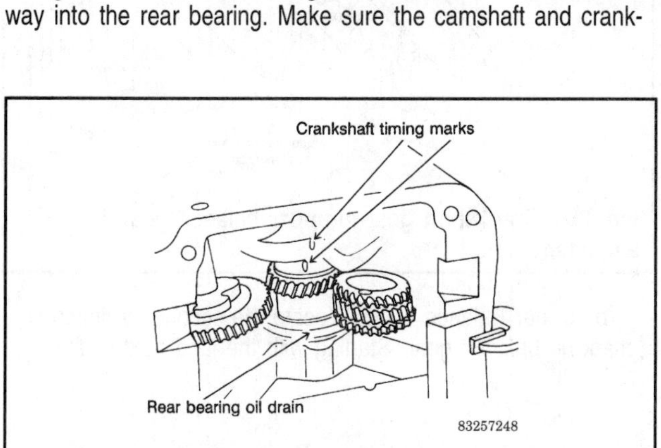

Fig. 247 Aligning the crankshaft and balance gears

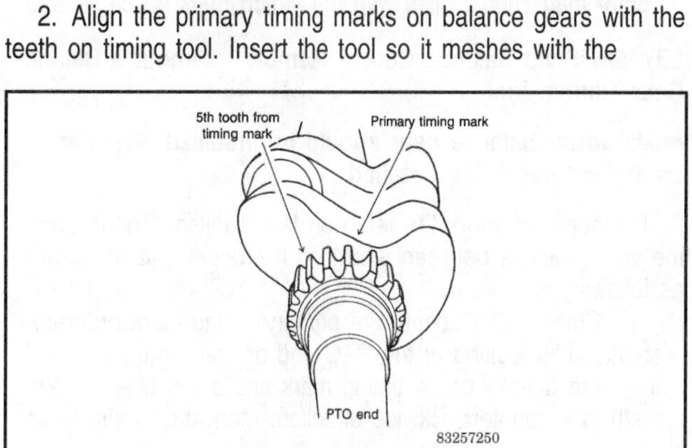

Fig. 249 Marking a tooth on the crankshaft gear

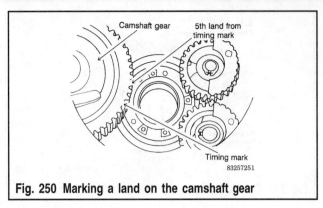

Fig. 250 Marking a land on the camshaft gear

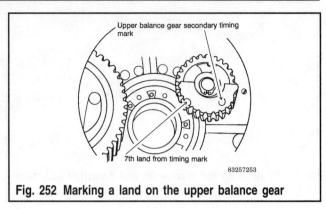

Fig. 252 Marking a land on the upper balance gear

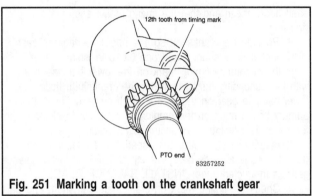

Fig. 251 Marking a tooth on the crankshaft gear

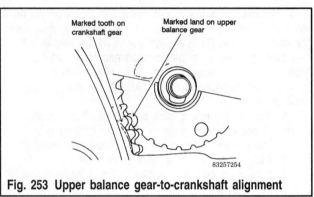

Fig. 253 Upper balance gear-to-crankshaft alignment

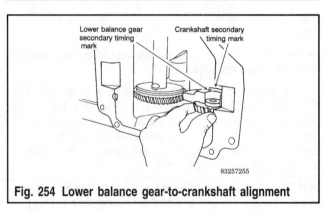

Fig. 254 Lower balance gear-to-crankshaft alignment

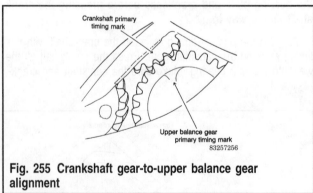

Fig. 255 Crankshaft gear-to-upper balance gear alignment

gears. Hold or clamp the tool against oil pan gasket surface of crankcase.

3. Lubricate the rear bearing surface of crankshaft. Insert the PTO end of crankshaft through the rear bearing. 'Straddle' the primary and secondary timing marks on crankshaft over the rear bearing oil drain. Press the crankshaft into the rear bearing until the crankshaft gear is just above the camshaft gear, but not in mesh with it. Do not remove the balance gear timing tool at this time.

4. Align the fifth (5th) land marked on camshaft gear with the fifth (5th) tooth marked on camshaft gear. Press the crankshaft all the way into the rear bearing. Make sure the camshaft and crankshaft gears mesh and the marks align while pressing.

5. Remove the balance gear timing tool. Check the timing of the crankshaft, camshaft, and balance gears. The primary timing mark on crankshaft should align with the secondary timing mark on lower balance gear. The primary timing mark on crankshaft should align with the timing mark on camshaft.

2B) New Style Balance Gear Assembly - without a Balance Gear Timing Tool

➡The lower balance gear should be installed after the crankshaft has been installed.

1. Count and mark the teeth on the crankshaft gear, and the land (notches between teeth) on the upper balance gear as follows:

   a. Crankshaft - Locate the primary timing mark on crankshaft. While looking at the PTO end of crankshaft, start with the tooth directly below timing mark and count twelve (12) teeth in a counterclockwise direction. Mark the twelfth tooth.

   b. Upper Balance Gear - Locate the secondary timing mark on balance gear. Starting with the land next to the timing mark, count seven (7) lands in a clockwise direction. Mark the seventh land.

2. Lubricate the rear bearing surface of crankshaft. Insert the PTO end of crankshaft through the rear bearing. Align the twelfth (12th) tooth marked on crankshaft gear with the seventh (7th) land marked on upper balance gear. Press the crankshaft into the rear bearing until the crankshaft gear is just above the camshaft gear, but not in mesh with it.

3. Align the timing mark on camshaft with the primary timing mark on crankshaft.

➡**Align the timing mark on camshaft with the primary timing mark on crankshaft. To align the marks, rotate the camshaft only - do not rotate the crankshaft. Rotating the crankshaft could cause the crankshaft gear to come out of mesh and the marks align while pressing.**

4. Install the ⅜ in. spacer and one (1) 0.010 in. shim spacer to the stub shaft for the lower balance gear.

5. Position the crankshaft so it is about 15 degrees past bottom dead center (BDC). Align the secondary timing mark on crankshaft. Install the lower balance gear to the stub shaft. If properly timed, the secondary timing mark on lower balance gear will now be aligned with the primary timing mark on crankshaft.

6. Secure the lower balance gear to stub shaft using one (1) 0.020 in. shim spacer and retaining ring (rounded edge towards gear). Check end play of lower balance gear as instructed in 'INSTALL BALANCE GEARS'.

7. Check the timing of the crankshaft, camshaft, and balance gears. The primary timing mark on crankshaft should align with the secondary timing mark on lower balance gear. The primary timing mark on crankshaft should align with the timing mark on camshaft. The primary timing mark on crankshaft should align with the primary timing mark on upper balance gear. If the marks do not align, the timing is incorrect and must be corrected.

Front Bearing Installation
▶ **See Figure 256**

Install the front bearing into the bearing plate using the #4747 handle and appropriate bearing installer. Make sure the bearing is bottomed fully, and straight and true in the bore.

CONTINUE ENGINE ASSEMBLY

1. Position the fuel line and wiring harness (if so equipped) to crankcase.

2. Adjust the fuel line and wiring harness to their final positions just before securing the bearing plate to the crankcase.

3. The installation of bearing plate and gaskets can be made considerably easier with the use of two simple, easy to make alignment guides. Using 2½ in. long bolts with the hexagon heads removed and screwdriver slots cut in the stem, screw the two headless bolts into the cylinder block diagonally from each other. Bolt thread sizes are ⅜-16 U.N.C.

4. Lubricate the bearing surface of crankshaft and bearing. Install the gasket, two or three 0.005 in. shims (as required), and bearing plate over studs.

➡**Crankshaft end play is determined by the thickness of the gasket and shims between crankcase and bearing plate. Check the end play after the bearing plate is installed.**

5. Install two hex cap screws and hand tighten. Remove the locating studs, and install the remaining two hex cap screws and hand tighten.

6. Tighten the screws evenly, drawing bearing plate to crankcase. Torque to 35 ft. lbs.

7. Check crankshaft end play between the inner bearing race and shoulder of crankshaft using a flat feeler gauge. Recommended total end play is 0.003-0.020 in.

If measured end play is not within limits, remove the bearing plate and, remove or install shims as necessary.

➡**Crankshaft end play is especially critical on gear reduction engines.**

8. Slide the appropriate seal sleeves over the crankshaft. Generously lubricate the lips of the oils seals with light grease. Slide the oil seals over the sleeves.

9. Use the #11795 handle and appropriate seal drivers to install the front oil seals to the following depths. Note that the front oil seal depth varies with engine model and type of bearing plate used-bearing plate configuration differs with type of ignition system used.

➡**For detailed piston inspection and piston ring installation procedures, refer to the 'Inspection And Repair/Reconditioning' section.**

10. Style A pistons: Install wrist pin and retainers.

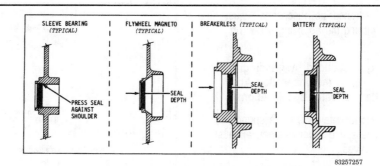

Fig. 256 Front oil seal location

11. Style 'C' And 'D' Pistons:

➡**Proper orientation of the piston to the connecting rod is extremely important. Improper orientation may cause extensive wear or damage.**

12. Orient piston and connecting rod so that the notch (Style 'C' piston) or Fly symbol (Style 'D' piston) on piston and the match mark on connecting rod are facing the same direction.

13. Install wrist pin and retainers.

➡**Proper orientation of the piston to the connecting rod is extremely important. Improper orientation may cause extensive wear or damage.**

14. Stagger the piston rings in their grooves until end gaps are 120 degrees apart.

15. Lubricate the piston and rings with engine oil. Install the piston ring compressor around piston.

16. Orient the notch (on style 'C' piston) or Fly symbol (Style 'D' piston) and match marks on connecting rod towards the flywheel end of crankshaft. Gently push the piston/connecting rod into bore - do not pound on piston.

17. Lubricate the crankshaft and connecting rod journal surfaces with engine oil. Install the connecting rod cap - make sure the match marks are aligned and the oil hole is towards the camshaft. It is important that marks on the connecting rod and cap line up and face flywheel end of engine.

18. Torque the capscrew to 20% over the nominal torque value listed below. Loosen cap screws to below the nominal value--do not leave overtorqued. Retorque bolts to the nominal value.

➡**To prevent damage to connecting rod and engine, do not overtorque-loosen--and retorque the hex nuts on Posi-Lock connecting rods. Torque nuts, in increments, directly to the specified value.**

19. Rotate the crankshaft until the piston is at top dead center in bore to protect the dipper on the connecting rod. If locking tabs are used, bend tabs to lock cap screws.

20. Install screws and oil drain plug as specified in 'General Information'.

21. Rotate the crankshaft until piston is at top dead center of compression stroke.

22. Install the valves and measure the valve-to-tappet clearance using a flat feeler gauge.

➡**Valve faces and seats must be lapped-in before checking /adjusting valve clearance. Refer to the 'Inspection And Repair/Reconditioning' section.**

23. Adjust valve-to-tappet clearance, as necessary. Adjust valve-to-tappet clearance by turning the adjusting screw on tappets.

➡**Large clearances can also be reduced by grinding the valves and/or valve seats. Refer to the 'Inspection And Repair/Reconditioning' section for valve specifications.**

24. Install the valve spring upper retainers, valve springs (close coils to top), intake valve spring lower retainer, exhaust valve rotator or retainer, and valves.

➡**Some models use a valve rotator on both valves.**

25. Compress springs using a valve spring compressor and install keepers.

26. On flywheel-magneto ignition systems, the magneto coil-core assembly ignition systems, the magneto coil-core assembly is secured in stationary position on the bearing plate. On the magneto-alternator systems, the coil is part of the stator assembly which is also secured to the bearing plate. Permanent magnets are affixed to the inside rim of the flywheel except in rotor type magneto systems. On these the magnet or rotor has a keyway and is press fitted on crankshaft. The magnet rotor is marked 'engine-side' for proper assembly.

27. After installing magneto components, run all leads out through holes provided (in 11 o'clock position) on bearing plate.

✳✳CAUTION

Damaging Crankshaft and Flywheel Could Cause Personal Injury!Using improper procedures to install the flywheel can crack or damage the crankshaft and/or flywheel. This not only causes extensive engine damage, but also is a serious threat to the safety of persons nearby, since broken fragments could be thrown from the engine. Always observe and use the following precautions and procedures when installing the flywheel.

➡**Before installing the flywheel, make sure the crankshaft taper and flywheel hub are clean dry and completely free of lubricants. The presence of lubricants can cause the flywheel to be overstressed and damaged when the cap screw is torqued to specification.**

Make sure square flywheel key is installed only in the flat area of keyway, not in the rounded area. The flywheel can become cracked or damaged if the key is installed in the rounded area of keyway.

Always use a flywheel strap wrench to hold flywheel when tightening flywheel fastener. Do not use any type of bar or wedge between the cooling fins or flywheel ring gear, as these parts could become cracked or damaged.

Do not use impact wrenches to install the flywheel retaining nut as this may overstress the nut and crack the flywheel hub.

Do not reuse a flywheel if it has been dropped or damaged in any way.

Do make a through visual inspection of the flywheel and crankshaft before installation to make sure they are in good condition and fee of cracks.

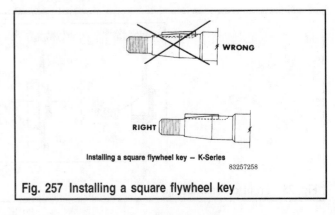

Installing a square flywheel key — K-Series

83257258

Fig. 257 Installing a square flywheel key

The old crankshaft design has a externally threaded end and uses a square key, plain washer, and marsden nut to align and secure the flywheel.

The new crankshaft design has an internally threaded end and uses a woodruff key, washer, and/or brushing, hex cap screw or hexnut.

28. Position key properly in keyway as shown, and carefully guide key slot in flywheel hub over the key while installing to avoid pushing the key inward.

29. On models w/Rope Start:

a. Install the nylon grass screen.

b. Install the spacer, rope pulley, plain washer, and hex cap screw (lubricate threads with oil). Hold the flywheel with a strap wrench and torque hex cap screw to 35-40 ft. lbs. If a hex nut is used, torque to 50-60 ft. lbs.

c. Install the wire mesh grass screen and grass screen retainer to rope pulley.

30. On models w/Retractable Start:

a. Install the grass screen.

b. Install the drive cup, plain washer, and hex cap screw (lubricate threads with oil). Hold the flywheel with a strap wrench and torque hex cap screw to 35-40 ft. lbs. If a hex nut is used, torque to 50-60 ft. lbs.

31. On models w/Electric Start:

a. Install the plain washer and hex cap screw (lubricate threads with oil). Hold the flywheel with a strap wrench and torque hex cap screw to 35-40 ft. lbs. If a hex nut is used, torque to 50-60 ft. lbs.

b. Install the grass screen.

32. For all models, torque the grass screen fasteners to 70-140 ft. lbs. for a metal grass screen and 20-30 inch lbs. for a plastic grass screen.

33. Install the spark plug lead and kill lead into the slots in the baffle.

34. Install the remaining self-tapping screws and the blower housing.

➡**On some models, the grass screen must be installed before installing the blower housing.**

35. Install push rod, breaker assembly, and breaker point lead.

36. Set breaker point gap at 0.020 in. full open.

37. Install gasket and breaker point cover.

38. Install the gasket and cylinder head. Always use a new gasket when head has been removed for service work.

39. Torque the hex cap screws and hex nuts (in increments) in the sequence and torques shown.

➡**The importance of torquing cylinder head bolts to specified values and following the recommended sequences cannot be overemphasized. Blown head gaskets and cylinder head distortion may result from improper torquing. Following is the recommended torquing procedure: 1. Lubricate the cylinder head bolts with oil before installation. 2. Initially torque each bolt to 10 ft. lbs. following the recommended torque sequence.**

40. Sequentially tighten each bolt in 10 ft. lbs. increments until the specified torque values are reached. After reaching the final torque value, run the engine for 15 minutes, stop, and allow to cool. Then, sequentially retorque the head bolts to the specified torque value.

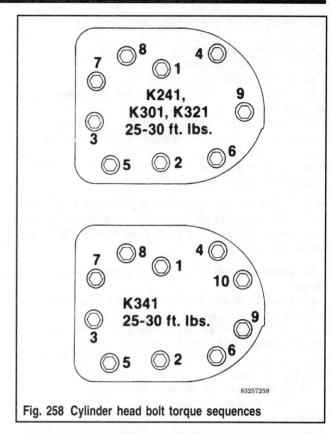

Fig. 258 Cylinder head bolt torque sequences

41. Make sure the spark plug is properly gapped. 5. Install the spark plug and torque it to 18-22 ft. lbs.

42. Install the stud, gasket, breather plate, reed, reed stop, seal, and filter. The accompanying illustrations show the correct order of assembly for two types of breather assemblies. Make sure reed valve is installed properly and that oil drain hole on breather plate is down.

43. Install the gasket, breather cover, and pawlnut.

44. Install the starter side air baffle, plain washer, and hex cap screws. Leave the screws loose.

45. Install the carburetor side air baffle, plain washer, and hex cap screws. Leave the screws loose.

46. Install the cylinder head baffle, plain washer, and hex cap screws. Leave the screws loose.

47. Tighten the screws securely when all pieces are in position.

➡**Shorter screws go into lower portion of blower housing.**

48. Install dipstick.

✳✳CAUTION

Gasoline may be present in the carburetor and fuel system. Gasoline is extremely flammable and it can explode if ignited. Keep sparks, open flames, and other sources of ignition away from the engine. Disconnect and ground the spark plug lead to prevent the possibility of sparks from the ignition system.

49. Install fuel tank with brackets.

50. Install fuel line on fuel tank outlet fitting.

51. Install electric starter.

52. Install hex cap screws which mount electric starter to engine.

53. Install key switch panel.

54. Connect lead to electrical starter.

55. Connect electrical connector(s).

56. Install the retractable starter and hex cap screws. Leave the screws slightly loose.

57. Pull the starter handle out 8-10 in. until the pawls engage in the drive cup. Hold the handle in this position and tighten screws securely.

58. Install the gasket, fuel pump, plain washers, and fillister head screws. Torque the screws to 37-45 inch lbs.

➡**Make sure the fuel pump lever is positioned above the camshaft. Damage to the fuel pump, and subsequent severe engine damage could result if the lever is positioned below the camshaft.**

59. Connect the fuel lines to fuel pump inlet and outlet fittings.

60. Install the throttle lever, bracket, spacer, plain washer and hex cap screw.

61. Install the governor spring to the governor arm. Install the governor arm to the cross shaft. Leave the pawlnut slightly loose as the governor arm and cross shaft will be adjusted after the carburetor and throttle linkage are installed.

62. Install the fuel line and hose clamps.

63. Install the gasket, carburetor, and slotted hex cap screws.

64. Install the throttle linkage into the nylon inserts in the governor arm and carburetor throttle lever.

65. Adjust the governor as instructed below.

66. Refer to 'Fuel System And Governor' section for carburetor adjustment procedure.

GOVERNOR ADJUSTMENT

The governor cross shaft/governor arm must be adjusted every time the governor arm is loosened or removed from cross shaft.

1. Pull the governor arm away from the carburetor as far as it will go.

2. Grasp end of cross shaft with pliers and turn counterclockwise as far as it will go. The governor shaft can be adjusted for end clearance by moving needle bearing in block. Set bearing to allow a slight back-and-forth movement of shaft.

3. Torque the pawlnut on governor arm to 15 inch lbs.

➡**Make sure there is at least $1/16$ in. clearance between the governor arm and the upper-left cam gear cover fastener to prevent interference.**

4. If the engine is equipped with a flat muffler, install muffler and gasket using cap screws. If equipped with a round muffler, install muffler and threaded exhaust pipe between the muffler and engine using a pipe wrench.

5. Install the base gasket, base, and air cleaner.

PREPARE THE ENGINE FOR OPERATION

Before operating the engine, be sure to do the following:
- Make sure all hardware is tightened securely and oil drain plugs are installed.
- Fill the crankcase with the right amount, weight, and type of oil.
- Fill the fuel tank with the proper type of gasoline and open fuel shut-off valve (if equipped).
- Adjust the carburetor main fuel needle, idle fuel needle, or idle speed adjusting screw as necessary. Refer to the 'Fuel System And Governor' section.
- Make sure the maximum engine speed does not exceed 3600 RPM. Adjust the high speed stop as necessary. Refer to the 'Fuel System And Governor' section.

RUN-IN PROCEDURES (RECONDITIONED ENGINES)

For overhauled engines or those rebuilt with a new short block or miniblock, use straight SAE 30, SG-quality oil for the first 5 hours of operation. Change the oil after this initial run-in period. Refill with SG-quality oil as specified under 'Oil Types'.

7

KOHLER
COMMAND

11 AND 12.5 HORSEPOWER
MODELS 7-30
14 HP MODELS 7-60
5 HORSEPOWER MODELS 7-2

5 HORSEPOWER MODELS

SPECIFICATIONS, TOLERANCES, AND SPECIAL TORQUE VALUES[1]

DESCRIPTION	C5 (5 Hp)
General Specifications	
Power (@ 3600 rpm, corrected to SAE J1349)	3.73 kW (5 hp)
Peak Torque (@ 2200 rpm)	11.4 N•m (8.4 lbf-ft)
Bore	67 mm (2.64 in)
Stroke	51 mm (2.01 in)
Displacement	180 cu cm (10.98 cu in)
Compression Ratio	8.5 : 1
Approx. Weight	16.33 kg (36 lb)
Approx. Oil Capacity	0.66 liter (0.7 U.S. qt)
Air Cleaner	
Base Nut Torque	6.8 N•m (58 lbf-in)
Angle Of Operation — Maximum (At Full Oil Level)	
Intermittent — All Directions	35°
Continuous — All Directions	20°
Camshaft	
End Play	0.15 – 0.55 mm (0.0059 – 0.0217 in)
Bore I.D. — Max. Wear Limit Crankcase	16.030 mm (0.6311 in)
Closure Plate	25.430 mm (1.0012 in)
Camshaft Bearing Surface O.D. — Max. Wear Limit Crankcase End	15.954 mm (0.6281 in)
Closure Plate End	25.350 mm (0.9980 in)
Carburetor	
Preliminary Low Idle Fuel Needle Setting	1 Turn
Fuel Bowl Retaining Screw Torque	9.8 N•m (87 lbf-in)

832590C1

Air Cleaner

This engine is equipped with a replaceable, high density paper air cleaner element and an oil-foam precleaner (which covers the paper element).

AIR CLEANER SERVICE

▶ **See Figures 1 and 2**

Check the air cleaner daily or before starting the engine. Check for and correct heavy buildup of dirt and debris, and loose or damaged components.

➡ Operating the engine with loose or damaged air cleaner components could allow unfiltered air into the engine, causing premature wear and failure.

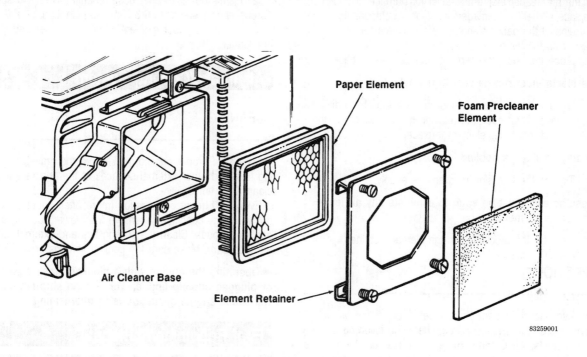

Fig. 1 Air cleaner element

Fig. 2 Removing the air cleaner cover

PRECLEANER SERVICE

Wash and re-oil the precleaner every 25 hours of operation (more often under dusty or dirty conditions).

1. Press and hold the tab at the bottom of the air cleaner cover.

➡**Choke control must be in the OFF position.**

2. Slide the air cleaner cover off of the air cleaner base (away from the retractable starter).

3. Remove the precleaner from the air cleaner element retainer.

4. Wash the precleaner in warm water with detergent. Rinse the precleaner thoroughly until all traces of detergent are eliminated. Squeeze out excess water (do not wring). Allow the precleaner to air-dry.

5. Saturate the precleaner with new engine oil. Squeeze out all excess oil.

6. Reinstall the precleaner in the element retainer.

7. Reinstall the air cleaner cover. Make sure the air cleaner cover latch snaps securely.

PAPER ELEMENT SERVICE

Every 100 hours of operation (more often under extremely dusty or dirty conditions), check the paper element. Replace the element as necessary.

1. Remove the air cleaner cover.

2. Remove the precleaner form the air cleaner element retainer.

3. Remove the element retainer and paper element from the air cleaner base as follows:

a. Loosen the four slot head screws in the element retainer.

➡**Do not remove screws.**

b. When screws are loosened sufficiently, unhook element retainer from bottom tabs on air cleaner base. Remove element retainer.

c. Remove paper element.

4. Replace a dirty, bent, or damaged element with a genuine Kohler element. Do not wash the paper element or use pressurized air, as this will damage the element. Handle new elements carefully; do not use if the sealing surfaces are bent or damaged.

5. When servicing the air cleaner, check the air cleaner base. make sure it is secured and not bent or damaged. Also

check the air cleaner element retainer for damage or improper fit. Replace all bent or damaged air cleaner components.

6. Reinstall the paper element , element retainer, and precleaner as follows:

a. Place the paper element into the air cleaner base.

➡**The pleats must run parallel to the cylinder.**

b. Unhook the element retainer into the top of the air cleaner base. Then hook the retainer into the bottom of the base to hold the paper element in place.

➡**Be sure retainer is hooked in tabs.**

c. Tighten the four slot head screws evenly.

➡**These screws must be snug to eliminate any air leaks around paper element.**

d. Reinstall the precleaner in the element retainer.

INSPECTION

Whenever the air cleaner cover is removed, or the paper element or precleaner are serviced, check the following areas/components: **Air Cleaner Base** — Make sure the base is secured and not cracked or damaged. Since the air cleaner base and carburetor are secured to the intake port with common hardware, it is extremely important that the nuts securing these components are tight at all times.**Breather Tube** — Make sure the tube is installed to both the air cleaner base and valve cover.

➡**Damaged, worn, or loose air cleaner components can allow unfiltered air into the engine causing premature wear and failure. Tighten or replace all loose or damaged components.**

DISASSEMBLY

The following procedure is for complete disassembly of all air cleaner components.

1. Remove the air cleaner cover.

2. Remove the precleaner from the air cleaner element retainer.

3. Remove the element retainer and paper element from the air cleaner base.

4. Remove the two air cleaner base mounting screws and the two air cleaner base mounting nuts.

5. Disconnect the breather tube from the air cleaner base, and remove the air cleaner base and gasket.

ASSEMBLY

Before reinstalling an air cleaner base that has been removed, make sure the four metal bushings, which reinforce the base mounting holes and maintain the proper torque of the mounting hardware, are in place.

1. Install the gasket and air cleaner vase, and connect the breather tube to the air cleaner base.

2. Install the air cleaner base mounting nuts and screws. Torque each fastener to 6.8 N-m (58 inch lbs.).

3. Install the paper element, element retainer and precleaner.

Air Intake/Cooling System

CLEANING

To ensure proper cooling, make sure the grass screen, cooling fins, and other external surfaces of the engine are kept clean at all times.

Every 100 hours of operation (more often under extremely dusty or dirty conditions), remove the blower housing and other cooling shrouds. Clean the cooling fins and external surfaces as necessary. Make sure the cooling shrouds are reinstalled.

➡**Operating the engine with a blocked grass screen, dirty or plugged cooling fins, and/or cooling shrouds removed, will cause engine damage due to overheating.**

Fuel Recommendations

✳✳CAUTION

Gasoline is extremely flammable and its vapors can explode if ignited. Before servicing the fuel system, make sure there are no sparks, open flames, or other sources of ignition nearby as these can ignite gasoline vapors. Disconnect and ground the spark plug lead to prevent the possibility of sparks from the ignition system.

For best results use only clean, fresh, regular grade, unleaded gasoline with a pump sticker octane rating of 87 or higher. In countries using the Research method, it should be 90 octane minimum.

Unleaded gasoline is recommended since it leaves less combustion chamber deposits. Regular grade, leaded gasoline can also be used; however be aware that the combustion chamber and cylinder head may require more frequent cleaning.

Use fresh gasoline to ensure it is blended for the season and to reduce the possibility of gum deposits forming which could clog the fuel system. Do not use gasoline left over from the previous season.

Do not add oil to the gasoline!

GASOLINE/ALCOHOL BLENDS

Up to 10% ethyl alcohol/90% unleaded gasoline can be used as fuel for Kohler engines. Do not use other gasoline/alcohol blends.

OPERATION

The typical fuel system and related components include the fuel tank, in-line fuel filter, fuel pump, carburetor, and interconnecting fuel lines.

The fuel from the tank is moved through the carburetor inlet, fuel moves by gravity feed from the fuel tank, through the fuel line(s) and in-line filter (if equipped), to the carburetor.

Fuel then enters the carburetor float bowl and is moved into the carburetor body. There, the fuel is mixed with air. This air-fuel mixture is then burned in the engine combustion chamber.

Fuel Filter

Some engines are equipped with an in-line fuel filter. Visually inspect the filter periodically, and replace with a genuine Kohler filter wen dirty.

Carburetor

▶ See Figure 3

✳✳CAUTION

Gasoline may be present in the carburetor and fuel system. Gasoline is extremely flammable and its vapors can explode if ignited. Keep sparks, open flames, and other sources of ignition away from the engine. Disconnect and ground the spark plug lead to prevent the possibility of sparks from the ignition system.

TROUBLESHOOTING

This engine is equipped with a fixed main jet carburetor. If engine troubles are experienced that appear to be fuel system related, check the following areas before adjusting or disassembling the carburetor.

Make sure the fuel tank is filled with clean, fresh gasoline.

➡**Make sure the fuel tank vent cap is not blocked and that it is operating properly.**

• Make sure fuel line(s) is unrestricted.
• Make sure the fuel tank filter screen is clean and unobstructed.
• If the fuel tank is equipped with a shutoff valve, make sure it is open and unobstructed.
• If the engine is equipped with an in-line fuel filter, make sure it is clean and unobstructed. Replace the filter if necessary.
• Make sure the air cleaner base and carburetor are securely fastened to the engine using gaskets in good condition.

• Make sure the air cleaner element is clean and all air cleaner components are fastened securely.
• Make sure the ignition system, governor system, exhaust system, and throttle and choke controls are operating properly.
• If, after checking the items listed above, engine starting/running problems still exist, it may be necessary to adjust or service the carburetor.

ADJUSTMENT

▶ See Figure 4

➡**Carburetor adjustments should be made only after the engine has warmed up.**

This engine is equipped with a fixed main jet carburetor. The carburetor is designed to deliver the correct fuel-to-air mixture to the engine under all operating conditions. The main fuel jet is calibrated at the factory and is not adjustable. The idle fuel adjusting needle is also set at the factory and normally does not need adjustment.

If, however, the engine is hard-starting or does not operate properly, it may be necessary to adjust or service the carburetor.

1. With the engine stopped, turn the low idle fuel adjusting needle in (clockwise) until it bottoms lightly.

➡**The tip of the idle fuel adjusting needle is tapered to critical dimensions. Damage to the needle and the seat in the carburetor body will result if the needle is forced.**

2. Preliminary Low Idle Fuel Needle Setting: Turn the low idle fuel adjusting needle out (counterclockwise) 1 full turn from lightly bottomed.

3. Start the engine and run at half throttle for five to ten minutes to warm up. the engine must be warm before making final settings (Steps 4,5, and 6).

4. Low Idle Speed setting: Place the throttle control into the idle or slow position. Set the low idle speed to 1200 rpm (± 75 rpm) by turning the low idle speed adjusting screw in or out. Check the speed using a tachometer.

➡**The actual low idle speed depends on the application; refer to equipment manufacturer's recommendations. The recommended low idle speed for basic engines is 1200 rpm. To ensure best results when setting the low idle fuel needle, the low idle speed must not exceed 1500 rpm.**

5. Low Idle Fuel Needle Setting: Place the throttle into the idle or slow position. Turn the low idle fuel adjusting needle

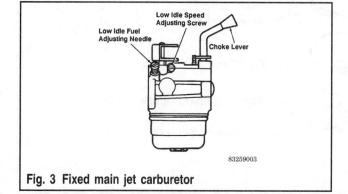

Fig. 3 Fixed main jet carburetor

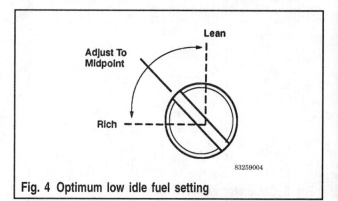

Fig. 4 Optimum low idle fuel setting

out (counterclockwise) from the preliminary setting until the engine speed decreases (rich). Note the position of the needle. Now turn the adjusting needle in (clockwise). The engine speed may increase, then it will decrease as the needle is turned in further (lean). Note the position of the needle. Set the adjusting needle midway between the rich and lean settings.

6. Recheck the idle speed using a tachometer. Readjust the speed as necessary.

DISASSEMBLY

▶ See Figure 5

1. Remove the bowl retaining screw, retaining screw gasket, and fuel bowl.
2. Remove the bowl gasket, float shaft, float, and fuel inlet needle.

3. Remove the low idle fuel adjusting needle and spring. Remove the low idle speed adjusting screw and spring.

➡**Further disassembly of the carburetor (removal of the welch plugs, fuel inlet seat, throttle plate and shaft, and choke plate and shaft) is recommended only if these parts are to be cleaned or replaced.**

Welch Plug Removal
▶ See Figure 6

In order to clean the off-idle ports and bowl vent thoroughly, remove the welch plugs covering these areas.

Use tool no. KO-1018, or equivalent, and the following procedure to remove the welch plugs.

1. Pierce the welch plug with the tip of the tool.

➡**To prevent damage to the carburetor, do not allow the tool to strike the carburetor body.**

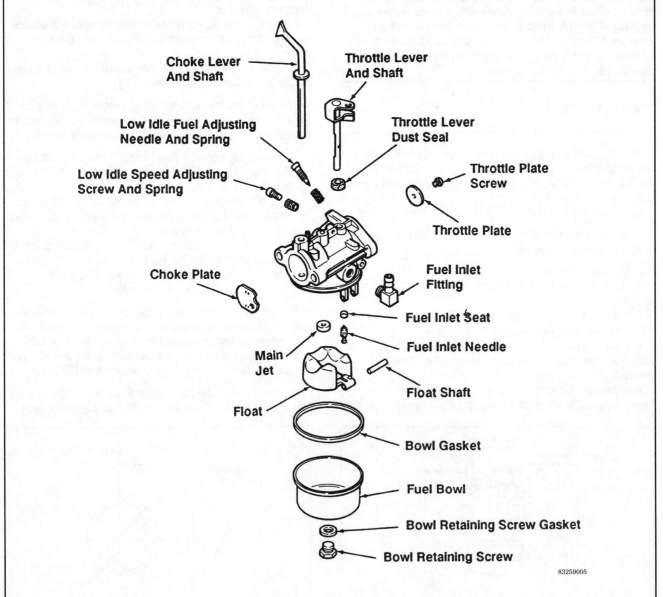

Choke Lever And Shaft

Throttle Lever And Shaft

Low Idle Fuel Adjusting Needle And Spring

Throttle Lever Dust Seal

Low Idle Speed Adjusting Screw And Spring

Throttle Plate Screw

Throttle Plate

Choke Plate

Fuel Inlet Fitting

Fuel Inlet Seat

Main Jet

Fuel Inlet Needle

Float Shaft

Float

Bowl Gasket

Fuel Bowl

Bowl Retaining Screw Gasket

Bowl Retaining Screw

83259005

Fig. 5 Carburetor exploded view

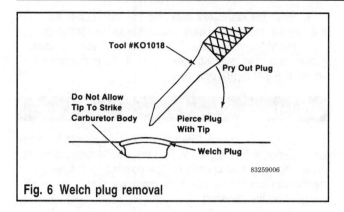

Fig. 6 Welch plug removal

2. Pry out the welch plug with the tip of the tool.

Fuel Inlet Set Removal

To remove the fuel inlet seat, pull it out of the carburetor body using a screw, drill bit, or similar tool.

➡**Always install a new fuel inlet seat. Do not reinstall a seat that has been removed.**

Choke Shaft Removal

1. Because the edges of the choke plate are beveled, mark the choke plate and carburetor body to ensure correct reassembly. Also take note of the choke plate position in the bore, and the position of the choke lever.
2. Grasp the choke plate with a pliers. Pull it out of the slot in the choke shaft.
3. Remove the choke shaft.

Throttle Shaft Removal

1. Because the edges of the throttle plate are beveled, mark the throttle plate and carburetor body to ensure correct reassembly. Also take not of the throttle plate position in the bore, and the position of the throttle lever.
2. Carefully and slowly remove the screw which secures the throttle plate to the throttle shaft. Remove the throttle plate.
3. File off any burrs which may have been left on the throttle shaft when the screw was removed. Do this before removing the throttle shaft from the carburetor body.
4. Remove the throttle lever/shaft assembly with the foam dust seal.

CLEANING

✳✳CAUTION

Carburetor cleaners and solvents are extremely flammable. Keep sparks, flames, and other sources of ignition away from the area. Follow the cleaner manufacturer's warnings and instructions on its proper and safe use. Never use gasoline as a cleaning agent.

All parts should be cleaned thoroughly using a carburetor cleaner (such as acetone). Make sure all gum deposits are removed from the following areas:**Carburetor body and bore** — especially the areas where the throttle plate, choke plate

and shafts are seated.**Idle fuel and off-idle ports** in carburetor bore, main jet, bowl vent, and fuel inlet needle and seat.

➡**These areas can be cleaned with a piece of fine wire in addition to cleaners. Be careful not to enlarge the ports, or break the wire inside the ports. Wearing proper eye protection, blow out all passages with compressed air.**

Float and float hinge.Fuel bowl.Throttle plate, choke plate, throttle shaft, and choke shaft.

➡**Do not submerge the carburetor in cleaner or solvent when fiber, rubber, or foam seals or gaskets are installed. The cleaner may damage these components.**

INSPECTION

Carefully inspect all components and replace those that are worn or damaged.
Inspect the carburetor body for cracks, holes, and other wear or damage.
Inspect the float for cracks, holes, and missing or damaged float tabs. Check the float hinge and shaft for wear or damage.
Inspect the fuel inlet needle and seat for wear or damage.
Inspect the tip of the low idle fuel adjusting needle for wear or grooves.
Inspect the throttle and choke shaft and plate assemblies for wear or excessive play.

REPAIR

Always use new gaskets when servicing or reinstalling carburetors. Repair kits are available which include new gaskets and other components.
Components such as the throttle and choke shaft assemblies, throttle plate, choke plate, low idle fuel needle, and others are available separately.

ASSEMBLY

Throttle Shaft Installation

1. Install the foam dust seal on the throttle shaft.
2. Insert the throttle lever/shaft assembly into the carburetor body. Position the cutout portion of the shaft so it faces the carburetor mounting flange.
3. Install the throttle plate to the throttle shaft. Make sure the plate is positioned properly in the bore as noted and marked during disassembly. Apply Loctite® no. 609 to the threads of the throttle plate retaining screw so that it is slightly loose.
4. Apply finger pressure to the throttle lever/shaft to keep it firmly seated against the pivot in the carburetor body. Rotate the throttle shaft until the throttle plate closes the bore around its entire perimeter; then tighten the screw.
5. Operate the throttle lever. Check for binding between the throttle plate and carburetor bore. Loosen the screw and adjust the throttle plate as necessary. Torque the screw to 0.9-1.4 N-m (8-12 inch lbs.).

Choke Shaft Installation

1. Insert the choke shaft into the carburetor body until the choke shaft detent collar touches the top of the detent spring on the carburetor.

2. Lift the detent spring away from the choke shaft detent collar using a small screwdriver, and insert the choke shaft further into the carburetor body until it bottoms. Spring should now engage with detents of collar.

3. Position the choke as noted and marked during disassembly. Insert the choke plate into the slot in the choke shaft. Make sure that the choke plate is inserted far enough that its locking tabs are positioned on each side of the choke shaft.

Fuel Inlet Seat Installation

Press the fuel inlet seat into the bore in the carburetor body until it bottoms.

Welch Plug Installation

▶ **See Figure 7**

Use tool no. KO-1017, or equivalent.

1. Position the carburetor body with the welch plug cavities to the top.

2. Place a new welch plug into the cavity with the raised surface up.

3. Use the end of the tool that is about the same size as the plug and flatten the plug. Do not force the plug below the surface of the cavity.

4. After the plugs are installed, seal them with fingernail polish or lacquer (or an equivalent sealant). Allow the sealant to dry completely.

CARBURETOR BODY REASSEMBLY

1. Install the low idle speed adjusting screw and spring.

2. Install the low idle fuel adjusting needle and spring. Turn the adjusting needle in (clockwise) until it bottoms lightly.

➡**The tip of the fuel adjusting needle is tapered to critical dimensions. Damage to the needle and the seat in the carburetor body will result if the needle is forced.**

3. Turn the low idle fuel adjusting needle out (counterclockwise) 1 full turn from lightly bottomed.

➡**Upon installation of the reassembled carburetor, follow the final adjustment procedures in this Section.**

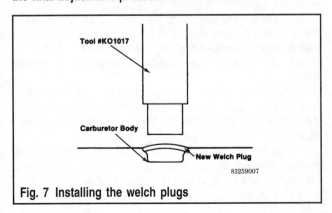

Fig. 7 Installing the welch plugs

4. Insert the fuel inlet needle into the float. Lower the float/needle into the carburetor body. Install the float shaft.

5. Install the bowl gasket, fuel bowl, bowl retaining screw gasket, and bowl retaining screw. Torque the bowl retaining screw to 9.8 N-m (87 inch lbs.).

Governor

This engine is equipped with a centrifugal flyweight mechanical governor. It is designed to hold the engine speed constant under changing load conditions. The governor gear/flyweight mechanism is mounted inside the crankcase and is driven off the gear on the camshaft.

OPERATION

Centrifugal force acting on the rotating governor gear assembly causes the flyweights to move outward as speed increases, and inward as speed decreases. As the flyweights move outward, they cause the regulating pin to extend from the governor gear assembly.

The regulating pin contacts the tab on the cross shaft, causing the shaft to rotate when the engine speed changes. One end of the cross shaft protrudes through the side of the crankcase. Through external linkage attached to the cross shaft, the rotating action is transmitted to the throttle lever of the carburetor.

When the engine is at rest, and the throttle control is in the **fast** position, the tension of the governor spring holds the throttle plate open. When the engine is operating (the governor gear assembly is rotating), the force applied by the regulating pin against the cross shaft tends to close the throttle plate. The governor spring tension and the force applied by the regulating pin are in equilibrium during operation, holding the engine speed constant.

When load is applied and the engine speed (and governor gear speed) decreases, the governor spring tension moves the governor lever to open the throttle plate wider. This allows more fuel into the engine, increasing engine speed. (This action occurs very rapidly, so a reduction in speed is hardly noticed.) As the speed reaches the governed setting, the governor spring tension and the force applied by the regulating pin will again be in equilibrium. This maintains the engine speed at a relatively constant level.

The governed speed setting is determined by the position of the throttle control. It can be variable or constant, depending on the application.

INITIAL ADJUSTMENT

Make this initial adjustment whenever the governor lever is loosened or removed from the cross shaft. To ensure proper setting, make sure the throttle linkage is connected to the governor lever and to the carburetor throttle lever.

1. Loosen the governor lever hex nut.

2. Pull and hold the governor lever away from the carburetor so that the carburetor throttle plate is in the wide open throttle position.

3. Grasp the governor cross shaft with a pliers and turn the shaft counterclockwise as far as it will go.

4. Tighten the hex nut securely.

HIGH IDLE SPEED ADJUSTMENT

The recommended maximum no-load high idle speed for this engine speed is 3600 rpm. The actual high idle speed depends on the application. Refer to the equipment manufacturer's instructions for specific information.

❋❋CAUTION

Do not tamper with the governor setting. Overspeed is hazardous and could cause personal injury.

The high idle speed is set by turning the high speed adjusting screw on the high idle speed/kill switch bracket assembly in or out.

➡**Although certain engine components have been removed for clarity of the photograph, never run the engine with the air cleaner assembly removed. Damage to the engine might otherwise result.**

1. Start the engine and allow it to warm up. Place the throttle control lever into the fast or high idle position.

2. Check the engine speed with a tachometer.

3. To increase the high idle speed, turn the high idle speed adjusting screw out (counterclockwise), while applying light pressure to (and thereby gradually moving) the throttle control lever in the high idle speed direction (toward the carburetor), until the desired speed is attained.

LOW IDLE SPEED ADJUSTMENT

The low idle speed is set by turning the low idle speed adjusting screw on the carburetor in or out. This setting must be made in conjunction with the idle fuel mixture setting.

SENSITIVITY ADJUSTMENT

Governor sensitivity is adjusted by repositioning the governor spring in the holes in the governor lever. If speed surging occurs with a change in load, the governor is set too sensitive. If a big drop in speed occurs when normal load is applied, the governor should be set for greater sensitivity.

The governor lever has five holes for use in adjusting governor sensitivity. For the least governor sensitivity, the governor spring should be inserted in the lever hole closest to the governor cross shaft. The lever holes become increasingly sensitive the farther they are from the cross shaft. For the greatest governor sensitivity, the governor spring should be inserted in the lever hole farthest from the governor cross shaft.

Lubrication System

OIL RECOMMENDATIONS

Using the proper type and weight of oil in the crankcase is extremely important. So is checking oil daily and changing oil regularly. Failure to use the correct oil, or using dirty oil, causes premature engine wear and failure.

Use high-quality detergent oil of API (American Petroleum Institute) service class SF or SG. Select the viscosity based on the air temperature at the time of operation as shown in the following table.

➡**Using other than service SF or SG oil or extending oil change intervals longer than recommended can cause engine damage.**

A logo or symbol on oil containers identifies the API service class and SAE viscosity grade.

OIL LEVEL CHECK

The importance of checking and maintaining the proper oil level in the crankcase cannot be overemphasized. Check oil BEFORE EACH USE as follows:

Engines With Extended Oil Fill Tube/Dipstick
▶ **See Figures 8 and 9**

1. Make sure the engine is stopped, level, and is cool so the oil has had time to drain into the sump.

2. To keep dirt, grass clippings, etc., out of the engine, clean the area around the oil fill cap/dipstick before removing it.

3. Unthread and remove the oil fill cap/dipstick; wipe oil off. Reinsert the dipstick into the tube and rest the oil fill cap on the tube. Do not thread the cap onto the tube.

4. Remove the dipstick and check the oil level. The oil level should be up to, but not over, the **F** mark on the dipstick.

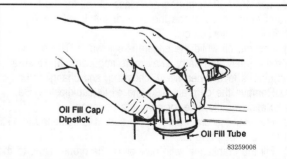

Oil Fill Cap/Dipstick

Oil Fill Tube

83259008

Fig. 8 Checking the oil level on engines with an extended oil fill/tube dipstick

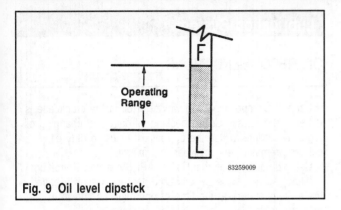

Fig. 9 Oil level dipstick

5. If the level is low, add oil of the proper type, up to the **F** mark on the dipstick. Always check the level with the dipstick before adding more oil.

➡ **To prevent extensive engine wear or damage, always maintain the proper oil level in the crankcase. Never operate the engine with the oil level below the L mark or over the F mark on the dipstick.**

OIL SENTRY

Some engines are equipped with an optional Oil Sentry oil level monitor. If the oil level gets low, Oil Sentry will either shut off the engine or activate a warning signal, depending on the application.

➡ **Make sure the oil level is checked BEFORE EACH USE and is maintained up to the F mark on the dipstick. This includes engines equipped with Oil Sentry.**

OIL CHANGE

◗ **See Figure 10**

For a new engine, change oil after the first 5 hours of operation. Thereafter, change oil after every 100 hours of operation.

For an overhauled engine or those rebuilt with a new short block, use 10W-30-weight service class SF oil for the first 5 hours of operation. Change the oil after this initial run-in period. Refill with service class SF.

Change the oil while the engine is still warm. The oil will flow more freely and carry away more impurities. Make sure the engine is level when filling, checking, and changing the oil.

1. Remove the oil drain plug and oil fill cap/dipstick. Be sure to allow ample time for complete drainage.

2. Reinstall the drain plug. Make sure it is tightened to 17.6 N-m (13 foot lbs.) torque.

3. Fill the crankcase, with new oil of the proper type, to the **F** mark on the dipstick. Always check the level with the dipstick before adding more oil.

4. Reinstall the oil fill cap/dipstick and tighten securely.

➡ **To prevent extensive engine wear or damage, always maintain the proper oil level in the crankcase. Never operate the engine with the oil level below the L mark or over the F mark on the dipstick.**

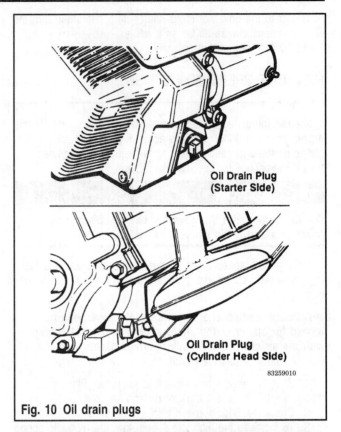

Fig. 10 Oil drain plugs

Retractable Starter

◗ **See Figure 11**

✳✳CAUTION

Retractable starters contain a powerful, flat wire recoil spring that is under tension. Do not remove the center screw from the starter until the spring tension is released. Removing the center screw before releasing spring tension, or improper starter disassembly, can cause the sudden and potentially dangerous release of the spring.

Always wear safety goggles when servicing retractable starters; full face protection is recommended.

➡ **To ensure personal safety, and proper starter disassembly and reassembly, follow the procedures in this section carefully.**

REMOVAL

1. Remove air cleaner cover.
2. Remove the four hex. flange screws securing the starter to blower housing.
3. Remove the starter.

INSTALLATION

1. Install the retractable starter and four hex. flange screws to blower housing. Leave the screws slightly loose.

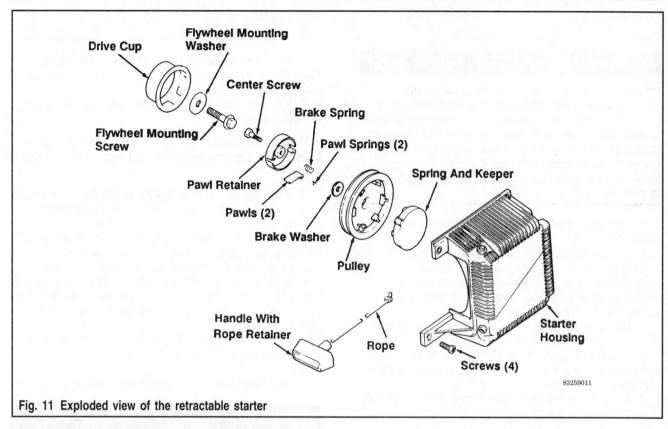

Fig. 11 Exploded view of the retractable starter

2. Pull the starter handle out until the pawls engage in the drive cup. Hold the handle in this position and tighten the screws securely.

3. Install the air cleaner cover.

ROPE REPLACEMENT

The rope can be replaced without complete starter disassembly.

1. Remove the starter from the engine blower housing.

2. Pull the rope out approx. 12 in. and tie a temporary (slip) knot in it to keep it from retracting into the starter.

3. Remove the rope retainer from inside the starter handle. Untie the single knot and remove the rope retainer and handle.

4. Hold the pulley firmly and untie the slip knot. Allow the pulley to rotate slowly as the spring tension is released.

5. When all spring tension on the starter pulley is released, remove the rope from pulley.

6. Tie a single knot in one end of the new rope.

7. Rotate the pulley counterclockwise (when viewed from pawl side of pulley) until the spring is tight. (Approx. 5 full turns of pulley.).

8. Rotate the pulley clockwise until the rope hole in pulley is aligned with rope guide bushing of starter housing.

➡**Do not allow the pulley/spring to unwind. Enlist the aid of a helper if necessary, or use a C-clamp to hold the pulley in position.**

9. Insert the new rope through the rope hole in starter pulley and rope guide bushing of housing.

10. Tie a slip knot approx. 12 in. from the free end of rope. Hold the pulley firmly and allow it to rotate slowly until the slip knot reaches the guide bushing of housing.

11. Slip the handle and rope retainer onto the rope. Tie a single knot at the end of the rope. Install the rope retainer into the starter handle.

12. Untie the slip knot and pull on the handle until the rope is fully extended. Slowly retract the rope into thew starter.

➡**When the spring is properly tensioned, the rope will retract fully and the handle will stop against the starter housing.**

PAWLS (DOGS) REPLACEMENT

The starter must be completely disassembled to replace the starter pawls. A pawl repair kit is available which includes the following components:

DISASSEMBLY

❋❋CAUTION

Do not remove the center screw from starter until the spring tension is released. Removing the center screw before releasing spring tension, or improper starter disassembly, can cause the sudden and potentially dangerous release of the spring. Follow these instructions carefully to ensure personal safety and proper starter disassembly. Make sure adequate face protection is worn by all persons in the area.

1. Release spring tension and remove the handle and starter rope.
2. Remove the center screw and pawl retainer.
3. Remove the brake spring and brake washer.
4. Carefully note the positions of the pawls and pawl springs before removing them. Remove the pawls and pawl springs from the starter pulley.
5. Rotate the pulley clockwise 2 full turns. This will ensure the spring is disengaged from the starter housing.
6. Hold the pulley into the starter housing. Invert the pulley/housing so the pulley is away from your face, and away from the others in the area.
7. Rotate the pulley slightly from side to side and carefully separate the pulley from the housing.

➡If the pulley and the housing do not separate easily, the spring could be engaged in the starter housing, or there is still tension on the spring. Return the pulley to the housing and repeat step 5 before separating the pulley and housing.

8. Note the position of the spring and keeper assembly in the pulley. Remove the spring and keeper assembly from the pulley as a package.

❋❋CAUTION

Do not remove the spring from the keeper. Severe personal injury could result from the sudden uncoiling of the spring.

INSPECTION AND SERVICE

1. Carefully inspect the rope, pawls, housing, center screw, and other components for wear or damage.
2. Replace all worn or damaged components.
3. Do not attempt to rewind a spring that has come out of the keeper. Order and install a new spring and keeper assembly.
4. Clean all old grease and dirt from the starter components. Generously lubricate the spring and center shaft with any commercially-available bearing grease.

REASSEMBLY

1. Make sure the spring is well-lubricated with grease. Place the spring and keeper assembly inside the pulley (with spring towards pulley).
2. Install the pulley with spring and keeper assembly into the starter housing.

➡Make sure the pulley is fully seated against the starter housing. Do not wind the pulley and recoil spring at this time.

3. Install the pawl springs and pawls into the starter pulley.
4. Place the brake washer in the recess in the starter housing hub.
5. Lubricate the brake spring sparingly with grease. Place the spring on the plain washer. (Make sure the threads in center shaft remain clean, dry, and free of grease and oil.).
6. Apply a small amount of Loctite® #271 to the threads of the center screw. Install the center screw, with retainer, to the center shaft. Torque the screw to 7.4-8.5 N-m (65-75 inch lbs.).
7. Tension the spring and install the rope and handle as instructed in steps 6 though 12 under Rope Replacement above.
8. Install the starter to the engine blower housing.

Spark Plug

Engine misfire or starting problems are often caused by a spark plug that is in poor condition or with an improper gap setting.

This engine is factory-equipped with the following spark plug:Type — Champion RC12YC (or equivalent)Gap — 0.030 in.

Thread Size :
14 mm Reach — $3/4$
Hex Size — $5/8$ in.

SPARK PLUG SERVICE

▶ **See Figures 12, 13, 14, 15 and 16**

Every 100 hours of operation, remove the spark plug, check its condition, and reset the gap or replace with a new plug as necessary.

1. Before removing the spark plug, clean the area around the base of the plug to keep dirt and debris out of the engine.
2. remove the plug and check its condition. Replace the plug if worn or if reuse is questionable.

➡Do not clean the spark plug in a machine which uses abrasive grit. Some grit could remain on the spark plug and enter the engine, causing extensive wear and damage.

3. Check the gap using a feeler gauge. Adjust the gap to 0.76 mm (0.030 in) by carefully bending the ground electrode.
4. reinstall the spark plug into the cylinder head. Torque the spark plug into the cylinder head. Torque the spark plug to 24.4-29.8 Nm (18-22 foot lbs.).

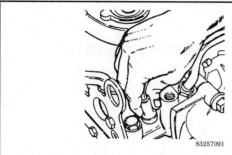

Fig. 12 Twist and pull on the boot, never on the plug wire

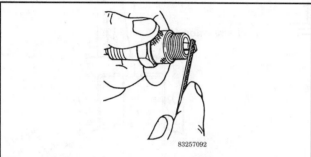

Fig. 13 Plugs that are in good condition can be filed and reused

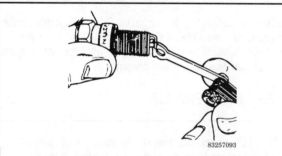

Fig. 14 Adjust the electrode gap by bending the side electrode

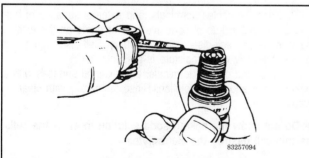

Fig. 15 Always used a wire gauge to check the plug gap

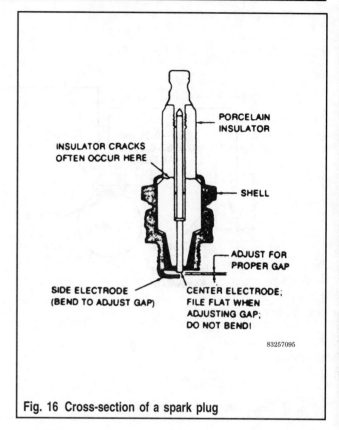

Fig. 16 Cross-section of a spark plug

INSPECTION

Inspect the spark plug as soon as it is removed from the cylinder head. the deposits on the tip are an indication of the internal condition of the piston rings, valves, and carburetor.

Electronic Magneto Ignition System

▶ **See Figure 17**

This engine is equipped with a dependable electronic magneto ignition system. The system consists of the following components:**A magneto assembly** — which is permanently affixed to the flywheel.**An electronic magneto ignition module** — which mounts on the engine crankcase.**A kill switch** — (or key switch) which grounds the module to stop the engine.**A spark plug.**

OPERATION

As the flywheel rotates and the magnet assembly moves past the ignition module, a low voltage is induced in the primary windings of the module. When the primary voltage is precisely at its peak, the module induces a high voltage in it s secondary windings. This high voltage creates a spark plug at the tip of the spark plug. this spark ignites the air-fuel mixture in the combustion chamber.

The timing of the spark is automatically controlled by the module. Therefore, other than periodically checking/replacing the spark plug, no maintenance, timing, or adjustments are necessary or possible with this system.

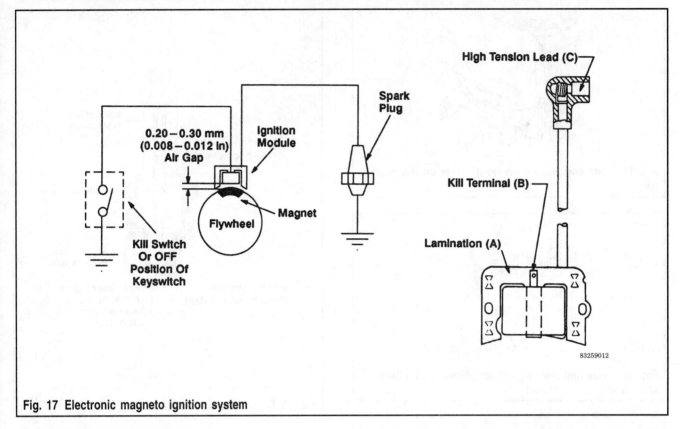

Fig. 17 Electronic magneto ignition system

IGNITION MODULE REMOVAL AND INSTALLATION

Refer to the Engine Disassembly and Reassembly Sections for complete ignition module removal and installation procedures.

Battery

A 12-volt battery with a rating of approximately 32-amp hours-250 cold cranking amps, is normally used.

If the battery charge is not sufficient to crank the engine, recharge the battery.

➡**Do not attempt to jump start the engine with another battery. Starting the engine with batteries larger than those recommended can burn out the starter motor.**

BATTERY CHARGING

✳✳CAUTION

Batteries contain sulfuric acid. To prevent acid burns, avoid contact with skin, eyes, and clothing. Batteries produce explosive hydrogen gas while being charged. To prevent a fire or explosion, charge batteries only in well ventilated areas. Keep sparks, open flames, and other sources of ignition away from the battery at all times. Keep batteries out of the reach of children. Remove all jewelry when servicing batteries.

Before disconnecting the negative (-) ground cable, make sure all switches are OFF. If ON, a spark will occur at the ground cable terminal which could cause an explosion if hydrogen gas or gasoline vapors are present.

BATTERY MAINTENANCE

1. Regularly check the level of electrolyte. Add distilled water as necessary to maintain the recommended level.

➡**Do not overfill the battery. Poor performance or early failure due to loss of electrolyte will result.**

2. Keep the cables, terminals, and external surfaces of battery clean. A build-up of corrosive acid or grime on the external surface can self-discharge the battery. Self-discharging happens rapidly when moisture is present.

3. Wash the cables, terminals, and external surfaces with a baking soda and water solution. Rinse thoroughly with clear water.

➡**Do not allow the baking soda solution to enter the cells as this will destroy the electrolyte.**

BATTERY TEST

Test the battery voltage by connecting D.C. voltmeter across the battery terminals — crank the engine. If the battery drops below 9 volts while cranking, the battery is discharged or faulty.

TROUBLESHOOTING

This engine is equipped with a 0.5 Amp unregulated battery charging system.

➡Observe the following guidelines to prevent damage to the electrical system and components.

Electric Start Engines 0.5 Amp Unregulated Battery Charging System

▶ See Figure 18

PRECAUTIONS

1. Make sure the battery polarity is correct. A negative (-) ground system is used.
2. Disconnect the stator lead, wiring harness, and any other electrical accessories in common ground with the engine before performing electric welding on the equipment powered by the engine.
3. Prevent the stator (AC) lead from touching or shorting while the engine is running. This could damage the stator.

TROUBLESHOOTING GUIDE

Minimum Output
Engine Speed (rpm) — Output (Volts)
1600 — 2.49-3.39
2000 — 4.19-5.57
2400 — 7.18-9.10
2800 — 9.68-11.90
3200 — 1.90-14.00
3600 — 13.29-14.97

If voltage is significantly lower than listed values replace stator.

➡**Voltmeter must have an integrating time (RMS) function to yield a true DC reading, and a high input impedance (>10 M ohms) such that the output voltage is not artificially lowered due to meter loading.**

Electric Starter

PRECAUTIONS

- Do not crank the engine for more than 10 seconds at a time. If the engine does not start, allow a 60-second cool-down period between starting attempts. Failure to follow these guidelines can burn out the starter motor.
- If the engine develops sufficient speed to disengage the starter but does not keep running (a false start), the engine rotation must be allowed to come to a complete stop before attempting to restart the engine. If the starter is engaged while the flywheel is rotating, the starter pinion and flywheel ring gear may clash, resulting in damage to the starter.
- If the starter does not crank the engine, shut off the starter immediately. Do not make further attempts to start the engine until the condition is corrected. Do not attempt to jump start the engine with another battery. Starting with batteries larger than those recommended can burn out the stator motor.
- Do not drop the starter or strike the starter frame. Doing so can damage the ceramic permanent magnets inside the starter frame.

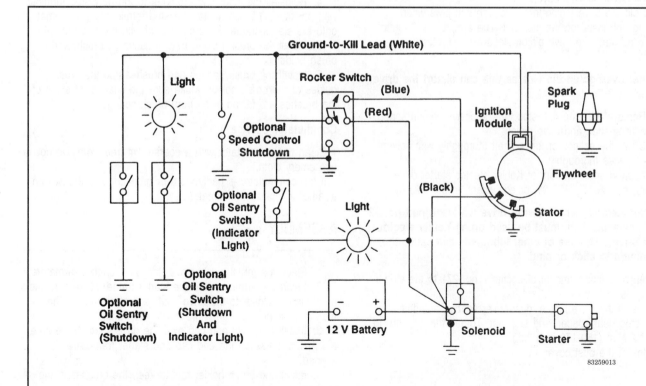

Fig. 18 Wiring diagram for electric start engines with 0.5 amp unregulated battery charging system

OPERATION

When power is applied to the starter, the armature rotates. As the armature rotates, the drive pinion moves out on the splined drive shaft and into mesh with flywheel ring gear. When the pinion reaches the end of the drive shaft, it rotates the flywheel and cranks the engine.

When the engine starts, the flywheel rotates faster than the starter armature and drive pinion. This moves the drive pinion out of mesh while the ring gear and into the retracted position. When power is removed from the starter, the armature stops rotating and the drive pinion is held in the retraced position by the anti-drift spring.

REMOVAL AND INSTALLATION

Refer to the Engine Disassembly and Assembly sections for starter removal and installation procedures.

STARTER DRIVE SERVICE

Every 100 hours of operation (or annually, whichever occurs first), clean and lubricate the splines on the starter drive shaft. If the drive pinion is worn, or has chipped or broken teeth, it must be replaced.

It is not necessary to completely disassemble the starter to service the drive components. Service the drive as follows:

1. Remove the starter from the engine.
2. Remove the dust cover.
3. Hold the drive pinion in a vice with soft jaws when removing and installing the stop nut. The armature will rotate with the nut until the drive pinion stops against internal spacers.

➡**Do not overtighten the vice as this can distort the drive pinion.**

4. Remove the stop nut, stop gear spacer, anti-drift spring, dust cover spacer, and drive pinion.
5. Clean the splines on drive shaft thoroughly with solvent. Dry the splines thoroughly.
6. Apply a small amount of Kohler electric starter drive lubricant, part no. 52 357 01, or equivalent, to the splines.

➡**Kohler electric electric starter drive lubricant, part no. 52 357 01, or equivalent, must be used on all Kohler electric starter drives. The use of other lubricants can cause the drive pinion to stick or bind.**

7. Apply a small amount of Loctite® no. 271 to the stop nut threads.
8. Install the drive pinion, dust cover spacer, anti-drift spring, stop gear spacer, and stop nut. Torque the stop nut to 17.0-19.2 N·m (150-170 inch lbs.).
9. Install the dust cover.

DISASSEMBLY

▶ **See Figure 19**

1. Remove the dust cover, stop nut, gear spacer, anti-drift spring, dust cover spacer, and drive pinion. Refer to Starter Drive Service above.
2. Scribe a small line on the drive end cap, opposite the line on the starter frame. These lines will serve as match marks when reassembling the starter.
3. Remove the through bolts.
4. Remove the commutator end cap with brushes and brush springs.

➡**The wiring lead of the positive (+) brush is attached to the insulated terminal on the starter frame. When the commutator end cap is removed, the positive (+) brush should be removed from the brush guide of the brush holder, and will remain attached to the starter frame insulated terminal.**

5. Remove the drive end cap.
6. Remove the armature and thrust washer from inside the starter frame.

Brush Replacement

1. Remove the brush springs from the brush guides of the brush holder.
2. Brush the brush holder screws, negative (-) brush, and plastic brush holder.
3. Remove the hex nuts from the stud terminal. Remove the stud terminal with positive (+) brush and rubber insulating grommet from the starter frame.
4. Reinstall the insulating grommet to the new stud terminal with positive (+) brush. Install the stud terminal with grommet onto the starter frame. Secure the stud with the hex nut.
5. Install the brush holder, new negative (-) brush, and brush holder screws.
6. Install the brush springs and brushes into the brush guides of the brush holder. Make sure the chamfered sides of the brushes are facing away from brush springs.

Commutator Service

Clean the commutator with a coarse, lint free cloth. Do not use emery cloth.

If the commutator is badly worn or grooved, turn it down on a lathe or replace the armature.

ASSEMBLY

1. Place the thrust washer over the drive shaft of armature.
2. Insert the armature into the starter frame. Make sure the magnets are closer to the drive shaft end of armature. The magnets will hold the armature inside the frame.
3. Install the drive end cap over the drive shaft. Make sure the match marks on the end cap and starter frame are aligned.
4. Install the brush holder tool to keep the brushes in the pockets of the commutator end cap.
5. Align the match marks on the commutator end cap and starter frame. Hold the drive end cap commutator end caps firmly to the starter frame. Remove the brush holder tool.

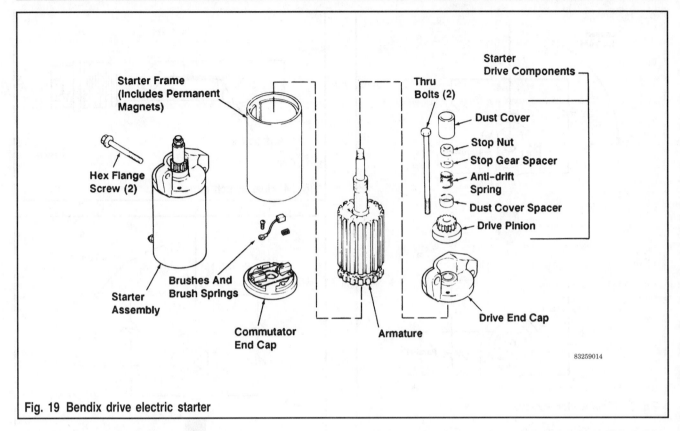

Fig. 19 Bendix drive electric starter

6. Install the through bolts and tighten securely.

7. Lubricate the drive shaft with Kohler electric starter drive lubricant. Install the pinion, dust cover spacer, anti-drift spring, stop gear spacer, stop nut, and dust cover. Refer to Starter Drive Service above.

Oil Sentry Oil Level Monitor

OPERATION

Some engines are equipped with optional Oil Sentry system. Oil sentry uses a float switch in the oil pan to detect a low engine oil level. On stationary or unattended applications (pumps, generators, etc.) the float switch can be used to ground the ignition module to stop the engine. On vehicular applications (garden tractors, mowers, etc.) and those equipped with a battery or electric start, the float switch can be used to activate a **low oil** warning light.

FLOAT SWITCH REMOVAL

▸ **See Figures 20, 21, 22, 23 and 24**

1. Make sure the engine/equipment is resting on a level surface.

2. Remove the oil drain plug and drain oil from crankcase.

3. Disconnect float switch leads.

4. Using a ⁹⁄₁₆ in. open end wrench, turn switch counterclockwise ¼ TURN to loosen. STOP turning switch when flat

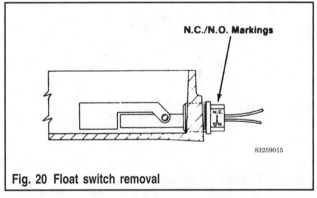

Fig. 20 Float switch removal

surface on float switch is in a horizontal position. (Flat surface parallel with base of oil pan and N.C./N.O. markings down.).

5. Turn the switch counterclockwise in ½ TURN INCREMENTS using a smooth, continuous action. Pause briefly between increments and keep the flat surface of float switch in a horizontal position (parallel with base of oil pan).

❊❊WARNING

To prevent damage to the float switch, and to enable you to feel if the float strikes the oil pan, REMOVE THE SWITCH BY HAND as soon as it is loose enough for you to do so.

➡When turning the float switch, use a smooth, continuous action for the ENTIRE ½ turn increment.

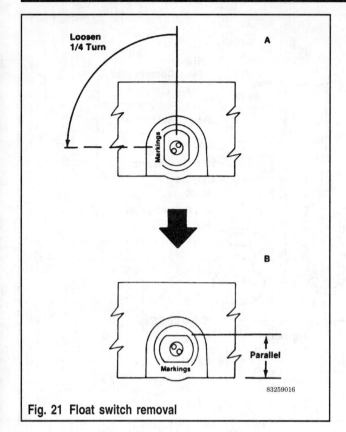

Fig. 21 Float switch removal

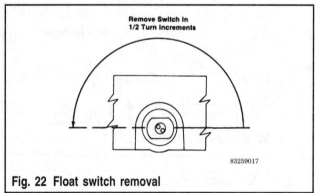

Fig. 22 Float switch removal

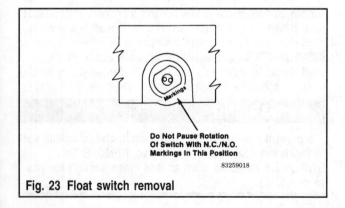

Fig. 23 Float switch removal

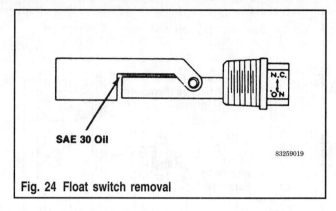

Fig. 24 Float switch removal

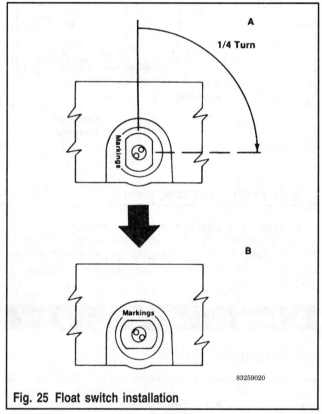

Fig. 25 Float switch installation

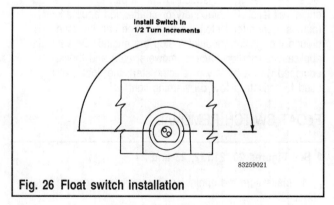

Fig. 26 Float switch installation

6. If the float does not strike the oil pan, STOP turning the switch, then use the following procedure:

a. Turn the switch clockwise until the flat surface is in a vertical position. (N.C./N.O. markings on left.) This will allow the float to return against the switch body.

b. Turn the switch counterclockwise ¼ TURN. STOP turning switch when flat surface is in a horizontal position. (Flat surface parallel with base of oil pan and N.C./N.O. markings down.

c. Turn the switch counterclockwise in ½ TURN INCREMENTS as instructed in step 5 above.

FLOAT SWITCH INSTALLATION

♦ **See Figures 26, 27, 28 and 29**

1. Make sure the engine/equipment is resting on a level surface.

2. Remove the oil drain plug and drain oil from crankcase.

3. When adding this switch as an accessory, remove and discard the ½ in. NPSF pipe plug from the location in oil pan where switch will be installed.

❋❋WARNING

To prevent damage to the float switch, and to enable you to feel if the float strikes the oil pan, INSTALL THE SWITCH BY HAND as long as it is loose enough for you to so.

4. Apply Loctite® No. 592 Teflon® sealant (or equivalent) to the entire thread area of switch.

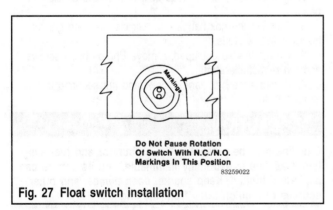

Do Not Pause Rotation
Of Switch With N.C./N.O.
Markings In This Position
83259022

Fig. 27 Float switch installation

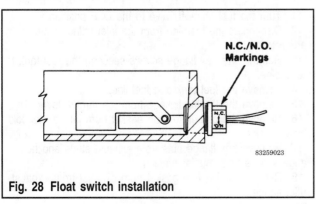

N.C./N.O.
Markings

83259023

Fig. 28 Float switch installation

5. Apply a thick film of clean SAE 30 oil to the float and switch body.

6. Hold the switch with the flat surface in a vertical position . (N.C./N.O. marking stop the left.) Insert the switch into the oil pan and turn the switch ¼ turn. STOP turning the switch when the flat surface on switch is in a horizontal position. Flat surface parallel with base of oil pan and N.C./N.O. markings up.).

➡**Several ½ turn increments may be required until the threads on switch engage in oil pan.   When turning the float switch, use a smooth continuous action for the entire ½ turn increment. Pausing the rotation of the switch may cause the float to strike the oil pan.**

7. If the float does not strike the oil pan, STOP turning the switch, then use the following procedure.

a. Turn the switch counterclockwise until the flat is in a vertical position, (N.C./N.O. markings on left.) This will allow the float to return against the switch body.

b. Turn the switch clockwise ¼ TURN. STOP turning switch when flat surface is in a horizontal position. (Flat surface parallel with base of oil pan and N.C./N.O. markings up.).

c. Turn the switch clockwise in ½ TURN INCREMENTS as instructed in step 6 above.

8. Turn in the switch approximately five (5) to six (6) full turns to obtain the proper position. Use a $9/16$ in. open end wrench to tighten the switch. The N.C. markings on switch will be at the top when the switch is positioned properly.

FLOAT SWITCH TEST

♦ **See Figure 29**

Test switch for continuity by placing an ohmmeter or continually test light across leads.

Switch Position A — No Continuity (Switch open); Switch Position B — Continuity (Switch closed)

Perform the following tests to ensure that the float switch is positioned and working properly before connecting the leads.

➡**These tests apply to engines equipped with a standard oil pan and dipstick. Special oil pan and/or dipstick arrangements can give inaccurate test results.**

1. Connect a continuity test light across float switch leads. The light should be off after oil is above the **L** mark on the dipstick.

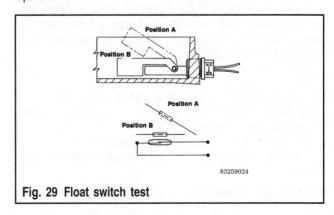

Position A

Position B

Position A

Position B

83259024

Fig. 29 Float switch test

2. If the float switch fails this test:

a. Make sure the switch is in the proper position with the N.C. markings at the top.

b. If switch is positioned properly, drain oil and remove switch (see Float Switch Removal). If the float is not attached to the switch body, the oil pan must be removed.

c. Replace a faulty or broken switch with a new one. (See Float Switch Installation.)

Operational Test

Reconnect the leads and perform the following test.

1. Make sure the oil level is up to, but not over the **F** mark on dipstick.

2. Start the engine. If the switch is wired as a low oil level shutdown, the engine should start. If the switch is wired to activate a low oil warning light, the light should be off.

3. Stop the engine. Drain the oil until the oil level is below the **L** mark on the dipstick. If properly wired, the engine will not start, or the light will be **on**.

4. If the test results of steps 2 and 3 are not as indicated, check for improper wiring and/or improper float installation.

Engine Mechanical Service

▶ See Figures 30, 31, 32, 33, 34, 35, 36, 37, 38, 39, 40, 41, 42, 43 and 44

DISASSEMBLY

❊❊CAUTION

Before servicing the engine or equipment, always disconnect the spark plug lad to prevent the engine from starting accidentally. Ground the lead to prevent sparks which could cause fires.

The following sequence is suggested for complete engine disassembly. This procedure can be varied to accommodate options or special equipment.

Clean all parts thoroughly as the engine is disassembled. Only clean parts can be accurately inspected and gauged for wear or damage. There are many commercially available cleaners that will quickly remove grease, oil, and grime from engine parts. When such a cleaner is used, follow the manufacturer's instructions and safety precautions carefully.

Make sure all traces of the cleaner are removed before the engine is reassembled and put into operation. Even small amounts of these cleaners can quickly break down the lubricating properties of engine oil.

1. Disconnect the spark plug.

2. Drain the oil.

3. Remove the oil drain plug and oil fill cap/dipstick.

4. Allow ample time for the oil to drain from the crankcase. Tip engine speed draining.

5. Remove the air cleaner cover as follows:

a. Press and hold the tab at the bottom of the air cleaner cover.

➡**Choke control must be in the OFF position.**

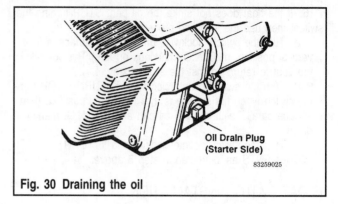

Fig. 30 Draining the oil

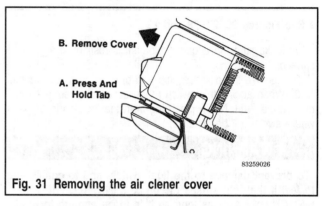

Fig. 31 Removing the air clener cover

b. Slide the air cleaner cover off of the air cleaner base (away from retractable starter).

6. Loosen the slot-head screws and remove the air cleaner element retainer.

7. Remove the Phillips head screws from the air cleaner base.

8. Remove the hex flange nuts, air cleaner, and gasket from the intake studs.

9. On models so equipped, remove Phillips head screws and the retractable starter.

10. Remove the hex flange screw and nut securing the oil fill tube to the fuel tank.

❊❊CAUTION

Gasoline may be present in the carburetor and fuel system. Gasoline is extremely flammable, and its vapors can explode if ignited. Keep sparks, open flames, and other sources of ignition away from the engine.

11. Turn the fuel shut-off valve to the OFF position.

12. Disconnect the fuel line from the inlet fitting of the carburetor.

13. Remove the hex flange screws securing the fuel tank to the engine.

14. Remove the fuel tank and fuel line.

15. Remove the throttle linkage from the throttle lever clip.

16. Remove the carburetor and gasket from intake manifold studs.

17. Remove hex flange nuts from exhaust studs and hex flange screws from muffler bracket.

18. Remove muffler and gasket from exhaust from exhaust outlet flange.

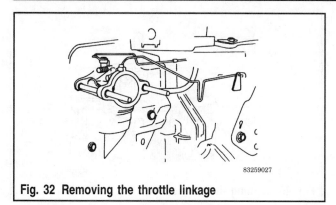

Fig. 32 Removing the throttle linkage

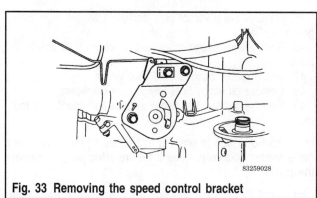

Fig. 33 Removing the speed control bracket

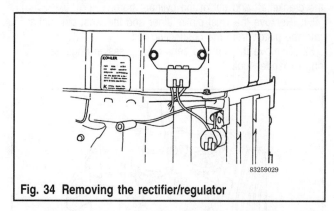

Fig. 34 Removing the rectifier/regulator

19. Loosen four hex flange screws and remove the cylinder head baffle. Remove the spark plug wire grommet from the cylinder head baffle.

20. Remove hex flange screws from the top cylinder barrel baffle. Remove the baffle and disconnect the kill-switch linkage from the throttle lever.

21. Remove the hex flange screws from the bottom cylinder baffle. Remove the baffle.

22. Remove the Phillips head screws from the blower housing. Remove the blower housing.

23. Remove the heat deflector and gaskets from the intake studs.

24. Loosen hex flange nut on governor lever and remove lever from governor shaft. Disconnect return spring from the governor lever.

25. Disconnect the kill-switch lead from the kill-switch.

26. Remove the hex flange screw from the throttle/kill-switch levers.

27. Remove throttle lever, bushing, and kill-switch lever from engine.

28. Remove the hex flange screws from the valve cover. Remove the valve cover from the cylinder head.

29. Remove the breather assembly and valve cover gasket from the cylinder head.

30. Remove the hex flange screws, cylinder head, push rods, and cylinder head gasket.

31. Remove the spark plug.

32. Remove the rocker arm adjusting nuts, rocker arms and balls, rocker arm studs, and pushrod guide plate.

33. Compress the valve springs by pushing down on the valve spring cap.

34. Remove the valve spring caps, valve springs, and valves.

35. Position the flywheel so the magnet is away from the ignition module.

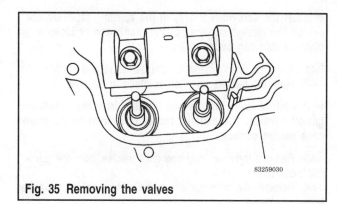

Fig. 35 Removing the valves

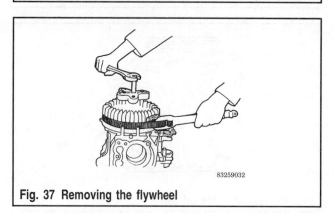

Fig. 36 Loosening the flywheel fastener

Fig. 37 Removing the flywheel

36. Remove the hex flange screws securing the ignition module to the crankcase.

➡**Always use a flywheel strap wrench to hold the flywheel when loosening or tightening the flywheel retaining fastener. Do not use any type of bar or wedge between the cooling fins as the fins could be cracked or damaged. Always use a puller to remove the flywheel from the crankshaft. Do not strike the flywheel or crankshaft, as these parts could be cracked or damaged.**

37. Remove the hex flange screw, plain washer, and drive cup.

38. Remove the flywheel from the crankshaft using a puller.

39. Remove the six hex flange screws securing the closure plate to the crankcase.

40. Locate the splitting tabs in the seam of the closure plate and crankcase. Pry the closure plate from the crankcase using a large, flat-blade screwdriver.

➡**Insert the screwdriver only in the splitting tabs. Do not pry on the gasket surfaces of the crankcase or closure plate as this can cause leaks.**

41. Align the timing marks on the camshaft and crankshaft.

42. Mark the tappets as whether intake or exhaust.

➡**The intake tappet is the one farthest from the crankcase gasket surface. The exhaust tappet is nearest to the crankcase gasket surface.**

43. Remove the snap ring and plain washer from the governor cross shaft.

44. Remove the governor cross shaft and small plain washer from the closure.

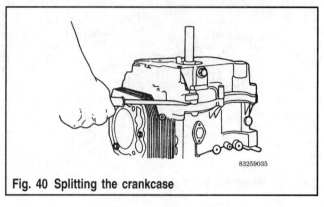

Fig. 40 Splitting the crankcase

45. Remove the governor gear and plain washer from the governor shaft.

46. Remove hex flange screw and oil sentry float switch baffle.

47. Use a rubber band to hold float switch.

48. Remove oil sentry float switch from crankcase.

49. Remove the two hex flange screws and connecting rod cap.

➡**If a carbon ridge is present at the top of the bore, use a ridge reamer tool to remove it before attempting to remove the piston.**

50. Carefully push the connecting rod and piston away from the crankshaft and out of the cylinder bore.

51. Remove the wrist pin retainer and the wrist pin. Separate the piston from the connecting rod.

52. Remove the top and center compression rings using a ring expander tool.

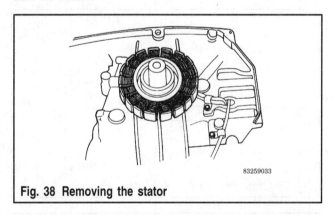

Fig. 38 Removing the stator

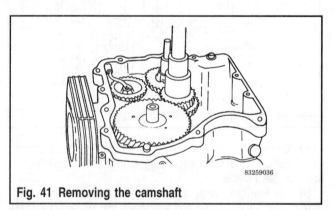

Fig. 41 Removing the camshaft

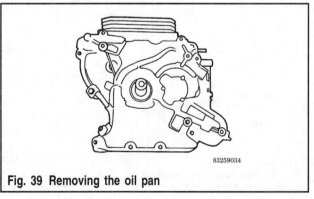

Fig. 39 Removing the oil pan

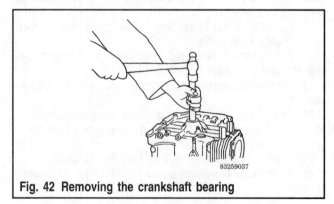

Fig. 42 Removing the crankshaft bearing

53. Remove the oil control ring rails, then remove the rails spacer.

54. Remove the woodruff key from the flywheel taper end of the crankshaft.

55. Press the crankshaft from the crankcase.

56. Remove the oil seals from the crankcase and closure plate.

57. Remove the bearings from the crankcase and closure plate.

CLEANING AND INSPECTION

Clean all parts thoroughly. Only clean parts can be accurately inspected and gauged for war or damage. There are many commercially available cleaners that will quickly remove grease, oil, and grime from engine parts. When such a cleaner is used, follow the manufacturer's instructions and safety precautions carefully.

Make sure all traces of the cleaner are removed before the engine is reassembled and placed into operation. Even small amounts of the these cleaners can quickly break down the lubricating properties of engine oil.

Camshaft Inspection and Service

Inspect the gear teeth of the camshaft. If the teeth are badly worn, chipped, or some are missing, replacement of the camshaft will be necessary.

Crankshaft Inspection and Service

▶ See Figure 43

Inspect the gear teeth of the crankshaft. If the teeth are badly worn, chipped, or some are missing, replacement of the crankshaft will be necessary.

Inspect the crankshaft bearings for scoring, grooving, etc. Do not replace bearings unless they show signs of damage or are out of running clearance specifications. If the crankshaft turns easily and noiselessly, and there is no evidence of scoring, grooving, etc., on the races or bearing surfaces, the bearings can be reused.

Inspect the crankshaft keyways. If worn or chipped, replacement of the crankshaft will be necessary.

Inspect the crankpin for score marks or metallic pickup. Slight score marks can be cleaned with crocus cloth soaked in oil. If wear limits are exceeded, it will either be necessary to replace the crankshaft or regrind the crankpin to 0.25 mm (0.010 in) undersize. If reground, a 0.25 mm (0.010 in) under-

size connecting rod (big end) must then be used to achieve proper running clearance. Measure the crankpin for size, taper, and out-of-round.

➡**If the crankpin is reground, visually check to insure that the fillet blends smoothly with the crankpin surface.**

Crankcase Inspection and Service

Check all gasket surfaces to make sure they are free of gasket fragments. Gasket surfaces must also be free of deep scratches or nicks.

Check the cylinder bore wall for scoring. In severe cases, unburned fuel can cause scuffing and scoring of the cylinder wall. It washes the necessary lubricating oils off the piston and cylinder wall. As raw fuel seeps down the cylinder wall, the piston rings make metal to metal contact with the wall. Scoring of the cylinder wall can also be caused by localized hot spots resulting from blocked cooling fins or from inadequate or contaminated lubrication.

If the cylinder bore is badly scored, excessively worn, tapered, or out of round, resizing i necessary. Use a measuring device (inside micrometer, etc.) to determine amount of wear, then select the nearest suitable oversize of either 0.25 mm (0.010 in) or 0.50 mm (0.020 in). Resizing to one of these oversizes will allow usage of the available oversize piston and ring assemblies. Initially, resize using a boring bar, then use the following procedures for honing the cylinder.

Honing

While most commercially available cylinder hones can be used with either portable drills or drill presses, the use of a low speed drill press is preferred as it facilitates more accurate alignment of the bore in relation to the crankshaft crossbore. Honing is best accomplished at a drill speed of about 250 RPM and 60 strokes per minute. After installing coarse stones in hone, proceed as follows:

1. Lower hone into bore and after centering, adjust so that the stones are in contact with the cylinder wall. Use of a commercial cutting-cooling agent is recommended.

2. With the lower edge of each stone positioned even with the lowest edge of the bore, start drill and honing process. Move the hone up and down while resizing to prevent the formation of cutting ridges. Check the size frequently.

➡**Measure the piston diameter and resize the bore to the piston to obtain the specified running clearances. Keep in mind the temperatures caused by honing may cause inaccurate measurements. Make sure the bore is cool when measuring.**

3. When the bore is within 0.064 mm (0.0025 in) of desired size, remove the coarse stones and replace with burnishing stones. Continue with the burnishing stones until within 0.013 mm (0.0005 in) of desired size and then use finish stones (220-280 grit) and polish to final size. A crosshatch should be observed if honing is done correctly. The crosshatch should intersect at approximately 23-33 degrees off the horizontal. Too flat of an angle could cause the rings to skip and wear excessively, too steep of an angle will result in high oil consumption.

4. After resizing, check the bore for roundness, taper, and size. Use am inside micrometer, telescoping gauge, or bore gauge to take measurements. The measurements should be taken at three locations in the cylinder — at the top, middle,

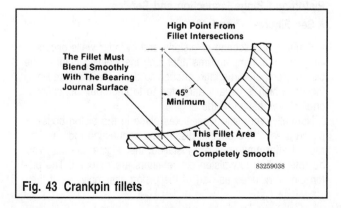

The Fillet Must Blend Smoothly With The Bearing Journal Surface

High Point From Fillet Intersections

45° Minimum

This Fillet Area Must Be Completely Smooth

83259038

Fig. 43 Crankpin fillets

and bottom. Two measurements should be taken (perpendicular to each other) at each of the three locations.

Measuring Piston-To-Bore Clearance

▶ **See Figure 44**

Before installing into the cylinder bore, it is necessary that the clearance be accurately checked. This step is often overlooked, and if the clearances are not within specifications, engine failure will usually result.

➡**Do not use a feeler gauge to measure piston-to-bore clearance — it will yield inaccurate measurements. Always use a micrometer.**

Use the following procedure to accurately measure the piston-to-bore clearance:

1. Use a micrometer and measure the diameter of the piston 6 mm (0.24 in) above the bottom of the piston skirt and perpendicular to the piston pin.
2. Use an inside micrometer, telescoping gauge, or bore gauge and measure the cylinder bore. Take the measurement approximately 40 mm (1.6 in) below the top of the bore and perpendicular to the piston pin.
3. Piston-to-bore clearance in the difference between the bore diameter and the piston diameter (step 2 minus step 1).

Flywheel Inspection

Inspect the flywheel for cracks, and the flywheel keyway for damage. Replace flywheel if cracked. Replace the flywheel, the crankshaft, and the key if flywheel key is sheared or the keyway damaged.

Inspect the ring gear for cracks or damage. Kohler does not provide ring gears as a serviceable part. Replace the flywheel if the ring gear is damaged.

Cylinder Head and Valves Inspection And Service

▶ **See Figure 45**

Carefully inspect the valve mechanism parts. Inspect the valve springs and related hardware for excessive war or distortion. Check the valves and valve seat area or inserts for evidence of deep pitting, cracks, or distortion. Check clearance of the valve stems in guides.

Hard stating, or loss of power accompanied by high fuel consumption may be symptoms of faulty valves. Although these symptoms could also be attributed to worn rings, remove and check the valves first. After removal, clean the valve heads, faces, and stems with a power brush. Then, carefully

inspect each valve for defects such as warped head, excessive corrosion, or worn stem end. Replace valves found to be in bad condition. A normal valve and valves in bad condition are shown in the accompanying illustrations.

VALVE GUIDES

If a valve guide is worn beyond specifications, it will not guide the valve in a straight line. This may result in burnt valve aces or seats, loss of compression, and excessive oil consumption.

To check valve guide-to-valve stem clearance, thoroughly clean the valve guide and, using a split-ball gauge, measure the inside diameter. Then, using an outside micrometer, measure the diameter of the valve stem at several points on the stem where it moves in the valve guide. Use the largest stem diameter to calculate the clearance. If the clearance exceeds 7.134 mm (0.2809 in) on intake valve or 7.159 mm (0.2819 in) on exhaust valve, determine whether the valve stem or the guide is responsible for the excessive clearance.

Maximum allowable inside diameter is 5.085 mm (0.2002 in) on the intake valve guide and 5.080 mm (0.2000 in) on the exhaust valve guide.

If the valve stem diameter is within specifications, then recondition the valve guide.

RECONDITIONING VALVE GUIDE

The valve guides in the cylinder head are not removable. Use a 0.25 mm (0.010 in) O/S reamer. Tool no. KO-1033, or equivalent.

VALVE SEAT INSERTS

▶ **See Figure 46**

The valve seats are not replaceable. If the seats become badly pitted, cracked, or distorted, the inserts can be reconditioned.

Use a standard valve seat cutter and cut seat to the required dimensions.

LAPPING VALVES

Reground or new valves must be lapped in, to provide fit. Use a hand valve grinder with suction cup for final lapping. Lightly coat valve face with fine grade of grinding compound, then rotate valve on seat with grinder. Continue grinding until smooth surface is obtained on seat and on valve face. Thoroughly clean cylinder head in soap and hot water to remove all traces of grinding compound. After drying cylinder head apply a light coating of SAE 10 oil to prevent rusting.

Piston and Rings Inspection and Service

▶ **See Figures 47 and 48**

Scuffing and scoring of pistons and cylinder walls occurs when internal temperatures approach the welding point of the piston. Temperatures high enough to do this are created by friction, which is usually attributed to to improper lubrication, and/or overheating of the engine.

Normally, very little wear takes place in the piston boss-piston area. If the original piston and connecting rod can be reused after new rings are installed, the original pin can also be reused but new piston pin retainers are required. The piston pin is included as part of the piston assembly — if the

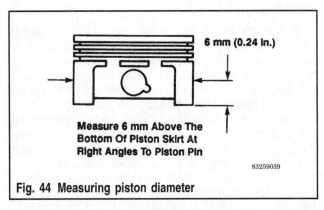

Fig. 44 Measuring piston diameter

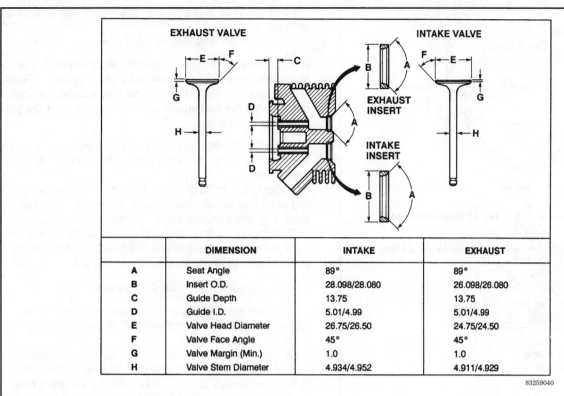

	DIMENSION	INTAKE	EXHAUST
A	Seat Angle	89°	89°
B	Insert O.D.	28.098/28.080	26.098/26.080
C	Guide Depth	13.75	13.75
D	Guide I.D.	5.01/4.99	5.01/4.99
E	Valve Head Diameter	26.75/26.50	24.75/24.50
F	Valve Face Angle	45°	45°
G	Valve Margin (Min.)	1.0	1.0
H	Valve Stem Diameter	4.934/4.952	4.911/4.929

83259040

Fig. 45 Valve details

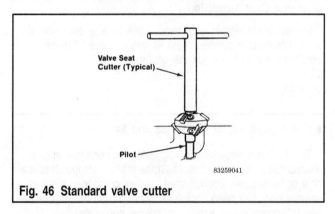

83259041

Fig. 46 Standard valve cutter

piston pin or the pin boss of the piston is worn or damaged, a new piston assembly is required.

Ring failure is usually indicated by excessive oil consumption and blue exhaust smoke. When rings fail, oil is allowed to enter the combustion chamber where it burned along with the fuel. High oil consumption can also occur when the piston ring end gap is incorrect because the ring cannot properly conform to the cylinder wall under this condition. Oil control is also lost when ring gaps are not staggered during installation.

When cylinder temperatures get too high, lacquer and varnish collect on pistons causing rings to stick, which results in rapid wear. A worn ring usually takes on a shiny or bright appearance. Scratches on rings and pistons are caused by abrasive material such as carbon, dirt, or pieces of hard metal.

Detonation damage occurs when a portion of the fuel charge ignites spontaneously from heat and pressure shortly after ignition. This creates two flame fronts which meet and explode to create extreme hammering pressures on a specific area of the piston. Detonation generally occurs from using fuels with too low an octane rating.

Preignition, or ignition of the fuel charge before the timed spark, can cause damage similar to detonation. Preignition

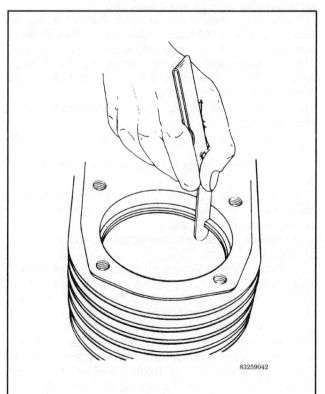

83259042

Fig. 47 Measuring ring end gap

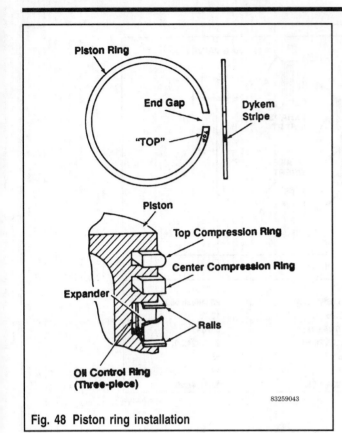

Fig. 48 Piston ring installation

damage is often more sever than detonation damage — often a hole is quickly burned right through the piston dome. Preignition is caused by a hot spot in the combustion chamber from sources such as: glowing carbon deposits, blocked fins, improperly seated valve, or wrong spark plug.

Replacement pistons are available in STD bore size, and in 0.25 mm (0.010 in) and 0.50 mm (0.20 in) oversizes. Replacement pistons include new piston ring sets and new piston pins.

Service replacement piston rings sets are also available separately for STD pistons, and for 0.25 mm (0.010 in) and 0.50 mm (0.020 in) oversized pistons.

Always use new piston rings when installing pistons. Never reuse old rings.

The cylinder bore must be deglazed before service ring sets are used.

Some important points to remember when servicing piston rings:

1. If the cylinder bore does not need reboring and if the old piston is within wear limits and free of score or scuff marks, the old piston may be reused.

2. Remove old rings and clean up grooves. Never reuse old rings.

3. Before installing the rings on piston, place the top two rings, each in turn, in its running area in cylinder bore and check end gap. This gap should be 0.75 mm (0.030 in) max. in a used cylinder bore and 0.25-0.45mm (0.010-0.018 in) in a new cylinder bore.

4. After installing the new rings on piston, check piston-to-ring side clearance. Maximum recommended side clearance is: Top ring — 0.040-0.085 mm (0.0016-0.0033 in); Middle ring — 0.040-0.072 mm (0.0016-0.0028 in); Oil control ring — 0.140-0.275 mm (0.0055-0.0108 in)

If side clearance is greater than specified, a new piston must be used.

To install piston rings, proceed as follows:

➡**Rings must be installed correctly. Ring installation instructions are usually included with new rings sets. Follow instructions carefully. Use a piston ring expander to install rings. Install the bottom (oil control) ring first and the top compression ring last.**

5. Oil Control Ring (Bottom Groove): Install expander and then the rails. Make sure the ends of expander are not overlapped.

6. Compression Ring (Center Groove): Install the center ring using a piston ring installation tool. Make sure the **top** mark is up and the PINK stripe is to the left of end gap.

7. Compression Ring (Top Groove): Install the top ring using a piston ring installation tool. Make sure the **top** mark is up and the BLUE stripe is to the left of end gap.

Connecting Rods Inspection And Service

Check bearing area (big end) for excessive wear, score marks, running and side clearances. Replace rod and cap if scored or excessively worn.

Service replacement connecting rods are available in STD crankpin size and 0.25 mm (0.010 in) undersize. The 0.25 mm (0.010 in) undersize rod can be identified by the drilled hole located in the lower end of the rod shank.

Governor Gear Inspection

Inspect the governor gear teeth. Look for any evidence of worn, chipped, or cracked teeth. If one or more of these problems is noted, replace the governor gear.

ASSEMBLY

◆ **See Figures 49, 50, 51, 52, 53 and 54**

The following sequence is suggested for complete engine reassembly. This procedure assumes that all components are new or have been reconditioned, and all components subassembly work has been completed. This procedure may be varied to accommodate options or special equipment.

➡**Make sure the engine is assembled using all specified torque values, tightening sequences, and clearances. Failure to observe specification could cause severe engine wear or damage. Always use new gaskets.**

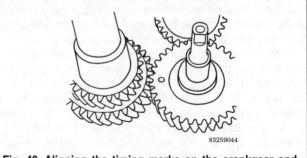

Fig. 49 Aligning the timing marks on the crankgear and balance shaft

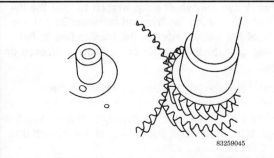

Fig. 50 Aligning the timing marks on the crankgear and camgear

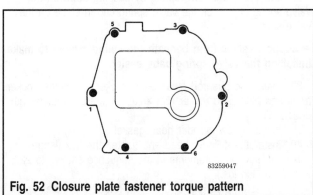

Fig. 51 Checking camshaft endplay

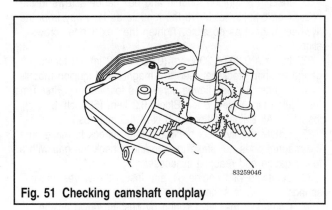

Fig. 52 Closure plate fastener torque pattern

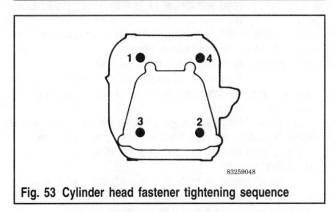

Fig. 53 Cylinder head fastener tightening sequence

1. Assemble the NU-12018 bearing installer to the NU-4747 handle, or equivalent.

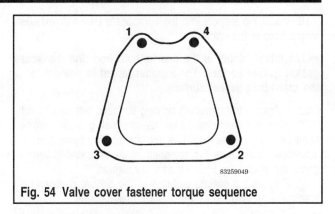

Fig. 54 Valve cover fastener torque sequence

2. Position the installer/bearing to the bearing bore of the crankcase or closure plate.

3. Drive the bearing into the bearing bore. Make sire the bearing is installed straight and true, and bottoms in the bore.

4. Lubricate the flywheel and bearing surface of the crankshaft.

5. Insert the crankshaft through the flywheel and bearing.

6. Assemble the piston, connecting rod, wrist pin and wrist pin retainers.

➡**The connecting rod must be assembled so the side with the cast numbers is opposite the FLY mark on the piston. Proper orientation of the piston/connecting rod inside the engine is extremely important. Improper orientation can cause extensive wear or damage.**

7. Stagger the piston rings in the grooves until the end gaps are 120 degrees apart.

8. Lubricate thew cylinder bore, piston and rings with engine oil. Compress the piston rings using a piston ring compressor.

9. Orient the FLY mark on piston towards the flywheel side of crankcase. Gently push the piston/connecting rod into bore. Do not pound on the piston.

10. Lubricate the crankshaft journal and connecting rod bearing surfaces with engine oil. Install the connecting rod cap to the connecting rod.

➡**The connecting rod cap must be installed with its match mark aligned with the connecting rod match mark. Improper installation can cause serious engine damage.**

11. Install the hex flange screws and torque in several increments to 9 N-m (80 inch lbs.).

12. Rotate the crankshaft until the piston is at top dead center in the cylinder bore.

13. Install oil sentry float switch in crankcase. Torque the float switch to 13.6 N-m (120 inch lbs.).

➡**The oil sentry float switch must be installed so the float arm is free to pivot toward the center of the crankcase.**

14. Install the oil sentry float switch baffle and hex flange screw.

15. Install plain washer and governor on crankcase stud.

16. Install small plain washer on cross shaft and install cross shaft (from inside closure plate) through bore in closure plate.

17. Install plain washer and hitch pin.

18. Lubricate the tappets and tappet bores in crankcase with engine oil.

19. Install the tappets into the appropriate intake or exhaust tappet bore in the crankcase.

➡ **The intake tappet is the one farthest from the crankcase gasket gasket surface. The exhaust tappet is nearest to the crankcase gasket surface.**

20. Lubricate the camshaft bearing surfaced with engine oil.
21. Align the timing marks on the camshaft gear and crankshaft gear. Lower the camshaft into the bearing surface in crankcase. Make sure the camshaft, crankshaft, and governor gears mesh and the timing marks are aligned.
22. Prepare the sealing surfaces of the closure plate and crankcase as directed by the sealant manufacturer.

➡ **Do not scrape the surfaces when cleaning as this will damage the surfaces and could cause leaks. The use of a gasket-removing solvent is recommended.**

23. RTV silicone sealant is used as a gasket between the closure plate and crankcase. GE Silmate® type RTV-1473 or RTV-108 silicone sealant (or equivalent) is recommended.
24. Apply a 1/16 in. bead of sealant to the closure plate.
25. Install the closure plate to the crankcase and install the six hex flange screws. Tighten the screws hand tight.

➡ **Turn governor cross shaft clockwise before installing the closure plate. Cross shaft must rest on governor for proper operation.**

26. Torque the fasteners, in several increments in the sequence shown, to 22.6 N-m (200 inch lbs.).
27. Slide the seat protector sleeve NU-12021, or equivalent, over the crankshaft. Generously lubricate the lips of the oil seal with light grease. Slide the oil seal over he sleeve.
28. Assemble handle and seal driver. Install the crankcase seal until the driver bottoms against the crankcase. Assemble handle and seal driver. Install the closure plate seal until the driver bottoms against the closure plate.

➡ **Oil Seal on PTO side of crankshaft should be installed to a depth of 5-7 mm (0.020-0.28 in) below lip of crankshaft bore.**

❈❈CAUTION

Damaging Crankshaft And Flywheel Can Cause Personal Injury! Using improper procedures to install the flywheel can crack or damage the crankshaft and/or flywheel. This not only causes extensive engine damage, but can also cause personal injury, since broken fragments could be thrown from the engine. Always observe and use the following precautions and procedures when installing the flywheel.

➡ **Before installing the flywheel, make sure the crankshaft taper and flywheel hub are clean, dry and completely free of lubricants. The presence of lubricants can case the flywheel to be over-stressed and damaged when the flange screw is torqued to specification.**

➡ **Make sure the flywheel key is installed properly in the keyway. The flywheel can become cracked or damaged if the key is not installed properly in the keyway.**

➡ **Always use a flywheel strap wrench to hold the flywheel when tightening the flywheel fastener. Do not use any type of bar wedge between the cooling fins or flywheel ring gear, as these parts could become cracked or damaged.**

29. Install the woodruff key into the keyway in the crankshaft.
30. Place the flywheel over the keyway/crankshaft. Install the drive cup, plain washer (flat side of washer towards the drive cup), and the hex flange screw.
31. Hold the flywheel with a strap wrench and torque the hex flange screw to 67.8 N-m (50 foot lbs.).
32. Install the ignition module and hex flange screws to the bosses on crankcase. Move the module as far away from the flywheel/magnet as possible. Tighten the hex flange screws slightly.
33. Insert a 0.203-0.305 mm (0.008-0.012 in) flat feeler gauge or shim stock between the magnet and ignition module.
34. Tighten the hex flange screws as follows: First Time Installation On A New Short Block: 6.2 N-m (55 inch lbs.).   All Reinstallations: 4.0 N-m (35 inch lbs.).
35. Rotate the flywheel back and forth; check to make sure the magnet does not strike the module. Check the gap with a feeler gauge and readjust if necessary.
36. Lubricate with engine oil, and install the valves, valve springs, and valve spring caps.
37. Compress the valve spring by pushing down on the valve spring cap. Lock the valve spring cap in place on the valve stem.

➡ **Support valves from beneath the cylinder head to make installing the valve spring caps easier.**

38. Install the pushrod guide plate, rocker arm studs, rocker arms and balls, and rocker arm adjusting nuts. Lubricate with engine oil.
39. Install a new cylinder head gasket.
40. Install the cylinder head and tighten the hex flange screws in several increments in the sequence shown, to 22.6 N-m (200 inch lbs.).
41. Set spark plug gap at 0.76 mm (0.030 in). Install spark plug in cylinder head and torque to 24.4-29.8 N-m (18-22 foot lbs.).
42. Install the push rods. Check that the push rods are seated on the tappets and rocker arms.
43. Adjust valve to tappet clearance as follows:
   a. Position the crankshaft so the piston is at the top of the compression stroke (the camshaft is not pushing the tappets and push rods).
   b. Insert a flat feeler gauge between the rocker arm and valve stem. The recommended valve to rocker arm clearance for both intake and exhaust is 0.038-0.051 mm (0.0015-0.0020 in).
   c. Adjust clearance by turning the adjusting nut clockwise to decrease valve to rocker arm clearance, counterclockwise to increase valve to rocker arm clearance.
44. Install a new valve cover gasket. Install the breather assembly on the cylinder head.
45. Install a valve cover. Torque the hex flange screws to 3.4 N-m (30 inch lbs.) using the sequence shown.
46. Install kill-switch lever, bushing, and throttle lever on the crankcase using hex flange screw.

47. Connect the kill-switch lead to the kill-switch.

48. Install the return spring in the first hole of the throttle lever (the hole nearest the end of the throttle lever).

49. Install the governor lever on the governor cross shaft. install the return spring in the third (middle) hole of the governor lever.

➡ **Leave all hardware slightly loose until all sheet metal parts are in position.**

50. Install the blower housing using Phillips head screws in the locations shown.

51. Install the heat detector and gaskets on the intake studs.

52. Install the bottom cylinder baffle using hex flange screws.

53. Connect the kill-switch linkage in the first hole of the throttle lever. Install the top cylinder barrel baffle using hex flange screws.

54. Install the spark plug wire and grommet into the cylinder head baffle. Install the cylinder head baffle onto the top and bottom cylinder baffles.

55. Tighten all hardware.

56. Install new gasket and muffler on exhaust outlet flange.

57. install hex flange nuts onto exhaust studs and hex flange screws in muffler bracket. Torque hex flange nuts to 22.6 N-m (200 inch lbs.).

58. Install new gasket and carburetor onto intake manifold studs.

59. Install the throttle linkage to the carburetor throttle lever using the linkage clip.

✳✳CAUTION

Gasoline may be present in the carburetor and fuel system. Gasoline is extremely flammable, and its vapors can explode if ignited. Keep sparks, flames, and other sources of ignition away from the engine.

60. Install the fuel tank and fuel line.

61. Secure the fuel tank to the engine using hex flange screws. Torque the hex flange screws to 17 N-m (150 inch lbs.).

62. Install the fuel line on the carburetor inlet fitting. Secure the fuel line with a hose clamp.

63. Make sure the two O-rings on the oil fill tube and the O-ring in the oil fill cap are in place.

64. Install oil fill tube into the hole in the crankcase.

65. Secure the oil fill tube to the fuel tank with the hex flange screw and lock nut.

66. Install the retractable starter and Phillips head screws to the blower housing.

67. Install gasket, air cleaner, and hex flange nuts on intake studs.

68. Install Phillips head screws to air cleaner.

69. Instal the precleaner and air cleaner element retainer. Be sure the retainer is hooked on tabs. Tighten the four slot head screws evenly.

70. Install the air cleaner cover as follows:

   a. Slide the air cleaner cover onto the air cleaner base.

   b. Be sure the tab at the bottom of the air cleaner cover snaps in place to lock the cover on the air cleaner base.

PREPARE THE ENGINE FOR OPERATION

The engine is now completely reassembled. Before starting or operating the engine, be sure to do the following:

1. Make sure all hardware is tightened securely.

2. Make sure the oil drain plug and oil sentry pressure switch are tightened securely.

3. Fill the crankcase with the correct amount, weight, and type of oil.

4. Adjust the governor. Refer to the Fuel System and Governor section.

5. Adjust the carburetor idle fuel needle or idle sped adjusting screw as necessary. Refer to the Fuel System and Governor section.

6. Make sure the maximum engine speed does not exceed 3600 rpm. Adjust the throttle and choke controls and the high speed stop as necessary. Refer to the Fuel System and Governor section.

Automatic Compression Release (ACR)

This engine is equipped with an Automatic Compression Release (ACR) mechanism. ACR lowers compression at cranking speeds to make starting easier.

OPERATION

The ACR mechanism consists of a lever and control pin assembly attached to the gear on the camshaft. At cranking speeds (700 RPM or lower), the control pin protrudes above the exhaust cam lobe. This pushes the exhaust valve off its seat during the first part of the compression stroke. The reduced compression results in an effective compression ratio of about 2:1 during cranking.

After starting, engine speed increases to over 700 RPM. Centrifugal force moves the lever, and the control pin drops into the recess in the exhaust valve and the engine operates at full power.

When the engine is stopped, the spring returns the lever and control pin assembly to the compression release position ready for the next start.

11 AND 12.5 HORSEPOWER MODELS

SPECIFICATIONS, TOLERANCES, AND SPECIAL TORQUE VALUES[1]

DESCRIPTION	Command 11, 12.5, 14 Hp
General Specifications	

Power (@ 3600 rpm, corrected to SAE J1349)
    Command 11 . . . . 8.20 kW (11 hp)
    Command 12.5 . . . . 9.33 kW (12.5 hp)
    Command 14 . . . . 10.50 kW (14 hp)
Peak Torque
    Command 11 . . . . 27.4 N•m (20.2 lbf•ft)
    Command 12.5 . . . . 27.8 N•m (20.5 lbf•ft)
    Command 14 . . . . 28.9 N•m (21.3 lbf•ft)
Bore . . . . 87 mm (3.43 in)
Stroke . . . . 67 mm (2.64 in)
Displacement . . . . 398 cm³ (24.3 in³)
Compression Ratio 8.5:1
Approx. Weight . . . . 36.3 kg (80 lb)
Approx. Oil Capacity . . . . 1.9 L (2.0 U.S. qt)

Air Cleaner

Base Nut Torque . . . . 9.9 N•m (88 lbf•in)

Angle Of Operation — Maximum (At Full Oil Level)

Intermittent — All Directions . . . . 35°
Continuous — All Directions . . . . 25°

Balance Shaft

End Play (Free) . . . . 0.0575/0.3625 mm (0.0023/0.0143 in)
Running Clearance . . . . 0.025/0.063 mm (0.0009/0.0025 in)
Bore I.D. — New . . . . 20.000/20.025 mm (0.7874/0.7884 in)
Bore I.D. — Max. Wear Limit . . . . 20.038 mm (0.7889 in)
Balance Shaft Bearing Surface O.D. — New . . . . 19.962/19.975 mm (0.7859/0.7864 in)
Balance Shaft Bearing Surface O.D. — Max. Wear Limit . . . . 19.959 mm (0.7858 in)

Camshaft

End Play (With Shims) . . . . 0.076/0.127 mm (0.003/0.005 in)
Running Clearance . . . . 0.025/0.063 mm (0.0010/0.0025 in)
Bore I.D. — New . . . . 20.000/20.025 mm (0.7874/0.7884 in)
Bore I.D. — Max. Wear Limit . . . . 20.038 mm (0.7889 in)
Camshaft Bearing Surface O.D. — New . . . . 19.962/19.975 mm (0.7859/0.7864 in)
Camshaft Bearing Surface O.D. — Max. Wear Limit . . . . 19.959 mm (0.7858 in)

832590C2

Tightening Torque: N•m (lbf•in) + or – 20%

Bolts, Screws, Nuts And Fasteners Assembled Into Cast Iron Or Steel			Grade 2 Or 5 Fasteners Into Aluminum
Grade 2	Grade 5	Grade 8	
Size			
8–32	2.3 (20)	2.8 (25)	———	2.3 (20)
10–24	3.6 (32)	4.5 (40)		3.6 (32)
10–32	3.6 (32)	4.5 (40)	———	
1/4–20	7.9 (70)	13.0 (115)	18.7 (165)	7.9 (70)
1/4–28	9.6 (85)	15.8 (140)	22.6 (200)	
5/16–18	17.0 (150)	28.3 (250)	39.6 (350)	17.0 (150)
5/16–24	18.7 (165)	30.5 (270)		
3/8–16	29.4 (260)	———		
3/8–24	33.9 (300)	———		

Tightening Torque N•m (lbf•ft) + or – 20%

Size	Grade 2	Grade 5	Grade 8	
5/16–24	———	———	40.7 (30)	———
3/8–16	———	47.5 (35)	67.8 (50)	
3/8–24		54.2 (40)	81.4 (60)	
7/16–14	47.5 (35)	74.6 (55)	108.5 (80)	
7/16–20	61.0 (45)	101.7 (75)	142.4 (105)	
1/2–13	67.8 (50)	108.5 (80)	155.9 (115)	
1/2–20	94.9 (70)	142.4 (105)	223.7 (165)	
9/16–12	101.7 (75)	169.5 (125)	237.3 (175)	
9/16–18	135.6 (100)	223.7 (165)	311.9 (230)	
5/8–11	149.2 (110)	244.1 (180)	352.6 (260)	
5/8–18	189.8 (140)	311.9 (230)	447.5 (330)	
3/4–10	199.3 (150)	332.2 (245)	474.6 (350)	
3/4–16	271.2 (200)	440.7 (325)	637.3 (470)	

83259050

Fig. 55 Fastener torque specifications for U.S. measure fasteners

Tightening Torque: N•m (lbf•in) + or – 10%

Size	Property Class					Noncritical Fasteners Into Aluminum
	4.8	5.8	8.8	10.9	12.9	
M4	1.2 (11)	1.7 (15)	2.9 (26)	4.1 (36)	5.0 (44)	2.0 (18)
M5	2.5 (22)	3.2 (28)	5.8 (51)	8.1 (72)	9.7 (86)	4.0 (35)
M6	4.3 (38)	5.7 (50)	9.9 (88)	14.0 (124)	16.5 (146)	6.8 (60)
M8	10.5 (93)	13.6 (120)	24.4 (216)	33.9 (300)	40.7 (360)	17.0 (150)

Tightening Torque: N•m (lbf•ft) + or – 10%

	Property Class					Noncritical Fasteners Into Aluminum
	4.8	5.8	8.8	10.9	12.9	
M10	21.7 (16)	27.1 (20)	47.5 (35)	66.4 (49)	81.4 (60)	33.9 (25)
M12	36.6 (27)	47.5 (35)	82.7 (61)	116.6 (86)	139.7 (103)	61.0 (45)
M14	58.3 (43)	76.4 (55)	131.5 (97)	184.4 (136)	219.7 (162)	94.9 (70)

83259051

Fig. 56 Fastener torque specifications for metric measure fasteners

Size	Into Cast Iron	Into Aluminum
1/8" NPT	--------------	4.5 (40 lbf•in)
1/4"	17.0 (150 lbf•in)	11.3 (100 lbf•in)
3/8"	20.3 (180 lbf•in)	13.6 (120 lbf•in)
1/2"	27.1 (20 lbf•ft)	17.6 (13 lbf•ft)
3/4"	33.9 (25 lbf•ft)	21.7 (16 lbf•ft)
X-708-1	27.1/33.9 (20/25 lbf•ft)	27.1/33.9 (20/25 lbf•ft)

Torque Conversions

N•m = lbf•in x 0.113
N•m = lbf•ft x 1.356
lbf•in = N•m x 8.85
lbf•ft = N•m x 0.737

83259052

Fig. 57 Oil pan drain plug torque specifications Nm (U.S.)

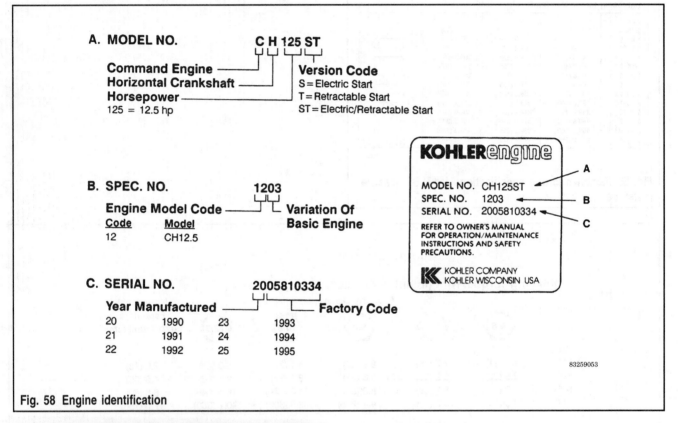

A. MODEL NO.   C H 125 ST

Command Engine
Horizontal Crankshaft
Horsepower
125 = 12.5 hp

Version Code
S = Electric Start
T = Retractable Start
ST = Electric/Retractable Start

B. SPEC. NO.   1203

Engine Model Code   Variation Of Basic Engine

Code	Model
12	CH12.5

C. SERIAL NO.   2005810334

Year Manufactured   Factory Code

20	1990	23	1993
21	1991	24	1994
22	1992	25	1995

KOHLER engine

MODEL NO. CH125ST — A
SPEC. NO. 1203 — B
SERIAL NO. 2005810334 — C

REFER TO OWNER'S MANUAL FOR OPERATION/MAINTENANCE INSTRUCTIONS AND SAFETY PRECAUTIONS.

KOHLER COMPANY
KOHLER WISCONSIN USA

83259053

Fig. 58 Engine identification

Air Cleaner

These engines are equipped with a replaceable, high-density paper air cleaner element. Some engines are also equipped with an oiled-foam precleaner which surrounds the paper element.

SERVICE

Check the air cleaner daily or before starting the engine. Check for and correct heavy buildup of dirt and debris, and loose components.

➡ **Operating the engine with loose or damaged air cleaner components could allow unfiltered air into the engine causing premature wear and failure.**

Precleaner Service

If so equipped, wash and reoil the precleaner every 25 hours of operation (more often under extremely dusty or dirty conditions).

1. Remove the precleaner from the paper element.
2. Wash the precleaner in warm water with detergent. Rinse the precleaner thoroughly until all traces of detergent are eliminated. Squeeze out excess water (do not wring). Allow the precleaner to air-dry.
3. Saturate the precleaner with new engine oil. Squeeze out all excess oil.
4. Reinstall the precleaner over the paper element.
5. Reinstall air cleaner cover, and air cleaner cover retaining knob. Make sure the knob is tightened securely.

Servicing the Paper Element

Every 100 hours of operation (more often under extremely dusty or dirty conditions), check the paper element. Replace the element as necessary.

1. Remove the precleaner (if so equipped) from the paper element.

2. Remove the wing nut, washer, element cover, and air cleaner element.

3. Do not wash the paper element or use pressurized air, as this will damage the element. Replace a dirty, bent, or damaged element with a new element. Handle new elements carefully; do not use if the sealing surfaces are bent or damaged.

4. Reinstall the paper element, element cover, washer, wing nut, precleaner, air cleaner cover, and air cleaner cover retaining knob. Make sure the knob is tightened securely.

Inspect Air Cleaner Components

Whenever the air cleaner cover is removed, or the paper element or precleaner are serviced, check the following areas/components:

• Covered Air Cleaner Element — Inspect the rubber grommet in the hole of the air cleaner element cover. Replace the grommet if it is worn or damaged.

• Air Cleaner Base — Make sure the base is secured and not cracked or damaged. Since the air cleaner base and carburetor are secured to the intake port with common hardware, it is extremely important that the nuts securing these components are tight at all times.

• Breather Tube — Make sure the tube is installed to both the air cleaner base and valve cover.

➡Damaged, worn, or loose air cleaner components can allow unfiltered air into the engine causing premature wear and failure. Tighten or replace all loose or damaged components.

DISASSEMBLY

1. Remove the air cleaner cover retaining knob and air cleaner cover.

2. If so equipped, remove the precleaner from paper element.

3. Remove the wing nut, washer, element cover, and air cleaner element.

4. Disconnect the breather hose from the valve cover.

5. Remove the air cleaner base mounting nuts, air cleaner base, and gasket.

6. If necessary, remove the self-tapping screws and elbow from air cleaner base.

ASSEMBLY

1. Install the elbow and self-tapping screws to air cleaner base.

2. Install the gasket, air cleaner base, and base mounting nuts. Torque the nuts to 9.9 N-m (88 inch lbs.).

3. Connect the breather hose to the air cleaner base (and valve cover). Secure with hose clamps.

4. If necessary, install the grommet into the cover of air cleaner element. Install the air cleaner element, element cover, washer, and wing nut.

5. If equipped, install the precleaner (washed and oiled) over the paper element.

6. Install the air cleaner cover and air cleaner cover retaining knob. Tighten the knob securely.

Air Intake/Cooling System

CLEANING

To ensure proper cooling, make sure the grass screen, cooling fins, and other external surfaces of the engine are kept clean at all times.

Every 100 hours of operation (more often under extremely dusty, dirty conditions), remove the blower housing and other cooling shrouds. Clean the cooling fins and external surfaces as necessary. Make sure the cooling shrouds are reinstalled.

➡Operating the engine with a blocked grass screen, dirty or plugged cooling fins, and/or cooling shrouds removed, will cause engine damage due to overheating.

Fuel Recommendations

✳✳CAUTION

Gasoline is extremely flammable and its vapors can explode if ignited. Before servicing the fuel system, make sure there are no sparks, open flames, or other sources of ignition nearby as these can ignite gasoline vapors. Disconnect and ground the spark plug lead to prevent the possibility of sparks from the ignition system.

GENERAL RECOMMENDATIONS

• Purchase gasoline in small quantities and store in clean, approved containers. A container with a capacity of 2 gallons or less with a pouring spout is recommended. Such a container is easier to handle and helps to eliminate spoilage during refueling.

• Do not use gasoline left over from the previous season, to minimize gum deposits in your fuel system and to ensure easy starting.

• Do not add oil to the gasoline.

• Do not overfill the fuel tank. Leave room for the fuel to expand.

FUEL TYPE

For best results, use only clean, fresh, unleaded gasoline with a pump sticker octane rating of 87 or higher. I countries using the Research method, it should be 90 octane minimum.

Unleaded gasoline is recommended, as it leaves less combustion chamber deposits. Leaded gasoline may be used is areas where unleaded is not available and exhaust emissions are not regulated. Be aware however, that the cylinder head will require more frequent service.

GASOLINE/ALCOHOL blends

Gasohol (up to 10% ethyl alcohol, 90% unleaded gasoline by volume) is approved as a fuel for Kohler engines. Other gasoline/alcohol blends are not approved.

GASOLINE/ETHER BLENDS

Methyl Tertiary Butyl Ether (MTBE) and unleaded gasoline blends (up to a maximum of 15% MTBE by volume) are approved as a fuel for Kohler engines. Other gasoline/ether blends are not approved.

FUEL SYSTEM OPERATION

The typical fuel system and related components include the fuel tank, in-line fuel filter, fuel pump, carburetor, and interconnecting fuel lines.

The fuel from the tank is moved through the in-line filter and fuel lines by the pump. On engines not equipped with a fuel pump, the fuel tank outlet is located above the carburetor inlet; gravity moves the fuel.

Fuel then enters the carburetor float bowl and is moved into the carburetor body. There, the fuel is then burned in the engine combustion chamber.

Fuel Filter

Some engines are equipped with an in-line fuel filter. Visually inspect the filter periodically, and replace when dirty with a genuine Kohler filter.

Fuel Pump

◆ **See Figure ?**

Some engines are equipped with an optical mechanically operated fuel pump. On applications using a gravity feed fuel system, the fuel pump mounting pad is covered with a metal plate.

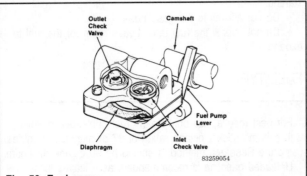

Fig. 59 Fuel pump

The fuel pump body is constructed of nylon. The nylon body insulates the fuel from the engine crankcase. This prevents the fuel from vaporizing inside the pump.

The mechanical pump is operated by a lever which rides on the engine camshaft. This lever transmits a pumping action to the diaphragm inside the pump body. On the downward stroke of the diaphragm, fuel is drawn in through the inlet check valve. On the upward stroke of the diaphragm, fuel is forced out through the outlet check valve.

REMOVAL

1. Disconnect the fuel lines from the inlet and outlet fittings of pump.
2. Remove the hex, flange screws, fuel pump, and gasket.
3. If necessary, remove the fittings from the pump body.

REPAIR

Nylon-bodied fuel pumps are not serviceable and must be replaced when faulty. Replacement pumps are available in kits that include the pump and mounting gasket.

INSTALLATION

1. Fittings - Apply a small amount of Permatex Aviation Perm-A-Gasket® (or equivalent) gasoline resistant thread sealant to the threads of fittings. Turn the fittings into the pump 6 full turns; continue turning the fittings in the same direction until the desired position is reached.
2. Install new gasket, fuel pump, and hex, flange screws.

➡**Make sure the fuel pump lever is positioned to the RIGHT of the camshaft (when looking at fuel pump mounting pad). Damage to the fuel pump, and subsequent severe damage could result if the lever is positioned to the left of the camshaft.**

Torque the hex, flange screws as follows:
• First Time Installation On A New Short Block — 9.0 N-m (80 inch lbs.)
• All Reinstallations — 7.3 N-m (65 inch lbs.)
3. Connect the fuel lines to the inlet and outlet fittings.

Carburetor

These engines are equipped with an adjustable main jet carburetor. This subsection covers the troubleshooting, idle adjustment, and service procedures for the carburetor.

✳✳CAUTION

Gasoline may be present in the carburetor and fuel system. Gasoline is extremely flammable and its vapors can explode if ignited. Keep sparks, open flames, and other sources of ignition away from the engine. Disconnect and ground the spark plug lead to prevent the possibility of sparks from the ignition system.

TROUBLESHOOTING

If engine troubles are experienced that appear to be fuel related, check the following area before adjusting or disassembling the carburetor.

- Make sure the fuel tank is filled with clean, fresh gasoline.
- Make sure the fuel tank cap vent is not blocked and that it is operating properly.
- Make sure fuel is reaching the carburetor. This includes checking the fuel shut-off valve, fuel tank filter screen, in-line fuel filter, fuel lines, and fuel pump for restrictions or faulty components as necessary.
- Make sure the air cleaner base and carburetor is securely fastened to the engine using gaskets in good condition.
- Make sure the air cleaner element is clean and all air cleaner components are fastened securely.
- Make sure the ignition system, governor system, exhaust system, and throttle and choke controls are operating properly.
- If the engine is hard-starting or runs roughly or stalls at low idle speed, it may be necessary to adjust or service the carburetor.

ADJUSTMENT

➡**Carburetor adjustments should only be made after the engine has warmed up.**

The carburetor is designed to deliver the correct fuel-to-air mixture to the engine under all operating conditions. The main fuel jet (power screw) is calibrated at the factory and is adjustable. The idle fuel adjusting needle is also set at the factory and is adjustable. The idle fuel adjusting needle is also set at the factory and normally does not need adjustment.

If the engine is hard-starting or runs roughly or stalls at low idle speed, it may be necessary to adjust or service the carburetor.

1. With the engine stopped turn the low idle and high idle fuel adjusting needles in (clockwise) until they bottom lightly.

➡**The tip of the idle fuel and high idle fuel adjusting needles are tapered to critical dimensions. Damage to the needles and the seats in carburetor body will result if the needles are forced.**

2. Preliminary settings: Turn the adjusting needles out (counterclockwise) from lightly bottomed to the positions in the chart.

3. Start the engine and run at half-throttle for 5 to 10 minutes to warm up. The engine must be warm before making final settings. Check that the throttle and choke plates can fully open.

4. High idle fuel needle setting: Place the throttle into the 'fast" position. If possible place the engine under load. Turn the high idle adjusting needle in (slowly) until engine speed decreases and then back out approximately ¼ turn for best high-speed performance.

5. Low idle speed setting: Place the throttle control into the 'idle" or 'slow" position. Set the low idle sped to 1500 rpm and

± 75 rpm by turning the low idle speed adjusting screw in or out. Check the speed using a tachometer.

➡**The actual low idle speed depends on the application - refer to equipment manufacturer's recommendations. The recommended low idle speed for basic engines is 1500 rpm. To ensure best results when setting the low idle speed should not exceed 1500 rpm ± 75 rpm.**

6. Low idle fuel needle setting: Place the throttle into the 'idle" or 'slow" position. Turn the low idle fuel adjusting needle in (slowly) until the engine speed decreases and then back out approximately ⅛-¼ turn to obtain best low speed performance.

7. Recheck the idle speed using a tachometer. Readjust the speed as necessary.

DISASSEMBLY

▶ **See Figure 60**

1. Remove the power screw, needle and spring, main jet, power screw gasket and fuel bowl.

2. Remove the bowl gasket, float shaft, shaft, float, and fuel inlet needle.

3. Remove the low idle fuel adjusting needle and spring. Remove the low idle speed adjusting screw and spring.

➡**Further disassembly to remove the welch plug, fuel inlet seat, throttle plate and shaft is recommended only if these parts are to be cleaned or replaced.**

Welch Plug Removal

▶ **See Figure 61**

In order to clean the 'off-idle" ports and bowl vent thoroughly, remove the welch plug covering these areas. Use tool no. KO-1018 and the following procedure to remove the welch plug.

1. Pierce the welch plug with the tip of the tool,

➡**To prevent damage to the carburetor, do not allow the tool to strike the carburetor body.**

2. Pry out the welch plug with the tip of the tool.

Fuel Inlet Seat Removal

To remove the fuel inlet seat, pull it out of the carburetor body using a screw, drill bit, or similar tool.

➡**Always install a new fuel inlet seat. Do not reinstall a seat that has been removed.**

Choke Shaft Removal

1. Because the edges of the choke plate are beveled, mark the choke plate and carburetor body to ensure correct reassembly. Also take note of the choke plate position in bore, and the position of the choke lever and return spring.

2. Grasp the choke plate with pliers. Pull it out the slot in the choke shaft.

3. Remove the choke shaft and choke return spring.

Throttle Shaft Removal

1. Because the edges of the throttle plate are beveled, mark the throttle plate and carburetor body to ensure correct

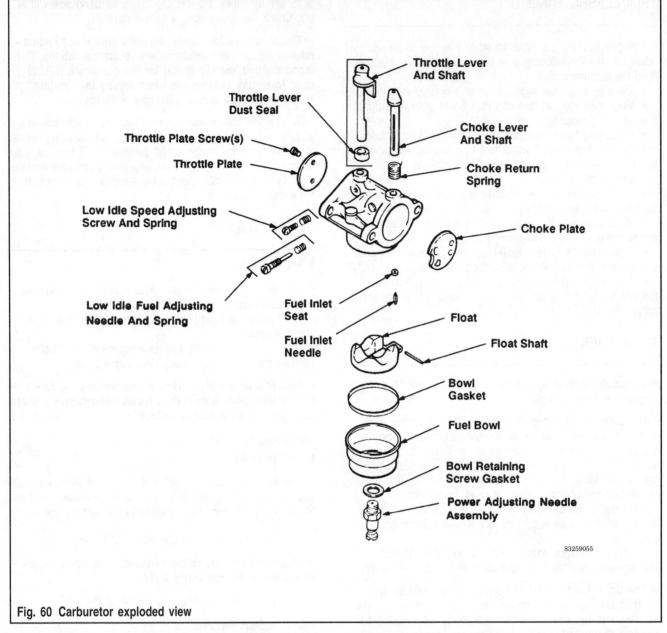

Throttle Lever And Shaft

Throttle Lever Dust Seal

Throttle Plate Screw(s)

Throttle Plate

Low Idle Speed Adjusting Screw And Spring

Low Idle Fuel Adjusting Needle And Spring

Choke Lever And Shaft

Choke Return Spring

Choke Plate

Fuel Inlet Seat

Fuel Inlet Needle

Float

Float Shaft

Bowl Gasket

Fuel Bowl

Bowl Retaining Screw Gasket

Power Adjusting Needle Assembly

83259055

Fig. 60 Carburetor exploded view

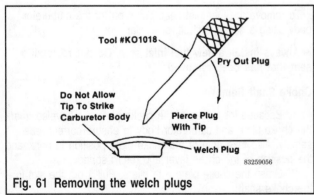

Tool #KO1018

Pry Out Plug

Do Not Allow Tip To Strike Carburetor Body

Pierce Plug With Tip

Welch Plug

83259056

Fig. 61 Removing the welch plugs

reassembly. Also take note of the throttle plate position in bore, and the position of the throttle lever.

2. Carefully and slowly remove the screws securing the throttle plate to the throttle shaft. Remove the throttle plate.

3. File off any burrs which may have been left on the throttle shaft when the screws were removed. Do this before removing the throttle shaft from the carburetor body.

4. Remove the throttle lever/shaft assembly with foam dust seal.

Cleaning

✳✳CAUTION

Carburetor cleaners and solvents are extremely flammable. Keep sparks, flames, and other sources of ignition away from the area. Follow the cleaner manufacturer's warnings and instructions on its proper and safe use. Never use gasoline as a cleaning agent.

All parts should be cleaned thoroughly using a carburetor cleaner (such as acetone). Make sure all gum deposits are removed from the following areas:

- Carburetor body and bore — especially the areas where the throttle plate, choke plate and shafts are seated.
- Idle fuel and "off-idle" ports in carburetor bore, power screw, bowl vent, and fuel inlet needle and seat.

➡**These areas can be cleaned with a fine piece of wire in addition to cleaners. Be careful not to enlarge the ports, or break the wire inside the ports. Blow out all passages with compressed air.**

- Float and float hinge.
- Fuel bowl.
- Throttle plate, choke plate, throttle shaft, and choke shaft.

➡**Do not submerge the carburetor in cleaner or solvent when fiber, rubber, or foam seals or gaskets are installed. The cleaner may damage these components.**

Inspection

- Carefully inspect all components and replace those that are worn or damaged.
- Inspect the carburetor body for cracks, holes, and other wear or damage.
- Inspect the float for cracks, holes, and other wear or damage.
- Inspect the fuel inlet needle and seat for wear or damage.
- Inspect the tip of the low idle fuel adjusting needle and power screw needle for wear or grooves.
- Inspect the throttle and choke shaft and plate assemblies for wear or excessive play.

Repair

Always use new gaskets when servicing or reinstalling carburetors. Repair kits are available which include new gaskets and other components. These kits are described below.

Components such as the throttle and choke shaft assemblies, throttle plate, choke plate, low idle fuel needle, power screw, and others, are available separately.

REASSEMBLY

Throttle Shaft Installation

1. Install the foam dust seal on the throttle shaft.
2. Insert the throttle/shaft assembly into the carburetor body. Position the cutout portion of the shaft so it faces the carburetor mounting flange.
3. Install the throttle plate to the throttle shaft. Make sure the plate is positioned properly in the bore as noted and marked during disassembly. Apply Loctite® no. 609 to the threads of the throttle plate retaining screws. Install the screws so they are slightly loose.
4. Apply finger pressure to the throttle lever/shaft to keep it firmly seated against the pivot in the carburetor body. Rotate the throttle shaft until the throttle shaft until the throttle plate closes the bore around its perimeter; then tighten the screws.
5. Operate the throttle lever. Check for binding between the throttle plate and carburetor bore. Loosen the screws and ad-

just the throttle plate as necessary. Torque the screws to 0.9/1.4 N-m (8/12 inch lbs.).

Choke Shaft Installation

1. Install the choke return spring to the choke shaft.
2. Insert the choke lever with return spring into the carburetor body.
3. Rotate the choke lever approximately 1/2 turn counterclockwise. Make sure the choke return spring hooks on the carburetor body.
4. Position the choke plate as noted and marked during disassembly. Insert the choke plate into the slot in the choke shaft. Make sure the choke shaft is locked between the tabs on the choke plate.

Fuel Inlet Seat Installation

Press the fuel inlet seat into the bore in carburetor body until it bottoms.

Welch Plug Installation

▸ **See Figure 62**

Use tool no. KO-1017 and install new plugs as follows:
1. Position the carburetor body with the welch plug cavity to the top.
2. Place a new welch plug into the cavity with the raised surface up.
3. Use the end of the tool that is about the same size as the plug and flatten the plug. Do not force the plug below the surface of the cavity.
4. After the plug is installed, seal it with sealant. Allow the sealant to dry.

➡**If a commercial sealant is not available, fingernail polish can be used.**

CARBURETOR REASSEMBLY

1. Install the low idle speed adjusting screw and spring.
2. Install the low idle fuel adjusting needle and spring. Turn the adjusting needle in (clockwise) until it bottoms lightly.

➡**The tip of the idle fuel adjusting needle is tapered to critical dimensions. Damage to the needle and the seat in carburetor body will result if the needle is forced.**

3. Turn the low idle fuel adjusting out (counterclockwise) as specified in the "Adjustment" portion of this section.

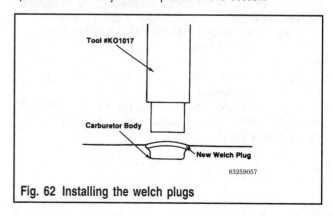

Fig. 62 Installing the welch plugs

4. Insert the fuel inlet needle into the float. Lower the float/needle into the carburetor body.

5. Install the float shaft.

6. Install the bowl gasket, fuel bowl, bowl retainer gasket, and power screw.

7. Torque the power screw to: 5.1/6.2 N-m (45/55 inch lbs.).

HIGH ALTITUDE OPERATION

When operating the engine at altitudes of 1830 m (6000 ft.) and above, the main fuel mixture tends to get overrich. An overrich mixture can cause conditions such as black, sooty exhaust smoke, misfiring, loss of speed and power, poor fuel economy, and poor or slow governor response.

To compensate for the effects of high altitude, a special high altitude main fuel jet can be installed. High altitude jets are sold in kits which include the jet and necessary gaskets.

Governor

These engines are equipped with a centrifugal flyweight mechanical governor. It is designed to hold the engine speed constant under changing load conditions. The governor gear/flyweight mechanism is mounted inside the crankcase and is driven off the gear on the camshaft.

Centrifugal force acting on the rotating governor gear assembly causes the flyweights to move outward as speed increases and inward as speed decreases. As the flyweights move outward, they cause the regulating pin to move outward.

The regulating pin contacts the tab on the cross shaft, causing the shaft to rotate when the engine speed changes. One end of the cross shaft protrudes through the side of the crankcase. Through external linkage attached to the cross shaft, the rotating action is transmitted to the throttle lever of the carburetor.

When the engine is at rest, and the throttle is in the "fast" position, the tension of the governor spring holds the throttle plate open. When the engine is operating (the governor gear assembly is rotating), the force applied by the regulating pin against the cross shaft tends to close the throttle plate. The governor spring tension and the force applied by the regulating pin are in "equilibrium" during operation, holding the engine speed constant.

When load is applied and the engine speed (and governor gear speed) decreases, the governor spring tension moves the governor arm to open the throttle plate wider. This allows more fuel into the engine; increasing engine speed. (This action takes place very rapidly, so a reduction in speed is hardly noticed.) As the speed reaches the governed setting, the governor spring tension and the force applied by the regulating pin will again be in equilibrium. This maintains the engine speed at a relatively constant level.

The governed speed setting is determined by the position of the throttle control. It can be variable or constant, depending on the application.

INITIAL ADJUSTMENT

Make this initial adjustment whenever the governor arm is loosened or removed from the cross shaft. To ensure proper setting, make sure the throttle linkage is connected to the governor arm and the throttle lever on the carburetor.

1. Pull the governor lever away from the carburetor (wide open throttle).

2. Insert a nail in the cross shaft hole or grasp the cross shaft with pliers and turn the shaft counterclockwise as far as it will go.

3. Tighten the hex nut securely.

SENSITIVITY ADJUSTMENT

Governor sensitivity is adjusted by repositioning the governor spring in the holes in the governor lever. If speed surging occurs with a change in load, the governor is set too sensitive. If a big drop in speed occurs when normal load is applied, the governor should be set for greater sensitivity.

REMOTE THROTTLE AND CHOKE ADJUSTMENT

1. Adjust the throttle lever. see this section.

2. Install remote throttle cable in hole in the throttle lever.

3. Install remote choke cable in hole in the choke lever.

4. Secure remote cables loosely with the cable clamps.

5. Position the throttle cable so that the throttle lever is against stop.

6. Tighten the throttle cable clamp.

7. Position the choke cable so that the carburetor choke plate is fully closed.

8. Tighten the choke cable clamp.

9. Check carburetor idle speed. See Adjust Carburetor in this section.

Lubrication System

Using the proper type and weight of oil in the crankcase is extremely important. So, is checking oil daily and changing oil regularly. Failure to use the correct oil, or using dirty oil, causes premature engine wear and failure.

OIL TYPE

Use a high-quality oil of API (American Petroleum Institute) service class SF or SG. Select the viscosity based on the air temperature at the time of operation as shown in the following table.

CHECK OIL LEVEL

The importance of checking and maintaining the proper oil level in the crankcase cannot be overemphasized. Check oil BEFORE EACH USE as follows:

1. Make sure the engine is stopped, level, and is cool so the oil has had time to drain into the sump.

➡**Using other than service class SF or SG oil or extending oil change intervals longer than recommended can cause engine damage.**

2. To keep dirt, grass clippings, etc., out of the engine, clean the area around the oil fill cap/dipstick before removing it.

3. Remove the oil fill cap/dipstick; wipe oil off. Reinsert the dipstick into the tube and seat the oil fill cap on the tube.

4. Remove the dipstick and check the oil level. The oil level should up to, but not over, the "F" mark on the dipstick.

5. If the oil level is low, add oil of the proper type, up to the "F" mark on the dipstick. Always check the level with the dipstick before adding more oil.

➡**To prevent extensive engine wear or damage, always maintain the proper oil level in the crankcase. Never operate the engine with the oil level below the "L" mark or over the "F" mark on the dipstick.**

Oil Sentry

Some engines are equipped with an optional Oil Sentry oil pressure monitor. If the oil pressure gets low, Oil Sentry will either shut off the engine or activate a warning signal, depending on the application.

➡**Make sure the oil level is checked BEFORE EACH USE and is maintained up to the "F" mark on the dipstick. This includes engines equipped with Oil Sentry.**

CHANGING OIL AND OIL FILTER

Changing Oil

For a new engine, change oil after the first 5 hours of operation. Thereafter, change oil after every 100 hours of operation.

For an overhauled engine or those rebuilt with a new short block, use 10W-30-weight service class SF oil for the first 5 hours of operation. Change the oil after this initial run-in period. Refill with service class SF oil as specified in the "Viscosity Grades" table.

Change the oil while the engine is still warm. The oil will flow freely and carry away more impurities. Make sure the engine is level when filling, checking, and changing the oil.

Change the oil as follows:

1. Remove the oil drain plug and oil fill cap/dipstick. Be sure to allow ample time for complete drainage.

2. Reinstall the drain plug. Make sure it is tightened to 13.6 N-m (10 inch lbs.) torque.

3. Fill the crankcase, with new oil of the proper type, to the "F" mark on the dipstick. Always check the level with the dipstick before adding more oil.

4. Reinstall the oil fill cap/dipstick.

➡**To prevent extensive engine wear or damage, always maintain the proper oil level in the crankcase. Never operate the engine with the oil level below the "L" mark over the "F" mark on the dipstick.**

Changing Oil Filter

Replace the oil filter after every oil change (every 200 hours of operation).

1. Drain the oil from the engine crankcase.

2. Allow the oil filter to drain.

3. Remove the old filter and wipe off the filter adapter.

4. Apply a thin coat of new oil to the rubber gasket on the replacement oil filter.

5. Install the replacement oil filter to the filter adapter. Turn the filter clockwise until the rubber gasket contacts the filter adapter, then tighten the filter an additional 1/2 turn.

6. Reinstall the drain plug. Torque the drain plug to 7.3/9.0 N-m (65/80 inch lbs.).

7. Fill the crankcase with new oil as instructed under "Change Oil." Add an additional 0.24 L (1/2 pint) of oil for the filter capacity.

8. Start the engine and check for oil leaks. Correct any leaks before placing the engine into service.

Oil Pump

SERVICE

The oil pump rotors can be serviced without removing the closure plate. Remove the oil pump cover on the PTO side of closure plate to service the rotors.

The closure plate must be removed to service the oil pickup and oil pressure relief valve.

Oil Sentry Oil Pressure Monitor

Some engines are equipped with an optional Oil Sentry oil pressure monitor. Oil Sentry will either stop the engine or activate a "low oil" warning light, if the pressure gets low. Actual Oil Sentry use will depend on the engine application.

The pressure switch is designed to break contact as the oil pressure increases and make contact as the oil pressure decreases. At oil pressure above approx. 3.0 to 5.0 psi., the switch contacts open. At oil pressures below approx. 3.0 to 5.0 psi., the switch contact close.

On stationary or unattended applications (pumps, generators, etc.), the pressure switch can be used the activate a "low oil" warning light.

➡**Oil sentry is not a substitute for checking the oil level BEFORE EACH USE. Make sure the oil level is maintained up to the "F" mark on the dipstick.**

INSTALLATION

The pressure switch is installed in the oil filter adapter in one of the main oil galleries of the closure plate. On engines not equipped with Oil Sentry, the installation hole is sealed with a 1/8-27 N.P.T.F. pipe plug.

To install the Oil Sentry switch to the oil filter adapter of closure plate:

1. Apply Loctite® #592 pipe sealant with Teflon® (or equivalent) to the threads of the switch.

2. Install the switch into the tapped hole in oil filter adapter. Torque the switch to 7/9 N-m (70 inch lbs.).

TESTING

The Oil Sentry pressure monitor is a normally closed type switch. It is calibrated to open (break contact) with increasing pressure and close (make contact) with decreasing pressure within the range of 3.0/5.0 psi.

Compressed air, a pressure regulator, pressure gauge and a continuity tester are required to test the switch.

1. Connect the continuity tester across the blade terminal and the metal case of switch. With 0 psi pressure applied to the switch, the tester should indicate continuity (switch closed).

2. Gradually increase the pressure to thew switch. The tester should indicate a change to no continuity (switch open) as the pressure increases through the range of 3.0/5.0 psi. The switch should remain open as the pressure is increased to 90 psi maximum.

3. Gradually decrease the pressure to the switch. The tester should indicate a change to continuity (switch closed) as the pressure decreases through the range of 3.0/5.0 psi; approaching 0 psi. If the switch does not operate as specified, replace the switch.

Retractable Starter

✳✳CAUTION

The spring is under tension! Retractable starters contain a powerful, flat wire recoil spring that is under tension. Do not remove the center screw from the starter until the spring is released. Removing the center screw before releasing spring tension, or improper starter disassembly, can cause the sudden and potentially dangerous release of the spring.

Always wear safety goggles when servicing retractable starters - full face protection is recommended.

To ensure personal safety and proper starter disassembly and reassembly, follow the procedures in this section carefully.

REMOVAL

▶ **See Figure 63**

1. remove the five hex flange screws securing the starter blower housing.

2. Remove the starter.

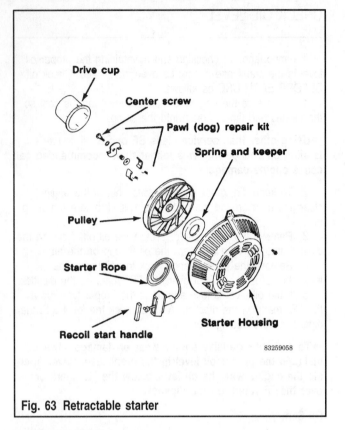

Fig. 63 Retractable starter

INSTALLATION

1. Install the retractable starter and five hex flange screws to blower housing. Leave the screws slightly loose.

2. Pull the starter handle out until the pawls engage in this position and tighten the screws securely.

ROPE REPLACEMENT

The rope can be replaced without complete starter disassembly.

1. Remove the starter from the engine blower housing.

2. Pull the rope out approx. 12″ and tie a temporary (slip) knot in it to keep it from retracting into the starter.

3. Remove the rope retainer from inside the starter handle. Untie the single knot and remove the rope retainer and handle.

4. Hold the pulley firmly and untie the slip knot. Allow the pulley to rotate slowly as the spring tension is released.

5. When all spring tension on the starter pulley is released, remove the rope from the pulley.

6. Tie a single knot in one end of the new rope.

7. Rotate the pulley counterclockwise (when viewed from pawl side of pulley) until the spring is tight. (Approximately. 6 full turns of pulley.)

8. Rotate the pulley counterclockwise until the rope hole in pulley is aligned with rope guide bushing of starter housing.

➡**Do not allow the pulley/spring to unwind. Enlist the aid of a helper if necessary, or use a C-clamp to hold the pulley in position.**

9. Insert the new rope through the rope hole in starter pulley and rope guide bushing of housing.

10. Tie a slip knot approx. 12" from the free end of rope. Hold the pulley firmly and allow it to rotate slowly until the slip knot reaches the guide bushing of housing.

11. Slip the handle and rope retainer onto the rope. Tie a single knot at the end of the rope. Install the rope retainer into the starter handle.

12. Untie the slip knot and pull on the handle until the rope is fully extended. Slowly retract the rope into the starter.   When the spring is properly tensioned, the rope will retract fully and the handle will stop against the starter housing.

PAWLS (DOGS) REPLACEMENT

The starter must be completely disassembled to replace the starter pawls. a pawl repair kit is available which includes the following components:

DISASSEMBLY

✳✳CAUTION

Spring Under Tension! Do not remove the center screw from starter until the spring tension is released. Removing the center screw before releasing spring tension, or improper starter disassembly, can cause the sudden and potentially dangerous release of the spring. Follow these instructions carefully to ensure personal safety and proper starter disassembly. Make sure adequate face protection is worn by all persons in the area.

1. Release spring tension and remove the handle and starter rope. (Refer to "Rope Replacement", steps 2 through 5 above.)

2. Remove the center screw, washer, and pawl retainer.

3. Remove the brake spring and brake washer.

4. Carefully note the positions of the pawls and pawl springs before removing them. Remove the pawls and pawl springs from the starter pulley.

5. Rotate the pulley clockwise 2 full turns. This will ensure the spring is disengaged from the starter housing.

6. Hold the pulley into the starter housing. Invert the pulley/housing so the pulley is away from your face, and away from others in the area.

7. Rotate the pulley slightly from side to side and carefully separate the pulley from the housing.

➡**If the pulley and the housing do not separate easily, the spring could be engaged in the starter housing, or there is still tension on the spring. Return the pulley to the housing and repeat step 5 before separating the pulley and housing.**

8. Note the position of the spring and keeper assembly in the pulley. Remove the spring and keeper assembly from the pulley as a package.

✳✳CAUTION

Do not remove the spring from the keeper. Severe personal injury could result from the sudden uncoiling of the spring.

INSPECTION AND SERVICE

1. Carefully inspect the rope, pawls, housing, center screw, and other components for wear or damage.

2. Replace all worn or damaged components.

3. Do not attempt to rewind a spring that has come out of the keeper. Order and install a new spring and keeper assembly.

4. Clean all old grease and dirt from the starter components. Generously lubricate the spring and center shaft with any commercially available bearing grease.

REASSEMBLY

1. Make sure the spring is well lubricated with grease. Place the spring and keeper assembly inside the pulley (with spring towards pulley).

2. Install the pulley with spring and keeper assembly into the starter housing. Make sure the pulley is fully seated against the starter housing. Do not wind the pulley and recoil spring at this time.

3. Install the pawls springs and pawls into the starter pulley.

4. Place the brake washer in the recess in starter pulley; over the center shaft.

5. Lubricate the brake spring sparingly with grease. Place the spring on the plain washer. (Make sure the threads in center shaft remain clean, dry, and free of grease and oil.)

6. Apply a small amount of Loctite® #271 to the threads of the center screw. Install the center screw, with washer and retainer, to the center shaft. Torque the screw 7.4/8.5 N-m (65/75 inch lbs.).

7. Tension the spring and install the rope and handle as instructed in steps 6 through 12 under "Rope Replacement" above.

8. Install the starter to the engine blower housing.

Spark Plug

Engine misfire or starting problems are often caused by a spark plug that is in poor condition or with an improper gap setting.

This engine is equipped with the following spark plug:
Type: Champion RC12YC (or equivalent)
Gap: 1.02mm (0.040 in.)
Thread Size: 14mm
Reach: 19.1mm (3/4 in.)
Hex Size: 15.9mm (5/8 in)

REMOVAL & INSTALLATION

▶ **See Figures 64, 65, 66, 67 and 68**

Every 100 hours of operation, remove the spark plug, check its condition, and reset the gap or replace with a new plug as necessary.

1. Before removing the spark plug, clean the area around the base of the plug to keep dirt and debris out of the engine.

2. Remove the plug and check its condition. Replace the plug if worn or reuse is questionable.

➡**Do not clean the spark plug in a machine using abrasive grit. Some grit could remain in the spark plug and enter the engine causing extensive wear and damage.**

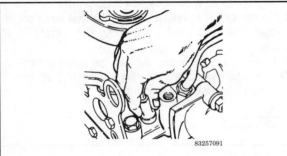

Fig. 64 Twist and pull on the boot, never on the plug wire

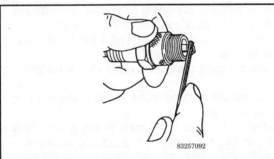

Fig. 65 Plugs that are in good condition can be filed and reused

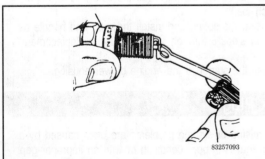

Fig. 66 Adjust the electrode gap by bending the side electrode

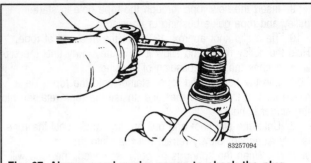

Fig. 67 Always used a wire gauge to check the plug gap

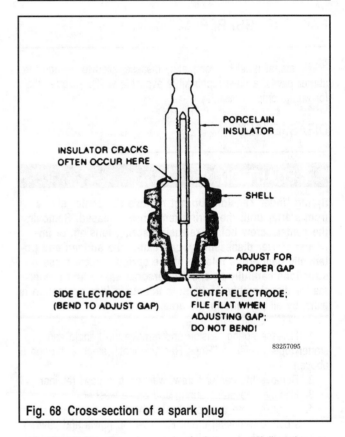

Fig. 68 Cross-section of a spark plug

3. Check the gap using a wire feeler gauge. Adjust the gap to 1.02 mm (0.040 in) by carefully bending the ground electrode.

4. Reinstall the spark plug into the cylinder head. Torque the spark plug to 38.0/43.4 N-m (28/32 ft. lbs.).

INSPECTION

Inspect the spark plug as soon as it is removed from the cylinder head. The deposits on the tip are an indication of the general condition of the piston rings, valves, and carburetors.

Battery

▸ **See Figure 69**

A 12-volt with a rating of approximately 32-amp hours/250 cold cranking amps, is normally used. Refer to the operating instructions of the equipment this engine powers for specific information.

If the battery charge is not sufficient to crank the engine, recharge the battery.

➡**Do not attempt to "jump start" the engine with another battery. Starting the engine with batteries larger than those recommended can burn out of the starter motor.**

BATTERY CHARGING

✳✳CAUTION

Batteries contain sulfuric acid. To prevent acid burns, avoid contact with skin, eyes, and clothing. Batteries produce explosive hydrogen gas while being charged.   To prevent a fire or explosion, charge batteries only in well ventilated areas. Keep sparks, open flames, and other sources of ignition away from the battery at all times.   Keep batteries out of the reach of children. Remove all jewelry when servicing batteries.   Before disconnecting the negative (-) ground cable, make sure all switches are OFF. If ON, a spark will occur at the ground cable terminal which could cause an explosion if hydrogen gas or gasoline vapors are present.

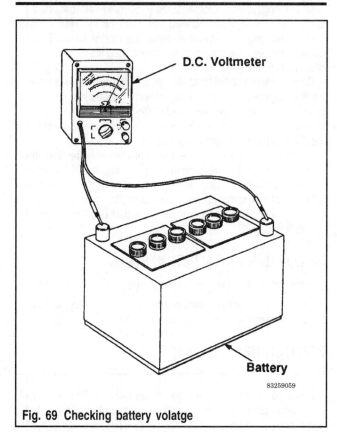

D.C. Voltmeter

Battery

83259059

Fig. 69 Checking battery volatge

BATTERY MAINTENANCE

1. Regularly check the level of electrolyte. Add distilled water as necessary to maintain the recommended level.

➡**Do not overfill the battery. Poor performance or early failure due to loss of electrolyte will result.**

2. Keep the cables, terminals, and external surfaces of battery clean. A build-up of corrosive acid or grime on the external surfaces can self-discharge the battery. Self-discharging happens rapidly when moisture is present.

3. Wash the cables, terminals, and external surfaces with a baking soda and water solution. Rinse thoroughly with clear water.

➡**Do not allow the baking soda solution to enter the cells as this will destroy the electrolyte.**

Battery Test

Test the battery voltage by connecting D.C. voltmeter across the battery terminals - crank the engine. If the battery drops below 9 volts while cranking, the battery is discharged or faulty.

Electronic Magneto Ignition System

▸ **See Figure 70**

These engines are equipped with a dependable electronic magneto ignition system. The system consists of the following components:

• A magnet assembly which is permanently affixed to the flywheel.

• An electronic magneto ignition module which mounts on the engine crankcase.

• A kill switch (or key switch) which grounds the module to stop the engine.

• A spark plug.

OPERATION

As the flywheel rotates and the magnet assembly moves past the ignition module, a low voltage is induced in the primary windings of the module. When the primary voltage id precisely at its peak, the module induces a high voltage in its secondary windings. This high voltage creates a spark at the tip of the spark plug. This spark ignites the fuel-air mixture in the combustion chamber.

The timing of the spark is automatically controlled by the module. Therefore, other than periodically checking/replacing the spark plug, no maintenance, timing, or adjustments are necessary or possible with this system.

➡**Use a low-voltage (2 volts or less) ohmmeter when ohmmeter is required. Always zero ohmmeter on each scale before testing to ensure accurate readings.**

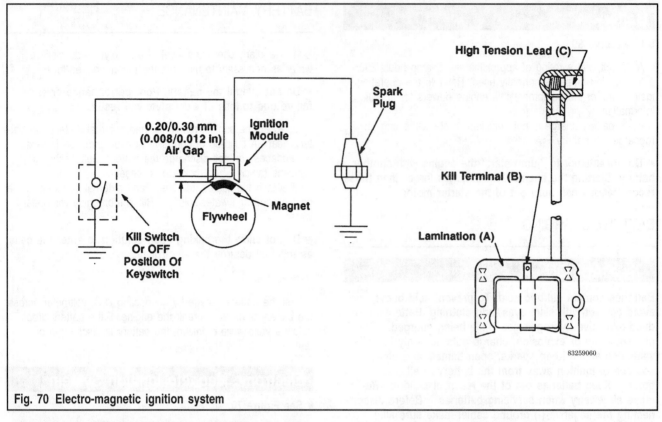

Fig. 70 Electro-magnetic ignition system

Battery Charging Systems

▶ **See Figures 71 and 72**

This engine is equipped with a 15 Amp regulated battery charging system.

➡ **Observe the following guidelines to prevent damage to the electrical system and components.**

1. Make sure the battery polarity is correct. A negative (-) ground system is used.

2. Disconnect the rectifier-regulator leads and/or wiring harness plug before doing electric welding on the equipment powered by engine. Also disconnect other electrical accessories in common ground with the engine.

3. Prevent the stator (AC) leads from touching or shorting while the engine is running. This could damage the stator.

Bendix Drive Electric Starter

▶ **See Figure 73**

• Do not crank the engine continuously for more than 10 seconds at a time. If the engine does not start, allow a 60-second cool-down period between starting attempts. Failure to follow these guidelines can burn out the starter motor.

• If the engine develops sufficient speed to disengage the starter but does not keep running (a false start), the engine rotation must be allowed to come to a complete stop before attempting to restart the engine. If the starter is engaged while the flywheel is rotating, the starter pinion and flywheel ring gear may clash, resulting in damage to the starter.

• If the starter does not crank the engine, shut off the starter immediately. Do not make further attempts to start the engine until the condition is corrected. Do not attempt to jump start the engine with another battery. Starting with batteries larger than those recommended can burn out the starter motor.

• Do not drop the starter or strike the starter frame. Doing so can damage the ceramic permanent magnets inside the starter frame.

When power is applied to the starter, the armature rotates. As the armature rotates, the drive pinion moves out on the splined drive shaft and into mesh with the flywheel ring gear. When the pinion reaches the end of the drive shaft, it rotates the flywheel and "cranks" the engine.

When the engine starts, the flywheel rotates faster than the starter armature and drive pinion. This moves the drive pinion out of mesh with the ring gear and into the retracted position. When power is removed from the starter, the armature stops and the drive pinion is held in the retracted position by the anti-drift spring.

REMOVAL & INSTALLATION

Refer to the Engine Disassembly and Reassembly sections for starter removal and installation procedures.

STARTER DRIVE SERVICE

Every 500 hours of operation (or annually, whichever occurs first), clean and lubricate the splines on the starter drive shaft. If the drive pinion is worn, or has chipped or broken teeth, it must be replaced.

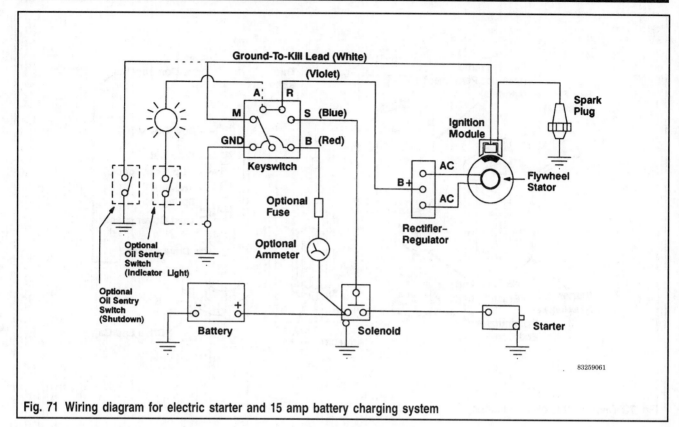

Fig. 71 Wiring diagram for electric starter and 15 amp battery charging system

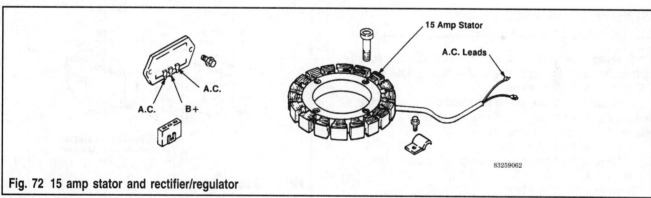

Fig. 72 15 amp stator and rectifier/regulator

It is not necessary to completely disassemble the starter to service the drive components. Service the drive as follows:

1. Remove the starter from the engine.
2. remove the dust cover.
3. Hold the drive pinion in a vice with soft jaws when removing and installing the stop nut. The armature will rotate with the nut until the drive pinion stops against internal spacers.

➡**Do not overtighten the vice as this can distort the drive pinion.**

4. Remove the stop nut, stop gear spacer, anti-drift spring, dust cover spacer, and drive pinion.

5. Clean the splines on drive shaft thoroughly with solvent. Dry the splines thoroughly.

6. Apply a small amount of Kohler electric starter drive lubricant to the splines.

➡**Kohler electric starter drive lubricant must be used on all Kohler electric starter drives. The use of other lubricants can cause the drive pinion to stick or bind.**

7. Apply a small amount of Loctite® no. 271 to the stop nut threads.

8. Install the drive pinion, dust cover spacer, anti-drift spring, stop gear spacer, and stop nut. Torque the stop nut to 17.0/19.2 N-m (135 inch lbs.).

9. Install the dust cover.

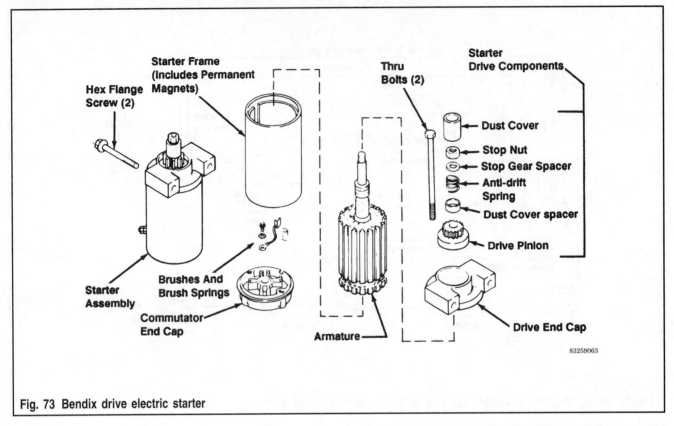

Fig. 73 Bendix drive electric starter

STARTER DISASSEMBLY

1. Remove the dust cover, stop nut, stop gear spacer, anti-drift spring, dust cover spacer, and drive pinion. Refer to "Starter Drive Service" above.

2. Scribe a small line on the drive end cap, opposite the line on the starter frame. These lines will service as match marks when reassembling the starter.

3. Remove the through bolts.

4. Remove the commutator end cap with brushes and brush springs.

5. remove the drive end cap.

6. remove the armature and thrust washer from inside the starter frame.

Brush Replacement

▶ **See Figure 74**

1. Remove the brush springs from the pockets in brush holder.

2. Remove the self-tapping screws, negative (-) brushes, and plastic brush holder.

3. Remove the hex flange nut and fiber washer from the stud terminal. Remove the stud terminal with positive (+) brushes and plastic insulating bushing from the end cap.

4. Reinstall the insulating bushing to the new stud terminal with positive (+) brushes. Install the stud terminal with bushing into the commutator end cap. Secure the stud with the fiber washer and hex flange screw.

5. Install the brush holder, new negative (-) brushes, and self-tapping screws.

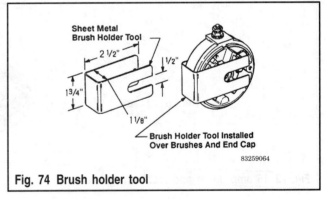

Fig. 74 Brush holder tool

6. install the brush springs and brushes into the pockets in brush holder. Make sure the chamfered sides of brushes are away from the brush springs.

➡**Use a brush holder tool to keep the brushes in the pockets. A brush holder tool can easily be made from thin sheet metal.**

Commutator Service

Clean the commutator with a coarse, lint free cloth. Do not use emery cloth.

If the commutator is badly worn or grooved, turn it down on a lathe or replace the armature.

STARTER REASSEMBLY

1. Place the thrust washer over the drive shaft of armature.

2. Insert the armature into the starter frame. Make sure the magnets are closer to the drive shaft end of armature. The magnets will hold the armature inside the frame.

3. Install the drive end cap over the drive shaft. Make sure the match marks on the end cap and starter frame are aligned.

4. Install the brush holder tool to keep the brushes in the pockets of the commutator end cap.

5. Align the match marks on the commutator end cap and starter frame. Hold the drive end and commutator end caps firmly to the starter frame. Remove the brush holder tool.

6. Install the through bolts and tighten securely.

7. Lubricate the drive shaft with Kohler electric starter drive lubricant. Install the drive pinion, dust cover spacer, anti-drift spring, stop gear spacer, stop nut, and dust cover. Refer to "Starter Drive Service" above.

Solenoid Shift Electric Starter

▶ **See Figure 75**

When power is applied to the starter the electric solenoid moves the drive pinion out onto the drive shaft and into mesh with the flywheel ring gear. When the pinion reaches the end of the drive shaft it rotates the flywheel and cranks the engine.

When the engine starts and the start switch is released the starter solenoid is deactivated, the drive lever moves back, and the drive pinion moves out of mesh with the ring gear into the retracted position.

STARTER REMOVAL & INSTALLATION

Refer to the Engine Disassembly and Assembly sections for starter removal and installation procedures.

STARTER DISASSEMBLY

1. Remove clip.

2. Remove cap screws and solenoid. Scribe alignment marks on caps and frame to aid assembly.

3. Remove through bolts, drive end cap, commutator end cap, and frame.

4. Remove drive lever.

5. Remove thrust washer and retainer to remove drive pinion from shaft.

Brush Replacement

See the procedure as explained in the Bendix Drive section, above.

Commutator Service

See the procedure as explained in the Bendix Drive section, above.

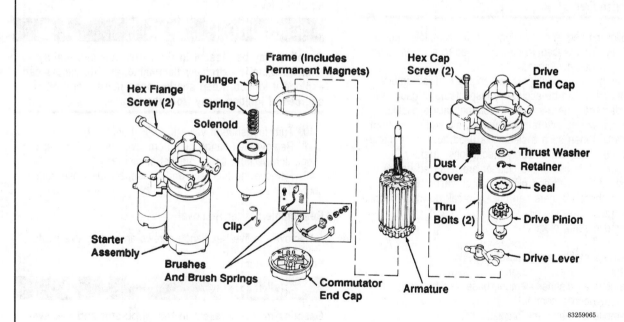

Fig. 75 Solenoid shift electric starter

STARTER REASSEMBLY

1. Slide frame over armature and place commutator end cap in position. Hold in position temporarily with tape.

➡ **Be sure alignment marks on caps and frame are in proper position.**

2. Place drive pinion (with seal), thrust washer and retainer on drive shaft.
3. Place lever in position on drive shaft.
4. Place solenoid plunger on drive lever and position drive end cap over drive shaft. (Be sure the rubber dust cover is in place at the drive lever.)
5. Fasten the end caps with the through bolts.
6. Place the spring in the solenoid and fasten solenoid to drive end cap using hex cap screws.
7. Replace the clip.

Engine Mechanical

ENGINE DISASSEMBLY

❋❋CAUTION

Before servicing the engine or equipment, always disconnect the spark plug lead to prevent the engine from starting accidentally. Ground the lead to prevent sparks which could cause fires.

The following sequence is suggested for complete engine disassembly. This procedure can be varied to accommodate options or special equipment.

Clean all parts thoroughly as the engine is disassembled. Only clean parts can be accurately inspected and gauged for wear or damage. There are many commercially available cleaners that will quickly remove grease, oil, and grime from engine parts. When such a cleaner is used, follow the manufacturer's instructions and safety precautions carefully.

Make sure all traces of the cleaner are removed before the engine is reassembled and placed into operation. Even small amounts of these cleaners can quickly break down the lubricating properties of engine oil.

1. Remove spark plug.
2. Drain oil.
3. Remove muffler and bracket.
4. Remove air cleaner cover.
5. Remove air cleaner element, base, and breather hose.
6. Remove choke control.
7. Remove fuel tank and bracket.
8. Remove retractable starter.
9. Remove fuel pump.
10. Remove starter cover - electrical starter.
11. Remove rectifier-regulator.
12. Remove Oil Sentry.
13. Remove throttle control bracket.
14. Remove carburetor,
15. Remove valve cover.
16. Remove cylinder head baffle.
17. Remove blower housing and baffles.
18. Remove ignition module.
19. Remove fuel line.
20. Remove cylinder head push rods/gasket.
21. Remove drive cup, grass screen, flywheel and fan.
22. Remove stator and wiring harness.
23. Remove oil fill tube - if necessary.
24. Remove closure plate.
25. Remove camshaft and hydraulic lifters.
26. Remove balance shaft.
27. Remove connecting rod.
28. Remove piston.
29. Remove crankshaft.
30. Remove main bearing.
31. Remove governor gear.
32. Remove governor cross shaft seal.
33. Disconnect spark plug lead.

➡ **Pull on boot only, to prevent damage to spark plug lead.**

34. Drain the oil and remove the filter.
35. Remove the muffler
36. Remove the knob and air cleaner cover.
37. Remove the wing nut, washer, element cover, element and precleaner.
38. Remove the hex flange nuts from the intake studs, and the air cleaner base and gasket from the studs.
39. Loosen the hose clamp and disconnect the breather hose from the rocker arm cover. Remove the air cleaner base from the studs and disconnect choke linkage from the carburetor choke lever.

❋❋CAUTION

Gasoline may be present in the carburetor and fuel system. Gasoline is extremely flammable, and its vapors can explode if ignited. Keep sparks, open flames, and other sources of ignition away from the engine.

40. Turn fuel shut-off valve to OFF (horizontal) position.
41. Remove hex flange nuts from lower bracket and hex flange screws from upper bracket of fuel tank.
42. Remove the fuel tank and disconnect fuel hose from shut-off valve.

Retractable Starter Removal

Remove the five hex flange screws and retractable starter.

Fuel Pump Removal

❋❋CAUTION

Gasoline may be present in the carburetor and fuel system. Gasoline is extremely flammable, and it's vapors can explode if ignited. Keep sparks, open flames, and other sources of ignition away from the engine.

1. Disconnect the fuel line from the outlet and inlet fittings of the fuel pump.
2. Remove the two hex flange screws, fuel pump, and gasket.

Electric Starter Removal

1. Disconnect the lead from the stud terminal. Disconnect both leads on Solenoid Shift Starter.
2. Remove the two hex flange screws and starter cover.
3. Remove the starter assembly and spacers from the studs.

Rectifier-Regulator Removal

1. Remove the wire connector from the rectifier-regulator.
2. Remove the two hex flange screws and rectifier-regulator.

Oil Sentry Removal

1. Disconnect the lead from the Oil Sentry switch.
2. Remove Oil Sentry switch from the oil filter adapter.

Throttle Control Bracket Removal

1. Remove two hex flange screws from throttle control bracket.
2. Remove governor lever spring from throttle control bracket.

Carburetor Removal

✳✳CAUTION

Gasoline may be present in the carburetor and fuel system. Gasoline is extremely flammable, and it's vapors can explode if ignited. Keep sparks, open flames, and other sources of ignition away from the engine.

1. Remove fuel line from carburetor inlet fitting.
2. Disconnect the throttle linkage from the bushing in carburetor governor lever.
3. Remove carburetor and gasket from intake studs.

Valve Cover Removal

Remove the five hex flange cover screws and valve cover from the cylinder head assembly.

➡**The valve cover is sealed to the cylinder head using RTV silicone sealant. When removing valve cover, use care not to damage the gasket surfaces of cover and cylinder head.**

Cylinder Head Baffle Removal

Remove the hex flange screws securing the cylinder head baffle to the cylinder head. Remove the baffle.

Blower Housing and Baffles Removal

Remove the hex flange screws from blower housing and baffles. Disconnect the wire harness from the key switch, if equipped. Remove the blower housing and baffles.

Ignition Module Removal

1. Disconnect the kill lead from the ignition module terminal.
2. Rotate flywheel magnet away from ignition module.

3. Remove the two hex flange screws and ignition module.

Fuel Line Removal

Remove the hex flange screw, clip, and fuel line.

Cylinder Head Removal

Remove the hex flange screws, spacer (from the screw by the exhaust port), cylinder head, push rods, and cylinder head gasket.

Cylinder Head Disassembly

1. Remove the spark plug.
2. Remove the hex flange screw, breather reed retainer, and breather reed.
3. Remove the rocker shaft (from the breather side of head), and rocker arms.
4. Remove the valves:
    a. Compress the valve springs using a valve spring compressor.
    b. Remove the keepers, valve spring caps, valve springs, exhaust valve rotator, intake valve spring seat, and intake valve stem seal.
5. Remove the two hex cap screws and rocker bridge.

Drive Cup, Grass Screen, Flywheel, and Fand Removal

➡**Always use the flywheel strap wrench to hold the flywheel when loosening or tightening the flywheel and fan retaining fasteners. Do not use any type of bar or wedge between the fins of cooling fan as the fins could become cracked or damaged.**

Always use a puller to remove the flywheel from the crankshaft. Do not strike the crankshaft or flywheel, as these parts could become cracked or damaged.

1. Remove the hex flange screw, plain washer, and driver cup.
2. Unsnap and remove the grass screen from fan.
3. Remove the flywheel from the crankshaft using a puller.
4. Remove the four hex flange screws and fan from flywheel.

Stator and Wiring Harness Removal

1. Remove the stator leads from connector body.
2. Remove the hex flange screw and clip securing the stator leads to the crankcase.
3. Remove the hex flange screw and clip securing the kill lead to the crankcase. Remove the four hex socket head screws and stator.

Closure Plate Removal

1. Remove the twelve hex flange screws securing the closure plate to the crankcase.
2. Locate the splitting notches in the seam of the closure plate and crankcase. Pry the closure plate from the crankcase using a large flat-blade screwdriver.

➡**Insert the screwdriver only in the splitting notches. Do not pry on the gasket surfaces of the closure plate or crankcase as this can cause leaks.**

Oil Pickup, Oil Pressure Relief Valve, Oil Pump, and Oil Seal

1. Remove the oil seal from the closure plate.
2. Remove the hex flange screw, clip, oil pickup, and O-ring seal.
3. Remove the hex socket screw, oil pressure relief bracket, relief valve body, piston, and spring.
4. Remove the three hex flange screws, oil pump cover, O-ring, and oil pump rotors.

Camshaft and Hydraulic Lifters Removal

1. Remove the camshaft and shim.
2. Mark or identify the hydraulic lifters as either intake or exhaust. Remove the lifters from the crankcase.

➡**The intake hydraulic lifter is farthest from the crankcase gasket surface. The exhaust hydraulic lifter is nearest to the crankcase gasket surface.**

Balance Shaft Removal

Remove the balance shaft from the crankcase.

Connecting Rod & Piston Removal

1. Remove the two hex flange screws and connecting rod cap.

➡**If a carbon ridge is present at the top of the bore, use a ridge reamer tool to remove it before attempting to remove the piston.**

2. Carefully push the connecting rod and the piston away from the crankshaft and out of the cylinder bore.
3. Remove the wrist pin retainer and wrist pin. Separate the piston from the connecting rod.
4. Remove the top and center compression rings using a ring expander tool.
5. Remove the oil control ring rails, then remove the rails spacer.

Crankshaft Removal

1. Remove the woodruff key from the flywheel taper end of crankshaft.
2. Remove the crankshaft from the crankcase.

Flywheel End Oil Seal and Bearing Removal

1. Remove the oil seal from crankcase.
2. Remove the bearing from the crankcase using handle #NU-4747 and bearing remover #KO-1029.

Governor Cross Shaft and Governor Gear Removal

1. Remove the hitch in and plain washer from governor cross shaft.
2. Remove the cross shaft and plain washer from the crankcase.
3. Remove the governor cross shaft oil seal from the crankcase.

4. If necessary, remove the governor gear and regulating pin.

➡**The governor gear is held onto the governor gear shaft by small molded tabs in the gear. When the gear is removed from the shaft these tabs are destroyed. This will require replacement of the gear; therefore, remove the gear only if absolutely necessary (such as when reboring, doing major engine rebuilding, etc.).**

ENGINE COMPONENT OVERHAUL

Clean all parts thoroughly. Only clean parts can be accurately inspected and gauged for wear or damage. There are many commercially available cleaners that will quickly remove grease, oil, and grime from engine parts. When such a cleaner is used, follow the manufacturer's instructions and safety precautions carefully.

Make sure all traces of the cleaner are removed before the engine is reassembled and placed into operation. Even small amounts of these cleaners can quickly break down the lubricating properties of engine oil.

Camshaft

Inspect the gear teeth of the camshaft. If the teeth are badly worn, chipped, or some are missing, replacement of the camshaft will be necessary.

Crankshaft

▶ **See Figure 76**

Inspect the gear teeth of the crankshaft. If the teeth are badly worn, chipped or some are missing, replacement of the crankshaft will be necessary.

Inspect the crankshaft bearings for scoring, grooving, etc. Do not replace bearings unless they show signs of damage or are out of running clearance specifications. If the crankshaft turns easily and noiselessly, and there is no evidence of scoring, grooving, etc., on the races of bearing surfaces, the bearings can be reused.

Inspect the crankshaft keyways. If worn or chipped, replacement of the crankshaft will be necessary.

Inspect the crankpin for score marks or metallic pickup. Slight score marks can be cleaned with crocus cloth soaked in oil. If wear limits, as stated in "Specifications and Tolerances" are exceeded, it will be necessary to either replace the crankshaft or regrind the crankpin to 0.25 mm (0.010 in) undersize.

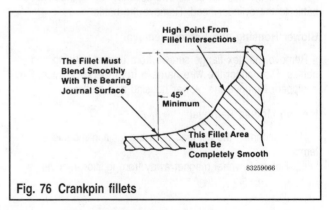

High Point From Fillet Intersections

The Fillet Must Blend Smoothly With The Bearing Journal Surface

45° Minimum

This Fillet Area Must Be Completely Smooth

83259066

Fig. 76 Crankpin fillets

If reground, a 0.25 mm (0.010 in) undersize connecting rod (big end) must then be used to achieve proper running clearance. Measure the crankpin for size, taper, and out-of-round.

➡**If the crankpin is reground, visually check to ensure that the fillet blends smoothly with the crankpin surface.**

When regrinding a crankshaft, grinding stone deposits can get caught in oil passages which could cause severe engine damage. Remove the sealing plug each time the crankshaft is ground to provide easy access for cleaning any grinding deposits that may collect in the oil passages.

Crankcase

Check all gasket surfaces to make sure they are free of gasket fragments. Gasket surfaces must also be free of deep scratches or nicks.

Check a the cylinder bore wall for scoring. In severe cases, unburned fuel can cause scuffing and scoring of the cylinder wall. It washes the necessary lubricating oils off the piston and cylinder wall. As raw fuel seeps down the cylinder wall, the piston rings make metal to metal contact with the wall. Scoring of the cylinder wall can also be caused by localized hot spots resulting from blocked cooling fins or from inadequate or contaminated lubrication.

If the cylinder bore is badly scored, excessively worn, tapered, or out of round, resizing is necessary. Use an inside micrometer to determine amount of wear then select the nearest suitable oversize of either 0.25 mm (0.010 in) or 0.50 mm (0.020 in). Resizing to one of these oversizes will allow usage of the available oversize piston and ring assemblies. Initially, resize using a boring bar, then use the following procedures for honing the cylinder.

HONING

While most commercially available cylinder hones can be used with either portable drills or drill presses, the use of a low speed drill press is preferred as it facilitates more accurate alignment of the bore in relation to the crankshaft crossbore. Honing is best accomplished at a drill speed of about 2500 RPM and 60 strokes per minute. After installing coarse stones in hone, proceed as follows:

1. Lower hone into bore and after centering, adjust so that the stones are in contact with the cylinder wall. Use of a commercial cutting-cooling agent is recommended.

2. With the lower edge of each stone positioned even with the lowest edge of the bore, start drill and honing process. Move the hone up and down while resizing to prevent the formation of cutting ridges. Check the size frequently.

➡**Measure the piston diameter and resize the bore to the piston to obtain the specified running clearances. Keep in mind the temperatures caused by honing may cause inaccurate measurements. make sure the bore is cool when measuring.**

3. When the bore is within 0.064 mm (0.025 in) of desired size, remove the coarse stones and replace with burnished stones. Continue with the burnishing stones until within 0.013 mm (0.0005 in) of desired size, and then use finish stones (220-280 grit) and polish to final size. A crosshatch should be observed if honing is done correctly. The crosshatch should intersect at approximately 23-33 degrees off the horizontal. Too

flat of an angle could cause the rings to skip and wear excessively, to steep of an angle will result in high oil consumption.

4. After resizing, check the bore for roundness, taper, and size. Use an inside micrometer, telescoping gauge, or bore gauge to take measurements. The measurements should be taken at three locations in the cylinder - at the top, middle, and bottom. Two measurements should be taken (perpendicular to each other) at each of the three locations.

MEASURING PISTON-TO-BORE CLEARANCE

▶ **See Figure 77**

Before installing the piston into the cylinder bore, it is necessary that the clearance be accurately checked. This step is often overlooked, and if the clearances are not within specifications, engine failure will usually result.

➡**Do not use a feeler gauge to measure piston-to-bore clearance - it will yield inaccurate measurements. Always use a micrometer.**

Use the following procedure to accurately measure the piston-to-bore clearance:

1. Use a micrometer and measure the diameter of the piston 6 mm (.024 in) above the bottom of the piston skirt and perpendicular to the piston pin.

2. Use an inside micrometer, telescoping gauge, or bore gauge and measure the cylinder bore. Take the measurement approximately 63.5 mm (2.5 in) below the top of the bore and perpendicular to the piston pin.

3. Piston-to-bore clearance is the difference between the bore diameter (step 2 minus step 1).

Flywheel

Inspect the flywheel for cracks, and the flywheel keyway for damage. Replace flywheel if cracked. Replace the flywheel, the crankshaft, and the key if flywheel key is sheared or the keyway damaged.

Inspect the ring gear for cracks or damage. Kohler does not provide ring gears as a serviceable part. Replace the flywheel if the ring gear is damaged.

Cylinder Head and Baffles

Carefully inspect the valve mechanism parts. Inspect the valve springs and related hardware for excessive wear or distortion. Check the valves and valve seat area or inserts for evidence of deep pitting, cracks, or distortion. Check clearance of the valve stems in guides.

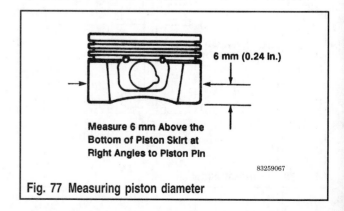

Measure 6 mm Above the
Bottom of Piston Skirt at
Right Angles to Piston Pin

6 mm (0.24 In.)

83259067

Fig. 77 Measuring piston diameter

Hard starting, or loss of power accompanied by high fuel consumption may be symptoms of faulty valves. Although these symptoms could also be attributed to worn rings, remove and check the valves first. After removal, clean the valve heads, faces, ad stems with a power wire brush. Then, carefully inspect each valve for defects such as warped head, excessive corrosion, or worn stem end. Replace valves found to be in bad condition. A normal valve and valves in bad condition are shown in the accompanying illustrations.

VALVE GUIDES

If a valve guide is worn beyond specifications, it will not guide the valve in a straight line. This may result id burnt valve faces or seats, loss of compression, and excessive oil consumption.

To check valve guide to valve stem clearance, thoroughly clean the valve guide and, using a split-ball gauge, measure the inside diameter. Then, using an outside micrometer, measure the diameter of the valve stem at several points in the stem where it moves in the valve guide. Use the largest stem diameter to calculate the clearance. If the clearance exceeds 7.134 mm (0.2809 in) on intake or 7.159 mm (0.2819 in) on exhaust valve, determine whether the valve stem or the guide is responsible for the excessive clearance.

If the valve stem diameter is within specifications, then recondition the valve guide.

The valve guides in the cylinder head are not removable. Use a 0.25 mm (0.010 in) O/S reamer. Tool no. KO-1026.

VALVE SEAT INSERTS
▶ **See Figure 78**

Intake valve seats are usually machined into the cylinder head, however, certain applications may specify hard alloy inserts. The valve seats are not replaceable. If the seats become badly pitted, cracked, or distorted, the inserts can be reconditioned.

Use a standard valve seat cutter and cut seat to dimensions shown.

LAPPING VALVES
▶ **See Figure 79**

Reground or new valves must be lapped in, to provide fit. Use a hand valve grinder with suction cup for final lapping. Lightly coat valve face with "fine" grade of grinding compound, then rotate valve on seat with grinder. Continue grinding until smooth surface is obtained on seat and on valve face. Thor-

oughly clean cylinder head in soap and hot water to remove all traces of grinding compound. After drying cylinder head, apply a light coating of SAE 10 oil to prevent rusting.

INTAKE VALVE STEM SEAL

These engines use valve stem seals on the intake valves. Always use a new seal when valves are removed from cylinder head. The seals should also be replaced if deteriorated or damaged in any way. Never reuse an old seal.

Pistons and Rings
▶ **See Figures 80 and 81**

Scuffing and scoring of pistons and cylinder walls occurs when internal temperature approach the welding point of the piston. Temperatures high enough to do this are created by friction, which is usually attributed to improper lubrication, and/or overheating of the engine.

Normally, very little wear takes place in the piston boss-piston pin area. If the original piston and connecting rod can be reused after new rings are installed, the original pin can also be reused but new piston pin retainers are required. The piston pin is included as part of the piston assembly - if the pin boss in piston or the pin, are worn or damaged, a new piston assembly is required.

Ring failure is usually indicated by excessive oil consumption and blue exhaust smoke. When rings fail, oil is allowed to enter the combustion chamber where it is burned along with the fuel. High oil consumption can also occur when the piston ring end gap is incorrect because the ring cannot properly conform to the cylinder wall under this condition. Oil control is also lost when ring gaps are not staggered during installation.

When cylinder temperatures get too high, lacquer and varnish collect on pistons causing rings to stick which results in rapid wear. A worn ring usually takes on a shiny or bright appearance.

Scratches on rings and pistons are caused by abrasive materials such as carbon dirt, or pieces of hard metal.

Detonation damage occurs when a portion of the fuel charge ignites spontaneously from heat and pressure shortly after ignition. This creates two flame fronts which meet and explode to create extreme hammering pressures on a specific area of the piston. Detonation generally occurs from using fuels with too low of an octane rating.

Preignition or ignition of the fuel charge before the timed spark can cause damage similar to detonation. Preignition damage is often more sever than detonation damage - often a hole is quickly burned right through the piston dome. Preignition is caused by a hot spot in the combustion chamber from sources such as: glowing carbon deposits, blocked fins, improperly seated valve, or wrong plug.

Replacement pistons are available in STD bore size and in .025 mm (0.010 in), and 0.50 mm (0.20 in), oversizes. Replacement pistons include new piston ring sets and new piston pins.

Service replacement piston ring sets are also available separately for STD, 0.25 mm (0.010 in), and 0.50 mm (0.020 in), oversized pistons. Always use new piston rings when installing pistons. Never reuse old rings.

The cylinder bore must be deglazed before service ring sets ar used.

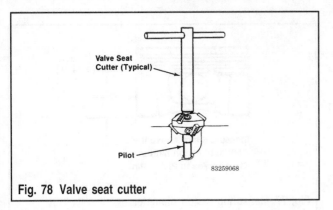

Valve Seat Cutter (Typical)

Pilot

83259068

Fig. 78 Valve seat cutter

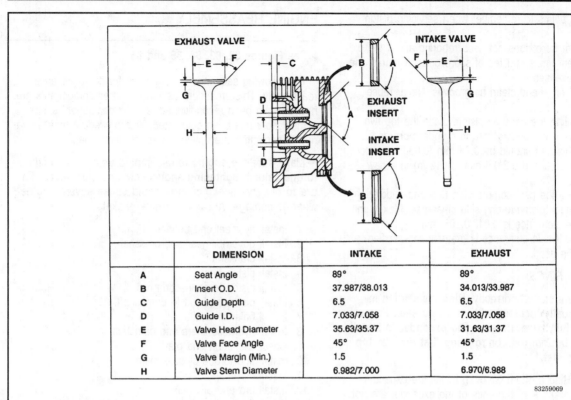

	DIMENSION	INTAKE	EXHAUST
A	Seat Angle	89°	89°
B	Insert O.D.	37.987/38.013	34.013/33.987
C	Guide Depth	6.5	6.5
D	Guide I.D.	7.033/7.058	7.033/7.058
E	Valve Head Diameter	35.63/35.37	31.63/31.37
F	Valve Face Angle	45°	45°
G	Valve Margin (Min.)	1.5	1.5
H	Valve Stem Diameter	6.982/7.000	6.970/6.988

83259069

Fig. 79 Valve details

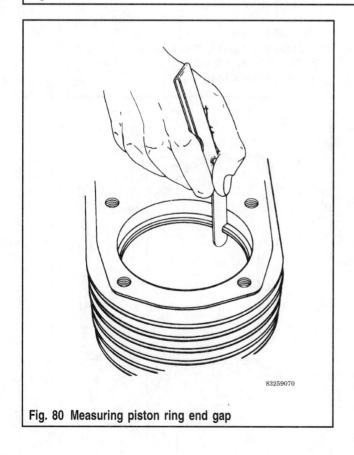

83259070

Fig. 80 Measuring piston ring end gap

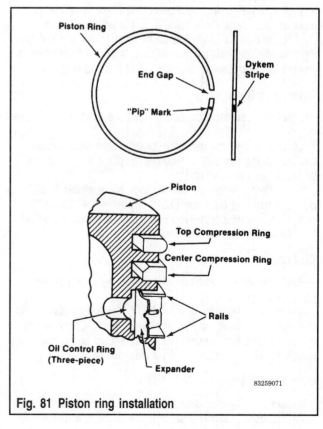

83259071

Fig. 81 Piston ring installation

Some important points to remember when servicing piston rings:

1. If the cylinder bore does not need reboring and if the old piston is within wear limits and fee of score or scuff marks, the old piston may be reused.

2. Remove old rings and clean up grooves. Never reuse old rings.

3. Before installing the rings on piston, place the top two rings, each in turn, in its running area in cylinder bore and check end gap. This gap should be 0.75 mm (0.030 in) max. in a used cylinder bore and 0.3/0.5 mm (0.012 in) in a new cylinder bore.

4. After installing the new compression (top and middle) rings on piston, check piston-to-ring side clearance. Maximum recommended side clearance is 0.040/0.105 mm (0.0016/0.0041 in). If side clearance is grater than specified, a new piston must be used.

INSTALL PISTON RINGS

➡Rings must be installed correctly. Ring installation instructions are usually included with new ring sets. Follow instructions carefully. Use a piston ring expander to install rings. Install the bottom (oil control) ring first and the top compression ring last.

1. Oil Control Ring (Bottom Groove): Install the expander and then the rail. Make sure the ends of the expander are not overlapped.

2. Compression Ring (Center Groove): Install the center ring using a piston ring installation tool. Make sure the "pip" mark is up and the PINK stripe is to the left of end gap.

3. Compression Ring (Top Groove): Instal the top ring using a piston ring installation tool. Make sure to "pip" mark is up and the BLUE stripe is to the left of the end gap.

Connecting Rods

Offset Stepped-cap Connecting Rods are used in all of these engines.

Check bearing area (big end) for excessive wear, score marks, running and side clearances. Replace rod and cap if scored or excessively worn.

Service replacement connecting rods are available in STD crankpin size and 0.25 mm (0.010 in.) undersize. The 0.25 mm (0.010 in) undersized rod can be identified by the drilled hole located in the lower end of the rod shank.

Oil Pump

Pump can be checked/replaced without removing closure plate.

Check oil pressure relief valve body, piston, and spring. Piston and body should be free of nicks or burrs. Check spring for ear or distortion. Spring free length should be approximately 0.992 in. Replace spring if distorted or worn.

Governor Gear

Inspect the governor gear teeth. Look for any evidence of worn, chipped, or cracked teeth. If one or more of these problems is noted, replace the governor gear.

The governor gear must be replaced once it is removed from the engine.

ENGINE REASSEMBLY

▶ **See Figures 82, 83, 84, 85 and 86**

The following sequence is suggested for complete engine reassembly. This procedure assumes that all components are new or have been reconditioned, and all component subassembly work has been completed. This procedure may be varied to accommodate options or special equipment.

➡**Make sure the engine is assembled using all specified torque values, tightening sequences, and clearances. Failure to observe specifications could cause severe engine wear or damage. Always use new gaskets.**

1. Install flywheel end bearing.
2. Install governor gear and crosshaft.
3. Install crankshaft.
4. Install piston rings.
5. Install piston to connecting rod.
6. Install piston and rod to crankshaft.
7. Install balance shaft.
8. Install hydraulic lifters and camshaft.
9. Check camshaft end play.
10. Install and torque closure plate.
11. Install oil pump.
12. Install pro end oil seal.
13. Install flywheel end oil seal.
14. Install stator and leads.
15. Install flywheel, grass screen and drive cup.
16. Install fuel line.
17. Install and adjust ignition module.
18. Assemble cylinder head.
19. Install cylinder head.

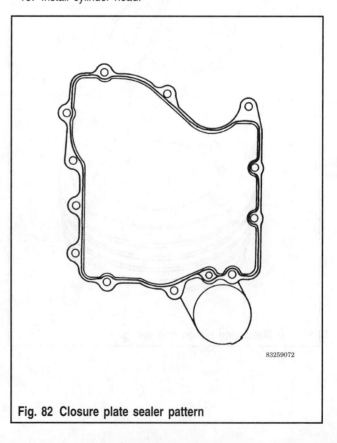

83259072

Fig. 82 Closure plate sealer pattern

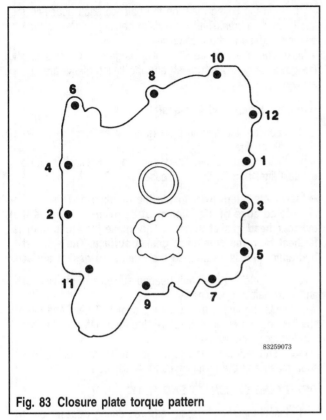

Fig. 83 Closure plate torque pattern

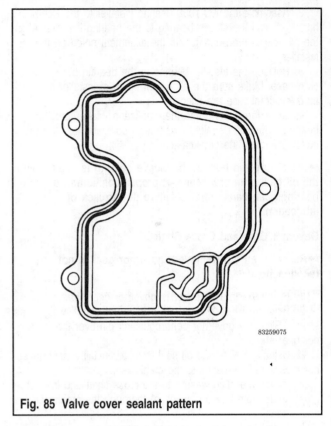

Fig. 85 Valve cover sealant pattern

Flywheel End Bearing

1. Mark the position of one of the crankcase bearing oil galleries on the crankcase.

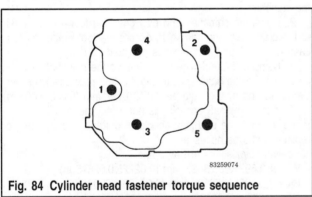

Fig. 84 Cylinder head fastener torque sequence

20. Install baffles and blower housing.
21. Install cylinder head baffle.
22. Install valve cover.
23. Install fuel pump.
24. Install electric starter and cover.
25. Install fuel tank.
26. Install rectifier-regulator.
27. Install carburetor.
28. Install and adjust governor arm.
29. Install throttle bracket.
30. Install choke and air cleaner base plate.
31. Install air cleaner element/precleaner and cover.
32. Install oil filter and Oil Sentry.
33. Install dipstick.
34. Install retractable starter.
35. Install muffler and bracket.

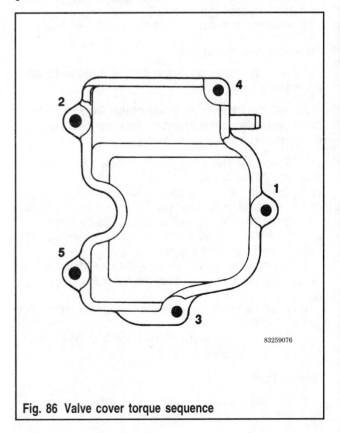

Fig. 86 Valve cover torque sequence

2. Assemble the KO-1028 bearing installer to the NU-4747 handle. Install the sleeve bearing to the bearing installer. Align the oil hole in the bearing with the alignment notch on the installer.

3. Position the installer/bearing to the bearing bore of crankcase. Make sure the alignment notch of installer and mark on crankcase are aligned.

4. Drive the bearing into the crankcase. Make sure the bearing is installed straight and true in bore and that the tool bottoms against the crankcase.

➡**Make sure the hole in the sleeve bearing is aligned with the oil gallery in crankcase. Improper positioning of the bearings can cause engine failure due to lack of lubrication.**

Governor Gear and Cross Shaft

➡**Reuse of an old (removed) governor gear is not recommended.**

1. Install the thrust washer to governor gear shaft.

2. Position the regulating pin to governor gear/flyweights as shown. Slide the governor gear/regulating pin over the governor gear shaft.

3. Using the KO-1030 oil seal installer, install a new governor cross shaft oil seal into the crankcase.

4. Install one plain washer to the cross shaft and insert the cross shaft (from inside crankcase) thorough the crankcase and oil seal.

5. Install one plain washer and hitch pin.

Crankshaft

1. Lubricate the flywheel end bearing surfaces of the crankshaft and crankcase with engine oil.

2. Insert the crankshaft through the flywheel end bearing.

Piston and Connecting Rod

1. Install the piston, connecting rod, piston pin, and piston pin retainers.

➡**Proper orientation of the piston/connecting rod inside the engine is extremely important. Improper orientation can cause extensive wear or damage.**

2. Stagger the piston rings in the grooves until the end gaps are 120 degrees apart.

3. Lubricate the cylinder bore, piston, and rings with engine oil. Compress the piston rings using a piston ring compressor.

4. Orient the "Fly" mark on piston towards the flywheel side of crankcase. Gently push the piston/connecting rod into bore. Do not pound on the piston.

5. Lubricate the crankshaft journal and connecting rod bearing surfaces with engine oil. Install the rod cap to connecting rod.

6. Install the hex flange screws and torque in increments to 22.6 N-m (200 inch lbs.).

7. Rotate the crankshaft until the piston is at the top dead center in the cylinder bore.

Balance Shaft

1. Lubricate the balance shaft bearing surfaces of crankshaft and balance shaft with engine oil.

2. Align the timing mark on the balance shaft gear and the larger gear on crankshaft. Lower the balance shaft into the bearing surface in the crankcase.

Make sure the balance shaft gear, large crankshaft gear and the governor gear teeth mesh and the timing marks are aligned.

Hydraulic Lifters and Camshaft

1. Lubricate the hydraulic lifters and lifter bores in crankcase with engine oil.

2. Install the hydraulic lifters into the appropriate intake or exhaust lifter bore in the crankcase.

➡**Install the lifters from inside the crankcase. The chamfered edge of the lifter must be inserted towards the cylinder head gasket surface. The intake hydraulic lifter is farthest from the crankcase gasket surface. The exhaust hydraulic lifter is nearest to the crankcase gasket surface.**

3. Lubricate the camshaft bearing surfaces of crankcase and camshaft with engine oil.

4. Align the timing marks on the camshaft gear and the smaller gear on crankshaft. Lower the camshaft into the bearing surface in crankcase.

Make sure the camshaft gear and smaller gear on crankshaft mesh and the timing marks are aligned.

DETERMINE CAMSHAFT END PLAY

1. Install the shim spacer, removed during disassembly, to the camshaft.

2. Install the camshaft end play checking tool no. KO-1031 to the crankcase and camshaft. secure the tool to the crankcase with the hex flange screws provided.

3. Using a flat feeler gauge, measure the camshaft end play between the shim spacer and the end play checking tool. Camshaft end play should be 0.076/0.127 mm (0.003/0.005 in).

4. If the camshaft end play is not within the specified range, remove the end play checking tool and add, remove, or replace shims as necessary.

Several color shims are available:
White: 0.69215/0.73025 mm (0.02725/0.02875 in)
Blue: 0.74295/0.78105 mm (0.02929/0.03075 in)
Red: 0.79375/0.83185 mm (0.03125/0.03275 in)
Yellow: 0.84455/0.88265 mm (0.03325/0.03475 in)
Green: 0.89535/0.99345 mm (0.03525/0.03675 in)
Gray: 0.94615/0.98425 mm (0.03725/0.03875 in)
Black: 0.99695/1.03505 mm (0.03925/0.04075 in)

5. Reinstall the end play checking tool and recheck end play.

6. Repeat steps 4 and 5 until the end play is within the specified range.

Oil Pressure Relief Valve

1. Place the relief valve body in the cavity of the closure plate.

2. Insert the piston and spring into the body.

3. Install the bracket and hex flange screw.

Oil Pickup

Install the oil pickup, O-ring, and hex flange screw.

➡**Lightly grease O-ring and install before oil pickup.**

Closure Plate to Crankshaft

RTV sealant is used as a gasket between the closure plate and crankcase. GE Silmate® type RTV-1473 or RTV-108 silicone sealant (or equivalent) is recommended.

1. Prepare the sealing surfaces of the crankcase and closure plate as directed by the sealant manufacture.

➡**Do not scrape the surfaces when cleaning as this will damage the surfaces. This could result in leaks. The use of a gasket removing solvent is recommended.**

2. Apply a 1/16″ bead of sealant to the closure plate as shown.

3. Install the closure plate to the crankcase and install the twelve hex flange screws. Tighten the screws hand tight.

4. Torque the fasteners, in the sequence shown to 24.4 N-m (216 inch lbs.).

Oil Pump

1. Lubricate the oil pump cavity and oil pump rotors with engine oil. Install the outer and inner oil pump rotors.

2. Install the O-ring in the groove in the closure plate.

3. Install the oil pump cover (machined side towards O-ring). Secure with three hex flange screws.

➡**Apply sealant to the oil pump cover hex flange screws to prevent leakage.**

4. Torque the screws as follows:
   • First Time Installation On A New Closure Plate: 6.2 N-m (55 inch lbs.).
   • Reinstallation On A Used Closure Plate: 4.0 N-m (35 inch lbs.)

Oil Seals

1. Slide the seal protector sleeve, no. KO-1037, over the crankshaft. Generously lubricate the lips of oil seal with light grease. Slide the oil seal over the sleeve.

2. Use handle no. KO-1036 and seal driver no. KO-1027. Install the seals until the driver bottoms against the crankcase of closure plate.

Stator and Wiring Harness

1. Position the stator leads towards the hole i the crankcase. Insert the stator leads through the hole to the outside of the crankcase.

2. Install the stator using four hex socket head screws. Torque the screws to 4.0 N-m (35 inch lbs.).

3. secure the stator leads to the crankcase with the clip and hex flange screw.

4. Install the connector body to the stator leads.

5. Secure the kill lead to the crankcase with the clip and hex flange screw.

Fan and Flywheel

❋❋CAUTION

Using improper procedures to install the flywheel can crack or damage the crankshaft and/or flywheel. This not only causes extensive engine damage, but can also cause

personal injury, since broken fragments could be thrown from the engine. Always observe and use the following precautions and procedures when installing the flywheel.

➡**Before installing the flywheel make sure the crankshaft taper and flywheel hub are clean, dry and completely free of lubricants. The presence of lubricants can cause the flywheel to be over-stressed and damaged when the flange screw is torqued to specification.**

➡**Make sure the flywheel key is installed properly in the keyway. The flywheel can become cracked or damaged if the key is not installed properly in the keyway.**

➡ **Always use a flywheel strap wrench to hold the flywheel when tightening the flywheel fastener. Do not use any type of bar wedge between the cooling fins or flywheel ring gear, as these parts could become cracked or damaged.**

1. Install the fan, spacers and hex flange screws to the flywheel. Torque the hex flange screws to 9.9 N-m (88 inch lbs.).

2. Install the woodruff key into the keyway in the crankshaft.

3. Place the flywheel over the keyway/crankshaft. Install grass screen, drive cup, plain washer (flat side of plain washer towards the drive cup), and the hex flange screw.

4. Hold the flywheel with a strap wrench and torque the hex flange screw to 66.4 N-m (491 inch lbs.).

Fuel Line

Install the fuel line, clamp and hex flange screw.

Ignition Module

1. Install the ignition module and hex flange screws to the bosses on crankcase. Move the module as far from the flywheel/magnet as possible. Tighten the hex flange screws slightly.

2. Insert a 10 mm (0.394 in) flat feeler gauge or shim stock between the magnet and ignition module. Loosen the hex flange screws so the magnet pulls the module against the feeler gauge.

3. Tighten the hex flange screws as follows:
   • First Time Installation On A New Short Block: 6.2 N-m (55 inch lbs.)
   • All Reinstallations: 4.0 N-m (35 inch lbs.).

4. Rotate the flywheel back and forth; check to make sure the magnet does not strike the module.

5. Check the gap with feeler gauge and readjust if necessary. Final Air Gap: 0.203/0.305 mm (0.008/0.012 in)

6. Connect the kill lead to the tab terminal on ignition module.

Cylinder Head Components

1. Install the rocker bridge to the cylinder head. Make sure the small (counterbored) hole is towards the exhaust port side of the cylinder head. Secure the rocker bridge with two hex cap screws.

2. Install the intake valve stem seal, intake valve, intake valve spring seat, intake valve spring, and valve spring cap.

Compress the valve spring using a valve spring compressor and install the keeper.

3. Position the rocker arms over the valve stems and rocker arm bridge. Insert the pin (from breather reed side) through the rocker bridge and rocker arms.

Cylinder Head

1. Install a new cylinder head gasket.

2. Install the cylinder head spacer (closest to the exhaust port) and hex flange screws. Torque the screws in increments of 10 ft. lbs. in the sequence shown to 40.7 N-m (30 inch lbs.).

3. Install the push rods and compress the valve springs. Snap the push rods underneath the rocker arms.

4. Install the spark plug into the cylinder head. Torque the spark plug to 38.0/43.4 N-m (28/32 inch lbs.).

5. Install the breather reed, breather reed retainer, and hex flange screw.

Baffles and Blower Housing

➡ **Leave all hardware slightly loose until all sheet metal pieces are in position.**

1. Install the heat deflector, intake manifold, and gaskets to the cylinder head intake port using two hex socket screws. Torque the hex socket screws to 9.9 N-m (88 inch lbs.).

2. Install the grommet around the high tension lead. Insert the grommet into the slot in the blower housing. Install the blower housing and baffles using hex flange screws.

3. Install cylinder head baffle to the cylinder head using hex flange screws.

4. Tighten all hardware.

Valve Cover and Muffler Bracket

RTV silicone sealant is used as a gasket between the valve cover and crankcase. GE Silmate® type RTV-1473 or RTV-108 silicone sealant (or equivalent) is recommended.

1. Prepare the sealing surfaces of the cylinder head and valve cover as directed by the sealant manufacturer.

➡ **Do not scrape surfaces when cleaning as this will damage the surface and could cause leaks. The use of a gasket removing solvent is recommended.**

2. Apply a 1/16″ bead of sealant to the cylinder head as shown.

3. Install the valve cover, lift bracket (lifting hole towards flywheel), and two hex flange screws.

4. Torque the screws in the sequence shown, as follows:
- First Time Installation On A New Cylinder Head: 10.7 N-m (95 inch lbs.).
- All Reinstallation: 7.3 N-m (65 inch lbs.).

Fuel Pump

1. Install the rubber line and two hose clamps to the fuel pump end of the metal fuel line. Secure the rubber fuel line to the steel fuel line with on of the clamps.

2. Install the gasket, fuel pump, and two hex flange screws. Torque the screws as follows:
- First Time Installation On A New Short Block: 9.0 N-m (80 inch lbs.).
- All Reinstallations: 7.3 N-m (65 inch lbs.).

Electric Starter

ELECTRIC STARTER (BENDIX DRIVE OR SOLENOID SHIFT)

1. Install the starter and spacers on the mounting studs.
2. Install the starter cover and the two hex flange screws.
3. Connect the lead to the starter terminal(s).

Fuel Tank

1. Connect the fuel hose to the shut-off valve.
2. Install the hex flange screws to upper bracket of fuel tank. Install hex flange nuts to studs in lower bracket of fuel tank.

Rectifier-Regulator

1. Install the rectifier-regulator and hex flange screws.
2. Install the connector to the rectifier-regulator,

Carburetor and External Governor Components

1. Install the rubber fuel line and tow hose clamps to the metal fuel line. Secure the metal fuel line with one of the hose clamps.

2. Install the bushing and the throttle linkage to the carburetor throttle lever.

3. Install the gasket and carburetor over the intake studs. Install the free end of the rubber fuel line to the carburetor fuel inlet fitting as the carburetor is inserted over the studs. Secure the fuel line with the other hose clamp.

4. Install the throttle linkage and bushing to governor lever.

5. Install the governor lever to governor cross shaft. Do not tighten the hex nut on the governor lever until the lever is adjusted (step 6).

6. Adjust the governor lever/governor gear.
   a. Pull the governor lever away from the carburetor (wide open throttle).
   b. Insert a nail in the cross shaft hole or grasp the cross shaft with a pliers and turn the shaft counterclockwise as far as it will go,
   c. Tighten the hex nut securely.

Throttle Bracket

1. Install the throttle bracket assembly with two hex flange screws.

2. Install the governor spring in the appropriate hole in the governor arm and throttle control lever, as indicated in the chart. Note that hole positions are counted from the top of the lever. RPM should be checked with a tachometer.

Air Cleaner

1. Connect choke linkage to the carburetor choke lever. Install the base plate to the studs and connect the breather hose to the rocker arm cover.

2. Install the air cleaner base and gasket to the studs and torque the hex flange nuts to 9.9 N-m (88 inch lbs.).

3. Install the element and precleaner, element cover, washer, and wing nut.

4. Install the air cleaner cover and knob.

Retractable Starter

1. Install the retractable starter and five hex flange screws to blower housing. Leave the screws slightly loose.

2. Pull the starter handle out until the pawls engage in the drive cup. Hold the handle in this position and tighten the screws securely.

Muffler

1. Install the gasket, muffler, and hex flange nuts to the exhaust port studs. Leave the nuts slightly loose.

2. Secure the muffler bracket using the two hex flange screws.

3. Torque the hex flange nuts to 24.4 N-m (215 inch lbs.), screws to 9.9 N-m (88 inch lbs.).

14 HP MODELS

SPECIFICATIONS, TOLERANCES, AND SPECIAL TORQUE VALUES[1]

Air Cleaner

Base Nut Torque .... 9.9 N•m (88 lbf•in)

Angle Of Operation — Maximum
(At Full Oil Level)

Intermittent — All Directions .... 35°

Continuous — All Directions .... 25°

Balance Shaft

End Play (Free) .... 0.0575/0.3625 mm (0.0023/0.0143 in)

Running Clearance .... 0.025/0.063 mm (0.0009/0.0025 in)

Bore I.D. — New .... 20.000/20.025 mm (0.7874/0.7884 in)

Bore I.D. — Max. Wear Limit .... 20.038 mm (0.7889 in)

Balance Shaft Bearing Surface O.D. — New .... 19.962/19.975 mm (0.7859/0.7864 in)

Balance Shaft Bearing Surface O.D. — Max. Wear Limit .... 19.959 mm (0.7858 in)

Camshaft

End Play (With Shims) .... 0.076/0.127 mm (0.003/0.005 in)

Running Clearance .... 0.025/0.063 mm (0.0010/0.0025 in)

Bore I.D. — New .... 20.000/20.025 mm (0.7874/0.7884 in)

Bore I.D. — Max. Wear Limit .... 20.038 mm (0.7889 in)

Camshaft Bearing Surface O.D. — New .... 19.962/19.975 mm (0.7859/0.7864 in)

Camshaft Bearing Surface O.D. — Max. Wear Limit .... 19.959 mm (0.7858 in)

Carburetor

Preliminary Low Idle Fuel Needle Setting .... 1 Turn

Fuel Bowl Nut Torque .... 5.1/6.2 N•m (45/55 lbf•in)

Charging

Stator Mounting Screw Torque .... 4.0 N•m (35 lbf•in)

832590C3

Air Cleaner

These engines are equipped with a replaceable, high-density paper air cleaner element. Some engines are also equipped with an oiled-foam precleaner which surrounds the paper element.

SERVICE

Check the air cleaner daily or before starting the engine. Check for and correct heavy buildup of dirt and debris, and loose components.

➡**Operating the engine with loose or damaged air cleaner components could allow unfiltered air into the engine causing premature wear and failure.**

Precleaner Service

If so equipped, wash and reoil the precleaner every 25 hours of operation (more often under extremely dusty or dirty conditions).

1. Remove the precleaner from the paper element.
2. Wash the precleaner in warm water with detergent. Rinse the precleaner thoroughly until all traces of detergent are eliminated. Squeeze out excess water (do not wring). Allow the precleaner to air-dry.
3. Saturate the precleaner with new engine oil. Squeeze out all excess oil.
4. Reinstall the precleaner over the paper element.
5. Reinstall air cleaner cover, and air cleaner cover retaining knob. Make sure the knob is tightened securely.

Servicing the Paper Element

Every 100 hours of operation (more often under extremely dusty or dirty conditions), check the paper element. Replace the element as necessary.

1. Remove the precleaner (if so equipped) from the paper element.
2. Remove the wing nut, washer, element cover, and air cleaner element.
3. Do not wash the paper element or use pressurized air, as this will damage the element. Replace a dirty, bent, or damaged element with a new element. Handle new elements carefully; do not use if the sealing surfaces are bent or damaged.
4. Reinstall the paper element, element cover, washer, wing nut, precleaner, air cleaner cover, and air cleaner cover retaining knob. Make sure the knob is tightened securely.

Inspect Air Cleaner Components

Whenever the air cleaner cover is removed, or the paper element or precleaner are serviced, check the following areas/components:

- Covered Air Cleaner Element — Inspect the rubber grommet in the hole of the air cleaner element cover. Replace the grommet if it is worn or damaged.
- Air Cleaner Base — Make sure the base is secured and not cracked or damaged. Since the air cleaner base and carburetor are secured to the intake port with common hardware, it is extremely important that the nuts securing these components are tight at all times.
- Breather Tube — Make sure the tube is installed to both the air cleaner base and valve cover.

➡**Damaged, worn, or loose air cleaner components can allow unfiltered air into the engine causing premature wear and failure. Tighten or replace all loose or damaged components.**

DISASSEMBLY

1. Remove the air cleaner cover retaining knob and air cleaner cover.

2. If so equipped, remove the precleaner from paper element.
3. Remove the wing nut, washer, element cover, and air cleaner element.
4. Disconnect the breather hose from the valve cover.
5. Remove the air cleaner base mounting nuts, air cleaner base, and gasket.
6. If necessary, remove the self-tapping screws and elbow from air cleaner base.

ASSEMBLY

1. Install the elbow and self-tapping screws to air cleaner base.
2. Install the gasket, air cleaner base, and base mounting nuts. Torque the nuts to 9.9 N-m (88 inch lbs.).
3. Connect the breather hose to the air cleaner base (and valve cover). Secure with hose clamps.
4. If necessary, install the grommet into the cover of air cleaner element. Install the air cleaner element, element cover, washer, and wing nut.
5. If equipped, install the precleaner (washed and oiled) over the paper element.
6. Install the air cleaner cover and air cleaner cover retaining knob. Tighten the knob securely.

Air Intake/Cooling System

CLEANING

To ensure proper cooling, make sure the grass screen, cooling fins, and other external surfaces of the engine are kept clean at all times.

Every 100 hours of operation (more often under extremely dusty, dirty conditions), remove the blower housing and other cooling shrouds. Clean the cooling fins and external surfaces as necessary. Make sure the cooling shrouds are reinstalled.

➡**Operating the engine with a blocked grass screen, dirty or plugged cooling fins, and/or cooling shrouds removed, will cause engine damage due to overheating.**

Fuel System and Governor

✳✳CAUTION

Gasoline is extremely flammable and its vapors can explode if ignited. Before servicing the fuel system, make sure there are no sparks, open flames, or other sources of ignition nearby as these can ignite gasoline vapors. Disconnect and ground the spark plug lead to prevent the possibility of sparks from the ignition system.

FUEL RECOMMENDATIONS

General Recommendations

• Purchase gasoline in small quantities and store in clean, approved containers. A container with a capacity of 2 gallons or less with a pouring spout is recommended. Such a container is easier to handle and helps to eliminate spoilage during refueling.

• Do not use gasoline left over from the previous season, to minimize gum deposits in your fuel system and to ensure easy starting.

• Do not add oil to the gasoline.

• Do not overfill the fuel tank. Leave room for the fuel to expand.

Fuel Type

For best results, use only clean, fresh, unleaded gasoline with a pump sticker octane rating of 87 or higher. I countries using the Research method, it should be 90 octane minimum.

Unleaded gasoline is recommended, as it leaves less combustion chamber deposits. Leaded gasoline may be used is areas where unleaded is not available and exhaust emissions are not regulated. Be aware however, that the cylinder head will require more frequent service.

Gasoline/Alcohol blends

Gasohol (up to 10% ethyl alcohol, 90% unleaded gasoline by volume) is approved as a fuel for Kohler engines. Other gasoline/alcohol blends are not approved.

Gasoline/Ether blends

Methyl Tertiary Butyl Ether (MTBE) and unleaded gasoline blends (up to a maximum of 15% MTBE by volume) are approved as a fuel for Kohler engines. Other gasoline/ether blends are not approved.

Fuel System Operation

The typical fuel system and related components include the fuel tank, in-line fuel filter, fuel pump, carburetor, and interconnecting fuel lines.

The fuel from the tank is moved through the in-line filter and fuel lines by the pump. On engines not equipped with a fuel pump, the fuel tank outlet is located above the carburetor inlet; gravity moves the fuel.

Fuel then enters the carburetor float bowl and is moved into the carburetor body. There, the fuel is then burned in the engine combustion chamber.

Fuel Filter

Some engines are equipped with an in-line fuel filter. Visually inspect the filter periodically, and replace when dirty with a genuine Kohler filter.

Fuel Pump

▶ **See Figure 87**

Some engines are equipped with an optical mechanically operated fuel pump. On applications using a gravity feed fuel system, the fuel pump mounting pad is covered with a metal plate.

The fuel pump body is constructed of nylon. The nylon body insulates the fuel from the engine crankcase. This prevents the fuel from vaporizing inside the pump.

The mechanical pump is operated by a lever which rides on the engine camshaft. This lever transmits a pumping action to the diaphragm inside the pump body. On the downward stroke of the diaphragm, fuel is drawn in through the inlet check valve. On the upward stroke of the diaphragm, fuel is forced out through the outlet check valve.

REMOVAL

1. Disconnect the fuel lines from the inlet and outlet fittings of pump.
2. Remove the hex, flange screws, fuel pump, and gasket.
3. If necessary, remove the fittings from the pump body.

REPAIR

Nylon-bodied fuel pumps are not serviceable and must be replaced when faulty. Replacement pumps are available in kits that include the pump and mounting gasket.

INSTALLATION

▶ **See Figure 88**

1. Fittings - Apply a small amount of Permatex Aviation Perm-A-Gasket® (or equivalent) gasoline resistant thread sealant to the threads of fittings. Turn the fittings into the pump 6 full turns; continue turning the fittings in the same direction until the desired position is reached.

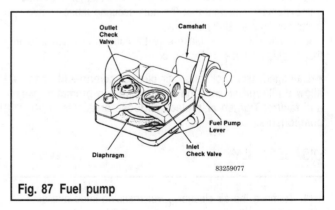

Fig. 87 Fuel pump

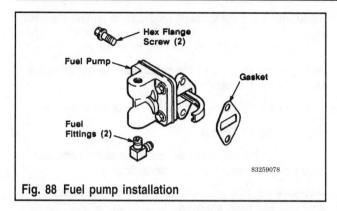

Fig. 88 Fuel pump installation

2. Install new gasket, fuel pump, and hex, flange screws.

➡**Make sure the fuel pump lever is positioned to the RIGHT of the camshaft (when looking at fuel pump mounting pad). Damage to the fuel pump, and subsequent severe damage could result if the lever is positioned to the left of the camshaft.**

Torque the hex, flange screws as follows:
- First Time Installation On A New Short Block — 9.0 N-m (80 inch lbs.)
- All Reinstallations — 7.3 N-m (65 inch lbs.)
3. Connect the fuel lines to the inlet and outlet fittings.

Carburetor

▶ **See Figure 89**

These engines are equipped with an adjustable main jet carburetor. This subsection covers the troubleshooting, idle adjustment, and service procedures for the carburetor.

✳✳CAUTION

Gasoline may be present in the carburetor and fuel system. Gasoline is extremely flammable and its vapors can explode if ignited. Keep sparks, open flames, and other sources of ignition away from the engine. Disconnect and ground the spark plug lead to prevent the possibility of sparks from the ignition system.

TROUBLESHOOTING

If engine troubles are experienced that appear to be fuel related, check the following area before adjusting or disassembling the carburetor.
- Make sure the fuel tank is filled with clean, fresh gasoline.
- Make sure the fuel tank cap vent is not blocked and that it is operating properly.
- Make sure fuel is reaching the carburetor. This includes checking the fuel shut-off valve, fuel tank filter screen, in-line fuel filter, fuel lines, and fuel pump for restrictions or faulty components as necessary.
- Make sure the air cleaner base and carburetor is securely fastened to the engine using gaskets in good condition.
- Make sure the air cleaner element is clean and all air cleaner components are fastened securely.

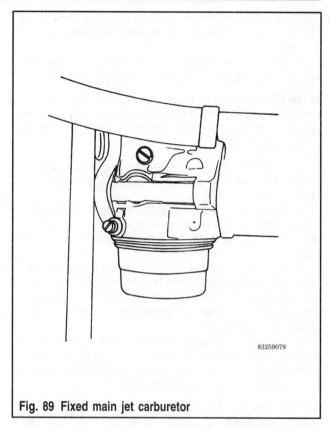

Fig. 89 Fixed main jet carburetor

- Make sure the ignition system, governor system, exhaust system, and throttle and choke controls are operating properly.
- If the engine is hard-starting or runs roughly or stalls at low idle speed, it may be necessary to adjust or service the carburetor.

ADJUSTMENT

▶ **See Figure 90**

➡**Carburetor adjustments should only be made after the engine has warmed up.**

The carburetor is designed to deliver the correct fuel-to-air mixture to the engine under all operating conditions. The main fuel jet (power screw) is calibrated at the factory and is adjustable. The idle fuel adjusting needle is also set at the factory and is adjustable. The idle fuel adjusting needle is also set at the factory and normally does not need adjustment.

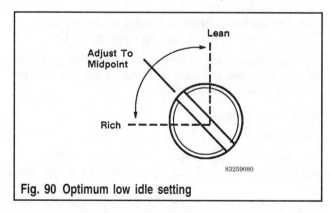

Fig. 90 Optimum low idle setting

If the engine is hard-starting or runs roughly or stalls at low idle speed, it may be necessary to adjust or service the carburetor.

1. With the engine stopped turn the low idle and high idle fuel adjusting needles in (clockwise) until they bottom lightly.

➡ **The tip of the idle fuel and high idle fuel adjusting needles are tapered to critical dimensions. Damage to the needles and the seats in carburetor body will result if the needles are forced.**

2. Preliminary settings: Turn the adjusting needles out (counterclockwise) from lightly bottomed to the positions in the chart.

3. Start the engine and run at half-throttle for 5 to 10 minutes to warm up. The engine must be warm before making final settings. Check that the throttle and choke plates can fully open.

4. High idle fuel needle setting: Place the throttle into the "fast" position. If possible place the engine under load. Turn the high idle adjusting needle in (slowly) until engine speed decreases and then back out approximately ¼ turn for best high-speed performance.

5. Low idle speed setting: Place the throttle control into the "idle" or "slow" position. Set the low idle sped to 1500 rpm and ± 75 rpm by turning the low idle speed adjusting screw in or out. Check the speed using a tachometer.

➡ **The actual low idle speed depends on the application - refer to equipment manufacturer's recommendations. The recommended low idle speed for basic engines is 1500 rpm. To ensure best results when setting the low idle speed should not exceed 1500 rpm ± 75 rpm.**

6. Low idle fuel needle setting: Place the throttle into the "idle" or "slow" position. Turn the low idle fuel adjusting needle in (slowly) until the engine speed decreases and then back out approximately ⅛-¼ turn to obtain best low speed performance.

7. Recheck the idle speed using a tachometer. Readjust the speed as necessary.

DISASSEMBLY

▶ **See Figure 91**

1. Remove the power screw, needle and spring, main jet, power screw gasket and fuel bowl.
2. Remove the bowl gasket, float shaft, shaft, float, and fuel inlet needle.
3. Remove the low idle fuel adjusting needle and spring. Remove the low idle speed adjusting screw and spring.

➡ **Further disassembly to remove the welch plug, fuel inlet seat, throttle plate and shaft is recommended only if these parts are to be cleaned or replaced.**

Welch Plug Removal

▶ **See Figure 92**

In order to clean the "off-idle" ports and bowl vent thoroughly, remove the welch plug covering these areas. Use tool

no. KO-1018 and the following procedure to remove the welch plug.

1. Pierce the welch plug with the tip of the tool,

➡ **To prevent damage to the carburetor, do not allow the tool to strike the carburetor body.**

2. Pry out the welch plug with the tip of the tool.

Fuel Inlet Seat Removal

To remove the fuel inlet seat, pull it out of the carburetor body using a screw, drill bit, or similar tool.

➡ **Always install a new fuel inlet seat. Do not reinstall a seat that has been removed.**

Choke Shaft Removal

1. Because the edges of the choke plate are beveled, mark the choke plate and carburetor body to ensure correct reassembly. Also take note of the choke plate position in bore, and the position of the choke lever and return spring.
2. Grasp the choke plate with pliers. Pull it out the slot in the choke shaft.
3. Remove the choke shaft and choke return spring.

Throttle Shaft Removal

1. Because the edges of the throttle plate are beveled, mark the throttle plate and carburetor body to ensure correct reassembly. Also take note of the throttle plate position in bore, and the position of the throttle lever.
2. Carefully and slowly remove the screws securing the throttle plate to the throttle shaft. Remove the throttle plate.
3. File off any burrs which may have been left on the throttle shaft when the screws were removed. Do this before removing the throttle shaft from the carburetor body.
4. Remove the throttle lever/shaft assembly with foam dust seal.

Cleaning

✳✳CAUTION

Carburetor cleaners and solvents are extremely flammable. Keep sparks, flames, and other sources of ignition away from the area. Follow the cleaner manufacturer's warnings and instructions on its proper and safe use. Never use gasoline as a cleaning agent.

All parts should be cleaned thoroughly using a carburetor cleaner (such as acetone). Make sure all gum deposits are removed from the following areas:

• Carburetor body and bore — especially the areas where the throttle plate, choke plate and shafts are seated.
• Idle fuel and "off-idle" ports in carburetor bore, power screw, bowl vent, and fuel inlet needle and seat.

➡ **These areas can be cleaned with a fine piece of wire in addition to cleaners. Be careful not to enlarge the ports, or break the wire inside the ports. Blow out all passages with compressed air.**

• Float and float hinge.
• Fuel bowl.
• Throttle plate, choke plate, throttle shaft, and choke shaft.

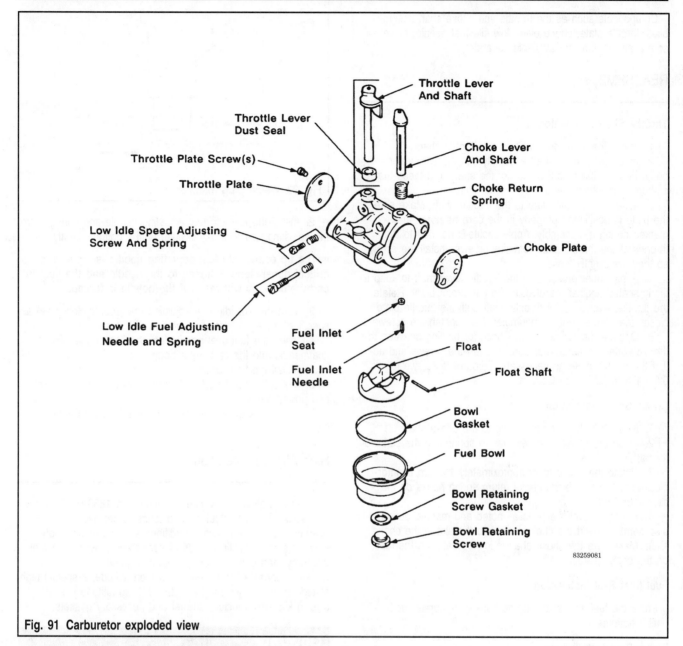

Fig. 91 Carburetor exploded view

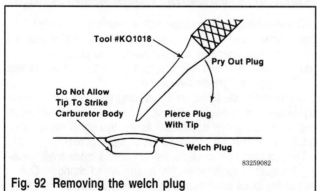

Fig. 92 Removing the welch plug

➡Do not submerge the carburetor in cleaner or solvent when fiber, rubber, or foam seals or gaskets are installed. The cleaner may damage these components.

Inspection

- Carefully inspect all components and replace those that are worn or damaged.
- Inspect the carburetor body for cracks, holes, and other wear or damage.
- Inspect the float for cracks, holes, and other wear or damage.
- Inspect the fuel inlet needle and seat for wear or damage.
- Inspect the tip of the low idle fuel adjusting needle and power screw needle for wear or grooves.
- Inspect the throttle and choke shaft and plate assemblies for wear or excessive play.

Repair

Always use new gaskets when servicing or reinstalling carburetors. Repair kits are available which include new gaskets and other components. These kits are described below.

Components such as the throttle and choke shaft assemblies, throttle plate, choke plate, low idle fuel needle, power screw, and others, are available separately.

REASSEMBLY

Throttle Shaft Installation

1. Install the foam dust seal on the throttle shaft.
2. Insert the throttle/shaft assembly into the carburetor body. Position the cutout portion of the shaft so it faces the carburetor mounting flange.
3. Install the throttle plate to the throttle shaft. Make sure the plate is positioned properly in the bore as noted and marked during disassembly. Apply Loctite® no. 609 to the threads of the throttle plate retaining screws. Install the screws so they are slightly loose.
4. Apply finger pressure to the throttle lever/shaft to keep it firmly seated against the pivot in the carburetor body. Rotate the throttle shaft until the throttle shaft until the throttle plate closes the bore around its perimeter; then tighten the screws.
5. Operate the throttle lever. Check for binding between the throttle plate and carburetor bore. Loosen the screws and adjust the throttle plate as necessary. Torque the screws to 0.9/1.4 N-m (8/12 inch lbs.).

Choke Shaft Installation

1. Install the choke return spring to the choke shaft.
2. Insert the choke lever with return spring into the carburetor body.
3. Rotate the choke lever approximately 1/2 turn counterclockwise. Make sure the choke return spring hooks on the carburetor body.
4. Position the choke plate as noted and marked during disassembly. Insert the choke plate into the slot in the choke shaft. Make sure the choke shaft is locked between the tabs on the choke plate.

Fuel Inlet Seat Installation

Press the fuel inlet seat into the bore in carburetor body until it bottoms.

Welch Plug Installation
▶ **See Figure 93**

Use tool no. KO-1017 and install new plugs as follows:
1. Position the carburetor body with the welch plug cavity to the top.
2. Place a new welch plug into the cavity with the raised surface up.
3. Use the end of the tool that is about the same size as the plug and flatten the plug. Do not force the plug below the surface of the cavity.
4. After the plug is installed, seal it with sealant. Allow the sealant to dry.

➡**If a commercial sealant is not available, fingernail polish can be used.**

Carburetor Reassembly

1. Install the low idle speed adjusting screw and spring.

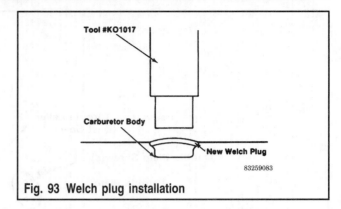

Fig. 93 Welch plug installation

2. Install the low idle fuel adjusting needle and spring. Turn the adjusting needle in (clockwise) until it bottoms lightly.

➡**The tip of the idle fuel adjusting needle is tapered to critical dimensions. Damage to the needle and the seat in carburetor body will result if the needle is forced.**

3. Turn the low idle fuel adjusting out (counterclockwise) as specified in the "Adjustment" portion of this section.
4. Insert the fuel inlet needle into the float. Lower the float/needle into the carburetor body.
5. Install the float shaft.
6. Install the bowl gasket, fuel bowl, bowl retainer gasket, and power screw.
7. Torque the power screw to: 5.1/6.2 N-m (45/55 inch lbs.).

High Altitude Operation

When operating the engine at altitudes of 1830 m (6000 ft.) and above, the main fuel mixture tends to get overrich. An overrich mixture can cause conditions such as black, sooty exhaust smoke, misfiring, loss of speed and power, poor fuel economy, and poor or slow governor response.

To compensate for the effects of high altitude, a special high altitude main fuel jet can be installed. High altitude jets are sold in kits which include the jet and necessary gaskets.

Governor

These engines are equipped with a centrifugal flyweight mechanical governor. It is designed to hold the engine speed constant under changing load conditions. The governor gear/flyweight mechanism is mounted inside the crankcase and is driven off the gear on the camshaft.

Centrifugal force acting on the rotating governor gear assembly causes the flyweights to move outward as speed increases and inward as speed decreases. As the flyweights move outward, they cause the regulating pin to move outward.

The regulating pin contacts the tab on the cross shaft, causing the shaft to rotate when the engine speed changes. One end of the cross shaft protrudes through the side of the crankcase. Through external linkage attached to the cross shaft, the rotating action is transmitted to the throttle lever of the carburetor.

When the engine is at rest, and the throttle is in the "fast" position, the tension of the governor spring holds the throttle plate open. When the engine is operating (the governor gear assembly is rotating), the force applied by the regulating pin against the cross shaft tends to close the throttle plate. The governor spring tension and the force applied by the regulating pin are in "equilibrium" during operation, holding the engine speed constant.

When load is applied and the engine speed (and governor gear speed) decreases, the governor spring tension moves the governor arm to open the throttle plate wider. This allows more fuel into the engine; increasing engine speed. (This action takes place very rapidly, so a reduction in speed is hardly noticed.) As the speed reaches the governed setting, the governor spring tension and the force applied by the regulating pin will again be in equilibrium. This maintains the engine speed at a relatively constant level.

The governed speed setting is determined by the position of the throttle control. It can be variable or constant, depending on the application.

INITIAL ADJUSTMENT

Make this initial adjustment whenever the governor arm is loosened or removed from the cross shaft. To ensure proper setting, make sure the throttle linkage is connected to the governor arm and the throttle lever on the carburetor.

1. Pull the governor lever away from the carburetor (wide open throttle).
2. Insert a nail in the cross shaft hole or grasp the cross shaft with pliers and turn the shaft counterclockwise as far as it will go.
3. Tighten the hex nut securely.

SENSITIVITY ADJUSTMENT

Governor sensitivity is adjusted by repositioning the governor spring in the holes in the governor lever. If speed surging occurs with a change in load, the governor is set too sensitive. If a big drop in speed occurs when normal load is applied, the governor should be set for greater sensitivity.

REMOTE THROTTLE AND CHOKE ADJUSTMENT

1. Adjust the throttle lever. see this section.
2. Install remote throttle cable in hole in the throttle lever.
3. Install remote choke cable in hole in the choke lever.
4. Secure remote cables loosely with the cable clamps.
5. Position the throttle cable so that the throttle lever is against stop.
6. Tighten the throttle cable clamp.
7. Position the choke cable so that the carburetor choke plate is fully closed.
8. Tighten the choke cable clamp.
9. Check carburetor idle speed. See Adjust Carburetor in this section.

Lubrication System

♦ See Figure 94

OIL TYPE

Using the proper type and weight of oil in the crankcase is extremely important. So, is checking oil daily and changing oil regularly. Failure to use the correct oil, or using dirty oil, causes premature engine wear and failure.

Use a high-quality oil of API (American Petroleum Institute) service class SF or SG. Select the viscosity based on the air temperature at the time of operation as shown in the following table.

CHECK OIL LEVEL

The importance of checking and maintaining the proper oil level in the crankcase cannot be overemphasized. Check oil BEFORE EACH USE as follows:

1. Make sure the engine is stopped, level, and is cool so the oil has had time to drain into the sump.

➡ **Using other than service class SF or SG oil or extending oil change intervals longer than recommended can cause engine damage.**

2. To keep dirt, grass clippings, etc., out of the engine, clean the area around the oil fill cap/dipstick before removing it.
3. Remove the oil fill cap/dipstick; wipe oil off. Reinsert the dipstick into the tube and seat the oil fill cap on the tube.
4. Remove the dipstick and check the oil level. The oil level should up to, but not over, the "F" mark on the dipstick.
5. If the oil level is low, add oil of the proper type, up to the "F" mark on the dipstick. Always check the level with the dipstick before adding more oil.

➡ **To prevent extensive engine wear or damage, always maintain the proper oil level in the crankcase. Never operate the engine with the oil level below the "L" mark or over the "F" mark on the dipstick.**

Oil Sentry

Some engines are equipped with an optional Oil Sentry oil pressure monitor. If the oil pressure gets low, Oil Sentry will

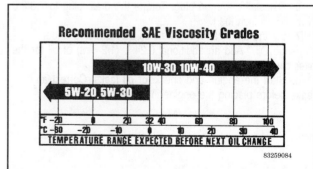

Recommended SAE Viscosity Grades

10W-30, 10W-40

5W-20, 5W-30

TEMPERATURE RANGE EXPECTED BEFORE NEXT OIL CHANGE

83259084

Fig. 94 Engine oil viscosity chart

either shut off the engine or activate a warning signal, depending on the application.

➡ **Make sure the oil level is checked BEFORE EACH USE and is maintained up to the "F" mark on the dipstick. This includes engines equipped with Oil Sentry.**

CHANGING OIL AND OIL FILTER

Changing Oil

For a new engine, change oil after the first 5 hours of operation. Thereafter, change oil after every 100 hours of operation.

For an overhauled engine or those rebuilt with a new short block, use 10W-30-weight service class SF oil for the first 5 hours of operation. Change the oil after this initial run-in period. Refill with service class SF oil as specified in the "Viscosity Grades" table.

Change the oil while the engine is still warm. The oil will flow freely and carry away more impurities. Make sure the engine is level when filling, checking, and changing the oil.

Change the oil as follows:

1. Remove the oil drain plug and oil fill cap/dipstick. Be sure to allow ample time for complete drainage.

2. Reinstall the drain plug. Make sure it is tightened to 13.6 N-m (10 inch lbs.) torque.

3. Fill the crankcase, with new oil of the proper type, to the "F" mark on the dipstick. Always check the level with the dipstick before adding more oil.

4. Reinstall the oil fill cap/dipstick.

➡ **To prevent extensive engine wear or damage, always maintain the proper oil level in the crankcase. Never operate the engine with the oil level below the "L" mark over the "F" mark on the dipstick.**

Changing Oil Filter

Replace the oil filter after every oil change (every 200 hours of operation).

1. Drain the oil from the engine crankcase.

2. Allow the oil filter to drain.

3. Remove the old filter and wipe off the filter adapter.

4. Apply a thin coat of new oil to the rubber gasket on the replacement oil filter.

5. Install the replacement oil filter to the filter adapter. Turn the filter clockwise until the rubber gasket contacts the filter adapter, then tighten the filter an additional 1/2 turn.

6. Reinstall the drain plug. Torque the drain plug to 7.3/9.0 N-m (65/80 inch lbs.).

7. Fill the crankcase with new oil as instructed under "Change Oil." Add an additional 0.24 L (1/2 pint) of oil for the filter capacity.

8. Start the engine and check for oil leaks. Correct any leaks before placing the engine into service.

Oil Pump

SERVICE

The oil pump rotors can be serviced without removing the closure plate. Remove the oil pump cover on the PTO side of closure plate to service the rotors.

The closure plate must be removed to service the oil pickup and oil pressure relief valve.

Oil Sentry Oil Pressure Monitor

Some engines are equipped with an optional Oil Sentry oil pressure monitor. Oil Sentry will either stop the engine or activate a "low oil" warning light, if the pressure gets low. Actual Oil Sentry use will depend on the engine application.

The pressure switch is designed to break contact as the oil pressure increases and make contact as the oil pressure decreases. At oil pressure above approx. 3.0 to 5.0 psi., the switch contacts open. At oil pressures below approx. 3.0 to 5.0 psi., the switch contact close.

On stationary or unattended applications (pumps, generators, etc.), the pressure switch can be used the activate a "low oil" warning light.

➡ **Oil sentry is not a substitute for checking the oil level BEFORE EACH USE. Make sure the oil level is maintained up to the "F" mark on the dipstick.**

INSTALLATION

The pressure switch is installed in the oil filter adapter in one of the main oil galleries of the closure plate. On engines not equipped with Oil Sentry, the installation hole is sealed with a 1/8-27 N.P.T.F. pipe plug.

To install the Oil Sentry switch to the oil filter adapter of closure plate:

1. Apply Loctite® #592 pipe sealant with Teflon® (or equivalent) to the threads of the switch.

2. Install the switch into the tapped hole in oil filter adapter. Torque the switch to 7/9 N-m (70 inch lbs.).

TESTING

The Oil Sentry pressure monitor is a normally closed type switch. It is calibrated to open (break contact) with increasing pressure and close (make contact) with decreasing pressure within the range of 3.0/5.0 psi.

Compressed air, a pressure regulator, pressure gauge and a continuity tester are required to test the switch.

1. Connect the continuity tester across the blade terminal and the metal case of switch. With 0 psi pressure applied to the switch, the tester should indicate continuity (switch closed).

2. Gradually increase the pressure to thew switch. The tester should indicate a change to no continuity (switch open) as the pressure increases through the range of 3.0/5.0 psi. The

switch should remain open as the pressure is increased to 90 psi maximum.

3. Gradually decrease the pressure to the switch. The tester should indicate a change to continuity (switch closed) as the pressure decreases through the range of 3.0/5.0 psi; approaching 0 psi. If the switch does not operate as specified, replace the switch.

Retractable Starter

✳✳CAUTION

The spring is under tension! Retractable starters contain a powerful, flat wire recoil spring that is under tension. Do not remove the center screw from the starter until the spring is released. Removing the center screw before releasing spring tension, or improper starter disassembly, can cause the sudden and potentially dangerous release of the spring.

Always wear safety goggles when servicing retractable starters - full face protection is recommended.

To ensure personal safety and proper starter disassembly and reassembly, follow the procedures in this section carefully.

REMOVAL

▶ **See Figure 95**

1. remove the five hex flange screws securing the starter blower housing.
2. Remove the starter.

INSTALLATION

1. Install the retractable starter and five hex flange screws to blower housing. Leave the screws slightly loose.
2. Pull the starter handle out until the pawls engage in this position and tighten the screws securely.

ROPE REPLACEMENT

The rope can be replaced without complete starter disassembly.
1. Remove the starter from the engine blower housing.
2. Pull the rope out approx. 12″ and tie a temporary (slip) knot in it to keep it from retracting into the starter.
3. Remove the rope retainer from inside the starter handle. Untie the single knot and remove the rope retainer and handle.
4. Hold the pulley firmly and untie the slip knot. Allow the pulley to rotate slowly as the spring tension is released.
5. When all spring tension on the starter pulley is released, remove the rope from the pulley.
6. Tie a single knot in one end of the new rope.

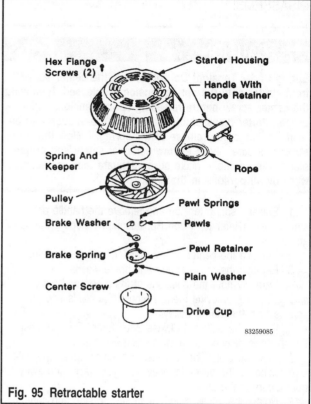

Fig. 95 Retractable starter

7. Rotate the pulley counterclockwise (when viewed from pawl side of pulley) until the spring is tight. (Approximately. 6 full turns of pulley.)
8. Rotate the pulley counterclockwise until the rope hole in pulley is aligned with rope guide bushing of starter housing.

➡**Do not allow the pulley/spring to unwind. Enlist the aid of a helper if necessary, or use a C-clamp to hold the pulley in position.**

9. Insert the new rope through the rope hole in starter pulley and rope guide bushing of housing.
10. Tie a slip knot approx. 12″ from the free end of rope. Hold the pulley firmly and allow it to rotate slowly until the slip knot reaches the guide bushing of housing.
11. Slip the handle and rope retainer onto the rope. Tie a single knot at the end of the rope. Install the rope retainer into the starter handle.
12. Untie the slip knot and pull on the handle until the rope is fully extended. Slowly retract the rope into the starter.   When the spring is properly tensioned, the rope will retract fully and the handle will stop against the starter housing.

PAWLS (DOGS) REPLACEMENT

The starter must be completely disassembled to replace the starter pawls. a pawl repair kit is available which includes the following components:

DISASSEMBLY

> ※※**CAUTION**
>
> **Spring Under Tension! Do not remove the center screw from starter until the spring tension is released. Removing the center screw before releasing spring tension, or improper starter disassembly, can cause the sudden and potentially dangerous release of the spring. Follow these instructions carefully to ensure personal safety and proper starter disassembly. Make sure adequate face protection is worn by all persons in the area.**

1. Release spring tension and remove the handle and starter rope. (Refer to "Rope Replacement", steps 2 through 5 above.)
2. Remove the center screw, washer, and pawl retainer.
3. Remove the brake spring and brake washer.
4. Carefully note the positions of the pawls and pawl springs before removing them. Remove the pawls and pawl springs from the starter pulley.
5. Rotate the pulley clockwise 2 full turns. This will ensure the spring is disengaged from the starter housing.
6. Hold the pulley into the starter housing. Invert the pulley/housing so the pulley is away from your face, and away from others in the area.
7. Rotate the pulley slightly from side to side and carefully separate the pulley from the housing.

➡ **If the pulley and the housing do not separate easily, the spring could be engaged in the starter housing, or there is still tension on the spring. Return the pulley to the housing and repeat step 5 before separating the pulley and housing.**

8. Note the position of the spring and keeper assembly in the pulley. Remove the spring and keeper assembly from the pulley as a package.

> ※※**CAUTION**
>
> **Do not remove the spring from the keeper. Severe personal injury could result from the sudden uncoiling of the spring.**

INSPECTION AND SERVICE

1. Carefully inspect the rope, pawls, housing, center screw, and other components for wear or damage.
2. Replace all worn or damaged components.
3. Do not attempt to rewind a spring that has come out of the keeper. Order and install a new spring and keeper assembly.
4. Clean all old grease and dirt from the starter components. Generously lubricate the spring and center shaft with any commercially available bearing grease.

REASSEMBLY

1. Make sure the spring is well lubricated with grease. Place the spring and keeper assembly inside the pulley (with spring towards pulley).
2. Install the pulley with spring and keeper assembly into the starter housing. Make sure the pulley is fully seated against the starter housing. Do not wind the pulley and recoil spring at this time.
3. Install the pawls springs and pawls into the starter pulley.
4. Place the brake washer in the recess in starter pulley; over the center shaft.
5. Lubricate the brake spring sparingly with grease. Place the spring on the plain washer. (Make sure the threads in center shaft remain clean, dry, and free of grease and oil.)
6. Apply a small amount of Loctite® #271 to the threads of the center screw. Install the center screw, with washer and retainer, to the center shaft. Torque the screw 7.4/8.5 N-m (65/75 inch lbs.).
7. Tension the spring and install the rope and handle as instructed in steps 6 through 12 under "Rope Replacement" above.
8. Install the starter to the engine blower housing.

Spark Plug

▶ **See Figures 96, 97, 98, 99 and 100**

Engine misfire or starting problems are often caused by a spark plug that is in poor condition or with an improper gap setting.

This engine is equipped with the following spark plug:
Type: Champion RC12YC (or equivalent)
Gap: 1.02mm (0.040 in.)
Thread Size: 14mm
Reach: 19.1mm (3/4 in.)
Hex Size: 15.9mm (5/8 in)

REMOVAL & INSTALLATION

Every 100 hours of operation, remove the spark plug, check its condition, and reset the gap or replace with a new plug as necessary.

1. Before removing the spark plug, clean the area around the base of the plug to keep dirt and debris out of the engine.

83257091

Fig. 96 Twist and pull on the boot, never on the plug wire

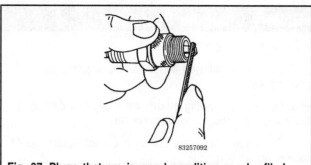

Fig. 97 Plugs that are in good condition can be filed and reused

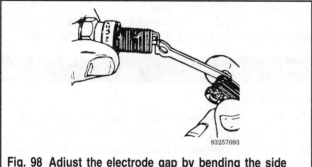

Fig. 98 Adjust the electrode gap by bending the side electrode

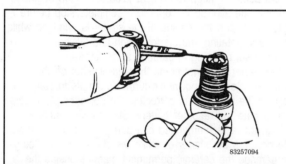

Fig. 99 Always used a wire gauge to check the plug gap

2. Remove the plug and check its condition. Replace the plug if worn or reuse is questionable.

➡**Do not clean the spark plug in a machine using abrasive grit. Some grit could remain in the spark plug and enter the engine causing extensive wear and damage.**

3. Check the gap using a wire feeler gauge. Adjust the gap to 1.02 mm (0.040 in) by carefully bending the ground electrode.

4. Reinstall the spark plug into the cylinder head. Torque the spark plug to 38.0/43.4 N-m (28/32 ft. lbs.).

INSPECTION

Inspect the spark plug as soon as it is removed from the cylinder head. The deposits on the tip are an indication of the general condition of the piston rings, valves, and carburetors.

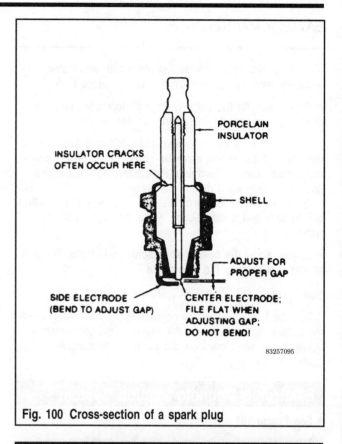

Fig. 100 Cross-section of a spark plug

Battery

A 12-volt with a rating of approximately 32-amp hours/250 cold cranking amps, is normally used. Refer to the operating instructions of the equipment this engine powers for specific information.

If the battery charge is not sufficient to crank the engine, recharge the battery.

➡**Do not attempt to "jump start" the engine with another battery. Starting the engine with batteries larger than those recommended can burn out of the starter motor.**

BATTERY CHARGING

✳✳CAUTION

Batteries contain sulfuric acid. To prevent acid burns, avoid contact with skin, eyes, and clothing. Batteries produce explosive hydrogen gas while being charged. To prevent a fire or explosion, charge batteries only in well ventilated areas. Keep sparks, open flames, and other sources of ignition away from the battery at all times.   Keep batteries out of the reach of children. Remove all jewelry when servicing batteries. Before disconnecting the negative (-) ground cable, make sure all switches are OFF. If ON, a spark will occur at the ground cable terminal which could cause an explosion if hydrogen gas or gasoline vapors are present.

BATTERY MAINTENANCE

1. Regularly check the level of electrolyte. Add distilled water as necessary to maintain the recommended level.

➡**Do not overfill the battery. Poor performance or early failure due to loss of electrolyte will result.**

2. Keep the cables, terminals, and external surfaces of battery clean. A build-up of corrosive acid or grime on the external surfaces can self-discharge the battery. Self-discharging happens rapidly when moisture is present.

3. Wash the cables, terminals, and external surfaces with a baking soda and water solution. Rinse thoroughly with clear water.

➡**Do not allow the baking soda solution to enter the cells as this will destroy the electrolyte.**

Battery Test

Test the battery voltage by connecting D.C. voltmeter across the battery terminals - crank the engine. If the battery drops below 9 volts while cranking, the battery is discharged or faulty.

Electronic Magneto Ignition System

▶ **See Figure 101**

These engines are equipped with a dependable electronic magneto ignition system. The system consists of the following components:
- A magnet assembly which is permanently affixed to the flywheel.
- An electronic magneto ignition module which mounts on the engine crankcase.
- A kill switch (or key switch) which grounds the module to stop the engine.
- A spark plug.

OPERATION

As the flywheel rotates and the magnet assembly moves past the ignition module, a low voltage is induced in the primary windings of the module. When the primary voltage id precisely at its peak, the module induces a high voltage in its secondary windings. This high voltage creates a spark at the tip of the spark plug. This spark ignites the fuel-air mixture in the combustion chamber.

The timing of the spark is automatically controlled by the module. Therefore, other than periodically checking/replacing the spark plug, no maintenance, timing, or adjustments are necessary or possible with this system.

➡**Use a low-voltage (2 volts or less) ohmmeter when ohmmeter is required. Always zero ohmmeter on each scale before testing to ensure accurate readings.**

Battery Charging Systems

▶ **See Figures 102 and 103**

This engine is equipped with a 15 Amp regulated battery charging system.

➡**Observe the following guidelines to prevent damage to the electrical system and components.**

1. Make sure the battery polarity is correct. A negative (-) ground system is used.
2. Disconnect the rectifier-regulator leads and/or wiring harness plug before doing electric welding on the equipment powered by engine. Also disconnect other electrical accessories in common ground with the engine.
3. Prevent the stator (AC) leads from touching or shorting while the engine is running. This could damage the stator.

Electric Starter

▶ **See Figure 104**

- Do not crank the engine continuously for more than 10 seconds at a time. If the engine does not start, allow a 60-second cool-down period between starting attempts. Failure to follow these guidelines can burn out the starter motor.
- If the engine develops sufficient speed to disengage the starter but does not keep running (a false start), the engine rotation must be allowed to come to a complete stop before attempting to restart the engine. If the starter is engaged while the flywheel is rotating, the starter pinion and flywheel ring gear may clash, resulting in damage to the starter.
- If the starter does not crank the engine, shut off the starter immediately. Do not make further attempts to start the engine until the condition is corrected. Do not attempt to jump start the engine with another battery. Starting with batteries larger than those recommended can burn out the starter motor.
- Do not drop the starter or strike the starter frame. Doing so can damage the ceramic permanent magnets inside the starter frame.

Bendix Drive Electric Starter

When power is applied to the starter, the armature rotates. As the armature rotates, the drive pinion moves out on the splined drive shaft and into mesh with the flywheel ring gear. When the pinion reaches the end of the drive shaft, it rotates the flywheel and "cranks" the engine.

When the engine starts, the flywheel rotates faster than the starter armature and drive pinion. This moves the drive pinion out of mesh with the ring gear and into the retracted position. When power is removed from the starter, the armature stops and the drive pinion is held in the retracted position by the anti-drift spring.

REMOVAL & INSTALLATION

Refer to the Engine Disassembly and Reassembly sections for starter removal and installation procedures.

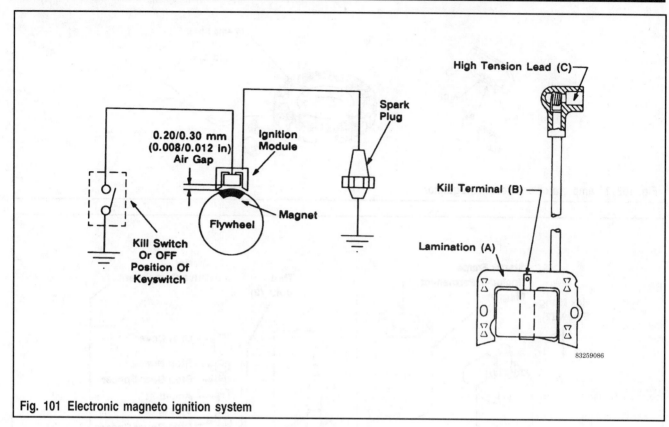

Fig. 101 Electronic magneto ignition system

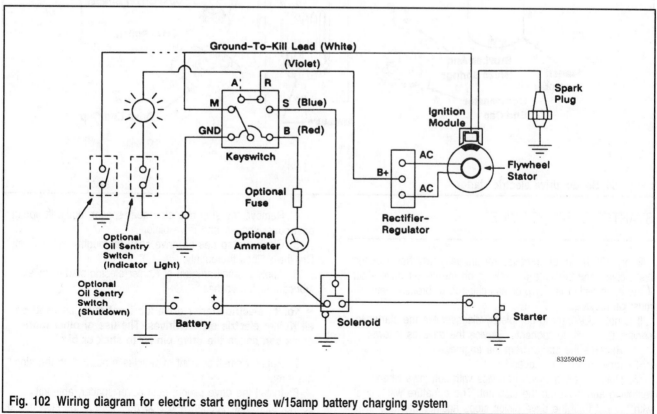

Fig. 102 Wiring diagram for electric start engines w/15amp battery charging system

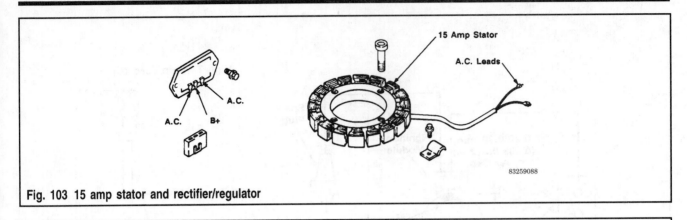

Fig. 103 15 amp stator and rectifier/regulator

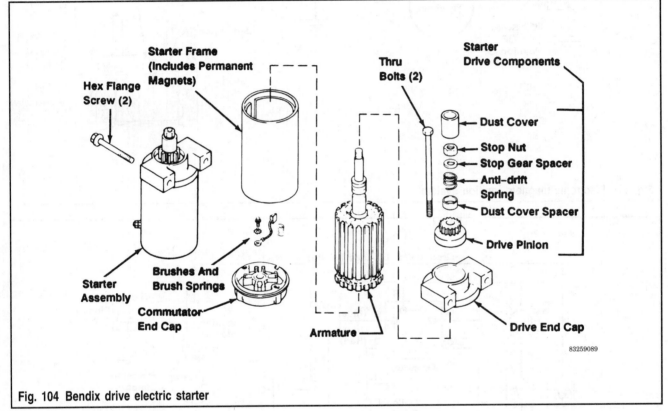

Fig. 104 Bendix drive electric starter

STARTER DRIVE SERVICE

Every 500 hours of operation (or annually, whichever occurs first), clean and lubricate the splines on the starter drive shaft. If the drive pinion is worn, or has chipped or broken teeth, it must be replaced.

It is not necessary to completely disassemble the starter to service the drive components. Service the drive as follows:

1. Remove the starter from the engine.
2. remove the dust cover.
3. Hold the drive pinion in a vice with soft jaws when removing and installing the stop nut. The armature will rotate with the nut until the drive pinion stops against internal spacers.

➡**Do not overtighten the vice as this can distort the drive pinion.**

4. Remove the stop nut, stop gear spacer, anti-drift spring, dust cover spacer, and drive pinion.
5. Clean the splines on drive shaft thoroughly with solvent. Dry the splines thoroughly.
6. Apply a small amount of Kohler electric starter drive lubricant to the splines.

➡**Kohler electric starter drive lubricant must be used on all Kohler electric starter drives. The use of other lubricants can cause the drive pinion to stick or bind.**

7. Apply a small amount of Loctite® no. 271 to the stop nut threads.
8. Install the drive pinion, dust cover spacer, anti-drift spring, stop gear spacer, and stop nut. Torque the stop nut to 17.0/19.2 N·m (135 inch lbs.).
9. Install the dust cover.

STARTER DISASSEMBLY

1. Remove the dust cover, stop nut, stop gear spacer, anti-drift spring, dust cover spacer, and drive pinion. Refer to "Starter Drive Service" above.

2. Scribe a small line on the drive end cap, opposite the line on the starter frame. These lines will service as match marks when reassembling the starter.

3. Remove the through bolts.

4. Remove the commutator end cap with brushes and brush springs.

5. remove the drive end cap.

6. remove the armature and thrust washer from inside the starter frame.

Brush Replacement

▶ **See Figure 105**

1. Remove the brush springs from the pockets in brush holder.

2. Remove the self-tapping screws, negative (-) brushes, and plastic brush holder.

3. Remove the hex flange nut and fiber washer from the stud terminal.   Remove the stud terminal with positive (+) brushes and plastic insulating bushing from the end cap.

4. Reinstall the insulating bushing to the new stud terminal with positive (+) brushes. Install the stud terminal with bushing into the commutator end cap. Secure the stud with the fiber washer and hex flange screw.

5. Install the brush holder, new negative (-) brushes, and self-tapping screws.

6. install the brush springs and brushes into the pockets in brush holder. Make sure the chamfered sides of brushes are away from the brush springs.

➡**Use a brush holder tool to keep the brushes in the pockets. A brush holder tool can easily be made from thin sheet metal.**

Commutator Service

Clean the commutator with a coarse, lint free cloth. Do not use emery cloth.

If the commutator is badly worn or grooved, turn it down on a lathe or replace the armature.

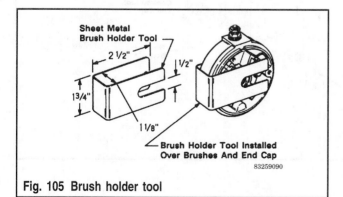

Fig. 105 Brush holder tool

STARTER REASSEMBLY

1. Place the thrust washer over the drive shaft of armature.

2. Insert the armature into the starter frame. Make sure the magnets are closer to the drive shaft end of armature. The magnets will hold the armature inside the frame.

3. Install the drive end cap over the drive shaft. Make sure the match marks on the end cap and starter frame are aligned.

4. Install the brush holder tool to keep the brushes in the pockets of the commutator end cap.

5. Align the match marks on the commutator end cap and starter frame. Hold the drive end and commutator end caps firmly to the starter frame. Remove the brush holder tool.

6. Install the through bolts and tighten securely.

7. Lubricate the drive shaft with Kohler electric starter drive lubricant. Install the drive pinion, dust cover spacer, anti-drift spring, stop gear spacer, stop nut, and dust cover. Refer to "Starter Drive Service" above.

Solenoid Shift Electric Starter

▶ **See Figures 106 and 107**

When power is applied to the starter the electric solenoid moves the drive pinion out onto the drive shaft and into mesh with the flywheel ring gear. When the pinion reaches the end of the drive shaft it rotates the flywheel and cranks the engine.

When the engine starts and the start switch is released the starter solenoid is deactivated, the drive lever moves back, and the drive pinion moves out of mesh with the ring gear into the retracted position.

STARTER REMOVAL & INSTALLATION

Refer to the Engine Disassembly and Assembly sections for starter removal and installation procedures.

STARTER DISASSEMBLY

1. Remove clip.

2. Remove cap screws and solenoid. Scribe alignment marks on caps and frame to aid assembly.

3. Remove through bolts, drive end cap, commutator end cap, and frame.

4. Remove drive lever.

5. Remove thrust washer and retainer to remove drive pinion from shaft.

Brush Replacement

See the procedure as explained in the Bendix Drive section, above.

Commutator Service

See the procedure as explained in the Bendix Drive section, above.

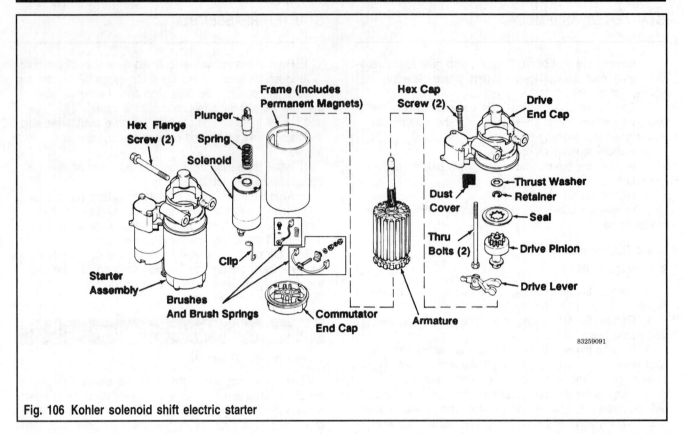

Fig. 106 Kohler solenoid shift electric starter

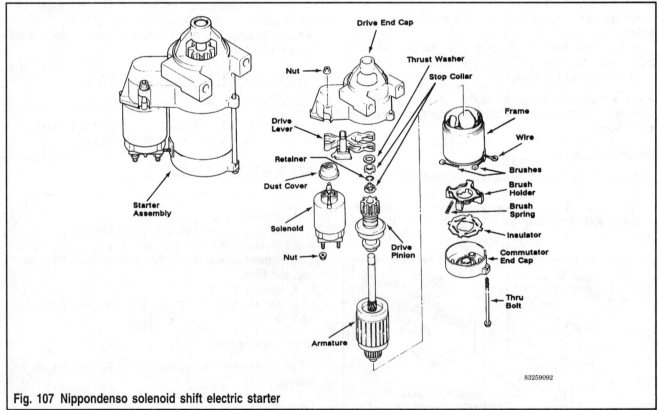

Fig. 107 Nippondenso solenoid shift electric starter

STARTER REASSEMBLY

1. Slide frame over armature and place commutator end cap in position. Hold in position temporarily with tape.

➡**Be sure alignment marks on caps and frame are in proper position.**

2. Place drive pinion (with seal), thrust washer and retainer on drive shaft.
3. Place lever in position on drive shaft.
4. Place solenoid plunger on drive lever and position drive end cap over drive shaft. (Be sure the rubber dust cover is in place at the drive lever.)
5. Fasten the end caps with the through bolts.
6. Place the spring in the solenoid and fasten solenoid to drive end cap using hex cap screws.
7. Replace the clip.

Engine Mechanical

ENGINE DISASSEMBLY

▶ **See Figures 108 and 109**

✳✳CAUTION

Before servicing the engine or equipment, always disconnect the spark plug lead to prevent the engine from starting accidentally. Ground the lead to prevent sparks which could cause fires.

The following sequence is suggested for complete engine disassembly. This procedure can be varied to accommodate options or special equipment.

Clean all parts thoroughly as the engine is disassembled. Only clean parts can be accurately inspected and gauged for wear or damage. There are many commercially available cleaners that will quickly remove grease, oil, and grime from engine parts. When such a cleaner is used, follow the manufacturer's instructions and safety precautions carefully.

Make sure all traces of the cleaner are removed before the engine is reassembled and placed into operation. Even small amounts of these cleaners can quickly break down the lubricating properties of engine oil.

1. Remove spark plug.
2. Drain oil.
3. Remove muffler and bracket.
4. Remove air cleaner cover.
5. Remove air cleaner element, base, and breather hose.
6. Remove choke control.
7. Remove fuel tank and bracket.
8. Remove retractable starter.
9. Remove fuel pump.
10. Remove starter cover - electrical starter.
11. Remove rectifier-regulator.
12. Remove Oil Sentry.
13. Remove throttle control bracket.
14. Remove carburetor,

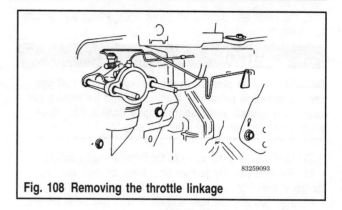

Fig. 108 Removing the throttle linkage

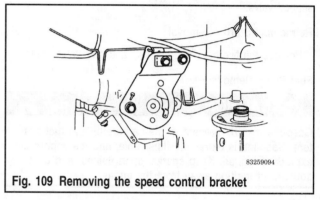

Fig. 109 Removing the speed control bracket

15. Remove valve cover.
16. Remove cylinder head baffle.
17. Remove blower housing and baffles.
18. Remove ignition module.
19. Remove fuel line.
20. Remove cylinder head push rods/gasket.
21. Remove drive cup, grass screen, flywheel and fan.
22. Remove stator and wiring harness.
23. Remove oil fill tube - if necessary.
24. Remove closure plate.
25. Remove camshaft and hydraulic lifters.
26. Remove balance shaft.
27. Remove connecting rod.
28. Remove piston.
29. Remove crankshaft.
30. Remove main bearing.
31. Remove governor gear.
32. Remove governor cross shaft seal.
33. Disconnect spark plug lead.

➡**Pull on boot only, to prevent damage to spark plug lead.**

34. Drain the oil and remove the filter.
35. Remove the muffler
36. Remove the knob and air cleaner cover.
37. Remove the wing nut, washer, element cover, element and precleaner.
38. Remove the hex flange nuts from the intake studs, and the air cleaner base and gasket from the studs.
39. Loosen the hose clamp and disconnect the breather hose from the rocker arm cover. Remove the air cleaner base

from the studs and disconnect choke linkage from the carburetor choke lever.

❊❊CAUTION

Gasoline may be present in the carburetor and fuel system. Gasoline is extremely flammable, and its vapors can explode if ignited. Keep sparks, open flames, and other sources of ignition away from the engine.

40. Turn fuel shut-off valve to OFF (horizontal) position.
41. Remove hex flange nuts from lower bracket and hex flange screws from upper bracket of fuel tank.
42. Remove the fuel tank and disconnect fuel hose from shut-off valve.

Retractable Starter Removal

Remove the five hex flange screws and retractable starter.

Fuel Pump Removal

❊❊CAUTION

Gasoline may be present in the carburetor and fuel system. Gasoline is extremely flammable, and it's vapors can explode if ignited. Keep sparks, open flames, and other sources of ignition away from the engine.

1. Disconnect the fuel line from the outlet and inlet fittings of the fuel pump.
2. Remove the two hex flange screws, fuel pump, and gasket.

Electric Starter Removal

1. Disconnect the lead from the stud terminal. Disconnect both leads on Solenoid Shift Starter.
2. Remove the two hex flange screws and starter cover.
3. Remove the starter assembly and spacers from the studs.

Rectifier-Regulator Removal

♦ **See Figure 110**

1. Remove the wire connector from the rectifier-regulator.
2. Remove the two hex flange screws and rectifier-regulator.

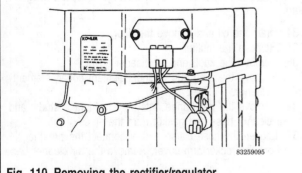

83259095

Fig. 110 Removing the rectifier/regulator

Oil Sentry Removal

1. Disconnect the lead from the Oil Sentry switch.
2. Remove Oil Sentry switch from the oil filter adapter.

Throttle Control Bracket Removal

1. Remove two hex flange screws from throttle control bracket.
2. Remove governor lever spring from throttle control bracket.

Carburetor Removal

❊❊CAUTION

Gasoline may be present in the carburetor and fuel system. Gasoline is extremely flammable, and it's vapors can explode if ignited. Keep sparks, open flames, and other sources of ignition away from the engine.

1. Remove fuel line from carburetor inlet fitting.
2. Disconnect the throttle linkage from the bushing in carburetor governor lever.
3. Remove carburetor and gasket from intake studs.

Valve Cover Removal

Remove the five hex flange cover screws and valve cover from the cylinder head assembly.

➡ **The valve cover is sealed to the cylinder head using RTV silicone sealant. When removing valve cover, use care not to damage the gasket surfaces of cover and cylinder head.**

Cylinder Head Baffle Removal

Remove the hex flange screws securing the cylinder head baffle to the cylinder head. Remove the baffle.

Blower Housing and Baffles Removal

Remove the hex flange screws from blower housing and baffles. Disconnect the wire harness from the key switch, if equipped. Remove the blower housing and baffles.

Ignition Module Removal

1. Disconnect the kill lead from the ignition module terminal.
2. Rotate flywheel magnet away from ignition module.
3. Remove the two hex flange screws and ignition module.

Fuel Line Removal

Remove the hex flange screw, clip, and fuel line.

Cylinder Head Removal

Remove the hex flange screws, spacer (from the screw by the exhaust port), cylinder head, push rods, and cylinder head gasket.

Cylinder Head Disassembly

♦ **See Figure 111**

1. Remove the spark plug.

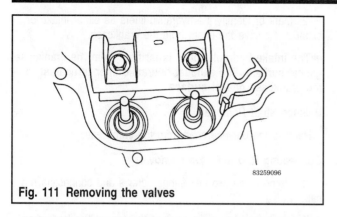

Fig. 111 Removing the valves

2. Remove the hex flange screw, breather reed retainer, and breather reed.

3. Remove the rocker shaft (from the breather side of head), and rocker arms.

4. Remove the valves:

a. Compress the valve springs using a valve spring compressor.

b. Remove the keepers, valve spring caps, valve springs, exhaust valve rotator, intake valve spring seat, and intake valve stem seal.

5. Remove the two hex cap screws and rocker bridge.

Drive Cup, Grass Screen, Flywheel, and Fand Removal

▶ **See Figures 112 and 113**

➡**Always use the flywheel strap wrench to hold the flywheel when loosening or tightening the flywheel and fan retaining fasteners. Do not use any type of bar or wedge between the fins of cooling fan as the fins could become cracked or damaged.**

Always use a puller to remove the flywheel from the crankshaft. Do not strike the crankshaft or flywheel, as these parts could become cracked or damaged.

1. Remove the hex flange screw, plain washer, and driver cup.

2. Unsnap and remove the grass screen from fan.

3. Remove the flywheel from the crankshaft using a puller.

4. Remove the four hex flange screws and fan from flywheel.

Stator and Wiring Harness Removal

▶ **See Figure 114**

1. Remove the stator leads from connector body.

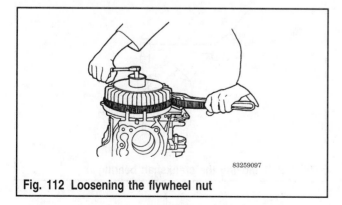

Fig. 112 Loosening the flywheel nut

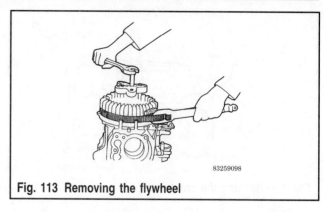

Fig. 113 Removing the flywheel

2. Remove the hex flange screw and clip securing the stator leads to the crankcase.

3. Remove the hex flange screw and clip securing the kill lead to the crankcase. Remove the four hex socket head screws and stator.

Closure Plate Removal

▶ **See Figures 115 and 116**

1. Remove the twelve hex flange screws securing the closure plate to the crankcase.

2. Locate the splitting notches in the seam of the closure plate and crankcase. Pry the closure plate from the crankcase using a large flat-blade screwdriver.

➡**Insert the screwdriver only in the splitting notches. Do not pry on the gasket surfaces of the closure plate or crankcase as this can cause leaks.**

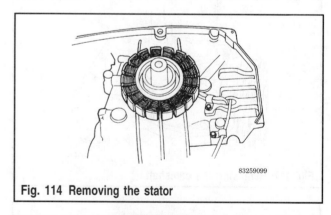

Fig. 114 Removing the stator

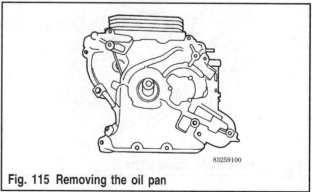

Fig. 115 Removing the oil pan

Fig. 116 Splitting the crankcase

Oil Pickup, Oil Pressure Relief Valve, Oil Pump, and Oil Seal

1. Remove the oil seal from the closure plate.
2. Remove the hex flange screw, clip, oil pickup, and O-ring seal.
3. Remove the hex socket screw, oil pressure relief bracket, relief valve body, piston, and spring.
4. Remove the three hex flange screws, oil pump cover, O-ring, and oil pump rotors.

Camshaft and Hydraulic Lifters Removal
▶ **See Figures 117 and 118**

1. Remove the camshaft and shim.

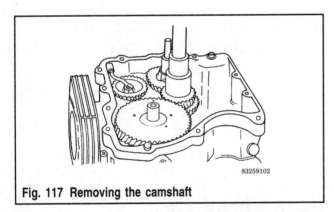

Fig. 117 Removing the camshaft

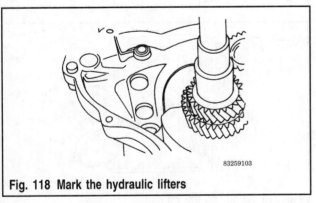

Fig. 118 Mark the hydraulic lifters

2. Mark or identify the hydraulic lifters as either intake or exhaust. Remove the lifters from the crankcase.

➡**The intake hydraulic lifter is farthest from the crankcase gasket surface. The exhaust hydraulic lifter is nearest to the crankcase gasket surface.**

Balance Shaft Removal

Remove the balance shaft from the crankcase.

Connecting Rod & Piston Removal

1. Remove the two hex flange screws and connecting rod cap.

➡**If a carbon ridge is present at the top of the bore, use a ridge reamer tool to remove it before attempting to remove the piston.**

2. Carefully push the connecting rod and the piston away from the crankshaft and out of the cylinder bore.
3. Remove the wrist pin retainer and wrist pin. Separate the piston from the connecting rod.
4. Remove the top and center compression rings using a ring expander tool.
5. Remove the oil control ring rails, then remove the rails spacer.

Crankshaft Removal
▶ **See Figure 119**

1. Remove the woodruff key from the flywheel taper end of crankshaft.
2. Remove the crankshaft from the crankcase.

Flywheel End Oil Seal and Bearing Removal

1. Remove the oil seal from crankcase.
2. Remove the bearing from the crankcase using handle #NU-4747 and bearing remover #KO-1029.

Governor Cross Shaft and Governor Gear Removal
▶ **See Figure 120**

1. Remove the hitch in and plain washer from governor cross shaft.
2. Remove the cross shaft and plain washer from the crankcase.
3. Remove the governor cross shaft oil seal from the crankcase.

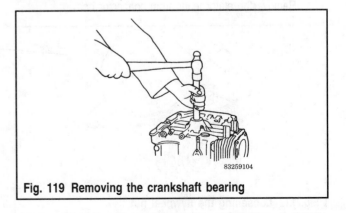

Fig. 119 Removing the crankshaft bearing

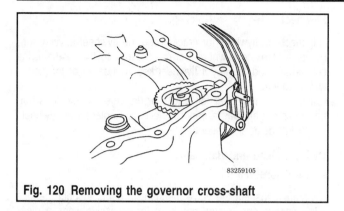

Fig. 120 Removing the governor cross-shaft

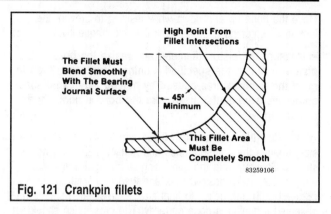

Fig. 121 Crankpin fillets

4. If necessary, remove the governor gear and regulating pin.

➡The governor gear is held onto the governor gear shaft by small molded tabs in the gear. When the gear is removed from the shaft these tabs are destroyed. This will require replacement of the gear; therefore, remove the gear only if absolutely necessary (such as when reboring, doing major engine rebuilding, etc.).

ENGINE COMPONENT OVERHAUL

Clean all parts thoroughly. Only clean parts can be accurately inspected and gauged for wear or damage. There are many commercially available cleaners that will quickly remove grease, oil, and grime from engine parts. When such a cleaner is used, follow the manufacturer's instructions and safety precautions carefully.

Make sure all traces of the cleaner are removed before the engine is reassembled and placed into operation. Even small amounts of these cleaners can quickly break down the lubricating properties of engine oil.

Camshaft

Inspect the gear teeth of the camshaft. If the teeth are badly worn, chipped, or some are missing, replacement of the camshaft will be necessary.

Crankshaft

▶ See Figure 121

Inspect the gear teeth of the crankshaft. If the teeth are badly worn, chipped or some are missing, replacement of the crankshaft will be necessary.

Inspect the crankshaft bearings for scoring, grooving, etc. Do not replace bearings unless they show signs of damage or are out of running clearance specifications. If the crankshaft turns easily and noiselessly, and there is no evidence of scoring, grooving, etc., on the races of bearing surfaces, the bearings can be reused.

Inspect the crankshaft keyways. If worn or chipped, replacement of the crankshaft will be necessary.

Inspect the crankpin for score marks or metallic pickup. Slight score marks can be cleaned with crocus cloth soaked in oil. If wear limits, as stated in "Specifications and Tolerances" are exceeded, it will be necessary to either replace the crank-

shaft or regrind the crankpin to 0.25 mm (0.010 in) undersize. If reground, a 0.25 mm (0.010 in) undersize connecting rod (big end) must then be used to achieve proper running clearance. Measure the crankpin for size, taper, and out-of-round.

➡If the crankpin is reground, visually check to ensure that the fillet blends smoothly with the crankpin surface.

When regrinding a crankshaft, grinding stone deposits can get caught in oil passages which could cause severe engine damage. Remove the sealing plug each time the crankshaft is ground to provide easy access for cleaning any grinding deposits that may collect in the oil passages.

Crankcase

Check all gasket surfaces to make sure they are free of gasket fragments. Gasket surfaces must also be free of deep scratches or nicks.

Check a the cylinder bore wall for scoring. In severe cases, unburned fuel can cause scuffing and scoring of the cylinder wall. It washes the necessary lubricating oils off the piston and cylinder wall. As raw fuel seeps down the cylinder wall, the piston rings make metal to metal contact with the wall. Scoring of the cylinder wall can also be caused by localized hot spots resulting from blocked cooling fins or from inadequate or contaminated lubrication.

If the cylinder bore is badly scored, excessively worn, tapered, or out of round, resizing is necessary. Use an inside micrometer to determine amount of wear then select the nearest suitable oversize of either 0.25 mm (0.010 in) or 0.50 mm (0.020 in). Resizing to one of these oversizes will allow usage of the available oversize piston and ring assemblies. Initially, resize using a boring bar, then use the following procedures for honing the cylinder.

HONING

While most commercially available cylinder hones can be used with either portable drills or drill presses, the use of a low speed drill press is preferred as it facilitates more accurate alignment of the bore in relation to the crankshaft crossbore. Honing is best accomplished at a drill speed of about 2500 RPM and 60 strokes per minute. After installing coarse stones in hone, proceed as follows:

1. Lower hone into bore and after centering, adjust so that the stones are in contact with the cylinder wall. Use of a commercial cutting-cooling agent is recommended.

2. With the lower edge of each stone positioned even with the lowest edge of the bore, start drill and honing process.

Move the hone up and down while resizing to prevent the formation of cutting ridges. Check the size frequently.

➡**Measure the piston diameter and resize the bore to the piston to obtain the specified running clearances. Keep in mind the temperatures caused by honing may cause inaccurate measurements. make sure the bore is cool when measuring.**

3. When the bore is within 0.064 mm (0.025 in) of desired size, remove the coarse stones and replace with burnished stones. Continue with the burnishing stones until within 0.013 mm (0.0005 in) of desired size, and then use finish stones (220-280 grit) and polish to final size. A crosshatch should be observed if honing is done correctly. The crosshatch should intersect at approximately 23-33 degrees off the horizontal. Too flat of an angle could cause the rings to skip and wear excessively, to steep of an angle will result in high oil consumption.

4. After resizing, check the bore for roundness, taper, and size. Use an inside micrometer, telescoping gauge, or bore gauge to take measurements. The measurements should be taken at three locations in the cylinder - at the top, middle, and bottom. Two measurements should be taken (perpendicular to each other) at each of the three locations.

MEASURING PISTON-TO-BORE CLEARANCE

▶ **See Figure 122**

Before installing the piston into the cylinder bore, it is necessary that the clearance be accurately checked. This step is often overlooked, and if the clearances are not within specifications, engine failure will usually result.

➡**Do not use a feeler gauge to measure piston-to-bore clearance - it will yield inaccurate measurements. Always use a micrometer.**

Use the following procedure to accurately measure the piston-to-bore clearance:

1. Use a micrometer and measure the diameter of the piston 6 mm (.024 in) above the bottom of the piston skirt and perpendicular to the piston pin.

2. Use an inside micrometer, telescoping gauge, or bore gauge and measure the cylinder bore. Take the measurement approximately 63.5 mm (2.5 in) below the top of the bore and perpendicular to the piston pin.

3. Piston-to-bore clearance is the difference between the bore diameter (step 2 minus step 1).

Flywheel

Inspect the flywheel for cracks, and the flywheel keyway for damage. Replace flywheel if cracked. Replace the flywheel, the crankshaft, and the key if flywheel key is sheared or the keyway damaged.

Inspect the ring gear for cracks or damage. Kohler does not provide ring gears as a serviceable part. Replace the flywheel if the ring gear is damaged.

Cylinder Head and Baffles

▶ **See Figure 123**

Carefully inspect the valve mechanism parts. Inspect the valve springs and related hardware for excessive wear or distortion. Check the valves and valve seat area or inserts for evidence of deep pitting, cracks, or distortion. Check clearance of the valve stems in guides.

Hard starting, or loss of power accompanied by high fuel consumption may be symptoms of faulty valves. Although these symptoms could also be attributed to worn rings, remove and check the valves first. After removal, clean the valve heads, faces, ad stems with a power wire brush. Then, carefully inspect each valve for defects such as warped head, excessive corrosion, or worn stem end. Replace valves found to be in bad condition. A normal valve and valves in bad condition are shown in the accompanying illustrations.

VALVE GUIDES

If a valve guide is worn beyond specifications, it will not guide the valve in a straight line. This may result id burnt valve faces or seats, loss of compression, and excessive oil consumption.

To check valve guide to valve stem clearance, thoroughly clean the valve guide and, using a split-ball gauge, measure the inside diameter. Then, using an outside micrometer, measure the diameter of the valve stem at several points in the stem where it moves in the valve guide. Use the largest stem diameter to calculate the clearance. If the clearance exceeds 7.134 mm (0.2809 in) on intake or 7.159 mm (0.2819 in) on exhaust valve, determine whether the valve stem or the guide is responsible for the excessive clearance.

If the valve stem diameter is within specifications, then recondition the valve guide.

The valve guides in the cylinder head are not removable. Use a 0.25 mm (0.010 in) O/S reamer. Tool no. KO-1026.

VALVE SEAT INSERTS

▶ **See Figure 124**

Intake valve seats are usually machined into the cylinder head, however, certain applications may specify hard alloy inserts. The valve seats are not replaceable. If the seats become badly pitted, cracked, or distorted, the inserts can be reconditioned.

Use a standard valve seat cutter and cut seat to dimensions shown.

LAPPING VALVES

Reground or new valves must be lapped in, to provide fit. Use a hand valve grinder with suction cup for final lapping. Lightly coat valve face with "fine" grade of grinding compound, then rotate valve on seat with grinder. Continue grinding until

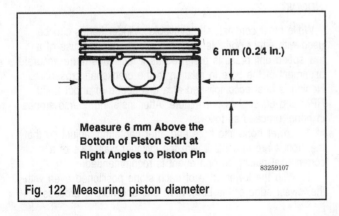

6 mm (0.24 in.)

Measure 6 mm Above the Bottom of Piston Skirt at Right Angles to Piston Pin

83259107

Fig. 122 Measuring piston diameter

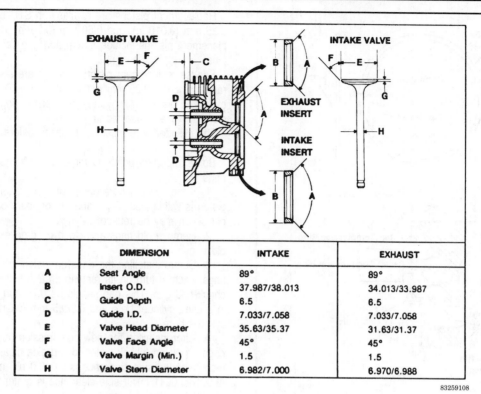

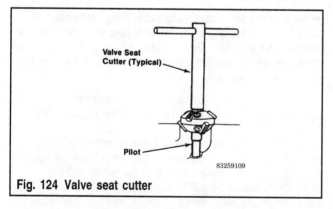

	DIMENSION	INTAKE	EXHAUST
A	Seat Angle	89°	89°
B	Insert O.D.	37.987/38.013	34.013/33.987
C	Guide Depth	6.5	6.5
D	Guide I.D.	7.033/7.058	7.033/7.058
E	Valve Head Diameter	35.63/35.37	31.63/31.37
F	Valve Face Angle	45°	45°
G	Valve Margin (Min.)	1.5	1.5
H	Valve Stem Diameter	6.982/7.000	6.970/6.988

83259108

Fig. 123 Valve details

Valve Seat Cutter (Typical)

Pilot

83259109

Fig. 124 Valve seat cutter

smooth surface is obtained on seat and on valve face. Thoroughly clean cylinder head in soap and hot water to remove all traces of grinding compound. After drying cylinder head, apply a light coating of SAE 10 oil to prevent rusting.

INTAKE VALVE STEM SEAL

These engines use valve stem seals on the intake valves. Always use a new seal when valves are removed from cylinder head. The seals should also be replaced if deteriorated or damaged in any way. Never reuse an old seal.

Pistons and Rings

▶ See Figures 125 and 126

Scuffing and scoring of pistons and cylinder walls occurs when internal temperature approach the welding point of the piston. Temperatures high enough to do this are created by

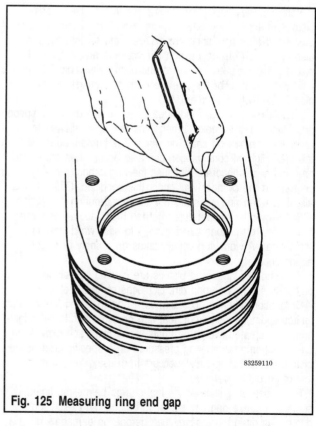

83259110

Fig. 125 Measuring ring end gap

friction, which is usually attributed to improper lubrication, and/or overheating of the engine.

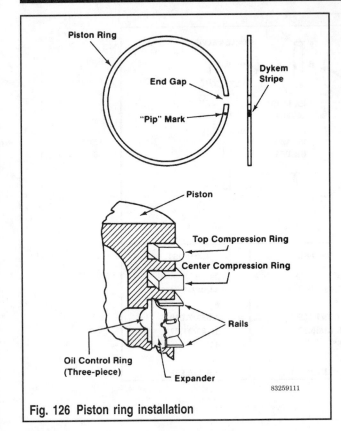

Fig. 126 Piston ring installation

Normally, very little wear takes place in the piston boss-piston pin area. If the original piston and connecting rod can be reused after new rings are installed, the original pin can also be reused but new piston pin retainers are required. The piston pin is included as part of the piston assembly - if the pin boss in piston or the pin, are worn or damaged, a new piston assembly is required.

Ring failure is usually indicated by excessive oil consumption and blue exhaust smoke. When rings fail, oil is allowed to enter the combustion chamber where it is burned along with the fuel. High oil consumption can also occur when the piston ring end gap is incorrect because the ring cannot properly conform to the cylinder wall under this condition. Oil control is also lost when ring gaps are not staggered during installation.

When cylinder temperatures get too high, lacquer and varnish collect on pistons causing rings to stick which results in rapid wear. A worn ring usually takes on a shiny or bright appearance.

Scratches on rings and pistons are caused by abrasive materials such as carbon dirt, or pieces of hard metal.

Detonation damage occurs when a portion of the fuel charge ignites spontaneously from heat and pressure shortly after ignition. This creates two flame fronts which meet and explode to create extreme hammering pressures on a specific area of the piston. Detonation generally occurs from using fuels with too low of an octane rating.

Preignition or ignition of the fuel charge before the timed spark can cause damage similar to detonation. Preignition damage is often more sever than detonation damage - often a hole is quickly burned right through the piston dome. Preignition is caused by a hot spot in the combustion chamber from sources such as: glowing carbon deposits, blocked fins, improperly seated valve, or wrong plug.

Replacement pistons are available in STD bore size and in .025 mm (0.010 in), and 0.50 mm (0.20 in), oversizes. Replacement pistons include new piston ring sets and new piston pins.

Service replacement piston ring sets are also available separately for STD, 0.25 mm (0.010 in), and 0.50 mm (0.020 in), oversized pistons. Always use new piston rings when installing pistons. Never reuse old rings.

The cylinder bore must be deglazed before service ring sets ar used.

Some important points to remember when servicing piston rings:

1. If the cylinder bore does not need reboring and if the old piston is within wear limits and fee of score or scuff marks, the old piston may be reused.

2. Remove old rings and clean up grooves. Never reuse old rings.

3. Before installing the rings on piston, place the top two rings, each in turn, in its running area in cylinder bore and check end gap. This gap should be 0.75 mm (0.030 in) max. in a used cylinder bore and 0.3/0.5 mm (0.012 in) in a new cylinder bore.

4. After installing the new compression (top and middle) rings on piston, check piston-to-ring side clearance. Maximum recommended side clearance is 0.040/0.105 mm (0.0016/0.0041 in). If side clearance is grater than specified, a new piston must be used.

INSTALL PISTON RINGS

➡**Rings must be installed correctly. Ring installation instructions are usually included with new ring sets. Follow instructions carefully. Use a piston ring expander to install rings. Install the bottom (oil control) ring first and the top compression ring last.**

1. Oil Control Ring (Bottom Groove): Install the expander and then the rail. Make sure the ends of the expander are not overlapped.

2. Compression Ring (Center Groove): Install the center ring using a piston ring installation tool. Make sure the "pip" mark is up and the PINK stripe is to the left of end gap.

3. Compression Ring (Top Groove): Instal the top ring using a piston ring installation tool. Make sure to "pip" mark is up and the BLUE stripe is to the left of the end gap.

Connecting Rods

Offset Stepped-cap Connecting Rods are used in all of these engines.

Check bearing area (big end) for excessive wear, score marks, running and side clearances. Replace rod and cap if scored or excessively worn.

Service replacement connecting rods are available in STD crankpin size and 0.25 mm (0.010 in.) undersize. The 0.25 mm (0.010 in) undersized rod can be identified by the drilled hole located in the lower end of the rod shank.

Oil Pump

Pump can be checked/replaced without removing closure plate.

Check oil pressure relief valve body, piston, and spring. Piston and body should be free of nicks or burrs. Check spring for ear or distortion. Spring free length should be approximately 0.992 in. Replace spring if distorted or worn.

Governor Gear

Inspect the governor gear teeth. Look for any evidence of worn, chipped, or cracked teeth. If one or more of these problems is noted, replace the governor gear.

The governor gear must be replaced once it is removed from the engine.

ENGINE REASSEMBLY

The following sequence is suggested for complete engine reassembly. This procedure assumes that all components are new or have been reconditioned, and all component subassembly work has been completed. This procedure may be varied to accommodate options or special equipment.

➡**Make sure the engine is assembled using all specified torque values, tightening sequences, and clearances. Failure to observe specifications could cause severe engine wear or damage. Always use new gaskets.**

1. Install flywheel end bearing.
2. Install governor gear and crosshaft.
3. Install crankshaft.
4. Install piston rings.
5. Install piston to connecting rod.
6. Install piston and rod to crankshaft.
7. Install balance shaft.
8. Install hydraulic lifters and camshaft.
9. Check camshaft end play.
10. Install and torque closure plate.
11. Install oil pump.
12. Install pro end oil seal.
13. Install flywheel end oil seal.
14. Install stator and leads.
15. Install flywheel, grass screen and drive cup.
16. Install fuel line.
17. Install and adjust ignition module.
18. Assemble cylinder head.
19. Install cylinder head.
20. Install baffles and blower housing.
21. Install cylinder head baffle.
22. Install valve cover.
23. Install fuel pump.
24. Install electric starter and cover.
25. Install fuel tank.
26. Install rectifier-regulator.
27. Install carburetor.
28. Install and adjust governor arm.
29. Install throttle bracket.
30. Install choke and air cleaner base plate.
31. Install air cleaner element/precleaner and cover.
32. Install oil filter and Oil Sentry.
33. Install dipstick.
34. Install retractable starter.
35. Install muffler and bracket.

Flywheel End Bearing

▶ **See Figure 127**

1. Mark the position of one of the crankcase bearing oil galleries on the crankcase.
2. Assemble the KO-1028 bearing installer to the NU-4747 handle. Install the sleeve bearing to the bearing installer. Align the oil hole in the bearing with the alignment notch on the installer.
3. Position the installer/bearing to the bearing bore of crankcase. Make sure the alignment notch of installer and mark on crankcase are aligned.
4. Drive the bearing into the crankcase. Make sure the bearing is installed straight and true in bore and that the tool bottoms against the crankcase.

➡**Make sure the hole in the sleeve bearing is aligned with the oil gallery in crankcase. Improper positioning of the bearings can cause engine failure due to lack of lubrication.**

Governor Gear and Cross Shaft

➡**Reuse of an old (removed) governor gear is not recommended.**

1. Install the thrust washer to governor gear shaft.
2. Position the regulating pin to governor gear/flyweights as shown. Slide the governor gear/regulating pin over the governor gear shaft.
3. Using the KO-1030 oil seal installer, install a new governor cross shaft oil seal into the crankcase.
4. Install one plain washer to the cross shaft and insert the cross shaft (from inside crankcase) thorough the crankcase and oil seal.
5. Install one plain washer and hitch pin.

Crankshaft

1. Lubricate the flywheel end bearing surfaces of the crankshaft and crankcase with engine oil.
2. Insert the crankshaft through the flywheel end bearing.

Piston and Connecting Rod

1. Install the piston, connecting rod, piston pin, and piston pin retainers.

➡**Proper orientation of the piston/connecting rod inside the engine is extremely important. Improper orientation can cause extensive wear or damage.**

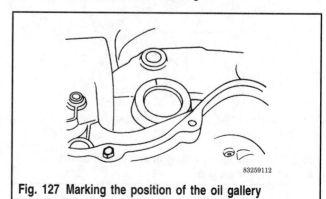

83259112

Fig. 127 Marking the position of the oil gallery

2. Stagger the piston rings in the grooves until the end gaps are 120 degrees apart.

3. Lubricate the cylinder bore, piston, and rings with engine oil. Compress the piston rings using a piston ring compressor.

4. Orient the "Fly" mark on piston towards the flywheel side of crankcase. Gently push the piston/connecting rod into bore. Do not pound on the piston.

5. Lubricate the crankshaft journal and connecting rod bearing surfaces with engine oil. Install the rod cap to connecting rod.

6. Install the hex flange screws and torque in increments to 22.6 N-m (200 inch lbs.).

7. Rotate the crankshaft until the piston is at the top dead center in the cylinder bore.

Balance Shaft

▶ **See Figure 128**

1. Lubricate the balance shaft bearing surfaces of crankshaft and balance shaft with engine oil.

2. Align the timing mark on the balance shaft gear and the larger gear on crankshaft. Lower the balance shaft into the bearing surface in the crankcase.

Make sure the balance shaft gear, large crankshaft gear and the governor gear teeth mesh and the timing marks are aligned.

Hydraulic Lifters and Camshaft

▶ **See Figure 129**

1. Lubricate the hydraulic lifters and lifter bores in crankcase with engine oil.

2. Install the hydraulic lifters into the appropriate intake or exhaust lifter bore in the crankcase.

➡**Install the lifters from inside the crankcase. The chamfered edge of the lifter must be inserted towards the cylinder head gasket surface. The intake hydraulic lifter is farthest from the crankcase gasket surface. The exhaust hydraulic lifter is nearest to the crankcase gasket surface.**

3. Lubricate the camshaft bearing surfaces of crankcase and camshaft with engine oil.

4. Align the timing marks on the camshaft gear and the smaller gear on crankshaft. Lower the camshaft into the bearing surface in crankcase.

Make sure the camshaft gear and smaller gear on crankshaft mesh and the timing marks are aligned.

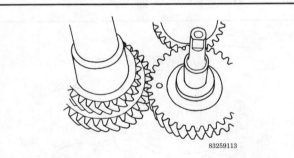

Fig. 128 Aligning the timing marks on the crankgear and balance shaft gear

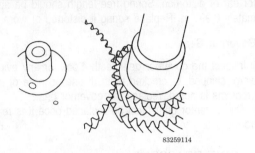

Fig. 129 Aligning the timing marks on the crankgear and camgear

DETERMINE CAMSHAFT END PLAY

▶ **See Figure 130**

1. Install the shim spacer, removed during disassembly, to the camshaft.

2. Install the camshaft end play checking tool no. KO-1031 to the crankcase and camshaft. secure the tool to the crankcase with the hex flange screws provided.

3. Using a flat feeler gauge, measure the camshaft end play between the shim spacer and the end play checking tool. Camshaft end play should be 0.076/0.127 mm (0.003/0.005 in).

4. If the camshaft end play is not within the specified range, remove the end play checking tool and add, remove, or replace shims as necessary.

Several color shims are available:
White: 0.69215/0.73025 mm (0.02725/0.02875 in)
Blue: 0.74295/0.78105 mm (0.02929/0.03075 in)
Red: 0.79375/0.83185 mm (0.03125/0.03275 in)
Yellow: 0.84455/0.88265 mm (0.03325/0.03475 in)
Green: 0.89535/0.99345 mm (0.03525/0.03675 in)
Gray: 0.94615/0.98425 mm (0.03725/0.03875 in)
Black: 0.99695/1.03505 mm (0.03925/0.04075 in)

5. Reinstall the end play checking tool and recheck end play.

6. Repeat steps 4 and 5 until the end play is within the specified range.

Oil Pressure Relief Valve

1. Place the relief valve body in the cavity of the closure plate.

2. Insert the piston and spring into the body.

3. Install the bracket and hex flange screw.

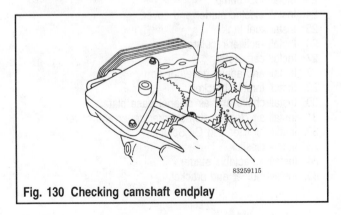

Fig. 130 Checking camshaft endplay

Oil Pickup

Install the oil pickup, O-ring, and hex flange screw.

➡ **Lightly grease O-ring and install before oil pickup.**

Closure Plate to Crankshaft

▶ **See Figures 131 and 132**

RTV sealant is used as a gasket between the closure plate and crankcase. GE Silmate® type RTV-1473 or RTV-108 silicone sealant (or equivalent) is recommended.

1. Prepare the sealing surfaces of the crankcase and closure plate as directed by the sealant manufacture.

➡ **Do not scrape the surfaces when cleaning as this will damage the surfaces. This could result in leaks. The use of a gasket removing solvent is recommended.**

2. Apply a 1/16″ bead of sealant to the closure plate as shown.

3. Install the closure plate to the crankcase and install the twelve hex flange screws. Tighten the screws hand tight.

4. Torque the fasteners, in the sequence shown to 24.4 N-m (216 inch lbs.).

Oil Pump

1. Lubricate the oil pump cavity and oil pump rotors with engine oil. Install the outer and inner oil pump rotors.

2. Install the O-ring in the groove in the closure plate.

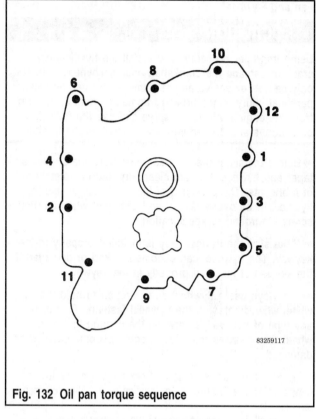

Fig. 132 Oil pan torque sequence

3. Install the oil pump cover (machined side towards O-ring). Secure with three hex flange screws.

➡ **Apply sealant to the oil pump cover hex flange screws to prevent leakage.**

4. Torque the screws as follows:
   • First Time Installation On A New Closure Plate: 6.2 N-m (55 inch lbs.).
   • Reinstallation On A Used Closure Plate: 4.0 N-m (35 inch lbs.)

Oil Seals

1. Slide the seal protector sleeve, no. KO-1037, over the crankshaft. Generously lubricate the lips of oil seal with light grease. Slide the oil seal over the sleeve.

2. Use handle no. KO-1036 and seal driver no. KO-1027. Install the seals until the driver bottoms against the crankcase of closure plate.

Stator and Wiring Harness

1. Position the stator leads towards the hole i the crankcase. Insert the stator leads through the hole to the outside of the crankcase.

2. Install the stator using four hex socket head screws. Torque the screws to 4.0 N-m (35 inch lbs.).

3. secure the stator leads to the crankcase with the clip and hex flange screw.

4. Install the connector body to the stator leads.

5. Secure the kill lead to the crankcase with the clip and hex flange screw.

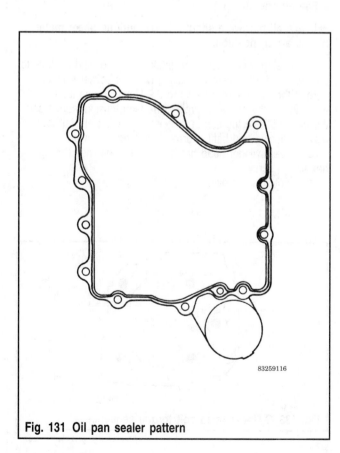

Fig. 131 Oil pan sealer pattern

Fan and Flywheel

✳✳CAUTION

Using improper procedures to install the flywheel can crack or damage the crankshaft and/or flywheel. This not only causes extensive engine damage, but can also cause personal injury, since broken fragments could be thrown from the engine. Always observe and use the following precautions and procedures when installing the flywheel.

➡**Before installing the flywheel make sure the crankshaft taper and flywheel hub are clean, dry and completely free of lubricants. The presence of lubricants can cause the flywheel to be over-stressed and damaged when the flange screw is torqued to specification.**

➡**Make sure the flywheel key is installed properly in the keyway. The flywheel can become cracked or damaged if the key is not installed properly in the keyway.**

➡ **Always use a flywheel strap wrench to hold the flywheel when tightening the flywheel fastener. Do not use any type of bar wedge between the cooling fins or flywheel ring gear, as these parts could become cracked or damaged.**

1. Install the fan, spacers and hex flange screws to the flywheel. Torque the hex flange screws to 9.9 N-m (88 inch lbs.).
2. Install the woodruff key into the keyway in the crankshaft.
3. Place the flywheel over the keyway/crankshaft. Install grass screen, drive cup, plain washer (flat side of plain washer towards the drive cup), and the hex flange screw.
4. Hold the flywheel with a strap wrench and torque the hex flange screw to 66.4 N-m (491 inch lbs.).

Fuel Line

Install the fuel line, clamp and hex flange screw.

Ignition Module

1. Install the ignition module and hex flange screws to the bosses on crankcase. Move the module as far from the flywheel/magnet as possible. Tighten the hex flange screws slightly.
2. Insert a 10 mm (0.394 in) flat feeler gauge or shim stock between the magnet and ignition module. Loosen the hex flange screws so the magnet pulls the module against the feeler gauge.
3. Tighten the hex flange screws as follows:
   • First Time Installation On A New Short Block: 6.2 N-m (55 inch lbs.)
   • All Reinstallations: 4.0 N-m (35 inch lbs.).
4. Rotate the flywheel back and forth; check to make sure the magnet does not strike the module.
5. Check the gap with feeler gauge and readjust if necessary. Final Air Gap: 0.203/0.305 mm (0.008/0.012 in)
6. Connect the kill lead to the tab terminal on ignition module.

Cylinder Head Components

1. Install the rocker bridge to the cylinder head. Make sure the small (counterbored) hole is towards the exhaust port side of the cylinder head. Secure the rocker bridge with two hex cap screws.
2. Install the intake valve stem seal, intake valve, intake valve spring seat, intake valve spring, and valve spring cap. Compress the valve spring using a valve spring compressor and install the keeper.
3. Position the rocker arms over the valve stems and rocker arm bridge. Insert the pin (from breather reed side) through the rocker bridge and rocker arms.

Cylinder Head

▶ **See Figure 133**

1. Install a new cylinder head gasket.
2. Install the cylinder head spacer (closest to the exhaust port) and hex flange screws. Torque the screws in increments of 10 ft. lbs. in the sequence shown to 40.7 N-m (30 inch lbs.).
3. Install the push rods and compress the valve springs. Snap the push rods underneath the rocker arms.
4. Install the spark plug into the cylinder head. Torque the spark plug to 38.0/43.4 N-m (28/32 inch lbs.).
5. Install the breather reed, breather reed retainer, and hex flange screw.

Baffles and Blower Housing

▶ **See Figure 134**

➡**Leave all hardware slightly loose until all sheet metal pieces are in position.**

1. Install the heat deflector, intake manifold, and gaskets to the cylinder head intake port using two hex socket screws. Torque the hex socket screws to 9.9 N-m (88 inch lbs.).
2. Install the grommet around the high tension lead. Insert the grommet into the slot in the blower housing. Install the blower housing and baffles using hex flange screws.
3. Install cylinder head baffle to the cylinder head using hex flange screws.
4. Tighten all hardware.

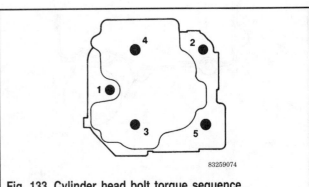

83259074

Fig. 133 Cylinder head bolt torque sequence

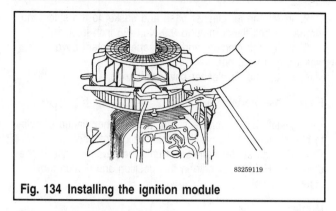

Fig. 134 Installing the ignition module

Valve Cover and Muffler Bracket

▶ See Figures 135 and 136

RTV silicone sealant is used as a gasket between the valve cover and crankcase. GE Silmate® type RTV-1473 or RTV-108 silicone sealant (or equivalent) is recommended.

1. Prepare the sealing surfaces of the cylinder head and valve cover as directed by the sealant manufacturer.

➡**Do not scrape surfaces when cleaning as this will damage the surface and could cause leaks. The use of a gasket removing solvent is recommended.**

2. Apply a 1/16″ bead of sealant to the cylinder head as shown.

3. Install the valve cover, lift bracket (lifting hole towards flywheel), and two hex flange screws.

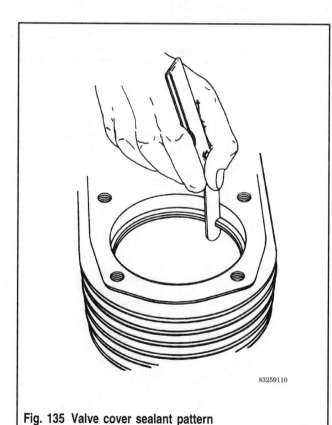

Fig. 135 Valve cover sealant pattern

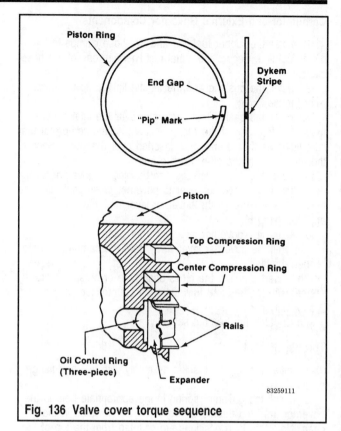

Fig. 136 Valve cover torque sequence

4. Torque the screws in the sequence shown, as follows:
   • First Time Installation On A New Cylinder Head: 10.7 N-m (95 inch lbs.).
   • All Reinstallation: 7.3 N-m (65 inch lbs.).

Fuel Pump

1. Install the rubber line and two hose clamps to the fuel pump end of the metal fuel line. Secure the rubber fuel line to the steel fuel line with on of the clamps.

2. Install the gasket, fuel pump, and two hex flange screws. Torque the screws as follows:
   • First Time Installation On A New Short Block: 9.0 N-m (80 inch lbs.).
   • All Reinstallations: 7.3 N-m (65 inch lbs.).

Electric Starter

ELECTRIC STARTER (BENDIX DRIVE OR SOLENOID SHIFT)

1. Install the starter and spacers on the mounting studs.
2. Install the starter cover and the two hex flange screws.
3. Connect the lead to the starter terminal(s).

Fuel Tank

1. Connect the fuel hose to the shut-off valve.
2. Install the hex flange screws to upper bracket of fuel tank. Install hex flange nuts to studs in lower bracket of fuel tank.

Rectifier-Regulator

1. Install the rectifier-regulator and hex flange screws.
2. Install the connector to the rectifier-regulator,

Carburetor and External Governor Components

1. Install the rubber fuel line and tow hose clamps to the metal fuel line. Secure the metal fuel line with one of the hose clamps.

2. Install the bushing and the throttle linkage to the carburetor throttle lever.

3. Install the gasket and carburetor over the intake studs. Install the free end of the rubber fuel line to the carburetor fuel inlet fitting as the carburetor is inserted over the studs. Secure the fuel line with the other hose clamp.

4. Install the throttle linkage and bushing to governor lever.

5. Install the governor lever to governor cross shaft. Do not tighten the hex nut on the governor lever until the lever is adjusted (step 6).

6. Adjust the governor lever/governor gear.

   a. Pull the governor lever away from the carburetor (wide open throttle).

   b. Insert a nail in the cross shaft hole or grasp the cross shaft with a pliers and turn the shaft counterclockwise as far as it will go,

   c. Tighten the hex nut securely.

Throttle Bracket

1. Install the throttle bracket assembly with two hex flange screws.

2. Install the governor spring in the appropriate hole in the governor arm and throttle control lever, as indicated in the chart. Note that hole positions are counted from the top of the lever. RPM should be checked with a tachometer.

Air Cleaner

▶ See Figure 137

1. Connect choke linkage to the carburetor choke lever. Install the base plate to the studs and connect the breather hose to the rocker arm cover.

2. Install the air cleaner base and gasket to the studs and torque the hex flange nuts to 9.9 N-m (88 inch lbs.).

3. Install the element and precleaner, element cover, washer, and wing nut.

4. Install the air cleaner cover and knob.

Retractable Starter

1. Install the retractable starter and five hex flange screws to blower housing. Leave the screws slightly loose.

2. Pull the starter handle out until the pawls engage in the drive cup. Hold the handle in this position and tighten the screws securely.

Muffler

1. Install the gasket, muffler, and hex flange nuts to the exhaust port studs. Leave the nuts slightly loose.

2. Secure the muffler bracket using the two hex flange screws.

3. Torque the hex flange nuts to 24.4 N-m (215 inch lbs.), screws to 9.9 N-m (88 inch lbs.).

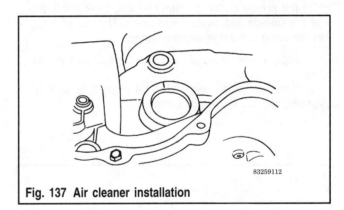

83259112

Fig. 137 Air cleaner installation

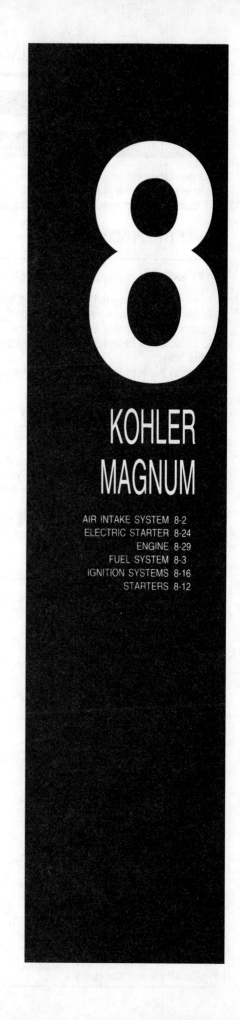

8

KOHLER
MAGNUM

AIR INTAKE SYSTEM

Air Cleaner

These engines are equipped with a high-density paper air cleaner element. Some models may also be equipped with an oiled foam precleaner which surrounds the paper element.

ELEMENT REPLACEMENT

▶ **See Figure 1**

1. Remove the wing nut and air cleaner cover.
2. Remove the precleaner (if so equipped), element cover nut, element cover, paper element, and seal.
3. Remove the base screws, air cleaner base, gasket, and breather hose.
4. Install the breather hose, gasket, air cleaner base, and screws.

➡**Make sure breather hose seals tightly in air cleaner base and breather cover to prevent unfiltered air from entering engine.**

5. Install the seal, paper element, element cover, and element cover nut. Tighten nut to 50 in. lbs. torque.
6. if so equipped, install the precleaner (cleaned and oiled) over the paper element.
7. Install the air cleaner cover and wing nut. Tighten wing nut until it is snug against cover — do not overtighten.

SERVICE

Precleaner

1. If so equipped, wash and re-oil the precleaner every 25 operating hours (more often under extremely dusty or dirty conditions).
2. Rinse the precleaner thoroughly until all traces of detergent are eliminated. Squeeze out excess water (do not wring). Allow precleaner to air dry.
3. Saturate the precleaner in clean, fresh engine oil. Squeeze out excess oil.
4. Reinstall the precleaner over paper element.

Paper Element

Every 100 operating hours (more often under extremely dusty or dirty conditions) check the paper element. Replace the element as follows:

1. Remove the precleaner (if so equipped), element cover nut, element cover, and paper element.
2. Replace a dirty, bent or damaged element with a new element. Handle new elements carefully; do not use if surfaces are bent or damaged.

➡**Do not wash the paper element or use compressed air as this will damage the element.**

3. Reinstall the paper element, element cover, and element cover nut. Make sure nut is tightened securely and element is sealed tightly against the element cover and air cleaner base.

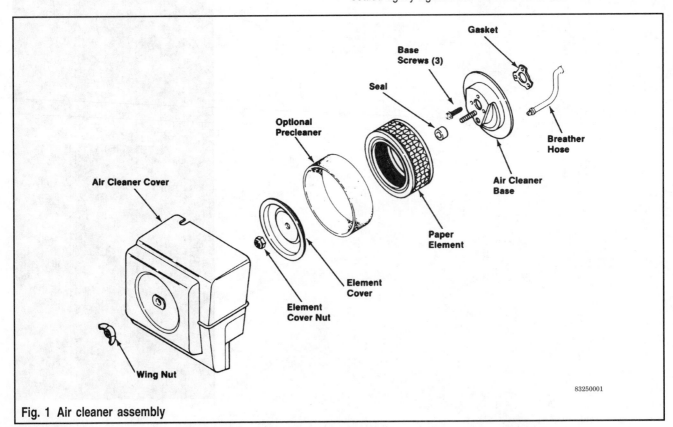

83250001

Fig. 1 Air cleaner assembly

4. Install the precleaner (cleaned and oiled) over the paper element.

5. Install the air cleaner cover and wing nut. Tighten wing nut until it is snug against cover — do not overtighten.

Inspect Air Cleaner Components

Whenever the air cleaner cover is removed, or servicing the element or precleaner, check the following components: **Air Cleaner Base** — Make sure it is secured tightly to carburetor and is not bent or damaged.

Element Cover and Element Cover Nut — Make sure element cover is not bent or damaged. Make sure element cover nut is secured tightly to seal element between air cleaner base and element cover. Tighten nut to 50 in. lbs. torque.

➡ **Breather Tube — Make sure it is sealed tightly in air cleaner base and breather cover.**

➡**Damaged, worn, or loose air cleaner components could allow unfiltered air into the engine causing premature wear and failure. Replace all damaged or worn components.**

FUEL SYSTEM

Carburetor

The typical fuel system and related components include the fuel tank with vented cap, shutoff valve with screen, in-line fuel filter, fuel pump, carburetor, and interconnecting fuel line.

OPERATION

The fuel from the tank is moved through the screen and shutoff valve, in-line filter, and fuel lines by the fuel pump. Fuel then enters the carburetor float bowl and is moved into the carburetor body where it is mixed with air. This fuel-air mixture is then burned in the engine combustion chamber.

❊❊CAUTION

Gasoline is extremely flammable and its vapors can explode if ignited. Before troubleshooting the fuel system, make sure there are no sources of heat, flames, or sparks nearby as these can ignite gasoline vapors. Disconnect and ground the spark plug lead to eliminate the possibility of sparks from the ignition system.

Fuel Tank

The fuel tank is made of a tough, impact resistant material. If the tank does become cracked or damaged, it is not repairable and must be replaced.

Refer to the 'Engine Disassembly and Reassembly' sections for complete fuel tank removal and installation procedures.

Fuel Shutoff Valve

Some engines are equipped with a fuel shutoff valve with a wire mesh screen. On engines without the shutoff valve, a straight outlet fitting is installed. The shutoff valve or outlet fitting are installed into the bottom of the fuel tank with a rubber grommet.

REMOVAL

1. Grasp the shutoff valve or outlet fitting and pull from tank using a side-to-side twisting motion.

2. Remove the grommet from tank, or from shutoff valve or outlet fitting.

INSTALLATION

1. Install the grommet into the bottom of the fuel tank.
2. Press the shutoff valve or outlet fitting securely into the grommet.

Isolation Mounts

Isolation mounts are used on models with side mounted tanks. Install isolation mounts into fuel tank hand tight.

Fuel Filter

Some engines are equipped with an in-line fuel filter. Visually inspect the filter periodically. Replace when dirty with a new filter.

Fuel Pump

◗ **See Figure 2**

Most Magnum engines are equipped with a mechanically operated fuel pump. On applications using a gravity feed fuel system, the fuel pump is not used and the pump mounting pad on the crankcase is covered.

The fuel pump body is constructed of a nylon material. The nylon body insulates the fuel from the hot engine crankcase and prevents fuel from vaporizing inside the pump.

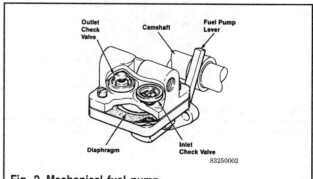

Fig. 2 Mechanical fuel pump

The mechanical fuel pump is operated by a lever which rides on the engine camshaft. The lever transmits a pumping action to the diaphragm inside the pump body. This pumping action draws fuel in through the inlet check valve on the downward stroke of the diaphragm. On the upward stroke of the diaphragm, the fuel is forced out through the outlet check valve.

REMOVAL

1. Disconnect the fuel lines from the inlet and outlet fittings of the pump.
2. Remove the fillister head screws, plain washers, fuel pump, and gasket.
3. If necessary, remove the fittings from pump body.

REPAIR AND INSTALLATION

▶ **See Figure 3**

Nylon-bodied fuel pumps are not serviceable and must be replaced when faulty. Replacement pumps are available in kits which include the pump, mounting gasket, and plain washers.

1. Apply a small amount of Permatex Aviation Perm-A-Gasket® (or equivalent) gasoline resistant thread sealant to fittings. Turn fittings into pump 6 full turns; continue turning fittings in same direction until desired position is reached.
2. Install new gasket, fuel pump, plain washers, and fillister head screws.

✳✳WARNING

Make sure the fuel pump lever is positioned above the camshaft. Damage to the fuel pump, and subsequent severe engine damage could result if the lever is positioned below the camshaft.

Make sure the plain washers are installed next to the mounting flange to prevent damage from the lockwasher.

3. Torque screws 40-45 inch lbs.
4. Connect fuel lines to inlet and outlet fittings.

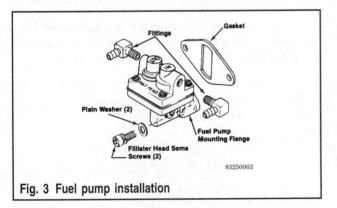

Fig. 3 Fuel pump installation

Carburetor

✳✳CAUTION

Gasoline may be present in the carburetor and fuel system. Gasoline is extremely flammable and its vapors can explode if ignited. Keep sparks, open flame, and other sources of ignition away from the engine. Wipe up spilled fuel immediately.

ADJUSTMENT

▶ **See Figures 4 and 5**

Turning the adjusting needles in (clockwise) decreases the supply of fuel to the carburetor. This gives a leaner fuel/air mixture. Turning the adjusting needles out (counterclockwise) increases the supply of fuel to the carburetor. This gives a richer fuel/air mixture.

✳✳WARNING

Incorrect settings can cause a fouled spark plug, overheating, excessive valve wear, and other problems. To ensure correct settings, make sure the following adjustment procedures are used.

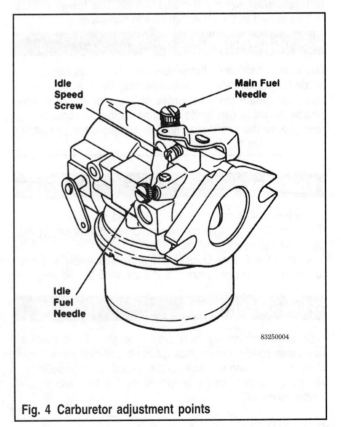

Fig. 4 Carburetor adjustment points

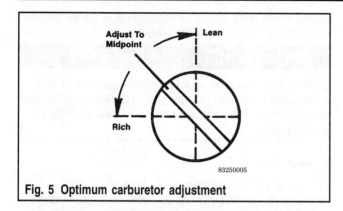

Fig. 5 Optimum carburetor adjustment

Make carburetor adjustments after the engine has warmed.

1. Stop the engine. Turn the main fuel and idle fuel adjusting needles in (clockwise) until they bottom lightly.

✳✳WARNING

The ends of the main fuel and idle fuel adjusting needles are tapered to critical dimensions. Damage to needles and seats will result if needles are forced.

2. Preliminary Settings: Turn the main fuel and idle fuel adjusting needles out (counterclockwise) from lightly bottomed as follows: Start the engine and run at half-throttle for 5-10 minutes to warm up. Engine must be warm before making final settings (steps 4-6).

3. Final Setting — Main Fuel: Place throttle in wide open position; and if possible, place engine under load. Turn main fuel adjusting needle out (counterclockwise) from preliminary setting until the engine speed decreases (rich). Note the position of the needle. Now turn the adjusting needle in (clockwise). The engine speed may increase, then it will decrease as the needle is turned in (lean). Note the position of the needle.

4. Final Setting — Idle Fuel: Place throttle into idle or slow position. Set idle fuel adjusting needle using the same procedure as in step 4.

➡**To ensure best results when setting idle fuel mixture, the idle speed must not exceed 1500 RPM. Typical idle speed is 1200 RPM. See step 6.**

5. Idle Speed Setting: Place throttle into idle or slow position. Set idle speed to 1200 rpm ± 75 RPM by turning the idle speed adjusting screw in or out.

➡**The actual idle speed depends on the application. Refer to the equipment manufacturer's instructions for specific idle speed settings.**

DISASSEMBLY

▶ **See Figure 6**

1. Remove the bowl retaining screw, retaining screw gasket, and fuel bowl.

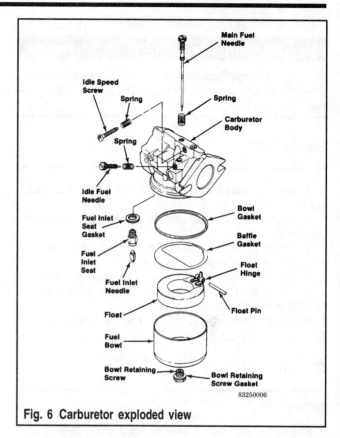

Fig. 6 Carburetor exploded view

2. Remove the float pin, float fuel inlet needle, baffle gasket, and bowl gasket.

3. Remove the fuel inlet seat and inlet seat gasket. Remove the idle fuel and main fuel adjusting needles and springs. Remove the idle speed adjusting screw and spring.

4. Further disassembly to remove the throttle and choke shafts is recommended only if these parts are to be replaced. Refer to 'Throttle And Choke Shaft Replacement.'

CLEANING

✳✳CAUTION

Carburetor cleaners and solvents are extremely flammable. Keep sparks, flames, and other sources of ignition away from area. Follow the cleaner manufacturer's warnings and instructions on its proper and safe use. Never use gasoline as a cleaning agent.

All parts should be carefully cleaned using a carburetor cleaner (such as acetone). Be sure all gum deposits are removed from the following areas: **Carburetor body and bore:** especially the areas where throttle plate, choke plate, and shafts are seated.

Float and float hinge. Fuel bowl. Idle fuel and 'off-idle' ports in carburetor bore, ports in main fuel adjusting needle, and main fuel seat. These areas can be cleaned a piece of fine wire in addition to cleaners. Be careful not to enlarge the ports, or break the cleaning wire within ports.

Blow out all passages with compressed.

INSPECTION

1. Carefully inspect all components and replace those that are worn or damaged.
2. Inspect the carburetor body for cracks, holes, and other wear or damage.
3. Inspect the float for dents or holes. Check the float hinge for wear, and missing or damaged float tabs.
4. Inspect the inlet needle and seat for wear or grooves.
5. Inspect the tips of the main fuel and idle fuel adjusting needles for wear or grooves.
6. Inspect the throttle and choke shaft and plate assemblies for wear or excessive play.

REPAIR

▶ **See Figure 7**

Always use new gaskets when servicing and reinstalling carburetors. Several repair kits are available which include the gaskets and other components.

Several kits are available for replacement of the throttle and choke shafts in Kohler-built carburetors.

Sub-assembly

1. To ensure correct reassembly, mark choke plate and carburetor body with a marking pen. Also take note of choke plate position in bore, and choke lever position.
2. Carefully and slowly remove the screws securing choke plate to choke shaft. Remove and save the choke plate as it will be reused.
3. File off any burrs which may have been left on choke shaft when screws were removed. Place carburetor on work bench with choke side down. Remove choke shaft; the detent ball and spring will drop out.
4. Note the position of the choke lever with respect to the cutout portion of choke shaft.
5. Carefully grind or file away the riveted portion of shaft. Remove and save choke lever; discard old choke shaft.
6. Install choke lever to new choke shaft from kit. Make sure lever is installed correctly as noted in step 4. Secure lever to choke shaft. Apply Loctite® to threads of (1) #3-48x$^{7}/_{32}$ in. brass screw; secure lever to shaft.
7. To ensure correct reassembly, mark throttle plate and carburetor body with a marking pen. Also take note of throttle plate position in bore, and throttle lever position.

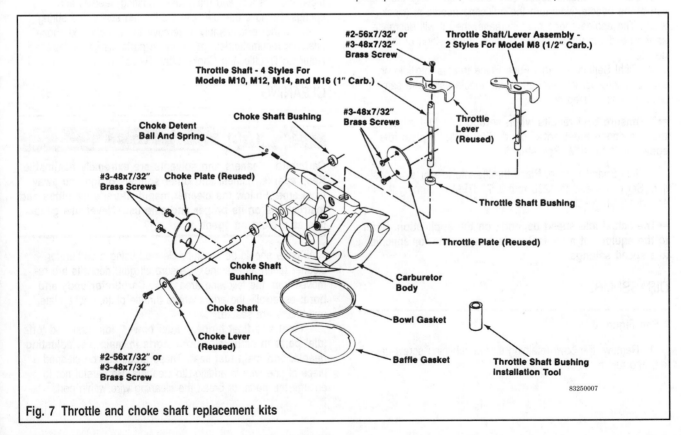

Fig. 7 Throttle and choke shaft replacement kits

8. Carefully and slowly remove the screws securing the throttle plate to throttle shaft. Remove and save the throttle plate as it will be reused.

9. File off any burrs which may have been left on throttle shaft when screws were removed.

✳✳WARNING

Failure to remove burrs from the throttle shaft may cause permanent damage to carburetor body when shaft is removed.

10. Remove the throttle shaft from carburetor body. Remove and discard the foam rubber dust seal from throttle shaft.

a. Note the position of the throttle lever with respect to the cutout portion of throttle shaft.

b. Carefully grind or file away the riveted portion of shaft; remove throttle lever.

c. Carefully compare the old shaft to the new shafts from kit. Select the appropriate new shaft and discard the old shaft.

d. Install throttle lever to throttle shaft. Make sure lever is installed correctly as noted in step a.

e. Apply Loctite® to threads of (1) #2-56x$7/32$ in. brass screw (use #3-48x$7/32$ in. screw with $2^{49}/64$ in. shaft); secure lever to shaft.

11. Mount the carburetor body in a drill press vise. Keep vice slightly loose.

12. Install a drill of the following specified size in drill press chuck. Lower drill (not rotating) through both choke shaft bores; then tighten vice. This ensures the carburetor body and drill are perpendicular and in correct alignment. Use a $1/4$ in. dia drill.

13. Install a $19/64$ in. dia. drill in chuck. Set drill press speed to a low speed suitable for aluminum. Feed drill slowly to obtain a good finish to holes.

14. Ream the choke shaft bores to a final size of $5/16$ in.. For best results use a piloted $5/16$ in. reamer.

15. Blow out all metal chips using compressed air. Thoroughly clean the carburetor body in a carburetor cleaner.

Carburetor Assembly
▶ **See Figures 8, 9 and 10**

1. Install screws into the tapped holes that enter the choke shaft bores until the screws bottom lightly.

2. Coat the outside surface of choke shaft bushings with Loctite® from kit. Carefully press bushings into carburetor body using a smooth-jawed vice. Stop pressing when bushings bottom against screws.

3. Allow Loctite® to 'set' for 5-10 minutes then remove screws.

4. Install new choke shaft in bushings. Rotate shaft and check for binding.

➡**If binding occurs, locate and correct the cause before proceeding. Use choke shaft to align bushings if necessary.**

5. Remove choke shaft and allow Loctite® to 'set' for an additional 30 minutes before proceeding.

6. Wipe away all excess Loctite® from bushings and choke shaft.

7. Make sure the dust seal counterbore in carburetor is thoroughly clean and free of chips and burrs.

8. Install a throttle shaft (without throttle lever) into carburetor body to use as a pilot. Use one of the remaining new throttle shafts from kit.

9. Coat the outside surface of throttle shaft bushing with Loctite® from kit. Slip bushing over shaft. Using installation tool from kit and vice, press bushing into counterbore until it bottoms in carburetor body.

10. Allow Loctite® to 'set' for 5-10 minutes then remove throttle shaft.

11. Install new throttle shaft with lever into carburetor body. Rotate shaft and check for binding.

➡**If binding occurs, locate cause and correct before proceeding. Use throttle shaft to align bushing if necessary.**

12. Remove shaft and allow Loctite® to 'set' for an additional 30 minutes before proceeding.

13. Wipe away all excess Loctite® from bushing and throttle shaft.

14. Install new detent spring and ball into carburetor body in the side opposite choke lever.

15. Compress detent ball and spring and insert choke shaft through bushings. Make sure the choke lever is on the correct side of carburetor body.

16. Install choke plate to choke shaft. Make sure marks are aligned and plate is positioned properly in bore. Apply Loctite® to threads of (2) #3-48x$7/32$ in. screws. Install screws so they are slightly loose.

17. Operate choke lever. Check for binding between choke plate and carburetor bore. Loosen screws and adjust plate as necessary; then tighten screws securely.

18. Install throttle shaft into carburetor with cutout portion of shaft facing out.

19. Install throttle plate to throttle shaft. Make sure marks are aligned and plate is positioned properly in bore. Apply Loctite® to threads of (2) #3-48x$7/32$ in. screws. Install screws so they are slightly loose.

20. Apply finger pressure to throttle shaft to keep it firmly seated against pivot in carburetor body. Rotate the throttle shaft until throttle plate fully closes the bore around its entire perimeter; then tighten screws.

21. Operate throttle lever and check for binding between throttle plate and carburetor bore. Loosen screws and adjust plate as necessary; then tighten screws securely.

22. Install the fuel inlet seal gasket and fuel inlet seat into carburetor body. Torque seat to 35-45 inch lbs.

23. Install the fuel inlet needle into inlet seat. Install Float and slide float pin through float hinge and float hinge towers on carburetor body.

24. Set Float Level: Invert carburetor so the float tab rests on the fuel inlet needle. There should be $11/64$ in. ± $1/32$ in. clearance between the machined surface of body and the free end of float. Behind the float tab with a small screwdriver to adjust.

25. Set float drop: Turn the carburetor over to its normal operating position and allow float to drop to its lowest level. The float drop should be limited to $1^1/32$ in. between the machined surface of the body and the bottom of the free end of float. Bend the float tab with a small screwdriver to adjust.

26. Check float-to-float hinge tower clearance: Invert the carburetor so the float tab rests on the fuel inlet needle. Insert a

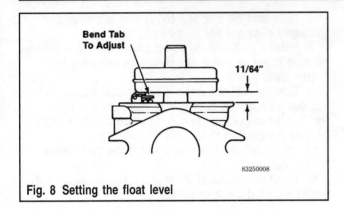

Fig. 8 Setting the float level

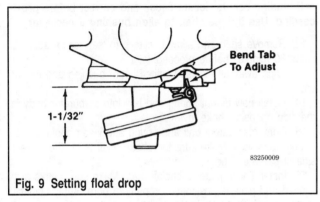

Fig. 9 Setting float drop

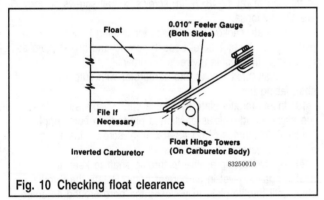

Fig. 10 Checking float clearance

0.010 in. feeler gauge between float and float hinge towers. If the feeler gauge cannot be inserted, or there is interference between between the float and towers, file the towers to obtain the proper clearance.

27. Install the bowl gasket and baffle gasket. Position baffle gasket so the inner edge is against the float hinge towers.

28. Install the fuel bowl so it is centered on the baffle gasket. Make sure the baffle gasket and bowl are positioned properly to ensure a good seal.

29. Install the bowl retaining screw gasket and bowl retaining screw. Torque screw to 50-60 inch lbs.

30. Install the idle speed adjusting screw and spring. Install the idle fuel and main fuel adjusting needles and springs. Turn the adjusting needles clockwise until they bottom lightly.

✳✳WARNING

The ends of adjusting needles are tapered to critical dimensions. Damage to needles and seats will result if needles are forced.

31. Reinstall the carburetor to the engine using a new gasket.

32. Adjust the carburetor as outlined under the 'Adjustment' portion of this section.

Governor

▶ **See Figure 11**

Magnum engines are equipped with a centrifugal flyweight mechanical governor. It is designed to hold the engine speed constant under changing load conditions. The governor gear/flyweight mechanism is mounted within the crankcase and is driven off the gear on the camshaft.

Centrifugal force acting on the rotating governor gear assembly causes the flyweights to move outward as speed increases and inward as speed decreases. As the flyweights move outward they force the regulating pin of the assembly to move outward. The regulating pin contacts the tab on the cross shaft, causing the shaft to rotate with changing speed. One end of the cross shaft protrudes through the side of the crankcase. Through external linkage attached to the cross shaft, the rotating action is transmitted to the throttle plate of carburetor.

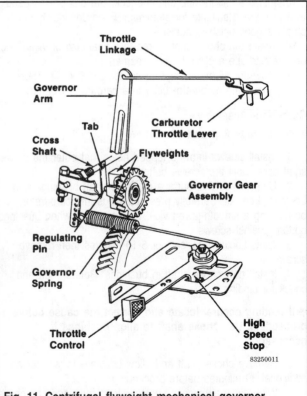

Fig. 11 Centrifugal flyweight mechanical governor

When the engine is at rest and the throttle is in the 'fast' position, the tension of the governor spring holds the throttle valve open. When the engine is operating (governor gear assembly is rotating), the force applied by the regulating pin against the cross shaft tends to close the throttle valve. The governor spring tension and the force applied by the regulating pin are in 'equilibrium' during operation, holding the engine speed constant.

When a load is applied and the engine speed (and governor speed) decreases, the governor spring tension moves the governor arm to open the throttle plate wider. This admits more fuel and restores engine speed. (This action takes place very rapidly, so a reduction in speed is hardly noticed.) As the speed reaches the governed setting, the governor spring tension and the force applied by the regulating pin will again be in equilibrium. This maintains engine speed at a relatively constant level.

Governed speed may be at a fixed point as on constant speed applications, or variable as determined by a throttle control lever.

ADJUSTMENT

✳✳CAUTION

The maximum allowable speed for these engines is 3600 RPM, no load. Never tamper with the governor setting to increase the maximum speed. Severe personal injury and damage to the engine or equipment can result if operated at speeds above maximum.

Initial Adjustment

▶ **See Figure 12**

Make this initial adjustment whenever the governor arm is loosened or removed from cross shaft. Make sure the throttle linkage is connected to governor arm and throttle lever on carburetor to ensure proper setting.

1. Pull the governor arm away from the carburetor as far as it will go.
2. Grasp the end of cross shaft with pliers and turn counterclockwise as far as it will go.
3. Tighten the pawlnut on governor arm to 15 inch lbs. torque. Make sure there is at least 1/16 in. clearance between governor arm and upper left cam gear cover fastener to prevent interference.

High Speed Adjustment

▶ **See Figure 13**

The maximum allowable speed is 3600 RPM, no load. The actual high speed setting depends on the application. Refer to the equipment manufacturer's instructions for specific high speed settings. Check the operating speed with a tachometer; do not exceed the maximum. To adjust high speed stop:

1. Loosen the lock nut on high speed adjusting screw.
2. Turn the adjusting screw in or out until desired speed is reached. Tighten the lock nut.
3. Recheck the speed with the tachometer; readjust if necessary.

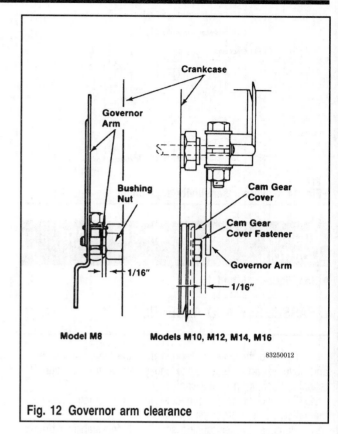

Fig. 12 Governor arm clearance

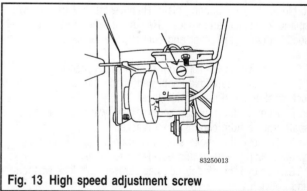

Fig. 13 High speed adjustment screw

Sensitivity Adjustment

▶ **See Figure 14**

Governor sensitivity is adjusted by repositioning the governor spring in the holes in governor arm. If set too sensitive, speed surging will occur with a change in load. If a big drop in speed occurs when normal load is applied, the governor should be set for greater sensitivity.

The standard spring position is in the sixth hole from the cross shaft. The position can vary, depending on the engine application. Therefore, make a note of (or mark) the spring position before removing it from the governor arm.

To increase sensitivity, increase the governor spring tension by moving the spring towards the cross shaft.

To decrease sensitivity, and allow broader control, decrease spring tension by moving the spring away from the cross shaft.

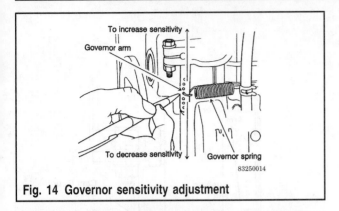

Fig. 14 Governor sensitivity adjustment

Engine-Mounted Throttle and Choke Controls

▶ **See Figures 15 and 16**

DISASSEMBLY AND ASSEMBLY

1. Remove the governor spring from throttle control lever and governor arm. Remove the choke linkage from linkage retaining bushing in choke shaft.

2. Remove the wiring connector from key switch. Remove the kill switch lead from kill switch.

3. Remove the self-tapping screw securing stabilizer bracket to blower housing/bearing plate. Remove the hex cap screw, split lock washer, and spacer. Remove the hex cap screws securing cam gear cover and control panel bracket to crankcase.

4. Remove key switch from control panel bracket.

5. Remove the choke knob, pan head screws, internal tooth lock washers, and panel with decal.

6. Remove the hex cap screws, plain washers, and choke shaft bracket from control panel bracket.

7. Remove the choke shaft with e-ring. Remove hex cap screws, plain washers, and bowden wire bracket.

8. Remove the slotted hex cap screws, split lock washers, and bowden wire bracket.

9. Remove the bowden wire clamps, choke shaft bushings, and linkage retaining bushings as necessary.

10. Remove the hex cap screw and throttle control bracket assembly from control panel bracket.

11. Remove the hex cap screw, plain washer, and stabilizer bracket from throttle control bracket.

12. Remove the hex lock nut, locking tab, plain washer, throttle control lever, and wave washer from throttle control bracket.

13. Remove bowden wire clamp, linkage retaining bushing, high speed adjusting screw, and nut as necessary.

To assemble:

14. Install the bowden wire clamp to throttle control bracket.

15. Install the linkage retaining bushing, high speed adjusting screw, and nut to throttle control lever.

16. Install the wave washer, throttle control lever, plain washer, locking tab, and hex lock nut to throttle control bracket. Make sure the locking tab goes into the hole in throttle control bracket. Torque hex lock nut to 15-25 inch lbs.

17. Install the stabilizer bracket, plain washer, and hex cap screw to throttle control bracket.

18. Install the throttle control bracket assembly and hex cap screw to control panel bracket.

19. Install the linkage retaining bushings to choke shaft.

20. Install the choke shaft bushings to choke shaft bracket.

21. Install the bowden bracket wire bracket, split lock washers, and slotted hex cap screws to choke shaft bracket.

22. Install the bowden wire bracket, plain washers, and hex cap screws to choke shaft bracket.

23. Install choke shaft with e-ring into bushings. Install the choke shaft bracket assembly, plain washers, and hex cap screw to control panel bracket.

24. Install the panel with decal, internal tooth lock washers, and pan head screws. Install choke knob to choke shaft.

25. Install key switch to control panel bracket.

26. Secure the control panel bracket and cam gear cover to crankcase using new gaskets and the hex cap screws removed previously. Install the spacer, split lock washer, and hex cap screw. Secure the stabilizer bracket to blower housing/bearing plate with the self-tapping screw removed previously.

27. Install the wiring connector to key switch. Install the kill switch lead to kill switch.

28. Install the choke linkage to linkage retaining bushing in choke shaft. Install the governor spring to throttle control lever and governor arm.

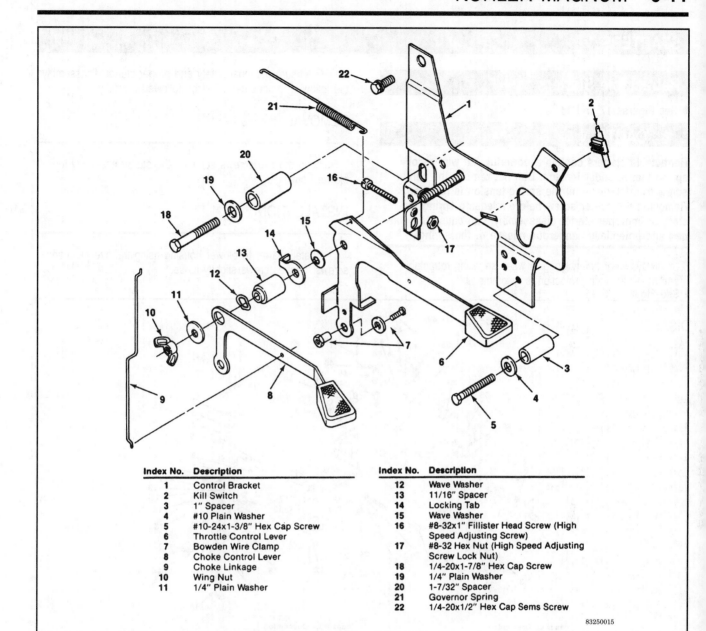

Index No.	Description
1	Control Bracket
2	Kill Switch
3	1" Spacer
4	#10 Plain Washer
5	#10-24x1-3/8" Hex Cap Screw
6	Throttle Control Lever
7	Bowden Wire Clamp
8	Choke Control Lever
9	Choke Linkage
10	Wing Nut
11	1/4" Plain Washer

Index No.	Description
12	Wave Washer
13	11/16" Spacer
14	Locking Tab
15	Wave Washer
16	#8-32x1" Fillister Head Screw (High Speed Adjusting Screw)
17	#8-32 Hex Nut (High Speed Adjusting Screw Lock Nut)
18	1/4-20x1-7/8" Hex Cap Screw
19	1/4" Plain Washer
20	1-7/32" Spacer
21	Governor Spring
22	1/4-20x1/2" Hex Cap Sems Screw

83250015

Fig. 15 Engine mounted throttle and choke controls — M8

STARTERS

Retractable Starters

▶ See Figures 17 and 18

✳✳CAUTION

Retractable starters contain a powerful, flat wire recoil spring that is under tension. Don not remove the center screw from starter until the spring tension is released. Removing the center screw before releasing spring tension, or improper starter disassembly, can cause the sudden and potentially dangerous release of the spring.

Always wear safety goggles when servicing retractable starters — full face protection is recommend

To ensure personal safety and proper starter disassembly, the following procedures must be followed carefully.

REMOVAL

Remove the five screws securing the starter assembly to blower housing.

INSTALLATION

1. Install starter to blower housing using the five mounting screws. Leave screws slightly loose.

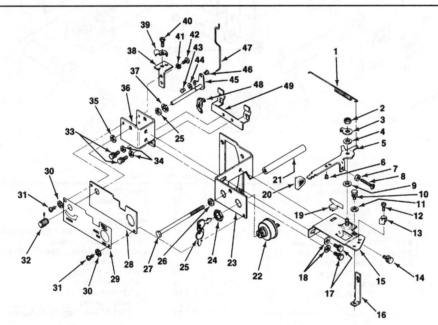

Index No.	Description
1	Governor Spring
2	1/4-20 Hex Lock Nut
3	Locking Tab
4	1/4" Plain Washer
5	Throttle Control Lever
6	Linkage Retaining Bushing (Black Plastic)
7	#10-24 Hex Nut (High Speed Adjusting Screw Lock Nut)
8	#10-24x3/8" Fillister Head Screw (High Speed Adjusting Screw)
9	1/4" Wave Washer
10	1/4-20x3/8" Hex Cap Sems Screw
11	1/4" Plain Washer
12	#10-24x3/8" Sltd. Hex Head Screw
13	Bowden Wire Clamp
14	1/4-20x3/8" Hex Cap Sems Screw
15	Throttle Control Bracket
16	Stabilizer Bracket
17	1/4-20x5/8" Hex Cap Sems Screw (2)
18	1/4" Plain Washer (2)
19	Kill Switch
20	Knob
21	4-1/2" Spacer
22	Keyswitch
23	Control Panel Bracket
24	Hex Nut
25	Keys

Index No.	Description
26	1/4" Split Lock Washer
27	1/4-20x5-1/16" Hex Cap Screw
28	Panel
29	Decal
30	#10 Internal Tooth Lock Washer (2)
31	#10-24x3/8" Pan Head Screw (2)
32	Choke Knob
33	#10-24x1/2" Hex Cap Sems Screw (2)
34	1/4" Plain Washer (2)
35	Choke Shaft Bushing (2)
36	Choke Shaft Bracket
37	E-Ring
38	Bowden Wire Bracket
39	Bowden Wire Clamp
40	#10-24x3/8" Sltd. Hex Head Screw
41	#10 Split Lock Washer (2)
42	#10-24x3/8" Sltd. Hex Head Screw (2)
43	Linkage Retaining Bushing (White Plastic)
44	3/16" Plain Washer
45	Choke Shaft
46	Linkage Retaining Bushing (Black Plastic)
47	Choke Linkage
48	Bowden Wire Clamp (2)
49	Bowden Wire Bracket

83250016

Fig. 16 Engine mounted throttle and choke controls — M10, 12, 14, 16

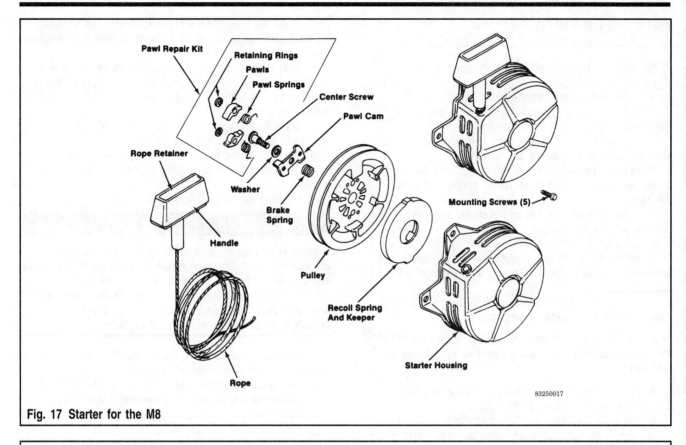

Fig. 17 Starter for the M8

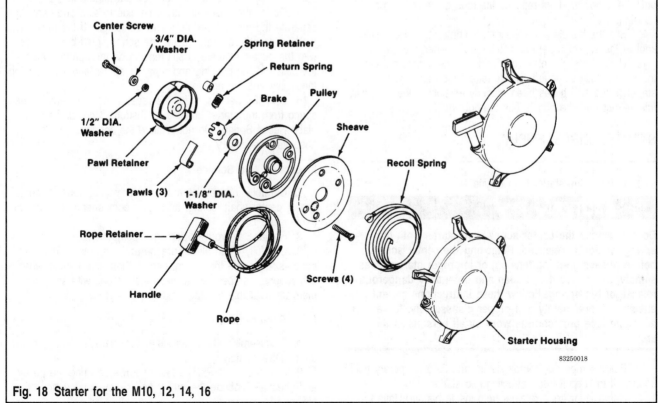

Fig. 18 Starter for the M10, 12, 14, 16

2. Pull the handle out approx. 8 to 10 in. until the pawls engage in the drive cup. Hold the handle in this position and tighten the screws securely.

ROPE REPLACEMENT

The rope can be replaced without complete starter disassembly.

1. Remove the starter from engine.

2. Pull the rope out approx. 12 in. and tie a temporary (slip) knot in it to keep it from retracting into starter.

3. Remove the rope retainer from inside handle. Untie the knot and remove the retainer and handle.

4. Hold the pulley firmly with thumb and untie the slip knot. Allow the pulley to rotate slowly as the spring tension is released.

5. When all spring tension on the starter pulley is released, remove old rope from pulley.

6. Tie a single knot in one end of new rope.

7. Rotate the pulley counterclockwise (when viewed from pawl side of pulley) until the spring is tight. (Approx. 6 full turns of pulley).

8. Rotate the pulley clockwise until the rope pocket is aligned with the rope guide bushing housing.

➡ **Do not allow pulley/spring to unwind. Enlist the aid of a helper if necessary, or use a c-clamp to hold pulley in position.**

9. Insert the new rope into the rope pocket of pulley and through rope guide bushing in housing.

10. Tie a slip knot approx. 12 in. from the free end of rope. Hold pulley firmly with thumb and allow pulley to rotate slowly until the temporary knot reaches the rope guide bushing in housing.

11. Slip the handle and rope retainer onto rope. Tie a single knot at the end of rope and install rope retainer into handle.

12. Untie the slip knot in rope and pull the handle out until the rope is fully extended. Slowly retract the rope into the starter. If the spring has been properly tensioned, the rope will fully retract until the handle hits the housing.

STARTER DISASSEMBLY

1. Remove the starter from engine.

❋❋CAUTION

Do not remove the center screw from starter until the spring tension is released. Removing the center screw before releasing spring tension, or improper starter disassembly, can cause the sudden and potentially dangerous release of the spring. Follow these instructions carefully to ensure personal safety and proper disassembly. Make sure adequate face protection is worn by all persons in the area.

2. Pull the rope out approx. 12 in. and tie a temporary (slip knot) in it to keep it from retracting into starter.

3. Remove the rope retainer from inside handle. Untie the knot and remove retainer and handle.

4. Rotate the pulley counterclockwise until the notch in pulley is next to the rope guide bushing.

5. Hold the pulley firmly to keep it from turning. Untie the slip knot and pull the rope through the bushing.

6. Place the rope into the notch in pulley. This will keep the rope from interfering with the starter housing leg reinforcements as the pulley is rotated (step 7).

7. Hold the housing and pulley with both hands. Release pressure on the pulley and allow it to rotate slowly as the spring tension is released. Be sure to keep the rope in the notch.

8. Make sure the spring tension is fully released. (The pulley should rotate easily in either direction.).

9. When all spring tension on the pulley is released, remove the center screw, ¾ in. DIA. washer, and ½ in. DIA. washer.

10. Carefully lift the pawl retainer from pulley.

❋❋CAUTION

A small return spring and nylon spring retainer (spacer) are located under the pawl retainer. These parts are fragile and can easily be lost or damaged. If necessary, use a small screwdriver to loosen the spring retainer from the post on pulley. Replace the spring if it is broken, stretched, or shows other signs of damage.

11. Remove the 1⅛ in. DIA. thrust washer, brake, return spring, nylon spring retainer, and pawls.

12. Rotate the pulley clockwise 2 full turns. There should be no resistance to this rotation. This will ensure the pulley is disengaged from the recoil spring.

13. Hold the pulley into the starter housing and invert starter so the pulley is away from your face and others in the area.

14. Rotate the pulley slightly from side to side and carefully separate the pulley from the starter housing. If the pulley and housing do not separate easily, the spring could be engaged with the pulley, or there is still tension on the spring. Return the pulley to the housing and repeat step 12 before separating the pulley and the housing.

15. Only if it necessary for the repair of starter, remove the spring from the starter housing as instructed under 'To Replace Recoil Spring.' Do not remove the spring unless it is absolutely necessary.

Inspection And Service

1. Carefully inspect the rope, starter pawls, housing, center screw, center shaft, spring, and other components for wear or damage.

2. Replace all worn or damaged components.

3. Carefully clean all oil, grease and dirt from starter components. Lubricate the spring, center shaft, and certain other components as specified in these instructions with any commercially available bearing grease.

Rope Replacement

1. Disassemble starter as instructed in steps 2 through 14 under 'Disassembly.'

2. Remove the 4 Phillips head screws securing the pulley and sheave. Separate the pulley and sheave and remove the old rope.

3. Position the new rope in the notch in the pulley and around the rope lock post.

➡ **Rope of the incorrect diameter and/or type will not lock properly in the pulley.**

4. Install the sheave on the pulley and install the 4 Phillips head screws. Use care not to strip or cross-thread the threads in pulley.

5. Inspect the pulley to make sure the sheave is securely joined to the pulley. Pull firmly on the rope to make sure it is securely retained in the pulley.

Recoil Spring Replacement

❋❋CAUTION

Do not attempt to pull or pry the recoil spring or the housing. Doing so can cause the sudden and potentially dangerous release of the spring from the housing. Follow these instructions carefully to ensure personal safety and proper spring replacement. Make sure adequate face protection is worn throughout the following procedure.

1. Carefully note the position of the spring in the housing. The new spring must be installed in the proper position — it is possible to install it backwards in the housing.

2. Place the housing on a flat wooden surface with the recoil spring and center shaft down and away from you.

3. Grasp the housing by the top so that your fingers are protected. Do not wrap your fingers around the edge of the hosing.

4. Lift the housing and rap it firmly against the wooden surface. Repeat this procedure until the spring is released from the spring pocket in housing.

5. Discard the old spring.

❋❋CAUTION

Do not attempt to rewind or reinstall a spring once it has been removed from the starter housing. Severe personal injury could result from the sudden uncoiling of the spring. Always order and install a new spring which is held in a specially designed 'C-ring' spring retainer.

6. Thoroughly clean the starter housing removing all old grease and dirt.

7. Carefully remove the masking tape surrounding the new spring/C-ring.

8. Position the spring/C-ring to the housing so the spring hook is over the post in the housing. make sure the spring is coiled in the correct direction.

9. Using Seal Installer #11791 and Handle #11795, or equivalent:

    a. Hook the spring hook over the post in housing.

    b. Make sure the spring/C-ring is centered over the spring pocket in housing.

    c. Drive the spring out of the C-ring and into the spring pocket using the seal installer and handle.

10. Make sure all of the spring coils are bottomed against ribs in spring pocket. Use the seal installer and handle to bottom the coils, as necessary.

11. Lubricate the spring moderately with wheel bearing before reassembling the starter.

STARTER ASSEMBLY

1. Install the recoil spring into the starter housing as instructed under 'Recoil Spring.'

2. Sparingly lubricate the center shaft of starter with wheel bearing grease.

3. Make sure the rope is in good condition. If necessary, replace the rope as instructed under 'Rope Replacement.' Ready the pulley and rope for assembly by unwinding all of the rope from pulley. Place the rope into the notch in pulley. This will keep the rope from interfering with the starter housing leg reinforcements as the pulley is rotated later during reassembly.

4. Install the pulley onto the center shaft. If the pulley does not seat fully, it is resting on the inner spring coil. Rotate the pulley slightly from side to side while exerting slight downward pressure. This should move the inner spring coil out of the way and allow the pulley to drop into position. The pulley is in position when the center shaft is flush with the face of the pulley. Do not wind the pulley and recoil spring at this time.

5. Install the starter pawls into the appropriate pockets in the pulley.

6. Sparingly lubricate the underside of the 1⅛ in. DIA. washer with grease and install it over the center shaft. Make sure the threads in center shaft remain clean, dry, and free of grease or oil.

7. Sparingly lubricate the insides of the 'legs' of the brake spider with grease. Install the brake to the retainer.

8. Install the small return ring to the pawl retainer. Make sure it is positioned properly.

9. Position the pawl retainer and return spring next to the small post on pulley. Install the free loop of the return spring over the post. Install the nylon spring retainer over the post.

10. Invert the pawl retainer over the pawls and center hub of pulley. Take great care not to damage or unhook the return spring. Make sure the pawls are positioned in the slots of pawl retainer.

11. As a test, rotate the pawl retainer slightly clockwise. Pressure from the return spring should be felt. In addition, the pawl retainer should return to its original position when released. If no spring pressure is felt or the retainer does not return, the spring is damaged, unhooked, or improperly assembled. Repeat steps 8, 9, and 10 to correct the problem.

12. Sparingly lubricate the ½ in. DIA. washer and ¾ in. DIA. washer with grease. Install the ½ in. DIA. washer then the ¾ in. DIA. washer in the center of pawl retainer. Make sure the threads in center shaft remain clean, dry, and free of grease or oil.

13. Apply a small amount of Loctite® #271 to the threads of center screw. Install the center screw to center shaft. Torque screw to 55-70 inch lbs.

14. Rotate the pulley counterclockwise (when viewed from the pawl side of pulley) until the spring is tight. (Approx. 4 full turns of pulley.) Make sure the fully extended rope is held in the notch in pulley to prevent interference with the housing leg reinforcements.

15. Rotate the pulley clockwise until the notch is aligned with the rope guide bushing of housing.

➡**Do not allow the pulley/spring to unwind. Enlist the aid of a helper, or use a c-clamp to hold pulley in position.**

16. Insert the free end of rope through rope guide bushing. Tie a temporary (slip) knot approx. 12 in. from the free end of rope.

17. Hold the pulley firmly with thumbs and allow the pulley to rotate slowly until the slip knot reaches the rope guide bushing of housing.

18. Slip the handle and rope retainer onto rope. Tie a single knot at the end of rope and install rope retainer into handle.

19. Untie the slip knot and pull the handle out until the rope is fully extended. Slowly retract the rope into the starter. If the spring has been properly tensioned, the rope will fully retract until the handle hits the housing.

IGNITION SYSTEMS

Electronic Magneto Ignition System

▶ See Figure 19

This engine is equipped with a state-of-the-art electronic magneto ignition system. This system consists of the following components:

• A magnet assembly, which is PERMANENTLY affixed to the flywheel.

• An electronic magneto ignition module, which is mounted to the engine bearing plate.

• A kill switch or key switch which stops the engine by grounding the ignition module.

OPERATION

As the flywheel rotates and the magnet assembly moves past the ignition module, a low voltage is induced in the primary windings of the module. When the primary voltage is precisely at its peak, the module induces a high voltage in its necessary windings. This high voltage creates a spark at the tip of the spark plug, igniting the fuel-air mixture in the combustion chamber. The timing of the spark is automatically controlled by the module. Therefore, no ignition timing adjustments are necessary or possible with this system.

❋❋WARNING

Do not connect 12 volts to the ignition system or to any wire connected to the ignition module.

The ignition system operates independently of the battery, starting charging, and other auxiliary electrical systems. Connecting 12 volts to the ignition module can cause the module to burn out. This type of damage is not covered by the engine warranty.

A break-before-make type key switch is required to prevent damage to the ignition module.

IGNITION MODULE REPLACEMENT

▶ See Figures 20, 21, 22, 23 and 24

1. Remove the high-tension lead and kill lead from slots in air baffle. Remove the kill lead from kill terminal of module.

2. Remove the hex cap screws, plain washers, module, and air baffle.

3. Separate the module and air baffle.

4. Install the module, plain washers, and hex cap screws. Move the modules as far from flywheel/magnet as possible — tighten the screws slightly.

5. Insert a 0.018 in. flat feeler gauge (or shim stock) between the magnet and module.

6. Loosen the hex cap screws so the magnet pulls module down. Tighten the screws to 32 inch lbs.

7. Remove the feeler gauge or shim stock. Due to the pull of the magnet, the bearing plate will flex lightly. The magnet-to-module air gap should be within the range of 0.012-0.016 in..

8. Rotate the flywheel back and forth; check to make sure the magnet does not strike the module. Check gap with feeler gauge and readjust if necessary.

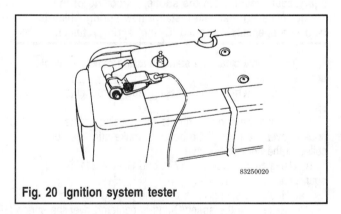

Fig. 20 Ignition system tester

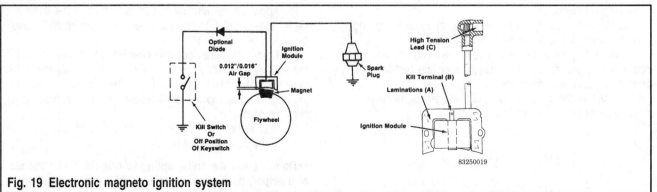

Fig. 19 Electronic magneto ignition system

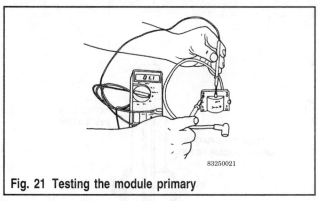

Fig. 21 Testing the module primary

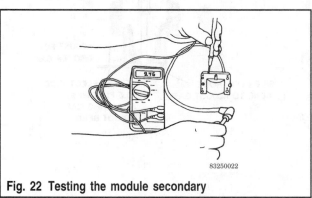

Fig. 22 Testing the module secondary

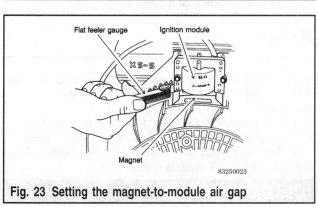

Fig. 23 Setting the magnet-to-module air gap

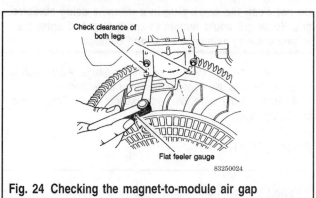

Fig. 24 Checking the magnet-to-module air gap

9. Install the air baffle over module. Install the kill lead to terminal of module. Install high-tension lead and kill lead to slots in baffle.

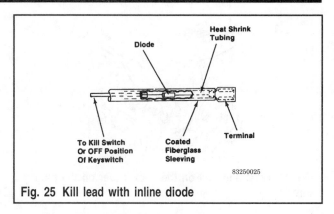

Fig. 25 Kill lead with inline diode

KILL LEAD WITH OPTIONAL DIODE

▶ See Figure 25

An optional in-line diode is installed in the kill lead of some Magnum engines. This diode protects the module from burning out, in the event voltage is applied to the kill lead.

The diode is rated such that failure (and subsequent module burn out) is highly unlikely. In the event a module with a diode protected kill lead does burn out, the diode should be tested.

Diode Test

Use an ohmmeter (or continuity tester) to test the diode.

1. Disconnect the kill lead terminals from the kill switch and ignition module.

2. Place the meter leads (or tester leads) across the kill lead. In one direction, the resistance should be infinity ohms (open circuit — no continuity). Reverse the test leads; some resistance should be measured (closed circuit — continuity).

3. If the resistance is infinity ohms in both directions (no continuity), the kill lead or diode is open.

4. Cut the protective tubing to expose the leads of diode. Perform the resistance (or continuity) test in step 2 to the diode leads. This will confirm if the lead or the diode is at fault.

5. If the resistance is 0 ohms in both directions (continuity), the diode is shorted.

Kill Lead/Diode Replacement

When servicing the kill lead, the entire lead can be replaced or, just the portion containing the diode.

Replacing the entire lead usually requires removing the bearing plate (refer to the 'Disassembly and Reassembly' sections). Use the following procedure to replace just the portion of lead with diode.

1. Cut off the diode portion of kill lead approximately 4¾ in. from terminal.

2. Strip ¼ in. of insulation from kill lead.

3. Crimp the 'insulating' connector of replacement diode/lead assembly to kill lead.

Spark Plug

▶ See Figures 26, 27, 28, 29 and 30

Engine misfire or starting problems are often caused by a spark plug in poor condition or with improper gap settings.

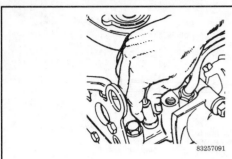

Fig. 26 Twist and pull on the boot, never on the plug wire

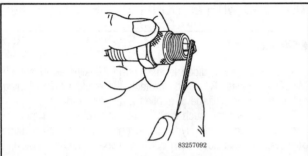

Fig. 27 Plugs that are in good condition can be filed and reused

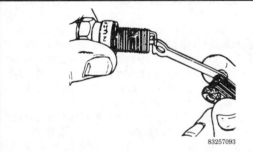

Fig. 28 Adjust the electrode gap by bending the side electrode

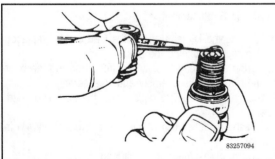

Fig. 29 Always used a wire gauge to check the plug gap

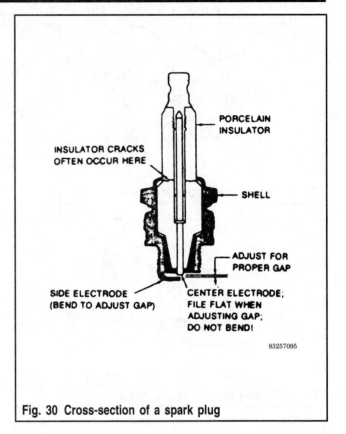

Fig. 30 Cross-section of a spark plug

SERVICE

Every 100 operating hours remove the spark plug, check its condition, and reset gap or replace with new plug as necessary.

1. Before removing the spark plug, clean the area around the base of plug to keep dirt and debris out of the engine.

2. Remove the plug and check its condition. Replace the plug if it is worn or if reuse is questionable.

✳✳CAUTION

Do not clean the spark plug in a machine using abrasive grit. Some grit could remain in spark plug and enter the engine causing extensive wear and damage.

3. Check the gap using a wire feeler gauge. Adjust gap to 0.025 in. by carefully bending the ground electrode.

4. Reinstall the spark plug into cylinder head. Torque plug to 18-22 ft. lbs.

INSPECTION

Inspect the spark plug as soon as is removed from the cylinder head. The deposits on the tip are an indication of the general condition of piston rings, valves, and carburetor.

Battery

▶ **See Figures 31, 32, 33, 34, 35, 36, 37 and 38**

Batteries are supplied by the equipment manufacturer. A 12-volt battery with a rating of at least 32 amp. hr.; 250 is recommended. Refer to the equipment manufacturer's instructions for specific information.

BATTERY TEST

If the battery charge is not sufficient to crank the engine, recharge the battery.

✳✳CAUTION

Do not attempt to jump start the engine with another battery. Starting the engine with batteries larger than those recommended can burn out the starter motor.

Test the battery voltage by connecting D.C. voltmeter across the battery terminals — crank the engine. If the battery drops below 9 volts while cranking, the battery is discharged or faulty.

BATTERY CHARGING

✳✳CAUTION

Batteries contain sulfuric acid. To prevent acid burns, avoid contact with skin, eyes, and clothing. Batteries produce explosive hydrogen gas while being charged. Charge the battery in well ventilated areas. Keep cigarettes, sparks, open flame, and other sources of ignition away from the battery at all times. To prevent accidental shorting and the resulting sparks, remove all jewelry when servicing the battery. When disconnecting battery cables, always disconnect the negative (-) (ground) cable first. When connecting battery cables, always connect the negative cable last. Before disconnecting the negative (-) ground cable, make sure all switches are OFF. If ON, a spark will occur at the ground cable terminal which could cause an explosion if hydrogen gas or gasoline vapors are present. Keep batteries and acid out of the reach of children.

BATTERY MAINTENANCE

1. Regularly check the level of electrolyte. Add distilled water as necessary to maintain the recommended level.

✳✳WARNING

Do not overfill the battery. Poor performance or early failure due to loss of electrolyte will result.

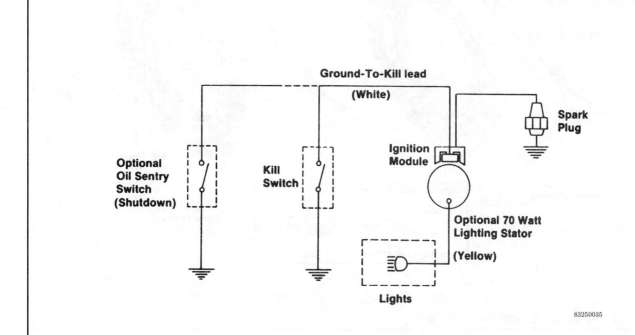

Fig. 31 Wiring diagram for manual start engines w/70 watt lighting

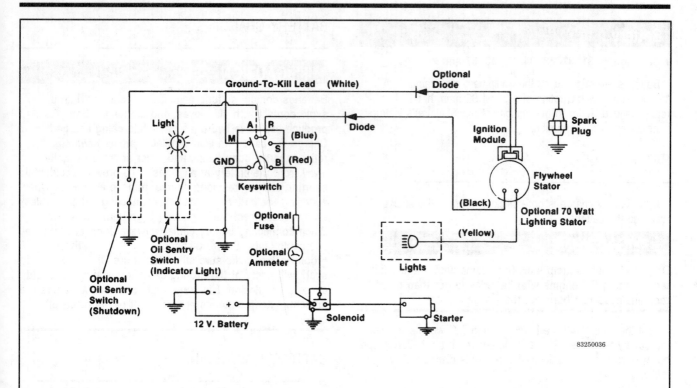

Fig. 32 Wiring diagram for electric start engines w/1.25 amp or 3 amp unregulated battery charging systems and 70 watt lighting

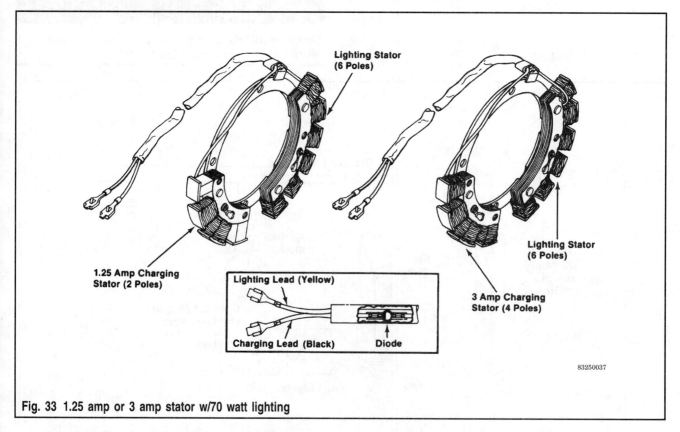

Fig. 33 1.25 amp or 3 amp stator w/70 watt lighting

2. Keep the cables, terminals, and external surfaces of battery clean. A build-up of corrosive acid or grime on the exter- nal surfaces can self- discharge the battery. Self-discharging happens rapidly when moisture is present.

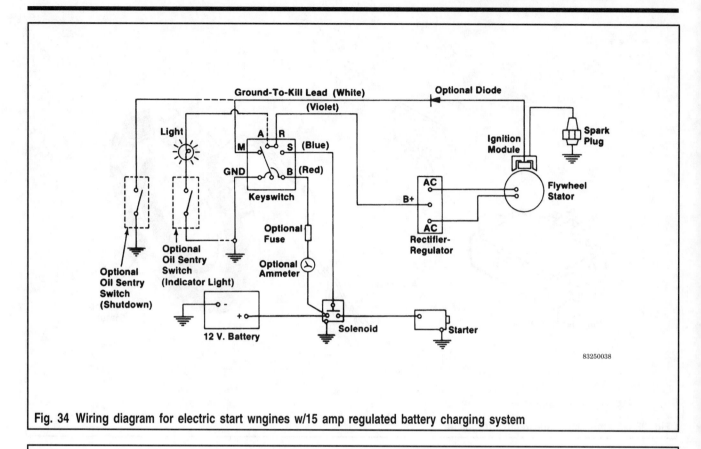

Fig. 34 Wiring diagram for electric start wngines w/15 amp regulated battery charging system

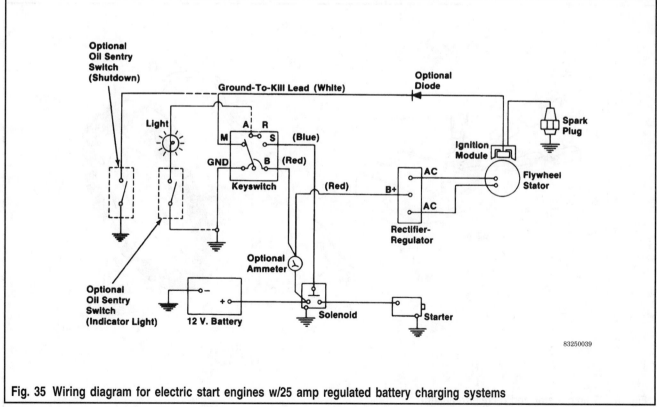

Fig. 35 Wiring diagram for electric start engines w/25 amp regulated battery charging systems

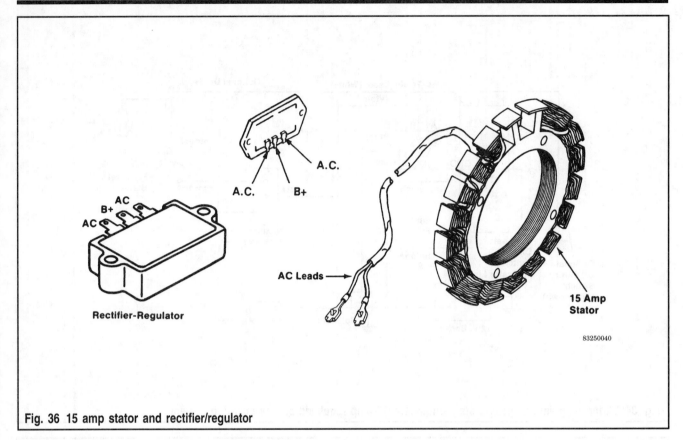

Fig. 36 15 amp stator and rectifier/regulator

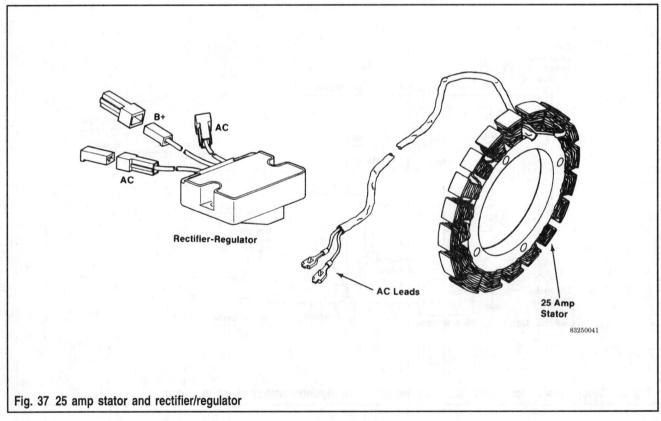

Fig. 37 25 amp stator and rectifier/regulator

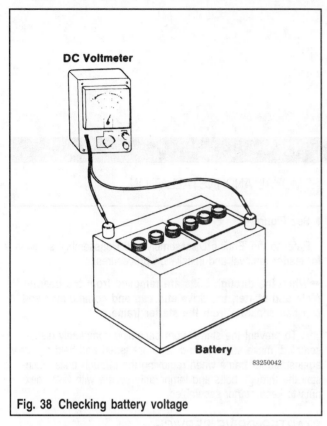

Fig. 38 Checking battery voltage

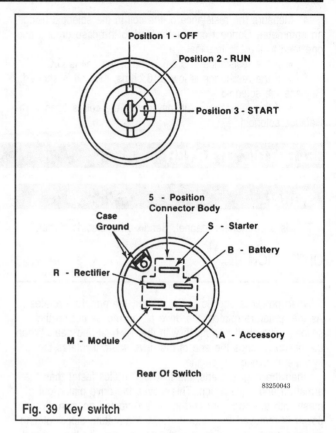

Fig. 39 Key switch

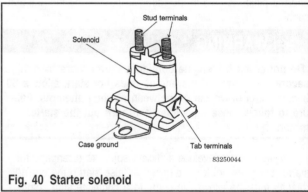

Fig. 40 Starter solenoid

3. Wash the cables, terminals, and external surfaces with a baking soda and water solution. Rinse thoroughly with clean water.

✳✳WARNING

Do not allow the baking soda solution to enter the cells as this will destroy the electrolyte.

Key Switch

▶ **See Figure 39**

A key switch is used on Magnum engines equipped with instrument panels. It is a three position (OFF, RUN, START), break-before-make type switch.

TESTING

Test the switch for continuity using an ohmmeter or continuity test light. For each switch position, continuity should be present across the terminals listed in the table below.

Starter Solenoid

▶ **See Figure 40**

A solenoid is used on electric start engines equipped with an instrument panel or key switch. The solenoid is an electrically-actuated normally open switch designed for heavy current loads.

The solenoid is used to switch the heavy current required by the starter using the key switch (designed for low current loads).

TESTING

1. Connect an ohmmeter or continuity tester across the stud terminals of solenoid.

2. Apply 12 volts DC across the tab terminal and case ground of solenoid and observe ohmmeter or tester.

➡ **Apply positive (+) of voltage supply to tab terminal; negative (-) to case ground.**

3. The ohmmeter or tester should indicate continuity as long as voltage is applied. If there is no continuity, the solenoid is probably faulty and should be replaced.

4. Measure the resistance of the coil in the solenoid using an ohmmeter. Connect one meter lead to the case ground and one lead to the tab terminal.

- If the resistance is 5.2-6.3 ohms, the coil is OK.
- If the resistance is low or 0 ohms, the coil is shorted. Replace the solenoid.
- If the resistance is infinity ohms, the coil is open. Replace solenoid.

ELECTRIC STARTER

Starter

This is a permanent magnet, Bendix-drive electric starter.

OPERATION

When power is applied to the starter, the armature rotates. As the armature rotates, the drive pinion moves out on the splined drive shaft into mesh with the flywheel ring gear. When the pinion reaches the end of the drive shaft, it rotates the flywheel cranking the engine.

When the engine starts, the flywheel rotates faster than the armature and drive pinion. This moves the drive pinion out of mesh with the ring gear and into the retracted position. When power is removed from the starter, the armature stops rotating and the pinion is held in the retracted position by the anti-drift spring.

✳✳WARNING

Do not crank the engine continuously for more than 10 seconds at a time. If the engine does not start, allow a 60 second cool down period between starting attempts. Failure to follow these guidelines can burn out the starter motor.

If the engine develops sufficient speed to disengage the starter but does not keep running (a 'false start'), the engine rotation must be allowed to come to a complete stop before attempting to restart the engine. If the starter is engaged while the flywheel is rotating, the starter pinion and flywheel ring gear may clash. This can damage the start

➡ **If the starter does not crank the engine, shut off the starter immediately. Do not make further attempts to start the engine until the condition is corrected. Do not jump start using another battery. Using batteries larger than those recommended can burn out the starter motor.**

Do not drop the starter or strike the starter frame. Doing so can damage the ceramic permanent magnets.

REMOVAL AND INSTALLATION

▶ **See Figure 41**

Refer to the 'Engine Disassembly and Reassembly' sections for starter removal and installation procedures.

➡**When the through bolts are removed from the bearing plate and starter, the drive end cap and commutator end cap can separate from the starter frame.**

To prevent the starter from becoming completely disassembled, make sure the end caps are taped and held securely against starter frame when removing the through bolts. Reinstall the through bolts and temporarily secure with ¼-20 hex nuts to keep starter assembled.

STARTER DRIVE SERVICE

Every 500 operating hours or annually (whichever occurs first), clean and lubricate the drive splines of the starter. If the drive pinion is badly worn, or has chipped or broken teeth, it must be replaced.

It is not necessary to disassemble the starter to service the drive components. Service the drive as follows:

1. Hold the drive pinion in a vice with soft jaws when removing and installing the stop nut. The armature will rotate with the nut only until the drive pinion stops against internal spacers.

➡**Do not overtighten the vice as this can distort the drive pinion.**

2. Remove the dust cover, stop nut, stop gear spacer, anti-drift spring, dust cover spacer, and drive pinion.
3. Clean the drive shaft splines with solvent. Dry the splines thoroughly.
4. Apply a small amount of Loctite® No. 271 to stop nut threads.
5. Reinstall the drive pinion, dust cover spacer, anti-drift spring, stop gear spacer, and stop nut. Torque stop nut to 160 inch lbs. Install the dust cover.

DISASSEMBLY

▶ **See Figure 42**

1. Remove the dust cover, stop nut, stop gear spacer, anti-drift spring, dust cover spacer, and drive pinion. Refer to 'Starter Drive Service.'

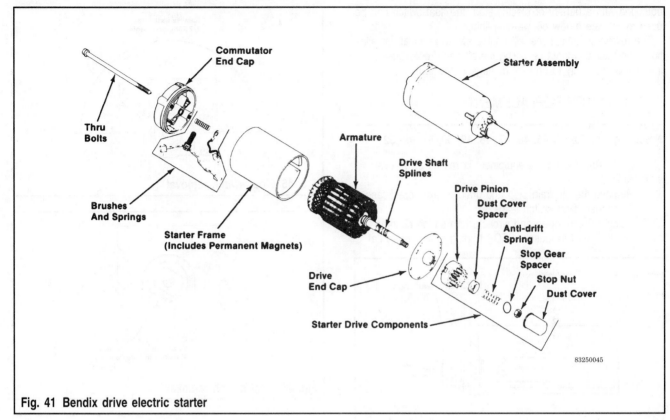

Fig. 41 Bendix drive electric starter

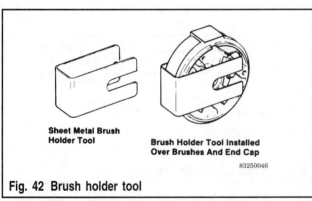

Fig. 42 Brush holder tool

2. Remove the temporary nuts and through bolts.

3. Remove the drive end cap and thrust washer.

4. Remove the commutator end cap with brushes and springs.

5. Remove the armature from inside the starter frame.

6. Install the brush springs and brushes into the pockets in brush holder. Make sure the chamfered sides of brushes are away from the springs.

➡**Use a brush holder tool to keep the brushes in the pockets. A brush holder tool can easily be made from a thin sheet metal.**

Commutator Service

Clean the commutator with a coarse, lint free cloth. Do not use emery cloth. If the commutator is badly worn or grooved, turn down on a lathe, or replace the armature.

ASSEMBLY

1. Insert the armature into the starter frame. Make sure the magnets are closer to the drive shaft end of armature. The magnets will hold the armature inside the frame.

2. Install the thrust washer and drive end cap. Make sure the match marks on end cap and frame are aligned.

3. Install the brush holder tool to keep the brushes in the pockets of commutator end cap.

4. Install the commutator end cap to armature and starter frame. Firmly hold the drive end cap and commutator end cap to the starter frame. Remove the brush holder tool.

5. Make sure the match marks on end cap are aligned. Install the through bolts and temporary nuts to keep the starter assembled.

6. Install the drive pinion, dust cover spacer, anti-drifting spring, stop gear spacer, stop nut, and dust cover. Refer to 'Starter Drive Service.'

Oil Sentry Oil Level Monitor

OPERATION

Some engines are equipped with optional Oil Sentry system. Oil Sentry uses a float switch in the oil pan to detect a low engine oil level. On stationary or unattended applications (pumps, generators, etc.) the float switch can be used to ground the ignition module to stop the engine. On vehicular applications (garden tractors, mowers, etc.) and those equipped with a battery or electric start, the float switch can be used to activate a 'low oil' warning light.

The following instructions will enable switch removal, installation, and testing without removing the oil pan. Follow these instructions carefully to prevent damage to the switch.

FLOAT SWITCH REPLACEMENT

▶ **See Figures 43, 44, 45, 46, 47, 48, 49, 50, 51 and 52**

1. Make sure the engine/equipment is resting on a level surface.
2. Remove the oil drain plug and drain oil from crankcase.
3. Disconnect float switch leads.
4. Using a ⁹/₁₆ in. open end wrench, turn switch counterclockwise ¼ TURN to loosen. STOP turning switch when flat

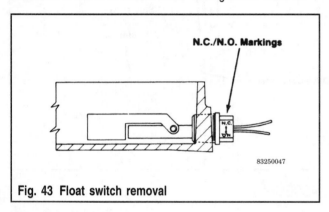

Fig. 43 Float switch removal

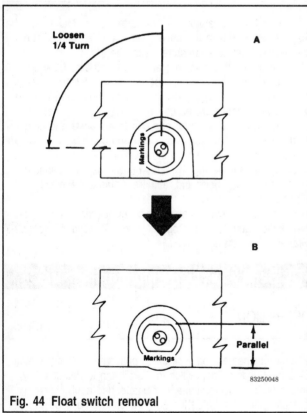

Fig. 44 Float switch removal

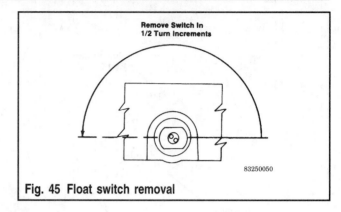

Fig. 45 Float switch removal

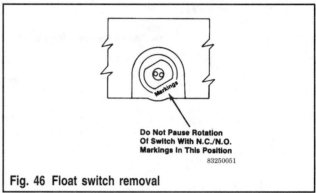

Fig. 46 Float switch removal

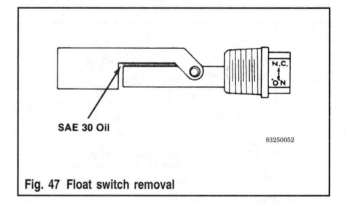

Fig. 47 Float switch removal

surface on float switch is in a horizontal position. (Flat surface parallel with base of oil pan and N.C./N.O. markings down.).

5. Turn the switch counterclockwise in ½ TURN INCREMENTS using a smooth, continuous action. Pause briefly between increments and keep the flat surface of float switch in a horizontal position (parallel with base of oil pan).

✳✳WARNING

To prevent damage to the float switch, and to enable you to 'feel' if the float strikes the oil pan, REMOVE THE SWITCH BY HAND as soon as it is loose enough for you to do so.

When turning the float switch, use a smooth, continuous action for the ENTIRE ½ turn increment. Pausing the rotation of the switch in the position shown will cause the float to strike

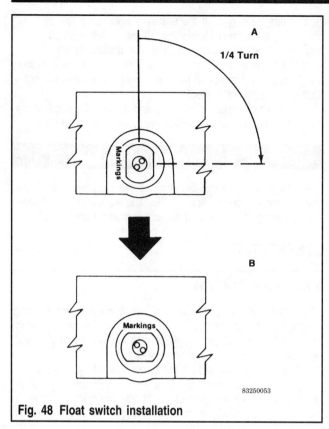

Fig. 48 Float switch installation

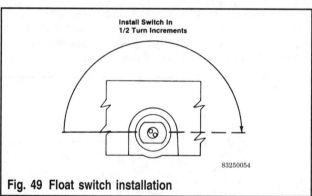

Fig. 49 Float switch installation

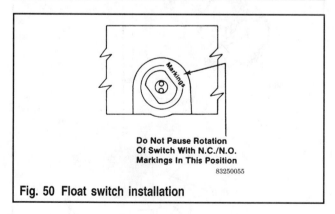

Fig. 50 Float switch installation

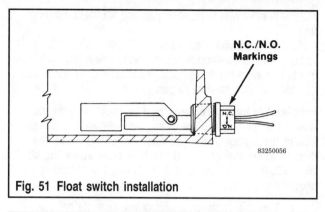

Fig. 51 Float switch installation

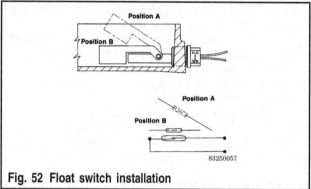

Fig. 52 Float switch installation

the oil pan. If the float does not strike the oil pan, STOP turning the switch, then use the following procedure.

6. Turn the switch clockwise until the flat surface is in a vertical position as shown. (N.C./N.O. markings on left.) This will allow to float to return against the switch body.

7. Turn the switch counterclockwise ¼ TURN. STOP turning switch when flat surface is in a horizontal position. (Flat surface parallel with base of oil pan and N.C./N.O. markings down.).

8. Turn the switch counterclockwise in ½ TURN INCREMENTS as instructed in step 5 above.

9. Make sure the engine/equipment is resting on a level surface.

10. Remove the oil drain plug and drain oil from crankcase.

11. When adding this switch as an accessory, remove and discard the ½ in. NPSF pipe plug from the location in oil pan where switch will be installed.

✳✳WARNING

To prevent damage to the float switch, and to enable you to 'feel' if the float strikes the oil pan, INSTALL THE SWITCH BY HAND as long as it is loose enough for you to do so.

12. Apply Loctite® No. 592 Teflon® sealant (or equivalent) to the entire thread area of switch.

13. Apply a thick film of clean SAE 30 oil to the float and switch body as shown.

14. Hold the switch with the flat surface in a vertical position. (N.C./N.O. markings to the left.) Insert the switch into the oil pan and turn the switch ¼ turn. STOP turning the switch when the flat surface on switch is in a horizontal position. Flat

surface parallel with base of oil pan and N.C./N.O. markings up.).

15. Turn the switch clockwise in ½ TURN INCREMENTS using a smooth continuous action. Pause briefly between increments and keep the flat surface of switch in a horizontal position (parallel with base of oil pan.).

➡**Several ½ turn increments may be required until the threads on switch engage in oil pan. Pausing the rotation of the switch in the position shown will cause the float to strike the oil pan. If the float does strike the oil pan, STOP turning the switch, then use the following procedure:**

a. Turn the switch counterclockwise until the flat surface is in a vertical position, (N.C./N.O. markings on left.) This will allow the float to return against the switch body.

b. Turn the switch clockwise ¼ turn. STOP turning switch when flat surface is in a horizontal position. (Flat surface parallel with base of oil pan and N.C./N.O. markings up.).

c. Turn the switch clockwise in ½ turn increments as instructed in step 7 above.

16. Turn in the switch approximately five (5) to six (6) full turns to obtain the proper position. Use a ⁹/₁₆ in. open end wrench to tighten the switch. The 'N.C.' markings on switch will be at the top when the switch is positioned properly.

FLOAT SWITCH TEST

Test switch for continuity by placing a ohmmeter of continuity test light across leads.
- Switch Position A: No Continuity (switch open)
- Switch Position B: Continuity (switch closed)

Perform the following tests to ensure that the float switch is positioned and working properly before connecting the leads.

➡**These tests apply to engines equipped with a standard oil pan and dipstick. Special oil pan and/or dipstick arrangements can give inaccurate test results.**

1. Connect a continuity test light across float switch leads. The light should be 'on'.
2. Install oil drain plug and refill crankcase with oil. The light should be 'off' after oil is above the 'L' mark on the dipstick.
3. If the float switch fails this test:
a. Make sure the switch is in the proper position with the 'N.C.' markings at the top.
b. If switch is positioned properly, drain oil and remove switch (see 'Float Switch Removal'). If the float is not attached to the switch body, the oil pan must be removed.
c. replace a faulty or broken switch with a new one. (See 'Float Switch Installation'.)

OPERATIONAL TEST

Reconnect the leads and perform the following test.
1. Make sure the oil level is up to , but not over 'F' mark on dipstick.

2. Start the engine. If the switch is wired as a low oil level shutdown, the engine should start. If the switch is wired to activate a low oil warning light, the light should be 'off.'
3. Stop the engine. Drain the oil until the oil level is below the 'L' mark on dipstick. If properly wired, the engine will not start or the light will be 'on.'
4. If the test results of steps 2 and 3 are not as indicated, check for improper wiring and/or improper float installation.

Automatic Compression Release

All Magnum single cylinder engines are equipped with Automatic Compression Release (ACR). The ACR mechanism lowers compression at cranking speeds to make starting easier.

OPERATION

▶ **See Figures 53 and 54**

The ACR mechanism consists of two flyweights and a spring attached to the gear on camshaft. When the engine is rotating at low cranking speeds (600 RPM or lower) the flyweights are held by the spring in the position shown.

In this position, the tab on the larger flyweight protrudes above the exhaust cam lobe. This lifts the exhaust valve off of its seat during the first part of the compression stroke. The reduced compression results in an effective compression ratio of about 2:1 during cranking.

After the engine speed increases to about 600 RPM, centrifugal force moves the flyweights to the position shown. In this position the tab on the larger flyweight drops into the recess in the exhaust cam lobe. When in recess, the tab has

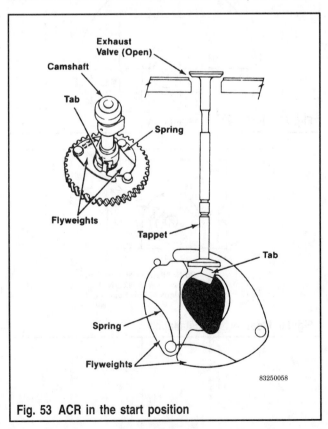

Fig. 53 ACR in the start position

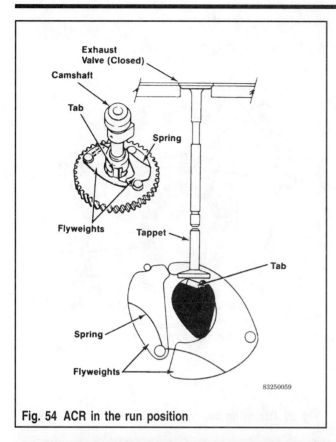

Fig. 54 ACR in the run position

no effect on the exhaust valve and the engine operates at full compression and full power.

When the engine is stopped, the spring returns the flyweights to the position shown, ready for the next start.

INSPECTION AND SERVICE

1. Check exhaust valve to tappet clearance and adjust as necessary to specification.

2. Remove cylinder head and turn the crankshaft clockwise by hand and observe the exhaust valve carefully. When the piston is approx. ⅔ of the way up the cylinder during the compression stroke, the exhaust valve should lift off the seat slightly. If the exhaust valve does not lift, the ACR spring may be unhooked or broken. To service the spring, remove the oil pan and rehook the spring or replace it. The camshaft does not have to be removed.

The flyweights are not serviceable. If they are stuck or worn excessively, the camshaft must be replaced.

❄❄WARNING

The tab on the flyweights is hardened and is not adjustable. Do not attempt to bend the tab — it will break and a new camshaft will be required.

ENGINE

Engine Mechanical Service

COMPRESSION TESTING

Because of the ACR mechanism, it is difficult to obtain an accurate compression reading.

To check the condition of the combustion chamber, and related mechanisms, physical inspection and a crankcase vacuum test are recommended.

DISASSEMBLY

▶ **See Figures 55, 56, 57, 58, 59, 60, 61, 62, 63, 64, 65 and 66**

❄❄CAUTION

Before servicing the engine or equipment, always remove the spark plug lead to prevent the engine fro starting accidentally. Ground the lead to prevent sparks that could cause fires.

The following sequence is suggested for complete engine disassembly. This procedure may have to be varied slightly to accommodate options or special equipment.

Clean all parts thoroughly as the engine is disassembled. Only clean parts can be accurately inspected and gauged for wear or damage. There are many commercially available cleaners that quickly remove grease, oil, and grime from engine parts. When such a cleaner is used, follow the manufacturer's instructions carefully. Make sure all traces of the cleaner are removed before the engine is reassembled and placed in operation — even small amounts of these cleaners quickly break down the lubricating properties of engine oil.

1. Disconnect the spark plug.
2. Drain the oil.
3. Remove the wing nut and air cleaner cover.
4. Remove the precleaner, element cover nut, element cover, paper element, and seal.

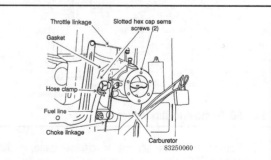

Fig. 55 Removing the choke linkage, throttle linkage and carburetor

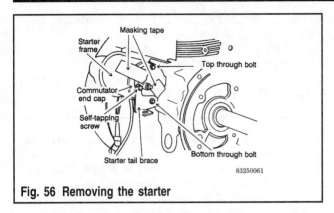

Fig. 56 Removing the starter

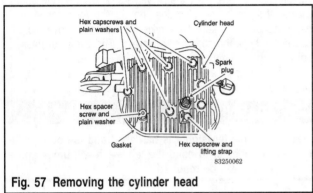

Fig. 57 Removing the cylinder head

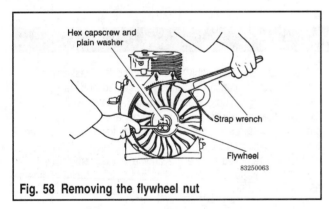

Fig. 58 Removing the flywheel nut

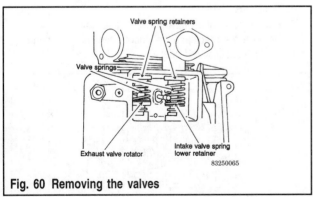

Fig. 60 Removing the valves

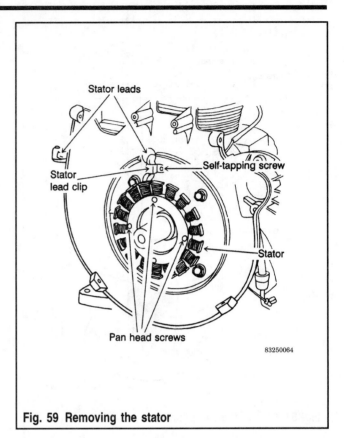

Fig. 59 Removing the stator

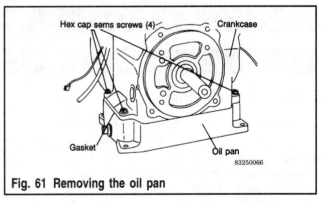

Fig. 61 Removing the oil pan

5. Remove the three screws, air cleaner base, gasket, and breather hose.

6. Remove the muffler and threaded exhaust pipe.

7. Remove two screws and the rectifier-regulator.

8. Remove the electrical connector from the rectifier-regulator.

✳✳CAUTION

Gasoline may be present in carburetor and fuel system. Gasoline is extremely flammable, and its vapors can explode if ignited. Keep cigarettes, sparks, open flames, and other sources of ignition away from the engine. Wipe up spilled fuel immediately.

9. Remove the throttle linkage from the nylon inserts in governor arm and carburetor throttle lever. Remove the choke linkage from the nylon insert in the choke control lever, then from the carburetor choke lever.

10. Close the fuel shut-off valve at fuel tank (if so equipped) or drain fuel from tank.

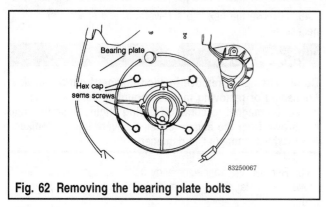

Fig. 62 Removing the bearing plate bolts

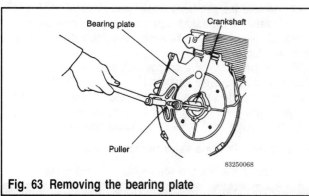

Fig. 63 Removing the bearing plate

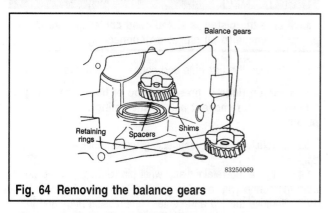

Fig. 64 Removing the balance gears

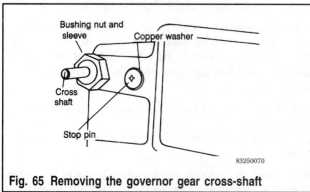

Fig. 65 Removing the governor gear cross-shaft

11. Loosen the hose clamp and remove fuel line from the carburetor inlet.

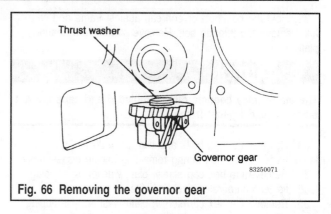

Fig. 66 Removing the governor gear

12. Remove two slotted hex cap screws, the carburetor, and gasket.

13. Note the position of the governor spring in governor arm.

14. Loosen pawlnut and remove governor arm from cross shaft.

➡Loosening pawlnut or removing governor arm will disrupt governor arm to cross shaft adjustment. Readjustment will be required upon reassembly.

15. Remove the governor spring from the governor arm.

16. Remove the electrical connector from the back of key switch.

17. Remove the hex cap screw, plain washer, and spacer. Remove the self-tapping screw securing the bracket to blower housing/bearing plate. Remove four hex cap screws, cam gear cover, gasket, throttle and choke control assembly, and gasket.

18. Remove pawlnut, breather cover, and gasket.

19. Remove the filter, seal, reed stop, reed, breather plate, gasket, and stud.

20. Disconnect the fuel line from the fuel pump inlet fitting.

21. Remove the fillister head screws, plain washers, fuel pump, and gasket.

22. Remove five screws and the starter assembly.

✳✳CAUTION

Gasoline may be present in fuel tank and fuel system. Gasoline is extremely flammable, and its vapors can explode if ignited. Keep cigarettes, sparks, and open flames, and other sources of ignition away from engine. Wipe up spilled fuel immediately.

23. Disconnect leads from starter and solenoid.

24. Remove the self-tapping screws and solenoid from the fuel tank lower bracket.

25. Remove the fuel line from tank outlet fitting.

26. Remove the self-tapping screw securing starter tail brace to fuel tank bracket and remove the two acorn nuts and washers at the top of fuel tank.

27. Remove two hex cap screws and fuel tank with bracket.

28. Remove pawl nuts, fuel tank lower bracket, and isolation mounts from fuel tank, if necessary.

29. Remove the dipstick, fillister head screws, oil fill/dipstick tube, and gasket.

30. Place across the commutator end cap and starter frame.

31. Remove the self-tapping screw, bottom through bolt, and starter tail brace.

32. Hold the commutator end cap against frame and carefully remove top through bolt. Remove starter from bearing plate.

✳✳CAUTION

The starter may become disassembled if end caps are not taped or held against frame.

33. Reinstall through bolts and two (2) ¼-20 hex nuts to keep starter assembled during remaining engine disassembly.
34. Remove the hex cap screw, plain washer, spark plug lead clip, and carburetor side air baffle.
35. Remove the hex cap screw, plain washer, self-tapping screw, and starter side air baffle.
36. Remove the truss head screw and cylinder head baffle.
37. Remove the remaining self-tapping screws and the blower housing.
38. Remove the spark plug, hex cap screws, lifting strap, hex spacer screw, plain washers, cylinder head, and gasket.
39. Remove the hex cap screws, plain washers, ignition module, and air baffle.
40. Remove the spark plug lead and kill lead from the slots in the baffle. Separate the ignition module and air baffle, if necessary.

✳✳CAUTION

Always use a flywheel strap wrench to hold the flywheel when loosening or tightening flywheel and fan retaining fasteners. Do not use any type of bar or wedge between fins of cooling fan, as the fins could become cracked or damaged. Always use a puller to remove flywheel from crankshaft. Do not strike the crankshaft or flywheel, as these parts could become cracked or damaged.

41. With rope start models:
    a. Remove the grass screen retainer and wire mesh grass screen from rope pulley.
    b. Hold the flywheel with a strap wrench and loosen the hex cap screw. Remove the hex cap screw, plain washer, rope pulley, and spacer. Remove the nylon grass screen from the fan.
42. With retractable start models:
    a. Hold the flywheel with a strap wrench and loosen hex cap screw securing flywheel to crankshaft. Remove the hex cap screw, plain washer, and drive cup.
    b. Remove the grass screen from the fan.
43. With electric start models:
    a. Remove the grass screen from the fan.
    b. Hold the flywheel with a strap wrench and loosen hex cap screw securing flywheel to crankshaft. Remove the hex cap screw and plain washer.
44. Remove the flywheel from the crankshaft using a puller.

45. Remove the hex cap screws, fan, and spacers, if necessary.

✳✳CAUTION

Do not attempt to remove ignition magnet from flywheel. Loosening or removing magnet mounting screws could cause the magnet to come off during engine operation and be thrown from the engine causing severe injury. Replace the flywheel if magnet is damaged.

46. Remove the connector body from the stator leads. Remove the self-tapping screw and stator lead clip from bearing plate.
47. Remove the pan head screws and stator.
48. Rotate the crankshaft until the piston is at top dead center of compression stroke (both valves closed and piston flush with top of bore).
49. Compress the valve springs with a valve spring compressor and remove the keepers.
50. Remove the valve spring compressor, then remove the valves, intake valve spring lower retainer, exhaust valve rotator, valve springs, and valve spring upper retainers.

➡Some models use a valve rotator on both valves.

51. Remove the hex cap screws, oil pan, and gasket.

✳✳WARNING

Make sure the piston is at top dead center in bore to prevent damage to oil dipper on connecting rod.

52. Remove the hex nuts and connecting rod cap.

➡If a carbon ridge is present at top of bore, use a ridge reamer tool to remove it before attempting to remove piston.

53. Carefully push the connecting rod and piston out top of bore.
54. Remove the retainer and wrist pin. Separate the piston from the connecting rod.
55. Remove the top and center compression rings and the oil control ring spacer using a ring expander tool.

➡To make reassembly easier, mark the positions of the fuel line and wiring harness before removing the bearing plate. Mark the locations where they exit from between the crankcase and bearing plate, on both the carburetor and starter sides.

56. Remove the hex cap screws securing the bearing plate to crankcase.
57. Remove the bearing plate from the crankshaft using a puller.

➡The front bearing may remain either in the bearing plate or on the crankshaft when the bearing plate is removed.

58. Press the crankshaft out of the crankcase from the PTO side.

➡If the repair does not require separating the bearing plate from the crankshaft, the crankshaft and bearing plate can be pressed out as an assembly.

59. Drive the camshaft pin (and cup plug on bearing plate side) out of the crankcase from the PTO side.

60. Remove the camshaft pin, camshaft, and shim(s) (on bearing plate side of camshaft).

61. Mark the tappets as being either intake or exhaust. Remove the tappets from the crankcase.

➡**The intake valve tappet is closest to the bearing plate side of crankcase. The exhaust valve tappet is closest to the PTO side of crankcase.**

62. Remove the retaining rings, shims, balance gears with needle bearings, shims, and spacers.

63. Remove the stop pin, copper washer, governor gear, and thrust washer.

64. remove bushing nut and sleeve. Remove cross shaft from inside crankcase.

65. Remove the oil seals from the crankcase and bearing plate.

66. Press the bearings out of the bearing plate and crankcase.

➡**If the bearings have remained on the crankshaft, remove bearing by using a puller.**

INSPECTION AND REPAIR/CONDITIONING

All parts should be thoroughly cleaned — dirty parts cannot be accurately gauged or inspected properly for wear or damage. There are many commercially available cleaners that quickly remove grease, oil and grime accumulation from engine parts. If such a cleaner is used, follow the manufacturer's instructions carefully, and make sure that all of the cleaner is removed before the engine is reassembled and placed in operation. Even small amounts of these cleaners quickly break down the lubricating properties of engine oils.

Flywheel Inspection

Inspect the flywheel for cracks, and the flywheel keyway for damage. Replace flywheel if cracked. replace the flywheel, the crankshaft, and the key if flywheel key is sheared or the keyway damaged.

Inspect ring gear for cracks or damage. Kohler no longer provides ring gears as a serviceable part. Replace flywheel if the ring gear is damaged.

Cylinder Head Inspection

Blocked cooling fins often cause localized 'hot spots' which can result in a 'blown' cylinder head gasket. If the gasket fails, high temperature gases can burn away portions of the aluminum alloy head. A cylinder head in this condition must be replaced.

If the cylinder head appears in good condition, use a block of wood or plastic scraper to scrape away carbon deposits. Be careful not to nick or scratch the aluminum, especially in gasket seating area.

The cylinder head should also be checked for flatness. Use a feeler gauge and a surface plate or a piece of plate glass to make this check. Cylinder head flatness should not vary more than 0.003 in.; if it does, replace the cylinder head.

➡**Measure cylinder head flatness between each cap screw hole.**

In cases where the head is warped or burned, it will also be necessary to replace the head screws. The high temperatures that warped or burned the head could have made the screws ductile which will cause them to stretch when tightened.

Cylinder Block Inspection and Reconditioning

Check all gasket surfaces to make sure they are free of gasket fragments. gasket surfaces must also be free of deep scratches or nicks.

Scoring of the Cylinder Wall: Unburned fuel, in severe cases, can cause scuffing and scoring of the cylinder wall. As raw fuel seeps down the cylinder wall, it washes the necessary lubricating oils off the piston and cylinder wall so that the piston rings make metal to metal contact with the wall. Scoring of the cylinder wall can also be caused by localized hot spots resulting from blocked cooling fins or from inadequate or contaminated lubrication.

If the cylinder bore is badly scored, excessively worn, tapered, or out of round, resizing is necessary. Use an inside micrometer to determine amount of wear. Resizing to one of these oversizes will allow usage of the available oversize piston and ring assemblies. Initially, resize using a boring bar, then use the following procedures for honing the cylinder:

HONING

While most commercially available cylinder hones can be used with either portable drills or drill presses, the use of a low speed drill press is preferred as it facilitates more accurate alignment of the bore in relation to the crankshaft crossbore. Honing is best accomplished at a drill speed of about 250 RPM and 60 strokes per minute. After installing coarse stones in hone, proceed as follows:

1. Lower hone into bore and after centering, adjust so that stones are in contact with the cylinder wall. Use of a commercial cutting-cooling agent is recommended.

2. With the lower edge of each stone positioned even with the lowest edge of the bore, start drill and honing process. Move hone up and down while resizing to prevent formation of cutting ridges. Check size frequently.

➡**Keep in mind the temperatures caused by honing may cause inaccurate measurements. Make sure the block is cool when measuring.**

3. When bore is within 0.0025 in. of desired size, remove coarse stones and replace with burnishing stones. Continue with burnishing stones until within 0.0005 in. of desired size and then use finish stones (220-280 grit) and polish to final size. A cross hatch should be observed if honing is done correctly. The cross hatch should intersect at approximately 23-33 degrees off the horizontal. Too flat an angle could cause the rings to skip and wear excessively, too high an angle will result in high oil consumption.

4. After resizing, check the bore for roundness, taper, and size. Use an inside micrometer, telescoping gauge, or bore gauge to take measurements. The measurements should be taken at three locations in the cylinder — at the top, middle,

and bottom. Two measurements should be taken (perpendicular to each other) at each of the three locations.

5. Thoroughly clean cylinder wall with soap and hot water. Use a scrub brush to remove all traces of boring/honing debris. Dry thoroughly and apply a light coat of SAE 10 oil to prevent rust.

Measuring Piston-to-Bore Clearance

▶ **See Figure 67**

Before installing the piston into the cylinder bore, it is necessary that the clearance be accurately checked. This step is often overlooked, and if the clearances are not within specifications, generally engine failure will result.

➡**Do not use a feeler gauge to measure piston-to-bore clearance — it will yield inaccurate measurements. Use a micrometer.**

The following procedures should be used to accurately measure the piston to bore clearance.

1. Use a micrometer and measure the diameter of the piston.

2. Use an inside micrometer, telescoping gauge, or bore gauge and measure the cylinder bore. Take the measurement approximately $2\frac{1}{2}$ in. below the top of the bore and perpendicular to the piston pin.

3. Piston-to-bore clearance is the difference between the bore and the piston diameter (step 2 minus step 1). Clearance should be: 0.007-0.010 in. (style A piston), 0.003-0.005 in. (style C piston), 0.0034-0.0051 in. (style D piston).

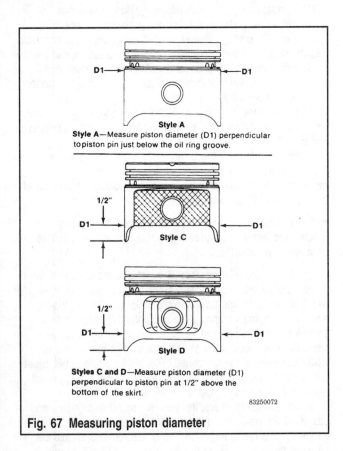

Style A—Measure piston diameter (D1) perpendicular to piston pin just below the oil ring groove.

Style C

Style D

Styles C and D—Measure piston diameter (D1) perpendicular to piston pin at 1/2" above the bottom of the skirt.

83250072

Fig. 67 Measuring piston diameter

Valve Inspection and Service

▶ **See Figure 68**

Carefully inspect valve mechanism parts. Inspect valve springs and related hardware for excessive wear or distortion. Valve spring free height should be at approximately the dimension given in the chart below. Check valves and valve seat area or inserts for evidence of deep pitting, cracks or distortion. Check clearance of valve stems in guides.

Hard starting, or loss of power accompanied by high fuel consumption may be symptoms of faulty valves. Although these symptoms could also be be attributed to worn rings, remove and check valves first. After removal, clean valve head, face and stem with power wire brush and then carefully inspect for defects such as warped valve head, excessive corrosion or worn stem end. Replace valves found to be in bad condition. A normal valve and valves in bad condition are shown in the accompanying illustrations.

Valve Guides

▶ **See Figure 69**

If a valve guide is worn beyond specifications, it will not guide the valve in a straight line. This may result in a burnt valve face or seat, loss of compression, and excessive oil consumption.

To check valve guide to stem clearance, thoroughly clean the valve guide and, using a split-ball gauge, measure the inside diameter. Then, using an outside micrometer, measure the diameter of the valve stem at several points on the stem where it moves in the valve guide. Use the largest stem diameter to calculate the clearance. If the clearance exceeds 0.006 on intake or 0.008 on exhaust valves, determine whether the valve stem or the guide is responsible for the excessive clearance.

➡**The exhaust valves on these engines have a slightly tapered valve stem to help prevent sticking. Because of the taper, the valve stem must be measured in two places to determine if valve stem is worn. If the valve stem diameter is within specifications, replace the valve guide.**

VALVE GUIDE REPLACEMENT

The valve guides are a tight press fit in the cylinder block. A valve guide removal tool is recommended to remove the guides. To remove valve guide, proceed as follows:

1. Install $\frac{5}{16}$-18 NC nut on coarse threaded end of $3\frac{1}{2}$ in. long stud.

2. Insert other end of stud through valve guide bore and install $\frac{5}{16}$-24 NF nut. Tighten both nuts securely.

➡**Valve guide must be held firmly by the stud assembly so that all side hammer force will act on the guide.**

3. Assemble the valve guide removal adapter to the stud and then the slide hammer to the adapter.

4. use the slide hammer to pull the guide out.

To Install:

5. Make sure valve guide bore is clean and free of nicks or burrs.

6. Using valve guide driver, align and then press guide in until valve guide driver bottoms on valve guide counterbore.

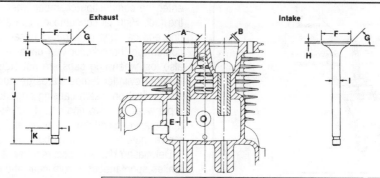

Dimension		M8		M10, M12, M14		M16	
		Intake	Exhaust	Intake	Exhaust	Intake	Exhaust
A	Seat Angle	89°	89°	89°	89°	89°	89°
B	Seat Width	.037/.045	.037/.045	.037/.045	.037/.045	.037/.045	.037/.045
C	Insert O. D.	—	1.2525/1.2535	—	1.2525/1.2535	1.5035/1.5045	1.5035/1.5045
D	Guide Depth	1.281	1.312	1.996*	1.996*	1.996*	1.996*
E	Guide I.D.	.312/.313	.312/.313	.312/.313	.312/.313	.312/.313	.312/.313
F	Valve Head Diameter	1.380/1.370	1.130/1.120	1.380/1.370	1.130/1.120	1.380/1.370	1.380/1.370
G	Valve Face Angle	45°	45°	45°	45°	45°	45°
H	Valve Margin (Min.)	.031	.031	.031	.031	.031	.031
I	Valve Stem Dia.	.3103/.3110	—	.3103/.3110	—	.3103/.3110	—
	Valve Stem Dia. @ J	—	.3074/.3081	—	.3074/.3081	—	.3074/.3081
	Valve Stem Dia. @ K	—	.3084/.3091	—	.3084/.3091	—	.3084/.3091
J	(I) Measurement Location	—	2.530	—	3.060	—	3.060
K	(I) Measurement Location	—	.535	—	.835	—	.835

*Approximate. Should be flush with top of valve guide counterbore.

83250073

Fig. 68 Valve details

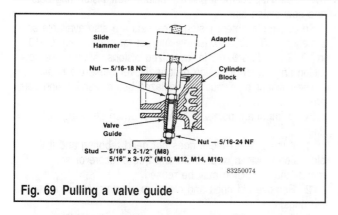

Fig. 69 Pulling a valve guide

7. Valve guides are often slightly compressed during insertion. Use a piloted reamer and then a finishing reamer to resize the guide bore to 0.3125 in..

Valve Seat Inserts

The intake valve seat is usually machined into the cylinder block, however, certain applications may specify a hard alloy insert. The exhaust valve seat is a replaceable alloy insert. If the seat becomes badly pitted, cracked, or distorted, the insert must be replaced.

The insert is a tight press fit in the cylinder block. A valve seat removal tool is recommended for this job. Since insert removal causes loss of metal in the insert bore area, use only service replacement inserts, which are slightly larger to provide proper retention in the cylinder block. Make sure new insert is properly started and pressed into bore to prevent cocking of the insert.

VALVE SEAT INSERT REPLACEMENT

◆ See Figures 70 and 71

1. Install valve seat puller on forcing screw and lightly secure with washer and nut.
2. Center the puller assembly on valve seat insert.
3. Hold forcing screw with a hex wrench to prevent turning and slowly tighten nut.

➡Make sure sharp lip on puller (see insert) engages in joint between bottom of valve seat insert and cylinder block counterbore, all the way around.

4. Continue to tighten nut until puller is tight against valve seat insert.
5. Assemble adapter to valve seat puller forcing screw and slide hammer to adapter.
6. Use slide hammer to remove valve seat insert.
To install:
7. Make sure valve seat insert bore is clean and free of nicks or burrs.
8. Align valve seat insert in counterbore and using valve seat installer and driver, press seat until bottomed.
9. Use a standard valve seat cutter and cut seat to dimensions shown.

Reground or new valves must be lapped in to provide proper fit. Use a hand valve grinder with suction cup for final lapping. Lightly coat valve face with 'fine' grade of grinding compound, then rotate valve on seat with grinder. Continue grinding until smooth surface is obtained on seat and on valve face. Thoroughly clean cylinder block in soap and hot water to remove all traces of grinding compound. After drying cylinder block apply a light coating of SAE 10 oil to prevent rusting.

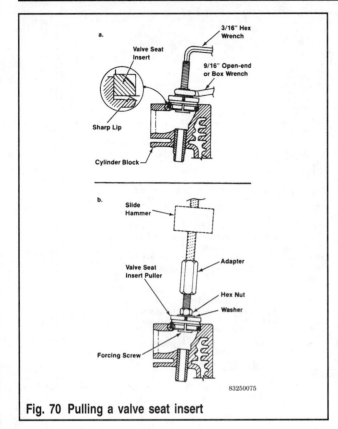

Fig. 70 Pulling a valve seat insert

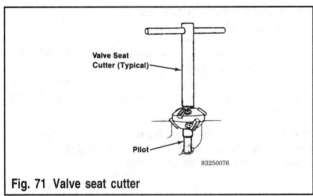

Fig. 71 Valve seat cutter

Piston and Rings

INSPECTION

Scuffing and scoring of piston and cylinder wall occurs when internal temperatures approach the melting point of the piston. Temperatures high enough to do this are created by friction, which is usually attributed to improper lubrication, and/or overheating of the engine.

Normally, very little wear takes place in the piston boss-piston pin area. If the original piston and connecting rod can be reused after new rings are installed, the original pin can also be reused but new piston pin retainers are required. The piston pin is included as part of the piston assembly — if the pin boss in piston, or the pin are worn or damaged, a new piston assembly is required.

Ring failure is usually indicted by excessive oil consumption and blue exhaust smoke. When rings fail, oil is allowed to enter the combustion chamber where it is burned along with the fuel. High oil consumption can also occur when the piston ring end cap gap is incorrect because the ring cannot properly conform to the cylinder wall under this condition. Oil control is also lost when ring gaps are not staggered during installation.

When cylinder temperatures get too high, lacquer and varnish collect on piston causing rings to stick which results in rapid wear. A worn ring usually takes on a shiny or bright appearance. Scratches on rings and piston are caused by abrasive material such as carbon, dirt, or pieces of hard metal.

Detonation damage occurs when a portion of the fuel charge ignites spontaneously from neat and pressure shortly after ignition. This creates two flame fronts which meet and explode to create extreme hammering pressures on a specific area of the piston. Detonation generally occurs from using fuels with too low of an octane rating.

Preignition or ignition of the fuel charge before the timed spark can cause damage similar to detonation. Preignition damage is often more quickly burned right through the piston dome. Preignition is caused by a hot spot in the combustion chamber from its sources such as: Glowing carbon deposits, blocked fins, improperly seated valves or wrong spark plug.

Service

▶ See Figures 72 and 73

Magnum service replacement pistons are available in STD bore size, and in 0.003 in., 0.010 in., 0.020 in., and 0.030 in. oversizes. Replacement pistons include new piston ring sets and new piston pins.

Service replacement piston rings sets are also available separately for STD/.003 in. (same ring set for both sizes), 0.010 in., 0.020 in. and 0.030 in. oversized pistons. Always use new piston rings when installing pistons. Never reuse old rings.

The cylinder bore must be deglazed before service ring sets are used.

Some important points to remember when servicing piston rings:

1. If the cylinder block does not need reboring and if the old piston is within wear limits and free of score or scuff marks, the old piston may be reused.

2. Remove old rings and clean up grooves. Never reuse old rings.

3. Before installing new rings on piston, place top two rings, each in turn, in its running area in cylinder bore and check end cap.

4. After installing the new compression (top and middle) rings on piston, check piston-to-ring side clearance. Maximum recommended side clearance is 0.006 in.. If side clearance is greater than 0.006 in., a new piston must be used.

PISTON RINGS

Rings must be installed correctly. Ring installation instructions are usually included with new ring sets. Follow instructions carefully. Use a piston ring expander to install rings. Install the bottom (oil control) ring first and the top compression ring last.

Posi-Lock Connecting Rods

Posi-Lock connecting rods are used in all Magnum engines.

Style A

1. Oil Control Ring - (Bottom Groove): Install the expander spring, spacer (use a piston ring installation tool), and rails.

2. Compression Ring - (Center Groove): Install the expander spring and then the ring (use a piston ring installation tool). Make sure "pip" mark on ring is up.

3. Compression Ring - (Top Groove): Install the chrome compression ring (use a piston ring installation tool). Make sure "pip" mark on ring is up.

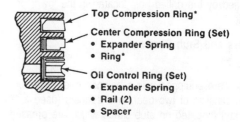

Top Compression Ring*

Center Compression Ring (Set)
• Expander Spring
• Ring*

Oil Control Ring (Set)
• Expander Spring
• Rail (2)
• Spacer

Style C

1. Oil Control Ring - (Bottom Groove): Install the expander spring, spacer (use a piston ring installation tool), and rails.

2. Compression Ring - (Center Groove): Install the expander, ring (use a piston ring installation tool), and rail. Make sure "pip" mark on ring is up.

3. Compression Ring - (Top Groove): Install the chrome compression ring (use a piston ring installation tool). Make sure "pip" mark on ring is up.

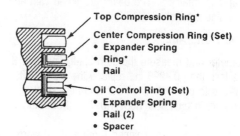

Top Compression Ring*

Center Compression Ring (Set)
• Expander Spring
• Ring*
• Rail

Oil Control Ring (Set)
• Expander Spring
• Rail (2)
• Spacer

Style D

1. Oil Control Ring - (Bottom Groove): Install the expander and rails. Make sure the ends of expander are not overlapped.

2. Compression Ring - (Center Groove): Install the beveled ring using a piston ring installation tool. Make sure the "pip" mark is up and PINK dykem stripe is to the left of end gap.

3. Compression Ring - (Top Groove): Install the top ring using a piston ring installation tool. Make sure the "pip" mark is up and BLUE dykem stripe is to the left of end gap.

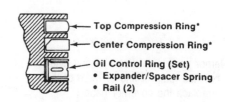

Top Compression Ring*

Center Compression Ring*

Oil Control Ring (Set)
• Expander/Spacer Spring
• Rail (2)

*Install with "pip" mark facing up.

83250077

Fig. 72 Piston ring installation

Model	End Gap	
	New	Used
M8	.007/.017	.007/.027
M10 M12 M14 M16	.010/.020	.010/.030

83250078

Fig. 73 Measuring ring end gap

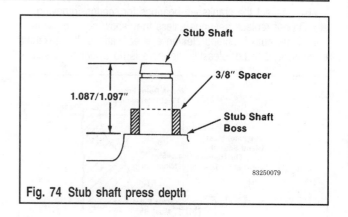

Stub Shaft

3/8" Spacer

Stub Shaft Boss

1.087/1.097"

83250079

Fig. 74 Stub shaft press depth

INSPECTION AND SERVICE

Check bearing area (big end) for excessive wear, score marks, running and side clearances. Replace rod and cap if scored or excessively worn.

Service replacement connecting rods are available in STD crankpin size and 0.010 undersize. The 0.010 in. undersize rod can be identified by the drilled hole located in the lower end of the rod shank.

Balance Gears and Stub Shafts

▶ See Figure 74

Most engines are equipped with a balance gear system.

The system consists of two gears and spacers (used to control end play) mounted on stub shafts which are pressed into the crankcase. The gears and spacers are held on the shafts with snap-ring retainers. The gears are timed with and driven by the engine crankshaft.

INSPECTION AND REPAIR

Use a micrometer and measure the stub shaft diameter. If the diameter is less than 0.4996 in., replace the stub shaft. use an arbor press to push old shaft out and new shaft in. Press the new shaft in until it is 1.087-1.097 in. from stub shaft boss.

Inspect the gears fro worn or chipped teeth and for worn needle bearings. Use an arbor press and driver to replace bearings, if required.

Governor Gear

INSPECTION

Inspect the governor gear teeth. Look for any evidence of worn, chipped or cracked teeth. If one or more of these problems is noted, replace the governor gear.

Camshaft and Crankshaft

INSPECTION AND SERVICE

▶ See Figure 75

Inspect the gear teeth on both the crankshaft and camshaft. If the teeth are badly worn, chipped or some are missing, replacement of the damaged components will be necessary.

Also, inspect the crankshaft bearings for scoring, grooving, etc. Do not replace bearings unless they show signs of damage or are out of running clearance specifications. If crankshaft turns easily and noiselessly, and there is no evidence of scor-

ing, grooving, etc., on the races or bearing surfaces, the bearings can be reused.

Check crankshaft keyways. If worn or chipped, replacement of the crankshaft will be necessary. Also inspect the crankpin for score marks or metallic pickup. Slight score marks can be cleaned with crocus cloth soaked in oil. If wear limits are exceeded, it will be necessary to either replace the crankshaft or regrind the crankpin to 0.010 in. undersize. If reground, a 0.010 in. undersize connecting rod (big end) must then be used to achieve proper running clearance. Measure the crankpin for size, taper and out-of-round.

➡ If the crankpin is reground, visually check to insure that the fillet blends smoothly with the crankpin surface.

ASSEMBLY

▶ See Figures 76, 77, 78, 79, 80, 81, 82, 83, 84, 85, 86, 87, 88 and 89

The following sequence is suggested for complete engine reassembly. This procedure assumes that all components are new or have been reconditioned, and all component subassembly work has been completed. This procedure may have to varied slightly to accommodate options or special equipment.

✳✳WARNING

Make sure the engine is assembled using all specified torque values, tightening sequences, and clearances. Failure to observe specifications could cause severe engine wear or damage.

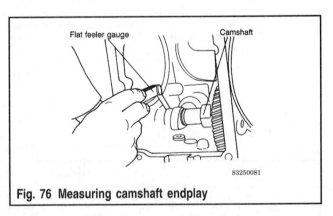

Fig. 76 Measuring camshaft endplay

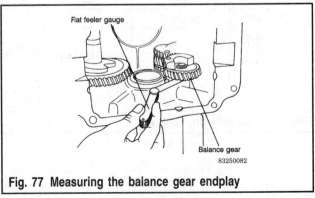

Fig. 77 Measuring the balance gear endplay

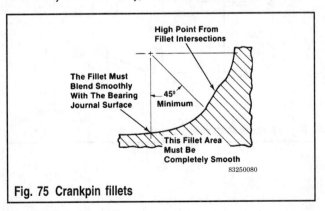

Fig. 75 Crankpin fillets

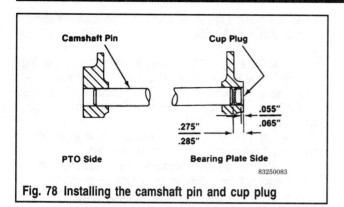

Fig. 78 Installing the camshaft pin and cup plug

1. Install the rear bearing into crankcase using the #4747 handle and appropriate bearing installer. Make sure the bearing is bottomed fully, and is straight and true in bore.

2. Install the cross shaft, sleeve, and bushing nut. Torque bushing nut to 100-120 inch lb.

3. Install the thrust washer, governor gear, copper washer, and stop pin.

4. Install the intake valve tappet and exhaust valve tappet into crankcase. (Intake valve tappet towards bearing plate side; exhaust valve tappet towards PTO side of crankcase.).

5. Install the camshaft, one 0.010 in. shim spacer, and the camshaft pin (from bearing plate side). Do not drive the camshaft pin into its final position at this time.

6. Measure the camshaft end play between the spacer and crankcase boss using a flat feeler gauge. Recommended camshaft end play is 0.005-0.010 in.. Add or subtract 0.005 in. and/or 0.010 in. shim spacers as necessary to obtain the proper end play.

7. Drive the camshaft pin into the PTO side of crankcase until it is 0.300-0.330 in. from machined bearing plate gasket surface.

8. Apply Loctite® #290 (or equivalent) to cup plug. Install the cup plug into bore in bearing plate mounting surface. Plug should be FLUSH to 0.030 in. below mounting surface.

➡ **The balance gears must be timed to the crankshaft whenever the crankshaft is installed. Use a balance gear timing tool to simplify this procedure. If the balance gears must be timed without using the tool, do not install the lower balance gear (closest to oil pan) until after the crankshaft has been installed.**

9. On engines with balance gears: Install the ³⁄₈ in. spacer, one 0.010 in. shim spacer, balance gear, one 0.020 in. shim spacer, and retaining ring (rounded edge towards balance gear). Check end play with a flat feeler gauge. Recommended end play is 0.002-0.010 in.. If end play is not within range, install or remove 0.005 in. and 0.010 in. spacers, as necessary. Balance gear spacer kit, Kohler Part 47 755 01, contains enough ³⁄₈ in., 0.005 in., 0.010 in., and 0.020 in. spacers to obtain correct end play for both balance gears.

10. On Engines Without Balance Gears:

   a. Lubricate the crankshaft rear bearing surface. Insert the crankshaft through the rear bearing.

➡ **If the crankshaft and bearing plate have not been separated, position the fuel line and wiring harness between the bearing plate and crankcase before pressing the crankshaft all the way in.**

b. Align the primary timing mark on crankshaft with the timing mark on camshaft. Press the crankshaft into rear bearing. Make sure the camshaft and crankshaft gears mesh and that the timing marks remain aligned while pressing.

11. On Engines With Balance Gears:

METHOD 1 — WITH BALANCE GEAR TIMING TOOL

   a. Align the primary timing marks of balance gears with the teeth on timing tool. Insert tool so it meshes with gears. Hold or clamp tool against oil pan gasket surface.

   b. Lubricate the crankshaft rear bearing surface. Insert the PTO end of crankshaft through rear bearing. 'Straddle' the primary and secondary timing marks on crankshaft over the rear bearing oil drain. Press the crankshaft into rear bearing

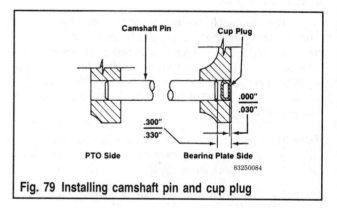

Fig. 79 Installing camshaft pin and cup plug

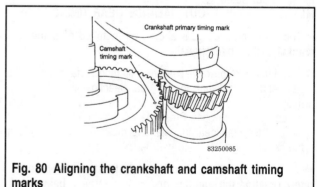

Fig. 80 Aligning the crankshaft and camshaft timing marks

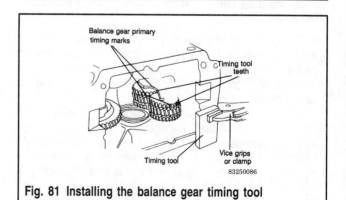

Fig. 81 Installing the balance gear timing tool

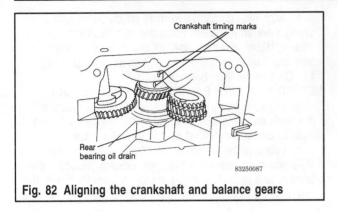

Fig. 82 Aligning the crankshaft and balance gears

until the crankgear is just above the camshaft gear but but not in mesh with it.

→If the crankshaft and bearing plate have not been separated, position the fuel line and wiring harness between the bearing plate and crankcase before pressing the crankshaft all the way in.

c. Remove the balance gear timing tool and align the primary timing mark on the crankshaft with the timing mark on the camshaft gear. Press the crankshaft all the way into the rear bearing. Make sure the camshaft and crankshaft gears mesh and that the timing marks align while pressing.

d. Check the timing of the crankshaft, camshaft, and balance gears:
- The primary timing mark on crankshaft should align with the secondary timing mark on lower balance gear.
- The primary timing mark on crankshaft should align with the timing mark on camshaft.

→If the marks do not align, the timing is incorrect and must be corrected.

METHOD 2 — WITHOUT BALANCE GEAR TIMING TOOL

→The lower balance gear should be installed after the crankshaft has been installed.

e. Lubricate the crankshaft rear bearing surface. Insert the PTO end of crankshaft through rear bearing. Align the primary timing mark on crankshaft with the primary timing mark on upper balance gear. Press the crankshaft into rear bearing until the crankgear just starts to mesh (about 1/16 in.) with the center ring of balance gear teeth.

→If the crankshaft and bearing plate have not been separated, position the fuel line and wiring harness between the bearing plate and crankcase before pressing the crankshaft all the way in.

f. Align the primary timing mark on the crankshaft with the timing mark on the camshaft gear. Press the crankshaft all the way into the rear bearing. Make sure the camshaft and crankshaft gears mesh and that the timing marks align while pressing.

g. Position the crankshaft so it is about 15 degrees past BDC. Install 3/8 in. spacer, and one 0.010 in. shim spacer. Align the secondary timing mark on the lower balance gear with the secondary timing timing mark on the crankshaft. Install the lower balance gear on the stub shaft. If properly

timed, the primary timing mark on the crankshaft will now be aligned with the secondary timing mark on the lower balance gear.

h. Install one (1) 0.020 in. shim spacer and retaining ring (rounded edge towards gear). Check end play of lower balance gear as instructed under 'INSTALL BALANCE GEARS'.

i. Check the timing of the crankshaft, camshaft, and balance gears:
- The primary mark on crankshaft should align with the primary timing mark on upper balance gear.
- The primary mark on crankshaft should align with the secondary timing mark on lower balance gear.
- The primary mark on crankshaft should align with the timing mark on camshaft.

→If the marks do not align, the timing is incorrect and must be corrected.

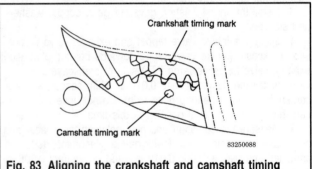

Fig. 83 Aligning the crankshaft and camshaft timing marks

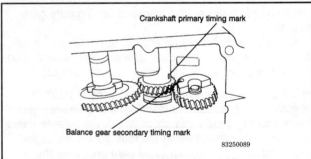

Fig. 84 Checking the crankshaft and balance gear alignment

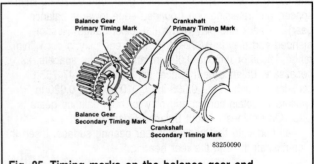

Fig. 85 Timing marks on the balance gear and crankshaft

12. Install the front bearing into the bearing plate using the #4747 handle and appropriate bearing installer. Make sure the bearing is bottomed fully, and straight and true in the bore.

13. Position the fuel line and wiring harness (if so equipped) to crankcase.

14. Adjust the fuel line and wiring harness to their final positions just before securing the bearing plate to the crankcase.

15. Install ⅜-16 studs into two of the bearing plate mounting holes. The studs ease locating and assembly of shims, gasket, and bearing plate.

16. Lubricate the bearing surface of crankshaft and bearing. Install the gasket, two or three 0.005 in. shims (as required)*, and bearing plate over studs.

➡Crankshaft end play is determined by the thickness of the gasket and shims between crankcase and bearing plate. Check the end play after the bearing plate is installed.

17. Install two hex cap screws and hand tighten. remove the locating studs, and install the remaining two hex cap screws and hand tighten.

18. Position the wiring harness and fuel line in their final positions as marked during disassembly.

19. Tighten the screws evenly, drawing bearing plate to crankcase. Torque screws to 35 ft. lbs.

20. Check crankshaft end play between the inner bearing race and shoulder of crankshaft using a flat feeler gauge. Recommended total end play is 0.003-0.020 in.. If measured end play is not within limits, remove the bearing plate and, remove or install shims as necessary.

21. Slide the appropriate seal sleeves over the crankshaft. Generously lubricate the lips of the oil seals with light grease. Slide the oil seals over the sleeves.

22. Use the #11795 handle and appropriate seal drivers to install the oil seals to the following depths:
- Front Oil Seal (Bearing Plate) — ¹/₁₀₀₀ in.
- Rear Oil Seal (Crankcase PTO End) — ⅛ in.
- Tolerance on seal position — +³/₁₀₀ in.; -¹/₁₀₀ in.

23. With 'Style A' Piston — Install wrist pin and retainers.

24. With 'Style C' Piston

❊❊WARNING

Proper orientation of the piston to the connecting rod is extremely important. Improper orientation may cause extensive wear or damage.

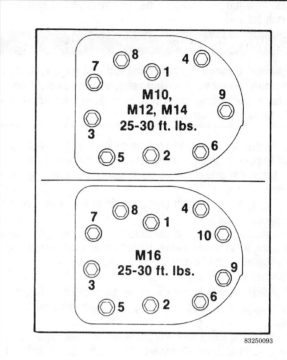

Fig. 88 Cylinder head torque sequence — M10, 12, 14, 16

25. Orient piston and connecting rod so that the notch on piston and the match mark on connecting rod are facing the same direction.

26. Install wrist pin and retainers.

27. Stagger the piston rings in their grooves until end gaps are 120 degrees apart.

28. Lubricate the piston and rings with engine oil. Install the piston ring compressor around piston.

29. Orient the notch (on 'Style C' piston) and match marks on connecting rod towards the flywheel end of crankshaft. Gently push the piston/connecting rod into bore — do not pound on piston.

30. Lubricate the crankshaft and connecting rod journal surfaces with engine oil. Install the connecting rod cap — make sure the match marks are aligned and the oil hole is towards the camshaft. Install hex nuts and torque in increments as

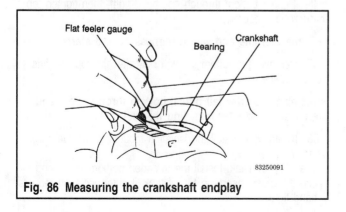

Fig. 86 Measuring the crankshaft endplay

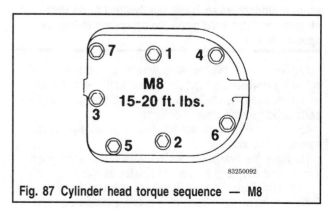

Fig. 87 Cylinder head torque sequence — M8

follows: New rod — 260 inch lbs.; Used/Reinstalled Rod — 200 inch lbs.

✳✳WARNING

To prevent damage to connecting rod and engine, do not overtorque — loosen — and retorque the hex nuts on Posi-Lock connecting rods. Torque nuts, in increments, directly to the specified value.

31. Rotate the crankshaft until the piston is at top dead center in bore to protect the dipper on the connecting rod.

32. Install the gasket, oil pan, and hex cap screws. Tighten screws securely.

33. Rotate the crankshaft until piston is at top dead center of compression stroke.

34. Install the valves and measure the valve-to-tappet clearance using a flat feeler gauge. Valve-to-tappet cold clearance: Intake Valve — 0.008-0.010 in.; Exhaust valve — 0.017-0.019 in.

35. Adjust valve-to-tappet clearance, as necessary. Adjust valve-to-tappet clearance by turning the adjusting screw on tappets.

36. Install the valve spring upper retainers, valve springs (close coils to top), intake valve spring lower retainer, exhaust valve rotator, and valves.

➡Some models use a valve rotator on both valves.

37. Compress springs using a valve spring compressor and install keepers.

38. Route the leads through the hole in bearing plate. Install stator and pan head screws. Install stator lead clip and self-tapping screw.

39. insert stator leads into outer positions of connector body.

✳✳CAUTION

Damaging Crankshaft and Flywheel Could Cause Personal Injury! Using improper procedures to install the flywheel can crack or damage the crankshaft and/or flywheel. This not only causes extensive engine damage, but also is a serious threat to the safety of persons nearby, since broken fragment could be thrown from the engine. Always observe and use the following precautions and procedures when installing the flywheel: Before installing the flywheel, make sure the crankshaft taper and flywheel hub are clean, dry, and completely free of lubricants. the presence of lubricants can cause the flywheel to be overstressed and damaged when the cap screw is torqued to specification.

Always use a flywheel strap wrench to hold flywheel when tightening flywheel fastener. Do not use any type of bar or wedge between the cooling fins or flywheel ring gear, as these parts could become cracked or damaged.

40. install the spacers, fan and hex flange screws. Torque screws to 115 inch lbs.

41. Place the flywheel on the crankshaft. Install the grass screen, and drive cup or rope start pulley as follows:

a. With Rope Start — Install the nylon grass screen to the fan. Install the spacer, rope pulley, plain washer, and hex cap screw (lubricate threads with oil). Hold the flywheel with a strap wrench and torque hex cap screw to 40-45 ft. lbs. Install the grass screen to the fan.

b. With Retractable Start — Install the grass screen to the fan. Install the drive cup, plain washer, and hex cap screw (lubricate threads with oil). Hold the flywheel with a strap wrench and torque hex cap screw to 40-45 ft. lbs.

c. With Electric Start — Install the plain washer and hex cap screw (lubricate threads with oil). Hold the flywheel with a strap wrench and torque hex cap screw to 40-45 ft. lbs. Install the grass screen to the fan.

42. install the module, plain washers, and hex cap screws. Move the module as far from the flywheel/magnet as possible — tighten the hex cap screws slightly.

43. insert a 0.018 in. flat feeler gauge or shim stock between magnet and module. Loosen hex cap screws so magnet pulls module down. Tighten hex cap screws to 32 inch lbs. Remove feeler gauge or shim stock. Due to the pull of the magnet the bearing plate will flex slightly. The magnet to module air gap should be within the final range of 0.012-0.016 in..

44. Rotate the flywheel back and forth; check to make sure the magnet does not strike module. Check gap with a feeler gauge and readjust, if necessary.

45. Install the air baffle over the module. Install the kill lead and spark plug into the baffle.

46. Install the gasket, cylinder head, plain washers, hex spacer screw, lifting strap, and hex cap screws.

47. Torque the hex cap screws and hex spacer screw (in increments) in the sequence shown, to 25-30 ft. lbs.

48. Install the blower housing and two lower self-tapping screws. Leave the screws loose.

49. Install the cylinder head baffle and truss head screw. leave the screw loose.

50. Install the starter side air baffle, plain washer, hex cap screw, and self-tapping screw. Leave the screws loose.

51. Install the carburetor side air baffle, spark plug lead clip, plain washer, and hex cap screw. Leave the screw loose.

52. Tighten the screws securely when all pieces are in position.

53. Remove the 1/4-20 hex nuts and through bolts from electric starter.

➡Be sure to hold the commutator end cap and drive end cap against the starter frame to prevent the starter from becoming disassembled.

54. install the starter to bearing plate and install the through bolts. Tighten bottom through bolt hand tight, then tighten top through bolt securely.

➡Route fuel line and wiring harness behind starter.

55. Remove the bottom through bolt after the top bolt has been tightened.

➡Install the starter tail brace, bottom through bolt, and starter solenoid after installing the fuel tank.

56. If removed, install the isolation mounts to the fuel tank as follows:

a. Top of Tank: Install the threaded portion, 1/2 in. long, into the brass inserts in top of tank. Tighten isolation mounts hand tight.

b. Bottom of Tank: Install the threaded portion, ⅜ in. long, into the brass inserts in bottom of tank. Tighten isolation mounts hand tight.

57. Install bracket and pawl nuts to bottom isolation mounts. Tighten pawl nuts securely.

58. Insert the threaded portion of top isolation mounts through holes in cylinder head baffle. Install the plain washers and acorn nuts. Tighten the acorn nuts securely.

59. Install the plain washers and hex cap screws through bracket and into crankcase. Leave the screws slightly loose.

60. Adjust the position of the fuel tank until the top of tank is even with top of blower housing. Tighten hex cap screws securely.

61. Install the fuel filter and fuel line to tank outlet.

62. Install the starter tail brace, bottom through bolt, plain washer, and self-tapping screw. Tighten all screws securely when all pieces are in position.

63. Install the solenoid and self-tapping screws to fuel tank bracket. Install leads to solenoid and starter.

64. Install the gasket, oil fill/dipstick tube, fillister head screws, and dipstick.

65. Install the retractable starter and hex cap screws. Leave the screws slightly loose.

66. Pull the starter handle out 8-10 in. until the pawls engage in the drive cup. Hold the handle in this position and tighten screws securely.

67. Install the gasket, fuel pump, plain washers, and fillister head screws.

✳✳WARNING

Make sure the fuel pump lever is positioned above the camshaft. Damage to the fuel pump, and subsequent severe engine damage could result if the lever is positioned below the camshaft. Torque the screws to 40-45 inch lbs.

68. Connect the fuel line to fuel pump inlet fitting.

69. Install the stud, gasket, breather plate, reed, reed stop, seal, and filter.

70. Install the long leg of governor spring to throttle control lever.

71. Install the gasket, throttle and choke control assembly, gasket, cam gear cover, plain washer, hex cap screws, and self-tapping screw. Torque cam gear cover screws to 115 inch lbs. Install the spacer, plain washer, and hex cap screw.

72. Install the connector/wiring harness to back of key switch.

73. Install the governor spring to the governor arm. Install the governor arm to the cross shaft. Leave the pawlnut slightly loose as the governor arm and cross shaft will be adjusted after the carburetor and throttle linkage are installed.

74. Install the fuel line and hose clamps.

75. Install the gasket, carburetor, and slotted hex cap screws.

76. Install the throttle linkage into the nylon inserts in the governor arm and carburetor throttle lever.

77. Adjust the governor as instructed below.

78. Install the choke linkage to the carburetor choke lever, and then into the nylon insert in the choke control lever.

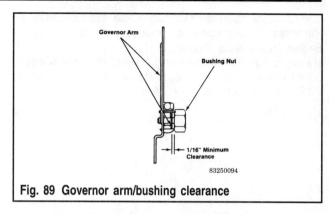

Fig. 89 Governor arm/bushing clearance

79. The governor cross shaft/governor arm must be adjusted every time the governor arm is loosened or removed from cross shaft.

   a. Pull the governor arm away from the carburetor as far as it will go.

   b. Grasp end of cross shaft with pliers and turn counterclockwise as far as it will go.

   c. Torque the pawlnut on governor arm to 15 inch lbs.

➡**Make sure there is at least ¹⁄₁₆ in. clearance between the governor arm and the upper-left cam gear cover fastener to prevent interference.**

80. Insert the B+ lead into the center position of connector. Install the stator leads/connector to the rectifier-regulator.

81. Install the rectifier-regulator and hex cap screws.

82. Install the threaded exhaust pipe and muffler.

83. Install the breather hose, gasket, air cleaner base, and screws.

✳✳WARNING

Make sure that the breather hose seals tightly in the air cleaner base and the breather cover to prevent unfiltered air from entering the engine.

84. Install the seal, paper element, element cover, and element cover nut. Torque the nut to 50 inch lbs.

85. If equipped, install the optional foam, precleaner (cleaned and oiled) over the paper element.

86. Install the air cleaner and wing nut. Tighten the wing nut until it is snug. Do not overtighten.

PREPARE THE ENGINE FOR OPERATION

The engine is now completely reassembled. Before operating the engine, be sure to do the following:
- Make sure all hardware is tightened securely and oil drain plugs are installed.
- Fill the crankcase with the correct amount, weight, and type of oil.
- Fill the fuel tank with the proper type of gasoline and open fuel shut-off valve (if equipped).

- Adjust the carburetor main fuel needle, idle fuel needle, or idle speed adjusting screw as necessary. Refer to the 'Fuel System And Governor' section.
- Make sure the maximum engine speed does not exceed 3600 RPM. Adjust the high speed stop as necessary. Refer to the 'Fuel System And Governor' section.

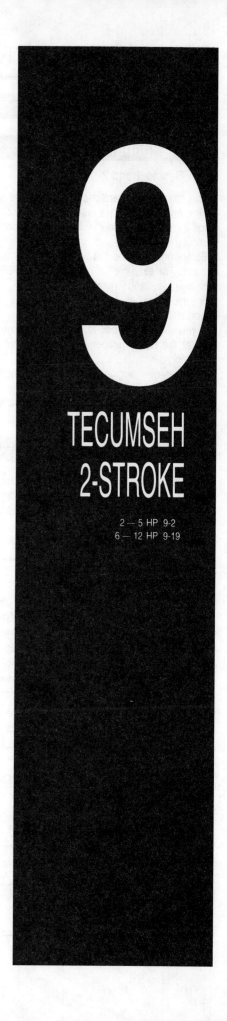

9

TECUMSEH 2-STROKE

2 — 5 HP

Engine Identification

▶ **See Figures 1 and 2**

The identification taps may be located at a variety of places on the engine. The type number is the most important number since it must be included with any correspondence about a particular engine.

Early engines listed the type number as a suffix of the serial number. For example on number 123456789 P 234, 234 is the type number. In the number 123456789 H 104-02B; 104-02B is the type number. In either case the type number is important.

If you use short block to repair the engine, be sure that you transfer the serial number and type number tag to the new short block.

On the newer engines, reference is sometimes made to the model number. The model number tells the number of cylinders, the design (vertical or horizontal) and the cubic inch displacement.

Air Cleaners

The instructions below detail the procedures involved in cleaning the various types of elements. See the illustrations for exploded views to aid disassembly and assembly.

POLYURETHANE AIR CLEANER

▶ **See Figure 3**

1. Wash the element in a solvent or detergent and water solution by squeezing similar to a sponge.
2. Clean the air cleaner housing and cover with the same solution. Dry thoroughly.
3. Dry the element by squeezing or with compressed air if available.
4. Apply a generous quantity of oil to the element sides and open ends. Squeeze vigorously to distribute oil and to remove excess oil.

ALUMINUM FOIL AIR CLEANER

▶ **See Figure 4**

1. Dip the aluminum foil filter in solvent. Flush out all dirt particles.
2. Shake out the filter thoroughly to remove all solvent, then dip the filter element in oil. Allow the oil to drain from the filter. Clean the screens and filter body.

➡**The concave screen and retainer cover or ring are not used on later models. They are replaced with a clip which rolls into the groove in the lip of the body.**

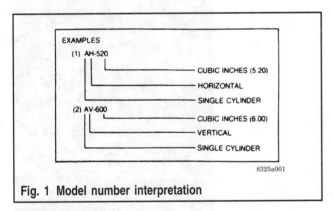

Fig. 1 Model number interpretation

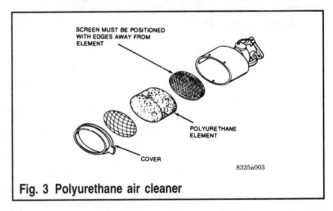

Fig. 3 Polyurethane air cleaner

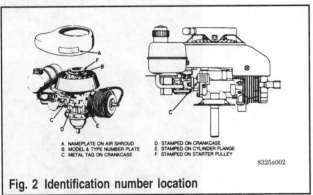

Fig. 2 Identification number location

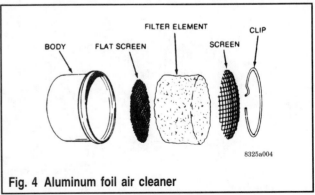

Fig. 4 Aluminum foil air cleaner

FELT TYPE AIR CLEANERS

♦ See Figure 5

1. To clean felt air cleaners, merely blow compressed air through the element in the reverse direction to normal air flow. Felt elements may also be washed in nonflammable solvents or soapy water. Blow dry with compressed air.

➡**Power Products type numbers 641 and up, use a gasket between the element and the base. The gasket is used only with this element; earlier versions did not have a gasket.**

FIBER ELEMENT AIR CLEANER

1. Remove the filter and place the cover in a normal position on the filter. With the filter element down to semi-seal it, blow compressed air through the cover hole to reverse air flow, forcing dirt particles out.
2. Clean the cover mounting bracket with a damp cloth.

DRY PAPER AIR CLEANER

♦ See Figure 6

1. Tap the element on a workbench or any solid object to dislodge larger particles of dirt.
2. Wash the element in soap and water. Rinse from the inside until it is thoroughly flushed and the water coming through is free of soap.

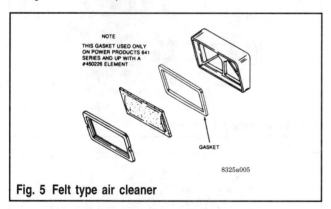

NOTE
THIS GASKET USED ONLY ON POWER PRODUCTS 641 SERIES AND UP WITH A #450226 ELEMENT

GASKET

8325a005

Fig. 5 Felt type air cleaner

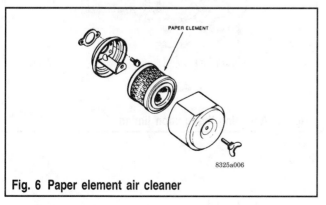

PAPER ELEMENT

8325a006

Fig. 6 Paper element air cleaner

3. Allow the element to dry completely or use low pressure compressed air blown from the inside to speed the process.
4. Inspect the element for cracks or holes, and replace if necessary.

Lubrication

OIL AND FUEL RECOMMENDATIONS

Power Products 2 cycle engines are mist-lubricated by oil mixed with the gasoline. For the best performance, use regular grade, leaded fuel, with 2 cycle or outboard oil rated SAE 30 or SAE 40. Regular grade unleaded fuel is an acceptable substitute. The terms 2 cycle or outboard are used by various manufacturers to designate oil they have designed for use in 2 cycle engines. Multiple weight oil such as all season 10W-30, are not recommended.

If you have to mix the gas and oil when the temperature is below 35°F (1.6°C), heat up the oil first, then mix it with the gas. Oil will not mix with gas when the temperature is approaching freezing. However, if you use oil that has been warmed first it will not be affected by low temperatures.

The proportion of oil to fuel is absolutely critical to two-stroke operation. If too little oil is used, overheating and damage to engine parts will occur (this can even result from running the engine too lean). If excessive oil is used, spark plug fouling, smoke in the exhaust, and even misfire can occur. Mix carefully and precisely. Follow Power Products recommendations for your particular engine, and disregard fuel container labels.

Fuel/oil mix must be clean and fresh. Fuel deteriorates enough to form troublesome gum and varnish after more than a month. Dirt in the fuel can cause clogging of carburetor passages and even engine wear.

Tune-Up Specifications

All spark plugs are gapped at 0.035 in. (0.90mm). Because of the great number of individual models of Power Products engines that exist, an individual chart of Tune-Up specifications is impractical. Refer to the charts at the end of this section for breaker point gap and timing dimension specifications.

Spark Plugs

♦ See Figures 7, 8, 9, 10 and 11

Spark plugs should be removed and cleaned of deposits frequently, especially in two-stroke engines because they burn the lubricating oil right with the fuel. Carefully inspect the plug for severely eroded electrodes or a cracked insulator, and replace the plug if either condition exists or if deposits cannot be adequately removed. Set the gap to 0.035 in. (0.90mm) with a wire type feeler gauge, and install the plug, torquing it to 18-22 ft. lbs.

Make sure to replace the plug with one of the same type and heat range. If the plug is fouled, poor quality or old fuel, a rich mixture, or the wrong fuel/oil mix may be at fault. Also, make sure the engine's exhaust ports are not clogged.

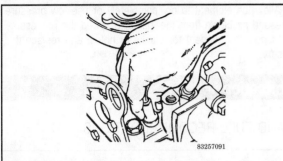

Fig. 7 Twist and pull on the boot, never on the plug wire

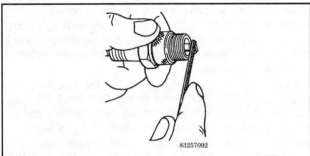

Fig. 8 Plugs that are in good condition can be filed and reused

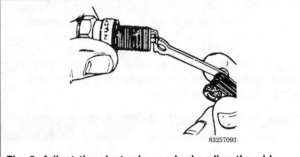

Fig. 9 Adjust the electrode gap by bending the side electrode

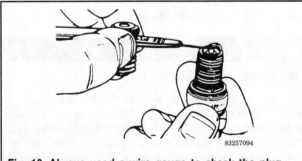

Fig. 10 Always used a wire gauge to check the plug gap

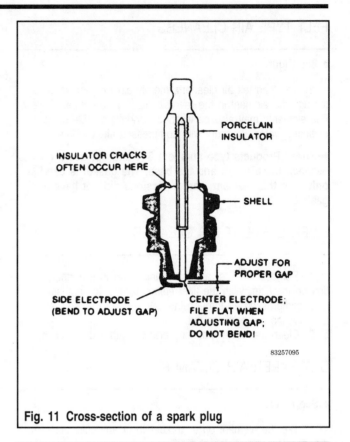

Fig. 11 Cross-section of a spark plug

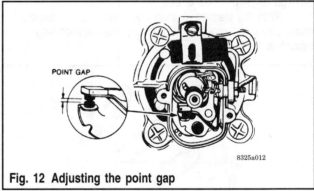

Fig. 12 Adjusting the point gap

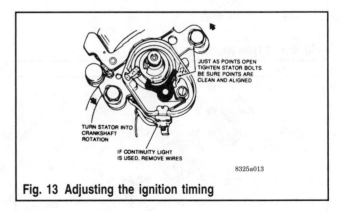

Fig. 13 Adjusting the ignition timing

Breaker Points

REMOVAL & INSTALLATION

1. First remove the flywheel as described below:

   a. Remove the screws, engine shroud, and starter. Determine the direction of rotation of the flywheel nut by looking at the threads. Then, place a box wrench on the nut and tap with a soft hammer in the proper direction.

   b. The flywheel is removed with a special puller or a special knock off tool. The knock off tool must not be used on 660, 670, or 1500 ball bearing models. To use the knock off tool, screw it onto the crankshaft until it is within $^1/_{16}$ in. (1.6mm) of the flywheel. Hold the flywheel firmly and rap the top of the puller sharply with a hammer to jar it loose. Pull the flywheel off.

   c. If a puller is being used, the flywheel will have three cored holes into which a set of self-tapping screws are turned. The handle which operates the bolt at the center of the puller's collar is then turned to pull the flywheel off the crankshaft. If this is not adequate to do the job, heat the center of the alloy flywheel with a butane torch to expand it before turning the puller handle.

2. Remove the nuts that hold the electrical leads to the screw on the movable breaker point spring. Remove the movable breaker point from the stud.

3. Remove the screw and stationary breaker point. Put a new stationary breaker point on breaker plate; install the screw, but do not tighten it fully.

4. Position a new movable breaker point on the post.

5. Check that the new points contact each other properly and remove all grease, fingerprints, and dirt from the points.

ADJUSTMENT

▶ See Figure 12

1. If necessary (as when checking the gap of old points), loosen the screw which mounts the stationary breaker contact. Rotate the crankshaft until the contact cam follower rests right on the highest point of the cam.

2. Using a flat feeler gauge of the dimension shown under 'point gap' in the charts at the rear of this section, check the dimension of the gap and, if incorrect, move the breaker base in the appropriate direction by wedging a screwdriver between the dimples on the base plate and the notch in the breaker plate. When gap is correct, tighten the screw. Recheck the gap and, if necessary, reset it.

Adjusting Ignition Timing

▶ See Figure 13

1. Remove the spark plug and install a special timing tool or thin ruler. With the tool lockscrew loose or the ruler riding on the piston, rotate the crankshaft back and forth to find Top Dead Center. Tighten the tool lockscrew or use a straightedge across the cylinder head, if you're using a ruler, to measure

Top Center. Look up the timing dimension in the specifications at the back of this section.

2. Turn the crankshaft backwards so the piston descends. Then, reset the position of the special tool downward the amount of the dimension and tighten the lockscrew, or move the ruler down that amount.

3. Very carefully bring the piston upward by turning the crankshaft until the top of the piston just touches the tool or ruler.

4. Install a piece of cellophane between the contact surfaces. Loosen the ignition stator lockscrews and turn the stator until the cellophane is clamped tightly between the contact surfaces. Turn the stator until the cellophane can just be pulled out (as contacts start to open), and then tighten the stator lockscrews.

Magneto Armature Air Gap

▶ See Figures 14 and 15

There are two types of magnetos. Compare appearance of your engine with each illustration to determine whether it looks like the type which employs a gap of 0.005-0.008 in. (0.13-0.20mm) or the type with a gap of 0.015 in. (0.38mm).

1. Loosen the two screws which hold the laminations and coil to the block. Turn the flywheel around so the magnets line up directly with the ends of the laminations. Pull the laminations/coil assembly upward.

2. Insert a gap gauge of 0.005-0.008 in. (0.13-0.20mm) between the flywheel and laminations on either side. On units with a gap of 0.015 in. (0.38mm), use two of the above numbered parts or a 0.015 in. (0.38mm) gauge. Allow the attrac-

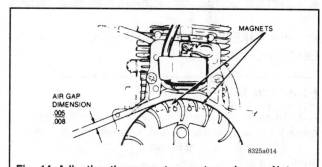

Fig. 14 Adjusting the magneto armature air gap. Note the magnets on the flywheel

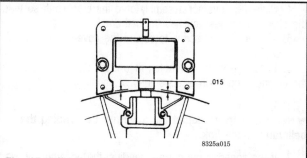

Fig. 15 Adjusting the magneto armature air gap. Note the wider air gap

tion of the magnets to pull the coil/laminations assembly to-ward the flywheel.

3. On the magneto type with a 0.005-0.008 in. (0.13-0.20mm) dimension, use Loctite® Grade A on the screws and torque to 35-45 inch lbs. On the other type magneto, torque the screws to 20-30 inch lbs. Recheck the gap and readjust, if necessary.

Mixture Adjustment

1. If the engine will not start or the carburetor has recently been disassembled, both idle and main mixture screws may be turned in very gently until they just bottom, and then turned out exactly one turn. In this case, back out the idle speed screw until throttle is free, then turn screw in until it just contacts the throttle. Turn it one turn more exactly.

2. Allow the engine to warm up to normal running temperature. With the engine running at maximum recommended rpm, loosen the main metering screw until the engine rolls, then tighten the screw until the engine starts to cut out. Note the number of turns from one extreme to the other. Loosen the screw to a point midway between the extremes.

3. Set the throttle to idle speed and repeat Step 2 for that adjusting screw.

➡Some carburetors do not have a main mixture adjustment. Others employ a drilled mixture screw which provides about the right mixture when fully screwed in. In some weather conditions, operation may be improved by setting this screw just slightly off the seat for smoother running.

Governor Adjustments

POWER TAKEOFF END MECHANICAL GOVERNOR

▶ See Figure 16

1. To adjust the governor, remove the outboard bearing housing. Use a ³/₃₂ in. allen wrench to loosen the set screw.

2. Squeeze the top and bottom governor rings, fully compressing the governor spring.

3. Hold the upper arm of the bell crank parallel to the crankshaft and insert a ³/₃₂ in. allen wrench between the upper ring and the bell crank.

4. Slide the governor assembly onto the crankshaft so that the allen wrench just touches the bell crank.

5. Tighten the set screw to secure the governor to the crankshaft.

6. Install the bearing adapter, mount the engine, and check the engine speed with a tachometer. It should be about 3200-3400 rpm.

➡Never attempt to adjust the governor by bending the bellcrank or the link.

7. If the speed is not correct, readjust the governor by moving the assembly toward the crankcase to increase speed, and away from the crankcase to decreased speed.

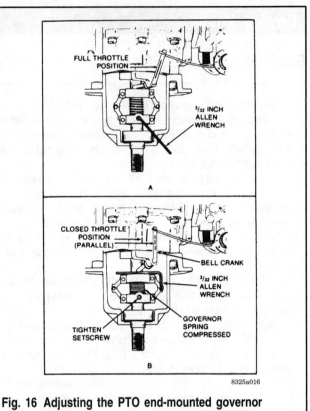

Fig. 16 Adjusting the PTO end-mounted governor

MECHANICAL FLYWHEEL TYPE GOVERNOR ADJUSTMENT

▶ See Figure 17

As engine speed increases, the links are thrown outward, compressing the link springs. The links apply a thrust against the slide ring, moving it upward and compressing the governor spring. As the slide ring moves away from the thrust block of the bellcrank assembly, the throttle spring causes the thrust block to maintain engagement and close the throttle slightly.

As the throttle closes and engine speed decreases, force on the slide ring decreases so that it moves downward, pivots the bellcrank outward to overcome the force of the throttle spring, and opens the throttle to speed up the engine. In this manner, the operating speed of the engine is stabilized to the adjusted governor setting.

1. To adjust the governor, loosen the bracket screw and slide the governor bellcrank assembly toward or away from the

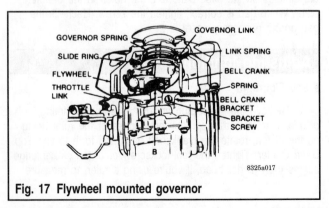

Fig. 17 Flywheel mounted governor

flywheel. Move the bellcrank toward the flywheel to increase speed and away from the flywheel to decrease speed.

2. Tighten the screw to secure the bracket.

3. Make minor speed adjustments by bending the throttle link at the bend in the center of the link.

➡**Do not lubricate the governor assembly or the governor bellcrank assembly of flywheel mounted governors.**

ADJUSTING 2-CYCLE AIR VANE GOVERNORS

▶ **See Figure 18**

1. Loosen the self-locking nut that holds the governor spring bracket to the engine crankcase.

2. Adjust the spring bracket to increase or decrease the governor spring tension. Increasing spring tension increases speed and decreasing spring tension decreases speed.

3. After adjusting, the spring bracket should not be closer than 1/16 in. (1.6mm) to the crankcase.

4. Tighten the self-locking nut.

Carburetor

REMOVAL & INSTALLATION

▶ **See Figure 19**

1. Remove the air cleaner. Drain the fuel tank. Disconnect the carburetor fuel lines.

2. If necessary, remove any shrouding or control panels to gain access to the carburetor.

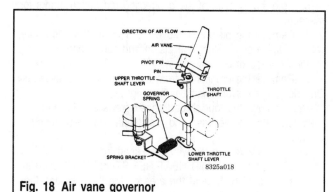

Fig. 18 Air vane governor

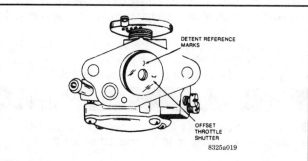

Fig. 19 Position the detent marks on the throttle shutter as shown

3. Disconnect the choke or throttle control wires at the carburetor.

4. Remove the cap screws, or nuts and lockwashers and remove the carburetor from the engine.

5. To install, reverse the removal procedure, using new gaskets.

OVERHAUL

Tecumseh two-and four-stroke engines employ a common series of carburetors. Refer to the Four-Stroke Tecumseh Engine section for specific carburetor overhaul procedures.

Idle Governor

SERVICE

1. Remove the shutter fastener and allow the shutter to drop out of the air horn.

2. Note location of the spring end in the disc-shaped throttle lever. The spring should be placed into the same hole during reassembly.

3. Remove the retainer clip and lift out the throttle shaft.

4. Replace all worn parts and reassemble in reverse order.

➡**Note the position of the throttle shutter, as shown. The reference marks must be positioned as shown when shutter is installed.**

Fuel Pump

SERVICE

Float Type Carburetor With Integral Pump
▶ **See Figure 20**

1. If the engine runs, but roughly, make both carburetor mixture adjustments.

2. Make sure the fuel supply is adequate and the tank is in the proper position.

3. Make sure the fuel tank valve is open.

4. Make sure the pick-up tube is not cracked.

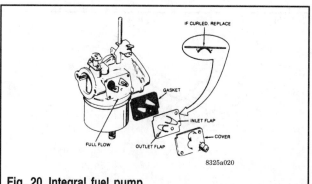

Fig. 20 Integral fuel pump

5. Remove the carburetor and make sure the pulsation passage is properly aligned.

6. Check for air leaks at the gasket surface.

7. Remove the cover and check the condition of the inlet and outlet flaps — if curled, replace the flap leaf.

Engine Overhaul

DISASSEMBLY

Split Crankcase Engines

1. Remove the shroud and fuel tank if so equipped.

2. Remove the flywheel and the ignition stator.

3. Remove the carburetor and governor linkage. Carefully note the position of the carburetor wire links and springs for reinstallation.

4. Lift off the reed plate and gasket if present and inspect them. They should not bend away from the sealing surface plate more than 0.010 in. (0.25mm).

5. Remove the spark plug and inspect it.

6. Remove the muffler. Be sure that the muffler and the exhaust ports are not clogged with carbon. Clean them if necessary.

7. Remove the transfer port cover and check for a good seal.

8. Remove the cylinder head, if so equipped.

➡**Some models utilize a locking compound on the cylinder head screws. Removing the screws on such engines can be difficult. This is especially true with screws having slotted head for a straight screwdriver blade. The screws can be removed if heat is applied to the head of the screw with an electric soldering iron.**

9. On engines having a governor mounted on the power take-off end of the crankshaft, remove the screws that hold the outboard bearing housing to the crankcase. Clean the PTO end of the crankshaft and remove the outboard bearing housing and bearing. Loosen the set screw that holds the governor assembly to the crankshaft, slide the entire governor assembly from the crankshaft. Remove the screw that holds the governor bellcrank bracket to the crankshaft. Remove the governor bellcrank and bracket.

10. Make match marks on the cylinder and crankcase. Remove the four nuts and lockwashers that hold the cylinder to the crankcase.

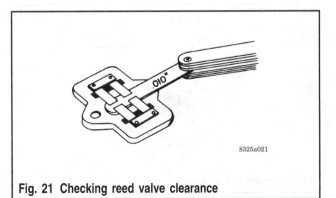

8325a021

Fig. 21 Checking reed valve clearance

11. Remove the cylinder by pulling it straight out from the crankcase.

12. To separate the two crankcase halves, remove all of the screws that hold the crankcase halves together.

13. With the crankcase in a vertical position, grasp the top half of the crankcase and hold it firmly. Strike the top end of the crankcase with a rawhide mallet, while holding the assembly over a bench to prevent damage to parts when they fall. The top half of the crankcase should separate from the remaining assembly.

14. Invert the assembly and repeat the procedure to remove the other casting half from the crankcase on ball bearing units.

15. Each time the crankshaft is removed from the crankcase, seals at the end of the crankcase should be replaced. To replace the seals, use a screwdriver or an ice pick to remove the seal retainers and remove and discard the old seals. Install the seals in the bores of the crankcase halves. The seals must be inserted into the bearing well with the channel groove toward the internal side of the crankcase. Retain the seal with the retainer. Seat the retainer spring into the spring groove.

Uniblock Engines

1. Remove the shroud and fuel tank. Note the condition of the air vane governor, if so equipped.

2. Remove the starter cap and flywheel nut, noting the position of the belleville washer.

3. Remove the flywheel — see 'Breaker Points Removal and Installation,' above.

4. Remove the head. Save the old head gasket for use when replacing the piston, but procure a new gasket for use in final assembly.

5. Remove the cylinder block cover plate to gain access to the connecting rod bolts.

6. Note the location of the connecting rod match marks for reassembly.

7. Remove the piston. Remove the ridge first, if necessary, with a ridge reamer. Push the piston and connecting rod through the top of cylinder.

8. Remove the crankshaft from the cylinder block assembly. On engines with crankshaft ball bearings:

   a. Remove the four shroud base screws and tap the shroud base so the base and crankshaft can be removed together.

   b. To remove bearing with crankshaft from base: USE SAFETY GLASSES AND HEAT RESISTANT GLOVES. Using a propane torch, heat the area on the base around the outside of the bearing until there is enough expansion to remove the base from the bearing on the crankshaft. Now remove and discard the seal retainer ring, seal retainer and seal.

   c. To remove the bearing race, remove the retainer ring on the crankshaft with snap ring pliers, and with the use of a bearing splitter or arbor press, remove the ball bearing.

❊❊CAUTION

Support the crankshaft's top counterweight to prevent bending. Also, bearing is to be pressed on via the inner race only.

9. On other engines:

a. When equipped with a sleeve or needle bearing, use a seal protector and lift the crankshaft out of the cylinder. Be careful not to lose the bearing needles.

b. When equipped with a ball bearing, use a mallet to strike the crankshaft on the P.T.O. end while holding the block in your hand.

10. In assembly, bear the following points in mind:

a. Use a ring compressor to install piston. Be careful not to allow the rings to catch on the recess for the head gasket. Use the old head gasket to take up the space in the recess. Do not force the piston into the cylinder, or damage to rings or piston could occur.

b. To install the ball bearing on the crankshaft, slide the bearing on the crankshaft and fit it on the shaft by tapping using a mallet and tool, part number 670258 or press the ball bearing on the crankshaft with an arbor press. Install the retainer ring.

c. To install the crankshaft with a ball bearing, heat the shroud base to expand the bearing seat and drop the ball bearing into the seat of the base shroud. Allow it to cool. Install a new seal retainer ring, seal retainer, and seal.

11. After the shroud base and flywheel are back in place, adjust the air gap between the coil core and flywheel as described above under 'Magnet Air Gap Adjustment.'

CONNECTING ROD SERVICE

▶ **See Figures 22 and 23**

1. For engines using solid bronze or aluminum connecting rods, remove the two self-locking cap screws which hold the connecting rod to the crankshaft and remove the rod cap. Note the match marks on the connecting rod and cap. These marks must be reinstalled in the same position to the crankshaft.

2. Engines using steel connecting rods are equipped with needle bearings at both crankshaft and piston pin end. Remove the two set screws that hold the connecting rod and cap to the crankshaft, taking care not to lose the needle bearings during removal.

3. Needle bearings at the piston pin end of steel rods are caged and can be pressed out as an assembly if damaged.

4. Check the connecting rod for cracks or distortion. Check the bearing surfaces for scoring or wear. Bearing diameters should be within the limits indicated in the table of specifications located at the end of this section.

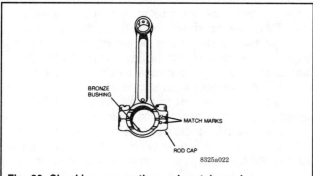

Fig. 22 Checking connecting rod match marks

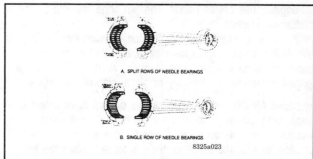

A. SPLIT ROWS OF NEEDLE BEARINGS

B. SINGLE ROW OF NEEDLE BEARINGS

8325a023

Fig. 23 Needle bearing arrangement. Double rows of bearings are placed with the tapered edges facing out

5. There are two basic arrangements of needles supplied with the connecting rod crankshaft bearing: split rows of needles and a single row of needles. Service needles are supplied with a beeswax coating. The beeswax holds the needles in position.

6. To install the needle bearings, first make sure that the crankshaft bearing journal and the connecting rod are free from oil and dirt.

7. Place the needle bearings with the beeswax onto a cool metallic surface to stiffen the beeswax. Body temperature will melt the wax, so avoid handling.

8. Remove the paper backing on the bearings and wrap the needles around the crankshaft journal. The beeswax will hold the needles onto the journal. Position the needles uniformly onto the crankpin.

➡ **When installing the split row of needles, wrap each row of needles around the journal and try to seal them together with gentle but firm pressure to keep the bearings from unwinding.**

9. Place the connecting rod onto the journal, position the rod cap, and secure it with the capscrews. Tighten the screw to the proper specifications.

10. Force solvent (lacquer thinner) into the needles just installed to remove the beeswax, then force 30W oil into the needles for proper lubrication.

PISTON AND RINGS SERVICE

1. Clean all carbon from the piston and ring grooves.

2. Check the piston for scoring or other damage.

3. Check the fit of the piston in the cylinder bore. Move the piston from side-to-side to check clearance. If the clearance is not greater than 0.0003 in. (0.0076mm) and the cylinder is not scored or damaged, then the piston need not be replaced.

4. Check the piston ring side clearance to make sure it is within the limits recommended.

5. Check the piston rings for wear by inserting them into the cylinder about ½ in. (13mm) from the top of the cylinder. Check at various places to make sure that the gap between the ends of the ring does not exceed the dimensions recommended in the specifications table at the end of this section. Bore wear can be checked in the same way, except that a new ring is used to measure the end gap.

6. If replacement rings have a beveled or chamfered edge, install them with the bevel up toward the top of the piston. Not

all engines use beveled rings. The two rings installed on the piston are identical.

7. When installed, the offset piston used on the AV600 and the AV520 engines must have the **V** stamped in the piston head (some have hash marks) facing toward the right as the engine is viewed from the top or piston side of the engine.

➡**Some AV520 and AV600 engines do not have offset pistons. Only offset pistons will have the 'V' or the hash marks on the piston head. Domed pistons must be installed so that the slope of the piston is toward the exhaust port.**

CRANKSHAFT SERVICE

▶ **See Figure 24**

1. Use a micrometer to check the bearing journals for out-of-roundness. The main bearing journals should not be more than 0.0005 in. (0.0127mm) out-of-round. Connecting rod journals should not be more than 0.001 in. (0.025mm) out-of-round. Replace a crankshaft that is not within these limits.

➡**Do not attempt to regrind the crankshaft since undersize parts are not available.**

2. Check the tapered portion of the crankshaft (magneto end), keyways, and threads. Damaged threads may be restored with a thread die. If the taper of the shaft is rusty, it indicates that the engine has been operating with a loose flywheel. Clean the rust off the taper and check for wear. If the taper or keyway is worn, replace the crankshaft.

3. Check all of the bearing journal diameters. They should be within the limits indicated in the specifications table at the end of this section.

4. Check the crankshaft for bends by placing it between two pivot points. Position the dial indicator feeler on the crankshaft bearing surface and rotate the shaft. The crankshaft should not be more than 0.002-0.004 in. (0.05-0.10mm) out-of-round.

BEARING SERVICE

1. Do not remove the bearings unless they are worn or noisy. Check the operation of the bearings by rotating the bearing cones with your fingers to check for roughness, bind-ing, or any other signs of unsatisfactory operation. If the bearings do not operate smoothly, remove them.

2. To remove the bearings from the crankcase, the crankcase must be heated. Use a hot plate to heat the crankcase to no more than 400°F (204°C). Place a 1/8 in. (3mm) steel plate over the hot plate to prevent overheating. At this temperature, the bearings should drop out with a little tapping of the crankcase.

3. The replacement bearing is left at room temperature and dropped into the heated crankcase. Make sure that the new bearing is seated to the maximum depth of the cavity.

➡**Do not use an open flame to heat the crankcase halves and do not heat the crankcase halves to more than 400°F (204°C). Uneven heating with an open flame or excessive temperature will distort the case.**

4. The needle bearings will fall out of the bearing cage with very little urging. Needles can be reinstalled easily by using a small amount of all-purpose grease to hold the bearings in place.

5. Cage bearings are removed and replaced in the same manner as ball bearings.

6. Sleeve bearings cannot be replaced. Both crankcase halves must be discarded if a bearing is worn excessively.

ASSEMBLY

1. The gasket surface where the crankcase halves join must be thoroughly clean before reassembly. Do not buff or use a file or any other abrasive that might damage the mating surfaces.

➡**Crankcase halves are matched. If one needs to be replaced, then both must be replaced.**

2. Place the PTO half of the crankcase onto the PTO end of the crankshaft. Use seal protectors where necessary.

3. Apply a thin coating of sealing compound to the contact surface of one of the crankcase halves.

4. Position one crankcase half on the other. The fit should be such that some pressure is required to bring the two halves together. If this is not the case, either the crankcase halves and/or the crankshaft must be replaced.

5. Secure the halves with the screws provided, tightening the screws alternately and evenly. Before tightening the screws, check the union of the crankcase halves on the cylinder mounting side. The halves should be flat and smooth at the union to provide a good mounting face for the cylinder. If necessary, realign the halves before tightening the screws.

6. The sleeve tool should be placed into the crankcase bore from the direction opposite the crankshaft. Insert the tapered end of the crankshaft through the half of the crankcase to which the magneto stator is mounted. Remove the seal tools after installing the crankcase halves.

7. Stagger the ring ends on the piston and check for the correct positioning on domed piston models.

8. Place the cylinder gasket on the crankcase end.

9. Place the piston into the cylinder using the chamfer provided on the bottom edge of the cylinder to compress the rings.

10. Secure the cylinder to the crankcase assembly.

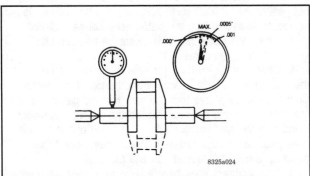

Fig. 24 Checking the crankshaft for out-of-round

Type Number-to-Letter Cross Reference Chart

Type No.	Column Letter	Type No.	Column Letter	Type No.	Column Letter
1 thru 44	A	611	F	1001	D
46 thru 68	B	614–01 thru 614–04	E	1002 thru 1002D	G
69 thru 77	D	614–05 thru 614–06A	F	1003	C
78 thru 80	C	615–01 thru 615–01A	E	1004 thru 1004A	H
81 thru 83	D	615–04	F	1005 thru 1005A	C
84 thru 85	C	615–05 thru 615–10A	E	1006	H
86 thru 87	B	615–11 thru 615–18	F	1007 thru 1008B	C
89	C	615–19 thru 615–27A	E	1009 thru 1010B	G
91	D	615–28	F	1011 thru 1019E	H
93	D	615–29 thru 615–39	E	1020	G
94 thru 98	H	616–01 thru 616–11	E	1021	C
99	C	616–12 and 616–14	F	1022 thru 1022C	D
		616–16 thru 616–40	E	1023 thru 1026A	H
201 thru 208	A	617A thru 617–01	E	1027 thru 1027B	C
209 thru 244	B	617–02 thru 617–03A	F	1028 thru 1030C	H
245	D	617–04 thru 617–04A	E	1031B thru 1031C	G
246 thru 248	B	617–05 thru 617–05A	F	1032	C
249 thru 251	D	617–06	E	1033 thru 1033F	G
252	B	618 thru 618–16	E	1034 thru 1034G	H
253 thru 255	D	618–17	F	1035 thru 1036	C
256 thru 261	B	618–18	E	1037 thru 1039C	D
262 thru 265A	D	619 thru 619–03	E	1040 thru 1041A	D
266 thru 267	B	621 thru 621–11	E	1042 thru 1042C	G
268 thru 272	D	621 thru 621–12A	F	1042D thru 1042E	H
273 thru 275	B	621–13 thru 621–15	E	1042F	G
276	D	622 thru 622–07	E	1042G and 1042H	H
277 thru 279	B	623 and 623A	E	1042I thru 1043B	G
280 thru 281	D	623–01	F	1043C thru 1043F	H
282	B	623–02 thru 623–33E	E	1043G thru 1044E	G
283 thru 284	D	623–34	F	1045 thru 1045F	B
285 thru 286	B	623–35 and 623–36	E	1046 thru 1051B	H
287	D	624 thru 630–09	J	1052	G
288 thru 291B	B	632–02A	K	1053 thru 1054C	H
292 thru 294	D	632–04 and 632–05A	K	1055 thru 1056B	G
295 thru 296	B	633 thru 634–09	J	1057	H
297	F	635A and B	K	1058 thru 1058A	G
298 thru 299	B	635–03B	K	1059	D
		635–04B	K	1060 thru 1060C	H
301 thru 375		635–06A thru 635–10	K	1061 and 1062	C
These are twin		636 thru 636–11	J	1063A thru 1063C	H
cylinder units –		637 thru 637–16	J	1064	G
out of production		638 thru 638–100	F	1065 thru 1067C	D
401 thru 402	L	639	M	1068 and 1068A	G
403	M	640	O	1069	H
404 thru 406B	L	641	K	1070 thru 1070B	C
407 thru 407C	M	642	I	1071 thru 1075D	H
408 thru 410B	L	643	J	1076 thru 1084	H
411 thru 411B	M	650	N	1085 and 1085A	G
412 thru 423	L	670	H	1086 thru 1140A	H
424 thru 427	M				
				1141 thru 1142	G
501 thru 509A	E	701 thru 701–1C	G	1143 thru 1145	H
		701–2 thru 701–2D	H	1146 thru 1149A	I
601 thru 601–01	F	701–3 thru 701–3C	G	1149A thru 1153B	H
602 thru 602–02B	F	701–4 thru 701–4C	H	1158 thru 1159	G
603A thru 603–23	E	701–5 thru 701–5C	G	1160 thru 1161A	I
604 thru 604–25	E	701–6 thru 701–6C	H	1162 and 1162A	G
605 thru 605–18	E	701–7 thru 701–7C	G	1163 thru 1163A	H
606 thru 606–05B	E	701–8 thru 701–11	H	1165 thru 1168A	G
606–06 thru 606–07	F	701–12 thru 701–13A	G	1169 thru 1174	H
606–08 thru 606–13	E	701–14 thru 701–17	H	1175 and 1175A	I
608 thru 608–C	E	701–18 thru 701–19	G	1176 thru 1177A	H
610–A thru 610–14	E	701–19A	H	1179 thru 1179A	I
610–15 thru 610–16	F	701–20 thru 701–22A	G	1180 thru 1181	G
610–19 thru 610–20	E	701–24	H	1182 and 1182A	I

8325a0C1

Type Number-to-Letter Cross Reference Chart (cont.)

Type No.	Column Letter	Type No.	Column Letter	Type No.	Column Letter
1183 thru 1185	G	1345	J	1441	P
1186A thru 1186C	J	1348 and 1348A	G	1442 thru 1442B	G
1187 thru 1192	G	1350 thru 1350C	I	1443	I
1192A thru 1196B	J			1444 and 1444A	G
1197 thru 1197A	G	1550A	P	1445	I
1198 and 1198A	H			1446 and 1446A	A
1199 and 1199A	J	1351 thru 1351B	K	1447	G
		1352 thru 1352B	I	1448 thru 1450	F
1206 thru 1208B	J	1353	K	1450A thru 1450B	F
1210 thru 1215C	K	1354 thru 1355A	I	1450C thru 1450E	F
1216 thru 1216A	H	1356 and 1356B	K	1453	K
1217 thru 1220A	J	1357 thru 1358	I	1454 and 1454A	A
1221	H	1359 thru 1362B	K	1455 thru 1456	K
1222 thru 1223	J	1363 thru 1369A	G	1459	G
1224 thru 1225A	H	1372 thru 1375A	G	1460 thru 1460F	A
1226 thru 1227B	K	1376 and 1376A	J	1461	K
1228 thru 1229A	J	1377 thru 1378	K	1462	A
1230 thru 1231	H	1379 thru 1379A	H	1463	K
1232	G	1380 thru 1380B	K	1464 thru 1464B	L
1233 thru 1237B	K	1381	G	1465	A
1238 thru 1238C	H	1382	H	1466 thru 1466A	F
1239 thru 1243	J	1383 thru 1383B	K	1467 thru 1468	K
1244 thru 1245A	K	1384 thru 1385	G	1471 thru 1471B	E
1246 thru 1247	J	1386	I	1472 thru 1472C	L
1248	H	1387 thru 1388	K	1473 thru 1473B	A
1249 thru 1251A	J	1389	G	1474	L
1252 thru 1254A	K	1390 thru 1390B	H	1475 thru 1476	A
1255	J	1391	K	1477	G
1256 thru 1262B	K	1392	J	1478	J
1263	J	1393	G	1479	G
1264 thru 1265E	K	1394 thru 1395A	P	1482 and 1482A	F
1266 thru 1267	G	1396	K	1483	F
1269 thru 1270D	K	1397	H	1484 thru 1484D	A
1271 thru 1271B	G	1398 thru 1399	K	1485	G
1272 thru 1275	K			1486	D
1276	H	1400	K	1487	J
1277 thru 1279D	K	1401 thru 1401F	F	1488 thru 1488D	A
1280	J	1402 and 1402B	G	1489 thru 1490B	C
1283 thru 1284D	K	1403A and 1403B	K	1491	L
1286 thru 1286A	H	1404 and 1404A	K	1493 and 1493A	G
1287	K	1405 thru 1406A	H	1494 and 1495A	B
1288	J	1407 thru 1408	K	1496	G
1289 thru 1289A	G	1409A	J	1497	A
1290 thru 1293A	K	1410 thru 1412A	I	1498	E
1294 thru 1295	J	1413 thru 1416	I	1499	F
1296 thru 1298	K	1417 and 1417A	K		
		1418 and 1418A	P	1500	E
1300 thru 1303	K	1419 and 1419A	K	1501A thru 1501E	A
1304 thru 1307	J	1420 and 1420A	H	1503 thru 1503D	L
1308 thru 1316A	K	1421 and 1421A	K	1506 thru 1507	F
1317 thru 1317B	J	1422 thru 1423	P	1508	G
1318 thru 1320B	K	1424	K	1509	C
1321 thru 1322	J	1425	G	1510	L
1323	K	1426 thru 1426B	P	1511	C
1325	G	1427 and 1427A	H	1512 and 1512A	B
1326 thru 1326F	K	1428 thru 1429A	P	1513	L
1327 thru 1327B	H	1430A	G	1515 thru 1516C	C
1328B and 1328C	K	1431	P	1517	E
1329	J	1432 and 1432A	G	1518	D
1330	H	1433	K	1519 thru 1521	A
1331	J	1434 thru 1435A	P	1522	L
1332	J	1436 and 1436A	I	1523	A
1333	I	1437	J	1524	B
1334	I	1439	I	1527	C
1343 thru 1344A	H	1440 thru 1440D	A	1528	A

Type Number-to-Letter Cross Reference Chart (cont.)

Type No.	Column Letter	Type No.	Column Letter	Type No.	Column Letter
1529A and 1529B	C	2045 thru 2046B	F	40069 thru 40075	N
1530 thru 1530B	A	2047 thru 2048	D		
1531 thru 1535B	C	2049 thru 2049A	F	710101 thru 710116	E
1536	L	2050 thru 2050A	D	710124 thru 710130	E
1537	A	2051 thru 2052A	D	710131 thru 710137	E
1538 thru 1541A	L	2053 thru 2055	F	710138 thru 710149	E
1542	E	2056	D	710154	M
1543 thru 1546	A	2057	F	710201 thru 710209	E
1547	C	2058 thru 2058B	D	710210 thru 710218	E
1549	C	2059 thru 2063	F	710219 thru 710227	E
		2064 thru 2064A	D	710228	M
S-1801 thru 1822	F	2065	F	710150	G
1823 and 1824	E	2066	D	710151	H
		2067 thru 2071B	F	710155	L
2001 thru 2003	B			710157	H
2004 thru 2006	D	2200 thru 2201A	D	710229	H
2007 thru 2007B	B	2202 thru 2204	E	710230	N
2008 thru 2008B	D	2205 thru 2205A	D	710234	N
2009	B	2206 thru 2206A	G		
2010 thru 2011	D	2207	D		
2012	F	2208	G	200.183112	F
2013 thru 2014	B	2768	C	200.18322	F
2015 thru 2018	D			200.193132	F
2020	F	40001 thru 40028	N	200.193142	F
2021	D	40029 thru 40032C	O	200.193152	G
2022 thru 2022B	F	40033	N	200.193162	G
2023 thru 2026	D	40034 thru 40045A	O	200.203172	H
2027	B	40046	N	200.203182	H
2028 thru 2029	D	40047 thru 40052	O	200.203192	H
2030 thru 2031	B	40053 thru 40054A	N	200.213112	H
2032 thru 2033A	D	40056 thru 40060B	O	200.213122	H
2034 thru 2035C	F	40061 thru 40062B	N	200.503111	F
2036 thru 2037	B	40063 thru 40064	O	200.583111	F
2038	F	40065 thru 40066	N	200.593121	F
2039 thru 2044	C	40067 thru 40068	O	200.613111	F

8325a0C3

Specifications for 2 Cycle Engines with Split Crankcases

	A	B	C	D	E	F	G	H
Bore	1.500 1.5005	1.6253 1.6258	1.7503 1.7508	1.7503 1.7508	2.000	2.000	2.000	2.000
Stroke	1.375	1.50	1.50	1.50	1.50	1.50	1.50	1.50
Displacement Cubic Inches	2.43	3.10	3.60	3.60	4.70	4.70	4.70	4.70
Point Gap	.020	.020	.020	.020	.020	.020	.015	.015
Timing B.T.D.C. Before Top Dead Center	$1/16''$ or .0625	$5/32''$ or .1562	$5/32''$ or .1562	$1/4''$ or .250	$5/32''$ or .1562	$5/32''$ or .1562	$5/32''$	$1/4''$ or .250 $11/64''$ for "super"
Spark Plug Gap	.030	.030	.030	.030	.030	.030	.030	.030
Piston Ring End Gap	.003 .008	.005 .010	.005 .010	.005 .010	.006 .011	.006 .011	.006 .011	.006 .011
Piston Diameter	1.4966 1.4969	1.6216 1.6219	1.7461 1.7464	1.7461 1.7464	1.9948 1.9951	1.9948 1.9951	1.9948 1.9951	1.9949 1.9955
Piston Ring Groove Width	.095 .096	.095 .096	.095 .096	.095 .096	.095 .096	.095 .096	.095 .096	.095 .096
Piston Ring Width	.093 .0935	.093 .0935	.093 .0935	.093 .0935	.093 .0935	.093 .0935	.093 .0935	.093 .0935
Piston Pin Diameter	.3750 .3751	.3750 .3751	.3750 .3751	.3750 .3751	.3750 .3751	.3750 .3751	.3750 .3751	Early .3750 .3761 Late .4997 .4999
Connecting Rod Diameter Crank Bearing	.6869 .6874	.6869 .6874	.6869 .6874	.6986 .6989 w/o needles	.6869 .6874	.6869 .6874	.6869 .6874	.6935 .6939 w/o needles
Crankshaft Rod Needle Diameter				.0653 .0655				.0653 .0655
Crank Pin Journal Diameter	.6860 .6865	.6860 .6865	.6860 .6865	.5615 .5618	.6860 .6865	.6860 .6865	.6860 .6865	.5615 .5618
Crankshaft P.T.O. Side Main Brg. Dia.	.6689 .6693	.6689 .6693	.6690 .6694	.6689 .6693	.9995 1.0000	.6689 .6693	.9995 1.0000	.9995 1.0000
Crankshaft Magneto Side Main Brg. Dia.	.6689 .6693	.6689 .6693	.6689 .6693	.6690 .6694	.7495 .7500	.6689 .6693	.7495 .7500	.7495 .7500
Crankshaft End Play	.003 .008	.003 .008	.003 .008	.003 .008	.009 .022	.009 .022	.009 .022	.009 .022

8325a0C4

Specifications for 2 Cycle Engines with Split Crankcases (cont.)

I	J	K	L	M	N	O	P	
2.000	2.000	2.093 2.094	2.2505 2.2510	2.2505 2.2510	2.5030 2.5035	2.5030 2.5035	2.000	Bore
1.50	1.625	1.63	2.00	2.00	1.625	1.680	1.50	Stroke
4.70	5.10	5.80	8.00	8.00	7.98	8.25	4.70	Displacement Cubic Inches
.015	.020	.015	.020	.020	.020	.020	.015	Point Gap
11/64" or .175	11/64" or .175	3/32" or .095	1/8" or .125	3/32" or .095	11/64" or .175	11/64" or .175	.90	Timing B.T.D.C. Before Top Dead Center
.035	.030	.030	.030	.030	.030	.030	.035	Spark Plug Gap
.006 .011	.006 .011	.006 .011	.007 .015	.007 .015	.005 .013	.005 .013	.006 .011	Piston Ring End Gap
1.9948 1.9951	1.9951 1.9948	2.0880 2.0883	2.2460 2.2463	2.2460 2.2463	2.4960 2.4963	2.4960 2.4963	1.9948 1.9951	Piston Diameter
.095 .096	.095 .096	T.0655 .0665 L.0645 .0655	.095 .096	.095 .096	T.0655 .0665 L.0645 .0655	T.0655 .0665 L.0645 .0655	.095 .096	Piston Ring Groove Width
.093 .0935	.093 .0935	.0615 .0625	.093 .0935	.093 .0935	.0615 .0625	.0615 .0625	.093 .0935	Piston Ring Width
.4997 .4999	.3750 .3751	.4997 .4999	.5000 .5001	.3000 .5001	.4997 .4999	.4997 .4999	.4997 .4999	Piston Pin Diameter
.6941 .6944	.6869 .6874	.9407 .9412	1.000 1.0004	1.000 1.0004	.9407 .9412	.9407 .9412	.6941 .6944	Connecting Rod Diameter Crank Bearing
.0653 .0655		.0943 .0945	.0943 .0945	.0943 .0945	.0943 .0945	.0943 .0945	.0653 .0655	Crankshaft Rod Needle Diameter
.6860 .6865	.6860	.7499 .7502	.8096 .8099	.8096 .8099	.7499 .7502	.7499 .7502	.6860 .6865	Crank Pin Journal Diameter
.9995 1.0000	.9995 1.0000	.6990 .6994	.9839 .9842	.9839 .9842	.7871 .7875	.7871 .7875	.9995 1.0000	Crankshaft P.T.O. Side Main Brg. Dia.
.7495 .7500	.7495 .7500	.6990 .6994	.9839 .9842	.9839 .9842	.7498 .7501	.7498 .7501	.7495 .7500	Crankshaft Magneto Side Main Brg. Dia.
.009 .022	.009 .022	.003 .008	.003 .008	.003 .008	.008 .013	.008 .013	.009 .022	Crankshaft End Play

8325a0C5

Uniblock Cross Reference Chart

Type No. Vertical Crankshaft Engines	Column No.	Type No. Horizontal Crankshaft Engines	Column No.	Type No. Horizontal Crankshaft Engines	Column No.
638 thru 638–100	6	1401 thru 1401F	16	1508	7
		1401G, H	17	1509	3
642–01, A	9A	1401J	27	1510	12
642–02, A, B, C, D	9A	1402 and 1402B	7	1511	3
642–02E, F	9B	1425	7	1512 and 1512A	2
642–03, A, B	9A	1430A	7	1513	12
642–04, A, B, C	9A	1432 and 1432A	7	1515 thru 1516C	3
642–05, A, B	9A	1440 thru 1440D	1	1517	5
642–06, A	9A	1442 thru 1442B	7	1518	4
642–07, A, B	9A	1444 and 1444A	7	1519 thru 1521	1
		1448 thru 1450	16	1522	12
642–07C	9B	1450A thru 1450B	16	1523	1
642–08	9B	1450C thru 1450E	16	1524	2
642–08A, B	9A	1450F	17	1525A	16
642–09 thru 642–14	9A	1454 and 1454A	1	1527	3
642–13A, 14A, 14B	9B	1459	7	1528	1
642–15 thru 642–22	9B	1460 thru 1460F	1	1529A and 1520B	3
642–24 thru 642–30	9C	1462	1	1530 thru 1530B	1
		1464 thru 1464B	12	1531	3
670–01 thru 670–101	8	1465	1	1534A	17
		1466 thru 1466A	16	1535B	3
200–183112	6	1471 thru 1471B	5	1536	12
200–183122	6	1472 thru 1472C	12	1537	1
200–193132	6	1473 thru 1473B	1	1538 thru 1541A	12
200–193142	6	1474	12	1542	5
200–193152	7	1475 thru 1476	1	1543 thru 1546	1
200–193162	7	1479	7	1517	3
200–203172	8	1482 and 1482A	16	1519	3
200–203182	8	1483	16	1551	16
200–203192	8	1484 thru 1484D	3	1552	20
200–213112	8	1485	7	1553	16
200–213122	8	1486	4	1554 and 1554A	3
200–243112	8	1488 thru 1488D	1	1555 and 1556	16
200–283012	8	1489 thru 1490B	3	1561	19
		1491	12	1572	2
		1493 and 1493A	7	1573	3
		1494 and 1495A	2	1574 thru 1577	23
		1496	7	1575	24
		1497	1	1578	25
		1498	5	1581 thru 1582A	23
		1499	16	1583 thru 1595	26
		1500	5	200–503111	16
		1501A thru 1501E	1	200–583111	16
		1503 thru 1503D	12	200–593121	16
		1506	16	200–613111	16
		1506B	17	200–672102	26
		1507	16	200–682102	26

8325a0C6

Torque Specifications for 2-Cycle Engines

Application	Torque	Application	Torque
Carburetor and Reed Plates		Cylinder head to cylinder (635 type engine)	45–50 in. pounds
Carburetor to crankcase, Carburetor to adapter, or Carburetor adapter to crankcase	70–75 in. pounds	Cable clip and transfer port cover to cylinder	25–30 in. pounds
Carburetor to snow blowers cover	30–35 in. pounds	Cable clip to cylinder Stop lever to cylinder	25–30 in. pounds
Carburetor outlet fitting	40 in. pounds	Spark plug	18–22 ft. pounds
Reed and cover plates 639 type engines	50–60 in. pounds with Loctite, type A	Transfer cover cross port engines	25–30 in. pounds
Reed to plate 635 type engines	12–18 in. pounds	**Crankshaft and Connecting Rods**	
Crankcase and Cylinder		Aluminum and bronze rods to rod cap	40–50 in. pounds
Crankcase to crankcase cover	23–30 in. pounds	Steel rod to rod cap	70–80 in. pounds
Crankcase to crankcase cover screws	35–40 in. pounds	Flywheel nut on tapered end of crankshaft: Aluminum hub on iron or steel shaft.	18–25 ft. pounds
Mounting cylinder or carburetor to crankcase studs	50 in. pounds	Steel hub flywheel on iron shaft	18–25 ft. pounds
Base to crankcase	240–250 in. pounds	Steel hub flywheel on steel shaft	30 ft. pounds
Cylinder to crankcase nuts	70–75 in. pounds	**Governor and Bell Crank Parts**	
In cylinder	100–110 in. pounds	Power take-off end governor lower ring to crankshaft setscrew	30–35 in. pounds
Spark plug stop lever and head shroud to head	50–60 in. pounds	Bell crank bracket to crankcase	20–25 in. pounds
Cylinder head to cylinder	30–40 in. pounds	Governor cover to crankcase	50–60 in. pounds
Cylinder head to cylinder	50–60 in. pounds 80–90 in. pounds		

8325a0C7

Specifications Chart

		1	2	3	4	5	6	7	8	9A	9B	9C	12
Bore		2.093 2.094	2.093 2.094	2.093 2.094	2.093 2.094	2.093 2.094	2.093 2.094	2.093 2.094	2.093 2.094	2.093 2.094	2.093 2.094	2.093 2.094	2.093 2.094
Stroke		1.250	1.410	1.410	1.410	1.410	1.500	1.500	1.500	1.500	1.500	1.500	1.410
Cu. In. Displacement		4.40	4.80	4.80	4.80	4.80	5.20	5.20	5.20	5.20	5.20	5.20	4.80
Point Gap		.017	.017	.017	.017	.017	.018	.017	.020	.018	.020	.020	.017
Timing B.T.D.C.		.122″	.100″	.135″	.100″	.135″	.100″	.185″	.070″	.110″	.085″ See Note 1	.078″ See Note 2	.135″
Spark Plug Gap		.035	.035	.035	.035	.035	.035	.035	.035	.035	.035	.035	.035
Piston Ring End Gap		.007 .017	.007 .017	.006 .011	.006 .014	.006 .011	.006 .014	.007 .017	.006 .016	.007 .017	.006 .016	.006 .016	.007 .017
Piston Diameter		2.0080 2.0870	2.0880 2.0870	2.0885 2.0875	2.0885 2.0875	2.0885 2.0875	2.0880 2.0870	2.0880 2.0870	2.0880 2.0870	2.0880 2.0870	2.0880 2.0820	2.0880 2.0870	2.0880 2.0870
Piston Ring Groove Width	(Top)	.0655 .0665	.0655 .0665	.0655 .0665	.0975 .0985	.0655 .0665	.0975 .0985	.0655 .0665	.0655 .0665	.0655 .0665	.0655 .0665	.0655 .0665	.0655 .0665
	(Bot.)	.0645 .0655	.0645 .0655	.0645 .0655	.0955 .0965	.0645 .0655	.0955 .0965	.0645 .0655	.0645 .0655	.0645 .0655	.0645 .0655	.0645 .0655	.0645 .0655
Piston Ring Width		.0625 .0615	.0625 .0615	.0625 .0615	.0925 .0935	.0625 .0615	.0925 .0935	.0625 .0615	.0625 .0615	.0625 .0615	.0625 .0615	.0625 .0615	.0625 .0615

8325a0C8

Piston Pin Diameter	.4999 / .4997	.4999 / .4997	.4999 / .4997	.3750 / .3751	.4999 / .4997	.3750 / .3751	.4999 / .4997	.4999 / .4997	.4999 / .4997	.4999 / .4997	.4999 / .4997
Connecting Rod Diameter Crank Bearing	—	—	—	.6886 / .6879 Dowels	—	—	—	1.0053 / 1.0023 Liner Dia.	—	1.0053 / 1.0023 Liner Dia.	1.0053 / 1.0023 Liner Dia
Crankshaft Rod Needle Dia.	.0655 / .0653	.0655 / .0653	.0655 / .0653	—	.0655 / .0653	—	.0655 / .0653	.0781 / .0780	—	.0781 / .0780	.0781 / .0780
Crank Pin Journal Diameter	.5618 / .5611	.5621 / .5614	.5621 / .5614	.6865 / .6857	.5618 / .5611	.6865 / .6857	.5618 / .5611	.8450 / .8442	.6865 / .6857	.8450 / .8442	.8450 / .8442
Crankshaft P.T.O. Side Main Brg. Dia.	.6695 / .6691	.6695 / .6691	.6695 / .6691	.6695 / .6691	.6695 / .6691	.8750 / .8745	.6694 / .6690	1.0003 / .9998	.8750 / .8745	1.0003 / .9998	1.0003 / .9998
Crankshaft Magneto Side Main Brg. Dia.	.6695 / .6691	.6695 / .6691	.6695 / .6691	.6695 / .6691	.6695 / .6691	.7500 / .7495	.6694 / .6690	.6695 / .6691	.7500 / .7495	.7503 / .7498	.6695 / .6691
Crankshaft End Play	None	None	None	None	None	.003 / .016	None	None	.003 / .016	.003 / .016	None

NOTE 1: 642-08 14A, 14B B.T.D.C. = .110"
642-16D, 19A, 20A, 21 22 B.T.D.C. = .078"
NOTE 2: 642-24, 26, 29 B.T.D.C. = .085"
643-2a, 25, 26 B.T.D.C. = .085"
NOTE 3: 643-13 B.T.D.C. = .095"
NOTE 4: 643-03A, 05A, 13, 14 = .020"

8325a0C9

Engine Identification

◆ **See Figures 25 and 26**

The identification taps may be located at a variety of places on the engine. The type number is the most important number since it must be included with any correspondence about a particular engine.

Early engines listed the type number as a suffix of the serial number. For example on number 123456789 P 234, 234 is the type number. In the number 123456789 H 104-02B; 104-02B is the type number. In either case the type number is important.

If you use short block to repair the engine, be sure that you transfer the serial number and type number tag to the new short block.

On the newer engines, reference is sometimes made to the model number. The model number tells the number of cylinders, the design (vertical or horizontal) and the cubic inch displacement.

Air Cleaners

The instructions below detail the procedures involved in cleaning the various types of elements. See the illustrations for exploded views to aid disassembly and assembly.

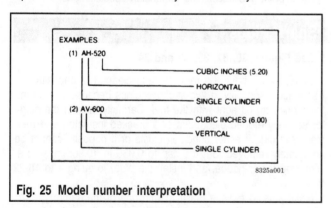

Fig. 25 Model number interpretation

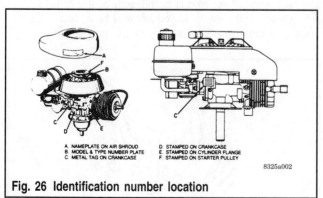

Fig. 26 Identification number location

POLYURETHANE AIR CLEANER

1. Wash the element in a solvent or detergent and water solution by squeezing similar to a sponge.
2. Clean the air cleaner housing and cover with the same solution. Dry thoroughly.
3. Dry the element by squeezing or with compressed air if available.
4. Apply a generous quantity of oil to the element sides and open ends. Squeeze vigorously to distribute oil and to remove excess oil.

ALUMINUM FOIL AIR CLEANER

◆ **See Figure 27**

1. Dip the aluminum foil filter in solvent. Flush out all dirt particles.
2. Shake out the filter thoroughly to remove all solvent, then dip the filter element in oil. Allow the oil to drain from the filter. Clean the screens and filter body.

➡ **The concave screen and retainer cover or ring are not used on later models. They are replaced with a clip which rolls into the groove in the lip of the body.**

FELT TYPE AIR CLEANERS

◆ **See Figure 28**

1. To clean felt air cleaners, merely blow compressed air through the element in the reverse direction to normal air flow. Felt elements may also be washed in nonflammable solvents or soapy water. Blow dry with compressed air.

➡ **Power Products type numbers 641 and up, use a gasket between the element and the base. The gasket is used only with this element; earlier versions did not have a gasket.**

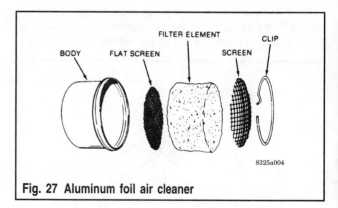

Fig. 27 Aluminum foil air cleaner

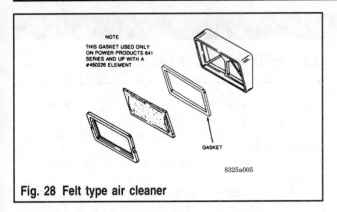

Fig. 28 Felt type air cleaner

FIBER ELEMENT AIR CLEANER

1. Remove the filter and place the cover in a normal position on the filter. With the filter element down to semi-seal it, blow compressed air through the cover hole to reverse air flow, forcing dirt particles out.
2. Clean the cover mounting bracket with a damp cloth.

DRY PAPER AIR CLEANER

▶ **See Figure 29**

1. Tap the element on a workbench or any solid object to dislodge larger particles of dirt.
2. Wash the element in soap and water. Rinse from the inside until it is thoroughly flushed and the water coming through is free of soap.
3. Allow the element to dry completely or use low pressure compressed air blown from the inside to speed the process.
4. Inspect the element for cracks or holes, and replace if necessary.

Lubrication

OIL AND FUEL RECOMMENDATIONS

Power Products 2 cycle engines are mist-lubricated by oil mixed with the gasoline. For the best performance, use regular grade, leaded fuel, with 2 cycle or outboard oil rated SAE 30

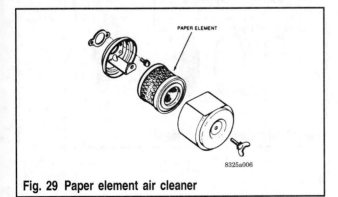

Fig. 29 Paper element air cleaner

or SAE 40. Regular grade unleaded fuel is an acceptable substitute. The terms 2 cycle or outboard are used by various manufacturers to designate oil they have designed for use in 2 cycle engines. Multiple weight oil such as all season 10W-30, are not recommended.

If you have to mix the gas and oil when the temperature is below 35°F (1.6°C), heat up the oil first, then mix it with the gas. Oil will not mix with gas when the temperature is approaching freezing. However, if you use oil that has been warmed first it will not be affected by low temperatures.

The proportion of oil to fuel is absolutely critical to two-stroke operation. If too little oil is used, overheating and damage to engine parts will occur (this can even result from running the engine too lean). If excessive oil is used, spark plug fouling, smoke in the exhaust, and even misfire can occur. Mix carefully and precisely. Follow Power Products recommendations for your particular engine, and disregard fuel container labels.

Fuel/oil mix must be clean and fresh. Fuel deteriorates enough to form troublesome gum and varnish after more than a month. Dirt in the fuel can cause clogging of carburetor passages and even engine wear.

Tune-Up Specifications

All spark plugs are gapped at 0.035 in. (0.90mm). Because of the great number of individual models of Power Products engines that exist, an individual chart of Tune-Up specifications is impractical. Refer to the charts at the end of this section for breaker point gap and timing dimension specifications.

Spark Plugs

▶ **See Figures 30, 31, 32, 33 and 34**

Spark plugs should be removed and cleaned of deposits frequently, especially in two-stroke engines because they burn the lubricating oil right with the fuel. Carefully inspect the plug for severely eroded electrodes or a cracked insulator, and replace the plug if either condition exists or if deposits cannot be adequately removed. Set the gap to 0.035 in. (0.90mm) with a wire type feeler gauge, and install the plug, torquing it to 18-22 ft. lbs.

Make sure to replace the plug with one of the same type and heat range. If the plug is fouled, poor quality or old fuel, a rich mixture, or the wrong fuel/oil mix may be at fault. Also, make sure the engine's exhaust ports are not clogged.

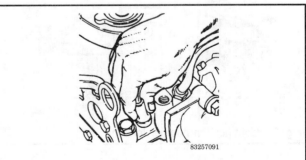

Fig. 30 Twist and pull on the boot, never on the plug wire

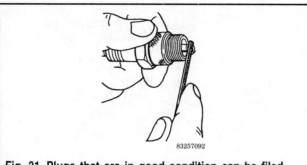

Fig. 31 Plugs that are in good condition can be filed and reused

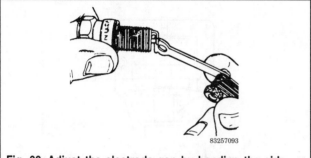

Fig. 32 Adjust the electrode gap by bending the side electrode

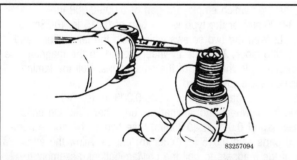

Fig. 33 Always used a wire gauge to check the plug gap

Breaker Points

REMOVAL & INSTALLATION

1. First remove the flywheel as described below:
   a. Remove the screws, engine shroud, and starter. Determine the direction of rotation of the flywheel nut by looking at the threads. Then, place a box wrench on the nut and tap with a soft hammer in the proper direction.
   b. The flywheel is removed with a special puller or a special knock off tool. The knock off tool must not be used on 660, 670, or 1500 ball bearing models. To use the knock off tool, screw it onto the crankshaft until it is within 1/16 in. (1.6mm) of the flywheel. Hold the flywheel firmly and rap the top of the puller sharply with a hammer to jar it loose. Pull the flywheel off.

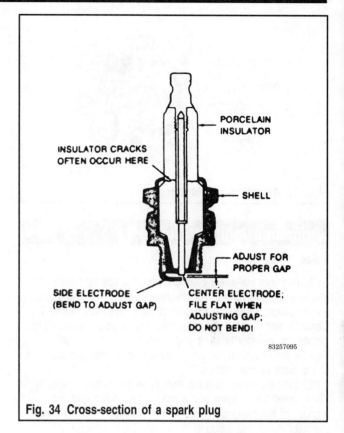

Fig. 34 Cross-section of a spark plug

   c. If a puller is being used, the flywheel will have three cored holes into which a set of self-tapping screws are turned. The handle which operates the bolt at the center of the puller's collar is then turned to pull the flywheel off the crankshaft. If this is not adequate to do the job, heat the center of the alloy flywheel with a butane torch to expand it before turning the puller handle.
2. Remove the nuts that hold the electrical leads to the screw on the movable breaker point spring. Remove the movable breaker point from the stud.
3. Remove the screw and stationary breaker point. Put a new stationary breaker point on breaker plate; install the screw, but do not tighten it fully.
4. Position a new movable breaker point on the post.
5. Check that the new points contact each other properly and remove all grease, fingerprints, and dirt from the points.

ADJUSTMENT

▶ **See Figure 35**

1. If necessary (as when checking the gap of old points), loosen the screw which mounts the stationary breaker contact. Rotate the crankshaft until the contact cam follower rests right on the highest point of the cam.
2. Using a flat feeler gauge of the dimension shown under 'point gap' in the charts at the rear of this section, check the dimension of the gap and, if incorrect, move the breaker base in the appropriate direction by wedging a screwdriver between the dimples on the base plate and the notch in the breaker plate. When gap is correct, tighten the screw. Recheck the gap and, if necessary, reset it.

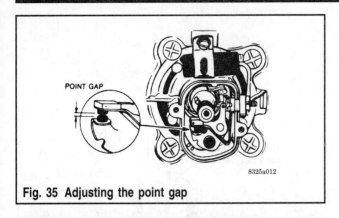

Fig. 35 Adjusting the point gap

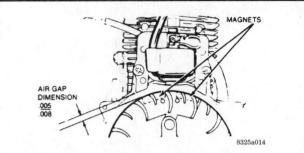

Fig. 37 Adjusting the magneto armature air gap. Note the magnets on the flywheel

Adjusting Ignition Timing

▶ **See Figure 36**

1. Remove the spark plug and install a special timing tool or thin ruler. With the tool lockscrew loose or the ruler riding on the piston, rotate the crankshaft back and forth to find Top Dead Center. Tighten the tool lockscrew or use a straightedge across the cylinder head, if you're using a ruler, to measure Top Center. Look up the timing dimension in the specifications at the back of this section.

2. Turn the crankshaft backwards so the piston descends. Then, reset the position of the special tool downward the amount of the dimension and tighten the lockscrew, or move the ruler down that amount.

3. Very carefully bring the piston upward by turning the crankshaft until the top of the piston just touches the tool or ruler.

4. Install a piece of cellophane between the contact surfaces. Loosen the ignition stator lockscrews and turn the stator until the cellophane is clamped tightly between the contact surfaces. Turn the stator until the cellophane can just be pulled out (as contacts start to open), and then tighten the stator lockscrews.

Magneto Armature Air Gap

▶ **See Figures 37 and 38**

There are two types of magnetos. Compare appearance of your engine with each illustration to determine whether it looks

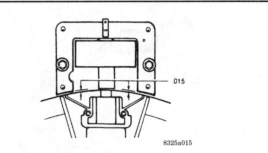

Fig. 38 Adjusting the magneto armature air gap. Note the wider air gap

like the type which employs a gap of 0.005-0.008 in. (0.13-0.20mm) or the type with a gap of 0.015 in. (0.38mm).

1. Loosen the two screws which hold the laminations and coil to the block. Turn the flywheel around so the magnets line up directly with the ends of the laminations. Pull the laminations/coil assembly upward.

2. Insert a gap gauge of 0.005-0.008 in. (0.13-0.20mm) between the flywheel and laminations on either side. On units with a gap of 0.015 in. (0.38mm), use two of the above numbered parts or a 0.015 in. (0.38mm) gauge. Allow the attraction of the magnets to pull the coil/laminations assembly toward the flywheel.

3. On the magneto type with a 0.005-0.008 in. (0.13-0.20mm) dimension, use Loctite® Grade A on the screws and torque to 35-45 inch lbs. On the other type magneto, torque the screws to 20-30 inch lbs. Recheck the gap and readjust, if necessary.

Mixture Adjustment

1. If the engine will not start or the carburetor has recently been disassembled, both idle and main mixture screws may be turned in very gently until they just bottom, and then turned out exactly one turn. In this case, back out the idle speed screw until throttle is free, then turn screw in until it just contacts the throttle. Turn it one turn more exactly.

2. Allow the engine to warm up to normal running temperature. With the engine running at maximum recommended rpm, loosen the main metering screw until the engine rolls, then tighten the screw until the engine starts to cut out. Note the number of turns from one extreme to the other. Loosen the screw to a point midway between the extremes.

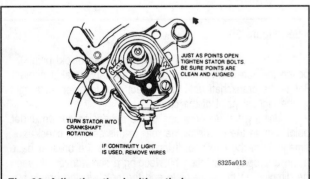

Fig. 36 Adjusting the ignition timing

3. Set the throttle to idle speed and repeat Step 2 for that adjusting screw.

➡**Some carburetors do not have a main mixture adjustment. Others employ a drilled mixture screw which provides about the right mixture when fully screwed in. In some weather conditions, operation may be improved by setting this screw just slightly off the seat for smoother running.**

Governor Adjustments

POWER TAKEOFF END MECHANICAL GOVERNOR

▶ **See Figure 39**

1. To adjust the governor, remove the outboard bearing housing. Use a ³⁄₃₂ in. allen wrench to loosen the set screw.
2. Squeeze the top and bottom governor rings, fully compressing the governor spring.
3. Hold the upper arm of the bell crank parallel to the crankshaft and insert a ³⁄₃₂ in. allen wrench between the upper ring and the bell crank.
4. Slide the governor assembly onto the crankshaft so that the allen wrench just touches the bell crank.
5. Tighten the set screw to secure the governor to the crankshaft.

6. Install the bearing adapter, mount the engine, and check the engine speed with a tachometer. It should be about 3200-3400 rpm.

➡**Never attempt to adjust the governor by bending the bellcrank or the link.**

7. If the speed is not correct, readjust the governor by moving the assembly toward the crankcase to increase speed, and away from the crankcase to decreased speed.

MECHANICAL FLYWHEEL TYPE GOVERNOR ADJUSTMENT

▶ **See Figure 40**

As engine speed increases, the links are thrown outward, compressing the link springs. The links apply a thrust against the slide ring, moving it upward and compressing the governor spring. As the slide ring moves away from the thrust block of the bellcrank assembly, the throttle spring causes the thrust block to maintain engagement and close the throttle slightly.

As the throttle closes and engine speed decreases, force on the slide ring decreases so that it moves downward, pivots the bellcrank outward to overcome the force of the throttle spring, and opens the throttle to speed up the engine. In this manner, the operating speed of the engine is stabilized to the adjusted governor setting.

1. To adjust the governor, loosen the bracket screw and slide the governor bellcrank assembly toward or away from the flywheel. Move the bellcrank toward the flywheel to increase speed and away from the flywheel to decrease speed.
2. Tighten the screw to secure the bracket.
3. Make minor speed adjustments by bending the throttle link at the bend in the center of the link.

➡**Do not lubricate the governor assembly or the governor bellcrank assembly of flywheel mounted governors.**

ADJUSTING 2-CYCLE AIR VANE GOVERNORS

▶ **See Figure 41**

1. Loosen the self-locking nut that holds the governor spring bracket to the engine crankcase.
2. Adjust the spring bracket to increase or decrease the governor spring tension. Increasing spring tension increases speed and decreasing spring tension decreases speed.

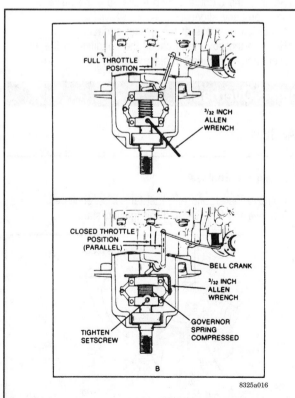

8325a016

Fig. 39 Adjusting the PTO end-mounted governor

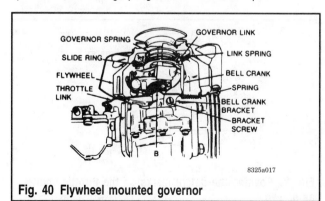

8325a017

Fig. 40 Flywheel mounted governor

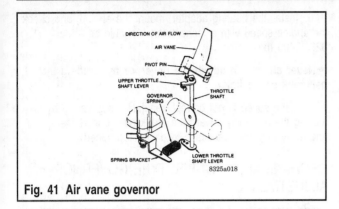

Fig. 41 Air vane governor

3. After adjusting, the spring bracket should not be closer than 1/16 in. (1.6mm) to the crankcase.

4. Tighten the self-locking nut.

Carburetor

REMOVAL & INSTALLATION

▶ See Figure 42

1. Remove the air cleaner. Drain the fuel tank. Disconnect the carburetor fuel lines.

2. If necessary, remove any shrouding or control panels to gain access to the carburetor.

3. Disconnect the choke or throttle control wires at the carburetor.

4. Remove the cap screws, or nuts and lockwashers and remove the carburetor from the engine.

5. To install, reverse the removal procedure, using new gaskets.

OVERHAUL

Tecumseh two-and four-stroke engines employ a common series of carburetors. Refer to the Four-Stroke Tecumseh Engine section for specific carburetor overhaul procedures.

Idle Governor

SERVICE

1. Remove the shutter fastener and allow the shutter to drop out of the air horn.

2. Note location of the spring end in the disc-shaped throttle lever. The spring should be placed into the same hole during reassembly.

3. Remove the retainer clip and lift out the throttle shaft.

4. Replace all worn parts and reassemble in reverse order.

➡**Note the position of the throttle shutter, as shown. The reference marks must be positioned as shown when shutter is installed.**

Fuel Pump

SERVICE

Float Type Carburetor With Integral Pump

▶ See Figure 43

1. If the engine runs, but roughly, make both carburetor mixture adjustments.

2. Make sure the fuel supply is adequate and the tank is in the proper position.

3. Make sure the fuel tank valve is open.

4. Make sure the pick-up tube is not cracked.

5. Remove the carburetor and make sure the pulsation passage is properly aligned.

6. Check for air leaks at the gasket surface.

7. Remove the cover and check the condition of the inlet and outlet flaps — if curled, replace the flap leaf.

Engine Overhaul

DISASSEMBLY

Split Crankcase Engines

1. Remove the shroud and fuel tank if so equipped.

2. Remove the flywheel and the ignition stator.

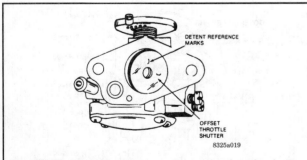

Fig. 42 Position the detent marks on the throttle shutter as shown

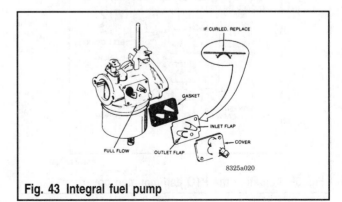

Fig. 43 Integral fuel pump

3. Remove the carburetor and governor linkage. Carefully note the position of the carburetor wire links and springs for reinstallation.

4. Lift off the reed plate and gasket if present and inspect them. They should not bend away from the sealing surface plate more than 0.010 in. (0.25mm).

5. Remove the spark plug and inspect it.

6. Remove the muffler. Be sure that the muffler and the exhaust ports are not clogged with carbon. Clean them if necessary.

7. Remove the transfer port cover and check for a good seal.

8. Remove the cylinder head, if so equipped.

➡**Some models utilize a locking compound on the cylinder head screws. Removing the screws on such engines can be difficult. This is especially true with screws having slotted head for a straight screwdriver blade. The screws can be removed if heat is applied to the head of the screw with an electric soldering iron.**

9. On engines having a governor mounted on the power take-off end of the crankshaft, remove the screws that hold the outboard bearing housing to the crankcase. Clean the PTO end of the crankshaft and remove the outboard bearing housing and bearing. Loosen the set screw that holds the governor assembly to the crankshaft, slide the entire governor assembly from the crankshaft. Remove the screw that holds the governor bellcrank bracket to the crankcase. Remove the governor bellcrank and bracket.

10. Make match marks on the cylinder and crankcase. Remove the four nuts and lockwashers that hold the cylinder to the crankcase.

11. Remove the cylinder by pulling it straight out from the crankcase.

12. To separate the two crankcase halves, remove all of the screws that hold the crankcase halves together.

13. With the crankcase in a vertical position, grasp the top half of the crankcase and hold it firmly. Strike the top end of the crankcase with a rawhide mallet, while holding the assembly over a bench to prevent damage to parts when they fall. The top half of the crankcase should separate from the remaining assembly.

14. Invert the assembly and repeat the procedure to remove the other casting half from the crankcase on ball bearing units.

15. Each time the crankshaft is removed from the crankcase, seals at the end of the crankshaft should be replaced. To replace the seals, use a screwdriver or an ice pick to remove the seal retainers and remove and discard the old seals. Install

the seals in the bores of the crankcase halves. The seals must be inserted into the bearing well with the channel groove toward the internal side of the crankcase. Retain the seal with the retainer. Seat the retainer spring into the spring groove.

Uniblock Engines

1. Remove the shroud and fuel tank. Note the condition of the air vane governor, if so equipped.

2. Remove the starter cap and flywheel nut, noting the position of the belleville washer.

3. Remove the flywheel — see 'Breaker Points Removal and Installation,' above.

4. Remove the head. Save the old head gasket for use when replacing the piston, but procure a new gasket for use in final assembly.

5. Remove the cylinder block cover plate to gain access to the connecting rod bolts.

6. Note the location of the connecting rod match marks for reassembly.

7. Remove the piston. Remove the ridge first, if necessary, with a ridge reamer. Push the piston and connecting rod through the top of cylinder.

8. Remove the crankshaft from the cylinder block assembly. On engines with crankshaft ball bearings:

   a. Remove the four shroud base screws and tap the shroud base so the base and crankshaft can be removed together.

   b. To remove bearing with crankshaft from base: USE SAFETY GLASSES AND HEAT RESISTANT GLOVES. Using a propane torch, heat the area on the base around the outside of the bearing until there is enough expansion to remove the base from the bearing on the crankshaft. Now remove and discard the seal retainer ring, seal retainer and seal.

   c. To remove the bearing race, remove the retainer ring on the crankshaft with snap ring pliers, and with the use of a bearing splitter or arbor press, remove the ball bearing.

✳✳CAUTION

Support the crankshaft's top counterweight to prevent bending. Also, bearing is to be pressed on via the inner race only.

9. On other engines:

   a. When equipped with a sleeve or needle bearing, use a seal protector and lift the crankshaft out of the cylinder. Be careful not to lose the bearing needles.

   b. When equipped with a ball bearing, use a mallet to strike the crankshaft on the P.T.O. end while holding the block in your hand.

10. In assembly, bear the following points in mind:

   a. Use a ring compressor to install piston. Be careful not to allow the rings to catch on the recess for the head gasket. Use the old head gasket to take up the space in the recess. Do not force the piston into the cylinder, or damage to rings or piston could occur.

   b. To install the ball bearing on the crankshaft, slide the bearing on the crankshaft and fit it on the shaft by tapping using a mallet and tool, part number 670258 or press the ball bearing on the crankshaft with an arbor press. Install the retainer ring.

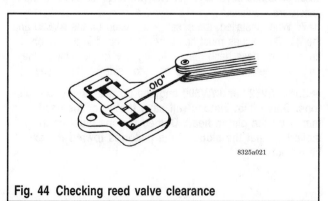

8325a021

Fig. 44 Checking reed valve clearance

c. To install the crankshaft with a ball bearing, heat the shroud base to expand the bearing seat and drop the ball bearing into the seat of the base shroud. Allow it to cool. Install a new seal retainer ring, seal retainer, and seal.

11. After the shroud base and flywheel are back in place, adjust the air gap between the coil core and flywheel as described above under 'Magnet Air Gap Adjustment.'

CONNECTING ROD SERVICE

▶ **See Figures 45 and 46**

1. For engines using solid bronze or aluminum connecting rods, remove the two self-locking cap screws which hold the connecting rod to the crankshaft and remove the rod cap. Note the match marks on the connecting rod and cap. These marks must be reinstalled in the same position to the crankshaft.

2. Engines using steel connecting rods are equipped with needle bearings at both crankshaft and piston pin end. Remove the two set screws that hold the connecting rod and cap to the crankshaft, taking care not to lose the needle bearings during removal.

3. Needle bearings at the piston pin end of steel rods are caged and can be pressed out as an assembly if damaged.

4. Check the connecting rod for cracks or distortion. Check the bearing surfaces for scoring or wear. Bearing diameters should be within the limits indicated in the table of specifications located at the end of this section.

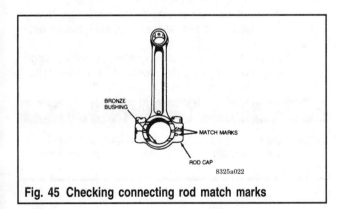

Fig. 45 Checking connecting rod match marks

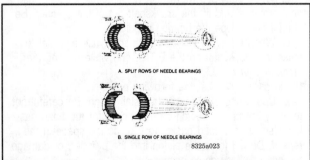

Fig. 46 Needle bearing arrangement. Double rows of bearings are placed with the tapered edges facing out

5. There are two basic arrangements of needles supplied with the connecting rod crankshaft bearing: split rows of needles and a single row of needles. Service needles are supplied with a beeswax coating. The beeswax holds the needles in position.

6. To install the needle bearings, first make sure that the crankshaft bearing journal and the connecting rod are free from oil and dirt.

7. Place the needle bearings with the beeswax onto a cool metallic surface to stiffen the beeswax. Body temperature will melt the wax, so avoid handling.

8. Remove the paper backing on the bearings and wrap the needles around the crankshaft journal. The beeswax will hold the needles onto the journal. Position the needles uniformly onto the crankpin.

➡**When installing the split row of needles, wrap each row of needles around the journal and try to seal them together with gentle but firm pressure to keep the bearings from unwinding.**

9. Place the connecting rod onto the journal, position the rod cap, and secure it with the capscrews. Tighten the screw to the proper specifications.

10. Force solvent (lacquer thinner) into the needles just installed to remove the beeswax, then force 30W oil into the needles for proper lubrication.

PISTON AND RINGS SERVICE

1. Clean all carbon from the piston and ring grooves.
2. Check the piston for scoring or other damage.
3. Check the fit of the piston in the cylinder bore. Move the piston from side-to-side to check clearance. If the clearance is not greater than 0.0003 in. (0.0076mm) and the cylinder is not scored or damaged, then the piston need not be replaced.
4. Check the piston ring side clearance to make sure it is within the limits recommended.
5. Check the piston rings for wear by inserting them into the cylinder about ½ in. (13mm) from the top of the cylinder. Check at various places to make sure that the gap between the ends of the ring does not exceed the dimensions recommended in the specifications table at the end of this section. Bore wear can be checked in the same way, except that a new ring is used to measure the end gap.
6. If replacement rings have a beveled or chamfered edge, install them with the bevel up toward the top of the piston. Not all engines use beveled rings. The two rings installed on the piston are identical.
7. When installed, the offset piston used on the AV600 and the AV520 engines must have the **V** stamped in the piston head (some have hash marks) facing toward the right as the engine is viewed from the top or piston side of the engine.

➡**Some AV520 and AV600 engines do not have offset pistons. Only offset pistons will have the 'V' or the hash marks on the piston head. Domed pistons must be installed so that the slope of the piston is toward the exhaust port.**

CRANKSHAFT SERVICE

▶ **See Figure 47**

1. Use a micrometer to check the bearing journals for out-of-roundness. The main bearing journals should not be more than 0.0005 in. (0.0127mm) out-of-round. Connecting rod journals should not be more than 0.001 in. (0.025mm) out-of-round. Replace a crankshaft that is not within these limits.

➡ **Do not attempt to regrind the crankshaft since undersize parts are not available.**

2. Check the tapered portion of the crankshaft (magneto end), keyways, and threads. Damaged threads may be restored with a thread die. If the taper of the shaft is rusty, it indicates that the engine has been operating with a loose flywheel. Clean the rust off the taper and check for wear. If the taper or keyway is worn, replace the crankshaft.

3. Check all of the bearing journal diameters. They should be within the limits indicated in the specifications table at the end of this section.

4. Check the crankshaft for bends by placing it between two pivot points. Position the dial indicator feeler on the crankshaft bearing surface and rotate the shaft. The crankshaft should not be more than 0.002-0.004 in. (0.05-0.10mm) out-of-round.

BEARING SERVICE

1. Do not remove the bearings unless they are worn or noisy. Check the operation of the bearings by rotating the bearing cones with your fingers to check for roughness, binding, or any other signs of unsatisfactory operation. If the bearings do not operate smoothly, remove them.

2. To remove the bearings from the crankcase, the crankcase must be heated. Use a hot plate to heat the crankcase to no more than 400°F (204°C). Place a ⅛ in. (3mm) steel plate over the hot plate to prevent overheating. At this temper-

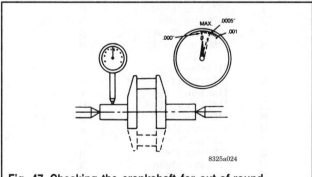

8325a024

Fig. 47 Checking the crankshaft for out-of-round

ature, the bearings should drop out with a little tapping of the crankcase.

3. The replacement bearing is left at room temperature and dropped into the heated crankcase. Make sure that the new bearing is seated to the maximum depth of the cavity.

➡ **Do not use an open flame to heat the crankcase halves and do not heat the crankcase halves to more than 400°F (204°C). Uneven heating with an open flame or excessive temperature will distort the case.**

4. The needle bearings will fall out of the bearing cage with very little urging. Needles can be reinstalled easily by using a small amount of all-purpose grease to hold the bearings in place.

5. Cage bearings are removed and replaced in the same manner as ball bearings.

6. Sleeve bearings cannot be replaced. Both crankcase halves must be discarded if a bearing is worn excessively.

ASSEMBLY

1. The gasket surface where the crankcase halves join must be thoroughly clean before reassembly. Do not buff or use a file or any other abrasive that might damage the mating surfaces.

➡ **Crankcase halves are matched. If one needs to be replaced, then both must be replaced.**

2. Place the PTO half of the crankcase onto the PTO end of the crankshaft. Use seal protectors where necessary.

3. Apply a thin coating of sealing compound to the contact surface of one of the crankcase halves.

4. Position one crankcase half on the other. The fit should be such that some pressure is required to bring the two halves together. If this is not the case, either the crankcase halves and/or the crankshaft must be replaced.

5. Secure the halves with the screws provided, tightening the screws alternately and evenly. Before tightening the screws, check the union of the crankcase halves on the cylinder mounting side. The halves should be flat and smooth at the union to provide a good mounting face for the cylinder. If necessary, realign the halves before tightening the screws.

6. The sleeve tool should be placed into the crankcase bore from the direction opposite the crankshaft. Insert the tapered end of the crankshaft through the half of the crankcase to which the magneto stator is mounted. Remove the seal tools after installing the crankcase halves.

7. Stagger the ring ends on the piston and check for the correct positioning on domed piston models.

8. Place the cylinder gasket on the crankcase end.

9. Place the piston into the cylinder using the chamfer provided on the bottom edge of the cylinder to compress the rings.

10. Secure the cylinder to the crankcase assembly.

Type Number-to-Letter Cross Reference Chart

Type No.	Column Letter	Type No.	Column Letter	Type No.	Column Letter
1 thru 44	A	611	F	1001	D
46 thru 68	B	614–01 thru 614–04	E	1002 thru 1002D	G
69 thru 77	D	614–05 thru 614–06A	F	1003	C
78 thru 80	C	615–01 thru 615–01A	E	1004 thru 1004A	H
81 thru 83	D	615–04	F	1005 thru 1005A	C
84 thru 85	C	615–05 thru 615–10A	E	1006	H
86 thru 87	B	615–11 thru 615–18	F	1007 thru 1008B	C
89	C	615–19 thru 615–27A	E	1009 thru 1010B	G
91	D	615–28	F	1011 thru 1019E	H
93	D	615–29 thru 615–39	E	1020	G
94 thru 98	H	616–01 thru 616–11	E	1021	C
99	C	616–12 and 616–14	F	1022 thru 1022C	D
		616–16 thru 616–40	E	1023 thru 1026A	H
201 thru 208	A	617A thru 617–01	E	1027 thru 1027B	C
209 thru 244	B	617–02 thru 617–03A	F	1028 thru 1030C	H
245	D	617–04 thru 617–04A	E	1031B thru 1031C	G
246 thru 248	B	617–05 thru 617–05A	F	1032	C
249 thru 251	D	617–06	E	1033 thru 1033F	G
252	B	618 thru 618–16	E	1034 thru 1034G	H
253 thru 255	D	618–17	F	1035 thru 1036	C
256 thru 261	B	618–18	E	1037 thru 1039C	D
262 thru 265A	D	619 thru 619–03	E	1040 thru 1041A	D
266 thru 267	B	621 thru 621–11	E	1042 thru 1042C	G
268 thru 272	D	621 thru 621–12A	F	1042D thru 1042E	H
273 thru 275	B	621–13 thru 621–15	E	1042F	G
276	D	622 thru 622–07	E	1042G and 1042H	H
277 thru 279	B	623 and 623A	E	1042I thru 1043B	G
280 thru 281	D	623–01	F	1043C thru 1043F	H
282	B	623–02 thru 623–33E	E	1043G thru 1044E	G
283 thru 284	D	623–34	F	1045 thru 1045F	B
285 thru 286	B	623–35 and 623–36	E	1046 thru 1051B	H
287	D	624 thru 630–09	J	1052	G
288 thru 291B	B	632–02A	K	1053 thru 1054C	H
292 thru 294	D	632–04 and 632–05A	K	1055 thru 1056B	G
295 thru 296	B	633 thru 634–09	J	1057	H
297	F	635A and B	K	1058 thru 1058A	G
298 thru 299	B	635–03B	K	1059	D
		635–04B	K	1060 thru 1060C	H
301 thru 375		635–06A thru 635–10	K	1061 and 1062	C
These are twin		636 thru 636–11	J	1063A thru 1063C	H
cylinder units–		637 thru 637–16	J	1064	G
out of production		638 thru 638–100	F	1065 thru 1067C	D
401 thru 402	L	639	M	1068 and 1068A	G
403	M	640	O	1069	H
404 thru 406B	L	641	K	1070 thru 1070B	C
407 thru 407C	M	642	I	1071 thru 1075D	H
408 thru 410B	L	643	J	1076 thru 1084	H
411 thru 411B	M	650	N	1085 and 1085A	G
412 thru 423	L	670	H	1086 thru 1140A	H
424 thru 427	M				
				1141 thru 1142	G
501 thru 509A	E	701 thru 701–1C	G	1143 thru 1145	H
		701–2 thru 701–2D	H	1146 thru 1149A	I
601 thru 601–01	F	701–3 thru 701–3C	G	1149A thru 1153B	H
602 thru 602–02B	F	701–4 thru 701–4C	H	1158 thru 1159	G
603A thru 603–23	E	701–5 thru 701–5C	G	1160 thru 1161A	I
604 thru 604–25	E	701–6 thru 701–6C	H	1162 and 1162A	G
605 thru 605–18	E	701–7 thru 701–7C	G	1163 thru 1163A	H
606 thru 606–05B	E	701–8 thru 701–11	H	1165 thru 1168A	G
606–06 thru 606–07	F	701–12 thru 701–13A	G	1169 thru 1174	H
606–08 thru 606–13	E	701–14 thru 701–17	H	1175 and 1175A	I
608 thru 608–C	E	701–18 thru 701–19	G	1176 thru 1177A	H
610–A thru 610–14	E	701–19A	H	1179 thru 1179A	I
610–15 thru 610–16	F	701–20 thru 701–22A	G	1180 thru 1181	G
610–19 thru 610–20	E	701–24	H	1182 and 1182A	I

8325a0C1

Type Number-to-Letter Cross Reference Chart (cont.)

Type No.	Column Letter	Type No.	Column Letter	Type No.	Column Letter
1183 thru 1185	G	1345	J	1441	P
1186A thru 1186C	J	1348 and 1348A	G	1442 thru 1442B	G
1187 thru 1192	G	1350 thru 1350C	I	1443	I
1192A thru 1196B	J			1444 and 1444A	G
1197 thru 1197A	G	1550A	P	1445	I
1198 and 1198A	H			1446 and 1446A	A
1199 and 1199A	J	1351 thru 1351B	K	1447	G
		1352 thru 1352B	I	1448 thru 1450	F
1206 thru 1208B	J	1353	K	1450A thru 1450B	F
1210 thru 1215C	K	1354 thru 1355A	I	1450C thru 1450E	F
1216 thru 1216A	H	1356 and 1356B	K	1453	K
1217 thru 1220A	J	1357 thru 1358	I	1454 and 1454A	A
1221	H	1359 thru 1362B	K	1455 thru 1456	K
1222 thru 1223	J	1363 thru 1369A	G	1459	G
1224 thru 1225A	H	1372 thru 1375A	G	1460 thru 1460F	A
1226 thru 1227B	K	1376 and 1376A	J	1461	K
1228 thru 1229A	J	1377 thru 1378	K	1462	A
1230 thru 1231	H	1379 thru 1379A	H	1463	K
1232	G	1380 thru 1380B	K	1464 thru 1464B	L
1233 thru 1237B	K	1381	G	1465	A
1238 thru 1238C	H	1382	H	1466 thru 1466A	F
1239 thru 1243	J	1383 thru 1383B	K	1467 thru 1468	K
1244 thru 1245A	K	1384 thru 1385	G	1471 thru 1471B	E
1246 thru 1247	J	1386	I	1472 thru 1472C	L
1248	H	1387 thru 1388	K	1473 thru 1473B	A
1249 thru 1251A	J	1389	G	1474	L
1252 thru 1254A	K	1390 thru 1390B	H	1475 thru 1476	A
1255	J	1391	K	1477	G
1256 thru 1262B	K	1392	J	1478	J
1263	J	1393	G	1479	G
1264 thru 1265E	K	1394 thru 1395A	P	1482 and 1482A	F
1266 thru 1267	G	1396	K	1483	F
1269 thru 1270D	K	1397	H	1484 thru 1484D	A
1271 thru 1271B	G	1398 thru 1399	K	1485	G
1272 thru 1275	K			1486	D
1276	H	1400	K	1487	J
1277 thru 1279D	K	1401 thru 1401F	F	1488 thru 1488D	A
1280	J	1402 and 1402B	G	1489 thru 1490B	C
1283 thru 1284D	K	1403A and 1403B	K	1491	L
1286 thru 1286A	H	1404 and 1404A	K	1493 and 1493A	G
1287	K	1405 thru 1406A	H	1494 and 1495A	B
1288	J	1407 thru 1408	K	1496	G
1289 thru 1289A	G	1409A	J	1497	A
1290 thru 1293A	K	1410 thru 1412A	I	1498	E
1294 thru 1295	J	1413 thru 1416	I	1499	F
1296 thru 1298	K	1417 and 1417A	K		
		1418 and 1418A	P	1500	E
1300 thru 1303	K	1419 and 1419A	K	1501A thru 1501E	A
1304 thru 1307	J	1420 and 1420A	H	1503 thru 1503D	L
1308 thru 1316A	K	1421 and 1421A	K	1506 thru 1507	F
1317 thru 1317B	J	1422 thru 1423	P	1508	G
1318 thru 1320B	K	1424	K	1509	C
1321 thru 1322	J	1425	G	1510	L
1323	K	1426 thru 1426B	P	1511	C
1325	G	1427 and 1427A	H	1512 and 1512A	B
1326 thru 1326F	K	1428 thru 1429A	P	1513	L
1327 thru 1327B	H	1430A	G	1515 thru 1516C	C
1328B and 1328C	K	1431	P	1517	E
1329	J	1432 and 1432A	G	1518	D
1330	H	1433	K	1519 thru 1521	A
1331	J	1434 thru 1435A	P	1522	L
1332	J	1436 and 1436A	I	1523	A
1333	I	1437	J	1524	B
1334	I	1439	I	1527	C
1343 thru 1344A	H	1440 thru 1440D	A	1528	A

8325a0C2

Type Number-to-Letter Cross Reference Chart (cont.)

Type No.	Column Letter	Type No.	Column Letter	Type No.	Column Letter
1529A and 1529B	C	2045 thru 2046B	F	40069 thru 40075	N
1530 thru 1530B	A	2047 thru 2048	D		
1531 thru 1535B	C	2049 thru 2049A	F	710101 thru 710116	E
1536	L	2050 thru 2050A	D	710124 thru 710130	E
1537	A	2051 thru 2052A	D	710131 thru 710137	E
1538 thru 1541A	L	2053 thru 2055	F	710138 thru 710149	E
1542	E	2056	D	710154	M
1543 thru 1546	A	2057	F	710201 thru 710209	E
1547	C	2058 thru 2058B	D	710210 thru 710218	E
1549	C	2059 thru 2063	F	710219 thru 710227	E
		2064 thru 2064A	D	710228	M
S–1801 thru 1822	F	2065	F	710150	G
1823 and 1824	E	2066	D	710151	H
		2067 thru 2071B	F	710155	L
2001 thru 2003	B			710157	H
2004 thru 2006	D	2200 thru 2201A	D	710229	H
2007 thru 2007B	B	2202 thru 2204	E	710230	N
2008 thru 2008B	D	2205 thru 2205A	D	710234	N
2009	B	2206 thru 2206A	G		
2010 thru 2011	D	2207	D	200.183112	F
2012	F	2208	G	200.18322	F
2013 thru 2014	B	2768	C	200.193132	F
2015 thru 2018	D			200.193142	F
2020	F	40001 thru 40028	N	200.193152	G
2021	D	40029 thru 40032C	O	200.193162	G
2022 thru 2022B	F	40033	N	200.203172	H
2023 thru 2026	D	40034 thru 40045A	O	200.203182	H
2027	B	40046	N	200.203192	H
2028 thru 2029	D	40047 thru 40052	O	200.213112	H
2030 thru 2031	B	40053 thru 40054A	N	200.213122	H
2032 thru 2033A	D	40056 thru 40060B	O	200.503111	F
2034 thru 2035C	F	40061 thru 40062B	N	200.583111	F
2036 thru 2037	B	40063 thru 40064	O	200.593121	F
2038	F	40065 thru 40066	N	200.613111	F
2039 thru 2044	C	40067 thru 40068	O		

8325a0C3

Specifications for 2 Cycle Engines with Split Crankcases

	A	B	C	D	E	F	G	H
Bore	1.500 1.5005	1.6253 1.6258	1.7503 1.7508	1.7503 1.7508	2.000	2.000	2.000	2.000
Stroke	1.375	1.50	1.50	1.50	1.50	1.50	1.50	1.50
Displacement Cubic Inches	2.43	3.10	3.60	3.60	4.70	4.70	4.70	4.70
Point Gap	.020	.020	.020	.020	.020	.020	.015	.015
Timing B.T.D.C. Before Top Dead Center	$1/16$" or .0625	$5/32$" or .1562	$5/32$" or .1562	$1/4$" or .250	$5/32$" or .1562	$5/32$" or .1562	$5/32$"	$1/4$" or .250 $11/64$" for "super"
Spark Plug Gap	.030	.030	.030	.030	.030	.030	.030	.030
Piston Ring End Gap	.003 .008	.005 .010	.005 .010	.005 .010	.006 .011	.006 .011	.006 .011	.006 .011
Piston Diameter	1.4966 1.4969	1.6216 1.6219	1.7461 1.7464	1.7461 1.7464	1.9948 1.9951	1.9948 1.9951	1.9948 1.9951	1.9949 1.9955
Piston Ring Groove Width	.095 .096	.095 .096	.095 .096	.095 .096	.095 .096	.095 .096	.095 .096	.095 .096
Piston Ring Width	.093 .0935	.093 .0935	.093 .0935	.093 .0935	.093 .0935	.093 .0935	.093 .0935	.093 .0935
Piston Pin Diameter	.3750 .3751	.3750 .3751	.3750 .3751	.3750 .3751	.3750 .3751	.3750 .3751	.3750 .3751	Early .3750 .3761 Late .4997 .4999
Connecting Rod Diameter Crank Bearing	.6869 .6874	.6869 .6874	.6869 .6874	.6986 .6989 w/o needles	.6869 .6874	.6869 .6874	.6869 .6874	.6935 .6939 w/o needles
Crankshaft Rod Needle Diameter				.0653 .0655				.0653 .0655
Crank Pin Journal Diameter	.6860 .6865	.6860 .6865	.6860 .6865	.5615 .5618	.6860 .6865	.6860 .6865	.6860 .6865	.5615 .5618
Crankshaft P.T.O. Side Main Brg. Dia.	.6689 .6693	.6689 .6693	.6690 .6694	.6689 .6693	.9995 1.0000	.6689 .6693	.9995 1.0000	.9995 1.0000
Crankshaft Magneto Side Main Brg. Dia.	.6689 .6693	.6689 .6693	.6689 .6693	.6690 .6694	.7495 .7500	.6689 .6693	.7495 .7500	.7495 .7500
Crankshaft End Play	.003 .008	.003 .008	.003 .008	.003 .008	.009 .022	.009 .022	.009 .022	.009 .022

8325a0C4

Specifications for 2 Cycle Engines with Split Crankcases (cont.)

I	J	K	L	M	N	O	P	
2.000	2.000	2.093 2.094	2.2505 2.2510	2.2505 2.2510	2.5030 2.5035	2.5030 2.5035	2.000	Bore
1.50	1.625	1.63	2.00	2.00	1.625	1.680	1.50	Stroke
4.70	5.10	5.80	8.00	8.00	7.98	8.25	4.70	Displacement Cubic Inches
.015	.020	.015	.020	.020	.020	.020	.015	Point Gap
11/64" or .175	11/64" or .175	3/32" or .095	1/8" or .125	3/32" or .095	11/64" or .175	11/64" or .175	.90	Timing B.T.D.C. Before Top Dead Center
.035	.030	.030	.030	.030	.030	.030	.035	Spark Plug Gap
.006 .011	.006 .011	.006 .011	.007 .015	.007 .015	.005 .013	.005 .013	.006 .011	Piston Ring End Gap
1.9948 1.9951	1.9951 1.9948	2.0880 2.0883	2.2460 2.2463	2.2460 2.2463	2.4960 2.4963	2.4960 2.4963	1.9948 1.9951	Piston Diameter
.095 .096	.095 .096	T.0655 .0665 L.0645 .0655	.095 .096	.095 .096	T.0655 .0665 L.0645 .0655	T.0655 .0665 L.0645 .0655	.095 .096	Piston Ring Groove Width
.093 .0935	.093 .0935	.0615 .0625	.093 .0935	.093 .0935	.0615 .0625	.0615 .0625	.093 .0935	Piston Ring Width
.4997 .4999	.3750 .3751	.4997 .4999	.5000 .5001	.3000 .5001	.4997 .4999	.4997 .4999	.4997 .4999	Piston Pin Diameter
.6941 .6944	.6869 .6874	.9407 .9412	1.000 1.0004	1.000 1.0004	.9407 .9412	.9407 .9412	.6941 .6944	Connecting Rod Diameter Crank Bearing
.0653 .0655		.0943 .0945	.0943 .0945	.0943 .0945	.0943 .0945	.0943 .0945	.0653 .0655	Crankshaft Rod Needle Diameter
.6860 .6865	.6860	.7499 .7502	.8096 .8099	.8096 .8099	.7499 .7502	.7499 .7502	.6860 .6865	Crank Pin Journal Diameter
.9995 1.0000	.9995 1.0000	.6990 .6994	.9839 .9842	.9839 .9842	.7871 .7875	.7871 .7875	.9995 1.0000	Crankshaft P.T.O. Side Main Brg. Dia.
.7495 .7500	.7495 .7500	.6990 .6994	.9839 .9842	.9839 .9842	.7498 .7501	.7498 .7501	.7495 .7500	Crankshaft Magneto Side Main Brg. Dia.
.009 .022	.009 .022	.003 .008	.003 .008	.003 .008	.008 .013	.008 .013	.009 .022	Crankshaft End Play

8325a0C5

Uniblock Cross Reference Chart

Type No.	Column No.	Type No.	Column No.	Type No.	Column No.
Vertical Crankshaft Engines		**Horizontal Crankshaft Engines**		**Horizontal Crankshaft Engines**	
638 thru 638-100	6	1401 thru 1401F	16	1508	7
		1401G, H	17	1509	3
642-01, A	9A	1401J	27	1510	12
642-02, A, B, C, D	9A	1402 and 1402B	7	1511	3
642-02E, F	9B	1425	7	1512 and 1512A	2
642-03, A, B	9A	1430A	7	1513	12
642-04, A, B, C	9A	1432 and 1432A	7	1515 thru 1516C	3
642-05, A, B	9A	1440 thru 1440D	1	1517	5
642-06, A	9A	1442 thru 1442B	7	1518	4
642-07, A, B	9A	1444 and 1444A	7	1519 thru 1521	1
		1448 thru 1450	16	1522	12
642-07C	9B	1450A thru 1450B	16	1523	1
642-08	9B	1450C thru 1450E	16	1524	2
642-08A, B	9A	1450F	17	1525A	16
642-09 thru 642-14	9A	1454 and 1454A	1	1527	3
642-13A, 14A, 14B	9B	1459	7	1528	1
642-15 thru 642-22	9B	1460 thru 1460F	1	1529A and 1520B	3
642-24 thru 642-30	9C	1462	1	1530 thru 1530B	1
		1464 thru 1464B	12	1531	3
670-01 thru 670-101	8	1465	1	1534A	17
		1466 thru 1466A	16	1535B	3
200-183112	6	1471 thru 1471B	5	1536	12
200-183122	6	1472 thru 1472C	12	1537	1
200-193132	6	1473 thru 1473B	1	1538 thru 1541A	12
200-193142	6	1474	12	1542	5
200-193152	7	1475 thru 1476	1	1543 thru 1546	1
200-193162	7	1479	7	1517	3
200-203172	8	1482 and 1482A	16	1519	3
200-203182	8	1483	16	1551	16
200-203192	8	1484 thru 1484D	3	1552	20
200-213112	8	1485	7	1553	16
200-213122	8	1486	4	1554 and 1554A	3
200-243112	8	1488 thru 1488D	1	1555 and 1556	16
200-283012	8	1489 thru 1490B	3	1561	19
		1491	12	1572	2
		1493 and 1493A	7	1573	3
		1494 and 1495A	2	1574 thru 1577	23
		1496	7	1575	24
		1497	1	1578	25
		1498	5	1581 thru 1582A	23
		1499	16	1583 thru 1595	26
		1500	5	200-503111	16
		1501A thru 1501E	1	200-583111	16
		1503 thru 1503D	12	200-593121	16
		1506	16	200-613111	16
		1506B	17	200-672102	26
		1507	16	200-682102	26

8325a0C6

Torque Specifications for 2-Cycle Engines

Application	Torque	Application	Torque
Carburetor and Reed Plates		**Cylinder head to cylinder** (635 type engine)	45–50 in. pounds
Carburetor to crankcase, Carburetor to adapter, or Carburetor adapter to crankcase	70–75 in. pounds	Cable clip and transfer port cover to cylinder	25–30 in. pounds
Carburetor to snow blowers cover	30–35 in. pounds	Cable clip to cylinder Stop lever to cylinder	25–30 in. pounds
Carburetor outlet fitting	40 in. pounds	Spark plug	18–22 ft. pounds
Reed and cover plates 639 type engines	50–60 in. pounds with Loctite, type A	Transfer cover cross port engines	25–30 in. pounds
Reed to plate 635 type engines	12–18 in. pounds	**Crankshaft and Connecting Rods**	
Crankcase and Cylinder		Aluminum and bronze rods to rod cap	40–50 in. pounds
Crankcase to crankcase cover	23–30 in. pounds	Steel rod to rod cap	70–80 in. pounds
Crankcase to crankcase cover screws	35–40 in. pounds	Flywheel nut on tapered end of crankshaft: Aluminum hub on iron or steel shaft.	18–25 ft. pounds
Mounting cylinder or carburetor to crankcase studs	50 in. pounds	Steel hub flywheel on iron shaft	18–25 ft. pounds
Base to crankcase	240–250 in. pounds	Steel hub flywheel on steel shaft	30 ft. pounds
Cylinder to crankcase nuts	70–75 in. pounds	**Governor and Bell Crank Parts**	
In cylinder	100–110 in. pounds	Power take-off end governor lower ring to crankshaft setscrew	30–35 in. pounds
Spark plug stop lever and head shroud to head	50–60 in. pounds	Bell crank bracket to crankcase	20–25 in. pounds
Cylinder head to cylinder	30–40 in. pounds	Governor cover to crankcase	50–60 in. pounds
Cylinder head to cylinder	50–60 in. pounds 80–90 in. pounds		

8325a0C7

10

TECUMSEH
4-STROKE

General Engine Specifications
Vertical Crankshaft Engines

Model	Bore & Stroke	Displace- ment	Horse- power
LAV25	2.3125 × 1.8438	7.75	2½
LAV30	2.3125 × 1.8438	7.75	3
TVS75	2.3125 × 1.8438	7.75	3
LV35	2.5000 × 1.8438	9.06	3½
LAV35	2.5000 × 1.8438	9.06	3½
TVS90	2.5000 × 1.8438	9.06	3½
LAV40	2.6250 × 1.9375	10.5	4
TVS105	2.6250 × 1.9375	10.5	4
V40, V40B	2.5000 × 2.2500	11.04	4
VH40	2.5000 × 2.2500	11.04	4
LAV50	2.812 × 1.9375	12.0	5
TVS120	2.812 × 1.9375	12.0	5
V50	2.625 × 2.2500	12.17	5
VH50	2.625 × 2.2500	12.17	5

8325b0C1

General Engine Specifications
Horizontal Crankshaft Engines

Model	Bore & Stroke (in.)	Displace- ment (cu in.)	Horse- power
H25	2.3125 × 1.8438	7.75	2½
H30	2.3125 × 1.8438	7.75	3
H35	2.5000 × 1.8438	9.06	3½
H40	2.5000 × 2.2500	11.04	4
HH40	2.5000 × 2.2500	11.04	4
HS40	2.6250 × 1.9375	10.5	4
H50	2.6250 × 2.2500	12.17	5
HH50	2.6250 × 2.2500	12.17	5
HS50	2.8120 × 1.9375	12.0	5

8325b0C2

Engine Identification

▶ See Figures 1 and 2

Lauson 4 cycle engines are identified by a model number stamped on a nameplate. The nameplate is located on the crankcase of vertical shaft models and on the blower housing of horizontal shaft models.

A typical model number appears on the illustration showing the location of the nameplate for vertical crankshaft engines. This number is interpreted as follows:

- V — vertical shaft engine
- 60 — 6.0 horsepower
- 70360J — the specification number. The last three numbers (360) indicate that this particular engine is a variation on the basic model line.
- 2361J — serial number
- 2 — year of manufacture
- 361 — the calendar day of manufacture
- J — line and shift location at the factory.

Air Cleaner

SERVICE

Dry Paper Element
▶ See Figure 3

The Tecumseh treated paper element type air cleaner consists of a pleated paper element encased in a metal housing and must be replaced as a unit. A flexible tubing and hose clamps connect the remotely mounted air filter to the carburetor.

Clean the element by lightly tapping it. Do not distort the case. When excessive carburetor adjustment or loss of power results, inspect the air filter to see if it is clogged. Replacing a severely restricted air filter should show an immediate performance improvement.

A plain paper element is also used. It should be removed every 10 hours, or more often if the air is dusty. Tap or blow out the dirt from the inside with low pressure air. This type should be replaced at 50 hours. If clogged sooner, it may be washed in soap and water and rinsed by flushing from the

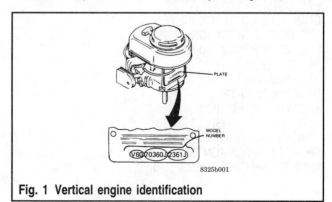

8325b001

Fig. 1 Vertical engine identification

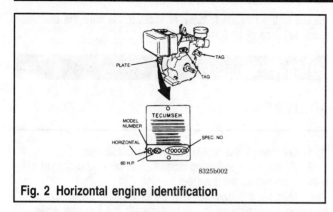

Fig. 2 Horizontal engine identification

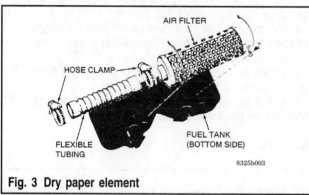

Fig. 3 Dry paper element

inside until the water is clear. Blow dry with low pressure compressed air.

To service the KLEEN-AIRE system, remove the element, wash it in soap and mild detergent, pat dry, and then coat with oil. Squeeze the oil to distribute it evenly and remove the excess. Make sure all mounting surfaces are tight to prevent leakage.

Oil/Foam Air Cleaners

Clean and re-oil the air cleaner element every 25 hours of operation under normal operating conditions. The capacity of the oil/foam air cleaner is adequate for a full season's use without cleaning. Under very dusty conditions, clean the air cleaner every few hours of operation.

The oil/foam air cleaner is serviced in the following manner:

1. Remove the screw that holds the halves of the air cleaner shell together and retains it to the carburetor.

2. Remove the air cleaner carefully to prevent dirt from entering the carburetor.

3. Take the air cleaner apart (split the two halves).

4. Wash the foam in kerosene or liquid detergent and water to remove the dirt.

5. Wrap the foam in a clean cloth and squeeze it dry.

6. Saturate the foam in clean engine oil and squeeze it to remove the excess oil.

7. Assemble the air cleaner and fasten it to the carburetor with the attaching screw.

Oil Bath Air Cleaner

▶ See Figure 4

Pour the old oil out of the bowl. Wash the element thoroughly in solvent and squeeze it dry. Clean the bowl and refill it with the same type of oil used in the crankcase.

Lubrication

OIL AND FUEL RECOMMENDATIONS

Use fresh (less than one month old) gasoline, of 'Regular" grade. Unleaded fuel is preferred, but leaded fuel is acceptable.

Use oil having SG classification. Use these viscosities for aluminum engines:

• Summer — above 32°F (0°C): S.A.E. 30 (S.A.E. 10W-30 or 10W-40 are acceptable substitutes).

• Winter — Below 32°F (0°C): S.A.E. 5W-30 (S.A.E. 10W is an acceptable substitute). (Including Snow King Snow Blower Engines)

• Winter — Below 0°F (-18°C) only: S.A.E. 10W diluted with 10% kerosene is an acceptable substitute. (Including Snow King Snow Blower Engines)Use these viscosities for cast iron engines:

• Summer — Above 32°F (0°C): S.A.E. 30

• Winter — Above 32°F (0°C): S.A.E. 10W

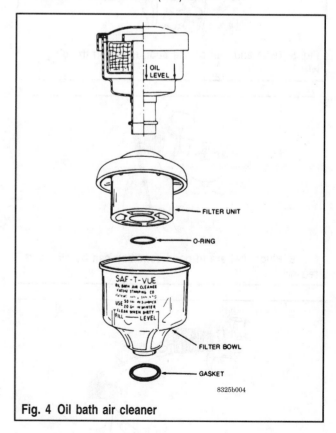

Fig. 4 Oil bath air cleaner

Tune-Up Specifications

The following basic specifications apply to all the engines covered in this section:
- Spark Plug Gap: 0.030 in. (0.76mm)
- Ignition Point Gap: 0.020 in. (0.50mm)
- Valve Clearance: 0.010 in. (0.25mm) for both intake and exhaustFor timing dimension, which varies from engine to engine.

Spark Plug Service

▶ **See Figures 5, 6, 7, 8 and 9**

Spark plugs should be removed, cleaned, and adjusted periodically. Check the electrode gap with a wire feeler gauge and adjust the gap. Replace the plugs if the electrodes are pitted and burned or the porcelain is cracked. Apply a little graphite grease to the threads to prevent sticking. Be sure the cleaned plugs are free of all foreign material.

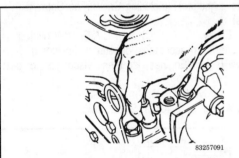

Fig. 5 Twist and pull on the boot, never on the plug wire

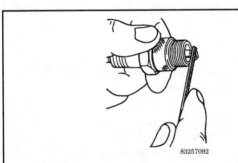

Fig. 6 Plugs that are in good condition can be filed and reused

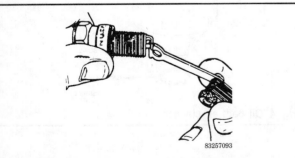

Fig. 7 Adjust the electrode gap by bending the side electrode

Breaker Points

ADJUSTMENT

1. Disconnect the fuel line from the carburetor.
2. Remove the mounting screws, fuel tank, and shroud to provide access to the flywheel.
3. Remove the flywheel with either a puller (over 3.5 hp) or by using a screwdriver to pry underneath the flywheel while tapping the top lightly with a soft hammer.
4. Remove the dust cover and gasket from the magneto and crank the engine over until the breaker points of the magneto are fully opened.
5. Check the condition of the points and replace them if they are burned or pitted.
6. Check the point gap with a feeler gauge. Adjust them, if necessary, as per the directions on the dust cover.

REPLACEMENT

1. Gain access to the points and inspect them as described above. If the points are badly pitted, follow the remaining steps to replace them.
2. Remove the nuts that hold the electrical leads to the screw on the movable breaker point spring. Remove the movable breaker point from stud.
3. Remove the screw and stationary breaker point. Put a new stationary breaker point on the breaker plate; install the screw, but do not tighten. This point must be moved to make the proper air gap when the points are adjusted.
4. Position a new movable breaker point on the stud.
5. Adjust the breaker point gap with a flat feeler gauge and tighten the screw.
6. Check the new point contact pattern and remove all grease, finger-prints, and dirt from contact surfaces.
7. Adjust the timing as described below.

Ignition Timing

ADJUSTMENT

1. Remove the cylinder head bolts, and move the head (with gasket in place) so that the spark plug hole is centered over the piston.
2. Using a ruler (through the spark plug hole) or special plunger type tool, carefully turn the engine back and forth until the piston is at exactly Top Dead Center. Tighten the thumbscrew on the tool.
3. Find the timing dimension for your engine in the specifications at the rear of the manual. Then, back off the position of the piston until it is about halfway down in the bore. Lower the ruler (or loosen the thumbscrew and lower the plunger, if using the special tool) exactly the required amount (the amount of the timing dimension). Then, hold the ruler in place (or

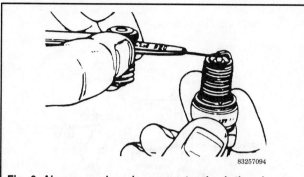

83257094

Fig. 8 Always used a wire gauge to check the plug gap

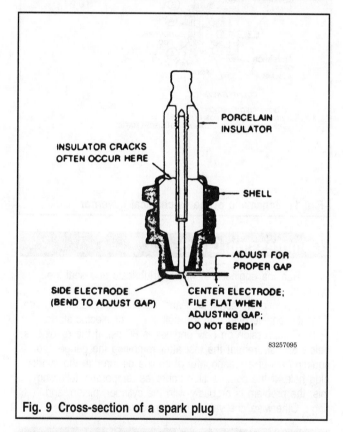

PORCELAIN INSULATOR

INSULATOR CRACKS OFTEN OCCUR HERE

SHELL

ADJUST FOR PROPER GAP

SIDE ELECTRODE (BEND TO ADJUST GAP)

CENTER ELECTRODE; FILE FLAT WHEN ADJUSTING GAP; DO NOT BEND!

83257095

Fig. 9 Cross-section of a spark plug

tighten the special tool thumbscrew) and, finally, carefully rotate the engine forward until the piston just touches the ruler or tool plunger.

4. Install a timing light or place a very thin piece of cellophane between the contact points. Loosen and rotate the stator just until the timing light shows a change in current flow or the cellophane pulls out of contact gap easily. Then, tighten stator bolts to specified torque.

5. Install the leads, point cover, flywheel, and shrouding.

SOLID STATE IGNITION SYSTEM CHECKOUT

The only on-engine check which can be made to determine whether the ignition system is working, is to separate the high tension lead from the spark plug and check for spark. If there is a spark, then the unit is all right and the spark plug should be replaced. No spark indicates that some other part needs replacing.

Check the individual components as follows:
• High Tension Lead: Inspect for cracks or indications of arcing. Replace the transformer if the condition of the lead is questionable.
• Low Tension Leads: Check all leads for shorts. Check the ignition cut-off lead to see that the unit is not grounded. Repair the leads, if possible, or replace them.
• Pulse Transformer: Replace and test for spark.
• Magneto: Replace and test for spark. Time the magneto by turning it counterclockwise as far as it will go and then tighten the retaining screws.
• Flywheel: Check the magnets for strength. With the flywheel off the engine, it should attract a screwdriver that is held 1 in. (25mm) from the magnetic surface on the inside of the flywheel. Be sure that the key locks the flywheel to the crankshaft.

Carburetor Mixture Adjustments

Chart of Initial Carburetor Adjustments

Adjustment	For Engines Built Prior to 1977	For Engines Built After 1977
Main Adjustment	V50-60-1¼ H50-60-1¼	Same
Idle Adjustment	V50-60-70-1 Turn H50-60-70-1 Turn	Same
Idle Speed (Top of Carburetor) Regulating screw	Back out screw, then turn in until screw just touches throttle lever and continue 1 turn more (if idle RPM is given set final idle speed with a tachometer)	

8325b0C3

1. If the carburetor has been overhauled, or the engine won't start, make initial mixture screw adjustments as specified in the chart.

2. Start the engine and allow it to warm up to normal running temperature. With the engine running at maximum recommended rpm, loosen the main adjustment screw until engine rpm drops off, then tighten the screw until the engine starts to cut out. Note the number of turns from one extreme to the other. Loosen the screw to a point midway between the extremes.

➡**Some carburetors have fixed jets. If there is no main adjusting screw and receptacle, no adjustment is needed.**

3. After the main system is adjusted, move the speed control lever to the idle position and follow the same procedure for adjusting the idle system.

4. Test the engine by running it under a normal load. The engine should respond to load pickup immediately. An engine that 'dies" is too lean. An engine which ran roughly before picking up the load is adjusted too rich.

Governor

ADJUSTMENT

Air Vane Type
▶ **See Figure 10**

1. Operate the engine with the governor adjusting lever or panel control set to the highest possible speed position and check the speed. If the speed is not within the recommended limits, the governed speed should be adjusted.

2. Loosen the locknut on the high speed limit adjusting screw and turn the adjusting screw out to increase the top engine speed.

Mechanical Type
▶ **See Figure 11**

1. Set the control lever to the idle position so that no spring tension affects the adjustment.

2. Loosen the screw so that the governor lever is loose in the clamp.

3. Rotate both the lever and the clamp to move the throttle to the full open position (away from the idle speed regulating screw).

4. Tighten the screw when no end-play exists in the direction of open throttle.

5. Move the throttle lever to the full speed setting and check to see that the control linkage opens the throttle.

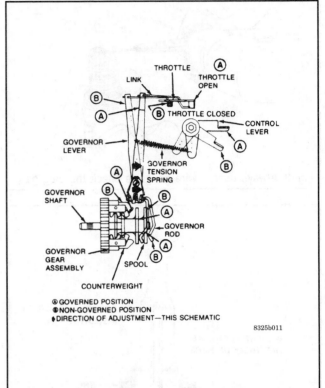

Fig. 11 Schematic of the mechanical governor

Compression Check

1. Run the engine until warm to lubricate and seal the cylinder.

2. Remove the spark plug and install a compression gauge. Turn the engine over with the pull starter or electric starter.

3. Compression on new engines is 80 psi. If the reading is below 60 psi., repeat the test after removing the gauge and squirting about a teaspoonful of engine oil through the spark plug hole. If the compression improves temporarily following this, the problem is probably with the cylinder, piston, and rings. Otherwise, the valves require service.

Carburetor

▶ **See Figures 12, 13, 14, 15, 16, 17, 18, 19 and 20**

➡ **Four-cycle Tecumseh engines use float or diaphragm type carburetors.**

REMOVAL & INSTALLATION

1. Drain the fuel tank. Remove the air cleaner and disconnect the carburetor fuel lines.

2. If necessary, remove any shrouding or control panels to provide access to carburetor.

3. Disconnect the choke or throttle control wires at the carburetor.

4. Remove the cap screws, or nuts and lockwashers that hold the carburetor to the engine; remove the carburetor.

5. Secure the carburetor on to engine.

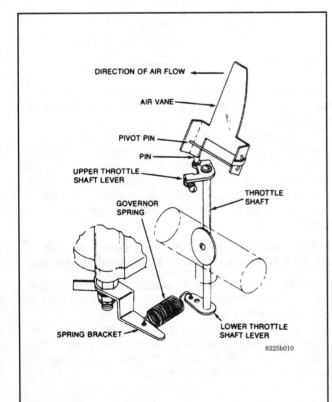

Fig. 10 Schematic of the air vane governor

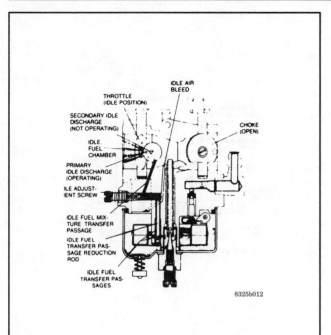

Fig. 12 The idle operation of a float feed carburetor. The throttle plate closes, restricting the flow of fuel and air, forcing the engine to run on a reduced volume of fuel and air

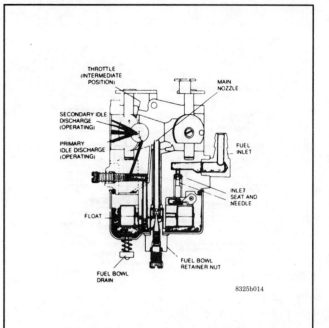

Fig. 14 The intermediate operation of a float feed carburetor. The throttle plate opens slightly to reduce restriction and the engine runs on an increased volume of fuel and air

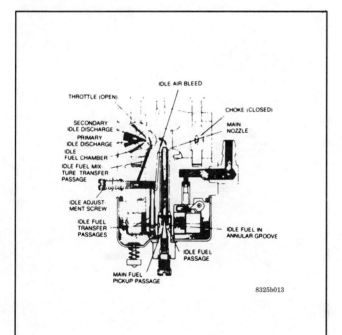

Fig. 13 The choke position of a float feed carburetor. The closed choke plate restricts air, creating a richened mixture by drawing in a greater proportion of fuel

6. Install the shrouding or control panels. Connect the choke and throttle control wires.

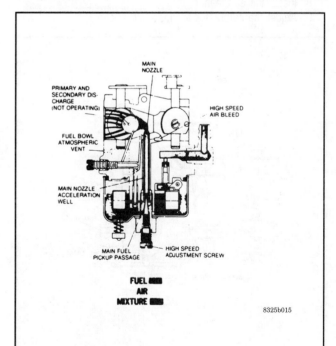

Fig. 15 The high speed operation of a float feed carburetor. The air venturi replaces the throttle plate as the restricting device and the engine runs on its greatest volume of fuel and air

7. Position the control panel to carburetor. Connect the carburetor fuel lines.

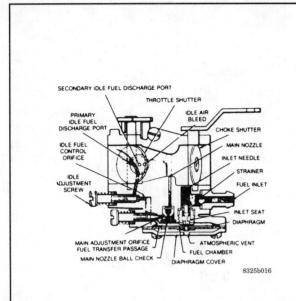

Fig. 16 The choke position of a diaphragm carburetor. The closed choke plate restricts air, creating a richened mixture by drawing in a greater proportion of fuel

8. Install the air cleaner.
9. Adjust the carburetor as described above.

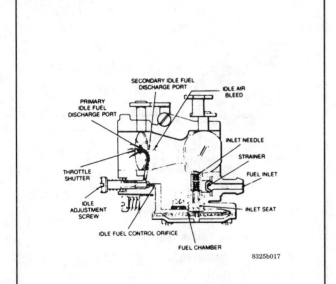

Fig. 17 The idle operation of a diaphragm carburetor. The throttle plate closes, restricting the flow of fuel and air, forcing the engine to run on a reduced volume of fuel and air

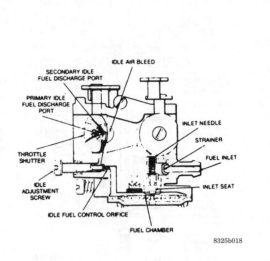

Fig. 18 The intermediate operation of a diaphragm carburetor. The throttle plate opens slightly to reduce restriction and the engine runs on an increased volume of fuel and air

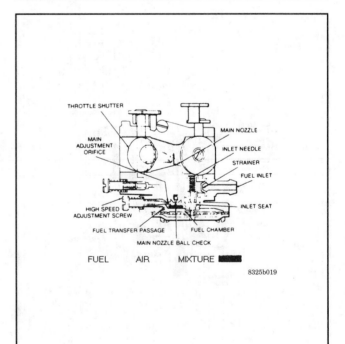

Fig. 19 The high speed operation of a diaphragm carburetor. The air venturi replaces the throttle plate as the restricting device and the engine runs on its greatest volume of fuel and air

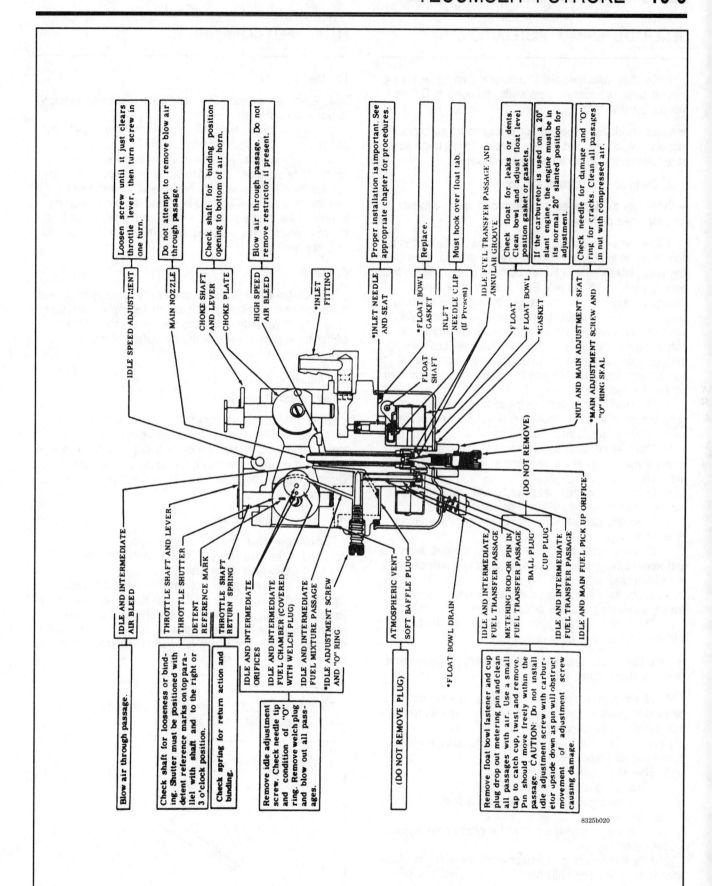

Fig. 20 Service hints for float feed carburetors

IDLE SPEED ADJUSTMENT — Loosen screw until it just clears throttle lever, then turn screw in one turn.

MAIN NOZZLE — Do not attempt to remove blow air through passage.

CHOKE SHAFT AND LEVER — Check shaft for binding position opening to bottom of air horn.

CHOKE PLATE

HIGH SPEED AIR BLEED — Blow air through passage. Do not remove restrictor if present.

*INLET FITTING

*INLET NEEDLE AND SEAT — Proper installation is important. See appropriate chapter for procedures.

*FLOAT BOWL GASKET — Replace.

INLET NEEDLE CLIP (If Present) — Must hook over float tab.

IDLE FUEL TRANSFER PASSAGE AND ANNULAR GROOVE

FLOAT SHAFT

*FLOAT — Clean float for leaks or dents. Clean bowl and adjust float level position gasket or gaskets.

FLOAT BOWL — If the carburetor is used on a 20° slant engine, the engine must be in its normal 20° slanted position for adjustment.

*GASKET

NUT AND MAIN ADJUSTMENT SEAT

*MAIN ADJUSTMENT SCREW AND "O" RING SEAL — Check needle for damage and "O" ring for cracks. Clean all passages in nut with compressed air.

IDLE AND INTERMEDIATE AIR BLEED — Blow air through passage.

THROTTLE SHAFT AND LEVER — Check shaft for looseness or binding. Shutter must be positioned with detent reference marks on top parallel with shaft and to the right or 3 o'clock position.

THROTTLE SHUTTER

DETENT REFERENCE MARK

THROTTLE SHAFT RETURN SPRING — Check spring for return action and binding.

IDLE AND INTERMEDIATE ORIFICES

IDLE AND INTERMEDIATE FUEL CHAMBER (COVERED WITH WELCH PLUG)

IDLE AND INTERMEDIATE FUEL MIXTURE PASSAGE

*IDLE ADJUSTMENT SCREW AND "O" RING — Remove idle adjustment screw. Check needle tip and condition of "O" ring. Remove welch plug and blow out all passages.

ATMOSPHERIC VENT

SOFT BAFFLE PLUG — (DO NOT REMOVE PLUG)

*FLOAT BOWL DRAIN — (DO NOT REMOVE PLUG)

IDLE AND INTERMEDIATE FUEL TRANSFER PASSAGE

METERING ROD OR PIN IN FUEL TRANSFER PASSAGE — Remove float bowl fastener and cup plug drop out metering pin and clean all passages with air. Use a small tap to catch cup, twist and remove. Pin should move freely within the passage. CAUTION: Do not install idle adjustment screw with carburetor upside down as pin will obstruct movement of adjustment screw causing damage.

BALL PLUG — (DO NOT REMOVE)

CUP PLUG

IDLE AND INTERMEDIATE FUEL TRANSFER PASSAGE

IDLE AND MAIN FUEL PICK UP ORIFICE

8325b020

GENERAL OVERHAUL INSTRUCTIONS

1. Carefully disassemble the carburetor removing all non-metallic parts, i.e.; gaskets, viton seats and needles, O-rings, fuel pump valves, etc.

➡**Nylon check balls used in some diaphragm carburetor models may or may not be serviceable. Check to be sure of serviceability before attempting removal.**

2. Clean all metallic parts with solvent.

➡**Nylon can be damaged if subjected to harsh cleaners for prolonged periods.**

3. The large O-rings sealing the fuel bowl to the carburetor body must be in good condition to prevent leakage. If the O-ring leaks, interfering with the atmospheric pressure in the float bowl, the engine will run rich. Foreign material can enter through the leaking area and cause blocking of the metering orifices. This O-ring should be replaced after the carburetor has been disassembled for repair. Lubricate the new O-ring with a small amount of oil to allow the fuel bowl to slide onto the O-ring properly. Hold the carburetor body in an inverted position and place the O-ring on the carburetor body and then position the fuel bowl.

4. The small O-rings used on the carburetor adjustment screws must be in good condition or a leak will develop and cause improper adjustment of carburetor.

5. Check all adjusting screws for wear. The illustration shows a worn screw and a good screw. Replace screws that are worn.

6. Check the carburetor inlet needle and seat for wear, scoring, or other damage. Replace defective parts.

7. Check the carburetor float for dents, leaks, worn hinge or other damage.

8. Check the carburetor body for cracks, clogged passages, and worn bushings. Clean clogged air passages with clean, dry compressed air.

9. Check the diaphragms on diaphragm carburetors for cracks, punctures, distortion, or deterioration.

10. Check all shafts and pivot pins for wear on the bearing surfaces, distortion, or other damage.

➡**Each time a carburetor is disassembled, it is good practice to install a repair kit.**

11. Where there is excessive vibration, a damper spring may be used to assist in holding the float against the inlet needle thus minimizing the flooding condition. Two types of springs are available; the float shaft (hinge pin) type and the inlet needle mounted type.

12. Float shaft spring positioning:

    a. The spring is slipped over the shaft.

    b. The rectangular shaped spring end is hooked onto the float tab.

    c. The shorter angled spring end is placed onto the float bowl gasket support.

13. Note that on late model carburetors, the spring clip fastened to the inlet needle has been revised to provide a damping effect. The clip fastens to the needle and is hooked over the float tab.

FLOAT FEED CARBURETOR

Throttle

1. Examine the throttle lever and plate prior to disassembly. Replace any worn parts.

2. Remove the screw in the center of the throttle plate and pull out the throttle shaft lever assembly.

3. When reassembling, it is important that the lines on the throttle plate are facing out when in the closed position. Position the throttle plates with the two lines at 12 and 3 o'clock. The throttle shaft must be held in tight to the bottom bearing to prevent the throttle plate from riding on the throttle bore of the body which would cause excessive throttle plate wear and governor hunting.

Choke

Examine the choke lever and shaft at the bearing points and holes into which the linkage is fastened and replace any worn parts. The choke plate is inserted into the air horn of the carburetor in such a way that the flat surface of the choke is toward the fuel bowl.

Idle Adjusting Screw
▶ **See Figure 21**

Remove the idle screw from the carburetor body and examine the point for damage to the seating surface on the taper. If damaged, replace the idle adjusting needle. Tension is maintained on the screw with a coil spring and sealed with an O-ring. Examine and replace the O-ring if it is worn or damaged.

High Speed Adjusting Jet

Remove the screw and examine the taper. If the taper is damaged at the area where it seats, replace the screw and fuel bowl retainer nut as an assembly.

The fuel bowl retainer nut contains the seat for the screw. Examine the sealing O-ring on the high speed adjusting screw. Replace the O-ring if it indicates wear or cuts. During the reassembly of the high speed adjusting screw, position the coil spring on the adjusting screw, followed by the small brass washer and the O-ring seal.

Fuel Bowl

To remove the fuel bowl, remove the retaining nut and fiber washer. Replace the nut if it is cracked or worn.

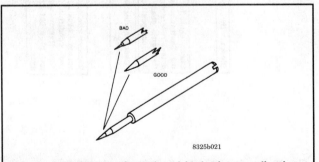

Fig. 21 Appearance of good and bad mixture adjusting screws

The retaining nut contains the transfer passage through which fuel is delivered to the high speed and idle fuel system of the carburetor. It is the large hole next to the hex nut end of the fitting. If a problem occurs with the idle system of the carburetor, examine the small fuel passage in the annular groove in the retaining nut. This passage must be clean for the proper transfer of fuel into the idle metering system.

The fuel bowl should be examined for rust and dirt. Thoroughly clean it before installing it. If it is impossible to properly clean the fuel bowl, replace it.

Check the drain valve for leakage. Replace the rubber gasket on the inside of the drain valve if it leaks.

Examine the large O-ring that seals the fuel bowl to the carburetor body. If it is worn or cracked, replace it with a new one, making sure the same type is used (square or round).

Inlet Needle and Seat

▶ **See Figures 21 and 23**

1. The inlet needle sits on a rubber seat in the carburetor body instead of the usual metal fitting.

2. Remove it, place a few drops of heavy engine oil on the seat, and pry it out with a short piece of hooked wire.

3. The grooved side of the seat is inserted first. Lubricate the cavity with oil and use a flat faced punch to press the inlet seat into place.

4. Examine the inlet needle for wear and rounding off of the corners. If this condition does exist, replace the inlet needle.

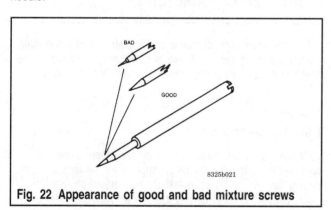

Fig. 22 Appearance of good and bad mixture screws

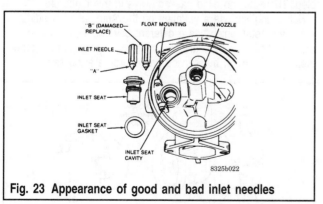

Fig. 23 Appearance of good and bad inlet needles

Float

▶ **See Figures 24, 25, 26 and 27**

1. Remove the float from the carburetor body by pulling out the float axle with a pair of needle nose pliers. The inlet needle will be lifted off the seat because it is attached to the float with an anchoring clip.

2. Examine the float for damage and holes. Check the float hinge for wear and replace it if worn.

3. The float level is checked by positioning a #4 drill bit across the rim between the center leg and the unmachined surface of the index pad, parallel to the float axle pin. If the index pad is machined, the float setting should be made with a #9 drill bit.

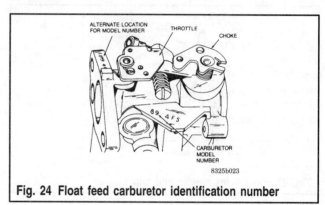

Fig. 24 Float feed carburetor identification number

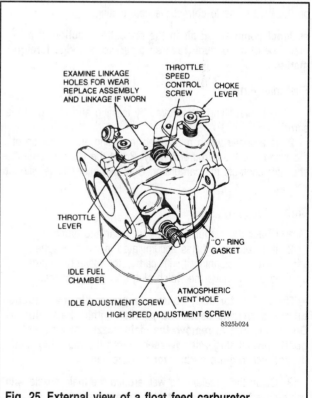

Fig. 25 External view of a float feed carburetor

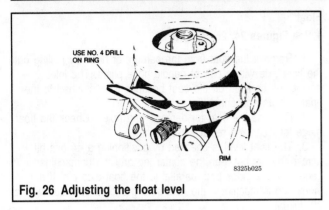

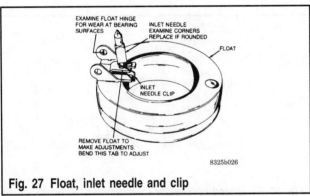

Fig. 26 Adjusting the float level

Fig. 27 Float, inlet needle and clip

4. Remove the float to make an adjustment. Bend the tab on the float hinge to correct the float setting.

➡**Direct compressed air in the opposite direction of normal flow of air or fuel (reverse taper) to dislodge foreign matter.**

Fuel Inlet Fitting

1. The inlet fitting is removed by twisting and pulling at the same time.

2. Use sealer when reinstalling the fitting. Insert the tip of the fitting into the carburetor body. Press the fitting in until the shoulder contacts the carburetor. Only use inlet fittings without screens.

Carburetor Body

1. Check the carburetor body for wear and damage.

2. If excessive dirt has accumulated in the atmospheric vent cavity, try cleaning it with carburetor solvent or compressed air. Remove the welch plug only as a last resort.

➡**The carburetor body contains a pressed-in main nozzle tube at a specific depth and position within the venturi. Do not attempt to remove the main nozzle. Any change in nozzle positioning will adversely affect the metering quality and will require carburetor replacement.**

3. Clean the accelerating well around the main nozzle with compressed air and carburetor cleaning solvents.

4. The carburetor body contains two cup plugs, neither of which should be removed. A cup plug located near the inlet seat cavity, high up on the carburetor body, seals off the idle bleed. This is a straight passage drilled into the carburetor

throat. Do not remove this plug. Another cup plug is located in the base where the fuel bowl nut seals the idle fuel passage. Do not remove this plug or the metering rod.

5. A small ball plug located on the side of the idle fuel passage seals this passage. Do not remove this ball plug.

6. The welch plug on the side of the carburetor body, just above the idle adjusting screw, seals the idle fuel chamber. This plug can be removed for cleaning of the idle fuel mixture passage and the primary and secondary idle fuel discharge ports. Do not use any tools that might change the size of the discharge ports, such as wire or pins.

Resilient Tip Needle

Replace the inlet needle. Do not attempt to remove or replace the seat in the carburetor body.

Viton Seat

Using a 10-24 or 10-32 tap, turn the tap into the brass seat fitting until it grasps the seat firmly. Clamp the tap shank into a vise and tap the carburetor body with a soft hammer until the seat slides out of the body.

To replace the viton seat, position the replacement over the receptical with the soft rubber like seat toward the body. Use a flat punch and a small hammer to drive the seat into the body until it bottoms on the shoulder.

TECUMSEH AUTOMATIC NON-ADJUSTABLE FLOAT FEED CARBURETOR

This carburetor has neither a choke plate nor idle and main mixture adjusting screws. There is no running adjustment. The float adjustment is the standard Tecumseh float setting of 0.210 in. (5.3mm).

Cleaning
▶ **See Figures 28 and 29**

Remove all non-metallic parts and clean them using a procedure similar to that for the other carburetors. Never use wires through any of the drilled holes. Do not remove the baffling welch plug unless it is certain there is a blockage under the plug. There are no blind passageways in this carburetor.

Some engines use a variation on the Automatic Nonadjustable carburetor which has a different bowl hold-on nut and main jet orifices. There are two main jet orifices and a deeper fuel reserve cavity, but service procedures are the same.

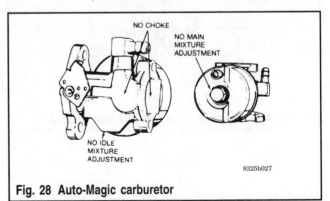

Fig. 28 Auto-Magic carburetor

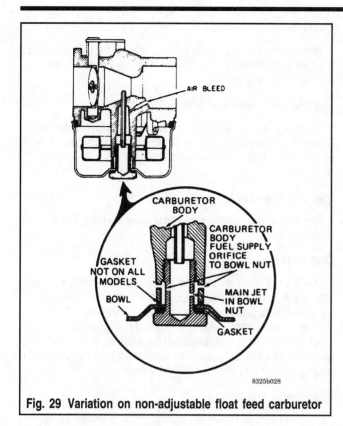

Fig. 29 Variation on non-adjustable float feed carburetor

WALBRO AND TILLOTSON FLOAT FEED CARBURETORS

▶ See Figure 30

Procedures are similar to those for the Tecumseh float carburetor with the exceptions noted below.

Main Nozzle

The main nozzle in Wallbro carburetors is cross drilled after it is installed in the carburetor. Once removed, it cannot be reinstalled, since it is impossible to properly realign the cross drilled holes. Grooved service replacement main nozzles are available which allow alignment of these holes.

Float Shaft Spring

Carefully position the float shaft spring on models so equipped. The spring dampens float action when properly assembled. Use needlenosed pliers to hook the end of the spring over the float hinge and then insert the pin as far as possible before lifting the spring from the hinge into position. Leaving the spring out or improper installation will cause unbalanced float action and result in a touchy adjustment.

Float Adjustment

1. To check the float adjustment, invert the assembled float carburetor body. Check the clearance between the body and the float, opposite the hinge. Clearance should be 1/8 in. ± 1/64 in. (3mm ± 0.4mm).
2. To adjust the float level, remove the float shaft and float. Bend the lip of the float tang to correct the measurement.
3. Assemble the parts and recheck the adjustment.

TILLOTSON E FLOAT FEED TYPE CARBURETOR

▶ See Figure 31

The following adjustments are different for this carburetor.

Running Adjustment

1. Start the engine and allow it to warm up to operating temperatures. Make sure the choke is fully opened after the engine is warmed up.
2. Run the engine at a constant speed while slowly turning the main adjustment screw in until the engine begins to lose speed; then slowly back it out about 1/8-1/4 of a turn until maximum speed and power is obtained (4000 rpm). This is the correct power adjustment.
3. Close the throttle and cause the engine to idle slightly faster than normal by turning the idle speed regulating screw in. Then turn the idle adjustment speed screw in until the engine begins to lose speed; then turn it back 1/4-1/2 of a turn until the engine idles smoothly. Adjust the idle speed regulating screw until the desired idling speed is acquired.
4. Alternately open and close the throttle a few times for an acceleration test. If stalling occurs at idle speeds, repeat the adjustment procedures to get the proper idle speed.

Float Level Adjustment

1. Remove the carburetor float bowl cover and float mechanism assembly.
2. Remove the float bowl cover gasket and, with the complete assembly in an upside down position and the float lever tang resting on the seated inlet needle, a measurement of 1⁵/64 in. (27.4mm) should be maintained from the free end flat rim,

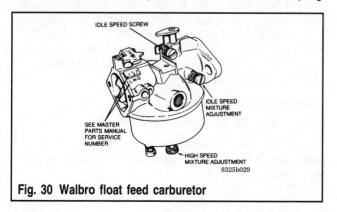

Fig. 30 Walbro float feed carburetor

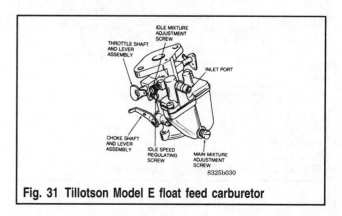

Fig. 31 Tillotson Model E float feed carburetor

or edge of the cover, to the toe of the float. Measurement can be checked with a standard straight rule or depth gauge.

3. If it is necessary to raise or lower the float lever setting, remove the float lever pin and the float, then carefully bend the float lever tang up or down as required to obtain the correct measurement.

DIAPHRAGM CARBURETORS

◆ See Figures 32 and 33

Diaphragm carburetors have a rubber-like diaphragm that is exposed to crankcase pressure on one side and to atmospheric pressure on the other side. As the crankcase pressure decreases, the diaphragm moves against the inlet needle allowing the inlet needle to move from its seat which permits fuel to flow through the inlet valve to maintain the correct fuel level in the fuel chamber.

An advantage of this type of system over the float system, is that the engine can be operated in any position.

➡ In rebuilding, use carburetor cleaner only on metal parts, except for the main nozzle in the main body.

Throttle Plate

Install the throttle plate with the short line that is stamped in the plate toward the top of the carburetor, parallel with the throttle shaft, and facing out when the throttle is closed.

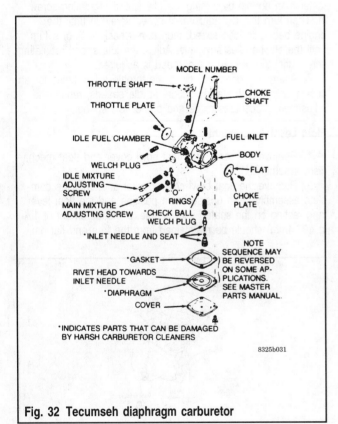

Fig. 32 Tecumseh diaphragm carburetor

Choke Plate

Install the choke plate with the flat side of the choke toward the fuel inlet side of the carburetor. The mark faces in and is parallel to the choke shaft.

Idle Mixture Adjustment Screw

There is a neoprene O-ring on the needle. Never soak the O-ring in carburetor solvent. Idle and main mixture screws vary in size and design, so make sure that you have the correct replacement.

Idle Fuel Chamber

The welch plug can be removed if the carburetor is extremely dirty.

Diaphragms

Diaphragms are serviced and replaced by removing the four retaining screws from the cover. With the cover removed, the diaphragm and gasket may be serviced. Never soak the diaphragm in carburetor solvent. Replace the diaphragm if it is cracked or torn. Be sure there are no wrinkles in the diaphragm when it is replaced. The diaphragm rivet head is always placed facing the inlet needle valve.

Inlet Needle and Seat

The inlet seat is removed by using either a slotted screwdriver (early type) or a 9/32 in. socket. The inlet needle is spring loaded, so be careful when removing it.

Fuel Inlet Fitting

All of the diaphragm carburetors have an integral strainer in the inlet fitting. To clean it, either reverse flush it or use compressed air after removing the inlet needle and seat. If the strainer is lacquered or otherwise unable to be cleaned, replace the fitting.

CRAFTSMAN FUEL SYSTEMS

Changes In Late Model Carburetors

The newest Craftsman carburetors incorporate the following changes:
- The cable form of control is replaced by a control knob.
- The fuel pickup is longer and has a collar machined into it which must be installed tight against the carburetor body.
- The fuel pickup screen is pressed onto the ends of the fill tubes on both models, but the measured depth has changed.
- The cross-drilled passages have been eliminated, as has the O-ring on the body. There are no cup plugs.
- The fuel tank and reservoir tube have been revised-the reservoir tube being larger.

Disassembly and Service
◆ See Figures 34, 35, 36, 37, 38, 39 and 40

1. Remove the air cleaner assembly and remove the four screws on the top of the carburetor body to separate the fuel tank from the carburetor.

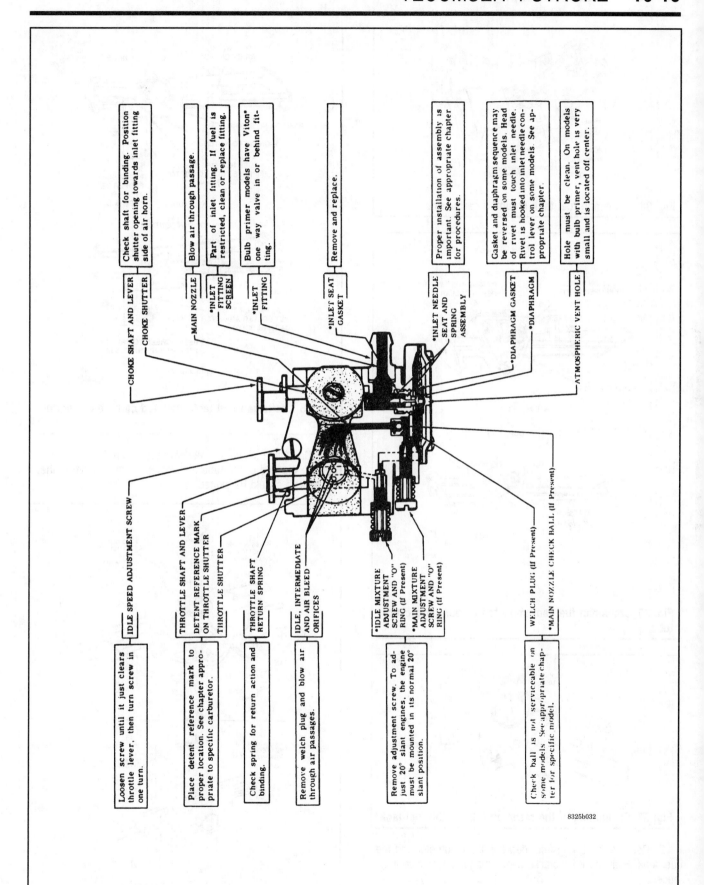

Check shaft for binding. Position shutter opening towards inlet fitting side of air horn.

Blow air through passage.

Part of inlet fitting. If fuel is restricted, clean or replace fitting.

Bulb primer models have Viton* one way valve in or behind fitting.

Remove and replace.

Proper installation of assembly is important. See appropriate chapter for procedures.

Gasket and diaphragm sequence may be reversed on some models. Head of rivet must touch inlet needle. Rivet is hooked into inlet needle control lever on some models. See appropriate chapter.

Hole must be clean. On models with bulb primer, vent hole is very small and is located off center.

CHOKE SHAFT AND LEVER
CHOKE SHUTTER
*MAIN NOZZLE
*INLET FITTING SCREEN
*INLET FITTING
*INLET SEAT GASKET
*INLET NEEDLE SEAT AND SPRING ASSEMBLY
*DIAPHRAGM GASKET
*DIAPHRAGM
ATMOSPHERIC VENT HOLE

Loosen screw until it just clears throttle lever, then turn screw in one turn.

Place detent reference mark to proper location. See chapter appropriate to specific carburetor.

Check spring for return action and binding.

Remove welch plug and blow air through air passages.

Remove adjustment screw. To adjust 20° slant engines, the engine must be mounted in its normal 20° slant position.

Check ball is not serviceable on some models. See appropriate chapter for specific model.

IDLE SPEED ADJUSTMENT SCREW
THROTTLE SHAFT AND LEVER
DETENT REFERENCE MARK ON THROTTLE SHUTTER
THROTTLE SHUTTER
THROTTLE SHAFT RETURN SPRING
IDLE, INTERMEDIATE AND AIR BLEED ORIFICES
*IDLE MIXTURE ADJUSTMENT SCREW AND "O" RING (If Present)
*MAIN MIXTURE ADJUSTMENT SCREW AND "O" RING (If Present)
*WELCH PLUG (If Present)
*MAIN NOZZLE CHECK BALL (If Present)

8325b032

Fig. 33 Service hints for the diaphragm carburetor

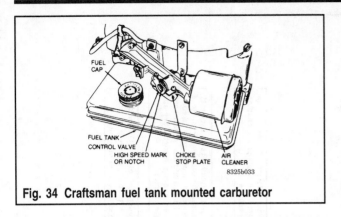

FUEL
CAP

FUEL TANK
CONTROL VALVE

HIGH SPEED MARK   CHOKE   AIR
OR NOTCH   STOP PLATE   CLEANER

8325b033

Fig. 34 Craftsman fuel tank mounted carburetor

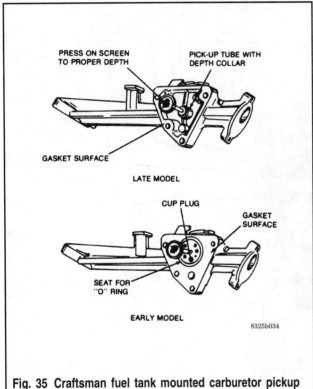

PRESS ON SCREEN   PICK-UP TUBE WITH
TO PROPER DEPTH   DEPTH COLLAR

GASKET SURFACE

LATE MODEL

CUP PLUG

GASKET
SURFACE

SEAT FOR
"O" RING

EARLY MODEL

8325b034

Fig. 35 Craftsman fuel tank mounted carburetor pickup tube

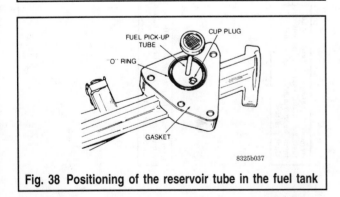

FUEL PICK-UP   CUP PLUG
TUBE

"O" RING

GASKET

8325b037

Fig. 38 Positioning of the reservoir tube in the fuel tank

2. Remove the O-ring from between the carburetor and the fuel tank. Examine it for cracks and damage and replace it if necessary.

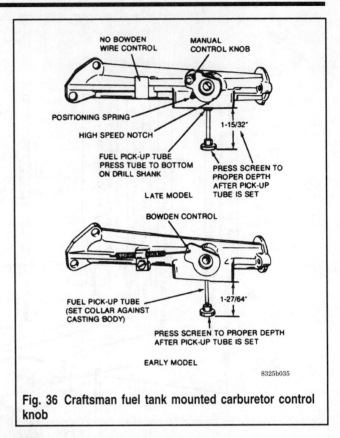

NO BOWDEN   MANUAL
WIRE CONTROL   CONTROL KNOB

POSITIONING SPRING

HIGH SPEED NOTCH

FUEL PICK-UP TUBE   1-15/32"
PRESS TUBE TO BOTTOM
ON DRILL SHANK   PRESS SCREEN TO
                                     PROPER DEPTH
LATE MODEL   AFTER PICK-UP
                                     TUBE IS SET

BOWDEN CONTROL

FUEL PICK-UP TUBE   1-27/64"
(SET COLLAR AGAINST
CASTING BODY)
                 PRESS SCREEN TO PROPER DEPTH
                 AFTER PICK-UP TUBE IS SET

EARLY MODEL

8325b035

Fig. 36 Craftsman fuel tank mounted carburetor control knob

3. Carefully remove the reservoir tube from the fuel tank. Observe the end of the tube that rested on the bottom of the fuel tank. It should be slotted.

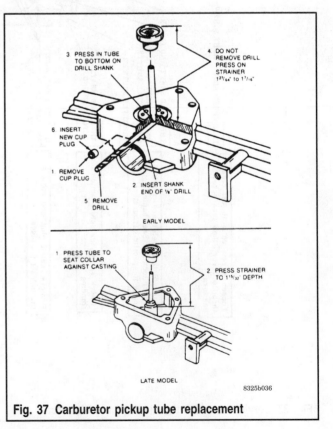

3 PRESS IN TUBE   4 DO NOT
  TO BOTTOM ON   REMOVE DRILL
  DRILL SHANK   PRESS ON
                              STRAINER
                              1²⁷/₆₄" to 1⁷/₁₆"

6 INSERT
  NEW CUP
  PLUG

1 REMOVE
  CUP PLUG

                2 INSERT SHANK
                  END OF ⅛" DRILL

5 REMOVE
  DRILL

                              EARLY MODEL

1 PRESS TUBE TO
  SEAT COLLAR
  AGAINST CASTING
                              2 PRESS STRAINER
                                TO 1¹⁵/₃₂" DEPTH

LATE MODEL   8325b036

Fig. 37 Carburetor pickup tube replacement

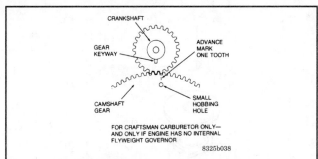

Fig. 39 Timing an engine with the Craftsman fuel tank-mounted carburetor

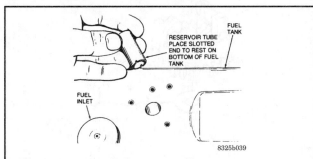

Fig. 40 The fuel pickup tube and O-ring on early model Craftsman carburetors

4. Remove the control valve by turning the valve clockwise until the flange is clear of the retaining boss. Pull the valve straight out and examine the O-ring seal for damage or wear. If possible, use a new O-ring when reassembling.

5. Examine the fuel pick up tube. There are no valves or ball checks that may become inoperative. These parts can normally be cleaned with carburetor solvent. If it is found that the passage cannot be cleared, the fuel pick up tubes can be replaced. Carefully remove the old ones so as not to enlarge the opening in the carburetor body. If the pickup tube and screen must be replaced, follow the directions shown in the illustration for the type of carburetor (early or late model) on which you are working.

6. Assemble the carburetor in reverse order of disassembly. Use new O-rings and gaskets. When assembling reservoir tube, hold the carburetor upside down and place the reservoir tube over the pickup tube with the slotted end up. Make sure the intake manifold gasket is correctly positioned-it can be assembled blocking the intake passage partially.

Adjustments

1. Move the carburetor control valve to the high speed position. The mark on the face of the valve should be in alignment with the retaining boss on the carburetor body.

2. Move the operator's control on the equipment to the high speed position.

3. Insert the bowden wire into the hole of the control valve. Clamp the bowden wire sheath to the carburetor body.

➡**If the engine was disassembled and the camshaft removed, be sure that the timing marks on the camshaft gear and the keyway in the crankshaft gear are aligned**

when reinserting the camshaft. Then lift the camshaft enough to advance the camshaft gear timing mark to the right (clockwise) ONE tooth, as viewed from the power take-off end of the crankshaft.

CRAFTSMAN FLOAT TYPE CARBURETORS

▶ **See Figures 41 and 42**

Craftsman float type carburetors are serviced in the same manner as the other Tecumseh float type carburetors. The throttle control valve has three positions: stop, run, and start.

When the control valve is removed, replace the O-ring. If the engine runs sluggishly, consider the possibility of a leaky O-ring.

To remove the fuel pickup tube, clamp the tube in a vise and then twist the carburetor body. Check the small jet in the air horn while the tube is out. In replacement, position the tube squarely and then press in on the collar until the collar seats.

In assembly, install the gasket and bolt through the bowl, position the centering spacer onto the bolt, and then attach the parts to the carburetor body.

The camshaft timing mark must be advanced one tooth in relation to crankshaft gear timing mark, as shown in the illustration above. However, when this type carburetor is used with a float bowl reservoir and variable governor adjustment, time it as for other Tecumseh engines — with the camshaft and crankshaft timing marks aligned.

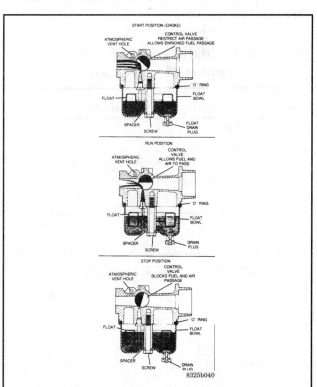

Fig. 41 Operating positions of a Craftsman float type carburetor

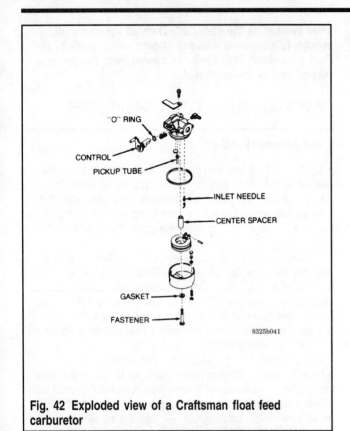

Fig. 42 Exploded view of a Craftsman float feed carburetor

Governor

▶ See Figures 43 and 44

Shaft Installed Dimension/ Engine Model Chart

Engine Model	"A" Exposed Shaft Length (see figure)
LAV30-50 H25-35 HS40-50	$1^{5}/_{16}$"
V50	$1^{19}/_{32}$"
H50 HH40	$1^{7}/_{16}$"

8325b0C4

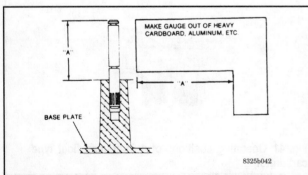

Fig. 43 Measuring governor shaft installed dimensions

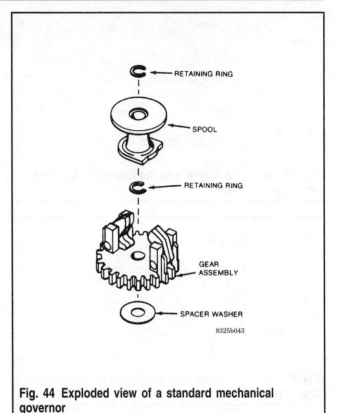

Fig. 44 Exploded view of a standard mechanical governor

The mechanical governor is located inside the mounting flange. See engine disassembly instructions, below. To disassemble the governor, see the illustration, and: remove the retaining ring, pull off the spool, remove the second retaining ring, and then pull off the gear assembly and retainer washer.

Check for wear on all moving surfaces, but especially gear teeth, the inside diameter of the gear where it rides on the shaft, and the flyweights where they work against the spool.

If the governor shaft must be replaced, it should be started into the boss with a few taps using a soft hammer, and then pressed in with a press or vise. The shaft can be installed by positioning a wooden block on top and tapping the upper surface of the block, but the use of a vise or press is much preferred.

The shaft must be pressed in until just the required length is exposed, as measured from the top of the shaft boss to the upper end of the shaft. See the chart.

The governor is installed in reverse of the removal procedure. Connect the linkage and then adjust as described in the Tune-Up section.

Engine Overhaul

TIMING GEARS

▶ See Figures 45, 45 and 45

Correctly matched camshaft gear and crankshaft gear timing marks are necessary for the engine to perform properly.

On all camshafts the timing mark is located in line with the center of the hobbing hole (small hole in the face of the gear).

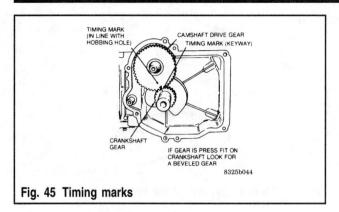

Fig. 45 Timing marks

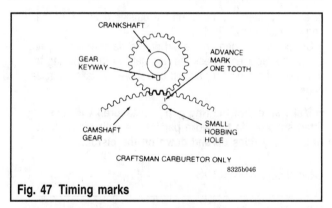

Fig. 46 Timing marks

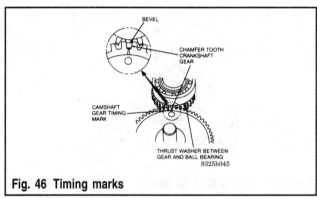

Fig. 47 Timing marks

If no line is visible, use the center of the hobbing hole to align with the crankshaft gear marked tooth.

On crankshafts where the gear is held on by a key, the timing mark is the tooth in line with the keyway.

On crankshafts where the gear is pressed onto the crankshaft, a tooth is bevelled to serve as the timing mark.

On engines with a ball bearing on the power take-off end of the crankshaft, look for a bevelled tooth which serves as the crankshaft gear timing mark.

If the engine uses a Craftsman type carburetor, the camshaft timing mark must be advanced clockwise one tooth ahead of the matching timing mark on the crankshaft, the exception being the Craftsman variable governed fuel systems.

➡️**If one of the timing gears, either the crankshaft gear or the camshaft gear, is damaged and has to be replaced, both gears should be replaced.**

CRANKSHAFT INSPECTION

Inspect the crankshaft for worn or crossed threads that can't be redressed; worn, scratched, or damaged bearing surfaces; misalignment; flats on the bearing surfaces. Replace the shaft if any of these problems are in evidence — do not try to straighten a bent shaft.

In replacement, be sure to lubricate the bearing surfaces and use oil seal protectors. If the camshaft gear requires replacement, replace the crankshaft gear, also.

LAV35-LAV50 and H35-HS50 crankshafts have a press fit gear. If the camshaft gear requires replacement on these engines, the crankshaft must be replaced, as the gear cannot be replaced separately.

PISTONS

▶ **See Figures 48, 49, 50 and 51**

When removing the pistons, clean the carbon from the upper cylinder bore and head. The piston and pin must be replaced in matched pairs.

A ridge reamer must be used to remove the ridge at the top of the cylinder bore on some engines.

Clean the carbon from the piston ring groove. A broken ring can be used for this operation.

Some engines have oversize pistons which can be identified by the oversize engraved on the piston top.

There is a definite piston-to-connecting rod-to-crankshaft arrangement which must be maintained when assembling these

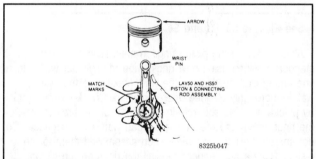

Fig. 48 Piston-to-rod relationship for the LAV50 and HS50 engines

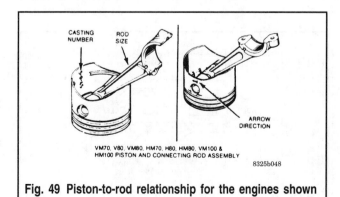

Fig. 49 Piston-to-rod relationship for the engines shown

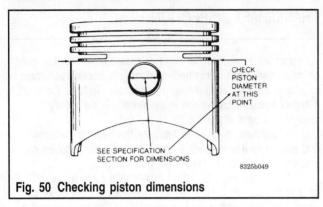

Fig. 50 Checking piston dimensions

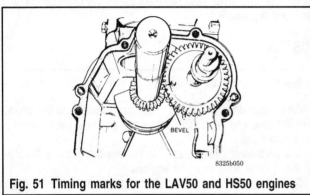

Fig. 51 Timing marks for the LAV50 and HS50 engines

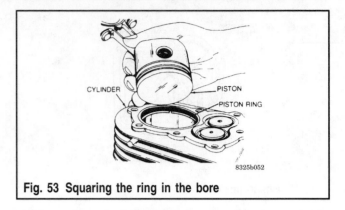

Fig. 53 Squaring the ring in the bore

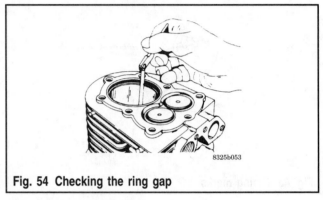

Fig. 54 Checking the ring gap

parts. If the piston is assembled in the bore 180° out of position, it will cause immediate binding of the parts.

PISTON RINGS

▶ See Figures 52, 53 and 54

Always replace the piston rings in sets. Ring gaps must be staggered. When using new rings, wipe the cylinder wall with fine emery cloth to deglaze the wall. Make sure the cylinder wall is thoroughly cleaned after deglazing. Check the ring gap by placing the ring squarely in the center of the area in which the rings travel and measuring the gap with a feeler gauge. Do not spread the rings too wide when assembling them to the pistons. Use a ring leader to install the rings on the piston.

The top compression ring has an inside chamfer. This chamfer must go UP. If the second ring has a chamfer, it must also

face UP. If there is a notch on the outside diameter of the ring, it must face DOWN.

Check the ring gap on the old ring to determine if the ring should be replaced. Check the ring gap on the new ring to determine if the cylinder should be rebored to take oversize parts.

➡Make sure that the ring gap is measured with the ring fitted squarely in the worn part of the cylinder where the ring usually rides up and down on the piston.

CONNECTING RODS

▶ See Figures 55, 56 and 57

Be sure that the match marks align when assembling the connecting rods to the crankshaft. Use new self-locking nuts. Whenever locking tabs are included, be sure that the tabs lock the nuts securely. NEVER try to straighten a bent crankshaft

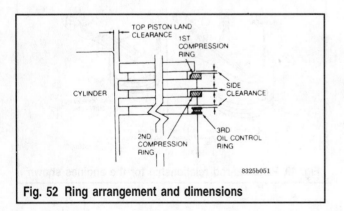

Fig. 52 Ring arrangement and dimensions

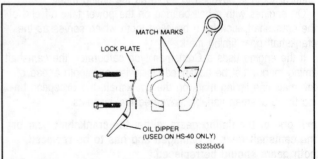

Fig. 55 Connecting rod for the LAV40 and HS40 engines

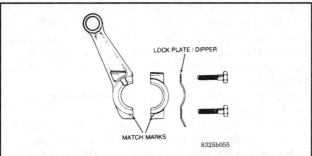

Fig. 56 Connecting rod for the V70, V80, H70, and H80 engines

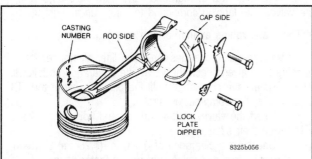

Fig. 57 Connecting rod and piston assembly for the V80 and H80 engines

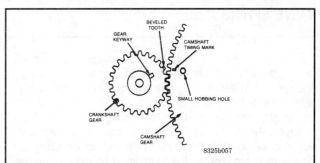

Fig. 58 Timing mark alignment for camshaft removal on VM80 and VM100 engines

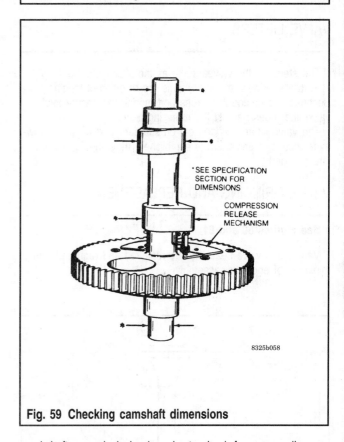

Fig. 59 Checking camshaft dimensions

or connecting rod. Replace them if necessary. When replacing either the piston, rod, crankshaft, or camshaft, liberally lubricate all bearings with engine oil before assembly.

The following engines have offset connecting rods: LAV40, LAV50, HS40, HS50. The LAV40, LAV50, HS40, HS50 engines have the caps fitted from opposite to the camshaft side of the engine.

On engine with Durlock rod bolts, torque the bolts as follows:

- LAV25-50, H25, 35, HS40-50, 110 inch lbs.
- V50, H50, VH50, HH50, 150 inch lbs.

➡**Early type caps can be distorted if the cap is not held to the crankpin while threading the bolts tight. Undue force should not be used.**

Later rods have serrations which prevent distortion during tightening. They also have match marks which must face out when assembling the rod. On the LAV50 and the HS50 engines, the piston must be fitted to the rod with the arrow on the top of the piston pointing to the right and the match marks on the rod facing you when the piston is pinned to the rod.

CAMSHAFT

▶ **See Figures 58 and 59**

Before removing the camshaft, align the timing marks to relieve the pressure on the valve lifters, on most engines.

In installation, align the gears for this type of camshaft as they were right before removal. After installation, turn the crankshaft gear clockwise in order to check for proper alignment of timing marks.

Clean the camshaft in solvent, then blow the oil passages dry with compressed air. Replace the camshaft if it shows wear of evidence of scoring. Check the cam dimensions against those in the chart.

If the engine has a mechanical fuel pump, it may have to be removed to properly reinstall the camshaft. If the engine is equipped with the Insta-matic Ezee-Start Compression Release, and any of the parts have to be replaced due to wear or damage, the entire camshaft must be replaced. Be sure that the oil pump (if so equipped) barrel chamber is toward the fillet of the camshaft gear when assembled.

➡**If a damaged gear is replaced, the crankshaft gear should also be replaced.**

VALVE SPRINGS

The valve springs should be replaced whenever an engine is overhauled. Check the free length of the springs. Comparing one spring with the other can be a quick check to notice any differences. If a difference is noticed, carefully measure the free length, compression length, and strength of each spring.

Some valve springs use dampening coils — coils that are wound closer together than most of the coils of the spring. Where these are present, the spring must be mounted so the dampening coils are on the stationary (upper) end of the spring.

VALVE LIFTERS

The stems of the valves serve as the lifters. On the 4 hp light frame models, the lifter stems are of different lengths. Because this engine is a cross port model, the shorter intake valve lifter goes nearest the mounting flange.

The valve lifters are identical on standard port engines. However once a wear pattern is established, they should not be interchanged.

VALVE GRINDING AND REPLACEMENT

▶ **See Figures 60 and 61**

Valves and valve seats can be removed and reground with a minimum of engine disassembly.

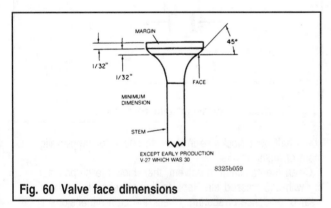

Fig. 60 Valve face dimensions

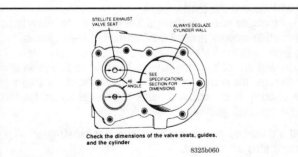

Fig. 61 Checking the dimensions of the valve seats, guides and cylinder

Remove the valves as follows:

1. Raise the lower valve spring caps while holding the valve heads tightly against the valve seat to remove the valve spring retaining pin. This is best achieved by using a valve spring compressor. Remove the valves, springs, and caps from the crankcase.

2. Clean all parts with a solvent and remove all carbon from the valves.

3. Replace distorted or damaged valves. If the valves are in usable condition, grind the valve faces in a valve refacing maching and to the angle given in the specifications chart at the end of this section. Replace the valves if the faces are ground to less than 1/32 in. (0.8mm).

4. Whenever new or reground valves are installed, lap in the valves with lapping compound to insure an air-tight fit.

➡**There are valves available with oversize stems.**

5. Valve grinding changes the valve lifter clearance. After grinding the valves check the valve lifter clearance as follows:

   a. Rotate the crankshaft until the piston is set at the TDC position of the compression stroke.

   b. Insert the valves in their guides and hold the valves firmly on their seats.

   c. Check for a clearance of 0.010 in. (0.25mm) between each valve stem and valve lifter with a feeler gauge.

   d. Grind the valve stem in a valve resurfacing machine set to grind a perfectly square face with the proper clearance.

6. Install valves as follows on Early Models:

   a. Position the valve spring and upper and lower valve spring caps under the valve guides for the valve to be installed.

   b. Install the valves in the guides, making sure that the valve marked **EX** is inserted in the exhaust port. The valve stem must pass through the valve spring and the valve spring caps.

   c. Insert the blade of a screwdriver under the lower valve spring cap and pry the spring up.

   d. Insert the valve pin through the hole in the valve stem with a long nosed pliers. Make sure the valve pin is properly seated under the lower valve spring cap.

7. Install the valves as follows on Later Models:

   a. Position the valve caps and spring in the valve compartment.

   b. Install the valves in guides with the valve marked **EX** in exhaust port. The valve stem must pass through the upper valve cap and spring. The lower cap should sit around the valve lifter exposed end.

   c. Compress the valve spring so that the shank is exposed. DO NOT TRY TO LIFT THE LOWER CAP WITH THE SPRING.

   d. Lift the lower valve cap over the valve stem shank and center the cap in the smaller diameter hole.

   e. Release the valve spring tension to lock the cap in place.

REBORING THE CYLINDER

1. First, decide whether to rebore for 0.010 in. (0.25mm) or 0.020 in. (0.50mm).

2. Use any standard commercial hone of suitable size. Chuck the hone in the drill press with the spindle speed of about 600 rpm.

3. Start with coarse stones and center the cylinder under the press spindle. Lower the hone so the lower end of the stones contact the lowest point in the cylinder bore.

4. Rotate the adjusting nut so that the stones touch the cylinder wall and then begin honing at the bottom of the cylinder. Move the hone up and down at a rate of 50 strokes a minute to avoid cutting ridges in the cylinder wall. Every fourth or fifth stroke, move the hone far enough to extend the stones 1 in. (25mm) beyond the top and bottom of the cylinder bore.

5. Check the bore size and straightness every thirty or forty strokes. If the stones collect metal, clean them with a wire brush each time the hone is removed.

6. Hone with coarse stones until the cylinder bore is within 0.002 in. (0.05mm) of the desired finish size. Replace the coarse stones with burnishing stones and continue until the bore is to within 0.0005 in. (0.0127mm) of the desired size.

7. Remove the burnishing stones and install finishing stones to polish the cylinder to the final size.

8. Clean the cylinder with solvent and dry it thoroughly.

9. Replace the piston and piston rings with the correct oversize parts.

REBORING VALVE GUIDES

The valve guides are permanently installed in the cylinder. However, if the guides wear, they can be rebored to accommodate a $\frac{1}{32}$ in. (0.8mm) oversize valve stem. Rebore the valve guides in the following manner:

1. Ream the valve guides with a standard straight shanked hand reamer or a low speed drill press.

2. Redrill the upper and lower valve spring caps to accommodate the oversize valve stem.

3. Reassemble the engine, installing valves with the correct oversize stems in the valve guides.

REGRINDING VALVE SEATS

The valve seats need regrinding only if they are pitted or scored. If there are no pits or scores, lapping in the valves will provide a proper valve seat. Valve seats are not replaceable. Regrind the valve seats as follows:

1. Use a grinding stone or a reseater set to provide the proper angle and seal face dimensions.

2. If the seat is over $\frac{3}{32}$ in. (2.4mm) wide after grinding, use a 15° stone or cutter to narrow the face to the proper dimensions.

3. Inspect the seats to make sure that the cutter or stone has been held squarely to the valve seat and that the same dimension has been held around the entire circumference of the seat.

4. Lap the valves to the reground seats.

Engine Bearings

▶ See Figures 62, 63, 64, 65, 66, 67, 68 and 69

LIGHTWEIGHT ALUMINUM BEARING REPLACEMENT

The aluminum bearing must be cut out using the rough cut reamer and the procedures shown. Follow illustrated steps to install bronze bushings in place of the aluminum bearings.

LONG LIFE AND CAST IRON ENGINES WITHOUT BALL BEARINGS

The worn bronze bushing must be driven out before the new bushing can be installed. Follow illustrated steps to replace the main bearings on these units.

LONG LIFE AND CAST IRON ENGINES WITH BALL BEARING ON THE P.T.O. (POWER TAKE-OFF) SIDE OF THE CRANKSHAFT

The side cover containing the ball bearing must be removed and a substitute cover with either a new bronze bushing or aluminum bearing must be used instead. Follow illustrated steps.

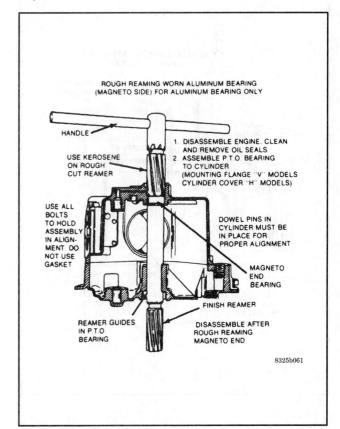

WORN BUSHING REMOVAL (MAGNETO END)
FOR BRONZE BUSHING ONLY

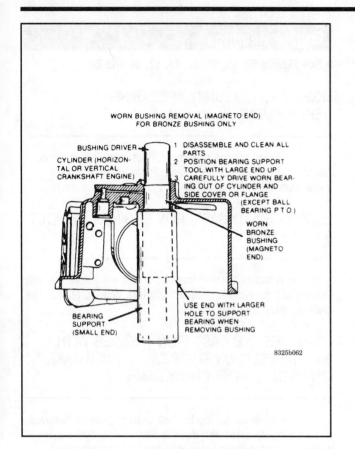

BUSHING DRIVER

CYLINDER (HORIZON-
TAL OR VERTICAL
CRANKSHAFT ENGINE)

1 DISASSEMBLE AND CLEAN ALL
  PARTS
2 POSITION BEARING SUPPORT
  TOOL WITH LARGE END UP
3 CAREFULLY DRIVE WORN BEAR-
  ING OUT OF CYLINDER AND
  SIDE COVER OR FLANGE
  (EXCEPT BALL
  BEARING P.T.O.)

WORN
BRONZE
BUSHING
(MAGNETO
END)

USE END WITH LARGER
HOLE TO SUPPORT
BEARING WHEN
REMOVING BUSHING

BEARING
SUPPORT
(SMALL END)

8325b062

FINISH REAMING NEW MAGNETO BUSHING

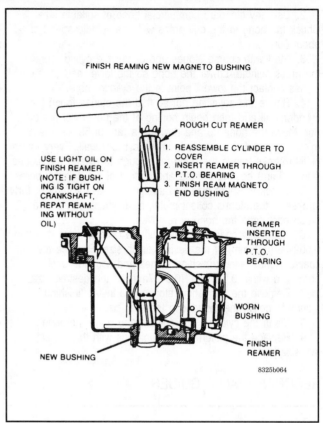

ROUGH CUT REAMER

USE LIGHT OIL ON
FINISH REAMER.
(NOTE: IF BUSH-
ING IS TIGHT ON
CRANKSHAFT,
REPAT REAM-
ING WITHOUT
OIL)

1. REASSEMBLE CYLINDER TO
   COVER
2. INSERT REAMER THROUGH
   P.T.O. BEARING
3. FINISH REAM MAGNETO
   END BUSHING

REAMER
INSERTED
THROUGH
P.T.O.
BEARING

WORN
BUSHING

FINISH
REAMER

NEW BUSHING

8325b064

INSTALLING NEW BRONZE BUSHING
(MAGNETO SIDE)

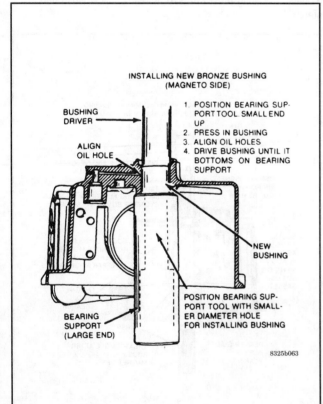

BUSHING
DRIVER

ALIGN
OIL HOLE

1. POSITION BEARING SUP-
   PORT TOOL SMALL END
   UP
2. PRESS IN BUSHING
3. ALIGN OIL HOLES
4. DRIVE BUSHING UNTIL IT
   BOTTOMS ON BEARING
   SUPPORT

NEW
BUSHING

BEARING
SUPPORT
(LARGE END)

POSITION BEARING SUP-
PORT TOOL WITH SMALL-
ER DIAMETER HOLE
FOR INSTALLING BUSHING

8325b063

ROUGH REAMING WORN ALUMINUM BEARING
(P.T.O. END) FOR ALUMINUM BEARING ONLY

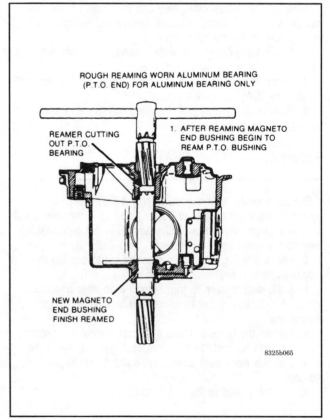

REAMER CUTTING
OUT P.T.O.
BEARING

1. AFTER REAMING MAGNETO
   END BUSHING BEGIN TO
   REAM P.T.O. BUSHING

NEW MAGNETO
END BUSHING
FINISH REAMED

8325b065

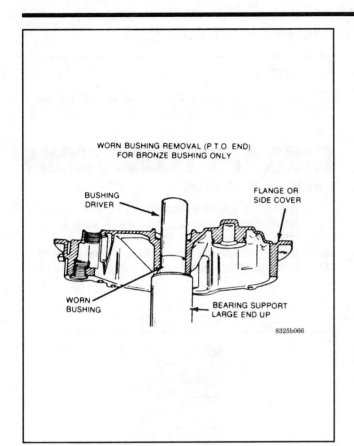

WORN BUSHING REMOVAL (P.T.O. END)
FOR BRONZE BUSHING ONLY

BUSHING DRIVER

FLANGE OR SIDE COVER

WORN BUSHING

BEARING SUPPORT LARGE END UP

8325b066

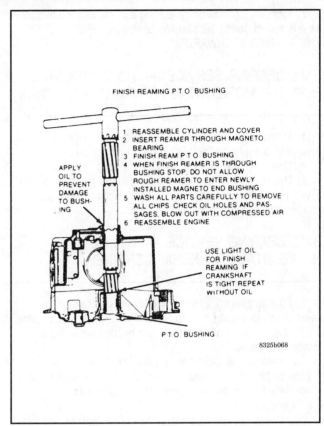

FINISH REAMING P.T.O. BUSHING

APPLY OIL TO PREVENT DAMAGE TO BUSHING

1 REASSEMBLE CYLINDER AND COVER
2 INSERT REAMER THROUGH MAGNETO BEARING
3 FINISH REAM P.T.O. BUSHING
4 WHEN FINISH REAMER IS THROUGH BUSHING STOP, DO NOT ALLOW ROUGH REAMER TO ENTER NEWLY INSTALLED MAGNETO END BUSHING
5 WASH ALL PARTS CAREFULLY TO REMOVE ALL CHIPS CHECK OIL HOLES AND PASSAGES. BLOW OUT WITH COMPRESSED AIR
6 REASSEMBLE ENGINE

USE LIGHT OIL FOR FINISH REAMING IF CRANKSHAFT IS TIGHT REPEAT WITHOUT OIL

P.T.O. BUSHING

8325b068

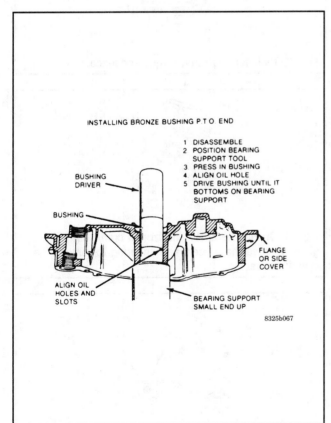

INSTALLING BRONZE BUSHING P.T.O. END

BUSHING DRIVER

BUSHING

ALIGN OIL HOLES AND SLOTS

1 DISASSEMBLE
2 POSITION BEARING SUPPORT TOOL
3 PRESS IN BUSHING
4 ALIGN OIL HOLE
5 DRIVE BUSHING UNTIL IT BOTTOMS ON BEARING SUPPORT

FLANGE OR SIDE COVER

BEARING SUPPORT SMALL END UP

8325b067

SPECIAL TOOLS

The task of main bearing replacement is made easier by using one of two Tecumseh main bearing tool kits. Kit 670161 is used to replace main bearings on the lightweight engines except HS, LAV40 and 50 models. Kit 670165 is used to replace main bearings on the medium weight engines except HS, LAV40 and 50 models.

GENERAL NOTES ON BUSHING REPLACEMENT

1. Your fingers and all parts must be kept very clean when replacing bushings.

2. On splash lubricated horizontal engines, the oil hole in the bushing is to be lined up with the oil hole that leads into the slot in the original bearing.

3. In the event it is necessary to replace the mounting flange or cylinder cover, the magneto end bearing must be rebushed. The P.T.O. bearing should also be rebushed to assure proper alignment.

4. Oil should be used to finish-ream the bushings. In the event the crankshaft does not rotate freely repeat the finish-reaming operation without oil.

5. Kerosene should be used as a cutting lubricant while rough-reaming.

6. Be sure that the dowel pins are in the cylinder block when assembling the mounting flange or cylinder cover. Use all bolts to hold the assembly together.

7. Remove the reamer by rotating it in the same direction as it is turned during the reaming operation. DO NOT TURN THE REAMER BACKWARDS.

BALL BEARING SERVICE — H20 THROUGH HS50 H.P. HORIZONTAL CRANKSHAFT ENGINES

1. Remove the crankshaft P.T.O. end oil seal. Drive an awl or similar tool into the metal seal body and pry out.
2. Use snap ring pliers to remove the snap ring.
3. Reassembly is in reverse order. Secure the cylinder cover, install the snap ring and oil seal. Protect the oil seal to prevent damage during installation.

BALL BEARING SERVICE — H40, 50 HORIZONTAL CRANKSHAFT ENGINES

1. Prior to attempting removal of the cylinder cover, observe the area around the crankshaft P.T.O. oil seal. Compare it with the illustration, and if there are bearing locks, follow instructions below:

a. Remove the locking nuts using the proper socket wrench. Note fiber washer located under nut; this must be reinstalled. Lift side cover from cylinder after removing the side cover bolts.

b. Install the bearing retainer bolts, fiber washer and locking nuts in the proper sequence in the cover.

2. Also note the following points:

a. On some engines, a locking type retainer bolt is used. To release the bolt, merely loosen the locking nut and turn the retainer bolt counterclockwise to the unlocked position with needle nose pliers to permit the side cover to be removed. Note that the flats on the retainer bolts must be turned so they face the crankshaft to be relocked upon installation. Don't force them! Torque the locking nuts only to 15-22 inch lbs.

b. The ball bearing used in horizontal crankshaft engines has a restricted fit. The bearing is heated and put onto the cold crankshaft. As the bearing cools it grasps the crankshaft tightly and must be removed cold. Remove the ball bearing with a bearing splitter (separator) and a puller. The bearing may be heated by placing it into a container with a sufficient amount of oil to cover the bearing. The bearing should not rest on the bottom of the container. Suspend the bearing on a wire or set the bearing onto a spacer block of wood or wire mesh. Heat the oil and bearing carefully until the oil smokes, quickly remove the bearing and slide it onto the crankshaft.

c. The bearing must seat tightly against the thrust washer which in turn rests tightly against the crankshaft gear.

d. When a ball bearing is used it is not possible to see the keyway in the crankshaft gear which is normally used for timing. Because of this, one tooth of the crankshaft gear is chamfered. This chamfered tooth of the crankshaft gear is positioned opposite the timing mark on the camshaft gear. The use of a ball bearing requires the removal of the crankshaft when it is necessary to remove the camshaft. When replacing the crankshaft and camshaft, mate the timing marks and insert it into the cylinder block as an assembly.

TORQUING THE CYLINDER HEAD

▶ See Figure 70

Torque the cylinder head to 200 inch lbs. in 4 equal stages of 50 inch lbs. Follow the sequence shown for each tightening stage.

Barrel and Plunger Oil Pump System

▶ See Figures 71, 72, 73 and 74

This system is driven by an eccentric on the camshaft. Oil is drawn through the hollow camshaft from the oil sump on its intake stroke. The passage from the sump through the camshaft is aligned with the pump opening. As the camshaft continues rotation (pressure stroke), the plunger force the oil

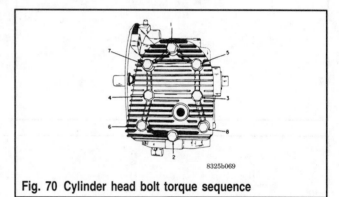

Fig. 70 Cylinder head bolt torque sequence

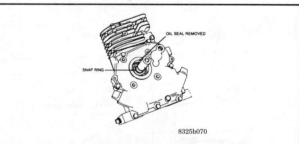

Fig. 71 On H20-HS50 H.P. horizontal crankshaft engines, remove the oil seal and snapring to remove the cylinder cover

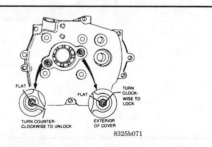

Fig. 72 On H40-HM100 H.P. horizontal crankshaft engines, locking and unlocking the bearing from the outside

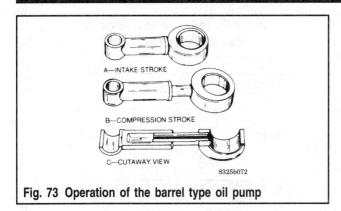

Fig. 73 Operation of the barrel type oil pump

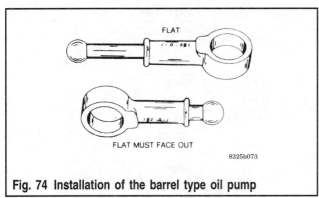

Fig. 74 Installation of the barrel type oil pump

out. The other port in the camshaft is aligned with the pump, and directs oil out of the top of the camshaft.

At the top of the camshaft, oil is forced through a crankshaft passage to the top main bearing groove which is aligned with the drilled crankshaft passage. Oil is directed through this passage to the crankshaft connecting rod journal and then spills from the connecting rod to lubricate the cylinder walls. Splash is used to lubricate the other parts of the engine.

A pressure relief port in the crank case relieves excessive pressures when the oil viscosity is extremely heavy due to cold temperatures, or when the system is plugged or damaged. Normal pressure is 7 psi.

SERVICE

▶ See Figures 75 and 76

Remove the mounting flange or the cylinder cover, whichever is applicable. Remove the barrel and plunger assembly and separate the parts.

Clean the pump parts in solvent and inspect the pump plunger and barrel for rough spots or wear. If the pump plunger is scored or worn, replace the entire pump.

Before reassembling the pump parts, lubricate all of the parts in engine oil. Manually operate the pump to make sure the plunger slides freely in the barrel.

Lubricate all the parts and position the barrel on the camshaft eccentric. If the oil pump has a chamfer only on one side, that side must be placed toward the camshaft gear. The flat goes away from the gear, thus out to work against the flange oil pickup hole.

Install the mounting flange. Be sure the plunger ball seats in the recess in the flange before fastening it to the cylinder.

Spray Mist Lubrication

Late model LAV40, LAV30, and LAV35 engines have a spray mist lubrication system. This system is the same as the barrel and plunger oil pump system except that (1) the pressure relief port is changed to a calibrated spray mist orifice and (2) the crankshaft is not rifle-drilled from the top main to the crank pin. Lubrication is sprayed to the narrow rod cap area through the spray mist hole.

Splash Lubrication
▶ See Figure 75

Some engines utilize the splash type lubrication system. The oil dipper, on some engines, is cast onto the lower connecting rod bearing cap. It is important that the proper parts are used to ensure the longest engine life.

Gear Type Oil Pump System
▶ See Figure 76

The gear type lubrication pump is a crankshaft driven, positive displacement pump. It pumps oil from the oil sump in the engine base to the camshaft, through the drilled camshaft passage to the top main bearing, through the drilled crankshaft, to the connecting rod journal on the crankshaft.

Spillage from the connecting rod lubricates the cylinder walls and normal splash lubricates the other internal working parts. There is a pressure relief valve in the system.

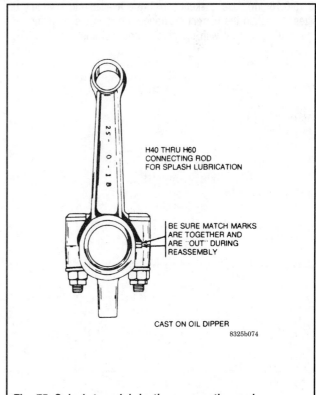

Fig. 75 Splash type lubrication connecting rod

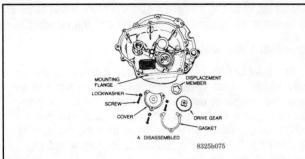

Fig. 76 Disassembled view of the gear type lubrication system

SERVICE

Disassemble the pump as follows: remove the screws, lockwashers, cover, gear, and displacement member.

Wash all of the parts in solvent. Inspect the oil pump ddive gear and displacement member for worn or broken teeth, scoring, or other damage. Inspect the shaft hole in the drive gear for wear. Replace the entire pump if cracks, wear, or scoring is evident.

To replace the oil pump, position the oil pump displacement member and oil pump gear on the shaft, then flood all the parts with oil for priming during the initial starting of the engine.

The gasket provides clearance for the drive gear. With a feeler gauge, determine the clearance between the cover and the oil pump gear. The clearance desired is 0.006-0.007 in. (0.15-0.18mm). Use gaskets, which are available in a variety of sizes, to obtain the correct clearance. Position the oil pump cover and secure it with the screws and lockwashers.

Cross Reference for Vertical Crankshaft Engines

Size	Model	Column
2½ HP	LAV25	1
3HP	LAV30	1
	TVS75	1
3½ HP	LV35	2
	LAV35	2
	TVS90	2
4 HP	LAV40	3
	TVS105	3
	V40 thru V40B	7
	VH40	7
10.0 CI	ECV100	4
	TNT100	14
10.5 CI	ECV105	5
11.0 CI	ECV110	12
5 HP	LAV50	6
	TVS120	6
	V50	9
	VH50	9
12.0 CI	ECV120	13
	TNT120	15

8325b0C5

Cross Reference Chart for Horizontal Crankshaft Engines

Size	Model	Column
2½ HP	H25	1
3 HP	H30	1
3½ HP	H35	2
9.0 CI	ECH90	2
4 HP	H40	7
	HH40	7
	HS40	3
5 HP	H50	9
	HH50	9
	HS50	6

8325b0C6

Craftsman Engines Cross Reference Chart

Craftsman Engine Models	See Column	Craftsman Engine Models	See Column	Craftsman Engine Models	See Column
143.50040	7	143.196042–143.196072	10	143.224022	
143.50045		143.196082	8	143.224032	4
				143.224062	2
143.131022–143.131102	1	143.197012	3	143.224072	4
143.135012–143.135112	9	143.197022	5	143.224092–143.224132	2
143.136012–143.136052	8	143.197032		143.224142	1
143.137012	7	143.197042–143.197072	3	143.224162–143.224222	2
143.137032		143.197082	5	143.224232	4
				143.224242	
143.141012–143.143032	1	143.201032–143.203012	1	143.224252–143.224282	2
143.145012–143.145072	9	143.204022	4	143.224292	4
143.146012	8	143.204032–143.204052	2	143.224302	
143.146022		143.204062	4	143.224312–143.224342	2
143.147012–143.147032	7	143.204072–143.204092	2	143.224352	4
		143.204102	4	143.224362	
143.151012–143.153032	1	143.204132		143.224372–143.224422	2
143.154012–143.154142	2	143.204142–143.204192	2	143.224432	4
143.155012–143.155062	9	143.204202	4	143.225012	13
143.156012	8	143.205022	9	143.225022	
143.156022		143.206012	8	143.225032–143.225052	9
143.157012–143.157032	7	143.206022	10	143.225062	13
		143.206032	8	143.225072	
143.161012–143.163062	1	143.207012–143.207052	3	143.225082–143.225102	9
143.164012–143.164202	2	143.207062	5	143.226012	8
143.165012–143.165052	9	143.207072	3	143.226032	
143.166012–143.166052	8	143.207082	5	143.226072	10
143.167012–143.167042	7			143.226082	
		143.213012–143.213042	1	143.226092–143.226122	11
143.171012–143.171172	1	143.214012–143.214032	2	143.226131–143.226182	8
143.171202	2	143.214042–143.214072	4	143.226192	11
143.171212–143.173042	1	143.214082–143.214252	2	143.226202	10
143.174012–143.174292	2	143.214262–143.214282	4	143.226212	
143.175012–143.175072	9	143.214292	2	143.226222–143.226262	8
143.176012–143.176092	8	143.214302		143.226272	10
143.177012–143.177072	7	143.214312	4	143.226282	
		143.214322		143.226292	11
143.181042–143.183042	1	143.214332	2	143.226302	10
143.184012–143.184212	2	143.214342		143.226312	11
143.184232–143.184252	4	143.214352	4	143.226322	8
143.184262–143.184402	2	143.216012–143.216032	10	143.226332	
143.185012–143.185052	9	143.216042–143.216062	3	143.226342	10
143.186012	8	143.216072–143.216092	10	143.226352	11
143.186022–143.186042	10	143.216122	8	143.227012–143.227072	12
143.186052	8	143.216132	11		
143.186062		143.216142	8	143.233012	1
143.186072–143.186112	10	143.216152	11	143.233032	
143.186122	8	143.216162		143.233042	
143.187022–143.187102	3	143.216172		143.234022–143.234052	2
		143.216182	8	143.234062–143.234092	4
143.191012–143.191052	1	143.217012–143.217032	5	143.234102–143.234162	2
143.194012–143.194052	2	143.217042–143.217072	3	143.234192	
143.194062	4	143.217092	5	143.234202	
143.194072–143.194092	2	143.217102	3	143.234212–143.234232	4
143.194102	4			143.234242–143.234262	2
143.194112–143.194142	2	143.223012–143.223052	1	143.235012	13
143.195012	9	143.224012	2		
143.195022					
143.196012–143.196032	8				

8325bC17

Craftsman Engines Cross Reference Chart (cont.)

Craftsman Engine Models	See Column	Craftsman Engine Models	See Column	Craftsman Engine Models	See Column
143.235022		143.246242	10	143.264422	2
143.235032	6	143.246252	11	143.264432–143.264482	4
143.235042	13	143.246262	10	143.264492	2
143.235052		143.246272–143.247292	11	143.264502	
143.235062	9	143.246302	10	143.264512	4
143.235072	6	143.246312		143.264522	2
143.236012	8	143.246322	11	143.264542	
143.236022–143.236042	11	143.246332		143.264562–143.264672	4
143.236052	8	143.246342	10	143.264682	2
143.236062	11	143.246352	8	143.265012–143.265192	6
143.236072		143.246362	16	143.266012	11
143.236082	8	143.246382		143.266022	
143.236092	10	143.246392	8	143.266032	8
143.236102	8			143.266042	10
143.236112		143.254012–143.254052	2	143.266052	
143.236122	10	143.254062	4	143.266062	8
143.236132	8	143.254072–143.254122	2	143.266082	
143.236142	11	143.254142–143.254192	4	143.266092–143.266132	10
143.236152	8	143.254212	2	143.266142–143.266242	11
143.237012	12	143.254222		143.266252	8
143.237022		143.254232–143.254292	4	143.266262	11
143.237032	5	143.254302	2	143.266272–143.266302	10
143.237042	3	143.254312		143.266312	11
		143.254322	4	143.266322	
143.244032	2	143.254332	2	143.266332	10
143.244042	4	143.254342	4	143.266342	11
143.244052		143.254352		143.266352	10
143.244062		143.254362	2	143.266362	11
143.244072–143.244112	2	143.254372	4	143.266372–143.266412	8
143.244122–143.244142	4	143.254382		143.266422	11
143.244202	2	143.254392	2	143.266432–143.266452	8
143.244212	4	143.254402	4	143.266462	16
143.244222	2	143.254412		143.266472	
143.244232		143.254432	2	143.266482	11
143.244242	4	143.254442	4	143.267012–143.267042	3
143.244252	4	143.254452	2		
143.244262–143.244282	2	143.254462	4	143.274022–143.274072	4
143.244292–143.244332	4	143.254472	2	143.274092–143.274132	2
143.245012	6	143.254482		143.274142	4
143.245042	9	143.254492	4	143.274152	
143.245052–143.245072	13	143.254502–143.254532	2	143.274162–143.274182	2
143.245082		143.255012–143.255112	6	143.274192–143.274242	4
143.245092	6	143.256012	11	143.274252	2
143.245102–143.245132	13	143.256022	8	143.274262	4
143.245142	6	143.256032	10	143.274272–143.274322	2
143.245152		143.256042	11	143.274402–143.274482	4
143.245162	13	143.256052	8	143.275012–143.275052	6
143.245172	6	143.256062	11	143.276022	4
143.245182		143.256072		143.276032	11
143.245192	13	143.256082	8	143.276042	
143.246012	8	143.256092		143.276052	16
143.246022	11	143.256102	10	143.276062–143.276162	11
143.246032		143.256112	11	143.276182	8
143.246042	8	143.256122	8	143.276192	10
143.246052–143.246072	10	143.256132	10	143.276202	8
143.246082	11	143.257012–143.257072	3	143.276222	10
143.246092				143.276242	11
143.246102	10	143.264012–143.264042	2	143.276252	8
143.246112		143.264052–143.264082	4	143.276262	11
143.246122	11	143.264092	2	143.276272	11
143.246132	10	143.264102	4	143.276282	10
143.246142		143.264232–143.264342	2	143.276292	11
143.246152–143.246212	11	143.264352–143.264372	4	143.276302	11
143.246222	10	143.264382	2	143.276322–143.276342	10
143.246232	11	143.264392–143.264412	4	143.276352	11

8325bC18

Craftsman Engines Cross Reference Chart (cont.)

Craftsman Engine Models	See Column	Craftsman Engine Models	See Column	Craftsman Engine Models	See Column
143.276362	16	143.531052	1	143.596022	
143.276372 – 143.276392	10	143.531082		143.596042	8
143.276402	16	143.531122		143.596052	10
143.276412	8	143.531132		143.596072 – 143.596122	8
143.276422	10	143.531142	2	143.597012 – 143.597032	3
143.276432 – 143.276472	11	143.531152	1		
143.276482	16	143.531172		143.601022 – 143.601062	1
143.277012	3	143.531182		143.604012	2
143.277022		143.534012 – 143.534072	2	143.604022	4
		143.535012 – 143.535062	9	143.604032	2
143.284012	2	143.536012 – 143.536062	8	143.604042	
143.284022	1	143.537012	7	143.604052	4
143.284032	2			143.604062	2
143.284042	4	143.541012	1	143.604072	
143.284052	2	143.541042 – 143.541062	1	143.605012	9
143.284062		143.541112 – 143.541152	1	143.605022	
143.284072	4	143.541172 – 143.541202	1	143.605052	
143.284082	2	143.541222	1	143.606012 – 143.606052	10
143.284092		143.541282 – 143.541302	1	143.606092	8
143.284102	4	143.544012 – 143.544042	2	143.606102	10
143.284112	2	143.545012 – 143.545042	9	143.607012 – 143.607032	3
143.284142		143.546012 – 143.546022	8	143.607042 – 143.607062	3
143.284152		143.547012 – 143.547032	7		
143.284162				143.611012 – 143.611112	1
143.284182		143.551012	1	143.614012 – 143.614032	4
143.284212	4	143.551032		143.614042	2
143.284312	2	143.551052 – 143.551192	1	143.614052	4
143.284322		143.554012 – 143.554082	2	143.614062 – 143.614162	2
143.284332	4	143.555012 – 143.555052	9	143.615012 – 143.615092	9
143.284342		143.556012 – 143.556282	8	143.616012	10
143.284352		143.557012 – 143.557082	7	143.616022 – 143.616112	8
143.284372	16			143.616122	10
143.284382	4	143.565022	9	143.616132	8
143.284402	2	143.566002 – 143.566202	8	143.616142	
143.284412		143.566212	9	143.617012 – 143.617182	3
143.284432	2	143.566222 – 143.566252	8		
143.284362		143.567012 – 143.567042	7	143.621012 – 143.621092	1
143.284392	2			143.624012 – 143.624112	2
143.284442		143.571002 – 143.571122	1	143.625012 – 143.625132	9
143.284482		143.571152	2	143.626012	10
143.284452	4	143.571162	1	143.626022	8
143.284472		143.571172		143.626032	10
143.285012	6	143.574022 – 143.574102	2	143.626042	8
143.285022		143.575012 – 143.575042	9	143.626052 – 143.626122	10
143.285032		143.576002 – 143.576202	8	143.626132	8
143.286102	16			143.626142	10
143.286022	17	143.581002 – 143.581102	1	143.626152	
143.286032	10	143.584012 – 143.584142	2	143.626162	8
143.286072 – 143.286092	11	143.585012 – 143.585042	9	143.626172	10
143.286112		143.586012 – 143.586042	8	143.626182	8
143.286122		143.586052 – 143.586062	10	143.626192	10
143.286132	10	143.586072 – 143.586082	8	143.626202	8
143.286142	11	143.586112 – 143.586142	10	143.626212	10
143.286152		143.586152	8	143.626222 – 143.626262	8
143.286162		143.586162	10	143.626282	11
143.286172		143.586172 – 143.586242	8	143.626292	10
		143.586252	10	143.626302	8
143.505010	8	143.586262 – 143.586282	8	143.626312	10
143.505011		143.587012 – 143.587042	3	143.626322	
143.521081	9			143.627012 – 143.627042	3
143.525021	9	143.591012 – 143.591142	1		
143.526011		143.594022 – 143.594082	2	143.631012 – 143.631092	1
143.526021	8	143.594092	2	143.634012	2
143.526031	8	143.594102		143.634032	
		143.595012	9	143.635012	9
		143.595042		143.635022	
		143.596012	10		

8325bC19

Craftsman Engines Cross Reference Chart (cont.)

Craftsman Engine Models	See Column	Craftsman Engine Models	See Column	Craftsman Engine Models	See Column
143.635032	6	143.656202	8	143.674012	2
143.635052	9	143.656212	11	143.675012	6
143.636012	11	143.656222		143.675022	
143.636022		143.656232	10	143.675032	9
143.636032	10	143.656242	11	143.675042	6
143.636042	11	143.656252	8	143.676012	11
143.636052	8	143.656262	10	143.676022	
143.636062	10	143.656272		143.676032	10
143.636072	11	143.656282	11	143.676042	11
143.637012	3	143.657012–143.657052	3	143.676052	
				143.676062	16
143.641012–143.641062	1	143.661012–143.661062	1	143.676072	
143.641072	2	143.664012–143.664332	2	143.676102	10
143.644012–143.644082	2	143.665012–143.665082	6	143.676112	8
143.645012–143.645032	6	143.666012	10	143.676122	10
143.646012–143.646032	10	143.666022		143.676132	8
143.646042	11	143.666032	11	143.676142	11
143.646052		143.666042–143.666072	10	143.676152	16
143.646072–143.646102	10	143.666082	11	143.676162	
143.646112	8	143.666092		143.676172	10
143.646122	10	143.666102–143.666142	8	143.676182	11
143.646132		143.666152	11	143.676192	10
143.646142	11	143.666162		143.676202	11
143.646152	10	143.666172	8	143.676212	16
143.646162	11	143.666202		143.676222	11
143.646172	10	143.666222	10	143.676232	8
143.646182		143.666232	11	143.676242	8
143.646192	8	143.666242	8	143.676252	11
143.646202	10	143.666252	10	143.676262	16
143.646212–143.646232	11	143.666272	8	143.677012	3
143.647012–143.647062	3	143.666282	10	143.677022	
		143.666292	8		
143.651012–143.651072	1	143.666302	10		
143.654022–143.654322	2	143.666312		143.686012	17
143.655012	6	143.666322	11	143.686022	
143.655032		143.666332	16	143.686032	11
143.656012–143.656052	8	143.666342	10	143.686042	
143.656062	10	143.666352	11	143.686052	
143.656082	11	143.666362	16	143.686062	10
143.656092	8	143.666372	8	143.686072	9
143.656102	10	143.666382	10	143.687012	3
143.656112	8	143.667012	3		
143.656122–143.656152	10	143.667022			
143.656162–143.656182	8	143.667032	6	143.694126	4
143.656192	10	143.667042–143.667082	3	143.694132	2
				143.694134	11

8325bC20

Engine Specifications

Reference Column	1	2	3	4	5	6	7	8	9	10	11	12
Displacement	7.75	9.06	10.5	10.0	10.5	12.0	11.04	12.17	11.5	12.0	10.0	12.0
Stroke	1²⁷/₃₂"	1²⁷/₃₂"	1¹⁵/₁₆"	1²⁷/₃₂"	1¹⁵/₁₆"	1¹⁵/₁₆"	2¼"	2¼"	1¹⁵/₁₆"	1¹⁵/₁₆"	1²⁷/₃₂"	1¹⁵/₁₆"
Bore	2.3125 2.3135	2.5000 2.5010	2.625 2.626	2.625 2.626	2.625 2.626	2.812 2.813	2.5000 2.5010	2.625 2.626	2.750 2.751	2.812 2.813	2.625 2.626	2.812 2.813
Timing Dimension Before Top Dead Center for Vertical Engines	V.060 .070	V.065	V.035	.035	.035	V.040 .060	V.050	H.050	V.035	V.035	V.035	V.035
Timing Dimension Before Top Dead Center for Horizontal Engines	H.060 .070	H.030 .040	H.035			H.055	H.050	H.050				
Point Setting	.020	.020	.020	.020	.020	.020	.020	.020	.020	.020	.020	.020
Spark Plug Gap	.030	.030	.030	.030	.030	.030	.030	.030	.030	.030	.030	.030
Valve Clearance	.010 Both	.010 Both	.010 Both	.010 Both	.010 Both	.010 Both	.010 Both	.010 Both	.010 Both	.010 Both	.010 Both	.010 Both
Valve Seat Angle	46°	46°	46°	46°	46°	46°	46°	46°	46°	46°	46°	46°
Valve Spring Free Length	1.135"	1.135"	1.135"	1.135"	1.135"	1.135"	1.562"	1.462"	1.135"	1.135"	1.135"	1.135"
Valve Guides Over-Size Dimensions	.2805 .2815	.2805 .2815	.2805 .2815	.2805 .2815	.2807 .2817	In. .280 Ex. .278	.3432 .3442	.343 .344	.2805 .2815	.2805 .2815	.2805 .2815	.2805 .2815
Valve Seat Width	.035 .045	.035 .045	.035 .045	.035 .045	.035 .045	.035 .045	.042 .052	.042 .052	.035 .045	.035 .045	.035 .045	.035 .045
Crankshaft End Play	.005 .027	.005 .027	.005 .027	.005 .027	.005 .027	.005 .027	.005 .027	.005 .027	.005 .027	.005 .027	.005 .027	.005 .027
Crankpin Journal Diameter	.8610 .8615	.8610 .8615	.9995 1.0000	.8610 .8615	.9995 1.0000	.9995 1.0000	1.0615 1.0620	1.0615 1.0620	.9995 1.0000	.9995 1.0000	.8610 .8615	.9995 1.0000
Cylinder Main Bearing Dia.	.8755 .8760	.8755 .8760	1.0005 1.0010	.8755 .8760	1.0005 1.0010	1.0005 1.0010	1.0005 1.0010	1.0005 1.0010	1.0005 1.0010	1.0005 1.0010	.8755 .8760	1.0005 1.0010

8325bC11

Specification												
Cylinder Cover Main Bearing Dia.	.8755/.8760	.8755/.8760	1.0005/1.0010	.8755/.8760	1.2010/1.2020	1.0005/1.0010	1.0005/1.0010	1.0005/1.0010	1.0005/1.0010	1.0005/1.0010	.8755/.8760	1.0005/1.0010
Conn. Rod. Dia. Crank Bearing	.8620/.8625	.8620/.8625	1.0005/1.0010	.8620/.8625	1.0005/1.0010	1.0005/1.0010	1.0630/1.0635	1.0630/1.0635	1.0005/1.0010	1.0005/1.0010	.8620/.8625	1.0005/1.0010
Piston Diameter	2.3090/2.3095	2.4950/2.4955	2.6260/2.6205	2.6200/2.6205	2.604/2.608	2.8070/2.8075	2.492/2.4945	2.6210/2.6215	2.7450/2.7455	2.8070/2.8075	2.6200/2.6205	2.8070/2.8075
Piston Pin Diameter	.5629/.5631	.5629/.5631	.5629/.5631	.5629/.5631	.5631/.5635	.5629/.5631	.6248/.6250	.6248/.6250	.5629/.5631	.5629/.5631	.5629/.5631	.5629/.5631
Width of Comp. Ring Groove	.0955/.0977	.0955/.0975	.0925/.0935	.0955/.0975	.0955/.0975	.0955/.0975	.0955/.0975	.0955/.0975	.0795/.0815	.0955/.0975	.0955/.0975	.0955/.0975
Width of Oil Ring Grooves	.125/.127	.125/.127	.156/.158	.156/.158	.156/.158	.156/.158	.156/.158	.156/.158	.1565/.1585	.1565/.1585	.1565/.1585	.1565/.1585
Side Clearance of Ring Groove (Top) Comp.	.002/.005		.002/.004	.002/.005	.002/.005	.003/.004	.002/.003	.003/.004	.002/.004	.003/.004	.003/.0045	.0028/.0039
Side Clearance of Ring Groove (Bot.) Oil		.002/.003	.001/.004	.001/.004	.001/.004	.002/.003	.002/.003	.002/.004	.001/.002	.001/.002	.0010/.0030	.0018/.0038
Ring End Gap	.007/.020	.007/.020	.007/.020	.007/.020	.007/.020	.007/.020	.007/.020	.007/.020	.007/.020	.007/.020	.007/.020	.007/.020
Top Piston Land Clearance	.0015/.0145	.015/.018	.0165/.0215	.017/.022	.017/.022	.017/.022	.015/.018	.017/.020	.024/.027	.018/.021	.017/.022	.017/.022
Piston Skirt Clearance	.0025/.0040	.0045/.0060	.0045/.0060	.0045/.0060	.0050/.0065	.0045/.0060	.0055/.0070	.0035/.0050	.0045/.0060	.0045/.0060	.0045/.0060	.0045/.0060
Camshaft Bearing Dia.	.4975/.4980	.4975/.4980	.4975/.4980	.4975/.4980	.505/.513	.4975/.4980	.6230/.6235	.6230/.6235	.4975/.4980	.4975/.4980	.4975/.4980	.4975/.4980
Dia. of Crankshaft Mag. Main Brg.	.8735/.8740	.8735/.8740	.9985/.9990	.8735/.8740	.9985/.9990	.9985/.9990	.9985/.9990	.9985/.9990	.9990/.9995	.9990/.9995	.8735/.8740	.9985/.9990
Dia. of Crankshaft P.T.O. Main Brg.	.8735/.8740	.8735/.8740	.9985/.9990	.8735/.8740	.9985/.9990	.9985/.9990	.9985/.9990	.9985/.9990	.9985/.9990	.9985/.9990	.8735/.8740	.9985/.9990

A. For VM80 & HM80 engines only - Displacement is 19.41"

B. For VM80 & HM80 engines only - Bore is 3.125" (3⅛").
   3.126"

Note C. For VM80 & HM80 engines only - Piston Diameter is 3.1205"
   3.1195"

6-12 HP

Engine Identification

▶ See Figures 77 and 78

General Engine Specifications
Vertical Crankshaft Engines

Model	Bore & Stroke	Displacement	Horsepower
V60	2.625 × 2.5000	13.53	6
VH60	2.625 × 2.5000	13.53	6
V70	2.750 × 2.5313	15.0	7
VH70	2.750 × 2.5313	15.0	7
VM70	2.750 × 2.5313	15.0	7
V80	3.062 × 2.5313	18.65	8
VM80	3.125 × 2.5313	19.41	8
VM100	3.187 × 2.5313	17.16	8
ECV100	2.625 × 1.8438	10.0	—
TNT100	2.625 × 1.8434	20.2	—
ECV105	2.625 × 1.9375	10.5	—
ECV110	2.750 × 1.9375	11.5	—
ECV120	2.812 × 1.9375	12.0	—
TNT120	2.812 × 1.9375	12.0	—

8325bC13

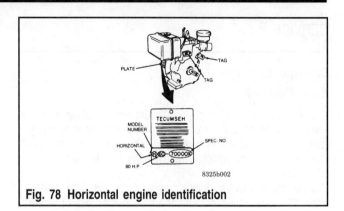

Fig. 78 Horizontal engine identification

General Engine Specifications
Horizontal Crankshaft Engines

Model	Bore & Stroke (in.)	Displacement (cu in.)	Horsepower
H60	2.6250 × 2.5000	13.53	6
HH60	2.6250 × 2.5000	13.53	6
H70	2.7500 × 2.5313	15.0	7
HH70	2.7500 × 2.5313	15.0	7
HM70	2.9375 × 2.5313	17.16	7
H80	3.0620 × 2.5313	18.65	8
HM80	3.0620 × 2.5313	18.65	8
HM100	3.1870 × 2.5313	20.2	10
ECH90	2.5000 × 1.8438	9.06	—

8325bC14

Lauson 4 cycle engines are identified by a model number stamped on a nameplate. The nameplate is located on the crankcase of vertical shaft models and on the blower housing of horizontal shaft models.

A typical model number appears on the illustration showing the location of the nameplate for vertical crankshaft engines. This number is interpreted as follows:

- V — vertical shaft engine
- 60 — 6.0 horsepower
- 70360J — the specification number. The last three numbers (360) indicate that this particular engine is a variation on the basic model line.
- 2361J — serial number
- 2 — year of manufacture
- 361 — the calendar day of manufacture
- J — line and shift location at the factory.

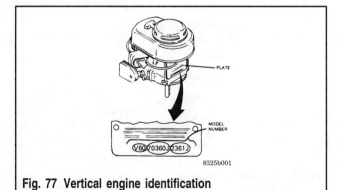

8325b001

Fig. 77 Vertical engine identification

Air Cleaner Service

SERVICE

Dry Paper Element
▶ **See Figure 79**

The Tecumseh treated paper element type air cleaner consists of a pleated paper element encased in a metal housing and must be replaced as a unit. A flexible tubing and hose clamps connect the remotely mounted air filter to the carburetor.

Clean the element by lightly tapping it. Do not distort the case. When excessive carburetor adjustment or loss of power results, inspect the air filter to see if it is clogged. Replacing a severely restricted air filter should show an immediate performance improvement.

A plain paper element is also used. It should be removed every 10 hours, or more often if the air is dusty. Tap or blow out the dirt from the inside with low pressure air. This type should be replaced at 50 hours. If clogged sooner, it may be washed in soap and water and rinsed by flushing from the inside until the water is clear. Blow dry with low pressure compressed air.

To service the KLEEN-AIRE system, remove the element, wash it in soap and mild detergent, pat dry, and then coat with oil. Squeeze the oil to distribute it evenly and remove the excess. Make sure all mounting surfaces are tight to prevent leakage.

Oil/Foam Air Cleaners

Clean and re-oil the air cleaner element every 25 hours of operation under normal operating conditions. The capacity of the oil/foam air cleaner is adequate for a full season's use without cleaning. Under very dusty conditions, clean the air cleaner every few hours of operation.

The oil/foam air cleaner is serviced in the following manner:

1. Remove the screw that holds the halves of the air cleaner shell together and retains it to the carburetor.

2. Remove the air cleaner carefully to prevent dirt from entering the carburetor.

3. Take the air cleaner apart (split the two halves).

4. Wash the foam in kerosene or liquid detergent and water to remove the dirt.

5. Wrap the foam in a clean cloth and squeeze it dry.

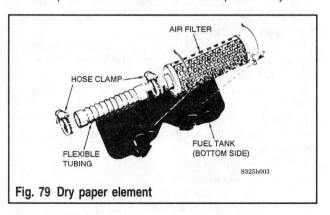

Fig. 79 Dry paper element

6. Saturate the foam in clean engine oil and squeeze it to remove the excess oil.

7. Assemble the air cleaner and fasten it to the carburetor with the attaching screw.

Oil Bath Air Cleaner
▶ **See Figure 80**

Pour the old oil out of the bowl. Wash the element thoroughly in solvent and squeeze it dry. Clean the bowl and refill it with the same type of oil used in the crankcase.

Lubrication

OIL AND FUEL RECOMMENDATIONS

Use fresh (less than one month old) gasoline, of 'Regular" grade. Unleaded fuel is preferred, but leader fuel is acceptable.

Use oil having SG classification. Use these viscosities for aluminum engines:

- Summer — above 32°F (0°C): S.A.E. 30 (S.A.E. 10W-30 or 10W-40 are acceptable substitutes)
- Winter — Below 32°F (0°C): S.A.E. 5W-30 (S.A.E. 10W is an acceptable substitute). (Including Snow King Snow Blower Engines)
- Winter — Below 0°F (-18°C) only: S.A.E. 10W diluted with 10% kerosene is an acceptable substitute. (Including Snow King Snow Blower Engines)Use these viscosities for cast iron engines:
- Summer — Above 32°F (0°C): S.A.E. 30
- Winter — Above 32°F (0°C): S.A.E. 10W

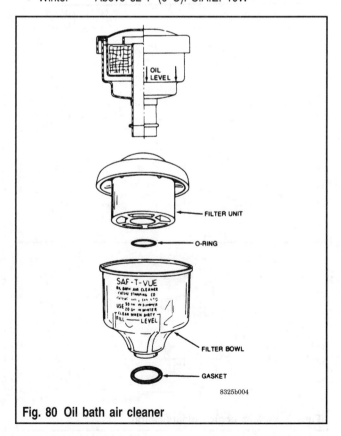

Fig. 80 Oil bath air cleaner

Tune-Up Specifications

The follow basic specifications apply to all the engines covered in this section:
- Spark Plug Gap: 0.030 in. (0.76mm)
- Ignition Point Gap: 0.020 in. (0.50mm)
- Valve Clearance: 0.010 in. (0.25mm) for both intake and exhaust. For timing dimension, which varies from engine to engine.

Spark Plug Service

▶ **See Figures 81, 82, 83, 84 and 85**

Spark plugs should be removed, cleaned, and adjusted periodically. Check the electrode gap with a wire feeler gauge and adjust the gap. Replace the plugs if the electrodes are pitted and burned or the porcelain is cracked. Apply a little graphite

83257091

Fig. 81 Twist and pull on the boot, never on the plug wire

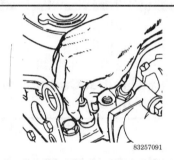

83257092

Fig. 82 Plugs that are in good condition can be filed and reused

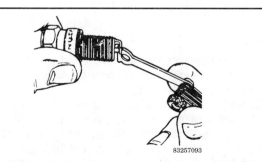

83257093

Fig. 83 Adjust the electrode gap by bending the side electrode

grease to the threads to prevent sticking. Be sure the cleaned plugs are free of all foreign material.

Breaker Points

ADJUSTMENT

1. Disconnect the fuel line from the carburetor.
2. Remove the mounting screws, fuel tank, and shroud to provide access to the flywheel.
3. Remove the flywheel with either a puller (over 3.5 hp) or by using a screwdriver to pry underneath the flywheel while tapping the top lightly with a soft hammer.
4. Remove the dust cover and gasket from the magneto and crank the engine over until the breaker points of the magneto are fully opened.
5. Check the condition of the points and replace them if they are burned or pitted.
6. Check the point gap with a feeler gauge. Adjust them, if necessary, as per the directions on the dust cover.

REPLACEMENT

1. Gain access to the points and inspect them as described above. If the points are badly pitted, follow the remaining steps to replace them.
2. Remove the nuts that hold the electrical leads to the screw on the movable breaker point spring. Remove the movable breaker point from stud.
3. Remove the screw and stationary breaker point. Put a new stationary breaker point on the breaker plate; install the screw, but do not tighten. This point must be moved to make the proper air gap when the points are adjusted.
4. Position a new movable breaker point on the stud.
5. Adjust the breaker point gap with a flat feeler gauge and tighten the screw.
6. Check the new point contact pattern and remove all grease, finger-prints, and dirt from contact surfaces.
7. Adjust the timing as described below.

Ignition Timing

ADJUSTMENT

1. Remove the cylinder head bolts, and move the head (with gasket in place) so that the spark plug hole is centered over the piston.
2. Using a ruler (through the spark plug hole) or special plunger type tool, carefully turn the engine back and forth until the piston is at exactly Top Dead Center. Tighten the thumbscrew on the tool.
3. Find the timing dimension for your engine in the specifications at the rear of the manual. Then, back off the position of the piston until it is about halfway down in the bore. Lower the ruler (or loosen the thumbscrew and lower the plunger, if using the special tool) exactly the required amount (the amount of the timing dimension). Then, hold the ruler in place (or

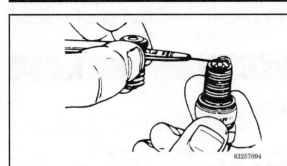

83257094

Fig. 84 Always used a wire gauge to check the plug gap

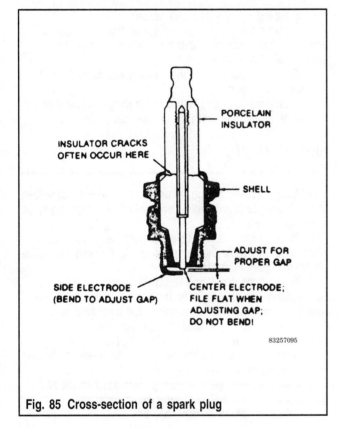

PORCELAIN INSULATOR

INSULATOR CRACKS OFTEN OCCUR HERE

SHELL

ADJUST FOR PROPER GAP

SIDE ELECTRODE (BEND TO ADJUST GAP)

CENTER ELECTRODE; FILE FLAT WHEN ADJUSTING GAP; DO NOT BEND!

83257095

Fig. 85 Cross-section of a spark plug

tighten the special tool thumbscrew) and, finally carefully rotate the engine forward until the piston just touches the ruler or tool plunger.

4. Install a timing light or place a very thin piece of cellophane between the contact points. Loosen and rotate the stator just until the timing light shows a change in current flow or the cellophane pulls out of contact gap easily. Then, tighten stator bolts to specified torque.

5. Install the leads, point cover, flywheel, and shrouding.

SOLID STATE IGNITION SYSTEM CHECKOUT

The only on-engine check which can be made to determine whether the ignition system is working, is to separate the high tension lead from the spark plug and check for spark. If there is a spark, then the unit is all right and the spark plug should be replaced. No spark indicates that some other part needs replacing.

Check the individual components as follows:

• High Tension Lead: Inspect for cracks or indications of arcing. Replace the transformer if the condition of the lead is questionable.

• Low Tension Leads: Check all leads for shorts. Check the ignition cut-off lead to see that the unit is not grounded. Repair the leads, if possible, or replace them.

• Pulse Transformer: Replace and test for spark.

• Magneto: Replace and test for spark. Time the magneto by turning it counterclockwise as far as it will go and then tighten the retaining screws.

• Flywheel: Check the magnets for strength. With the flywheel off the engine, it should attract a screwdriver that is held 1 in. (25mm) from the magnetic surface on the inside of the flywheel. Be sure that the key locks the flywheel to the crankshaft.

Carburetor Mixture Adjustments

Chart of Initial Carburetor Adjustments

Adjustment	For Engines Built Prior to 1977	For Engines Built After 1977
Main Adjustment Up to 7 HP	V-60-70-1¼ H-60-70-1¼	Same
Main Adjustment VM70-80-100 & HM70-80-100	1¼	1½
Idle Adjustment Up to 7 HP	V-60-70-1 Turn H-60-70-1 Turn	Same
Idle Adjustment VM70-80-100 & HM70-80-100	1½	1¼
Idle Speed (Top of Carburetor) Regulating screw	Back out screw, then turn in until screw just touches throttle lever and continue 1 turn more (if idle RPM is given set final idle speed with a tachometer)	

8325bC15

1. If the carburetor has been overhauled, or the engine won't start, make initial mixture screw adjustments as specified in the chart.

2. Start the engine and allow it to warm up to normal running temperature. With the engine running at maximum recommended rpm, loosen the main adjustment screw until engine rpm drops off, then tighten the screw until the engine starts to cut out. Note the number of turns from one extreme to the

other. Loosen the screw to a point midway between the extremes.

➡**Some carburetors have fixed jets. If there is no main adjusting screw and receptacle, no adjustment is needed.**

3. After the main system is adjusted, move the speed control lever to the idle position and follow the same procedure for adjusting the idle system.

4. Test the engine by running it under a normal load. The engine should respond to load pickup immediately. An engine that 'dies" is too lean. An engine which ran roughly before picking up the load is adjusted too rich.

Governor

ADJUSTMENT

Air Vane Type
◗ See Figure 86

1. Operate the engine with the governor adjusting lever or panel control set to the highest possible speed position and check the speed. If the speed is not within the recommended limits, the governed speed should be adjusted.

2. Loosen the locknut on the high speed limit adjusting screw and turn the adjusting screw out to increase the top engine speed.

Mechanical Type
◗ See Figure 87

1. Set the control lever to the idle position so that no spring tension affects the adjustment.

2. Loosen the screw so that the governor lever is loose in the clamp.

3. Rotate both the lever and the clamp to move the throttle to the full open position (away from the idle speed regulating screw).

4. Tighten the screw when no end-play exists in the direction of open throttle.

5. Move the throttle lever to the full speed setting and check to see that the control linkage opens the throttle.

Compression Check

1. Run the engine until warm to lubricate and seal the cylinder.

2. Remove the spark plug and install a compression gauge. Turn the engine over with the pull starter or electric starter.

3. Compression on new engines is 80 psi. If the reading is below 60 psi., repeat the test after removing the gauge and squirting about a teaspoonful of engine oil through the spark plug hole. If the compression improves temporarily following this, the problem is probably with the cylinder, piston, and rings. Otherwise, the valves require service.

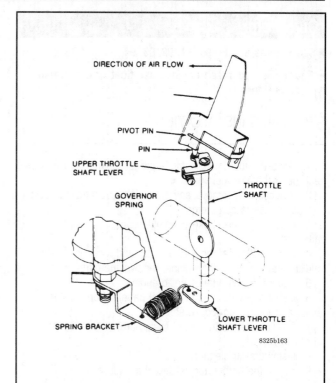

Fig. 86 Schematic of the operation of an air vane governor

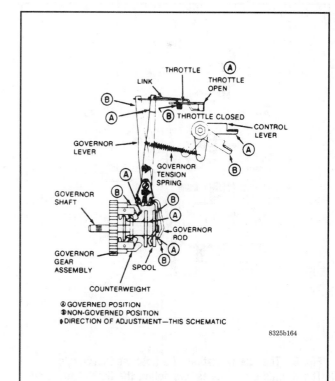

Fig. 87 Schematic of the operation of a mechanical governor

Carburetor

▶ See Figures 88, 89, 90, 91, 92, 93, 94, 95 and 96

➡Four-cycle Tecumseh engines use float or diaphragm type carburetors.

REMOVAL & INSTALLATION

1. Drain the fuel tank. Remove the air cleaner and disconnect the carburetor fuel lines.

2. If necessary, remove any shrouding or control panels to provide access to carburetor.

3. Disconnect the choke or throttle control wires at the carburetor.

4. Remove the cap screws, or nuts and lockwashers that hold the carburetor to the engine; remove the carburetor.

5. Secure the carburetor on to engine.

6. Install the shrouding or control panels. Connect the choke and throttle control wires.

7. Position the control panel to carburetor. Connect the carburetor fuel lines.

8. Install the air cleaner.

9. Adjust the carburetor as described above.

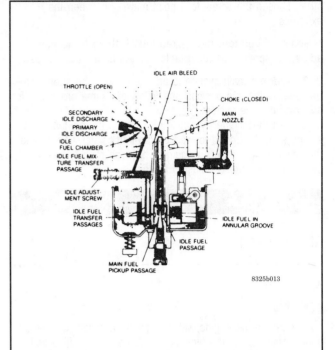

Fig. 89 The choke position of a float feed carburetor. The closed choke plate restricts air, creating a richened mixture by drawing in a greater proportion of fuel

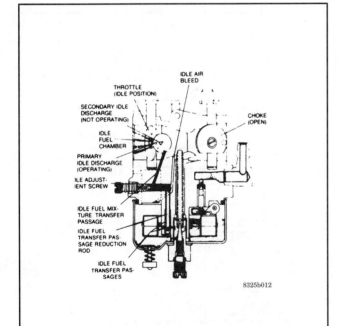

Fig. 88 The idle operation of a float feed carburetor. The throttle plate closes, restricting the flow of fuel and air, forcing the engine to run on a reduced volume of fuel and air

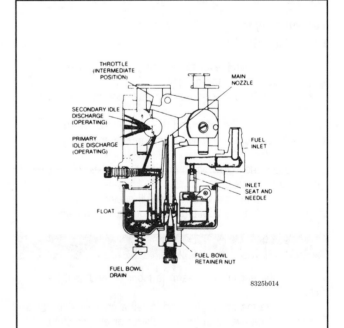

Fig. 90 The intermediate operation of a float feed carburetor. The throttle plate opens slightly to reduce restriction and the engine runs on an increased volume of fuel and air

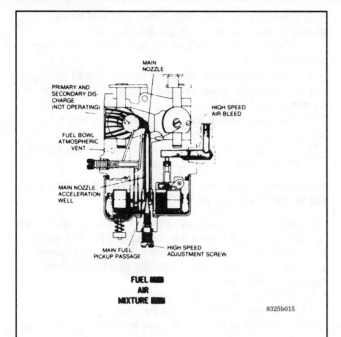

Fig. 91 The high speed operation of a float feed carburetor. The air venturi replaces the throttle plate as the restricting device and the engine runs on its greatest volume of fuel and air

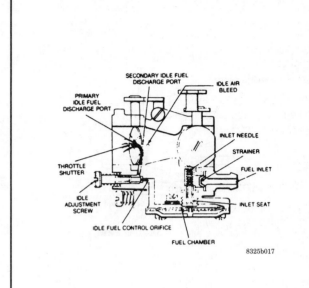

Fig. 93 The idle operation of a diaphragm carburetor. The throttle plate closes, restricting the flow of fuel and air, forcing the engine to run on a reduced volume of fuel and air

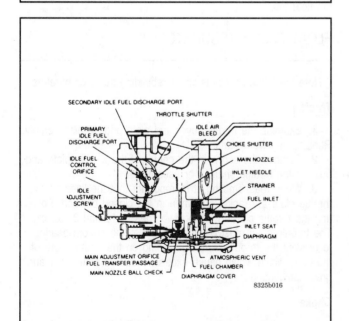

Fig. 92 The choke position of a diaphragm carburetor. The closed choke plate restricts air, creating a richened mixture by drawing in a greater proportion of fuel

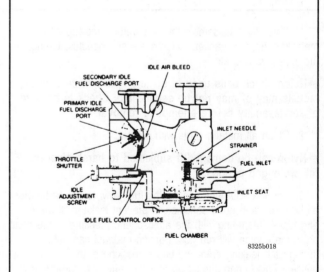

Fig. 94 The intermediate operation of a diaphragm carburetor. The throttle plate opens slightly to reduce restriction and the engine runs on an increased volume of fuel and air

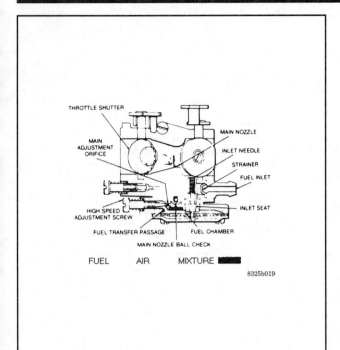

THROTTLE SHUTTER

MAIN NOZZLE

MAIN ADJUSTMENT ORIFICE

INLET NEEDLE

STRAINER

FUEL INLET

INLET SEAT

HIGH SPEED ADJUSTMENT SCREW

FUEL TRANSFER PASSAGE

FUEL CHAMBER

MAIN NOZZLE BALL CHECK

FUEL   AIR   MIXTURE

8325b019

Fig. 95 The high speed operation of a diaphragm carburetor. The air venturi replaces the throttle plate as the restricting device and the engine runs on its greatest volume of fuel and air

GENERAL OVERHAUL INSTRUCTIONS

1. Carefully disassemble the carburetor removing all non-metallic parts, i.e.; gaskets, viton seats and needles, O-rings, fuel pump valves, etc.

➡**Nylon check balls used in some diaphragm carburetor models may or may not be serviceable. Check to be sure of serviceability before attempting removal.**

2. Clean all metallic parts with solvent.

➡**Nylon can be damaged if subjected to harsh cleaners for prolonged periods.**

3. The large O-rings sealing the fuel bowl to the carburetor body must be in good condition to prevent leakage. If the O-ring leaks, interfering with the atmospheric pressure in the float bowl, the engine will run rich. Foreign material can enter through the leaking area and cause blocking of the metering orifices. This O-ring should be replaced after the carburetor has been disassembled for repair. Lubricate the new O-ring with a small amount of oil to allow the fuel bowl to slide onto the O-ring properly. Hold the carburetor body in an inverted position and place the O-ring on the carburetor body and then position the fuel bowl.

4. The small O-rings used on the carburetor adjustment screws must be in good condition or a leak will develop and cause improper adjustment of carburetor.

5. Check all adjusting screws for wear. The illustration shows a worn screw and a good screw. Replace screws that are worn.

6. Check the carburetor inlet needle and seat for wear, scoring, or other damage. Replace defective parts.

7. Check the carburetor float for dents, leaks, worn hinge or other damage.

8. Check the carburetor body for cracks, clogged passages, and worn bushings. Clean clogged air passages with clean, dry compressed air.

9. Check the diaphragms on diaphragm carburetors for cracks, punctures, distortion, or deterioration.

10. Check all shafts and pivot pins for wear on the bearing surfaces, distortion, or other damage.

➡**Each time a carburetor is disassembled, it is good practice to install a repair kit.**

11. Where there is excessive vibration, a damper spring may be used to assist in holding the float against the inlet needle thus minimizing the flooding condition. Two types of springs are available; the float shaft (hinge pin) type and the inlet needle mounted type.

12. Float shaft spring positioning:
    a. The spring is slipped over the shaft.
    b. The rectangular shaped spring end is hooked onto the float tab.
    c. The shorter angled spring end is placed onto the float bowl gasket support.

13. Note that on late model carburetors, the spring clip fastened to the inlet needle has been revised to provide a damping effect. The clip fastens to the needle and is hooked over the float tab.

FLOAT FEED CARBURETOR

Note the following points when rebuilding these carburetors.

Throttle

1. Examine the throttle lever and plate prior to disassembly. Replace any worn parts.

2. Remove the screw in the center of the throttle plate and pull out the throttle shaft lever assembly.

3. When reassembling, it is important that the lines on the throttle plate are facing out when in the closed position. Position the throttle plates with the two lines at 12 and 3 o'clock. The throttle shaft must be held in tight to the bottom bearing to prevent the throttle plate from riding on the throttle bore of the body which would cause excessive throttle plate wear and governor hunting.

Choke

Examine the choke lever and shaft at the bearing points and holes into which the linkage is fastened and replace any worn parts. The choke plate is inserted into the air horn of the carburetor in such a way that the flat surface of the choke is toward the fuel bowl.

Idle Adjusting Screw

▶ **See Figure 97**

Remove the idle screw from the carburetor body and examine the point for damage to the seating surface on the taper. If damaged, replace the idle adjusting needle. Tension is

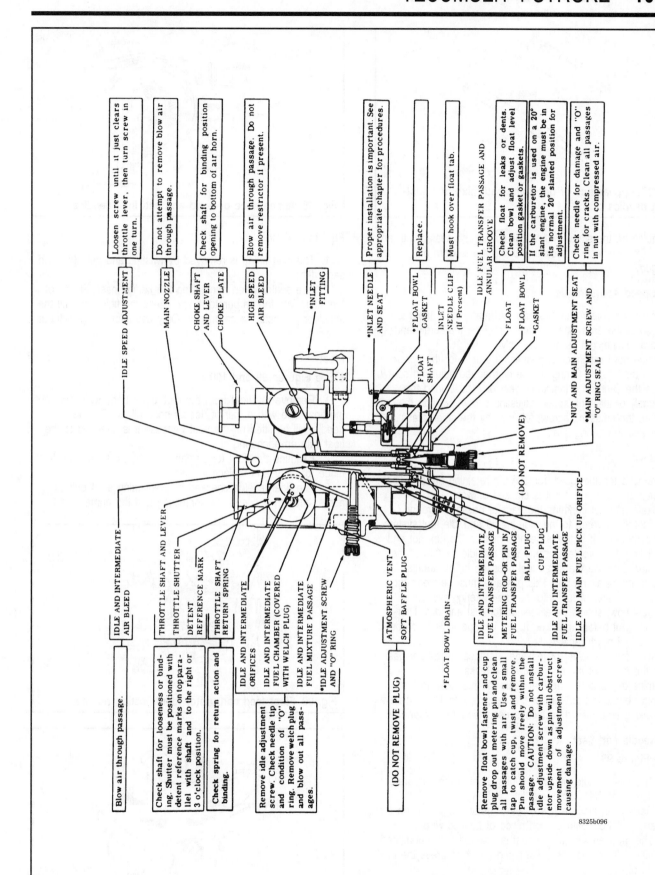

Loosen screw until it just clears throttle lever, then turn screw in one turn.

Do not attempt to remove blow air through passage.

Check shaft for binding position opening to bottom of air horn.

Blow air through passage. Do not remove restrictor if present.

Proper installation is important. See appropriate chapter for procedures.

Replace.

Must hook over float tab.

Check float for leaks or dents. Clean bowl and adjust float level position gasket or gaskets.

If the carburetor is used on a 20° slant engine, the engine must be in its normal 20° slanted position for adjustment.

Check needle for damage and "O" ring for cracks. Clean all passages in nut with compressed air.

IDLE SPEED ADJUSTMENT

MAIN NOZZLE

CHOKE SHAFT AND LEVER

CHOKE PLATE

HIGH SPEED AIR BLEED

*INLET FITTING

*INLET NEEDLE AND SEAT

*FLOAT BOWL GASKET

FLOAT SHAFT

INLET NEEDLE CLIP (If Present)

IDLE FUEL TRANSFER PASSAGE AND ANNULAR GROOVE

FLOAT

FLOAT BOWL

*GASKET

(DO NOT REMOVE)

NUT AND MAIN ADJUSTMENT SEAT

*MAIN ADJUSTMENT SCREW AND "O" RING SEAL

Blow air through passage.

Check shaft for looseness or binding. Shutter must be positioned with detent reference marks on top parallel with shaft and to the right or 3 o'clock position.

Check spring for return action and binding.

Remove idle adjustment screw. Check needle tip and condition of "O" ring. Remove welch plug and blow out all passages.

(DO NOT REMOVE PLUG)

Remove float bowl fastener and cup plug drop out metering pin and clean all passages with air. Use a small tap to catch cup, twist and remove. Pin should move freely within the passage. CAUTION: Do not install idle adjustment screw with carburetor upside down as pin will obstruct movement of adjustment screw causing damage.

IDLE AND INTERMEDIATE AIR BLEED

THROTTLE SHAFT AND LEVER

THROTTLE SHUTTER

DETENT REFERENCE MARK

THROTTLE SHAFT RETURN SPRING

IDLE AND INTERMEDIATE ORIFICES

IDLE AND INTERMEDIATE FUEL CHAMBER (COVERED WITH WELCH PLUG)

IDLE AND INTERMEDIATE FUEL MIXTURE PASSAGE

*IDLE ADJUSTMENT SCREW AND "O" RING

ATMOSPHERIC VENT

SOFT BAFFLE PLUG

*FLOAT BOWL DRAIN

IDLE AND INTERMEDIATE FUEL TRANSFER PASSAGE

METERING ROD OR PIN IN FUEL TRANSFER PASSAGE

BALL PLUG

CUP PLUG

IDLE AND INTERMEDIATE FUEL TRANSFER PASSAGE

IDLE AND MAIN FUEL PICK UP ORIFICE

8325b096

Fig. 96 Service hints for float feed carburetors

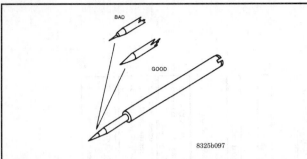

Fig. 97 Appearance of good and bad mixture adjustment needles

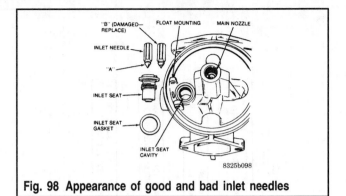

Fig. 98 Appearance of good and bad inlet needles

maintained on the screw with a coil spring and sealed with an O-ring. Examine and replace the O-ring if it is worn or damaged.

High Speed Adjusting Jet

Remove the screw and examine the taper. If the taper is damaged at the area where it seats, replace the screw and fuel bowl retainer nut as an assembly.

The fuel bowl retainer nut contains the seat for the screw. Examine the sealing O-ring on the high speed adjusting screw. Replace the O-ring if it indicates wear or cuts. During the reassembly of the high speed adjusting screw, position the coil spring on the adjusting screw, followed by the small brass washer and the O-ring seal.

Fuel Bowl

To remove the fuel bowl, remove the retaining nut and fiber washer. Replace the nut if it is cracked or worn.

The retaining nut contains the transfer passage through which fuel is delivered to the high speed and idle fuel system of the carburetor. It is the large hole next to the hex nut end of the fitting. If a problem occurs with the idle system of the carburetor, examine the small fuel passage in the annular groove in the retaining nut. This passage must be clean for the proper transfer of fuel into the idle metering system.

The fuel bowl should be examined for rust and dirt. Thoroughly clean it before installing it. If it is impossible to properly clean the fuel bowl, replace it.

Check the drain valve for leakage. Replace the rubber gasket on the inside of the drain valve if it leaks.

Examine the large O-ring that seals the fuel bowl to the carburetor body. If it is worn or cracked, replace it with a new one, making sure the same type is used (square or round).

Inlet Needle and Seat

▶ See Figure 98

1. The inlet needle sits on a rubber seat in the carburetor body instead of the usual metal fitting.
2. Remove it, place a few drops of heavy engine oil on the seat, and pry it out with a short piece of hooked wire.
3. The grooved side of the seat is inserted first. Lubricate the cavity with oil and use a flat faced punch to press the inlet seat into place.
4. Examine the inlet needle for wear and rounding off of the corners. If this condition does exist, replace the inlet needle.

Fuel Inlet Fitting

1. The inlet fitting is removed by twisting and pulling at the same time.
2. Use sealer when reinstalling the fitting. Insert the tip of the fitting into the carburetor body. Press the fitting in until the shoulder contacts the carburetor. Only use inlet fittings without screens.

Float

▶ See Figures 99 and 100

1. Remove the float from the carburetor body by pulling out the float axle with a pair of needle nose pliers. The inlet needle will be lifted off the seat because it is attached to the float with an anchoring clip.
2. Examine the float for damage and holes. Check the float hinge for wear and replace it if worn.
3. The float level is checked by positioning a #4 drill bit across the rim between the center leg and the unmachined

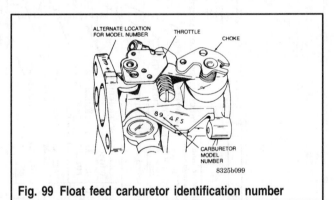

Fig. 99 Float feed carburetor identification number

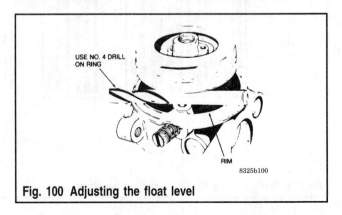

Fig. 100 Adjusting the float level

surface of the index pad, parallel to the float axle pin. If the index pad is machined, the float setting should be made with a #9 drill bit.

4. Remove the float to make an adjustment. Bend the tab on the float hinge to correct the float setting.

➡**Direct compressed air in the opposite direction of normal flow of air or fuel (reverse taper) to dislodge foreign matter.**

Carburetor Body

1. Check the carburetor body for wear and damage.

2. If excessive dirt has accumulated in the atmospheric vent cavity, try cleaning it with carburetor solvent or compressed air. Remove the welch plug only as a last resort.

➡**The carburetor body contains a pressed-in main nozzle tube at a specific depth and position within the venturi. Do not attempt to remove the main nozzle. Any change in nozzle positioning will adversely affect the metering quality and will require carburetor replacement.**

3. Clean the accelerating well around the main nozzle with compressed air and carburetor cleaning solvents.

4. The carburetor body contains two cup plugs, neither of which should be removed. A cup plug located near the inlet seat cavity, high up on the carburetor body, seals off the idle bleed. This is a straight passage drilled into the carburetor throat. Do not remove this plug. Another cup plug is located in the base where the fuel bowl nut seals the idle fuel passage. Do not remove this plug or the metering rod.

5. A small ball plug located on the side of the idle fuel passage seals this passage. Do not remove this ball plug.

6. The welch plug on the side of the carburetor body, just above the idle adjusting screw, seals the idle fuel chamber. This plug can be removed for cleaning of the idle fuel mixture passage and the primary and secondary idle fuel discharge ports. Do not use any tools that might change the size of the discharge ports, such as wire or pins.

Resilient Tip Needle

Replace the inlet needle. do not attempt to remove or replace the seat in the carburetor body.

Viton Seat

Using a 10-24 or 10-32 tap, turn the tap into the brass seat fitting until it grasps the seat firmly. Clamp the tap shank into a vise and tap the carburetor body with a soft hammer until the seat slides out of the body.

To replace the viton seat, position the replacement over the receptacle with the soft rubber like seat toward the body. Use a flat punch and a small hammer to drive the seat into the body until it bottoms on the shoulder.

TECUMSEH AUTOMATIC NON-ADJUSTABLE FLOAT FEED CARBURETOR

This carburetor has neither a choke plate nor idle and main mixture adjusting screws. There is no running adjustment. The float adjustment is the standard Tecumseh float setting of #4 drill bit.

Cleaning

▶ **See Figures 101, 102 and 103**

Remove all non-metallic parts and clean them using a procedure similar to that for the other carburetors. Never use wires through any of the drilled holes. Do not remove the baffling welch plug unless it is certain there is a blockage under the plug. There are no blind passageways in this carburetor.

Some engines use a variation on the Automatic Nonadjustable carburetor which has a different bowl hold-on nut and main jet orifices. There are two main jet orifices and a deeper fuel reserve cavity, but service procedures are the same.

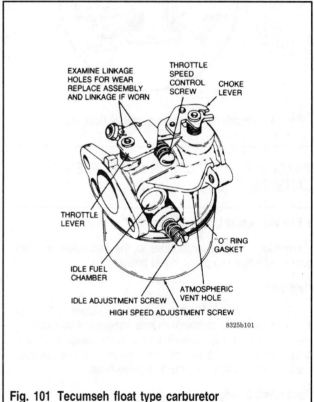

Fig. 101 Tecumseh float type carburetor

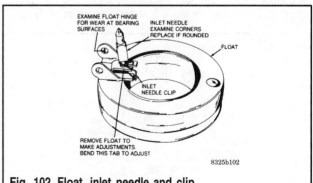

Fig. 102 Float, inlet needle and clip

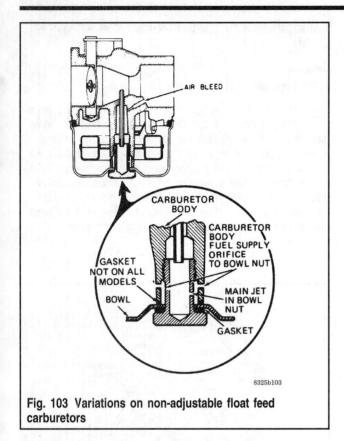

Fig. 103 Variations on non-adjustable float feed carburetors

WALBRO AND TILLOTSON FLOAT FEED CARBURETORS

▶ See Figures 104, 105, 106 and 107

Procedures are similar to those for the Tecumseh float carburetor with the exceptions noted below.

Main Nozzle

The main nozzle in Wallbro carburetors is cross drilled after it is installed in the carburetor. Once removed, it cannot be reinstalled, since it is impossible to properly realign the cross drilled holes. Grooved service replacement main nozzles are available which allow alignment of these holes.

Float Shaft Spring

Carefully position the float shaft spring on models so equipped. The spring dampens float action when properly as-

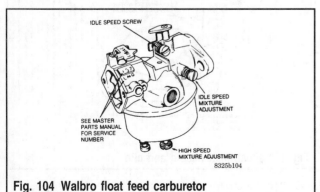

Fig. 104 Walbro float feed carburetor

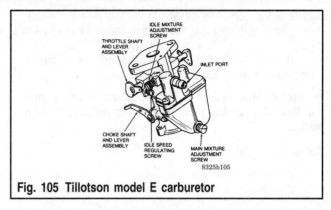

Fig. 105 Tillotson model E carburetor

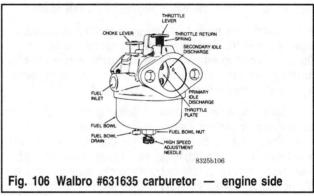

Fig. 106 Walbro #631635 carburetor — engine side

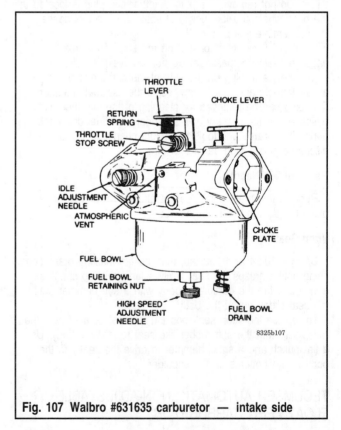

Fig. 107 Walbro #631635 carburetor — intake side

sembled. Use needlenosed pliers to hook the end of the spring over the float hinge and then insert the pin as far as possible before lifting the spring from the hinge into position. Leaving the spring out or improper installation will cause unbalanced float action and result in a touchy adjustment.

Float Adjustment

1. To check the float adjustment, invert the assembled float carburetor body. Check the clearance between the body and the float, opposite the hinge. Clearance should be $\frac{1}{8}$ in. $\pm \frac{1}{64}$ in. (3mm $\pm$ 0.4mm).

2. To adjust the float level, remove the float shaft and float. Bend the lip of the float tang to correct the measurement.

3. Assemble the parts and recheck the adjustment.

TILLOTSON E FLOAT FEED TYPE CARBURETOR

The following adjustments are different for this carburetor.

Running Adjustment

1. Start the engine and allow it to warm up to operating temperatures. Make sure the choke is fully opened after the engine is warmed up.

2. Run the engine at a constant speed while slowly turning the main adjustment screw in until the engine begins to lose speed; then slowly back it out about $\frac{1}{8}$-$\frac{1}{4}$ of a turn until maximum speed and power is obtained (4000 rpm). This is the correct power adjustment.

3. Close the throttle and cause the engine to idle slightly faster than normal by turning the idle speed regulating screw in. Then turn the idle adjustment speed screw in until the engine begins to lose speed; then turn it back $\frac{1}{4}$-$\frac{1}{2}$ of a turn until the engine idles smoothly. Adjust the idle speed regulating screw until the desired idling speed is acquired.

4. Alternately open and close the throttle a few times for an acceleration test. If stalling occurs at idle speeds, repeat the adjustment procedures to get the proper idle speed.

Float Level Adjustment

1. Remove the carburetor float bowl cover and float mechanism assembly.

2. Remove the float bowl cover gasket and, with the complete assembly in an upside down position and the float lever tang resting on the seated inlet needle, a measurement of $\frac{15}{64}$ in. (6mm) should be maintained from the free end flat rim, or edge of the cover, to the toe of the float. Measurement can be checked with a standard straight rule or depth gauge.

3. If it is necessary to raise or lower the float lever setting, remove the float lever pin and the float, then carefully bend the float lever tang up or down as required to obtain the correct measurement.

WALBRO CARBURETORS FOR V80, VM80, H80, AND HM80 ENGINES

Adjustment

The following initial carburetor adjustments are to be used to start the engine. For proper carburetion adjustment, the atmospheric vent must be open. Examine and clean it if necessary.

1. Idle adjustment — $1\frac{1}{4}$ turn from its seat.

2. High speed adjustment — $1\frac{1}{2}$ turn form its seat.

3. Throttle stop screw — 1 turn after contacting the throttle lever.

After the engine reaches normal operating temperature, make the final adjustments for best idle and high speed within the following ranges. Recommended speeds: Idle: 1800-2300 rpm. High speed: 3450-3750 rpm.

Rebuilding Notes

♦ See Figures 108, 109 and 110

1. The throttle plate is installed with the lettering (if present) facing outward when closed. The throttle plate is installed on the throttle level with the lever in the closed position. If there is binding after the plate is in position, loosen the throttle plate and reposition it.

2. Before removing the fuel bowl nut, remove the high speed adjusting needle. Use a $\frac{7}{16}$ in. (11mm) box wrench or socket to remove the fuel bowl nut. When replacing the fuel

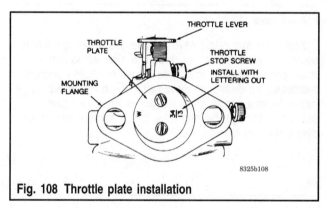

Fig. 108 Throttle plate installation

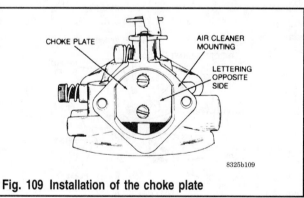

Fig. 109 Installation of the choke plate

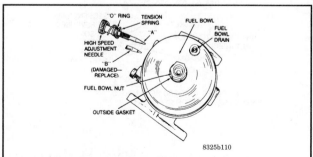

Fig. 110 Installation of the fuel bowl and the high speed adjustment needle

bowl nut, be sure to position the fiber gasket under the nut and tighten it securely.

3. Examine the high speed needle tip and, if it appears to be worn, replace it. When the high speed jet is replaced, the main nozzle, which includes the jet seat, should also be replaced. The original main nozzle cannot be used. There are special replacement nozzles available.

4. The inlet needle valve is replaceable if it appears to be worn. The inlet valve seat is also replaceable and should be replaced if the needle valve is replaced.

Float Adjustment

1. The Float setting for this carburetor is $5/64$-$7/64$ in. (2.0-2.8mm).

2. The float is set in the traditional manner, at the opposite end of the float from the float hinge and needle valve.

3. Bend the adjusting tab to adjust the float level.

DIAPHRAGM CARBURETORS

▶ **See Figure 111**

Diaphragm carburetors have a rubber-like diaphragm that is exposed to crankcase pressure on one side and to atmospheric pressure on the other side. As the crankcase pressure decreases, the diaphragm moves against the inlet needle allowing the inlet needle to move from its seat which permits fuel to flow through the inlet valve to maintain the correct fuel level in the fuel chamber.

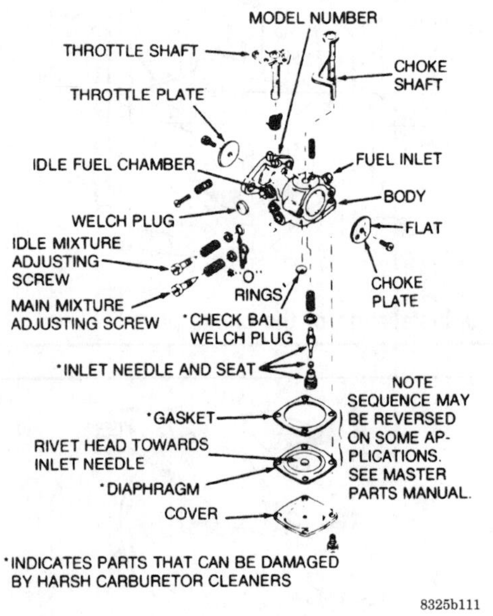

Fig. 111 Tecumseh diaphragm carburetor

An advantage of this type of system over the float system, is that the engine can be operated in any position.

➥**In rebuilding, use carburetor cleaner only on metal parts, except for the main nozzle in the main body.**

Throttle Plate

Install the throttle plate with the short line that is stamped in the plate toward the top of the carburetor, parallel with the throttle shaft, and facing out when the throttle is closed.

Choke Plate

Install the choke plate with the flat side of the choke toward the fuel inlet side of the carburetor. The mark faces in and is parallel to the choke shaft.

Idle Mixture Adjustment Screw

There is a neoprene O-ring on the needle. Never soak the O-ring in carburetor solvent. Idle and main mixture screws vary in size and design, so make sure that you have the correct replacement.

Idle Fuel Chamber

The welch plug can be removed if the carburetor is extremely dirty.

Diaphragms

Diaphragms are serviced and replaced by removing the four retaining screws from the cover. With the cover removed, the diaphragm and gasket may be serviced. Never soak the diaphragm in carburetor solvent. Replace the diaphragm if it is cracked or torn. Be sure there are no wrinkles in the diaphragm when it is replaced. The diaphragm rivet head is always placed facing the inlet needle valve.

Inlet Needle and Seat

The inlet seat is removed by using either a slotted screwdriver (early type) or a $9/32$ in. socket. The inlet needle is spring loaded, so be careful when removing it.

Fuel Inlet Fitting

All of the diaphragm carburetors have an integral strainer in the inlet fitting. To clean it, either reverse flush it or use compressed air after removing the inlet needle and seat. If the strainer is lacquered or otherwise unable to be cleaned, replace the fitting.

CRAFTSMAN FUEL SYSTEMS

▶ **See Figures 112, 113 and 114**

Changes in Late Model Carburetors

The newest Craftsman carburetors incorporate the following changes:
- The cable form of control is replaced by a control knob.
- The fuel pickup is longer and has a collar machined into it which must be installed tight against the carburetor body.
- The fuel pickup screen is pressed onto the ends of the fill tubes on both models, but the measured depth has changed.

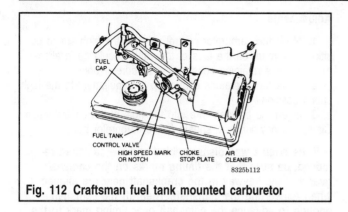

Fig. 112 Craftsman fuel tank mounted carburetor

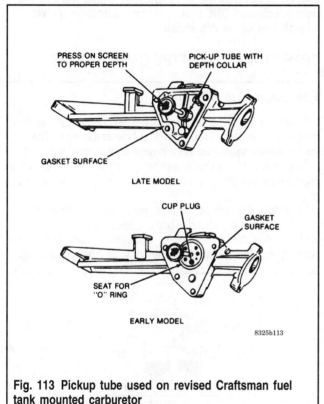

Fig. 113 Pickup tube used on revised Craftsman fuel tank mounted carburetor

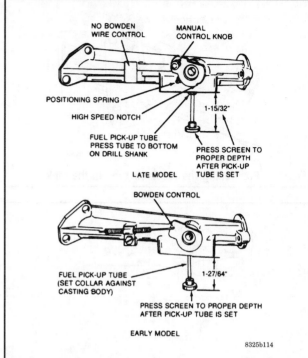

Fig. 114 Manual control knob and several other changes incorporated on revised Craftsman fuel tank mounted carburetor

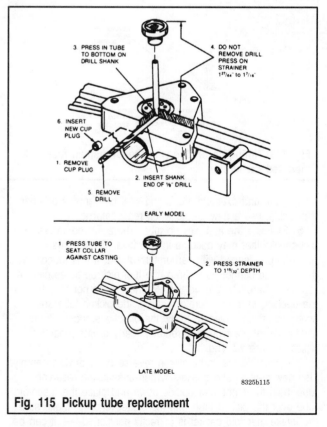

Fig. 115 Pickup tube replacement

• The cross-drilled passages have been eliminated, as has the O-ring on the body. There are no cup plugs.
• The fuel tank and reservoir tube have been revised — the reservoir tube being larger.

Disassembly and Service

▶ See Figures 115, 116, 117 and 118

1. Remove the air cleaner assembly and remove the four screws on the top of the carburetor body to separate the fuel tank from the carburetor.

2. Remove the O-ring from between the carburetor and the fuel tank. Examine it for cracks and damage and replace it if necessary.

3. Carefully remove the reservoir tube from the fuel tank. Observe the end of the tube that rested on the bottom of the fuel tank. It should be slotted.

4. Remove the control valve by turning the valve clockwise until the flange is clear of the retaining boss. Pull the valve

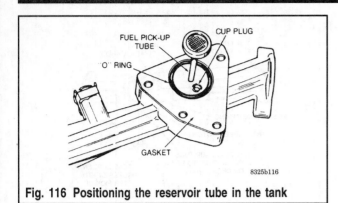

Fig. 116 Positioning the reservoir tube in the tank

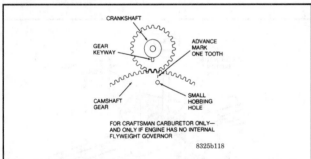

Fig. 117 The fuel pickup tube and O-ring on early Craftsman carburetors

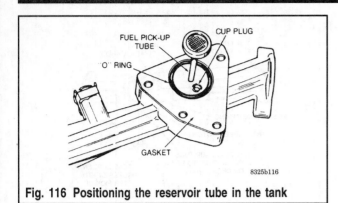

Fig. 118 Timing an engine with the Craftsman fuel tank mounted carburetor

straight out and examine the O-ring seal for damage or wear. If possible, use a new O-ring when reassembling.

5. Examine the fuel pick up tube. There are no valves or ball checks that may become in-operative. These parts can normally be cleaned with carburetor solvent. If it is found that the passage cannot be cleared, the fuel pick up tubes can be replaced. Carefully remove the old ones so as not to enlarge the opening in the carburetor body. If the pickup tube and screen must be replaced, follow the directions shown in the illustration for the type of carburetor (early or late model) on which you are working.

6. Assemble the carburetor in reverse order of disassembly. Use new O-rings and gaskets. When assembling reservoir tube, hold the carburetor upside down and place the reservoir tube over the pickup tube with the slotted end up. Make sure the intake manifold gasket is correctly positioned — it can be assembled blocking the intake passage partially.

Adjustments

1. Move the carburetor control valve to the high speed position. The mark on the face of the valve should be in alignment with the retaining boss on the carburetor body.

2. Move the operator's control on the equipment to the high speed position.

3. Insert the bowden wire into the hole of the control valve. Clamp the bowden wire sheath to the carburetor body.

➡**If the engine was disassembled and the camshaft removed, be sure that the timing marks on the camshaft gear and the keyway in the crankshaft gear are aligned when reinserting the camshaft. Then lift the camshaft enough to advance the camshaft gear timing mark to the right (clockwise) ONE tooth, as viewed from the power take-off end of the crankshaft.**

CRAFTSMAN FLOAT TYPE CARBURETORS

▶ **See Figures 119 and 120**

Craftsman float type carburetors are serviced in the same manner as the other Tecumseh float type carburetors. The throttle control valve has three positions: stop, run, and start.

When the control valve is removed, replace the O-ring. If the engine runs sluggishly, consider the possibility of a leaky O-ring.

To remove the fuel pickup tube, clamp the tube in a vise and then twist the carburetor body. Check the small jet in the air horn while the tube is out. In replacement, position the tube squarely and then press in on the collar until the collar seats.

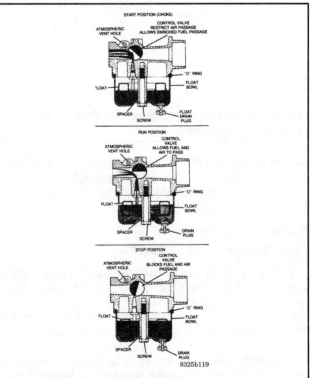

Fig. 119 Start, run and stop throttle positions of a Carftsman float type carburetor

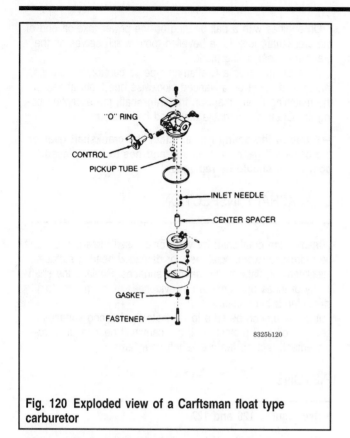

Fig. 120 Exploded view of a Carftsman float type carburetor

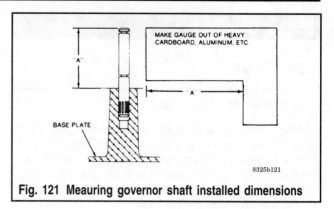

Fig. 121 Meauring governor shaft installed dimensions

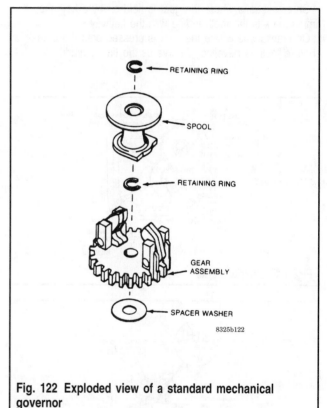

Fig. 122 Exploded view of a standard mechanical governor

In assembly, install the gasket and bolt through the bowl, position the centering spacer onto the bolt, and then attach the parts to the carburetor body.

The camshaft timing mark must be advanced one tooth in relation to crankshaft gear timing mark, as shown in the illustration above. However, when this type carburetor is used with a float bowl reservoir and variable governor adjustment, time it as for other Tecumseh engines — with the camshaft and crankshaft timing marks aligned.

Governor

♦ See Figures 121 and 122

Shaft Installed Dimension/ Engine Model Chart

Engine Model	"A" Exposed Shaft Length (see figure)
TNT100-120 ECV100-105-110-120 ECH90 TVS75-90-105-120	$1^5/_{16}$"
V60 VM70-80-100	$1^{19}/_{32}$"
H60 HH60-70	$1^7/_{16}$"
HM70-80-100	$1^{13}/_{32}$"

8325bC16

The mechanical governor is located inside the mounting flange. See engine disassembly instructions, below. To disassemble the governor, see the illustration, and: remove the re-taining ring, pull off the spool, remove the second retaining ring, and then pull off the gear assembly and retainer washer.

Check for wear on all moving surfaces, but especially gear teeth, the inside diameter of the gear where it rides on the shaft, and the flyweights where they work against the spool.

If the governor shaft must be replaced, it should be started into the boss with a few taps using a soft hammer, and then pressed in with a press or vise. The shaft can be installed by positioning a wooden block on top and tapping the upper surface of the block, but the use of a vise or press is much preferred.

The shaft must be pressed in until just the required length is exposed, as measured from the top of the shaft boss to the upper end of the shaft. See the chart.

The governor is installed in reverse of the removal procedure. Connect the linkage and then adjust as described in the Tune-Up section.

Engine Overhaul

TIMING GEARS

▶ **See Figures 123, 124 and 125**

Correctly matched camshaft gear and crankshaft gear timing marks are necessary for the engine to perform properly.

On all camshafts the timing mark is located in line with the center of the hobbing hole (small hole in the face of the gear). If no line is visible, use the center of the hobbing hole to align with the crankshaft gear marked tooth.

On crankshafts where the gear is held on by a key, the timing mark is the tooth in line with the keyway.

On crankshafts where the gear is pressed onto the crankshaft, a tooth is bevelled to serve as the timing mark.

On engines with a ball bearing on the power take-off end of the crankshaft, look for a bevelled tooth which serves as the crankshaft gear timing mark.

If the engine uses a Craftsman type carburetor, the camshaft timing mark must be advanced clockwise one tooth ahead of the matching timing mark on the crankshaft, the exception being the Craftsman variable governed fuel systems.

➡**If one of the timing gears, either the crankshaft gear or the camshaft gear, is damaged and has to be replaced, both gears should be replaced.**

CRANKSHAFT INSPECTION

Inspect the crankshaft for worn or crossed threads that can't be redressed; worn, scratched, or damaged bearing surfaces; misalignment; flats on the bearing surfaces. Replace the shaft if any of these problems are in evidence — do not try to straighten a bent shaft.

In replacement, be sure to lubricate the bearing surfaces and use oil seal protectors. If the camshaft gear requires replacement, replace the crankshaft gear, also.

PISTONS

▶ **See Figures 126 and 127**

When removing the pistons, clean the carbon from the upper cylinder bore and head. The piston and pin must be replaced in matched pairs.

A ridge reamer must be used to remove the ridge at the top of the cylinder bore on some engines.

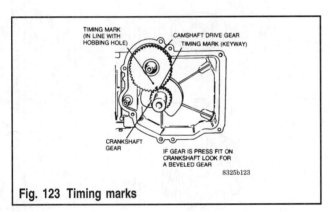

Fig. 123 Timing marks

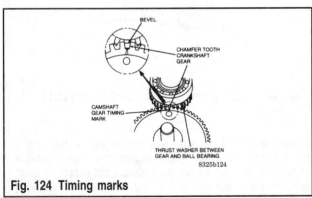

Fig. 124 Timing marks

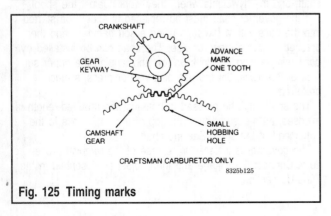

Fig. 125 Timing marks

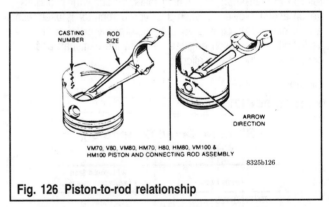

Fig. 126 Piston-to-rod relationship

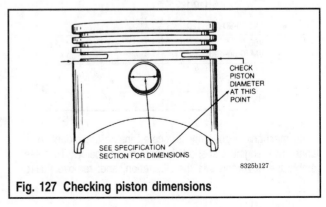

Fig. 127 Checking piston dimensions

Clean the carbon from the piston ring groove. A broken ring can be used for this operation.

Some engines have oversize pistons which can be identified by the oversize engraved on the piston top.

There is a definite piston-to-connecting rod-to-crankshaft arrangement which must be maintained when assembling these parts. If the piston is assembled in the bore 180° out of position, it will cause immediate binding of the parts.

PISTON RINGS

▶ **See Figures 128, 129 and 130**

Always replace the piston rings in sets. Ring gaps must be staggered. When using new rings, wipe the cylinder wall with fine emery cloth to deglaze the wall. Make sure the cylinder wall is thoroughly cleaned after deglazing. Check the ring gap

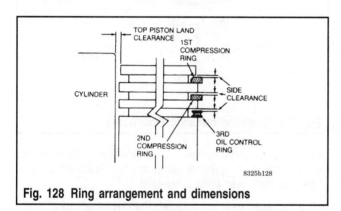
Fig. 128 Ring arrangement and dimensions

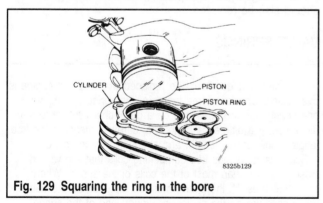
Fig. 129 Squaring the ring in the bore

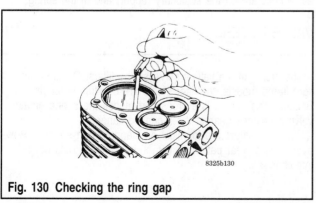
Fig. 130 Checking the ring gap

by placing the ring squarely in the center of the area in which the rings travel and measuring the gap with a feeler gauge. Do not spread the rings too wide when assembling them to the pistons. Use a ring leader to install the rings on the piston.

The top compression ring has an inside chamfer. This chamfer must go UP. If the second ring has a chamfer, it must also face UP. If there is a notch on the outside diameter of the ring, it must face DOWN.

Check the ring gap on the old ring to determine if the ring should be replaced. Check the ring gap on the new ring to determine if the cylinder should be rebored to take oversize parts.

➡**Make sure that the ring gap is measured with the ring fitted squarely in the worn part of the cylinder where the ring usually rides up and down on the piston.**

CONNECTING RODS

▶ **See Figures 131 and 132**

Be sure that the match marks align when assembling the connecting rods to the crankshaft. Use new self-locking nuts. Whenever locking tabs are included, be sure that the tabs lock the nuts securely. NEVER try to straighten a bent crankshaft or connecting rod. Replace them if necessary. When replacing either the piston, rod, crankshaft, or camshaft, liberally lubricate all bearings with engine oil before assembly.

The following engines have offset connecting rods: V70, VH70, VM70, V80, VM80, H70, HH70, HM70, H80, HM80. ECV105, ECV110, ECM120, and TNT120 engines have the caps fitted from opposite to the camshaft side of the engine. The following engines have the cap fitted from the camshaft

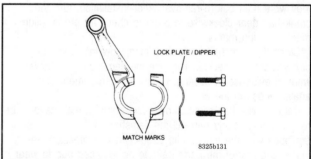
Fig. 131 Connecting rod assembly for the V70, H70 engines

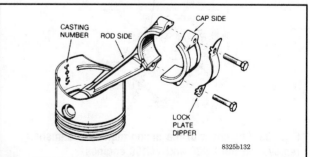
Fig. 132 Connecting rod assembly for the V80, H80 engines

side: V70, VH70, VM70, V80, VM80, H70, HH70, HM70, H80, HM80, VM100 and HM100.

On H70, HH70, HM70, H80, HM80, V70, VH70, VM70, V80, VM80, VM100 and HM100 engines, a dipper is stamped into the lockplate. Use a new lockplate whenever the rod cap is removed.

On engine with Durlock rod bolts, torque the bolt as follows:
- TVS75, 90, 105, 120, ECH90, TNT100, 120, ECV100, 105, 110 and 120-110 inch lbs.
- V60-80, H60-80, VH60-70, HH60-70, VM70-100 and HM70-100-150 inch lbs.

➥**Early type caps can be distorted if the cap is not held to the crank pin while threading the bolts tight. Undue force should not be used.**

Later rods have serrations which prevent distortion during tightening. They also have match marks which must face out when assembling the rod. On the V80 and H80 engines, the piston and rod must fit so that the number inside the casting is on the rod side of the rod/cap combination.

CAMSHAFT

▶ **See Figures 133 and 134**

Before removing the camshaft, align the timing marks to relieve the pressure on the valve lifters, on most engines.

On VM80 and VM100 engines, while the basic timing is the same, the crankshaft must be rotated so that the timing mark is located 3 teeth further counterclockwise (referring to crankshaft gear rotation). This clears the camshaft of the compression release mechanism for easier removal.

In installation, align the gears for this type of camshaft as they were right before removal. After installation, turn the crankshaft gear clockwise in order to check for proper alignment of timing marks.

Clean the camshaft in solvent, then blow the oil passages dry with compressed air. Replace the camshaft if it shows wear of evidence of scoring. Check the cam dimensions against those in the chart.

If the engine has a mechanical fuel pump, it may have to be removed to properly reinstall the camshaft. If the engine is equipped with the Insta-matic Ezee-Start Compression Release, and any of the parts have to be replaced due to wear

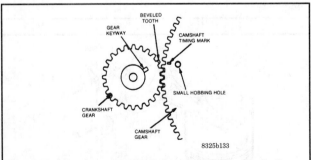

Fig. 133 Alignment of the timing marks for camshaft removal on the VM80 and VM100 engines

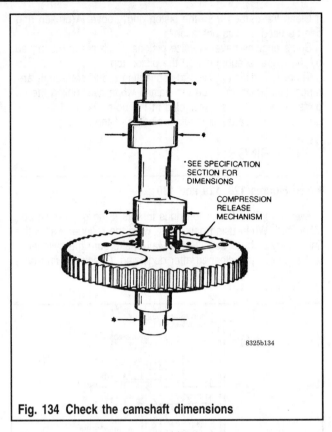

Fig. 134 Check the camshaft dimensions

or damage, the entire camshaft must be replaced. Be sure that the oil pump (if so equipped) barrel chamber is toward the fillet of the camshaft gear when assembled.

➥**If a damaged gear is replaced, the crankshaft gear should also be replaced.**

VALVE SPRINGS

The valve springs should be replaced whenever an engine is overhauled. Check the free length of the springs. Comparing one spring with the other can be a quick check to notice any difference. If a difference is noticed, carefully measure the free length, compression length, and strength of each spring.

Some valve springs use dampening coils that are wound closer together than most of the coils of the spring. Where these are present, the spring must be mounted so the dampening coils are on the stationary (upper) end of the spring.

VALVE LIFTERS

The stems of the valves serve as the lifters. On the 4 hp light frame models, the lifter stems are of different lengths. Because this engine is a cross port model, the shorter intake valve lifter goes nearest the mounting flange.

The valve lifters are identical on standard port engines. However once a wear pattern is established, they should not be interchanged.

VALVE GRINDING AND REPLACEMENT

▶ **See Figure 135**

Valves and valve seats can be removed and reground with a minimum of engine disassembly.

Remove the valves as follows:

1. Raise the lower valve spring caps while holding the valve heads tightly against the valve seat to remove the valve spring retaining pin. This is best achieved by using a valve spring compressor. Remove the valves, springs, and caps from the crankcase.

2. Clean all parts with a solvent and remove all carbon from the valves.

3. Replace distorted or damaged valves. If the valves are in usable condition, grind the valve faces in a valve refacing machine and to the angle given in the specifications chart at the end of this section. Replace the valves if the faces are ground to less than 1/32 in. (0.8mm).

4. Whenever new or reground valves are installed, lap in the valves with lapping compound to insure an air-tight fit.

➡**There are valves available with oversize stems.**

5. Valve grinding changes the valve lifter clearance. After grinding the valves check the valve lifter clearance as follows:

  a. Rotate the crankshaft until the piston is set at the TDC position of the compression stroke.

  b. Insert the valves in their guides and hold the valves firmly on their seats.

  c. Check for a clearance of 0.010 in. (0.25mm) between each valve stem and valve lifter with a feeler gauge.

  d. Grind the valve stem in a valve resurfacing machine set to grind a perfectly square face with the proper clearance.

6. Install valves as follows on Early Models:

  a. Position the valve spring and upper and lower valve spring caps under the valve guides for the valve to be installed.

  b. Install the valves in the guides, making sure that the valve marked **EX** is inserted in the exhaust port. The valve stem must pass through the valve spring and the valve spring caps.

  c. Insert the blade of a screwdriver under the lower valve spring cap and pry the spring up.

  d. Insert the valve pin through the hole in the valve stem with a long nosed pliers. Make sure the valve pin is properly seated under the lower valve spring cap.

7. Install the valves as follows on Later Models:

  a. Position the valve caps and spring in the valve compartment.

  b. Install the valves in guides with the valve marked **EX** in exhaust port. The valve stem must pass through the upper valve cap and spring. The lower cap should sit around the valve lifter exposed end.

  c. Compress the valve spring so that the shank is exposed. DO NOT TRY TO LIFT THE LOWER CAP WITH THE SPRING.

  d. Lift the lower valve cap over the valve stem shank and center the cap in the smaller diameter hole.

  e. Release the valve spring tension to lock the cap in place.

REBORING THE CYLINDER

▶ **See Figure 136**

1. First, decide whether to rebore for 0.010 in. (0.25mm) or 0.020 in. (0.50mm).

2. Use any standard commercial hone of suitable size. Chuck the hone in the drill press with the spindle speed of about 600 rpm.

3. Start with coarse stones and center the cylinder under the press spindle. Lower the hone so the lower end of the stones contact the lowest point in the cylinder bore.

4. Rotate the adjusting nut so that the stones touch the cylinder wall and then begin honing at the bottom of the cylinder. Move the hone up and down at a rate of 50 strokes a minute to avoid cutting ridges in the cylinder wall. Every fourth or fifth stroke, move the hone far enough to extend the stones 1 in. (25mm) beyond the top and bottom of the cylinder bore.

5. Check the bore size and straightness every thirty or forty strokes. If the stones collect metal, clean them with a wire brush each time the hone is removed.

6. Hone with coarse stones until the cylinder bore is within 0.002 in. (0.05mm) of the desired finish size. Replace the coarse stones with burnishing stones and continue until the bore is to within 0.0005 in. (0.0127mm) of the desired size.

7. Remove the burnishing stones and install finishing stones to polish the cylinder to the final size.

8. Clean the cylinder with solvent and dry it thoroughly.

9. Replace the piston and piston rings with the correct oversize parts.

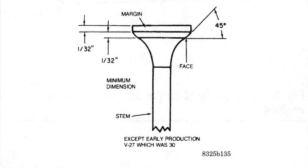

Fig. 135 Dimensions of the valve face

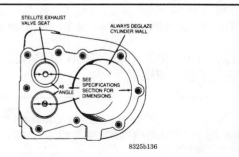

Fig. 136 Checking the dimensions of the valve seats, guides and the cylinder

REBORING VALVE GUIDES

The valve guides are permanently installed in the cylinder. However if the guides wear, they can be rebored to accommodate a 1/32 in. (0.8mm) oversize valve stem. Rebore the valve guides in the following manner:

1. Ream the valve guides with a standard straight shanked hand reamer or a low speed drill press.

2. Redrill the upper and lower valve spring caps to accommodate the oversize valve stem.

3. Reassemble the engine, installing valves with the correct oversize stems in the valve guides.

REGRINDING VALVE SEATS

The valve seats need regrinding only if they are pitted or scored. If there are no pits or scores, lapping in the valves will provide a proper valve seat. Valve seats are not replaceable. Regrind the valve seats as follows:

1. Use a grinding stone or a reseater set to provide the proper angle and seal face dimensions.

2. If the seat is over 3/64 in. (1.2mm) wide after grinding, use a 15° stone or cutter to narrow the face to the proper dimensions.

3. Inspect the seats to make sure that the cutter or stone has been held squarely to the valve seat and that the same dimension has been held around the entire circumference of the seat.

4. Lap the valves to the reground seats.

Torquing Cylinder Head

▶ See Figures 137 and 138

Torque the cylinder head to 200 inch lbs. in 4 equal stages of 50 inch lbs. Follow the sequence shown in the appropriate illustration for each tightening stage.

Bearing Service

▶ See Figures 139, 140, 141, 142, 143, 144, 145 and 146

LIGHTWEIGHT ALUMINUM BEARING REPLACEMENT

The aluminum bearing must be cut out using the rough cut reamer and the procedures shown. Follow illustrated steps to install bronze bushings in place of the aluminum bearings.

LONG LIFE AND CAST IRON ENGINES WITHOUT BALL BEARINGS

The worn bronze bushing must be driven out before the new bushing can be installed. Follow illustrated steps to replace the main bearings on these units.

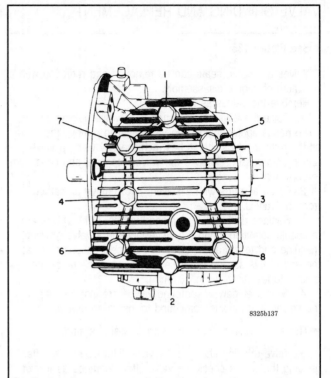

Fig. 137 Cylinder head torque sequence for all engines except 8 hp

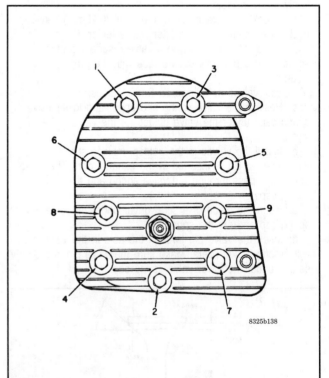

Fig. 138 Cylinder head torque sequence for 8 hp engines

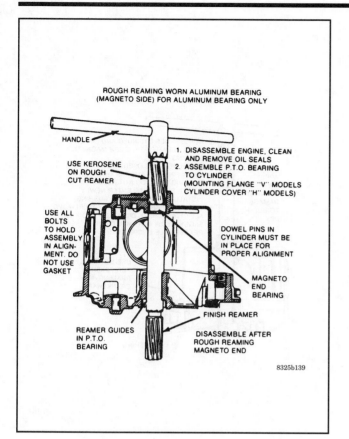

ROUGH REAMING WORN ALUMINUM BEARING
(MAGNETO SIDE) FOR ALUMINUM BEARING ONLY

HANDLE

USE KEROSENE
ON ROUGH
CUT REAMER

1. DISASSEMBLE ENGINE, CLEAN
   AND REMOVE OIL SEALS
2. ASSEMBLE P.T.O. BEARING
   TO CYLINDER
   (MOUNTING FLANGE "V" MODELS
   CYLINDER COVER "H" MODELS)

USE ALL
BOLTS
TO HOLD
ASSEMBLY
IN ALIGN-
MENT. DO
NOT USE
GASKET

DOWEL PINS IN
CYLINDER MUST BE
IN PLACE FOR
PROPER ALIGNMENT

MAGNETO
END
BEARING

REAMER GUIDES
IN P.T.O.
BEARING

FINISH REAMER

DISASSEMBLE AFTER
ROUGH REAMING
MAGNETO END

8325b139

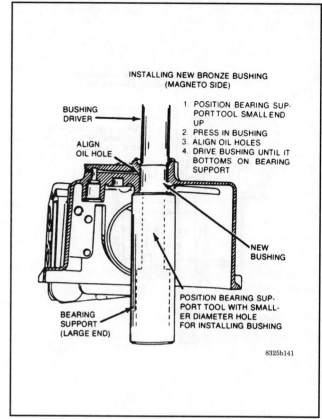

INSTALLING NEW BRONZE BUSHING
(MAGNETO SIDE)

BUSHING
DRIVER

ALIGN
OIL HOLE

1. POSITION BEARING SUP-
   PORT TOOL, SMALL END
   UP
2. PRESS IN BUSHING
3. ALIGN OIL HOLES
4. DRIVE BUSHING UNTIL IT
   BOTTOMS ON BEARING
   SUPPORT

NEW
BUSHING

BEARING
SUPPORT
(LARGE END)

POSITION BEARING SUP-
PORT TOOL WITH SMALL-
ER DIAMETER HOLE
FOR INSTALLING BUSHING

8325b141

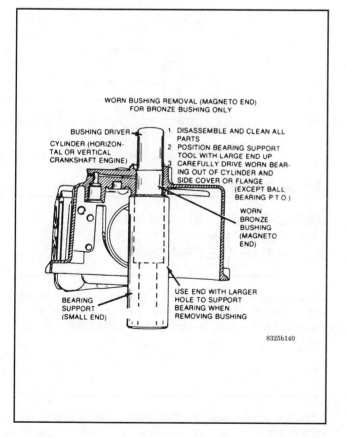

WORN BUSHING REMOVAL (MAGNETO END)
FOR BRONZE BUSHING ONLY

BUSHING DRIVER

CYLINDER (HORIZON-
TAL OR VERTICAL
CRANKSHAFT ENGINE)

1. DISASSEMBLE AND CLEAN ALL
   PARTS
2. POSITION BEARING SUPPORT
   TOOL WITH LARGE END UP
3. CAREFULLY DRIVE WORN BEAR-
   ING OUT OF CYLINDER AND
   SIDE COVER OR FLANGE
   (EXCEPT BALL
   BEARING P.T.O.)

WORN
BRONZE
BUSHING
(MAGNETO
END)

BEARING
SUPPORT
(SMALL END)

USE END WITH LARGER
HOLE TO SUPPORT
BEARING WHEN
REMOVING BUSHING

8325b140

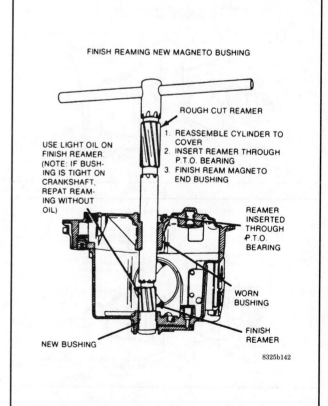

FINISH REAMING NEW MAGNETO BUSHING

ROUGH CUT REAMER

USE LIGHT OIL ON
FINISH REAMER.
(NOTE: IF BUSH-
ING IS TIGHT ON
CRANKSHAFT,
REPAT REAM-
ING WITHOUT
OIL)

1. REASSEMBLE CYLINDER TO
   COVER
2. INSERT REAMER THROUGH
   P.T.O. BEARING
3. FINISH REAM MAGNETO
   END BUSHING

REAMER
INSERTED
THROUGH
P.T.O.
BEARING

WORN
BUSHING

FINISH
REAMER

NEW BUSHING

8325b142

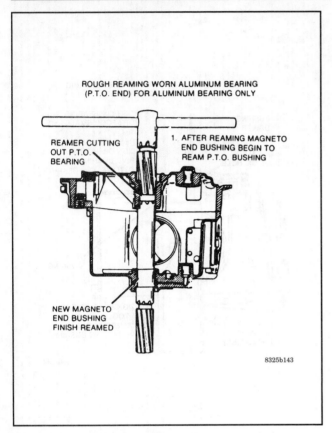

ROUGH REAMING WORN ALUMINUM BEARING
(P.T.O. END) FOR ALUMINUM BEARING ONLY

REAMER CUTTING
OUT P.T.O.
BEARING

1. AFTER REAMING MAGNETO
END BUSHING BEGIN TO
REAM P.T.O. BUSHING

NEW MAGNETO
END BUSHING
FINISH REAMED

8325b143

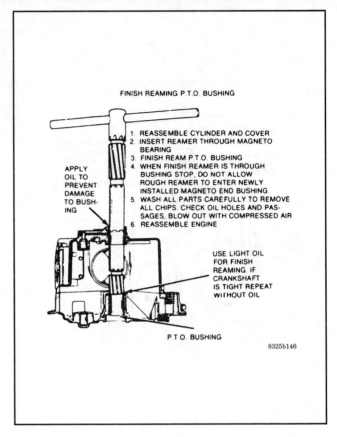

FINISH REAMING P.T.O. BUSHING

APPLY
OIL TO
PREVENT
DAMAGE
TO BUSH-
ING

1. REASSEMBLE CYLINDER AND COVER
2. INSERT REAMER THROUGH MAGNETO
   BEARING
3. FINISH REAM P.T.O. BUSHING
4. WHEN FINISH REAMER IS THROUGH
   BUSHING STOP, DO NOT ALLOW
   ROUGH REAMER TO ENTER NEWLY
   INSTALLED MAGNETO END BUSHING
5. WASH ALL PARTS CAREFULLY TO REMOVE
   ALL CHIPS. CHECK OIL HOLES AND PAS-
   SAGES, BLOW OUT WITH COMPRESSED AIR
6. REASSEMBLE ENGINE

USE LIGHT OIL
FOR FINISH
REAMING. IF
CRANKSHAFT
IS TIGHT REPEAT
WITHOUT OIL

P.T.O. BUSHING

8325b146

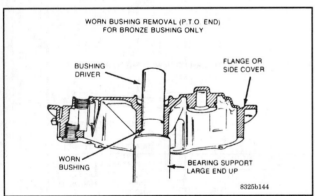

WORN BUSHING REMOVAL (P.T.O. END)
FOR BRONZE BUSHING ONLY

BUSHING
DRIVER

FLANGE OR
SIDE COVER

WORN
BUSHING

BEARING SUPPORT
LARGE END UP

8325b144

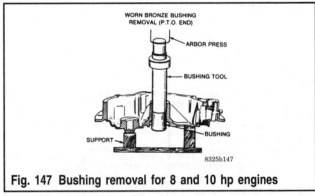

WORN BRONZE BUSHING
REMOVAL (P.T.O. END)

ARBOR PRESS

BUSHING TOOL

SUPPORT

BUSHING

8325b147

Fig. 147 Bushing removal for 8 and 10 hp engines

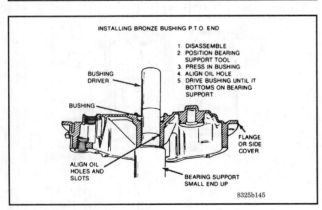

INSTALLING BRONZE BUSHING P.T.O. END

BUSHING
DRIVER

BUSHING

1. DISASSEMBLE
2. POSITION BEARING
   SUPPORT TOOL
3. PRESS IN BUSHING
4. ALIGN OIL HOLE
5. DRIVE BUSHING UNTIL IT
   BOTTOMS ON BEARING
   SUPPORT

ALIGN OIL
HOLES AND
SLOTS

FLANGE
OR SIDE
COVER

BEARING SUPPORT
SMALL END UP

8325b145

LONG LIFE AND CAST IRON ENGINES WITH BALL BEARING ON THE P.T.O. (POWER TAKE-OFF) SIDE OF THE CRANKSHAFT

The side cover containing the ball bearing must be removed and a substitute cover with either a new bronze bushing or aluminum bearing must be used instead. Follow illustrated steps.

SPECIAL TOOLS

The task of main bearing replacement is made easier by using one of two Tecumseh main bearing tool kits. Kit 670161 is used to replace main bearings on the lightweight engines except HS models. Kit 670165 is used to replace main bearings on the medium weight engines except HS models.

GENERAL NOTES ON BUSHING REPLACEMENT

1. Your fingers and all parts must be kept very clean when replacing bushings.

2. On splash lubricated horizontal engines, the oil hole in the bushing is to be lined up with the oil hole that leads into the slot in the original bearing.

3. In the event it is necessary to replace the mounting flange or cylinder cover, the magneto end bearing must be rebushed. The P.T.O. bearing should also be rebushed to assure proper alignment.

4. Oil should be used to finish-ream the bushings. In the event the crankshaft does not rotate freely repeat the finish-reaming operation without oil.

5. Kerosene should be used as a cutting lubricant while rough-reaming.

6. Be sure that the dowel pins are in the cylinder block when assembling the mounting flange or cylinder cover. Use all bolts to hold the assembly together.

7. Remove the reamer by rotating it in the same direction as it is turned during the reaming operation. DO NOT TURN THE REAMER BACKWARDS.

REMOVAL AND INSTALLATION OF CRANKSHAFT BUSHING FOR 8 AND 10 HORSEPOWER ENGINES

▶ **See Figures 147, 148 and 149**

The illustrations show the mounting flange for a vertical engine. Procedures also apply to the cylinder cover for a horizontal engine. Use tool No. 670247 removal end and arbor press to press bushing from P.T.O. bearing end. Note the position of oil slots in the bushing which must align with the oil slots in the mounting flange.

To install, insert a new bushing on the installation end of tool No. 670247. Position the slots so they properly align with the oil slots in the cover and press the bushing in with an arbor press.

After the new bushing is installed, use a light coating of oil and finish reaming with reamer, part No. 670248 (handle 690160). Assemble the P.T.O. mounting flange to the cylinder. Use all bolts with dowel pins to hold the assembly in alignment. Insert the tool through the bushing and cylinder crankshaft magneto end bearing as shown. Rotate the cutting edge clockwise in the P.T.O. bushing. Remove the tool in the same direction of rotation. Do not allow the cutting edge of reamer to touch the magneto end of the cylinder.

BALL BEARING SERVICE H60 THROUGH HM100 H.P. HORIZONTAL CRANKSHAFT ENGINES

▶ **See Figure 150**

1. Prior to attempting removal of the cylinder cover, observe the area around the crankshaft P.T.O. oil seal. Compare

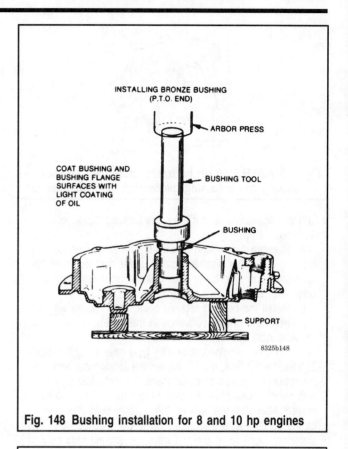

Fig. 148 Bushing installation for 8 and 10 hp engines

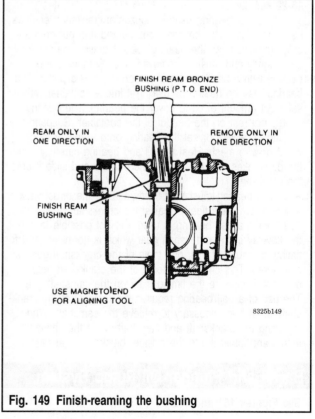

Fig. 149 Finish-reaming the bushing

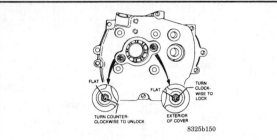

Fig. 150 H60-HM100 HP horizontal crankshaft engines — locking and unlocking the bearing from the outside

it with the illustration, and if there are bearing locks, follow instructions below:

a. Remove the locking nuts using the proper socket wrench. Note fiber washer located under nut: this must be reinstalled. Lift side cover from cylinder after removing the side cover bolts.

b. Install the bearing retainer bolts, fiber washer and locking nuts in the proper sequence in the cover.

2. Also note the following points:

a. On some engines, a locking type retainer bolt is used. To release the bolt, merely loosen the locking nut and turn the retainer bolt counterclockwise to the unlocked position with needle nose pliers to permit the side cover to be removed. Note that the flats on the retainer bolts must be turned so they face the crankshaft to be relocked upon installation. Don't force them! Torque the locking nuts only to 15-22 inch lbs.

b. The ball bearing used in horizontal crankshaft engines has a restricted fit. The bearing is heated and put onto the cold crankshaft. As the bearing cools it grasps the crankshaft tightly and must be removed cold. Remove the ball bearing with a bearing splitter (separator) and a puller. The bearing may be heated by placing it into a container with a sufficient amount of oil to cover the bearing. The bearing should not rest on the bottom of the container. Suspend the bearing on a wire or set the bearing onto a spacer block of wood or wire mesh. Heat the oil and bearing carefully until the oil smokes, quickly remove the bearing and slide it onto the crankshaft.

c. The bearing must seat tightly against the thrust washer which in turn rests tightly against the crankshaft gear.

d. When a ball bearing is used it is not possible to see the keyway in the crankshaft gear which is normally used for timing. Because of this, one tooth of the crankshaft gear is chamfered. This chamfered tooth of the crankshaft gear is positioned opposite the timing mark on the camshaft gear. The use of a ball bearing requires the removal of the crankshaft when it is necessary to remove the camshaft. When replacing the crankshaft and camshaft, mate the timing marks and insert it into the cylinder block as an assembly.

Barrel and Plunger Oil Pump System

▶ **See Figures 151 and 152**

This system is driven by an eccentric on the camshaft. Oil is drawn through the hollow camshaft from the oil sump on its intake stroke. The passage from the sump through the camshaft is aligned with the pump opening. As the camshaft

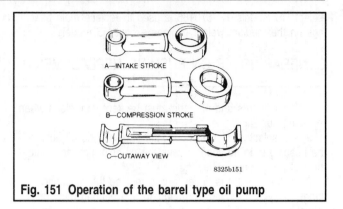

Fig. 151 Operation of the barrel type oil pump

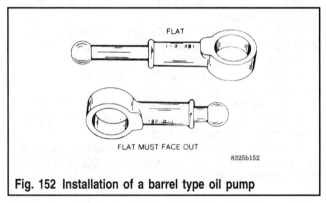

Fig. 152 Installation of a barrel type oil pump

continues rotation (pressure stroke), the plunger force the oil out. The other port in the camshaft is aligned with the pump, and directs oil out of the top of the camshaft.

At the top of the camshaft, oil is forced through a crankshaft passage to the top main bearing groove which is aligned with the drilled crankshaft passage. Oil is directed through this passage to the crankshaft connecting rod journal and then spills from the connecting rod to lubricate the cylinder walls. Splash is used to lubricate the other parts of the engine.

A pressure relief port in the crankcase relieves excessive pressures when the oil viscosity is extremely heavy due to cold temperatures, or when the system is plugged or damaged. Normal pressure is 7 psi.

SERVICE

Remove the mounting flange or the cylinder cover, whichever is applicable. Remove the barrel and plunger assembly and separate the parts.

Clean the pump parts in solvent and inspect the pump plunger and barrel for rough spots or wear. If the pump plunger is scored or worn, replace the entire pump.

Before reassembling the pump parts, lubricate all of the parts in engine oil. Manually operate the pump to make sure the plunger slides freely in the barrel.

Lubricate all the parts and position the barrel on the camshaft eccentric. If the oil pump has a chamfer only on one side, that side must be placed toward the camshaft gear. The flat goes away from the gear, thus out to work against the flange oil pickup hole.

Install the mounting flange. Be sure the plunger ball seats in the recess in the flange before fastening it to the cylinder.

Splash Lubrication

▶ **See Figure 153**

Some engines utilize the splash type lubrication system. The oil dipper, on some engines, is cast onto the lower connecting rod bearing cap. It is important that the proper parts are used to ensure the longest engine life.

Gear Type Oil Pump System

▶ **See Figure 154**

The gear type lubrication pump is a crankshaft driven, positive displacement pump. It pumps oil from the oil sump in the engine base to the camshaft, through the drilled camshaft passage to the top main bearing, through the drilled crankshaft, to the connecting rod journal on the crankshaft.

Spillage from the connecting rod lubricates the cylinder walls and normal splash lubricates the other internal working parts. There is a pressure relief valve in the system.

SERVICE

Disassemble the pump as follows: remove the screws, lockwashers, cover, gear, and displacement member.

Wash all of the parts in solvent. Inspect the oil pump drive gear and displacement member for worn or broken teeth, scoring, or other damage. Inspect the shaft hole in the drive gear for wear. Replace the entire pump if cracks, wear, or scoring is evident.

To replace the oil pump, position the oil pump displacement member and oil pump gear on the shaft, then flood all the parts with oil for priming during the initial starting of the engine.

The gasket provides clearance for the drive gear. With a feeler gauge, determine the clearance between the cover and the oil pump gear. The clearance desired is 0.006-0.007 in. (0.15-0.18mm). Use gaskets, which are available in a variety of sizes, to obtain the correct clearance. Position the oil pump cover and secure it with the screws of lockwashers.

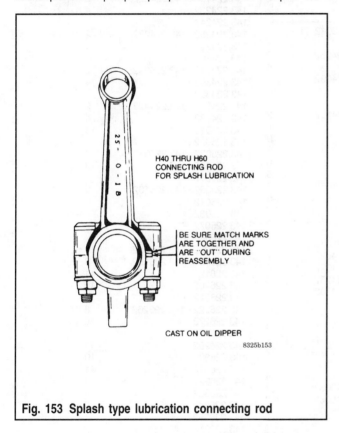

Fig. 153 Splash type lubrication connecting rod

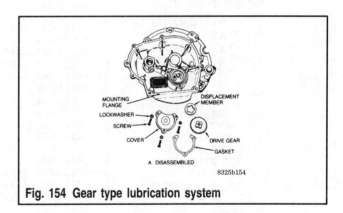

Fig. 154 Gear type lubrication system

Craftsman Engines Cross Reference Chart

Craftsman Engine Models	See Column	Craftsman Engine Models	See Column	Craftsman Engine Models	See Column
143.50040	7	143.196042–143.196072	10	143.224022	
143.50045		143.196082	8	143.224032	4
				143.224062	2
143.131022–143.131102	1	143.197012	3	143.224072	4
143.135012–143.135112	9	143.197022	5	143.224092–143.224132	2
143.136012–143.136052	8	143.197032		143.224142	1
143.137012	7	143.197042–143.197072	3	143.224162–143.224222	2
143.137032		143.197082	5	143.224232	4
				143.224242	
143.141012–143.143032	1	143.201032–143.203012	1	143.224252–143.224282	2
143.145012–143.145072	9	143.204022	4	143.224292	4
143.146012	8	143.204032–143.204052	2	143.224302	
143.146022		143.204062	4	143.224312–143.224342	2
143.147012–143.147032	7	143.204072–143.204092	2	143.224352	4
		143.204102	4	143.224362	
143.151012–143.153032	1	143.204132		143.224372–143.224422	2
143.154012–143.154142	2	143.204142–143.204192	2	143.224432	4
143.155012–143.155062	9	143.204202	4	143.225012	13
143.156012	8	143.205022	9	143.225022	
143.156022		143.206012	8	143.225032–143.225052	9
143.157012–143.157032	7	143.206022	10	143.225062	13
		143.206032	8	143.225072	
143.161012–143.163062	1	143.207012–143.207052	3	143.225082–143.225102	9
143.164012–143.164202	2	143.207062	5	143.226012	8
143.165012–143.165052	9	143.207072	3	143.226032	
143.166012–143.166052	8	143.207082	5	143.226072	10
143.167012–143.167042	7			143.226082	
		143.213012–143.213042	1	143.226092–143.226122	11
143.171012–143.171172	1	143.214012–143.214032	2	143.226131–143.226182	8
143.171202	2	143.214042–143.214072	4	143.226192	11
143.171212–143.173042	1	143.214082–143.214252	2	143.226202	10
143.174012–143.174292	2	143.214262–143.214282	4	143.226212	
143.175012–143.175072	9	143.214292	2	143.226222–143.226262	8
143.176012–143.176092	8	143.214302		143.226272	10
143.177012–143.177072	7	143.214312	4	143.226282	
		143.214322		143.226292	11
143.181042–143.183042	1	143.214332	2	143.226302	10
143.184012–143.184212	2	143.214342		143.226312	11
143.184232–143.184252	4	143.214352	4	143.226322	8
143.184262–143.184402	2	143.216012–143.216032	10	143.226332	
143.185012–143.185052	9	143.216042–143.216062	3	143.226342	10
143.186012	8	143.216072–143.216092	10	143.226352	11
143.186022–143.186042	10	143.216122	8	143.227012–143.227072	12
143.186052	8	143.216132	11		
143.186062		143.216142	8	143.233012	1
143.186072–143.186112	10	143.216152	11	143.233032	
143.186122	8	143.216162		143.233042	
143.187022–143.187102	3	143.216172		143.234022–143.234052	2
		143.216182	8	143.234062–143.234092	4
143.191012–143.191052	1	143.217012–143.217032	5	143.234102–143.234162	2
143.194012–143.194052	2	143.217042–143.217072	3	143.234192	
143.194062	4	143.217092	5	143.234202	
143.194072–143.194092	2	143.217102	3	143.234212–143.234232	4
143.194102	4			143.234242–143.234262	2
143.194112–143.194142	2	143.223012–143.223052	1	143.235012	13
143.195012	9	143.224012	2		
143.195022					
143.196012–143.196032	8				

8325bC17

Craftsman Engines Cross Reference Chart (cont.)

Craftsman Engine Models	See Column	Craftsman Engine Models	See Column	Craftsman Engine Models	See Column
143.235022		143.246242	10	143.264422	2
143.235032	6	143.246252	11	143.264432-143.264482	4
143.235042	13	143.246262	10	143.264492	2
143.235052		143.246272-143.247292	11	143.264502	
143.235062	9	143.246302	10	143.264512	4
143.235072	6	143.246312		143.264522	2
143.236012	8	143.246322	11	143.264542	
143.236022-143.236042	11	143.246332		143.264562-143.264672	4
143.236052	8	143.246342	10	143.264682	2
143.236062	11	143.246352	8	143.265012-143.265192	6
143.236072		143.246362	16	143.266012	11
143.236082	8	143.246382		143.266022	
143.236092	10	143.246392	8	143.266032	8
143.236102	8			143.266042	10
143.236112		143.254012-143.254052	2	143.266052	
143.236122	10	143.254062	4	143.266062	8
143.236132	8	143.254072-143.254122	2	143.266082	
143.236142	11	143.254142-143.254192	4	143.266092-143.266132	10
143.236152	8	143.254212	2	143.266142-143.266242	11
143.237012	12	143.254222		143.266252	8
143.237022		143.254232-143.254292	4	143.266262	11
143.237032	5	143.254302	2	143.266272-143.266302	10
143.237042	3	143.254312		143.266312	11
		143.254322	4	143.266322	
143.244032	2	143.254332	2	143.266332	10
143.244042	4	143.254342	4	143.266342	11
143.244052		143.254352		143.266352	10
143.244062		143.254362	2	143.266362	11
143.244072-143.244112	2	143.254372	4	143.266372-143.266412	8
143.244122-143.244142	4	143.254382		143.266422	11
143.244202	2	143.254392	2	143.266432-143.266452	8
143.244212	4	143.254402	4	143.266462	16
143.244222	2	143.254412		143.266472	
143.244232		143.254432	2	143.266482	11
143.244242	4	143.254442	4	143.267012-143.267042	3
143.244252	4	143.254452	2		
143.244262-143.244282	2	143.254462	4	143.274022-143.274072	4
143.244292-143.244332	4	143.254472	2	143.274092-143.274132	2
143.245012	6	143.254482		143.274142	4
143.245042	9	143.254492	4	143.274152	
143.245052-143.245072	13	143.254502-143.254532	2	143.274162-143.274182	2
143.245082		143.255012-143.255112	6	143.274192-143.274242	4
143.245092	6	143.256012	11	143.274252	2
143.245102-143.245132	13	143.256022	8	143.274262	4
143.245142	6	143.256032	10	143.274272-143.274322	2
143.245152		143.256042	11	143.274402-143.274482	4
143.245162	13	143.256052	8	143.275012-143.275052	6
143.245172	6	143.256062	11	143.276022	4
143.245182		143.256072		143.276032	11
143.245192	13	143.256082	8	143.276042	
143.246012	8	143.256092		143.276052	16
143.246022	11	143.256102	10	143.276062-143.276162	11
143.246032		143.256112	11	143.276182	8
143.246042	8	143.256122	8	143.276192	10
143.246052-143.246072	10	143.256132	10	143.276202	8
143.246082	11	143.257012-143.257072	3	143.276222	10
143.246092				143.276242	11
143.246102	10	143.264012-143.264042	2	143.276252	8
143.246112		143.264052-143.264082	4	143.276262	11
143.246122	11	143.264092	2	143.276272	11
143.246132	10	143.264102	4	143.276282	10
143.246142		143.264232-143.264342	2	143.276292	11
143.246152-143.246212	11	143.264352-143.264372	4	143.276302	11
143.246222	10	143.264382	2	143.276322-143.276342	10
143.246232	11	143.264392-143.264412	4	143.276352	11

8325bC18

Craftsman Engines Cross Reference Chart (cont.)

Craftsman Engine Models	See Column	Craftsman Engine Models	See Column	Craftsman Engine Models	See Column
143.276362	16	143.531052	1	143.596022	
143.276372–143.276392	10	143.531082		143.596042	8
143.276402	16	143.531122		143.596052	10
143.276412	8	143.531132		143.596072–143.596122	8
143.276422	10	143.531142	2	143.597012–143.597032	3
143.276432–143.276472	11	143.531152	1		
143.276482	16	143.531172		143.601022–143.601062	1
143.277012	3	143.531182		143.604012	2
143.277022		143.534012–143.534072	2	143.604022	4
		143.535012–143.535062	9	143.604032	2
143.284012	2	143.536012–143.536062	8	143.604042	
143.284022	1	143.537012	7	143.604052	4
143.284032	2			143.604062	2
143.284042	4	143.541012	1	143.604072	
143.284052	2	143.541042–143.541062	1	143.605012	9
143.284062		143.541112–143.541152	1	143.605022	
143.284072	4	143.541172–143.541202	1	143.605052	
143.284082	2	143.541222	1	143.606012–143.606052	10
143.284092		143.541282–143.541302	1	143.606092	8
143.284102	4	143.544012–143.544042	2	143.606102	10
143.284112	2	143.545012–143.545042	9	143.607012–143.607032	3
143.284142		143.546012–143.546022	8	143.607042–143.607062	3
143.284152		143.547012–143.547032	7		
143.284162				143.611012–143.611112	1
143.284182		143.551012	1	143.614012–143.614032	4
143.284212	4	143.551032		143.614042	2
143.284312	2	143.551052–143.551192	1	143.614052	4
143.284322		143.554012–143.554082	2	143.614062–143.614162	2
143.284332	4	143.555012–143.555052	9	143.615012–143.615092	9
143.284342		143.556012–143.556282	8	143.616012	10
143.284352		143.557012–143.557082	7	143.616022–143.616112	8
143.284372	16			143.616122	10
143.284382	4	143.565022	9	143.616132	8
143.284402	2	143.566002–143.566202	8	143.616142	
143.284412		143.566212	9	143.617012–143.617182	3
143.284432	2	143.566222–143.566252	8		
143.284362		143.567012–143.567042	7	143.621012–143.621092	1
143.284392	2			143.624012–143.624112	2
143.284442		143.571002–143.571122	1	143.625012–143.625132	9
143.284482		143.571152	2	143.626012	10
143.284452	4	143.571162	1	143.626022	8
143.284472		143.571172		143.626032	10
143.285012	6	143.574022–143.574102	2	143.626042	8
143.285022		143.575012–143.575042	9	143.626052–143.626122	10
143.285032		143.576002–143.576202	8	143.626132	8
143.286102	16			143.626142	10
143.286022	17	143.581002–143.581102	1	143.626152	
143.286032	10	143.584012–143.584142	2	143.626162	8
143.286072–143.286092	11	143.585012–143.585042	9	143.626172	10
143.286112		143.586012–143.586042	8	143.626182	8
143.286122		143.586052–143.586062	10	143.626192	10
143.286132	10	143.586072–143.586082	8	143.626202	8
143.286142	11	143.586112–143.586142	10	143.626212	10
143.286152		143.586152	8	143.626222–143.626262	8
143.286162		143.586162	10	143.626282	11
143.286172		143.586172–143.586242	8	143.626292	10
		143.586252	10	143.626302	8
143.505010	8	143.586262–143.586282	8	143.626312	10
143.505011		143.587012–143.587042	3	143.626322	
143.521081	9			143.627012–143.627042	3
143.525021	9	143.591012–143.591142	1		
143.526011		143.594022–143.594082	2	143.631012–143.631092	1
143.526021	8	143.594092	2	143.634012	2
143.526031	8	143.594102		143.634032	
		143.595012	9	143.635012	9
		143.595042		143.635022	
		143.596012	10		

8325bC19

Craftsman Engines Cross Reference Chart (cont.)

Craftsman Engine Models	See Column	Craftsman Engine Models	See Column	Craftsman Engine Models	See Column
143.635032	6	143.656202	8	143.674012	2
143.635052	9	143.656212	11	143.675012	6
143.636012	11	143.656222		143.675022	
143.636022		143.656232	10	143.675032	9
143.636032	10	143.656242	11	143.675042	6
143.636042	11	143.656252	8	143.676012	11
143.636052	8	143.656262	10	143.676022	
143.636062	10	143.656272		143.676032	10
143.636072	11	143.656282	11	143.676042	11
143.637012	3	143.657012–143.657052	3	143.676052	
				143.676062	16
143.641012–143.641062	1	143.661012–143.661062	1	143.676072	
143.641072	2	143.664012–143.664332	2	143.676102	10
143.644012–143.644082	2	143.665012–143.665082	6	143.676112	8
143.645012–143.645032	6	143.666012	10	143.676122	10
143.646012–143.646032	10	143.666022		143.676132	8
143.646042	11	143.666032	11	143.676142	11
143.646052		143.666042–143.666072	10	143.676152	16
143.646072–143.646102	10	143.666082	11	143.676162	
143.646112	8	143.666092		143.676172	10
143.646122	10	143.666102–143.666142	8	143.676182	11
143.646132		143.666152	11	143.676192	10
143.646142	11	143.666162		143.676202	11
143.646152	10	143.666172	8	143.676212	16
143.646162	11	143.666202		143.676222	11
143.646172	10	143.666222	10	143.676232	8
143.646182		143.666232	11	143.676242	8
143.646192	8	143.666242	8	143.676252	11
143.646202	10	143.666252	10	143.676262	16
143.646212–143.646232	11	143.666272	8	143.677012	3
143.647012–143.647062	3	143.666282	10	143.677022	
		143.666292	8		
143.651012–143.651072	1	143.666302	10		
143.654022–143.654322	2	143.666312		143.686012	17
143.655012	6	143.666322	11	143.686022	
143.655032		143.666332	16	143.686032	11
143.656012–143.656052	8	143.666342	10	143.686042	
143.656062	10	143.666352	11	143.686052	
143.656082	11	143.666362	16	143.686062	10
143.656092	8	143.666372	8	143.686072	9
143.656102	10	143.666382	10	143.687012	3
143.656112	8	143.667012	3		
143.656122–143.656152	10	143.667022		143.694126	4
143.656162–143.656182	8	143.667032	6	143.694132	2
143.656192	10	143.667042–143.667082	3	143.694134	11

8325bC20

Torque Specifications

Model/Part	Inch Pounds	Ft. Pounds
Cylinder Head Bolts	160–200	13–16
Connecting Rod Bolts	65–75	5.5–6
ECH90, ECV100, TNT100	75–80	6.2–6.7
TVS120, 5 H.P. Small Frame (Durlok Rod Bolts)	110–130	9.1–10.8
6 H.P. Medium Frame	86–110	7.1–9.1
6 H.P. Medium Frame (Durlok Rod Bolts)	130–150	10.8–12.5
ECV105, ECV110, ECV120, TNT120	80–95	6.6–7.9
7, 8 & 10 Medium Frame	106–130	8.8–10.8
7, 8 & 10 Medium Frame (Durlok Rod Bolts)	150–170	12.5–14.1
Cylinder Cover or Flange-to-Cylinder	65–110	5.5–9
Cylinder Cover 6–7 H.P. Medium Frame, H Models	100–140	8.3–11.6
Flywheel Nut	360–396	30–33
Spark Plug	180–360	15–30
Magneto Stator to Cylinder	40–90	3.3–7.5
Starter to Blower Housing or Cylinder	40–60	3.5–5
Housing Baffle to Cylinder	48–72	4–6
Breather Cover (Top Mount ECV)	40–50	3.3–4.1
Breather Cover	20–26	1.7–2.1
Intake Pipe to Cylinder	72–96	6–8
Carburetor to Intake Pipe	48–72	4–6
Air Cleaner to Carburetor (Plastic)	8–12	1
Tank Plate to Bracket (Plastic)	100–144	9–12
Tank to Housing	45–65	3.7–5
Muffler Bolts to Cylinder		
1–5 H.P. Small Frame	30–45	2.5–3.5
4–5 H.P. Medium Frame	90–150	8–12
6:1 Gear Reduction Cover to Housing	100–144	8.5–12
Gear Reduction Cover to Housing	65–110	5–9
Oil Drain Plug		
1/8—27	35–50	1.1–4.1
1/4—18	65–85	4.5–7
3/8—18	80–100	6.6–9
5/8—18	90–150	7.5–12.5
1/2–14	80–100	6.6–9
Ball Bearing Retainer 2.5		
2.5–5 H.P. Small Frame	45–60	3.7–5
5–10 H.P. Medium Frame	15–22	1.5
Craftsman Exclusive Fuel System to Cylinder	72–96	6–8
Electric Starter-to-Cylinder	50–60	4–5

8325bC21

Cross Reference for Vertical Crankshaft Engines

Size	Model	Column
6 HP	V60	8
	VH60	8
7 HP	V70	10
	VH70	10
	VM70	17
8 HP	V80	11
	VM80	11
10 HP	VM100	16

8325bC22

Cross Reference Chart for Horizontal Crankshaft Engines

Size	Model	Column
6 HP	H60	8
	HH60	8
7 HP	H70	10
	HH70	10
	HM70	17
8 HP	H80	11
	HM80	11
10 HP	HM100	16

8325bC23

Engine Specifications

Reference Column	1	2	3	4	5	6	7	8	9	10	11	12
Displacement	7.75	9.06	10.5	10.0	10.5	12.0	11.04	12.17	11.5	12.0	10.0	12.0
Stroke	1²⁷⁄₃₂"	1²⁷⁄₃₂"	1¹⁵⁄₁₆"	1²⁷⁄₃₂"	1¹⁵⁄₁₆"	1¹⁵⁄₁₆"	2¼"	2¼"	1¹⁵⁄₁₆"	1¹⁵⁄₁₆"	1²⁷⁄₃₂"	1¹⁵⁄₁₆"
Bore	2.3125 2.3135	2.5000 2.5010	2.625 2.626	2.625 2.626	2.625 2.626	2.812 2.813	2.5000 2.5010	2.625 2.626	2.750 2.751	2.812 2.813	2.625 2.626	2.812 2.813
Timing Dimension Before Top Dead Center for Vertical Engines	V.060 .070	V.065	V.035	.035	.035	V.040 .060	V.050	H.050	V.035	V.035	V.035	V.035
Timing Dimension Before Top Dead Center for Horizontal Engines	H.060 .070	H.030 .040	H.035			H.055	H.050	H.050				
Point Setting	.020	.020	.020	.020	.020	.020	.020	.020	.020	.020	.020	.020
Spark Plug Gap	.030	.030	.030	.030	.030	.030	.030	.030	.030	.030	.030	.030
Valve Clearance	.010 Both	.010 Both	.010 Both	.010 Both	.010 Both	.010 Both	.010 Both	.010 Both	.010 Both	.010 Both	.010 Both	.010 Both
Valve Seat Angle	46°	46°	46°	46°	46°	46°	46°	46°	46°	46°	46°	46°
Valve Spring Free Length	1.135"	1.135"	1.135"	1.135"	1.135"	1.135"	1.562"	1.462"	1.135"	1.135"	1.135"	1.135"
Valve Guides Over-Size Dimensions	.2805 .2815	.2805 .2815	.2805 .2815	.2805 .2815	.2807 .2817	In .280 Ex .278	.3432 .3442	.343 .344	.2805 .2815	.2805 .2815	.2805 .2815	.2805 .2815
Valve Seat Width	.035 .045	.035 .045	.035 .045	.035 .045	.035 .045	.035 .045	.042 .052	.042 .052	.035 .045	.035 .045	.035 .045	.035 .045
Crankshaft End Play	.005 .027	.005 .027	.005 .027	.005 .027	.005 .027	.005 .027	.005 .027	.005 .027	.005 .027	.005 .027	.005 .027	.005 .027
Crankpin Journal Diameter	.8610 .8615	.8610 .8615	.9995 1.0000	.8610 .8615	.9995 1.0000	.9995 1.0000	1.0615 1.0620	1.0615 1.0620	.9995 1.0000	.9995 1.0000	.8610 .8615	.9995 1.0000
Cylinder Main Bearing Dia.	.8755 .8760	.8755 .8760	1.0005 1.0010	.8755 .8760	1.0005 1.0010	1.0005 1.0010	1.0005 1.0010	1.0005 1.0010	1.0005 1.0010	1.0005 1.0010	.8755 .8760	1.0005 1.0010

8325bC11

Specification												
Cylinder Cover Main Bearing Dia.	.8755 .8760	.8755 .8760	1.0005 1.0010	.8755 .8760	1.2010 1.2020	1.0005 1.0010	1.0005 1.0010	1.0005 1.0010	1.0005 1.0010	1.0005 1.0010	.8755 .8760	1.0005 1.0010
Conn. Rod. Dia. Crank Bearing	.8620 .8625	.8620 .8625	1.0005 1.0010	.8620 .8625	1.0005 1.0010	1.0005 1.0010	1.0630 1.0635	1.0630 1.0635	1.0005 1.0010	1.0005 1.0010	.8620 .8625	1.0005 1.0010
Piston Diameter	2.3090 2.3095	2.4950 2.4955	2.6260 2.6205	2.6200 2.6205	2.604 2.608	2.8070 2.8075	2.492 2.4945	2.6210 2.6215	2.7450 2.7455	2.8070 2.8075	2.6200 2.6205	2.8070 2.8075
Piston Pin Diameter	.5629 .5631	.5629 .5631	.5629 .5631	.5629 .5631	.5631 .5635	.5629 .5631	.6248 .6250	.6248 .6250	.5629 .5631	.5629 .5631	.5629 .5631	.5629 .5631
Width of Comp. Ring Groove	.0955 .0977	.0955 .0975	.0925 .0935	.0955 .0975	.0955 .0975	.0955 .0975	.0955 .0975	.0955 .0975	.0795 .0815	.0955 .0975	.0955 .0975	.0955 .0975
Width of Oil Ring Grooves	.125 .127	.125 .127	.156 .158	.156 .158	.156 .158	.156 .158	.156 .158	.156 .158	.1565 .1585	.1565 .1585	.1565 .1585	.1565 .1585
Side Clearance of Ring Groove (Top) Comp.	.002 .005	.002 .004	.002 .004	.002 .005	.002 .005	.003 .004		.002 .004	.002 .004	.003 .004	.003 .0045	.0028 .0039
(Bot.) Oil		.001 .004	.001 .004	.001 .004	.001 .004	.002 .003	.002 .003	.002 .004	.001 .002	.001 .002	.0010 .0030	.0018 .0038
Ring End Gap	.007 .020	.007 .020	.007 .020	.007 .020	.007 .020	.007 .020	.007 .020	.007 .020	.007 .020	.007 .020	.007 .020	.007 .020
Top Piston Land Clearance	.0015 .0145	.015 .018	.0165 .0215	.017 .022	.017 .022	.017 .022	.015 .018	.017 .020	.024 .027	.018 .021	.017 .022	.017 .022
Piston Skirt Clearance	.0025 .0040	.0045 .0060	.0045 .0060	.0045 .0060	.0050 .0065	.0045 .0060	.0055 .0070	.0035 .0050	.0045 .0060	.0045 .0060	.0045 .0060	.0045 .0060
Camshaft Bearing Dia.	.4975 .4980	.4975 .4980	.4975 .4980	.4975 .4980	.505 .513	.4975 .4980	.6230 .6235	.6230 .6235	.4975 .4980	.4975 .4980	.4975 .4980	.4975 .4980
Dia. of Crankshaft Mag. Main Brg.	.8735 .8740	.8735 .8740	.9985 .9990	.8735 .8740	.9985 .9990	.9985 .9990	.9985 .9990	.9985 .9990	.9990 .9995	.9990 .9995	.8735 .8740	.9985 .9990
Dia. of Crankshaft P.T.O. Main Brg.	.8735 .8740	.8735 .8740	.9985 .9990	.8735 .8740	.9985 .9990	.9985 .9990	.9985 .9990	.9985 .9990	.9985 .9990	.9985 .9990	.8735 .8740	.9985 .9990

A. For VM80 & HM80 engines only - Displacement is 19.41"

B. For VM80 & HM80 engines only - Bore is 3.125" (3⅛"). 3.126"

Note C. For VM80 & HM80 engines only - Piston Diameter is 3.1205" 3.1195"

8325bC12

13-20 HP

Engine Identification

▶ **See Figures 155 and 156**

Lauson 4 cycle engines are identified by a model number stamped on a nameplate. The nameplate is located on the crankcase of vertical shaft models and on the blower housing of horizontal shaft models.

A typical model number appears on the illustration showing the location of the nameplate for vertical crankshaft engines. This number is interpreted as follows:

- V — vertical shaft engine
- 130 — 13.0 horsepower
- 70360J — the specification number. The last three numbers (360) indicate that this particular engine is a variation on the basic model line.
- 2361J — serial number
- 2 — year of manufacture
- 361 — the calendar day of manufacture
- J — line and shift location at the factory.

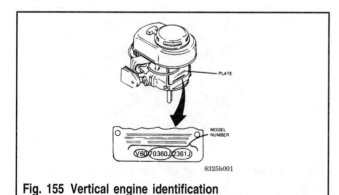

Fig. 155 Vertical engine identification

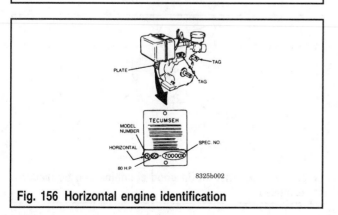

Fig. 156 Horizontal engine identification

Air Cleaner Service

SERVICE

Dry Paper Element
▶ **See Figure 157**

The Tecumseh treated paper element type air cleaner consists of a pleated paper element encased in a metal housing and must be replaced as a unit. A flexible tubing and hose clamps connect the remotely mounted air filter to the carburetor.

Clean the element by lightly tapping it. Do not distort the case. When excessive carburetor adjustment or loss of power results, inspect the air filter to see if it is clogged. Replacing a severely restricted air filter should show an immediate performance improvement.

A plain paper element is also used. It should be removed every 10 hours, or more often if the air is dusty. Tap or blow out the dirt from the inside with low pressure air. This type should be replaced at 50 hours. If clogged sooner, it may be washed in soap and water and rinsed by flushing from the inside until the water is clear. Blow dry with low pressure compressed air.

To service the KLEEN-AIRE system, remove the element, wash it in soap and mild detergent, pat dry, and then coat with oil. Squeeze the oil to distribute it evenly and remove the excess. Make sure all mounting surfaces are tight to prevent leakage.

Oil/Foam Air Cleaners

Clean and re-oil the air cleaner element every 25 hours of operation under normal operating conditions. The capacity of the oil/foam air cleaner is adequate for a full season's use without cleaning. Under very dusty conditions, clean the air cleaner every few hours of operation.

The oil/foam air cleaner is serviced in the following manner:
1. Remove the screw that holds the halves of the air cleaner shell together and retains it to the carburetor.
2. Remove the air cleaner carefully to prevent dirt from entering the carburetor.
3. Take the air cleaner apart (split the two halves).

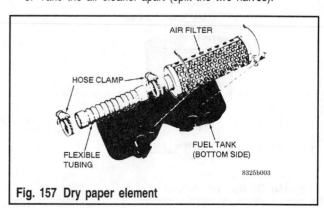

Fig. 157 Dry paper element

4. Wash the foam in kerosene or liquid detergent and water to remove the dirt.

5. Wrap the foam in a clean cloth and squeeze it dry.

6. Saturate the foam in clean engine oil and squeeze it to remove the excess oil.

7. Assemble the air cleaner and fasten it to the carburetor with the attaching screw.

Oil Bath Air Cleaner

♦ See Figure 158

Pour the old oil out of the bowl. Wash the element thoroughly in solvent and squeeze it dry. Clean the bowl and refill it with the same type of oil used in the crankcase.

Lubrication

OIL AND FUEL RECOMMENDATIONS

Use fresh (less than one month old) gasoline, of 'Regular" grade. Unleaded fuel is preferred, but leader fuel is acceptable.

Use oil having SG classification. Use these viscosities for aluminum engines:

• Summer — above 32°F (0°C): S.A.E. 30 (S.A.E. 10W-30 or 10W-40 are acceptable substitutes)

• Winter — Below 32°F (0°C): S.A.E. 5W-30 (S.A.E. 10W is an acceptable substitute). (Including Snow King Snow Blower Engines)

• Winter — Below 0°F (-18°C) only: S.A.E. 10W diluted with 10% kerosene is an acceptable substitute. (Including Snow King Snow Blower Engines)Use these viscosities for cast iron engines:

• Summer — Above 32°F (0°C): S.A.E. 30
• Winter — Above 32°F (0°C): S.A.E. 10W

Tune-Up Specifications

The follow basic specifications apply to all the engines covered in this section:

• Spark Plug Gap: 0.030 in. (0.76mm)
• Ignition Point Gap: 0.020 in. (0.50mm)
• Valve Clearance: 0.010 in. (0.25mm) for both intake and exhaust.For timing dimension, which varies from engine to engine.

Spark Plug Service

♦ See Figures 159, 160, 161, 162 and 163

Spark plugs should be removed, cleaned, and adjusted periodically. Check the electrode gap with a wire feeler gauge and adjust the gap. Replace the plugs if the electrodes are pitted and burned or the porcelain is cracked. Apply a little graphite grease to the threads to prevent sticking. Be sure the cleaned plugs are free of all foreign material.

Breaker Points

ADJUSTMENT

1. Disconnect the fuel line from the carburetor.

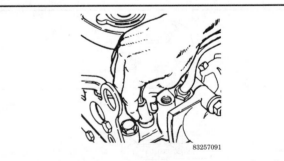

Fig. 159 Twist and pull on the boot, never on the plug wire

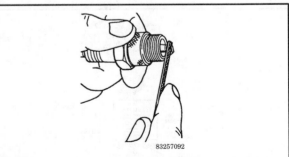

Fig. 160 Plugs that are in good condition can be filed and reused

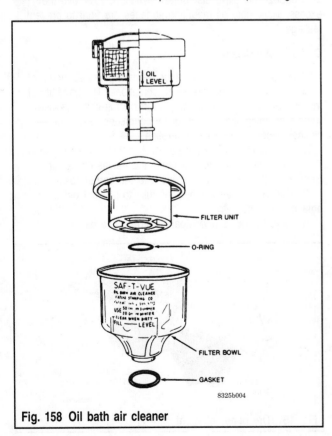

Fig. 158 Oil bath air cleaner

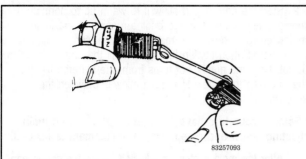

Fig. 161 Adjust the electrode gap by bending the side electrode

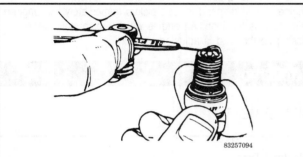

Fig. 162 Always used a wire gauge to check the plug gap

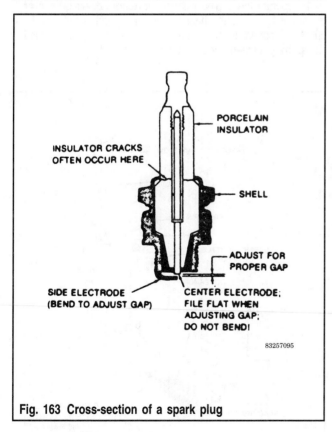

Fig. 163 Cross-section of a spark plug

2. Remove the mounting screws, fuel tank, and shroud to provide access to the flywheel.

3. Remove the flywheel with either a puller (over 3.5 hp) or by using a screwdriver to pry underneath the flywheel while tapping the top lightly with a soft hammer.

4. Remove the dust cover and gasket from the magneto and crank the engine over until the breaker points of the magneto are fully opened.

5. Check the condition of the points and replace them if they are burned or pitted.

6. Check the point gap with a feeler gauge. Adjust them, if necessary, as per the directions on the dust cover.

REPLACEMENT

1. Gain access to the points and inspect them as described above. If the points are badly pitted, follow the remaining steps to replace them.

2. Remove the nuts that hold the electrical leads to the screw on the movable breaker point spring. Remove the movable breaker point from stud.

3. Remove the screw and stationary breaker point. Put a new stationary breaker point on the breaker plate; install the screw, but do not tighten. This point must be moved to make the proper air gap when the points are adjusted.

4. Position a new movable breaker point on the stud.

5. Adjust the breaker point gap with a flat feeler gauge and tighten the screw.

6. Check the new point contact pattern and remove all grease, finger-prints, and dirt from contact surfaces.

7. Adjust the timing as described below.

Ignition Timing

ADJUSTMENT

1. Remove the cylinder head bolts, and move the head (with gasket in place) so that the spark plug hole is centered over the piston.

2. Using a ruler (through the spark plug hole) or special plunger type tool, carefully turn the engine back and forth until the piston is at exactly Top Dead Center. Tighten the thumbscrew on the tool.

3. Find the timing dimension for your engine in the specifications at the rear of the manual. Then, back off the position of the piston until it is about halfway down in the bore. Lower the ruler (or loosen the thumbscrew and lower the plunger, if using the special tool) exactly the required amount (the amount of the timing dimension). Then, hold the ruler in place (or tighten the special tool thumbscrew) and, finally carefully rotate the engine forward until the piston just touches the ruler or tool plunger.

4. Install a timing light or place a very thin piece of cellophane between the contact points. Loosen and rotate the stator just until the timing light shows a change in current flow or the cellophane pulls out of contact gap easily. Then, tighten stator bolts to specified torque.

5. Install the leads, point cover, flywheel, and shrouding.

SOLID STATE IGNITION SYSTEM CHECKOUT

The only on-engine check which can be made to determine whether the ignition system is working, is to separate the high tension lead from the spark plug and check for spark. If there is a spark, then the unit is all right and the spark plug should be replaced. No spark indicates that some other part needs replacing.

Check the individual components as follows:

• High Tension Lead: Inspect for cracks or indications of arcing. Replace the transformer if the condition of the lead is questionable.

• Low Tension Leads: Check all leads for shorts. Check the ignition cut-off lead to see that the unit is not grounded. Repair the leads, if possible, or replace them.

• Pulse Transformer: Replace and test for spark.

• Magneto: Replace and test for spark. Time the magneto by turning it counterclockwise as far as it will go and then tighten the retaining screws.

• Flywheel: Check the magnets for strength. With the flywheel off the engine, it should attract a screwdriver that is held 1 in. (25mm) from the magnetic surface on the inside of the flywheel. Be sure that the key locks the flywheel to the crankshaft.

Carburetor Mixture Adjustments

Chart of Initial Carburetor Adjustments

Adjustment	For Engines Built Prior to 1977	For Engines Built After 1977
Main Adjustment Up to 7 HP	V-60-70-1¼ H-60-70-1¼	Same
Main Adjustment VM70-80-100 & HM70-80-100	1¼	1½
Idle Adjustment Up to 7 HP	V-60-70-1 Turn H-60-70-1 Turn	Same
Idle Adjustment VM70-80-100 & HM70-80-100	1½	1¼
Idle Speed (Top of Carburetor) Regulating screw	Back out screw, then turn in until screw just touches throttle lever and continue 1 turn more (if idle RPM is given set final idle speed with a tachometer)	

8325bC27

1. If the carburetor has been overhauled, or the engine won't start, make initial mixture screw adjustments as specified in the chart.

2. Start the engine and allow it to warm up to normal running temperature. With the engine running at maximum recommended rpm, loosen the main adjustment screw until engine rpm drops off, then tighten the screw until the engine starts to cut out. Note the number of turns from one extreme to the other. Loosen the screw to a point midway between the extremes.

➥**Some carburetors have fixed jets. If there is no main adjusting screw and receptacle, no adjustment is needed.**

3. After the main system is adjusted, move the speed control lever to the idle position and follow the same procedure for adjusting the idle system.

4. Test the engine by running it under a normal load. The engine should respond to load pickup immediately. An engine that 'dies" is too lean. An engine which ran roughly before picking up the load is adjusted too rich.

Governor

ADJUSTMENT

Air Vane Type
♦ **See Figure 165**

1. Operate the engine with the governor adjusting lever or panel control set to the highest possible speed position and check the speed. If the speed is not within the recommended limits, the governed speed should be adjusted.

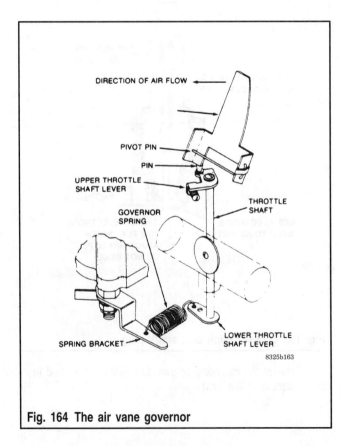

Fig. 164 The air vane governor

2. Loosen the locknut on the high speed limit adjusting screw and turn the adjusting screw out to increase the top engine speed.

Mechanical Type
▶ **See Figure 165**

1. Set the control lever to the idle position so that no spring tension affects the adjustment.
2. Loosen the screw so that the governor lever is loose in the clamp.
3. Rotate both the lever and the clamp to move the throttle to the full open position (away from the idle speed regulating screw).
4. Tighten the screw when no end-play exists in the direction of open throttle.
5. Move the throttle lever to the full speed setting and check to see that the control linkage opens the throttle.

Compression Check

1. Run the engine until warm to lubricate and seal the cylinder.
2. Remove the spark plug and install a compression gauge. Turn the engine over with the pull starter or electric starter.
3. Compression on new engines is 80 psi. If the reading is below 60 psi., repeat the test after removing the gauge and squirting about a teaspoonful of engine oil through the spark plug hole. If the compression improves temporarily following

this, the problem is probably with the cylinder, piston, and rings. Otherwise, the valves require service.

Carburetor

▶ **See Figures 166, 167, 168, 169, 170, 171, 172, 173 and 174**

➡**Four-cycle Tecumseh engines use float or diaphragm type carburetors.**

REMOVAL & INSTALLATION

1. Drain the fuel tank. Remove the air cleaner and disconnect the carburetor fuel lines.
2. If necessary, remove any shrouding or control panels to provide access to carburetor.
3. Disconnect the choke or throttle control wires at the carburetor.
4. Remove the cap screws, or nuts and lockwashers that hold the carburetor to the engine; remove the carburetor.
5. Secure the carburetor on to engine.
6. Install the shrouding or control panels. Connect the choke and throttle control wires.
7. Position the control panel to carburetor. Connect the carburetor fuel lines.
8. Install the air cleaner.
9. Adjust the carburetor as described above.

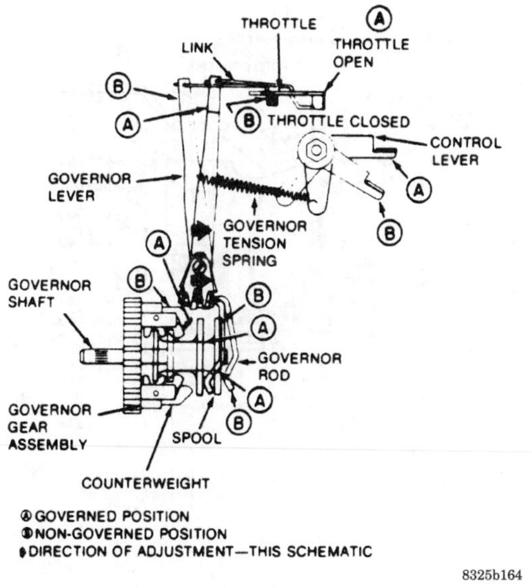

Fig. 165 The mechanical governor

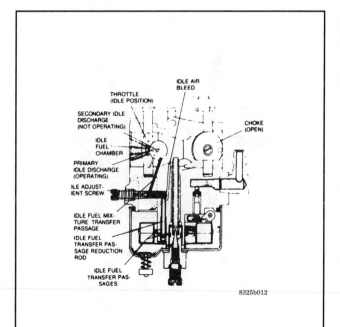

Fig. 166 The idle operation of a float feed carburetor. The throttle plate closes, restricting the flow of fuel and air, forcing the engine to run on a reduced volume of fuel and air

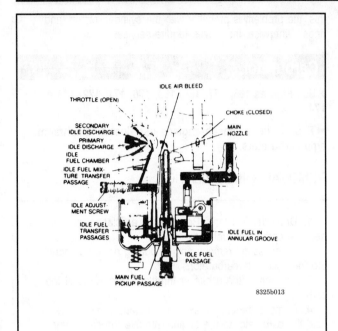

Fig. 167 The choke position of a float feed carburetor. The closed choke plate restricts air, creating a richened mixture by drawing in a greater proportion of fuel

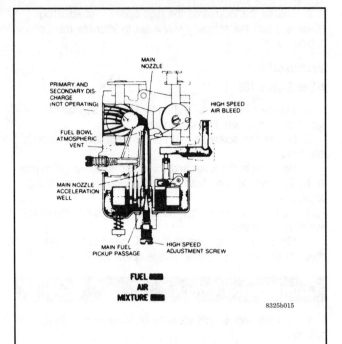

Fig. 169 The high speed operation of a float feed carburetor. The air venturi replaces the throttle plate as the restricting device and the engine runs on its greatest volume of fuel and air

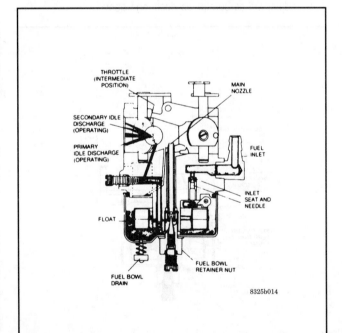

Fig. 168 The intermediate operation of a float feed carburetor. The throttle plate opens slightly to reduce restriction and the engine runs on an increased volume of fuel and air

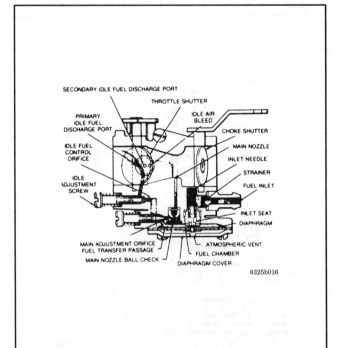

Fig. 170 The choke position of a diaphragm carburetor. The closed choke plate restricts air, creating a richened mixture by drawing in a greater proportion of fuel

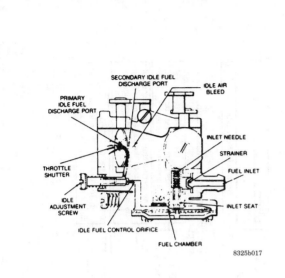

Fig. 171 The idle operation of a diaphragm carburetor. The throttle plate closes, restricting the flow of fuel and air, forcing the engine to run on a reduced volume of fuel and air

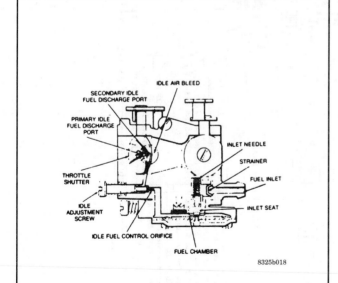

Fig. 172 The intermediate operation of a diaphragm carburetor. The throttle plate opens slightly to reduce restriction and the engine runs on an increased volume of fuel and air

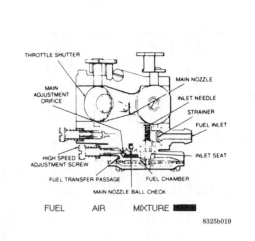

Fig. 173 The high speed operation of a diaphragm carburetor. The air venturi replaces the throttle plate as the restricting device and the engine runs on its greatest volume of fuel and air

GENERAL OVERHAUL INSTRUCTIONS

1. Carefully disassemble the carburetor removing all non-metallic parts, i.e.; gaskets, viton seats and needles, O-rings, fuel pump valves, etc.

➡**Nylon check balls used in some diaphragm carburetor models may or may not be serviceable. Check to be sure of serviceability before attempting removal.**

2. Clean all metallic parts with solvent.

➡**Nylon can be damaged if subjected to harsh cleaners for prolonged periods.**

3. The large O-rings sealing the fuel bowl to the carburetor body must be in good condition to prevent leakage. If the O-ring leaks, interfering with the atmospheric pressure in the float bowl, the engine will run rich. Foreign material can enter through the leaking area and cause blocking of the metering orifices. This O-ring should be replaced after the carburetor has been disassembled for repair. Lubricate the new O-ring with a small amount of oil to allow the fuel bowl to slide onto the O-ring properly. Hold the carburetor body in an inverted position and place the O-ring on the carburetor body and then position the fuel bowl.

4. The small O-rings used on the carburetor adjustment screws must be in good condition or a leak will develop and cause improper adjustment of carburetor.

5. Check all adjusting screws for wear. The illustration shows a worn screw and a good screw. Replace screws that are worn.

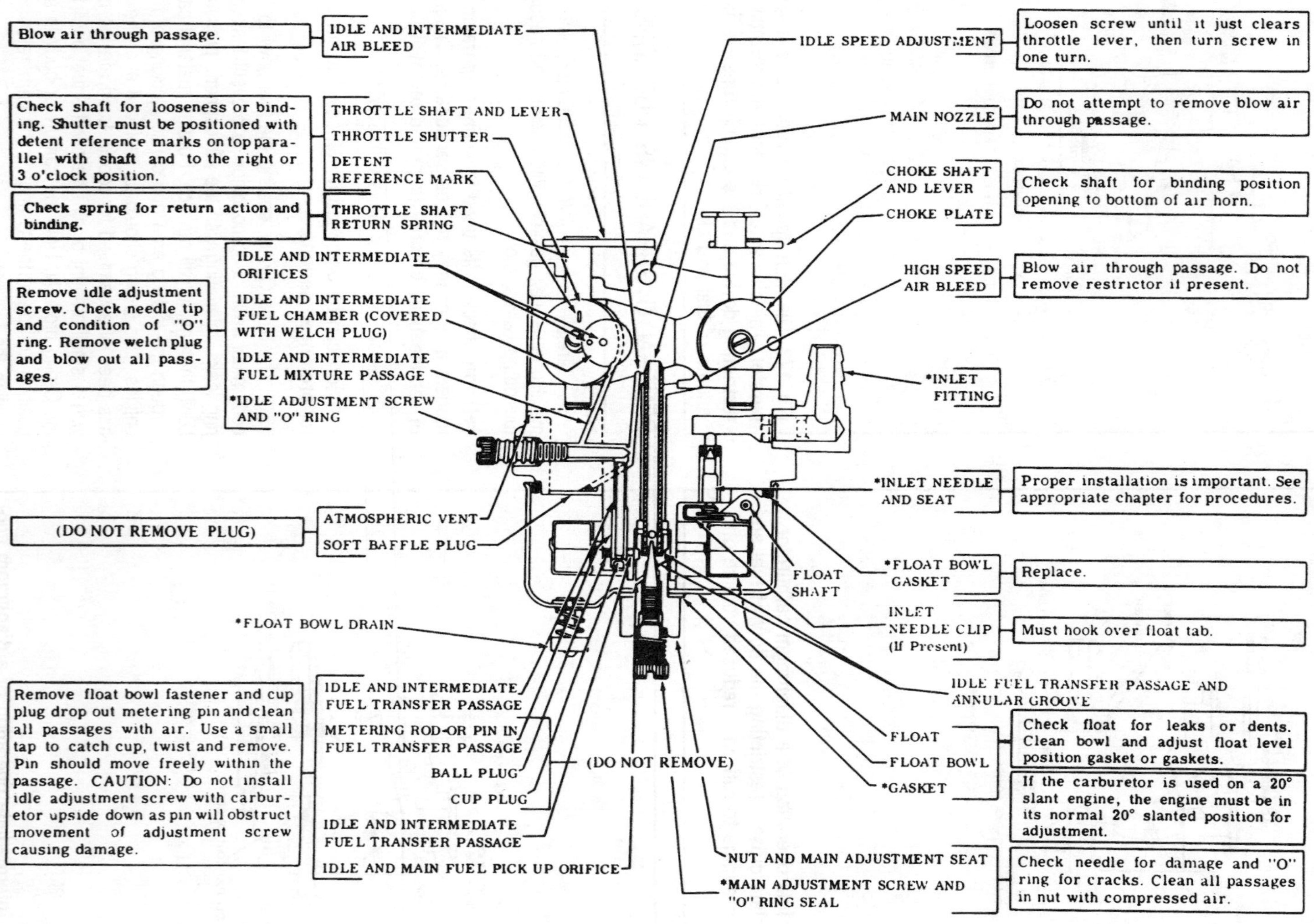

Loosen screw until it just clears throttle lever, then turn screw in one turn.

Do not attempt to remove blow air through passage.

Check shaft for binding position opening to bottom of air horn.

Blow air through passage. Do not remove restrictor if present.

Proper installation is important. See appropriate chapter for procedures.

Replace.

Must hook over float tab.

Check float for leaks or dents. Clean bowl and adjust float level position gasket or gaskets.

If the carburetor is used on a 20° slant engine, the engine must be in its normal 20° slanted position for adjustment.

Check needle for damage and "O" ring for cracks. Clean all passages in nut with compressed air.

IDLE SPEED ADJUSTMENT

MAIN NOZZLE

CHOKE SHAFT AND LEVER

CHOKE PLATE

HIGH SPEED AIR BLEED

*INLET FITTING

*INLET NEEDLE AND SEAT

*FLOAT BOWL GASKET

INLET NEEDLE CLIP (If Present)

IDLE FUEL TRANSFER PASSAGE AND ANNULAR GROOVE

FLOAT

FLOAT BOWL

*GASKET

NUT AND MAIN ADJUSTMENT SEAT

*MAIN ADJUSTMENT SCREW AND "O" RING SEAL

FLOAT SHAFT

(DO NOT REMOVE)

IDLE AND INTERMEDIATE AIR BLEED

THROTTLE SHAFT AND LEVER

THROTTLE SHUTTER

DETENT REFERENCE MARK

THROTTLE SHAFT RETURN SPRING

IDLE AND INTERMEDIATE ORIFICES

IDLE AND INTERMEDIATE FUEL CHAMBER (COVERED WITH WELCH PLUG)

IDLE AND INTERMEDIATE FUEL MIXTURE PASSAGE

*IDLE ADJUSTMENT SCREW AND "O" RING

ATMOSPHERIC VENT

SOFT BAFFLE PLUG

*FLOAT BOWL DRAIN

IDLE AND INTERMEDIATE FUEL TRANSFER PASSAGE

METERING ROD·OR PIN IN FUEL TRANSFER PASSAGE

BALL PLUG

CUP PLUG

IDLE AND INTERMEDIATE FUEL TRANSFER PASSAGE

IDLE AND MAIN FUEL PICK UP ORIFICE

Blow air through passage.

Check shaft for looseness or binding. Shutter must be positioned with detent reference marks on top parallel with shaft and to the right or 3 o'clock position.

Check spring for return action and binding.

Remove idle adjustment screw. Check needle tip and condition of "O" ring. Remove welch plug and blow out all passages.

(DO NOT REMOVE PLUG)

Remove float bowl fastener and cup plug drop out metering pin and clean all passages with air. Use a small tap to catch cup, twist and remove. Pin should move freely within the passage. CAUTION: Do not install idle adjustment screw with carburetor upside down as pin will obstruct movement of adjustment screw causing damage.

8325b096

Fig. 174 Service hints for float feed carburetors

6. Check the carburetor inlet needle and seat for wear, scoring, or other damage. Replace defective parts.

7. Check the carburetor float for dents, leaks, worn hinge or other damage.

8. Check the carburetor body for cracks, clogged passages, and worn bushings. Clean clogged air passages with clean, dry compressed air.

9. Check the diaphragms on diaphragm carburetors for cracks, punctures, distortion, or deterioration.

10. Check all shafts and pivot pins for wear on the bearing surfaces, distortion, or other damage.

➡**Each time a carburetor is disassembled, it is good practice to install a repair kit.**

11. Where there is excessive vibration, a damper spring may be used to assist in holding the float against the inlet needle thus minimizing the flooding condition. Two types of springs are available; the float shaft (hinge pin) type and the inlet needle mounted type.

12. Float shaft spring positioning:

a. The spring is slipped over the shaft.

b. The rectangular shaped spring end is hooked onto the float tab.

c. The shorter angled spring end is placed onto the float bowl gasket support.

13. Note that on late model carburetors, the spring clip fastened to the inlet needle has been revised to provide a damping effect. The clip fastens to the needle and is hooked over the float tab.

FLOAT FEED CARBURETOR

Note the following points when rebuilding these carburetors.

Throttle

1. Examine the throttle lever and plate prior to disassembly. Replace any worn parts.

2. Remove the screw in the center of the throttle plate and pull out the throttle shaft lever assembly.

3. When reassembling, it is important that the lines on the throttle plate are facing out when in the closed position. Position the throttle plates with the two lines at 12 and 3 o'clock. The throttle shaft must be held in tight to the bottom bearing to prevent the throttle plate from riding on the throttle bore of the body which would cause excessive throttle plate wear and governor hunting.

Choke

Examine the choke lever and shaft at the bearing points and holes into which the linkage is fastened and replace any worn parts. The choke plate is inserted into the air horn of the carburetor in such a way that the flat surface of the choke is toward the fuel bowl.

Idle Adjusting Screw
▶ **See Figure 175**

Remove the idle screw from the carburetor body and examine the point for damage to the seating surface on the taper. If damaged, replace the idle adjusting needle. Tension is

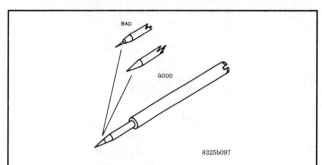

Fig. 175 Appearance of good and bad mixture adjustment needles

maintained on the screw with a coil spring and sealed with an O-ring. Examine and replace the O-ring if it is worn or damaged.

High Speed Adjusting Jet

Remove the screw and examine the taper. If the taper is damaged at the area where it seats, replace the screw and fuel bowl retainer nut as an assembly.

The fuel bowl retainer nut contains the seat for the screw. Examine the sealing O-ring on the high speed adjusting screw. Replace the O-ring if it indicates wear or cuts. During the reassembly of the high speed adjusting screw, position the coil spring on the adjusting screw, followed by the small brass washer and the O-ring seal.

Fuel Bowl

To remove the fuel bowl, remove the retaining nut and fiber washer. Replace the nut if it is cracked or worn.

The retaining nut contains the transfer passage through which fuel is delivered to the high speed and idle fuel system of the carburetor. It is the large hole next to the hex nut end of the fitting. If a problem occurs with the idle system of the carburetor, examine the small fuel passage in the annular groove in the retaining nut. This passage must be clean for the proper transfer of fuel into the idle metering system.

The fuel bowl should be examined for rust and dirt. Thoroughly clean it before installing it. If it is impossible to properly clean the fuel bowl, replace it.

Check the drain valve for leakage. Replace the rubber gasket on the inside of the drain valve if it leaks.

Examine the large O-ring that seals the fuel bowl to the carburetor body. If it is worn or cracked, replace it with a new one, making sure the same type is used (square or round).

Inlet Needle and Seat
▶ **See Figure 176**

1. The inlet needle sits on a rubber seat in the carburetor body instead of the usual metal fitting.

2. Remove it, place a few drops of heavy engine oil on the seat, and pry it out with a short piece of hooked wire.

3. The grooved side of the seat is inserted first. Lubricate the cavity with oil and use a flat faced punch to press the inlet seat into place.

4. Examine the inlet needle for wear and rounding off of the corners. If this condition does exist, replace the inlet needle.

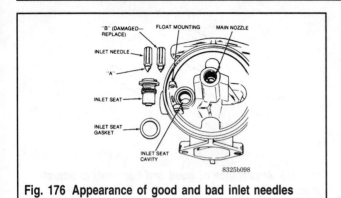

Fig. 176 Appearance of good and bad inlet needles

Fuel Inlet Fitting

1. The inlet fitting is removed by twisting and pulling at the same time.

2. Use sealer when reinstalling the fitting. Insert the tip of the fitting into the carburetor body. Press the fitting in until the shoulder contacts the carburetor. Only use inlet fittings without screens.

Float

♦ **See Figures 177 and 178**

1. Remove the float from the carburetor body by pulling out the float axle with a pair of needle nose pliers. The inlet needle will be lifted off the seat because it is attached to the float with an anchoring clip.

2. Examine the float for damage and holes. Check the float hinge for wear and replace it if worn.

3. The float level is checked by positioning a #4 drill bit across the rim between the center leg and the unmachined

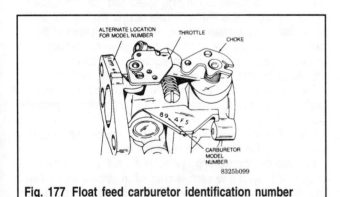

Fig. 177 Float feed carburetor identification number

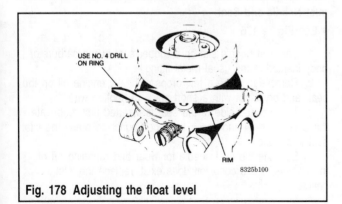

Fig. 178 Adjusting the float level

surface of the index pad, parallel to the float axle pin. If the index pad is machined, the float setting should be made with a #9 drill bit.

4. Remove the float to make an adjustment. Bend the tab on the float hinge to correct the float setting.

➡ **Direct compressed air in the opposite direction of normal flow of air or fuel (reverse taper) to dislodge foreign matter.**

Carburetor Body

1. Check the carburetor body for wear and damage.

2. If excessive dirt has accumulated in the atmospheric vent cavity, try cleaning it with carburetor solvent or compressed air. Remove the welch plug only as a last resort.

➡ **The carburetor body contains a pressed-in main nozzle tube at a specific depth and position within the venturi. Do not attempt to remove the main nozzle. Any change in nozzle positioning will adversely affect the metering quality and will require carburetor replacement.**

3. Clean the accelerating well around the main nozzle with compressed air and carburetor cleaning solvents.

4. The carburetor body contains two cup plugs, neither of which should be removed. A cup plug located near the inlet seat cavity, high up on the carburetor body, seals off the idle bleed. This is a straight passage drilled into the carburetor throat. Do not remove this plug. Another cup plug is located in the base where the fuel bowl nut seals the idle fuel passage. Do not remove this plug or the metering rod.

5. A small ball plug located on the side of the idle fuel passage seals this passage. Do not remove this ball plug.

6. The welch plug on the side of the carburetor body, just above the idle adjusting screw, seals the idle fuel chamber. This plug can be removed for cleaning of the idle fuel mixture passage and the primary and secondary idle fuel discharge ports. Do not use any tools that might change the size of the discharge ports, such as wire or pins.

Resilient Tip Needle

Replace the inlet needle. do not attempt to remove or replace the seat in the carburetor body.

Viton Seat

Using a 10-24 or 10-32 tap, turn the tap into the brass seat fitting until it grasps the seat firmly. Clamp the tap shank into a vise and tap the carburetor body with a soft hammer until the seat slides out of the body.

To replace the viton seat, position the replacement over the receptacle with the soft rubber like seat toward the body. Use a flat punch and a small hammer to drive the seat into the body until it bottoms on the shoulder.

TECUMSEH AUTOMATIC NON-ADJUSTABLE FLOAT FEED CARBURETOR

This carburetor has neither a choke plate nor idle and main mixture adjusting screws. There is no running adjustment. The float adjustment is the standard Tecumseh float setting of #4 drill bit.

Cleaning

▶ **See Figures 179, 180 and 181**

Remove all non-metallic parts and clean them using a procedure similar to that for the other carburetors. Never use wires through any of the drilled holes. Do not remove the baffling welch plug unless it is certain there is a blockage under the plug. There are no blind passageways in this carburetor.

Some engines use a variation on the Automatic Nonadjustable carburetor which has a different bowl hold-on nut and main jet orifices. There are two main jet orifices and a deeper fuel reserve cavity, but service procedures are the same.

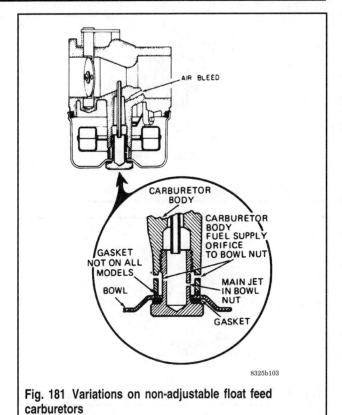

8325b103

Fig. 181 Variations on non-adjustable float feed carburetors

WALBRO AND TILLOTSON FLOAT FEED CARBURETORS

▶ **See Figures 182, 183, 184 and 185**

Procedures are similar to those for the Tecumseh float carburetor with the exceptions noted below.

Main Nozzle

The main nozzle in Wallbro carburetors is cross drilled after it is installed in the carburetor. Once removed, it cannot be reinstalled, since it is impossible to properly realign the cross drilled holes. Grooved service replacement main nozzles are available which allow alignment of these holes.

Float Shaft Spring

Carefully position the float shaft spring on models so equipped. The spring dampens float action when properly as-

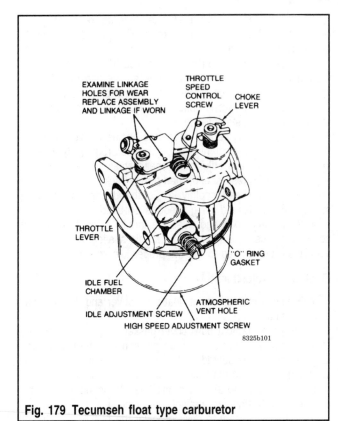

8325b101

Fig. 179 Tecumseh float type carburetor

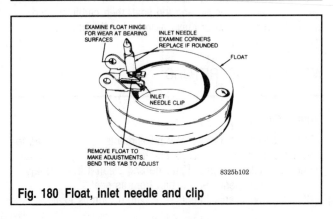

8325b102

Fig. 180 Float, inlet needle and clip

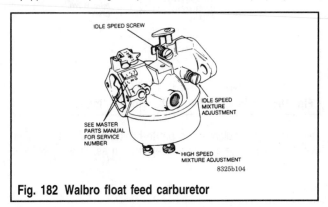

8325b104

Fig. 182 Walbro float feed carburetor

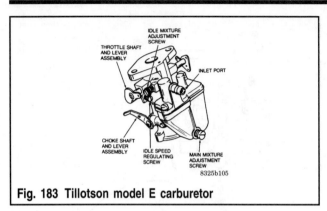

Fig. 183 Tillotson model E carburetor

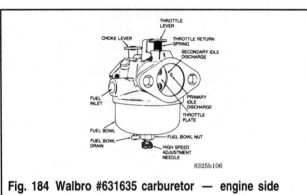

Fig. 184 Walbro #631635 carburetor — engine side

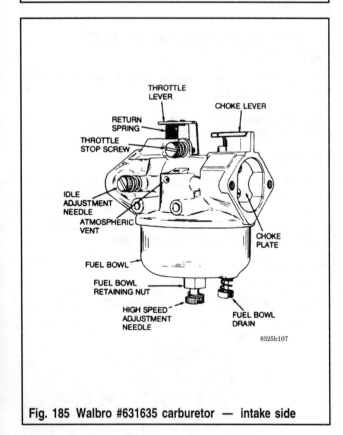

Fig. 185 Walbro #631635 carburetor — intake side

sembled. Use needlenosed pliers to hook the end of the spring over the float hinge and then insert the pin as far as possible

before lifting the spring from the hinge into position. Leaving the spring out or improper installation will cause unbalanced float action and result in a touchy adjustment.

Float Adjustment

1. To check the float adjustment, invert the assembled float carburetor body. Check the clearance between the body and the float, opposite the hinge. Clearance should be 1/8 in. ± 1/64 in. (3mm ± 0.4mm).

2. To adjust the float level, remove the float shaft and float. Bend the lip of the float tang to correct the measurement.

3. Assemble the parts and recheck the adjustment.

TILLOTSON E FLOAT FEED TYPE CARBURETOR

The following adjustments are different for this carburetor.

Running Adjustment

1. Start the engine and allow it to warm up to operating temperatures. Make sure the choke is fully opened after the engine is warmed up.

2. Run the engine at a constant speed while slowly turning the main adjustment screw in until the engine begins to lose speed; then slowly back it out about 1/8-1/4 of a turn until maximum speed and power is obtained (4000 rpm). This is the correct power adjustment.

3. Close the throttle and cause the engine to idle slightly faster than normal by turning the idle speed regulating screw in. Then turn the idle adjustment speed screw in until the engine begins to lose speed; then turn it back 1/4-1/2 of a turn until the engine idles smoothly. Adjust the idle speed regulating screw until the desired idling speed is acquired.

4. Alternately open and close the throttle a few times for an acceleration test. If stalling occurs at idle speeds, repeat the adjustment procedures to get the proper idle speed.

Float Level Adjustment

1. Remove the carburetor float bowl cover and float mechanism assembly.

2. Remove the float bowl cover gasket and, with the complete assembly in an upside down position and the float lever tang resting on the seated inlet needle, a measurement of 15/64 in. (6mm) should be maintained from the free end flat rim, or edge of the cover, to the toe of the float. Measurement can be checked with a standard straight rule or depth gauge.

3. If it is necessary to raise or lower the float lever setting, remove the float lever pin and the float, then carefully bend the float lever tang up or down as required to obtain the correct measurement.

WALBRO CARBURETORS

Adjustment

The following initial carburetor adjustments are to be used to start the engine. For proper carburetion adjustment, the atmospheric vent must be open. Examine and clean it if necessary.

1. Idle adjustment — 1 1/4 turn from its seat.
2. High speed adjustment — 1 1/2 turn form its seat.

3. Throttle stop screw — 1 turn after contacting the throttle lever.

After the engine reaches normal operating temperature, make the final adjustments for best idle and high speed within the following ranges. Recommended speeds: Idle: 1800-2300 rpm. High speed: 3450-3750 rpm.

Rebuilding Notes

▶ **See Figures 186, 187 and 188**

1. The throttle plate is installed with the lettering (if present) facing outward when closed. The throttle plate is installed on the throttle level with the lever in the closed position if there is binding after the plate is in position, loosen the throttle plate and reposition it.

2. Before removing the fuel bowl nut, remove the high speed adjusting needle. Use a $^7/_{16}$ in. (11mm) box wrench or

socket to remove the fuel bowl nut. When replacing the fuel bowl nut, be sure to position the fiber gasket under the nut and tighten it securely.

3. Examine the high speed needle tip and, if it appears to be worn, replace it. When the high speed jet is replaced, the main nozzle, which includes the jet seat, should also be replaced. The original main nozzle cannot be used. There are special replacement nozzles available.

4. The inlet needle valve is replaceable if it appears to be worn. The inlet valve seat is also replaceable and should be replaced if the needle valve is replaced.

Float Adjustment

1. The Float setting for this carburetor is $^5/_{64}$-$^7/_{64}$ in. (2.0-2.8mm).

2. The float is set in the traditional manner, at the opposite end of the float from the float hinge and needle valve.

3. Bend the adjusting tab to adjust the float level.

DIAPHRAGM CARBURETORS

▶ **See Figure 189**

Diaphragm carburetors have a rubber-like diaphragm that is exposed to crankcase pressure on one side and to atmospheric pressure on the other side. As the crankcase pressure decreases, the diaphragm moves against the inlet needle allowing the inlet needle to move from its seat which permits fuel to flow through the inlet valve to maintain the correct fuel level in the fuel chamber.

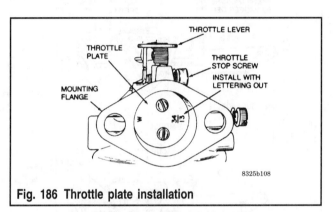

Fig. 186 Throttle plate installation

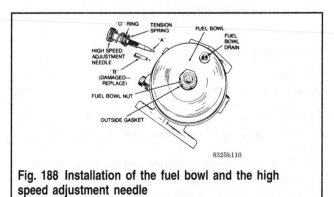

Fig. 187 Installation of the choke plate

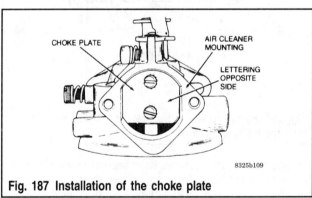

Fig. 188 Installation of the fuel bowl and the high speed adjustment needle

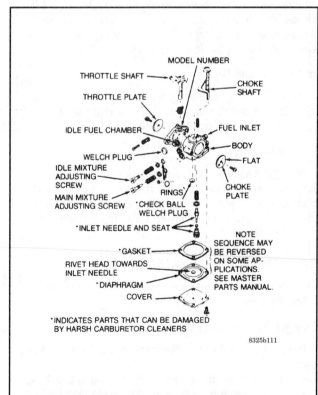

Fig. 189 Tecumseh diaphragm carburetor

An advantage of this type of system over the float system, is that the engine can be operated in any position.

→In rebuilding, use carburetor cleaner only on metal parts, except for the main nozzle in the main body.

Throttle Plate

Install the throttle plate with the short line that is stamped in the plate toward the top of the carburetor, parallel with the throttle shaft, and facing out when the throttle is closed.

Choke Plate

Install the choke plate with the flat side of the choke toward the fuel inlet side of the carburetor. The mark faces in and is parallel to the choke shaft.

Idle Mixture Adjustment Screw

There is a neoprene O-ring on the needle. Never soak the O-ring in carburetor solvent. Idle and main mixture screws vary in size and design, so make sure that you have the correct replacement.

Idle Fuel Chamber

The welch plug can be removed if the carburetor is extremely dirty.

Diaphragms

Diaphragms are serviced and replaced by removing the four retaining screws from the cover. With the cover removed, the diaphragm and gasket may be serviced. Never soak the diaphragm in carburetor solvent. Replace the diaphragm if it is cracked or torn. Be sure there are no wrinkles in the diaphragm when it is replaced. The diaphragm rivet head is always placed facing the inlet needle valve.

Inlet Needle and Seat

The inlet seat is removed by using either a slotted screwdriver (early type) or a $^9/_{32}$ in. socket. The inlet needle is spring loaded, so be careful when removing it.

Fuel Inlet Fitting

All of the diaphragm carburetors have an integral strainer in the inlet fitting. To clean it, either reverse flush it or use compressed air after removing the inlet needle and seat. If the strainer is lacquered or otherwise unable to be cleaned, replace the fitting.

CRAFTSMAN FUEL SYSTEMS

▶ **See Figures 190, 191 and 192**

Changes in Late Model Carburetors

The newest Craftsman carburetors incorporate the following changes:
- The cable form of control is replaced by a control knob.
- The fuel pickup is longer and has a collar machined into it which must be installed tight against the carburetor body.
- The fuel pickup screen is pressed onto the ends of the fill tubes on both models, but the measured depth has changed.

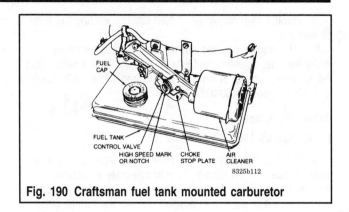

Fig. 190 Craftsman fuel tank mounted carburetor

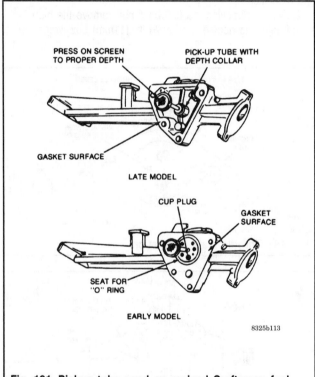

Fig. 191 Pickup tube used on revised Craftsman fuel tank mounted carburetor

- The cross-drilled passages have been eliminated, as has the O-ring on the body. There are no cup plugs.
- The fuel tank and reservoir tube have been revised — the reservoir tube being larger.

Disassembly and Service

▶ **See Figures 193, 194, 195 and 196**

1. Remove the air cleaner assembly and remove the four screws on the top of the carburetor body to separate the fuel tank from the carburetor.

2. Remove the O-ring from between the carburetor and the fuel tank. Examine it for cracks and damage and replace it if necessary.

3. Carefully remove the reservoir tube from the fuel tank. Observe the end of the tube that rested on the bottom of the fuel tank. It should be slotted.

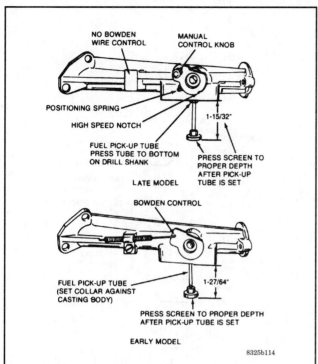

Fig. 192 Manual control knob and several other changes incorporated on revised Craftsman fuel tank mounted carburetor

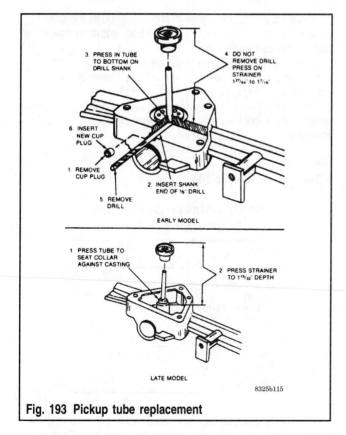

Fig. 193 Pickup tube replacement

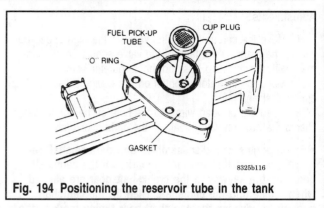

Fig. 194 Positioning the reservoir tube in the tank

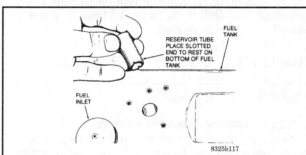

Fig. 195 The fuel pickup tube and O-ring on early Craftsman carburetors

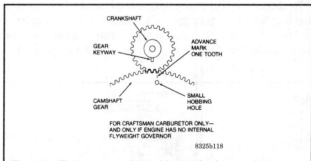

Fig. 196 Timing an engine with the Craftsman fuel tank mounted carburetor

straight out and examine the O-ring seal for damage or wear. If possible, use a new O-ring when reassembling.

5. Examine the fuel pick up tube. There are no valves or ball checks that may become in-operative. These parts can normally be cleaned with carburetor solvent. If it is found that the passage cannot be cleared, the fuel pick up tubes can be replaced. Carefully remove the old ones so as not to enlarge the opening in the carburetor body. If the pickup tube and screen must be replaced, follow the directions shown in the illustration for the type of carburetor (early or late model) on which you are working.

6. Assemble the carburetor in reverse order of disassembly. Use new O-rings and gaskets. When assembling reservoir tube, hold the carburetor upside down and place the reservoir tube over the pickup tube with the slotted end up. Make sure the intake manifold gasket is correctly positioned — it can be assembled blocking the intake passage partially.

4. Remove the control valve by turning the valve clockwise until the flange is clear of the retaining boss. Pull the valve

Adjustments

1. Move the carburetor control valve to the high speed position. The mark on the face of the valve should be in alignment with the retaining boss on the carburetor body.

2. Move the operator's control on the equipment to the high speed position.

3. Insert the bowden wire into the hole of the control valve. Clamp the bowden wire sheath to the carburetor body.

➡**If the engine was disassembled and the camshaft removed, be sure that the timing marks on the camshaft gear and the keyway in the crankshaft gear are aligned when reinserting the camshaft. Then lift the camshaft enough to advance the camshaft gear timing mark to the right (clockwise) ONE tooth, as viewed from the power take-off end of the crankshaft.**

CRAFTSMAN FLOAT TYPE CARBURETORS

▶ **See Figures 197 and 198**

Craftsman float type carburetors are serviced in the same manner as the other Tecumseh float type carburetors. The throttle control valve has three positions: stop, run, and start.

When the control valve is removed, replace the O-ring. If the engine runs sluggishly, consider the possibility of a leaky O-ring.

To remove the fuel pickup tube, clamp the tube in a vise and then twist the carburetor body. Check the small jet in the air horn while the tube is out. In replacement, position the tube squarely and then press in on the collar until the collar seats.

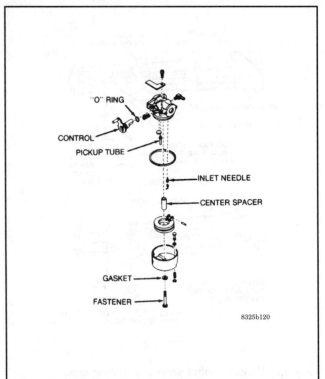

Fig. 198 Exploded view of a Carftsman float type carburetor

In assembly, install the gasket and bolt through the bowl, position the centering spacer onto the bolt, and then attach the parts to the carburetor body.

The camshaft timing mark must be advanced one tooth in relation to crankshaft gear timing mark, as shown in the illustration above. However, when this type carburetor is used with a float bowl reservoir and variable governor adjustment, time it as for other Tecumseh engines — with the camshaft and crankshaft timing marks aligned.

Governor

▶ **See Figures 199 and 200**

Shaft Installed Dimension/ Engine Model Chart

Engine Model	"A" Exposed Shaft Length (see figure)
TNT100-120 ECV100-105-110-120 ECH90 TVS75-90-105-120	$1^5/_{16}$"
V60 VM70-80-100	$1^{19}/_{32}$"
H60 HH60-70	$1^7/_{16}$"
HM70-80-100	$1^{13}/_{32}$"

8325bC16

The mechanical governor is located inside the mounting flange. See engine disassembly instructions, below. To disassemble the governor, see the illustration, and: remove the retaining ring, pull off the spool, remove the second retaining ring, and then pull off the gear assembly and retainer washer.

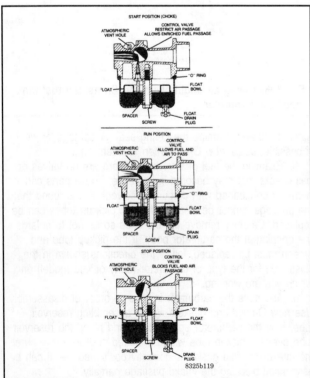

Fig. 197 Start, run and stop throttle positions of a Carftsman float type carburetor

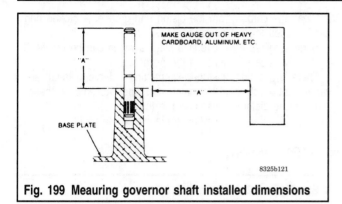

Fig. 199 Meauring governor shaft installed dimensions

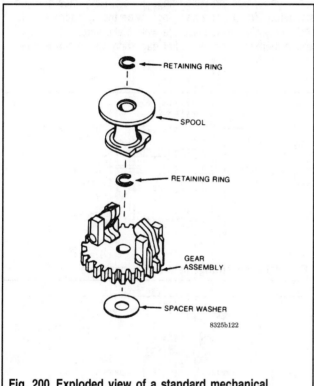

Fig. 200 Exploded view of a standard mechanical governor

Check for wear on all moving surfaces, but especially gear teeth, the inside diameter of the gear where it rides on the shaft, and the flyweights where they work against the spool.

If the governor shaft must be replaced, it should be started into the boss with a few taps using a soft hammer, and then pressed in with a press or vise. The shaft can be installed by positioning a wooden block on top and tapping the upper surface of the block, but the use of a vise or press is much preferred.

The shaft must be pressed in until just the required length is exposed, as measured from the top of the shaft boss to the upper end of the shaft. See the chart.

The governor is installed in reverse of the removal procedure. Connect the linkage and then adjust as described in the Tune-Up section.

Engine Overhaul

TIMING GEARS

▶ See Figures 201, 202 and 203

Correctly matched camshaft gear and crankshaft gear timing marks are necessary for the engine to perform properly.

On all camshafts the timing mark is located in line with the center of the hobbing hole (small hole in the face of the gear). If no line is visible, use the center of the hobbing hole to align with the crankshaft gear marked tooth.

On crankshafts where the gear is held on by a key, the timing mark is the tooth in line with the keyway.

On crankshafts where the gear is pressed onto the crankshaft, a tooth is bevelled to serve as the timing mark.

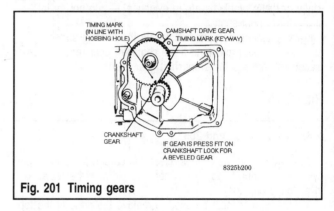

Fig. 201 Timing gears

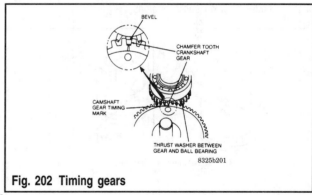

Fig. 202 Timing gears

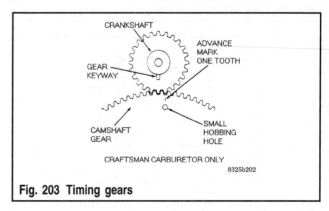

Fig. 203 Timing gears

On engines with a ball bearing on the power take-off end of the crankshaft, look for a bevelled tooth which serves as the crankshaft gear timing mark.

If the engine uses a Craftsman type carburetor, the camshaft timing mark must be advanced clockwise one tooth ahead of the matching timing mark on the crankshaft, the exception being the Craftsman variable governed fuel systems.

➡**If one of the timing gears, either the crankshaft gear or the camshaft gear, is damaged and has to be replaced, both gears should be replaced.**

CRANKSHAFT INSPECTION

Inspect the crankshaft for worn or crossed threads that can't be redressed; worn, scratched, or damaged bearing surfaces; misalignment; flats on the bearing surfaces. Replace the shaft if any of these problems are in evidence — do not try to straighten a bent shaft.

In replacement, be sure to lubricate the bearing surfaces and use oil seal protectors. If the camshaft gear requires replacement, replace the crankshaft gear, also.

PISTONS

▶ **See Figures 204 and 205**

When removing the pistons, clean the carbon from the upper cylinder bore and head. The piston and pin must be replaced in matched pairs.

A ridge reamer must be used to remove the ridge at the top of the cylinder bore on some engines.

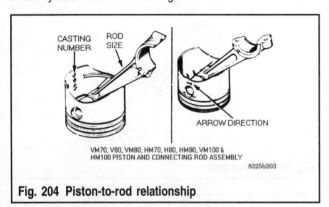
Fig. 204 Piston-to-rod relationship

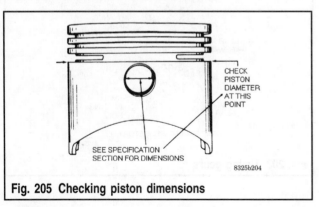
Fig. 205 Checking piston dimensions

Clean the carbon from the piston ring groove. A broken ring can be used for this operation.

Some engines have oversize pistons which can be identified by the oversize engraved on the piston top.

There is a definite piston-to-connecting rod-to-crankshaft arrangement which must be maintained when assembling these parts. If the piston is assembled in the bore 180° out of position, it will cause immediate binding of the parts.

PISTON RINGS

▶ **See Figures 206, 207 and 208**

Always replace the piston rings in sets. Ring gaps must be staggered. When using new rings, wipe the cylinder wall with fine emery cloth to deglaze the wall. Make sure the cylinder wall is thoroughly cleaned after deglazing. Check the ring gap

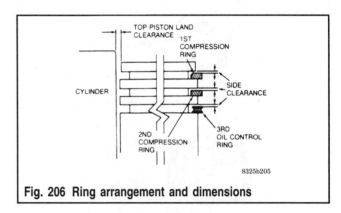
Fig. 206 Ring arrangement and dimensions

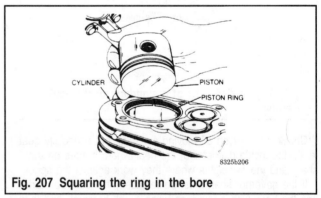
Fig. 207 Squaring the ring in the bore

Fig. 208 Checking the ring gap

by placing the ring squarely in the center of the area in which the rings travel and measuring the gap with a feeler gauge. Do not spread the rings too wide when assembling them to the pistons. Use a ring leader to install the rings on the piston.

The top compression ring has an inside chamfer. This chamfer must go UP. If the second ring has a chamfer, it must also face UP. If there is a notch on the outside diameter of the ring, it must face DOWN.

Check the ring gap on the old ring to determine if the ring should be replaced. Check the ring gap on the new ring to determine if the cylinder should be rebored to take oversize parts.

➡**Make sure that the ring gap is measured with the ring fitted squarely in the worn part of the cylinder where the ring usually rides up and down on the piston.**

CONNECTING RODS

▶ **See Figure 209**

Be sure that the match marks align when assembling the connecting rods to the crankshaft. Use new self-locking nuts. Whenever locking tabs are included, be sure that the tabs lock the nuts securely. NEVER try to straighten a bent crankshaft or connecting rod. Replace them if necessary. When replacing either the piston, rod, crankshaft, or camshaft, liberally lubricate all bearings with engine oil before assembly.

The following engines have offset connecting rods: V70, VH70, VM70, V80, VM80, H70, HH70, HM70, H80, HM80. ECV105, ECV110, ECM120, and TNT120 engines have the caps fitted from opposite to the camshaft side of the engine. The following engines have the cap fitted from the camshaft side: V70, VH70, VM70, V80, VM80, H70, HH70, HM70, H80, HM80, VM100 and HM100.

On H70, HH70, HM70, H80, HM80, V70, VH70, VM70, V80, VM80, VM100 and HM100 engines, a dipper is stamped into the lockplate. Use a new lockplate whenever the rod cap is removed.

On engine with Durlock rod bolts, torque the bolt as follows:
- TVS75, 90, 105, 120, ECH90, TNT100, 120, ECV100, 105, 110 and 120-110 inch lbs.
- V60-80, H60-80, VH60-70, HH60-70, VM70-100 and HM70-100-150 inch lbs.

➡**Early type caps can be distorted if the cap is not held to the crank pin while threading the bolts tight. Undue force should not be used.**

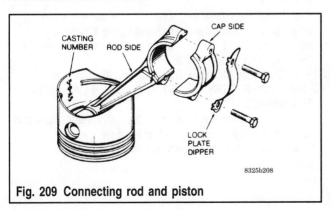

Fig. 209 Connecting rod and piston

Later rods have serrations which prevent distortion during tightening. They also have match marks which must face out when assembling the rod. On the V80 and H80 engines, the piston and rod must fit so that the number inside the casting is on the rod side of the rod/cap combination.

CAMSHAFT

▶ **See Figures 210 and 211**

Before removing the camshaft, align the timing marks to relieve the pressure on the valve lifters, on most engines.

On VM80 and VM100 engines, while the basic timing is the same, the crankshaft must be rotated so that the timing mark is located 3 teeth further counterclockwise (referring to crankshaft gear rotation). This clears the camshaft of the compression release mechanism for easier removal.

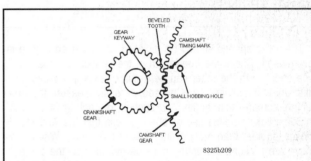

Fig. 210 Alignment of timing marks for camshaft removal

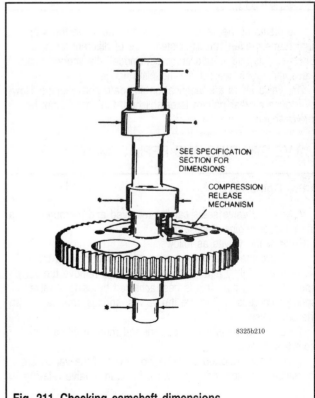

Fig. 211 Checking camshaft dimensions

In installation, align the gears for this type of camshaft as they were right before removal. After installation, turn the crankshaft gear clockwise in order to check for proper alignment of timing marks.

Clean the camshaft in solvent, then blow the oil passages dry with compressed air. Replace the camshaft if it shows wear of evidence of scoring. Check the cam dimensions against those in the chart.

If the engine has a mechanical fuel pump, it may have to be removed to properly reinstall the camshaft. If the engine is equipped with the Insta-matic Ezee-Start Compression Release, and any of the parts have to be replaced due to wear or damage, the entire camshaft must be replaced. Be sure that the oil pump (if so equipped) barrel chamber is toward the fillet of the camshaft gear when assembled.

➡**If a damaged gear is replaced, the crankshaft gear should also be replaced.**

VALVE SPRINGS

The valve springs should be replaced whenever an engine is overhauled. Check the free length of the springs. Comparing one spring with the other can be a quick check to notice any difference. If a difference is noticed, carefully measure the free length, compression length, and strength of each spring.

Some valve springs use dampening coils that are wound closer together than most of the coils of the spring. Where these are present, the spring must be mounted so the dampening coils are on the stationary (upper) end of the spring.

VALVE LIFTERS

The stems of the valves serve as the lifters. On the 4 hp light frame models, the lifter stems are of different lengths. Because this engine is a cross port model, the shorter intake valve lifter goes nearest the mounting flange.

The valve lifters are identical on standard port engines. However once a wear pattern is established, they should not be interchanged.

VALVE GRINDING AND REPLACEMENT

▶ **See Figure 212**

Valves and valve seats can be removed and reground with a minimum of engine disassembly.

Remove the valves as follows:

1. Raise the lower valve spring caps while holding the valve heads tightly against the valve seat to remove the valve spring retaining pin. This is best achieved by using a valve spring compressor. Remove the valves, springs, and caps from the crankcase.

2. Clean all parts with a solvent and remove all carbon from the valves.

3. Replace distorted or damaged valves. If the valves are in usable condition, grind the valve faces in a valve refacing

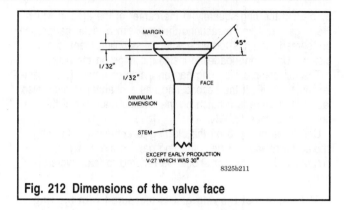

Fig. 212 Dimensions of the valve face

machine and to the angle given in the specifications chart at the end of this section. Replace the valves if the faces are ground to less than $\frac{1}{32}$ in. (0.8mm).

4. Whenever new or reground valves are installed, lap in the valves with lapping compound to insure an air-tight fit.

➡**There are valves available with oversize stems.**

5. Valve grinding changes the valve lifter clearance. After grinding the valves check the valve lifter clearance as follows:

a. Rotate the crankshaft until the piston is set at the TDC position of the compression stroke.

b. Insert the valves in their guides and hold the valves firmly on their seats.

c. Check for a clearance of 0.010 in. (0.25mm) between each valve stem and valve lifter with a feeler gauge.

d. Grind the valve stem in a valve resurfacing machine set to grind a perfectly square face with the proper clearance.

6. Install valves as follows on Early Models:

a. Position the valve spring and upper and lower valve spring caps under the valve guides for the valve to be installed.

b. Install the valves in the guides, making sure that the valve marked **EX** is inserted in the exhaust port. The valve stem must pass through the valve spring and the valve spring caps.

c. Insert the blade of a screwdriver under the lower valve spring cap and pry the spring up.

d. Insert the valve pin through the hole in the valve stem with a long nosed pliers. Make sure the valve pin is properly seated under the lower valve spring cap.

7. Install the valves as follows on Later Models:

a. Position the valve caps and spring in the valve compartment.

b. Install the valves in guides with the valve marked **EX** in exhaust port. The valve stem must pass through the upper valve cap and spring. The lower cap should sit around the valve lifter exposed end.

c. Compress the valve spring so that the shank is exposed. DO NOT TRY TO LIFT THE LOWER CAP WITH THE SPRING.

d. Lift the lower valve cap over the valve stem shank and center the cap in the smaller diameter hole.

e. Release the valve spring tension to lock the cap in place.

REBORING THE CYLINDER

1. First, decide whether to rebore for 0.010 in. (0.25mm) or 0.020 in. (0.50mm).

2. Use any standard commercial hone of suitable size. Chuck the hone in the drill press with the spindle speed of about 600 rpm.

3. Start with coarse stones and center the cylinder under the press spindle. Lower the hone so the lower end of the stones contact the lowest point in the cylinder bore.

4. Rotate the adjusting nut so that the stones touch the cylinder wall and then begin honing at the bottom of the cylinder. Move the hone up and down at a rate of 50 strokes a minute to avoid cutting ridges in the cylinder wall. Every fourth or fifth stroke, move the hone far enough to extend the stones 1 in. (25mm) beyond the top and bottom of the cylinder bore.

5. Check the bore size and straightness every thirty or forty strokes. If the stones collect metal, clean them with a wire brush each time the hone is removed.

6. Hone with coarse stones until the cylinder bore is within 0.002 in. (0.05mm) of the desired finish size. Replace the coarse stones with burnishing stones and continue until the bore is to within 0.0005 in. (0.0127mm) of the desired size.

7. Remove the burnishing stones and install finishing stones to polish the cylinder to the final size.

8. Clean the cylinder with solvent and dry it thoroughly.

9. Replace the piston and piston rings with the correct oversize parts.

REBORING VALVE GUIDES

▶ See Figure 213

The valve guides are permanently installed in the cylinder. However if the guides wear, they can be rebored to accommodate a 1/32 in. (0.8mm) oversize valve stem. Rebore the valve guides in the following manner:

1. Ream the valve guides with a standard straight shanked hand reamer or a low speed drill press.

2. Redrill the upper and lower valve spring caps to accommodate the oversize valve stem.

3. Reassemble the engine, installing valves with the correct oversize stems in the valve guides.

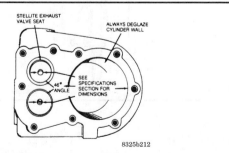

Fig. 213 Checking the dimensions of the valve seats, guides, and the cylinder

REGRINDING VALVE SEATS

The valve seats need regrinding only if they are pitted or scored. If there are no pits or scores, lapping in the valves will provide a proper valve seat. Valve seats are not replaceable. Regrind the valve seats as follows:

1. Use a grinding stone or a reseater set to provide the proper angle and seal face dimensions.

2. If the seat is over 3/64 in. (1.2mm) wide after grinding, use a 15° stone or cutter to narrow the face to the proper dimensions.

3. Inspect the seats to make sure that the cutter or stone has been held squarely to the valve seat and that the same dimension has been held around the entire circumference of the seat.

4. Lap the valves to the reground seats.

Torquing Cylinder Head
▶ See Figure 214

Torque the cylinder head to 200 inch lbs. in 4 equal stages of 50 inch lbs. Follow the sequence shown in the appropriate illustration for each tightening stage.

Bearing Service

▶ See Figures 215, 216, 217, 218, 219, 220, 221 and 222

LIGHTWEIGHT ALUMINUM BEARING REPLACEMENT

The aluminum bearing must be cut out using the rough cut reamer and the procedures shown. Follow illustrated steps to install bronze bushings in place of the aluminum bearings.

LONG LIFE AND CAST IRON ENGINES WITHOUT BALL BEARINGS

The worn bronze bushing must be driven out before the new bushing can be installed. Follow illustrated steps to replace the main bearings on these units.

LONG LIFE AND CAST IRON ENGINES WITH BALL BEARING ON THE P.T.O. (POWER TAKE-OFF) SIDE OF THE CRANKSHAFT

The side cover containing the ball bearing must be removed and a substitute cover with either a new bronze bushing or aluminum bearing must be used instead. Follow illustrated steps.

SPECIAL TOOLS

The task of main bearing replacement is made easier by using one of two Tecumseh main bearing tool kits. Kit 670161 is used to replace main bearings on the lightweight engines

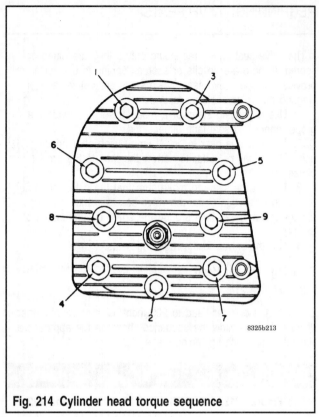

Fig. 214 Cylinder head torque sequence

8325b213

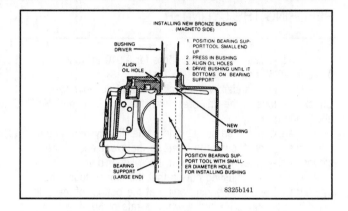

INSTALLING NEW BRONZE BUSHING (MAGNETO SIDE)

8325b141

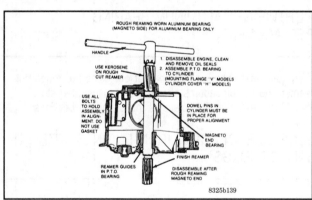

ROUGH REAMING WORN ALUMINUM BEARING (MAGNETO SIDE) FOR ALUMINUM BEARING ONLY

8325b139

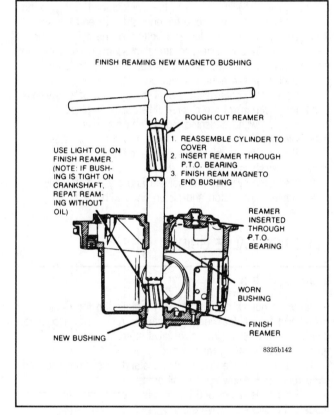

FINISH REAMING NEW MAGNETO BUSHING

8325b142

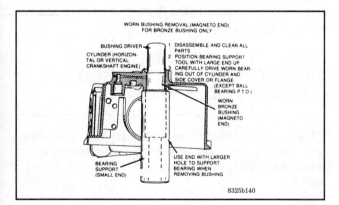

WORN BUSHING REMOVAL (MAGNETO END) FOR BRONZE BUSHING ONLY

8325b140

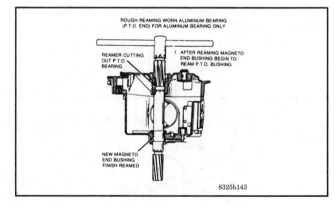

ROUGH REAMING WORN ALUMINUM BEARING (P.T.O. END) FOR ALUMINUM BEARING ONLY

8325b143

except HS models. Kit 670165 is used to replace main bearings on the medium weight engines except HS models.

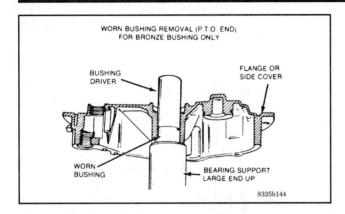

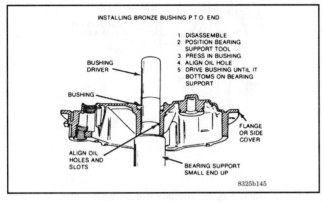

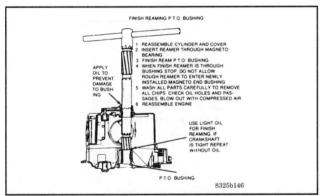

GENERAL NOTES ON BUSHING REPLACEMENT

1. Your fingers and all parts must be kept very clean when replacing bushings.

2. On splash lubricated horizontal engines, the oil hole in the bushing is to be lined up with the oil hole that leads into the slot in the original bearing.

3. In the event it is necessary to replace the mounting flange or cylinder cover, the magneto end bearing must be rebushed. The P.T.O. bearing should also be rebushed to assure proper alignment.

4. Oil should be used to finish-ream the bushings. In the event the crankshaft does not rotate freely repeat the finish-reaming operation without oil.

5. Kerosene should be used as a cutting lubricant while rough-reaming.

6. Be sure that the dowel pins are in the cylinder block when assembling the mounting flange or cylinder cover. Use all bolts to hold the assembly together.

7. Remove the reamer by rotating it in the same direction as it is turned during the reaming operation. DO NOT TURN THE REAMER BACKWARDS.

Barrel and Plunger Oil Pump System

▶ **See Figures 223 and 224**

This system is driven by an eccentric on the camshaft. Oil is drawn through the hollow camshaft from the oil sump on its intake stroke. The passage from the sump through the camshaft is aligned with the pump opening. As the camshaft continues rotation (pressure stroke), the plunger force the oil out. The other port in the camshaft is aligned with the pump, and directs oil out of the top of the camshaft.

At the top of the camshaft, oil is forced through a crankshaft passage to the top main bearing groove which is aligned with the drilled crankshaft passage. Oil is directed through this passage to the crankshaft connecting rod journal and then spills from the connecting rod to lubricate the cylinder walls. Splash is used to lubricate the other parts of the engine.

A pressure relief port in the crankcase relieves excessive pressures when the oil viscosity is extremely heavy due to cold temperatures, or when the system is plugged or damaged. Normal pressure is 7 psi.

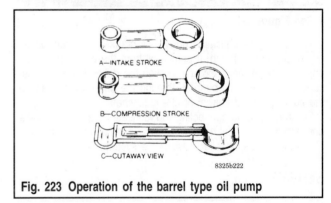

Fig. 223 Operation of the barrel type oil pump

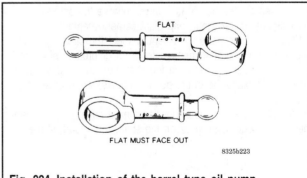

Fig. 224 Installation of the barrel type oil pump

SERVICE

Remove the mounting flange or the cylinder cover, whichever is applicable. Remove the barrel and plunger assembly and separate the parts.

Clean the pump parts in solvent and inspect the pump plunger and barrel for rough spots or wear. If the pump plunger is scored or worn, replace the entire pump.

Before reassembling the pump parts, lubricate all of the parts in engine oil. Manually operate the pump to make sure the plunger slides freely in the barrel.

Lubricate all the parts and position the barrel on the camshaft eccentric. If the oil pump has a chamfer only on one side, that side must be placed toward the camshaft gear. The flat goes away from the gear, thus out to work against the flange oil pickup hole.

Install the mounting flange. Be sure the plunger ball seats in the recess in the flange before fastening it to the cylinder.

Splash Lubrication

▶ **See Figure 225**

Some engines utilize the splash type lubrication system. The oil dipper, on some engines, is cast onto the lower connecting rod bearing cap. It is important that the proper parts are used to ensure the longest engine life.

Gear Type Oil Pump System

▶ **See Figure 226**

The gear type lubrication pump is a crankshaft driven, positive displacement pump. It pumps oil from the oil sump in the engine base to the camshaft, through the drilled camshaft passage to the top main bearing, through the drilled crankshaft, to the connecting rod journal on the crankshaft.

Spillage from the connecting rod lubricates the cylinder walls and normal splash lubricates the other internal working parts. There is a pressure relief valve in the system.

SERVICE

Disassemble the pump as follows: remove the screws, lockwashers, cover, gear, and displacement member.

Wash all of the parts in solvent. Inspect the oil pump drive gear and displacement member for worn or broken teeth, scoring, or other damage. Inspect the shaft hole in the drive gear for wear. Replace the entire pump if cracks, wear, or scoring is evident.

To replace the oil pump, position the oil pump displacement member and oil pump gear on the shaft, then flood all the

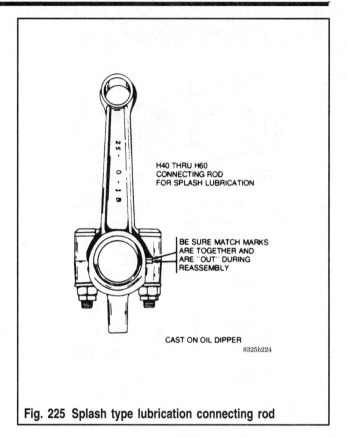

H40 THRU H60 CONNECTING ROD FOR SPLASH LUBRICATION

BE SURE MATCH MARKS ARE TOGETHER AND ARE "OUT" DURING REASSEMBLY

CAST ON OIL DIPPER

8325b224

Fig. 225 Splash type lubrication connecting rod

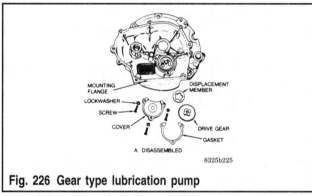

MOUNTING FLANGE
LOCKWASHER
SCREW
COVER
DISPLACEMENT MEMBER
DRIVE GEAR
GASKET
A. DISASSEMBLED
8325b225

Fig. 226 Gear type lubrication pump

parts with oil for priming during the initial starting of the engine.

The gasket provides clearance for the drive gear. With a feeler gauge, determine the clearance between the cover and the oil pump gear. The clearance desired is 0.006-0.007 in. (0.15-0.18mm). Use gaskets, which are available in a variety of sizes, to obtain the correct clearance. Position the oil pump cover and secure it with the screws of lockwashers.

TECUMSEH VECTOR

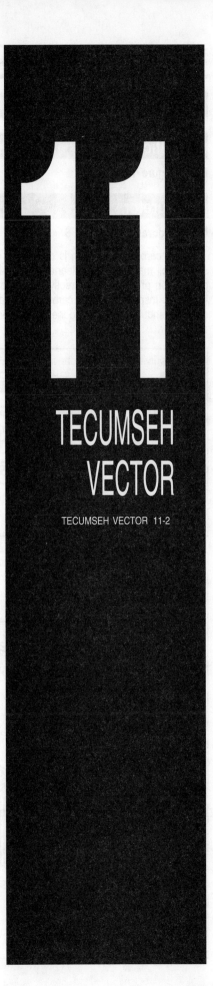

11

TECUMSEH
VECTOR

TECUMSEH VECTOR

Identification

▶ **See Figure 1**

Carburetor

▶ **See Figures 2, 3, 4 and 5**

Proper carburetor function is dependent on clean fresh fuel and a well maintained air cleaner system. Most causes of carburetion problems are directly related to stale fuel and dirt ingestion. Inspection of the carburetor for dirt wear and fuel deposits should always be done before servicing the carburetor.

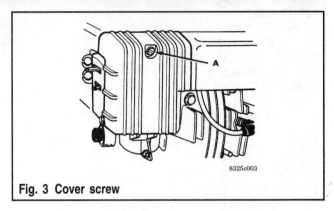

Fig. 3 Cover screw

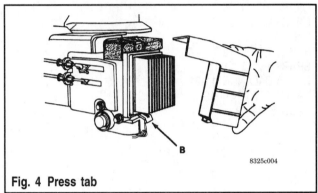

Fig. 4 Press tab

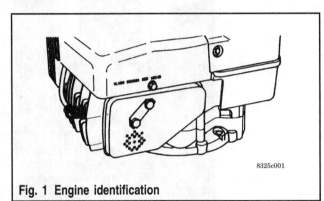

Fig. 1 Engine identification

Fig. 5 Filter

IDENTIFICATION

▶ **See Figure 6**

Tecumseh carburetors are identified by a model number and date code stamped on the carburetor as shown. When servicing carburetors, use the engine model number or the model number on the carburetor to find repair parts.

The Vector carburetor is a float feed, non-adjustable carburetor, with a 1 piece extruded aluminum body. The float bowl, float, nozzle, and venturi are nonmetallic, eliminating the corrosion and varnishing problems associated with similar metallic parts. Common service areas of the carburetor are contained in the fuel bowl. These areas are the float, needle, seat and

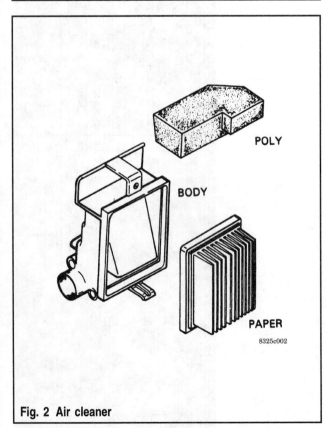

Fig. 2 Air cleaner

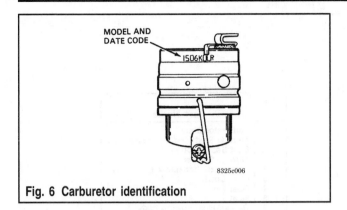

MODEL AND
DATE CODE

1506K R

8325c006

Fig. 6 Carburetor identification

main nozzle. All of these parts can be serviced without removing the carburetor body from the engine.

FLOAT BOWL SERVICE

1. Disconnect and plug the fuel line.
2. Remove the bowl drain screw.
3. Remove the float bowl by snapping the bale spring towards the throttle end of the carburetor.

➡**If a screwdriver or similar tool is used to aid in the bail removal, care must be taken not to permanently bend the retainer.**

4. Pull out the main nozzle and spring. Inspect the man nozzle for deposits. Be sure to check the cross holes on the body of the nozzle and the main jet orifice in the bottom of the nozzle. Use compressed air or monofilament fishing line to remove any deposits ion the main jet or cross holes. Replace the O-ring if damaged.

5. The float is held in the float bowl by the float pin which is pressed into tabs on the top of the float support towers. To remove the float:

    a. Insert a screwdriver into the rectangular hole in the float at the hinge.

    b. Carefully pry the hinge shaft out of the tabs in the float bowl.

    c. Carefully lift the float out of the float bowl and inspect for damage or deposits.

    d. Clean the idle passageway with compressed air, or probe with tag wire.

➡**The inlet needle is attached to the float and shield also be inspected for damage or deposits.**

6. The inlet seal can be removed with a small wire hook or forced out with compressed air. Inspect the float bowl and main nozzle area for sediment and deposits. Use a mild carburetor cleaner to loosen and remove deposits and sediment.

7. Install a new inlet seat into the float bowl. The grooved side of the inlet seat goes into the float bowl first. Place a drop of oil on the seat and press it in with a flat punch until it seats. Do not scratch the inlet bore.

8. Slide the inlet needle drops into the fuel inlet.

9. Snap the float shaft into the tabs in the float bowl. It is not necessary to adjust the float height even if the float has been replaced.

10. Drop the main nozzle spring into the main nozzle well in the float bowl. Put a small amount of oil on the main nozzle O-ring and push the nozzle into the main nozzle well, O-ring end first.

11. Place a new gasket on top of the float bowl (the gasket will only fit onto the float bowl one way.) Hold the float bowl to the carburetor body and snap the retainer into position. Reinstall the bowl drain screw, do not overtighten, reattach the fuel line.

➡**Bowl service is all that is normally required for routine carburetor maintenance.**

CARBURETOR OVERHAUL

◆ **See Figures 7, 8, 9, 10, 11, 12, 13, 14, 15 and 16**

To rebuild the carburetor body it is necessary to remove the carburetor from the engine.

1. Remove the speed control plate.
2. Remove the air cleaner body from the carburetor body.
3. Disconnect and plug the fuel line.
4. Remove the carburetor mounting studs.
5. Remove the governor link.
6. Drain the carburetor float bowl.
7. Disassemble the float bowl (see Bowl Service).

➡**Before disassembling the carburetor body, check the throttle shaft and body for excessive wear. Remove the shutter screw and discard. Inspect the shutter for damage; discard if damaged.**

Cleaning

To properly clean the carburetor body, the welch plugs should be removed to expose drilled passages. To remove welch plug, sharpen a small chisel to a sharp wedge point. Drive the chisel into the welch plug, push down on chisel and pry plug out of position.

After the welch plug is removed from the carburetor it can be soaked in a commercial carburetor cleaner no longer than 30 minutes. Be sure to follow the directions on the container.

After the carburetor has been soaked, all passages mat be probed with monofilament fishing line and compressed air to open plugged or restricted passages.

Assembly

1. Install a new welch plug over the idle fuel chamber with the raised portion up. Use a punch equal to the size of the plug, to fasten the plug. Do not dent or drive the center of the plug below the top surface of the carburetor.

2. Install the throttle and shutter (use a new shutter screw and dust seal). The scribe mark on the shutter must be in the 12 o'clock position.

➡**If the scribe mark is out of position the shutter may stick.**

3. To rebuild the Float Bowl, refer to the previous section on float bowl service.

4. Install the carburetor to the engine using a new gasket. The primer passage in the air cleaner body should be cleaned before it is reinstalled over the carburetor.

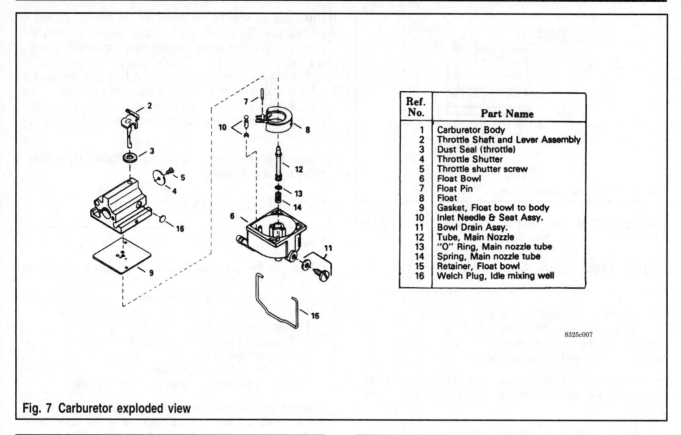

Ref. No.	Part Name
1	Carburetor Body
2	Throttle Shaft and Lever Assembly
3	Dust Seal (throttle)
4	Throttle Shutter
5	Throttle shutter screw
6	Float Bowl
7	Float Pin
8	Float
9	Gasket, Float bowl to body
10	Inlet Needle & Seat Assy.
11	Bowl Drain Assy.
12	Tube, Main Nozzle
13	"O" Ring, Main nozzle tube
14	Spring, Main nozzle tube
15	Retainer, Float bowl
16	Welch Plug, Idle mixing well

8325c007

Fig. 7 Carburetor exploded view

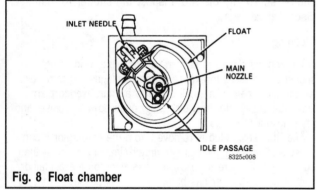

8325c008

Fig. 8 Float chamber

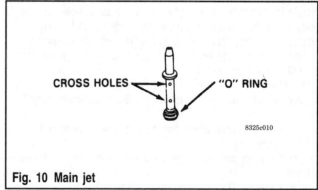

8325c010

Fig. 10 Main jet

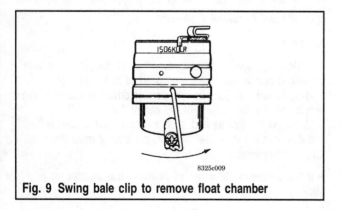

8325c009

Fig. 9 Swing bale clip to remove float chamber

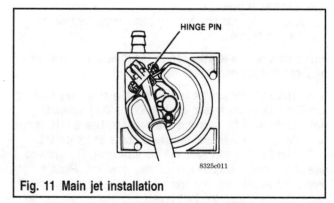

8325c011

Fig. 11 Main jet installation

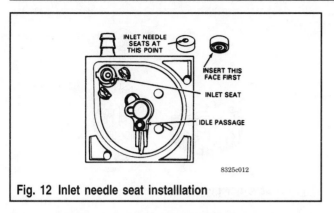

Fig. 12 Inlet needle seat installlation

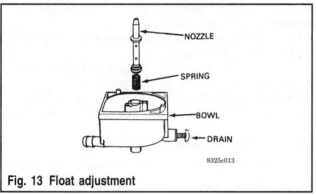

Fig. 13 Float adjustment

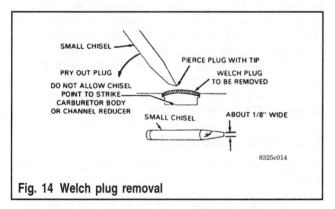

Fig. 14 Welch plug removal

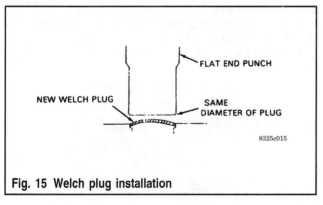

Fig. 15 Welch plug installation

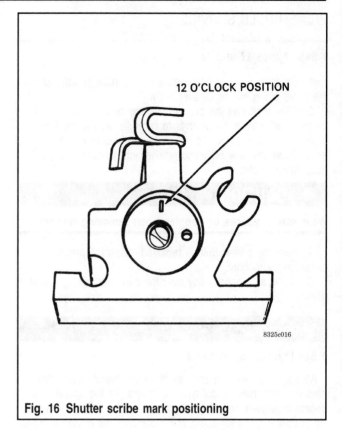

Fig. 16 Shutter scribe mark positioning

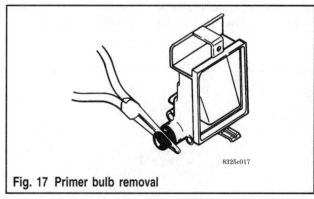

Fig. 17 Primer bulb removal

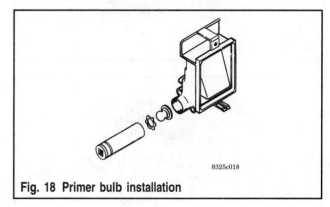

Fig. 18 Primer bulb installation

PRIMER BULB SERVICE

▶ **See Figures 17 and 18**

1. Grasp the primer bulb with a needle nose pliers and roll the pliers along the air cleaner body.
2. After removing the primer bulb, the retaining ring must be removed. Use a screwdriver to carefully pry the retainer out of the air cleaner body. Do not reuse old bulb or retainer.
3. Press the new bulb and retainer into position using a deep reach socket.

❊❊CAUTION

Wear safety glasses or goggles when removing retainer.

4. After the primer bulb is removed, clean the primer passages thoroughly.
5. Install air cleaner body over the carburetor using a new gasket.

Governors

▶ **See Figures 19, 20 and 21**

All Tecumseh 4-cycle engines of recent manufactures are equipped with mechanical type governors. As the speed of an engine increases, centrifugal force moves the weights outward, lifting up the governor spool which contacts the governor shaft; this in turn closes the throttle. As engine speed decreases, the weights are pulled inward by the spring which opens the throt-

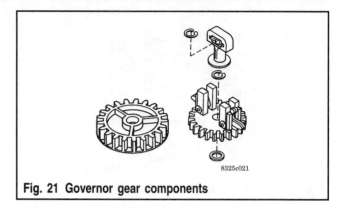

Fig. 21 Governor gear components

tle. Thus, the engine speed controls the throttle opening and maintains a certain governed speed.

The governor gear on the Vector engine is driven by the crankshaft through an idler gear.

The idler gear is captured in the mounting flange by both the crankshaft gear and the governor gear.

The governor and idler shafts are pressed into the flange or cover to a specific dimension and is serviced as an assembly.

LINKAGE INSTALLATION

The solid link is always connected from the throttle lever on the carburetor to the lower hole on the governor lever. The shorter bend has to be toward the governor. The governor extension spring is connected with the spring and hooked into the upper hole of the governor lever and the extension end hooked through the speed control lever. To remove the governor spring, carefully twist the extension end counterclockwise to unhook the extension spring at the speed control lever. Do not bend or distort governor extension spring.

GOVERNOR ADJUSTMENT

▶ **See Figure 22**

1. With engine stopped, loosen the screw holding the governor clamp and lever.
2. Turn the clamp clockwise, then push governor lever connected to the throttle to a full wide open position.
3. Hold the lever and clamp in this position and tighten the screw.

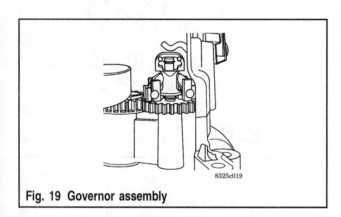

Fig. 19 Governor assembly

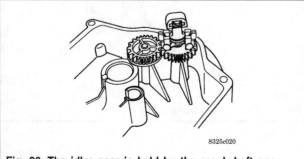

Fig. 20 The idler gear is held by the crankshaft gear and governor

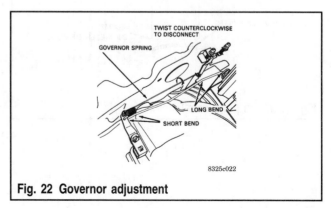

Fig. 22 Governor adjustment

SPEED CONTROLS

▶ **See Figure 23**

The Vector engine has an adjustable speed control. Never exceed the manufacturer's recommended speeds.

Rewind Starter

DISASSEMBLY

▶ **See Figure 24**

1. After removing the rewind from the engine blower housing, release the tension on the rewind spring.
2. Place a ¾ in. deep reach socket inside the retainer pawl. Set the rewind on a bench, supported on the socket.
3. Using a 5/16 in. roll punch, drive out the center pin.
4. All components that are in need of service should be replaced.

➡**Care must be used when handling the pulley because the rewind spring and cover is held by the bosses in the pulley.**

ASSEMBLY

1. Reverse the above listed keeping in mind that the starter dogs with the dog springs must snap back to the center of the pulley.
2. Always replace the center pin with a new pin upon reassembly. Also place the two new plastic washers between the center lag and retainer pawl. Disacard old plastic washer. The new plastic washers will be provided along with the new center pin.
3. Check retainer pawl. It is worn, bent or damaged in any manner replace upon reassembly.
4. Tap the new center pin in until it is within ⅛ in. of the top of the starter.

➡**Driving the center pin in too far will cause the retainer pawl to bend and the starter dogs will not engage starter cup.**

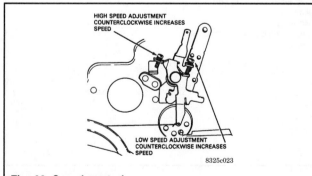

HIGH SPEED ADJUSTMENT
COUNTERCLOCKWISE INCREASES
SPEED

LOW SPEED ADJUSTMENT
COUNTERCLOCKWISE INCREASES
SPEED

8325c023

Fig. 23 Speed controls

Electric Starters

▶ **See Figures 25, 26 and 27**

REMOVAL

1. Remove face plate, air cleaner assembly and gas tank.
2. Compress plastic grommet and pull it out of the blower housing.
3. Slide the wire through slot, being careful not to cut the wire insulation.
4. Remove blower housing, remove flywheel (see flywheel section) and inspect ring gear for wear or damage. Replace if necessary (see flywheel section).
5. Remove nuts on both sides of pinion. Drop starter out of back plate and remove ground wire.

DRIVE ASSEMBLY SERVICE

Pinion gear parts should be checked for damage or wear. If the gear does not engage or slips, it should be washed in solvent (rubber parts cleaned with soap and water) to remove dirt and grease, and dried thoroughly. If damaged replace parts.

Remove, inspect and replace as necessary. Assembly is the reverse of disassembly.

➡**For ease of assembly, assemble armature into brush end frame first. Place a small amount of light grease such as Lubriplate® between the drive nut and helix on armature shaft. DO NOT apply lubricant to pinion driver.**

BRUSHES INSPECTION

Before removing armature, check brushes for wear. Make sure brushes are not worn to the point where brush wire bottoms out in the slot of brush holder. Brush springs must have enough strength to keep tension on the brushes and hold them against the commutator. If brushes are in need if change, replace the entire end cap assembly.

ARMATURE CHECK

If commutator bars are glazed or dirty, they can be turned down in a lather. While rotating, hold a strip of 00 sandpaper lightly on the commutator, moving it back and forth. Do not use emery cloth. Recut grooves between commutator bars to depth equal to the width of the insulators.

Using a continuity tester to make certain no continuity exists between the commutator (copper) and the iron of the armature, rotate armature and check out all commutator bars.

The armature can be thoroughly checked with a growler if available.

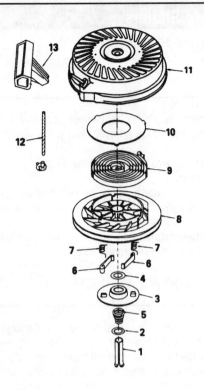

Ref. No.	Part Name
	Starter, Rewind
1	Pin, Spring (Incl. No. 4)
2	Washer
3	Retainer
4	Washer
5	Spring, Brake
6	Dog, Starter
7	Spring, Dog
8	Pulley
9	Spring, Rewind
10	Cover, Spring
11	Housing Assy., Starter
12	Rope, Starter (Length 98" & 9/64" dia.)
13	Handle, Starter

8325c024

Fig. 24 Rewind starter

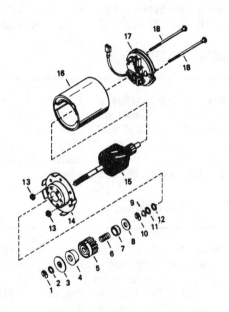

1. Retainer ring
2. Dust washer
3. Drive nut
4. Pinion driver
5. Gear
6. Anti-drift spring
7. Spring retainer (spring collapses into retainer)
8. Cup washer (cup washer cupped over retainer spring)
9. Washer (metal)
10. Retainer ring
11. Thrust washer (metal)
12. Washer (plastic)
13. Lock nuts
14. Cap assembly drive end
15. Armature
16. Housing
17. Cap assembly commutator end
18. Bolts

8325c025

Fig. 25 Electric starter

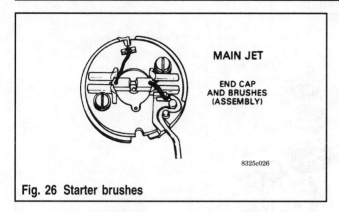

Fig. 26 Starter brushes

STARTER ASSEMBLY

1. Attach the ground wire prior to assembling the electric starter to the baffle.

2. Attach the black ground wire to the electric starter through bolt so that the wire extends between the two adjacent end cap prongs.

3. Place the starter into the black plate with the ground wire bolt away from the carburetor. Note that the throttle linkage is routed around starter while the governor spring is routed through the end cap springs.

4. Tighten nuts on starter bolts.

5. Place blower housing on engine and slide wires through slot making sure not to cut insulation.

6. Press plastic grommet into hole.

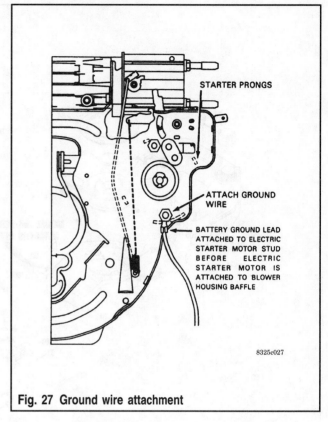

Fig. 27 Ground wire attachment

7. Reassemble gas tank, air cleaner assembly and face plate.

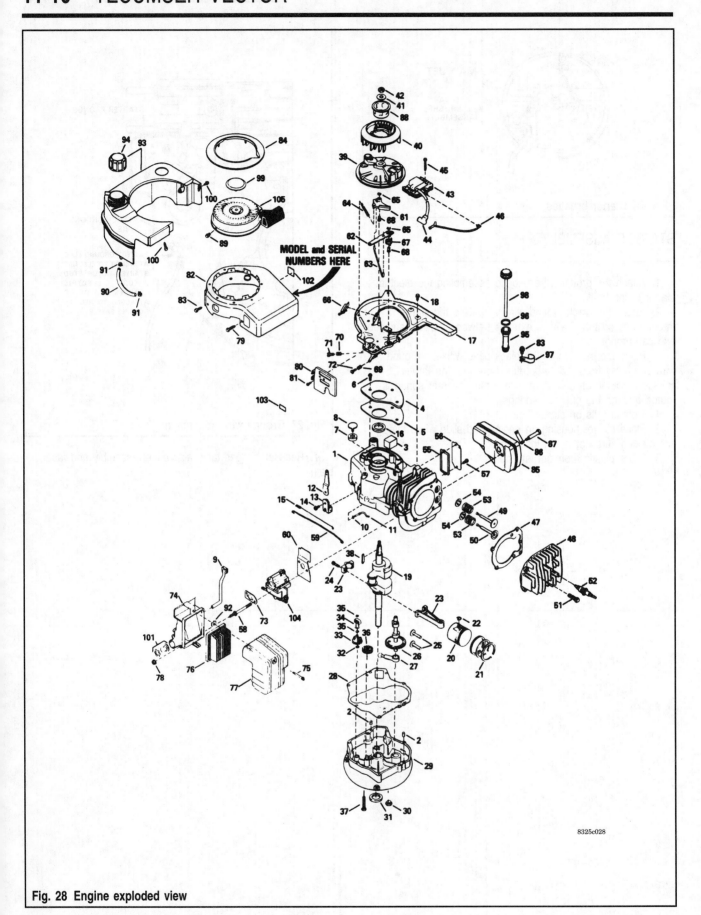

MODEL and SERIAL NUMBERS HERE

Fig. 28 Engine exploded view

8325c028

Ref. No.	Part Name	Ref. No.	Part Name
1	Cylinder Assy.	54	Cap, Valve spring
2	Pin, Dowel	55	Gasket, Valve cover
3	Element, Breather	56	Cover, Valve spring box
4	Cover, Breather	57	Screw, Hex hd. Sems, 10-24 x 1/2
5	Gasket, Breather cover	58	Stud, Carburetor mounting
6	Screw, Hex washer hd., thread forming, 10-24 x 1/2	59	Link, Governor
		60	Spacer, Carburetor mounting
7	Body, Breather valve	61	Lever Assy., Brake
8	Valve, Breather check	62	Lever, Brake control
9	Breather Tube	63	Link, Brake control lever
10	Washer, Flat	64	Spring, Brake
11	Rod, Governor	65	Ring, Retaining
12	Lever, Governor	66	Terminal
13	Clamp, Governor lever	67	Spring, Brake control lever
14	Screw, Hex washer hd., 8-32 x 5/16	68	Bushing, Brake control lever & brake lever
15	Spring, Governor extension	69	Spring, Compression
16	Seal, Oil	70	Spring, Compression
17	Baffle, Blower housing (Incl. No. 195)	71	Screw, Fil. hd., 5-40 x 7/16
18	Screw, Hex washer hd. taptite, 1/4-20 x 5/8	72	Screw, Fil. hd., 6-32 x 21/32
		73	Gasket, Carburetor to air cleaner
19	Crankshaft Assy.	74	Body, Air cleaner (Incl. Nos. 239, 299 & 350)
20	Piston & Pin Assy.		
21	Ring Set, Piston	75	Screw, Fil hd. Sems, 10-32 x 2-3/32
22	Ring, Piston pin retaining	76	Filter, Air cleaner (Paper)
23	Rod Assy., Connecting	77	Cover, Air cleaner
24	Bolt, Connecting rod	78	Nut, Lock, 1/4-20
25	Lifter, Valve	79	Screw, Hex washer hd., taptite, 1/4-20 x 11/16
26	Camshaft Assy.		
27	Pump Assy., Oil	80	Plate, Control Assy. cover
28	Gasket, Mounting Flange	81	Screw, Hex washer hd., taptite, 8-32 x 1/2
29	Flange, Mounting		
30	Plug, Oil drain	82	Housing, Blower
31	Seal, Oil	83	Screw, Hex washer hd. taptite, 1/4-20 x 1/2
32	Washer, Flat		
33	Gear Assy., Governor	84	Ring, Starter
34	Spool, Governor	85	Muffler
35	Ring, Retaining	86	Plate, Muffler locking
36	Gear, Idler	87	Screw, Hex hd. shoulder, 5/16-18 x 2-11/32
37	Screw, Hex washer hd., 1/4-20 x 1-9/16		
		88	Cup, Starter
38	Key, Flywheel	89	Screw, Hex washer hd., 8-32 x 21/64
39	Flywheel	90	Line, Fuel
40	Fan, Flywheel	91	Clamp, Fuel line
41	Washer, Belleville	92	Clip, "U" Type Nut, 10-32
42	Nut, Flywheel	93	Tank Assy., Fuel
43	Solid State Assy.	94	Cap, Fuel
44	Cover, Spark plug	95	Tube, Oil fill
45	Screw, Hex washer hd. Sems, 10-24 x 1	96	"O" Ring
		97	Clip, Fill tube
46	Wire Assy., Ground	98	Dipstick, Oil
47	Gasket, Cylinder head	99	Plug, Starter
48	Head, Cylinder	100	Screw, Hex washer hd. 10-32 x 35/64
49	Valve, Exhaust	101	Primer
50	Valve, Intake	102	Decal, Instruction
51	Screw, Hex flange hd., 5/16-18 x 1-1/2	103	Decal, Primer
52	Spark Plug (Champion RJ-19LM or equivalent)	104	Carburetor
53	Spring, Valve	105	Starter, Rewind

8325c28a

Flywheels

▶ See Figures 29, 30 and 31

The Vector engine uses one of two types of flywheels. The first type is a cast iron high inertia flywheel. This type of flywheel will have a pressed on ring gear if the engine is equipped with an electric starter. The other is an aluminum type.

REMOVAL

1. Disconnect the battery from the engine before servicing.
2. Remove the ignition module.
3. Remove the brake pressure from the flywheel.

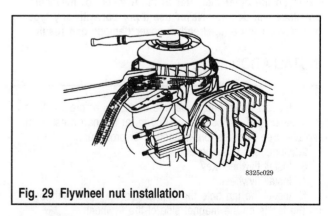

8325c029

Fig. 29 Flywheel nut installation

4. To remove the flywheel nut, use a flywheel strap wrench to hold the flywheel, while turning the flywheel nut counterclockwise.

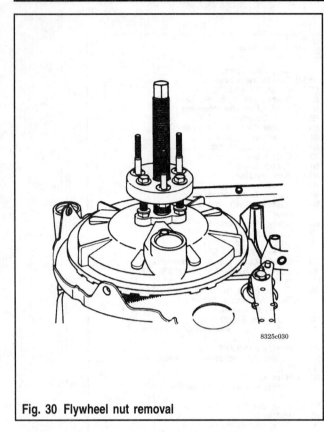

Fig. 30 Flywheel nut removal

5. Lift the starter cup and fan off of the flywheel. (Aluminum flywheel only).

6. Remove the flywheel using a flywheel puller or knock-off tool.

➠On engines with cored holes (not tapped) use flywheel puller Part No. 670306.

7. Screw a knock-off tool down until it touches the flywheel, then back off 1 turn. Using a large screwdriver, pry upward under the flywheel (side opposite the brake) and tap sharply and squarely on the knock-off tool to break the flywheel loose. If necessary rotate flywheel a half turn and repeat until it loosens.

➠Do not attempt to remove flywheel using a jaw type puller on the outer diameter of the flywheel or flywheel breakage will occur. Never use a pry bar with any type of curve on the end. Breather cover damage can result.

INSTALLATION

1. Inspect brake pad to be free of dirt, oil or grease. If pad is contaminated, or less than 0.060 in. at the narrowest point, replace. See Flywheel brake section for procedure.

2. Compress brake lever.

3. Install flywheel key.

4. Install flywheel.

5. Install the fan onto the flywheel so the Tecumseh logo on the fan is on the magnet side of the flywheel.

6. Place starter cup into position and torque nut to specification. Use a strap wrench to hold the flywheel.

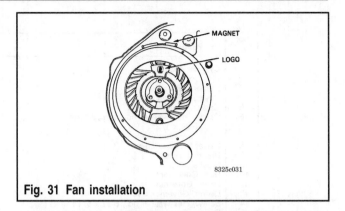

Fig. 31 Fan installation

Flywheel Brake System

OPERATION

▶ See Figures 32, 33 and 34

Use of the blade brake clutch in conjunction with either a top or side mounted recoil starter or 2 volt electric starter. The blade stops within three seconds after the operator lets go of the blade control bail at the operator position and the engine continues to run. Starter rope handle is on the engine.

Use of a recoil starter (top or side mounted) with the rope handle on the engine as opposed to within 24 inches of the operator position. This method is acceptable if the mower deck passes the 360 degree foot probe test. A specified foot probe must not contact the blade when applied completely around the entire blade housing. This alternative can be used with

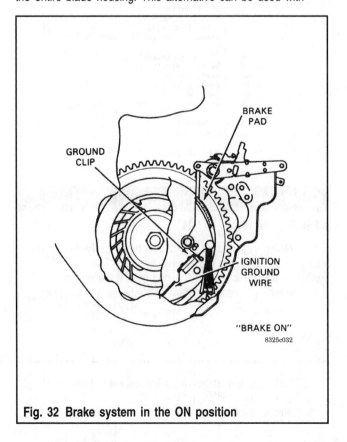

Fig. 32 Brake system in the ON position

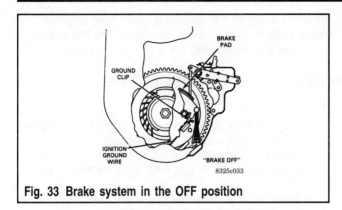

Fig. 33 Brake system in the OFF position

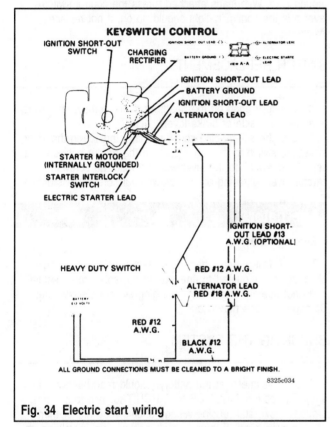

Fig. 34 Electric start wiring

engine mounted brake systems and typical bail controls. The blade stops within three seconds after the operator lets go of the blade control bail at the operator position and the engine is stopped.

The Brake Starter Mechanism may be used with either of two options for starting:

1. Manual Rope Start
2. 12 Volt Starter System

Each system requires the operator to start unit behind mower handle in operator zone area. The electric start system also provides a charging system for battery recharge when engine is running. **TO STOP THE ENGINE** In the stop position the brake pad is applied to the inside edge of the flywheel; at the same time the ignition system is grounded out. **TO START THE ENGINE** In order to restart the engine, the control must be applied. This action pulls the brake pad

away from the inside edge of the flywheel and opens the ignition ground switch.

On electric start systems the starter is energized to start the engine.

On non-electric start systems, recoil started rope must be pulled to start engine.

BRAKE PAD REPLACEMENT

▶ **See Figure 35**

1. If equipped with electric starter, locate wire routing through blower housing. Compress the grommet and pull out of the blower housing. Carefully slide wires through the slot. DO NOT cut the wire insulation on the blower housing.
2. Remove flywheel.
3. Remove pad lever E-clip. Lift pad lever, and unhook spring and link.
4. Attach link to pad lever install pad lever and E-clip.
5. Attach spring to lever first. Use a needle nose pliers to hook the spring into the baffle.

➡**It is important to attach the pad lever spring with the short hook on the pad lever and the long hook to the blower housing baffle.**

BRAKE CONTROL LEVER REPLACEMENT

▶ **See Figures 36 and 37**

1. Mark hole that spring is installed into baffle.
2. Remove E-clip from brake lever shaft.

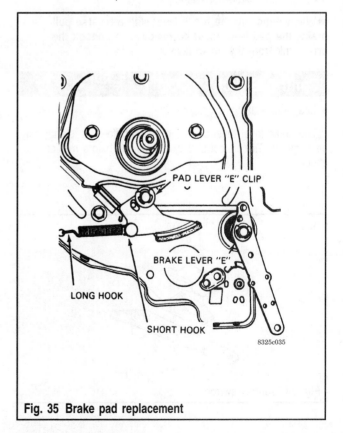

Fig. 35 Brake pad replacement

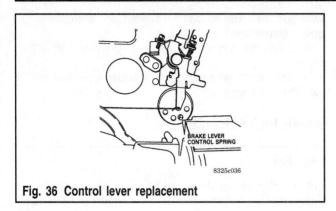

Fig. 36 Control lever replacement

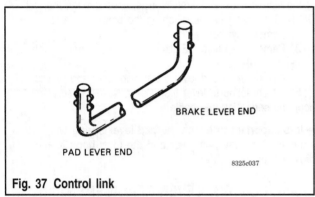

Fig. 37 Control link

3. Lift brake control lever and unhook link. Replace with new lever and reassemble in reverse order.

4. Replacement springs must be the same size and color.

5. Be sure control lever spring is in proper hole in blower housing baffle before reassembly.

➡ **When removing the brake lever with a reverse pull brake, the pad lever must be removed to unhook the brake link from the brake lever.**

Control Switch

▶ **See Figure 38**

The brake lever must close the switch before the starter can be engaged. Disconnect battery from circuit before making check.

Engines equipped with an electric starter have a control switch which is attached to the brake lever. The brake lever must close the switch before the starter can be engaged.

CHECKING THE CONTROL SWITCH

1. Disconnect the battery from the circuit. Use a continuity light or meter to check control switch operation.

2. Disconnect the wire harness at the engine.

3. Attach one continuity light lead to the electric starter lead.

4. Attach the other continuity light lead to the battery ground lead. With leads attached, press the control switch lever and the continuity light should go on, if not replace switch.

SWITCH REPLACEMENT

1. Carefully grind off the heads of rivets, remove the rivets from the back side of brake lever.

2. Use the self-tapping screw to make threads in the lever, install the switch to the brake lever in the proper position and secure the switch to the lever with the machine screws. Be careful, over-tightening of the screws could break the switch.

Alternator

▶ **See Figure 39**

The 350 Milliampere charging system consists of a single alternator coil mounted to one side of the solid state module.

Do not operate engine with charging system disconnected. Damage to diode may occur.

CHECKING THE SYSTEM

Connect voltmeter at the battery (should read battery voltage). The battery MUST BE IN CIRCUIT for test to perform properly. next, start engine-voltage should read higher than when engine is off. If there is a change upward in voltage, the charging system is working. If there is no change in voltage, the alternator should be replaced. Set volt/ohmmeter to 0-20 volt D.C. scale for test.

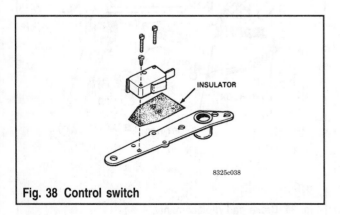

Fig. 38 Control switch

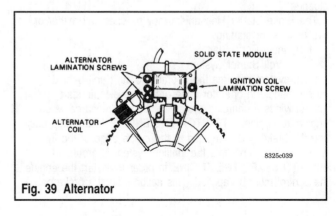

Fig. 39 Alternator

Ignition System

♦ **See Figures 40 and 41**

Tecumseh's solid state capacitor discharge ignition (CDI) is an all electronic ignition system and is encapsulated in epoxy for protection against dirt and moisture.

As the magnets in the flywheel rotate past the charge coil, electrical energy is produced in the module. This energy is transferred to a capacitor where it is stored until it is needed to fire the spark plug.

The magnet continues rotating past a trigger coil where a low voltage signal is produced and closes an electronic switch (SCR).

The energy which was stored in the capacitor is now transferred through the switch (SCR) to a transformer where the voltage is increased form 200 volts to 25,000 volts. This voltage is transferred by means of the high tension lead to th

spark plug, where it arcs across the electrode of the spark plug and ignited the fuel-air mixture.

Spark Plug

♦ **See Figures 42, 43, 44, 45 and 46**

SERVICE

Spark plugs should be replaced periodically. Check electrode gap with wire feeler gauge and adjust gap. Replace if electrode is pitted, burned or the porcelain is cracked. Be sure cleaned plugs are free of all foreign material. Use a spark plug tester to check for spark.

If spark plug fouls frequently, check for the following condition:

1. Incorrect spark plug

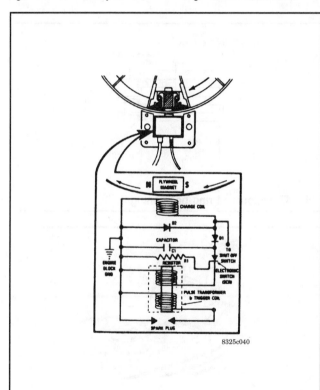

Fig. 40 Ignition module

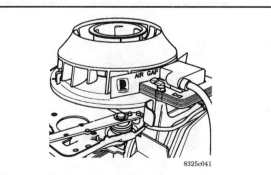

Fig. 41 Module air gap replacement

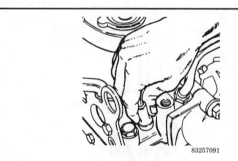

Fig. 42 Twist and pull on the boot, never on the plug wire

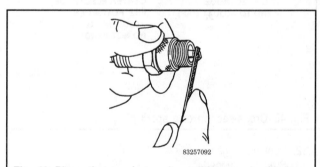

Fig. 43 Plugs that are in good condition can be filed and reused

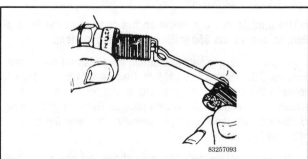

Fig. 44 Adjust the electrode gap by bending the side electrode

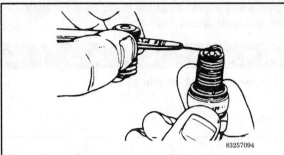

Fig. 45 Always used a wire gauge to check the plug gap

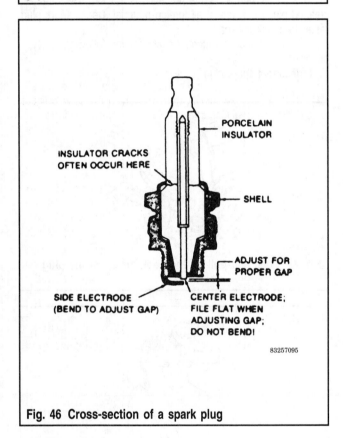

Fig. 46 Cross-section of a spark plug

2. Poor grade gasoline
3. Breather plugged
4. Oil level too high
5. Engine using excessive oil
6. Clogged air cleaner

➡ **The module has two holes on the one leg of the lamination, to secure the 350 milliamp charging system.**

The proper air gap setting between magnets and the laminations on CDI systems is 0.0125 in. Place a 0.0125 in. gauge between the magnets and laminations and tighten mounting screws to a torque of 30-40 inch pounds. Recheck gap setting to make certain there is proper clearance between the magnets and laminations.

➡ **Due to variations between pole shoes, air gap may vary from 0.005-0.020 in. when flywheel is rotated. There is no further timing adjustment on external laminations systems.**

Piston

▶ **See Figures 47, 48, 49 and 50**

Before removing piston, clean any carbon from the top of the cylinder bore to prevent ring breakage when removing the piston. Push the rod and piston out through the top of the cylinder.

Oversize pistons are identified by the size imprinted on the piston as shown. Check the piston for wear by measuring at the bottom of the skirt 90 degrees from the wrist pin hole. Clean the carbon from the piston ring grooves, install new rings and measure side clearance.

Replace rings in sets and always stagger ring gaps. When installing new rings, deglaze cylinder wall, using a commercially available deglazing tool.

Use a ring expander to remove and replace rings. Do not spread the rings too wide or breakage will result.

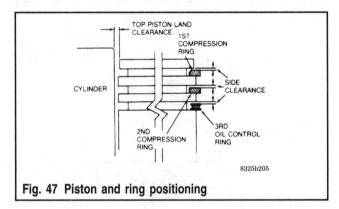

Fig. 47 Piston and ring positioning

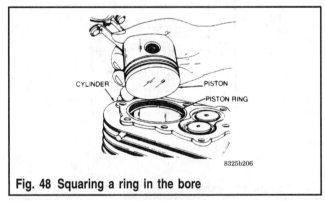

Fig. 48 Squaring a ring in the bore

Fig. 49 Checking the ring gap

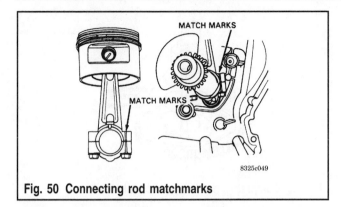

Fig. 50 Connecting rod matchmarks

Both compression rings will have a chamfer on the inside edge. The rings must be installed with the chamfer up.

To check ring end gap, place ring squarely in center of ring travel area. Using the piston to push the ring down into the cylinder at least one inch.

Check ring gap on new ring to determine if cylinder should be rebored to take oversize parts.

Connecting Rods

▶ **See Figure 51**

Match marked on connecting rods must always align and must face outward toward the mechanic when installed in an engine.

A new piston can be installed on to the connecting rod in either direction. It is necessary to replace the connecting rod be sure to mark the calve side of the piston.

Install the piston to the connecting rod so that the piston will be in the same position when reinstalled in the engine.

The connecting rod bolts use fine threads to insure torque retention. For the proper torque specification see the table of specifications.

Cylinder

INSPECTION

Check cylinder for dirty, broken or cracked fins, worn or scored bearings or scored cylinder bore surface, and warped head mounting surfaces.

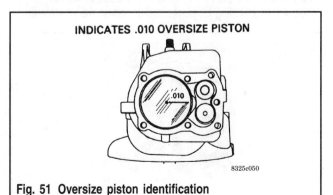

Fig. 51 Oversize piston identification

The cylinder must be replaced if the cylinder head mounting surface is warped extensively.

If cylinder bore is worn more tan 0.005 in. oversize, out-of-round or scored, it should be replaced or rebored to 0.010 in. or 0.020 in. oversize. In some cases engines are built with an oversize cylinder; in these instances, they are identified with the oversize value imprinted in the cylinder as pictured.

REBORING CYLINDER

To rebore cylinder use a commercial hone of suitable size chucked in a drill press with a spindle speed of approximately 600 R.P.M.

Start with coarse stones and center cylinder under press spindle. Lower them so lower end of stones contacts lowest point in cylinder bore.

Rotate adjusting nut so that stones touch cylinder wall and begin honing at bottom of cylinder. Move hone up or down at rate of 50 strokes per minute to avoid cutting ridges in cylinder wall. Every fourth or fifth stroke, move hone far enough to extend the stones one inch beyond top and bottom of cylinder bore.

Check bore every thirty or forty strokes for size and straightness. If stones collect metal, clean with a wire brush each time hone is removed.

Hone with coarse stones until cylinder bore is within 0.002 inch of desired finish size. Replace coarse stones with burnishing stones and continue until bore is to within 0.0005 inch of desired size.

Remove burnishing stones and install finishing stones to polish cylinder to final size.

Clean cylinder with soap and water, and dry thoroughly.

Replace piston and piston rings with correct oversize parts.

Cylinder Head

▶ **See Figure 52**

Check cylinder heads for warpage by placing on a flat surface. If warped extensively, replace head. Always replace head gasket and torque head bolts in 50 inch lbs. increments in the numbered sequence to a torque of 180-220 inch lbs.

Crankshaft

INSPECTION

Inspect crankshaft for worn, scratched or damaged bearing surfaces, out-of-round or flat spots on the journal area, or a bent P.T.O. end.

➡**Never try to straighten a bent crankshaft.**

When installing a crankshaft, lubricate all bearing surfaces and use oil seal protector part no. 670327 or equivalent

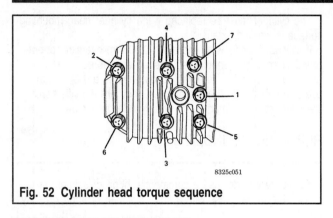

Fig. 52 Cylinder head torque sequence

CRANKSHAFT TIMING MARK

▶ **See Figures 53, 54 and 55**

The crankshaft has a pressed on timing gear. This gear has a small dimple punched on one side of the teeth on this gear. This dimple is a timing mark. With the crankpin at top dead center, the timing mark should be in the 2:30 position.

The camshaft has an aligning mark in line with the timing hole on the camshaft gear. Line this mark up with the dimple on the crankshaft gear.

Timing marks on crankshaft gear and camshaft gear must be aligned for proper valve timing.

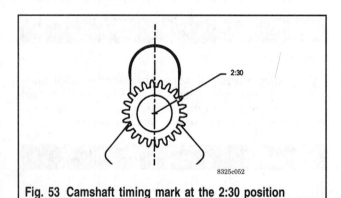

Fig. 53 Camshaft timing mark at the 2:30 position

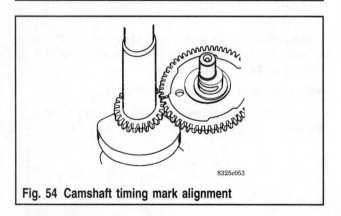

Fig. 54 Camshaft timing mark alignment

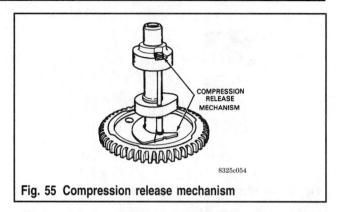

Fig. 55 Compression release mechanism

Camshaft

REMOVAL

Align timing marks to relieve train pressure. Lift out camshaft.

The camshaft has a mechanical compression release mechanism. A pin which runs through both cam lobes extends past the exhaust lobe and lifts the valve to relieve compression for easier starting. When the engine starts, centrifugal force moves the flyweights outward, moving the pin below the lobe, allowing full compression. The compression release mechanism is non-serviceable (replace camshaft assembly. if damaged or worn.)

Oil Pump

▶ **See Figures 56, 57, 58, 59, 60 and 61**

All Tecumesh Vector engines use a positive displacement plunger oil pump to pump oil from the crankcase, up through the camshaft to a passage in the breather box to the top of the crankshaft main bearing, and ultra balance bearings.

Oil is pressure sprayed out of a small hole between the crankshaft and ultra-balance bearing, to lubricate the connecting rod journal area. If a heavy leakage is noted from the breather cover check for plugged mist hole.

➡**Not all Vector engines are equipped with ultra-balance.**

An eccentric on the camshaft works the plunger in the barrel back and forth, forcing oil up the center of the camshaft. A ball on the end of the plunger locates in a recess in the flange cover. When installing oil pump. make certain the chamfered

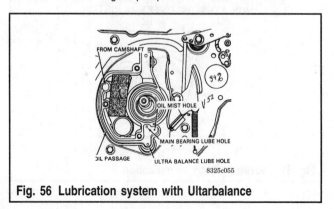

Fig. 56 Lubrication system with Ultarbalance

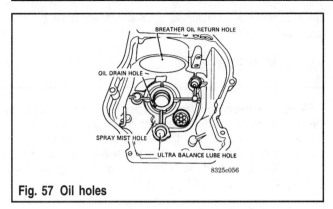

Fig. 57 Oil holes

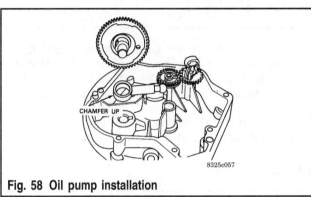

Fig. 58 Oil pump installation

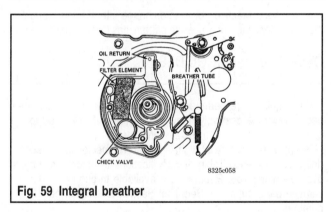

Fig. 59 Integral breather

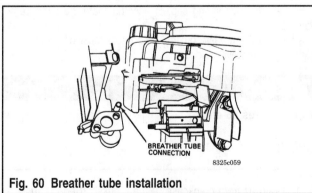

Fig. 60 Breather tube installation

side of the pump faces the camshaft, and the plunger ball seats in the recess of the flange cover.

The Vector engine has a top mounted integral breather.

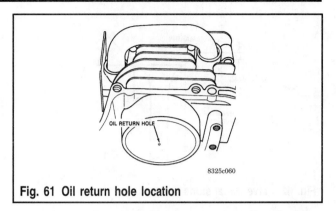

Fig. 61 Oil return hole location

The breather compartment is located under the flywheel. A check valve allows excess crankcase pressure to be vented through the element and out the breather tube. The breather tube is connected to the air cleaner body. When reassembling the breather, DO NOT pinch the filter element under the breather cover or leak may occur.

Condensed oil vapors are returned to the crankcase by means of the oil return hole. The oil return hole is opened and closed in the cylinder by the piston.

The breather filter element can be cleaned using solvent.

When reinstalling the check valve, apply oil to aid in assembly. A new breather valve body can be pressed into the block to replace a damaged breather valve body.

Valves

Valves must be in good condition, properly sealing and the proper gap must be maintained for full power, easy starting and efficient operation.

VALVE REMOVAL

1. To remove valves, raise the lower valve spring caps, while holding the valve head tightly against the seat.

2. Move the lower cap, so it will slip off the end of the valve.

3. Clean all parts and remove carbon from valve heads and stems. If valves are is usable condition, grind the valve faces to a 45 degree angle. Replace valves if they are damaged, distorted or if the margin is ground to less than 1/32 in.

Valve Seats

▶ See Figures 62, 63, 64 and 65

Valve seats are not replaceable. If they are burned or pitted, they can be reground using a grinding stone or valve reseater. Seats are ground at an angle of 46 degrees, to a width of 3/64 in.

The recommended procedure to properly cut a valve seat is to use the Neway Cutting System, which consists of three different degree-cutters.

1. Use the 60 degree cutter to clean and narrow the seat from the bottom toward the center.

2. Use the 31 degree cutter to clean and narrow the seat from the top toward the center.

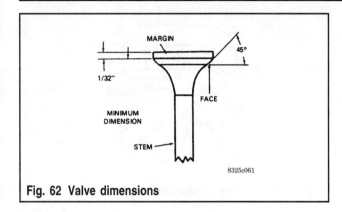

Fig. 62 Valve dimensions

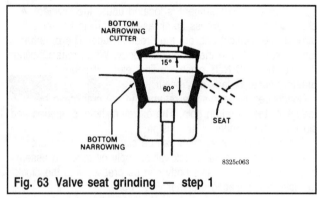

Fig. 63 Valve seat grinding — step 1

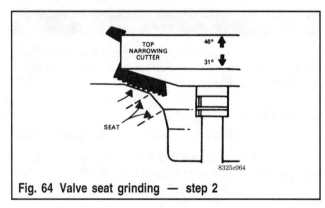

Fig. 64 Valve seat grinding — step 2

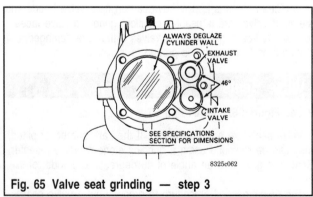

Fig. 65 Valve seat grinding — step 3

Valves are not identical. Make sure the valve marked **EX** or **X** is installed in the exhaust valve location, and the valve marked **I** is installed in the intake valve location. If the valves are unmarked, the nonmagnetic valve is installed in the exhaust valve location.

VALVE ADJUSTMENT

Clearance between the valve stem and lifter must be set to the recommended specifications when the engine is cold. Check these clearances with the piston T.D.C. on the compression stroke. Grind end of valve stem with a valve grinder, or use a **V** block to hold the valve square on grinding wheel, grinding to the proper 0.008 in. clearance.

VALVE INSTALLATION

1. Position valve caps and spring in the valve compartment.
2. Install valves in guides with valve marked **I** in the intake port. The valve stem must pass through the spring. The valve spring cap should sit around the valve lifter exposed end.
3. Use a valve spring compressor to compress the valve spring. Position the valve spring cap onto the valve stem and release valve spring tension to lock cap in place.

VALVE LIFTERS

It is a good practice not to interchange lifters, even though they are identical, once a wear pattern has been established.

OVERSIZE VALVE GUIDES

Valve guides are permanently installed in the cylinder. If they become worn excessively, they can be reamed oversized to accommodate a $^1/_{32}$ in. oversize valve stem.

Ream guides with a straight shanked reamer or low speed drill press. Refer to Table of Specifications to determine correct oversize dimension. Reamers are available through your Tecumseh parts suppliers. See Tool Section for correct part numbers.

Redrill the valve spring cap, to accommodate the oversize valve stems.

After oversizing valve guides - the seats must be recut to align with the valve guides.

Oil Seal

▶ See Figures 66 and 67

SERVICE

1. Drain oil from crankcase. If the crankshaft end is rusty or pitted, polish the crankshaft with emery cloth so it will not damage the bearings when the cover is removed.
2. Remove mounting bolts and slide seal protector-driver tool, Part No 670327 or equivalent, into the oil seal. If necessary, tap edge of flange or cover lightly with a soft hammer to remove cover.

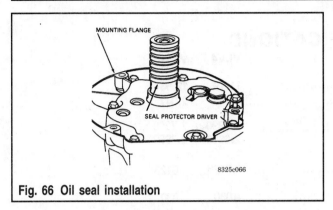

Fig. 66 Oil seal installation

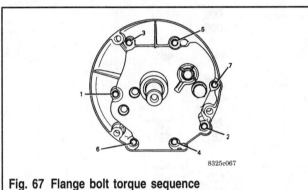

Fig. 67 Flange bolt torque sequence

3. Clean and inspect the cover for wear and scoring of bearings. Inspect crankshaft bearings. Replace any worn or damaged parts.

4. If crankshaft is out of engine, remove old oil seal by tapping them out with a screwdriver or punch from the inside, To remove a seal with the crankshaft in the engine, insert a screwdriver between the seal and the crankshaft and pry the seal out.

5. Lubricate the outside of the new oil seal with oil prior to installation.

6. Use seal driver-protector tool Part No. 670327, or equivalent. Place oil seal over the driver-protector and place over crankshaft, driving it into position using universal driver

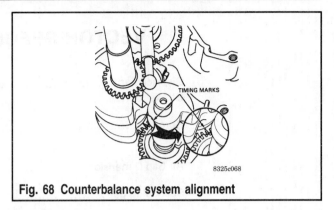

Fig. 68 Counterbalance system alignment

No. 670272. The seal will automatically be driven into the proper depth.

7. Torque flange bolts in numerical order as shown in illustration.

Ultra-Balance

▶ **See Figures 68 and 69**

The single shaft counterbalance is driven by a gear on the crankshaft, to counteract the imbalance caused by the counterweights on the crankshaft.

To correctly align the counterbalance system, rotate the piston to 90 past TDC and insert the single-shaft counterbalance into its boss in the crankcase with the punch mark on the gear in alignment with the crankshaft punch mark.

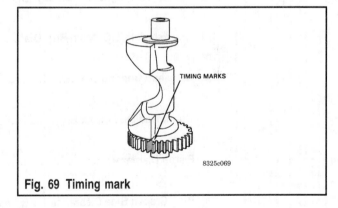

Fig. 69 Timing mark

VECTOR SPECIFICATIONS

	VLV4	VLV5
Displacement	11.19	12.56
Stroke	2.047	2.047
Bore	2.6390	2.7960
	2.6380	2.7950
Air Gap Dimension	.0125	.0125
Spark plug gap	.030	.030
Valve Clearance	.008	
	.008	
Valve Seat Angle	46°	46°
Valve Seat Width	.035	.035
	.045	.045
Valve Guide Oversize Dia.	.2807	.2807
	.2817	.2817
Crankshaft End Play	.005	.005
	.027	.027
Crankpin Journel Dia.	1.0235	1.0235
	1.0230	1.0230
Crankshaft Magneto Main Brg. Dia	1.0242	1.0242
	1.0237	1.0237
Crankshaft P.T.O. Main Brg. Dia	1.0242	1.0242
	1.0237	1.0237
Camshaft Bearing Dia.	.4980	.4980
	.4975	.4975
Conn. Rod Dia. Crank Brg.	1.0246	1.0246
	1.0240	1.0240
Piston Diameter	2.6340	2.7910
	2.6330	2.7900
Ring Groove Side Clearance 1st & 2nd Comp.	.002	.002
	.005	.005
Side Clearance (Bot.) Oil	.0005	.0005
	.0035	.0035
Piston Skirt Clearance	.004	.004
	.006	.006
Ring End Gap	.007	.007
	.020	.020
Cylinder Main Brg. Dia	1.0262	1.0262
	1.0257	1.0257
Cylinder Cover/Flange Main Bearing Diameter	1.0262	1.0262
	1.0257	1.0257

8325c0C2

VECTOR TORQUE SPECIFICATIONS

	Inch-Lbs.
Governor Rod Clamp to Lever	7-12
Breather Cover	40-50
Valve Box Cover	25-35
Connecting Rod	95-110
Cylinder Head	180-220
Mounting Flange or Cylinder Cover	100-130
Housing Baffle to Cylinder	80-120
Solid State Ignition to Cylinder	30-50
Alternating Coil Assembly to Lamination	25-35
Flywheel Nut	400-440
Housing to Baffle	35-45
Carburetor Stud to Cylinder	50-75
A/C Hex Nut to Stud	35-45
Control Face Plate to Baffle	30-40
Starter - Top Mounting	20-30
Electric Starter to Baffle	10-15
5/8-18 Plug (Hex Flange)	90-150
Plastic Tank to Housing	12-20
Threaded Fill Tube (Plastic)	45-65
Large Diameter Oil Fill Plug	Hand Tight
Muffler Mounting (Shoulder Screw)	100-165
A/C Body & Housing to Baffle	35-45
Conduit Clip Screw	5-15
Muffler Deflector	10-25

8325c0C3

GLOSSARY

AIR/FUEL RATIO: The ratio of air to gasoline by weight in the fuel mixture drawn into the engine.

AIR INJECTION: One method of reducing harmful exhaust emissions by injecting air into each of the exhaust ports of an engine. The fresh air entering the hot exhaust manifold causes any remaining fuel to be burned before it can exit the tailpipe.

ALTERNATOR: A device used for converting mechanical energy into electrical energy.

AMMETER: An instrument, calibrated in amperes, used to measure the flow of an electrical current in a circuit. Ammeters are always connected in series with the circuit being tested.

AMPERE: The rate of flow of electrical current present when one volt of electrical pressure is applied against one ohm of electrical resistance.

ANALOG COMPUTER: Any microprocessor that uses similar (analogous) electrical signals to make its calculations.

ARMATURE: A laminated, soft iron core wrapped by a wire that converts electrical energy to mechanical energy as in a motor or relay. When rotated in a magnetic field, it changes mechanical energy into electrical energy as in a generator.

ATMOSPHERIC PRESSURE: The pressure on the Earth's surface caused by the weight of the air in the atmosphere. At sea level, this pressure is 14.7 psi at 32{248}F (101 kPa at 0{248}C).

ATOMIZATION: The breaking down of a liquid into a fine mist that can be suspended in air.

AXIAL PLAY: Movement parallel to a shaft or bearing bore.

BACKFIRE: The sudden combustion of gases in the intake or exhaust system that results in a loud explosion.

BACKLASH: The clearance or play between two parts, such as meshed gears.

BACKPRESSURE: Restrictions in the exhaust system that slow the exit of exhaust gases from the combustion chamber.

BAKELITE: A heat resistant, plastic insulator material commonly used in printed circuit boards and transistorized components.

BALL BEARING: A bearing made up of hardened inner and outer races between which hardened steel balls roll.

BALLAST RESISTOR: A resistor in the primary ignition circuit that lowers voltage after the engine is started to reduce wear on ignition components.

BEARING: A friction reducing, supportive device usually located between a stationary part and a moving part.

BIMETAL TEMPERATURE SENSOR: Any sensor or switch made of two dissimilar types of metal that bend when heated or cooled due to the different expansion rates of the alloys. These types of sensors usually function as an on/off switch.

BLOWBY: Combustion gases, composed of water vapor and unburned fuel, that leak past the piston rings into the crankcase during normal engine operation. These gases are removed by the PCV system to prevent the buildup of harmful acids in the crankcase.

BRAKE PAD: A brake shoe and lining assembly used with disc brakes.

BRAKE SHOE: The backing for the brake lining. The term is, however, usually applied to the assembly of the brake backing and lining.

BUSHING: A liner, usually removable, for a bearing; an anti-friction liner used in place of a bearing.

CALIPER: A hydraulically activated device in a disc brake system, which is mounted straddling the brake rotor (disc). The caliper contains at least one piston and two brake pads. Hydraulic pressure on the piston(s) forces the pads against the rotor.

CAMSHAFT: A shaft in the engine on which are the lobes (cams) which operate the valves. The camshaft is driven by the crankshaft, via a belt, chain or gears, at one half the crankshaft speed.

CAPACITOR: A device which stores an electrical charge.

CARBON MONOXIDE (CO): A colorless, odorless gas given off as a normal byproduct of combustion. It is poisonous and extremely dangerous in confined areas, building up slowly to toxic levels without warning if adequate ventilation is not available.

CARBURETOR: A device, usually mounted on the intake manifold of an engine, which mixes the air and fuel in the proper proportion to allow even combustion.

CATALYTIC CONVERTER: A device installed in the exhaust system, like a muffler, that converts harmful byproducts of combustion into carbon dioxide and water vapor by means of a heat-producing chemical reaction.

CENTRIFUGAL ADVANCE: A mechanical method of advancing the spark timing by using flyweights in the distributor that react to centrifugal force generated by the distributor shaft rotation.

CHECK VALVE: Any one-way valve installed to permit the flow of air, fuel or vacuum in one direction only.

CHOKE: A device, usually a moveable valve, placed in the intake path of a carburetor to restrict the flow of air.

CIRCUIT: Any unbroken path through which an electrical current can flow. Also used to describe fuel flow in some instances.

CIRCUIT BREAKER: A switch which protects an electrical circuit from overload by opening the circuit when the current flow exceeds a predetermined level. Some circuit breakers must be reset manually, while most reset automatically

COIL (IGNITION): A transformer in the ignition circuit which steps up the voltage provided to the spark plugs.

COMBINATION MANIFOLD: An assembly which includes both the intake and exhaust manifolds in one casting.

COMBINATION VALVE: A device used in some fuel systems that routes fuel vapors to a charcoal storage canister instead of venting them into the atmosphere. The valve relieves fuel tank pressure and allows fresh air into the tank as the fuel level drops to prevent a vapor lock situation.

COMPRESSION RATIO: The comparison of the total volume of the cylinder and combustion chamber with the piston at BDC and the piston at TDC.

CONDENSER: 1. An electrical device which acts to store an electrical charge, preventing voltage surges.
2. A radiator-like device in the air conditioning system in which refrigerant gas condenses into a liquid, giving off heat.

CONDUCTOR: Any material through which an electrical current can be transmitted easily.

CONTINUITY: Continuous or complete circuit. Can be checked with an ohmmeter.

COUNTERSHAFT: An intermediate shaft which is rotated by a mainshaft and transmits, in turn, that rotation to a working part.

CRANKCASE: The lower part of an engine in which the crankshaft and related parts operate.

CRANKSHAFT: The main driving shaft of an engine which receives reciprocating motion from the pistons and converts it to rotary motion.

CYLINDER: In an engine, the round hole in the engine block in which the piston(s) ride.

CYLINDER BLOCK: The main structural member of an engine in which is found the cylinders, crankshaft and other principal parts.

CYLINDER HEAD: The detachable portion of the engine, fastened, usually, to the top of the cylinder block, containing all or most of the combustion chambers. On overhead valve engines, it contains the valves and their operating parts. On overhead cam engines, it contains the camshaft as well.

DEAD CENTER: The extreme top or bottom of the piston stroke.

DETONATION: An unwanted explosion of the air/fuel mixture in the combustion chamber caused by excess heat and compression, advanced timing, or an overly lean mixture. Also referred to as "ping".

DIAPHRAGM: A thin, flexible wall separating two cavities, such as in a vacuum advance unit.

DIESELING: A condition in which hot spots in the combustion chamber cause the engine to run on after the key is turned off.

DIFFERENTIAL: A geared assembly which allows the transmission of motion between drive axles, giving one axle the ability to turn faster than the other.

DIODE: An electrical device that will allow current to flow in one direction only.

DISC BRAKE: A hydraulic braking assembly consisting of a brake disc, or rotor, mounted on an axle, and a caliper assembly containing, usually two brake pads which are activated by hydraulic pressure. The pads are forced against the sides of the disc, creating friction which slows the vehicle.

DISTRIBUTOR: A mechanically driven device on an engine which is responsible for electrically firing the spark plug at a predetermined point of the piston stroke.

DOWEL PIN: A pin, inserted in mating holes in two different parts allowing those parts to maintain a fixed relationship.

DRUM BRAKE: A braking system which consists of two brake shoes and one or two wheel cylinders, mounted on a fixed backing plate, and a brake drum, mounted on an axle, which revolves around the assembly.

DWELL: The rate, measured in degrees of shaft rotation, at which an electrical circuit cycles on and off.

ELECTRONIC CONTROL UNIT (ECU): Ignition module, module, amplifier or igniter. See Module for definition.

ELECTRONIC IGNITION: A system in which the timing and firing of the spark plugs is controlled by an electronic control unit, usually called a module. These systems have no points or condenser.

ENDPLAY: The measured amount of axial movement in a shaft.

ENGINE: A device that converts heat into mechanical energy.

EXHAUST MANIFOLD: A set of cast passages or pipes which conduct exhaust gases from the engine.

FEELER GAUGE: A blade, usually metal, of precisely predetermined thickness, used to measure the clearance between two parts.

FIRING ORDER: The order in which combustion occurs in the cylinders of an engine. Also the order in which spark is distributed to the plugs by the distributor.

FLOODING: The presence of too much fuel in the intake manifold and combustion chamber which prevents the air/fuel mixture from firing, thereby causing a no-start situation.

FLYWHEEL: A disc shaped part bolted to the rear end of the crankshaft. Around the outer perimeter is affixed the ring gear. The starter drive engages the ring gear, turning the flywheel, which rotates the crankshaft, imparting the initial starting motion to the engine.

FOOT POUND (ft.lb. or sometimes, ft. lbs.): The amount of energy or work needed to raise an item weighing one pound, a distance of one foot.

FUSE: A protective device in a circuit which prevents circuit overload by breaking the circuit when a specific amperage is present. The device is constructed around a strip or wire of a lower amperage rating than the circuit it is designed to protect. When an amperage higher than that stamped on the fuse is present in the circuit, the strip or wire melts, opening the circuit.

GEAR RATIO: The ratio between the number of teeth on meshing gears.

GENERATOR: A device which converts mechanical energy into electrical energy.

HEAT RANGE: The measure of a spark plug's ability to dissipate heat from its firing end. The higher the heat range, the hotter the plug fires.

HUB: The center part of a wheel or gear.

HYDROCARBON (HC): Any chemical compound made up of hydrogen and carbon. A major pollutant formed by the engine as a byproduct of combustion.

HYDROMETER: An instrument used to measure the specific gravity of a solution.

INCH POUND (in.lb. or sometimes, in. lbs.): One twelfth of a foot pound.

INDUCTION: A means of transferring electrical energy in the form of a magnetic field. Principle used in the ignition coil to increase voltage.

INJECTOR: A device which receives metered fuel under relatively low pressure and is activated to inject the fuel into the engine under relatively high pressure at a predetermined time.

INPUT SHAFT: The shaft to which torque is applied, usually carrying the driving gear or gears.

INTAKE MANIFOLD: A casting of passages or pipes used to conduct air or a fuel/air mixture to the cylinders.

JOURNAL: The bearing surface within which a shaft operates.

KEY: A small block usually fitted in a notch between a shaft and a hub to prevent slippage of the two parts.

MANIFOLD: A casting of passages or set of pipes which connect the cylinders to an inlet or outlet source.

MANIFOLD VACUUM: Low pressure in an engine intake manifold formed just below the throttle plates. Manifold vacuum is highest at idle and drops under acceleration.

MASTER CYLINDER: The primary fluid pressurizing device in a hydraulic system. In automotive use, it is found in brake and hydraulic clutch systems and is pedal activated, either directly or, in a power brake system, through the power booster.

MODULE: Electronic control unit, amplifier or igniter of solid state or integrated design which controls the current flow in the ignition primary circuit based on input from the pick-up coil. When the module opens the primary circuit, the high secondary voltage is induced in the coil.

NEEDLE BEARING: A bearing which consists of a number (usually a large number) of long, thin rollers.

OHM:(Ω) The unit used to measure the resistance of conductor to electrical flow. One ohm is the amount of resistance that limits current flow to one ampere in a circuit with one volt of pressure.

OHMMETER: An instrument used for measuring the resistance, in ohms, in an electrical circuit.

OUTPUT SHAFT: The shaft which transmits torque from a device, such as a transmission.

OVERDRIVE: A gear assembly which produces more shaft revolutions than that transmitted to it.

OVERHEAD CAMSHAFT (OHC): An engine configuration in which the camshaft is mounted on top of the cylinder head and operates the valve either directly or by means of rocker arms.

OVERHEAD VALVE (OHV): An engine configuration in which all of the valves are located in the cylinder head and the camshaft is located in the cylinder block. The camshaft operates the valves via lifters and pushrods.

OXIDES OF NITROGEN (NOx): Chemical compounds of nitrogen produced as a byproduct of combustion. They combine with hydrocarbons to produce smog.

OXYGEN SENSOR: Used with the feedback system to sense the presence of oxygen in the exhaust gas and signal the computer which can reference the voltage signal to an air/fuel ratio.

PINION: The smaller of two meshing gears.

PISTON RING: An open ended ring which fits into a groove on the outer diameter of the piston. Its chief function is to form a seal between the piston and cylinder wall. Most automotive pistons have three rings: two for compression sealing; one for oil sealing.

PRELOAD: A predetermined load placed on a bearing during assembly or by adjustment.

PRIMARY CIRCUIT: Is the low voltage side of the ignition system which consists of the ignition switch, ballast resistor or resistance wire, bypass, coil, electronic control unit and pick-up coil as well as the connecting wires and harnesses.

PRESS FIT: The mating of two parts under pressure, due to the inner diameter of one being smaller than the outer diameter of the other, or vice versa; an interference fit.

RACE: The surface on the inner or outer ring of a bearing on which the balls, needles or rollers move.

REGULATOR: A device which maintains the amperage and/or voltage levels of a circuit at predetermined values.

RELAY: A switch which automatically opens and/or closes a circuit.

RESISTANCE: The opposition to the flow of current through a circuit or electrical device, and is measured in ohms. Resistance is equal to the voltage divided by the amperage.

RESISTOR: A device, usually made of wire, which offers a preset amount of resistance in an electrical circuit.

RING GEAR: The name given to a ring-shaped gear attached to a differential case, or affixed to a flywheel or as part a planetary gear set.

ROLLER BEARING: A bearing made up of hardened inner and outer races between which hardened steel rollers move.

ROTOR: 1. The disc-shaped part of a disc brake assembly, upon which the brake pads bear; also called, brake disc.
2. The device mounted atop the distributor shaft, which passes current to the distributor cap tower contacts.

SECONDARY CIRCUIT: The high voltage side of the ignition system, usually above 20,000 volts. The secondary includes the ignition coil, coil wire, distributor cap and rotor, spark plug wires and spark plugs.

SENDING UNIT: A mechanical, electrical, hydraulic or electromagnetic device which transmits information to a gauge.

SENSOR: Any device designed to measure engine operating conditions or ambient pressures and temperatures. Usually electronic in nature and designed to send a voltage signal to an on-board computer, some sensors may operate as a simple on/off switch or they may provide a variable voltage signal (like a potentiometer) as conditions or measured parameters change.

SHIM: Spacers of precise, predetermined thickness used between parts to establish a proper working relationship.

SLAVE CYLINDER: In automotive use, a device in the hydraulic clutch system which is activated by hydraulic force, disengaging the clutch.

SOLENOID: A coil used to produce a magnetic field, the effect of which is produce work.

SPARK PLUG: A device screwed into the combustion chamber of a spark ignition engine. The basic construction is a conductive core inside of a ceramic insulator, mounted in an outer conductive base. An electrical charge from the spark plug wire travels along the conductive core and jumps a preset air gap to a grounding point or points at the end of the conductive base. The resultant spark ignites the fuel/air mixture in the combustion chamber.

SPLINES: Ridges machined or cast onto the outer diameter of a shaft or inner diameter of a bore to enable parts to mate without rotation.

TACHOMETER: A device used to measure the rotary speed of an engine, shaft, gear, etc., usually in rotations per minute.

THERMOSTAT: A valve, located in the cooling system of an engine, which is closed when cold and opens gradually in response to engine heating, controlling the temperature of the coolant and rate of coolant flow.

TOP DEAD CENTER (TDC): The point at which the piston reaches the top of its travel on the compression stroke.

TORQUE: The twisting force applied to an object.

TORQUE CONVERTER: A turbine used to transmit power from a driving member to a driven member via hydraulic action, providing changes in drive ratio and torque. In automotive use, it links the driveplate at the rear of the engine to the automatic transmission.

TRANSDUCER: A device used to change a force into an electrical signal.

TRANSISTOR: A semi-conductor component which can be actuated by a small voltage to perform an electrical switching function.

TUNE-UP: A regular maintenance function, usually associated with the replacement and adjustment of parts and components in the electrical and fuel systems of a vehicle for the purpose of attaining optimum performance.

TURBOCHARGER: An exhaust driven pump which compresses intake air and forces it into the combustion chambers at higher than atmospheric pressures. The increased air pressure allows more fuel to be burned and results in increased horsepower being produced.

VACUUM ADVANCE: A device which advances the ignition timing in response to increased engine vacuum.

VACUUM GAUGE: An instrument used to measure the presence of vacuum in a chamber.

VALVE: A device which control the pressure, direction of flow or rate of flow of a liquid or gas.

VALVE CLEARANCE: The measured gap between the end of the valve stem and the rocker arm, cam lobe or follower that activates the valve.

VISCOSITY: The rating of a liquid's internal resistance to flow.

VOLTMETER: An instrument used for measuring electrical force in units called volts. Voltmeters are always connected parallel with the circuit being tested.

WHEEL CYLINDER: Found in the automotive drum brake assembly, it is a device, actuated by hydraulic pressure, which, through internal pistons, pushes the brake shoes outward against the drums.

BRIGGS & STRATTON 1970-84 13-20 HP MODELS

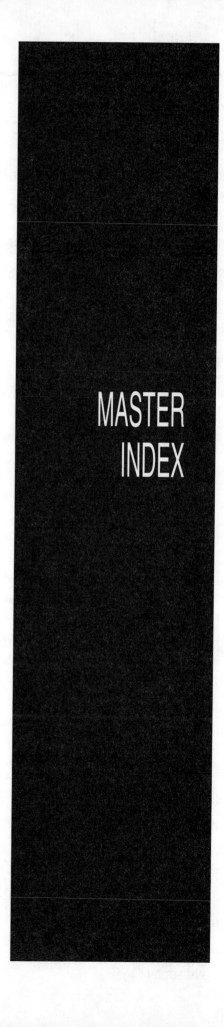

MASTER
INDEX

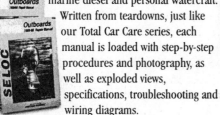

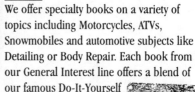